D1667218

Handbook of Analytical Techniques

edited by Helmut Günzler and
Alex Williams

Volume I

Handbook of Analytical Techniques

edited by Helmut Günzler and
Alex Williams

Volume I

Chapter 1
to
Chapter 18

Weinheim · New York · Chichester · Brisbane · Singapore · Toronto

Prof. Dr. Helmut Günzler
Bismarckstr. 4
D-69469 Weinheim
Germany

Alex Williams
19 Hamesmoor Way, Mytchett
Camberley, Surrey GU16 6JG
United Kingdom

Library of Congress Card No. applied for.

British Library Cataloguing-in-Publication Data:
A catalogue record for this book is available from the British Library.

Deutsche Bibliothek – CIP Cataloguing-in-Publication-Data:
A catalogue record for this publication is available from Die Deutsche Bibliothek.

ISBN 3-527-30165-8

Printed on acid-free paper.

Composition: Rombach GmbH, D-79115 Freiburg
Printing: Strauss Offsetdruck GmbH, D-69509 Mörlenbach
Bookbinding: Wilhelm Osswald & Co., D-67433 Neustadt (Weinstraße)
Cover Design: Gunter Schulz, D-67136 Fußgönheim
Printed in the Federal Republic of Germany.

Preface

The broad spectrum of analytical techniques available today is covered in this handbook. It starts with general articles on purpose and procedures of analytical chemistry, quality assurance, chemometrics, sampling and sample preparation followed by articles on individual techniques, including not only chromatographic and spectrometric techniques but also e. g. immunoassays, activation analysis, chemical and biochemical sensors, and techniques for DNA-analysis.

Most of the information presented is a thoroughly updated version of that included in the 5th edition of the 36-volume "Ullmann's Encyclopedia of Industrial Chemistry", the last edition that is available in print format. Some chapters were completely rewritten. The wealth of material in that Encyclopedia provides the user with both broad introductory information and in-depth detail of utmost importance in both industrial and academic environments. Due to its sheer size, however, the unabridged Ullmann's is inaccessible to many potential users, particularly individuals, smaller companies, or independent analytical laboratories. In addition there have been significant developments in analytical techniques since the last printed edition of the Encyclopedia was published, which is currently available in its 6th edition in electronic formats only. This is why all the information on analytical techniques has been revised and published in this convenient two-volume set.

Users of the "Handbook of Analytical Techniques" will have the benefit of up-to-date professional information on this topic, written and revised by acknowledged experts. We believe that this new handbook will prove to be very helpful to meet the many challenges that analysts in all fields are facing today.

Weinheim, Germany
Camberley, United Kingdom
January 2001

Helmut Günzler
Alex Williams

Contents

Volume I

5. Sampling ... 71

6. Sample Preparation for Trace Analysis ... 77

7. Trace Analysis ... 109

8. Radionuclides in Analytical Chemistry ... 127

9. Enzyme and Immunoassays ... 147

10. Basic Principles of Chromatography ... 173

Symbols and Units

Symbols and units agree with SI standards. The following list gives the most important symbols used in the handbook. Articles with many specific units and symbols have a similar list as front matter.

Symbol	Unit	Physical Quantity
a_B		activity of substance B
A_r		relative atomic mass (atomic weight)
A	m^2	area
c_B	mol/m^3, mol/L (M)	concentration of substance B
C	C/V	electric capacity
c_p, c_v	$J\ kg^{-1}K^{-1}$	specific heat capacity
d	cm, m	diameter
d		relative density (ϱ/ϱ_{water})
D	m^2/s	diffusion coefficient
D	Gy (= J/kg)	absorbed dose
e	C	elementary charge
E	J	energy
E	V/m	electric field strength
E	V	electromotive force
E_A	J	activation energy
f		activity coefficient
F	C/mol	Faraday constant
F	N	force
g	m/s^2	acceleration due to gravity
G	J	Gibbs free energy
h	m	height
h	$W \cdot s^2$	Planck constant
H	J	enthalpy
I	A	electric current
I	cd	luminous intensity
k	(variable)	rate constant of a chemical reaction
k	J/K	Boltzmann constant
K	(variable)	equilibrium constant
l	m	length
m	g, kg, t	mass
M_r		relative molecular mass (molecular weight)
n_D^{20}		refractive index (sodium D-line, 20 °C)
n	mol	amount of substance
N_A	mol^{-1}	Avogadro constant ($6.023 \times 10^{23}\ mol^{-1}$)
p	Pa, bar *	pressure
Q	J	quantity of heat
r	m	radius
R	$J\ K^{-1}mol^{-1}$	gas constant
R	Ω	electric resistance
S	J/K	entropy
t	s, min, h, d, month, a	time
t	°C	temperature
T	K	absolute temperature
u	m/s	velocity
U	V	electric potential
U	J	internal energy
V	m^3, L, mL	volume
w		mass fraction
W	J	work
x_B		mole fraction of substance B
Z		proton number, atomic number
α		cubic expansion coefficient

Symbol	Unit	Physical Quantity
α	W $m^{-2}K^{-1}$	heat-transfer coefficient (heat-transfer number)
α		degree of dissociation of electrolyte
$[\alpha]$	10^{-2}deg cm^2g^{-1}	specific rotation
η	Pa · s	dynamic viscosity
θ	°C	temperature
κ		c_p/c_v
λ	W $m^{-1}K^{-1}$	thermal conductivity
λ	nm, m	wavelength
μ		chemical potential
ν	Hz, s^{-1}	frequency
ν	m^2/s	kinematic viscosity (η/ϱ)
π	Pa	osmotic pressure
ϱ	g/cm^3	density
σ	N/m	surface tension
τ	Pa (N/m^2)	shear stress
φ		volume fraction
χ	Pa^{-1} (m^2/N)	compressibility

* The official unit of pressure is the pascal (Pa).

1. Analytical Chemistry: Purpose and Procedures

HANS KELKER formerly Hoechst AG, Frankfurt, Federal Republic of Germany

GÜNTHER TÖLG, formerly Institut für Spektrochemie und Angewandte Spektroskopie, Dortmund, Federal Republic of Germany

HELMUT GÜNZLER, Weinheim, Federal Republic of Germany

ALEX WILLIAMS, Mytchett, Camberley, UK.

1.1. The Evolution of Analytical Chemistry

"Analytical chemistry" (more simply: *analysis*) is understood today as encompassing any examination of chemical material with the goal of eliciting information regarding its constituents: their *character* (form, quality, or pattern of chemical bonding), *quantity* (concentration, content), *distribution* (homogeneity, but also distribution with respect to internal and external boundary surfaces), and *structure* (spatial arrangement of atoms or molecules). This goal is pursued using an appropriate combination of chemical, physical, and biological methods [1] – [6]. From a strategic standpoint the challenge is to solve the analytical problem in question as completely and reliably as possible with the available methods, and then to interpret the results correctly. Sometimes it becomes apparent that none of the methods at hand are in fact suitable, in which case it is the methods themselves that must be improved, perhaps the most important rationale for intensive basic research directed toward the increased effectiveness of problem-oriented analysis in the future.

More comprehensive contemporary definitions of analytical chemistry have been proposed [7], [8], underscoring above all the complexity of the discipline — which the authors of this introduction were also forced to confront.

Consistent with its close historical ties to chemical synthesis, modern analysis is still firmly embedded within the broader framework of chemistry in general. This is inevitably the case, because systematic analysis depends absolutely upon a solid, factual knowledge of matter. This point is as valid now as it was in 1862 when C. R. FRESENIUS stated in his classic *Introduction to Qualitative Chemical Analysis* [9]: "Chemical analysis is based directly on general chemistry, and it cannot be practiced without a knowledge thereof. At the same time it must be regarded as one of the fundamental pillars upon which the entire scientific edifice rests; for analysis is of almost equal importance with respect to all the branches of chemistry, the theoretical as well as the applied, and its usefulness to doctors, pharmacists, mineralogists, enlightened farmers, technologists, and others requires no discussion."

The *tools* of modern analysis are nevertheless based largely on physical principles. Mathematical techniques related to information theory, systems theory, and chemometrics are also making increasingly important inroads. It would in fact no longer be presumptuous to go so far as to describe "analytical science" as an independent discipline in its own right.

The pathway leading to the present exalted place of analysis within the hierarchy of chemistry specifically and the natural sciences generally has not always been a straight one, however. Indeed, from earliest times until well into the eighteenth century the very concept of "analysis" was purely implicit, representing only one aspect of the work of the alchemists and various practitioners of the healing arts (iatrochemists). Some more tangible objective always served as the driving force in an investigation, and "to analyze" was almost synonymous with the broader aim: a quest for precious metals, a desire to establish the content of something in a particular matrix, or a demonstration of pharmacological activity. Only after the time of LAVOISIER and with the emergence of a separate chemical science—a science largely divorced from external goals—is one able to discern what would today be regarded as typical "analytical" activity. The term "analysis" appears explicitly for the first time around the turn of the nineteenth century in the title of the book, *Handbuch zur chemischen Analyse der Mineralkörper* ("Handbook for the Chemical Analysis of Minerals") by W. A. LAMPADIUS. Further information regarding the history of analysis is available from the monograph by SZABADVARY [10].

Many of the greatest discoveries in chemistry could fairly be described as classic examples of successful analyses, including the discovery of oxygen, the halogens, and several other elements. Well into the nineteenth century, discovering a new chemical element was regarded as the highest and most prestigious achievement possible for an academic chemist, as documented, for example, by desperate attempts to gain further insight into the "rare earths," or to detect the elusive (but accurately predicted) homologues of lanthanum and cerium. MOSANDER in fact devoted his entire life to the latter search.

C. REMIGIUS FRESENIUS once again deserves credit for noting, toward the middle of the nineteenth century, that new analytical techniques invariably lead to fresh sets of discoveries. Whereas the element germanium was found on the basis of "classical" methods (CLEMENS WINKLER, 1886), FRESENIUS' observation clearly applies to the discovery of the alkali metals rubidium and cesium (by ROBERT W. BUNSEN after he and G. R. KIRCHHOFF first developed emission spectroscopy in 1861). Other relevant examples include the discoveries of radium and polonium (by Madame CURIE), hafnium (HEVESY and COSTER, 1922), and rhenium (I. TACKE and W. NODDACK, 1925), all with the aid of newly introduced X-ray spectrometric techniques. This is also an appropriate point to mention the discovery of nuclear fission by OTTO HAHN and FRITZ STRASSMANN (1938), another accomplishment with strongly analytical characteristics [10].

ROBERT BUNSEN is rightfully acknowledged as the harbinger of modern analysis, but much of the discipline's distinctive scientific character was provided by WILHELM OSTWALD [11] building on the activities of J. H. VAN'T HOFF and WALTHER NERNST.

Analytical chemistry in these early decades was often accorded the secondary status of a faithful servant, but even the few examples cited here demonstrate quite convincingly that it also pursued its own unique set of principles—and for its own sake, with a strictly scientific orientation. The principles themselves were shaped by BERZELIUS and WÖHLER; experiment rather than theoretical speculation was the starting point and source of inspiration in this era characterized largely by chemical reactions. Readers of the present essay should in fact take the time to examine the third edition of *Ullmann's* [12] and discover there what the expression "analytical procedure" actually meant even as late as the end of World War II. There can be no mistaking the fact that "purely chemical" methods were still domi-

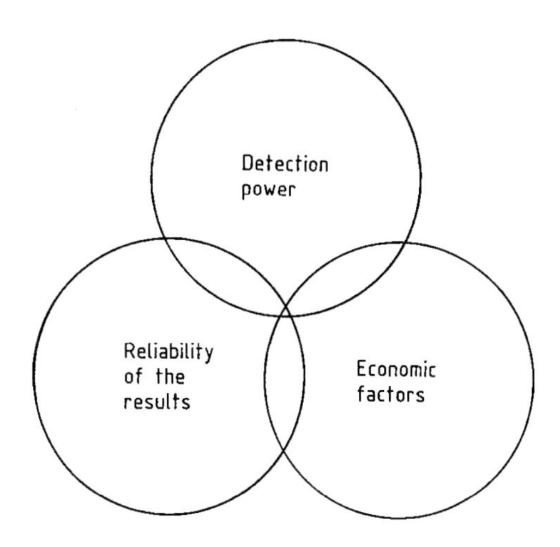

Figure 1. Qualitative criteria for use in evaluating analytical procedures

nant, and that the synthetic process constituted the model, particularly in the field of organic analysis.

Analytical chemistry has been responsible for many important contributions to our basic understanding of matter (e.g., the existence of the various elements, gas theory, stoichiometry, atomic theory, the law of mass action, nuclear fission, etc.), but the growth and development of a separate chemical industry ushered in a phase during which the scientific aspects of analysis suffered serious decline. The demand for analytical services shifted markedly in the direction of routine quality control, particularly with respect to synthetic organic products; indeed, significant resources were invested in the effort to dismember, resolve, and decompose synthetic substances into their simpler constituents (e.g., the chemical elements) — in strict conformity with the original meaning of the word "analysis" (*αναλυσια*, *resolution*). For many years organic elementary analysis was virtually the only analytical approach available for characterizing synthetic organic reaction products. The denigration suffered by analysis at that time relative to synthesis (and production) continues to exert a negative influence even today on the university training of analytical chemists.

Elemental analysis in certain other quarters enjoyed a climate much more congenial to further development, especially in the metalworking industry and geochemistry. The indispensable contributions of analysis were recognized here much earlier, particularly with respect to optimizing product characteristics (e.g., of steels and other alloys), and to providing detailed insight into the composition of the Earth's crust to facilitate the extraction of valuable raw materials. Geochemistry and the steel industry were particularly receptive to BUNSEN'S new methods of spectral analysis, for example, which in turn provided a powerful stimulus for the development of other modern instrumental techniques. These techniques encouraged the exploitation of new and innovative technologies, first in the fields of semiconductors and ultrapure metals, then optical fibers and superconductors, and, most recently, in high-temperature and functional ceramics. Extraordinarily stringent demands were imposed upon the various analytical methods with respect to detection limits, extending to the outermost limits what was possible, especially in the attempt to characterize impurities responsible for altering the properties of particular materials. At the same time, the information acquired was expected to reflect the highest possible standards of reliability — and to be available at an affordable price. These three fundamental quality criteria are in fact closely interrelated, as indicated in Figure 1.

The increasing effectiveness of analytical techniques in general led ultimately to progress in the area of organic materials as well, especially with the rapid development of chromatographic and molecular spectroscopic methods. At the same time it also became necessary to acknowledge that technological advances inevitably bring with them new safety and health risks. For this reason analysis today plays an essential role not only in supporting technological progress but also in detection and minimization of the associated risks.

Just as FRESENIUS predicted, analysis has advanced rapidly toward becoming a science in its own right, with interdisciplinary appeal and subject to intense interest extending far beyond the bounds of chemistry itself: to the geological and materials sciences, the biosciences, medicine, environmental research, criminology — even research into the history of civilization, to mention only a few of the most important areas of application. The chemical industry today is the source of only a relatively small fraction of the samples subject to analysis. Rocks, soils, water, air, and biological matrices, not to mention mankind itself and a wide array of consumer goods, together with raw materials and sources of energy constitute the broad spectrum of analytical samples in the modern era (Fig. 2).

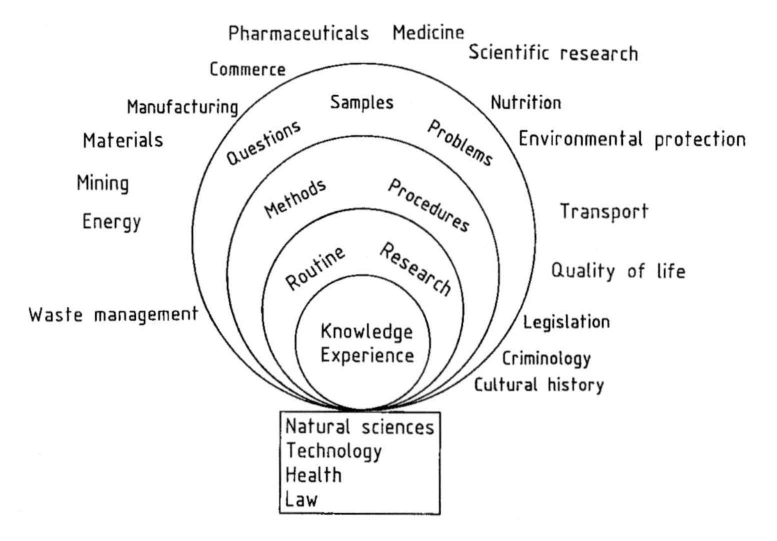

Figure 2. Overall task structure associated with analytical chemistry

Given this diversity of appeal the question has frequently been raised as to whether analysis really is an independent discipline, or if it should not instead be regarded simply as a service activity. The question is of course unrealistic, because analysis by its very nature is clearly both. Equally clear is the crucial importance of analysis to modern society. While the service function is undoubtedly more widely appreciated than other activities characterized by a "strictly scientific" focus, the latter also have an indispensable part to play in future progress.

The diversity characterizing the beneficiaries of analysis has actually remained fairly constant in recent decades, though immediate priorities have undergone a steady shift, particularly during the last 20 years with respect to ecology. Such a "paradigm shift" (THOMAS KUHN), marked by profound changes over time in both motivation and methodology, can occasionally assume revolutionary proportions. It remains an open question whether external change induces analysts to adapt and further develop their methodologies, or if the methodology itself provides the driving force. Here as elsewhere, however, there can be little doubt that "necessity is the mother of invention," capable of mobilizing forces and resources to an extent unimaginable in the absence of pressing problems.

Change also provides an incentive for deeper reflection: should we perhaps reformulate our understanding of the overall significance of analysis, lift it out of its customary chemico-physical framework and broaden its scope to include, for example, KANT'S "analytical judgments," or even psychoanalysis? Some would undoubtedly dismiss the questions as pointless or exaggerated, but from the perspective of the theory of learning they nevertheless provoke a considerable amount of interest and fascination [13], [14].

1.2. The Functional Organization of Analytical Chemistry

Attempting to summarize analytical chemistry in a single comprehensive schematic diagram is a major challenge, one that can only be addressed in an approximate way, and only after considerable simplification (Fig. 2) [5]. The fundamentals supporting the analysis must ultimately be the individual analyst's own store of knowledge, including the basic principles and laws of science and mathematics, together with the scope—and limitations—of existing analytical methods and procedures. Indispensable prerequisites to the successful resolution of an analytical problem include experience, a certain amount of intuition, and thorough acquaintance with a wide variety of modern analytical techniques. Familiarity with the extensive technical literature is also important (including the sources cited at the end of this article), an area in which modern systems of documentation can be of considerable assistance. For example, an astonishing level of perfection can almost be taken for granted with respect to computer-based systems for locating spectra. Another essential component of the analyst's information base is knowledge regarding the source of each analytical sample — whether it comes from industry, the environment, or from medicine. After all, only the analyst is in a position to provide an overview of the analytical data themselves when the time comes for critical interpretation of the experimental results.

Immediately adjacent to "knowledge" in the functional diagram characterizing analytical chemistry (Fig. 2) is a region occupied by two parallel lines of endeavor: routine analysis on one hand, and research and development on the other, with the latter directed toward new methods and procedures. Both are subject to initiatives and incentives from outside, including other branches of science, medicine, regulatory agencies, commerce, and industry, all of which encourage and foster innovative developments within analysis itself.

Figure 2 also underscores the fact that an analyst's primary activities are of a problem-oriented nature, determined largely by the needs of others. The problems themselves, represented here by the outermost circle, might originate almost anywhere within the material world. Analysis can even play a significant role in the very definition of a scientific investigation. Consider the case of archaeology, for example, a considerable part of which is now "archaeometry," simply a specialized type of analysis.

With respect to the development of new products — such as materials, semiconductors, pharmaceuticals, crop protection agents, or surfactants — analysis plays a companion role at every stage in the progression from research laboratory to market. Studies related to physiological and ecological behavior demand comprehensive analytical efforts as well as intimate knowledge of the materials in question.

1.3. Analysis Today

Figure 3 provides a representative sample of methods to be found in the arsenal of the modern analyst. The figure also highlights the rapid pace of developments in analytical chemistry during the twentieth century [15]. Continued success in meeting present and future analytical challenges involves more than simply the tools, however, most of which have already been perfected to the point of commercialization. Appropriate strategies are required as well, just as a hammer, a chisel, and a block of marble will not suffice to produce a sculpture. Analytical strategies are at least as important as the methods, and the strategies must themselves be devised by qualified analysts, because every complex analytical problem demands its own unique strategic approach.

It is this context that establishes the urgent need for reactivating as quickly as possible the long-neglected training of qualified analysts. New analytical curricula must also be devised in which special emphasis is placed on the close symbiotic relationship in modern analysis between chemistry and physics [6].

Figure 4 depicts in a generalized way the multileveled complex of pathways constituting a typical analytical process and linking a particular problem with its solution. From the diagram it becomes immediately apparent that the "analytical measurement," which is the focal point of most modern physical methods, in fact represents only a very small part of the whole, despite the fact that the treatise to which this essay serves as a preface focuses almost exclusively on the principles of instrumental methods and their limitations.

Physical methods clearly occupy the spotlight at the moment, but *chemical* methods of analysis are just as indispensable today as in the past. Especially when combined with physical methods, chemical techniques frequently represent the only means to achieving a desired end. This is generally the case in extreme trace analysis [16], for example, where attaining maximum sensitivity and reliability usually requires that the element or compound of interest first be isolated from an accompanying matrix and then concentrated within the smallest possible target area or solution volume prior to the physical excitation that leads ultimately to an analytical signal. Combination approaches involving both chemical and physical methods are today commonly referred to as *multistep procedures* (see Section 1.6.11), where some chemical step (e.g., digestion, or enrichment) often precedes an instrumental measurement, or an analysis is facilitated by preliminary chromatographic separation. Chromatographic separation in turn sometimes requires some type of prior chemical transformation [17], as in the gas-chromatographic separation of organic acids, which is usually preceded by esterification.

pre-1850	1850-1900	1900-1925	1925-1950	1950-1975	1975-1992 (selection)
Titrimetry	Electrogravimetry	Raman spectroscopy	γ-Spectroscopy	Neutron spectroscopy	Atomic force microscopy
Gravimetry	Coulometry	X-Ray emission spectroscopy	Electron spin resonance	Laser spectroscopy	Scanning tunnel microscopy
Gas-volumetry	Optical emission spectroscopy (OES)	X-Ray diffraction	NMR	Electron tunneling spectroscopy	Atom probe
		Mass spectroscopy	Gas chromatography	HPLC	Analytical electron microscopy
		IR spectroscopy	Paper chromatography	Candoluminescence	FT spectroscopy
		Isotope-dilution analysis	Distribution chromatography	MECA spectroscopy	ICR mass spectroscopy
		Radiochemical analysis	High-frequency titration	RHEED	TXRFA
		Adsorption chromatography	Electron diffraction	LEED	IR microscopy
		Conductometry	Activation analysis	Desorptiometry	Scanning auger electron spectroscopy
			Electrophoresis	SIMS	Dynamic SIMS
			Polarography	Mößbauer spectroscopy	CARS
			RBS	ESMA	Photoacoustic spectroscopy
				NQR	SERS
				EELS	FANES
				Auger electron spectroscopy	MONES
				Photoelectron spectroscopy	Appearance-potential spectroscopy
				Ionography	IBLE
				FIM	FAB-MS
				FEM	GD-AES
				EDX	GD-MS
				Neutron diffraction	Ion chromatography
				β-Backscattering	Ion neutralization spectroscopy
				Atomic absorption spectroscopy	EXAFS
				Reflection IRS	SNMS
				Microwave spectroscopy	UPS, XPS
				PIXE	Capillary gas chromatography
					Gel chromatography
					Supercritical-fluid chromatography
					HPTLC
					Scanning DC
					Plasma spectroscopy
					ICP-OES, MIP-OES
					ETA-AAS
					2D-NMR
					Multinuclear NMR
					Time-resolution spectroscopy
					GC-MS
					MS-MS-(MS)
					AAS-GC
					ICP-MS
					ICP-GC
					Laser ICP-MS
					Thermospray MS, Electrospray MS
					HPLC-MS
					C-13 NMR
					Charged-particle activation analysis
					Raman microprobe
					RP chromatography
					Solid phase AAS
					Headspace GC
					PIGE
					Static TOFSIMS
					LAMMA

Figure 3. Chronological summary of the arsenal of experimental methods currently available to analytical chemists; based on [15]

The terms "preanalysis" and "postanalysis" have been coined for characterizing steps that precede or follow a "true" analytical operation. Unfortunately, classification in this way tends to denigrate the importance of an operation like sampling or the evaluation of a set of final results, suggesting that these are secondary and relatively peripheral activities—reason enough for exercising considerable caution in use of the terms.

There can be no justification whatsoever for dismissing the importance of chemical reactions in analysis, as "superprogressive" instrumental analysts occasionally tend to do, treating chemical methods as relics of an outmoded past. Chemical reactions still have a crucial part to play in many operations: sometimes as useful adjuncts, but often enough at the very heart of the determination. It is worth recalling in this context that gravimetry—together with the volumetric methods to which it gave birth—remains virtually the only viable approach to direct and reliable absolute determination (i.e., to *calibration-free analysis*). Such anal-

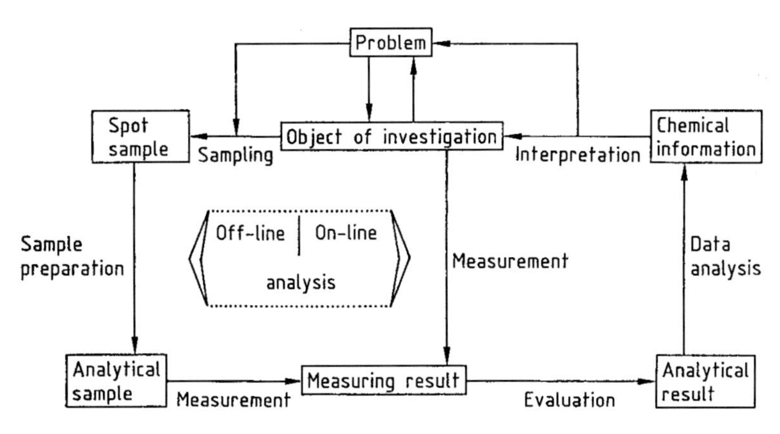

Figure 4. Schematic diagram of the analytical process; based on [15]

yses rely on "stoichiometric factors," which were painstakingly compiled over the course of decades in conjunction with the equally arduous and prolonged quest for an exact set of atomic masses.

Most physical methods, especially those associated with spectroscopy, lead only to *relative* information acquired through a comparison of two signals. This in turn presupposes a procedure involving a *calibration standard*, or reference sample of known composition. The only exceptions to this generalization—at least theoretically—are instrumental activation analysis (which involves the counting of activated atomic nuclei), isotope dilution (especially IDMS—isotope dilution mass spectrometry), and coulometry (assuming the strict validity of Faraday's Law). In view of quality assurance (see → Quality Assurance in Instrumentation), the named methods, jointly with gravimetry, volumetric analysis, and thermoanalysis, were recently designated as *primary methods of measurement* [18]–[21]. They play an important role in achieving traceable results in chemical measurements.

Some may feel that the foregoing observations direct excessive attention to the virtues of classical analytical chemistry. If so, the justification is a continuing need to emphasize the fact that optimal results are achieved when there is a close coupling between chemical and physical methods, and this despite antagonisms that persist between champions devoted to one approach or the other. Even today, classical principles— appropriately adapted—often constitute the most reliable guide.

1.4. Computers

A few remarks are necessary at this point on the subject of *electronic data processing* and the vital supportive role computers now play in analysis.

Developments in this area began with the central mainframe computer, to which a wide variety of isolated analytical devices could be connected. In recent years the trend has shifted strongly toward preliminary data processing via a minicomputer located directly at the site of data collection, followed in some cases by network transfer of the resulting information to a central computing facility. Often, however, the central computer is dispensed with entirely, with all data evaluation occurring on the spot. The powerful impact of electronic data processing on modern analysis dictates that it be addressed elsewhere in the present treatise in greater detail (→ Chemometrics).

The benefit of computers in modern analysis has been clearly established for some time. Computers now provide routine management and control support in a wide variety of analytical operations and procedures, and they are an almost indispensable element in data interpretation, processing, and documentation. Indeed, the lofty goals of "good laboratory practice" (GLP) would probably be beyond reach were it not for the assistance of computers. Computers also have a key role in such wide-ranging activities as automated sample introduction and the control of calibration steps (*robotics*). Process-independent tasks closely related to the ongoing work of a laboratory have long been delegated to computers, including the storage, retrieval, and management of data.

Nevertheless, the claim that we have entered an age of "computer-based analytical chemistry" (COBAC) is inappropriate and overly optimistic; "computer-aided" analysis would be a more satisfactory description, and one more consistent with terminology adopted in other disciplines. "Artificial intelligence," so-called expert systems [22], neural networks, and genetic algorithms will undoubtedly be increasingly important in the analytical chemistry of the future, but in most cases probably in the context of relatively complex routine investigations supported by extensive previous experience. It is unlikely that such methods will prove optimal even in the long term with respect to analytical research in uncharted waters, especially if results are required near the limit of detectability.

1.5. Analytical Tasks and Structures

1.5.1. Formulating the Analytical Problem

Generally speaking, problem-oriented analytical tasks can best be defined with reference to criteria most easily expressed as questions:

1) How has the problem at hand already been stated? Is the problem as stated truly relevant? If so, what is the maximum expenditure that can be justified for its solution, considering both material and economic resources? (Note that not every problem warrants the pursuit of an optimal analytical solution!)
2) What type and size of sample is available? What content range is predicted with respect to the analyte, and what mass of sample would be required to produce an answer?
3) What analytical strategy (including choice of a particular method) is most appropriate within the context set by considerations (1) and (2)?
4) Will critical assessment of the analytical results be possible, with evaluation of an uncertainty budget aiming to determine an *expanded uncertainty* of the analytical result [29], [30]? (see Section 1.6.2)

Ensuring the correctness of a set of results is extremely important, because nothing is more wasteful than acquiring a wrong answer, especially when account is taken of the subsequent interpretation and application of analytical data with respect to matters of safety, health, and the environment. The ultimate validity of an analytical result can be placed in serious jeopardy as early as the sampling stage, since inappropriate sampling can be a very significant source of error.

Such mathematical tools as statistical tests and uncertainty evaluation are prerequisite to the practical application of an analytical result. In any situation involving verification of compliance with conventions, agreements, regulations, or laws, analysis is expected to provide the meaningful and objective criteria required for assessing the material facts. This means that observed analytical values must be supplemented with quality criteria applicable to the analytical procedure itself, such as the limit of detection, limit of determination, standard deviation, and measurement uncertainty.

1.5.2. Research and Application

Two major branches of analytical chemistry can be distinguished by the types of challenges they address. The first is the problem-oriented service sector, or *routine analysis*. Here one is usually in a position to rely on existing and proven methods and procedures, though some adaptation may be required to accommodate a method to the particular task at hand.

The second area, *basic analytical research*, is the key to resolving an increasingly complex set of problems today and in the future—problems not subject to attack with tools that are currently available, or amenable only to unsatisfactory solutions (with appropriate regard for economic factors). This underscores the high degree of innovative scientific character associated with analysis as a discipline, innovation that often approaches revolutionary proportions. It is perfectly possible for epochal developments to emerge from basic principles that are themselves already well established. A striking example is provided by the path leading from organic elementary analysis as first introduced by JUSTUS LIEBIG, starting with rather large samples, via the work of F. EMICH and F. PREGL, and culminating in today's highly perfected micro techniques, a path that runs parallel to the development of the analytical balance.

It is also interesting to consider in this context the source of some of today's most important innovations, which increasingly result from a close symbiotic relationship between university research centers on one hand, and commercial instrument manufacturers on the other (where the latter often

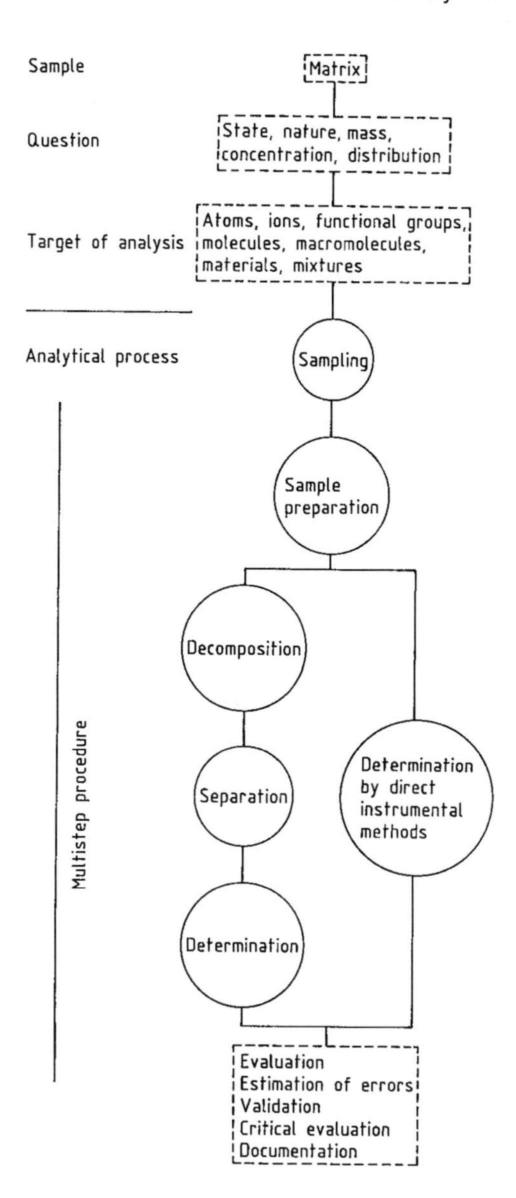

Figure 5. Strategic organization of the analytical process

have access to extensive in-house research facilities of their own, and may be in a position to introduce important independent initiatives). The reason for the collaborative trend is obvious: continued progress has been accompanied by a disproportionate increase in costs, and the resulting burden can no longer be borne by universities alone. Collaboration between industry and higher education is certainly to be welcomed, but not to the point that technical shortcomings still evident at the conclusion of a joint commercialization venture are suppressed or trivialized in the interest of profit, as has unfortunately occurred on more than one occasion.

1.5.3. An Organogram

The two complementary branches of analytical chemistry rely on a common foundation of structure and content, illustrated in the "organogram" of Figure 5.

Starting with an analytical sample (the matrix), and proceeding via the formulation of a specific question regarding the state, nature, mass, concentration, or distribution of that sample, as well as a definition (or at least partial definition) of the true target of the analysis (atoms, ions, molecules, etc.), two different paths might in principle be followed in pursuit of the desired objective. Both commence with the extremely critical steps of sampling (→ Sampling) and sample preparation (→ Sample Preparation for Trace Analysis), which must again be recognized as potential sources of significant error. Under certain conditions it may then be possible to embark immediately on qualitative and/or quantitative analysis of the relevant target(s) through direct application of a physical method in the form of an "instrumental" analysis (e.g., a spectroscopic determination following excitation of the sample with photons, electrons, other charged particles, or neutrons). Such instrumental methods can be subdivided into *simultaneous* and *sequential* methods, according to whether several analytes would be determined at the same time (as in the case of multichannel optical emission spectrometry) or one after another (with the help of a monochromator).

Immediate application of a direct instrumental method (e.g., atomic spectroscopy in one of its many variants) usually represents the most economical approach to elemental analysis provided the procedure in question is essentially unaffected by the sample matrix, or if one has access to appropriate reference materials similar in composition to the substance under investigation [23] – [26]. The alternative is an analytical method consisting of multiple operations separated by either space or time, often referred to as a *multistep procedure*, as indicated on the left in Figure 5. The possibility of combining two or more discrete techniques adds a whole new dimension to chemical analysis, although there is a long tradition of observing a formal distinction between "separation" and true "determination." Separation in this sense has often been understood to include chemical reactions undertaken for the purpose of preparing a new, more readily separable compound — as a solid phase, for example — together with the actual separation step itself (e.g., filtration or extraction), although the term is sometimes interpreted more literally and limited to the latter activity alone. Cases also come to mind in which individual "separation" and "determination" steps cannot be clearly differentiated (e.g., in chromatography).

A separation step might be preceded by some preliminary treatment of the sample, such as a prechromatographic operation [17], and this might also warrant special attention. *Trace enrichment* is typical of the fields in which prechromatographic techniques have much to offer.

Particularly in trace analysis, and in the absence of standard samples for calibration purposes, there still is no satisfactory alternative to relying at least initially on "wet-chemical" multistep procedures. This entails a detour consisting of sample decomposition with subsequent separation and enrichment of the analyte(s) of interest relative to interfering matrix constituents. A suitable form of the analyte(s) is then subjected to the actual determination step, which may ultimately involve one or more of the direct instrumental methods of analysis.

Multistep procedures are even more indispensable in the analysis of organic substances, where a chromatographic separation is often closely coupled with the actual method of determination, such as IR or mass spectrometry. Separations based on chemical reactions designed to generate new phases for subsequent mechanical isolation (e.g., precipitation, liquid–liquid partition) have also not been completely supplanted in elemental and molecular analysis.

Recent progress in analytical chemistry is marked by dramatic developments in two areas: (1) an enormous increase in the number of available analytical methods and opportunities for applying them in combination, and (2) new approaches to mathematical evaluation (chemometrics). As a result, most matrices are now subject to characterization with respect to their components both in terms of the bulk sample and at such internal and external phase interfaces as grain boundaries and surfaces — extending in some cases even into the extreme trace range. As in the past, the safest course of action entails separating the component(s) of interest in weighable form, or taking an indirect route via gravimetry or titrimetry as a way of establishing a state indicative of complete reaction.

Many modern methods of separation and determination result in the generation of some type of "signal", whereby an appropriate sensor or detector is expected to react in response to concentration or mass flow — perhaps as a function of time, and at least ideally in a linear fashion throughout the range of practical interest. Devices such as photocells, secondary electron multipliers, Golay cells, thermal conductivity cells, thermocouples, and flame ionization detectors convey information related to concentration changes. This information takes the form of an electrical signal (either a voltage or a current), which is fed to some type of measuring system, preferably at a level such that it requires no amplification. Sensor development is an especially timely subject, warranting extensive discussion elsewhere (→ Chemical and Biochemical Sensors).

Further processing of an analytical signal may have any of several objectives, including:

1) Incorporation of a "calibration function" that permits direct output of a concentration value
2) Establishing feedback control as one way of managing the data-acquisition process (e.g., in a *process computer*)
3) Recasting the primary signal to reflect more clearly the true analytical objective (e.g., "on-line" Fourier transformation, a common practice now in both IR and NMR spectroscopy)

1.5.4. Physical Organization of the Analytical Laboratory

Depending on the situation, assignments with which a particular analytical team is confronted might be linked organizationally and physically with the source of the samples in various ways. The following can be regarded as limiting cases:

1) Direct physical integration of the analytical function into the production or organization process, where "on-line" analysis represents the extreme
2) Strict physical separation of the sample source from subsequent analytical activities

It would be pointless to express a general preference for one arrangement or the other, but a few relevant considerations are worth examining.

Analysis "on the spot" eliminates the complications of sample transport, and it offers the potential for saving a considerable amount of time. This in turn facilitates rapid processing, an especially important factor when process control is dependent upon analytical data (e.g., in blast furnace operation). Analysis of this type is always associated with a very specific objective, usually involving a single analytical method and a single specialized type of instrumentation, and its economic viability must be critically evaluated on a case-by-case basis. Costs related to acquisition, amortization, and the repair of expensive equipment must all be considered, as must demands for personnel — who are likely to require special skills and training.

The obvious alternative to integrated analysis is a physically separate, central analytical facility like that traditionally maintained by a large chemical corporation. A laboratory of this sort typically reflects an interest in analysis in the most general sense, with provisions for the utilization of as many as possible — preferably all — of the conventional and fashionable analytical methods in anticipation of a very broad spectrum of assignments. Routine analysis in such a setting can conveniently be combined with the innovative development of new methods and procedures, thereby assuring optimal utilization of equipment that is becoming increasingly sophisticated and expensive. Considering the rapid pace of developments in major instrumentation, and the risks entailed in implementing modern approaches to automation, data processing, and laboratory operations generally, it often becomes apparent that centralization is the only economically justifiable course of action.

Similar considerations underscore the critical importance of continuing education for laboratory personnel, who must of necessity adapt to any changes in hardware. This perspective also sheds additional light on the independent scientific character of analysis, both in the industrial sphere and in academia. The problems encountered are essentially scientific in nature, the questions are fundamental, and the tools engaged in their solution reflect a complex development process that is technically demanding in the extreme.

1.5.5. The Target of Analysis

One of the fields in Figure 5 (the diagram singling out various stages in an analytical procedure) bears the label "Target of Analysis," and its structure deserves closer scrutiny. Until relatively recently the "target of an analysis" was always a list of constituent elements, together with the corresponding overall composition. An arduous trail of analytical research leads from the dualistic theory of matter (BERZELIUS and his contemporaries) to an understanding of the fine structure and conformation of molecules in the solid and liquid (dissolved) states, culminating in direct proof of the existence of atoms. In planning an analysis today it is almost self-evident that the first question to be addressed concerns the particular level in the hierarchically ordered concept "target" at which the investigation is to be conducted.

One important property of this hierarchy is that every higher level of order implies a specific set of properties at each of the lower levels. The reverse is not true, however, since the lower stages are independent and do not presuppose any higher degree of structure. Thus, in order to conduct a molecular structure determination on an organic substrate it is first necessary to ascertain the corresponding elemental composition. Needless to say, analysis at any level in the object hierarchy depends upon the availability of suitable procedures.

Atoms. As shown schematically in Figure 6, the hierarchy of targets begins with *atoms* (and the various *isotopes* of the elements) as the smallest fundamental units with analytical chemical relevance. This is already a rather profound observation in the case of certain geochemical questions, for example, since it is well known that the isotope ratios for such isotopically mixed elements as sulfur or uranium are by no means constant, and an isotope ratio (of chlorine, perhaps) can also be a useful or even indispensable parameter in the practice of mass spectrometry (→ Mass Spectrometry).

Molecules. *Ions* and *functional groups* have been assigned to a level of their own, located between that of the atoms and that of the *molecules*, which represents the next formal stage in our hierarchical scheme. Chemical reactions long constituted the sole basis for analysis at the molecular level, and together with the methods of atomic and molecular spectroscopy they continue to serve as the foundation of modern analysis.

Macromolecular Species. The transition from high molecular mass substances (*macromolecules*) to the highly ordered macroscopic crystalline

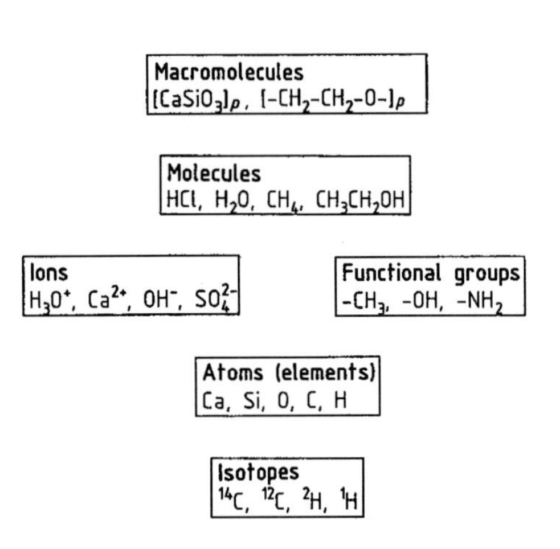

Figure 6. Hierarchical ordering of the various possible targets of analysis

range is somewhat indistinct, and it is also here that the concept of the molecule loses its validity (ignoring for the moment the notion of molecular crystals). Qualitatively distinct properties emerge, and these can in turn be exploited for purposes of analytical characterization: external form (crystalline habit), symmetry, ionic lattice structure (as well as molecular lattices and the structures of mixed forms), mechanical and electrical characteristics, specific gravity, phenomena associated with phase formation and conversion, melting point, etc. In the course of analysis, a solid material is degraded to a *matrix*, which often turns out to be a troublesome source of interference. At this stage the solid itself is regarded as homogeneous with respect to its composition, although the structure, surface properties, and microdistribution in terms of atoms or other components might also be subjected to analysis.

Heterogeneous multiphase systems (*mixtures*) lay claim to a special status, constituting the highest level in the hierarchical classification system. The prototype for this category is a rock: granite, for example, which is composed of three homogeneous substances. Precise identification at the mixture level typically requires the use of special techniques drawn from other disciplines, such as mineralogy or petrography.

Analysis of the Various Targets. A formal distinction can be made today between the detection and determination of free atoms, molecules, or ions on the one hand (via mass or optical spectroscopy in the gaseous state), and the analysis of molecules or ions in solution on the other (using such techniques as UV–VIS spectrophotometry, electrochemistry, and NMR or ESR spectroscopy). Mention should also be made of such "colligative" methods as ebullioscopy, cryoscopy, and gas-density determination. The overall importance of colligative methods has declined somewhat, but they continue to provide useful information regarding molecular mass and various dissociation phenomena. Their principal field of application today is the analysis of high polymers (macromolecules), the level immediately above molecules in our hierarchy of targets.

Polymeric molecules and ions have been assigned to a level of their own, primarily because of a unique set of methods that has been developed specifically for dealing with analytical targets in this size range. The most important are ultracentrifugation, certain types of liquid chromatography, light scattering, and—not to be overlooked—chemical approaches to the determination of end groups. Certain colligative methods are also very important here because of their extremely high sensitivity.

With respect to the optical analytical methods it could be claimed (cum grano salis) that *dispersive methods*—which today extend into the short-X-ray region—are increasingly being supplemented by *image-forming methods* (e.g., microscopy, polarizing microscopy, stereoscanning, electron microscopy, ultrasound microscopy). Image-forming methods appear to have reached a plateau (at least for the time being) with the quasimechanical/optical principle as manifested in scanning tunneling microscopy, which has succeeded for the first time in making atomic structures "visible." Methods for the study of surfaces [e.g., Auger spectrometry, X-ray photoelectron spectroscopy (XPS), and secondary ion mass spectroscopy (SIMS)] must also now be incorporated under the heading of general structural analysis (→ Surface Analysis).

Major advances have occurred in recent years in direct instrumental approaches in the *bulk analysis* of condensed matter, providing integrated insight into the various components comprising a complex sample; examples of such methods include X-ray fluorescence, atomic emission, atomic absorption, and atomic fluorescence spectrometry. Especially in optical atomic emission spectrometry and mass spectrometry, traditional sources of excitation such as arcs and sparks are increasingly giving way to alternative techniques: direct-current, high-frequency, and microwave plasmas; glow discharges; and lasers (→ Laser Analytical Spectroscopy).

Information regarding the qualitative and quantitative distribution of elements within individual phases and at phase interfaces (grain boundaries) is the primary goal of *microdistribution analysis*, in which special probe techniques involving electrons, ions, photons, and neutrons are used to excite the sample under investigation. In the organic realm, biological cells might be singled out as the prototypical analytical substrate. A fundamental distinction must of course be made here between "nondestructive" methods and methods that concentrate on the sample as a whole.

1.6. Definitions and Important Concepts

Terminology plays an important role in analysis, and several technical terms have in fact already appeared in preceding sections of this essay. Here we consider explicitly a select subset of these terms, mainly ones whose widespread usage is relatively recent, as well as a few that are often utilized incorrectly.

1.6.1. Sensitivity, Limit of Detection, and Detection Power

These three expressions tend to be used very loosely, even among analysts—despite the fact that each is subject to very precise definition [1].

The *sensitivity* E of an analytical method expresses the dependency of a measured response on the analytical value of primary interest. It is defined as the first derivative of the measurement function:

$$F(y)E = F'(y) = \mathrm{d}x/\mathrm{d}y$$

In the case of a linear calibration function of the type $x = a_x \cdot y + b_x$, the sensitivity is equal to the slope of the calibration line; i.e.,

$$E = \Delta_x/\Delta_y = a_x$$

The *limit of detection* of an individual analytical procedure is the lowest amount of an analyte in a sample which can be detected but not necessarily quantified as an exact value [27]. Expressed as a concentration c_L or a quantity q_L the *limit of detection* is derived from the smallest signal x_L which can be detected with reasonable certainty for a given analytical procedure. The value of x_L is given by the equation

$$x_L = x_{bl} + ks_{bl}$$

where x_{bl} is the mean of the blank measurements, s_{bl} is the standard deviation of the blank measurements, and k is a numerical factor chosen according to the level of confidence required. For many purposes, the limit of detection is taken to be $3\,s_{bl}$ or 3 x signal to noise ratio, but other values of k may be used if appropriate. For a more detailed discussion see section 7.4.3. The concept "limit of detection" is applicable only to a particular analytical procedure, one that can be precisely defined with respect to all its parameters, whereas the *detection power* is a crude estimate associated with an idealized analysis, in which external interfering factors are largely ignored. This term is therefore reserved for characterizing an analytical *principle*.

The demand for analytical procedures with ever-increasing detection power is especially acute in the context of biologically relevant trace elements because of the ubiquitous concentrations of these materials in all natural matrices. It is the environmental concentrations that effectively establish lowest levels of the corresponding elements that are subject to determination in any biotic matrix. In most cases these levels are in the range >0.1 ng/g, and thus within a region that could today be regarded as practically accessible—at least in principle. Exceptions include the concentrations of certain elements in Antarctic or Arctic ice samples, for example, or samples from research involving ultrapure substances, where the goal is to prepare materials (e.g., metals) of the highest possible purity. Relevant impurity concentrations in cases such as these may fall in the pg/g range, often far below the environmental background concentrations of the elements in question.

Establishing the trace-element content for a bulk material typically requires a sample at the upper end of the milligram or lower end of the gram range, but increasing importance is being attached to the acquisition of detailed information regarding the distribution and bonding states of elements within specific microregions (see Sections 1.6.5, 1.6.7, 1.6.14).

Studies related to construction materials and other solids frequently rely on information regarding the distribution of elements at external and internal boundary surfaces, including grain boundaries. With biotic matrices the attention may be focused on tissue compartments, individual cells, or even cell membranes, whereas an environmental analysis might be concerned with individual aerosol particles. Investigations in areas such as these — especially projects involving in situ microdistribution analysis — serve to further promote the ongoing quest for ever lower absolute limits of detection. Assume, for example, one wished to determine the elemental distribution in various protein fractions from blood serum. Blood-serum background levels are in the low ng/g range, so an analytical method would be required with a detection power 10 – 100 times greater. Attempting to determine quantitatively an elemental concentration on the order of 1 ng/g in a sample weighing only 1 μg presupposes an absolute detection power in the femtogram range ($1\ \mathrm{fg} = 10_{-15}\ \mathrm{g}$).

An interesting question in this context is the minimum mass of an element that would theoretically be required for a successful determination with a given statistical degree of certainty. If the statistical error is not to exceed 1 %, for example, then the root-*N* law of error analysis specifies that at least 10 000 atoms must be present for the determination to succeed. For the element zinc (atomic mass 60) this would correspond to a mass of only one attogram (10^{-18} g) [28]. Given the potential of laser spectroscopy (→ Laser Analytical Spectroscopy), attaining such a goal is no longer considered utopian.

The question "Is it really necessary that analysis continue to strive for greater detection power?" must therefore be answered with an unequivocal "Yes." It is naive to suggest (as some have) that the blame for the many problems with which we are today confronted lies exclusively with the availability of increasingly powerful analytical tools. Such a biased perspective reflects at best uncritical — indeed, irresponsible — misuse of the powers of analysis, although misuse of this type can never be ruled out completely.

1.6.2. Reliability — Measurement Uncertainty

One of the first problems faced by analysts is whether a method will provide a result that is fit for its intended purpose, i.e., whether it will produce a result of the required "accuracy". A quantitative indication of the accuracy is required if the user of the result is to make any judgement on the confidence to be placed in it, or to compare it in a rational way with the results of other analyses. The statement of a result is not complete without information about the "accuracy" or the "uncertainty".

There will always be an uncertainty about the correctness of a stated result, even when all the known or suspected components of error have been evaluated and the appropriate correction factors applied, since there is an uncertainty in the value of these correction factors. In addition, there will the uncertainty arising from random effects.

Recent developments have led to the formulations of consistent and quantitative procedures for evaluating and reporting the uncertainty of a result, which are applicable in all areas of measurement. These procedures have been set out in the ISO Guide to the Expression of Uncertainty in Measurement [29] and their application to analytical chemistry is described in the EURACHEM (Cooperation for Analytical Chemistry in Europe) Guide Quantifying Uncertainty in Analytical Measurement [30].

The approach set out in the ISO Guide treats all sources of uncertainty in a consistent manner and thus avoids the difficulties encountered in some previous approaches to the evaluation of uncertainty, which treated the uncertainty arising from systematic effects in a different manner to that arising from random effects.

In essence, for chemical analysis the ISO definition of uncertainty is:

> A parameter associated with the result of an analysis that characterizes the dispersion of the values that could reasonably be attributed to the concentration of the analyte.

Thus the analyst when reporting the result of an analysis is also being asked to provide a parameter that gives a quantitative indication of the range of the values that could reasonably be attributed to the concentration of the analyte. The ISO Guide recommends that this parameter should be reported as either:

A *standard uncertainty* defined as:

> uncertainty of the result of a measurement expressed as standard deviation (→ Chemometrics).

or as *expanded uncertainty* defined as:

> a quantity defining an interval about the result of a measurement that may be expected to encompass a large fraction of the distribution of values that could be attributed to the measurand (concentration of the analyte) and which is obtained by multiplying the *standard uncertainty* by a *coverage factor*, which in practice is typically in the range 2–3.

It is common practice to report the standard uncertainty using a value of 2 since this gives an interval with confidence level of approximately 95 %.

The evaluation of uncertainty requires a detailed examination of the measurement procedure. The first step is to identify the possible sources of uncertainty. The next step is to evaluate the size of the contribution from each source, or the combined contribution from a number of sources, expressed as a standard deviation. These contributions are then combined to give the standard uncertainty. Detailed examples are given in the EURACHEM Guide [30].

1.6.3. Elemental Analysis

A formal distinction between "elemental analysis" and "elementary analysis" (Section 1.6.4) is seldom carefully observed in English. *Elemental analysis* in the present context is understood to mean a determination of essentially all the elements present in a sample, irrespective of the type of bonding involved or the constitution of the matrix. Means toward that end include not only the classical methods (gravimetric analysis, titrimetry, spectrophotometry, electrochemical and kinetic methods, etc.) but also atomic spectrometric and radioanalytical methods, some of which are essentially nondestructive. From the standpoint of reliability, classical chemical methods are rarely surpassed by instrumental methods, though the latter typically do provide lower limits of detectability, and they are faster and more economical, generally offering the added potential for simultaneous multielement determination.

1.6.4. Elementary Analysis

Elementary analysis in the classical sense refers to quantitative determination of the constituent elements in an organic compound, especially carbon, hydrogen, oxygen, nitrogen, sulfur, the halogens, and phosphorus, although the definition would today be expanded to cover determination of any element present in an organic structure.

Precise characterization of organic substances became possible for the first time as a result of elementary analysis techniques developed by BERZELIUS, LIEBIG, DUMAS, KJELDAHL, and many others. The fundamental principle, which relies essentially on combustion of the material under investigation with subsequent determination of the combustion products, was adapted to microscale analysis by F. PREGL and later developed by other analysts to such a point that "microelementary analysis" now requires only a few milligrams of sample. This has led not only to shorter analysis times, but also the possibility of analyzing extremely small amounts of valuable natural substances.

Microelementary analysis has now been carried to a high degree of perfection as well as almost complete automation. Despite the advent of modern physical methods (MS, NMR, etc.) the classical techniques have lost little of their significance for synthetic chemists and biochemists, thanks mainly to their considerable advantage of providing reliable *absolute* values of actual mass proportions [31], [32].

There has been no lack of attempts to reduce requisite sample sizes even further into the lower microgram range [33], [34], but the associated techniques are often quite time-consuming, and efforts in this direction have been largely superseded by mass spectrometry, especially in its high-resolution mode. Nevertheless, work in the field of "ultramicroelementary analysis" is still of interest, especially since it has produced impressive evidence of the limits of classical microanalysis.

1.6.5. Microanalysis and Micro Procedures

"Microanalysis" is a term originally associated with classical analytical techniques capable of providing very accurate results from as little as ca. 1 mg of substance (relative error $\leq 1\,\%$; see the discussion of organic microelementary analysis in Section 1.6.4).

The term "micro procedure" is one of several now defined with respect to mass range by a DIN standard (DIN 32 630):

	Sample size
macro procedure	> 100 mg
semimacro procedure (also known as a semimicro procedure)	1 – 100 mg
micro procedure	1 µg – 1 mg
submicro procedure	< 1 µg

Progress in analytical chemistry in recent decades has been so extensive that most procedures could today be included in the "micro" or even "ultramicro" category, especially ones involving chromatography or spectroscopy. Under these circumstances the term "microanalysis" should now be restricted exclusively to the field of elementary analysis. "Chemically" oriented techniques for ultramicro elemental analysis in the microgram range [28] are based on extremely small reaction volumes (e.g., as little as ca. 1 µL) and correspondingly designed facilities for sample manipulation (ultramicro balances, microscopes, capillaries, and extremely tiny tools). Such methods have now been largely superseded by total reflection X-ray fluorescence (TRXRF), which permits rapid, simultaneous, quantitative determination of most elements with atomic numbers > 11 at levels extending into the lower picogram range [35], a striking example of the impressive advance in instrumental microbulk analysis (→ X-Ray Fluorescence Spectrometry).

1.6.6. Stereochemical and Topochemical Analysis

In this case the goal is to describe the target object (an atom, a molecule, or some other component of a solid phase or a solution) with respect to its spatial orientation in its surroundings. Sometimes the frame of reference is an external surface or an internal boundary surface (grain boundary). *Structural analysis* has as its ultimate objective describing all aspects of the overall structure of a particular phase, including the conformations of individual structural elements, perhaps also as a function of time. *Constitutional analysis* produces information regarding relative and absolute arrangements of atoms or atomic groupings (functional groups) within a molecule.

1.6.7. Microdistribution Analysis

Microdistribution analysis is a special type of topological analysis directed toward establishing lateral and depth distributions of the various elements making up a solid—preferably with explicit reference to the ways in which these elements are bonded, all expressed with the highest possible degree of positional resolution. Distribution with respect to phase boundaries may be important here as well. The designated goal is approached with the aid of techniques that permit beams of rays (e.g., laser photons, electrons, ions, or neutrons) to be focused extremely sharply, with a typical cross section of 1 µm or less. Alternative techniques maintain such spatial relationships as may exist within a series of signals, employing a multidimensional detector to transform the crude data into images.

Every effort is made here to achieve the highest possible absolute power of detection. Microdistribution analysis represents the primary field of application for *microprobe* techniques based on beams of laser photons, electrons, or ions, including electron microprobe analysis (EPMA), electron energy-loss spectrometry (EELS), particle-induced X-ray spectrometry (PIXE), secondary ion mass spectrometry (SIMS), and laser vaporization (laser ablation). These are exploited in conjunction with optical atomic emission spectrometry and mass spectrometry, as well as various forms of laser spectrometry that are still under development, such as laser atomic absorption spectrometry (LAAS), resonance ionization spectrometry (RIS), resonance ionization mass spectrometry (RIMS), laser-enhanced ionization (LEI) spectrometry, and laser-induced fluorescence (LIF) spectrometry [36] – [44].

1.6.8. Surface Analysis

Surface analysis is in turn a specialized form of microdistribution analysis, one that provides information on the coverage, distribution, and content of components either at the surface of a solid or in discrete layers located near the surface. Elemental analysis in this context is conducted in a single plane with no attempt at lateral resolution, utilizing, for example, total reflection X-ray fluorescence (TRXRF), glow-discharge mass spectrometry (GDMS), or secondary neutron mass spectrometry (SNMS). Positional resolution can also be achieved with probe techniques such as Auger electron spectrometry (AES) or secondary ion mass spectrometry (SIMS), and to a limited extent with Rutherford back-scattering (RBS) and X-ray photoelectron spectroscopy (XPS) [45] – [48] (→ Surface Analysis).

1.6.9. Trace Analysis

Here we encounter another term associated with a range whose definition has changed considerably with time. A "trace" was once understood to be a no longer determinable but nonetheless observable concentration of some undesired companion substance (impurity) within a matrix. In the meantime, trace analysis has become an important and very precise field of inquiry — subject to certain restrictions with respect to the achievable reliability, but indispensable in a number of disciplines (→ Trace Analysis). There is little point in attempting to express a "trace" in terms of absolute mass units; data should instead be reported on the basis of content (concentration) in a form such as "µg/kg" (mass proportion) or "µg/L" (mass concentration). These units are to be used in preference to the very popular abbreviations "ppm," "ppb," and "ppt", which need an additional indication to the respective unit (mass, volume, amount of substance).

A warning is in order against the practice of emphasizing wide disparities in content through inappropriate comparisons that make a sensational impression at the expense of reality. Phrases like "a Prussian in Munich," "a needle in a haystack," or "a grain of wheat in a hundredweight of rye" used as metaphors for "ppm" are inconsistent with the fact that what is really at issue is a trace *proportion*; that is, a homogeneous distribution or quasicontinuum within which single individuals can be isolated only hypothetically, and only at the molecular level.

Trace constituents at levels as low as a few micrograms per ton (i.e., in a mass ratio of 1 to 10^{12}) are today subject to meaningful analytical determination thanks to highly developed methods of separation and enrichment (multistep procedures, see Chap. 1.3 and Section 1.6.11) [49] – [52]. The principal challenge facing the trace analyst is that diminishing concentration leads to a rapid increase in systematic error [16], [52]. Extreme trace analysis with respect to the elements is therefore subject to large systematic deviations from "true" content, even though results obtained with a particular method may be quite reproducible.

Trace determinations based on atomic spectroscopy are usually matrix-dependent, *relative* methods, requiring the availability of standard reference samples for calibration. Unfortunately, no such standards yet exist for extreme ranges, so one is forced to rely instead on multistep procedures whereby trace amounts of elements of interest are excited in isolated form and within the smallest possible volume of analyte. "Limits of detectability" reported in the literature for methods of trace elemental analysis are almost invariably extrapolations based on determinations actually carried out at fairly high concentration. In no sense can these limits be regarded as reflecting real conditions owing to the problem of systematic error. Systematic errors are extremely difficult to detect, so it is advisable that one verify the validity of data acquired at each stage of the work using an alternative analytical approach. Only when data from different procedures agree within the appropriate statistical limits of error should one speak in terms of "reliable" results. Success in identifying and eliminating all the sources of error in an extreme trace analysis therefore presupposes a considerable amount of experience and a well-developed capacity for self-criticism.

Preferred methods in trace determination of the elements include atomic absorption spectrometry (AAS), optical emission spectrometry (OES) with any of a wide variety of excitation sources [e.g., sparks, arcs, high-frequency or microwave plasmas (inductively coupled plasma, ICP; microwave induced plasma, MIP; capacitively coupled microwave plasma, CMP), glow discharges (GD), hollow cathodes, or laser vaporization (laser ablation)], as well as mass spectrometry (again in combination with the various excitation sources listed), together with several types of X-ray fluorescence (XRF) analysis [51].

A special place is reserved for methods of *activation analysis*, involving slow and fast neutrons, charged particles, or photons, applied either directly or in combination with some type of radiochemical separation (Section 1.6.13). These methods quickly became almost indispensable, especially in extreme trace analysis of the elements, owing to a low risk of contamination and detection levels in at least some cases that are exceptionally favorable (→ Activation Analysis).

Electrochemical methods continue to be important as well, including inverse voltammetry, coulometry, amperometry, and potentiometry (→ Analytical Voltammetry and Polarography); indeed, their overall role has actually been expanded with the development of such chemical techniques as ion chromatography and chelate HPLC.

Problems associated with extreme trace analysis of the elements also affect extreme trace analyses of organic compounds, although background levels tend to be less relevant in this case [53]. All the separation methods most commonly applied to organic substances are chromatographic in nature, including thin layer chromatography (TLC; → Thin Layer Chromatography), high-performance (or high-pressure) liquid chromatography (HPLC; → Liquid Chromatography), gas chromatography (GC; → Gas Chromatography), and electrophoresis (more recently: capillary electrophoresis; → Electrophoresis), preferably combined with on-line mass or infrared spectrometry.

1.6.10. Trace Elements

The term "trace element" was first introduced in biochemistry after it became apparent during the 1920s that very low levels of certain elements in food can be important to life. Nine such elements had been identified by 1959, whereas today more than twenty different elements are regarded as essential, including several previously recognized only as toxic (e.g., arsenic, lead, and cadmium). Ambivalent physiological characteristics have now been ascribed to many elements, where toxicity may be manifested at high concentrations, but a low concentration is an absolute requirement, since a concentration even lower — or complete absence of such an element — leads directly to symptoms of illness [54]. A more appropriate descriptive term applicable in a nonbiological context might be "elemental traces."

1.6.11. Multistep Procedures

Situations frequently arise in which direct instrumental methods of analysis are inapplicable, perhaps because the corresponding detection power is insufficient, or in the case of a matrix-dependent method because no suitable calibration standards are available to correct for systematic errors. The best recourse is then a *multistep procedure*, in which actual determination is preceded by sample preparation, digestion, separation, or preconcentration steps. Individual operations within such a procedure must be linked as closely as possible, as in a "one-pot method" or one of the flow-injection or continuous-flow techniques that lend themselves so readily to automation. The goal is to concentrate an analyte from a rather large volume of solution (on the order of milliliters) for subsequent analysis on the microliter scale. Systematic errors can be minimized with on-line procedures, permitting highly reliable analysis at the picogram-per-milliliter level.

A search will also continue for elemental analysis techniques based on direct instrumental methods with enhanced powers of detection and more or less complete matrix independence. The motivation for this search goes beyond mere economic factors: direct methods are also less likely to be held hostage to blank readings, because physical sources of excitation (e.g., photons, electrons, charged particles, neutrons), which H. Malissa and M. Grasserbauer [55] characterize as "physical reagents," are essentially free of material contamination.

The trace analysis of organic substances is especially dependent on multistep procedures. In this case losses due to adsorption and vaporization are more worrisome potential sources of systematic error than elevated blank values.

1.6.12. Hyphenated Methods

This unfortunate piece of terminology is intended to emphasize the fact that multiple techniques, usually of an instrumental nature, often lend themselves to direct physical coupling, resulting in combinations whose formal designations contain hyphens (e.g., GC – MS). In contrast to the multistep procedures discussed previously, this type of combination involves a "real time" connection and true physical integration. The greatest challenge is to develop satisfactory interfaces for joining the various separation and detection systems. This particular problem is one that has long plagued the otherwise promising HPLC – MS combination.

1.6.13. Radioanalytical Methods and Activation Analysis

Methods based on nuclear reactions are restricted to laboratories specially equipped for handling radioactive substances (radionuclides) under the close supervision of trained personnel. The first important breakthroughs in trace analysis of the elements (e.g., in semiconductor applications) accompanied the development of activation analysis, which was originally based on excitation with slow or fast neutrons but later broadened to encompass the use of charged particles and photons as well. Activation methods were long held in exceptionally high esteem in the field of trace analysis, although competition eventually surfaced in the form of atomic spectroscopy. The drawbacks of activation methods (long analysis times, high cost, and current exaggerated fears with respect to radioactivity) are now perceived by many to outweigh the advantages (high detection power for many elements and relatively high reliability due to minimal complications from matrix effects or contamination), and the activation technique has recently been demoted to the status of one approach among many. This actually increases the need for stressing that activation analysis must still be regarded as an indispensable technique. Sometimes it in fact represents the only viable solution to a problem in extreme trace elemental analysis (e.g., in high-purity substance studies), whether applied directly (*instrumental* activation analysis) or—more often—in combination with radiochemical methods of separation (*radiochemical* activation analysis). Activation analysis also plays an essential role in the preparation of standard reference samples because of the fact that it is so reliable.

The same considerations apply with even greater force to the use of radioactive tracers in elemental and especially molecular analysis (→ Radionuclides in Analytical Chemistry).

1.6.14. Species Analysis (Speciation)

In biology the term *species* is used to describe a population of organisms with hereditary features that survive even after cross-breeding. The related chemical term *species analysis* was first introduced by biochemists, where a "chemical species" is understood to be a particular molecular form (configuration) of atoms of a single element or a cluster of atoms of different elements. Biologists thus define the term "species" very clearly, whereas chemists apply it in various ways:

1) For the *analysis* of a species, leading to its identification and quantification within some defined region of a sample (*speciation*)
2) For describing the *abundance* or *distribution* of various species of an element within a particular volume
3) In conjunction with the *reactivity* of a given species
4) With respect to the *transformation* of one species into another

Accordingly, various *categories* of chemical species can also be distinguished, including "original" or "conceptional" species, "matrix" species, and "analyte" species.

A species is said to be *original* or *conceptional* if it is resistant to change in contact with other matrices. If a chemical change does occur as a result of such contact, the material is called a *matrix* species. The third category refers to a species that undergoes a change during the course of an analytical procedure, in which case it is some new species (the *analyte* species) that becomes the subject of analytical detection. Consider, for example, the original (conceptional) species CH_3Hg^+, which in soil forms the matrix species CH_3Hg-humic acid, but is subsequently determined as the analyte species CH_3HgCl. Transformations of this type are a major source of the considerable challenge posed by species analysis relative to determining the total content of an element. Transformation may well begin as early as the sampling process, continuing throughout the period of sample storage and at the time of the analysis itself, so that original species present in situ and in vivo are never actually detected [56], [57]. Above all, in both environmental analysis and toxicology, detection of the original species is becoming more and more important (→ Sample Preparation for Trace Analysis).

1.6.15. Chemometrics

Chemometrics is the field encompassing those aspects of chemical analysis associated directly with measurement techniques, especially principles underlying the various types of detection. Opinions differ with respect to the meaning, purpose, and limitations of this discipline, but a relatively clear set of ideas is beginning to prevail. According to K. Doerffel et al. [58], chemometrics is concerned with evaluation of observed analytical data with the aid of mathematics, especially statistical methods. Chemometric methods facilitate the extraction of useful information even when the noise level of a signal is high, as well as the establishment of relationships linking multiple observations even when the results themselves seem widely divergent. A good example is provided by the mathematical resolution of partially overlapping signals ("peaks") in a chromatogram.

The tools of chemometrics encompass not only the familiar (univariant) methods of statistics, but especially the various multivariant methods, together with a package of "pattern-recognition" methods for time-series analyses and all the known models for signal detection and signal processing. Chemometric methods of evaluation have now become an essential part of environmental analysis, medicine, process analysis, criminology, and a host of other fields.

Chemometric methods have also been adapted to the development of labor-saving analytical strategies—the establishment, for example, of the ideal sampling frequency in a process analysis, or simplification of a multicomponent analysis so that it reflects only the truly relevant features. In addition, chemometrics plays an important part in quality-assurance programs directed toward analytical investigations. In the future, chemometrics should make a valuable contribution to the design of "legally binding" analyses with statistically assured results [59]–[63] (→ Chemometrics; see also Chap. 1.7).

1.6.16. DNA Analysis

DNA techniques are already being applied in a number of areas of analysis such as human health, identification of sex in certain species, personal identification, environmental and food analysis. The Polymerase Chain Reaction (PCR) is one of the most powerful methods of producing material for analysis from very small samples and can achieve up to a 10^6-fold increase in the target DNA. For example, with this technique it is possible to detect a range of pathogenic micro-organisms with a sensitivity which is orders of magnitude greater than previously achievable, and that is beyond the limits required for public health. It is also a very powerful technique for checking for food adulteration and food speciation. Applications of this technique are growing very rapidly and it could be one of the most important of the recent developments in analytical science.

1.7. "Legally Binding Analytical Results"

The recently coined phrase "legally binding analytical results" has been enthusiastically adopted by numerous authorities in response to problems raised in the administration of justice by such statistically sophisticated concepts as "confidence coefficient," "confidence interval," and the like. Reproducibility has become the primary criterion applied to analytical test results in a legal setting, not necessarily "correctness." This may appear to be an unscientific development, but it probably must be tolerated, at least within reasonable limits. In any case, this problem has been a subject of intense debate in recent years, offering the promise of welcome changes in the foreseeable future ("the theory of legal substantiation").

1.8. References

[1] K. Danzer, E. Than, D. Molch, L. Küchler: *Analytik-Systematischer Überblick,* 2nd ed., Akademische Verlagsgesellschaft Geest & Portig K.-G., Leipzig 1987.
[2] R. Bock: *Methoden der Analytischen Chemie,* **vol. 1: "Trennungsmethoden, " vol. 2: parts 1, 2, and 3: "Nachweis- und Bestimmungsmethoden,"** VCH Verlagsgesellschaft, Weinheim 1974–1987.
[3] H. Kienitz et al. (eds.): *Analytiker-Taschenbuch* **vols. 1–21,** Springer-Verlag, Berlin 1980–1999.
[4] G. Svehla (ed.): *Wilson and Wilson's Comprehensive Analytical Chemistry,* **vols. 1–28,** Elsevier, Amsterdam 1959–1991.
[5] *Ullmann's Encyclopedia of Industrial Chemistry,* 5th ed., **Vol. B5, B6,** Wiley-VCH, Weinheim 1994.
[6] R. Kellner, J.-M. Mermet, M. Otto, H. M. Widmer (eds.): *Analytical Chemistry,* Wiley-VCH, Weinheim 1998.
[7] K. Cammann, *Fresenius J. Anal. Chem.* **343** (1992) 812–813.
[8] M. Valcarcel, *Fresenius J. Anal. Chem.* **343** (1992) 814–816.
[9] C. R. Fresenius: *Anleitung zur qualitativen chemischen Analyse,* 12th ed., Vieweg u. Sohn, Braunschweig 1866, p. 4.
[10] F. Szabadvary: *Geschichte der analytischen Chemie,* Vieweg u. Sohn, Braunschweig 1966.
[11] W. Ostwald: *Die wissenschaftlichen Grundlagen der analytischen Chemie,* Leipzig 1894.
[12] Ullmann, 3rd ed., vol. 2/I.
[13] H. Malissa, *Fresenius J. Anal. Chem.* **337** (1991) 159.
[14] H. Malissa, *Fresenius J. Anal. Chem.* **343** (1992) 836.

[15] K. Danzer, *Mitteilungsblatt der Fachgruppe Analytische Chemie der GDCh* **4/1992** M 104 – M 110.
[16] G. Tölg, *Naturwissenschaften* **63** (1976) 99.
[17] W. Dünges: *Prächromatographische Mikromethoden,* Hüthig-Verlag, Heidelberg 1979.
[18] W. Richter in *Report on the Comité Consultatif pour la Quantité de Matière*, 1st meeting, Paris 1995.
[19] X. R. Pan, *Accred. Qual. Assur.* **1** (1996) 181–185.
[20] P. De Bièvre, *Accred. Qual. Assur.* **3** (1998) 481.
[21] W. Wegscheider, *Accred. Qual. Assur.* **4** (1999) 478–479.
[22] J. W. A. Klaessens, G. Kateman, B. G. M. Vanderginste, *TrAC Trends Anal. Chem.* **4** (1985) 114.
[23] B. Griepink, *Fresenius J. Anal. Chem.* **337** (1990) 812.
[24] Ph. Quevauviller, B. Griepink: Reference Materials in Quality Assurance. In H. Günzler (ed.): *Accreditation and Quality Assurance in Analytical Chemistry,* Springer, Berlin, Heidelberg 1996.
[25] B. Griepink, *Fresenius J. Anal. Chem.* **338** (1990) 360–362.
[26] B. Griepink, E. A. Maier, Ph. Quevauviller, H. Muntau, *Fresenius J. Anal. Chem.* **339** (1991) 599–603.
[27] J. Fleming, H. Albus, B. Neidhart, W. Wegscheider, *Accred. Qual. Assur.* **2** (1997) 51–52.
[28] G. Tölg in G. Svehla (ed.): *Wilson and Wilson's Comprehensive Analytical Chemistry,* **vol. III,** Elsevier, Amsterdam 1975, pp. 1 – 184.
[29] *ISO Guide to the Expression of Uncertainty in Measurement,* ISO, Geneva, Switzerland, 1993.
[30] The Quantification of Uncertainty in Chemical Analysis. Available for down loading from the EURACHEM web site (www.vtt.fi/ket/eurachem).
[31] F. Ehrenberger, S. Gorbach: *Quantitative organische Elementaranalyse,* VCH Verlagsgesellschaft, Weinheim 1991.
[32] T. S. Ma, R. C. Rittner: *Modern Organic Elemental Analysis,* Marcel Dekker, New York 1979.
[33] G. Tölg: *Ultramicro Elemental Analysis,* Wiley-Interscience, New York 1970.
[34] W. J. Kirsten: *Organic Elemental Analysis – Ultramicro, Micro, and Trace Methods,* Academic Press, New York 1983.
[35] G. Tölg, R. Klockenkämper, *Spectrochim. Acta Part B* **48 B** (1993) 111 – 127.
[36] K. Kiss: *Problem Solving with Microbeam Analysis,* Elsevier, Amsterdam 1988.
[37] E. Fuchs, H. Oppolzer, H. Rehme: *Particle Beam Microanalysis, Fundamentals, Methods and Applications,* VCH Verlagsgesellschaft, Weinheim 1990.
[38] J. C. Vickerman, A. E. Brown, N. M. Reed: *Secondary Ion Mass Spectrometry: Principles & Applications,* Oxford University Press, Oxford 1990.
[39] A. Benninghoven et al. (eds.): "Secondary Ion Mass Spectrometry SIMS VII," *Proceedings of the 7th International Conference on Secondary Ion Mass Spectrometry,* J. Wiley & Sons, Chichester 1990.
[40] M. Grasserbauer, H. W. Werner (eds.): *Analysis of Microelectronic Materials and Devices,* J. Wiley & Sons, Chichester 1991.
[41] L. Moenke-Blankenburg: "Laser Micro Analysis," *Chemical Analysis,* vol. 105, J. Wiley & Sons, New York 1989.
[42] K. Niemax in: *Analytiker-Taschenbuch,* vol. 10, Springer-Verlag, Heidelberg 1991, pp. 1 – 28.
[43] J. Uebbing, A. Ciocan, K. Niemax, *Spectrochim. Acta Part B* **47 B** (1992) 601.
[44] C. M. Miller, J. E. Parks (eds.): "Resonance Ionization Spectroscopy 1992," *Inst. Phys. Conf. Ser.* **128,** Institute of Physics Publishing, Bristol 1992.
[45] J. M. Watts: *Methods of Surface Analysis,* C.K.P., Cambridge 1989.
[46] M. Grasserbauer, *Philos. Trans. R. Soc. London* **A 333** (1990) 113.
[47] J. C. Riviere: *Surface Analytical Techniques,* Oxford University Press, Oxford 1990.
[48] D. Briggs, M. P. Seah: *Practical Surface Analysis,* 2nd ed., J. Wiley & Sons, Chichester, "Auger and X-Ray Photoelectron Spectroscopy," vol. 1, 1990; "Ion and Neutral Spectroscopy," vol. 2, 1992.
[49] A. Mizuike: *Enrichment Techniques for Inorganic Trace Analysis,* Springer-Verlag, Heidelberg 1983.
[50] J. Minczewski, J. Chwasstowska, R. Dybczinski: "Separation and Preconcentration Methods," in *Inorganic Trace Analysis,* Ellis Horwood Ltd., Chichester 1982.
[51] G. Tölg, *Anal. Chem. Acta* **238** (1993) 3 – 18.
[52] G. Tölg, P. Tschöpel: "Systematic Errors in Trace Analysis," in Z. B. Alfassi (ed.): *Determination of Trace Elements,* VCH, Weinheim, 1994.
[53] K. Beyermann: "Organische Spurenanalyse," in H. Hulpke, H. Hartkamp, G. Tölg (eds.): *Analytische Chemie für die Praxis,* Thieme-Verlag, Stuttgart 1982.
[54] G. Tölg, in H. Malissa, M. Grasserbauer, R. Belcher (eds.): *Nature, Aim and Methods of Microchemistry,* Springer-Verlag, Wien 1981, p. 203.
[55] M. Grasserbauer, *Angew. Chem.* **93** (1981) 1059.
[56] M. Bernhard, F. F. Brinckman, K. J. Irgolic: "The importance of Chemical `Speciation' in Environmental Processes," in M. Bernhard, F. E. Brinckman, P. J. Sadler (eds.): *Dahlemkonferenzen 1984,* Springer-Verlag, Heidelberg 1986.
[57] Group Report, Importance and Determination of Chemical Species in Biological Systems in: "The Importance of Chemical `Speciation' in Environmental Processes," in M. Bernhard, F. E. Brinckman, P. J. Sadler (eds.): *Dahlemkonferenzen 1984,* Springer-Verlag, Heidelberg 1986, pp. 17 – 38.
[58] K. Doerffel, K. Danzer, G. Ehrlich, M. Otto, *Mitteilungsbl. Chem. Ges. DDR* **31** (1984) 3.
[59] D. L. Massart et al.: *Chemometrics: A Textbook,* Elsevier, Amsterdam 1988.
[60] M. A. Sharaf, D. L. Illman, B. R. Kowalski: *Chemometrics,* J. Wiley, New York 1986.
[61] R. G. Brereton: *Chemometrics, Applications of Mathematics and Statistics to Laboratory Systems,* Ellis Horwood, Chichester 1990.
[62] B. G. M. Vandeginste, *Fresenius J. Anal. Chem.* **337** (1990) 786.
[63] St. J. Haswell: *Practical Guide to Chemometrics,* Marcel Dekker, New York 1992.

2. Quality Assurance in Instrumentation

Ludwig Huber, Agilent Technologies GmbH, P.O. Box 1280, D-76337 Waldbronn, Germany

2.1. Introduction

Analytical instruments play a major role in the process to achieve high quality and reliable analytical data. Thus everyone in the analytical laboratory should be concerned about the quality assurance of equipment. Quality standards usually applied in analytical laboratories, such as the ISO/IEC 17025 [1] EN 45001 [2], and the NAMAS [3] accreditation standard, stipulate that all instruments used must be adequately designed, well maintained, calibrated, and tested. Furthermore, regulations, principles and directives concerning laboratory work, such as Good Laboratory Practice (GLP) principles and regulations [4], [5] and Good Manufacturing Practice (GMP) [6] directives and regulations, include chapters that specifically deal with equipment.

Unfortunately quality standards and regulations are not specific enough to give clear guidelines on what to do on a day-to-day basis. Owing to this lack of clarity, most laboratories found they had to interpret these standards and regulations themselves. Private and public auditors and inspectors experienced similar problems and there have also been situations whereby the regulations have been interpreted differently by different inspectors.

Nevertheless, there has been some improvement. Inspection guides have been developed by regulatory agencies for use by inspectors, thus gaining a common understanding from the regulatory side. Interpretation guides have been developed by private organizations, providing a common understanding amongst accreditation bodies and users of equipment. Examples of such documents are *The Development and Application of Guidance on Equipment Qualification of Analytical Instruments* [7] and the U.S. *FDA Guide to Inspection of Pharmaceutical Quality Control Laboratories* [8].

This chapter discusses aspects of quality assurance of equipment as used in analytical laboratories. It provides guidelines on how to select a vendor and for installation and operational qualifications, ongoing performance control, maintenance, and error detection and handling that contribute to assuring the quality of analytical laboratory data. It refers mainly to an automated chromatography system as an example, but similar principles can be applied to other instrumentation.

It is not the scope of this chapter to discuss quality measures as applied during development and manufacturing of equipment hardware and software. This cannot be directly influenced by the user. Details on this topic can be found in

We herewith certify that the product/system

Product No:	79855 A/B
Serial No:	1456G3345

successfully passed all our production quality tests.

During final instrument performance verification the following functional characteristics were individually tested for conformance with our internal specifications:

Leak sensor function:	[√]
Solvent leaks:	[√]
Air leaks:	[√]
Sampling precision:	[√]
Operating test:	[√]

Date:	Nov. 20. 1992
Test Technician:	Schmidt

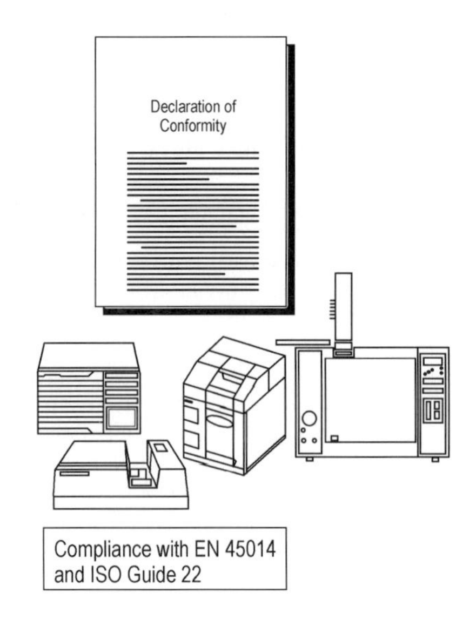

Figure 1. A declaration of conformity according to EN 45014 should be shipped with analytical instruments that show evidence of compliance to documented manufacturing specifications (from reference 9)

published literature [9], [10] and more recently online on the Internet [11].

2.2. Selecting a Vendor

For instruments purchased from a vendor, the quality process starts with the definition of specifications and the selection of the vendor. It is recommended to select vendors recognized as having quality processes in place for instrument design, development, manufacturing, testing, service, and support: for example ISO 9001 registration. Other criteria are the capability of the vendor to provide help in meeting the quality standards' requirements in the users laboratory. As examples, the vendor should provide operating procedures for maintenance and documentation should be available with guidelines on how to test the equipment in the user's environment. For more complex equipment the vendor should provide preventative maintenance and performance verification services in the user's laboratory. Instruments should be selected which have built-in features calibration, for self-diagnosis and for on-site and remote troubleshooting. Software should be available to do the required ongoing performance qualification automatically.

The user should get assurance from the vendor that software has been validated during its development process and that documented quality principles have been applied during manufacturing and testing of the equipment.

A "Declaration of Conformity" should be shipped with all instruments to document that the instrument operated within specification when it was shipped from the factory (Fig. 1). The "Declaration of Conformity" should be an extract from detailed and comprehensive test documentation and include the following information:

- The name and the address of the supplier
- Clear identification of the product (name, type and model number)
- Place and date the declaration was issued
- Name and signature of the supplier's authorized person
- Listing of tested items and check boxes with pass/fail information

Computer systems and software products should be supplied with declarations documenting the evidence of software development validation. The user should also get assurance that development validation procedures and documents can be made available to the user. Critical formulae used in the analytical process should be documented in the user's operating manual.

2.3. Installation and Operation of Equipment

To put equipment in routine operation requires three steps:

1) Preparing the site for installation
2) Installation of hardware and software
3) Operational, acceptance, and performance testing

It is important to do both operational and acceptance tests in the user's environment, even if these tests have been done before installation at the vendors location either as individual modules or as a complete system. Before and during routine use of the system, the performance and suitability of the complete system for the intended use should be verified.

2.3.1. Setting Specifications

Setting the right specifications should ensure that instruments have all the necessary functions and performance criteria that will enable them to be successfully implemented for the intended application and to meet business requirements. This process is also called Design Qualification (DQ). Errors in DQ can have a tremendous technical and business impact, and therefore a sufficient amount of time and resources should be invested in the DQ phase. For example, setting wrong operational specifications can substantially increase the workload for testing.

While IQ (Installation Qualification), OQ (Operational Qualification) and PQ are being performed in most regulated laboratories, DQ is a relatively new concept to many laboratories. It is rarely performed in those cases where the equipment is planned to be used not for a specific but for multiple applications.

This phase should include:

- Description of the analysis problem
- Description of the intended use of the equipment
- Description of the intended environment
- Preliminary selection of the functional and performance specifications (technical, environmental, safety)
- Instrument tests (if the technique is new)
- Final selection of the equipment
- Development and documentation of final functional and operational specifications

To set the functional and performance specifications, the vendor's specification sheets can be used as guidelines. However, it is not recommended to simply write up the vendor's specifications because compliance to the functional and performance specifications must be verified later on in the process during operational qualification and performance qualification. Specifying too many functions and setting the values too stringently, will significantly increase the workload for OQ.

2.3.2. Preparing for Installation

Before the instrument arrives at the user's laboratory, serious thought must be given to its location and space requirements. A full understanding of the new equipment has to be obtained from the vendor well in advance: required bench or floor space and environmental conditions such as humidity and temperature, and in some cases utility needs such as electricity, compressed gases for gas chromatographs, and water. Care should be taken that all these environmental conditions and electrical grounding are within the limits as specified by the vendor and that correct cables are used. Any special safety precautions should be considered, for example for radioactivity measurement devices and the location should also be checked for any devices generating electromagnetic fields nearby. For a summary of all points needing to be considered, see Table 1.

2.3.3. Installation

Once the instrument arrives, the shipment should be checked, by the user for completeness. It should be confirmed that the equipment ordered is what was in fact received. Besides the equipment hardware, other items should be checked for example correct cables, other accessories, and documentation. A visual inspection of the entire hardware should follow to detect any physical damage. For more complex instrumentation, wiring diagrams should be generated, if not obtained from the vendor. An electrical test of all modules and the system should follow. The impact of electrical devices close to the computer system should be considered and evaluated if a need arises. For example, when small voltages are sent between sensors and integrators or computers, electromagnetic energy emitted by poorly shielded nearby

Table 1. Steps towards routine use of instruments (from reference 9)

Before Installation	
●	Obtain manufacturer's recommendations for installation site requirements
●	Check the site for compliance with the manufacturer's recommendations (space, environmental conditions, utilities such as electricity, water, and gases)
●	Allow sufficient shelf space for SOPs, operating manuals, and disks
Documents	
☑	Manufacturer's recommended site preparation document
☑	Check-list for site preparation
Installation (installation qualification)	
●	Compare equipment as received with purchase order (including software, accessories, spare parts, and documentation such as operating manuals and SOPs)
●	Check equipment for any damage
●	Install hardware (computer, equipment, fittings and tubings for fluid connections, columns in HPLC and GC, cables for power, data flow and instrument control cables)
●	Install software on computer following the manufacturer's recommendation
●	Make back-up copy of software
●	Configure peripherals, e.g., printers and equipment modules
●	Evaluate electrical shielding (are their sources for electromagnetic fields nearby?)
●	Identify and make a list of all hardware
●	Make a list of all software installed on the computer
●	List equipment manuals and SOPs
●	Prepare installation reports
●	Train operator
Documents	
☑	Copy of original purchase order
☑	System schematics and wiring diagrams
☑	Equipment identification forms (in-house identification, name and model, manufacturer, serial number, firmware and software revision, location, date of installation)
☑	List of software programs with software revisions, disk storage requirements and installation date
☑	Accessory and documentation checklist
☑	Installation protocol
Pre-operation	
Examples:	
●	Passivate gas chromatograph if necessary
●	Flush HPLC fluid path with mobile phase
●	Verify wavelength accuracy of HPLC UV/visible detectors and recalibrate, if necessary
Documentation	
☑	Notebook and/or logbook entries
Operation (acceptance testing, operational qualification)	
●	Document anticipated functions and operational ranges of modules and systems
●	Perform basic functions of the application software, for example, integration, calibration, and reporting using data files supplied on disk (for details see reference 9)
●	Perform basic instrument control functions from both the computer and from the instrument's keyboard, for example, for an HPLC system switch on the detector lamp and the pump and set different wavelengths and flow rates
●	Test the equipment hardware for proper functioning
●	Document all the operation tests
●	Sign the installation and operation protocol
Documentation	
☑	Specifications on intended use and anticipated operating ranges
☑	Test procedures of the computer system with limits for acceptance criteria and templates with entry fields for instrument serial number, test results, corrective actions in case the criteria are not met and for printed name and signatures of the test engineer
☑	Procedures with templates for operational testing of equipment hardware describing the test details, acceptance criteria, and the definition of operational limits within which the system is expected to operate and against which the testing should be performed
☑	Signed installation and operational protocols. If the installation is made by a vendor firm's representative, the protocol should be signed by the vendor and the user firm's representative
After installation	
●	Affix a sticker on the instrument with the user firm's asset number, manufacturer's model and serial number, and firmware revision
●	Develop a procedure and a schedule for an on-going preventative maintenance, calibration, and performance verification
●	Prepare and maintain a logbook with entries of instrument problems

Table 1. continued

Documentation	
☑	Instrument sticker
☑	Operating procedures and schedules for preventative maintenance, calibration, and performance verification and procedures for error handling
☑	Notebook and/or logbook entries
On-going performance control (performance qualification, system suitability testing, analytical quality control) for routine analysis	
●	· Combine instrumentation with analytical methods, columns, reference material into an analysis system suitable to run the unknown samples
●	Define type and frequency of system suitability tests and/or analytical quality control (AQC) checks
●	Perform the tests as described above and document results
●	Develop procedures for definition of raw data and verification of raw data and processed data (for GLP/GMP)
Documentation	
☑	Test protocols
☑	Data sheets with acceptance criteria for system performance test results
☑	System performance test documents
☑	Quality control charts
☑	SOPs for definition of raw data and verification of processed data (for GLP/GMP)

Table 2. Form for computer system identification

Computer hardware
Manufacturer
Model
Serial number
Processor
Coprocessor
Memory (RAM)
Graphics adapter
Video memory
Mouse
Hard-disk
Installed drives
Space requirement
Printer
Manufacturer
Model
Serial number
Space requirement
Operating software
Operating system (version)
User interface (version)
Application software
Description
Product number (version)
Required disk space

fluorescent lamps or by motors can interfere with the transmitted data.

The installation should end with the generation and sign off of the installation report, in pharmaceutical manufacturing referred to as the Installation Qualification (IQ) document. It is recommended to follow documented procedures with checklists for installation and to use pre-printed forms for the installation report.

When the installation procedure is finished the hardware and software should be well documented with model, serial, and revision numbers.

For larger laboratories with lots of equipment this should be preferably a computer based data base (Table 2). Entries for each instruments should include:

- In-house identification number
- Name of the item of equipment
- The manufacturers name, address, and phone number for service calls, service contract number, if there is any
- Serial number and firmware revision number of equipment
- Software with product and revision number
- Date received
- Date placed in service
- Current location
- Size, weight
- Condition, when received, for example, new, used, reconditioned List with authorized users and responsible person

It is recommended to make copies of all important documentation: one should be placed close to the instrument, the other one should be kept in a safe place. A sticker should be put on the instrument with information on the instrument's serial number and the companies asset number.

2.3.4. Logbook

A bound logbook should be prepared for each instrument in which operators and service technicians record all equipment related activities in a chronological order. Information in the logbook may include:

HPLC instrument verification report

Test method: C\HPCEM\1\VERIF\Check.M
Data file directory: C\HPCEM\1\VERIF\Result.D
Original operator: Dr. Watson

Test item	**User limit**	**Actual**	**Com**
DAD noise	<5 x 10-5 AU	1 x 10-5 AU	pass
Baseline drift	<2 x 10-3 AU/hr	1.5 x 10-4 AU/hr	pass
DAD WL calibration	± 1 nm	± 1 nm	pass
DAD linearity	1.5 AU	2.2 AU	pass
Pump performance	<0.3% RSD RT	0.15% RSD RT	pass
Temp. stability	±0.15°C	±0.15°C	pass
Precision of peak area	<0.5°C RSD	0.09% RSD	pass

Verification test overall results pass
HP 1100 series system, Friday, November 14, 1997
Test engineer
Name: Signature:

Figure 2. Result of an automated HPLC hardware test. The user selects test items from a menu and defines acceptance limits. The instrument then performs the tasks and prints the actual results together with the limits, as specified by the user

- Logbook identification (number, valid time range)
- Instrument identification (manufacturer, model name/number, serial number, firmware revision, date received, service contact)

Column entry fields for dates, times, and events, for example, initial installation and calibration, updates, column changes, errors, repairs, performance tests, quality control checks, cleaning and maintenance plus fields for the name and signature of the technician making the entry.

2.3.5. Operation

After the installation of hardware and software an operational test should follow, a process which is referred in pharmaceutical manufacturing as Operational Qualification (OQ). The goal is to demonstrate that the system operates "as intended" in the user's environment.

For an automated GC or GC/MS system operational testing can mean, for example, verifying correct communication between the computer and the equipment but also checking the detector response and the precision of the retention times and peak areas. Vendors should provide Operating Procedures for the tests, limits for acceptance criteria, and recommendations in case these criteria cannot be met.

The words "intended ranges" in the definition of operational qualification is important. This means that the instrument does not need to be tested for all functions or the instrument manufacturers specifications, if not all functions of the instrument will be used or if it will not be used within all the instrument's limits. This also means that the user should specify the "intended range" before the tests for the operational qualification begin. For example, an HPLC pump may have been purchased with a 4 channel proportioning valve to switch automatically between different mobile phases. Even though in this case the pump is designed to perform gradient analysis, the gradient performance does not need to be tested if the pump will always be used for isocratic runs. Or if an HPLC UV/visible detector will always be used to measure relatively high concentrations, it is not required to measure the performance close to the detection limit of the instrument.

This also has an impact on the required maintenance efforts. To keep the baseline noise of a UV/visible detector within the manufacturers specification will require frequent cleaning of the flow cell, it will require ultra-pure mobile phases and the lamp may have to be changed frequently. For applications not requiring the greatest sensitivity this effort can be reduced if the acceptance limit for the baseline noise is set higher.

For the systems which consist of several hardware modules such as HPLC systems, operational testing may be done for the entire system or for each module (Fig. 2). Complete system testing is referred as holistic testing, testing each distinct, important part a computerized analytical system is called modular testing [12].

For this discussion it must be noted that for testing most of the items such as the flow rate precision, injection volume precision or the detector's linearity, more than one HPLC module is required. For example, in practice a pumping system and a detector are always needed for testing the precision of the autosampler. On the other hand, an injection system is required for the testing of the pumps precision, which is usually measured by injecting a series of standards and measuring the standard deviation of the peaks retention time. Some important characteristics are influenced by several parts of the system. An example is the precision of the peak area. This is influenced by the repeatability of the injection volume and by the stability of the solvent delivery system. In some instances especially when using a standard yielding a peak signal to noise ratio of 100 and below, the integration repeatability may determine the system precision. Because of these interdependencies testing using the holistic approach is preferred.

If the system does not perform as expected, individual modules may be interchangeably used to identify the source of the system problem, which means modular testing is recommended for troubleshooting purposes.

The installation and operational tests of equipment purchased from a vendor can be done by either a representative of the vendor's firm or by the user's firm. In any case, the installation should follow written protocols and should be done following the manufacturers recommendations regarding installation and testing.

2.4. Qualification of Software and Computer Systems

Correct function of software loaded on a computer system should be checked in the user's laboratory under typical operating conditions. Usually software and computer systems provide more functionality than those functions of the system which are used in the laboratory. Only those functions of the system which will be used should be tested. Therefore as the first step the intended functions of the system should be documented. Other documentation of such testing should include test conditions, acceptance criteria, summary of results, and names and signatures of persons who performed the tests.

Preferably such test and documentation of test results should be done automatically using always the same set of test files. In this way users are encouraged to perform the tests more frequently and user specific errors are eliminated. In some cases vendors provide test files and automated test routines for verification of a computer system's performance in the user's laboratory. For example, Agilent provides for the data systems such files and test automated test routines.

The procedure can be used to verify the performance of data systems for formal acceptance testing, operational qualification, or requalification:

- At installation
- After any change to the system (computer hardware, software updates)
- After hardware repair
- After extended use

Successful execution of the procedure ensures that

- Executed program files are loaded correctly on the hard disk
- The actual computer hardware is compatible with the software
- The actual version of the operating system and user interface software is compatible with the data system software

The method tests key functions of the system such as peak integration, quantitation, printout, and file storage and retrieval.

Test chromatograms derived from standards or real samples are stored on disk as the master file. Chromatograms are supplied as part of the software package or can be recorded by the user. This *master data file* goes through normal data evaluation from integration to report generation. Results are stored on the hard disk. The same results should always be obtained when using the same data file and method for testing purposes. The software includes a routine to carry out the performance tests and verification automatically.

To generate the master file, the user selects and defines a *master chromatogram* from a file menu and defines a method for integration and calibration. The instrument will perform the integration of the peaks using the specified method and stores the results in a checksum-protected, binary data file for use as a reference in later verifications. For

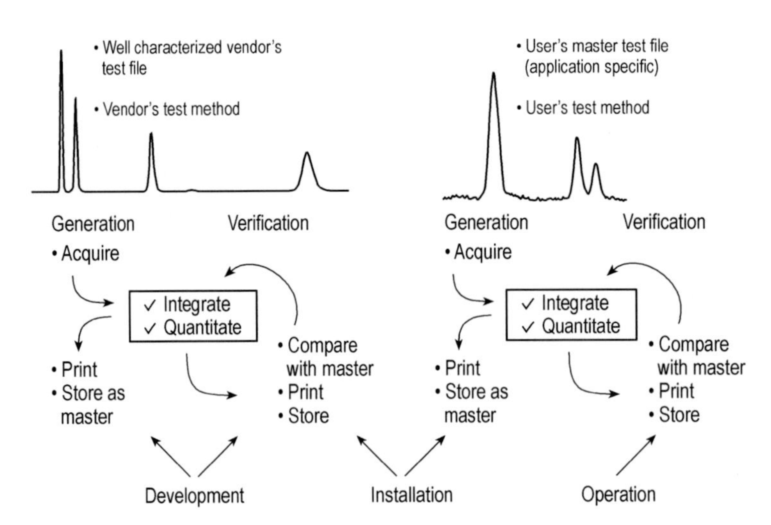

Figure 3. Verification process of chromatographic software. A standard chromatogram is generated by the vendor and well characterized and documented while the software is being developed. The results are verified at installation to make sure that the system works as intended during installation. The user can then generate his/her own application-specific "master" chromatograms to verify the ongoing performance of the system (from reference 9)

performance verification, the user again selects the same master chromatogram and the same method. The program integrates and evaluates peak data, compares it with the master set stored on disk, then prints a report informing the user of the successful verification of the data system (Fig. 3).

The advantages of this method are several fold.

1) A user can select one or more data files which are representative of the laboratory's samples. For example, one file may be a standard with nicely separated peaks which cover a wide range of calibration concentrations, useful to verify the accuracy and the linearity of the integrator. Second or third chromatograms can be used to verify the integrator's capability to integrate real-life samples reproducibly. Chromatograms may be selected with:
   - Peaks close to the detection limits
   - Peaks with shoulders
   - Poorly resolved peaks
   - Peaks on drifting baselines
   - Peaks with tailing
2) The entire process is fully automated, avoiding erroneous results and encouraging users to perform the test more frequently.
3) Results of the verification are documented in such a way that the documentation can be used directly for internal reviews and external audits, assessments and inspections, see for example Figure 4.

2.5. Routine Maintenance and Ongoing Performance Control

When the installation is finished and the equipment and the computer system are proven to operate well, the computerized system is put into routine analysis. Procedures should exist which show that *"... it will continue to do what it purports to do"*.

The characteristics of analytical hardware changes over time due to contamination, and normal usage of parts. Examples are the contamination of a flow cell of a UV detector, the abbreviation of the piston seal of a pump or the loss of light intensity of a UV detector. These changes will have a direct impact on the performance of analytical hardware. Therefore the performance of analytical instruments should be verified during the entire lifetime of the instrument.

Each laboratory should have a quality assurance program which should be well understood and used by individuals as well as by laboratory organizations to prevent, detect, and correct problems. The purpose is to ensure that the results have a high probability of being of acceptable quality. Ongoing activities may include preventative instrument maintenance, performance verification and calibration, system suitability testing, analysis of blanks and quality control samples, and ensuring system security. A plan should be set up to

===========================
Data System Verification Report
===========================

Tested configuration

Component	Revision	Serial number
Diode-array detector	1.0	3148G00859
HPLC 3D data system	Rev. A.02.00	N/A
Microsoft Windows	3.10 (enhanced mode)	N/A
MS-DOS	5.0	N/A
Processor	i486	N/A
CoProcessor	yes	N/A

Data system *verification test details*

Test name	: C:\HPCHEM\1\VERIFY\CHECK01.VAL
Data file	: C:\HPCHEM\1\VERIFY\CHECK01.VAL\DEMODAD.D
Method	: C:\HPCHEM\1\VERIFY\CHECK01.VAL\DEMODAD.M
Original acquisition Method	: PNASTD
Original operator	: Hans Obbens
Original injection date	: 08/02/1985
Original sample name	: AIRTEST

Signals tested

Signal 1 : DAD B, SIG=305, 190 Ref=550,100 of DEMODAD.D
Signal 1 : DAD B, SIG=270, 4 Ref=550,100 of DEMODAD.D
Signal 1 : DAD B, SIG=310, 4 Ref=550,100 of DEMODAD.D

Data system *verification test results*

Test module	Selected for test	Test result
Digital electronics test	yes	pass
Integration test	yes	pass
Quantification test	yes	pass
Print analytical report	yes	pass

Data system verification test overall results: pass

HP 1050 LC System, Friday, June 18, 1993 12:16:55 PM by ____________________

Page 1 of 1

Figure 4. A data system verification report

properly calibrate and verify the performance of the complete system after system updates and after repairs and to check the suitability of the system "for it's intended use" at regular intervals. In the following paragraphs different mechanisms to prove that systems perform as expected for their

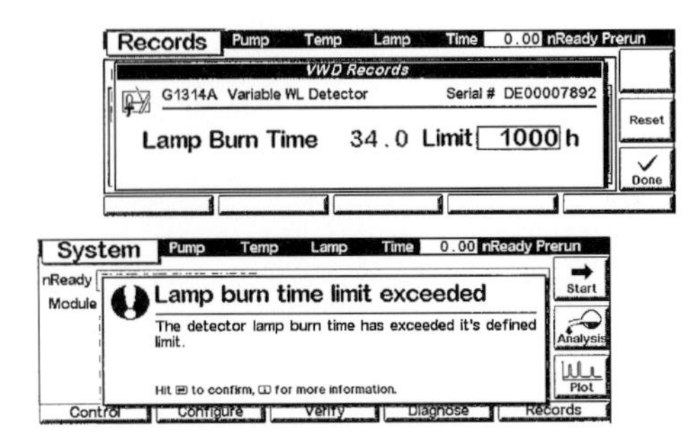

Figure 5. A typical early maintenance feedback system (EMF) informs the user when usage limits are reached, here being the burn time and limits of the detector lamp

intended application will be described. This does not mean that it is recommended to apply all the described procedures. The type and frequency of performance control activities depend on the instrument itself, the application, and on the required performance of the analytical method, as specified as part of the method validation process, or between the laboratory and its clients.

2.5.1. Preventative Maintenance

Operating Procedures for maintenance should be in place for every system component that requires periodic calibration or/and preventative maintenance. Critical parts should be listed and should be available at the user's site. The procedure should describe what should be done, when and what the qualifications of the engineer performing the tasks should be. System components should be labeled with the date of the last and next maintenance. All maintenance activities should be documented in the instrument's log book. Suppliers of equipment should provide a list of recommended maintenance activities and procedures (SOPs) on how to perform the maintenance. Some suppliers also offer maintenance contracts with services for preventative maintenance at scheduled time intervals.

Traditionally maintenance parts are replaced on a time base. For example, an HPLC pump seal every 2 months, a detector's lamp every 3 months or so. This is not economical for the laboratory and not environmentally friendly because frequently a replacement of the parts would not yet be necessary. A better way is to exchange maintenance parts on a usage basis. The user can enter limits for the lamp, the solvent pumped through and the number of injections. The instruments record the time usage and when the limits are exceeded the user is informed through the user interface. This allows timely exchange of the maintenance parts before the instrument performance goes under the acceptable limit.

All maintenance activities should be recorded in a maintenance logbook. To make this convenient, modern equipment includes electronic maintenance logbooks where the user enters the type of maintenance and the equipment records this activity together with the date and time. An example of such a system is shown in Figure 5.

2.5.2. Calibration

Operating devices may be miscalibrated after a while, for example the temperature accuracy of a GC column oven or the wavelength accuracy the optical unit of a UV/visible detector. This can have an impact on the performance of an instrument. Therefore a calibration program should be in place to recalibrate critical instrument items. All calibrations should follow documented procedures and the results should be recorded in the instrument's logbook. The system components should be labeled with the date of the last and next calibration. The label on the instrument should include the initials of the test engineer, the form should include his/her printed name and the full signature.

2.5.3. Performance Qualification

During performance qualification (PQ) or verification (PV) critical performance parameters of an analysis system are thoroughly tested. Examples are precision of retention times, migration times and peak areas of a chromatographic or electrophoresis system and the baseline noise of an HPLC UV/visible detector.

The performance is tested independently from a specific method using generic test conditions and test samples usually supplied by the instrument vendors. The supplier should also provide recommendations for performance limits (acceptance criteria) and recommended actions in case the criteria cannot be met.

The performance verification should follow documented procedures and the results should be recorded in the instrument's logbook. It is recommended to use templates for the results. An example is shown in Figure 6. The performance verification can be either performed by the user or

Instrument: HP 1050 Series VW detector
Serial number: 1448J3450
Test: Baseline noise
User's specification: 1.0 x 10^{-4} AU (HP's recommended limit 1.5 x 10 AU^{-5})
Test frequency: Every 12 months (HP's recom. every 12 months)

Date	Measured value	Corrective action	Final value	Test engineer	
				name	signature
2/3/93	1.4 x 10-5			Hughes	

Figure 6. Example for a performance verification template

by the supplier or any other third party on behalf of the user. Whoever does the performance verification should have documented evidence that she/he is qualified to do it. This may be certificates of successful participation in training courses.

A frequently asked question is on what performance characteristics should be verified and how often. The frequency of performance checks for a particular instrument depends on the "acceptable limits" as specified by the user. The more stringent the limits are, the sooner the instrument will drift out of the limits, the shorter is the frequency of the performance checks. The time intervals for the checks should be identified and documented for each set up equipment.

Reference 10 lists frequencies and parameters to be checked for chromatographs including liquid and gas chromatographs, electrodes, and for heating/cooling apparatus including freeze dryers, freezers, furnaces, hot air sterilizers, incubators, for spectrometers, autosamplers, balances, volumetric glassware, hydrometers, barometers, and thermometers.

Pre-printed forms should be available for:

- Schedules for preventative maintenance, such as cleaning, and replacement of spare parts
- Maintenance activities
- Calibration activities
- Performance verifications
- LogBooks

2.5.4. Analytical Quality Control (AQC) with Control Samples and Control Charts

The analysis of quality control samples with construction of quality control charts has been suggested as a way to build in quality checks on results as they are being generated. Such tests can then flag those values which may be erroneous for, for example, any of the following reasons:

- Operator is not qualified
- Reagents are contaminated
- GC carrier gas is impure
- HPLC mobile phase is contaminated
- Instrument characteristics have changed over time

For an accurate quality check, Quality Control (QC) samples are interspersed among actual samples at intervals determined by the total number of samples and the precision and reproducibility of the method. The control sample frequency will depend mainly on the known stability of the measurement process, a stable process requiring only occasional monitoring. HUBER [10] recommends that 5 % of sample throughput should consist of quality control samples for routine analysis and 20 – 50 % for more complex procedures.

Control samples should have a high degree of similarity to the actual samples analyzed, otherwise one cannot draw reliable conclusions on the performance of the measurement system. Control samples must be so homogeneous and stable that

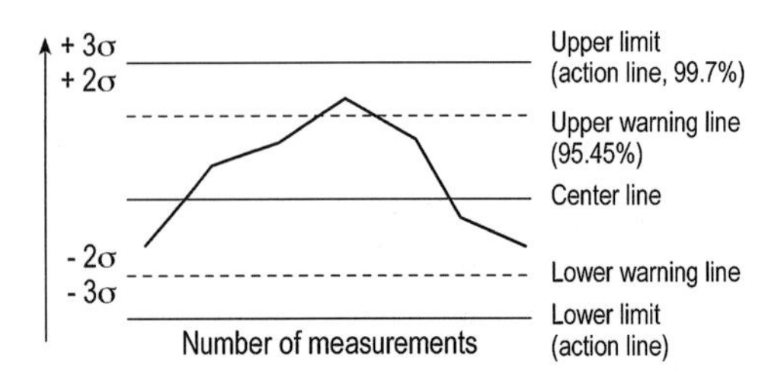

Figure 7. Quality control chart with warning lines and control lines

individual increments measured at various times will have less variability than the measurement process itself. QC samples are prepared by adding known amounts of analytes to blank specimens. They can be purchased as certified reference materials (CRMs) or may be prepared in-house. In the latter case, sufficient quantities should be prepared to allow the same samples to be used over a long period of time. Their stability over time should be proven and their accuracy verified, preferably through interlaboratory tests or by other methods of analysis.

The most widely used procedure for the ongoing equipment through QC samples involves the construction of control charts for quality control (QC) samples (Fig. 7). These are plots of multiple data points versus the number of measurements from the same QC samples using the same processes. Measured concentrations of a single measurement or the average of multiple measurements are plotted on the vertical axis and the sequence number of the measurement on the horizontal axis. Control charts provide a graphical tool to demonstrate statistical control, monitor a measurement process, diagnose measurement problems, and document measurement uncertainty. Many schemes for the construction of such control charts have been presented [10]. The most commonly used control charts are X-charts and R-charts as developed by Shewart. X-charts consist of a central line representing either the known concentration or the mean of 10–20 earlier determinations of the analyte in control material (QC sample). The standard deviation has been determined during method validation and is used to calculate the control lines in the control chart. Control limits define the bounds of virtually all values produced by a system under statistical control.

Control charts often have a center line and two control lines with two pairs of limits: a warning line at $\pm 2\,\sigma$ and an action line at $\pm 3\,\sigma$. Statistics predict that 95.45 and 99.7 % of the data will fall within the areas enclosed by the $\pm 2\,\sigma$ and $\pm 3\,\sigma$ limits, respectively. The center line is either the mean or the true value. In the ideal case, where unbiased methods are being used, the center line would be the true value. This would apply, for example, to precision control charts for standard solutions.

When the process is under statistical control, the day to day results are normally distributed about the center line. A result outside the warning line indicates that something is wrong. Such a result need not be rejected but documented procedures should be in place for suitable action. Instruments and sampling procedures should be checked for errors. Two successive values of the QC sample falling outside the action line indicate that the process is no longer under statistical control. In this case the results should be rejected and the process investigated for its unusual behavior. Further analyses should be suspended until the problem is resolved.

2.6. Handling of Defective Instruments

Clear instructions should be available to the operator on what to do in case the instrument breaks down or fails to function properly. Recommendations should be given on when the operator should try to fix the problem and when to call for service from the instrument vendor. For each instrument there should be a list of common and un-common failures. In cases of malfunction, it is not sufficient to repair the instrument on-site and to continue the measurements. The failure should be classified into a common or an un-common problem. A common problem like a defective lamp of a UV/visible detector requires short term action. The lamp should be replaced, and after a functional test the instrument can be used for further analyses. The failure and repair, and the result of the functional test should be entered into the instrument's logbook. In cases where there is an un-common failure which cannot be easily classified and repaired by the operator, several steps are required:

- The problem should be reported to the laboratory supervisor or to the person who is responsible for the instrument. The supervisor or the responsible person should decide on further action.

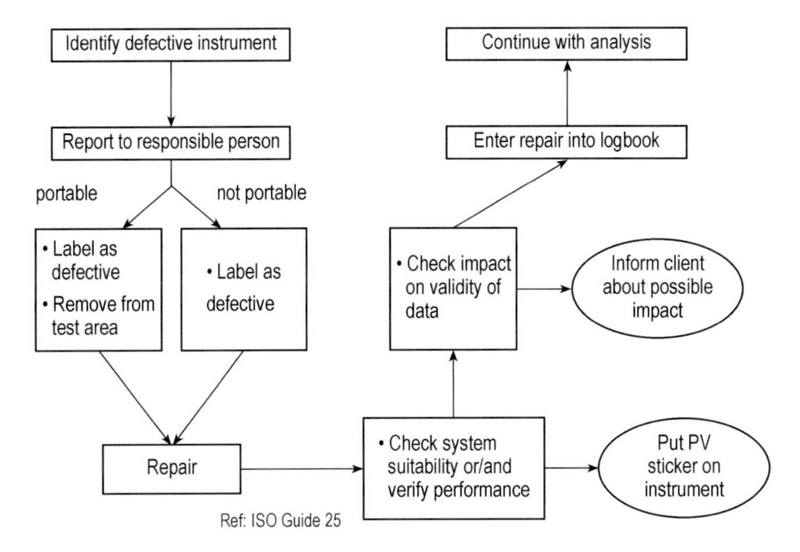

Figure 8. Handling of defective instruments

- The instrument should be taken out of the test area and stored at a specified place, or if this is not practical because of the size it should be clearly labeled as being defective. For example, small sized portable equipment such as a pH-meter can be taken out of the test area. Larger equipment such as a GC or an ICP-MS system should be labeled as being “out of service”.
- After repair, proper functioning should be verified by tests. The type of tests depend on the particular failure and possible impact on the system.
- Effects of the defect on previous test results have to be examined.
- Clients should be informed on any effect the defects could have had on the validity of test data.
- An entry on the defect, repair, and performance verification should be made in the equipment's logbook.

These procedures are summarized in Figure 8.

2.7. References

[1] ISO/IEC Guide 25: *General Requirements for the Competence of Calibration and Testing Laboratories,* International Organization for Standardization, Geneve, Switzerland 2000.

[2] EN 45001: *General Criteria for the Operation of Testing Laboratories, CEN/CENELEC,* The Joint European Standards Institution, Bruessel, Belgium 1989.

[3] NAMAS Accreditation Standard: *General Criteria for Calibration and Testing Laboratories,* M10 of the NAMAS Executive, National Physical Laboratory Teddington, Middlesex 1989.

[4] Food and Drug Administration: *Non-Clinical Laboratory Studies, Good Laboratory Practice Regulations,* U.S. Federal Register. vol. 41, No. 225, November 19, 1976, pp. 51206–51226 (Proposed Regulations) and vol. 43, No. 247, December 22, 1978, pp. 59986–60020, (Final Rule).

[5] OECD Series on Principles of Good Laboratory Practice and Compliance monitoring, Number 1: GLP Consensus Document: *The OECD Principles of Good Laboratory Practice,* Environment Monograph No. 45, Paris 1992.

[6] EC Guide to Good Manufacturing Practice for Medicinal Products in *The Rules Governing Medicinal Products in the European Community,* **vol. IV,** Office for Official Publications for the European Communities, Luxembourg, 1992.

[7] P. Bedson, M. Sargent, *Accred. Qual. Assur.* **1** (1996) 265–274.

[8] *FDA Guide to Inspection of Pharmaceutical Quality Control Laboratories,* The Division of Field Investigations Office of Regional Operations, Office of Regulatory Affairs U.S. Food & Drug Administration, July 1993.

[9] L. Huber: *Validation of Computerized Analytical Instruments,* Interpharm, Buffalo Grove, USA May 1995, 267 pages.

[10] L. Huber: *Validation and Qualification in Analytical Laboratories,* Interpharm, Buffalo Grove, USA Nov. 1998.

[11] Http://www.labcompliance.com, Global Online Resource for Validation and Compliance in Analytical Laboratories, 1999.

[12] ISO/IEC Guide 22: *Information on Manufacturer's Declaration of Conformity with Standards or Other Technical Specifications,* 1st ed., International Organization for Standardization, Geneve, Switzerland 1982.

3. Chemometrics

René Henrion, Weierstrass Institute of Applied Analysis and Stochastics, Berlin, Federal Republic of Germany

Günther Henrion, Humboldt University Berlin, Berlin, Federal Republic of Germany

In addition to the symbols listed in the front matter of this volume the following symbols are used:

b sensitivity ($=dy/dx$)
d distance
$\boldsymbol{D}$ distance matrix
n number of measurements (or objects, Chap. 3.8)
n_j number of parallel measurements
P probability
r correlation coefficient
$\boldsymbol{R}$ correlation matrix
s estimate of standard deviation (sdv)
$t(P,f)$ integration limit for Student's distribution
$\boldsymbol{v}_j$ eigenvectors
x independent variable (mostly: concentration)
x_d decision limit
x_D detection limit
x_i i-th value (predicting variable, Chap. 3.11)
$\bar{x}$ estimate of mean value
$\boldsymbol{X}$ matrix
$\boldsymbol{X}^T$ transpose of the matrix
y dependent variable (mostly: signal)
λ_j eigenvalues
μ mean value
σ standard deviation

3.1. Introduction

Using a widely accepted definition, *chemometrics* can be understood as "the chemical discipline that uses mathematical, statistical, and other methods employing formal logic (a) to design or select optimal measurement procedures and experiments, and (b) to provide maximum relevant chemical information by analyzing chemical data" [1]. Defined in this way, chemometrics shares the impact of mathematical modeling with many other disciplines carrying the suffix "metrics", such as biometrics, psychometrics, and econometrics. Nevertheless, the application to problems of chemistry puts emphasis on particular issues not present or less important in these other sciences. For example, keywords like calibration and signal resolution can be mentioned. However, chemometrics is not just the recipient of mathematical progress, it also stimulates the development and the foundation of new mathematical methods. For instance, the use of modern hyphenated analytical methods producing huge amounts of data with increasing complexity of structure, has become a driving force for algorithms in multimodal statistics. Over time, chemometric methods have become indispensable tools, e.g., for quality control, environmental and forensic analysis, medicine, and process control. Chemometrics, however, does not only interpret data or find optimal strategies for analytical work. It also provides a basic for theoretical treatment of analytical chemistry and establishes the field of analytical science as an independent discipline of chemistry. For comprehensive introductions to the field of chemometrics, the reader is referred to the basic monographs [1] and [2].

3.2. Measurements and Statistical Distributions

3.2.1. Measurements

One of the final aims of chemical analysis is the quantitative determination of species concentrations based on appropriate measurements. Usually these measurements exploit a functional relationship between concentrations and a physical (optical, electrochemical, etc.) signal. For a given sample under investigation, the obtained signal will never be constant when repeating the measurement. The reason is that a variety of conditions, such as temperature, conductivity, pH, change slightly over time and their superimposed impact on the measurement is out of control in the sense of a functional model. Therefore, it is useful to understand a measurement as a random variable x which can have many different outcomes or realizations x_i. In a fixed experiment, the set of realizations $x_1,...,x_N$ is called a sample of x with sample size N. Usually, a sample is characterized by its mean value $\bar{x}$ and its empirical standard deviation s:

$$\bar{x} = \frac{1}{N}\sum_{i=1}^{N} x_i ; \quad s = \sqrt{\frac{1}{N-1}\sum_{i=1}^{N}(x_i - \bar{x})^2}$$

Example 2.1. The analysis of an element concentration by repeated ($N = 16$) measurements of one laboratory gave the following results:

30.3, 29.9, 32.0, 32.0, 30.0, 31.0, 30.9, 30.3, 30.9, 30.1, 29.4, 29.6, 30.4, 30.9, 28.0, 29.6.

$N = 16$, $\bar{x} = 30.33$, $s = 0.99$.

3.2.2. Statistical Distributions

Typically, the outcomes of random variables are not arranged uniformly but show regions of differing density. The points corresponding to the measurements in Example 2.1 are arranged on a straight line in Fig. 1 a. Their density can be illustrated by means of so-called frequency histograms, where the range of measurements is subdivided into an appropriate number of subintervals of equal length. The number of measurements x_i falling inside a given interval is represented by a rectangle of corresponding height on top of the interval. With increasing sample size N, the histograms take on a more regular shape, which is well approximated by a smooth curve $f(t)$, as in Figure 1 b, where f is the density function of x. Clearly, the realizations of x are most likely to fall inside regions with high density values of f. More precisely, the probability of an outcome of x falling into a set A is given by the integral

$$P(A) = \int_A f(t)\mathrm{d}t \tag{2.1}$$

Of particular interest are sets given by left open intervals: $A = (-\infty, \tau)$. In this case, τ is called a $P(\mathrm{A})$ quantile of the distribution of x, which means that the probability for an outcome of x not to exceed τ equals $P(\mathrm{A})$. The density function is always normalized so as to render the integral over the whole real line equal to one. In this way, a density function completely determines the distribution of a random variable. The most prominent representative of probability distributions is the

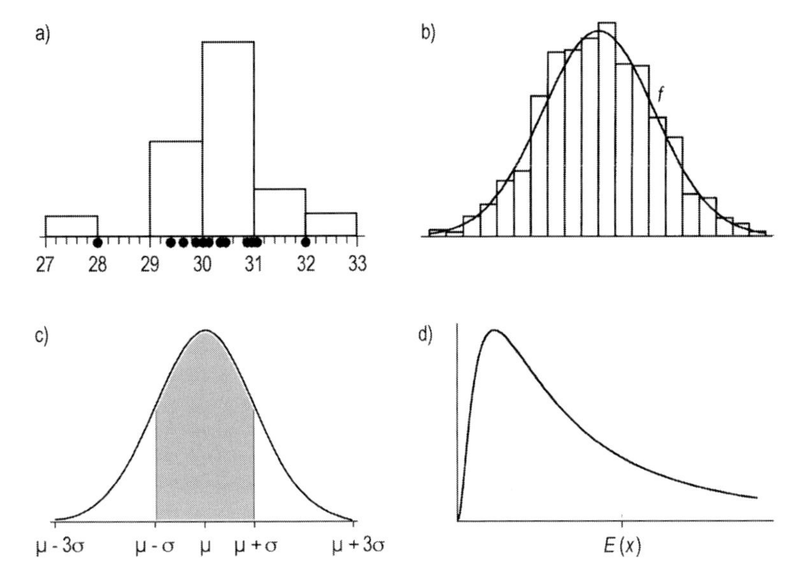

Figure 1. Illustration of densities for probability distributions.
a) Histogram for a small concrete measurement series; b) Histogram for a large series and limiting density function; c) Normal density function with symmetric interval of 68.3 % probability around the mean; d) Density and expected value for the logarithmic normal distribution

normal or Gaussian distribution with density function

$$f(t) = \frac{1}{\sigma\sqrt{2\pi}} e^{-\frac{1}{2}\left(\frac{t-\mu}{\sigma}\right)^2}$$

The bell-shaped profile of this function is shown in Figure 1 c. The parameters μ and σ determine the position of the maximum of f and the distance of the inflection points of f from this maximum. Actually, μ corresponds to the expected value E(x), and σ to the standard deviation σ(x) of the normal distribution. For general distributions, these quantities are defined as

$$E(x) = \int (t)\, t \mathrm{d}t$$

$$\sigma(x) = \sqrt{\int (t)[t - E(x)]^2\, \mathrm{d}t}$$

The square of the standard deviation $\sigma^2(x)$ is referred to as the variance of x. The expected value, in some sense, represents the most typical outcome of a random variable, whereas the standard deviation measures the average deviation of outcomes from the expected value. For the normal distribution, the probability of a realization falling inside the interval $\mu \pm \sigma$ — that is, closer to the expected value than one standard deviation — is 68.3 %. According to Equation (2.1), this value corresponds to the area portion under the density function for this interval (Fig. 1 c). Taking the larger interval $\mu \pm 3\,\sigma$ gives a probability of 99.7 %. This justifies the 3 σ rule, which states that realizations of normally distributed random variables almost surely fall inside this interval.

The importance of the normal distribution relies on the so-called central limit theorem, which roughly states that the distribution of the sum of a large number of independent random variables tends to be a normal one. In particular, measurements which are influenced by a large number of small independent errors are well approximated by a normal distribution with appropriate parameters μ and σ. Nevertheless, other random variables may follow distributions different from the normal one. Counting methods (e.g., X-ray fluorescence) must be described by a Poisson distribution, whereas concentrations in trace analysis are better modeled by a logarithmically normal distribution, that is, the logarithm of the measurements is normally distributed. The latter fact is intuitively clear, since concentrations cannot have negative values, so the density function should be zero on the negative axis (Fig. 1 d). The log-normal distribution also shows that the expected value does not coincide, in general, with the maximum of the density. Other important statistical distributions arise from

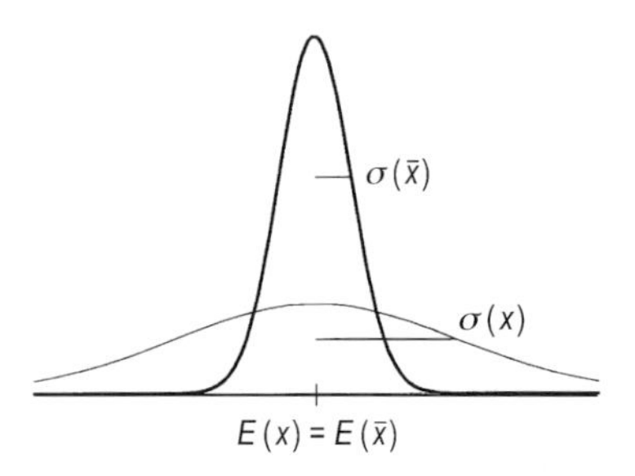

Figure 2. Densities for the normal distribution of a measurements x and of associated mean value $\bar{x}$

transformations of normally distributed random variables, e.g., the t, F and x^2 distributions.

3.2.3. Estimates

It is important to distinguish between parameters of statistical distributions such as density function, expected value, and standard deviation on the one hand, and corresponding sample characteristics such as histograms, mean value and empirical standard deviation on the other. The latter are calculated from actual samples and will converge — in a stochastic sense — with increasing sample size towards the distribution parameters which are the usually unknown characteristics of the abstract probability distribution behind infinitely many samples of one and the same experiment. The sample-based quantities are also called estimates of the corresponding statistical parameters.

Estimates are random variables themselves. For instance, repeating the same experiment as in Example 2.1., one would arrive at a different measurement series with slightly changed values of $\bar{x}$ and s. Repeating this experiment many times, one arrives at a distribution for $\bar{x}$ and s similar to that obtained before for x. Not surprisingly, the expected value of $\bar{x}$ stays the same as that of x: $E(\bar{x}) = E(x)$. The standard deviation, however, decreases by a factor of $\sqrt{N}$, where N is the sample size: $\sigma(\bar{x}) = \sigma(x) / \sqrt{N}$. Hence, the distribution of $\bar{x}$ is much narrower than that of x, (Fig. 2). Increasing the sample size of some measurement series by a factor of nine will therefore decrease the standard deviation of the associated mean value by a factor of three.

3.2.4. Error

Assume that c is the true value of some analytical quantity to be determined. For a given measurement series as in Example 2.1, the error is $|\bar{x}-c|$. We already know that $\bar{x} \to E(x)$ for $N \to \infty$. In the ideal case, where there is no bias, $E(x) = c$; hence, the error will become zero with increasing sample size. Otherwise, the determination has some bias $|E(x) - c| > 0$, and the error can only be reduced to the value of this bias. A statistical test for presence of bias is discussed in Section 3.3.2. The concept of precision refers to the scattering of measurements and is given by the (empirical) standard deviation of a measurement series. As long as precision refers to the results of a single laboratory, it is identified with the concept of *repeatability*. Precision of results provided by different laboratories is identified with the concept of *reproducibility*. For a quantification of both concepts, see Chapter 3.4.

3.3. Statistical Tests

Statistical tests serve the purpose of verifying hypotheses on parameters of distributions. As these parameters are usually unknown, the tests have to rely on estimates which are random variables, hence their outcome has to be interpreted in a probabilistic way.

3.3.1. General Procedure

The general procedure of a test may roughly be sketched as follows:

1) Formulate a "null hypothesis" H_0 usually involving an equality statment about a parameter of interest.
2) Fix a significance level P (usually $P = 95\,\%$ or $P = 99\,\%$).
3) Choose an appropriate "test statistic" T which is a random variable depending on the measurement sample and has a well-known distribution under validity of H_0.
4) Select some set A such that the probability of T having values in A equals P.
5) For a given sample, check if the value of T falls outside A. If yes, then reject H_0; otherwise, accept H_0.

Note that, according to this scheme, the probability of rejecting H_0 although H_0 holds true (error of the first kind) is less than $1-P$ (say 5 % or 1 %), which in this case gives the probability of the observed fact that T falls outside A. However, nothing is said about the probability of accepting H_0 although H_0 is false (error of the second kind) if we observe that T takes a value that falls in A. The realization of step 4 requires knowledge of quantiles for the distribution of T. Because of limited space, the necessary values for the different tests are not given here; reference is made to standard monographs on applied statistics or handbooks containing the corresponding data tables and also giving a more detailed introduction to statistical theory (e.g., [3], [4]). All tests presented in the following are based on normally distributed data. If this assumption, which should be satisfied in most cases, fails to hold, then one can use nonparametric tests as an alternative (see Section 3.3.3).

Test for Mean with Known Standard Deviation. As an illustration, consider a test for bias in the mean of a set of N laboratory results, when the value of σ is known. As null hypothesis, we formulate coincidence of the expected value behind the measurements with some known true value c:

$$H_0 : E(x) = c$$

A significance level of $P=95\,\%$ is fixed and the test statistic $T=\bar{x}$ is chosen. From Section 3.2.3, we know that, $E(T)=E(x)$ and $\sigma(T) = \sigma(x)/\sqrt{N}$. Furthermore, T is normally distributed whenever x is (which one has good reason to assume). It is convenient to select the set A, as required in step 4, in a symmetric way around the expected value (see shaded area in Fig. 1 c). For normal distributions, we already know that the intervals $[\mu-\sigma, \mu+\sigma]$ and $[\mu-3\,\sigma, \mu+3\,\sigma]$ have probabilities of 68.3 and 99.7 %, respectively. To realize the chosen probability of $P=95\,\%$, one would have to consider the interval $[\mu-1.96\,\sigma, \mu+1.96\,\sigma]$, where the value of 1.96 can be read off from appropriate data tables. Since we have to relate the parameters μ and σ here to the expected value and standard deviation of the distribution of $\bar{x}$ rather than that of x, the appropriate choice is (taking into account that $E(T)=E(x)=c$ for validity of H_0)

$$A = \left[c - 1.96\sigma(x)/\sqrt{N}, \quad c + 1.96\sigma(x)/\sqrt{N}\right] \qquad (3.1)$$

Now a decision according to step 5 can be made on the basis of an actual sample and a resulting value for T. Note, however, that the described procedure requires knowledge of the distribution parameter $\sigma(x)$, which is rarely available. A related test avoiding this assumption is described in Section 3.3.2.

Example 3.1. For the data of Example 2.1., it shall be checked if there is some bias to an assumed true value of $c=30$. We also assume that the standard deviation of the (abstract) distribution of the measurement results is known: $\sigma(x)=0.7$ (e.g., $\sigma(x)$ may have been approximated by empirical standard deviations s on the basis of a large sample size). The test interval calculates as $A=30\pm1.96\times0.7/4=30\pm0.343$. Since the mean value $T=\bar{x}=30.33$ is contained in this interval, we accept H_0 which means that — up to unavoidable random errors — the laboratory meets the true value (no bias). Recall, that nothing is said about the risk of a false statement here. If, in contrast, the true value were $c=29.5$, then T is not in A, hence we would reject H_0 and deduce the presence of some bias with a probability of error below $1-P=5\,\%$.

Confidence Intervals. The preceding derivations have shown that the mean value $\bar{x}$ belongs to the interval $E(x) \pm 1.96\sigma(x)/\sqrt{N}$ with probability 95 %. But, equivalently, $E(x)$ belongs to the interval $\bar{x} \pm 1.96\sigma(x)/\sqrt{N}$ with the same probability. Hence, given a mean value from a set of measurement results, the unknown distribution parameter $E(x)$ can be included in a symmetric interval around $\bar{x}$ with a given probability. This is called a P % confidence interval for the statistical parameter. For the data of Example 2.1. one obtains the 95 % confidence interval 30.33 ± 0.343. This is the usual way of indicating the result for an analytical concentration etc. For higher probabilities, the interval enlarges accordingly. Choosing, for instance the 99 % level, one would have to replace the above factor of 1.96 by 2.58 yielding the interval 30.33 ± 0.452.

One-Sided Tests. Frequently, some additional prior information on the considered statistical parameter is available. There may be, for instance, some danger of bias to the true concentration value as a consequence of sample preparation. From the logical background it might be clear, for instance, that the concentration of the analyte is increased if at all. Then, in the above terminology, we know that $E(x)\geq c$. Endowed with such one-sided information, it is reasonable to test the null hypothesis formulated above with a set A chosen as a one-

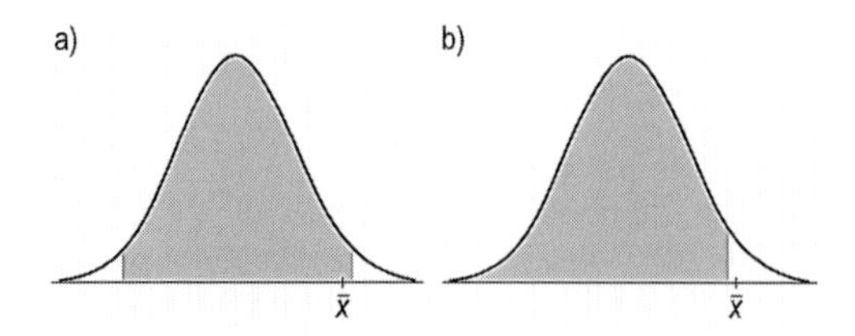

Figure 3. Symmetric (a) and one-sided (b) intervals around the expected value, both having equal probability

sided open interval rather than using the symmetric interval as before. Indeed, here we do not expect our mean value to be much below c, and it is the large values of the mean which are of potential interest now. In this way, we gain information on the crucial side of the inequality to be tested (here: $E(x) > c$) since the test set A is shifted to the left, and this makes it easier to detect excessive deviations on the right. Figure 3 shows that, although both intervals represent 95 % probability, a specific mean value $\bar{x}$ may fall outside the one-sided interval whereas it is inside the two-sided interval. This gives a gain in decision power when exploiting one-sided information. Therefore, we fix A in the form $(-\infty, \tau]$, where τ is chosen such that $P(A) = 95\,\%$. Hence, τ has to be selected as the 95 % quantile (see Section 3.2.2) of the normal distribution with parameters $\mu = E(x)$ and $\sigma = \sigma(x)$. This can be read off from data tables as the value $\tau = 1.65$. Accordingly, for the data of Example 3.1, the mean value now falls outside the interval $A = 30 \pm 1.65 \times 0.7/4 = 30 \pm 0.289$. That is why, in contrast to Example 3.1, the null hypothesis now has to be rejected, and it can be stated that there is a 95 % significant overestimation in the determinations as compared to the true value. The opposite case of knowing that $E(x) \leq c$ is treated analogously by exploiting the symmetry of the normal distribution.

3.3.2. Tests on Parameters of One or Two Measurement Series

Table 1 lists tests on statistical parameters of one or two measurement series. The test intervals are formulated for two-sided tests with probability $P = 95\,\%$. The occuring quantities refer to quantiles of corresponding distributions (F, t, χ^2) which can be found in data tables (e.g., [4]). Indices involving sample sizes refer to so-called degrees of freedom of the corresponding distribution. For instance, $t_{N-1;0.975}$ refers to the 97.5 % quantile of the t- distribution with N − 1 degrees of freedom (note that in the literature quantiles are frequently indicated in the complementary way as critical values, so that 97.5 % correspond to 2.5 %, and vice versa). A two-sided symmetric interval of probability 95 % would then be given by $[t_{N-1;0.025}, t_{N-1;0.975}]$ (since 97.5 % − 2.5 % = 95 %). Owing to the symmetry of the t distribution, $t_{N-1;0.025} = -t_{N-1;0.975}$, whence the interval as indicated in Table 1. There is no problem in changing to other probability levels or one-sided tests (see examples below).

Example 3.2. Using the data of Example 2.1, we want to test the hypotheses $E(x) = 30.8$ and $\sigma(x) = 2$ using the t- and the chi sqd distributions respectively. The 97.5 % quantile of the t-distribution with 15 degrees of freedom equals 2.13, hence $A = [-2.13, 2.13]$. The test statistic calculates for the test on the expected value to $T = (30.33 - 30.8) \times 4/0.99 = -1.89$. Since this value belongs to A, H_0 is accepted and there is no reason to assume a bias in the determinations with respect to the (hypothetical) true result of 30.8. If, however, one knows in advance that the expected value cannot exceed the true one, i.e., $E(x) \leq 30.8$, then A can be chosen as a right-open interval of the same probability $P = 95\,\%$: $A = [t_{N-1;0.05}, \infty) = [-t_{N-1;0.95}, \infty)$ (owing to the symmetry of the t-distribution). For the actual data with $N - 1 = 15$, one gets $A = [-1.75, \infty)$. Now, the value calculated above falls outside the test interval, hence, with the one-sided extra information one can deduce that there is a negative bias to 30.8 in the determinations of the laboratory at the probability level of 95 %. Note that we have just dealt with a one-sided test in the opposite direction to that considered in Section 3.3.1. Another important difference to that simpler test is that now the (theoretical) standard deviation need not be known. Instead, its empirical estimate can be used. To test the hypothesis on the standard deviation, using the chi sqd test gives $A = [6.26, 27.5]$ and $T = 3.68$; hence, the standard deviation behind the given measurement series must be considered different at the 95 % level from the value of 2 which was to be tested. For one-sided tests with additional a priori information that $\sigma(x) \leq 2$ or $\sigma(x) \geq 2$, one would fix the test intervals as $[\chi^2_{N-1;0.05}, \infty)$ and $(-\infty, \chi^2_{N-1;0.95}]$, respectively. Finally, with the given data, confidence intervals for the statistical parameters are easily constructed along the same line as described in Section 3.3.1. For instance, at the 95 % level, one obtains the interval $\bar{x} \pm 2.13$ for covering $E(x)$.

Frequently, it is not the statistical parameters of a single measurement series which have to be tested against specific values, but rather different series are to be tested against each other. For instance, one may ask if the mean values and the standard deviations of the results of two laboratories participating in the same experiment are in agreement. This can be answered by the last two tests recorded in Table 1.

Example 3.3. Consider the measurement data of Example 4.1 below. It shall be tested whether the expected values and standard deviations of the five measurements provided by

Table 1. Sample information, null hypothesis, test statistic, and test interval for common tests on parameters of one or two measurement series

Test	Sample data	H_0	T	A (two-sided, 95 %)
Comparison with fixed standard deviation	s, N	$\sigma(x) = \sigma_0$	$(N - 1)s^2 / \sigma_0^2$	$[\chi^2_{N-1;0.025}, \chi^2_{N-1;0.975}]$
Comparison with fixed expected value	$\bar{x}, s, N$	$E(x) = \mu_0$	$(\bar{x} - \mu_0)\sqrt{N} / s$	$[-t_{N-1;0.975}, t_{N-1;0.975}]$
Comparison of two standard deviations	s_1, N_1, s_2, N_2 $(s_1 \geq s_2)$	$\sigma(x_1) = \sigma(x_2)$	$\frac{s_1^2}{s_2^2}$	$[0, F_{N_1-1,N_2-1;0.975}]$
Comparison of two expected values	$\bar{x}_1, s_1, N_1$	$E(x_1) = E(x_2)$	$\frac{\bar{x}_1 - \bar{x}_2}{s'}\sqrt{\frac{N_1 N_2}{N_1 + N_2}}$	$[-t_{N-1;0.975}, t_{N-1;0.975}]$
$\sigma(x_1) = \sigma(x_2)$	$\bar{x}_2, s_2, N_2$		$s' = \sqrt{\frac{(N_1 - 1)s_1^2 + (N_2 - 1)s_2^2}{N_1 + N_2 - 2}}$	

laboratories A and B are different. Starting with the test on standard deviations, the measurement series with the higher empirical standard deviation is set to be the first one (with respect to index) by definition. So, the symbols s_1, N_1 in Table 1 refer to lab B and s_2, N_2 to lab A. As there is no a-priori information on how the two standard deviations relate to each other, a two-sided test is performed. Accordingly, using the 97.5 % quantile of the F-distribution with $N_1 - 1 = N_2 - 1 = 4$ degrees of freedom which is 9.60, one calculates $T = 0.09^2/0.08^2 = 1.27$ which is clearly inside the test interval. Consequently, there is no reason to assume differences between the underlying theoretical standard deviations of the two labs. For a one-sided test at the same 95 % probability level, the 97.5 % quantile of the F-distribution would have to be replaced by the 95 % quantile.

As for the comparison of expected values, one has to take into account first that the application of this test requires the theoretical standard deviations of the two labs to coincide (more sophisticated tests exist in case this assumption is violated). From the foregoing test, we have no reason to doubt about this coincidence for the data of labs A and B. Following the recipe in Table 1, we calculate $s' = \sqrt{(4 \times 0.09^2 + 4 \times 0.08^2) / 8} = 0.085$ and $T = (45.32 - 45.21)\sqrt{5 \times 5 / (5 + 5)} / 0.085 = 2.04$.

With $t_{4;0.975} = 2.78$, we see that T remains inside the test interval. Hence, we cannot deduce 95 % significant differences in the mean values of the results of labs A and B.

3.3.3. Outliers, Trend and Nonparametric Tests

Before extracting statistical characteristics from a measurement series, such as mean value, standard deviation, confidence interval, etc., it has to be checked whether the data are proper with respect to certain criteria. For instance, the sample may contain extremely deviating measurements owing to a gross error (e.g., simply a typing error or improper measurement conditions). Such values are called *outliers* and must be removed form further consideration as they strongly falsify the characteristics of the sample. The same argumentation holds true for the presence of a *trend* in measurements. This might be caused by directed changes of experimental conditions (e.g., temperature). This leads to a continuous shift of the mean value and increase of the empirical standard deviation with growing sample size, so there is no chance of approaching the theoretical parameters. The presence of outliers and trends can be tested as follows: the test statistic $T = d/s$, where s is the empirical standard deviation and d refers to the maximum absolute deviation of a single value x_i in the measurement series from its mean value, is checked against the values recorded in Table 2 (Grubbs test). If T exceeds the tabulated value, then the measurement x_i is considered as an outlier at the 99 % significance level (smaller levels should not be used, in general, for outlier testing). Similarly, the Trend test according to Neumann and Moore calculates the test statistic

$$T = \sum_{i=2}^{N} (x_i - x_{i-1})^2 / ((N - 1)s^2)$$

If it is smaller than the corresponding value in Table 2, a trend is evident at the 95 % level of probability.

Example 3.4. For the data of Example 2.1., the most outlying measurement with respect to the mean value is $x_{15} = 28.0$, hence $d = 2.33$ and $T = 2.33/0.99 = 2.35$, which is less than the tabulated value 2.75 for $N = 16$. Hence, the series can be considered outlier-free. The trend test statistic for the same data becomes $T = 1.59$ which is smaller than the tabulated value of 1.23. Hence, the series is also free of trends at the 95 % level.

All tests presented so far relied on the assumption of normally distributed measurements. Sometimes, strong deviations from this assumption can

Table 2. Critical values for the Grubbs test and Neumann/Moore test on the presence of outliers and trend, respectively

N	Outlier (99 %)	Trend (95 %)
3	1.16	
4	1.49	0.78
5	1.75	0.82
6	1.94	0.89
7	2.10	0.94
8	2.22	0.98
9	2.32	1.02
10	2.41	1.06
11	2.48	1.10
12	2.55	1.13
13	2.61	1.16
14	2.66	1.18
15	2.71	1.21
16	2.75	1.23

lead to incorrect conclusions from the outcome of these tests. If there is doubt about normality of the data (which itself can be tested as well) then it is recommended to apply so-called nonparametric counterparts of these tests. Nonparametric means that these tests are not based on a distribution. In this way, they are robust with respect to deviations from normality, although of course , less efficient in the presence of normality. Typically, nonparametric tests evaluate some ranking of specifically arranged measurements.

3.4. Comparison of Several Measurement Series

Up to now, we have dealt with statistical characterizations of single measurement series or pairwise comparisons. Frequently, a larger group of measurement series must be analyzed with respect to variance and mean. In the case of commercial products, for example, quality is tested in different laboratories. The different working conditions in individual laboratories provide an additional source of random error. In such situations, it is of interest to analyze the homogeneity of data with respect to standard deviations and mean values. In case that homogeneity can be assumed, much sharper confidence intervals for the precision and accuracy can be obtained from the pooled data rather than from a single lab's results. In the opposite case, the standard deviations relating to repeatability and reproducibility (see Section 3.2.4) can be estimated from *analysis of variance*. Testing for homogeneous data is not only useful, however, in the context of interlaboratory comparisons but also for the important question of representative sampling of materials to be analyzed.

3.4.1. Homogeneity of Variances

Assume that we are given a group of p measurement series with $N_j, \bar{x}_j, s_j$ denoting sample size, mean value and empirical standard deviation of series j. Further we set $N = \sum_{j=1}^{p} N_j$. We want to check whether all series have equal precision, i.e., H_0:

$$\sigma(x_1) = \ldots = \sigma(x_p)$$

This may be realized by the Bartlett test, in which, the following averaged standard deviation within the series is first determined:

$$s_g = \sqrt{\frac{\sum_{j=1}^{p}(N_j - 1)s_j^2}{N - p}} \tag{4.1}$$

Now, a test statistic is calculated from the data according to

$$T = \frac{2.306}{c}\left((N - p)\log s_g^2 - \sum_{j=1}^{p}(N_j - 1)\log s_j^2\right)$$

(where "log" = decadic logarithm) and

$$c = 1 + \frac{\left(\sum_{j=1}^{p}\frac{1}{N_j - 1} - \frac{1}{N - p}\right)}{3(p - 1)}$$

H_0 will then be rejected at probability level 95 % if $T > \chi^2_{p-1;0.95}$.

Example 4.1. A sample of FeSi was analyzed in $m = 7$ laboratories with the following results (% Si)

Laboratory	A	B	C	D	E	F	G
	45.09	45.20	45.37	45.23	45.40	45.63	44.93
	45.19	45.27	45.45	45.26	45.40	45.65	44.95
	45.22	45.30	45.48	45.31	45.45	45.73	44.95
	45.25	45.40	45.60	45.39	45.60	45.85	45.14
	45.31	45.43	45.62	45.44	45.60	45.85	45.17
$\bar{x}_j$	45.21	45.32	45.50	45.33	45.49	45.74	45.03
s_j	0.08	0.09	0.11	0.09	0.10	0.11	0.12

For the data from this table, one obtains the results

$p = 7, N_j = 5\,(j = 1, \ldots, p), N = 35, s_g = 0.10, c = 1.10, T = 0.64, \chi^2_{6;0.95} = 12.6$

Consequently, the null hypothesis cannot be rejected at the 95 % level, and one can assume homogeneous variances among the seven different measurement series.

3.4.2. Equality of Expected Values

The hypothesis for homogeneity of expected values is H_0:

$$E(x_1) = \ldots = E(x_p)$$

This hypothesis is tested by analysis of variance in which — apart from the already determined standard deviation within laboratories (Equation 3) — a so-called standard deviation of laboratory means (standard deviation between laboratories) figures as an important ingredient:

$$s_z = \frac{1}{p-1} \sum_{j=1}^{p} N_j (\bar{x}_j - = x)^2 \qquad (4.2)$$

where

$$= x = \frac{\sum_{j=1}^{p} N_j \bar{x}_j}{N} \qquad (4.3)$$

is the weighted mean of all single mean values or, equivalently, the overall mean of all measurements in the data table. H_0 will be rejected at probability level 95 % if $T = s_z^2 / s_g^2 > F_{p-1,N-p;0.95}$. The reason for using the one-sided test interval here in contrast to the two-sided one in Section 3.3.2 is that the theoretical standard deviations σ_g, σ_z behind their empirical estimates s_g, s_z can be shown always to satisfy the one-sided relation $\sigma_g \leq \sigma_z$, where equality holds exactly in the case that H_0 is true. If H_0 can be rejected on the basis of this test, then the standard deviations in Equations (4.1) and (4.2) represent the repeatability and reproducibility, respectively, of the given laboratory data. Otherwise, the expected values can be considered as being equal, and there is no reason to assume between laboratory bias in the given determinations. As a mean over means, the overall mean of Equation (4.3) has a much smaller variance than the single means (which in turn have smaller variances than the single measurements). This gives rise to very narrow confidence intervals for the mean and explains why interlaboratory homogeneity comparisons are so valuable in producing reference materials.

Similar to the test on equality of two expected values described in Section 3.3.2, its generalization to the multiple case requires homogeneity of variances, this means a Bartlett test must be performed first.

Example 4.2. For Example 4.1, the Bartlett test gave no reason to doubt the homogeneity of variances, so the test on multiple equality of expected values can be carried out. For the data of the example, one calculates

$s_z = 0.51, = x = 45.37, T = 26.58, F_{6.28;0.95} = 2.44$

Consequently, the expected values behind the single measurement series must be considered different at the 95 % level. In other words, there exists a strong laboratory bias which can be characterized by the strongly different values for s_z and s_g.

3.5. Regression and Calibration

3.5.1. Regression Analysis

Regression is a tool for modelling a set of N observed data pairs (x_i, y_i) by means of a functional relationship $y = f(x)$, where f belongs to a specified family of functions (e.g., linear functions, polynomials, bi-exponential functions, etc.). Usually, it is assumed that x is an independent variable which can be fixed without errors, whereas the variable y is dependent on x and is subject to random errors. In analytical chemistry, the role of y is frequently played by measured signals responding to some concentration x or recorded at some wavelength x. The main applications of regression in analytical chemistry are calibration (see Section 3.5.2) and signal resolution. Figure 4 a illustrates a calibration problem, while Figure 4 c shows the decomposition of a signal profile into a sum of two Gaussian profiles.

The family to which f is supposed to belong to is defined via some set of parameters which are coupled in a particular way with x. Therefore, it is reasonable to extend the above mentioned functional relationship to $y = f(x,p)$, where the parameter p is variable. Some prominent families of regression models are:

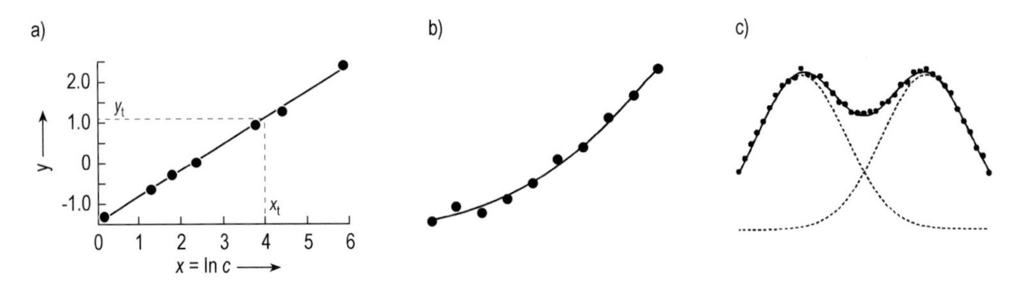

Figure 4. Illustration of different regression models. Linear model a), polynomial model b) and nonlinear model

1) Linear model	$f(x,p) = p_1 + p_2 x$
2) Polynomial model	$f(x,p) = p_1 + p_2 x + p_3 x^2 + \ldots + p_{n+1} x^n$
3) Multi-exponential model	$f(x,p) = p_1 e^{p_2 x} + \ldots + p_{2n-1} e^{p_{2n} x}$

Of course, the linear model is a special case of the polynomial model. Generally, a model is called *quasilinear* when f is a linear function of p. This does not exclude the case that f is nonlinear in x. In particular, the polynomial model is quasilinear although the functional dependence on x may be quadratic, as in Fig. 4 b. Given the data pairs (x_i, y_i), the parameter p yielding the best approximation of f to all these data pairs is found by minimizing the sum of the squares of the deviations between the measured values y_i and their modeled counterparts $f(x_i)$:

$$\mathrm{p} \rightarrow \text{minimize} \quad \sum_{i=1}^{N} (y_i - f(x_i, p))^2$$

The advantage of quasilinear models is that the exact solution for p is easily obtained by solving a system of linear equations (see Section 3.11.1). In the simple case of linear models, one can even directly indicate the explicit solution:

$$p_1 = \frac{s_{xx} s_y - s_x s_{xy}}{d}; \quad p_2 = \frac{N s_{xy} - s_x s_y}{d} \tag{5.1}$$

where

$$s_x = \sum_{i=1}^{N} x_i; \quad s_y = \sum_{i=1}^{N} y_i; \quad s_{xy} = \sum_{i=1}^{N} x_i y_i;$$
$$s_{xx} = \sum_{i=1}^{N} x_i^2; \quad d = N s_{xx} - (s_x)^2$$

The coefficients p_1 and p_2 refer to the intercept and slope of the straight line fitting the data points (see Fig. 4 a). Being based on the random variables x and y, the coefficients p_1 and p_2 are outcomes of random variables themselves. The true coefficients can be covered by the following 95 % confidence intervals:

$$p_1 \pm t_{N-2;0.975} s_R \sqrt{\frac{1}{N} + \frac{\bar{x}^2}{s_Q}}; \quad p_2 \pm \frac{t_{N-2;0.975} s_R}{\sqrt{s_Q}} \tag{5.2}$$

where

$$s_R = \sqrt{\frac{\sum_{i=1}^{N} (y_i - p_1 - p_2 x_i)^2}{N - 2}}$$

and

$$s_Q = s_{xx} - \frac{(s_x)^2}{N}.$$

In contrast to the polynomial model, the multi-exponential model is not quasilinear. This makes the determination of the parameters p more complicated. Then iterative methods have to be employed in order to approach a solution. This issue falls into the framework of nonlinear regression analysis. For more details, see [5].

3.5.2. Calibration

The purpose of quantitative chemical analysis is to determine concentrations of certain analytes on the basis of concentration-dependent instrumental responses or signals. The relation between signals (or functions thereof) and concentrations (or functions thereof) is established by calibration. Calibration consists of two steps: in the first step a functional dependence $y = f(x)$ is modeled on the basis of so-called calibration samples with known concentrations (e.g. reference materials) and corresponding signals, altogether giving a set of data pairs (x_i, y_i). To these data points, an appropriate function is fitted as described in Section 3.5.1. In the second step, the obtained model is applied to a test sample with unknown concentration on the basis of a measured signal. Here, the inverse function is used: $x = f^{-1}(y)$.

Example 5.1. The following calibration results were obtained for the determination of Ni in WO_3 by optical emission spectroscopy (OES):

sample	1	2	3	4	5	6	7
c/ppm	1	3.5	6.0	11.0	51.0	101	501
$x = \ln c$	0.00	1.25	1.79	2.40	3.93	4.62	6.22
y	–1.31	–0.66	–0.33	–0.04	0.84	1.15	2.26

The signal is in good linear relation with the logarithm of concentrations. The data are plotted in Fig. 4 a. From the data table one calculates the values required in Equation (5.1) as

$$s_x = 20.21; \quad s_y = 1.91; \quad s_{xy} = 21.16; \quad s_{xx} = 86.0;$$
$$d = 193.6; \quad p_1 = -1.36; \quad p_2 = 0.57$$

For a test sample, a signal with value $y_t = 1.0$ was measured. Starting from the obtained model $y = -1.36 + 0.57x$, one calculates the inverse function as $x = (y + 1.36)/0.57$; hence, $x_t = 2.36/0.57 = 4.14$. Finally, this last result is transformed back to concentrations via $x = \ln c$. This gives $c_t = e^{x_t} = 62.8$ as the concentration of Ni in the test sample.

In linear calibration, the intercept p_1 corresponds to the signal of the blank, i.e., the signal occuring in absence of the analyte. The slope p_2 is called the *sensitivity* of the signal. For nonlinear calibration curves, this sensitivity changes and has to be calculated as the first derivative d*y*/d*x* at a given concentration value x.

The methodology presented so far corresponds to the *classical calibration*. A different approach is *inverse calibration,* where the functional relationship is directly modeled in the form required for prediction of concentrations in test samples: $x = f(y)$. This approach has proven useful in the context of mutlivariate calibration (see Chap 3.11).

3.6. Characterization of Analytical Procedures

An analytical procedure is characterized by its range of concentration (including calibration), precision (random error), trueness (systematic error), selectivity, and principal limitations.

Trueness [6], [7]. A systematic error can occur as an *additive error* (e.g., an undetected blank) or a *multiplicative error* (e.g., an incorrect titer). Systematic errors are detected by analyzing a short series of m samples with "known" contents x_i and "found" contents y_i as the results. Evaluation by linear regression (Chap. 3.5) yields $y = a + bx$. An intercept $a \neq 0$ is due to an additive systematic error, whereas a slope $b \neq 1$ indicates a multiplicative error. The significance of a and b is tested by verifying that 0 and 1 do not belong to the 95 % confidence intervals around a and b, see Equation (5.2).

If samples with known concentration are not available (e.g., very often in the case of bioproducts), the multiplicative error can be detected in the following way. The sample (solution) is divided into two halves. One half is analyzed directly (result x_1). To the other half an exactly defined amount of the analyte x_+ is added (result x_2). Then the "recovery rate" b is given by

$$b = (x_2 - x_1)/x_+$$

This is performed for m samples, resulting in

$$\bar{b} = \frac{1}{m}\sum b_i; \quad s_b = \sqrt{\sum (b_i - \bar{b})^2/(m-1)}$$

A multiplicative systematic error is proved with 95 % probability if

$$t_b = \|1 - \bar{b}\|\sqrt{m}/s_b > t_{m-1;0.975}$$

Example 6.1. Trueness of determination of arsenic levels in yeast by hydride atomic absorption spectroscopy (AAS) should be tested. However, samples with known contents were not available. Therefore, the recovery rate (with $x_+ = 30$ µg As) was determined from $m = 5$ unknown samples.

Sample	1	2	3	4	5
x_1, µg As	5.8	13.8	30.0	43.1	66.8
x_2, µg As	35.2	43.3	59.6	72.4	96.5
b_i	0.9800	0.9833	0.9867	0.9767	0.9900

$\bar{b} = 0.9833$, $s_b = 0.005266$
$t_b = 7.09$; $t_{4;0.975} = 2.78$

Since $t_b > t_{4;0.975}$ the recovery rate differs undoubtedly from $b = 1$. The results gained by hydride-AAS are on average about 2 % too low.

Table 3. Multifactor plan according to PLACKETT and BURMAN for n = 7 components *

Experiment no.	Components B	C	D	E	F	G	H	Result
1	+	+	+	–	+	–	–	y_1
2	+	+	–	+	–	–	+	y_2
3	+	–	+	–	–	+	+	y_3
4	–	+	–	–	+	+	+	y_4
5	+	–	–	+	+	+	–	y_5
6	–	–	+	+	+	–	+	y_6
7	–	+	+	+	–	+	–	y_7
8	–	–	–	–	–	–	–	y_8

* Scheme for a multifactor plan with $n > 7$ see [6].

Selectivity [6]. In a multicomponent system consisting of an analyte A and other components (B, C, ..., N), each component contributes to the signal of the analyte. Selectivity is given if the signal for A is only randomly influenced by B, C, ... This can be tested by a simple multifactor plan according to PLACKETT and BURMAN [8] (see Table 3) with m measurements for $n = m - 1$ components ($m = 8$, 12, 16, etc.). All the m measurements need a solution of the analyte of concentration x_A. Then, B, C, ... are added in the concentration x_B^+, x_C^+, ... according to the sign (+) in the plan. All of these concentrations x_B^+, x_C^+, ..., x_j^+ must be similiar to the composition of the samples that are to be analyzed later. Each of the $i = 1, \ldots, m$ measurements gives a result y_i (e.g., extinction). The influence W of B, C, ... on the signal of the analyte A is given by

$$W_j = \left(\sum y_i^+ - \sum y_i^-\right)/m/2 \qquad (6.1)$$

$j = \mathrm{B, C}, \ldots$

Subsequently, the sensitivities b_j for B, C, ...

$$b_j = W_j / x_j^+ \qquad (6.2)$$

and b_A for the analyte A

$$b_A = y_m / x_A \qquad (6.3)$$

are calculated, and the desired polynomial for selectivity is given by

$$y = b_A x_A + b_B x_B + b_C x_C + \ldots \qquad (6.4)$$

where y is the intensity of the measured analytical signal. In this way the influence of all accompanying components B, C, ... can be evaluated (for details, see [9]).

Example 6.3. The selectivity for the determination of sodium ($x_{Na} = 100$ mg/L) by flame photometry in presence of Ca^{2+}, K^+, Mg^{2+}, SO_4^{2-}, Cl^-, and PO_4^{3-} had to be tested. According to the plan (Table) these components were added for the step (+) of the factors B ... G in the following concentration (mg/L):

100 mg Ca^{2+} (B); 100 mg K^+ (C); 100 mg Mg^{2+} (D);
100 mg SO_4^{2-} (E); 1000 mg Cl^- (F); 100 mg PO_4^{3-} (G)

Results y_i of the $m = 8$ measurements (scale reading)

$y_1 = 132$; $y_2 = 137$; $y_3 = 119$; $y_4 = 111$
$y_5 = 125$; $y_6 = 114$; $y_7 = 93$; $y_8 = 129$

Influences W_j (Eq. 6.1)
$W_B = (132+137+119+125)-(111+114+93+129)/4 = 16.5$;
$\rightarrow b_B = W_B/100 = 0.165$
$W_C = -3.5$; $W_D = -11.0$; $W_E = -5.5$; $W_F = 1.0$;
$W_G = -16.0$
Then follow all b_j ($j = \mathrm{B, C}, \ldots$) according to Equation (6.2). The sensitivity for the analyte is finally given by Equation (6.3) as $b_A = 129/100 = 0.129$. Then the desired polynomial of selectivity (Eq. 6.4) is
$y = 0.129\, x_{Na} + 0.165\, x_{Ca} - 0.035\, x_K - 0.110\, x_{Mg} - 0.055\, x_{SO^4} + 0.001\, x_{Cl} - 0.160\, x_{PO^4}$. The determination of Na is strongly influenced by Ca^{2+}, Mg^{2+}, and PO_4^{3-}, less by K^+ and SO_4^{2-}, the influence of Cl^- can be ignored. The analytical signal is enhanced by Ca^{2+} and depressed by Mg^{2+}, PO_4^{3-}, and SO_4^{2-}.

Principle Limitations. The analysis of traces is often influenced by the impurities of the reagents. Then a blank $\bar{y}_{bl} \neq 0$ (standard deviation σ_{bl}) is measured at the concentration $x = 0$ of the analyte. From Equation (3.1) we we know that an analytical signal $\bar{y}_a$ (n_j parallels) is different from the blank with 95 % probability if

$$(\bar{y}_a - \bar{y}_{bl}) > 1.96 \frac{\sigma_{bl}}{\sqrt{n_j}}$$

Extending the factor 1.96 to the value 3 would then even yield a difference with 99.7 % probability according to the 3σ-rule (see Section 3.2.2). Now, taking into account that the expected value of the signal of some analyte is never smaller than the blank, one can sharpen the statement above: the analytical signal is different from the blank with 99.85 % probability if

$$\bar{y}_a > \bar{y}_{bl} + 3 \frac{\sigma_{bl}}{\sqrt{n_j}} = y_c$$

where y_c is the critical value. The improvment in probability here is due to using one-sided additional information.

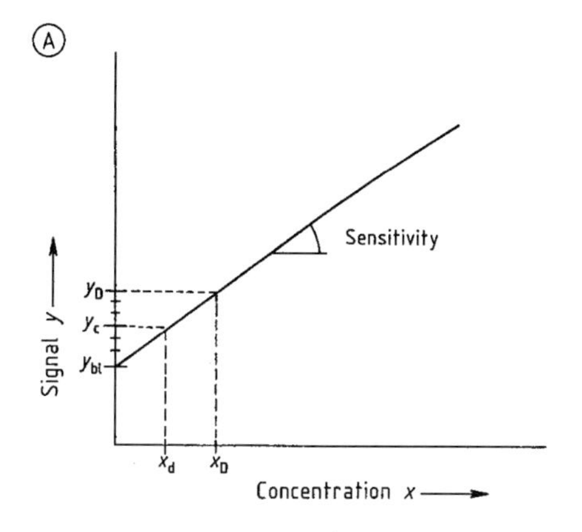

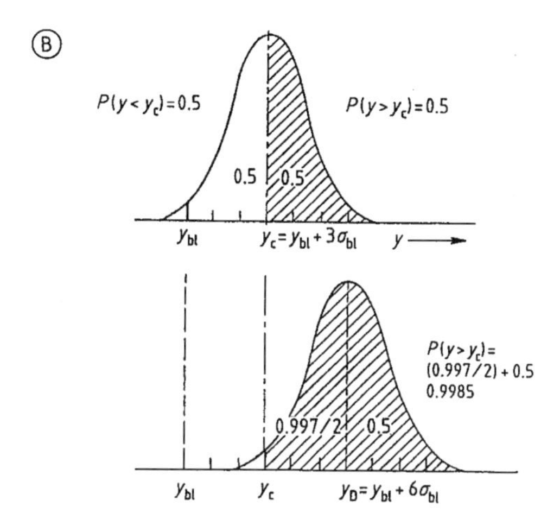

Figure 5. Limit of decision x_d and of detection x_D
A) Calibration function in presence of a blank; B) Probabilities for the detection of an impurity at limit of decision (upper curve) and at limit of detection (lower curve)

In the case of a blank the calibration function takes the form (see Fig. 5 A)

$$y = a + bx = \bar{y}_{bl} + bx$$

For $y = y_c$ the evaluation function will be

$$x = (y_c - a)/b = x_d$$

x_d—*the limit of decision* [10]—is due to the minimum concentration of the analyte that gives a significant analytical signal.

A signal $y = y_c$ results from a concentration $x = x_d$. The frequently repeated analysis of such a sample yields a signal $y < y_c$ as often as $y > y_c$ caused by the random error of the blank (see Fig. 5 B). Expressed as probabilities gives

$$P(y<y_c) = P(y>y_c) = 0.5$$

Therefore the value x will be interpreted as "analyte present" → (+) as often as "only blank → analyte absent" → (–), or again expressed as probabilities

$$P_x^+ = P_x^- = 0.5$$

Therefore the limit of decision x_d never can be used as guarantee of purity. A sufficiently high reliability for such a guarantee is given for the signal value

$$y_D = \bar{y}_{bl} + \frac{6\sigma_{bl}}{\sqrt{n_j}}$$

or the associated concentration

$$x_D = \frac{6\sigma_{bl}}{b\sqrt{n_j}}$$

x_D is termed the *detection limit* [10]. Then,

$$P_x^+ = P(y>y_c) = 0.9985$$
$$P_x^- = P(y<y_c) = 0.0015 \qquad (6.5)$$

An impurity can be detected now at a very high probability level. Therefore, only this limit of detection allows the characterization of high-purity material.

Example 6.4. The photometric determination of iron with triazin (absorptivity $A = 2.25 \times 10^3$ m^2/mol) yielded $\bar{y}_{bl}$ ($= E_{bl}$) = 0.08 and $\sigma_{bl} = 0.02$. For the analysis with duplicates, the critical value is

$$y_c\ (= E_c) = 0.08 + 3 \times 0.02/\sqrt{2} = 0.122$$

The evaluation function gives $x_d = (0.122 - 0.080)/2.25 \times 10^3 = 1.87 \times 10^{-5}$ mol/L and the limit of detection x_D follows from

$$y_D = 0.08 + 2 \times 3 \times 0.02/\sqrt{2} = 0.165$$

$$x_D = (0.165 - 0.080)/2.25 \times 10^3 = 3.78 \times 10^{-5} \text{ mol/L}$$

3.7. Signal Processing

Recording instrumental responses to the presence of some analyte is at the heart of qualitative

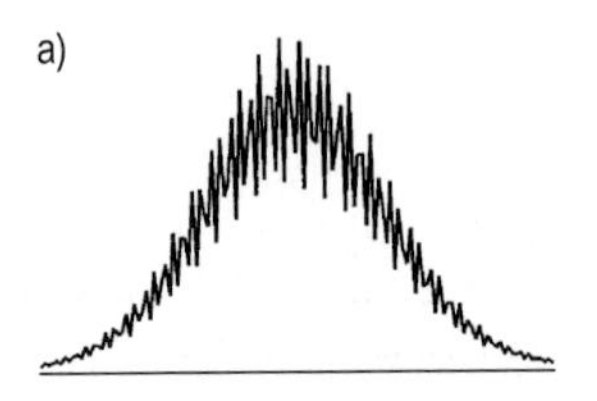

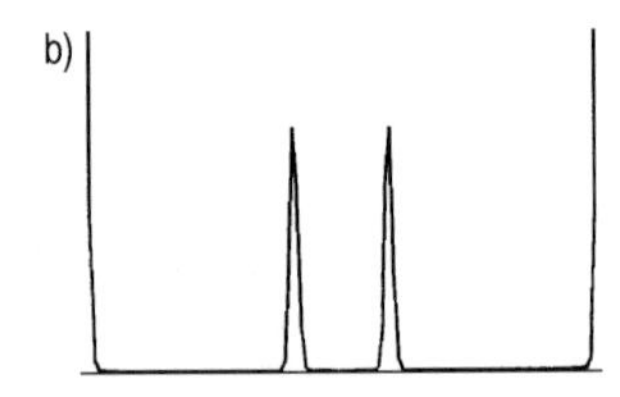

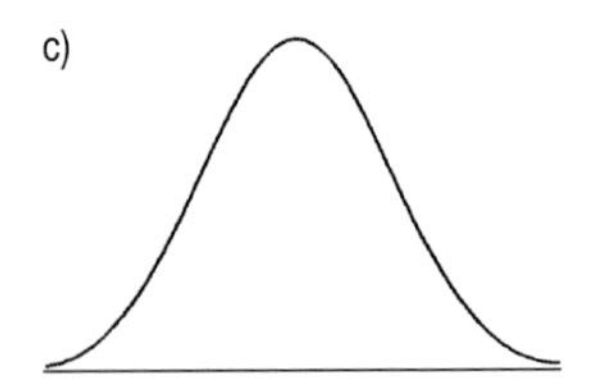

Figure 6. Illustration of Fourier transform
a) Noisy Gaussian signal peak; b) Resulting Fourier transform; c) Filtered signal

and quantitative chemical analysis. The physical nature of these responses or signals may be different according to appropriateness with respect to the given analyte (e.g. optical, electrochemical etc.). In the simplest case, just one signal is recorded, e.g. absorption at some wavelength. Modern instrumentation, however, allows signals to be monitored over a range of physical quantities such as wavelength or time or both of them coupled. Signals are always subjected to perturbations of different kind and origin. Such perturbations can be induced by noise related to the analyzing technique and its apparatus. Furthermore, signals of a certain analyte under investigation are frequently perturbed by interference with signals of different species, which leads to falsifications both in qualitative and quantitative analysis.

3.7.1. Fourier Transform

Contamination of analytical signals by high-frequency noise (e.g., 50 Hz modulated noise caused by the power line) can be detected and corrected by Fourier transformation. Figure 6 a shows a Gaussian signal (e.g. spectroscopic peak over the wavelength domain) containing sinoidal noise. For a discretized signal f with n values $f(0),...,f(n-1)$ the (discretized) Fourier transformed signal is defined as

$$F(u) = \frac{1}{n}\sum_{t=0}^{n-1} f(t)e^{-2\pi uti/n} \quad (u = 0, ..., n-1)$$

where i is the imaginary unit and n is the discretization number. As the Fourier transform is complex, one usually plots its absolute value as shown in Figure 6 b. The symmetry of the Fourier transform around the center point means that only the first half of the spectrum contains substantial information. Peaks in the Fourier transform indicate the presence of periodic contributions in the original signal. The higher the frequency of this contribution, the more this peak shifts to the center. Hence, each of the symmetrically arranged two peaks in the center of Figure 6 b corresponds to the presence of a high-frequency contribution in the original signal, whereas the two peaks on the boundary represent the (low-frequency) signal itself. In this way, noise and signal are clearly separated in the Fourier transform. To arrive at such separation in the original representation of the signal, one may exploit so-called filtering: in a first step, the Fourier transformed signal is convoluted with a so-called kernel which typically is some simple function having a peak in the center of the frequency domain; the convoluted signal is transformed back then to the original domain of the signal by Inverse Fourier transformation. The result is shown in Fig. 6 c. A clear improvement of the signal-to-noise ratio becomes evident when compared with Fig. 6 a.

3.7.2. Data Smoothing

Not all kinds of noise are efficiently removed by Fourier transformation. This is true, in particular, for nonperiodic perturbations such as single spikes. Then, direct data smoothing may be helpful. The simplest method is the moving average, in which each data point of the original signal is replaced by the average of its k nearest neighbors. The appropriate choice of the parameter k is crucial. Enhancing k improves the smoothing effect, which is desirable as far as the reduction of noise is concerned. If k is chosen too large, then information of the signal itself will be smoothed away. Figure 7 a shows an originally measured fluorescence spectrum (267 data points) with clear contamination by noise, whereas Figure 7 b plots a version smoothed by moving averages with $k=20$. More advanced techniques are based on polynomial smoothing.

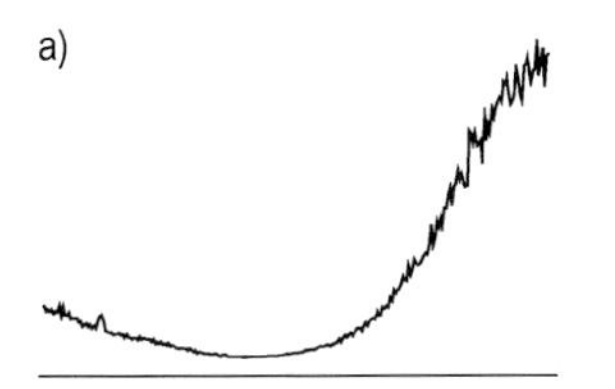

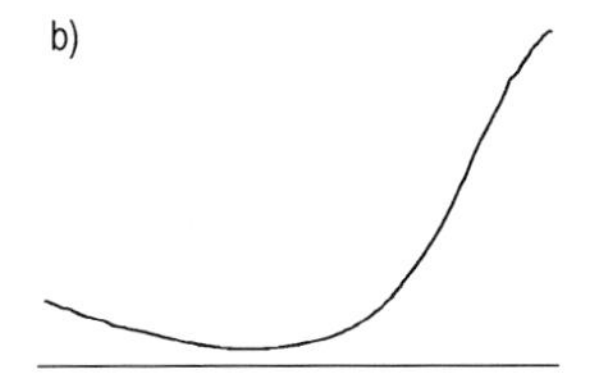

Figure 7. Signal smoothing
a) Noisy fluorescence spectrum; b) Spectrum smoothed by using moving average

Table 4. Small example of a fictive data table (arbitrary units)

Sample no.	[Cd]	[Pb]	[Cu]
1	3	3	1
2	1	2	3
3	4	3	3
4	2	2	0

3.7.3. Signal Resolution

Frequently, the analyte cannot be completely separated by chemical means from other species whose signals interfer with the signal of interest. Then, the resulting superposed signal may be distorted so much that quantitative analysis becomes impossible without additional mathematical tools. In the worst case, even the presence of an interfering component is hidden. Figure 8 a shows two idealized Gaussian peaks the superposition of which results in a single peak only. Such junction of peaks occurs when the distance between them is below a critical value. In this case, signal sharpening by differentiation is a possible remedy. Fig. 8 b shows the superposed single peak of Fig. 8 a in a shrinked domain along with its second derivative. Now, the presence of two underlying components is clearly evident again. In principle one could enhance the order of differentiation, but the gain in signal resolution quickly becomes negligible in comparison with the increase in noise.

3.8. Basic Concepts of Multivariate Methods

3.8.1. Objects, Variables, and Data Sets

The increasing performance of modern instruments allows the characterization of certain objects under investigation by a whole set of properties, which yield a typical fingerprint for each object. These properties are usually statistical variables, whose realization is measured experimentally.

Example 8.1. Several water samples (objects) are analyzed with respect to concentrations of a set of relevant trace elements (variables). Collecting the resulting data gives a constellation as in Table 4. In such data tables or data sets the rows correspond to objects and the columns to variables.

Real life data sets may be generated by very large numbers of objects or variables, depending on the problem. From a more technical point of view a distinction can be made between experimental and instrumental data sets. *Instrumental data sets* are the direct output of instruments, such as digitized spectra, chromatograms, or images in image processing. Here, the number of objects or variables may reach some ten or hundred thousands; therefore, data processing and evaluation generally are automated. *Experimental data* result from single observations (e.g., physico-chemical parameters of natural water at different locations) or from condensation of instrumental data (e.g., one concentration value for a sample instead of a digitized spectrum). As a consequence, acquisition of experimental data is much more expansive, leading to smaller data sets (in the range of several hundreds of objects or variables). As a rule, experimental data tables can be neither processed nor evaluated automatically but require consideration of specific circumstances in the treated problem.

Compared to classical measurement series restricted to a single variable, the multivariate approach yields a considerably greater amount of information for object characterization. This advantage, however, is connected with a loss of direct interpretability. If only two variables were measured, the whole data structure might be still visualized in a diagram, but this approach fails for larger problems. That is why the increase in information must be accompanied by some efficient tool of *data reduction* that enables interpretation

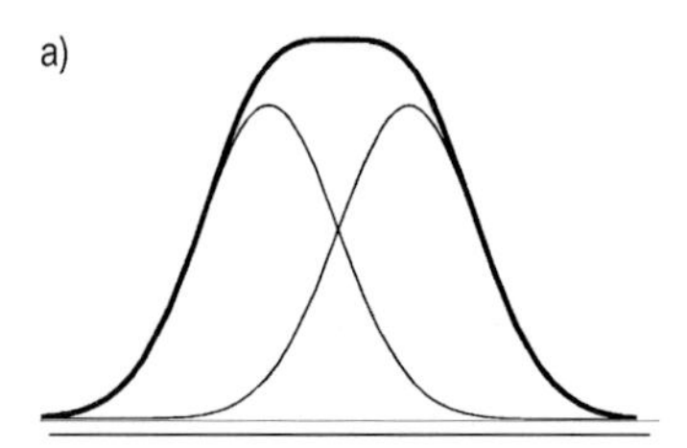

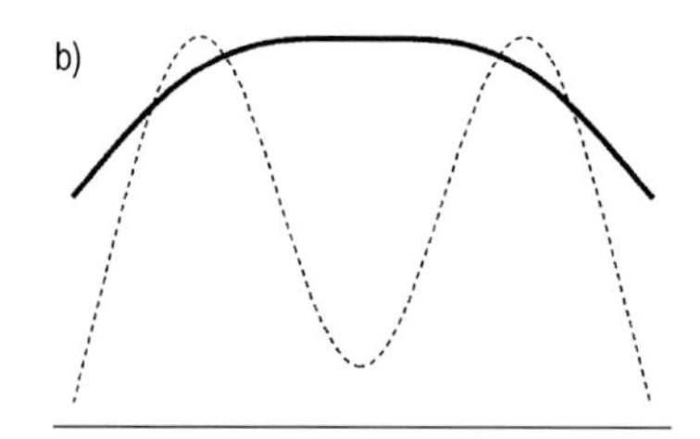

Figure 8. Signal sharpening by differentiation
a) Single peak as a superposition of two close Gaussian peaks; b) Second differential of the superposed signal, yielding two resolved peaks again

or visualization of the essential part of the data structure, thereby separating it from unessential noise. In this sense, data reduction can be understood as the main goal of multivariate methods in chemometrics. Whereas theoretical foundations were developed at the beginning of this century, practical applications of these procedures have revived considerably only during the last few decades (from the 1970s onward) in connection with modern computational equipment (for an introductory survey of multivariate statistics see [11]).

3.8.2. Correlation and Distance Matrices

An important task of multivariate data analysis is classification of objects and variables (i.e., subdivision of the whole data set into homogeneous groups of similar objects or variables, respectively). Similarity of variables is usually measured by their correlation coefficient, whereas similarity of objects is expressed in terms of the geometric distance. The *correlation coefficient* r of two variables x_1, x_2 is computed according to

$$r(x_1, x_2) = \frac{\sum_{i=1}^{n}(x_{i1} - \bar{x}_1)(x_{i2} - \bar{x}_2)}{\sqrt{\sum_{i=1}^{n}(x_{i1} - \bar{x}_1)^2} \cdot \sqrt{\sum_{i=1}^{n}(x_{i2} - \bar{x}_2)^2}} \qquad (8.1)$$

where n is the number of realizations (i.e., number of objects or rows of the data table); x_{ij} is the ith realization of the jth variable (i.e., entry of the data table that is located in the ith row and jth column); and $\bar{x}_j$ is the mean of the jth variable.

For the fictive example of Table 4, for instance, the following coefficients r are obtained r(Cd, Pb) = 0.89, r(Cd, Cu) = 0.09, and r(Pb, Cu) = 0.19. All information on correlation can conveniently be collected in the correlation matrix $\boldsymbol{R}$, which is the basis of many latent variable methods such as principal components, factor, or discriminant analysis:

$$\begin{array}{c} \\ \text{Cd} \\ \text{Pb} \\ \text{Cu} \end{array} \begin{array}{c} \begin{array}{ccc} \text{Cd} & \text{Pb} & \text{Cu} \end{array} \\ \begin{pmatrix} 1.00 & 0.89 & 0.09 \\ 0.89 & 1.00 & 0.19 \\ 0.09 & 0.19 & 1.00 \end{pmatrix} \end{array}$$

Sometimes, instead of correlation the covariance of two variables is used, which is just the numerator of Equation (8.1).

To recognize similarities among objects, they can best be treated as points in p-dimensional Euclidean space, where p is the number of variables (columns) of the data set. If in Table 4 only the concentrations of Cd and Pb had been measured, then all samples could be represented in a diagram as in Figure 9 (i.e., in a two-dimensional space or plane). Although for technical reasons, such representations cannot be realized if the number of variables exceeds 3, the mathematical concept of p-dimensional space is nevertheless useful, for example, when defining the distance of two objects. In a planar display, distances are easily computed by the Pythagorean theorem by taking the square root of the sum of squared coordinate differences between two points (e.g., the distance between samples 4 and 3 equals $\sqrt{5}$). Obviously, the more similar two objects are in all their coordinates the smaller their computed distance is. In this way, homogeneous classes of similar objects may be detected as those that have only small distances between them. Of course, computation of distances is not restricted to the plane but is extended to p-dimensional constellations in a straightforward way:

$$d(x_a, x_b) = \sqrt{\sum_{j=1}^{p}(x_{aj} - x_{bj})^2} \qquad (8.2)$$

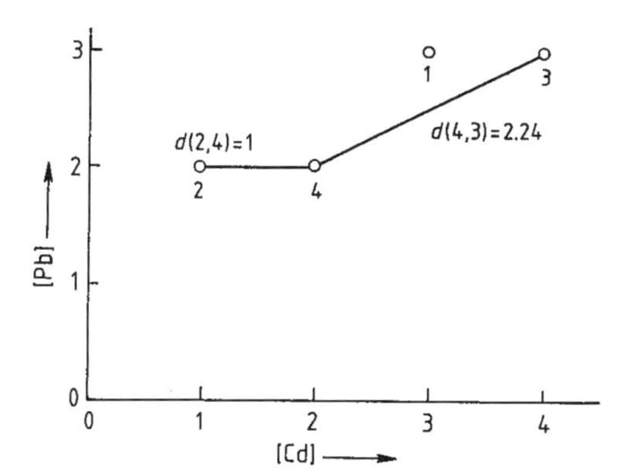

Figure 9. Plot of four samples from Table 4 in an elementary diagram

where x_a, x_b are the objects a and b; and x_{aj} is the entry of the data table in row a and column b. Taking all three coordinates of Table 4 into account gives, for instance:

$$d(4,3) = \sqrt{(2-4)^2 + (2-3)^2 + (0-3)^2} = 3.74$$

All pairwise distances are again collected to give the distance matrix $\boldsymbol{D}$, which forms the basis of many computational approaches including cluster analysis or multidimensional scaling:

$$\begin{array}{c} \\ 1 \\ 2 \\ 3 \\ 4 \end{array} \begin{array}{c} \begin{array}{cccc} 1 & 2 & 3 & 4 \end{array} \\ \begin{pmatrix} 0.00 & 3.00 & 2.24 & 1.73 \\ 3.00 & 0.00 & 3.16 & 3.16 \\ 2.24 & 3.16 & 0.00 & 3.74 \\ 1.73 & 3.16 & 3.74 & 0.00 \end{pmatrix} \end{array}$$

In this matrix the most similar pair of (different) objects is (1,4), while the most divergent pair is (3,4). Apart from the classical Euclidean distance defined by Equation 8.2, some further relevant measures exist such as Mahalanobis or Manhattan distance. The Mahalanobis distance, for instance, which is important in classification (see Chapter 3.10), is computed according to

$$d(\boldsymbol{x}_a, \boldsymbol{x}_b) = \sqrt{(\boldsymbol{x}_a - \boldsymbol{x}_b)^{\mathrm{T}} \boldsymbol{S}^{-1} (\boldsymbol{x}_a - \boldsymbol{x}_b)} \quad (8.3)$$

where $\boldsymbol{S}$ is the matrix of all pairwise covariances between variables.

3.8.3. Data Scaling

Data pretreatment is a necessary condition of multivariate analysis. Besides appropriate transformation of data according to their type of distribution (e.g., taking logarithms for concentration values in the range of the detection limit is recommended), data scaling is necessary as a first step of analysis for almost all methods. Scaling means that each variable is linearly transformed to have zero mean and unit standard deviation. In this way the fact of different variables being measured in different units (e.g., concentrations in parts per million or percent is circumvented, to avoid an arbitrary influence in the total variance or the pairwise object distances. After scaling, all variables have equal a priori influence.

3.9. Factorial Methods

3.9.1. Principal Components Analysis

Principal components analysis (PCA) as a statistical method was introduced by HOTELLING in 1933. Principal components are linear combinations of the original variables with optimal features: the first PC defines maximum variance among all possible linear combinations; the second PC defines maximum variance among all linear combinations uncorrelated with the first one, etc. In this way a small set of a few PCs (generally much fewer than the number of original variables) will suffice to represent the greatest part of the total data variation. Furthermore, in contrast to the original variables, different PCs are always uncorrelated, which makes them useful for many statistical applications. Geometrically, PCs may be identified as new coordinate axes that point to directions of large variance. Using the coordinate system defined by the first two PCs is a very popular approach for visualizing data structure in a diagram. From the computational viewpoint, PCs are obtained by solving the eigenvalue problem

$$\boldsymbol{R} \cdot \boldsymbol{v}_j = \lambda_j \cdot \boldsymbol{v}_j \quad (j = 1, \ldots, p) \quad (9.1)$$

where $\boldsymbol{R}$ is the correlation matrix (see Chap. 3.8); $\boldsymbol{v}_j$ are the eigenvectors; and λ_j are the eigenvalues of $\boldsymbol{R}$.

The number of linearly independent eigenvectors is at most p. Each $\boldsymbol{v}_j$ consists of p components v_{ij}, which are the coefficients (weights) of the original variables. The ith realization of the jth PC is (with x_{ik} being an entry of the scaled data table)

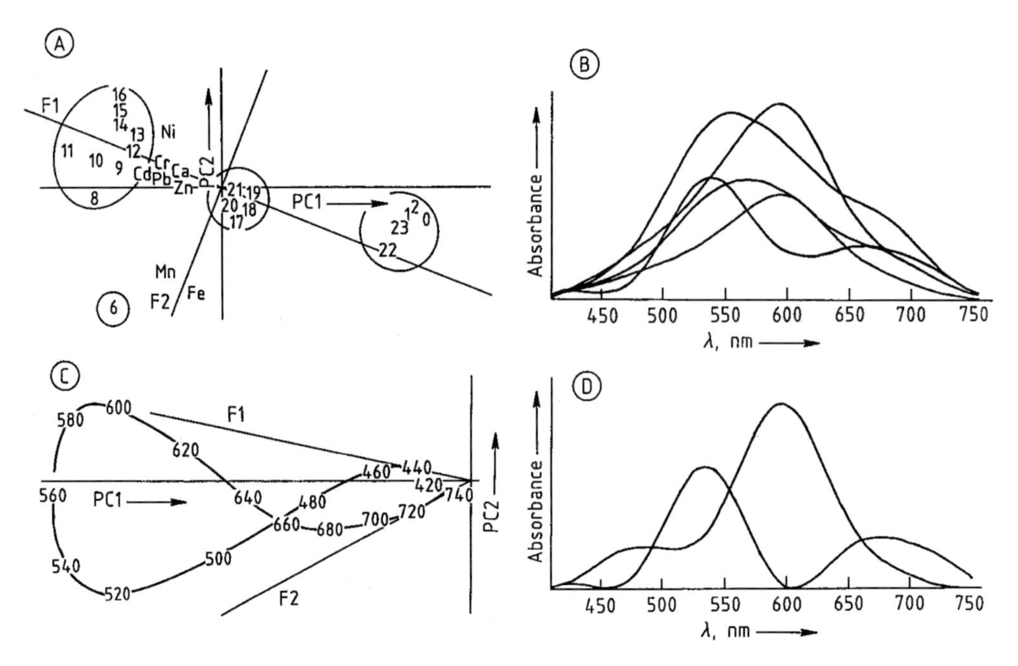

Figure 10. Examples of PCA and FA
A) PC diagram for trace element characterization of wastewater samples; B) Simulated (noise-added) absorbance spectra of five mixtures of two pure components; C) PC wavelength plot for B) including the axes F1 and F2 of pure components; D) Pure component spectra related to B) using the self-modeling method

$$PC_{ij} = \sum_{k=1}^{p} x_{ik} \cdot v_{kj} \qquad (9.2)$$

By using just the first two PCs, a diagram plot for the *objects* is obtained with PC_{i1}, PC_{i2} as coordinates of the *i*th object. These coordinates are frequently called scores. Similarly, in the same plot a representation of *variables* may be given by using v_{i1}, v_{i2} as coordinates (which are called loadings) for the *i*th variable. Finally, the eigenvalues contain information on the amount of data structure covered by a PC: λ_j is the variance of the *j*th PC. Frequently, indication of the variance percentage is preferred by relating one eigenvalue to the sum of all eigenvalues, which in the case of scaled data simply equals *p* (the number of original variables).

Example 9.1. Figure 10 A provides the PC plot (PC1 versus PC2) for a data set containing analytical results of eight trace elements (variables) in 20 samples (objects) of urban wastewater. The numbers in the plot refer to hours of sampling time in the course of a day. The element patterns of samples do not change continuously, but rather abruptly; four clusters of sampling time are visible: (6 h), (8 – 16 h), (17 – 21 h), and (22 – 2 h). The plot of variables reveals two major groups of correlating elements: (Fe, Mn), and (Ni, Cr, Cd, Cu, Pb, Zn). Even a joint interpretation of scores and loadings with respect to influence of variable groups in object groups is possible. This aspect is reconsidered below. In this example, the first two of eight eigenvalues were $\lambda_1 = 4.78$, $\lambda_2 = 1.49$; hence, ca. 78 % of total data variation is exhausted by the first two PCs, which is much more than by using any two of the original variables (recall that due to equal weighting by scaling, each pair of original variables covers a percentage of 2/8 = 25 % of the total variance).

An important kind of chemometric application of PCA is to find out the number of (unknown) components determining the spectra of a set of mixtures of these components.

Example 9.2. Figure 10 B shows simulated and noise-added spectra of five mixtures as an example. Each of these spectra may be digitized (in the example, 680 equidistant wavelengths in the range between 410 and 750 nm were used) to yield a data table with wavelengths as objects and absorbances of the mixtures at these wavelengths as variables. The total variance within such a data set is composed by a systematic part due to changing proportions of the underlying components within the mixtures and by a generally much smaller part due to noise. Consequently, an eigenvalue analysis should provide a group of large eigenvalues corresponding to the principal components, which reflect chemical composition, and a group of very small eigenvalues relating to PCs, which reflect only noise. Various methods exist for objective separation of both groups: graphical (e.g., scree-test), statistical (e.g., Bartlett's test on sphericity or cross-validation) or empirical procedures (e.g., Malinowskis indicator function). In the given artificial example the eigenvalue sequence is 4.453, 0.540, 0.004, 0.002, 0.000, making clear, even without any objective evaluation, that the number of underlying chemical components is probably two.

A review on PCA from the chemometric viewpoint may be found in [12].

3.9.2. Factor Analysis

Factor analysis (FA), developed in 1947 by THURSTONE, is a statistical method with different model assumptions compared to PCA. In chemometric applications, however, FA is frequently considered as an additional step after PCA, with the aim of making the results of the latter easier to interpret (a chemometric approach to FA is found in [13]). One important feature of FA is rotation of PCA solutions. In general, PCs are abstract variables without physical meaning. Therefore, an attempt is made to rotate the axes corresponding to the first few significant PCs so as to make the new axes better fit certain groups of variables and facilitate their interpretation as physically meaningful latent factors. The variety of rotations is subdivided into orthogonal (rotations in the true sense) and oblique transformations. *Orthogonal transformations* leave unchanged the variance percentage covered by the rotated axes. *Oblique transformations*, in contrast, do not provide uncorrelated factors but have more degrees of freedom for fitting groups of variables. From the computational viewpoint a successful transformation is achieved by optimizing an appropriate goal function (e.g., Kaisers varimax criterion, compare [13]). In Example (9.1) (Fig. 10 A) the optimal pair of orthogonal factors is denoted by F1, F2, respectively. It obviously fits the variable (and object) structure better than the PCs. Now, the above-mentioned joint interpretation of objects and variables may be given. Obviously, F1 is determined by (Ni, Cr, Cd, Pd, Cu, Zn) and it discriminates three major groups of sampling times with decreasing concentrations in the mentioned six elements from the beginning to the end of a day. This makes it reasonable to identify F1 as a factor in industrial pollution of wastewater. F2 consists of the (Fe, Mn) group, which is related to the 6-h sample. High concentrations of Fe and Mn in wastewater in the early morning might result from water pipes in private households. With some caution, this could represent a second, independent factor in wastewater exposure.

Another, frequently employed aspect of FA is target testing. This procedure is very useful for deriving new hypotheses from the given data set (for details of the method and applications, particularly in chromatography, see [13]). In Example (9.2) a specific chemical component might be suspected to be present in the mixtures. To verify this, the vector of digitized spectrum of this component could be considered a target to be tested. If the presence of the fixed component is probable, its digitized spectrum vector should be a linear combination of the score vectors of significant PCs since these are, in turn, linear combinations of the spectra of the unknown pure chemical components. The test is performed by multiple linear regresssion (see Chap. 3.11.1), indicating whether the target spectrum may be successfully predicted (within the range of experimental error) by a suitable linear combination of the PC spectra.

A powerful method of FA is *self-modeling of mixture spectra* [14], which allows, under a given assumption, the spectra of underlying pure components to be isolated even if these cannot be modeled by a certain band shape such as Gaussian or Lorentzian. This is a good alternative to conventional least-squares fit. Figure 10 C shows the PC score plot (680 points for different wavelengths in the range 410 – 750 nm), for Example (9.2). According to Beer's law, absorbances of mixtures are positive combinations of the absorbances of pure components. Therefore, the wavelength points of mixture spectra in a PC plot have positive coordinates with respect to factor axes representing the pure components. Consequently, these pure component axes must define some cone that contains all the wavelength points. This is easily illustrated if, as in the example, the number of pure components equals two. Of course, some degree of freedom exists for choosing the axes F1 and F2. Under the specific assumption, however, that each pure components absorbs radiation uniquely at least at one wavelength (within the considered range), the cone defined by F1 and F2 must be a minimal one (i.e., both axes must pass the wavelength loop tangentially; at 440 and 730 nm in the example). Having found F1 and F2 in this way, the pure component spectra are easily established: The absorbances of the first and second pure components, respectively, are given by projecting corresponding wavelength points of the PC plot onto F1 and F2, respectively. Projection

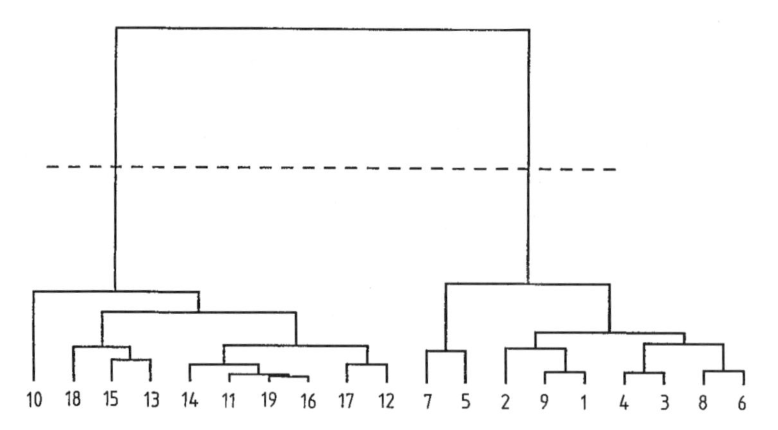

Figure 11. Dendrogram for hierarchic clustering of 19 tungsten powder samples

must be carried out parallel to both axes. The result is illustrated in Figure 10 D.

3.10. Classification Methods

3.10.1. Cluster Analysis

The classification of objects by their patterns in the data set plays a central role in chemometrics. Depending on the prior knowledge concerning the given problem such classification may be performed in a supervised or nonsupervised way. The main tool of nonsupervised or automatic classification is cluster analysis, which produces a partition of the set of objects into several homogeneous subgroups (for an introduction to this field, compare [15]). From a rough viewpoint, cluster analysis may be subdivided into hierarchic and nonhierarchic clustering. *Hierarchic clustering* generates a sequence of partitions that satisfy the definition of a hierarchy; i.e., if any two classes of possibly different partitions are compared, either these classes are disjoint or at least one class is contained in the other (overlapping is excluded). This sequence may be visualized by a so-called dendrogram similar to genealogical trees in taxonomic classification.

Example 10.1. Consider Figure 11, which resulted from hierarchic clustering of 19 tungsten powders being characterized by 11 trace elements. Partitions with different degrees of refinement are obtained by horizontal cuts of the dendrogram. At the lower end, all objects occur as separated classes; at the upper end, all objects are fused in a single class. The level at which certain subclasses are fused (vertical direction) may be interpreted as their dissimilarity. As a consequence, homogeneous groups are found below long branches of the tree. In the example, a bipartition (1 – 9 and 10 – 19) of the object set is suggested. In fact, the two groups of trace element patterns are in coincidence with two different chemical treatments of tungsten raw materials.

Depending on whether the sequence of partitions is generated by starting at the lower end (successive fusion) or at the uper end (successive splitting), agglomerative and divisive procedures can be distinguished. Most of the well-known methods are *agglomerative* and obey the following type of general algorithm:

1) Given a data set X compute the distance matrix $\boldsymbol{D}$ (see Chap. 3.8) and define each object to form a single class.
2) Find out the most similar pair of classes [e.g., C_1, C_2 $(C_1 \neq C_2)$], realizing the smallest distance $d_{1,2}$ in $\boldsymbol{D}$. Fuse all elements of C_1 and C_2 to yield a new, larger class $C_{1,2}$. Fix $d_{1,2}$ as the index of fusion (height in the dendrogram). Update $\boldsymbol{D}$ (with a reduced number of classes) by recomputing all distances between the new class $C_{1,2}$ and all old classes that differ from C_1 and C_2. Repeat step 2 until all objects are fused in a single class.

In fact, agglomerative procedures differ only by the way of updating the new distance matrix in each step. As an essential drawback of hierarchic clustering, "mistakes" of fusion at a low level of aggregation are retained, due to hierarchy, in the whole sequence of partitions.

If different partitions are obtained in an independent way by varying some parameter, *nonhierarchic clustering* is used. In so-called partition-making algorithms, such as k-means or Forgy's

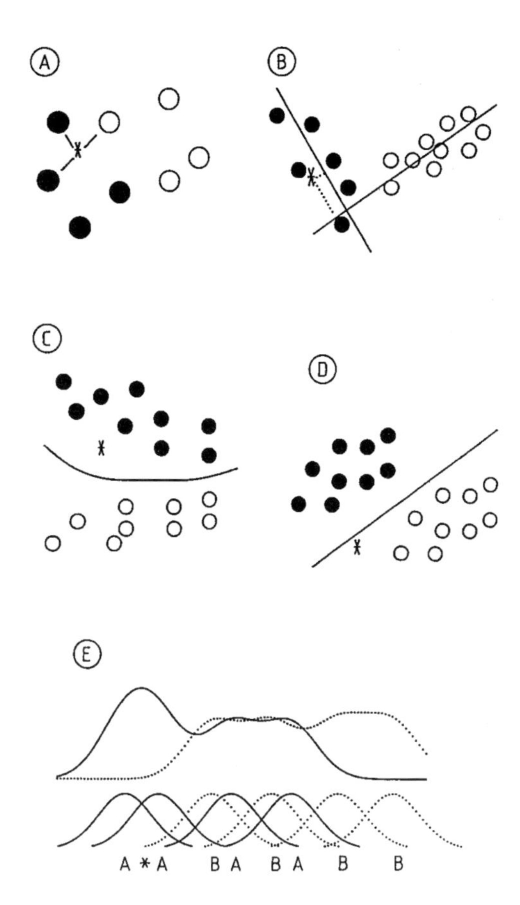

Figure 12. Illustration of supervised classification methods (filled and unfilled circles represent learning objects of different classes; asterisk indicates a test object)
A) The k-nearest neighbor (KNN) method; B) SIMCA method; C) Quadratic Bayesian classification; D) Linear Bayesian classification; E) ALLOC method

method, the number of classes expected must be predefined. Different numbers of classes yield different partitions that might have some overlapping; this would indicate an uncertainty in the class structure. Various criteria are used to decide which number of classes or which partition is appropriate. Other important representatives of nonhierarchic clustering are fuzzy clustering and potential clustering (the nonsupervised version of the ALLOC method discussed below).

3.10.2. Supervised Classification

The general principle of supervised classification may be sketched as follows: For specific predefined classes of interest (e.g., different labels of brandies in food chemistry) a number of samples can be characterized by certain properties (e.g., areas of relevant gas-chromatographic peaks). These samples are called learning objects because their origin is known. Based on the data set of learning objects, rules are derived for subdividing the space of all possible patterns into regions corresponding to the classes. Given any additional objects of unknown origin, they can then — after the same set of properties are determined for test as for learning objects — be assigned to the most probable class (according to the region into which the test object's pattern falls).

An important feature of supervised learning is estimation of the probability of misclassification which has to be kept as small as possible. Since a decision on correctness of classification cannot be made with real test objects having unknown origin, the risk of misclassification must be estimated from the learning objects. The simplest way of doing so would be to plug the data on learning objects into the classification rules that were derived before by the same objects. This method, which is called *resubstitution*, is easily and quickly applicable but it suffers from an underestimation of the real risk of errors. A useful alternative to resubstitution is the *leave-one-out method*, which deletes each learning object one at a time, derives the classification rule by means of the remaining objects, and applies this rule to the variable pattern of the left-out object. The number of misclassifications is much more realistic in this case. Formally applied, the leave-one-out method is very time consuming, but for most classification methods, efficient updating formulas exist to give reasonable computing times. The leave-one-out method, is a special type of cross-validation, where not one but several objects are left out simultaneously.

Some of the most frequently used classification methods are illustrated in Figure 12. For the purpose of visualization the objects are arranged in a plane (e.g., samples that are characterized by just two trace elements or peak areas, etc.). For simplicity, the discussion is restricted to the two-class problem (filled and unfilled circles represent different classes of learning objects). Figure 12 A relates to the *k-nearest neighbor* (*KNN*) *method.* Given the pattern (coordinates) of any test object (asterisk) the Euclidean distances can be computed between this test object and all the learning objects by using Equation (8.2). For a certain number k (e.g., $k = 3$) the test object is assigned to the class holding the majority under the k smallest distances. In Figure 12 A the test object would be assigned to the filled-circle class. Given a data set, the appropriate number for k is determined to yield minimum misclassification rate by the leave-one-out procedure. The main disadvantage of the KNN method is that all distances must be recomputed for each new test object, which makes this very time consuming for larger data sets.

Figure 12 B illustrates the soft independent modeling of class analogy (*SIMCA*) *method* [16], where each class of learning objects is modeled separately by a principal components approximation. In the figure, each of the two learning classes is sufficiently described—up to noise—by a straight line or one principal component, respectively. Now, a test object is classified according to its distance (orthogonal projection) to the approximating linear subspaces. In the figure the asterisk is assigned to the filled-circle group. The optimal number of principal components for each class is determined by cross-validation.

A classical method based on the assumption of normal distributions is *Bayesian classification* (Figs. 12 C and D). Roughly speaking (with some simplification), a test object is assigned to the class the centroid of which is nearest in the sense of Mahalanobis distance (see Eq. 8.3). Here, centroid refers to an average object having mean values of a certain class in all variables. The Mahalanobis distance is essentially based on the covariance matrix or, in illustrative terms, the shape of the class distribution. In the case of spherical distribution the Mahalanobis distance coincides with the conventional Euclidean distance (Eq. 8.2). In this special case the test object would be assigned to the class with nearest (in the sense of conventional distance) centroid. This trivial classification rule, however, is not appropriate for nonspherical distributions occurring due to correlations among variables. In contrast to this, the Mahalanobis distance yields an optimal classification rule, provided multivariate normality can be assumed. If the Mahalonobis distance is computed for the covariance matrices of each class separately (by assuming different shapes of distribution), a nonlinear decision curve (surface for more than two variables) is generally obtained (see Fig. 12 C, where the test object is assigned to the filled-circle class). This situation is referred to as quadratic classification. Under the assumption of equal shapes of distribution as in Figure 12 D, decision lines are obtained (hyperplanes for greater dimensions). Therefore this method is called linear classification. Here the test object would be assigned to the unfilled-circle class. A drawback of quadratic classification is the need to estimate many parameters (different class covariance matrices). If this estimation is not supplied with enough learning objects, the decision surface or classification rule, respectively, will become rather uncertain. This is why linear classification is often used even if the shapes of the distribution are unequal. Then an equalized covariance matrix is generated by averaging.

Another type of classification is the allocation of test objects *(ALLOC) method* [17], where the densities of class distributions are estimated by potential functions resulting from so-called density kernels. These kernels, usually Gaussian shaped, are constructed around each of the learning objects. Summing them up pointwise for each class separately yields the desired density estimations. Now a test object is assigned to the class having maximum "influence" in it; this means attaining the highest value of density function at the position of the object. In Figure 12 E, where objects are arranged in one dimension for simplicity, the test object would be assigned to class A. Here, solid curves (Gaussian kernels and summed-up potential function) refer to class A and dashed curves to class B. As for KNN (optimal k) and SIMCA (optimal number of PCs), some parameter must be adjusted in ALLOC, namely, the smoothing parameter, which determines the flatness of the kernels. Obviously, neither too sharp nor too flat kernels are desirable for optimal class separation. The best decision can be evaluated once again by some cross-validation technique.

3.11. Multivariate Regression

3.11.1. Multiple Linear Regression

Multiple linear regression (MLR) is a method that estimates the coefficients $c_1 \ldots c_p$ in a linear dependence

$$y = c_1 \cdot x_1 + \cdot \cdot \cdot + c_p \cdot x_p \tag{11.1}$$

of some response variable $\boldsymbol{y}$ on several predicting variables $\boldsymbol{x}_i$. If both predicting variables and responses are observed n times, the resulting data can be collected in a data set $\boldsymbol{X}$ (n rows and p columns) and a vector $\boldsymbol{y}$ (n components). The optimal choice of the coefficient vector c in Equation 11.1 (p components) is obtained as the solution of the linear equation $X^T Xc = X^T y$ (where $X^T$is the transpose of the matrix). By using the resulting coefficients c_j, new responses of y can be predicted from new constellations of the x_i by means of Equation 11.1. Besides statistical evaluations of the computed results (e.g., confidence intervals for coefficients and for predicted values) an essential task of MLR is selecting a significant subset among the possibly numerous predictors [5]. A typical application of MLR arises in multicomponent calibration: Given the spectrum of a sample that is a mixture of p components, the concentrations of these components in the mixture sample need to be determined. Digitizing the spectra of the mixture and of the pure components, respectively, by n distinct wavelengths yields vectors y and x_i, respectively, which are related according to Equation (11.1) by Beer's law, where the coefficients c_i represent the desired concentrations. The pure components need not even be known exactly. It would be sufficient to start computations with a large set of pure components that are possibly included. In an ideal situation, the correct subset could then be detected by some variable selection technique as mentioned above. A serious disadvantage of MLR is the multicollinearity problem. This problem occurs when all or some of the predicting variables are highly correlated. As a consequence, the estimation of regression coefficients in Equation (11.1) becomes more and more unreliable. The fact may be illustrated by the extreme situation in which all predicting variables are identical and, hence, the response variable may be described by arbitrary coefficients. In the example of spectral data, high correlations are generated by components having similar spectra (strong overlapping). Such situations would lead to unreliable computed concentrations.

3.11.2. Latent Variable Regression

The multicollinearity problem can be circumvented by application of *principal components regression* (PCR), which, roughly speaking, is MLR applied not to the original set of predicting variables but to a smaller set of principal components of these variables. Since PCs are always uncorrelated (see Chap. 3.9), multicollinearities may be avoided. The deleted components, however, although describing only a smaller amount of variance in the set of predictors, might account for a great part of prediction of the response variable that will be lost. A procedure for selecting latent variables as preditors by taking both multicollinearity and predictive ability into consideration is called latent root regression [18].

For multivariate calibration in analytical chemistry, the *partial least squares* (PLS) method [19], is very efficient. Here, the relations between a set of predictors and a set (not just one) of response variables are modeled. In multicomponent calibration the known concentrations of l components in n calibration samples are collected to constitute the response matrix Y (n rows, l columns). Digitization of the spectra of calibration samples using p wavelengths yields the predictor matrix X (n rows, p columns). The relations between X and Y are modeled by latent variables for both data sets. These latent variables (PLS components) are constructed to exhaust maximal variance (information) within both data sets on the one hand and to be maximally correlated for the purpose of good prediction on the other hand. From the computational viewpoint, solutions are obtained by a simple iterative procedure. Having established the model for calibration samples, component concentrations for future mixtures can be predicted from their spectra. A survey of multicomponent regression is contained in [20].

3.12. Multidimensional Arrays

The increasing complexity of experimental designs and hyphenated methods requires a generalization of data tables to higher dimensional arrays. Frequently, a fixed classical objects/variables setting is observed several times under different conditions, yielding a separate data table for each condition. Piecing all these together, one arrives at a cubic or three-dimensional data array, as sketched in Figure 13. In this data example, a set of six trace element concentrations was determined in five different fodder plants for 32 samples (eight parallel detrminations in each of four different sampling regions labeled a to d. While the general entry of a data table may be referred to as x_{ij} (see Chapter 3.8), it must be denoted by x_{ijk} for a three-dimensional array. As before, the indices relate to the corresponding ordered item in each of the three dimensions. In the array of Figure 13, for example, x_{123} represents the concentration of Cd determined in grass for the first sample of region a, where the order of indices

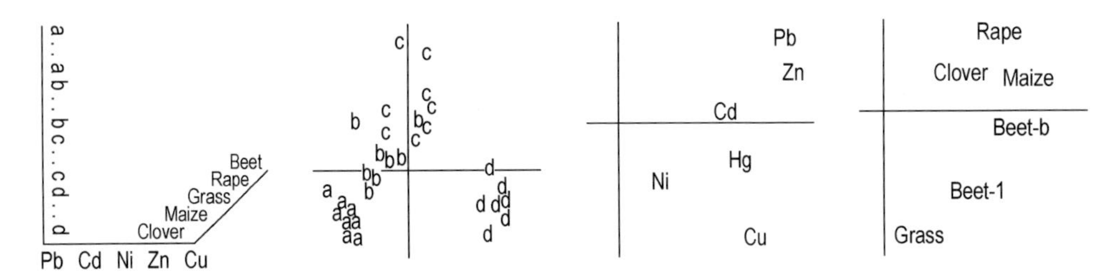

Figure 13. Example of a three-dimensional data array from environmental chemistry (left) and diagram plots for the first two factors of each of the three dimensions (to the right)

is: objects (samples), variables (elements), and "conditions" (fodder plants).

Such arrays raise the question of more generalizations of the table-oriented techniques presented in Chapters 3.9 to 3.11. The most prominent representatives of factorial methods are the so-called Tucker3 [21] and PARAFAC (parallel factor analysis) [22] models. For three-way arrays, the Tucker3 model is expressed as

$$x_{ijk} \approx \sum_{p=1}^{P} \sum_{q=1}^{Q} \sum_{r=1}^{R} c_{pqr}\, e_{ip}\, g_{jq}\, h_{kr} \qquad (12.1)$$

where, P, Q, R denote relatively small numbers of components for each of the three dimensions, the e_{ip}, g_{jq}, h_{kr} are entries of so-called component matrices (frequently required to be orthonormal), and the c_{pqr} refer to elements of a so-called core matrix. The approximation sign in Equation (12.1) is to be understood in a least-squares sense. The component matrices contain the scores or loadings of the factors for objects, variables, and conditions, and as such are generalizations of the principal components for data tables indeed, Eq. (9.2) relates to a special case of Eq.(12.1) when reducing the dimension of the array from three to two). In particular, choosing $P=Q=R=2$, the component matrices contain the coordinates for diagrams of objects, variables, and conditions. In this way, generalizations of plots like Figure 10 a can be obtained for three dimensions. The core elements indicate how single factors of different dimensions are linked with each other. Their squared values c_{pqr}^2 provide the amount of variation in the data explained by jointly interpreting factors p, q, and r, from objects, variables, and conditions, respectively. As an illustration, one may consult the diagram plots for samples, elements and fodder plants collected in Figure 13. For a brief interpretation of a first factor, one recognizes that all elements and fodder plants are given some signed (positive) scores on the first axis, which at the same time clearly differentiates between sampling regions in the order d > c, b > a. Hence, this first axis can be interpreted as one of general pollution exposure (not differentiating between elements and plants) with regions affected by pollution according to the just-stated order. A second factor then would differentiate between plants (e.g., grass as opposed to rape) or elements (Pb, Zn as opposed to Cu).

The PARAFAC model is slightly more restrictive, hence simpler, than Tucker 3:

$$x_{ijk} \approx \sum_{p=1}^{P} e_{ip}\, g_{jp}\, h_{kp} \qquad (12.2)$$

This model is particularly important for analysis of data from hyphenated methods. If, for instance, mixed samples of P chemical components are chararcterized by spectroscopic-chromatographic measurements, then the signal intensity x_{ijk} may be approximated as in Equation (12.2) by means of the spectral and chromatographic profiles g_{jp}, h_{kp} at wavelength j and retention time k of component p at unit concentration and the concentration e_{ip} of component p in mixture i. A peculiarity of the PARAFAC model is that its decomposition according to Equation (12.2) is unique, in contrast to the accordingly reduced decomposition of two-dimensional tables: whereas for data tables the factors can be rotated without changing the result of approximation, this is not possible for three-dimensional arrays. As a consequence, it is not necessary, for the decomposition to be followed by an appropriate rotation to arrive at chemically meaningful factors (e.g., spectra with nonnegative intensities or nonnegative concentration values). The PARAFAC model based on three-dimensional arrays is in principle able to find the correct and unique decomposition of the given data array in a single step. This fact highlights the importance of the model for qualitative (e.g., identification of spectral and chromatographic profiles of unknown pure components in the mixtures) and quantitative (e.g., concentrations of pure components in the mixture) analysis at the same time.

The Tucker 3 and the PARAFAC models are easily generalized to arrrays of arbitrarily high dimension by using tensorial notation (for a review, see [23])

3.13. References

[1] D. L. Massart, B. G. M. Vandeginste, S. N. Deming, Y. Michotte, L. Kaufman: *Chemometrics: a textbook,* Elsevier, Amsterdam 1988.

[2] M. A. Sharaf, D. L. Illmann, B. R. Kowalski: *Chemometrics,* John Wiley & Sons, New York 1986.

[3] D. C. Montgomery, G. C. Runger: *Applied Statistics and Probability for Engineers,* John Wiley & Sons, New York 1999.

[4] L. Sachs: *Angewandte Statistik,* Springer, Berlin 1999.

[5] S. Chatterjee, B. Price: *Regression Analysis by Example,* Wiley-Interscience, New York.

[6] K. Doerffel: *Statistische Methoden in der analytischen Chemie,* 5th ed., Deutscher Verlag für Grundstoffindustrie, Leipzig 1990.

[7] W. J. Youden: "Technique for Testing Accuracy of Analytical Data," *Anal. Chem.* **19** (1947) 946–950.

[8] R. Plackett, J. P. Burman, *Biometrika* **33** (1946) 305–310.

[9] K. Doerffel: "Selektivitätsprüfung von Analysenverfahren mit Hilfe unvollständiger Faktorpläne," *Pharmazie* **49** (1994) 216–218.

[10] J. C. Miller, J. N. Miller: *Statistics for Analytical Chemistry,* 2nd ed., Ellis Horwood Ltd., Chichester 1989.

[11] R. A. Johnson, D. W. Wichern: *Applied Multivariate Statistical Analysis,* Prentice Hall, New Jersey 1982.

[12] S. Wold, K. Esbensen, P. Geladi, *Chemom. Intell. Lab. Syst.* **2** (1987) 37–52.

[13] E. R. Malinowski, D. G. Howery: *Factor Analysis in Chemistry,* J. Wiley and Sons, New York 1980.

[14] W. H. Lawton, E. A. Sylvestre, *Technometrics* **13** (1971) 617.

[15] M. R. Anderberg: *Cluster Analysis for Applications,* Academic Press, New York 1973.

[16] S. Wold, *Pattern Recogn.* **8** (1975) 127–139.

[17] J. Hermans, J. D. F. Habbema: *Manual for the ALLOC Discriminant Analysis Programs,* University of Leiden, Leiden 1976.

[18] R. F. Gunst, R. L. Mason: *Regression Analysis and its Applications,* Marcel Dekker, New York 1980.

[19] H. Wold in K. G. Jöreskog, H. Wold (eds.): *Systems under Indirect Observation,* North Holland Publ., Amsterdam 1981.

[20] H. Martens, T. Næs: *Multivariate Calibration,* J. Wiley and Sons, Chichester 1989.

[21] L. R. Tucker, *Psychometrika* **31** (1966) 279–311.

[22] R. A. Haeshman, *UCLA Working Papers in Phonetics* **16** (1970) 1–84.

[23] R. Henrion, *Chemom. Intell. Lab. Syst.* **25** (1994) 1–23.

4. Weighing

Walter E. Kupper, Madison, New Jersey 07940, United States

4.1. Introduction

The field of mass measurement, commonly called weighing, is conveniently structured according to the degree of numerical precision involved. The classification and nomenclature of modern laboratory balances follows a decimal pattern based on the step size d of the associated digital display:

1) Precision balances: $d=1$, 0.1, 0.01, or 0.001 g
2) (Macro) analytical balances: $d=0.1$ mg
3) Semimicro balances: $d=0.01$ mg
4) Microbalances: $d=0.001$ mg
5) Ultramicrobalances: $d=0.0001$ mg

Weighing capacities range from 5 g for an ultramicrobalance to >60 kg for the largest precision balances (Figs. 1–5). Precision weighing instruments of greater capacity, known as industrial precision scales and not generally used in the laboratory, are available in a weighing range up to 6 t and $d=0.1$ kg.

The words "balance" and "scale" are often used interchangeably. "Balance" is derived from the Latin *bilanx*, having two pans [1]. "Scale" and "scales" are short forms of "a pair of scales" (old English, meaning dishes or plates). Modern weighing instruments no longer have two weighing pans, but the words "balance" and "scale" survive. The term "balance" is preferred for the more precise weighing instruments found in a laboratory, while "scale" is appropriate for all other weighing equipment.

4.2. The Principle of Magnetic Force Compensation

Analytical and precision balances are based on the principle of electromagnetic force compensation, where the weight of a sample is counterbalanced by a corresponding electromagnetic force. The measuring sensor or transducer in a balance is best described as a linear motor; i.e., it is an electromechanical aggregate that generates a straight-line force and motion on the basis of an electric current.

The operating principle of a loudspeaker (Fig. 6) provides a useful analogy. In a loudspeaker, an oscillatory, voice-modulated current is fed through a coil (c) to create a small magnetic field that interacts with the larger field from a surrounding permanent magnet (d) to generate an oscillatory force that in turn moves a diaphragm or voice cone (a), thus producing sound. In a balance sensor, direct current through a coil (c) similarly produces a static force of precisely the magnitude

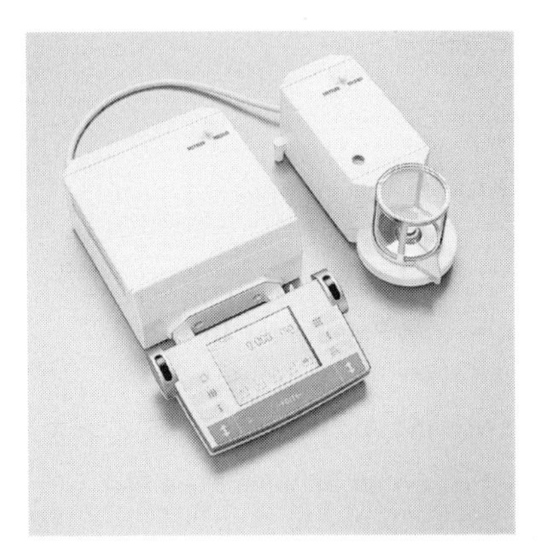

Figure 1. Microbalance Mettler MX5, 5100 mg × 1 μg
Automatic self-calibration with built-in reference masses.

Figure 2. Analytical balance Mettler AX201, 220 g × 0.01 mg
Automatic self-calibration with built-in reference masses.

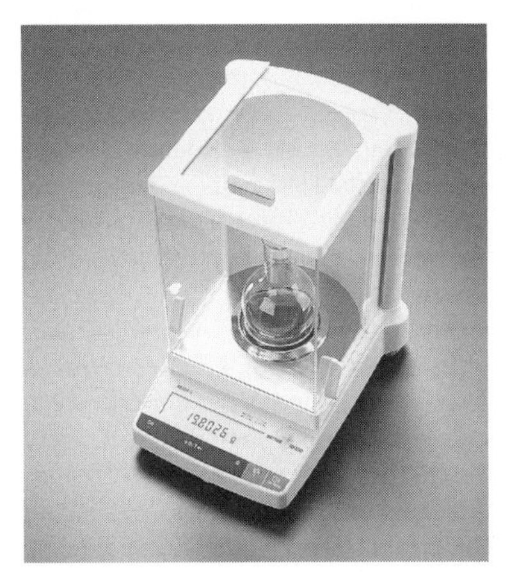

Figure 3. Analytical balance Mettler AB204-S, 210 g × 0.1 mg
Self-calibration with built-in reference mass.

Figure 4. Precision balance Mettler PR8002, DeltaRange, 8100 g × 0.01 g
Self-calibration with built-in reference mass.

Figure 5. Precision balance Mettler SR32001, 32 100 × 0.1 g
Self-calibration with built-in reference mass.

required to counterbalance the weight of a sample load (b). A position-maintaining feedback circuit (e) controls this current in such a way as to keep the overall magnetic force in equilibrium with the weight. The strength of the required current in this case provides a measure of the weight; i.e., the balance establishes a mass in grams on the basis of a current measured in milliamperes.

A typical electronic balance incorporates a system of links and levers like that shown schematically in Figure 7. The weighing pan (a) is guided by two parallel horizontal links and counterbalanced by the force coil (g) acting through a lever (c). The pivots (b) in this linkage are metallic leaf springs, called *flexures*. In the weighing process, the pan is initially depressed by the sample, after which it is restored to its null level by the servoac-

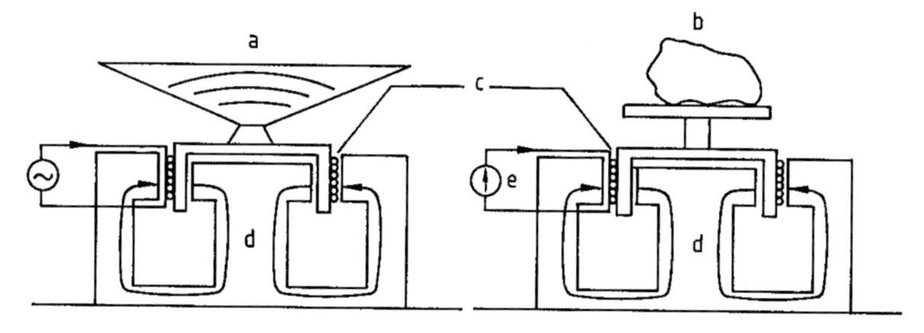

Figure 6. Analogy between a loudspeaker and a laboratory balance, as explained in the text
a) Diaphragm; b) Load; c) Coils; d) Magnets; e) Feedback circuit

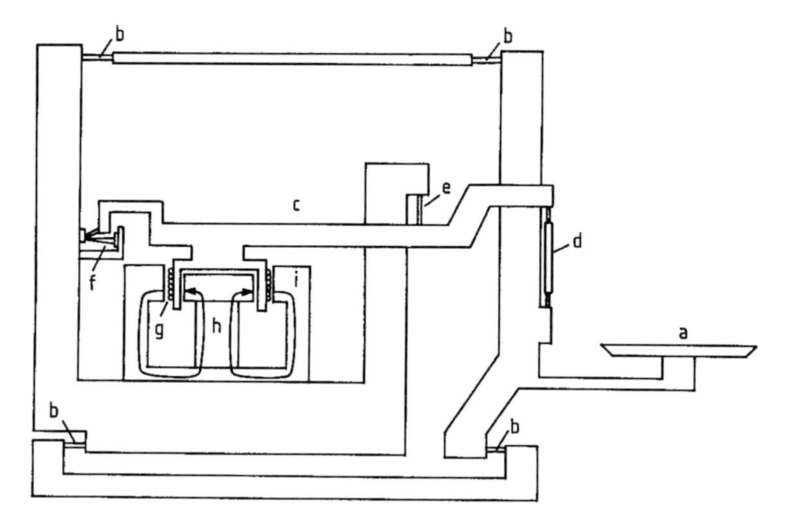

Figure 7. Schematic cross-section through an analytical balance
a) Weighing pan, cantilevered from a parallel-motion linkage; b) Flexure pivots of the parallel-motion linkage; c) Main lever or balance beam; d) Coupling between linkage and beam; e) Main pivot or fulcrum; f) Photoelectric position sensor; g) Force coil; h) Cylindrical permanent magnet; i) Cylindrical soft-iron shell, channeling the magnetic flux

tion of the force motor. This balancing process is controlled with sub-micron sensitivity by a photoelectric position sensor (f).

4.3. Automatic and Semiautomatic Calibration

The electronic signal-processing part of a balance is comparable to a digital voltmeter, except that the display is calibrated in units of mass. The *accuracy* (correspondence between a displayed value and the true mass on the balance) of a modern professional-grade balance is ensured by automatic or semiautomatic self-calibration. In a self-calibration cycle, a reference mass is either deposited on the balance internally or set on the pan manually by the operator, after which the balance's microprocessor updates the value of a calibration factor stored in a nonvolatile (power-independent) memory. This calibration function represents a direct relationship between the weighing object and the displayed value, in that the weighing process is referenced directly to a mass. A more sophisticated form of automatic calibration employs two built-in reference masses to assure both calibration at full capacity and linearity at midrange of the balance (see Figs. 1 and 2).

4.4. Processing and Computing Functions

Thanks to the capabilities of microprocessors, a number of special computation routines are common to most balances:

1) Pushbutton tare for subtracting container weights
2) Automatic zero tracking to keep the display zeroed when no weight is on the pan
3) Vibration filtering to make displayed values more immune to disturbances
4) Stability detection to prevent premature reading or transfer of transient or fluctuating results

In addition to these signal processing features, some balances also provide useful convenience options such as conversion to nonmetric units, parts counting, weight statistics, percentage calculation, and animal weighing.

4.5. Balance Performance

A weighing instrument is commonly tested by loading it with calibrated mass standards and then ascertaining whether the displayed weight values are accurate, meaning that each result corresponds to the value of the test standard within some predetermined tolerance limits. In the most elementary check of balance accuracy, a single test mass is placed once on the balance. This is a test that is often performed for extra assurance in critical applications.

A complete balance test consists of a systematic set of weighings designed to evaluate all aspects of balance performance, including repeatability, eccentric loading errors, linearity, and span calibration. For a balance to be declared in proper working condition, all test results must fall within the tolerance limits specified by the manufacturer. These tolerances normally apply only to new or newly serviced equipment. In other cases a so-called in-service tolerance — equal to twice the original tolerance — is customarily allowed.

Repeatability for successive weighings of the same load is calculated as standard deviation from ten weighings of the same mass, where the balance is reset to zero as necessary before each weighing. The procedure is usually conducted with a mass of approximately one-half the weighing capacity of the balance. This test measures short-term fluctuations and noise of the type present to some degree in every electronic instrument. Excessive standard deviation may be a result of adverse environmental conditions, such as building vibrations or air drafts. A factor equal to three times the standard deviation is customarily called the *uncertainty* of the balance, meaning that virtually all observed values (theoretically, 99.7 %) would fall within this limit on either side of the average.

In the *eccentric load test* a weight equal to one-half the capacity of the balance is placed at the center of the weighing pan and the balance is reset to zero. The weight is then moved, in turn, half-way to the left, right, front, and rear of the pan. Weight readings from the four eccentric positions, if different from zero, are called eccentric loading errors. If not specified separately, the tolerance limit for such errors is of the same order of magnitude as that for nonlinearity. Larger errors indicate that the mechanical geometry of the balance is out of adjustment, usually as a result of abusive treatment. A correctly adjusted balance is practically insensitive to eccentric placement of the load, but any residual errors will be minimized if samples and containers are approximately centered on the balance pan.

Nonlinearity refers to deviations from a mathematically straight line in a graph of display readings versus actual weights, where the line is drawn from the zero point to the full-capacity endpoint. Linearity errors in a balance are caused by practical imperfections in the theoretically linear scheme of electromagnetic force compensation. Nonlinearity manifests itself as a variation in results when the same object is weighed together with differing amounts of tare, and this is, in fact, how the nonlinearity of a balance is tested. A weight equal to one-half the capacity is weighed once without tare, then with a tare of the same magnitude as the weight itself. The mid-range linearity error of the balance is equal to one-half the difference between the two results. This test detects the most common type of balance nonlinearity, an error curve of parabolic shape with maximum positive or negative error at mid range. Higher-order components in the error curve can be detected with four weighings of a test weight equivalent to one-fourth the capacity, using tares equal to 0, 25, 50, and 75 % of the balance capacity. The advantage of such a procedure is that it does not require calibrated weights; the test weight as well as the tares represent nominal values only.

The *calibration error*, also known as sensitivity error or span error, is a small relative factor (a

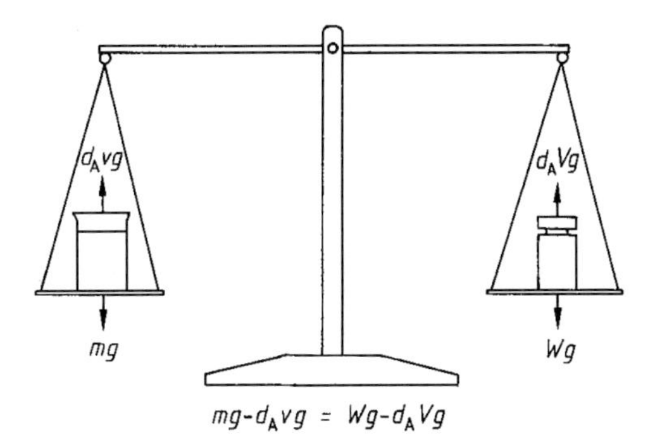

Figure 8. Gravity and buoyancy forces acting on a two-pan balance
v = sample volume. V = volume of the mass standard.
g = acceleration due to gravity.

constant percentage of the applied load) by which all weight readings may be biased. For a balance with a self-calibration device this error is limited to the error tolerance of the reference weight, which is of the order of 0.0001 % (1 ppm) in the case of an analytical balance. If the balance lacks self-calibration, calibration errors may develop:

1) With large changes in room temperature
2) When the balance is moved to a different location
3) As a long-term aging effect in the electromagnetic force sensor

Calibration errors should be ascertained with a certified test weight of substantially lower uncertainty than the calibration tolerance of the balance (insofar as this is possible or necessary in practice). In the finest analytical balances, the achievable calibration-error tolerance is dictated by the accuracy of the available mass standards. However, this limitation is not relevant at the level of 0.1 % to 0.01 % accuracy applicable to most laboratory work.

The temperature dependence of the calibration error (temperature drift) is tested in new balances at the factory. Test weighings are made in an environmental test chamber at various temperatures (e.g., 10, 20, and 30 °C). A typical tolerance for the span drift (sensitivity drift) as a function of temperature for an analytical balance is 1 ppm/ °C, also expressed as 10^{-6}/ °C. With a test weight of 100 g, this means that a correct reading of 100.0000 g at 20 °C could change to 99.9999 g or 100.0001 g (or anywhere in between) if the temperature were to change to either 19 or 21 °C.

4.6. Fitness of a Balance for Its Application

To select an appropriate balance for a given application, one needs to answer two basic questions:

1) How heavy are the samples to be investigated?
2) How accurately must they be weighed?

For example, if the accuracy objective is 0.1 %, then a balance should be selected for which the sample weights would represent at least a thousand times the value of the uncertainty, or three thousand times the standard deviation. Adherence to this rule may be encouraged by placing an appropriate label on the balance, such as “Minimum net sample weight 60 milligrams” for a balance with a standard deviation of 0.02 mg. To weigh smaller samples, one would select a balance with a smaller standard deviation. This requirement of 0.1 % accuracy and its interpretation in terms of the standard deviation are mandated for pharmaceutical weighing according to the *United States Pharmacopeia* [2].

4.7. Gravity and Air Buoyancy

All weighing relies ultimately on the concept of equilibrium with a symmetrical lever (Fig. 8). A *weight*, in the common usage of the term, is represented by the total value of calibrated weigh-masses required to put the lever in balance. According to international convention, the reference masses are to be fabricated from stainless steel with a density of 8 g/cm^3 [3]. The forces in equilibrium arise primarily from the downward pull of gravity on the two sides of the balance, but small buoyant forces from the surrounding air must also be considered. In Figure 8, the gravity and buoyancy forces are symbolized by arrows. The equation under the diagram expresses the equilibrium in mathematical terms. Solving the equilibrium equation for m leads to the conversion from weight to mass.

$$m = W \cdot (1 - d_A/D)/(1 - d_A/d) = W \cdot k \qquad (1)$$

m sample mass, expressed in the same metric unit as that applicable to W

Table 1. Weights at sea level (W_S) and in Denver, Colorado (W_D), for samples with a mass of one gram and various densities d

Sample density d, g/cm^3	Weight at sea level W_S, g	Weight in Denver W_D, g	Difference $(W_D - W_S)/W_S$, %
0.5	0.997749	0.998124	0.0376
1.0	0.998949	0.999124	0.0175
2.0	0.999549	0.999624	0.0075
4.0	0.999849	0.999874	0.0025
8.0	1.000000	1.000000	0.0000
20.0	1.000090	1.000074	−0.0017

W sample weight, defined in terms of the counterbalancing steel mass
d_A density of ambient air, g/cm^3
D density of the mass standard, normally 8 g/cm^3
d density of the sample, g/cm^3
k weight-to-mass conversion ratio for the sample

According to this expression, there are two cases in which sample mass equals sample weight ($k = 1$): at zero air density (vacuum), or if the sample density d equals the density D of the mass standard.

Typical values for air density are approximately 1.2 mg/cm^3 at sea level and 1.0 mg/cm^3 in Denver, Colorado (altitude 1600 m). The air density may also vary by as much as 6 % with changes in the weather. Given that the ambient air density is variable, it follows from Equation (1) that the same mass will produce slightly different weights at different times and places, as illustrated in Table 1. This variation is too small to be relevant in commercial applications, however, and the same is true for most laboratory weighings as well.

The conclusions derived above for a two-pan balance also apply to modern electronic balances. Weighing still entails comparison of the test object or sample with a calibrated steel mass. However, in the case of a single-pan electronic balance, the standard is placed on the pan at a particular point in time in order to calibrate the balance, and the weighing process occurs on the same pan at a subsequent time. Thus, weighing on an electronic balance involves a sequential rather than simultaneous comparison, which has one important implication: an electronic balance is the functional equivalent of a classical two-pan balance only if the surrounding gravity field and atmosphere are the same during weighing as when the balance was calibrated. Therefore, electronic balances must always be calibrated at the place where they are to be used, particularly since the force of gravity varies by more than 0.1 % over a territory as limited as the continental United States.

4.8. The Distinction Between Mass and Weight

Weight, in the primary definition of the term, means the direct result of a weighing, without any correction for air buoyancy. Mass, in the context of weighing, is understood as an absolute, buoyancy-corrected quantity. A problem arises if a weighing result is reported as a mass, when in fact no buoyancy correction has been made. The discrepancy amounts to about 0.1 % for a material with a density of 1 g/cm^3. In science there is a general desire to avoid the term "weight" in the sense of a weighing result, because scientific nomenclature treats weight as a *force*, properly expressed in dynes or newtons, whereas the gram is a unit of mass. To prevent misunderstandings, if weighing results are reported as "mass" then the context should make it clear whether or not stated values have been subjected to an air-buoyancy correction.

4.9. Qualitative Factors in Weighing

A proper environment and sufficient technical skill on the part of the operator are critical for achieving satisfactory weighing results. Precision balances weighing to 0.1 g perform well in almost any situation, but with analytical and microbalances a number of factors can be detrimental to accurate weighing, including air drafts, direct sunlight and other forms of radiant heat, building vibrations, and electrostatic and magnetic forces. A stable indoor climate is generally sufficient, but balances should be kept away from doors, windows, ovens, and heat and air conditioning outlets, and they should never be placed under ceiling outlets. A mid-range level of relative humidity is best, since this will prevent the buildup of electrostatic charge from excessively dry air as well as moisture absorption by samples and containers from air that is too humid. The balance table should rest on a solid foundation so the balance will not be affected if the operator moves or leans on the table. There should be no elevators or vibrating machinery nearby. An operator may find it necessary to conduct several experiments in a search for appropriate procedures when weighing hygroscopic, volatile, or ferromagnetic samples. Samples and containers should always be acclimated to room temperature to prevent air convection around the sample that might interfere with weighing.

4.10. Governmental Regulations and Standardization

Weighing is subject to governmental regulation in situations where errors or fraud might result in material or health hazards. Weighing in trade and commerce is commonly supervised by a governmental "Weights and Measures" agency through some system of certification and inspection of scales and weights [4], [5]. In areas other than commerce, the nature and scope of government control over weighing varies depending upon the field, as in the case of pharmaceuticals [2], nuclear materials [6], or military defense contracts [7]. In the United States, critical weighing applications are subject to a system of government-supervised self-control. Any organization conducting weighing processes subject to regulatory control must have adopted formal mass assurance procedures as part of an overall quality assurance system. The government exercises its oversight function through periodic audits. The main elements of a typical mass assurance program are as follows:

1) Balances available to the staff must conform to the accuracy required for the intended use
2) Balances must be maintained and calibrated on a regular cycle, with records kept on file
3) All balance operators must be instructed in correct weighing technique, and training records must be kept on file
4) Formal and documented weighing procedures must be in place
5) Orderly records must be maintained for all weighings that fall within the scope of the regulated activity

The most widely accepted guideline for establishing a system of mass assurance is found in International Standard ISO 9001, Section 4.11., "Inspection, Measuring and Test Equipment" [8]. Under ISO 9001, the quality system of an organization is to be audited and certified by independent quality assessors, analogous to the audits performed by the government on regulated organizations.

4.11. References

[1] B. Kisch: *Scales and Weights: A Historical Outline,* Yale University Press, New Haven 1965.

[2] *The United States Pharmacopeia,* 22nd ed., United States Pharmacopeia Convention, Inc., Rockville, Md., 1990.

[3] OIML Recommendation R-33, Conventional Value of the Result of Weighing in Air, Bureau International de Métrologie Légale, Paris, France, 1973.

[4] OIML International Recommendation R 76-1, Non-automatic Weighing Instruments. Part 1: Metrological and Technical Requirements—Tests, Bureau International de Métrologie Légale, Paris, France, 1993.

[5] NIST Handbook 44, U.S. Dept. of Commerce, National Institute of Standards and Technology, Gaithersburg, Md., 1993.

[6] MIL-STD-45662A, Calibration Systems Requirements, Department of Defense, Washington, DC 20360, 1988.

[7] Nuclear Regulatory Commission, Codes 10 CFR 21 and 10 CFR 50, 1989.

[8] ISO 9001,*Quality Systems—Model for Quality Assurance in Design/Development, Production, Installation and Servicing,* 1994.

5. Sampling

JAMES N. MILLER, Department of Chemistry, Loughborough University, Loughborough, United Kingdom

5.1. Introduction and Terminology

The determination of the concentrations of one or more components of a chemical material is a multistage process. Research in analytical chemistry understandably gives emphasis to methods of pretreating materials prior to analysis (e.g., dissolution, concentration, removal of interferences); to the production of standard materials and methods, such as those used to check for bias in measurement; to the development of analytical techniques demonstrating ever-improving selectivity, sensitivity, speed, cost per sample, etc.; and to new methods of optimization and data handling which enhance the information generated by these techniques. Many of these improvements are in vain, however, if the very first stage of the analytical process is not planned and managed properly: this is the *sampling* stage, which should ensure that specimens subjected to analysis are fully representative of the materials of interest. Repeated investigations show that much of the variation encountered when (for example) similar analyses are performed by different laboratories, arises at the sampling stage. This article outlines the general and statistical aspects of the sampling process. Detailed coverage of the chemical aspects of sampling in particular applications is to be found in reviews and monographs [1], [2].

The word *sample* is used in two rather different, but related senses in analytical science. In each case, the general implication is clear—a sample is a small and hopefully representative portion of a larger object. But some differences are also clear. Statisticians use the term to mean a small number of measurements which are assumed representative of a theoretically infinite *population* of measurements. Chemists and others use the term to mean a small number of specimens, taken in cases where the materials of interest are too large or too numerous to be examined in their entirety. Thus, a water analyst might take ten specimens of water from a particular stretch of river, knowing or assuming that they form a representative sample of the whole of that part of the river at a particular time. Laboratory staff might then make five replicate pH measurements on each specimen: in each case, these five pH values are a sample representing a potentially infinite number of pH values—the population. (It is conventional to use Greek letters for population statistics, and the Roman alphabet for sample statistics.) In this example, the chemical and statistical uses of the word sample are to some degree distinct. In a different application area, an analyst might remove every hundredth capsule from a pharmaceutical production line, and analzye each capsule once to determine the level of the active component. The concentrations obtained would form a sample (in both chemical and statistical senses of the word) representing the behavior of the production line. The latter example also reminds us that sampling theory is closely related to *time series* statistics, and to *quality control* methods.

Once the overall objectives of an analytical process have been set, the sampling steps must be carefully planned: how many specimens (or *sample increments;* see Chap. 5.2) should be taken? How should they be selected? How big should they be (in cases of bulk samples)? One difficulty is that in some analyses the nature of the sample and the precision of the sampling and measurement steps may not initially be known. In such cases, simple preliminary sampling plans may be used to provide such information, and the resulting data can be used iteratively to improve the sampling procedures for long-term use. Good sampling plans should be available in written form as protocols which describe experimental procedures in detail, set criteria on the circumstances in which a sample can be rejected, and list the additional information that should be gathered when a sample is taken (temperature, weather, etc.). In principle, such protocols allow samples to be taken by untrained staff, but it is desirable (if not always practicable) for a sample to be taken by the analytical scientist who is to perform the subsequent measurements.

5.2. Probability Sampling

Most sampling procedures combine a suitable method of selecting sample items or increments (individual portions) with a knowledge of probability theory. From the analytical data conclusions are drawn on, for example, the risks involved in accepting a particular product. Such methods are known collectively as probability sampling methods. The first step in such a procedure is the selection of the sample items or increments: it is critically important that these are not biased in any conscious or unconscious way, otherwise they will not be representative of the corresponding population. Usually, each part or element of the population must have an equal chance of selection. (This is not absolutely necessary, but if the probabilities of selecting different parts of the population are not equal, those probabilities should be known.) The best method of ensuring such equal probabilities is to use *random sampling*. This method (definitely not the same as haphazard sampling!) involves using a table of random numbers to indicate the items or increments to be selected. The approach can be applied to bulk samples as well as itemized/numbered materials. For example, a truckload of coal could be regarded as being divided into numbered cells by dividing the truck's internal dimensions both horizontally and vertically: the cells to be sampled are then obtained from the random number table. Sampling normally proceeds without replacement, i.e., an item or cell once sampled is not available for further sampling.

An important component of the sampling process—and an example of how it can be developed and improved with time—is to analyze for any systematic relationships within these random samples. For example, samples taken late in the working day may give results that differ significantly from those taken early in the morning. Such a trend, which if unnoticed might bias the results, can be taken into account once identified. Alternative sampling approaches are less satisfactory than a properly controlled random sample. Simplicity suggests that it might be easier, for example, to sample every hundredth item from a production line: but if the manufacturing process has an undesirable periodicity such a regular sampling plan might conceal the problem.

In some applications it is possible to use a *composite sample,* produced by taking several separate increments from a bulk material, and blending them. This process requires care (especially when solids of different particle size, etc., are involved) to produce a single sample that is analyzed several times. This procedure assumes that it is only the average composition of the bulk material which is of interest. When replicate measurements are made on this composite sample the sampling variance is greater (i.e., worse) than would have been the case without blending, but there is a benefit in that fewer measurements are necessary—considerations of time and cost often arise in the development of sampling plans.

Some materials to be sampled (e.g., ores or minerals) are often heterogeneous, i.e., the components of interest are distributed through the bulk in a nonrandom way. In such cases it is clearly inappropriate to use simple random sampling, and a more complex approach is necessary. This involves dividing the bulk material into a sensible number of segments or strata, and then (using random numbers, as before) taking samples from each stratum. It is usual to use strata of equal size, in which case the sample size for each stratum is also the same, but if the strata are by nature unequal in size, the number of samples taken should be proportional to these sizes. Such *stratified random sampling* clearly involves a compromise between the number of strata chosen and the labor involved in the analyses. If too few strata are chosen, it is possible to take acceptable numbers of samples from each, but some inhomogeneities in the bulk material may remain concealed; if too many strata are chosen, the number of measurements necessary is large, if the sampling variance is not to be too high. This is another instance where prior knowledge and/or iteratively gained experience helps to formulate a sensible sampling plan.

A useful qualitative guide to the principles underlying probability sampling is provided by an ASTM standard [3].

In many cases sampling schemes are regulated by statutes, such as those published by individual governments, the European Community, the Codex Alimentarius Commission, and so on. Such schemes are particularly common in the analysis of foodstuffs, and may specify the mass and number of the sample increments to be taken as well as the statutory limits for the analyte(s) under study.

5.3. Basic Sampling Statistics

The overall random error of an analytical process, expressed as the variance s^2, can be regarded as the sum of two other variances, that due to sampling s_1^2, and that due to the remaining measurement components of the process s_0^2. These variances are *estimates* of the corresponding population variances σ^2 and its components σ_1^2 and σ_0^2. The overall standard deviation s is calculated as usual from single measurements on each of h sample increments, and the confidence limits of the true mean value of the population μ are obtained from:

$$\mu = \bar{x} \pm t\, s/\sqrt{h} \tag{1}$$

where $\bar{x}$ is the mean of the h measurements, and the t-value is at the appropriate confidence level. This first calculation of the mean and standard deviation allows a sampling plan to be initiated; further measurements then give refined values.

From Equation (1), the variance of $\bar{x}$ is s^2/h, used as an estimate of σ^2/h. To reduce, i.e., improve this variance, it is necessary to take n measurements of each of the h increments. This allows the separation of the sampling and measurement variances by one-way analysis of variance. The replication of the measurements is expected to reduce the measurement variance σ_0^2, and the overall variance of the mean becomes $\sigma_0^2/nh + \sigma_1^2/h$. This relationship provides general guidance on the best practical way of reducing the overall variance. The measurement variance can be minimized by using a high-precision analytical method or by using higher values of n, but in practice the sampling variance is often much larger than the measurement variance, so the choice of h is much more important (see below). If a composite sample is used, i.e., h sample increments blended to form one sample which is then measured n times (see above), the variance of the mean of such replicates is $\sigma_0^2/n + \sigma_1^2/h$. This variance is higher (worse) than when h increments are measured n times each, but the number of measurements is reduced (n as opposed to nh).

When sampling a bulk material, it is clearly important to come to decisions about: (1) how many increments should be taken; and (2) how large they should be. The minimum number of increments h necessary to obtain a given level of confidence is:

$$h = t^2 s_1^2/E^2 \tag{2}$$

where E is the largest permissible difference between the sample estimate $\bar{x}$ and the corresponding population value μ for the determinand (the equation assumes that each increment is measured once only). The t-statistic is used at the desired confidence level, e.g., 95 % or 99.7 %. Since the value of t depends on h, it is necessary to make an initial estimate of h, using the $h = \infty$ value of t (e.g., $t = 1.96$ for 95 % confidence), thus obtaining from Equation (2) a preliminary value of h: a new value of t can then be taken, and a reliable final value of h approached iteratively. The ASTM standard [4] recommends the use of 99.7 % confidence levels. It should be noted that Equation (2) is only applicable if the determinand is distributed according to the Gaussian, or "normal", distribution: different distributions require separate expressions for h.

When a bulk material is examined, the size of each increment is also of importance. Clearly, if each increment is too large, it may conceal the extent of variation within the bulk material: if it is too small, many increments are necessary to reveal the extent of the sampling variance. INGAMELLS [5] utilized the fact that s_1^2 decreased as the increment size increased to develop the equation:

$$W R^2 = K_S \tag{3}$$

where W is the increment weight, R is the relative standard deviation of the sample composition, and K_S is a constant. The definition of K_S depends on the user (e.g., it might be the weight necessary to

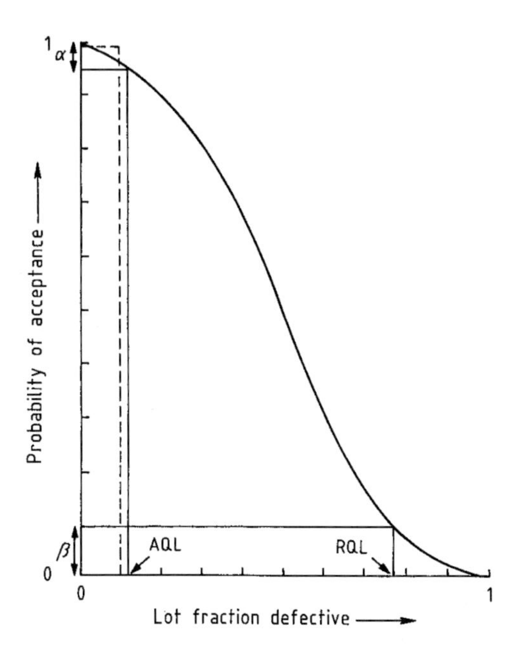

Figure 1. A typical OC curve
The broken line shows the result for an idealized sampling plan

limit the uncertainty in the composition of the material to 2 % with 95 % confidence); it is estimated by carrying out preliminary experiments to determine how s_1 and hence R vary with W. Once K_S is known, the minimum weight W required to produce a particular value of R is readily calculated.

Equations analogous to (2) and (3) can be derived for the more complex situations that arise in stratified random sampling [1], [6]. A computer program has been developed [7] to assist in the solution of sampling problems: it is especially directed at geochemical and other areas where the sizes and shapes of particulate solids may affect R^2.

5.4. Acceptance Sampling

This chapter deals with the problems associated with the acceptance or rejection of a product by a purchaser, on the basis of an examination of a *sample* taken from the whole *lot* of the product. The general aim of the methods summarized here is to protect both the manufacturer (e.g., of a pharmaceutical product which must contain a specified amount or range of the active ingredient), and also the consumer. The manufacturer is protected if good or valid lots of the product are (wrongly) rejected only infrequently; and the consumer is protected if bad or invalid lots are (wrongly) accepted only infrequently. Since the contents of the components of each lot will show variation (generally assumed to be Gaussian), the term "infrequently" has to be defined numerically (e.g., as 5 % of all occasions), and it is also necessary to give quantitative definition to "good/valid" and "bad/invalid". In some cases, lots are tested by means of a qualitative criterion (sampling by attributes), but more usually a quantitative measure is used (sampling by variables—this approach requires smaller samples than sampling by attributes to achieve a given degree of discrimination), and good and bad are thus defined numerically. Once these definitions are known, it is possible to develop a sampling plan, and describe it by an *operating characteristic* (OC) curve, sometimes known as a power curve, because of its relationship to the concept of the power of a statistical significance test.

An OC curve for sampling by attributes plots the *probability of acceptance* of a lot (y-axis) against the *lot fraction defective* (x-axis), i.e., the proportion of items in a lot which fail to meet the predetermined "good" criterion. A typical curve is shown in Figure 1. The success of a sampling plan in discriminating between good and bad lots depends on the shape of its curve. An ideal plan would give the shape shown in the figure as a broken line: the probability of acceptance would be 1.00 until the lot fraction defective reached a predetermined value, and would then fall abruptly to zero. In practice a curve is always obtained, so two points on it are particularly important. The first is called the *acceptance quality level* (AQL): this is the percentage of defective items that is (just) tolerable with a predetermined frequency, e.g., 5 %. A lot with this proportion of defectives will be rejected with a probability α. This is known as the *producer's risk.* It is often set at 5 %, but this is not universal; in some cases, plans are developed so that $\alpha = 10\,\%$ for small lots, but only 1 % for large lots. A second point of importance is the *rejectable quality level* (RQL), also known as the *lot tolerance percent defective* (LTPD), or the *limiting quality level* (LQL). If β is the probability of acceptance at the RQL, this is clearly the *consumer's risk,* i.e., the risk that the consumer will accept the lot, even though it has a more substantial portion of defective items. Quite commonly, $\beta = 10\,\%$.

To define a sampling plan when the sampling is by attributes, three numbers must be fixed. These are the sample size h, the *acceptance number*, and the *rejection number*. The acceptance number, sometimes given the symbol c, is the number which must not be exceeded by the number of defective items in the sample if the lot is to be accepted. The rejection number is the number at or above which the number of defectives in the sample leads to rejection of the lot. It might seem that, by definition, the rejection number is $c+1$, but this is not so, if the sampling plan is a *multiple* one, i.e., a plan in which a second or subsequent sample might be taken, depending on the results obtained for the first sample. In such cases, if the number of defectives in the first sample did not exceed c, the lot would be accepted; if the number of defectives exceeded the rejection number, the lot would be rejected; and if the number of defectives fell between these two extremes, one or more further samples might be taken. It is easy to see that multiple sampling plans might be highly discriminating, but the price paid is the much larger number of measurements needed, and such plans are not very widely used.

In the case of sampling by variables, the criteria needed to define the OC curve are inevitably different, and can best be illustrated by a simple example. Suppose that a manufactured fluorescent dye is of use to a laboratory analyst only if it contains at least 99 % of the active fluorophore. Any lot containing $\geq 99\,\%$ is acceptable, but the analyst is reluctant to buy the dye if the fluorophore content falls below 96 %. It is necessary to quantify these requirements: let us say that the manufacturer requires that a dye lot containing 99 % fluorophore must have a 95 % probability of acceptance, and the analyst requires that the probability of accepting a lot containing only 96 % fluorophore is only 1 %. It is also postulated that the sample size is h, and that the standard deviation σ for the fluorophore analysis, which reflects both sample inhomogeneities and measurement variation, is known exactly. (This is the case because of long experience of the production and testing of the dye.) It is necessary to estimate h and it is also stipulated that a dye lot will be accepted whenever the average fluorophore content over these h measurements equals or exceeds a value k, which is also to be estimated.

Suppose first that the true fluorophore content of a lot is 99 %. The sample average $\bar{x}$ is then normally distributed, with mean 99 and standard deviation σ/h. The probability of acceptance of the lot is the probability that $\bar{x} > k$; this is equal to the probability that the standard normal deviate:

$$z = (\bar{x} - 99)/\left(\sigma/\sqrt{h}\right) > (k - 99)/\left(\sigma/\sqrt{h}\right)$$

This probability has been defined as 0.95, so from the standard statistical tables for z, the result is:

$$(k - 99)/\left(\sigma/\sqrt{h}\right) = -1.64$$

In the alternative situation, suppose that the true fluorophore content of the dye lot is only 96 %. By exactly analogous arguments:

$$(k - 96)/\left(\sigma/\sqrt{h}\right) = +2.33$$

the latter number again being obtained from the tables for z at a probability of 0.99. Combining these results:

$$(k - 99)/(k - 96) = -(1.64/2.33) = -0.704$$

from which $k = 97.76$. If it is assumed that $\sigma = 2.0$, h can be obtained from either of the two above relationships. Thus:

$$(97.76 - 99)/\left(2.00/\sqrt{h}\right) = -1.64$$

from which h is 7, almost exactly. The sampling plan thus consists of taking seven measurements, and rejecting a lot whenever the average fluorophore content does not reach 97.76 %. A noteworthy feature of these calculations is that σ does not have to be known for k to be determined, so, as in other forms of sampling, a reasonable plan can be obtained, even when σ is initially unknown. It is possible to make a conservative estimate of σ, use it along with k to estimate h, and refine the estimate with increasing experience of the variance.

Sampling plans of this kind generate OC curves of the same form as those described above, although in cases of sampling by variables the x-axis is normally plotted as the mean determinant concentration (the fluorophore level in the example above). Other aspects of OC curves, of multiple sampling plans, and of *sequential* sampling plans, in which samples are taken one at a time, are dealt with in more advanced texts [8]–[10].

5.5. Conclusions

This brief survey of sampling and its application in various circumstances should indicate that the subject is of critical importance to the performance of the overall analytical process, whether the latter is laboratory based or process based. It is also a very good example of the principle that methods based on statistics and probability can, and should, be applied in the planning stage of an analysis — to use such methods only when data have been obtained is almost invariably too late!

5.6. References

[1] B. Kratochvil, J. K. Taylor, *Anal. Chem.* **53** (1981) 924 A – 938 A.
[2] D. T. E. Hunt, A. L. Wilson: *The Chemical Analysis of Water,* 2nd ed., Royal Society of Chemistry, London 1986.
[3] ASTM Standard E 105-58,*Standard Recommended Practice for Probability Sampling of Materials, ASTM, Philadelphia.*
[4] ASTM Standard E 122-72,*Standard Recommended Practice for Choice of Sample Size to Estimate the Average quality of a Lot or Process, ASTM, Philadelphia.*
[5] C. O. Ingamells, *Talanta* **21** (1974) 141 – 146.
[6] V. Barnett: *Elements of Sampling Theory,* Hodder and Stoughton, London 1974.
[7] P. Minkkinnen, *Anal. Chim. Acta* **196** (1987) 237 – 245.
[8] D. C. Montgomery: *Introduction to Statistical Quality Control,* Wiley, New York 1985.
[9] E. L. Crow, F. A. Davis, M. W. Maxfield: *Statistics Manual,* Dover, New York 1960.
[10] D. L. Massart, B. G. M. Vandeginste, L. M. Buydens, S. De Jong, P. J. Lewi, J. Smeyers-Verbeke, *Handbook of Chemometries & Dualmetries,* **Part 1,** Elsevier, Amsterdam 1997, Chapter 20.

6. Sample Preparation for Trace Analysis

Jutta Begerow, Medizinisches Institut für Umwelthygiene, Düsseldorf, Germany
Lothar Dunemann, Hygiene-Institut des Ruhrgebiets, Gelsenkirchen, Germany

6.1. Introduction

Modern analysis begins with a definition and outline of the problem and ends only after a detailed critical evaluation of the relevant analytical data is complete, permitting the presentation of a "result" (→ Analytical Chemistry, Purpose and Procedures). The analyst must therefore retain the ability to monitor a sample conscientiously and knowledgably throughout the analytical process. Only the analyst is in a position to assess the quality of a set of results and the validity of subsequent conclusions, although defining the problem and presenting the conclusions is almost always a cooperative multidisciplinary effort.

Analysis, and particularly trace analysis, thus entails more than the mere qualitative or quantitative detection of a particular element or chemical compound. For example, it presupposes knowledge on the part of the analyst with respect to the origin and structure of the sample matrix. The analyst must also possess a specialist's insight into analogous problems from other disciplines in order to assure the plausibility of the questions raised and critically evaluate and interpret the results, at least in a provisional way.

From these preliminary observations it will already have become clear that trace analysis, the focus of much concern in sample preparation, cannot be regarded as an end in itself. Rather, it is a very relevant and applications-oriented branch of analytical chemistry generally, one that, from a historical perspective, developed from within but became independent of chemical analysis as a whole, which was long regarded simply as a servant to the traditional subspecialities of chemistry and other disciplines (→ Analytical Chemistry, Purpose and Procedures). Nevertheless, the same tools, equipment, and methodological principles remain common to both general chemical analysis and modern trace analysis.

6.1.1. A Strategy Appropriate to Trace Analysis

Trace analytical efforts directed toward the determination of particular elements or compounds require the analyst steadfastly to pursue a single six-step strategy consisting of

1) Rigorous definition of the problem
2) Problem-oriented selection of a sample
3) Appropriate sample preparation
4) Quantitative determination of the analyte(s)
5) Plausibility tests with respect to the results, together with further evaluation
6) Professional interpretation and presentation of the findings

This strategy must be regarded as an indissoluble whole, with all the constituent parts remaining insofar as possible and practicable in the hands of a responsible trace analyst. Accordingly, it is absolutely essential that the analyst begins with a clear understanding of the problem—if necessary even formulating it. Only after the problem has been defined is it possible to establish the most appropriate approach to sampling and sample preparation. Sampling and sample preparation are followed by the determination itself in the strictest sense of the term, the outcome of which—an analytical result—must be checked for accuracy and plausibility before the findings can be presented. In many cases, the findings will require interpretation developed with the help of specialists from other disciplines, who in turn must familiarize themselves with the analyst's procedures.

Sample preparation plays a central role in the process, but it too often leads a "wall-flower" existence, with primary attention being directed to the determination step. This sense of priorities is reflected all too conspicuously in the equipment and investment planning of many analytical laboratories. However, a welcome trend in recent years points toward fuller recognition of the true importance of sample preparation in the quest for high-quality analytical results and valid conclusions [1], [2].

6.1.2. Avoidance of Systematic Errors

Trace analysis can satisfy the stringent requirements outlined above only by recognizing and minimizing possible sources of error. Unfortunately, the trace analyst almost always suffers from a lack of knowledge of the correct (or "true") analytical value. Nevertheless, it is precisely this fact that is the source of the analyst's peculiar responsibility with respect to the results in question. A purely theoretical examination of the dilemma provides little guidance for routine work, since it leads simply to a recognition that the "true" value represents an unachievable ideal.

At this point only a general consideration is feasible regarding possible sources of systematic errors that might influence the correctness of an analysis. Every sample preparation technique is associated with its own particular advantages as well as specific sources of error. In practice, the complete exclusion of systematic errors is possible only in rare cases, because optimization at one stage tends to cause problems elsewhere.

Generally speaking, any contact with vessel materials, reagents, or the ambient atmosphere, as well as any change in chemical or physical state, might result in systematic errors. It is therefore crucial that sample preparation be custom-tailored to the problem at hand with particular regard for the nature of the determination step.

Systematic errors arising from contact between a sample and various materials and chemicals are discussed in the context of the analyte (Sections 6.2.1 and 6.3.1), because different analytes require different types of treatment.

6.1.2.1. Trace Losses and Contamination

Trace *losses* are particularly common in the case of highly volatile analytes and as a consequence of adsorption effects, whereas trace *contamination* with otherwise prevalent elements and compounds may arise from laboratory air, vessels, chemicals, and various desorption effects.

The loss of trace analytes can be minimized by conducting all operations in a system that is almost completely sealed hermetically from the ambient atmosphere, and by using vessel materials characterized by a small effective surface area [3]. Surfaces should be free of fissures and preconditioned as necessary to minimize physical adsorption of the analyte (e.g., through ion exchange or hydrophobic interactions). It is also important to consider the possibility of diffusion into or through the vessel walls.

Addition of a radioactive tracer may make it possible to monitor losses that occur during the course of an analysis [4] (→ Radionuclides in Analytical Chemistry). Isotope dilution analysis is also useful for establishing the correctness of a result [5], [6].

Attention must be directed toward contamination in the early stages of laboratory planning and design to ensure that areas and rooms of equivalent sensitivity are situated adjacent to each other and shielded from less sensitive areas by appropriate structural and ventilation provisions. Analyses involving concentrations below the mg/kg or mg/L range, and especially in the ng or pg range, should of course be conducted in a clean-room environment. Precise requirements with respect to ventilation vary from case to case.

6.1.2.2. Uncertainty

The uncertainty on the result arises from both random and systematic effects but in trace analysis systematic effects largely determine the uncertainty of an analytical result. The search for and correction of systematic errors is therefore an important responsibility of every trace analyst. Even after correction for systematic errors the uncertainties on there corrections need to be evaluated and included in the overall uncertainty. Failure to correct for systematic errors leads to the considerable scattering frequently observed with collaborative analyses, and ultimately to inaccurate results. The uncertainty on the result increases disproportionately with decreasing amounts of analyte in the sample.

In trace analysis it is not sufficient simply to report a level of reproducibility for the actual determination of an analyte. Evaluating the quality of an analysis requires a knowledge of the reproducibility and the uncertainty arising from systematic effects (→ Chemometrics). Errors in sampling and/or sample preparation may be orders of magnitude greater than the standard deviation observed in several repetitions of a determination.

Whenever possible, reference materials should be included in the experimental plan as a way of checking for bias [7]. However, it is also important that the chemical state and environment of the analyte be as nearly identical as possible in both reference material and sample. The method of *standard addition* (where the spike should again be in the same chemical state as the native analyte) is particularly useful in a search for systematic errors, as is a comparison involving different analytical methods.

Internal and external quality control measures are a "must" in all trace analytical procedures, and an analytical result can only be accepted if all necessary actions in this respect have been carefully considered and the uncertainty on the result evaluated (Chapter 6.1). Quality control may therefore not only be a part of the determination step itself, but must also include the full analytical procedure including sample preparation. Internal quality control means the intra-laboratory procedure for quality control purposes including the control of uncertainty, whereas external quality control means the inter-laboratory comparison of the results of analyses performed on the same specimen by several independent laboratories (according to the definition of [8]).

6.2. Sample Preparation and Digestion in Inorganic Analysis

The object of sample preparation in inorganic analysis is to meet the requirements for a substantially trouble-free determination of the analyte. The most important of these with respect to trace analysis in any matrix include:

1) Conversion of the sample into a form consistent with the determination (dissolution)
2) Destruction of the matrix (digestion)
3) Isolation of the analyte from interfering substances that may be present (separation)
4) Enrichment of the sample with respect to trace analytes (concentration)

The dissolution step is designed to compensate for inhomogeneities in the sample. Dissolution and digestion also simplify the subsequent calibration step, ensuring that both the sample and the calibration solutions are in essentially the same chemical and physical state. The extent to which matrix constituents interfere in the determination process is significantly reduced by digestion, leading to a lower limit of detection for the determination. Digestion also facilitates concentration and separation steps.

6.2.1. Sample Treatment after the Sampling Process

6.2.1.1. Stabilization, Drying, and Storage

In trace analysis, sampling must always be followed by an appropriate stabilization step, with due regard for the nature of the matrix and the analyte.

Aqueous samples such as drinking water, surface water, and waste water but also beverages and urine samples, should always be acidified with mineral acids for stabilization purposes immediately after collection. This is especially true for the prevention of desorption processes during sampling and storage of samples in the course of trace metal analysis. Acidification reduces the tendency for ions to be adsorbed onto active sites at the surface of the containment vessel, and it also inhibits bacterial growth [9]). Glacial acetic acid and 65 % nitric acid have been shown to be suitable for this purpose. Approximately 1 mL of acid should be added per 100 mL of sample. Acidified urine in appropriate containers can be safely kept in a deep-frozen state for several months [10], [11]. Stored urine samples often deposit a sediment that may include considerable amounts of analyte (e.g., arsenic, antimony, copper, chromium, mercury, selenium, zinc) [10].

Loss of water from aqueous matrices (e.g., tissue, fruit, vegetables, and soil samples) may occur during storage. For this reason analytical results should always be reported in terms of dry mass to avoid false interpretations. Drying is best conducted immediately after sampling. Water removal can be accomplished by oven drying at elevated temperature, use of desiccating materials, or freeze drying (lyophilization). Freeze drying has been shown to be the most satisfactory procedure since it minimizes the loss of highly volatile elements and compounds. Drying at a temperature as low as 120 °C can result in the loss of up to 10 % of most elements, and losses for mercury, lead, and selenium may be considerably higher (e.g., 20 – 65 %) with certain matrices [12], [13].

Blood samples must be rendered incoagulable by the addition of an anticoagulant immediately after sampling. Suitable substances for this purpose include salts of ethylenediaminetetraacetic acid (EDTA), citric acid, oxalic acid, and heparin. A particular anticoagulant should be selected with reference to the analyte in question as well as the proposed analytical procedure. Particular attention should be directed to ensuring that the anticoagulant is not itself contaminated with the analyte. For recommendations concerning appropriate levels of anticoagulants see, for example, [14]. Blood sampling kits that already contain the aforementioned anticoagulants are now available commercially. Blood samples in suitable containers can be either refrigerated or deep-frozen prior to analysis. Such samples can normally be kept for several months at – 18 °C without measurable change in the concentration of an analyte. One exception is mercury; in this case reduction to elemental mercury leads to significant volatilization losses within as little as a few days [15]. If storage over several years is envisaged, the sample should either be dried at – 18 °C or quick-frozen in liquid nitrogen at – 196 °C and then stored at or below – 70 °C.

Determination of an analyte in serum or plasma requires that the analyte be isolated subsequent to sampling. This is no longer possible after deep-freezing because of the hemolytic nature of blood samples. In the case of elements at very low concentration it is preferable that an investigation be conducted on serum rather than plasma. This reduces the risk of contamination, because unlike blood plasma, serum is recovered without addition of an anticoagulant.

6.2.1.2. Homogenization and Aliquoting

In solid samples, elements are normally distributed in an inhomogeneous way. Trace-element determinations are usually restricted to relatively small samples, which requires that a fairly large sample be comminuted and homogenized prior to removal of an aliquot for analysis.

Comminution of the sample presents significant opportunity for contamination. Contamination of the sample from abrasion of the comminution equipment is fundamentally unavoidable, so efforts must be made to select the best possible equipment for each particular analytical task. Equipment is preferred in which the sample comes into contact only with surfaces fabricated from such high-purity plastics as polytetrafluoroethylene (PTFE), since this permits the sample to be used without restriction for the determination of a large number of elements. Friability can be increased by deep-freezing or drying the sample prior to comminution.

6.2.1.3. Requirements with Respect to Materials and Chemicals

Whenever samples, standard solutions, or reagents come into contact with containers, apparatus, or other materials, two opposing physicochemical processes occur at the interface: *adsorption* and *desorption* of ions and molecules.

Ion adsorption at active sites on a contact surface may lead to losses on the order of several nanograms per square centimeter. The importance of this factor therefore increases as the concentration of the analyte decreases. Adsorptive losses are a function of the nature and valence state of the analyte, the nature and concentration of accompanying ions in the solution, pH, temperature, and the duration of contact. Factors related to the containment vessel include the material from which it is constructed, its surface area, and the nature of any pretreatment.

Adsorption can be minimized through the use of quartz, PTFE, polypropylene (PP), perfluoroalkoxy resin (PFA), or glassy carbon vessels. There is no single ideal material, but quartz is preferred in the case of biological samples provided the use of hydrofluoric acid can be avoided. The surface area of the container and the contact time should be minimized, whereas the analyte concentration should be as high as possible. Cations are adsorbed less strongly from acidic solutions than from solutions that are neutral or basic because such ions are displaced from active sites on the container surface by the much more abundant protons . Adsorptive losses are also lower from solutions with high salt concentrations (e.g., urine) relative to those with low salt concentrations, such as standard solutions. Adsorption can be further reduced by adding a suitable complexing agent so long as this has no adverse affect on the subsequent analysis. For example, mercury can be stabilized as HgI_4^{2-}, and silver as $Ag(CN)_2^-$.

Containers, pipettes, and other laboratory implements contain traces of various elements both in their structure and adsorbed on the surface, and these are subject to desorption on contact with a sample, calibration solutions, or reagents, thereby contributing to a nonreproducible increase in blank results. Table 1 summarizes the inorganic impurities likely to be encountered with various vessel materials.

Careful and thorough cleaning of all vessels is extremely important in trace analysis both for inorganic and organic analytes. Materials must be specially selected with a specific view to the requirements of each task during sample handling and storage, with explicit attention being paid to trace purity. Thorough pretreatment (rinsing, boiling, steaming) with pure nitric, sulfuric, or hydrochloric acid in the case of metal analysis is strongly recommended [18]–[21]. Figure 1 illustrates a steaming apparatus useful for cleaning analytical vessels which are to be prepared for use in inorganic analyses, namely metals but also other inorganic analytes [22]. For organic analyses repeated cleaning of vessel surfaces with suitable solvents is essential.

Reagents for trace analytical applications must also satisfy stringent purity requirements, often exceeding those of commercially available reagents. Acids such as hydrochloric, nitric, sulfuric, hydrofluoric, and perchloric can be prepared relatively easily in the laboratory to a high standard of purity by distillation below the boiling point ("subboiling process"). An apparatus for this purpose is illustrated in Figure 2. Table 2 compares the elemental contents of commercially available

Table 1. Typical inorganic impurities in selected vessel materials [16], [17]

Element	Content, mg/kg				
	Quartz	GC [a]	PTFE [b]	PE [c]	BS [d]
V		4			< 2
Cr		0.5 – 5	10 – 30	15 – 300	< 3
Mn	0.01	0.1	60	10	< 6
Fe	0.2 – 0.8	1 – 15	10 – 35	600 – 2100	200
Co	< 0.001	< 0.01	0.3 – 1.7	0.1 – 5	< 0.1
Ni		0.5			< 2
Cu	< 0.01 – 0.07	0.2	20	4 – 15	1
Zn	< 0.01	< 0.3	8 – 10	25 – 90	2 – 4
As	0.0001 – 0.08	0.06			0.5 – 22
Cd		< 0.01			1
Sn		25 – 50			< 4
Sb	< 0.001	< 0.01	0.4	0.2 – 5	7 – 9
Hg	0.001	0.001			
Pb		0.4			

[a] Glassy carbon.
[b] Polytetrafluoroethylene.
[c] Polyethylene.
[d] Borosilicate glass.

Table 2. Elemental content of selected acids with varying degrees of purity [16]

Acid	Elemental content, µg/L				
	Cd	Pb	Cu	Fe	Zn
10 N HCl, analytically pure	0.1	0.5	1.0	100	8.0
10 N HCl, extra pure	0.03	0.13	0.2	11.0	0.3
2 N HCl, subboiled	0.01	< 0.05	0.07	0.6	0.2
15 N HNO_3, analytically pure	0.1	0.5	2.0	25.0	3.0
15 N, HNO_3, extra pure	0.06	0.7	3.0	14.0	5.0
15 N, HNO_3, subboiled	0.001	< 0.02	0.25	0.2	0.04

acids with those of acids prepared by the subboiling process.

All chemicals and reagents should be checked regularly for purity, and replaced as necessary. It is also important to remember the ever-present danger of contamination during reagent use as a result of interactions with container materials and the entrainment of impurities. For this reason reagents and standard solutions should always be prepared immediately before their use. As a general rule, the addition of reagents should be kept strictly to a minimum.

Possible contamination due to atmospheric dust and particle emission attributable to laboratory staff activities (smoking, use of cosmetics, etc.) can be avoided in trace analytical determinations at the µg/kg, µg/L, and µg/m^3 ranges by working under clean-room conditions. A laminar-flow "clean bench" ensures that particles with a diameter >0.1 µm will be removed with an efficiency of 99.9 % through high-efficiency, sub-micron-particulate air filtration. As indicated in Table 3, this can result in purification factors on the order of 100 with respect to such elements as manganese and vanadium [23].

6.2.2. Sample-Preparation Techniques; General Considerations

Sample-preparation techniques are classified, somewhat arbitrarily, according to external conditions. Categories include wet digestion, dry ashing, pressure digestion, microwave digestion, and cold-plasma ashing. However, these conventional designations are often imprecise or even misleading with respect to the actual mechanism of the process. Several very different names are sometimes applied to a single technique, which presents a considerable obstacle for anyone (particularly a nonspecialist) interested in acquiring a quick overview of systems applicable to a specific task. For

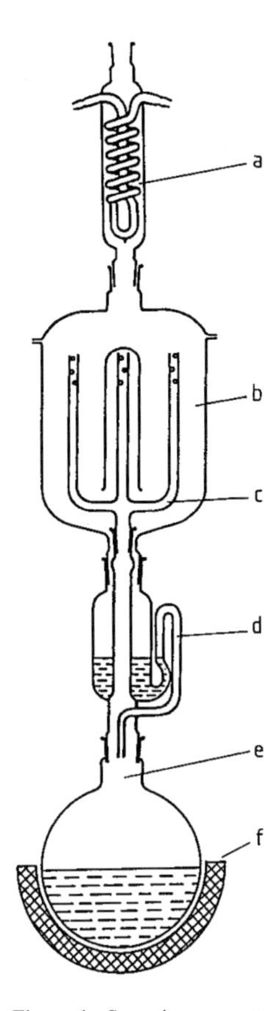

Figure 1. Steaming apparatus for cleaning analytical vessels with nitric acid, hydrochloric acid, or water (with permission from Kürner, Rosenheim/Oberbayern, Germany)
a) Reflux condenser; b) Steam chamber; c) Rack of quartz tubes; d) Overflow; e) Round-bottom flask (1 – 2 L capacity); f) Heating mantle

example, the expression "pressure digestion" is inappropriate, because it is a relatively high boiling temperature that ensures more effective digestion, not the associated high pressure. The conventional designation should nevertheless be retained in this case, if only because of its wide acceptance; attempting to rename the procedure now would introduce more confusion than clarity.

Another — very timely — example of incorrect terminology involves uncritical use of the expression "microwave digestion" for both acid digestion with microwave excitation and cold-plasma ashing. Although both techniques make use of microwave radiation, the direct effects of this radiation are of minor importance at most. Expressions like "microwave excitation," "microwave induction," or "microwave assistance" would be preferable.

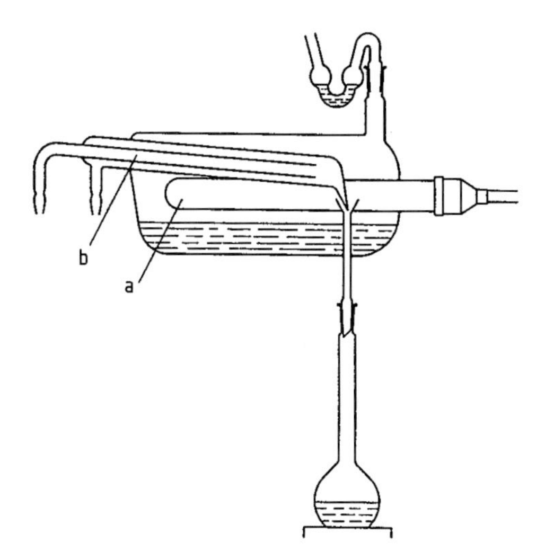

Figure 2. Quartz apparatus for purifying acids or water by subboiling distillation (with permission from Kürner, Rosenheim/Oberbayern, Germany); this device is also available in PFA for the subboiling distillation of hydrofluoric acid (with permission from Berghof Maassen, Eningen u. A., Germany)
a) Heater; b) Cold-finger condenser

Table 3. Typical levels of contamination by selected elements as a result of dust in the laboratory [23]

Element	Contamination level, ng cm^{-2} d^{-1}	
	Conventional laboratory	Clean workstation
Al	0.82	0.09
V	0.037	0.0003
Mn	0.063	0.002
Fe	1.5	0.2
Co	0.0036	0.0003
Cu	0.032	0.012
As	0.07	0.03
Sb	0.013	0.001

6.2.2.1. Special Factors Associated with Microwave-Assisted Digestion

Rapid, straightforward digestion processes have long been in demand in a wide variety of disciplines. The disparity between the several hours required for a conventional digestion process and the few seconds or minutes involved in elemental determination has become increasingly obvious in recent years with the increasing automation of analytical procedures. Recently introduced rapid digestion processes, which have profited greatly from modern microwave technology, can contribute to reducing this disparity.

The first published report on microwave ashing was that of KOIRTYOHANN et al., which appeared in 1975 [24]. Initially it attracted little attention because of the lack of appropriate expertise and apparatus.

Modern commercial microwave ovens operate at a frequency of 2.45 GHz. The corresponding energy is so low that such microwaves are incapable of rupturing molecular bonds directly, leading only to rotational excitation of dipoles and molecular motion associated with the migration of ions. No vibrational or electronic excitation occurs [25], [26]. It is for this reason that the designation "microwave-assisted digestion" has been suggested. Microwave energy introduced as radiation is dissipated (converted into heat) by ionic conduction and dipole rotation. Oscillating alignment and subsequent relaxation of dipoles occurs about 5×10^9 times per second at a frequency of 2.45 GHz, causing rapid heating of the sample.

Microwave radiation is generated by means of a magnetron (microwave diode). The radiation is then directed through a waveguide into the interior of the oven. Even distribution of the radiation is facilitated by a "mode stirrer" that prevents the development of standing waves, as well as a turntable and perhaps a rotating antenna located beneath the oven floor.

Starting in the late 1980s, microwave-based techniques became more and more available for analytical purposes [27]–[30]. Today in trace metal and non-metal analyses, microwave-assisted digestion has been successfully applied to the decomposition of broad range of organic matrix components in geological and biological samples as well as analysis and residue monitoring [22].

6.2.2.2. Safety Considerations

Necessary protective measures with respect to the laboratory workforce must be implemented for the handling of acids and other strongly oxidizing or poisonous chemicals. This includes protective clothing and an efficient fume extraction system. Perchloric acid as a digestion reagent offers advantages due to its powerful oxidizing action, but it should be used only when absolutely necessary since it entails extensive safety precautions (e.g., working in a special hood), and a definitive set of regulations must be scrupulously observed.

Special statutory regulations apply to operations involving pressure vessels, and pressure digestion techniques must be carried out with exceptional care and attention. Ashing an organic matrix can cause the pressure to rise to extremely high levels within a short period of time, and the reaction may become uncontrolled. The corresponding risk of a ruptured pressure vessel cannot be overemphasized. When developing a pressure digestion method it is important to begin with a small sample (≤ 100 mg dry mass; even less for pure carbon).

Additional safety considerations apply to the use of microwave technology. First, the equipment should be operated in strict accordance with instructions to ensure that no detectable amounts of microwave radiation will escape. Leak tests should be conducted on a regular basis by a trained technician. Vessels or other objects made of metal should never be placed in the oven chamber. Microwave equipment must also never be operated empty (i.e., in the absence of a sample solution, water, acid, etc.), since the magnetron might be destroyed by reflected radiation. The behavior of unknown samples should be tested on a small scale (see above).

6.2.3. Wet Digestion Techniques

Wet digestion with oxidizing acids is the most common sample-preparation procedure [31], [32]. This category can be extended to include processes involving bases or nonoxidizing acids as ashing reagents. Concentrated acids with the requisite high degree of purity are available commercially, but they can be purified further by subboiling distillation (see Section 6.2.1.3).

The most suitable acid for a digestion is a function of the sample matrix, the analyte, and the proposed determination method. Nitric acid is an almost universal digestion reagent, since it does not interfere with most determinations and is available commercially in sufficient purity. However, nitric acid has a rather low boiling point (122 °C), and its oxidizing power is often insufficient under atmospheric-pressure conditions. Hydrogen peroxide and hydrochloric acid can usefully be employed in conjunction with nitric acid as a way of improving the quality of a digestion. Hydrochloric acid and sulfuric acid may interfere with the determination of certain metals through the formation of stable compounds. As noted previously (Section 6.2.2.2), safety considerations are particularly important when using perchloric acid. Silicate samples require the further addition of hydrofluoric acid.

6.2.3.1. Wet Digestion at Atmospheric Pressure

This category includes all the various ambient-pressure wet digestion techniques involving concentrated acids (or bases). This very inexpensive technique is of inestimable value for routine analysis because it can easily be automated; all the relevant parameters (time, temperature, introduction of digestion reagents) lend themselves to straightforward control [33].

Systems of this type are limited by a low maximum digestion temperature, which cannot exceed the ambient-pressure boiling point of the corresponding acid or acid mixture. As noted previously, the oxidizing power of nitric acid with respect to many matrices is insufficient at such low temperatures. One possible remedy is the addition of sulfuric acid, which significantly increases the temperature of a digestion solution. Whether or not this expedient is practical depends on the matrix and the determination method. High-fat and high-protein samples are generally not subject to complete ashing at atmospheric pressure. Other disadvantages relate to the risk of contamination through laboratory air, the necessarily rather large amounts of required reagents, and the danger of trace losses. Nevertheless, systems operated at atmospheric pressure are preferred from the standpoint of workplace safety.

This category also includes the *solubilization* of simple matrices with low carbon content using special detergents or other solubilization reagents, as well as the enzymatic digestion techniques that are often used in food control and monitoring. Enzymatic digestion in trace analytical chemistry has been described in [34] for the example of the non-specific protease enzyme (pronase), which was used as an alternative sample preparation technique for the determination of trace elements in blood serum by ICP-MS. Using enzymatic digestion, a higher degree of instrument stability was achieved over 3 h period. The reported results are in good agreement with the certified values of the reference material [34]. Also "extraction – digestion" techniques belong to this category, e.g., using aqua regia at relatively low temperatures (e.g., 70 °C) over a long period of time (up to 5 d) as tool for the pseudo-total analysis of metals in soil samples [35]. These techniques are some importance as sample preparation techniques, but one should be aware that they do not lead to complete digestion of the organic matrix. Extraction – digestion techniques have become standard methods for the examination of water, waste water, sludge, and sediments using aqua regia for the subsequent determination of the acid-soluble portion of metals [36].

Thermally Convective Wet Digestion. The conventional approach to wet digestion, which has proven its worth over many years, entails a system equipped with heating units operating either at a fixed temperature or in response to a temperature program. Commonly employed digestion agents include nitric acid, sulfuric acid, hydrogen peroxide, hydrofluoric acid, and perchloric acid, as well as various combinations of these [33]. Most applications of wet digestion involve aqueous or organic matrices, such as surface waters, waste water, biological samples (tissue, body fluids, etc.), food samples, as well as soil and sewage sludge, coal, high-purity materials, and various technical materials [35].

Microwave-Assisted Wet Digestion. Microwave-assisted digestion in open vessels (atmospheric digestion) [37] is generally applicable only with simple matrices or strictly defined objectives, and the results are reproducible only if the specified ashing parameters are strictly observed. Losses may be encountered with mercury and possibly also with organometallic compounds (e.g., those containing arsenic, antimony, or tin). Addition of sulfuric acid is essential in order to achieve a sufficiently high digestion temperature in atmospheric-pressure equipment, where the boiling point of the acid establishes the maximum ashing temperature, although it is important to remember that the presence of sulfate interferes with many procedures for metal determination (e.g., graphite-furnace atomic absorption spectrometry, GF-AAS). Specific experimental conditions for microwave-assisted digestions have been summarized in [9], [22] as a function of the matrix to be ashed.

Ultraviolet Digestion (Photolysis). Ultraviolet (UV) digestion is utilized mainly in conjunction with uncontaminated or slightly contaminated waters, such as sea water and surface water. Liquids or slurries of solids are decomposed by UV radiation in the presence of small amounts of hydrogen peroxide and acids (e.g., beverages, wastewater, soil extracts). The corresponding digestion vessel should be placed in the closest possible proximity to the UV lamp to ensure a high light yield. In photolysis (see Fig. 3) the digestion mechanism can be characterized by the formation

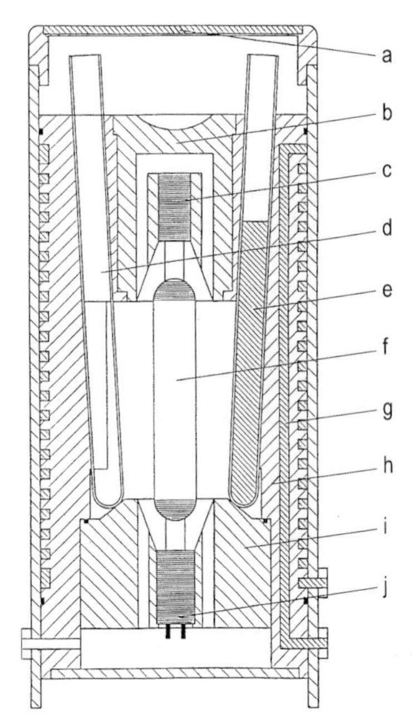

Figure 3. Apparatus for the UV digestion (photolysis) of liquid samples with low carbon content (with permission from Kürner, Rosenheim/Oberbayern, Germany)
a) UV protection glass; b) Lamp cooler; c) Lamp socket; d) Sample vessel; e) Sample solution; f) UV lamp; g) Coolant (water); h) Sample cooler; i) Lamp cooler; j) Lamp socket

of OH radicals from both water and hydrogen peroxide that is initialized by the aid of the UV radiation [9], [38]. These reactive radicals are able to oxidize the organic matrix of simple matrices containing up to about 100 mg/L of carbon to carbon dioxide and water. Effective cooling of the sample is essential, since losses might otherwise be incurred with highly volatile elements. Contamination can be minimized by the use of a nongas-tight stopper. Hydrogen peroxide addition may need to be repeated several times to produce a clear sample solution [39], [40]. Complete elimination of the matrix is of course possible only with very simple matrices (e.g., slightly contaminated water) or by combining photolysis with other digestion techniques [41], [42].

6.2.3.2. Pressure Digestion

Pressure digestion offers the advantage that the operation is essentially isolated from the laboratory atmosphere, thereby minimizing contamination, and digestion occurs at relatively high temperature due to boiling-point elevation effects. The pressure itself is in fact nothing more than an undesirable—but unavoidable—side effect. The principal argument in favor of this form of digestion is the vast amount of relevant experience acquired in recent decades. The literature is a treasure trove of practical information with respect to virtually every important matrix and a great number of elements (see, for example [43]). Pressure digestion is particularly suitable for trace and ultratrace analysis, especially when the supply of sample is limited.

Since the oxidizing power of a digestion reagent shows a marked dependence on temperature, a distinction must be made between "simple" pressure digestion and high-pressure digestion. Simple pressure digestions (< 20 bar) are limited to a temperature of ca. 180 °C, whereas with high-pressure apparatus (> 70 bar) the digestion temperature may exceed 300 °C.

6.2.3.2.1. Thermally Convective Pressure Digestion

Most sample containers for use in *thermally convective pressure digestion* are constructed from PTFE or PFA, although special quartz containers with PTFE holders are available for trace-analysis purposes. The sample container is mounted in a stainless-steel pressure vessel (autoclave) and then heated in a furnace to the desired temperature. Pressure digestions based on apparatus of the Tölg, Bernas, and Parr type are all feasible below ca. 180 °C, but above this temperature PTFE begins to "flow", rendering it unsuitable for use in high-pressure applications. Investigations with organic matrices have shown that at 180 °C as much as 10 % of the carbon present in high-fat and high-protein samples remains unashed [44]. Such samples must be processed further prior to analysis through evaporation of the digestion solution with perchloric acid.

All thermally initiated digestions have the disadvantage that a considerable amount of time is consumed in preheating the digestion solutions due to the need for heating the autoclave and sample vessels.

High-Pressure Digestion. The introduction of a high-pressure ashing (HPA) technique by Knapp [45] has not only reduced the effective ashing time to ca. two hours but also opened the way to digestion of extremely resistant materials [46]. High-

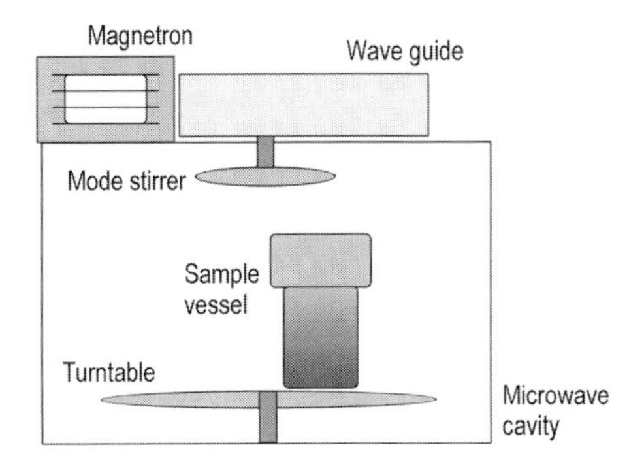

Figure 4. Apparatus for microwave-assisted pressure digestion

pressure ashing is conducted in quartz vessels, with a maximum ashing temperature as high as 320 °C at a pressure of ca. 100 bar. The quartz vessel is stabilized during the digestion process by subjecting it to an external pressure roughly equivalent to that developed within. Even thermally stable organic compounds undergo acid decomposition at such high temperatures. Essentially complete ashing can be accomplished with the vast majority of samples so far investigated. Nitric acid alone is a sufficiently powerful reagent in many cases.

6.2.3.2.2. Microwave-Assisted Pressure Digestion

The digestion of lightly contaminated water, many geological samples, and low-fat plant matrices can be accomplished successfully with a domestic microwave oven and relatively primitive sample holders. The recent introduction of complete commercial microwave systems specifically designed for sample ashing has led to important further developments. Such systems also offer special safety features and improved facilities for controlling and regulating the pressure and/or temperature [47]. Moreover, the distribution of microwave radiation inside the oven cavity tends to be significantly more homogeneous than with domestic devices, in part because the magnetron control is subject to shorter switching cycles.

Nitric acid alone does not always lead to satisfactory microwave-assisted ashing. Addition of hydrochloric acid or hydrogen peroxide is often helpful, and hydrofluoric acid must of course be present in the case of silicate-containing materials [39], [48]. The maximum attainable pressure seldom exceeds 10 bar.

A serious disadvantage of this simple approach is the frequent need for multistep ashing, in which a preliminary digestion is followed by a second treatment, perhaps with added hydrogen peroxide [40].

High-Pressure Microwave Digestion. High-pressure (or high-temperature) microwave-induced processes are designed to support digestion up to a pressure of perhaps 70 bar, opening the way to substantially higher temperatures (e.g., ca. 250 °C) without a corresponding increase in overall processing time. Temperature and digestion time ultimately determine the effectiveness of a digestion, with residual carbon content serving as a useful measure for quantitative assessment. The striking advantage of high-pressure, i.e., high-temperature (!) microwave digestion (see Fig. 4) is that almost complete dissolution can be accomplished with a single-stage process, even in the case of complex matrices. Examples found in the recent literature consist of new successful developments in the technique and apparatus [49] and applications to biological tissues [50] or technical products [51].

Examples of almost completely mineralizable matrices include resistant inorganic and high-fat organic samples. Nevertheless, limitations with respect to subsequent use of electrochemical methods of determination (Section 6.2.5.1) still represents a critical problem [52], and the results are often inferior to those from a conventional high-pressure asher (HPA).

6.2.4. "Dry" Digestion Techniques

The term "dry" digestion is intended to encompass all processes based on gaseous or solid digestion reagents. Such a distinction relative to wet digestion processes is not absolutely essential, but it does offer certain practical advantages [43].

6.2.4.1. Combustion in Air

Strictly speaking, "dry ashing" refers to the oxidation (combustion) of a substance in air at a temperature of several hundred degrees Celsius, often in a muffle furnace or similar apparatus.

Dry Ashing in a Muffle Furnace. Carbon-containing substances generally decompose satisfactorily under these conditions without the use of auxiliary agents. It is advantageous that reasona-

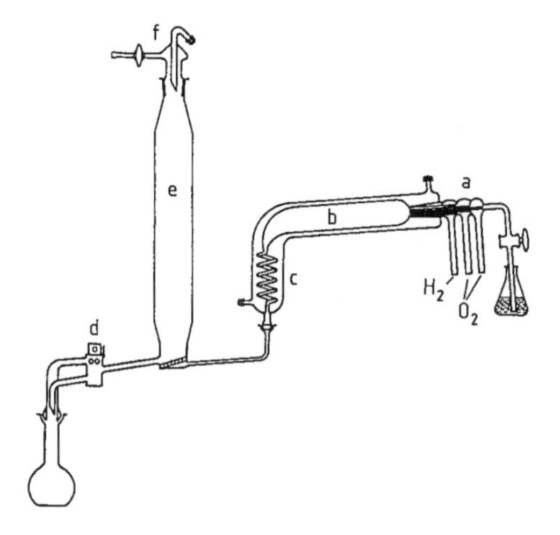

Figure 5. Wickbold combustion apparatus
a) Injection burner; b) Combustion chamber; c) Condenser; d) Multiport stopcock; e) Absorption chamber; f) Flushing port

bly large samples can be digested. In addition to routine biological samples (plants, foods), coal has also been subjected to dry ashing [53]. The importance of this technique has declined in recent years, primarily because it is so time-consuming; it also entails a high risk of contamination since the samples are kept in open vessels [54]. Losses must be anticipated in the case of volatile substances [55]. However, the literature provides numerous examples of decompositions that have been optimized through the addition of "ashing auxiliaries" (salt solutions or acids). Such additives often reduce the ashing time, especially for organic substances, and they may prevent the volatilization of element traces.

Dry Ashing in a Microwave Oven. The time required for dry ashing in air can be reduced by microwave heating of a special block containing the sample [25]. This block is fabricated from a material that is highly absorbent with respect to microwaves (e.g., silicon carbide), and is is surrounded by quartz insulation. At maximum power such a block can be heated to about 1000 °C within ca. 2 min. In addition to time saved during the heat-up phase this system offers the further advantages of a low energy requirement and a procedure that spares the technicians from exposure to intense heat when inserting and removing samples, in sharp contrast to work with a conventional muffle furnace [25].

6.2.4.2. Combustion in Oxygen

Combustion in oxygen is a decomposition procedure that eliminates many of the potential problems of sample contamination associated with digestion reagents [56]. The sole reagent here is gaseous oxygen, which is commercially available in very high purity. Related combustion techniques are based on the use of other oxidizing agents (e.g., sodium peroxide) from which free oxygen is released during the course of the operation. Analogous methods rely on oxidizing agents other than oxygen (e.g., chlorine [57], fluorine), but these have not been widely accepted, in part because they require working with very aggresive materials, often not available with the required level of purity and themselves capable of interfering with the determination process.

Atmospheric-Pressure Combustion. Although combustion in an open vessel is relatively easy to accomplish, it has the disadvantage of possible volatilization of element traces. Significant advantages are associated with the compromise of a quasi-closed quartz glass system consisting of a central combustion chamber, an ignition device (an infrared source), and a reflux condenser charged with liquid nitrogen. The analyte, together with inorganic combustion residues, is redissolved by refluxing with a small amount of acid. Such a system is applicable to almost all the elements, and it leads to nearly complete decomposition [56], [58]. Applications have been described with respect to a large number of organic matrices, including petroleum products, coal, and soils. Combustion systems also exist for analysis of nonmetals. SCHÖNIGER developed a particularly simple combustion process (now bearing his name) for general elemental analysis [59]. A special variant—combustion in an oxyhydrogen flame—is carried out in a Wickbold apparatus (Fig. 5) [60].

Combustion in a Closed Vessel. The risk of volatilization of trace analytes (e.g., mercury) can be reduced further by conducting the combustion in a special combustion bomb (e.g., a Parr bomb) [61]. Oxygen gas is introduced up to some predetermined pressure, after which the sample is subjected to explosive combustion with the aid of a detonation device (ignition wire). The analyte, together with inorganic residues, is again collected in a small volume of acid [62]. This technique is also applicable to the determination of nonmetals [63].

6.2.4.3. Cold-Plasma Ashing

Decomposition by cold-plasma ashing is accomplished with excited-state oxygen under a pressure of a few millibars. The plasma is formed at reduced pressure (0.01 – 1 mbar) in a high-frequency field established either with a magnetron (2.45 GHz) or with a semiconductor high-frequency generator. In the simplest application samples are ashed in flat dishes (e.g., Petri dishes) arranged in trays [64]. A quasi-closed system with a cold-finger condenser should of course be employed for trace analysis to minimize potential losses [65]. Oxidation of organic matrices occurs by way of reactive, short-lived radicals, leading to more complete matrix decomposition relative to most other approaches. This ashing technique is therefore particularly recommended as a precursor to electrochemical determinations (e.g., polarography). The oxygen plasma is a "cold plasma," which means that the sample temperature never exceeds ca. 150 °C. Risk with respect to the loss of volatile elements is therefore restricted to mercury. A reaction period of ca. 2 – 4 h is usually sufficient to disrupt the organic matrix to such an extent that the residue can be removed from the sample vessel and cold finger by refluxing with a small volume of concentrated acid (2 mL). The most important application of cold-plasma ashing is the mineralization of biological samples [66], although coal, graphite, and plastics (including PTFE) can also be ashed by this technique [65]. Samples as large as one gram or more can be accommodated.

6.2.4.4. Fusion

Fusion is the traditional approach to sample preparation for industrial and geological analyses [67], especially for the analysis of mineral samples (e.g., lithium metaborate melts [68]). The major problems associated with fusion include restricted means for purifying the required (solid) reagents and the presence of high salt concentrations in the resulting analyte solution. Especially for trace analysis, overloading of the matrix in this way is inadvisable.

6.2.5. Illustrative Examples

The sections that follow highlight a few particularly interesting examples of sample preparation, taking into account both the analytical method of choice and effective mineralization of specific matrices.

6.2.5.1. Sample Preparation as a Function of Analytical Method

In the case of *atomic spectrometry* [e.g., atomic absorption (AA), inductively coupled plasma – optical emission spectrometry (ICP –OES), inductively coupled plasma – mass spectrometry (ICP – MS)] matrix interference due to incomplete digestion may manifest itself as differences in the suction rates and aerosol yields for samples relative to calibrating solutions. Such differences may also reflect differences in the bonding states of the elements, which in turn leads to systematic errors and problems with calibration.

Electrochemical determination methods are particularly sensitive to incomplete sample mineralization. The result is matrix interference manifested in the case of inverse voltammetry, for example, as ghost peaks, signal suppression, and an enhanced hydrogen signal. With simple, aqueous samples such as seawater and river water, voltammetric determination of zinc, cadmium, lead, and copper can be effectively coupled with UV decomposition [69], [70]. The same determination method is applicable to these heavy metals in simple plant products such as sugars provided decomposition (ashing) is effected by combustion in a stream of oxygen (typical sample mass: 1.5 g) [71]. Combustion is also appropriate prior to the potentiometric determination of aluminum in organic samples [72]. Biological samples can be prepared for electrochemical analysis by cold-plasma or high-pressure ashing [41], [45], [56], [73].

An improved procedure has been described for atmospheric-pressure combustion of samples from high-purity metals [74]. High-pressure digestion with nitric and hydrofluoric acids [75], [76] and combustion techniques have both been shown to be suitable for the decomposition of samples of silicate-containing materials and fuels. Combustion in this case is carried out either in a stream of oxygen [77] or, to provide a closed system, in an oxygen bomb [78]. Mercury can be satisfactorily determined after prior Wickbold ashing [79].

Table 4. A comparison of the time required for microwave pressure digestion versus thermally convective digestion (for one sample and for six samples)

Operational step	Time, min			
	Microwave digestion (PMD)		Knapp digestion (HPA)	
	One sample	Six samples	One sample	Six samples
Preparation for digestion	10	30	10	30
Ashing, including warm-up	10	30 *	90–150	90–150
Cool-down	10	30	20	20
Preparation for analysis	5–10	30	5–10	30
Cleaning the equipment	10–20	20	10–20	20
Complete procedure	45–60	140	135–210	190–250

* Limited to two parallel samples.

The decomposition of inorganic substances can be accomplished with a mixture of hydrochloric and nitric acids in a closed vessel (pressure digestion). Silicate-containing minerals and glasses require the addition of hydrofluoric acid, and the presence of perchloric acid, phosphoric acid, or sulfuric acid is sometimes useful as well [80]. The resulting metal fluorides can be redissolved by adding a solution of boric acid. The same procedure is effective for ores and slags as well as quartz [81].

Cold-plasma ashing is useful as a way of excluding matrix interferences in the detection of elements with secondary-ion mass spectrometry (SIMS) [82].

6.2.5.2. Combined Use of Multiple Decomposition Techniques

A single ashing procedure is often insufficient for the complete decomposition of a complex matrix, leading some authors to recommend a combination of two or more techniques. One example will suffice to illustrate the principle: pressure ashing followed by UV photolysis. Thus, it has been shown that analysis of olive leaves for heavy metals by voltametric methods leads to distorted results after pressure digestion alone. Reliable data can be obtained only by supple-menting the digestion with UV irradiation to ensure adequate decomposition of the matrix [41].

6.2.5.3. Comparative Merits of the Various Sample-Preparation Techniques

High-fat samples present special problems with respect to decomposition due to the presence of highly resistant components. The result is often incomplete ashing. Moreover, there is always a risk of explosive decomposition caused by the formation of reactive radicals in the digestion solution. Unlike electrochemical methods of determination, which require the complete degradation of organic matrices, atomic absorption (particularly the flame technique) is compatible with a certain amount of residual carbon. Any digestion solution that is colorless and free of particles is a reasonable candidate for AAS analysis.

A systematic comparison of microwave-assisted decomposition techniques with cold-plasma ashing, conventional thermally convective pressure digestion, and high-pressure digestion is presented in [40]. Generally speaking, comparable results can be obtained with high-quality commercial ashing systems of all the common types, with the potential for nearly complete decomposition [e.g., the Paar microwave-assisted pressurized microwave decomposition (PMD) system, the Büchi high-pressure adapter, and the Knapp HPA system]. Microwave ashing of high-fat samples in an open vessel presents a serious risk of significant analyte loss despite the use of a reflux condenser, but losses with a microwave pressure-digestion system have been found to be no greater than with a corresponding convective heating system. Results also suggest that "memory effects" are less noticeable with a microwave system than in the case of a conventional heat source [40].

Table 4 provides a comparison of the time required for various steps in microwave versus convective thermal ashing, taking into account all the associated preparations, the cool-down phase, and cleaning of the apparatus, which has the effect of reducing somewhat the marked time advantage of the microwave-assisted ashing period. The threefold time advantage of the complete microwave-

Table 5. Overall ashing times associated with various procedures (for one sample)

Sample type	Ashing time, h				
	Pressure microwave *	Open microwave	Tölg **	Knapp **	Cold plasma
Plant matrix	0.75	0.92	4.5	3	10
Sunflower oil	0.75	0.92	4.5	3	20

* Single-stage digestion.
** Thermal convection heating.

assisted procedure is reduced even more if, for example, six samples are digested, because the PMD cannot process more than two samples at once. Table 5 shows that microwave-assisted pressure ashing is the most rapid of the techniques investigated, faster than the frequently used Tölg technique (Section 6.2.3.2.1) by a factor of six. Cold-plasma ashing takes even longer. It should be noted, however, that this comparison ignores the advantages that might be derived from parallel ashing of several samples, and it does not reflect special requirements associated with particular matrices or analytes.

Earlier investigations confirm these results. According to [41], however, complete degradation of many samples is achieved only through cold-plasma ashing or by a combination of wet ashing and UV irradiation. In a determination of aluminum in dialysis liquids it was observed that microwave-assisted ashing produced results comparable to those from conventional pressure ashing [52].

Five different digestion techniques are compared in [83] in an attempt to optimize the determination of mercury in soil via cold-vapor AAS. Only with closed systems was loss-free digestion assured. Open systems resulted either in mercury losses or incomplete recovery due to incomplete ashing. On the other hand, wet ashing in an open system with nitric and sulfuric acids was found to be preferable for the mineralization of biological samples, since it was easier to accomplish and led to lower blank readings [84].

Fusion with lithium metaborate is feasible for geological samples provided the melt is subsequently dissolved in acid in the course of a pressure- or microwave-assisted ashing process [85].

6.2.5.4. Decomposition Procedures for Determining Nonmetals

A pertinent example of an analysis for nonmetallic constituents is the digestion of water or food samples with peroxodisulfate, according to the Koroleff method [86], which was developed for the determination of all bound nitrogen in seawater, inland waters, rain water, groundwater, wastewater, or effluents. Measurable concentrations are in the range 0.02 – 4 mg of nitrogen per liter. The nitrogen-containing sample is first subjected to a pressure digestion in the course of which organic nitrogen compounds, nitrite ion, and ammonium ion are all oxidized to nitrate. This oxidation proceeds quantitatively only under alkaline conditions, achieved with the aid of a buffer system consisting of boric acid and sodium hydroxide. Reaction commences at pH 9.7 and ends at pH 5 – 6, at which point the nitrate exists in stable form in the acidic digestion solution. Nitrogen is not released in this way from certain five-membered heterocyclic systems [87], but the resulting error is probably negligible with most of the samples subject to investigation. The total nitrogen content in the digestion solutions can be established by photometric or ion-chromatographic methods. A Kjeldahl analysis can be used as a standard for comparison, using either the conventional technique or microwave-assisted heating. The accuracy of the results has been verified on the basis of standard reference material (IAEA A-11, full-cream milkpowder, in the case of foods) and by interlaboratory comparison (river water, wastewater) [88].

The fluoride content of silicates can be established after fusion with lithium metaborate [89], where fluoride is subsequently separated with the aid of superheated steam [90].

6.2.6. Evaluation Criteria

The maxim "one method is no method" is just as applicable to sample preparation as it is to quantitative determinations. A careful comparison of several decomposition techniques is the only way of assuring accurate results, particularly when little experience is available with respect to the decomposition of a specific matrix, or existing reports are contradictory.

6.2.6.1. Completeness

Decomposition in the strictest sense has as its goal a complete elimination of the matrix. Nevertheless, many decomposition procedures fall short of this ideal, permitting traces of the matrix to remain. Residual carbon content can be established with a total organic carbon (TOC) analyzer. This decreases with increasing oxidizing power of the decomposition reagent as well as with an increase in the duration and temperature of the digestion [43].

6.2.6.2. Uncertainty

As with all aspects of trace analysis, strict attention to possible sources of error is important in the successful application of a decomposition technique. The best determination is rendered completely worthless if the sample preparation step leads to deviations from the true value that exceed the reproducibility of the determination itself by several orders of magnitude. The risks are particularly great when one is working with very low concentrations. Even very small amounts of contamination introduced with the reagents, or "memory effects" from previous experiments, can generate utterly incorrect results. The laboratory atmosphere and materials present in the sample containers can also contribute to contamination and loss, and the potential for trace volatilization (of mercury, for example) at elevated temperature must always be taken into account. Sample preparation is an ideal place to apply the principles of "Good Laboratory Practice" (GLP), → Analytical Chemistry, Purpose and Procedures; even better, "Good *Analytical* Practice" (GAP). Construction materials for the equipment used in sample preparation must be selected very carefully, with appropriate consideration of possible adsorption–desorption effects. Steaming with concentrated acid is the most effective method for cleaning the equipment (Section 6.2.1.3).

The nature of any safety devices employed can also play a role with respect to uncertainty in the case of a pressure ashing system. For example, serious losses may be incurred if sample constituents are entrained during a pressure blow-off. With an irreversible safety device like a rupture disc such a problem may be immediately apparent, but the activation of a reversible device like a valve generally cannot be documented even though major losses may be excluded by automatic reclosure of the device. The optimal solution is almost always a function of the particular application in question.

6.2.6.3. Time Factors

Sample preparation is increasingly coming to be regarded as the "bottleneck" in analysis, because the actual determination methods are becoming increasingly rapid and ever more stringent demands are imposed with respect to precision and accuracy. The only remedies available are acceleration of the decomposition process and simultaneous preparation of multiple samples. However, the latter expedient inevitably leads to higher investment costs and increased operating expenses. Decomposition can be accelerated by working at higher temperature. In the case of wet and dry decompositions, heating can be accelerated by the use of microwave radiation. The subsequent cool-down phase also consumes less time, because the vessel walls are heated only indirectly via the hot sample, and there is no metal autoclave present, with its correspondingly high heat capacity.

Finally, an important parameter to consider in evaluating a routine sample preparation method is the number of staff involved. Several manufacturers have for some time been making a deliberate attempt to reduce staff requirements by introducing automation (or partial automation) into the regimen of sample preparation. The results have been quite satisfactory in the case of apparatus operating at atmospheric pressure, but the technical challenge is greater with systems operating under pressure, especially high pressure (see Section 6.2.8).

6.2.6.4. The Final Result

It is impossible to separate completely the consideration of a specific decomposition process from the subsequent analytical determination. Every decomposition must be adapted to suit a particular determination, which means there can be no such thing as an ideal decomposition method appropriate to all situations and applications. The preferred method must be established with reference to the following factors:

1) Nature of the analyte(s)
2) Concentration of the analyte
3) Matrix characteristics
4) Sample size

5) Required completeness of matrix degradation
6) Amount of time involved
7) Proposed method of determination

If the greatest possible degree of attention must be directed toward completeness in the trace range, decomposition in a closed system is preferred. Ensuring essentially complete decomposition requires the application of high temperature over a period of at least two hours. If sample preparation must be accomplished more rapidly (as in the case of food analysis, environmental studies, or quality control) tests should be conducted on the feasibility of a microwave-assisted process or direct analysis (see Section 6.2.8.2).

An investigation into the accuracy of an analysis should never be restricted to optimization of the determination itself: equal concern is warranted with respect to sample preparation. The key considerations listed above should be regarded only as rough starting points.

6.2.7. Concentration and Separation of Inorganic Trace Materials

Subsequent to any decomposition, but also in the case of liquid samples such as water and urine, the analytes of interest are generally present in dilute solution together with a large excess of foreign ions (e.g., alkali-metal and alkaline-earth cations). Separation and concentration of the analytes may be necessary to improve the limit of detection and exclude interference. Useful techniques in this regard include liquid–liquid extraction, solid-phase extraction, special precipitation reactions, and electrolytic deposition.

Liquid – liquid extraction is a widely used technique for separating and concentrating traces of various elements. For this purpose the analyte is first complexed with a suitable chelating agent, and the complex is then extracted into a water-immiscible solvent. Generally speaking, any complexing agent can be used provided it leads to stable, extractable complexes of the analyte in question. The complexing agent should be capable of separating the analyte as selectively and quantitatively as possible from accompanying ions. Such an extract can usually be used directly for a determination based on AAS. For example, nickel [91], cobalt [92], and cadmium [93] can all be separated and concentrated directly from urine after chelation with hexamethylene ammonium – hexamethylene dithiocarbamidate (HMA – HMDC), using the solvent mixture methyl isopropyl ketone – xylene (70: 30). The separation and enrichment of nickel and antimony in the form of HMA – HMDC complexes from digestion solutions of livers and kidneys has also been described, where copper and iron are removed in a preliminary step as the corresponding cupferron complexes [94]. Thallium can be extracted from a digestion solution after complexing with HMA – HMDC, diethyl dithiocarbamate, dithizone, and 8-mercaptoquinoline [95], or cupferron [96]. Antimony in urine has been extracted after nitric acid digestion as an ammonium pyrrolidine dithiocarbamate complex [97].

The advantage of this approach to concentration is that it permits very selective separation provided a complexing agent is selected with sufficient regard for accompanying ions. Its principal disadvantage is that the potential concentration factors tend to be rather small.

Metal complexes separable by liquid – liquid extraction can also be concentrated on and subsequently eluted from such adsorber resins as XAD, a reverse phase (e.g., C_{18}) [98], activated charcoal [99], or cellulose [100], leading to the technique of *solid-phase extraction*. Numerous reports describe the concentration of analytes on ion-exchange resins [101], [113] and such chelate-forming ion exchangers as Chelex [101], [114], [115] or Hyphan [116], [117]. Solid-phase extraction may result in concentration factors as high as 1000, particularly with dilute samples like drinking water or rain water.

Separation of element traces is also possible with *precipitation* reactions. This technique permits rapid and extensive concentration in a relatively uniform matrix, but it is not very selective. Aluminum hydroxide, magnesium hydroxide, iron hydroxide, and hydrogen sulfide have all been used for trapping trace amounts of various elements [101], [118] – [120], as have such organic precipitating agents as thionalide, cupferron, and dithiocarbamate [101], [121], [122]. A subsequent determination is carried out either directly on the separated precipitate or after restoring it to solution [118].

The *electrolytic deposition* of ions or complexes has also been described with such electrode materials as carbon, copper, or platinum. For example, this technique is utilized in the method of inverse voltammetric determination, where analyte is concentrated at the cathode [101].

New techniques and strategies have been introduced into sample preparation in the area of

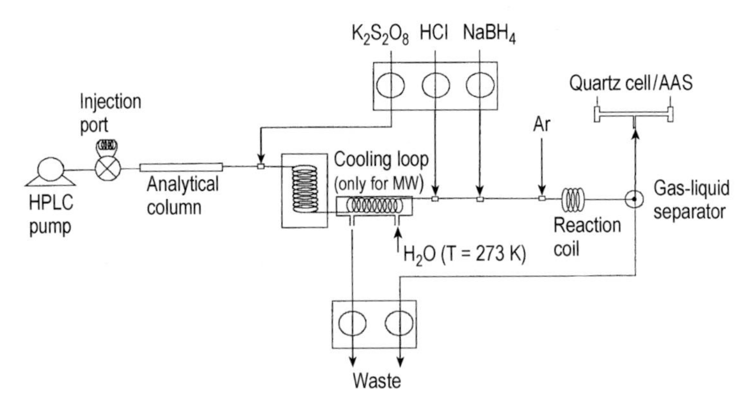

Figure 6. Schematic flow diagram of an on-line coupling of HPLC via UV or microwave-assisted reactors to hydride generation AAS for the determination of As species

preconcentration of inorganic analytes and the separation of such analytes from the matrix [102]–[104]. Very recently MANSFELDT and BIERNATH [105] described a micro-distillation apparatus (Fig. 6) for the determination of total cyanide in soils and sludge.

Another paper has suggested solid-phase extraction (SPE) for the preconcentration and speciation of Cr in waste water samples prior to AAS determination [106]. Interferences of di- and trivalent cations and diverse anions were studied. The detection limits for these species are about 1 µg/L. Other examples for SPE procedures are given in [107] for seawater analysis using on-line ICP-MS and in [108] for the separation and determination of precious metals.

New aspects of liquid–liquid extraction after a complexation step have been discussed for noble metals [109] and for rare earth elements [110].

Also, supercritical-fluid extraction (SFE) has been used for analyte enrichment and matrix separation in metal determinations [111]. The solubilities of metal dithiocarbamates in supercritical carbon dioxide have been characterized [112].

As an additional step in trace analysis, preconcentration techniques are resorted to only when either the sensitivity of the analytical determination is inadequate or severe matrix effects worsen the detection of the analytes [103].

6.2.8. Automation and Direct Analysis

6.2.8.1. Automation

Efforts have long been directed toward automation as a means of minimizing the labor and time involved in an analysis. The introduction of laboratory robots should make it possible to incorporate a significant degree of automation into the time-consuming, labor-intensive area of sample preparation as well, leading to more efficient, reliable, and reproducible sample work-up. An example of an automated system for microwave-assisted pressure ashing is presented in [123].

On-line procedures based on UV-photolysis or microwave-assisted dissolution have been established during recent years for the rapid digestion of relatively simple matrices such as water and waste water samples [130], [131]. In [124] on-line UV digestion with a segmented-flow device was applied for the determination of total cyanide prior to amperometric detection. The detection limit was found to be 0.2 µg/L. The throughput is about 30 analyses per hour.

Another example, that has been published recently by SUR et al. [125], is the on-line microwave and UV digestion of urine samples prior to hydride generation AAS (Fig. 7). The authors managed to determine as many as six arsenic species quantitatively [As(III), As(V), MMA, DMA, arsenobetaine, arsenocholine)]. For more information on species analysis see Section 6.2.9).

Ozone has also been used as a reagent with high quality criterions. For the determination of Hg in a batch cold-vapor system, ozone was used instead of permanganate or peroxidisulfate. The oxidation efficiency of ozone was found to be very high and the digestion was complete in less than 2 min [126].

Microwave decomposition at atmospheric pressure can also be used as an on-line technique, e.g., in a way that the sample passes through a PTFE coil with an inner diameter of about

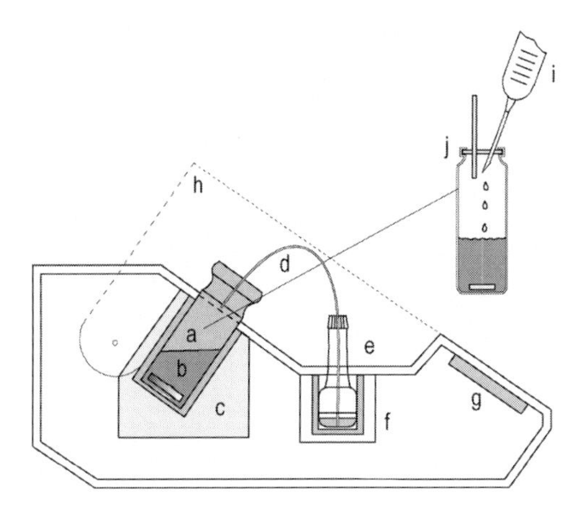

Figure 7. Apparatus for micro-distillation
a) Sample vessel; b) Frit; c) Heater; d) Capillary; e) Absorption vessel; f) Cooler; g) Keyboard and display; h) Protection hood; i) Syringe for reagent feeding; j) Septum (with permission from Dr. Tim Mansfeldt, Ruhr-Universität Bochum, Germany)

1 – 2 mm which is situated in a focused microwave-heated oven. A fully automated on-line sample pretreatment system which combines microwave digestion with sample preconcentration and matrix separation for heavy metals in blood samples is described in [127]. The digested sample solution was transferred on-line to a column packed with iminodiacetate resin for separation of matrix elements (alkaline, alkaline earth, nonmetals) that might interfere with the analyte mass signals in the ICP-MS detection method used. For Ni and Pb the authors reached detection limits in the upper ng/L range, for Fe, Cu, and Zn in the lower µg/L range. The sample throughput is given to be 6 samples/h.

PICHLER et al. [128] have shown that such an on-line microwave procedure can also be adapted to pressurized ashing. In this special case, the sample is injected into a constant stream of nitric acid which passes the microwave heating zone. Expansion of the coil due to the high vapor pressure is prevented by pressurizing its surroundings with nitrogen at up to 35 bar. The temperatures that can be reached in such a system may be as high as 230 °C or even higher.

An on-line SFE/HPLC procedure for the determination of Rh and Pd via chelates has been described in [129].

Flow-injection systems and continuous-flow systems can also make a contribution at the concentration and separation stages as a way of enhancing the potential advantages of automated atomic spectrometric determinations by the flame and graphite-furnace techniques [132], [133]. Another possible application of flow injection is in the conversion of an analyte into the particular oxidation state called for in some sample preparation or determination procedure, as in hydride generation or cold-vapor AAS [134], [135].

6.2.8.2. Direct Analysis

There are certain situations in which sample preparation can be completely or largely omitted, as in the case of graphite-furnace AAS analysis, which permits the direct introduction of solid samples. Here the "ashing step" has been incorporated into the determination itself, which is then referred to as a "direct process." This integration permits an analysis to be carried out more quickly, and the chance of errors is reduced by limiting the number of individual operations [136]. On the other hand, it is essential that the accuracy of such an analysis be confirmed by comparisons with standard reference materials.

Examples of the AAS analysis of powdered samples, as well as such solid foods as chocolate and flour, are presented in [137], [138]. The direct determination of mercury in soils, coal, and ash is discussed in [139]. ICP – OES (and ICP – MS) can also be transformed into a direct process through solid sample introduction and electrothermal pretreatment [140].

6.2.9. Analysis of Element Species

For the determination of the Hg species methylmercury, phenylmercury and Hg(II), ozone was used successfully in a batch cold-vapor system [126]. The preconcentration and speciation of Cr(III) and Cr(VI) in water sample can be performed using solid-phase extraction (SPE) [105]. An SPME (solid-phase microextraction) technique has been used as the sample preparation system for the innovative simultaneous multielement/multispecies determination of six different mercury, tin, and lead species in waters and urine with GC/MS-MS [141], [142]. The determination of arsenic species [As(III), As(V), MMA, DMA, arsenocholine, arsenobetaine)] has also been shown to work with an on-line digestion step prior to hydride generation AAS [125] (see Fig. 7).

6.3. Sample Preparation in Organic Analysis

Recent decades have witnessed significant advances in the efficiency and productivity of instrumental methods in the field of organic trace analysis. Chromatographic and spectroscopic methods in particular have improved greatly with respect to sensitivity. There has also been constant improvement in selectivity, to the point where some samples can now be subjected to analysis without prior preparation, although this is certainly not true in the majority of cases. The goal of sample preparation in organic trace analysis is to isolate the analyte from the sample matrix and then concentrate it and convert it into a form suitable for analysis by the selected method.

Microwave-based techniques have become versatile tools not only for total digestion of organic and inorganic matrices for use in inorganic analysis, but also in drying processes, solvent extraction, clean-up steps and specific reactions such as sample preparation steps in organic analysis [143].

6.3.1. Sample Treatment after the Sampling Process

6.3.1.1. Stabilization, Drying, and Storage

In the context of organic trace analysis, appropriate stabilization and storage precautions are a function of the nature of the analyte and its concentration. To avoid contamination by ambient air and dust, all operations should be conducted under clean-room conditions (clean-bench environment). It is of paramount importance to assure that contamination from vessels, covers, septa, and stabilizers is rigorously excluded.

Volatilization losses can pose a problem in the case of analytes with high vapor pressure. Evaporative loss can be largely avoided by completely filling each sample container and fitting it with an air-tight seal (water samples, biological fluids). This procedure ensures that there will be no oxidative loss due to oxygen present in the vapor space above the sample. For example, blood samples to be analyzed for volatile organic compounds (VOCs) are first treated with an anticoagulant, after which they can be stored for several months at +4 °C in Sovirel or Pyrex test tubes protected only by screw-cap closures and PTFE seals [144]. Similarly, VOCs adsorbed from air samples onto Tenax or activated charcoal can be sealed and stored for several months prior to analysis. Losses have been reported only with cyclic and aliphatic ketones (catalytic oxidation) and esters (hydrolysis) [145]. For these groups of substances, losses approach a maximum of 20 % after a four-week storage period. No losses were observed with benzene, toluene, and xylenes over a 24-month storage period [146]. Samples containing analytes of lower volatility can be stored in suitable containers for a few days at +4 °C, and for several months deep-frozen at – 18 °C or lower.

If an analyte is to be separated by Soxhlet extraction or supercritical fluid extraction, it is essential that the sample ist first dried. Standard practice is to add a drying agent like anhydrous sodium sulfate. Freeze-drying has also been recommended, but care must be exercised to prevent large nonreproducible losses of the analyte. For example, freeze-drying of milk leads to some loss of most PCB congeners. Losses may be as great as 50 – 74 % for monochlorobiphenyl, dichlorobiphenyl, and trichlorobiphenyl [147], [148].

6.3.1.2. Homogenization and Aliquoting

It is best to avoid homogenization prior to the determination of highly volatile analytes, since this often leads to nonreproducible losses. Liquid samples — preferably solids as well — should be transferred immediately after sampling to gas-tight sample vessels, followed by a direct determination on the basis of headspace gas chromatography (Section 6.3.3). If comminution is absolutely necessary, this should be accomplished in the deep-frozen state [149].

For relatively nonvolatile analytes, in such solid matrices as tissues, foods, or soil, the samples should be homogenized prior to aliquoting. This is often accomplished by trituration with sea sand, although use of a drying agent instead results in simultaneous removal of water, producing a homogenate suitable for immediate liquid extraction in a Soxhlet extractor (Section 6.3.2.3) or for supercritical fluid extraction (SFE; Section 6.3.2.4). Small amounts of sample can be homogenized in an ultrasound bath after the addition of water [150].

6.3.1.3. Requirements with Respect to Materials and Chemicals

Plastic containers are generally inappropriate for storing organic samples, since losses may be incurred through adsorption or migration through the vessel wall, and interfering substances may migrate into the sample from the container itself (plasticizers, antioxidants, monomers, etc.). In the case of analytes with high vapor pressures, volatilization losses must also be anticipated with plastic containers. Samples for organic trace analysis should always be stored in glass vessels with appropriate PTFE-sealed screw-cap closures or glass stoppers.

Materials and chemicals must be carefully checked for contamination before use, and if necessary cleaned and purified. Heating for several hours [151] or several days [152] is absolutely essential and very effective in the trace analysis of volatile analytes. Precleaning with boiling nitric acid may be necessary, as may a rinse with an organic solvent like acetone. Equipment for use in the determination of nonvolatile analytes is most effectively cleaned by rinsing with solvents that will be used subsequently in the course of the preparation process [153].

Glass vessels may also contribute to loss of an analyte, especially from aqueous solution, through adsorption on the vessel surface. This can be largely prevented by silanization of active sites on the glass. Alternatively, the vessel should be rinsed with solvent after removal of the sample, with the rinse solution then added to the sample itself [147].

Liquid – solid extraction procedures are accompanied by a serious risk of contamination. For example, blank tests with commercial extraction columns have revealed traces of plasticizers and antioxidants as well as various alkanes and alkenes [154]. For this reason certain manufacturers now offer solid-phase extraction columns made from glass. The extent of the po-tential contamination depends on the manufac-turer, the adsorbent load, conditioning steps, and the solvent used for the extraction [154], [155]. If no satisfactory commercial column is available, one can easily be prepared in the laboratory from a suitable adsorbent and an appropriate glass tube. For example, XAD can be purified to an extent sufficient for this purpose by Soxhlet extraction [156], [157], which reduces the level of impurities by a factor of 30. Silica gel and aluminum oxide can be purified by washing with solvents, followed by activation in a drying oven [152]. For example, unpurified silica gel has been shown to release as much as 160 ng/g of PCBs [155].

The risk of contamination must also be considered when a diffusive sampler is used to obtain air samples for determining volatile organic compounds. Thus, benzene, *n*-tetradecane, *n*-pentadecane, *n*-hexadecane, *n*-heptadecane, and dioctyl phthalate [158], along with trichloromethane, 1,1,1-trichloroethane, trichloroethene, and tetrachloroethene [159] have all been detected in extracts from unexposed diffusive samplers. The measured concentrations varied from batch to batch, ranging from 0.005 µg to 0.14 µg per diffusive sampler in the case of the chlorinated hydrocarbons [159], and from 0.1 – 2.6 µg per diffusion collector for the aliphatic hydrocarbons and dioctyl phthalate [158].

Ultrapure water can be prepared in the laboratory by filtration through an ion exchanger and activated charcoal. According to the manufacturer, one commercial water treatment system of this type reduces the residual hydrocarbon content of water to < 20 µg/L. Further reduction to < 5 µg/L is possible through UV irradiation (as with the "zero-water unit" manufactured by Gräntzel, Karlsruhe, Germany). If necessary, additional purification can be achieved with a stream of nitrogen at 80 °C [151], or by extraction with the solvent that will be used for extracting the analyte.

Most solvents are available commercially in sufficient purity, but blank test values should be established on a regular basis to ensure the absence of contaminants.

6.3.2. Separation of the Analyte

6.3.2.1. Hydrolysis

Many of the harmful substances and pollutants in biological samples, including their metabolites, are present wholly or in part in a conjugated (bound) form. This is the case, for example, with pentachlorophenol and phenol, which are excreted in the urine as glucuronide or glucuronide and sulfate, as well as with aromatic amines, present to some extent in the blood as hemoglobin conjugates. In such a situation the first step must accomplish release of the primary analytes. Conjugates can be cleaved by acid hydrolysis [160] – [163], basic hydrolysis [162], or enzymatic hydrolysis [162], [164]. Hydrolysis with the aid of enzymes, such as sulfatase and glucuronidase, has the significant advantage that bonds are cleaved extremely selectively, and information can also be obtained regarding the binding partners of the analyte. A disadvantage is the considerable amount of time required, typically several hours.

In many cases a portion of the matrix can be removed or converted into some more easily removable form before actual separation of the analytes is attempted. Relatively involatile, chemically stable analytes [e.g., organochlorine pesticides, PCBs, or polychlorinated dibenzo-*p*-dioxines (PCDDs)] can frequently be released from a matrix by acidic or alkaline hydrolysis with concentrated acid or base. Proteins can often be precipitated from biological samples by addition of acid, salt, or an organic solvent, or they can be broken down by enzymatic hydrolysis [165] – [167].

6.3.2.2. Liquid – Liquid Extraction

Extraction with organic solvents is the most common extraction technique for isolating an analyte from a liquid sample. The technique is based on distribution of the analyte between two immiscible liquid phases.

The decisive parameter with respect to extraction yield is the *distribution coefficient* for the analyte between the particular phases involved. A distribution coefficient can often be influenced advantageously by establishing a specific pH, thereby dividing the sample into strongly or weakly acidic, neutral, or basic fractions [168]. An example is provided by the extraction of aromatic amines from blood and urine [169]. Further possibilities include formation of ion pairs [170], complexation with metal salts, or salting-out of an aqueous phase.

If the distribution coefficient is sufficiently large, the simplest approach to liquid – liquid extraction is shaking the sample with an appropriate amount of an organic solvent. With smaller distribution coefficients or large sample volumes, continuous extraction or countercurrent extraction is required to achieve a complete separation. The apparatus for continuous extraction causes a liquid immiscible with the sample solution to circulate continuously and in finely divided form through the sample [171] – [173]. Extracted analytes are concentrated by distillation at appropriate times between individual extraction cycles.

Disadvantages of liquid – liquid extraction include the high dilution of the extract, which must subsequently be concentrated, and a high rate of solvent consumption. Both problems can be minimized by employing microextraction methods. Devices for extracting aqueous samples as large as 1 L with 200 μL of organic solvent have been described [174], [175].

Another variant of liquid – liquid extraction takes advantage of a liquid phase immobilized on a solid sorbent such as kieselguhr, Celite, Chromosorb W, or Chromosorb P [176]. The immobilized phase may be either aqueous or nonaqueous.

The formation of emulsions frequently presents problems in liquid – liquid extraction. Often such emulsions can be broken by centrifugation, freezing, or the addition of salt.

6.3.2.3. Soxhlet Extraction

Soxhlet extraction is chiefly applied to the separation of relatively nonvolatile analytes from solid samples. For example, this technique is useful for extracting pesticides, PCBs, and PCDD/PCDFs from various matrices, including fatty tissue, soil, and paper [177], [178]. The advantage of the method is high yield, achieved by continuous extraction of the sample with fresh extractant. Disadvantages include the considerable amount of time required (typically several hours to several days), the thermal stress to which analytes are subjected, and the fact that the analytes are obtained in very dilute form in a solvent, and must therefore be concentrated in a subsequent step. Any fat present in the sample is co-extracted, and this must also be removed separately.

6.3.2.4. Supercritical Fluid Extraction (SFE)

Recently, an extraction method long practiced in industry — supercritical fluid extraction (SFE) — has also been introduced into the analytical sector. As the name suggests, SFE employs supercritical fluids for extraction purposes in place of the organic solvents of conventional extraction.

A compound is said to be in a supercritical state when the critical pressure and critical temperature for that particular material are exceeded. Critical parameters for several relevant compounds are listed in Table 6. Supercritical fluids exhibit simultaneously properties associated with both gases and liquids (see Table 7). Thus, like gases, they are compressible, but they also display solvencies similar to those of liquids. Any increase in temperature at constant pressure reduces the

Table 6. Critical parameters for selected substances [179]

Substance	Critical temperature, °C	Critical pressure, MPa
Xe	16.6	5.83
CHF_3	25.9	4.83
$CClF_3$	28.8	3.92
CO_2	31.1	7.37
N_2O	36.4	7.24
NH_3	132.2	11.27
CH_3OH	239.4	8.09
H_2O	374.1	22.04

Table 7. Typical characteristics of gases, liquids, and supercritical fluids

Property	Gas	Liquid	Supercritical fluid
Density, g/mL	$10^{-4}-10^{-3}$	0.6 – 1.4	0.1 – 1
Diffusion coefficient, cm^2/s	10^{-1}	10^{-5}	$10^{-3}-10^{-4}$
Solvency	no	yes	yes
Compressibility	yes	no	yes

solvent power of a supercritical fluid, but it also leads to an increase in diffusion rate, which tends to lower the minimum required extraction time. Compared to conventional extractants, supercritical fluids have low viscosities and diffusion rates that are higher by a factor of 10 – 100, both of which contribute to reduced extraction times. With supercritical CO_2 and N_2O, which are gases under normal conditions, the extractant is separated by reducing the pressure to atmospheric levels, leading to simultaneous concentration of the extract.

Supercritical CO_2 is the most frequently used extractant for SFE. It has the advantage of being chemically rather inert, and its critical temperature is low, so it is valuable for the extraction of such thermolabile analytes as steroids and fragrances. The low critical pressure for CO_2 opens the way to a relatively broad range over which the solvency can be varied through adjustment of the pressure. Other advantages of CO_2 as an extractant include low toxicity, high purity, and low cost. The principal disadvantage of CO_2 is a relatively low polarity. Its solvent power with respect to polar analytes can be improved, however, by adding such polar *modifiers* as methanol, acetone, hexane, or dichloromethane.

Extraction with supercritical CO_2 has been recommended for separating a wide variety of analytes, including pesticides, PCBs, vitamins, and fragrances from meat, fish, baby food, and animal feed [180] – [185]. PAHs, PCBs, PCDO/PCDFs, and other substances have been extracted from soil, fly ash, sediment, air particles, polymers, and plants using supercritical CO_2 together with a modifier [186] – [196].

6.3.2.5. Solid-Phase Extraction (SPE)

Solid-phase extraction (SPE) is used for the selective separation and concentration of analytes from liquid samples. Extraction of the analytes is based in this case on the distribution of dissolved substances between a solid-phase surface and the sample liquid. Separation of various sample constituents may be a result of differing polarities, differences in molecular size, or differences with respect to ion-exchange capacity.

Many of the comments above regarding liquid – liquid extraction also apply to solid-phase extraction. However, the latter has the advantages that it can be accomplished more rapidly, requires less solvent, provides more highly concentrated extracts, and is relatively easy to automate. Laboratory robots and suitable column arrangement have rendered large parts of the sample preparation process automatic, thereby facilitating reproducibility and efficient operation [197], [198].

Many different adsorbents are applicable in this context (see Table 8). The most common adsorbents for solid-phase extraction are based on silica gel, the surface of which has been modified in some way.

Octadecyl surface phases (C_{18}) are used for the reversed-phase extraction of nonpolar substances from aqueous solutions. Typical applications include the extraction of organochlorine pesticides [199] – [201], organophosphorus pesticides [202], chlorinated and unchlorinated hydrocarbons, triazines [200], [201], PAHs and nitro-PAHs [200], carbamates [200], chlorophenols [200], [203], aflatoxins [204], plasticizers [205], vitamins [206], and medicaments such as barbiturates and antibiotics [207], [208].

Shorter octyl phases (C_8) are used for extracting substances of medium polarity. Substances that bind irreversibly to C_{18} phases can often be concentrated and re-eluted successfully with C_8 phases.

Normal phase materials, including unmodified silica gel, aluminum oxide, and Florisil, separate sample constituents into fractions of comparable polarity. They are often utilized to separate and concentrate pesticides [209] – [211], PCBs [147], [209], and PCDD/PCDFs [212] – [216] from such

Table 8. Important adsorption agents classified according to type

Nonpolar	Polar	Anion exchangers	Cation exchangers
Octadecyl	Cyano	Primary amine	Carboxylic acid
Octyl	Kieselguhr	Secondary amine	Sulfonic acid
Butyl	Silica gel	Quaternary ammonium salt	
Cyclohexyl	Florisil		
Phenyl	Aluminum oxide		
Amino			
Diol			

biological samples as blood, breast milk, fatty tissue, and foods. Normal phases are also used to extract polar sample constituents, such as amines, alcohols, phenols, dyes, medicaments, or vitamins [217].

Anion exchangers are mainly employed in the extraction of carbohydrates, peptides, nucleosides, and amino acids. Cation exchangers are useful for extracting amino acids and nucleosides [217].

Apart from modified silica gel, the most frequently used solid-phase adsorbents are activated charcoal, macroreticular resins (XAD), ordinary silica gel, aluminum oxide, and Florisil. Activated charcoal is a universal adsorbent for concentrating trace organic materials in aqueous solutions and air. XAD resins are also commonly employed for extracting organic trace constituents, such as alkoxyacetic acids [157], organochlorine pesticides [200], [218], carbamates [198], [202], and triazines [218] from aqueous matrices.

The activity of these solids must be accurately adjusted to ensure reproducible results and well-defined fractions, and sample solutions must always be carefully dried prior to extraction.

Exceptionally low limits of detectability can be achieved with automated on-line systems. Here the complete eluate is analyzed rather than simply an aliquot, which lowers the risk of evaporative losses and sample contamination. Automated sample preparation through solid-phase extraction is becoming increasingly important in extreme trace analysis, as in the determination of PCDDs in tissue or in the blood of uncontaminated persons [219].

A recently developed variant of solid-phase extraction is extraction with the aid of *extraction disks* [220], [221], membrane filter plates on which an appropriate solid-phase material has been immobilized. Advantages of extraction disks include higher flow rates and more rapid material exchange. Another variant involves the use of SPE for simultaneous extraction and derivatization. For this purpose the analyte is adsorbed onto a solid phase that has been impregnated with a derivatizing agent, permitting the separation and concentration of analytes that would otherwise not be adsorbed at all. Solid-supported reagents are useful in such applications as the determination of organic acids or cannabinoids in plasma [222], [223].

6.3.2.6. Solid-Phase Microextraction (SPME)

A new technique, which is applicable for sampling in air and liquids or in the headspace above a liquid or a solid sample, is solid-phase microextraction (SPME). The mechanism of SPME, which has been developed by Pawliszyn et al. [225], [226], is based on the partition equilibrium of the analytes between the sample or the headspace above the sample, respectively, and a fused silica fiber coated with a suitable stationary phase. The amount of analyte extracted by the fiber is proportional to the initial analyte concentration in the sample and depends on the type of fiber. After sampling, the fiber can be thermally desorbed directly into the injector of a gas chromatograph. SPME combines sampling, analyte enrichment, matrix separation, and sample introduction within one step [226]. Since its development, this innovative technique has found widespread use in environmental analysis. It has, for example, been applied in the determination of volatile organic compounds [227], [228], phenols [229], pesticides [230], polyaromatic hydrocarbons, and polychlorinated biphenyls [231] in water.

SPME fibers have been used as air sampling devices for volatile organic compounds in ambient and workplace air and give results that are in good agreement with traditional sampling methods [232]. Furthermore, grab sampling is used with stainless steel canisters or glass bulbs in combination with SPME [226], [233]. The dependence of the sampling rate on humidity and air temperature can be eliminated by correction factors [233]. The main drawback is the low storage stability of the samples, due to uncontrolled losses of analytes by adsorption on the walls of the canisters or by evaporation from the loaded fiber.

6.3.2.7. Miscellaneous Techniques

Microwave-assisted extraction (MAE) is an upcoming trend in rapid extraction techniques. Is has been applied recently, for example, to the extraction and determination of polycyclic hydrocarbons in marine sediments [234] and in wood samples [235] prior to an HPLC/UV determination.

An alternative approach to separating volatile compounds from liquid and solid samples is distillation. Simple distillation can accomplish the isolation of a volatile analyte from a nonvolatile residue, or separate multiple sample constituents with widely differing boiling points. A further application is separation of a sample into fractions with different boiling ranges.

Steam distillation is an effective way to separate from a matrix such steam-volatile analytes as phenol [224], alkylphenols [160], or formic acid [236]. Steam distillation offers the advantage that the analyte is recovered in a nearly matrix-free condition, although it is diluted with a large amount of water from which it must subsequently be separated and concentrated. Solid-phase extraction is particularly useful for this purpose (Section 6.3.2.5).

Another technique involving distillative concentration of sample constituents is *sweep co-distillation*. Here the sample is treated with a highly volatile solvent introduced with the aid of a stream of carrier gas. The solvent in turn transports soluble components of the sample to a cooled distillation receiver. The method is useful for such applications as the isolation of volatile pesticides from animal and plant fats [237], [238].

Membrane techniques, including dialysis, ultrafiltration, and reverse osmosis, are also applicable to sample preparation problems [239] – [241]. *Dialysis* separates analytes on the basis of their ability to diffuse through a membrane as a result of a concentration gradient. It is most frequently used for concentrating analytes in biological liquids. Automation and on-line coupling with liquid chromatography have both been reported [242], [243]. *Reverse osmosis* is very similar to dialysis, but in this case a pressure difference is used to cause the solvent to migrate through a membrane from a region of high analyte concentration to one of low analyte concentration. Reverse osmosis is frequently called upon for concentrating large volumes of very dilute solutions. *Ultrafiltration* is a method for concentrating large sample molecules present in dilute solution. Separation in this case depends directly on molecular size, and occurs when a sample solution is filtered through a membrane with an appropriate pore size. One possible application is in the separation of free molecules present in a biological fluid from similar molecules that are bound to receptors [244]; another is the desalination of a liquid sample.

6.3.3. Headspace Techniques

6.3.3.1. Static Headspace Technique

Highly volatile analytes can be separated for subsequent quantitative determination by one of several *headspace techniques*. All headspace techniques are based on the Henry – Dalton law, which states that, in a closed system, the vapor-space concentration of an analyte depends exclusively on the temperature and the corresponding analyte concentration in solution or on a solid surface. Consequently, raising the temperature causes volatile analytes to separate from their matrix and become concentrated in the surrounding vapor space. In the *static* headspace technique, a gas chromatographic determination is made of the analyte once equilibrium has been established. The separation efficiency characteristic of gas chromatography and the virtual absence of matrix-derived detector noise means that chlorinated, aromatic, and aliphatic hydrocarbons, along with alcohols, ketones, and esters, are all subject to determination essentially free of interference regardless of the matrix. Limits of detectability in the µg/L range permit volatile organic compounds to be determined in body fluids from persons occupationally exposed to contaminants [245], [246], as well as in foods [247] and water [248]. Nevertheless, the distribution coefficient does depend on the matrix, so a relatively complicated and time-consuming calibration is essential. Depending on the matrix, calibration can be based on matrix standards, standard addition, or internal standardization.

6.3.3.2. Dynamic Headspace Technique (Purge and Trap)

In the dynamic headspace technique, a carrier gas is passed continuously through the sample under investigation. Constant contact with fresh carrier gas causes volatile analytes to be removed gradually and almost completely from the sample. These are then concentrated on an adsorbent such as Tenax or Chromosorb. Quantitative transfer of the analytes to a gas chromatographic column is accomplished by thermodesorption, using either on-line or off-line techniques.

The concentration that results (typically by a factor of ca. 100) leads to limits of detectability in the ng/L range, permitting the simultaneous analysis of environmentally conditioned concentrations of as many as 48 volatile organic compounds in water and such body fluids as blood and urine [144], [249]–[251].

6.3.4. Determination of Trace Organic Materials in Air Samples

Air sampling for analysis of gases and vapors is accomplished either with the aid of a pump and subsequent concentration on a solid adsorbent (*active sampling*), or with a collector, in which case analytes reach the collecting phase via diffusion or permeation (*passive sampling*). Gaseous trace organic materials in air are often collected by adsorption on activated charcoal, Tenax, or XAD resin, although other adsorbents such as aluminum oxide or Florisil can also be used.

Analytes are usually desorbed from the adsorbent by washing with appropriate solvents, after which the eluates can be analyzed by gas chromatography. Carbon disulfide is the most commonly employed desorption agent [252]. For the desorption of polar analytes, small amounts of such polar solvents as methanol, 2-propanol, or 2-butanol can be added to the carbon disulfide as a way of increasing the desorption yield [253].

Analytes can also be desorbed thermally and analyzed on-line with a gas chromatograph. This expedient results in a significant increase in sensitivity, since the entire sample is utilized for the analysis. One disadvantage is that the binding power of activated charcoal (the most suitable adsorption material for most analytes) is too great to be overcome with conventional thermodesorbers, eliminating the possibility of using this combination for quantitative analysis. Quantitative desorption from charcoal can be accomplished with a microwave thermodesorption device, however, and this microwave thermodesorption is a mild process, suitable even for such labile analytes as fragrances [254].

Particle-bound substances like polycyclic aromatic hydrocarbons (PAH) are determined by collecting particulate air constituents on glass-fiber or PTFE filters, or in impactors, followed by extraction.

6.3.5. Analyte Concentration

Subsequent to one of the aforementioned extraction procedures (e.g., liquid–liquid extraction, steam distillation, Soxhlet extraction) the analytes are obtained in extremely dilute form in a large volume of solvent. Various approaches to concentration are possible depending on the nature of the analyte, the solvent, the initial volume, and the target volume.

Aqueous analytes, such as those from a steam distillation, can be concentrated by either solid-phase or liquid–liquid extraction (Sections 6.3.2.2 and 6.3.2.5). An analyte dissolved in a highly volatile solvent (as with liquid–liquid or Soxhlet extraction) is concentrated most effectively by evaporation of the solvent in a gas stream (e.g., nitrogen, helium), or with a rotary evaporator or a Kuderna–Danish concentrator. Losses must be anticipated with a rotary evaporator, however, especially of volatile analytes, though even less volatile constituents may also be lost through codistillation. Recoveries of volatile analytes are higher with a Kuderna–Danish concentrator. Concentrating an extract by solvent evaporation in a gas stream is appropriate only with relatively small volumes on account of the low vaporization rate. Analyte loss is possible through aerosol formation or evaporation.

6.3.6. Derivatization

Separation and concentration of an analyte must often be followed by some type of *derivatization*. Derivatization is conducted with one or more of the following objectives:

1) To facilitate chromatographic separation
2) To increase selectivity
3) To improve the limit of detection

The first objective applies mainly to gas chromatography, and involves the preparation of a more volatile or less polar form of the analyte. Improved selectivity is generally less important in this case due to the large number of theoretical plates available with capillary gas chromatography columns, in contrast to HPLC. However, the limit of detection plays an important role in both gas and liquid chromatography.

Most derivatizations for gas chromatography are esterifications or etherifications. For example, analytes containing carboxyl groups are often converted into methyl [156], [161], pentafluorobenzyl [255] – [258], or triethyl esters [259] – [261]. Analytes containing acidic hydroxyl groups (phenols, chlorophenols, glycol ethers) or amino groups (e.g., aniline) can be derivatized with such perfluorinated compounds as heptafluorobutyric anhydride, pentafluoropropionic anhydride [169], or pentafluorobenzoyl chloride [262].

Derivatization with halogenated compounds offers the advantage that the derivatives are subject to extremely sensitive detection with an electron capture detector (ECD). The disadvantage is that excess derivatization agent remains in the sample solution after the reaction, necessitating its removal prior to gas chromatographic determination because of the potential for interference during ECD detection.

Derivatizations for HPLC are designed mainly to improve the limit of detection, permitting the use of highly sensitive or selective detectors inapplicable to the analytes themselves. Enhanced absorption of UV/visible light is achieved by the introduction of chromophoric groups. Analytes can also be rendered fluorescent by the introduction of fluorophoric groups.

Carboxylic acids, including formic, acetic, lactic, propionic, malic, tartaric, and citric acids, can be transformed with benzyl, naphthacyl, phenacyl, or bromophenacyl bromides [263], [264]; *p*-nitrobenzyl bromide [265]; *p*-nitrophenacyl bromide; methoxyaniline; or *p*-nitrobenzyl-*N*,*N*′-diisopropylurea [266], [267] into esters that absorb UV or visible light. α-Keto acids (e.g., glycolic, glyoxylic acids) are detectable with UV light after derivatization with phenylhydrazine [268], [269].

Fluorescent compounds are obtained by reacting carboxylic acids with 4-bromomethyl-7-methoxycoumarin [270] – [273] or 4-bromomethyl-7-acetoxycoumarin [274], [275].

Analytes containing hydroxyl groups, such as phenols, glycols, and alcohols, can be converted with 3,5-dinitrobenzoyl chloride [276] or dabsyl chloride [277] – [279] into compounds that absorb UV or visible light. Fluorescent derivatives can be obtained with 7-(chlorocarbonyl) methoxy-4-methylcoumarin [280].

Derivatizations for HPLC purposes are accomplished either off-line or on-line. An on-line process may involve either pre-column or post-column reaction depending on the objective.

6.3.7. Coupled Techniques

Various coupled sample preparation and determination processes are increasingly utilized in trace organic analysis, whereby separated analytes pass directly in an on-line way into a chromatographic analysis system. The main advantages of such on-line procedures reflect the quantitative nature of analyte transfer from the sample preparation stage to the analytical determination, leading to an optimal limit of detection. This approach also minimizes the risk of sample loss and contamination. However, coupled techniques are relatively difficult to imple-ment, since parameters must be optimized not only with respect to sample preparation but also for the subsequent chromatographic separation. One of the most important coupled techniques at the present time is LC – GC. Automatic sample preparation is in this case accomplished via HPLC, permitting the separation of a complex matrix, or even a concentration of trace material. A subsequent capillary gas chromatographic analysis results in a high degree of resolution (with isomeric materials, for example) and the high signal-to-noise ratio essential for trace analysis thanks to the availability of such extremely sensitive and selective devices as electron capture detectors (ECD), flame photometry detectors (FPD), phosphorus – nitrogen-selective detectors (PND), and mass spectrometric detectors (MSD) [281] – [283]. Coupling is achieved with a sample loop (loop technique), which is used to transfer a small portion of the LC eluate to an uncoated GC precolumn (retention gap) [284]. Determinations of pesticide metabolites from maize [285] and organochlorine pesticides from fat [286] constitute two examples of the widespread application of this technique.

LC – MS coupling is also becoming an increasingly common analytical method now that effective interfaces have been developed (particle beam, thermospray, electrospray). Determinations of pesticides in sea water [220], carbamates and phenylurea derivatives in water samples [287], and organophosphorus pesticides and chlorophenols [288] are just a few examples of the use of this method.

LC–TLC coupling constitutes one possibility for carrying out multidimensional LC. Thus HPLC (usually reversed-phase chromatography) is coupled with thin layer (adsorption) chromatography in such a way that the eluate from the HPLC column is transferred via a capillary column to a mechanically transported TLC plate [289], [290].

The coupling of SFE with gas chromatography has also been described, as has chromatography with supercritical fluids (SFC). Examples include the separation and determination of PCBs, PAHs, and pesticides in such environmental samples as soil and sediments [291]–[294]. A frequently employed technique for determining traces of pesticide in aqueous samples is the on-line coupling of solid-phase extraction or dialysis with HPLC [242], [243], [295].

6.4. References

[1] P. Tschöpel: "Sample Treatment" in M. Stoeppler (ed.): *Hazardous Metals in the Environment,* vol. 12, Elsevier, Amsterdam 1992.

[2] O. Behne, *J. Clin. Chem. Clin. Biochem.* **19** (1981) 115.

[3] G. Knapp, *Int. J. Environ. Anal. Chem.* **22** (1985) 71.

[4] D. v. Renterghem, R. Cornelis, R. Vanholder, *Anal. Chim. Acta* **257** (1992) 1.

[5] J. Trettenbach, K. G. Heumann, *Fresenius Z. Anal. Chem.* **322** (1985) 306.

[6] L. B. Fischer, *Anal. Chem.* **58** (1986) 261.

[7] B. Grieping, H. Marchandise in: *Analytiker Taschenbuch,* vol. 6, Springer Verlag, Berlin 1986, p. 3.

[8] DFG Deutsche Forschungsgemeinschaft in J. Angerer, K. H. Schaller (eds.): *Analyses of Hazardous Substances in Biological Materials.* **vol. 5,** VCH Verlagsgesellschaft, Weinheim 1997, pp. XXI–XXII.

[9] J. Begerow, L. Dunemann, in L. Matter (ed.): *Elementspurenanalytik in biologischen Matrices,* Spektrum Akademischer Verlag, Heidelberg–Berlin–Oxford 1997, p. 27.

[10] D. Behne, G. V. Iyengar: "Spurenelementanalyse in biologischen Proben", *Analytiker-Taschenbuch,* vol. 6, Springer Verlag, Berlin 1986.

[11] M. Kiilunen, J. Järvisalo, O. Mäkitie, A. Aitio, *Int. Arch. Occup. Environ. Health* **59** (1987) 43.

[12] G. V. Iyengar, B. Sansoni: "Elemental Analysis of Biological Materials: Current Problems and Techniques with Special Reference to Trace Elements," *IAEA Technical Report 197,* Wien 1980, p. 73.

[13] G. V. Iyengar, K. Kasperek, L. E. Feindegen, *Sci. Total Environ.* **10** (1978) 1.

[14] J. Angerer: "Spezielle Vorbemerkungen," in D. Henschler (ed.): *Analysen in biologischem Material,* 10th suppl., VCH Verlagsgesellschaft, Weinheim 1991.

[15] J. C. Meranger, B. L. Hollebone, G. A. Blanchette, *J. Anal. Toxicol.* **33** (1981) 5.

[16] P. Tschöpel et al., *Fresenius Z. Anal. Chem.* **302** (1980) 1.

[17] K. Heydorn: *Neutron Activation Analysis for Clinical Trace Element Research,* **vol. 1,** CRC Press, Boca Raton, Fla. 1984, p. 42.

[18] J. R. Moody, P. Lindström, *Anal. Chem.* **47** (1977) 2264.

[19] S. P. Ericson et al., *Clin. Chem. (Winston-Salem N.C.)* **32** (1986) 1350.

[20] S. B. Adeloju, A. M. Bond, *Anal. Chem.* **57** (1985) 1728.

[21] R. Cornelis et al. in: *Nuclear Activation Techniques in the Life Sciences,* IAEA, Wien 1979, p. 165.

[22] J. Begerow, L. Dunemann in M. Stoeppler (ed.): *Sampling and Sample Preparation,* Springer-Verlag, Berlin–Heidelberg–New York 1997, p. 155.

[23] P. Lievens, J. Versieck, R. Cornelis, J. Hoste, *J. Radioanal. Chem.* **37** (1977) 483.

[24] A. Abu-Samra, J. S. Morris, S. R. Koirtyohann, *Anal. Chem.* **47** (1975) 1475.

[25] E. D. Neas, M. J. Collins in H. M. Kingston, L. B. Jassie (eds.): *Introduction to Microwave Sample Preparation,* Am. Chem. Soc., Washington, D.C., 1988, p. 7.

[26] R. A. Nadkarni, *Anal. Chem.* **56** (1984) 2233.

[27] Z. Sulcek, P. Povondra: *Methods of Decomposition in Inorganic Analysis,* CRC Press, Boca Raton, Fla. 1989.

[28] P. Aysola, P. Anderson, C. H. Langford, *Anal. Chem.* **59** (1987) 1582.

[29] R. T. White, Jr., G. E. Douthit, *J. Assoc. Off. Anal. Chem.* **68** (1985) 766.

[30] L.-Q. Xu, W.-X. Shen, *Fresenius Z. Anal. Chem.* **332** (1988) 45.

[31] G. Knapp, *Int. J. Environ. Anal. Chem.* **22** (1984) 71.

[32] G. Knapp, *Fresenius Z. Anal. Chem.* **317** (1984) 213.

[33] K. W. Budna, G. Knapp, *Fresenius Z. Anal. Chem.* **294** (1979) 122.

[34] F. R. Abou-Shakra, M. P. Rayman, N. I. Ward, V. Hotton, G. Bastian, *J. Anal. At. Spectrom.* **12** (1997) 429.

[35] H. Hodrejarv, A. Vaarmann, *Anal. Chim. Acta* **396** (1999) 293.

[36] M. Bettinelli, C. Baffi, G. M. Beone, S. Spezia, *At. Spectrosc.* **21** (2000) 50.

[37] A. C. Grillo, *Spectrosc. Int.* **1** (1989) 16.

[38] M. Kolb, P. Rach, J. Schäfer, A. Wild, *Fresenius J. Anal. Chem.* **342** (1992) 341.

[39] L. Dunemann in B. Welz (ed.): *5th Colloquium Atomspektrometrische Spurenanalytik,* Bodenseewerk Perkin-Elmer, Überlingen 1989, p. 593.

[40] L. Dunemann, M. Meinerling, *Fresenius J. Anal. Chem.* **342** (1992) 714.

[41] J. Hertz, R. Pani, *Fresenius Z. Anal. Chem.* **328** (1987) 487.

[42] F. Wahdat, R. Neeb, *Fresenius Z. Anal. Chem.* **335** (1989) 748.

[43] L. Dunemann, *Nachr. Chem. Tech. Lab.* **39** (1991) no. 10, M1.

[44] M. Würfels, E. Jackwerth, M. Stoeppler, *Fresenius Z. Anal. Chem.* **329** (1987) 459.

[45] P. Schramel, S. Haase, G. Knapp, *Fresenius Z. Anal. Chem.* **326** (1987) 142.

[46] P. Schramel, G. Lill, R. Seif, *Fresenius Z. Anal. Chem.* **326** (1987) 135.

[47] B. Zunk, *Anal. Chim. Acta* **236** (1990) 337.

[48] L. B. Fischer, *Anal. Chem.* **58** (1986) 261.

[49] H. Matusiewicz, *Anal. Chem.* **71** (1999) 3145.

[50] M. Deaker, W. Maher, *J. Anal. At. Spectrom.* **14** (1999) 1193.

[51] D. Merton, J. A. C. Broekaert, R. Brandt, N. Jakubowski, *J. Anal. At. Spectrom.* **14** (1999) 1093.

[52] R. A. Romero, J. E. Tahán, A. J. Moronta, *Anal. Chim. Acta* **257** (1992) 147.
[53] M. S. Chaudhary, S. Ahmad, A. Mannan, I. H. Qureshi, *J. Radioanal. Nucl. Chem.* **83** (1984) 387.
[54] M. Feinberg, C. Ducauze, *Anal. Chem.* **52** (1980) 207.
[55] D. Huljev, B. Huljev, Z. Rajkovic-Huljev, *Radiol. Iugosl.* **22** (1988) 403.
[56] G. Kaiser, G. Tölg, *Fresenius Z. Anal. Chem.* **325** (1986) 32.
[57] P. Barth, R. Caletka, V. Krivan, *Fresenius Z. Anal. Chem.* **319** (1984) 560.
[58] S. E. Raptis, G. Kaiser, G. Tölg, *Anal. Chim. Acta* **138** (1982) 93.
[59] J. Binkowski, P. Rutkowski, *Mikrochim. Acta* **1** (1987) 245.
[60] M. Kulke, F. Umland, *Fresenius Z. Anal. Chem.* **288** (1977) 273.
[61] M. Fujita et al., *Anal. Chem.* **40** (1968) 2042.
[62] S. S. Q. Hee, J. R. Boyle, *Anal. Chem.* **60** (1988) 1033.
[63] D. A. Levaggi, W. Qyung, M. Feldstein, *J. Air. Pollut. Control Assoc.* **21** (1971) 277.
[64] G. Schwedt, L. Dunemann, *LaborPraxis,* June 1990, 476.
[65] S. E. Raptis, G. Knapp, A. P. Schalk, *Fresenius Z. Anal. Chem.* **316** (1983) 482.
[66] D. Behne, P. A. Matamba, *Fresenius Z. Anal. Chem.* **274** (1975) 195.
[67] M. Gallorini, E. Orvini, A. Rolla, M. Burdisso, *Analyst (London)* **106** (1981) 328.
[68] C. Feldmann, *Anal. Chem.* **55** (1983) 2451.
[69] H. Brügmann, T. X. Gian, H. Berge, *Acta Hydrochim. Hydrobiol.* **16** (1988) 457.
[70] M. Weidenauer, K. H. Lieser, *Fresenius Z. Anal. Chem.* **320** (1985) 550.
[71] T. M. Karadahki, F. M. Najib, F. A. Mohammed, *Talanta* **34** (1987) 995.
[72] A. Campiglio, *Mikrochim. Acta* **3** (1987) 425.
[73] P. Ostapczuk, M. Froning, M. Stoeppler, *Fresenius Z. Anal. Chem.* **334** (1989) 61.
[74] K. Gretzinger, E. Grallath, G. Tölg, *Anal. Chim. Acta* **193** (1987) 1.
[75] H. Heinrichs, H. J. Brumsack, N. Loftfield, N. Kö-nig, *Z. Pflanzenernähr. Bodenkd.* **149** (1986) 350.
[76] S. Sprung, G. Kirchner, W. Rechenberg, *Zem.-Kalk-Gips, Ed. B* **37** (1984) 513.
[77] B. Morsches, G. Tölg, *Fresenius Z. Anal. Chem.* **219** (1966) 61.
[78] P. C. Lindahl, A. M. Bishop, *Fuel* **61** (1982) 658.
[79] R. Wickbold, *Angew. Chem.* **64** (1952) 134.
[80] J. Dolezal, J. Lenz, Z. Sulcek, *Anal. Chim. Acta* **47** (1969) 517.
[81] G. Tölg, *Pure Appl. Chem.* **44** (1975) 645.
[82] J. T. Brenna, G. H. Morrison, *Microbeam Anal.* **19th** (1984) 265.
[83] W. van Delft, G. Vos, *Anal. Chim. Acta* **209** (1988) 147.
[84] L. Vos, R. van Grieken, *Anal. Chim. Acta* **164** (1984) 83.
[85] M. Bettinelli, U. Baroni, N. Pastorelli, *J. Anal. At. Spectrom.* **2** (1987) 485.
[86] F. Koroleff, Baltic Intercalibration Workshop, Kiel March 7–14, 1977, p. 30.
[87] K. Grasshoff, M. Ehrhardt, K. Kremling (eds.): *Methods of Seawater Analysis,* Verlag Chemie, Weinheim 1983, p. 164.
[88] L. Dunemann, M. Meinerling, unpublished results (1992).
[89] W. Rechenberg, *Zem. Kalk Gips* **25** (1972) 410.
[90] F. Seel, E. Steigner, J. Burger, *Angew. Chem.* **76** (1964) 532.
[91] J. Angerer, R. Heinrich-Ramm, G. Lehnert, *Int. J. Environ. Anal. Chem.* **35** (1989) 81.
[92] E. Schumacher-Wittkopf, J. Angerer, *Int. Arch. Occup. Environ. Health* **49** (1981) 77.
[93] R. Heinrich, J. Angerer in P. Brätter, P. Schramel (eds.): *Trace Element-Analytical Chemistry in Medicine and Biology,* vol. 2, Walter De Gruyter, Berlin 1983.
[94] A. Dornemann, H. Kleist, *Fresenius Z. Anal. Chem.* **300** (1980) 197.
[95] B. Griepink, M. Sager, G. Tölg, *Pure Appl. Chem.* **60** (1988) 1425.
[96] M. Buratti et al., *Clin. Chim. Acta* **150** (1985) 53.
[97] R. Kobayashi, K. Imaizumi, *Anal. Sci.* **5** (1985) 61.
[98] H. Watanabe, *Anal. Chem.* **53** (1981) 738.
[99] H. Berndt, U. Harms, M. Sonneborn, *Fresenius Z. Anal. Chem.* **322** (1985) 329.
[100] P. Burba, P. G. Willmer, *Fresenius Z. Anal. Chem.* **329** (1987) 539.
[101] G. Schwedt: *Methoden der Spurenanreicherung anorganischer und organischer Stoffe aus Wässern,* Vogel-Verlag, Würzburg 1988.
[102] A. Taylor, S. Branch, H. M. Crews, D. F. Halls, L. M. Owen, M. White, *J. Anal. At. Spectrom.* **12** (1997) 119R.
[103] M. Hoenig, A.-M. de Kersabiec, *Spectrochim. Acta B* **51** (1996) 1297.
[104] J. Begerow, L. Dunemann in H. Günzler (ed.): *Analytiker-Taschenbuch,* **vol. 18,** Springer-Verlag, Berlin–Heidelberg–New York 1998, p. 67.
[105] T. Mansfeld, H. Biernath, *Anal. Chim. Acta* **406** (2000) 283.
[106] D. M. Adriá-Cerezo, M. Llobat-Estellés, A. M. Mauri-Aucejo, *Talanta* **51** (2000) 531.
[107] A. Seubert, G. Petzold, J. W. McLaren, *J. Anal. At. Spectrom.* **10** (1995) 371.
[108] I. Jarvis, M. M. Totland, K. E. Jarvis, *Analyst* **122** (1997) 19.
[109] J. Begerow, M. Turfeld, L. Dunemann, *Anal. Chim. Acta* **340** (1997) 277.
[110] V. K. Panday, K. Hoppstock, J. S. Becker, H.-J. Dietze, *At. Spectrosc.* **17** (1996) 98.
[111] K. E. Laintz, C. M. Wai, C. R. Yonker, R. D. Smith, *Anal. Chem.* **64** (1992) 2875.
[112] C. M. Wai, S. Wang, J.-J. Yu, *Anal. Chem.* **68** (1996) 3516.
[113] G. Schulze, O. Elsholz, *Fresenius Z. Anal. Chem.* **335** (1989) 724.
[114] M. Agarwal, R. B. Bennett, I. G. Stump, J. M.D'Auria, *Anal. Chem.* **47** (1975) 924.
[115] M. M. Kingston, I. L. Barnes, T. J. Brady, T. C. Rains, *Anal. Chem.* **50** (1978) 2064.
[116] P. Burba, K. H. Lieser, *Fresenius Z. Anal. Chem.* **297** (1979) 374.
[117] H. J. Fischer, K. H. Lieser, *Fresenius Z. Anal. Chem.* **335** (1989) 738.
[118] A. Disam, P. Tschöpel, G. Tölg, *Fresenius Z. Anal. Chem.* **295** (1979) 97.
[119] R. Chakrovorty, R. van Grieken, *Int. J. Environ. Anal. Chem.* **11** (1982) 67.
[120] B. Andresen, B. Salbu, *Radiochem. Radioanal. Lett.* **52** (1982) 19.
[121] C. L. Smith, J. M. Motooka, W. R. Willson, *Anal. Lett.* **17** (1984) 1715.

[122] J. Bandovskis, M. Vircavs, O. Veveris, A. Pelne, *Talanta* **34** (1987) 179.

[123] J. M. Labrecque in H. M. Kingston, L. B. Jassie (eds.): *Introduction to Microwave Sample Preparation,* Am. Chem. Soc., Washington D.C. 1988, p. 203.

[124] L. Solujic, E. B. Milosavljevic, M. R. Straka, *Analyst* **124** (1999) 1255.

[125] R. Sur, J. Begerow, L. Dunemann, *Fresenius' J. Anal. Chem.* **363** (1999) 526.

[126] K. Sasaki, G. E. Pacey, *Talanta* **50** (1999) 175.

[127] C.-C.Huang, M.-H. Yang, T.-S. Shih, *Anal. Chem.* **69** (1997) 3930.

[128] U. Pichler, A. Haase, G. Knapp, *Anal. Chem.* **71** (1999) 4050.

[129] B. W. Wenclawiak, T. Hees, C. E. Zöller, H.-P. Kabus, *Fresenius' J. Anal. Chem.* **358** (1997) 471.

[130] M. Burguera, J. L. Burguera, O. M. Alarcon, *Anal. Chim. Acta* **179** (1986) 351.

[131] V. Karanassios, F. H. Li, B. Liu, E. D. Salin, *J. Anal. At. Spectrom.* **6** (1991) 457.

[132] B. Welz, M. Schubert-Jacobs in B. Welz (ed.): *5th Colloquium Atomspektrometrische Spurenanalytik,* Bodenseewerk Perkin-Elmer, Überlingen 1989, p. 327.

[133] B. Welz, M. Sperling, X. Yin in B. Welz (ed.): *6th Colloquium Atomspektrometrische Spurenanalytik,* Bodenseewerk Perkin-Elmer, Überlingen 1992, p. 203.

[134] G. Schulze, H. Tessmer, O. Elsholz in B. Welz (ed.): *5th Colloquium Atomspektrometrische Spurenanalytik,* Bodenseewerk Perkin-Elmer, Überlingen 1989, p. 347.

[135] A. Meyer, L. Dunemann in B. Welz (ed.): *6th Colloquium Atomspektrometrische Spurenanalytik,* Bodenseewerk Perkin-Elmer, Überlingen 1991, p. 115.

[136] J. A. Broekaert, *Anal. Proc. (London)* **27** (1990) 336.

[137] I. Atsuya, K. Itoh, K. Akatsuka, *Fresenius Z. Anal. Chem.* **328** (1987) 338.

[138] P. Fecher, C. Malcherek in B. Welz (ed.): *5th Colloquium Atomspektrometrische Spurenanalytik,* Bodenseewerk Perkin-Elmer, Überlingen 1989, p. 501.

[139] K. H. Tobies, W. Großmann in B. Welz (ed.): *5th Colloquium Atomspektrometrische Spurenanalytik,* Bodenseewerk Perkin-Elmer, Überlingen 1989, p. 513.

[140] A. Sugimae, R. M. Barnes, *Anal. Chem.* **58** (1986) 785.

[141] L. Dunemann, H. Hajimiragha, J. Begerow, *Fresenius' J. Anal. Chem.* **363** (1999) 466.

[142] L. Moens, T. de Smaele, R. Dams, P. Van Den Broeck, P. Sandra, *Anal. Chem.* **69** (1997) 1604.

[143] Y. Jin, F. Liang, H. Zhang, L. Zhao, Y. Huan, D. Song, *Trends Anal. Chem.* **18** (1999) 479.

[144] F. Brugnone et al., *Int. Arch. Occup. Environ. Health* **64** (1992) 179.

[145] J. Rudling, E. Bjoerkholm, B.-O. Lundmark, *Ann. Occup. Hyg.* **30** (1986) 319.

[146] P. J. H. D. Verkoelen, M. W. F. Nielen, *J. High Resolut. Chromatogr. Chromatogr. Comm.* **11** (1988) 291.

[147] M. D. Erickson: *Analytical Chemistry of PCBs,* Ann Arbor Science Publishers, Stoneham, Mass. 1986.

[148] B. Bush, J. T. Snow, S. Connor, *J. Assoc. Off. Anal. Chem.* **66** (1983) 258.

[149] R. C. Entz, H. C. Hollifield, *J. Agric. Food Chem.* **30** (1982) 84.

[150] W. Wittfoht, W. J. Scott, H. Nau, *J. Chromatogr.* **448** (1988) 433.

[151] H. Hajimiragha, U. Ewers, R. Jansen-Rosseck, A. Brockhaus, *Int. Arch. Occup. Environ. Health* **58** (1986) 141.

[152] J. Angerer, G. Scherer, K. H. Schaller, J. Müller, *Fresenius J. Anal. Chem.* **339** (1991) 740.

[153] D. G. Patterson et al., *Anal. Chem.* **58** (1986) 705.

[154] G. A. Junk, M. J. Avery, J. J. Richard, *Anal. Chem.* **60** (1988) 1347.

[155] A. Bergmann, L. Reutegardh, M. Ahlman, *J. Chromatogr.* **291** (1984) 392.

[156] B. Wigilius et al., *J. Chromatogr.* **391** (1987) 169.

[157] J. Begerow, R. Heinrich-Ramm, J. Angerer, *Fresenius Z. Anal. Chem.* **331** (1988) 818.

[158] H. C. Shields, C. J. Weschler, *J. Air Pollut. Control Assoc.* **37** (1987) 1039.

[159] J. Begerow, E. Jermann, T. Keles, L. Dunemann, *Fresenius J. Anal. Chem.* (1994), in press.

[160] R. Heinrich, J. Angerer, *Fresenius Z. Anal. Chem.* **322** (1985) 766.

[161] G. Birner, H. G. Neumann, *Arch. Toxicol.* **62** (1988) 110.

[162] K. M. Engström, *Scand. J. Work Environ. Health* **10** (1984) 75.

[163] M. Balikova, J. Kohlicek, *J. Chromatogr.* **497** (1989) 159.

[164] E. R. Adlard, C. B. Milne, P. E. Tindle, *Chromatographia* **14** (1981) 507.

[165] J. A. F. de Silva, *J. Chromatogr.* **273** (1983) 345.

[166] K. G. Wahlund, T. Arvidson, *J. Chromatogr.* **282** (1983) 527.

[167] A. M. Rustum, *J. Chromatogr. Sci.* **27** (1989) 18.

[168] S. G. Colgrove, J. H. Svec, *Anal. Chem.* **53** (1981) 1737.

[169] J. Lewalter, U. Korallus, *Int. Arch. Occup. Environ. Health* **56** (1985) 179.

[170] M. Akerblom, G. Alex, *J. Assoc. Off. Anal. Chem.* **67** (1984) 653.

[171] J. Czuczwa et al., *J. Chromatogr.* **403** (1987) 233.

[172] T. L. Peters, *Anal. Chem.* **54** (1982) 1913.

[173] W. G. Jennings, A. Rapp: *Sample Preparation for Gas Chromatographic Analysis,* Hüthig Verlag, Heidelberg 1983.

[174] D. A. J. Murray, *J. Chromatogr.* **177** (1979) 135.

[175] J. F. J. van Rensberg, A. J. Hasset, *J. High Resolut. Chromatogr. Chromatogr. Comm.* **5** (1982) 574.

[176] N. F. Wood, *Analyst (London)* **94** (1969) 399.

[177] M. Teufel et al, *Arch. Environ. Contam. Toxicol.* **19** (1990) 646.

[178] S. Hashimoto, H. Ito, M. Morita, *Chemosphere* **25** (1992) 297.

[179] R. C. Reid, J. M. Prausnitz, B. E. Poling: *The Properties of Gases and Liquids,* 4th ed., McGraw-Hill, New York 1987.

[180] J. W. King, *J. Chromatogr. Sci.* **27** (1989) 355.

[181] K. S. Nam et al.: *Proceedings of the International Symposium on Supercritical Fluids,* French Chemical Society: Paris Index, France 1988, p. 743.

[182] J. W. King, J. H. Johnson, J. P. Friedrich, *J. Agric. Food Chem.* **37** (1989) 951.

[183] M. A. Schneiderman, A. K. Sharma, K. R. R. Mahanama, D. C. Locke, *J. Assoc. Off. Anal. Chem.* **71** (1988) 815.

[184] M. A. Schneiderman, A. K. Sharma, D. C. Locke, *J. Chromatogr. Sci.* **26** (1988) 458.

[185] K. Sugiyama, M. Saito, *J. Chromatogr.* **442** (1988) 121.

[186] P. Capriel, A. Haisch, S. U. Khan, *J. Agric. Food Chem.* **34** (1986) 70.
[187] V. Janda, G. Steenbeke, P. Sandra, *J. Chromatogr.* **479** (1989) 200.
[188] S. B. Hawthorne, D. J. Miller, *J. Chromatogr. Sci.* **24** (1986) 258.
[189] B. W. Wright, C. W. Wright, J. S. Fruchter, *Energy Fuels* **3** (1989) 474.
[190] S. B. Hawthorne, D. J. Miller, *Anal. Chem.* **59** (1987) 1705.
[191] M. M. Schantz, S. N. Chesler, *J. Chromatogr.* **363** (1986) 397.
[192] B. O. Brady, C. P. C. Kao, K. M. Dooley, F. C. Knopf, *Ind. Eng. Chem. Prod. Res. Dev.* **26** (1987) 261.
[193] N. Alexandrou, J. Pawliszyn, *Anal. Chem.* **61** (1989) 2770.
[194] F. I. Onuska, K. A. Terry, *J. High Resolut. Chromatogr. Chromatogr. Comm.* **12** (1989) 357.
[195] B. W. Wright, C. W. Wright, R. W. Gale, R. D. Smith, *Anal. Chem.* **59** (1987) 38.
[196] S. B. Hawthorne, M. S. Krieger, D. J. Miller, *Anal. Chem.* **61** (1989) 736.
[197] S. Forbes, *Anal. Chim. Acta* **196** (1987) 75.
[198] U. Juergens, *J. Chromatogr.* **371** (1986) 307.
[199] J. J. Richard, G. A. Junk, *Mikrochim. Acta* 1986 no. 1, 387.
[200] G. A. Junk, J. J. Richard, *Anal. Chem.* **60** (1988) 451.
[201] R. Bagnati, E. Benfenati, E. Davoli, R. Fanelli, *Chemosphere* **17** (1988) 59.
[202] J. Manes Vinuesa, J. C. Molto Cortes, C. Igualada Canas, G. Font Perez, *J. Chromatogr.* **472** (1989) 365.
[203] J. Angerer et al., *Fresenius J. Anal. Chem.* **342** (1992) 433.
[204] K. I. Tomlins, K. Jewers, R. D. Coker, *Chromatographia* **27** (1989) 67.
[205] M. R. Khan, C. P. Ong, S. F. Y. Li, H. D. Lee, *J. Chromatogr.* **513** (1990) 360.
[206] K. E. Savolainen, *J. Pharm. Sci.* **77** (1988) 802.
[207] R. Schmidt, P. Kupferschmidt, *Clin. Chem. (Winston-Salem N.C.)* **35** (1989) 1352.
[208] V. Marko, L. Soltes, K. Radova, *J. Chromatogr. Sci.* **28** (1990) 403.
[209] V. W. Burse et al., *J. Anal. Toxicol.* **14** (1990) 143.
[210] W. Butte, C. Fooken, *Fresenius J. Anal. Chem.* **336** (1990) 511.
[211] H. Bouwman, R. M. Coopan, A. J. Reinecke, *Chemosphere* **19** (1989) 1563.
[212] P. H. Cramer et al., *Chemosphere* **20** (1990) 821.
[213] A. K. D. Liem et al., *Chemosphere* **20** (1990) 843.
[214] P. Fuerst, H. A. Meemken, W. Groebel, *Chemosphere* **15** (1986) 1977.
[215] H. Beck et al., *Chemosphere* **16** (1987) 1977.
[216] R. G. Heath et al., *Anal. Chem.* **58** (1986) 463.
[217] H. F. Walton, R. D. Rocklin: *Ion Exchange in Analytical Chemistry,* CRC Press, Boca Raton, Fla. 1990.
[218] T. G. Kreindl, H. Malissa, K. Winsauer, *Mikro-chim. Acta* 1986 no. 1, 1.
[219] W. E. Turner, S. G. Isaacs, D. G. Patterson, Jr., *Chemosphere* **25** (1992) 805.
[220] D. Barceló, G. Durand, V. Bouvot, M. Nielen, *Environ. Sci. Technol.* **27** (1993) 271.
[221] E. R. Brouwer, H. Lingemann, U. A. T. Brinkmann, *Chromatographia* **29** (1990) 415.
[222] J. Rosenfeld, M. Mureika-Russell, S. Yeroushalmi, *J. Chromatogr.* **358** (1986) 137.
[223] J. M. Rosenfeld, M. Mureika-Russell, A. Phatak, *J. Chromatogr.* **283** (1984) 127.
[224] Z. Bardodej, *Arbeitsmed. Sozialmed. Arbeitshyg.* **3** (1968) 141.
[225] J. Pawliszyn: *Solid Phase Microextraction: Theory and Practice,* Wiley-VCH, New York 1997.
[226] C. Grote, J. Pawliszyn, *Anal. Chem.* **69** (1997) 587.
[227] I. Valor, C. Cortada, J. C. Moltó, *J. High Resolut. Chromatogr.* **19** (1996) 472.
[228] T. Nilsson, F. Pelusion, L. Montanarella, B. Larsen, S. Facchetti, J. Madsen, *J. High Resolut. Chromatogr.* **18** (1995) 617.
[229] M. Möder, S. Schrader, U. Franck, P. Popp, *Fresenius' J. Anal. Chem.* **357** (1997) 326.
[230] M. T. Sng, F. K. Lee, H. A. Lakso, *J. Chromatogr. A* **759** (1997) 225.
[231] D. W. Potter, J. Pawliszyn, *Environ. Sci. Technol.* **28** (1997) 298.
[232] P. A. Martos, J. Pawliszyn, *Anal. Chem.* **69** (1997) 206.
[233] F. Mangani, R. Cenciarini, *Chromatographia* **41** (1995) 678.
[234] V. Pino, J. H. Ayala, A. M. Afonso, V. Gonzáles, *J. Chromatogr. A* **869** (2000) 515.
[235] V. Pensado, C. Casais, C. Mejuto, R. Cela, *J. Chromatogr. A* **869** (2000) 505.
[236] J. Angerer: "Ameisensäure" in D. Henschler (ed.): *Analysen in biologischem Material,* 4. ed., Verlag Chemie, Weinheim 1980.
[237] A. B. Heath, R. R. Black, *J. Assoc. Off. Anal. Chem.* **70** (1987) 862.
[238] R. L. Brown, C. L. Farmer, R. G. Millar, *J. Assoc. Off. Anal. Chem.* **70** (1987) 442.
[239] R. A. Minear, L. H. Keith (eds.): *Water Analysis,* **vol. 3,** Academic Press, Orlando 1984, p. 84.
[240] T. N. Eisenberg, E. J. Middlebrooks: *Reverse Osmosis Treatment of Drinking Water,* Butterworths, London 1986.
[241] T. D. Brook: *Membrane Filtration: A User's Guide and Reference Manual,* Science Tech., Madison 1983.
[242] M. M. L. Aerts, W. M. J. Beek, U. A. T. Brinkmann, *J. Chromatogr.* **435** (1988) 613.
[243] D. C. Turnell, J. D. H. Cooper, *J. Chromatogr.* **395** (1987) 613.
[244] A. C. Metha, *Tr. AC Trends Anal. Chem. (Pers. Ed.)* **8** (1989) 107.
[245] K. Pekari, M. L. Riekkola, A. Aitio, *J. Chromatogr.* **491** (1989) 309.
[246] G. Machata, J. Angerer: "Gaschromatographie, Headspace-Technik" in D. Henschler (ed.): *Analysen in biologischem Material,* 7th suppl., Verlag Chemie, Weinheim 1983.
[247] V. C. Stein, R. S. Narang, *Arch. Environ. Contam. Toxicol.* **19** (1990) 593.
[248] S. L. Friant, I. H. Suffet, *Anal. Chem.* **51** (1979) 2167.
[249] L. Dunemann, H. Hajimiragha, *Anal. Chim. Acta* **283** (1993) 199.
[250] D. L. Ashley et al., *Anal. Chem.* **64** (1992) 1021.
[251] H. Hajimiragha, U. Ewers, A. Brockhaus, A. Boettger, *Int. Arch. Occup. Environ. Health* **61** (1989) 513.
[252] K. H. Pannwitz, *Drägerheft* **332** (1985) 10.
[253] K. H. Pannwitz, *Drägerheft* **327** (1983) 6.
[254] G. A. Reineccius, R. Liardon in R. G. Berger, S. Nitz, P. Schreier (eds.): "Topics in Flavor Research," Proceedings of *International Conference Freising-Weihenstephan,* 1985, p. 125.
[255] K. H. Schaller: "Mandelsäure" in: D. Henschler (ed.): *Analysen in biologischem Material,* 2nd suppl., Verlag Chemie, Weinheim 1978.

[256] J. Begerow, J. Angerer, *Fresenius J. Anal. Chem.* **366** (1990) 42.
[257] M. I. Daneshvar, J. B. Brooks, *J. Chromatogr.* **433** (1988) 248.
[258] H. B. Lee, T. E. Peart, J. M. Carron, *J. Chromatogr.* **498** (1990) 367.
[259] S. Jacobbson, A. Larsson, A. Arbin, A. Hagman, *J. Chromatogr.* **358** (1986) 137.
[260] R. v. Smith, S. Tsai, *J. Chromatogr.* **61** (1971) 29.
[261] P. Pfäffli, H. Savolainen, H. Keskinen, *Chromatographia* **27** (1989) 483.
[262] G. Johanson, H. Kronborg, P. Näslund, M. Byfalt Nordquist, *Scand. J. Work Environ. Health* **12** (1986) 594.
[263] Y. L. Emeillat, J. Menez, F. Berthou, L. Bardou, *J. Chromatogr.* **206** (1981) 89.
[264] R. Patience, J. Thomas, *J. Chromatogr.* **234** (1982) 225.
[265] E. Grushka, H. Durst, E. Kikta, *J. Chromatogr.* **112** (1975) 673.
[266] T. Jupile, *J. Chromatogr. Sci.* **17** (1979) 160.
[267] W. Steiner, E. Müller, D. Fröhlich, R. Battaglia, *Mitt. Geb. Lebensmitteluntersuch. Hyg.* **75** (1984) 37.
[268] M. Petrarulo, S. Pellegrino, M. Marangella, F. Linari, *J. Chromatogr.* **432** (1988) 37.
[269] M. Petrarulo et al., *J. Chromatogr.* **465** (1989) 87.
[270] W. Dünges, *Anal. Chem.* **49** (1977) 442.
[271] W. Dünges, *Chromatographia* **9** (1976) 624.
[272] E. Grushka, S. Lam, J. Chassin, *Anal. Chem.* **50** (1978) 1398.
[273] A. Crozier, J. Zaerr, R. Morris, *J. Chromatogr.* **238** (1982) 157.
[274] H. Tsuchiya, T. Hayastsi, H. Naruse, N. Takagi, *J. Chromatogr.* **234** (1982).
[275] H. D. Winkeler, K. Levsen, *Fresenius Z. Anal. Chem.* **333** (1989) 716.
[276] M. A. Carey, H. E. Perisinger, *J. Chromatogr. Sci.* **10** (1972) 537.
[277] J. K. Lin, J. Y. Yang, *Anal. Chem.* **47** (1975) 1634.
[278] J. Y. Chang, H. L. Creaser, *J. Chromatogr.* **116** (1976) 215.
[279] J. K. Lin, C. C. Lai, *Anal. Chem.* **52** (1980) 630.
[280] K. Karlson, M. Alsandro, M. Novotny, *Anal. Chem.* **57** (1985) 229.
[281] K. Grob, Jr., D. Fröhlich, B. Schilling, *J. Chromatogr.* **295** (1984) 55.
[282] K. Grob, Jr., C. Walder, B. Schilling, *J. High Resolut. Chromatogr. Chromatogr. Comm.* **9** (1986) 95.
[283] K. Grob, Jr., B. Schilling, *J. High Resolut. Chromatogr. Chromatogr. Comm.* **8** (1985) 726.
[284] K. Grob, Jr., G. Karrer, M. L. Riekkola, *J. Chromatogr.* **334** (1985) 129.
[285] H. J. Cortes, E. L. Olberding, J. H. Wetters, *Anal. Chim. Acta* **236** (1990) 173.
[286] R. Barcaralo, *J. High Resolut. Chromatogr. Chromatogr. Comm.* **518** (1990) 465.
[287] B. A. Anderson, *Am. Environ. Lab.* **1** (1989) 41.
[288] D. Barcelo et al., *Anal. Chem.* **62** (1990) 1696.
[289] S. A. Soper, K. L. Ratzlaff, T. Kuwana, *Anal. Chem.* **62** (1990) 1438.
[290] J. Strojek, S. A. Soper, K. L. Ratzlaff, T. Kuwana, *Anal. Sci.* **6** (1990) 121.
[291] S. B. Hawthorne, D. J. Miller, J. J. Langefeld, *J. Chromatogr. Sci.* **28** (1990) 2.
[292] J. R. Wheeler, M. E. McNally, *J. Chromatogr. Sci.* **27** (1989) 534.
[293] F. I. Onuska, K. A. Tery, *J. High Resolut. Chromatogr. Chromatogr. Commun.* **12** (1989) 527.
[294] M. R. Andersen, J. T. Swanson, N. C. Porter, B. E. Richter, *J. Chromatogr. Sci.* **27** (1989) 371.
[295] C. H. Marvin et al., *J. Chromatogr.* **518** (1990) 242.

7. Trace Analysis

HELMUT MÜLLER, Universität Halle-Wittenberg, Fachbereich Chemie, Merseburg, Federal Republic of Germany

HEINZ W. ZWANZIGER, Fachhochschule Merseburg, Fachbereich Chemie- und Umweltingenieurwesen, Merseburg, Federal Republic of Germany

JOHANNES FLACHOWSKY, Umweltforschungszentrum, Leipzig-Halle GmbH, Leipzig, Federal Republic of Germany

Symbols

1) Symbols of features (quantities, variables):

X	analyte feature (e.g., concentration)
Y	signal, response

2) Symbols of measured or calculated values of features:

x	concentration value
x_A	analytical value; analyte concentration
x_C	calibration value
$x_{decision}$	limit of decision
$x_{detection}$	limit of detection
$x_{quantitation}$	limit of quantitation
y	signal or response value
y_{blank}	blank value
y_{crit}	critical value

3) Symbols for running indices:

b	with respect to blank values
c	with respect to calibration
n	general running index

4) Symbols for terminal values of indices:

N	number of ...
N_a	... analytical values
N_b	... blank values
N_c	... calibration values

5) Symbols of statistical features and parameters:
symbols with a bar: mean values
symbols with a hat: calculated values according to a model

a_0	intercept in regression models
a_1	slope; sensitivity; regression coefficient
k	statistical factor (quantile of a distribution to be specified)
q	order of quantile
res	residual value (difference of measured and calculated values)
s	calculated ("sample") standard deviation
s^2	calculated ("sample") variance
sgn (cor_{XY})	sign of the correlation coefficient of X and Y
SS	read as "sum of squares" ... SS_{XX} ... of centered values of feature X SS_{YY} ... of centered values of feature Y
SS_{XY}	read as "sum of products of centered values of features X and Y
t, $t_q(\nu)$	values of Student's distribution (quantiles)
α	error probability (risk of first kind)

β	error probability (risk of second kind)
ν	degrees of freedom
σ	"population" standard deviation

7.1. Subject and Scope

The determination of analytes in very low concentration (traces) plays a fundamental role in many areas of science and technology.

A "trace" means a very small level of a substance that is present alongside a large excess of other components, which causes certain effects and must be detected (with qualitative information) and determined (with quantitative information).

A distinction is made between organic and inorganic "traces." Examples of *organic traces* are dioxins or furans in waste gases from refuse incineration plants, fluorochlorohydrocarbons (FCHCs) in the atmosphere, chlorinated hydrocarbons (CHCs) in water, and many others. *Inorganic traces* are referred to as trace elements, a concept that includes all elements (i.e., metals, semimetals, and nonmetals). Despite the different objects that must be monitored analytically, the methods and instruments in this field have so much in common that a special branch of analysis — trace analysis — has developed [1], [2].

For trace analysis, analytical methods with extreme powers of detection are required, which must often be coupled with enrichment steps. Another requirement follows from the fact that the traces occur alongside a large excess of other substances (matrix). The main components are present in $10^2 - 10^{10}$-fold excess. Therefore, methods must be developed that permit trace determinations without interference by the main component (i.e., specific methods); otherwise, separation of the trace and main components must be carried out before the determination.

The possibility of determining traces of elements arose in the 1920s with the development of spectrophotometry, spectroscopic methods, and polarography. Trace analysis was carried out very intensively and with great sophistication in connection with the development and production of the first atomic bomb. Later, it received an important stimulus from the development of novel materials, especially semiconductors.

Today, more areas than ever depend on the results of trace analysis: Nuclear energy, production of semiconductors and ultrapure substances, metallurgy, materials research and production, geology, mineralogy, oceanography, medicine, animal and plant physiology, criminology, ecology, and environmental research.

Currently, ecological research and routine analysis relevant to the environment provide trace analysis with imporant stimuli. At the same time, the centers of attention are shifting more and more from element trace analysis to organic trace analysis.

7.2. Fields of Work

Modern technology has many possibilities at its disposal for solving problems in trace analysis. A uniform characterization and assessment of trace-analytical methods requires the clearest possible definitions: For reporting a content G, mass relations — of which the simplest is percentage by mass — are unequivocal and independent of any additional information. The contents of the components to be determined can vary within wide limits. Very diverse proposals for the boundaries of these ranges are found. For practical purposes, the following classification has been established:

Main fraction	100 % – 10 %
Minor fraction	10 % – 0.1 %
Traces	less than 0.1 %
Micro fraction range	1000 – 1 μg/g
Nano fraction range	1000 – 1 ng/g

For nanofractions (contents below ng/g) the term "extreme trace analysis" has become generally accepted.

In addition to the specification of content ranges in parts by mass w, parts by volume σ, or parts by amount of substance x, in trace analysis the term ppm (parts per million; $1\ \text{ppm} = 1:10^6 = 10^{-4}\ \%$) is generally used for parts by mass w. Still lower concentrations are expressed in ppb (parts per billion); errors can arise because the intended reference quantity 10^9 ($1\ \text{ppb} = 10^{-7}\ \%$) is called "billion" mainly in the United States, but "milliard" in Europe. This quantity is sometimes represented as ppM (parts per milliard).

In the literature, ppm by mass is still encountered occasionally, and the ppm term is then used without further identification. Unfortunately, the terms ppm and ppb, which preferably should be used only for problems in solid analysis, are also used, instead of mass concentration β (or C) in milligrams or micrograms per liter, for specifying

Table 1. Specification of content units in trace analysis

Unit	Designation	Percentage
Parts by mass w		
mg/kg	1 ppm	10^{-4}
µg/kg	1 ppb	10^{-7}
ng/kg	1 ppt *	10^{-10}
pg/kg	1 ppq **	10^{-13}
Part by volume σ		
mL/m^3	ppm	10^{-4}
$µL/m^3$	ppb	10^{-7}
nL/m^3	ppt *	10^{-10}
pL/m^3	ppq **	10^{-13}

* ppt = Parts per trillion.
** ppq = Parts per quadrillion.

the concentration of trace contents in dissolved samples.

The content of trace components in gases is indicated as volume concentration σ or mass concentration β. Useful units are milliliters per cubic meter (ppm) and milligrams per cubic meter. The conversion is carried out according to

$$\sigma\left[\mathrm{mL/m^3}\right] = \frac{\text{molar volume [L/mol]}}{\text{molar mass [g/mol]}} \cdot \beta\left[\mathrm{mg/m^3}\right]$$

The molar volume is usually related to 25 °C and 101.3 kPa and is 22.47 L/mol. For dusts, the mass concentration β (mg/m^3) or the particle concentration C (particles/m^3 or fibers/m^3) is used.

Table 1 surveys the specification of content data in trace analysis. Limiting factors for the optimum choice of an analytical method for a trace-analytical problem are the portion of sample available and the expected mass concentration β of the analyte.

The sample portions or sample volumes most frequently used are in the range of a minimum of grams or milliliters to micrograms or microliters. Modern methods (e.g., capillary zone electrophoresis, CZE) require sample volumes only in the nanoliter range. This results in a high *mass sensitivity* that leads into the range of detection of individual molecules or atoms. By coupling capillary zone electrophoresis with laser-induced fluorescence detection, an absolute total of 600 molecules has even been detected.

Clearly distinguished from this is the "*concentration sensitivity*" of the individual methods. The current state of the art is ppm analysis of a wide range of analytes in a variety of matrices. The trend to ppb and in a few cases ppt (10^{-10} %) (parts per trillion, 10^{-12}) without enrichment is expected.

7.3. Methods of Modern Trace Analysis

Modern trace analysis is characterized by two different strategies, whose orientation is more physical or chemical, respectively:

1) Application of instrumental direct methods
2) Development of chemical-analytical compound methods

Instrumental Direct Methods. In instrumental direct methods, after sampling and possibly sample preparation, the sample is analyzed directly. Instrumental direct determination methods are usually matrix-dependent relative methods. A mathematical correction of the matrix effect is possible only in particular cases (e.g., X-ray fluorescence analysis, XRF). To compensate for systematic errors, therefore, standard reference materials are required, which must be very similar in composition to the sample to be analyzed (see Section 7.4.5). However, an optimum power of detection at high accuracy (freedom from systematic error) can be achieved only when the trace to be determined is present in isolated form in the highest possible mass concentration. In many cases of trace analysis, therefore, more complex multistage (compound) methods are required.

Compound Methods. In compound methods, after dissolution or digestion of the sample, the analyte is separated from the matrix, usually enriched, and then determined (Fig. 1). The calibration is relatively simple because the analyte components have been isolated. The only standards required are analyte solutions of known concentration, since the matrix can no longer interfere with the determination. The advantage of easy calibration is lessened, however, by the fairly large number of operations and the correspondingly greater risk of systematic errors (Fig. 1). This requires very careful and complicated optimization of compound methods to reduce such errors, which today are a central problem of extreme trace analysis. Compound methods are therefore often carried out in closed systems (digestion, enrichment, and/or separation in one apparatus) to prevent errors in the determination of trace contents resulting from the sample environment and sample preparation steps. The necessary trace-analytical hygiene is as important for samples as for standards, solvents, reagents, vessels, instruments, and the laboratory atmosphere. If the highest purity is required, particle emission by laboratory person-

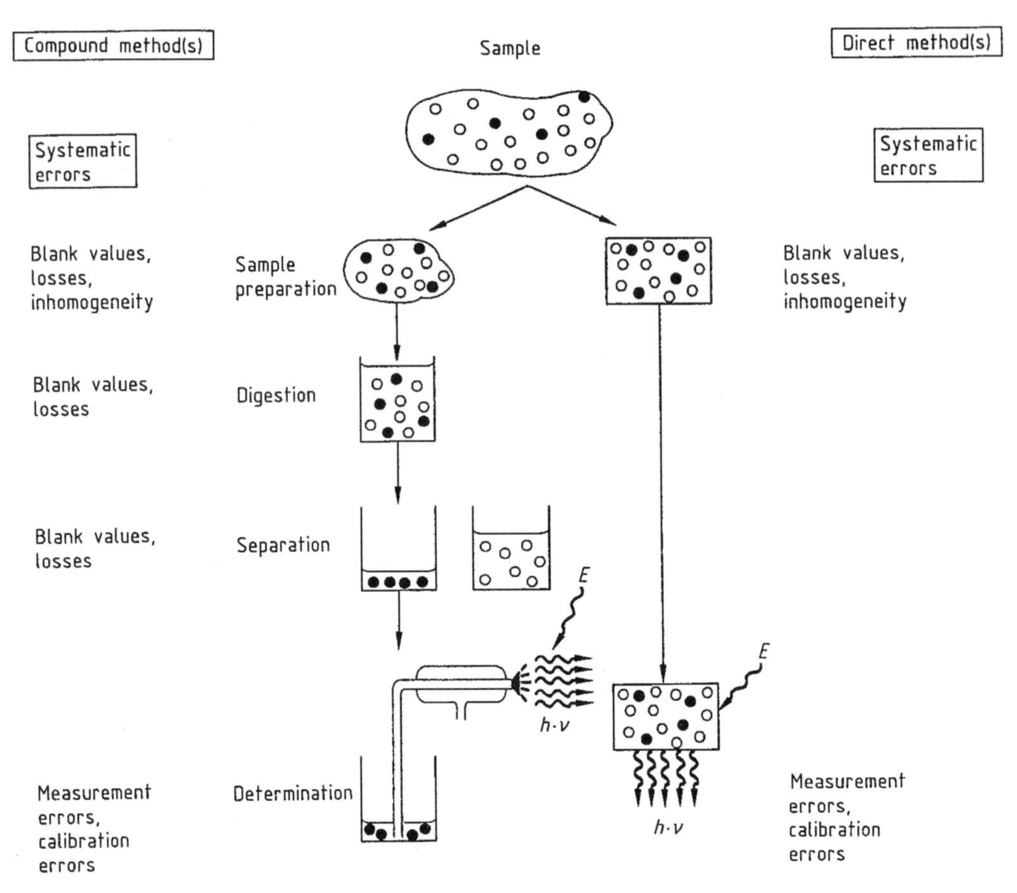

Figure 1. Systematic errors in direct and compound methods

nel must also be considered. Work is therefore carried out in clean rooms or clean boxes, through which laminar flow of filtered air occurs.

New operating techniques (e.g., flow methods), in combination with modern instrumental methods, enable the separation and enrichment processes to be automated in quasi-closed systems, and have contributed substantially to the increasing establishment of such composite methods in trace-analytical practice [3].

Despite the recognizable advantages of compound methods, work is increasingly being carried out on the (further) development of multielement methods with high powers of detection, especially for micro samples. Table 2 shows a survey of modern instrumental multianalyte methods.

Particularly for determining organic trace contents, analytical methods are used that combine systems with different measurement principles and therefore different selectivity (coupled or hyphenated techniques). Chromatographic methods, particularly gas chromatography, supercritical fluid chromatography (SFC), high-performance liquid chromatography, and capillary electrophoresis (CE) are often coupled with spectroscopic methods [7], [8]. The most important coupled systems in trace and environmental analysis are the following:

GC – MS, FTIR, AES/AAS, FTIR–MS, ICP–MS,
SFC – MS, FTIR, AES/AAS, FTIR–MS, ICP–MS,
HPLC – MS, FTIR, AES/AAS,
CE – MS, UV

Coupling of Chromatographic Methods
HPLC–GC, HPLC–GC–MS, SFC–GC–MS, HPLC–TLC–FTIR

The advantages of such coupled techniques lie in a frequently effective separation of the matrix (improvement of the accuracy), an increased

Table 2. Multianalyte methods [4]–[6]

Method	Concentration range, ppb	Matrix effects [a]	Simultaneous	Sequential
Mass spectrometry (MS)	1–10	++	+	
Spark ion source MS (SSMS)				
Secondary ion MS (SIMS)				
Inductively coupled plasma MS (ICP–MS)				
Glow discharge source MS (GD–MS)				
Laser-induced MS (LMS)				
Neutron activation analysis (NAA)	1–10	(+)	+	
X-ray fluorescence analysis (XRF)	10 [b] –1000	++	+	
Total reflection X-ray fluorescence (TRXRF)	10–100	++	+	
Atomic emission spectrometry (AES)	10–100	(+)		
Inductively coupled plasma AES (AES–ICP)	1–10	+++	+	+
Atomic fluorescence spectrometry (AFS)	100–1000	++		+
Gas chromatography (GC)	10–100 [c]		+	
Capillary GC	1–10 [c]		+	
High-performance liquid chromatography (HPLC)	10–100 [c]		+	
Capillary zone electrophoresis (CZE)	100–1000		+	

[a] + = low ... +++ = high.
[b] Wavelength-dispersive XRF (WDXRF).
[c] Dependent on detector principle.

power of detection, and a contribution to speciation analysis [9].

7.4. Calibration and Validation (see also → Chemometrics)

7.4.1. Conceptual Problems

A fully defined trace-analytical procedure is characterized by the concentration range, as well as the calibration function, the precision and accuracy, and the limits of decision, detection, and possibly quantitation. The use of the three last-mentioned terms in the literature unfortunately is not uniform and requires international standardization.

The *limit of decision* is an important criterion in the optimization of a trace-analytical procedure and the comparison of methods. It depends greatly on possible errors in individual steps. It can be determined in many different ways, which may vary in their complexity [10], [11] and whose general applicability on national level is still being elaborated [11], [12]. The precise specification of the means of calculating the limit of decision should therefore be a part of the analytical instructions. Determination of the limit of decision and determination of an analyte are based, except in the case of absolute methods (gravimetry, coulometry), not only on the actual course of the analysis but also essentially on the results of the calibration process.

7.4.2. Errors

Like any analytical procedure, trace analysis is subject to sources of error that can lead to systematic and random falsification of the observed values or test results. The reliability of results is therefore determined by the accuracy and precision. Measures of the precision are repeatability and reproducibility.

Observed values or results of measurement means a set $\{y_1, y_2, \ldots, y_N\}$ of N recorded values y_n (with $n = 1, 2, \ldots, N$) of a signal quantity Y (e.g., absorbance) that result from N given values x_n ($n = 1, 2, \ldots, N$) of a concentration quantity X or from which an analytical result x_A can be derived (calculated).

Systematic errors lead to one-sided displacement of results. Two cases exist. Case A: If the trace component to be determined is absent from a sample ($x_A = 0$), its presence can be simulated (e.g., if the value y_0 of the signal quantity Y specific to the analyte exceeds a critical level y_{crit}). For case B, in the presence of the analyte ($x_A > 0$), two extreme situations are conceivable: The "complete" loss of analyte (which leads to $y_A < y_{crit}$), or an analyte content that is no longer in the trace range ($y_A \gg y_{crit}$).

The risk of a wrong decision (in case A to accept the presence of the analyte, or in case B

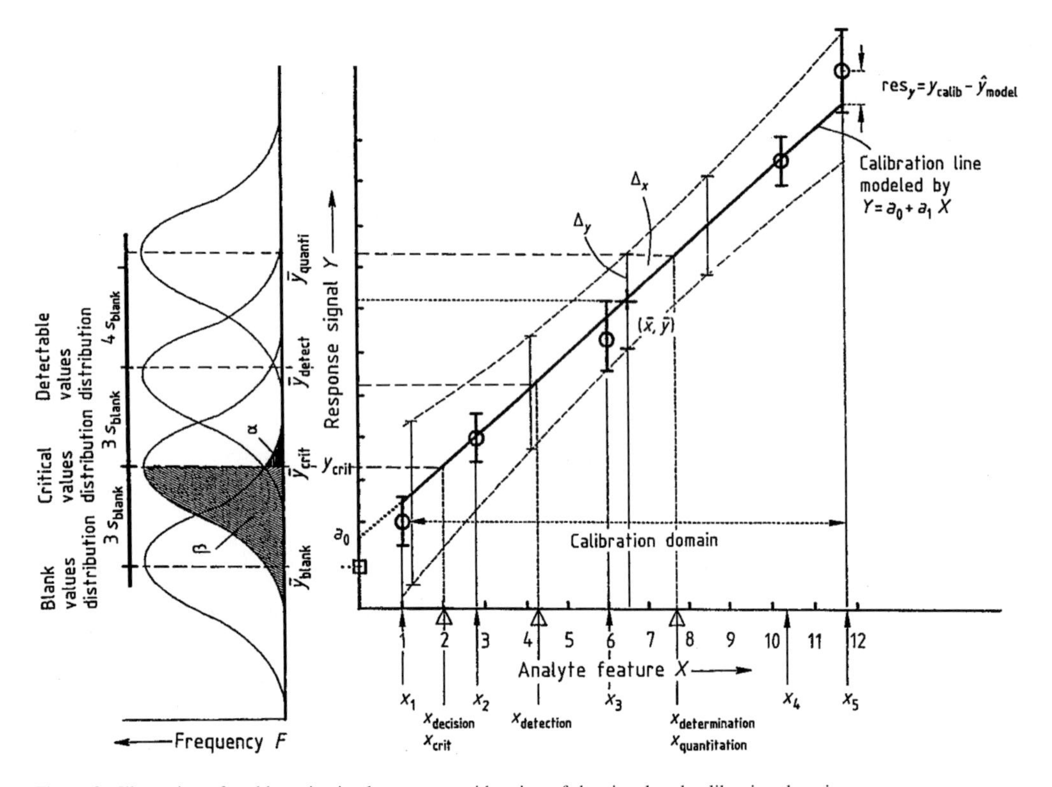

Figure 2. Illustration of problems in simultaneous consideration of the signal and calibration domains

to exclude its presence) can have quite different importance in the two cases. This is not discussed further here, since systematic sources of error can in principle be recognized and eliminated (see Section 7.4.4).

Unavoidable *random errors* cause a two-sided, random deviation of individual observed values from a particular value (target value, mean, median) and also result in different risks in the assessment of the signal or the reporting of analytical results.

7.4.3. The Critical Signal Value and Limits of the Procedure

The critical signal value y_{crit} plays a central part in determining the characteristics of the procedure. Via an experimentally supported or calculated calibration relation (e.g., $Y = a_0 + a_1 X$) it leads to the limit of decision $x_{\mathrm{crit}} = x_{\mathrm{decision}}$ (see Fig. 2).

The critical signal value is determined most simply from repeated analysis of several blank analytical portions or from repeated measurement of one blank analytical portion, the latter, however, provides less real characteristic quantities. Also, in the case of an absent analyte ($x = 0$), a total of N_b randomly determined signal values y_b ($b = 1, \ldots, N_b$), are measured, which must not be zero and whose important characteristics can be calculated easily.

Mean $\quad \bar{y}_{\mathrm{blank}} = \frac{1}{N_b} \sum_{b=1}^{N_b} y_b$

Variance $\quad s^2_{\mathrm{blank}} = \frac{1}{N_b - 1} \sum_{b=1}^{N_b} (y_b - \bar{y}_{\mathrm{blank}})^2$

Standard deviation $\quad s_{\mathrm{blank}} = \sqrt{s^2_{\mathrm{blank}}}$

The upper limit of the confidence interval of the mean of the blank value is then regarded as the critical value y_{crit} of the response value.

$$y_{\mathrm{crit}} = \bar{y}_{\mathrm{blank}} + s_{\mathrm{blank}} t_q(\nu) \sqrt{\frac{1}{N_b}}$$

It can be calculated by means of a tabulated value for Student's t distribution with $v = N_b - 1$ degrees of freedom and order $q = 1 - \alpha$ (one-sided values) (→ Chemometrics).

The notation for the t values is according to [13]. This notation assumes a symmetrical distribution of the observed value signals. However, for trace determinations, this idea, and particularly that of normally distributed values, are frequently unjustified. A logarithmic transformation may be necessary [14], [15].

If a signal value of any test portion falls above this critical value, one can assume, with a certain risk of error α, usually set at 5 % ($\alpha = 0.05$), that the signal value is not a blank value but results from the presence of analyte. At this point the qualitative decision about the presence of an analyte is made with respect to the signal.

The corresponding critical concentration $x_{\text{crit}} = x_{\text{decision}}$, the *limit of decision*, is obtained by combining the calibration relation with the expression for y_{crit}, by assuming that $\bar{y}_{\text{blank}} = a_0$.

$$x_{\text{decision}} = \frac{s_{\text{blank}}}{a_1} t_q(v) \sqrt{\frac{1}{N_b}}$$

Optimization of the limit of decision is thus possible by reducing the standard deviation of the blank value, increasing the sensitivity a_1, and increasing the number N_b of blank value measurements. The numerical value of the limit of decision can also be improved by taking future analytical steps into account (i.e., the number N_a of analyses of the analytical portion). If the standard deviation is assumed to be the same for the actual analysis as for the blank value measurements, then

$$x_{\text{decision}} = \frac{s_{\text{blank}}}{a_1} t_q(v) \sqrt{\frac{1}{N_b} + \frac{1}{N_a}}$$

A more sophisticated way to obtain the critical signal value is to calculate the upper confidence limit of the intercept a_0 of the calibration relation that was derived from N_c measurements of analytical calibration portions [11]:

$$y_{\text{crit}} = a_0 + s_{\text{res}} t_q(v = N_c - 2) \sqrt{\frac{1}{N_c} + \frac{1}{N_a} + \frac{\bar{x}_c^2}{SS_{XX}}}$$

(The residual standard deviation s_{res} of the calibration model and the sum of the squares of deviations SS_{XX} are defined in Section 7.4.4; $\bar{x}_c^2$ is the mean of the concentration in the calibration step.)

By inserting this expression in the calibration model, the limit of decision is again obtained. This limit could be lowered if the mean of the calibration range lay as close as possible to the limit of decision. In fact, the highest concentration value should be lower than $10 \times x_{\text{decision}}$. Figure 2 shows that the statistical distribution of observed values near the limit of decision provides values that fall with 50 % probability within the spread of the mean blank value. In other words, at this point the risk of wrongly accepting a blank signal as an analyte signal (an error of the first kind with a probability of α) is very low, but the risk of still rejecting an analyte signal as a blank value signal (an error of the second kind with a probability of β) is very high. An analyte signal is thus reliably detected when the last-mentioned risk is as small as possible. For $\alpha = \beta$, therefore, the *limit of detection* is twice the limit of decision (→ Chemometrics).

As can be seen in Figure 2, for a particular observed value $y_{\text{quantitation}}$ an acceptable analyte interval $\pm\Delta x$ (for $x_{\text{quantitation}}$) can be indicated only if $\Delta x \approx s_{\text{res}}/a_1$ is not too large. This quotient is a critical quantity for the practical application of an analytical method. In addition, the relative standard deviation of an analyte result must sometimes not exceed a specified maximum. These two requirements lead to the definition or pragmatic setting of so-called limits of *quantitation*. Since such limits depend on the particular value observed, they cannot be regarded as meaningful characteristics of the method.

Historically, the simple $k\sigma$ criteria used in Figure 2 are based on numerous simplifications. They are usually based on the standard deviation σ of blank value measurements: The limit of decision is defined with $k = 3$, the limit of detection with $k = 6$, and as one possibility, the limit of quantitation with $k = 10$ [16]. (The k values take into consideration the probability α of erroneous statistical and, therefore, erroneous analytical decisions. Thus, by fixing k or α, the purpose of the particular trace-analytical procedure can be taken into account.) The basic considerations in such definitions of method limits extend back to H. KAISER [14] and G. EHRLICH [17]. KAISER also tried to define characteristic quantities of methods for multicomponent analysis. For reasons of space, simultaneous multicomponent analysis cannot be discussed here [18], [19]. The previous discussion reveals how delicate results of trace analyses are in general. To achieve a responsible discussion in public it should be at least reported together with confidence limits (→ Chemometrics) and single values (means) should never be compared to critical target values.

7.4.4. Adequate Calibration Models

Calibration means the recording of observed values of response feature Y (signal quantity) typical of the method as a function of the analyte feature X (concentration or content) from N_c solid, liquid, or gaseous test portions. If possible, the analytical calibration portions should originate from different standard reference materials (SRMs), whose residual composition (matrix) corresponds largely to that of the expected analytical portions. If only one standard material or one standard portion is available, the analytical calibration portions should be prepared by spiking a blank test sample (standard addition) rather than by dilution of the standard (see Section 7.4.5). Here, analytical calibration portions for which the sampling error is negligible are assumed. (For nomenclature of sampling steps and of the aliquots arising from sample preparation steps, see [20].)

The objective of the calibration is to set up and statistically ensure the mathematical model of the functional dependence of the signal quantity Y on analyte content X. Linear models of the type

$$Y = a_0 + a_1 X$$

are most widely used.

Calibration in the narrower sense means determination of the sensitivity a_1. If N_c pairs of values (x_n, y_n) are available ($n = 1, 2, \ldots, N_c$), from which the arithmetic means ($\bar{x}$, $\bar{y}$) have been calculated, the sensitivity a_1 (regression coefficient, slope) can be calculated easily with the sum of squares of deviations from $n = 1$ to N_c, $SS_{XX} = (x_n - \bar{x})^2$ and $SS_{YY} = (y_n - \bar{y})^2$ and the sum of products of deviations $SS_{XY} = (x_n - \bar{x})(y_n - \bar{y})$ where $a_1 = SS_{XY}/SS_{XX}$.

Since all straight lines of calibration pass through the center of gravity of the data ($\bar{x}$, $\bar{y}$), the constant a_0 is obtained from the model equation.

Strictly speaking, a_0 and a_1 are estimates of the true but unknown characteristic quantities of the calibration model, which represents an unknown relationship between Y and X. For that reason, $\hat{a}_0$ and $\hat{a}_1$ are used. The $\hat{y}_n$ calculated by this model for every given x_n generally differs from the measured y_n by the residual $\text{res}_n = y_n - \hat{y}_n$.

This model is based on several assumptions:

1) The quantity X can be set at points x_n without error
2) The variance of Y (at points x_n) is constant in the calibration range ("homoscedastic", which can be tested with the Cochran test [13], [21]; if the assumption is not confirmed, weighted regression models should be used)
3) The quantity Y (at the points x_n) shows a normal distribution (see Fig. 2, left) and is outlier-free (the latter situation is tested with the Dixon test [13], [21]; if the assumption is not confirmed, a robust regression method can be used, for example)
4) The residuals res_n (see Figure 2) are normally distributed and do not correlate with the values of x_n (which can be tested qualitatively by simply plotting res_n against x_n; if the assumption is not met, a nonlinear analyte–signal relation rather than linear may be valid)

In general, a linear model is naturally assumed to be justified. If tests [for coefficients of determination, goodness of fit (GOF) and lack of fit (LOF); see below] show this to be improbable, a nonlinear calibration must be selected, in the simplest case by using quasi-linear models (consideration of higher powers of X; polynomials), or piecewise linear calibration can be performed [22]. (A survey of various basic regression models can be found in [23]).

In trace-analytical methods the nonconfirmation of item 1 above must be reckoned with. The characteristic quantities a_0 and a_1 can then still be calculated within the framework of a linear model:

1) $a_1 = \text{sgn}(\text{cor}_{XY}) \sqrt{SS_{YY}/SS_{XX}}$, if the error ratio (corresponding to the long-term standard deviations) of the two response features Y and X is equal to the slope.
2) $a_1 = \frac{d}{2} + \text{sgn}(\text{cor}_{XY}) \sqrt{\left(\frac{d}{2}\right)^2 + 1}$ where $d = \frac{SS_{YY} - SS_{XX}}{S_{XY}}$ if the errors of Y and X are equal (orthogonal regression)
3) $a_1 = SS_{XY}/(SS_{XX} - N_c \cdot \sigma^2{}_X)$, or
4) $a_1 = (SS_{YY} - N_c \cdot \sigma^2{}_Y)/SS_{XY}$, if the (long-term) variances σ^2 of the response features are known from previous experiments. For more details and alternatives see [24].

The characteristic quantities obtained for the straight lines must be tested for their statistical significance and should be quoted with a confidence interval. Their standard deviation is a function of the residual standard deviation s_{res} of the regression model. This is defined as:

$$s_{res} = \sqrt{\frac{1}{N_c - 2} \sum_{n=1}^{N_c} \mathrm{res}_n^2}$$

and also determines the width of the confidence band around the straight line of calibration.

If at the points x_n of the calibration range, only single measurements y_n are available, the quality of the calibration model can be evaluated only as a whole (sum of systematic model errors and random experimental errors) usually by testing the goodness of fit and by testing the coefficient of determination.

Only if replicate measurements of y_n have been made for at least one point x_n model errors and experimental errors can be weighed against each other by LOF and complete analysis of variances, ANOVA (→ Chemometrics). Replicate measurements also reduce the width of the confidence band.

Modeling, tests, calculation of the characteristic quantities of the method, and graphical visualization are supported by computer programs (e.g., [25]), which however usually consider only certain aspects of the calibration. An alternative method of calibration for imperfect response features and concentration quantities is to apply fuzzy theory [26].

7.4.5. Quality Assurance and Standard Reference Materials

Perhaps the greatest problem in trace analysis is assurance of the accuracy of the results (i.e., the avoidance of systematic errors). Systematic sources of error are possible in every step of an analytical process. The most reliable method for detecting systematic errors is continuous and comprehensive *quality assurance*, particularly by occasional analysis of (certified) standard reference materials. Strictly speaking, an analytical method cannot be calibrated if suitable (i.e., representative) standard reference materials adequately representing the matrix of the expected test samples are not available. However, *internal laboratory reference materials* can then usually be prepared, whose matrix largely resembles the matrix of the test portions expected. If problems occur in the preparation of such reference samples, the standard addition method (SAM) can be applied, in which internal laboratory standards are added stepwise to the test sample (analyte and matrix) and the sum of the two analyte contributions (from sample and standard) is determined.

In these cases, however, the danger exists that the speciation of the analyte in the added standard and that in the real sample are not identical.

Internal quality assurance can also be carried out by comparison of methods. In this case, two methods or procedures that are completely different in analytical principle should be applied to the same analytical problem. Comparison of one's own method with a certified method, is most useful. If the results of the two methods are plotted against each other in a scattergram, in the ideal case all values fall on a straight line with slope $a_1 = 1$ and intercept $a_0 = 0$. Here again, depending on the ratio of the random errors of the methods being compared, the characteristic quantities of the straight line corresponding to the actual data can be calculated as shown above, and interpreted and tested (statistically assured) as systematic additive (a_0) and multiplicative ($a_1 \cdot X$) error contributions.

External quality assurance, usually by participation in interlaboratory (round robin) tests, also contributes to the detection and avoidance of systematic errors.

However, the mere use of standard reference materials offers no guarantee of accurate, error-free calibration. Thus, many solid SRMs are useless for the calibration of solid microanalyses if the danger of microinhomogeneities is not recognized and if the minimum amount of test portion is not chosen accordingly [27].

This illustrates not only the difficulty of carrying out "perfect" quality assurance but once again the problems of reliable trace analysis in general.

7.5. Environmental Analysis

7.5.1. The Problem

The task of environmental analysis is the identification and quantification (screening and monitoring) of contaminants [28]. The analytical characterization and evaluation of dangerous wastes from the past are typical examples of applied environmental analysis. Traditionally, for the risk assessments of old waste deposits, analytical methods (in the form of costly laboratory analysis) are used remote from the site of investigation [29]. At abandoned waste deposits and industrial sites, contaminant distributions are extraordinarily het-

Table 3. Some activities of the US EPA in field analytical methods: number of sites by technology (EPA 542-R-97-011, November 1997)*

Technology	Number of Sites
Immunoassay	43
X-ray fluorescence	39
Cone penetrometer sensor	34
Gas chromatography	24
Fourier-transformed infrared spectrometry	3
Colorimetric test strip	3
Fiber-optic chemical sensor	3
Mercury vapor analyzer	2
Biosensor	1

* Some typical reports:
Cone penetrometer coupled with laser induced fluorescence probe
EPA/600/R97/019 (SCAPS-LIF), EPA/600/R97/020 (ROST)
Field portable gas chromatograph/ mass spectrometer
Bruker-Franzen EM 640 (EPA/600/R-97/149)
Viking SpectraTrak 672 (EPA/600/R-97/148)
Field portable X-ray fluorescence spectrometer (EPA/600/R-97-144,-145,-146)
Draft EPA Method 6200, March 1996: "Field Portable X-Ray Fluorescence Spectrometry for the Determination of Elemental Concentrations in Soil and Sediment".

erogeneous in relation to surface, depth, and composition. As a result, by inexpert and not directly verifiable sampling, errors of many hundred percent are caused that are strikingly disproportionate to the complexity of the analysis of the sample material [30]. This inconsistency between precise but cost-intensive single-sample analysis and an incorrect sample can be avoided by applying on-site measurement techniques with hand-held devices, portable devices, or mobile devices with full laboratory capability installed in a vehicle [31]. In the case of on-site work, the analytical results are mostly of limited precision, but since sampling, sample handling, sample preparation, and analytical work are combined directly at the site the information density is higher.

The cost of surveying and subsequent redevelopment has induced the EPA, within the framework of its SITE program, to start a monitoring and measurement technology program (MMTP). Part of this is the Environmental Technology Verification Program (ETV). The objective of this program is to organize the application of modern analytical techniques for field capability, to create new analytical methods for the field, and to apply these methods in addition to standard EPA specifications for contaminated sites [32].

The EPA has founded an online Field Analytic Technologies Encyclopedia (FATE), available on the Internet at http://fate.clu-in.org/. Summaries of applications of field analytical and site characterization technologies are available at http://clu-in.com. Some activities of the EPA are listed in Table 3. The newest results will be presented yearly at ON-SITE ANALYSIS conferences (http://www.infoscience.com). In Europe, too, there is some activity to establish such field screening methods [33].

7.5.2. Possibilities of Mobile Analysis

7.5.2.1. Definition and Requirements

Mobile analysis is a special form of on-site analysis in the environmental sphere. For this purpose the devices (sampling, sample preparation, measurement devices) are transported in a vehicle. The equipment can be integrated permanently into the car or operated hand-held.

In the screening mode, it needs methods which can identify unknown components and also quantify them. In the monitoring mode, the contaminant is known but not its distribution. In this case the analytical methods must be selective for the quantification of the contaminant. A typical example is the monitoring of organics (e.g., TNT, PAH) with immunoassay kits accepted by the EPA.

Mostly the field methods have higher detection limits than laboratory methods. Sample preparation and analytical procedure must be fast without extensive clean-up and separation steps.

The large amount of analytical information collected in the field is typical for mobile on-site analysis. The analytical techniques enable immediate analysis at the contaminated site, rapid change of site, and the necessary flexibility (surveying, site description, drill-core analysis, incident and damage analysis) in different measurement positions and at different sites.

Requirements for the use of field measurements are summarized below.

Analytical Parameter

Determination of field and sum parameters, toxic metals, anions and cations,
Organics: volatile organic compounds (VOC), total petroleum hydrocarbons (TPH), polynuclear aromatic hydrocarbons (PAH), polychlorinated biphenyls (PCB), pesticides
in soil, water, and soil vapor

Process Conditions

Sample preparation without or with only little chemical work

Linearity of the analytical signal over several orders of concentration

Large range of variation with small response of the analytical signal (robustness of the analytical method)

Multiparameter analysis with high selectivity

Simple procedures that require little knowledge of analytical chemistry

Low cost of the analytical procedure

Equipment Technology

Rugged design toward mechanical and climatic stress

Low-energy configuration

Minimum space requirement of the equipment with high operating convenience

Autonomous electrical energy generation

Availability of working surfaces and techniques for handling samples

Computer technique, modem connection, global position system device (GPS)

Geo-Environmental software (visualization and grid optimization)

Possible analytical methods for use in the field (ppm range or higher) are listed in what follows:

Field and global parameters

Hand-held devices for the water field parameters [T, pH, electroconductivity (EC), dissolved oxygen (DO), oxidation – reduction potential (ORP)]

Hand-held monitoring techniques for VOC and gases: flame ionization detection (FID), photoionization detector (PID), thermal conductivity detector (TCD), infrared sensor (IR), see Figure 3,

Transportable equipment for measuring adsorbable organic halogen compounds (AOX), chemical oxygen demand (COD), total organic carbon (TOC)

Anions/Cations

Ion-selective sensors, reaction kits, photometer, voltammeter for the determination of ions in water or leachates

Toxic Metals

EDXRF analyzer (hand-held or transportable, X-ray tube or radioactive isotope source) for elements in soils

Organics

GC, preferably with MS coupling and sample preparation [extraction, purge & trap, spray & trap, solid phase extraction (SPE), solid phase microextraction, headspace, thermodesorption, soil air trap], nondispersive infrared spectroscopy (NDIR), FT-IR, TLC, immunoassays and biosensors (for determination of defined substances)

7.5.2.2. Example of Equipment

Only in some cases are devices designed for work under field conditions. Generally, they are applicable for the determination of inorganic traces in water or leachate and also for the determination of gases and VOC in the soil air. Some special instruments (NDIR) exist for the determination of TPH in soils under field conditions. However, modern laboratory analytical devices are generally so safely transportable that they can be carried along, secured for transport, in vehicles and installed at the site of application. Part of the relevant technology has already been designed for mobile application. Examples include metal analysis by using EDXRFA (e.g., Spectrace QuanX, SpecTrace 6000, SpecTrace 9000) and the GC/MS EM 640 (membrane inlet and direct inlet) from Bruker Daltonik Bremen and Viking SpectraTrak 573 from Bruker Daltonics. The equipment is also fully suitable for work under laboratory conditions.

The equipment plan of a vehicle is determined largely by the planned task. For a widespread use of mobile on-site analysis, GC/MS and EDXRF are the base instruments. The devices work also with autosamplers. In the case of EDXRFA, the measuring results in the ppm range are obtained in 15 min, including sample preparation (drying, milling). In the case of GC/MS studies it is a function of the analytical problem and the quality, the amount, and the composition of the organics. The upper time limit for obtaining the analytical information is not less than one minute.

7.5.3. Selected Applications of Mobile Analysis

7.5.3.1. GC-MS Screening and IMS monitoring

A typical example of the application of GC-MS is the field screening of soil gas composition because of its simple sample preparation tech-

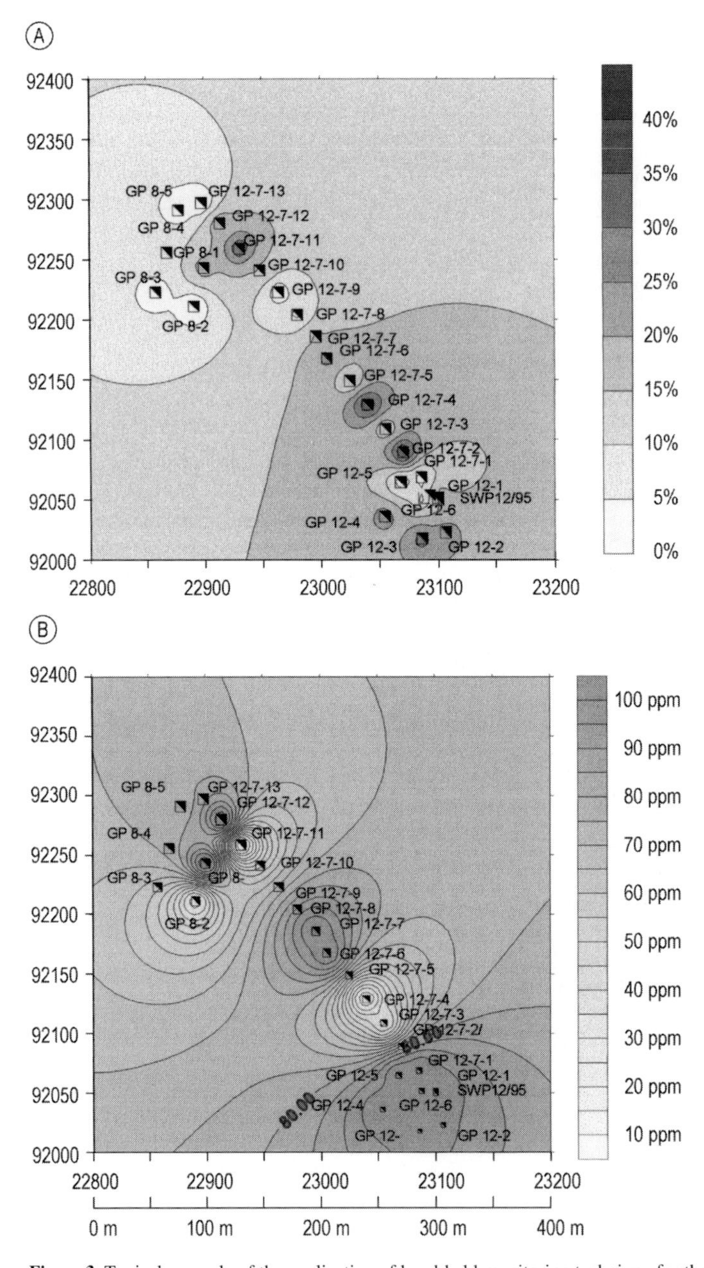

Figure 3. Typical example of the application of hand-held monitoring technique for the soil gas emission of a landfill site (depth of exploration of the drilling points: 1.50 – 1.80 m; visualisation by using SURFER software)
A) Determination of CH_4 distribution with an IR sensor; B) Determination of H_2S distribution with an electrochemical sensor

nique. A case study was to determine the VOC distribution, especially of tetrachloromethane (TCM), in a housing estate in the neighborhood of an old factory site. The first step of such studies is the determination of the soil gas composition and of the major components. The next step consists of choosing a fast, simple, and inexpensive measuring method for monitoring and controlling the major risk component, in this case TCM. Possible methods of TCM detection are gas reaction tubes, a hand-held photoionization detector (PID), a mobile gas chromatograph, a hand-held ion mo-

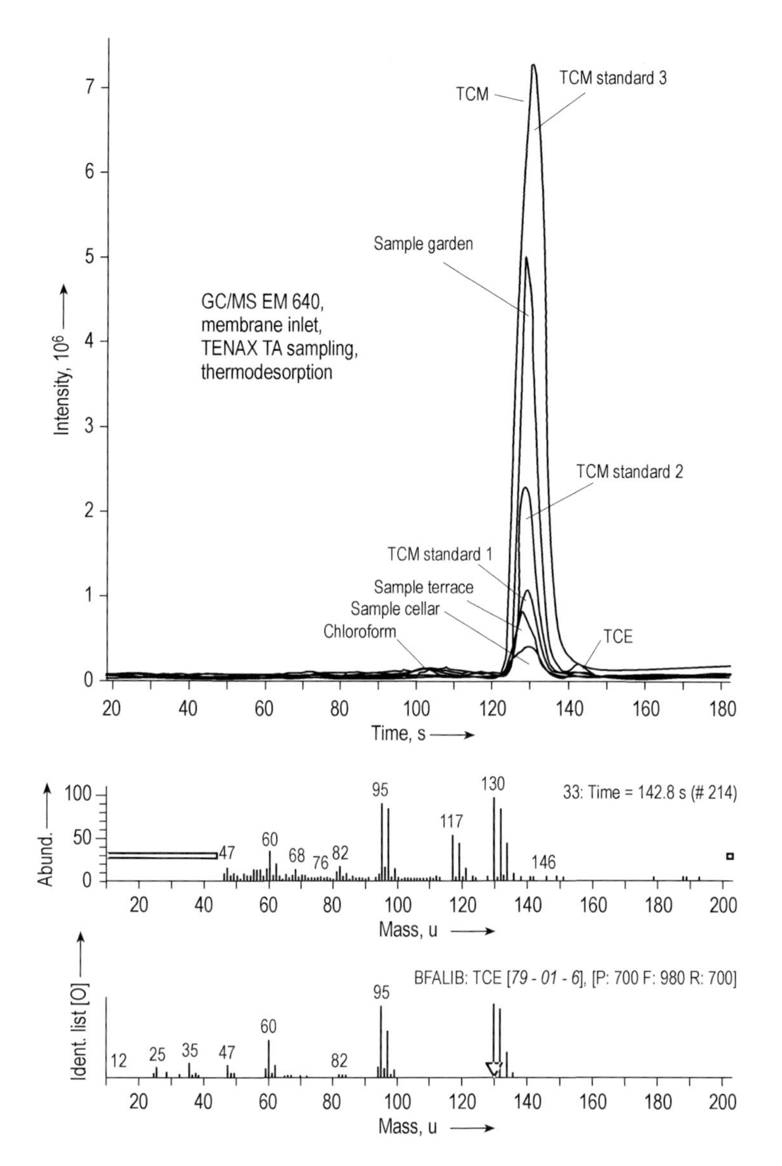

Figure 4. Presentation of some TIC runs of soil gas samples (garden, terrace; bore hole depth 2 m), gas standard and indoor air samples (cellar)
TCM = tetrachloromethane; TCE = trichloroethylene

bility spectrometer (IMS), and a mobile GC-MS. The application of gas reaction tubes is simple, but time consuming and expensive with poor reproducibility of results. The linear range of determination lies in an order of magnitude, and there is a pronounced nonconformity between the different tube types. The PID with the typically used 10.6 eV UV lamp is not suitable for TCM determination. The IMS is the fastest instrument (overall analysis time < 2 min). It gives only a sum signal for halogenated hydrocarbons in the negative operation mode (selective detection of a Cl peak) with high sensitivity (< 0.2 vol ppm), but with a small range of linearity of determination and the possibility of extension by using a dynamic gas dilution. The mobile GC-MS system is the best technique in all the cases of field screening of nonpolar compounds with a high dy-

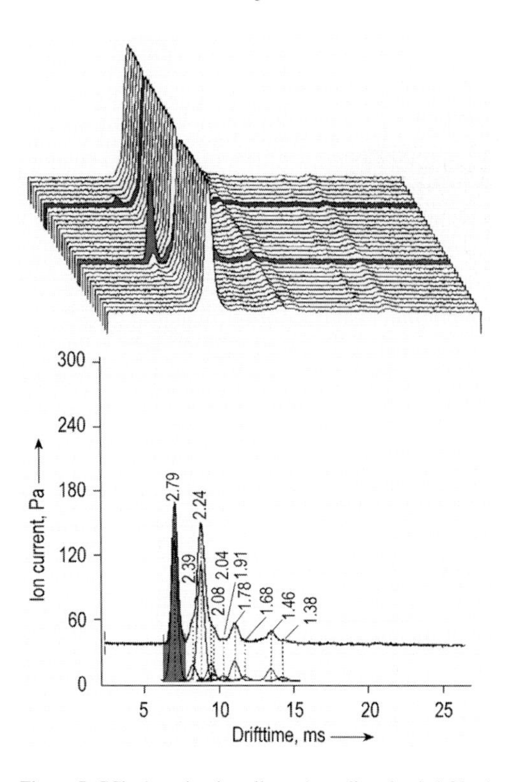

Figure 5. CCl_4 detection in soil gas (sampling depth 1.80 m) by using ion mobility spectrometry (8 bit IMS type RAID 1 from Bruker-Saxonia in the negative mode, push injection of 1 mL soil gas, measurement of the Cl-peak
red peak: sample taken near a walnut tree; Green peak: sample taken at the garden fence
Spectral range with two different sample injections (top) and a typical single spectrum (bottom) with a Cl peak (green)

namic range of more than three orders of magnitude under real field conditions and also allows the detection and identification of other contaminants. The system is expensive and its operation is not simple. A good compromise is to use GC-MS as a master instrument and IMS as a fast monitoring instrument. The accordance of results between GC-MS and IMS is better then 20 %. Figure 4 summarizes the application of GC-MS in this field. After collection (indoor air and soil gas of boreholes) of the volatile organic compounds on TENAX TA the organics are thermodesorped and separated by gas chromatography. The total ion chromatogram (TIC) is analyzed with the aid of a MS spectra data bank and analytical software specially set up for the system (in this case Bruker DataAnalysis software), and the substances are identified. The system is also calibrated for quantification and the concentrations of substances are calculated with the aid of the software.

Figure 5 shows the detection of TCM by ion mobility. The term ion mobility spectrometry refers to the principles, practice, and instrumentation for characterizing chemical substances by means of gas-phase ion mobilities [34]. Ion mobilities are determined from ion velocities that are measured in a drift tube with supporting electronics. Ion mobilities are characteristic of substances and can provide a means for detecting and identifying vapors. In practice, a vapor sample is introduced into the reaction region of a drift tube in which neutral molecules of the vapor undergo ionization, and the resultant ions, i.e., product ions, are injected into the drift region for mobility analysis. Mobility K is determined from the drift velocity attained by ions in a weak electric field of the drift region at atmospheric pressure. The normalized value K_0 is a characteristic constant of the product ion (in this case $K_0 = 2.79\ cm^2\ V^{-1}\ s^{-1}$ for the Cl peak).

By using a mass spectrometer with direct inlet (e.g., Bruker EM 640S) it is possible to inject directly the soil air samples and to report only the typical mass fragments of TCM in the SIM mode (Fig. 6). In this way the analytical procedure is faster and more sensitive. The TCM distribution using all results of IMS and GC-MS measuring can be visualized with special software (e.g., SURFER software from Golden Software Inc.). An example of TCM distribution is incorporated in Figure 6.

7.5.3.2. Elemental Analysis by Mobile EDXRF

Energy-dispersive X-ray fluorescence analysis (EDXRF) has proven to be ideal for on-site analysis [35]. With the availability of Peltier-cooled semiconductor detectors in EDXRF, its mobile application to field screening of toxic metals is unproblematic. In combination with geo-statistical methods or geographic information systems this technique provides important and meaningful information about the distribution of contaminants on hazardous waste sites. In Figure 7 this is illustrated by the determination of the metal distribution in a flood plain area in the neighborhood of a landfill site. A radioisotope is used to excite the characteristic fluorescent X-rays in a sample in hand-held devices. The instruments are rugged, fast, and easy to use. The sensitivity is lower than that of instruments equipped with adjustable X-ray tube sources, which have a detection limit in the ppm range. EDXRF can simultaneously determine 20 or more elements from

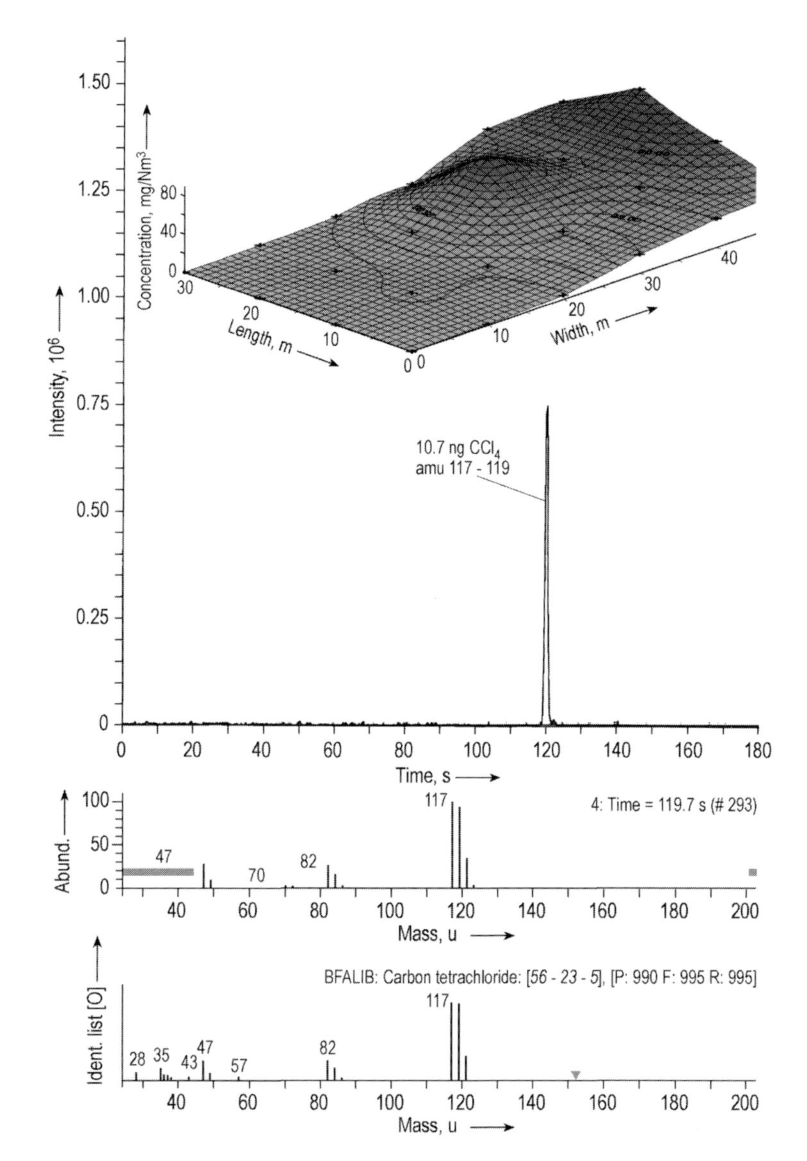

Figure 6. Selective detection of tetrachloromethane (TCM) contaminations by using SIM mode (measuring conditions: Bruker Daltonik GC/MS EM 640S with direct inlet; split 1 : 10; direct injection of 0.2 mL gas samples, 30 m HP 5 ms, $d_i = 0.25$ mm, $f_{th} = 0.25$ mm). The upper part of the figure incorporates the 3D visualisation of TCM distribution of soil gas (sampling depth 1.80 m; combination of IMS and GC-MS data) of part of the contaminated area in the neighborhood of the housing estate

sodium to uranium within only a few minutes. The instruments are also applicable in the laboratory. Some of these (e.g., the QuanX from Spectrace) also work with a 20-position sampler for automatic operation. The sample preparation is simple and fast and consists of drying and milling (particle size < 100 μm). The accuracy of the results depends on the quality of the used standards and the difference between the matrix of the standards and the sample (Fig. 8). Therefore, the results were calculated with the so-called Fundamental Parameter algorithm supplied with the spectrometer software. This algorithm corrects automatically for interelement and matrix effects. For this reason, no site-specific calibration is required, which is an enormous advantage over empirical calibration models.

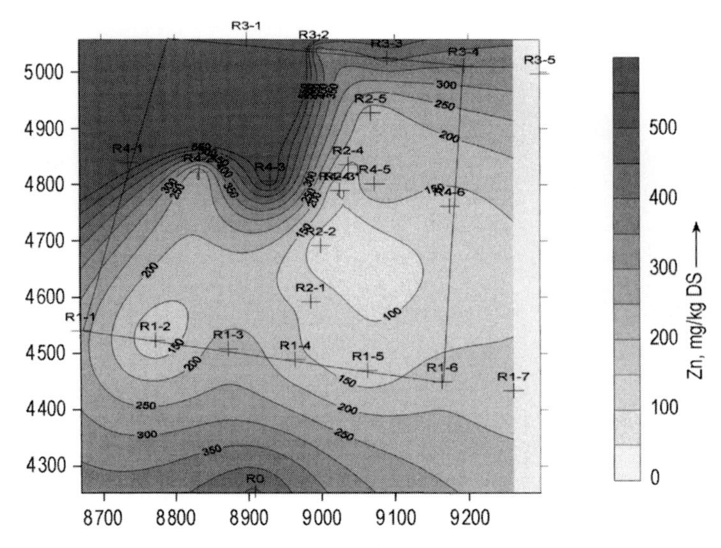

Figure 7. Determination of element distribution (e.g., Zn) in the subsoil of a landfill site by using mobile XRF (SPECTRACE 6000) and geostatistical methods

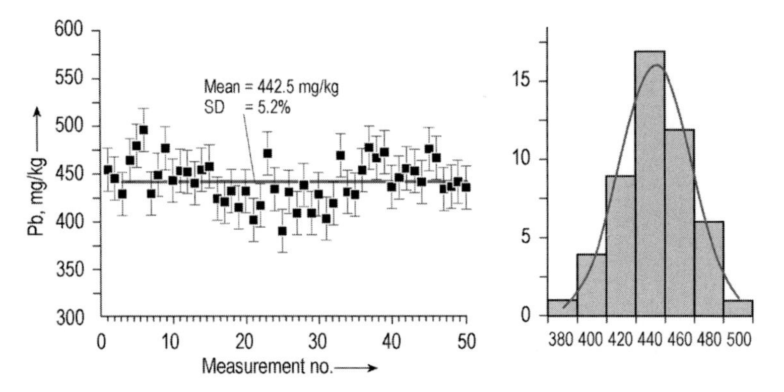

Figure 8. Reproducibility of the field screening EDXRF method [mean value, standard deviation (SD) and gaussian distribution of Pb content in excavated material from a waste disposal site]: determination of the repetition accuracy over a time scale of two months

Typical for the application of mobile EDRFA is the fast determination of element distributions in drilling cores or in finding hot spots in soil profiles (Fig. 9).

7.5.4. Conclusions

Field screening methods are used to determine the presence or absence of typical components at a given site. Needs for screening methods exist not only for site characterization, but also for determining the progress of remediation efforts. A driving force in mobile screening technology development and commercialization is the ever-increasing costs associated with environmental compliance. For the different contaminants, there many different field screening instruments. Many companies offer field test kits and a variety of detector tubes. Battery-operated spectrophotometers are available to support field analyses. Gas chromatography is one of the most mature technologies being applied in this field. The most recent advances involve the use of kits based on immunochemistry, GC-MS coupling, and XRFA. Attractive features include a high degree of analyte selectivity and ppb to ppm sensitivity in most matrices.

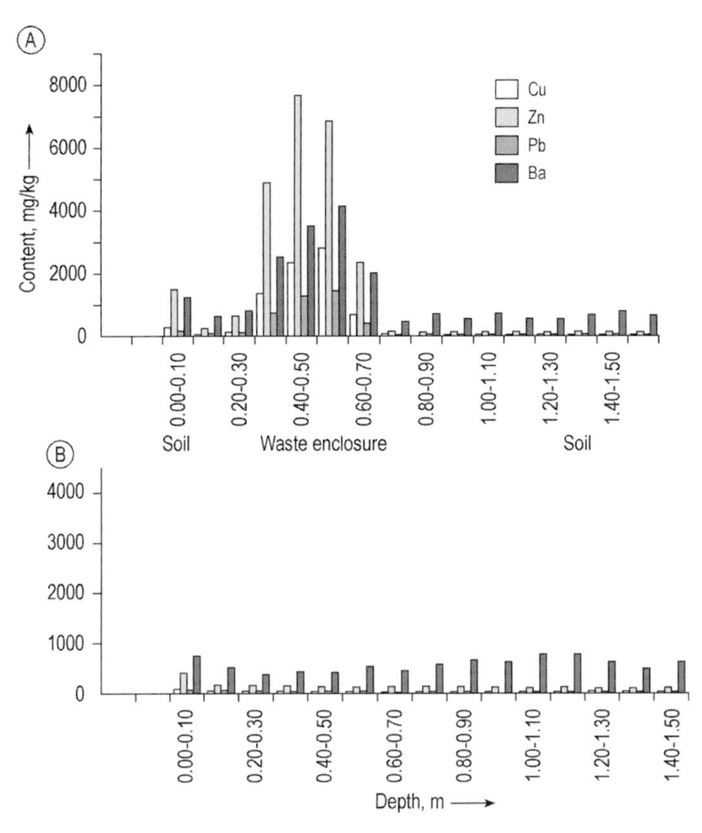

Figure 9. On-site determination of waste enclosure in a flood plain area in neighbourhood of a landfill site by using transportable EDXRF spectrometer (sampling by digging in the depth, waste contamination in profile 1)
A) Digging profile 1; B) Digging profile 2

7.6. References

[1] O. G. Koch, G. A. Koch-Dedic: *Handbuch der Spurenanalyse,* 2nd ed., **vol. 1,** 1, Springer Verlag, Berlin 1974.

[2] K. Beyermann: *Organische Spurenanalyse,* G. Thieme, Stuttgart 1982.

[3] G. Knapp, M. Michaelis: "Chemie in Labor und Biotechnik," *CLB Chem. Labor. Betr.* **44** (1993).

[4] G. Tölg, *Analyst (London)* **112** (1987) 365–376.

[5] B. Sansoni, *Pure Appl. Chem.* **59** (1987) 579–610.

[6] R. Kellner et al. (ed.): *Analytical Chemistry,* Wiley-VCH, Weinheim 1998.

[7] G. Schwedt, A. Meyer, *LaborPraxis* **16** (1992) 712–721.

[8] U. Thiele, *Lab. Trend.* 1999 10–14.

[9] L. Dunemann, J. Begerow: *Kopplungstechniken zur Elementspeziesanalytik,* VCH-Verlagsgesellschaft, Weinheim 1995.

[10] M. Luthardt, E. Than, H. Heckendorff, *Fresenius' Z. Anal. Chem.* **326** (1987) 331–339.

[11] DIN 32 645 (May 1994): *Chemische Analytik; Nachweis-, Erfassungs- und Bestimmungsgrenze; Ermittlung unter Wiederholbedingungen; Begriffe, Verfahren, Auswertung.*

[12] DIN 55 350, part 34 (Feb. 1991): *Begriffe der Qualitätssicherung und Statistik; Erkennungsgrenze, Erfassungsgrenze und Erfassungsvermögen.*

[13] *ISO Standards Handbook 3: Statistical Methods,* 3rd ed., ISO, Genf 1989.

[14] H. Kaiser, *Spectrochim. Acta* **3** (1947) 40–67.

[15] K. Doerffel, G. Michaelis, *Fresenius' Z. Anal. Chem.* **328** (1987) 226–227.

[16] G. L. Long, J. D. Winefordner, *Anal. Chem.* **55** (1983) 712A–724A.

[17] G. Ehrlich, *Fresenius' Z. Anal. Chem.* **232** (1967) 1–17.

[18] K. R. Beebe, B. R. Kowalski, *Anal. Chem.* **59** (1987) 1007A–1017A.

[19] C. W. Brown, R. J. Obremski, *Appl. Spectr. Rev.* **20** (1984) 373–418.

[20] W. Horwitz, *Pure Appl. Chem.* **62** (1990) 1193–1208.

[21] DIN ISO 5725 (April 1988): *Präzision von Meßverfahren; Ermittlung der Wiederhol- und Vergleichspräzision von festgelegten Meßverfahren durch Ringversuche.*

[22] L. M. Schwartz, *Anal. Chem.* **49** (1977) 2062–2068.

[23] J. D. Hwang, J. D. Winefordner, *Prog. Anal. Spectrosc.* **11** (1988) 209–249.

[24] C. Hartmann, J. Smeyers-Verbeke, D. L. Massart, *Analysis* **21** (1993) 125–132.

[25] J. Kramer: *STATCAL,* Umschau Verlag, Frankfurt/Main 1992.

[26] Y. Hu, J. Smeyers-Verbeke, D. L. Massart, *Chemom. Intell. Lab. Systems* **8** (1990) 143 – 155.

[27] J. Pauwels, C. Vandecasteele, *Fresenius J. Anal. Chem.* **345** (1993) 121 – 123.

[28] I. L. Marr, M. C. Cresser, L. J. O. Ottendorfer: *Umweltanalytik,* Thieme, Stuttgart 1988.

[29] V. Franzius, R. Stegmann, K. Wolf, E. Brandt: *Handbuch der Altlastensanierung,* **vol. 2,** R. v. Decker's Verlag, Heidelberg 1992 chap. 4.

[30] L. H. Keith (ed.): "Principles of Environmental Sampling," *ACS Professional Reference Book,* 2nd ed., American Chemical Society, Washington D.C., 1991.

[31] J. Flachowsky: "Mobile Umweltanalytik," in H. Günzler et al. (eds.): *Analytiker-Taschenbuch,* **vol. 18,** Springer-Verlag, Berlin – Heidelberg – New York 1998, pp. 143 – 180.

[32] Abstracts of the Eight International Conference ON-SITE ANALYSIS, Lake Las Vegas, NV, January 2000, www.ifpac.com.

[33] H. J. Gottlieb et al. (ed.): "Field Screening Europe," Proc. of the first international conference on strategies and techniques for the investigation and monitoring of contaminated sites, Kluwer Academic, Dordrecht 1997.

[34] G. A. Eiceman, Z. Karpas: *Ion Mobility Spectrometry,* CRC Press, Boca Raton – Ann Arbor – London – Tokyo 1994.

[35] G. Matz, W. Schröder, J. Flachowsky: "On-Site Investigation of Contaminated Soil by GC-MS and EDXRF-Techniques," in F. Arendt, G. J. Annokkée, R. Bosman, W. J. van den Brink (eds.): *Contaminated Soil '93, 4th Int. KfK/TNO Conf. on Contaminated Soil, Berlin May 1993,* vol. I, pp. 657 – 664, Kluwer Academic Publishers, Dordrecht 1993.

8. Radionuclides in Analytical Chemistry

Rainer P. H. Garten, Max-Planck-Institut für Metallforschung, Dortmund, Federal Republic of Germany

Juraj Tölgyessy, Department of Chemistry, Faculty of Natural Sciences, Matej Bel University, Banská Bystrica, Slovak Republic

Abbreviations

A	analyte
A_1	activity of component 1
AA	activation analysis
AAS	atomic absorption spectrometry
B	reagent
IDA	isotope dilution analysis
R*	labeled radioreagent
SRM	standard reference material

8.1. Introduction

8.1.1. Definition and Purpose

Radioanalytical chemistry covers the use of radioactive nuclides and nuclear radiation for analytical purposes:

1) In the analytical determination of any chemical species or element

2) In the development, improvement, and quality assurance of analytical procedures for the benefit of analytical principles

8.1.2. History

The use of radionuclide techniques in analytical chemistry was first reported in 1913 by G. HEVESY and F. PANETH in a study of the solubility of lead sulfide in water, using the natural lead isotope ^{210}Pb as indicator [67]. Isotope dilution analysis was introduced by O. HAHN in 1923 [68], using ^{231}Pa to determine the yield of ^{234}Pa. The development of radioreagent methods followed, and further development of radioanalytical chemistry has established a range of analytical methods and techniques [1] – [4], [61], [65], [87], [93], [95], [97]. These include the use of artificial radionuclides and labeled compounds, the principles of nuclear activation [4] – [10], [66] (→ Activation Analysis), and absorption and scattering of radiation [11], [12]. The most important procedures are shown in Table 1.

8.1.3. General Features

Radiochemical analysis is based on two outstanding features of radioactivity:

1) The high sensitivity and ease of measurement of radioactive radiation [13], [14]
2) The possibility of labeling chemical compounds with radioactive tracers [15] – [17], [69], [70]

Radionuclide techniques often have higher sensitivity than other analytical methods. The amounts of nuclides, correlated to an activity of 1000 Bq (see Table 2), can be derived from the law of radioactive decay. These amounts vary considerably, corresponding to the wide range of radioactive half-lifes. For 90 % of the commonly used nuclides [17], half-lifes range from several minutes to several years, so the corresponding masses are extremely low.

Radionuclides are often diluted with inactive isotopes, but specific activities (i.e., activity per total mass of element) are still very high. Since the background in nuclear spectroscopy is very low, and sensitivity is high, activities as low as 0.2 Bq can be readily detected (^{3}H, ^{125}I, ^{132}I), and even 0.01 Bq (several γ emitters, e.g. ^{24}Na, ^{38}Cl, ^{42}K, ^{46}Sc, ^{59}Fe, ^{60}Co, ^{65}Zn, ^{110m}Ag, ^{182}Ta, ^{187}W, ^{198}Au). Total activities of the order of 0.1 – 20 kBq (3 – 500 nCi) are often sufficient in analytical applications. The precision of the final result can be better than 2 %, if the counting error is kept below 1.5 %. This can be obtained from counting times between 100 s and 5 h.

Analytical applications of radionuclide techniques rely on the assumption that different isotopes of the same element exhibit the same properties in any macroscopic physical or chemical process, and that radioactive labeling does not influence the other properties of a chemical species. This is generally the case, with deviations below 1 % (with the exception of hydrogen isotopes) owing to isotopic fractionation or radiation effects. For analytical purposes, the radiotracer and the analyte must be present in the same chemical form. This is usually to achieve, but specialized preparative techniques may be necessary for radioactive labeling of more complex organic compounds.

8.1.4. Importance and Current Trends

Radionuclides are used in many subdivisions of analytical chemistry (see Table 1). Of major importance are radiotracers in methodological and pathway studies, isotope dilution analysis (IDA), radioimmunoassay, and nuclear activation analysis (AA) (→ Activation Analysis) [66]. They are all especially suited to analyze the extremely small amounts of substances encountered in ultratrace analysis or in trace analysis of microsamples.

Over the past two decades, the emphasis has shifted from high detection power in routine analysis toward an independent approach, applying this high detection power to the development of analytical procedures and reference materials.

Radiochemical methods for routine analysis have lost ground to other, primarily spectroscopic, methods (see Chap. 8.3) [18] – [21]. Nonradiochemical methods often yield highly reproducible results, but may involve systematic errors.

The need for increased reliability and analytical quality control has emphasized the usefulness of radiochemical methods for the certification of standard reference materials (SRM) [9], [22], [23]. Radioanalytical methods are often suitable for homogeneity testing and distribution analysis of traces in SRMs. Activation analysis, radiotracer techniques, and isotope dilution analysis are becoming increasingly important for assessment of analytical quality.

Table 1. The scope of radioanalytical chemistry

Division	Section	Examples
Tracer labeling in organic analysis	labeling with tritium, ^{14}C, and heteroatoms (^{32}P, ^{36}Cl, ^{35}S, ^{125}I, ^{131}I, etc.)	random molecular sites selective molecular sites specific molecular sites
Tracer techniques in inorganic and organic analysis (addition of radioactive compounds to the sample)	chemical methodological studies and quality control of fundamental analytical processes (multistage procedures)	combustion and decomposition
		equilibrium constants
		phase separation efficiencies
		extraction
		volatilization
		precipitation
		redissolving precipitates
		coprecipitation
		electrochemical separation
		stability of solutions
		evaluation of sources of errors in trace and ultra-trace processes
		adsorption
		desorption
		volatilization
		contamination
		yields of complex analytical procedures
	kinetic studies	diffusion and forced diffusion (including electromigration) in solutions and gases
		diffusion, forced diffusion, and self-diffusion [a]
		chemical reactions: from long reaction times down to the ms regime (μs in special cases); macroscopic reaction order and rate
		reactive intermediate studies
		reaction mechanisms
		metabolism in organims and organelles
		rates of emission, immission, and migration of chemical species in environmental systems: natural, anthropogenic, and intentionally introduced radionuclides
	isotope dilution analysis	stoichiometric
		substoichiometric
		sub- and superequivalence method
	radiotracer- and radioreagent based analytical methods and pathway studies	radiochromatography
		radioreagent methods
		radiorelease techniques
		radioimmunoassay
		radiometric titration
		isotope exchange
		autoradiography
		studies of metabolism
		pathways of tracers in the environment (generally short-lived, for safety reasons)
		wear and wear protection in engineering materials
		corrosion and corrosion protection in engineering, construction, and biomaterials
Activation analysis (AA), (production of radionuclides in the sample by nuclear reactions, their separation and measurement)	discriminating types of AA	Induced by
		thermal neutrons (reactors)
		epithermal neutrons (reactors)
		fast neutrons (generators, cyclotrons, isotope sources)
		protons (accelerators)
		α particles (sources, accelerators)

Table 1. (continued)

Division	Section	Examples
		fast ions of low to mid-range Z (^{6}Li, ^{9}Be, ^{12}C, ^{14}N, and others from accelerators)
		γ-radiation (reactors, sources, cyclotrons, synchrotrons)
		delayed (ordinary) AA vs. prompt [b] (prompt nuclear reaction analysis NRA) neutron- or charged particle-induced reactions
		fast [c] (spectroscopy of short-lived nuclides) vs. intermediate [d] and long lived [e] AA
		purely instrumental (mostly γ-spectrometric) vs. radiochemical i.e., group separations
		or specific element separations (including substoichiometric calibration techniques)
	applications	certification analysis of standard reference materials
		basic research in analytical science and technology
		routine control of high-purity materials
		trace elements in biomaterials (including medicine)
		routine production control of commercial high-performance materials
		on-line and in-line process control (e.g., bomb detection)
		trace elements from environmental materials (water, dust, soil, sludge)
		trace elements in plants (pollutants, uptake, storage, mobilization paths)
		trace elements in geochemical and cosmochemical matter
Determination of natural (H, C, K, Th, U, Ra, etc.) and anthropogenic (Cs, Sr, Tc, Np, Pu, etc.) radioactivity		levels and distribution of activity in environmental materials (air, water, soil, etc.)
		dating in archeology, fine art, historical, and environmental research (^{14}C, ^{3}H, ^{7}Be, ^{10}Be, ^{40}K, ^{232}Th, ^{235}U, ^{238}U)
Nuclide radiation sources in analysis (on-line and remote sensing devices)		radionuclide-induced X-ray fluorescence analysis (on-line product flow control, outdoor and remote sensing)
		resonance scattering of γ rays
		resonance absorption of γ rays
		absorption and backscattering of X, α, β and γ radiation (on-line control):
		mass density monitoring
		product mass flow control
		monitoring of coatings and films
		radiographic testing of materials and devices
		detection of radiation induced radicals
		radiation dosimetry
		particle size distribution
		absorption and moderation of neutrons for light element detection
		remote B- and Cd-sensing
		humidity measurement of product flows
		plasma desorption mass spectrometry (soft ionization technique induced by fast heavy fission nuclei from ^{252}Cf-source)
		prompt (n, γ) reactions (for prospecting, oil and gas well logging)
		α- and β-ionization detector devices (pressure gauges, gas flow metering, smoke sensors, electron capture detectors).

[a] Diffusion of a compound or element in a matrix containing an excess of the same species: (active and inactive) isotopic tracer methods are the only methods available. [b] $\Delta t < 10^{6}$ s. [c] half-lifes $\tau < 1$ h. [d] half-lifes $\tau < 3$ d. [e] Half-lives $\tau > 20$ d.

Table 2. Mass corresponding to 1 kBq for carrier-free radionuclides

Nuclide	Mass [a], ng	Half-live
^{36}Cl	820	3×10^{5} a
^{14}C	6.1	5.8×10^{3} a
^{63}Ni	0.57	120 a
^{85}Kr	6.8×10^{-2}	10.6 a
^{125}I	1.6×10^{-3}	60 d
^{3}H	2.8×10^{-3}	12.4 a
^{131}I	2.2×10^{-4}	8.04 d
^{35}S	6.3×10^{-5}	87.2 d
^{32}P	9.5×10^{-5} *	14.3 d
^{99m}Te	5.1×10^{-6}	6.02 h
^{18}F	2.8×10^{-7}	1.83 h

* Commercially available with maximum specific activity of 2×10^{14} Bq/g.

Radioanalytical methods are well suited to the determination of basic analytical data, such as equilibrium constants, or kinetic data (see Section 8.3.1). These data are important for development and optimization of new analytical procedures [61].

The most frequent analytical use of tracers is in the biomedical sciences. Radioimmunoassays, in the form of rapid diagnostic test kits have led to the development of nonradioactive assays based on similar principles [2], [24].

Many important tracer applications can be substituted by other methods. Even IDA and tracer applications in self-diffusion studies can be replaced by inactive isotope tracer methods using mass spectrometry and other methods for isotope ratio determination. However, because of the extremely high sensitivity of IDA, radioactive tracers are of unique usefulness in radioimmunoassays, radiorelease reagents, radiochromatography, AA, and for systematic studies in trace and ultratrace analysis, physiological chemistry, IDA, diffusion, isotope exchange, and physical chemistry of solids.

Radioanalytical chemistry will continue to occupy an important position, despite increasing criticism of nuclear technology.

Recently, there has been little innovative work in the field of methodology, but the applications field has continued to expand [7], [25], [66]. Highlights include application of isotope enrichment techniques in environmental studies [26], [71], [72], improved accuracy in the characterization of SRM [23], [27], [28], preirradiation chemistry with high-purity reagents under clean room conditions to permit speciation from AA [7], [26] (sometimes termed molecular AA [25]), derivative AA [7], accelerator-based dating methods combined with IDA, short-time and pulsed reactor activation, extension of the sub- and superequivalence method [73]–[75], in-vivo analysis by prompt-γ neutron AA [7] and by short-lived AA [8], [29], and extended use of the single comparator (k_0) standardization method in neutron AA [30].

8.2. Requirements for Analytical Use of Radionuclides

8.2.1. Safety and Operational Aspects

Safety and operational aspects involve the hazards of ingestion of radionuclides and exposure to radiation [76].

8.2.2. The Labeled Substance

Aspects of the labeled substance involve the following requirements [14], [20], [26], [31]–[33], [66]:

1) Radionuclide and radiochemical *purity* are essential. Radiochemical purity can be checked by reverse IDA (see Section 8.4.2). The labeled substance must react identically to the analyte.
2) Complete *homogenization* and *isotope exchange* must be achieved, requiring successive transformation of the active tracer as well as of the inactive analyte into all chemical species that are considered to occur in the complete system [20], [26], [34].
3) *Isotope effects* must be negligible.
4) For most elements, suitable radionuclides for analytical use are commercially available. There are only seven elements of interest (B, He, Li, N, Ne, O, Mg), for which this is not the case [17].
5) The *preparation* of a specific isotope, of an isotopically modified (i.e., mixture of labeled and unlabeled), isotopically substituted, or even site-specifically labeled compound is often mandatory in analytical applications.
6) *Synthesis* (including biosynthesis) of these labeled compounds often requires much more effort than the actual radioanalytical experiment [34]. A variety of labeled compounds and pharmaceuticals are available commerically from stock, and a further range

of compounds will be prepared on request by a number of suppliers [35]. In special cases, doubly and triply labeled molecules have been used to follow the reactions of different functional groups.

7) Costs of radionuclides for analytical applications are typically in the range \$ 150 – 400 per 40 MBq (= 1 mCi), and higher where preparation is difficult or time consuming (e.g., ^{54}Mn, ^{85}Sr, ^{36}Cl: \$ 700 – 4000 per 40 MBq). Labeled substances from stock usually cost between \$ 2000 (^{14}C) and \$ 200 (^{3}H) per 40 MBq; the price depends on the cost of preparation.

8.2.3. Activity Measurements

Commercially available instrumentation is generally used, with the exception of some research work and highly specialized routine analytical applications.

Quantitative measurement of radioactive substances is generally straightforward with the usual counting devices and relatively simple with γ ray, X ray, and hard β emitters. The *detection limit DL* of the concentration of the analyte is given [36], [77] by:

$$DL = \frac{f}{s} \sqrt{B} \quad (1)$$

where $f = 3$ is appropriate for detection at the 99.9 % confidence level, B is the blank level, including analytical blank and spectral background of the detection device, and s is the analytical sensitivity given by the slope of the analytical calibration function, i.e., under idealized conditions:

$$s = \varepsilon \eta \, t \, a \, m_{\text{sample}} \quad (2)$$

where ε is the detection efficiency, η is the emission probability, t is the counting time, a is the specific activity of the analyte, and m_{sample} is the mass.

Specimen preparation for radioactivity measurements aims, in general, to obtain the analyte:

1) As completely as possible and with the highest possible degree of purity from extraneous radioactivity and from excess matrix matter
2) Concentrated on the smallest possible area or volume of a solid, liquid, or gaseous (for H, C, S, Ar, Kr, Xe, Rn) target
3) On a specimen support that is compatible with the chemical nature of the species analyzed, with the specific radiation to be measured, and with the detector system, e.g., liquid scintillation cocktails [37] for weak β emitters, or aluminum trays for dried layers of hard β emitters [13].

These preparation requirements are compatible with the more general rules for specimen preparation in trace and ultra-trace analysis by any technique [18], [40], [45]. Requirements are different for measurements of: γ radiation of energy >50 keV; hard β radiation >200 keV; α radiation; weak β radiation below 200 keV; and X radiation below 70 keV. In this respect, specimen preparation follows the general outline of preparation procedures for small counting samples [13], [37].

8.2.4. Choice of Radionuclide

The choice of radionuclide involves:

1) Measurement, possibly including discrimination between different indicator nuclides or indicator and interfering nuclides (e.g., from byproducts or blank level impurities)
2) Suitable half-life
3) Chemical species

γ *Emitters* are preferred for multitracer experiments (most of all in nuclear AA) for applications where the tracer is dispersed in, or shielded by a material (e.g., in-vivo diagnostics in biology and medicine, measurement from bulky solid material), but also from tracer solutions.

Pure β *emitters* are especially useful where the shorter range of their radiation is utilized, e.g., for autoradiography [39], and for surface and thin film studies, including adsorption, corrosion, and catalysis. Measurements of β-emitting tracer solutions are performed with high efficiency by liquid scintillation.

α *Emitters* are only used as tracers in special cases, either where their very low range is important (e.g., surface and thin film techniques), or where they represent the only available nuclides of the element (e.g., actinides).

Imperfect tracers are often used when labeling of a specific chemical site is inappropriate with isotopic nuclides. Imperfect tracers are widely used to label gas streams with inert gases (^{85}Kr, ^{41}Ar, ^{133}Xe), water with dissolved anionic tracers (^{82}Br, ^{131}I, ^{24}Na), hydrocarbon fuels with dissolved ^{60}Co-naphthenate or 1,2-^{82}Br-dibromoethane. Similarly, loss of metal ions at ultra-trace levels due to adsorption on the vessel walls and by volatilization have been modeled by ^{195}Au, ^{203}Hg, ^{60}Co [18], [20], [40], [78], which are easily measured, and which are available at high specific activity.

8.2.5. Appraisal of Radionuclide Use in Analysis

8.2.5.1. Advantages

Utilization of radionuclides in analysis has the following advantages:

1) Cost effectiveness [4], [9], [40] – [44], [66] (Section 8.2.2).
2) Element specificity and chemical site specificity (Section 8.2.2).
3) High dynamic range in a single experimental run. Radionuclide blanks are often extraordinarily low, far below detectability. Multielement analysis by simultaneous use of different tracer nuclides, often without separation.
4) Quality and quantity of the signal-producing decay are independent of influences such as temperature, pressure, external fields, and the chemical environment of the nucleus.
5) Locally resolved identification. Distribution detection, e.g., from within an analytical decomposition apparatus, a growing plant, a thin-layer chromatogram, or sections from biological tissue or engineering materials. Lateral resolution as low as 10 μm by autoradiography [39].
6) Predetermination of counting precision [13], [49].
7) Following reaction schemes or metabolites: in vitro and in vivo with fast element detection, often from outside the analytical system at comparable levels of sensitivity and accuracy.
8) No interference with analysis by other methods (nondestructive, conservative).

8.2.5.2. Disadvantages

Disadvantages of radionuclides in analytical chemistry are as follows:

1) Synthesis of labeled compounds may be difficult
2) Safety considerations (see Section 8.2.1) are required to avoid incorporation of radiotoxic substances and radiation exposure, specialized laboratory and equipment, qualified and skilled staff, disposal of radioactive waste
3) Boron, aluminum, nitrogen, and oxygen are of analytical interest, but no nuclides are available with half-lifes longer than 10 min
4) Specific sources of error due to radiotracers (see Section 8.2.5.3) have to be controlled
5) No distinction between different chemical species without additional separation
6) In general, sample material cannot be returned to normal use after analysis. Safe handling may restrict use of some analytical processes

8.2.5.3. Sources of Error

In analytical applications of radionuclides, specific sources of error can generally be ascribed to either analytical and trace-analytical considerations, or particular features of radionuclides and their measurement.

Difficulties in handling sub-nanogram amounts of analyte are well known [40], [45]. Because of the very low concentrations considered, chemical techniques such as solvent extraction, electrodeposition, volatilization, coprecipiration, chromatography, and ion exchange may be necessary. Additionally, surface adsorption on vessel walls, colloidal matter, and dust particles in the system may be a significant source of error.

Errors may be prevented by blank experiments, high acid concentration above 1 mol/L (diminishing adsorption), selection and preconditioning of vessel materials [20], [40], [44] – [46]. Since these problems are not specific to radiotracers, they can be most effectively reduced by addition of an inactive isotopic carrier of the same chemical species.

Difficulties in Obtaining Representative Labeling. For a tracer experiment, it is necessary to process the tracer so that it is chemically identical to, and in the same phase as the unlabeled substance. Merely mixing the radiotracer with the analyte sample is generally not sufficient (see Section 8.2.2) [20], [26], [33], [34], [47].

Radiation-induced effects in labeled molecules occur with compounds of high specific activity, labeled with weak β emitters or even α emitters. Since autoradiolysis takes place mainly via secondary radical reactions of fragments formed from solvents or major components, radiolytic damage can be reduced by cooling and by adding a radical scavenger.

Impurities from decay products (i.e., from radioactive decay and from radiation-induced decay of chemical compounds) can interfere with the activity measurements, or with the chemical reactions employed to process the sample. Radionuclide purity is a function of time.

Limited radiochemical purity may result in detection of radioactivity from chemical species different from the intentionally labeled one. This is of prime importance in specifically labeled enantiomers. Labeled compounds should be purified immediately before use. Chromatography and reverse IDA (see Section 8.4.2) are important for checking radiochemical purity.

Isotope effects are important in reactions where tritium is involved in the rate-determining step [48], because it reacts more slowly than hydrogen. Similar effects, but to a much lesser degree, are reported for ^{14}C, ^{32}S, and other nuclides [4], [17], [48].

Errors Due to Measurement Setup. Some sources of error that are specific to radionuclides result from the measuring technique:

1) Errors due to radioactive decay are generally smaller than 0.2 % since half-lifes are accurately known [3], [13], [30], [49].
2) *Statistical uncertainty* results rom counting a limited number N of radioactive decay events. Mostly, there is no problem in obtaining $N \approx 10^4$ with precision at the 1 % level, as long as background is low.
3) *Geometric errors* result from uncertainties in the dimensions of the sample, detector, and the distance between them. They are often difficult to calculate on an absolute scale, and are therefore calibrated empirically. Depending on the calibration precision, on the type of nuclide, sample, and detector, the relative uncertainty can range from 50 to 0.2 %. Calibration on an absolute scale depends on the use of accurate standards, which are rarely available with a certified accuracy better than 2–3 %.
4) *Errors from absorption and backscattering* of radiation are generally corrected empirically as well.
5) *Errors from instrumental sources, quenching, and luminescence* can be kept very low (ca. 0.1 %) by control of temperature and humidity, uniform count rates from samples and standards to suppress differences in dead time and pulse pile-up, and uniform matrix of samples and standards in liquid scintillation counting [37]. Otherwise, errors as high as 40 % [37], [79] can be reduced to a few percent, either by utilizing fast detectors and electronics, or by empirically correcting for pulse pile-up losses, dead time, and spectrum instability [5], [13], [79].

8.3. Radiotracers in Methodological Studies

Radioanalytical chemistry offers a variety of accurate and highly sensitive methods. Even with the increased performance of other trace analysis methods [18]–[21], including electrothermal atomic absorption spectrometry, atomic emission, and fluorescence spectrometry (→ Atomic Spectroscopy), coupling methods, and high-performance sample introduction in organic and inorganic mass spectrometry (→ Mass Spectrometry), chromatographic separations (→ Gas Chromatography; → Liquid Chromatography), and neutron AA (→ Activation Analysis), there are still a number of areas where radionuclide applications are unsurpassed (see Section 8.1.4).

Picogram amounts of chemical species can be traced through a complete chemical process, and problems in process performance are often revealed and overcome rapidly and simply [18], [20], [22], [23], [28], [33], [40], [41], [45], [47], [50], [61], [78], [93], [97].

Important information can be obtained rapidly and cheaply by radioindicators. Specific procedures for quantitatie determinations based on radiotracers (see Chaps. 8.4, 8.5) involve additional principles.

8.3.1. Principles and Importance

Radioactive tracer analysis is based on the simple proportionality between the mass m_x of analyte x and the radioactivity A_1 initially added to the sample:

$$m_x = k \cdot A_1 \tag{3}$$

where k is a proportionality constant, measured for an individual experiment. This is based on the condition that the radioactive substance m_1 added to the sample does not increase the mass of the analyte, i.e., $m_x >> m_1$. Otherwise, for determination of m_x from A_1 the formalism of IDA (Chap. 8.4) is applied, but very often the simple proportionality between mass $(m_x + m_1)$ and A_1 is sufficient for a methodological investigation.

By adding a radioactive tracer before the individual process stage, and following it by radioactivity measurements, the recovery, distribution, and loss are measured independently of any additional risk of contamination. By means of nuclear activation, radioactive tracers can be introduced into solid samples and can be used to determine the recovery and loss during decomposition of solids, and separation from solid samples by volatilization, leaching, and solvent extraction [9], [10], [40], [51]–[64], [67]–[80], [93], [97].

A variety of examples from the general analytical field [2], [7], [17], [25] and from trace element analysis [20], [33], [44], [46] covers separate stages of combined analytical procedures such as decomposition [20], [44], [45], evaporation [81], coprecipitation, adsorption, and washout of precipitates; contamination and adsorption [18], [33], separation by chromatography, ion-exchange, liquid–liquid distribution, and electrodeposition [4], [33], [34], [38], [40], [41], [78].

Radioanalytical indicator methods (including the radioreagent method, see Chap. 8.5) experienced a revival in the 1960s and early 1970s. Modern direct instrumental methods [19] underemphasized the importance of radioanalytical methods, which are necessary to establish the accuracy of spectroscopic methods, to avoid systematic errors, and to assess trace content in SRMs. An important advantage is the ability to reveal accidental losses, even at the ultra-trace level.

The most common areas of analytical applications are:

1) Determination of equilibrium constants, including solubility and complex dissociation constants, phase distribution in solvent extraction, adsorption, ion exchange, coprecipitation; and kinetic data, such as rate constants
2) Tracing the distribution of a chemical species in an analytical system for decomposition, chromatography, evaporation, and separation
3) Determination of recovery, yield, and loss of a chemical species in a process

8.3.2. Control of Sampling

Inaccuracies in sampling procedures occur, owing to contamination, loss, and nonrepresentative sampling. Sampling, pretreatment, storage, etc., all affect the accuracy.

Extended radioindicator studies on trace element sampling (e.g., of Al, Cr, Mn, Fe, Co, Zn, Sr. Cs, As, Hg, Pb) have been reviewed [4], [5], [20], [32]. In environmental analysis (e.g., solids, water, aerosols, biological tissue), adequate sampling and sample treatment is of primary importance to account for variation in biological activity due to the distribution of chemical species between different phases and solid fractions [5], [16], [20], [22], [26], [32], [46]. Sample homogeneity may be a significant limitation to the quality of analytical results. It is of increasing importance with decreasing concentrations from the μg/g to the ng/g level, and it limits the amounts of subsamples and the number of replicates. Studies on sample homogeneity, and on sampling and storage of very dilute solutions, are conveniently performed with radiotracers for numerous trace elements [16], [18], [20], [33], [44], [45].

8.3.3. Control of Contamination and Loss

Contamination generally poses the most important problem in ultra-trace analysis. It causes fluctuations of the blank values, thus defining the lower limits of detection, but also introduces systematic inaccuracies. Tracer techniques can be used to study the sources of contamination: reagent blanks, vessel walls, airborne pollutants, etc.

Contamination cannot be overcome by radiochemical means. Purification procedures for reagents and vessels and a clean working environment are required and SRMs should be used. Tracer techiques are unique tools in its investigation and control (for reviews see [18], [34], [40], [44]), owing to the advantages mentioned in Chapter 8.2 and Section 8.3.1 [18], [44]. Radioindicator studies have been conducted on reagent purification by electrolysis, sublimation, sub-boiling distillation, liquid–liquid distribution, ion exchange, and on vessel cleaning by rinsing, leaching, and steaming.

In general, the same principles are applied to radiochemical investigations of losses by adsorption (i.e., during storage, pretreatment, precipitation, filtration), by volatilization (during decomposition, ashing, storage, digestion, drying), and by chemical reaction (complexation, ion exchange, photochemical and redox reaction) using radioindicators [20], [40], [45].

8.3.4. Separation Procedures

Separation stages and chemical reactions, e.g., for decomposition or phase transformations, are essential components of combined analytical procedures [52], [53]. They should be reproducible, quantitative, selective, and unequivocal.

Separation procedures are based on the principles of volatilization, liquid–liquid distribution, adsorption, diffusion, chromatography, ion exchange, electrophoresis, precipitation, coprecipitation, and electrodeposition. In all of these, radiotracers provide the best tool for methodological investigations, determination of equilibrium constants, kinetic data, and optimization of applied analytical data (yield, interference levels, etc.) [54]. Use of radiotracers in complex multielement separation schemes is reviewed in [4], [17], [20], [41], [54], radiochromatography is reviewed in [55], [61], [93], [97].

8.3.5. Control of the Determination Stage

Radionuclides in methodological studies of determination methods are sometimes useful. These are the measurement processes associated with the determination stage of a combined analytical procedure. Sources of systematic error in atomic absorption spectroscopy, optical emission spectrometry, and electrochemical methods. as well as optimization of the determination procedures have been examined.

Examples are the behavior of trace elements in graphite furnace atomizers [81] and investigations of the double layer structure on analytical electrodes and its exchange reactions with the solution [56].

8.4. Isotope Dilution Analysis

The principle of isotope dilution analysis (IDA) [31], [47], [90], [97], [98] involves measurement of the change in isotopic ratio when portions of a radiolabeled and nonlabeled form of the same chemical species are mixed. To perform a radioisotope IDA, an aliquot of a radioactive spike substance of known specific activity $a_i = A_i/m_i$ is added to the test sample containing an unknown analyte mass m_x. This analyte mass is then calculated from a determination of the specific activity a_x of the resulting mixture after complete homogenization. The important advantage of IDA is that the analyte need not be isolated quantitatively. This is often significant at the trace concentration level, where quantitative separation is not feasible or inconvenient, or if interferences occur, e.g., in the analysis of mixtures of chemically similar compounds. Only a portion of the analyte is separated, so that the separation reagent is not used in excess, and this generally improves selectivity [31], [54], [57]–[59]. Selective determination of *chemical species* in various states (e.g., tri- and pentavalent As, Sb, tri- and tetravalent Ru, Ir) by IDA is favored by the supplementary selectivity gained from the use of substoichiometric amounts of reagent. Losses during purification and other steps are taken into account. A suitable separation procedure is needed to isolate part of the homogenized sample [2], [31].

In more advanced variants of IDA, measurements of masses are substituted by volumetric and complementary activity measurements.

The same principle is also applied to inactive stable isotope IDA. Mass spectrometric determination of isotope ratios then replaces the activity measurements [82] (→ Mass Spectrometry).

8.4.1. Direct Isotope Dilution Analysis

In this most frequently used version of IDA (single IDA), the mass m_x of inactive analyte substance can be determined by using the labeled substance of mass m_1, radioactivity A_1, and the specific activity a_1:

$$a_1 = A_1/m_1 \tag{4}$$

After homogenization with the analyte mass m_x, the specific activity has decreased to:

$$a_2 = \frac{A_1}{m_1 + m_x} = \frac{A_2}{m_2} \tag{5}$$

The total radioactivity of the system is unchanged:

$$(m_1 + m_x)a_2 = m_1 a_1 \tag{6}$$

The unknown mass m_x is calculated from:

$$m_x = m_1\left(\frac{a_1}{a_2} - 1\right) \tag{7}$$

The specific activities a_1 and a_2 are obtained from activities A_1 and A_2 and masses m_1 and m_2 of the fractions isolated from the initial labeled substance (Eq. 4) and from the homogenized solution (Eq. 5).

Equation (7) shows that IDA depends on the relative change in specific activity due to isotope dilution. Low-yield separations may be sufficient. For high precision, a_1 should be high, with mass $m_1 \ll m_x$. Then Equation (7) reduces to

$$m_x \approx m_2 \frac{A_1}{A_2} = \frac{A_1}{a_2} \qquad (8)$$

Equation (8) shows that the accuracy of m_x is limited by the physicochemical determination of m_2.

Example. Determination of glycine in hydrolyzed protein [98]: 152.6 mg hydrolyzed protein homogenized with 5.07 mg ^{14}C-labeled glycine of specific activity $a_1 = (96.2 \pm 1.2)$ counts min^{-1} mg^{-1} (relative activities are sufficient). After separation of a portion of glycine from the mixture, specific activity of $a_2 = (51.3 \pm 0.9)$ counts min^{-1} mg^{-1}. Equation (7) gives:

$$m_x = 5.07\,\mathrm{mg}\left(\frac{96.2}{51.3} - 1\right) = (4.44 \pm 0.17)\,\mathrm{mg}$$

The percentage of glycine in the protein is:

$$\frac{4.44}{152.6} \cdot 100 = (2.91 \pm 0.11)\%$$

8.4.2. Reverse Isotope Dilution Analysis

Another application of IDA is the determination of the activity A_y of a radioactive substance y from a complex mixture of radioactive substances by adding a known mass m_c of an inactive carrier substance. This technique is also based on Equation (7), reversely solved for activity A_1. Homogenization and separation of a portion m_2 (activity A_2, specific activity a_2) gives:

$$A_y = a_2 \left(m_c + m_y\right) \qquad (9)$$

from the specific activity a_y of substance y. If the inactive carrier is used in significant excess (i.e., $m_c >> m_y$) the separation yield is:

$$A_y = A_2 \frac{m_c}{m_2} \qquad (10)$$

To obtain m_y, the specific activity a_y is required: $m_y = A_y/a_y$.

Reverse IDA (indirect IDA, or dilution with inactive isotopes) is particularly important in organic analysis and biochemistry [24], [32] to test for radiochemical purity and stability of labeled compounds. It is often used in radiochemical nuclear AA for the separation of activated element traces from a variety of interfering radionuclides in the analytical sample, for determination of the isolated activity, in comparison with that of a reference sample (i.e., to determine the radiochemical yield according to Eq. 10).

In contrast to direct IDA, reverse IDA is not limited by the sensitivity of the analytical determination of the isolated mass m_2, but merely by the specific activity a_y of the radioactive substance. By adjusting the carrier mass m_c, the sensitivity and accuracy can be increased, so reverse IDA is suitable for trace and microanalysis. Losses by absorption from carrier-free solutions must be taken into account. In separations from mixtures of radioactive substances, high-quality purification is important.

Double (Multiple) IDA. Since the specific activity a_y is often unknown, double IDA can be used to determine m_y. Two equal aliquots containing the same, but unknown mass m_y are diluted with different known carrier masses, m_c and m'_c. After separation of portions m_2, m'_2 and measurement of specific activities a_2, a'_2 one obtains from Equation (9):

$$a_2 = \frac{A_y}{m_c + m_y} = \frac{a_y m_y}{m_c + m_y} \qquad (11)$$

$$a'_2 = \frac{a_y m_y}{m'_c + m_y} \qquad (12)$$

Thus:

$$m_y = \frac{a'_2 m'_c - a_2 m_c}{a_2 - a'_2} \qquad (13)$$

$$a_y = \frac{(m_c - m'_c) a_2 a'_2}{a_2 m_c - a'_2 m'_c} \qquad (14)$$

Multiple (direct and reverse) IDA produces a calibration graph $m_x = f(1/a_2)$.

8.4.3. Derivative Isotope Dilution Analysis

If the preparation of labeled substances is too difficult, derivative IDA may be applied. This technique combines a radioreagent and IDA, preferably of low selectivity, to react with one or more analytes. It is mainly used with mixtures of complex organic compounds. The basic stages are:

1) Reaction of analyte A with radioreagent B*, of known specific activity, to form a radioactive product AB*. Excess B* may be removed in an optional purification step:

$$A + B^* \longrightarrow AB^* + B^*_{ex} \quad (15)$$

2) The mass of AB* is determined by mixture with a known mass of inactive AB by reverse IDA:

$$AB^* + B^*_{ex} + AB \longrightarrow (AB^* + AB) + B^*_{ex} \quad (16)$$

3) Separation of diluted product (AB* + AB) and determination of its specific activity.

The principal requirements for this technique are:

1) Reaction (Eq. 15) should be quantitative or of known yield.
2) Isotope exchange between substances AB and B* must be prevented. Since inorganic substances are quite amenable to such exchange, application of derivative IDA to inorganic analytes is impossible.
3) Purification from excess reagent must be quantitative to avoid significant bias, even from minor impurities.

As a variant, dilution with A* is applied prior to the first reaction (Eq. 15) with radioreagent B*.

Determination of the reaction yield and analysis of inorganic substances are possible with this variant technique [2]. Its main importance is in the determination of trace amounts in complex mixtures. It is routinely used in biochemistry and physiological chemistry [2], [31], [32], [75]. The simpler approaches of radioimmunoassays are generally preferred.

Further variants are based on combinations with double derivative IDA, or addition of an inactive derivative prior to direct IDA.

8.4.4. Substoichiometric Isotope Dilution Analysis

Utilization of conventional IDA for trace determinations is limited by the necessity to isolate macroscopic amounts of the analyte substance for determination of m_1 and m_2 by physicochemical methods. The principle of substoichiometry developed by Růžička and Starý [57] avoids this limitation.

Exactly equal (low) masses m_1 and m_2 are isolated from the spike solution of initial specific activity $a_1 = A_1/m_1$ and from the solution after homogenization with the analyte ($a_2 = A_2/m_2$). Equation (7) reduces to:

$$m_x = m_1 \left(\frac{A_1}{A_2} - 1 \right) \quad (17)$$

The masses m_1 and m_2 are isolated using the same amount of the separation reagent, which must be stoichiometrically insufficient and consumed quantitatively (or to exactly the same amount) in this reaction. It must form a product which can easily be separated from the excess of unreacted substance. Separation reagents should have high chemical stability, no tendency to adsorption at very low concentrations, and sufficient selectivity for the analyte. Separations with enrichment factors of $10^4 - 10^5$ are typical. Since the mass m_1 of the added radioactive substance must be accurately known, reverse IDA can be applied.

Error propagation is comprehensively discussed by Tölgyessy and Kyrš [2]. At trace levels below 1 µg, standard deviations of ca. 3 % and accuracies of ca. 6 % are typical; precision and accuracy can be 1.5 % or better at higher concentrations.

8.4.4.1. Substoichiometric Separation by Liquid – Liquid Distribution

The extraction of metal chelates is suitable for separating equal masses of metals from solutions of different concentrations. Extraction of a metal chelate MA from a metal M^{n+} by an organic reagent HA is generally described by [47], [57], [58]:

$$M^{n+} + n\,(HA)_{org} \rightleftharpoons (MA_n)_{org} + nH^+ \quad (18)$$

with the extraction constant K:

$$K = \frac{[MA_n]_{org}\,[H]^n}{[M^{n+}]\,[HA]^n_{org}} \quad (19)$$

where subscript org denotes the organic phase. In trace element analysis by substoichiometric IDA, concentrations of $10^{-6} - 10^{-9}$ mol/L are typically extracted from aqueous samples of volumes V in the range 10 mL to 100 µL into organic phases of smaller volumes (typically one-tenth of the sample volume) of $10^{-5}-10^{-8}$ mol/L solutions of chelating reagent.

The dependence of Equation (18) on the pH value of the solution makes it necessary to keep within an optimum pH range. The substoichiometric principle requires a higher than 99 % consumption of the initial concentration c_{iHA} in Equation (18), so that $[HA]_{org} < 0.01c_{iHA}$. Thus the solution must satisfy the condition:

$$pH > -\log(0.01c_{iHA}) - \frac{1}{n}\log K \quad (20)$$

With increasing pH, dissociation of acid HA (dissociation constant K_{HA}) in the aqueous phase increases. This dissociation is negligible at:

$$pH < pK_{HA} + \log q_{HA} + \log\frac{V_{org}}{V} \quad (21)$$

where q_{HA} is the distribution coefficient of the reagent, V is the sample volume, and V_{org} is the volume of the organic phase. Hydrolysis of the metal at increasing pH must not interfere with the separation.

The reagent (e.g., diphenylthiocarbazone, diethyldithiocarbamic acid (2-thenoxy)-3,3,3-trifluoroacetone, 8-hydroxyquinoline) must be stable to decomposition, even at very low concentrations in acid and neutral media, and must form an extractable chelate with a sufficiently high value of K (for further useful ligands see [38], [40], [58], [83]). Optimum conditions for substoichiometric extraction are calculated from K, K_{HA}, and q_{HA}. Selectivity of the separation can be improved by use of additional masking reagents [54], [57].

Substoichiometric extraction of ion-association complexes is generally restricted to higher concentration levels (e.g., determination of Na, P). Water-soluble complexes of more than 20 metals with ethylenediaminotetraacetic acid can be used for substoichiometric determination at concentrations below 1 ng/g, provided that the excess of unreacted metal is separated by liquid–liquid distribution, ion-exchange, carrier coprecipitation, or other suitable means. Selectivity is high and can be further increased by masking reagents.

8.4.4.2. Redox Substoichiometry

Oxidation or reduction by a substoichiometric mass of reagent is followed by separation, extraction, coprecipitation, etc. [73]. $KMnO_4$, $K_2Cr_2O_7$, and $FeSO_4$ have been used as substoichiometric redox reagents.

8.4.4.3. Displacement Substoichiometry

Complete extraction of an analyte metal M_1 with excess complexing reagent is followed by removal of the excess and subsequent displacement of M_1 by a substoichiometric amount of another metal M_2, provided that M_2 has a sufficiently higher extraction constant K_2. Further variants are discussed elsewhere [2].

8.4.4.4. Applications

Applications include trace element determination in rocks, soils, biogenic samples, and pure metals by direct IDA, i.e., added radiotracer [2], [54], [60]. Another field of increasing importance is in radiochemical AA for assessment and improvement of accuracy at trace and ultra-trace levels by determination of the radiochemical yield. This applies primarily to biomedical samples, high-purity materials, and environmental materials 7]–[9], [32], [84].

8.4.5. Sub- and Superequivalence Method

The sub- and superequivalence method is based on the isoconcentration principle. Three versions of this method exist: the direct method for the determination of nonradioactive substances [99], the reverse method for the determination of radioactive substances [100], and the universal isotope dilution method for determination of both [101].

Historically, the *reverse method* has been developed first [100]. In the reverse method, two series of aliquots are formed from a solution of the radioactive (labeled) analyte. The first series is formed by equal volumes of the analyte solution [containing x = (x,..,x,...,x) amounts of labeled analyte] while the (identical) aliquots of the second series contain ξ times higher volumes of the same solution (x = ξx,..., ξx,...., ξx). The members of the first series are isotopically diluted by successively adding fixed amounts of the same substance in inactive form $y=(y_1,\ldots,y_\xi,\ldots,y_a)$, and so the members of the first series contain $(x+y) = (x+y_1, \ldots, x+y_\xi,\ldots, x+y_a)$ amounts of the substance. After establishing the same separation conditions for all aliquots of both series, e.g., by adding masking and buffer solutions, and after dilution to the same volume, a fixed amount of reagent is added to all aliquots. Then the separation is performed at the same temperature and pressure.

The added amount of reagent is fixed well in the aliquot containing the highest total concentration of analyte (substoichiometric regime), but it is in excess in the aliquot containing the lowest total amount of substance. The ratio of the activities of the isolated fractions from the second series (which are theoretically identical) and the activities of the isolated fractions of particular members of the first series are plotted as a function of the dilution increment and the mass of analyte is determined from this plot [2], [73], [89].

In the *direct method* two series of aliquots are formed from a solution of the labeled counterpart of the substance to be analyzed (which is not radioactive). The first series is formed by increasing volumes of this solution containing ξy_1, ξy_2, $\ldots\xi y_a$ amounts of the labeled substance. The second series is formed by increasing, but ξ times lower amounts of the same solution. To the aliquots of the second series constant amounts of the substance to be analyzed (in nonradioactive form) are added. After establishing the same separation conditions for all aliquots of both series, the separation is performed. The evaluation of the analytical results is identical to that in the reverse method.

The *universal method* (two partially different procedures) is (are) the combination of both methods.

Besides a large variety of metals, several chemical species of specific interest, e.g., cyclic 3′,5-adenosine monophosphate, thyroxine and 3,5,3′-trioiodothyronine have been determined by sub- and superequivalence IDA and the results show higher accuracy of the method over that of radioimmunoassay [102], [103]. The principle of the method also allows the determination of certain physicochemical characteristics, e.g., estimation of the stability constants of complexes and the solubility of precipitates.

The method is a calibration curve approach. It does not require that all the separated amounts of the aliquots mentioned are exactly the same, as in substoichiometric IDA (Section 8.4.4). This is the reason why separation reactions can be used that are not quantitative, a sufficient prerequisite is reproducibility [2], [24], [74].

An alternative calibration curve approach for nonquantitative reactions is to determine the distribution coefficient by a concentration-dependent distribution method [2], [75], which is a variant of the radioreagent method (Section 8.5.1).

8.5. Radioreagent Methods

Radioreagent methods (RRM) are based on the use of a radioactive species in a quantitative reaction, and measurement of the change in activity of that species in the course of the reaction. The radioactive species may be a labeled reagent, the analyte, or a substance able to undergo an exchange reaction with some compound of the analyte.

After separation from excess reagent (by liquid-liquid distribution, chromatography, precipitation, etc.), the mass or concentration of this product is determined from activity measurement. The determination is based on a radioactive substance chemically different from the analyte substance (in contrast to IDA), therefore the chemical reaction is of prime importance. By variation of this key reaction, the principle can be adapted to various procedures. The superiority of radioreagent methods over classical separation techniques arises from the use of an inactive carrier and the high sensitivity of the activity measurements, which are not subject to interference by the carrier or other substances [1], [16], [24], [34], [60], [87], [92], [94], [95], [97].

Depending on the type of chemical interaction and relationship of the analyte to the radioactive substance measured, RRM can be divided into three basic group:

1) The radioactive substance is a typical reagent which is able to react with the compound to be determined.
2) The method is based on "nonequivalent" competition; i.e., the radioactive substance is able to exchange with the analyte.
3) The method is based on "equivalent" competition; i.e., the labeled substance is chemically identical with the substance to be determined.

From the practical point of view, it is convenient to divide RRM into the following groups:

1) Simple RRM in which the reactions take place quantitatively, i.e., either the analyte or the reagent is completely consumed and compounds of definite composition are formed.
2) Methods of concentration-dependent distribution utilizing reactions where products of unstable composition are formed, but in which the extent of reaction is determined by the corresponding equilibrium constant or reaction time.

Table 3. Useful concentration regime of the radioreagent method

Analyte	Radionuclide	Labeled reagent	Separation	Concentration regime
H^+	^{131}I	KI	extr.I_2 into pyridine/chloroform	> 0.1 µg/mL
H_3BO_3	^{18}F	HF	extr.HBF_4 into 1,2-dichloroethane[b]	≈0.1 µg B
NaLS[c]	^{59}Fe	$[Fe(II)(phen)_3]$[c]	extr.$Fe(phen)_3(LS)_2$ into $ChCl_3$	5 – 300 ng/mL Na
Cl^-, Br^-, I^-	^{203}Hg	$C_6H_5Hg^+$	extr.C_6H_5HgCl into benzene	0.5 – 15 mg/mL
Cl^- .I^-	^{110m}Ag	$AgNO_3$	pptn.AgCl; AgI	0.4 – 250 µg/mL
Cl^-	^{203}Hg	$HgNO_3$	pptn.Hg_2Cl_2	0.8 – 13 µg/mL
Bi	^{131}I	KI[d]	extr.$HBiI_4$ into *n*-butylacetate	0.04 – 4 µg
Cationic detergents	^{131}I	Rose Bengal	extr. ion-associate into $CHCl_3$	

[a] Extr. = solvent extraction; pptn. = precipitation. [b] In presence of methylene blue. [c] LS = lauryl sulfate; phen = 1,10 phenanthroline. [d] In presence of ascorbic acid + Na_2SO_3 in 0.6 mol/L H_2SO_4.

3) Isotope exchange methods based on the exchange of isotopes between two different compounds of one element, with a radioactive isotope in one of the compounds and a nonradioactive isotope in the other, usually the analyte.
4) Radiorelease methods comprising procedures in which the analyte reacts with a radioactive reagent, thus releasing an aliquot of the reagent activity, in most cases into the gas phase.
5) Radiometric titration usually involves use of radioactive reagents to determine the equivalence point, but there are variants that do not use a radioreagent; instead they are based on a change in the intensity of radiation caused by its absorption or scattering in a medium containing the analyte [2], [32], [87], [92], [95], [96], [97].

8.5.1. Simple Radioreagent Methods

Simple radioreagent methods (SRRM) utilize a reaction with a suitable reagent, and sometimes with a third substance; one of the reagents is radioactive and the reaction is quantitative [2], [32], [87], [95]. SRRM can be classified into the following three groups:

1) Determination with labeled reagents,
2) Determination with labeled analyte,
3) Determination with labeled competing substance

8.5.1.1. Determination with Labeled Reagents

An excess of radioactive reagent solution of known analytical concentration is usually used. The active product is separated from the excess of radioactive reagents by precipitation, formation of extractable chelates, or sorption.

The substance to be determined (n_x mol) reacts with a radioactive reagent (n_R mol), forming a precipitate, extractable compound, or other separable substance. After the separation of the product, the radioactivity of the product (A_P), that of the excess of unreacted reagent (A_E), or both are measured. The following relationship is valid:

$$\frac{A_R}{n_R} = \frac{A_P}{z\,n_x} = \frac{A_E}{n_{R-z\,n_x}}$$

Thus

$$n_x = \frac{n_R\,A_P}{z\,A_R} = \frac{n_R}{z}\left(1 - \frac{A_E}{A_R}\right)$$

where A_R is the radioactivity of the reagent and z denotes the stoichiometric ratio in the compound formed (the number of moles of the reagent interacting with 1 mol of the test substance). If n is not exactly known, a calibration curve approach can be used. Quantitative isolation of trace amounts of the reaction products poses another important problem (Table 3, [2], [41], [47], [60]).

In this way, the concentration of chloride in water was determined by precipitation with silver labeled with ^{110m}Ag ;^{14}CO was used for the determination of hemoglobin in blood; and ^{131}I for the determination of bismuth in the presence of an excess of iodide ions using extraction separation [2].

8.5.1.2. Determination with Labeled Analyte

The first step involves labeling of the analyte. To this end, the radioactive indicator is added to the unknown sample and isotope exchange is allowed to take place; then the reagent is added in a substoichiometric but known quantity. After the reaction is complete, the phases are separated and their respective activities determined. In those cases where the activity of the equilibrium solution is measured, the unknown concentration can be calculated from:

$$x + x_0 = \frac{A_{\text{init}} B}{(A_{\text{init}} - A_{\text{equilib}})}$$

where x is the quantity of analyte (equivalents), x_o the amount of added radioactive indicator (equivalents), B the amount of unlabeled reagent (equivalents), and A_{init} and A_{equilib} represent the total activity of the initial and equilibrium solutions, respectively. In this variant of the RRM, solvent extraction is frequently used.

8.5.1.3. Determination with Labeled Competing Substances

Analyte M_2 competes with the radioactively labeled substance $^*M_1$ in the formation of a compound with a reagent R, where the compound precipitates from the solution

$$^*M_1R\downarrow + M_2 \rightleftharpoons M_2R\downarrow + ^*M_1$$

$^*M_1R$ and M_2R occur in a different phase than either substance M_1 or M_2. The equilibrium constant of the reaction, as a rule is >1 so that a significant change takes place.

This method is used when the labeling of reagent R cannot be applied or when the reagent and its compound with the analyte are transferred into the same phase. Generally, a relatively soluble precipitate with one radioactively labeled component (e.g., $^{45}CaCO_3$) is brought into contact with a solution containing the analyte (Pb^{2+}), which displaces the radioactive component ($^{45}Ca^{2+}$) from the precipitate into the solution, and a less soluble precipitate ($PbCO_3$) is formed. The resulting radioactivity of the solution is proportional to the initial amount of lead.

8.5.2. Method of Concentration-Dependent Distribution

Concentration-dependent distribution (CDD) is based on the utilization of a calibration graph that shows the dependence of the distribution ratio of the substance to be determined in a two-phase system on the total concentration of the substance [2], [32], [87], [92], [95], [96]. Characteristic features of the method of CDD are as follows:

- The analysis is based on the distribution of a radioactive substance between two phases or between several parts of the systems used (paper chromatography, electrophoresis),
- The ratio of the activities in both phases or parts of the system depends strongly on the initial concentration of the unknown substance,
- The distribution is not given by the stoichiometric ratio of the reacting substances, i.e., not by total saturation of one phase or a nearly complete consumption of the reagent, but is given exclusively by the corresponding equilibrium or kinetic constants.

Example. Determination of chloride in water samples. The sample is shaken with an excess of labeled AgCl precipitate and the radioactivity of silver is measured in the equilibrium solution. The higher the chloride concentration the lower is the concentration of silver in solution (owing to the constancy of the AgCl solubility product). The Cl^- ions determined do not react with the AgCl precipitate, they only suppress its dissolution. The amount of chloride in the precipitate has no direct relationship to the concentration of Cl^- ions in the sample solution.

In CCD is useful to define two extreme types of determination: *saturation analysis* and *nonsaturation analysis*.

In *saturation analysis*, the cause of the change in the distribution of the radioactive substance between the phases is an increase in the total saturation of the reagent by the given substance. This term originally appeared in the literature in connection with the determination of biochemically important substances (steroids, hormones, vitamins) and is equivalent to the term radioimmunoassay.

In *nonsaturation analysis*, the cause of the change in the distribution coefficient is the shift of the chemical equilibrium due to an increase in the concentration of the substance to be determined, irrespective of the extent of saturation. The determination of some extractants can serve as a typical example. The number of procedures described that use nonsaturation analysis is limited.

8.5.3. Isotope Exchange Methods

Isotope exchange methods (IEM) are based on the exchange of isotopes between two different compounds of the element X, one of which (XA) is nonradioactive, the other (XB) being labeled with a radioactive isotope [2], [61], [87], [97]. After isotopic equilibrium is reached

$$XA + {}^{*}XB \rightleftharpoons {}^{*}XA + XB$$

the specific activities of the element X in both compounds are equal

$$A_1/m = A_2/m_x$$

where A_1 and A_2 are the equilibrium radioactivities of XB and XA, respectively, and m_x and m are the amounts of X in XA and XB, respectively. The value of m_x can be computed from:

$$m_x = mA_2/A_1$$

or from the calibration graph of $m_x = f(A_2/A_1)$, $m_x = f(A_2)$ or $m_x = f(A_1)$.

Isotopic exchange can be carried out in either a heterogeneous or a homogeneous system.

Example. Methylmercury and phenylmercury are determined down to 10 pg/mL by liquid-liquid extraction from 3 mol/L aqueous HCl into benzene, and addition of $K^{131}I$ to the separated organic phase. Inactive chloride is completely displaced by ^{131}I:

$$(CH_3HgCl)_{org} + {}^{131}I^- \longrightarrow (CH_3Hg^{131}I)_{org} + Cl^-$$

The method has been used for determination of methylmercury species in fish and in drinking water [61].

8.5.4. Radioimmunoassay

Radioimmunoassay is the widely used radioreagent method.

8.5.5. Radiorelease Methods

In radiorelease methods the analyte substance A reacts with a radioactive reagent, so that radioactive R^* is released into a second phase, without being replaced by an inactive analyte [62]. Applications involve either release of radioreagents from solids or liquids into the gas phase, or release from solids into a liquid. Radiorelease methods may be classified according to the type of the radioactive reagents employed, i.e. (1) radioactive kryptonates, (2) radioactive metals, and (3) radioactive salts and other substances.

8.5.5.1. Radioactive Kryptonates [2], [87], [92], [94], [95]

The term radioactive kryptonates is used for substances into which atoms or ions of krypton-85 (^{85}Kr) are incorporated (by diffusion of ^{85}Kr; by bombardment with accelerated krypton ions; by crystallization of the kryptonated substances from a melt, or by placing the solution in an atmosphere of ^{85}Kr, etc.). The radionuclide can be released from the solid lattice by any chemical or physical reaction that breaks down the lattice at the solid surface. The released inert gas is conveniently measured from its main β radiation (0.67 MeV). The half life of 10.27 a is suitable for long-term remote sensing devices.

Applications of Radioactive Kryptonates [32], [92], [94]–[96]. The determination of *oxygen* is performed by surface oxidation of copper or pyrographite kryptonate, at elevated temperature, resulting in destruction of the surface layer and release of ^{85}Kr proportional to the oxygen mass. Detection limits are at the 10 ng/m^3 level. Ozone oxidizes copper kryptonate at temperatures below 100 °C, whereas reaction with oxygen starts well above 200 °C, so this detector can detect O_3 and O_2 differentially. Determination of *ozone* in air is feasible over a concentration range of $10^{-7} - 10^{-3}$ g/m^3 with hydroquinone kryptonate:

$$[C_6H_4(OH)_2]_3[^{85}Kr] + O_3 \longrightarrow 3\,C_6H_4O_2 + 3\,H_2O + {}^{85}Kr\uparrow$$

Sulfur dioxide has also been determined by a method based on the mechanism of double release. In the first stage, sulfur dioxide reacts with sodium chlorate to release chlorine dioxide which is a strong oxidizing agent. The chlorine dioxide then oxidizes radioactive hydroquinone kryptonate and gaseous ^{85}Kr is released. The following reactions are invoved:

$$SO_2 + 2\,NaClO_3 \longrightarrow 2\,ClO_2 + Na_2SO_4$$

$$ClO_2 + [C_6H_4(OH)_2]_3[^{85}Kr] \longrightarrow {}^{85}Kr\uparrow$$

A radioactive kryptonate of silica has been suggested for the determination of *hydrogen fluoride* in air. *Oxygen dissolved in water* can be measured by the use of thallium kryptonate

$$4\,Tl[^{85}Kr] + O_2 + 2\,H_2O \longrightarrow Tl^+ + 4\,OH^- + {}^{85}Kr\uparrow$$

Thallium is oxidized by oxygen, and the amount of ^{85}Kr released or the decrease in the activity of kryptonated thallium is proportional to the dissolved oxygen concentration in the sample (at constant pH) down to 0.3 μg/mL.

8.5.5.2. Radioactive Metals

Oxidizing agents in solution can react at the surface of a labeled metal, releasing radioactive ions into the solutions which are used for determination of the oxidants. The decisive factors in the choice of the metal are the following: the metal should not react with water, but with oxidizing agents to yield ions which do not form precipitates in aqueous media. The metal should have a radionuclide with suitable nuclear properties. These conditions are met by thallium (^{204}Tl) and silver (^{110m}Ag) [54], [60], [61].

Dissolved *oxygen* has been determined in seawater, drinking water, and wastewater down to the ng/g level with metallic thallium labeled with ^{204}Tl. Selectivity requires removal or masking of other oxidizing agents. The *vanadate* ion is assayed by acidifying the corresponding sample (pH ≈ 3) and passing it over a column containing radioactive metallic ^{110m}Ag. The labeled silver is oxidized, dissolved, eluted from the column, and detected. The measurement of *dichromate* ion concentration in natural waters can be carried out similarly.

8.5.5.3. Radioactive Salts and Other Radioactive Substances [2], [32], [92], [96]

Sulfur dioxid is determined on the basis of the reaction:

$$5\,SO_2 + 2\,K^{131}IO_3 + 4\,H_2O \longrightarrow K_2SO_4 + 4\,H_2SO_4 + {}^{131}I_2\uparrow$$

This reaction takes place in an alkaline solution through which air containing sulfur dioxide is bubbled. After completion of the reaction, the solution is acidified and the iodine released is extracted. *Active hydrogen* in organic substances may be determined by reaction with lithium aluminum hydride labeled with tritium (^3H). The activity of released tritium is measured using a proportional counter.

A number of determinations are based on the formation of a soluble complex between the analyte in solution with a radioactively labeled precipitate. In this way it is possible to determine anions forming soluble complexes (e.g., CN^-, $S_2O_3^-$, I^-, F^-). The principle of the determination are given in the following equations:

$$2\,CN^- + {}^{203}Hg(IO_3)_2\downarrow \longrightarrow {}^{203}Hg(CN)_2 + 2\,IO_3^-$$
(> 0.50 µg/mL, ± 55 %)

$$2\,Cl^- + {}^{203}Hg(IO_3)_2\downarrow \longrightarrow {}^{203}HgCl_2 + 2\,IO_3^-$$
(> 1 µg/mL, ± 5 %)

$$n\,S_2O_3^{2-} + m\,{}^{110m}AgSCN\downarrow \longrightarrow {}^{110m}Ag_m(S_2O_3)_2^{(2n-m)-} + m\,SCN^-$$

$$2\,I^- + {}^{203}HgI_2\downarrow \longrightarrow {}^{203}HgI_4^{2-}$$
(> 25 µg/mL, ± 5 %)

8.5.6. Radiometric Titration

Radiometric titrations follow the relation between the radioactivity of one component or phase of the solution under analysis and the volume of added titrant. The compound formed during the titration must be easily separable from the excess of unreacted ions. This separation is directly ensured only in the case of precipitation reactions. In other types of reactions, the separation can be accomplished using an additional procedure. The endpoint is determined from the change in the activity of the residual solution or of the other phase.

According to the type of chemical reaction used, methods based on the formation of precipitates and methods based on complex formation can be distinguished. Because of the necessity for handling precipitates, precipitation radiometric titrations are difficult to apply to less than milligram amounts and, therefore, have no special advantages over other volumetric methods. The sensitivity of complexometric titrations is limited by the sensitivity of the determination of the endpoint. However, the use of radiometric detection can substantially increase the sensitivity of this type of determination. For the separation of the product from the initial component, liquid-liquid distribution, ion-exchange, electrophoresis, or paper chromatography are most often used [2], [63], [88], [93], [97].

This application of radiometric titrations has declined over the past three decades [64]. Their main advantage is where classical methods for detection of the endpoint are either impossible or subject to interference from the titration medium.

Radiopolarography offers highly increased sensitivity and selectivity over polarographic current measurement, without interference from major components of the solution. It measures the amount of labeled ions deposited in single drops in a dropping mercury electrode as a function of potential [85].

8.6. References

General References

[1] IUPAC Commission on Radiochemistry, Nuclear Techniques, Anal. Chem. Div., IUPAC, *Pure Appl. Chem.* **63** (1991) 1269–1306.
[2] J. Tölgyessy, M. Kyrš: *Radioanalytical Chemistry,* **vol. 1,** Ellis Horwood, Chichester 1989, p. 354 ff.
[3] J. Tölgyessy, M. Kyrš: *Radioanalytical Chemistry,* **vol. 2,** Ellis Horwood, Chichester 1989, p. 498 ff.
[4] H. A. Das, A. Faanhof, H. A. van der Sloot: "Radioanalysis in Geochemistry," *Dev. Geochem.* **5** (1989) 482 pp.
[5] P. Bode, H. T. Wolterbeek, *J. Trace Microprobe Tech.* **81** (1990) 121–138.
[6] D. De Soete, R. Gijbels, J. Hoste: *Neutron Activation Analysis,* Wiley-Interscience, London 1972.
[7] W. D. Ehmann, D. E. Vance, *CRC Crit. Rev. in Anal. Chem.* **20** (1989) 405–443.
[8] N. M. Spyrou, *J. Trace Microprobe Tech.* **6** (1988) 603–619.
[9] J. Hoste, F. De Corte, W. Maenhaut in P. J. Elving, V. Krivan, E. Kolthoff (eds.): *Treatise on Analytical Chemistry,* 2nd ed., **part I, vol. 14,** J. Wiley & Sons, New York 1986, p. 645.
[10] V. Krivan in W. Fresenius et al. (eds.): *Analytiker Taschenbuch,* **vol. 5,** Springer-Verlag, Berlin 1985, p. 36.
[11] Y. Suzuki, *Radioisotopes* **36** (1987) 414–421.
[12] C. G. Clayton, M. R. Wormald, J. S. Schweitzer in IAEA (ed.): *Proc. Int. Sympos. Nucl. Techn. Explor. Exploit, Energy Miner. Resourc.,* IAEA, Vienna 1991, p. 107.
[13] G. F. Knoll: *Radiation Detection and Measurement,* 2nd ed., J. Wiley, New York 1989, p. 754.
[14] M. F. L'Annunziata: *Radionuclide Tracers: Their Detection and Measurement,* Academic Press, New York 1987, p. 505.
[15] L. I. Wiebe, *Radiat. Phys. Chem.* **24** (1984) 365–372.
[16] H. Filthuth, *Chem. Anal. (N.Y.)* **108** (1990) 167–183.
[17] H. J. M. Bowen: "Nuclear Activation and Radioisotopic Methods of Analysis," in P. J. Elving, V. Krivan, E. Kolthoff (eds.): *Treatise on Analytical Chemistry,* 2nd ed., part I, Sect. K, vol. 14, J. Wiley & Sons, New York 1986, p. 193.
[18] G. Tölg, *Analyst (London)* **112** (1987) 365–376.
[19] J. A. C. Broekaert, G. Tölg, *Fresenius' Z. Anal. Chem.* **326** (1987) 495–509.
[20] V. Krivan: "Activation and Radioisotopic Methods of Analysis," in P. J. Elving, V. Krivan, I. M. Kolthoff (eds.): *Treatise on Analytical Chemistry,* part I, Sect. K, 2nd ed., vol. 14, J. Wiley & Sons, New York 1986, p. 340.
[21] M. Linscheid, *Nachr. Chem. Tech. Lab.* **39** (1991) 132–137.
[22] W. R. Wolf, M. Stoeppler: "Proc. 4th Int. Symp. Biol. Environm. Reference Mater.," *Fresenius' J. Anal. Chem.* **338** (1990) 359–581.
[23] L. L. Jackson et al., *Anal. Chem.* **63** (1991) 33 R–48 R.
[24] J. W. Ferkany, *Life Sci.* **41** (1987) 881–884.
[25] W. D. Ehmann, J. D. Robertson, St. W. Yates, *Anal. Chem.* **62** (1990) 50 R–70 R;**64** (1992) 1 R–22 R.
[26] R. Naeumann, E. Steinnes, V. P. Guinn, *J. radioanal. Nucl. Chem.* **168** (1993) 61–68.
[27] R. R. Greenberg, *J. Radioanal. Nucl. Chem.* **113** (1987) 233–247.
[28] V. P. Guinn, H. S. Hsia, N. L. Turglio, *J. Radioanal. Nucl. Chem.* **123** (1988) 249–257.
[29] L. Grodzins, *Nucl. Instrum. Methods Phys. Res. Sect. B* **B 56–B 57** (1991) 829–833.
[30] F. De Corte, A. Simonits, *J. Radioanal. Nucl. Chem.* **133** (1989) 43–130.
[31] J. Tölgyessy, T. Braun, M. Kyrš: *Isotope Dilution Analysis,* Pergamon Press, Oxford 1972.
[32] J. Tölgyessy, E. H. Klehr: *Nuclear Environmental Chemical Analysis,* Ellis Horwood, Chichester 1987.
[33] V. Krivan, *Sci. Total Environ.* **64** (1987) 21–40.
[34] J. R. Jones (ed.): *Isotopes – Essential Chemistry and Applications,* 2nd ed., Royal Society of Chemistry, CRC Press, Boca Raton 1988, p. 270.
[35] G. R. Choppin, J. Rydberg: *Nuclear Chemistry: Theory and Applications,* Pergamon Press, Oxford 1980, p. 75.
[36] J. Mandel in I. M. Kolthoff, P. J. Elving (eds.): *Treatise on Analytical Chemistry,* **part I, vol. 1,** J. Wiley & Sons, New York 1978, p. 243.
[37] C. T. Peng: *Sample Preparation for Liquid Scintillation Counting,* Amersham International, United Kingdom, 1977.
[38] K. Robards, P. Starr, E. Patsalides, *Analyst (London)* **116** (1991) 1247–1273.
[39] A. W. Rogers: *Techniques of Autoradiography,* 3rd ed., Elsevier, Amsterdam 1979.
[40] A. Mizuike: *Enrichment Techniques for Inorganic Trace Analysis,* Springer-Verlag, Berlin 1983, p. 144.
[41] A. Dyer, *Analyst (London)* **114** (1989) 265–267.
[42] V. Krivan, *Talanta* **29** (1982) 1041–1050.
[43] Yu. A. Zolotow, M. Grasserbauer, *Pure Appl. Chem.* **57** (1985) 1133–1152.
[44] J. Versieck, R. Cornelis: *Trace Elements in Human Plasma or Serum,* CRC Press, Boca Raton 1989, p. 23.
[45] P. Tschöpel, G. Tölg, *J. Trace Microprobe Tech.* **1** (1982) 1–77.
[46] R. Cornelis, *Mikrochim. Acta* 1991 (1991) no. III, 37–44.
[47] *Ullmann,* 4th ed., **5,** 709.
[48] L. Melander, W. H. Saunders: *Reaction Rates of Isotopic Molecules,* Wiley-Interscience, New York 1980.
[49] K. H. Lieser: "Activation and Radioisotopic Methods of Analysis," in P. J. Elving, V. Krivan, I. M. Kolthoff (eds.): *Treatise on Analytical Chemistry,* part I, Sect. K, 2nd ed., vol. 14, J. Wiley & Sons, New York 1986, p. 1.
[50] W. Schmid, V. Krivan, *Anal. Chem.* **57** (1985) 30–34.
[51] R. Cornelis, J. Versieck: "Activation and Radioisotopic Methods of Analysis," in P. J. Elving, V. Krivan, I. M. Kolthoff (eds.): *Treatise on Analytical Chemistry,* part I, Sect. K, 2nd ed., vol. 14, J. Wiley & Sons, New York 1986, p. 665.
[52] *Ullmann,* 4th ed., **5,** 1.
[53] G. Tölg et al., *Pure Appl. Chem.* **60** (1988) 1417–1424.
[54] J. Starý, J. Ružička in G. Svehla (ed.): *Wilson and Wilson's Comprehensive Analytical Chemistry,* vol. 7, Elsevier, Amsterdam 1976.
[55] D. M. Wieland, M. C. Tobes, T. J. Mangner: *Analytical and Chromatographic Techniques in Radiopharmaceutical Chemistry,* Springer Verlag, New York 1986, p. 300.
[56] B. B. Damaskin, O. A. Petrii, V. V. Batrakov in R. Parsons (ed.): *Adsorption of Organic Compounds on Electrodes,* Plenum Press, New York 1971.
[57] J. Ružička, J. Starý: *Substoichiometry in Radiochemical Analysis,* Pergamon Press, Oxford 1968.
[58] F. Umland, A. Janssen, D. Thierig, G. Wünsch: *Theorie und praktische Anwendung von Komplexbildnern,*

Methoden der Analyse in der Chemie, vol. 9, Akademische Verlagsgesellschaft, Frankfurt 1971, p. 759.
[59] H. Yoshioka, T. Kambara, *Talanta* **31** (1984) 509–513.
[60] J. Starý: "Activation and Radioisotopic Methods of Analysis," in P. J. Elving, V. Krivan, I. M. Kolthoff (eds.): *Treatise on Analytical Chemistry,* part I, Sect. K, 2nd ed., vol. 14, J. Wiley & Sons, New York 1986, p. 241.
[61] J. Tölgyessy, S. Varga: *Nuclear Analytical Chemistry,* vol. 1–3, University Park Press, London 1972.
[62] V. Balek, J. Tölgyessy: *Emanation Thermal Analysis and Other Radiometric Emanation Methods,* Elsevier, Amsterdam 1984.
[63] T. Braun, J. Tölgyessy: *Radiometrische Titrationen,* Hirzel Verlag, Stuttgart 1968.
[64] G. Kraft, J. Fischer: *Indikation von Titrationen,* De Gruyter, Berlin 1972, p. 304.
[65] S. J. Parry: *Activation Spectrometry in Chemical Analysis,* Wiley & Sons, New York 1991, p. 243.
[66] W. D. Ehrmann, D. E. Vance: *Radiochemistry and Nuclear Methods of Analyses,* Wiley & Sons, New York 1991, p. 531.

Specific References

[67] G. Hevesy, F. Paneth, *Z. Anorg. Chem.* **82** (1913) 322–328.
[68] O. Hahn, *Z. Phys. Chem. (Leipzig)* **103** (1923) 461.
[69] Amersham International plc, Amersham Life Science Products 1993, Amersham/Little Chalfont, Amersham 1992, p. 184.
[70] Du Pont de Nemours Biotechnol. Systems, NEN Research Products 1991, Bad Homburg 1991, p. 164.
[71] J. T. Van Elteren, H. A. Das, C. L. De Ligny, J. Agterdenbos, *Anal. Chim. Acta* **222** (1989) 159–167.
[72] V. Hodge, M. Stalland, M. Koide, E. D. Goldberg, *Anal. Chem.* **58** (1986) 616–620.
[73] H. Yoshioka, K. Hasegawa, T. Kambara, *J. Radioanal. Nucl. Chem.* **117** (1987) 47–59.
[74] J. Klas, Z. Korenova, J. Tölgyessy, *J. Radioanal. Nucl. Chem.* **102** (1986) 111–120; **109** (1987) 337–351.
[75] S. Banerjee, *Anal. Chem.* **60** (1988) 1626–1629.
[76] R. H. Clarke: The 1990 Recommendations of ICRP, Supplement to the Radiological Protection Bulletin No. 119, NRPB, Chilton 1991.
[77] L. A. Currie, *Anal. Chem.* **40** (1968) 586–593.
[78] K. Hoppstock, R. P. H. Garten, P. Tschöpel, G. Tölg, *Fresenius J. Anal. Chem.* **343** (1992) 778–781.
[79] G. P. Westphal, *J. Trace Microprobe Tech.* **2** (1984/85) 217–235.
[80] D. Gawlik, K. Berthold, F. Chisela, P. Brätter, *J. Radioanal. Nucl. Chem.* **112** (1987) 309–320.
[81] W. Schmid, V. Krivan, *Anal. Chem.* **57** (1985) 30–34.
[82] K. G. Heumann, *Fresenius Z. Anal. Chem.* **324** (1986) 601–611.
[83] M. Schuster, *Fresenius J. Anal. Chem.* **342** (1992) 791–794.
[84] K.-H. Thiemer, V. Krivan, *Anal. Chem.* **62** (1990) 2722–2727.
[85] C. F. Miranda, R. Muxart, J. Vernois, G. Zuppiroli, *Radiochim. Acta* **19** (1973) 153.
[86] K. Hoppstock, R. P. H. Garten, P. Tschöpel, G. Tölg, *Anal. Chim. Acta,* (1994) in press.
[87] J. Tölgyessy, E. Bujdosó: *Handbook of Radioanalytical Chemistry,* **vols. 1–2,** CRC Press, Boca Raton 1991.
[88] T. Braun, J. Tölgyessy: *Radiometric Titrations,* Pergamon Press, Oxford 1967.
[89] J. Klas, J. Tölgyessy, J. Lesný: *Sub- superekvivalentová izotopová zried'ovacia analýza (Sub-superequivalent Isotope Dilution Analysis)* (in Slovak), Veda, Bratislava 1985.
[90] J. Tölgyessy, T. Braun, M. Kyrš: *Analiz metodom izotopnogo razbavlenija (Isotope Dilution Analysis)* (in Russian), Atomizdat, Moscow 1975.
[91] J. Klas, J. Tölgyessy, E. H. Klehr, *Radiochem. Radioanal. Lett.* **18** (1974) 83–88.
[92] J. Tölgyessy, M. Harangozó: "Radio-reagent Methods," in *Encyclopedia of Analytical Science, Radiochemical Methods,* Academic Press, London 1995 pp. 4317–4323.
[93] J. Tölgyessy: *Nuclear Radiation in Chemical Analysis* (in Hungarian), Müszaki Kiadó, Budapest 1965, p. 432.
[94] J. Tölgyessy: *Radioaktivnye kriptonaty v nauke i technike (Radioactive Kryptonates in Science and Technology), in Buduschee nauki (The Future of Science)* (in Russian), Izdatelstvo Znanie, Moscow 1977.
[95] M. Kyrš et al.: *Novye metody radioanaliticheskoi chimii (New Methods of Radioanalytical Chemistry)* (in Russian), Atomizdat, Moscow 1982.
[96] J. Tölgyessy, Ju. V. Jakovlev, G. N. Bilimovitch: *Diagnostika okruchajushej sredy radioanaliticheskimi metodami (Diagnostics of the Environments with Radioanalytical Methods)* (in Russian), Energoatomizdat, Moscow 1985.
[97] J. Tölgyessy: *Radioanaliticheskaja chimija (Radioanalytical Chemistry* (in Russian), Energoatomizdat, Moscow 1987.
[98] S. Aronoff: *Techniques of Radiochemistry,* The Iowa State College Press Building, Ames, IA 1956.
[99] V. R. S. Rao, Ch. P. Rao, G. Tataiah, *Radiochem. Radioanal. Lett.* **30** (1977) 365.
[100] J. Klas, J. Tölgyessy, E. H. Klehr, *Radiochem. Radioanal. Lett.* **18** (1974) 83.
[101] J. Klas, J. Tölgyessy, J. Lesný, *Radiochem. Radioanal. Lett.* **31** (1977) 171–179.
[102] J. Lesný, J. Klas, Z. Koreňová, *J. Radioanal. Nucl. Chem., Lett.* **155** (1991) 145–153.
[103] J. Lesný, Z. Koreňová, J. Klas, *J. Radioanal. Nucl. Chem., Lett.* **155** (1991) 155–168.

9. Enzyme and Immunoassays

JAMES N. MILLER, Department of Chemistry, Loughborough University, United Kingdom (Chap. 9)

REINHARD NIESSNER, Technische Universität München, München, Federal Republic of Germany (Chap. 9.2)

DIETMAR KNOPP, Technische Universität München, München, Federal Republic of Germany (Chap. 9.2)

9.1. Enzymatic Analysis Methods

9.1.1. Introduction

Many analyses in biological and clinical chemistry, and in the environmental, food and forensic sciences, involve the study of exceedingly complex multicomponent sample matrices. Moreover, the analytes under study are frequently present in trace amounts. Analytical techniques which combine exceptional selectivity with high sensitivity are thus required. Preeminent among these methods are those where the necessary selectivity is provided by biomolecules with very specific activities. This article surveys two groups of methods, those based on the activities of enzyme and of antibodies (the latter methods are normally called immunoassays).

The catalytic activity of enzymes provides for an enormous range of analytical techniques. Analytes are not restricted to conventional organic molecules, but include virtually all chemical species, including gases and metal ions. The ability of a single enzyme molecule to catalyze the reaction of numerous substrate molecules also provides an amplification effect which enhances the sensitivity of the analyses. A further advantage is that most enzyme-catalyzed reactions can be followed by simple, widely available spectroscopic or electrochemical methods. Enzymes are normally active only in mild conditions — aqueous solutions at moderate temperatures and controlled pH — and this restricts the circumstances in which they can be used, but some enzymes have proved remarkably robust to, e.g., mixed aqueous – organic solvent systems or elevated temperature; immobilized (insolubilized) enzymes often show enhanced stability compared with their solution analogs (see Section 9.1.7).

Antibodies have been used in analysis for over 60 years, and offer an unexpectedly wide range of techniques and applications. In some cases, the specific combination of an antibody with the corresponding antigen or hapten can be detected directly (e.g., by nephelometry), but more often such reactions are monitored by a characteristic label such as a radioisotope, fluorophore, etc. Since antibody reactions do not have a built-in amplification effect, these labels are frequently necessary to provide sufficient analytical sensitivity. Antibodies of different classes vary greatly in stability, but some are relatively robust proteins, and this contributes significantly to the range of methods available.

Enzymes and antibodies are proteins, with the ability to bind appropriate ligands very strongly and specifically. Enzymes frequently need cofactors or metal ions for (full) activity. Antibodies are multifunctional molecules, whose ability to bind to cell surfaces and to other proteins, at sites distinct from the antigen/hapten-binding sites, extends the range of labeling and detection methods available in immunoassays.

The analytical potential of enzymes and antibodies can be combined. Enzyme immunoassays, in which enzymes act as label groups in antibody-based analyses, are very well established; they are probably the most widely used immunosassay methods. An important development is the production of single molecules which combine antibody and enzyme activities. A critical feature of enzyme activity is that the transition state of the substrate binds very tightly to the enzyme at a position near the amino acid side-chain groups that participate in the catalytic reaction. It is thus possible, by using haptens which are analogs of substrate transition states, to generate antibodies with catalytic activity. These "catalytic antibodies" are of great research interest, and their potential is emphasized by the possibility that they might be used to develop catalytic activities not present in living organisms.

This article can only attempt a general survey of enzyme and immunoassay methods, with references to more detailed reviews and books in specific areas. Many important clinical, veterinary, and forensic analyses in which enzymes themselves are the analytes are not discussed; this article considers only those methods where enzymes (and antibodies) are analytical reagents.

9.1.2. Enzymes: Basic Kinetics

The fundamental features of enzyme catalytic activity are described in detail in textbooks [1]. The simplest mechanism, outlined by MICHAELIS and MENTEN, proposes that the enzyme E reacts reversibly with the substrate S to form a complex, which subsequently decomposes to release the free enzyme and the product molecule P:

$$E + S \longrightarrow ES \longrightarrow E + P \quad (1)$$

In this model, the combination reaction of E and S, with a rate constant k_1, occurs at the active site of the enzyme. This is a relatively small region of the E molecule; often a cleft or depression accessible from the surrounding aqueous solution. Its conformation, polarity, and charge distribution are complementary to those of the S molecule. The reverse process, dissociation of the ES complex, has a rate constant k_2, and the decomposition of ES to give E and P has a rate constant k_3. These three rate constants can be used to calculate the Michaelis constant K_M:

$$K_M = (k_2 + k_3)/k_1 \quad (2)$$

The Michaelis constant clearly has units in mol/L, and in practice values between 10^{-1} and 10^{-7} mol/L. It is also convenient to define the turnover number of an enzyme, k_{cat}, which is the number of substrate molecules converted to product per unit time by a single enzyme molecule, in conditions where the enzyme is fully saturated with substrate. From the kinetic model $k_{cat} = k_3$, but if more complex models are used k_{cat} depends on several rate constants. Depending on the enzyme and the reaction conditions (see Section 9.1.3) k_{cat} varies between ca. 1 s^{-1} and 500 000 s^{-1}.

The kinetic model outlined above is clearly oversimplified (e.g., it should really include an

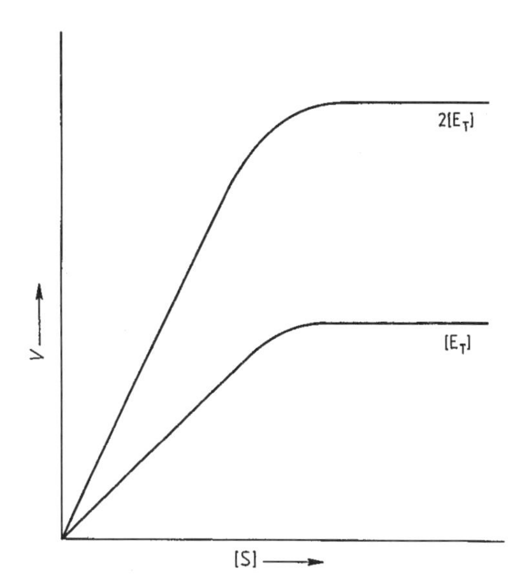

Figure 1. Relationship between reaction velocity V and substrate concentration [S] at different levels of enzyme concentration, $[E_T]$ and 2 $[E_T]$

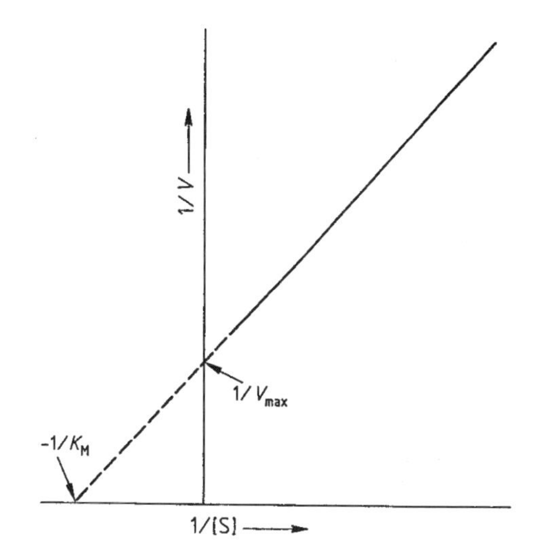

Figure 2. Double-reciprocal enzyme rate plot

intermediate complex EP, in which the product molecule is bound to the enzyme), but it describes adequately the kinetic behavior of most enzyme-catalyzed reactions. If the steady-state hypothesis is applied to the ES complex in this model, the initial velocity of the enzyme-catalyzed reaction is:

$$V = k_3[E_T][S]/([S] + K_M) \tag{3}$$

where squared brackets indicate concentrations, and $[E_T]$ is the total enzyme concentration, irrespective of whether it is combined with S. When $[S] >> K_M$, i.e., when the enzyme active sites are saturated with substrate, V is simply $k_3 [E_T]$. This is commonly called V_{max}, so Equation (3) becomes:

$$V = V_{max}[S]/([S] + K_M) \tag{4}$$

Equations (3) and (4) show that, when V is plotted against [S] at constant $[E_T]$ (Fig. 1), the reaction rate is approximately proportional to [S] at low substrate concentrations, but eventually reaches the plateau level V_{max}. So provided $[S] << K_M$, substrate concentrations can be determined directly from reaction rate measurements. Such analyses are widely used (Section 9.1.4). Equation (4) also shows that, when $[S] = K_M$, $V = V_{max}/2$. This provides a method of estimating K_M. In practice, K_M and V_{max} are usually determined by one of several transformations of Equation (4); the most common is the Lineweaver–Burk double-reciprocal method, in which $1/V$ is plotted against $1/[S]$ (Fig. 2) to give a straight line of slope K_M/V_{max} and x- and y-axis intercepts $-1/K_M$ and $1/V_{max}$. This leads to the following:

$$(1/V) = (1/V_{max}) + (K_M/[S]V_{max}) \tag{5}$$

An important feature of enzymes is that their active sites can often be occupied by, or react with, molecules other than the substrate, leading to inhibition of enzyme activity. Several inhibition mechanisms are known, but it is necessary only to distinguish between irreversible and reversible inhibition. Irreversible inhibition arises when the inhibitor molecule I dissociates very slowly or not at all from the enzyme active site. The best-known examples occur when I reacts covalently with a critical residue in the active site. Inhibition of cholinesterase enzymes by the reaction of organophosphorus compounds with a serine residue is a case in point. This type of inhibition is said to be noncompetitive—enzyme activity cannot be restored by addition of excess substrate. So although addition of I reduces V_{max}, K_M is unaffected. The double-reciprocal plot in such cases has the same x-axis intercept as the plot for the uninhibited enzyme, but greater slope.

Reversible inhibition occurs when the EI complex can dissociate rapidly, just as the ES complex does. The most common examples arise when S and I are chemically or sterically similar, and compete for the active site (competitive inhibition — but note that not all reversible inhibition mechanisms are competitive). Thus, alcohol dehydrogenase enzymes oxidize ethanol to acetaldehyde; but other alcohols, e.g., methanol or ethylene glycol, can act as competitive and reversible inhibitors. The extent of inhibition in such cases depends on the relative concentrations of I and S: if the latter is present in great excess, inhibition is negligible. In these cases, V_{max} is not affected, but K_M may be increased; the double-reciprocal plot has the same *y*-axis intercept, but greater slope than the plot for the uninhibited enzyme.

When steady-state theory is applied to these mechanisms, in both types of inhibition the initial rate of the enzyme-catalyzed reaction is inversely related to [I], provided that the latter is small. Again, the possibility of a simple quantitative analysis of I by rate measurements is apparent. Such analyses are not so numerous as determinations of substrates, but some are very important (Section 9.1.5).

A number of enzymes are inactive or weakly active in the absence of activators. The best known examples arise with metalloenzymes, whose active sites include a main-group or transition metal ion. When the active holoenzyme, i.e., the complete molecule, including the metal ion, is treated with a complexing agent to remove the metal ion, the resulting inactive apoenzyme may provide a sensitive reagent for determination of the metal, enzyme activity being directly proportional to the concentration of activator at low levels of the latter (Section 9.1.6).

9.1.3. Enzyme Assays: Practical Aspects

9.1.3.1. General Considerations

All enzymatic analyses of substrates, inhibitors, and activators involve the determination of reaction rates: since initial rates are required, experiments are usually quite short. It is possible in principle to mix the reactants and measure the rate directly, by recording an optical or electrochemical signal at regular intervals, the rate being the slope of the resulting signal plotted against time. When many samples are to be analyzed, such an approach is impracticable, and pseudo-rate methods are used. The reaction is allowed to proceed for a fixed time after the initial mixing; it is then stopped abruptly by a large change in pH or temperature (see below), or by addition of a specific enzyme inhibitor, and the extent of substrate depletion or product formation is measured. The signal recorded is then compared with the results of matching experiments on standard analyte solutions, and the test concentrations calculated by interpolation. In simple enzymatic assays, the calibration graphs are usually linear (see above), but in complex systems such as enzyme immunoassays, curved calibration plots are usual, requiring more complex statistical evaluation.

9.1.3.2. Spectrometric Methods

UV – Visible absorption spectrometry is the most commonly used method in modern enzymatic analysis. Suitable spectrometers are found in every laboratory, and many analyses use simple, robust, relatively cheap instruments. The availability of disposable polystyrene or acrylic cuvettes is a further advantage for conventional solution studies, but many analyses now use microtiter plates, the individual wells of which can be examined in a simple spectrometer. Continuous flow systems are also simple to set up. Most spectrometers are interfaced with personal computers, which offer a range of control and data handling options.

Several generic applications of UV – visible spectrometry are common in enzymology (Section 9.1.4). Many oxidoreductase enzymes involve NAD/NADH or NADP/NADPH as cofactor systems. Such reactions are readily followed by absorption measurements at 340 nm, where the absorption of NADH and NADPH is strong, but that of NAD and NADP negligible (Fig. 3). This approach is so convenient that it is often applied in the form of a coupled enzyme reaction: if neither substrate nor product in the main reaction of interest is readily determined, one of them may be transformed via a second enzyme-catalyzed reaction involving NAD(P)/NAD(P)H, with detection at 340 nm. This principle has been extended so that three or more reactions occur in sequence, the final one either generating or consuming the readily determined NAD(P)H. Such indirect detection of the primary reaction is much less complex than it seems: often, all the enzymes, and any other reagents needed, can be included in the initial reaction mixture, the specificities of the of enzyme-catalyzed reactions is the sequence ensuring that they do not interfere. Extra sensitivity, and the opportunity to work in the visible region, is

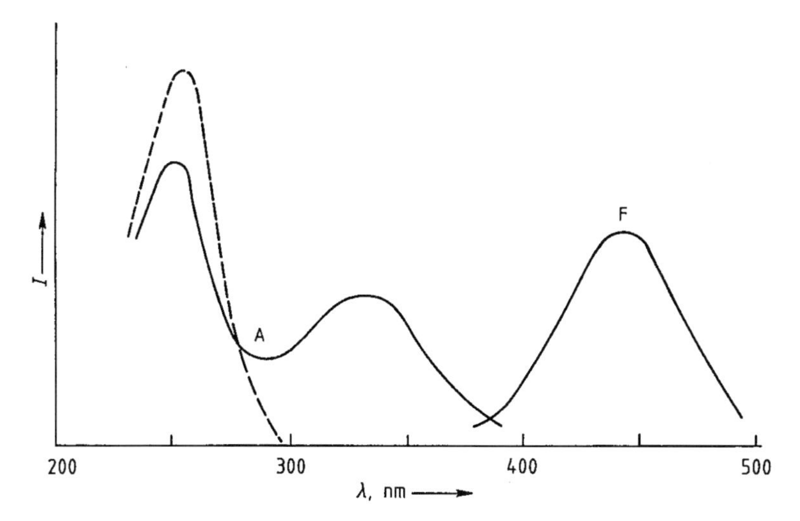

Figure 3. Absorption (A) and fluorescence (F) spectra NAD[P]H (——); and NAD[P] (- - -) (absorption only)

provided by using NAD(P)H to reduce colorless tetrazolium salts, in the presence of phenazine methasulfate (PMS), to intensely colored formazans, which absorb at ca. 500 nm. This further step is of particular value when a very rapid qualitative or semiquantitative result is required.

A further approach using UV–visible spectrometry involves colorigenic substrates. These molecules change color when hydrolyzed in the enzymatic reaction under study. A good example is provided by *p*-nitrophenol esters, which are colorless, but are hydrolzyed by appropriate enzymes to yellow *p*-nitrophenol, which can be determined at ca. 405 nm. Many substrates of this type are readily available, e.g., *p*-nitrophenyl phosphate as a phosphatase substrate, the corresponding sulfate as a substrate for aryl sulfatases, *N*-carbobenzoxy-L-tyrosine-*p*-nitrophenyl ester as a substrate for proteolytic enzymes such as chymotrypsin, etc.

Several analytically important enzymes catalyze reactions in which hydrogen peroxide is generated. Such reactions are easily followed colorimetrically, hydrogen peroxide being used in the presence of a second (peroxidase) enzyme to oxidize leuco-dyes to colored products. The same reactions can be used to monitor peroxidases themselves when they are used as labels in enzyme immunoassays; in modern systems, the enzyme-catalyzed reaction is halted by a large pH change, which simultaneously generates a distinctive color in the dye reaction product.

Fluorescence spectrometry is widely used in enzymology, usually because of its extra sensitivity compared with absorption methods. Fluorophores are characterized by two specific wavelengths, absorption and emission. The latter is the longer of the two, excited molecules having lost some energy between the processes of photon absorption and emission, and this allows the fluorescence to be determined (at 90° to the incident light beam in most instruments) against a dark background. This provides limits of detection unattainable by absorption spectrometry. A second important characteristic of fluorescence spectrometry is its versatility; it is just as easy to study concentrated solutions, suspensions, solid surfaces, flowing systems, etc., as it is to measure dilute solutions.

The generic approaches in absorption spectrometry are mirrored in fluorescence. Thus, NAD(P)H can be determined at an emission wavelength of ca. 460 nm with excitation at ca. 340 nm (Fig. 3). For work at longer wavelengths, NAD(P)H reduces nonfluorescent resazurin to the intensely yellow-fluorescent (ca. 580 nm) resorufin (this can also be determined colorimetrically at ca. 540 nm). Fluorigenic substrates are available for many hydrolytic enzymes, the best-known being those based on resorufin esters, and on esters of 4-methylumbelliferone; the esters are nonfluorescent, but are hydrolyzed to fluorescent products. Peroxidase activity can be monitored via the conversion of a resorcin derivative to a fluorescent product with hydrogen peroxide.

Recent years have seen an explosion of interest in chemiluminescence (CL) and bioluminescence

(BL) methods in enzymatic analyses. As in fluorescence, these methods measure photon emission from excited molecules, but here the latter are generated chemically, as reaction products. The reactions concerned are oxidations, often in mildly alkaline solutions in the presence of a catalyst. The advantages are:

1) No exciting light source is needed, so scattered light phenomena are minimized
2) It is usual to collect as many as possible of the emitted photons without the use of optical filters or monochromators. Screening devices using photographic detection have been developed. Sensitive Polaroid films allow the convenient qualitative or semiquantitative monitoring of chemiluminescent reactions.
3) Extreme sensitivities are possible
4) The reagents are generally free of hazard, despite offering limits of detection similar to those secured by radiochemical methods

Since the intensity of luminescence is proportional to the rate of generation of the excited reaction product, the principal precautions to be observed are those that apply when reaction rates are measured (control of temperature, pH, etc.), with the important additions that the order and the reproducibility of mixing of the reagents are critical, especially in conditions which yield short bright bursts of light. In such cases, fully automated CL/BL methods, such as those using flow injection analysis, are of obvious value. The common CL/BL techniques mirror those in UV–visible and fluorimetric measurements. Processes involving NAD(P) and NAD(P)H can be followed with the aid of coupled reactions (in which two or more enzymes catalyzing consecutive reactions are involved), and reactions in which hydrogen peroxide is generated are readily monitored by using the peroxide to oxidize a well-known luminescent compound, such as luminol. Luminogenic substrates, e.g., for phosphatase enzymes, have been developed, and these find particular use in CL immunoassays. Sensitive microtiter plate readers are now widely available for both fluorescence and chemiluminescence detection, so these spectroscopic methods are proving ever more popular in the monitoring of enzyme catalyzed reactions.

9.1.3.3. Other Methods

Although spectrometric methods still dominate the measurement systems used in biospecific analyses, other methods are important. Some of these are applied in specialized areas. The optical activity of saccharides can be used to follow reactions such as the inversion (hydrolysis) of sucrose, catalyzed by invertase. The products (D-glucose and fructose) have a combined optical rotation (e.g., at 589 nm) quite different from that of sucrose, so a simple polarimeter provides a good method of determining sucrose, the enzyme's substrate, and of other species which inhibit invertase activity. Radiolabeled substrates are often employed in research (^{14}C and ^{3}H are the most common isotopes), but safety precautions preclude the routine use of this approach. Since the radioactivity of the label is not changed in the enzymatic reaction, physical separation is necessary before the radiolabeled reaction product (or residual substrate) can be determined (see Section 9.2.4).

Electrochemical methods of following enzyme-catalyzed reactions are rapidly growing in popularity, mirroring the growth of these methods across analytical science in general. Particularly important are electrochemical enzyme sensors, portable or disposable devices which combine the specificity of an enzyme reaction with the simplicity and compactness of an electrical transducer. Selective electrodes have been adapted to form such sensors: for example, oxygen electrodes can monitor any enzyme-catalyzed reaction in which oxygen is consumed or produced. Other potentiometric and amperometric methods have achieved impressive successes, exemplified by the variety of glucose sensors now available. These devices use the glucose oxidase reaction (Section 9.1.4), and a variety of coupled reaction systems which provide an electrochemical response. In early applications, the hydrogen peroxide generated in the oxidation of glucose was used to oxidize hexacyanoferrate(II) to hexacyanoferrate(III), this second reaction being followed amperometrically. More sophisticated methods are now applied, and are generally applicable to all reactions in which peroxide is produced.

The rates of enzyme-catalyzed reactions can be followed, not only by a wide range of instrumental techniques, but also with the aid of a variety of chemical principles. In favorable cases, reaction rates can be determined directly by observing consumption of the substrate or appearance of the product, but if these compounds lack distinctive physical properties, indirect approaches, especially coupled reactions, can be used (see Section 9.1.3.2).

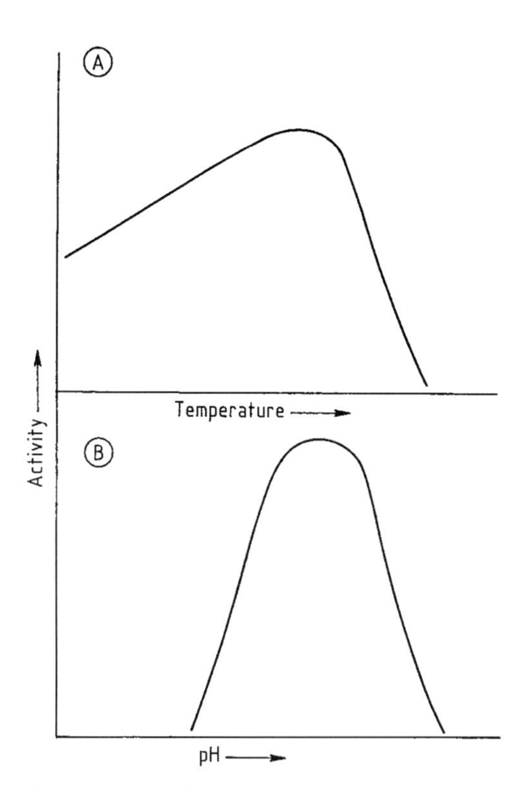

Figure 4. Effects of temperature (A), and pH (B) on the activity of a typical enzyme
The maxima and widths of the curves vary from one enzyme to another.

9.1.3.4. Effects of pH, Buffer, Composition, and Temperature

Virtually all enzymatic assays are carried out at 20 – 50 °C in aqueous buffers of known pH and controlled composition. Both temperature and buffer properties affect the rates of enzyme-catalyzed reactions markedly. The effects of temperature can usually be summarized by a bell-shaped curve (Fig. 4 A). At lower temperatures, reaction rates increase with temperature, but beyond a certain point, denaturation (unfolding) of the enzyme molecules begins, so they lose their ability to bind the substrate, and the reaction rate falls. The temperature giving maximum activity varies from one enzyme to another, according to the robustness of the molecule. In some cases, it may be convenient to use a temperature rather below this maximum, otherwise the rate becomes too high to measure precisely. The rates of many enzyme-catalyzed reactions increase by a factor of ca. 2 over a range of 10 °C in the region below the maximum of the curve; it is thus necessary to control the temperature to within ca. ± 0.1 °C.

The pH of the buffer solution in which the analysis takes place must also be controlled. Most enzymes have a pH dependence of their activity of the type shown in Figure 4 B, and pH should be controlled to within ± 0.02. The optimum pH, and the range over which the enzyme is active, vary widely from one enzyme to another. These phenomena arise from the effects of pH on the structure of the enzyme itself, on the affinity constant between the enzyme and the substrate, on V_{max}, and also on any coupled indicator reaction that is used. The choice of buffer recipe for any given pH may be important. The ionic strength of the buffer and the salts contained in it can influence the rate and mechanism of the main and coupled reactions, sometimes with unpredictable results. It may be necessary to choose a buffer whose properties represent a compromise between the ideal conditions for the main reaction, and the ideal conditions for the indicator reaction(s). Immobilized enzymes (Section 9.1.7) often have pH (and temperature) dependence significantly different from their soluble counterparts. This all points to the value of establishing and maintaining a well-defined buffer system for enzymatic analyses.

9.1.3.5. Sources and Activity of Enzymes

A wide range of enzymes is commercially available from numerous suppliers, so it is rarely necessary for the analyst to prepare such materials. Enzymes with the same name, but isolated from different species, may be quite distinct in their chemical and biological properties, and their activity (e.g., their requirements for cofactors). Enzymes are normally supplied as freeze-dried or crystallized proteins, and should be stored carefully at 4 °C or − 20 °C. Repeated freezing and thawing of protein solutions is not advisable, so it may be necessary to divide the dissolved enzyme into small aliquots, each of which is frozen and used just once.

The specificity of many enzyme-catalyzed reactions is a major reason for their use. Many organisms, or separate organs from a particular species, contain only one enzyme capable of catalyzing a given reaction. Thus, it may be unnecessary to use high-purity enzyme preparations; quite crudely purified material is often sufficient. In some cases, however, it is essential to ensure removal of a particular contaminant; enzyme preparations used in reactions involving hydrogen peroxide should be free of catalase, the enzyme which decomposes H_2O_2 to oxygen and water.

In practice many enzymatic analyses are performed with the aid of pre-packed kits. Such kits contain a number of vials which incorporate optimized mixtures of enzymes, cofactors, buffer salts, etc., as required, often in a dried form. Each vial provides a single analysis. The analyst needs only to reconstitute the reagents by the addition of water, add the sample, and perform the measurement after a given time period, recommended by the kit manufacturer.

Since an enzyme used as an analytical reagent may be quite impure, it is essential to know how much substrate a given weight of the enzyme preparation will convert to the corresponding product. This information can be derived from the activity of the material, given in International Units (IU). Unit enzyme activity converts one micromole of substrate to product per minute at 25 °C and optimal pH. Commercial enzyme preparations are described in terms of IU/mg or, in the case of an enzyme supplied in solution or suspension, IU/mL.

9.1.4. Enzyme Assays: Determination of Substrates

9.1.4.1. General Considerations

As shown in Section 9.1.2, the rate of an enzyme-catalyzed reaction is proportional to the concentration of substrate, if the latter is small compared with K_M. (The guideline $[S] < 0.2\,K_M$ is often used.) This provides a simple approach to substrate determinations, which certainly represent the largest class of enzymatic analyses in modern practice, e.g., the measurement of blood glucose or blood cholesterol. In many instances, the analyses are highly selective, but not absolutely specific for the target substrate.

9.1.4.2. Determination of Glucose

Several different enzymes have been used to determine β-D-glucose, but by far the most popular approach uses glucose oxidase. Enzymes are often known by their Enzyme Commission numbers: glucose oxidase is E.C. 1.1.3.4. The first digit defines the enzyme as an oxidoreductase, the second as an oxidoreductase in which the hydrogen or electron donor is an alcohol, and the third as one in which the hydrogen or electron acceptor is molecular oxygen. As there are several enzymes exhibiting these three characteristics, the fourth digit purely acts as a means of differentiating between similar substances. This enzyme, normally obtained from *Aspergillus niger,* has a molecular mass of ca. 80 000, uses flavin adenine dinucleotide as cofactor, and catalyzes, at pH 7 – 8, the reaction:

$$\text{D-Glucose} + O_2 \longrightarrow \text{D-Gluconolactone} + H_2O_2$$

This reaction is highly specific (α-D-glucose is oxidized at only 1 % of the rate of the β-anomer), but not perfectly so, as 2-deoxy-D-glucose (which does not ocur at significant levels in blood) is also readily oxidized. Scores of different methods have been used to follow the rate of this reaction, almost all based on the properties of the very reactive product, hydrogen peroxide. (Many of these methods can also be used to determine other substrates which are oxidized by oxidoreductase enzymes to yield, among other products, H_2O_2, e.g., cholesterol, xanthine, L- and D-amino acids, D-galactose etc.). Colorimetric methods are based on conversion of a chromogen [e.g., 2,2′-azino-di(3-ethylbenzothiazolinsulfonate) diammonium salt (ABTS)] to a colored product (for ABTS, $\lambda_{max} = 405$ nm) with the aid of a second enzyme, horseradish peroxidase. Analogous fluorimetric methods, such as the oxidation of nonfluorescent resorcin derivative to a fluorescent product, have also been used. The optimum pH for the coupled peroxidase step is ca. 10, but both enzymes show adequate activity at pH ca. 8.5, an example of a compromise pH. Glucose oxidase reactions can also be monitored by the chemiluminescence from luminol stimulated by peroxidase in the presence of hexacyanoferrate (III). Numerous electrochemical methods have also been developed. Many glucose tests are available which use the enzyme in immobilized form. A typical dipstick test for urine samples uses a cellulose matrix which contains immobilized glucose oxidase and peroxidase along with a chromogen. In the presence of glucose, the initially colorless matrix develops a color within a few seconds. (An exactly similar test is available using galactose oxidase to test for D-galactose.)

With such methods, glucose is routinely determined in blood (normal level ca. 5 mol/L) and foodstuffs. In all these analyses, it is assumed that molecular oxygen is present in excess, but it is also possible to use the enzyme in an excess of glucose to estimate levels of dissolved oxygen in aqueous or mixed aqueous – organic solvents. This unexpected application exemplifies the broad range of enzymatic analyses.

9.1.4.3. Determination of Ethanol

Ethanol can be determined with the aid of alcohol oxidase (E.C. 1.1.3.13), but a more usual approach uses alcohol dehydrogenase (ADH, E.C. 1.1.1.1. — the third digit indicates that this oxidoreductase has $NAD(P)^+$ as its hydrogen or electron acceptor). In neutral buffers, this zinc metalloenzyme, normally obtained from yeast, but also from mammalian liver, catalyzes the reaction:

$$CH_3CH_2OH + NAD^+ \longrightarrow CH_3CHO + NADH + H^+$$

The oxidation can be followed by monitoring the appearance of NADH by UV-absorption spectrometry, colorimetry, fluorimetry, or CL/BL methods (Section 9.1.3.2). Common sample matrices are blood plasma or various foods and beverages; as in the case of glucose, their analyte levels are often quite high, so few sensitivity problems arise. The catalytic action of this enzyme is not very specific: *n*-propanol and *n*-butanol, isobutanol, allyl alcohol, ethylene glycol, glycerol, and methanol (oxidized to formaldehyde, hence its toxicity) are among other alcohols that may interfere. This lack of specificity is one reason for the interest shown in alcohol oxidase as an alternative. The importance of routine alcohol determinations has encouraged the development of many methods using immobilized ADH in the form of "alcohol electrodes" or continuous flow analyzers.

9.1.5. Enzyme Assays: Determination of Inhibitors

9.1.5.1. Inhibitors of Cholinesterase Enzymes

Enzyme assays based on inhibition effects are not as commonly employed as substrate determinations, but one or two are very important. Preeminent is the determination of organophosphorus compounds by using their inhibitory effect on cholinesterase enzymes (E.C. 3.1.1.8—the first digit signifies a hydrolase enzyme, the second that the compounds hydrolyzed are esters, and the third that they are phosphoric monoesters). The latter catalyze the conversion of acylcholines to choline and the corresponding acid:

$$\text{Acylcholine} + H_2O \longrightarrow \text{Choline} + \text{Acid}$$

This reaction is crucial to many living systems, so the organophosphorus compounds (pesticides, chemical warfare agents) that inhibit it by irreversible binding to the active site are often highly toxic. A range of cholinesterase enzymes is available, the most common being the enzyme isolated from horse serum, which is a pseudocholinesterase or butyrylcholinesterase. The specificity of such an enzyme is broad, and a wide range of esters can be hydrolyzed. This is of value in inhibitor analyses as the "normal" substrate can be replaced by, e.g., a fluorigenic substrate to facilitate detection. Thus, resorufin butyrate, a nonfluorescent ester of resorufin, is hydrolyzed by the horse serum enzyme to the intensely fluorescent resorufin. In the presence of an inhibitor, the expected growth in fluorescence intensity is reduced. Since a free acid is produced in the hydrolysis reaction, electrochemical methods provide viable alternatives, and enzyme reactors using immobilized cholinesterases are commercially available.

Organophosphorus compounds are often used as mixtures in pesticide formulations, and an ingenious application allowed the identification of such mixtures by combining the fluorigenic enzyme assay with thin layer chromatography (TLC). The sample was separated using TLC, and the plate treated sequentially with the enzyme and the fluorigenic substrate. The result was a mostly fluorescent TLC plate, with dark areas marking the positions of the enzyme inhibitors. In addition to organohosphorus compounds, carbamates (also used as pesticides) and the calabar bean alkaloid physostigmine act as inhibitors of cholinesterases, and can also be determined.

9.1.5.2. Other Inhibition Methods

The literature on enzyme inhibition assays is very large, but many of the methods are of limited value because they lack specificity, and/or do not compete with other techniques. Thus, many enzymes are inhibited to a greater or lesser extent by transition metal ions, but the lack of specificity for individual ions and the preference for alternative techniques, e.g., ion chromatography, has reduced the value of such approaches.

One area of great current importance is the search for enzyme inhibitors, especially inhibitors of proteolytic enzymes, in high-throughput screening programs for new drugs. Combinatorial chemistry methods can generate candidate inhibitors at the rate of thousands per working day, so rapid and efficient screening procedures are essential. Most currently used methods employ microtiter plate formats with fluorescence detection.

9.1.6. Enzyme Assays: Determination of Activators

9.1.6.1. Determination of Metal Ions by Metalloenzymes

Activator analyses are the least common application of enzymatic methods, but, as in the case of inhibitor analyses, a small number of methods are extremely valuable. Of great interest are methods where incorporation of metal ions in the active site of an enzyme is critical. Removal of the metal from the holoenzyme (e.g., by mildly acidic solutions of EDTA) generates the apoenzyme, which lacks enzyme activity. This apoenzyme then becomes the reagent in the subsequent analysis, along with its natural substrate and any indicator reagents that may be required. In a sample containing the appropriate metal ion, the activity of the enzyme is wholly or partially restored. In contrast to the metal ion inhibition effects mentioned above, these analyses are often highly selective, as the active sites of metalloenzymes accommodate (with generation of activity) only a very few metal ions (ionic radius may be as important as ionic charge).

An example of such analyses is provided by isocitrate dehydrogenase (E.C. 1.1.1.42), normally isolated from pig heart. This enzyme catalyzes the oxidation of isocitrate ions to α-oxoglutarate ions; the reaction is NADP dependent, and results in a decarboxylation:

$$\text{Isocitrate} + \text{NADP} \longrightarrow \text{NADPH} + CO_2 + H^+ + \alpha\text{-Oxoglutarate}$$

The usual methods for following NAD(P)-dependent reactions are available here. The enzyme requires the presence of Mg^{2+} ions for activity, and extensive studies show that, apart from Mg^{2+}, the only other ions that combine with apo-ICDH (apo-isocitrate-dehydrogenase) to restore its activity are Mn^{2+} and to a lesser extent Zn^{2+} and Co^{2+}. This approach has been successfully applied to the determination of Mg^{2+} in blood and other samples at concentrations below 1 μmol/L, and to the analysis of Mn^{2+} at concentrations as low as 5 ppb. The advantage of this method of trace metal determination is that only a single oxidation state of the metal is determined—in the manganese analysis, only Mn(II), with no interference from Mn(VII) or other oxidation states.

9.1.6.2. Other Activator Analyses

The most obvious activator assays are those in which cofactors are analyzed: all the common cofactors can be determined in this way, and in many cases their own properties provide simple means of measuring reaction rates. The analyses are usually very selective. ATP in red blood cells and many other samples is often determined by using its specific participation in the well-known firefly bioluminescence reaction:

$$\text{Luciferin} + O_2 + \text{ATP} \longrightarrow \text{Oxyluciferin} + \text{ADP} + PO_4^{3-} + \text{Light}$$

This reaction is catalyzed by firefly luciferase, which requires Mg^{2+}, and can also be used to determine this ion. Even in simple instruments, ppb levels of magnesium can be detected.

9.1.7. Immobilized Enzymes

9.1.7.1. Introduction

Analysts have attempted to use immobilized enzymes for over 30 years. The original motivations were two-fold: to conserve enzymes that were expensive and difficult to isolate, and to incorporate enzymes in reusable sensors, such as enzyme electrodes. During such researches, other advantages and applications of immobilized enzymes were found, and immobilization technologies were extended to other areas of biospecific analysis such as immunoassays (see below) and affinity chromatography. Immobilized enzymes have also been used extensively in manufacturing processes and for therapeutic purposes.

Several distinct approaches to the problem of immobilizing an enzyme in or on a solid or gel, with optimum retention of biological activity, are available. These include covalent binding, surface adsorption, gel entrapment, encapsulation within a semipermeable membrane, and chemical cross-linking. The first is by far the most popular, and many enzymes are commercially available in covalently immobilized form. Common solid phases include particles of well-known chromatographic media, e.g., agarose, glass and silica, polystyrene, polyacrylamide and cellulose, modern perfusion chromatography phases, and nylon in the form of flow tubing or membranes. The best-known covalent linkage methods include the use of cyanogen bromide (CNBr) to bind $-NH_2$ groups to hydroxylic matrices such as agarose or cellulose, chloro-*sym*-triazinyl derivatives to perform a similar function, and carbodiimides to link amine and carboxylic acid functional groups between enzymes and solid phases. Such covalent linkages are not infinitely stable; even the best immobilized enzyme reactors suffer slow leakage of the enzyme molecules into solution. However, many derivatives remain stable on storage and in use over periods of several months.

The idea of entrapping enzyme molecules within a polymeric matrix of known (average) pore size is attractive and has been frequently studied, polyacrylamide gels being widely used. However the outcomes have often been disappointing. The polymerization process, carried out in the presence of a solution of the enzyme, may generate sufficient heat to denature some of the enzyme; leakage may occur from within the polymer matrix; enzyme activity with respect to large substrates may diminish sharply because of the molecular sieving effect of the polymeric gel; and even the activity to small molecules may be diminished because the enzyme molecules are less available than they are on the surface of a solid particle.

However, interest in this approach to enzyme immobilization has recently been renewed by the application of silica-based sol-gel entrapment matrices. These provide mild hydrophilic environments with controlled pore sizes. Moreover sol-gels can be formed into optically clear monoliths of varying dimensions, thin films, and fibers. Several applications including the determination of metal ions and of peroxidase have already been described.

Enzymes are readily adsorbed to the surfaces of ion exchange matrices, plastic microtiter plates, hydroxyapatite, even charcoal. Such immobilized molecules may be suitable for one-off assays (i.e., not for repeated use), but they are often readily desorbed by quite small changes in their environment, and enzymes immobilized in this way have not found great application in routine analyses, despite the ease with which adsorption is achieved. However, the adsorption of antibodies on plastic surfaces is routinely used in many immunoassays, including enzyme immunoassays (Section 9.2.4.3.1).

Bifunctional reagents, such as glutaraldehyde, can cross-link enzyme molecules to produce insoluble aggregates that retain at least some activity, but these cross-linked enzymes often have unsuitable mechanical properties, and they have found little practical use.

Microencapsulation of enzymes (and other biomolecules) has been extensively studied. The microcapsules have semipermeable walls that allow small substrate and product molecules to pass through freely, while large enzyme molecules are retained in solution within the capsules. However, the polymerization process used in forming the capsules generates heat that may damage the enzymes.

9.1.7.2. Properties of Immobilized Enzymes

Many experiments have shown that the advantages of (usually covalently) immobilized enzymes extend beyond their reusability; their physicochemical properties are often significant-ly different from those of the corresponding enzyme in free solution. The immobilized enzyme usually has somewhat less activity than the same mass of enzyme in solution. But unless activity losses are severe, this disadvantage is not serious, and can be compensated by using more of the immobilized preparation, knowing that it can be used repeatedly.

A significant advantage of immobilization is the increased thermal stability conferred on many enzymes, allowing their use for longer periods at higher temperatures than would be possible for the soluble molecule. An early experiment showed that immobilized papain (a protease derived from papaya latex) retained over half its room-temperature activity at 80 °C in conditions in which the soluble enzyme was almost entirely denaturated.

Also interesting and valuable are the changes in the pH – activity curves that often accompany immobilization. A common result is broadening of the curve, i.e., the enzyme is active over a wider pH range than its soluble counterpart. This is normally ascribed to the range of microenvironments of different enzyme molecules in or on the solid matrix. Depending on the charge properties of this matrix, the optimum pH may undergo significant shifts. The optimum pH for an enzyme bound to a negatively charged carrier such as carboxymethylcellulose is shifted to higher values, while immobilization on a cationic matrix such as DEAE-cellulose (diethylaminoethyl-cellulose) has the opposite effect. These effects are ascribed to the change in the enzyme's microenvironment brought about by neighboring charged groups. Immobilization on a neutral carrier is not expected to change the pH optimum. Since these pH effects can be controlled to some degree, immobilization of an enzyme may confer significant advantages, e.g., using two or more enzymes with different pH optima in solution (Section 9.1.4.2) may be facilitated.

9.1.7.3. Application of Immobilized Enzymes

Since most of the enzymes routinely used in analytical science are now commercially available in immobilized form, it might be expected that many analyses conventionally carried out in solution are also performed with the immobilized products. In practice, many of the important applications of immobilized enzymes are those in which the enzyme preparation is incorporated into a more specialized analytical system, such as an enzyme electrode or continuous enzyme reactor.

One of the first applications of immobilized enzymes was the construction of an enzyme electrode for glucose, in which a thin layer of glucose oxidase entrapped in a polyacrylamide gel was placed over a conventional Clark oxygen electrode. When the enzyme electrode was immersed in glucose solution, the glucose penetrated the enzyme layer, consuming oxygen and reducing the potential of the Clark electrode. Apart from concerns over the long-term stability of the enzyme layer, the principal drawbacks of this and many other similar devices were their response times (often minutes rather than seconds), and their recovery times, i.e., the rates at which the electrodes could be used with different samples. Some modern devices have largely overcome these problems, but although enzyme electrodes (most based on platinum, oxygen, or other established sensors) have been developed for numerous analytes (amino acids, urea, sugars, simple alcohols, penicillin, cholesterol, carboxylic acids, inorganic anions, etc.), only a very few seem robust enough to survive routine laboratory use over long periods.

On the other hand disposable "one-shot" sensors based on immobilized enzymes are now of major importance in several biochemical and environmental application areas.

A second obvious area of application is in continuous flow analysis or flow injection analysis systems, in which the immobilized molecules form reactors that can be readily inserted and replaced in a flow analysis manifold. The physical form of the enzymes varies widely; packed-bed reactors are often used, but open-tube wall reactors and membrane reactors have also been investigated. A principal advantage of all such systems is that they can use all the optical or electrochemical detectors routinely used in flow analysis. However, the problems of producing stable and robust immobilized enzyme reactors have proved more intractable than many researchers hoped, and other advances (e.g., the use of more sensitive detectors, improved availability of low-cost soluble enzymes) have minimized the advantages of using solid phase enzymes.

Nonetheless, much research continues in this area, particularly work on thin layers or membranes to which enzymes are attached. In such cases, the enzyme kinetics may be favorable, and, of the available detection procedures, fluorescence spectroscopy is particularly suitable for the study of solid surfaces, and is highly sensitive.

9.2. Immunoassays in Analytical Chemistry

9.2.1. Introduction

In recent years, analytical chemistry has grown tremendously, especially in the fields of environmental and process analysis. Modern measurement techniques have led to lower limiting values in the areas of water, soil, and air analysis. This has been followed by legislation requiring environmentally compatible production processes under constant control. As a result, less expensive (and often faster) monitoring techniques are necessary. The search for "chemical sensors," which, similar to optical and electrical sensors, can be integrated into production monitoring and quality control, is a logical consequence of this ecological and economic pressure.

Immunoassays (IAs) are based on the formation of a thermodynamically stable antigen–antibody complex. These methods play an important role, especially in clinical chemistry, being used for the fast and safe detection of proteins, hormones, and pharmaceutical agents. These techniques promise to close the gap between the cheapest chemical sensors and conventional, expensive, slower analytical methods.

On the other hand, classical analytical chemistry is just starting to accept immunoassays; the reasons for this hesitancy are to be found in the necessity of:

1) Synthesizing an immunogen and coating antigen (hapten linked to different carrier proteins)
2) Isolating an antiserum after immunization, usually of a vertebrate
3) Development, optimization, and synthesis of a labeled analyte (hapten) derivative after isolation and characterization of the first antibodies (AB)
4) Validation of both the antibodies obtained, and the entire test

5) Possibly having to start the immunization all over again, if the antibodies do not fulfill the selectivity and affinity requirements

Every analytically usable method must fulfill the criteria of selectivity, sensitivity, calibration ability, and reproducibility.

The development of immunoassays is frequently based on empirical findings which are an obstacle to certification. In particular, despite the fact that polyclonal antibodies often permit better detection tests, they are not reproducible in any given laboratory. Even the production of monoclonal antibodies, which in principle is easily reproducible, does not aid worldwide acceptance and recognition of immunoassays because of their extremely complex production processes, which are increasingly being patented.

Nevertheless, a steady rapid development of immunotests can be observed, especially in the field of environmental monitoring [1]–[3], where the question of costs has become so important that the establishment of an immunotest can be worthwhile in spite of all the facts mentioned above. In the development of chemical sensors, efforts are also being made to utilize the biochemical recognition principle by coupling with optical, electrochemical, or other transducer (signal transfer) [4]–[6]. However, the slower kinetics involved in molecular biological processes require stricter maintenance of the experimental protocol, e.g., the time of reaction between the antibody and the tracer molecule or analyte (hapten) should be set up in an exactly reproducible manner by means of flow-injection analysis [7].

Immunoassays become important when:

1) Fast measurement and evaluation are required
2) Highest possible detection strength is required
3) Large numbers of samples are to be expected
4) Only complex and expensive analytical methods are otherwise available

The greatest potential for the use of immunoassays in environmental analytical chemistry is in screening, i.e., for the selection of contaminated and uncontaminated samples for further validation analysis.

9.2.2. Polyclonal, Monoclonal, and Recombinant Antibodies

The isolation of highly selective (specific), high-affinity antibodies is the primary requirement for the successful development of an immune test [8], [9].

In analytical chemistry, especially in the environmental field, analytes of molecular mass < 1000 are often of interest, but molecules of this type (haptens) cannot alone induce the generation of antibodies.

For this reason, it is necessary to produce an immunogen by coupling the hapten to a high molecular carrier protein. The possibilities for immunogen synthesis are restricted by the structure of the hapten. The presence of amino or carboxyl groups, which can be directly coupled to the carrier protein by a peptide bond, is most favorable. If the hapten contains no reactive group, it must be introduced. An example is the reaction of haptens containing hydroxyl groups, which are converted to hemisuccinates with succinic anhydride or *N*-bromosuccinimide. Haptens with keto groups can be converted to oxime derivatives.

Modified haptens or haptens containing reactive groups are then coupled to a high molecular carrier protein. High yields and no interfering side products are usually obtainable from reactions at room temperature in aqueous media and at near-neutral pH. The carrier proteins commonly used are keyhole limpet hemocyanin human or bovine serum albumin, bovine gammaglobulin, or thyroglobulin. Coupling reactions comprise: (1) the carbodiimide-mediated conjugation, (2) the mixed anhydride procedure, (3) the NHS (*N*-Hydroxysuccinimide) ester methods, and (4) the glutaraldehyde condensation [10]. The degree of coupling, i.e., the number of hapten molecules per protein molecule, should be determined, using either radiolabeled hapten, UV and visible spectroscopy, matrix assisted laser desorption time-of-flight mass spectrometry (MALDI TOF MS), or indirect methods.

The choice of a suitable position of attachment to link the hapten to the carrier is important because it is responsible for recognition by the antibody produced. Especially in the case of parent substances which are rich in isomers, only some of which are of toxicological interest, e.g., PCBs or dioxins, the reactive positions used for conjugate synthesis must be carefully selected.

For the generation of polyclonal antibodies, the immunogen is applied to a vertebrate (rabbit, sheep, etc.), usually with the aid of adjuvants which enhance immunogenicity, by a mechanism that is not fully understood. The production of polyclonal antibodies is not very reproducible. For this reason, antibody concentration (titer), specificity (selectivity of the antibodies), and affinity must be continually monitored, and the ability of the isolated antiserum to bind to the hapten must be determined. Further, the cross-reactivity of the antibodies should also be determined as early as possible. Cross-reactivity refers to the reaction of an antibody with molecules which are structurally very similar to the desired analyte (hapten); it determines the sensitivity to interfering substances in the final test.

Polyclonal immune sera from an immunization of this type consist of a spectrum of antibodies, which are active against the diverse epitopes of the same antigen and, therefore, possess various affinity constants.

The acceptance of immunoassays depends on the availability of antibodies. Since the amount of polyclonal antibodies is always limited, and even repeated immunizations of the same experimental animal species results in nonreproducible characteristics (specificity, affinity), the production of suitable monoclonal antibodies (they are product of a single cell alone) is most important. The method was developed by KÖHLER and MILSTEIN and encompasses the immortalization of a single-antibody producing cell (B-lymphocyte) by fusion to a myeloma cell [11]. The resulting hybridoma produces just one species of antibody which can be grown indefinitely and, therefore, offers a means to advise a consistent product in unlimited quantities. Up to now, compared to polyclonal antibodies monclonal antibodies are less frequently applied in environmental analysis.

In recent years, genetic engineering processes have entered the field of antibody production. A number of functional recombinant single-chain antibody fragments (scAbs, recombinant antibodies) which recognizes and binds to small compounds has been expressed and characterized in microorganisms and transgenic plants [12] – [14]. However, these techniques are still at the development stage.

9.2.3. Sample Conditioning

Sample preparation is of tremendous importance in the application of an immunoassay. Apart from the usual criteria for representative sampling, special attention should be paid to the following influences:

1) pH value
2) Humic substances
3) Surfactants
4) Heavy metals
5) Organic solvents

If enzymatic tracers are used, the actual detection reaction is frequently disturbed by influences exerted on the reactive center of the enzyme, especially those due to interactions of the antibody with polyfunctional groups, e.g., with the humic substances in natural water samples [15]. This should be taken into account, especially in the use of immunoassays for soil screening. In competitive immunoassays, analytes that are sparingly soluble in water can be solubilized by the addition of surfactants [16]. On the other hand, the surfactant can alter the tertiary structure of the antibody, and possibly of the enzyme involved. The use of organic solvents is being intensively studied [17]. In some cases, immunoassays are known to tolerate more than 10 vol % of solvent.

The separation or complexation of humic substances and/or heavy metals is successful in favorable cases, e.g., by the addition of bovine serum albumin (BSA) in a trinitrotoluene enzyme-linked immunosorbent assay (ELISA) [18]. In the screening of soil samples for pesticides, persistent centrifugation at 17 000 rpm, and simple dilution with bidistilled water in the ratio 1 : 10 was sufficient [19].

In each application, the appropriate measurement must be established by independent validation analysis. Nonspecific interactions are very frequent in immunoassays, and usually lead to a slight overdetermination of the analyte.

9.2.4. Immunoassays

9.2.4.1. Radioimmunoassay

The development of radioimmunoassays (RIA), based on studies by YALOW and BERSON, has opened up an area of application especially in clinical chemistry [20], [21].

In RIAs, a radioactive label is used for detection of the formation of an antibody – antigen complex. Thus, a simple, specific, sensitive, and precise determination of the radioactive isotope is available. At present, however, immunoassays are increasingly carried out with nonisotopically labeled antigens or antibodies because of the legal restrictions on the use of radioactive material.

9.2.4.1.1. Isotopic Dilution Radioimmunoassay

In this classical RIA, a variable amount of unlabeled analyte (hapten) competes with a constant amount of a radiolabeled substance for a limited number of antibody binding sites:

$$Ag^* + Ag \begin{array}{l} \nearrow Ag^*\,AB(s) + Ag(b) \\ \searrow Ag\,AB(s) + Ag^*(b) \end{array}$$

where Ag is the antigen (hapten or analyte); Ag* the radiolabeled antigen; AB the antibody; and (b) and (s) indicate species bound or in solution.

The more analyte in the sample, the more labeled antigen is found in solution. To quantify this amount, a separation step must be carried out, usually by subjecting the adsorptively bound antibodies (e.g., on a microtiter plate) to a washing step.

In principle, standard and analyte should have the same ability to displace a tracer molecule from the binding site on the antibody.

9.2.4.1.2. Immunoradiometric Assay (IRMA)

This process uses radioactively labeled antibody AB*. Two antibodies are required which recognize different binding sites on the antigen; the first is fixed on a microtiter plate or in a test tube, and the second is labeled, e.g., with ^{125}I. Sequential incubation is carried out, first with the sample or the standard analyte, followed by a separation washing step. Subsequent incubation with radiolabeled antibody AB* shows the antigen binding density through sandwich coupling:

$$AB(s) + Ag \longrightarrow AB-Ag(s)$$

$$AB-Ag(s) + AB^* \longrightarrow AB-Ag-AB^*$$

Small analyte molecules (haptens) cannot be detected by this technique because of the lack of several binding sites (epitopes), so it cannot be used for environmental analysis.

9.2.4.2. Nonisotopic Homogeneous Immunoassay

For diverse reasons, such as restrictions on the use of radioactive substances, nonisotopic immunoassays have gained general acceptance, especially in trace analysis and environmental analysis 1]–[3].

Consequently, a sensitive measuring principle other than radioactive decay is required. Furthermore, a distinction must be made between tests that require a separation step to determine the bound component and those that do not. Nonisotopic, homogeneous immunoassay allows the direct observation of the hapten bound to the antibody.

9.2.4.2.1. Latex Particle Agglutination Immunoassay

A large number of latex agglutination immunoassays have recently been adopted from clinical chemistry. These assays are based on the visualization of antigen–antibody complexes by the attachment of latex particles or gold colloids. Entities of this type with dimensions in the nanometer or micrometer range can be quantified by turbidimetry, nephelometry, light scattering techniques, and particle counters [22]–[25].

9.2.4.2.2. Enzyme-Multiplied Immunoassay Technique (EMIT)

In the EMIT [26], [27], the analyte is covalently bound to the enzyme in spatial proximity to the active site and, consequently, the formation of the antibody–antigen complex inactivates the enzyme; addition of hapten results in a reduction of this inactivation.

9.2.4.2.3. Apoenzyme Reconstitution Immunoassay System (ARIS)

If, however, the antigen is covalently bound to the prosthetic group of an enzyme such as glucose oxidase (E.C. 1.1.3.4) and an aliquot of the coupled antigen to flavin–adenine dinucleotide) is added to determine an analyte, free antibodies prevent the reconstitution of the enzyme. The concentration of the free antibody naturally depends on the analyte concentration in the sample. Similar to the EMIT technique, the ARIS is used in automatic analyzer systems in clinical chemistry [28].

9.2.4.2.4. Fluorophore-Labeled Homogeneous Immunoassay

At first glance, fluorescent labeling appears to have a much higher detection strength compared to colorimetric detection, but this is not the case. First, the affinity constant generally limits the detection strength of a process. Second, fluorophores are exposed to many influences, such as quenching by impurities, or even adsorption of the fluorophore molecule. However, the fact that the detection can be repeated is advantageous, whereas a chemical reaction is irreversible [29].

The best-known variant is the substrate-labeled fluorescence immunoassay (SLFIA) test [30]. Galactose is linked to the antigen and to the fluorophore, methylumbelliferone. β-Galactosidase (E.C. 3.2.1.23) is capable of hydrolyzing and releasing the fluorophore, as long as the antigen linked to galactose is not stabilized in an antibody – antigen complex. If free antigen is present, the formation of galactose – antigen – fluorophore – antibody complex is reduced. The measuring signal is conveniently proportional to the concentration of free antigen.

In pesticide trace analysis, efforts have been made to observe directly the coupling of antigen – antibody in the quenching of the fluorophore bound to a tracer molecule, by the use of time-resolved, laser-induced fluorescence [31]. On formation of an Ag* – AB complex, a significant change in the fluorescence lifetime is observed. However, the fluorescence quantum yield suffers drastically, and this test has never been used.

9.2.4.2.5. Homogeneous Fluorescence Polarization Immunoassay (FPIA)

Direct observation of the formation of a hapten – (fluorescently labeled) antibody complex is also possibe in polarized light [32]. The presence of free hapten reduces the antibody-tracer complex concentration, and the degree of polarization is lowered. The detection strength of this test is in the μmol/L range and thus not yet high enough for environmental analysis.

9.2.4.2.6. Homogeneous Liposome Immunoassay

Owing to "cell wall formation" by multilayer membranes, liposomes may encapsulate up to several hundred fluorophore molecules.

The measuring principle is based on the fact that hundreds of hapten molecules are incorporated into the cell membrane of liposomes, e.g., by linkage of phosphatidylethanolamine (a functional part of the multilayer membrane). In the presence of an antibody for the formation of an immune complex, destabilization of the entire vesicle occurs. The contents of the liposome, e.g., fluorophores dyes, spin-labeled molecules, or enzymes, are released. Liposome stability is a general problem associated with this method [33].

9.2.4.3. Nonisotopic Heterogeneous Immunoassay

The most common type of immunoassay is the nonisotopically labeled heterogeneous test [1]. Either the antibody or the hapten (analyte) to be detected is fixed to a solid interface via a covalent bond or by adsorption. The other significant difference lies in the fact that no radioactive labeling is used for signal production.

9.2.4.3.1. Enzyme-Linked Immunosorbent Assay (ELISA)

This assay is the most important in trace and environmental analysis. First, antibodies are immobilized on a solid carrier. For this, it may be necessary to take suitable measures, such as ^{60}Co radiation, to make the carrier sorption active. Standard 96-well microtiter plates, polystyrene beads, or test tubes can be used.

In a typical ELISA test on a microtiter plate (Fig. 5), the carrier is first coated with analyte-binding antibody. The sample or standard is added after washing off excess antibody with a surfactant solution. The volume added is typically 100 – 200 μL. After (optional) preincubation, a constant quantity of tracer, e.g., enzyme-labeled hapten, is added to the sample or standard. This initiates a competitive reaction, because only a limited number of antibodies are available for binding. After the tracer incubation period, sample (or standard) and excess reagents are washed away. The bound tracer concentration is inversely proportional to that of the analyte. The amount of bound tracer can be determined via an enzyme substrate reaction with a chromogenic substrate. After a reaction time long enough to produce sufficient dye, the enzyme is denatured (enzyme reaction stopped) by addition of acid. Subsequently, the depth of color formed in the individual cavities of the microtiter plate can be automatically determined with a "plate reader" (photometer).

The application of a competitive heterogeneous immunoassay is very common. Some important immunoassays for trace amounts of environmentally relevant compounds are presented in Table 1. Especially for agrochemicals, a large number of determination procedures have been established, because of legislation on drinking

Table 1. Selected ELISA tests in environmental analysis

Substance	Detection limit	Reference
Pesticides		
Alachlor	1 ppb	[34]
Aldicarb	300 ppb	[35]
Atrazine	50 ppt	[36]
Benomyl	250 ppb	[37]
Bentazon	2 ppb	[38]
Diflubenzuron	1 ppb	[39]
Endosulfan	3 ppb	[40]
Fenpropimorph	13 ppt	[41]
Hydroxyatrazine	50 ppt	[36]
Metazachlor	10 ppt	[42]
Methabenzthiazuron	50 ppt	[43]
Molinate	30 ppb	[44]
Norflurazon	1 ppb	[45]
Terbuthylazin	60 ppt	[46]
2.4-D	100 ppb	[47]
Aromatic compounds		
Benzo[*a*]pyrene	100 ppt	[48]
1-Nitropyrene	50 ppt	[49]
Pyrene	35 ppt	[50]
Trinitrotoluene	50 ppt	[51]
Halogenated aromatic hydrocarbons		
Pentachlorophenol	30 ppb	[52]
2,3,7,8-Tetrachlorodibenzodioxin	5 ppb	[53]
Heavy metals		
Mercury(II)	500 ppt	[54]
Surfactants		
Triton X-100	2 ppb	[55]

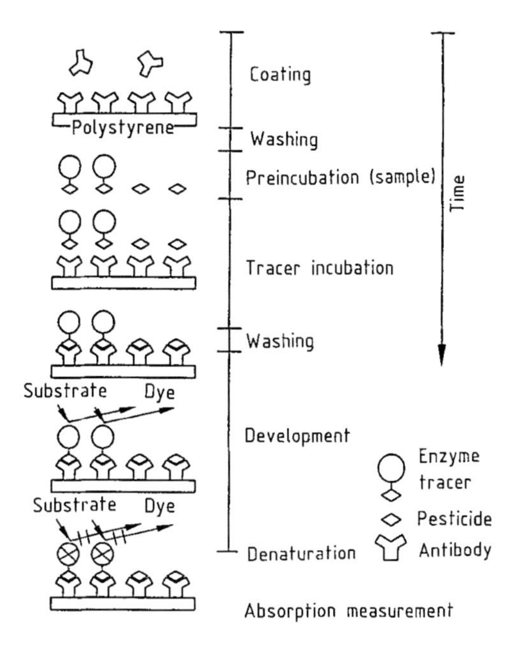

Figure 5. Principle of a direct competitive enzyme immunoassay (ELISA) with photometric detection

water which stipulates (as a precautionary measure) a limiting value of 100 ng/L for a single pesticide; this is not based on toxicological findings. The demand for sampling and pesticide analysis has given a sudden worldwide impetus to the development of immunoassays as a fast and cost-effective method of analysis.

The use of indirect competitive heterogeneous immunoassays (Fig. 6) is also common. Here, the hapten is bound to the surface of the carrier. Since hapten alone cannot be bound very reproducibly to a solid support by adsorption, a protein-linked hapten is used (e.g., to BSA). In principle, however, a conjugate (coating conjugate) different from the immunogen should be employed.

The advantage of this ELISA variant lies in the fact that (1) synthesis of a coating antigen is easier than that of a hapten-enzyme conjugate, (2) interfering matrix constituents are washed out before the addition of the enzyme-labeled second antibody, and (3) a prepared microtiter plate contains only slight amounts of a possibly toxic analyte [important when working with 2,3,7,8-tetrachlorodibenzodioxin (2,3,7,8-TCDD)]. After a washing step, sample and analyte-binding antibody are added. If larger amounts of free analyte are present in the added sample, only a few adsorptively bound analyte molecules react with AB 1. The AB – Ag complexes in solution are removed by a further washing step. The AB 1 molecules bound through hapten – BSA, which contain the analytical information, are detected by means of an enzyme-labeled second antibody. For this purpose, it is possible to use AB_2 that recognizes antibodies (IgG) from rabbits, from which AB 1 originates; these antibodies are commercially available. Signal production is again inversely proportional to the concentration of analyte. Enzymes commonly used are horseradish peroxidase (E.C. 1.11.1.7) and alkaline phosphatase (E.C. 3.1.3.1).

For environmental analysis, enzyme-labeled tracers are predominantly used because increasingly lower limiting values demand maximum sensitivity. The reason for this choice is based on the enormous turnover rate in chromophore or fluorophore formation by enzymes. In the most common tests, photometric evaluation predominates.

There has been no lack of attempts to reduce the detection limit further by the use of liposomes filled with enzymes [56] or fluorophores [57]. Unfortunately, liposomes are not very stable, and

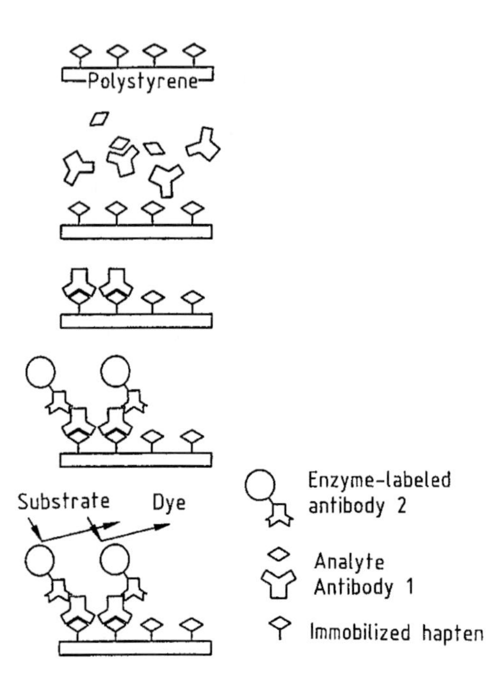

Figure 6. Principle of an indirect competitive enzyme immunoassay (ELISA)
The immobilized hapten corresponds to a hapten – BSA conjugate. AB 1 binds to the analyte, AB 2 binds to foreign antibodies and is covalently linked to an enzyme.

fluorophores with the highest quantum yields are extremely hydrophobic, and subject to quenching by adsorption to vessel walls.

Although they depend on special evaluation techniques, immunoassays based on signal production via chemi- and bioluminescence are successful [58]. In clinical chemistry, this type of signal production has already gained acceptance. This is, however, not the case in environmental analysis because of the often uncontrolled interaction of matrix components with the luminescence-producing enzyme.

Efforts are being made to improve the sensitivity of immunoassays by using long-lived fluorophore labeling and highly intense light sources (e.g., lasers). Labeling with rare earth chelates, which have fluorescence decay times in the range 600 – 1000 μs, permits an especially easy separation of scattered light from the signal [59] – [62]. The high cost of UV laser systems is presently an obstacle to its wider acceptance. However, the first applications at 780 nm, using less expensive semiconductor diode lasers, have already been published [63].

The application of electrically produced time-resolved luminescence without light source represents an extremely interesting variant [64].

9.2.4.3.2. Sandwich Immunoassay

This type of noncompetitive heterogeneous immunoassay (Fig. 7) is encountered especially in the case of larger antigens with several epitopes [65]. The antibody, immobilized on the microtiter plate in the first step, has the function of binding the analyte (antigen) or standard in the solution. After a washing step, the second antibody, which recognizes the second epitope of the analyte, is added. After another washing step, a labeled (in this case with an enzyme) third antibody, which binds to AB 2, is added. Apart from the large number of steps involved, all of which hinder reproducibility, problems are also posed by the necessity of suppressing nonspecific adsorption in the presence of added reagents. Furthermore, the color formation is not inversely proportional to the analyte concentration. Sandwich immunoassay is not applicable to environmentally relevant haptens because it requires an analyte molecule of a certain minimum size.

9.2.4.3.3. Immunoassays using Immobilized Affinity Ligands

New immunoassay concepts were described in which immobilized affinity ligands are used in unique flow-through systems (flow-injection immunoassay, FIIA). These systems are exclusively heterogeneous. Both immobilization of a specific antibody or a hapten on the solid support is possible. In most cases the scheme of the assay used is based on a sequential competitive enzyme immunoassay procedure [66], [67].

9.2.4.3.4. Immunoassays with Biotin and Avidin

Avidin (an egg yolk protein) exhibits an extremely strong affinity (ca. 10^{15} L/mol) for biotin (vitamin H). This irreversible interaction can be utilized in the design of heterogeneous immunoassays with biotinylated, antigen-specific antibodies [68], [69]. Biotinylation can be performed without interfering with antibody function. The antibodies are applied in the usual manner in a competitive immunoassay (Fig. 8). In the last step, enzyme-labeled avidin is added in excess. Since there are four possibilities for the binding of avidin

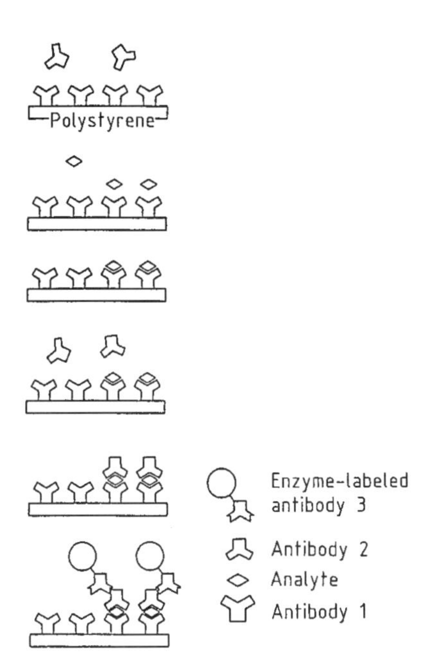

Figure 7. Principle of a noncompetitive immunoassay AB 1 is a catcher antibody, and binds to the analyte. AB 2 should be able to bind a second (identical or different) antigenic epitope. AB 3 binds to foreign antibodies (of type AB 2) and is covalently linked to an enzyme.

to biotin, labeled biotin can also be used to occupy binding sites of the bound avidin to amplify the recognition signal reaction. Streptavidin can be used instead of avidin.

9.2.4.3.5. Magnetic Particle Immunoassay

An especially elegant heterogeneous immunoassay format depends on coupling antibodies to magnetic particles. The assay is carried out in a test tube. The conjugate of analyte and peroxidase (tracer) is first added to the sample or standard and mixed well. An aliquot of an AB – magnetic particle suspension is then pipetted into the test tube and again mixed well. After a preselected incubation time, the particles are precipitated by a magnet. After pouring off the supernatant, the AB – magnetic particles are resuspended and re-precipitated (washing step). The depth of color, which is inversely proportional to the concentration of analyte in sample, is then determined in the usual manner via an enzyme – substrate reaction. Applications in the field of pesticide analysis have been published [70] – [72].

Instead of the tedious and destructive coupling of antibodies to magnetic particles, the utilization of natural bacterial magnetic particles is also possible [73].

9.2.5. Immunoaffinity Techniques

An immunosorbens represents an excellent agent for the specific preconcentration of analytes with the aid of antibodies (immunoextraction). Until now, this technique has been used mainly in clinical chemistry. Environmental applications started in 1994 [74]. The appropriate antibody is immobilized on a precolumn and, after a concentration phase, subjected to a desorption step. It is favorable if the eluent used for this purpose is the same as that employed in the subsequent liquid chromatographic separation.

The repeated use of the precolumn sometimes poses problems because desorption must be forcibly effected by denaturation of antibody, and is often reversible. The substantial band broadening in the desorption step represents another problem. The bands can be sharpened only if the analyte can be reconcentrated on a subsequent reversed phase column with a relatively weak eluent (low proportion of organic solvent).

Alternatively, immunoselective desorption can be carried out with a "displacer" having a still higher affinity for the immobilized antibody. In the subsequent chromatography, however, a reduction of the analytical "window" owing to the structural similarity between the displacer and the analyte is to be expected.

In recent years, a rather new technique — the sol-gel technology — was applied for the preparation of new immunosorbents (Fig. 9). Using this rather simple method, the antibodies are encapsulated in a silica network. They retain their activity and, in some cases gain even higher stability, and can react with ligands that diffuse into the highly porous matrix. Several papers were reported on the entrapment of polyclonal and monoclonal antibodies against environmental pollutants [75] – [78].

Preconcentration with immunosorbents, followed by quantification with an immunoassay, is not recommended because no increase in selectivity is achieved by using the same antibody in the preconcentration and in the test.

Immunosorbents are of interest in trace analytical problems which require optimum detection strength and selectivity.

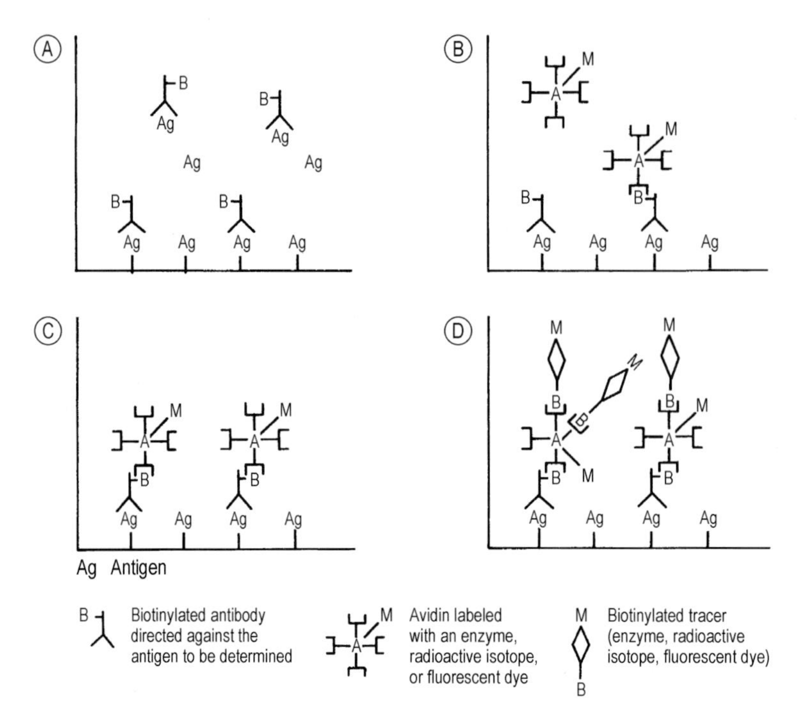

Figure 8. Principle of an immunoassay using the biotin – avidin system for labeling antibodies

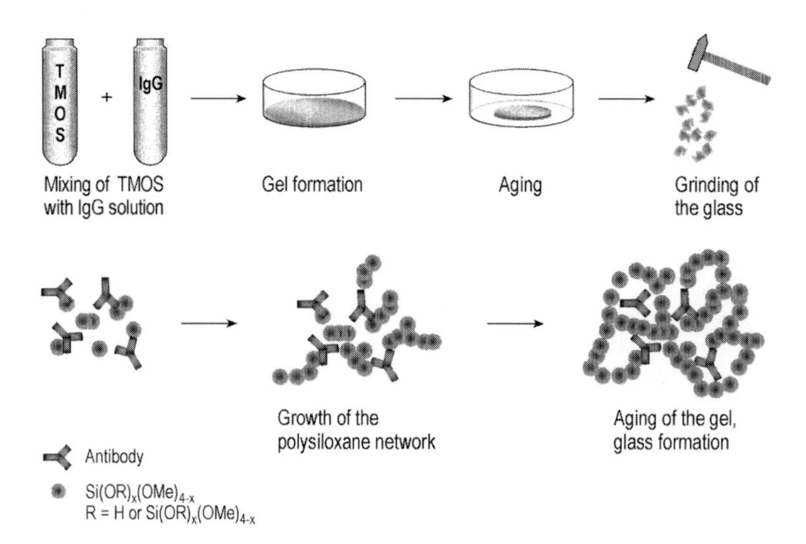

Figure 9. Encapsulation of antibodies in silicate glasses prepared by the sol – gel method. TMOS: tetramethoxysilane; IgG: immunoglobulin G

9.2.6. Immunoassays on Test Strips and other Planar Structures

Dry tests or test strips (also called lateral flow devices or immunomigration strips) are popular as quick tests in clinical chemistry. Immunological tests are especially interesting because they can make use of the high selectivity of AB – Ag complex formation [79], [80]. The objetive is to accommodate all the reagents required for a quanti-

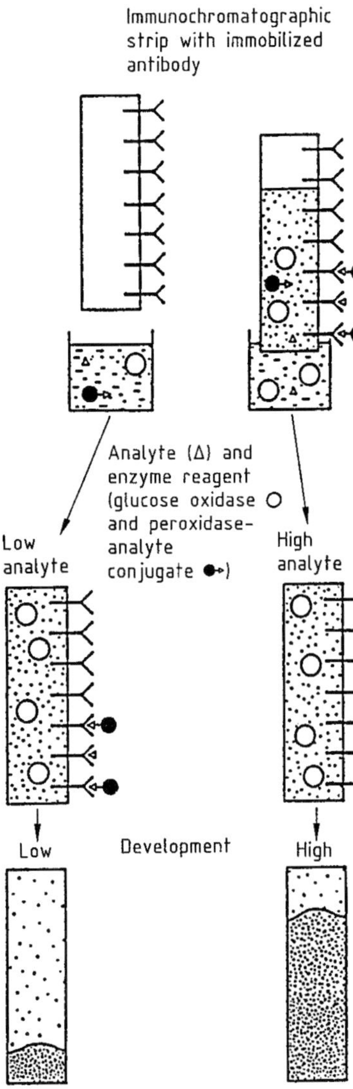

Figure 10. Principle of a test strip using enzyme immunochromatography

fiable test on a simple strip, filter, or capillary. It should be possible to dip the strip quickly into the liquid sample or to place a drop of sample on the carrier, and to determine the analyte concentration from the resulting depth of color or length of a colored band. Evaluation can be done visually or spots can be read-out by a pocket reflectometer. There should be tremendous possibilities for this test, especially for on-the-spot environmental analysis.

Two main variants are possible:

1) Accommodation of a homogeneous separation-free immunoassay on a dry matrix. This is possible for SLFIA and for ARIS. Other possibilities are labeling with gold particles or liposomes. The art lies in contact-free accommodation of all the reagents on the test strip, in such a way that no immune reaction takes place during storage.
2) The enzyme-channeling principle is especially suited to test strips. The test contains co-immobilized antibody and glucose oxidase. Two test strips must be used: one contains analyte-specific antibody, and the other is impregnated with peroxidase-specific antibody. The strips are dipped first into the sample solution and then into a color developer. The color developer contains a peroxidase tracer, glucose, and a chromogenic substrate (e.g., 4-chloro-1-naphthol). Channeling occurs by formation of H_2O_2 by glucose oxidase; this forms an insoluble dye on the surface of the test strip with 4-chloro-1-naphthol and peroxidase. In the case of the indicator test strip, the depth of color is influenced by competition with free analyte in the sample. The reference strip gives the standard value.

Enzyme immunochromatography (Fig. 10) represents an especially attractive variant [81]. A narrow strip of paper has the analyte-specific antibody immobilized on it. The sample is mixed with the tracer (analyte-labeled peroxidase and glucose oxidase) and the end of the test strip is dipped into this mixture. By capillary action, the solution is sucked up into the strip matrix. The sample is bound to the immobilized antibody as it travels up the strip, resulting in competition between the tracer and the analyte for the limited binding sites along the strip. A zone is formed which is proportional to the analyte concentration, and occupied by tracer molecules. After this mini-chromatography step, the entire strip is transferred to a color developing solution which consists of glucose and the peroxidase substrate, 4-chloro-1-naphthol.

A strong color is developed, owing to enzyme channeling only in places where the peroxidase tracer is bound, in addition to the glucose that is present everywhere. Thus, the occupied length, which is proportional to the concentration of analyte, can be visually quantified as in a thermometer.

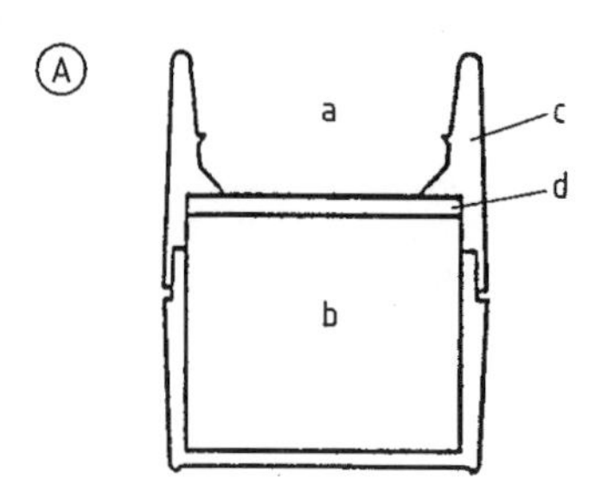

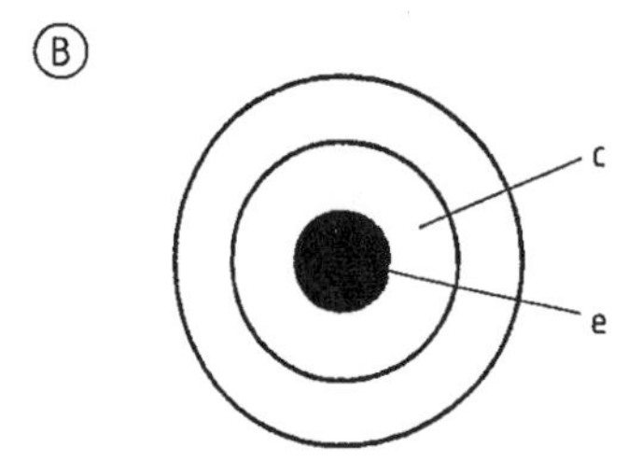

Figure 11. Cross section of the rapid immunofiltration test Hybritech ICON [72]
A) Side view; B) Top view
a) Fluid chamber; b) Absorbent reservoir; c) Membrane; d) Support disk; e) Area of immobilized antibody

Another heterogeneous immunoassay, developed for field use [82], [83], employs immunofiltration (Fig. 11). Antibodies are immobilized in the middle of a nylon membrane. The first few drops placed on this membrane are immediately sucked down into a reservoir by capillary action. A few drops of washing buffer are then added, followed by AP- (alkaline phosphatase) labeled antibody to the analyte and, subsequently, a few drops of indoxyl substrate. As in the case of the sandwich immunoassay, the depth of color of the spot formed in the middle is linearly proportional to the analyte concentration; i.e., an ELISA is obtained for determination of aflatoxin [84], atrazin [85], [86], [87], carbaryl [88] or polychlorinated biphenyls [89].

9.2.7. Characterization and Interference

Immunochemical determination seems to be especially suited to the screening of large numbers of samples (e.g., water and soil analyses). Immunoassays are well suited for the analysis of the smallest sample volumes (e.g., in fog water analysis) because they allow analysis without sample preparation in a few microliters [90]. In critical

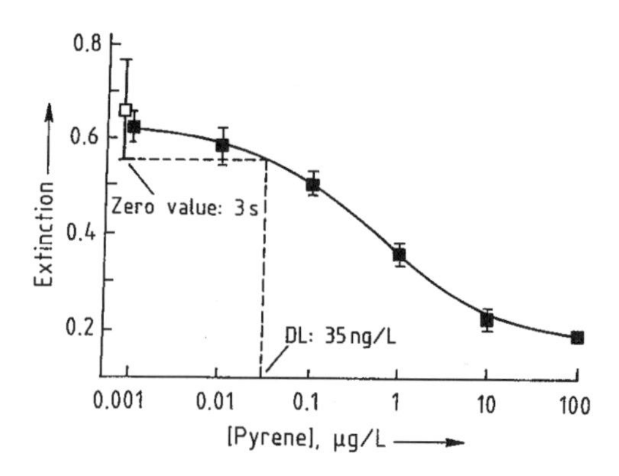

Figure 12. Typical calibration curve of a pyrene ELISA [50] The detection limit DL is calculated from the blank extinction value plus a three-fold standard deviation: 35 ppt. Test center point, C=0.583 µg/L; Margin of error: 1 Δ; n=12.

applications, immunoassays do not compete with, but rather supplement conventional trace analytical methods.

In contrast to applications in clinical chemistry, matrix interference must be expected when immunoassays are applied to environmental samples. Disturbances can be caused not only by the presence of high molecular, organic water components (humic acids, lignosulfonic acids), but also by inorganic ions (Al^{3+}, Fe^{3+}, other transition metals) and oxidizing agents (Cl_2, O_3, ClO_2, H_2O_2) used in water treatment.

An exceptional feature of competitive heterogeneous tests, which are at present the only tests that are sufficiently sensitive, is evaluation by photometrical readout of the microtiter plates.

In multiple determinations, the median is calculated for each calibration point (Fig. 12). The margins of error correspond to simple standard deviations. The medians, or the singly determined values, are fitted to a four-parameter function by a simplex process; each curve has seven calibration points. The zero value is not included in the mathematical evaluation, but is used only for control purposes:

$$Y = \frac{(A - D)}{\left[1 + \left(\frac{x}{C}\right)^R\right]}$$

where x is the concentration of analyte or hapten (µg/L); Y the extinction; A the maximal extinction (upper asymptote); D the minimal extinction

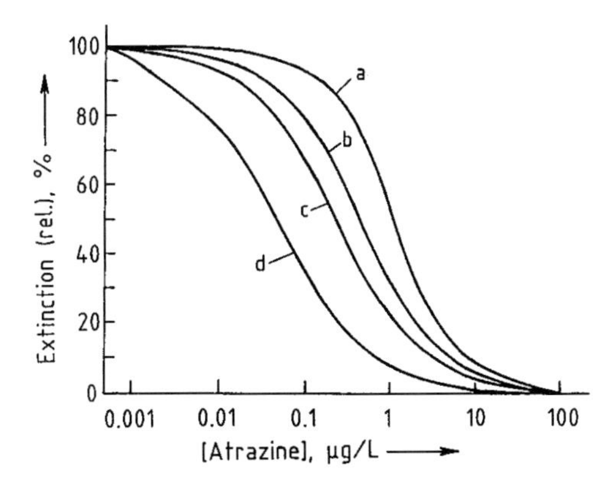

Figure 13. Calibration curves of an atrazine ELISA as a function of the incubation time of the tracer
a) 30 min; b) 10 min; c) 5 min;
Preincubation of unlabeled hapten: 60 min [91].

Table 2. Concentrations that cause 10 % inhibition of horseradish peroxidase (N = 12)

Ion	Concentration, mg/L (±2 *s*)
CN^-	0.001 ± 0.0002
SCN^-	0.02 ± 0.004
ClO_3^-	0.03 ± 0.004
N_3^-	0.04 ± 0.005
ClO_4^-	0.02 ± 0.003
F^-	0.15 ± 0.016
Cu^{2+}	35.40 ± 4.04
Ni^{2+}	8.70 ± 0.96
Co^{2+}	17.30 ± 2.11
Al^{3+}	43.10 ± 2.97
Fe^{3+}	33.10 ± 4.74
Fe^{2+}	1.30 ± 0.08
$[Fe(CN)_6]^{3-}$	0.02 ± 0.002
$[Fe(CN)_6]^{4-}$	0.17 ± 0.02
Sn^{2+}	18.00 ± 2.11

(lower asymptote); *C* the test center point (µ g/L); and *R* a curvature parameter.

The curves are normalized with respect to both *A* and *D*:

$$Y_N = \frac{1}{\left[1 + \left(\frac{x}{C}\right)^B\right]}$$

where Y_N approximately corresponds to the signal B/B_0 normalized to the zero value (*B* represents the extinction of the sample and B_0 the blank reading determined with the plate reader) and enforces convergence of all calibration curves for [analyte] $\rightarrow 0$ and [analyte] $\rightarrow \infty$. This standardization has the advantage that, unlike B/B_0, neither a zero value (associated with a relatively large error) nor an excess value (difficult to obtain owing to solubility or contamination problems) is required.

In the case of an ELISA, the tracer incubation time is important, as illustrated in Figure 13 for a triazine herbicide test [91]. The test center point *C* can be lowered by a factor of 25 by reducing the incubation time from 30 to 1 min; if it is increased to 40 h, a marked increase in the test center point *C* is obtained. Thus, the test covers atrazine concentrations of more than five powers of ten by simple variation of the incubation time. It is not known whether polyclonality or a kinetic effect (imbalance between antibody, hapten, and labeled hapten) is responsible for the observed shift of the calibration curves.

In immunoassays, antibodies react to a large number of chemically closely related species with varying sensitivity. This cross-sensitivity, also called cross-reactivity, is an indication of high structural similarity between haptens. Since cross-sensitivities cannot be exactly predicted, they must be experimentally determined for each antibody and each test design. Crossreactivities CR_{50} are normally determined by forming ratios of the 50 % values of the substances to be compared.

However, the actual influence of cross-reacting compounds can depend on their concentrations and, therefore, a single value for the cross-reactivity cannot contain all the information. For unknown samples, the influence of cross-reacting substances as a function of concentration should be determined in advance.

Water components can influence either the tracer (the tracer enzyme in ELISA), or the antibody, or the hapten – AB binding. The concentrations of substances that cause a 10 % inhibition of the horseradish peroxidase normally used as label in enzyme immunoassays are listed in Table 2.

Strongly oxidizing substances, such as ClO_3^- or complexing agents (F^-, SCN^-, CN^-) exert an inhibiting effect. The azide ion, which is used as a bacteriostatic agent in immunology, can also cause inhibition. If horseradish peroxidase is used as a tracer enzyme, NaN_3 can be employed only for the preservation of antibody solutions. No effect is exerted by the following ions: Hg^{2+}, $Cr_2O_7^{2-}$, Na^+, K^+, Li^+, Sr^{2+}, Mg^{2+}, Ba^{2+}, Ca^{2+}, Ag^+, Pb^{2+}, NH_4^+, Cl^-, NO_2^-, NO_3^-, and SO_4^{2-}.

In the presence of Ca^{2+}, the activity of horseradish peroxidase can be increased by up to 40 %, depending on the contact time and the Ca^{2+} concentration [15].

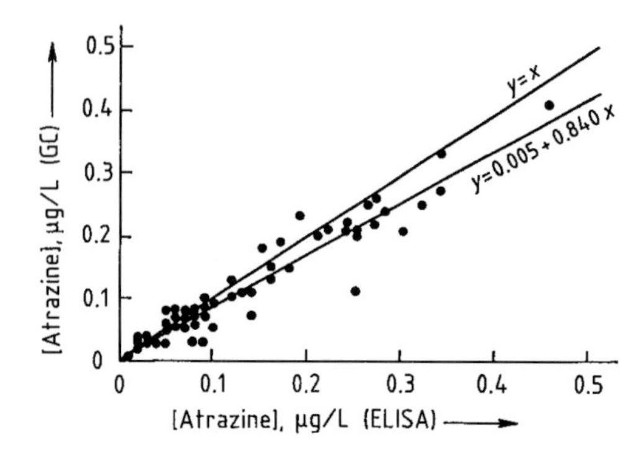

Figure 14. Correlation between two different analytical techniques (GC and ELISA) using atrazine as an example $n=89$, $r=0.956$ [15]

Sample components can interfere with antibodies in diverse ways. Evaluation is possible by means of calibration curves with the addition of various potentially interfering substances. Organic solvents (e.g., acetonitrile, methanol, and ethanol) in relatively low concentrations can affect the test. The maximal extinction falls increases on addition of solvents, the test center point is shifted by up to two powers of ten in the presence of 10 vol % of solvent [15]. In the immunological determination of trace substances (e.g., pesticides), the standard solutions should also be made up with the corresponding solvents.

An extremely important point in process characterization is validation with, e.g., classical methods, such as gas chromatography (GC) Fig. 14) or liquid chromatography. Atrazine in surface waters was measured by means of ELISA and GC, and the results compared. The results correlate well, and confirm the theory that immunological methods can give reliable values in practice. The results of both methods are equivalent with a probability of 95 %. No false negative values were obtained. The same applies to the use of immunoassays on soil extracts. In comparison with GC analyses, ELISA usually gives higher analyte concentrations, which is put down to the effects of humic acid.

9.2.8. Future Immunological Applications and Techniques

Apart from the extension of the tests already available to include analytes measurable only by expensive and tedious instrumental analytical techniques, the spectrum of applications is certain to widen.

Until now, water and body fluid (serum, urine) represent the most common test matrices. Soil analysis has already begun to profit tremendously from immunoassays. Very few applications exist in air monitoring, even though active and passive measuring procedures could profit increasingly from immunological techniques for the monitoring of diffuse sources.

The further development of immunolocalizing analytical techniques in conjunction with state-of-the-art optoelectronics and luminescence microscopy also seems worthwhile for the recognition of colloidal or cell-bound haptens.

In chemical sensors, the adaptation of immunological recognition techniques to optical and electrochemical transducers has been under investigation for several years.

An intense discussion presently goes on the usefulness of miniaturized analytical techniques, which may be considered part of the emerging nanotechnology complex. Immunological technology, after hyphenation with chromatographic and electrophoretic techniques, may be transferred to a chip format, obtaining a complete and very efficient "analytical-lab-on-a-chip". If a library of antibodies (an antibody array) is available, exhibiting a different pattern of affinity for a series of structurally related compounds, an appropriate statistical analysis (multivariate statistical techniques, parametric models) has the power to turn the problem of cross-reactivity into an advantage (multianalyte immunoassays). The breakthrough of this technology will mainly depend on the availability of improved solid supports and antibodies, and on the development of more simple instruments [92]–[95].

Increasing activities in validation and standardization of immunochemical tests, initiated by federal government and regulatory agencies of several countries and well-recognized national and international organizations will lead to more transparency and uniformity in method development and evaluation. This will help that the assays get rid of their image as "dubious biological tests", as they are sometimes called by analytical chemists [96]. Immunological methods cannot be assessed simply as "good" or "bad" but rather as suitable for an application or not.

9.3. References

[1] J. P. Sherry, *Crit. Rev. Anal. Chem.* **23** (1992) 217.
[2] J. P. Sherry, *Chemosphere* **34** (1997) 1011.
[3] E. P. Meulenberg et al., *Environ. Sci. Technol.* **29** (1995) 553.
[4] E. A. H. Hall (ed.): *Biosensoren,* Springer, Berlin 1990.
[5] B. Eggins (ed.): *Biosensors,* Wiley and Teubner, Chichester 1996.
[6] K. R. Rogers, A. Mulchandani (eds.): *Affinity Biosensors,* Humana Press, Totowa 1998.
[7] R. Schmid (ed.): *Flow Injection Analysis (FIA)* Based on Enzymes or Antibodies, VCH Verlagsgesellschaft, Weinheim 1991.
[8] P. Tijssen (ed.): *Practice and Theory of Enzyme Immunoassays,* Elsevier, Amsterdam 1985.
[9] J. Peters, H. Baumgarten (eds.): *Monoklonale Antikörper,* Springer, Berlin 1990.
[10] G. T. Hermanson (ed.): *Bioconjugate Techniques,* Academic Press, San Diego 1996.
[11] G. Köhler, C. Milstein, *Nature* **256** (1975) 52.
[12] H. A. Lee, M. R. A. Morgan, *Trends in Food Sci. Technol.* **4** (1993) 129.
[13] K. Kramer, *Anal. Letters* **31** (1998) 67.
[14] C. W. Bell et al., *ACS Symp. Ser.* **586** (1995) 413.
[15] T. Ruppert, L. Weil, R. Niessner, *Vom Wasser* **78** (1991) 387.
[16] P. Albro et al., *Toxicol. Appl. Pharmacol.* **50** (1979) 137.
[17] W. F. W. Stöcklein, A. Warsinke, F. W. Scheller, *ACS Symp. Ser.* **657** (1997) 373.
[18] C. Keuchel, L. Weil, R. Niessner, *SPIE* **1716** (1992) 44.
[19] R. Schneider, L. Weil, R. Niessner, *Int. J. Environ. Anal. Chem.* **46** (1992) 129.
[20] R. Yalow, S. Berson, *J. Clin. Invest.* **39** (1960) 1157.
[21] T. Chard (ed.): *An Introduction to Radioimmunoassay and Related Techniques,* Elsevier, Amsterdam 1990.
[22] P. Masson et al. in J. Langone, H. van Vunakis (eds.): *Immunochemical Techniques,* Part C, Academic Press, New York 1981, p. 106.
[23] Y. V. Lukin et al., *ACS Symp. Ser.* **657** (1997) 97.
[24] S. Obata et al., *Anal. Sci. Suppl.* **7** (1991) 1387.
[25] H. Weetall, A. Gaigalas, *Anal. Lett.* **25** (1992) 1039.
[26] E. Engvall in H. van Vunakis, J. Langone (eds.): *Methods in Enzymology,* **vol. 70,** Academic Press, New York 1970, p. 419.
[27] D. Morris et al., *Anal. Chem.* **53** (1981) 658.
[28] D. L. Morris, R. T. Buckler, *Meth. Enzymol.* **92** (1983) 413.
[29] E. Soini, I. Hemmilä, *Clin. Chem. (Winston-Salem, N.C.)* **25** (1979) 353.
[30] T. Ngo et al., *J. Immunol. Methods* **42** (1981) 93.
[31] R. Niessner et al.: "Application of Time-Resolved Laser-Induced Fluorescence for the Immunological Determination of Pesticides in the Water Cycle," in B. Hock, R. Niessner (eds.): *Immunological Detection of Pesticides and their Metabolites in the Water Cycle,* VCH Verlagsgesellschaft, Weinheim, in press.
[32] S. A. Eremin et al., *ACS Symp. Ser.* **586** (1995) 223.
[33] S. G. Reeves, S. T. A. Siebert, R. A. Durst, *ACS Symp. Ser.* **586** (1995) 210.
[34] J. Rittenburg et al. in M. Vanderlaan et al. (eds.): *Immunoassays for Trace Chemical Analysis,* **vol. 451,** American Chemical Society, Washington D.C., 1990, p. 28.
[35] J. Brady et al.: "Biochemical Monitoring for Pesticide Exposure," *ACS Symp. Ser.* **382** (1989) 262.
[36] J.-M. Schläppi, W. Föry, K. Ramsteiner, "Immunochemical Methods for Environmental Analysis," *ACS Symp. Ser.* **442** (1990) 199.
[37] H. Newsome, P. Collins, *J. Assoc. Off. Anal. Chem.* **70** (1987) 1025.
[38] Q. Li et al., *J. Agric. Food Chem.* **39** (1991) 1537.
[39] S. Wie, B. Hammock, *J. Agric. Food Chem.* **32** (1984) 1294.
[40] R. Dreher, B. Podratzki, *J. Agric. Food Chem.* **36** (1988) 1072.
[41] F. Jung et al., *J. Agric. Food Chem.* **37** (1989) 1183.
[42] H. Scholz, B. Hock, *Anal. Lett.* **24** (1991) 413.
[43] S. Kreissig et al., *Anal. Lett.* **24** (1991) 1729.
[44] Q. Li et al., *Anal. Chem.* **61** (1989) 819.
[45] B. Riggle, B. Dunbar, *J. Agric. Food Chem.* **38** (1990) 1922.
[46] P. Ulrich, R. Niessner, *Fres. Environ. Bull.* **1** (1992) 22.
[47] C. Hall et al., *J. Agric. Food. Chem.* **37** (1989) 981.
[48] A. Roda et al., *Environ. Technol. Lett.* **12** (1991) 1027.
[49] D. Knopp et al., *ACS Symp. Ser.* **657** (1997) 61.
[50] K. Meisenecker, D. Knopp, R. Niessner, *Analytical Methods and Instrumentation* **1** (1993) 114.
[51] C. Keuchel, L. Weil, R. Niessner, *Anal. Sci.* **8** (1992) 9.
[52] J. Van Emon, R. Gerlach, *Bull. Environ. Contam. Toxicol.* **48** (1992) 635.
[53] B. Watkins et al., *Chemosphere* **19** (1989) 267.
[54] D. Wylie et al., *Anal. Biochem.* **194** (1991) 381.
[55] S. Wie, B. Hammock, *Anal. Biochem.* **125** (1982) 168.
[56] T.-G. Wu, R. Durst, *Mikrochim. Acta* 1990, 187.
[57] S. Choquette, L. Locascio-Brown, R. Durst, *Anal. Chem.* **64** (1992) 55.
[58] A. Tsuji, M. Maeda, H. Arakawa, *Anal. Sci.* **5** (1989) 497.
[59] E. Diamandis, *Clin. Biochem.* **21** (1988) 139.
[60] E. Diamandis, T. Christopoulos, *Anal. Chem.* **62** (1990) 1149 A.
[61] T. Christopoulos, E. Diamandis, *Anal. Chem.* **64** (1992) 342.
[62] Y.-Y. Xu et al., *Analyst (London)* **116** (1991) 1155.
[63] T. Imasaka et al., *Anal. Chem.* **62** (1990) 2405.
[64] J. Kanakare et al., *Anal. Chim. Acta* **266** (1992) 205.
[65] Hybritech, US 4376110, 1983 (G. S. David, H. E. Greene).
[66] B. B. Kim et al., *Anal. Chim. Acta* **280** (1993) 1991
[67] P. M. Kraemer et al., *ACS Symp. Ser.* **657** (1997) 153.
[68] M. Wilchek, E. A. Bayer, *Meth. Enzymol.* **184** (1990) 14.
[69] F. Eberhard: "Entwicklung von Immunoassays und Markierungsmethoden für die Bestimmung von Schimmelpilzgiften mit immunooptischen Biosensoren," Dissertation, Universität Hannover 1991.
[70] F. Rubio et al., *Food Agric. Immun.* **3** (1991) 113.
[71] T. Lawruk et al., *Bull. Environ. Contam. Toxicol.* **48** (1992) 643.
[72] J. A. Itak et al., *ACS Symp. Ser.* **657** (1997) 261.
[73] N. Nakamura, T. Matsunaga, *Anal. Sci.* **Suppl. 7** (1991) 899.
[74] G. S. Rule, A. V. Mordehai, J. Henion, *Anal. Chem.* **66** (1994) 230.
[75] J. Zuehlke, D. Knopp, R. Niessner, *Fresenius J. Anal. Chem.* **352** (1995) 654.
[76] A. Turniansky et al., *J. Sol–Gel Sci. Technol.* **7** (1996) 135.
[77] M. Cichna et al., *Chem. Mater.* **9** (1997) 2640.

[78] T. Scharnweber, D. Knopp, R. Niessner, *Field Anal. Chem. Technol.* **4** (2000) 43.
[79] D. Morris, D. Ledden, R. Boguslaski, *J. Clin. Lab. Anal.* **1** (1987) 243.
[80] D. Litman in T. Ngo, H. Lenhoff (eds.): *Enzyme-Mediated Immunoassay,* Plenum Press, New York 1985, p. 155.
[81] R. Zuck et al., *Clin. Chem. (Winston-Salem, N.C.)* **31** (1985) 1144.
[82] G. Valkiro, R. Barton, *Clin. Chem. (Winston-Salem, N.C.)* **31** (1985) 1427.
[83] J. Rittenburg et al.: "Immunochemical Methods for Environmental Analysis," *ACS Symp. Ser.* **442** (1990) 28.
[84] E. Schneider et al., *Food Agric. Immun.* **3** (1991) 185.
[85] C. Morber, L. Weil, R. Niessner, *Fres. Environ. Bull.* **2** (1993) 151.
[86] C. Keuchel, R. Niessner, *Fresenius J. Anal. Chem.* **350** (1994) 538.
[87] A. Dankwardt, B. Hock, *Biosens. Bioelectron.* **8** (1993) XX.
[88] S. Morias, A. Maquieira, R. Puchades, *Anal. Chem.* **71** (1999) 1905.
[89] M. Del Carlo et al. in B. Hock et al. (eds.): *Biosensors and Environmental Diagnostics,* Chap. 2, Teubner Verlag, Stuttgart 1998.
[90] F. Trautner, K. Huber, R. Niessner, *J. Aerosol Sci.* **23** (1992) S 999.
[91] M. Weller, L. Weil, R. Niessner, *Mikrochim. Acta* **108** (1992) 29.
[92] R. Ekins, F. Chu, E. Biggart, *Anal. Chim. Acta* **227** (1989) 73.
[93] A. Brecht, G. Gauglitz, *Anal. Chim. Acta* **347** (1997) 219.
[94] M. G. Weller et al., *Anal. Chim. Acta* **393** (1999) 29.
[95] H. Kido, A. Maquieira, B. D. Hammock, *Anal. Chim. Acta* **411** (2000) 1.
[96] J. Rittenburg, J. Dautlick, *ACS Symp. Ser.* **586** (1995) 301.

10. Basic Principles of Chromatography

Georges Guiochon, Department of Chemistry, University of Tennessee, Knoxville, TN, 37996-1600, United States, Division of Chemical and Analytical Sciences, Oak Ridge National Laboratory, Oak Ridge, TN, 37831-6120, United States

Abbreviations

a	initial slope of the isotherm
C, C_m	mobile-phase concentration
C_s	stationary-phase concentration
d_f	film thickness
d_p	particle size (diameter)
D_a	axial dispersion coefficient
D_l	diffusion coefficient, stationary (liquid) phase
D_m	diffusion coefficient, mobile (gas) phase
F	phase ratio
F_v	mobile-phase flow rate
h	reduced plate height
H	height equivalent to a theoretical plate (HETP)
k_0	column permeability
k', k'_0	retention factor
K	Henry constant
L	column length
M_n	nth moment of a statistical distribution
N, N_c	number of plates
p	partial pressure
P	probability
P^0	vapor pressure
q	Langmuir isotherm
R	fraction of sample in the mobile phase; resolution
S_a	adsorbent surface area
S_i	detector response factor
t_R	retention time
u	mobile-phase velocity
v	reduced velocity
v_m	volume per plate of the mobile phase
v_s	volume per plate of the stationary phase
V_g	specific retention volume
V_G	volume of gas
V_l	volume of liquid
V_m	mobile-phase volume
V_N	correlated retention volume
V_R	retention volume
α	relative retention
γ^∞	activity coefficient at infinite dilution
ε_T	total packing porosity
η	mobile-phase viscosity
μ_n, $\bar{\mu}_n$	absolute and central moments
Φ	fractional loss of efficiency
σ	standard deviation
τ	time variance

10.1. Introduction

Chromatography is the most powerful separation technique available. For reasons that will be made clear later, it is easy to implement chromatographic techniques with small samples, and to carry out separations on an analytical scale. Separated components are usually detected on-line, and sometimes they are also characterized, after which they are generally discarded. In preparative applications, the purified components are collected for further investigation, although new difficulties may be encountered in an attempt to scale-up the equipment and operate at high concentrations.

Chromatography is a separation method based on differences in equilibrium constants for the components of a mixture placed in a diphasic system. A chromatographic system is one in which a fluid mobile phase percolates through a stationary phase. The stationary phase is often a bed of non-consolidated particles, but this is not essential. All that is in fact required is two phases in relative motion and excellent contact between these phases so that concentrations in the stationary phase are always very near their equilibrium values. For example, a tubular column through which the mobile phase flows and whose walls have been coated with a layer of stationary phase is a most suitable implementation of chromatography. In all cases the mobile phase is responsible for transporting the sample components through the stationary phase. The velocity of each component, hence its residence time, depends on the both mobile-phase velocity and the distribution equilibrium constant for that component. At least in principle, proper choice of the two phases constituting a chromatographic system permits selective adjustment of the relative migration rates of the mixture components and of the extent of their eventual separation.

The mobile phase is a fluid. The main criteria for its selection are:

1) Good solubility for the analytes
2) Low viscosity

All types of fluids have been used, but the chief types of chromatography are gas, dense-gas (more commonly called supercritical-fluid), and liquid chromatography. In *gas chromatography* (GC; → Gas Chromatography), a quasi-ideal gas is used; pressures are kept low, analyte solubility in the mobile phase depends only on its vapor pressure, and the role of the mobile phase is almost purely mechanical, with minor corrections due to nonideal behavior of the gas phase [1]. Retention adjustment is accomplished primarily by changing the temperature. With denser gases, an analyte's fugacity differs markedly from its vapor pressure, and analyte solubility depends strongly on the mobile-phase density as well as the temperature. Thus, retention adjustment in supercritical-fluid chromatography can be effected by changing not only the temperature but also the average pressure. In both gas and supercritical-fluid chromatography, the *selectivity* (i.e., the ability to achieve a separation) is relatively insensitive to temperature and density. In *liquid chromatography* (LC), the mobile phase is a mixture of solvents (→ Liquid Chromatography). Retention adjustments and selectivity changes are accomplished by changing the mobile-phase composition and introducing a variety of strong solvents or additives.

The stationary phase may be a solid (i.e., an adsorbent, with a relatively large specific surface area and very accessible pore channels) or a liquid coated on a solid support (to avoid stripping of the stationary phase by the streaming action of the mobile phase). Intermediate systems include immobilized polymers or surface-bonded shorter chemical species (e.g., $C_{18}H_{37}$) of widely varying chain length and molecular mass. Gas–solid, gas–liquid and liquid–solid chromatography are the most popular chromatographic methods. Ion-exchange resins in conjunction with ionic solvents, gels with pores comparable in size to the analyte molecules, and stationary phases containing suitable complexing agents or groups have also served as the basis for important methods, producing selectivities very different from those observed with more classical molecular interactions.

Another useful distinction involves the mechanism by which the mobile phase is transported through the stationary phase. The most popular approach is forced convection by pressurization, based on cylinders of compressed gases in gas chromatography, or a mechanical pump in liquid chromatography. In both cases the stationary phase is placed inside a tube, the *chromatographic column*. Other methods are available as well. For example, in thin layer chromatography (→ Thin Layer Chromatography) capillary forces suffice to draw a liquid stream through a dry, porous bed in the form of a paper sheet or a thin layer of adsorbent coated on a glass, metal, or plastic sheet. More recently, electro-osmosis has been used to force a liquid stream through a column [2]. Although the method is quite similar to standard column liquid chromatography, the resulting separations may be markedly superior.

10.2. Historical Development

It is not possible to review here the entire history of chromatography and do justice to all the scientists who have contributed to designing and developing the dozens of major separation techniques based on the principle of chromatography. The history of the origin of chromatography has been recounted by SAKODINSKY [3] and ETTRE [4]. Numerous other publications recall landmark contributions in the field.

Although a number of early publications introduced certain ideas related to chromatography, it is clear that the method was conceived and developed by the Russian chemist TSWETT at the beginning of this century [5]–[7]. TSWETT is best described as a biophysicist. He had a remarkably clear understanding of the interactions between the various phenomena involved in chromatography — some thermodynamic and others kinetic — and of the technical problems that would need to be solved to turn the chromatographic principle into a useful group of analytical methods. Unfortunately, living during the very early stages of the high-technology age, in a country soon to be ravaged by civil war, he was unable to succeed. His ideas remained untried until a group of German chemists who were informed of TSWETT's early work through WILLSTÄTTER [3] succeeded in using chromatography for the extraction and purification of plant pigments.

The original work of TSWETT is most interesting to read — at least the fragments available in the modern literature [3], [4], [8]. Many of the modern problems of high-performance liquid chromatography are already addressed in these early writings. TSWETT knew that small particles should be used for optimum performance, but he had no way to force a mobile-phase stream under pressure through a column. It is interesting also to note that the initial difficulties KUHN's group had to solve in developing their analytical schemes were due to an isomerization of carotenoid derivatives catalyzed by the silica gel used as stationary phase [3], [4]. TSWETT had deliberately used calcium carbonate for these separations, reserving silica gel for the separation of more stable compounds [7].

A. J. P. MARTIN has been associated with several critical developments in chromatography. His Nobel prize paper with SYNGE [9] represents the origin of both partition and thin layer chromatography (TLC). Originally carried out with paper sheets, TLC evolved later under the leadership of STAHL in the direction of consolidated layers of fine particles. Later, MARTIN, GORDON and CONSDEN [10] published the first two-dimensional chromatograms. Here a sample is introduced at one corner of a large, square paper sheet, and two successive developments are conducted in orthogonal directions, using two different solvent mixtures as mobile phases. This method leads to a dramatic increase in separation power. Finally, in cooperation with JAMES, MARTIN invented gas–liquid chromatography [11], a technique ideally suited to the incorporation of many developments in modern instrumentation that occurred at that time. Gas–liquid chromatography led first to a major new instrumentation technique, then to the rejuvenation of conventional liquid chromatography, and finally to the more recent development of capillary-zone electrophoresis.

10.3. Chromatographic Systems

Chromatography is based on phase-equilibrium phenomena. The components of the analyte sample are caused to equilibrate between two phases, a mobile phase and a stationary phase. Because the mobile phase percolates through the stationary phase, rapid mass transfer takes place, and the mobile phase carries the components through the column to a detector. The velocity of this transfer is related to the equilibrium constant. Hence, only compatible combinations of mobile and stationary phases can be used in practice.

10.3.1. Phase Systems Used in Chromatography

The mobile phase must be a fluid: A gas, a dense gas, or a liquid, where the rate of mass transfer through the mobile phase, characterized by the diffusion coefficient, decreases in the order listed. To compensate for this marked decrease, finer and finer particles are used as the mobile phase becomes more dense. The practical lack of compressibility of liquids supports this trade-off, since small particles demand higher pressures. However,viscous liquids (e.g.,glycerol,oligomers, or concentrated solutions of polymers) are excluded, because the associated pressure requirements would be unrealistic.

The stationary phase must be compatible with the mobile phase. Lack of solubility of the stationary phase in the mobile phase and a large interfacial area between the two phases are the two major requirements. Most adsorbents are applicable to any of the three types of mobile phase, giving rise, for example, to both gas–solid and liquid–solid chromatography. The specific surface area of the absorbent must be large, not only to enhance column efficiency but also to ensure sufficient retention [12]. This surface must be highly homogeneous as well to avoid the presence of active sites on which the adsorption energy is exceptionally high leading to excessive retention, and a poor band profile. Most of the adsorbents for chromatography have been developed to such a point that they display the required properties; alumina, silica, molecular sieves, and activated or graphitized carbons are the most popular.

Silica comes in two broad types: *Pure* or *underivatized silicas* (all of which have very similar properties, but among which chromatographers can recognize more varieties than there are commercial brands) and the *chemically bonded silicas* [13]. Virtually every conceivable type of chemical group—from alkyl derivatives to sugars and peptides—has been bonded to silica, leading to a large number of phases, each with its specific properties and applications. In liquid–solid chromatography, two types of systems are distinguished: *Normal-phase* systems based on a polar stationary phase and a nonpolar mobile phase (e.g., silica and dichloromethane), and *reversed-phase* systems, in which the stationary phase is nonpolar and the mobile phase polar (e.g., C_{18} silica and water–methanol solutions).

Liquids can also be used as the stationary phase in chromatography, but they must be immiscible with the mobile phase. In gas–liquid chromatography this means that the stationary phase must have a low vapor pressure to avoid its premature depletion by the mobile phase. Convection would be highly detrimental to the effectiveness of a separation, and this is avoided by coating the liquid phase on an inert solid. The support is usually a pulverulent solid consisting of porous particles. To avoid significant contributions from adsorption (i.e., mixed retention mechanisms), the solid should have a low specific surface area, but it should also be highly porous to permit the use of a large amount of liquid. Alternatively, a stationary liquid can be coated on a nonporous support (e.g., glass beads, or the inner wall of an empty tube). Considerable effort has been devoted to the development of reproducible methods for coating the inner wall of a quartz tube with a homogeneous, stable, thin film of liquid.

While gas–liquid chromatography has become the preferred implementation of gas chromatography because of the excessive adsorption energy associated with gas–solid equilibria, for most compounds but gases, liquid–solid chromatography is the preferred implementation of liquid chromatography. It is difficult to find a stationary liquid that is insoluble in the liquid mobile phase but at the same time not so different from this phase that the partition coefficients are either negligible or nearly infinite. The stability of a liquid coated on an inert support is low, and often the stationary phase is quickly stripped from the support in the form of a dilute emulsion.

Other important stationary phases for liquid chromatography are ion-exchange resins (*ion-exchange chromatography*, which uses anionic or cationic resins) and porous solids that do not adsorb the analytes but instead have pores so small that access to a fraction of the pore volume by some of the analyte molecules is more or less restricted depending on molecular size (*size-exclusion chromatography*).

Finally, chemical groups capable of forming charge transfer complexes or other types of labile complexes with specific components of a sample have also been used successfully as stationary phases. *Affinity chromatography* is one implementation of this concept, in which complex formation between a specific antibody and one analyte results in both exceptional selectivity for a particular component and an inherent lack of flexibility, two striking characteristics of this method.

10.3.2. Methods of Implementing Chromatography

There are three methods of implementing chromatography: Elution, displacement, and frontal analysis. An immense majority of analytical applications are carried out on the basis of elution. Equilibrium isotherms are often measured using frontal analysis. Finally, preparative separations, though most often accomplished by overloaded elution, also lend themselves to displacement. This has the advantage of producing more concentrated fractions, but it requires special development, choice of an appropriate displacer, and the extra step of column regeneration between the processing of successive batches.

In *elution*, a pulse of sample is injected into the mobile-phase stream before it enters the column. Under the linear conditions most often associated with analytical applications, each pulse of sample component behaves independently, migrating along the column at a velocity proportional to the mobile-phase velocity but dependent also on the equilibrium coefficient between the two phases. If the experimental conditions are properly chosen, pulses of the mixture components arrive separately at the column exit, and the chromatogram consists of a series of peaks, one for each component.

In *frontal analysis*, a stream of a solution of the sample in the mobile phase is abruptly substituted for the mobile-phase stream at the beginning of the experiment. If the components are present at very low concentrations, the front of each component "step" moves at the same velocity as the pulses in an elution experiment, resulting in a "staircase" in which the profile of each individual step can be regarded as an error function. However, in most applications of frontal analysis the component concentrations are in fact rather high, and the isotherms are no longer linear. Accordingly, they are not independent of each other, but are competitive; i.e., the concentration a given compound adsorbed at equilibrium depends on the concentrations of all the components present. The propagation of the various component steps results in breakthrough curves that are steep, and the concentrations of all the components in the eluent change when any one of them breaks through. As a consequence, frontal analysis can be used only to purify the first eluted component of a mixture, or to measure single-component or competitive (mainly binary) isotherms.

In *displacement* chromatography, a pulse of mixture is injected, and this is followed by a step of a single component called a *displacer*, which is adsorbed more strongly than any of the mixture components. After a certain period of time, during which the profiles of each pulse become reorganized, an isotachic pulse train is formed. In the isotachic train, each component forms its own concentration "boxcar," with a height that depends on the component and displacer isotherms as well as on the displacer concentration, and a width proportional to the amount of the corresponding component in the sample [14], [15].

10.4. Theory of Linear Chromatography

One can distinguish between two basic types of assumption common to all theoretical efforts to understand chromatography [16]. First, it may be assumed that the equilibrium isotherm is linear; alternatively, one might acknowledge that, under the experimental conditions selected, this isotherm is not linear. One can thus study models of either linear or nonlinear chromatography. Analytical applications almost always involve small or dilute samples, and under these conditions any curvature in the isotherm near the origin can be ignored and the isotherm can be regarded as linear. In preparative applications — more generally, in any case of concentrated solutions — this assumption is no longer tenable, and it becomes necessary to consider nonlinear chromatography (see Chap. 10.10).

Similarly, a decision must be made whether or not to take into account the influence on band profiles of such phenomena as axial dispersion (dispersion in the direction of the concentration gradient in the column) and resistance to mass transfer (i.e., the fact that equilibration between mobile and stationary phases is not instantaneous). These phenomena are responsible for the finite efficiency of actual columns. Neglecting them and assuming the column to have infinite efficiency leads to a model of *ideal* chromatography. Taking them into account results in one of the models of *nonideal* chromatography.

The combination of these two types of consideration leads to four theories of chromatography. The first, *linear, ideal chromatography*, is trivial: each component pulse moves without changes in its profile, and its motion is independent of other pulses. In *linear, nonideal chromatography* one studies the influence of kinetic phenomena on band profiles, assuming each component pulse to move independently of all the others, and at a constant velocity. The chromatogram of a mixture in this case is the sum of the chromatograms for each independent component. This is no longer true in *ideal, nonlinear chromatography*. Here one studies the influence of nonlinear phase-equilibrium thermodynamics on the band profile. The result depends on the amount of each component present in the mixture, and individual elution profiles for the various components cannot be obtained independently (see Chap. 10.10). Nevertheless, both linear, nonideal and ideal, nonlinear chromatography provide relatively simple solutions. In *nonlinear, nonideal chromatography* it is necessary to take into account both finite column efficiency and a nonlinear, nonindependent behavior of the isotherms, and only numerical solutions are possible.

10.4.1. Plate Models

These models assume that the chromatographic column can be divided into a series of a finite number of identical plates. Each plate contains volumes v_m and v_s of the mobile and stationary phases, respectively. The sample is introduced as a solution of known concentration in the mobile phase used to fill the required number of plates. Plate models are essentially empirical, and cannot be related to first principles. Depending upon whether one assumes continuous or batch operation, two plate models can be considered: The Craig model and the Martin and Synge model.

10.4.1.1. The Craig Model

In this model [17], mobile phase is transported through the column by withdrawing it from the last plate, moving the remaining mobile phase from each plate to the next one in succession, and refilling the first plate. The process is repeated as often as necessary to elute the material introduced. It is easily shown that the fraction R of the sample in the mobile phase is expressed by

$$R = \frac{v_m C_m}{v_m C_m + v_s C_s} = \frac{1}{1 + F a} = \frac{1}{1 + k_0'} \quad (1)$$

where C_m and C_s are the equilibrium concentrations in the mobile and stationary phases, respectively. F is the phase ratio, a is the initial slope of the isotherm, and k'_0 is the retention factor. After the moving process has been repeated r times, the probability P of finding a molecule in the plate of rank l is

$$P_{l,r} = \frac{r!}{l!(r-l)!} R^l (1-R)^{r-l} \quad (2)$$

and the elution profile (at the exit of the last plate, of rank N_c) is

$$f(r) = R \frac{r!}{N_c!(r-N_c)!} R^{N_c} (1-R)^{r-N_c} \quad (3)$$

When N_c exceeds about 50 plates, this profile cannot be distinguished from a Gaussian profile with a retention volume

$$V_R = V_m(1 + k_0') \quad (4)$$

where $V_m = N_c v_m$, and a standard deviation

$$\sigma_v = V_R\sqrt{\frac{1-R}{N_c}} \tag{5}$$

10.4.1.2. The Martin – Synge Model

This is a continuous plate model [9]. The mobile phase is transferred from one plate to the next, and each plate is regarded as a perfect mixer. Integration of the differential mass balance in every plate leads to a Poisson distribution inside the column, and to the elution profile:

$$f(v) = \frac{N}{V_R}\,e^{-\frac{Nv}{V_R}}\left(\frac{Nv}{V_R}\right)^{(N-1)}\frac{1}{(N-1)!} \tag{6}$$

As the plate number N increases, just as in the case of the Craig model the profile tends rapidly toward a Gaussian profile:

$$f(v) = \frac{1}{\sigma_v\sqrt{2\pi}}\exp\left[-\frac{(v-V_R)^2}{2\sigma_v^2}\right] \tag{7}$$

with a retention volume and standard deviation given by

$$V_R = N(v_m + a\,v_s) \tag{8a}$$

$$\sigma_v = (v_m + a\,v_s)\sqrt{N} \tag{8b}$$

Note that, although both models lead to the same profile, the resulting relationships between number of plates in the column and standard deviation differ. The conventional plate number as defined in chromatography is equal to N for the Martin and Synge model, and to $N_c(1+k'_0)/k'_0$ for the Craig model. In any discussion of column efficiency it is convenient to consider the *height equivalent to a theoretical plate* (or HETP).

$$H = \frac{L}{N} \tag{9}$$

This concept is developed further in Chapter 10.7.

10.4.2. Statistical Models

The *random walk* model has been used by GIDDINGS to relate various individual contributions in the mechanism of band broadening to the column efficiency [18]. This leads directly to a Gaussian band profile. The contributions of individual sources of band broadening to changes in the profile are additive. Thus, the column HETP is obtained as the sum of individual contributions

$$H = \sum h_i \tag{10}$$

which can be calculated for each identifiable source of band broadening.

GIDDINGS and EYRING [19] have introduced a more sophisticated, stochastic model that treats the chromatographic process as a Poisson distribution process. Assuming molecules undergo a series of random adsorption and desorption steps during their migration along the column, it is possible to derive a distribution for the residence times of the molecules injected in a pulse. This model has been extended by MCQUARRIE to the case of mixed mechanisms [20]. More recent investigations by DONDI et al. confirm that the model can account for both axial dispersion and mass-transfer resistance [21]. When the column efficiency is not too low, the resulting band profile tends toward a Gaussian distribution.

10.4.3. Mass-Balance Models

These models involve writing and integrating a differential mass-balance equation. The basic assumptions associated with writing such an equation are discussed in the literature [16]. There are two important models of this type, the equilibrium – dispersive model [22] and the lumped kinetic models, in addition to the general rate model. The differential mass balance equation is

$$\frac{\partial C}{\partial t} + F\frac{\partial q}{\partial t} + u\frac{\partial C}{\partial z} = D_a\frac{\partial^2 C}{\partial z^2} \tag{11}$$

where C and q are the mobile and stationary phase concentrations, respectively, F is the phase ratio [$F=(1-\varepsilon_T)/\varepsilon_T$, with ε_T the total packing porosity], u is the mobile-phase velocity, and D_a is the axial dispersion coefficient. The first two terms on the left of this equation are the accumulation terms, the third is the convective term. The term on the right is the dispersive term. This equation contains two functions of the time (t) and space (z) coordinates, C and q. These must be related by a second equation to fully determine the problem.

10.4.3.1. The Equilibrium – Dispersive Model

In this model, one assumes that any departure from equilibrium between the two phases of the chromatographic system is very small, and the stationary-phase concentration q is given practically everywhere by the equilibrium isotherm:

$$q = f(C) \tag{12}$$

For theoretical discussions, the Langmuir isotherm $q = a\,C/(1 + b\,C)$ is convenient. For a review of isotherms used in liquid chromatography, see [23]. With $D_a = 0$ this assumption would be equivalent to assuming that the column efficiency is infinite. However, the finite efficiency of an actual column can be taken into account by including in the axial dispersion coefficient the influences on band profiles due to both axial dispersion and the kinetics of mass transfer:

$$D_a = \frac{H\,u}{2} = \frac{L\,u}{2N} \tag{13}$$

where H is the column HEPT and L is the column length.

With a linear isotherm, Equation 11 becomes

$$(1 + k'_0)\frac{\partial C}{\partial t} + u\frac{\partial C}{\partial z} = D_a\frac{\partial^2 C}{\partial z^2} \tag{14}$$

where k'_0 is the retention factor. Solving this equation requires a choice of initial and boundary conditions [22]. In analytical chromatography, the initial condition is an empty column. Three types of boundary conditions have been studied:

1) An infinitely long column stretching from $z = -\infty$ to $z = +\infty$
2) An infinitely long column stretching from $z = 0$ to $z = +\infty$
3) A finite column from $z = 0$ to $z = L$

Depending on the boundary condition chosen, different solutions are obtained [16]. In practice, for values of N exceeding 30–50 plates there are no noticeable differences between the profiles. The first and second moments of these profiles are

1) $\left(1 + \frac{1}{N}\right)t_R$ and $\frac{t_R^2}{N}\left(1 + \frac{2}{N}\right)$
2) $\left(1 + \frac{1}{2N}\right)t_R$ and $\frac{t_R^2}{N}\left(1 + \frac{3}{4N}\right)$
3) t_R and $\frac{t_R^2}{N}\left[1 - \frac{1}{2N}\left(1 - e^{-2N}\right)\right]$

with $t_R = (1 + k'_0)\,t_0$, and $t_0 = L/u$.The differences relative to a Gaussian curve become rapidly negligible as N increases, ultimately reaching the values commonly achieved.

10.4.3.2. Lumped Kinetic Models

These models recognize that equilibrium between the mobile and stationary phases can never actually be achieved, and use a kinetic equation to relate $\partial q/\partial t$, the partial differential of q with respect to time, and the local concentrations, q and C. Several different kinetic models are possible depending on which step is assumed to be rate controlling. All are called "lumped kinetic" models because, for the sake of simplicity, the contributions of all other steps are lumped together with the one considered to be most important. These models are discussed and compared in detail in [24]. All such models are equivalent in linear chromatography, but not when the equilibrium isotherm is nonlinear [16]. The kinetic equation can be written as:

$$\frac{\partial q}{\partial t} = k_m(C_s - q) \tag{15}$$

where C_s is the stationary-phase concentration in equilibrium with the mobile-phase concentration C (and the isotherm is $C_s = a\,C$). A general solution of these models has been given by LAPIDUS and AMUNDSON [25]. Van DEEMTER et al. [26] have shown that the solution can be simplified considerably if one assumes reasonably rapid mass-transfer kinetics. The profile then becomes Gaussian, and the column HETP is given by:

$$H = \frac{2D_L}{u} + 2\left(\frac{k'_0}{1 + k'_0}\right)^2 \frac{u}{k'_0 k_m} \tag{16}$$

This is the well-known Van Deemter equation.

10.4.3.3. The Golay Equation

In the particular case of a cylindrical column the wall of which is coated with a layer of stationary phase, the system of partial differential equations of a simplified kinetic model can be solved analytically [27]. For a noncompressible mobile phase, GOLAY derived the following equation:

$$H = \frac{2\,D_m}{u} + \frac{1 + 6k'_0 + 11k'^2_0}{96\,(1 + k'_0)^2}\frac{u\,d_p^2}{D_m} + \frac{k'_0}{24\,(1 + k'_0)^2}\frac{u\,d_f^2}{D_l} \tag{17}$$

where d_p is the particle size, d_f is the thickness of the liquid film, D_m is the diffusion coefficient in the carrier gas, and D_l is the diffusion coefficient in the liquid phase.

This equation is in excellent agreement with all experimental results so far reported.

10.4.4. The General Rate Model

In this model, each step of the chromatographic process is analyzed in detail [16]. Separate mass-balance equations are written for the mobile phase that flows through the bed and for the stagnant mobile phase inside the particles. Separate kinetic equations are then written for the kinetics of adsorption/desorption and for mass transfer. Again, if these kinetics are not unduly slow, the band profile tends toward a Gaussian shape (in which case a simpler model is actually more suitable).

10.4.5. Moment Analysis

The *moments* of a statistical distribution provide a convenient description of this distribution. Since a band profile is the distribution of residence times for molecules injected into a chromatographic system, the use of moments to characterize these profiles is a natural approach. Furthermore, analytical solutions of the general rate model can be obtained in the Laplace domain. The inverse transform to the time domain is impossible to solve in closed form, but the moments of the solution can easily be calculated. Equations are available for the first five moments [28], [29].

The nth moment of a distribution is

$$M_n = \int_0^\infty C(t) t^n \, dt \tag{18}$$

It is more practical to use normalized distributions, so the absolute moments, μ_n, $n \geq 1$, and the central moments, $\bar{\mu}_n$, $n > 1$ are most com-monly employed:

$$\mu_n = \frac{M_n}{M_0} = \frac{\int_0^\infty C(t)\, t^n \, dt}{\int_0^\infty C(t) \, dt} \tag{19a}$$

$$\bar{\mu}_n = \frac{\int_0^\infty C(t)(t - \mu_1)^n \, dt}{\int_0^\infty C(t) \, dt} \tag{19b}$$

These moments have been used to study band profiles. Interpretations are straightforward for the zeroth (peak area), first (retention of the peak mass center), and second ($H = \bar{\mu}_2 L / \mu_1^2$) moments. Higher moments are related to deviations of the band profile from a Gaussian profile [30]. The use of such moments is limited, however, in large part because the accuracy of determination decreases rapidly with increasing order of a moment. Use of the first moment rather than the retention time of a band maximum has been suggested as a way of relating chromatographic and thermodynamic data for an unsymmetrical peak [30], but this is justified only with slow mass-transfer kinetics.

10.4.6. Sources of Band Asymmetry and Tailing in Linear Chromatography

The theory of chromatography shows that there should be no noticeable difference between a recorded peak shape and a Gaussian profile so long as the number of theoretical plates of the column exceeds 100 [16]. The difference is small even for 25 plates, and careful experiments would be needed to demonstrate that difference. However, it is common to experience in practice peak profiles that are not truly Gaussian.

There are basically four possible causes of unsymmetrical band profiles in chromatography. First, the injection profile may exhibit some tailing. Second, the column may be overloaded such that the isotherm of equilibrium between the two phases is no longer linear. Third, the stationary phase may be inhomogeneous, or mixed retention mechanisms may be operative. Finally, the column packing may not be homogeneous. Slow mass transfer is usually not responsible, contrary to common wisdom, because this would lead to an efficiency much lower than that actually observed.

Injections are rarely carried out under experimental conditions that permit the achievement of narrow, rectangular sample pulses. More often, the injection profile tails markedly. If the length of this tail is significant compared to the standard deviation of the band, the result is tailing of the analytical band. For an exponential injection profile with a time constant τ, the ratio τ/σ should be < 0.10 [31]. Such tailing is usually restricted to high-efficiency bands eluted early, often as a result of poorly designed injection ports or connecting tubes, or an insufficiently high injection-port temperature in gas chromatography. Direct connection of the injection system to the detector permits assessment of this particular source of tailing [32].

When the sample size is too large, the corresponding isotherm can no longer be considered identical to its initial tangent. If the isotherm is convex upward, the retention time decreases with increasing concentration, and the peak tails. Isotherms in gas chromatography are usually convex downward, and peaks from large samples tend to "front." However, peaks eluted early tend to have very sharp fronts and some tail. This is caused by a sorption effect, since the partial molar volume of the vapor is much larger than that of the dissolved solute [33].

In most cases, peak tailing can be explained by heterogeneity of the surface of the stationary phase [16]. Strongly adsorbing sites saturate before weak ones, and they have a much smaller saturation capacity. The efficiency increases when the sample size is reduced, but this observation is often of limited use, because detection limits establish a minimum for the acceptable sample size.

10.5. Flow Rate of the Mobile Phase

The velocity of the mobile phase is related to the pressure gradient, the mobile phase viscosity η, and the column permeability k_0 by the Darcy law [34]:

$$u = -\frac{k_0}{\eta}\frac{\mathrm{d}p}{\mathrm{d}z} \tag{20}$$

The minus sign here indicates that the flow is directed from high to low pressure. The differential equation can be integrated provided the relationship is known between local volume and pressure. Two equations of state can be used, one for gases (Boyle – Mariotte law), the other for liquids. For dense gases, the equation of state is far more complex [35].

Liquids can be considered as noncompressible for purposes of calculating a pressure gradient. In this case u is constant along the column, and the difference between the inlet and outlet pressures is

$$\Delta p = \frac{u\eta L}{k_0} \tag{21}$$

The retention time for an inert tracer is

$$t_0 = \frac{L}{u} \tag{22}$$

In gas chromatography the carrier gas behaves as an ideal gas from the standpoint of compressibility, so $p\,u = p_0\,u_0$, and the integration of Equation 20 gives

$$u_0 = \frac{k_0 p_0}{2\eta L}(p^2 - 1) \tag{23}$$

Because of gas-phase compressibility, the average gas velocity $\bar{u}$ is lower than the outlet velocity. It can be calculated that

$$\bar{u} = j\,u_0 = \frac{3\,(p^2 - 1)}{2\,(p^3 - 1)}\,u_0 \tag{24}$$

where j is the James and Martin compressibility factor [11]. The retention time for an inert tracer in this case is

$$t_0 = \frac{4\,\eta L^2\,(p_i^3 - p_0^3)}{3k_0\,(p_i^2 - p_0^2)} \tag{25}$$

10.5.1. Permeability and Porosity of the Packing

10.5.1.1. Column Porosity

There are three different porosities that must be considered: The *interparticle* or *external* porosity, the *intra particle* or *internal* porosity, and the *total* porosity. The external porosity is the fraction of the overall geometrical volume of the column available to mobile phase that flows between the particles. Chromatographers always strive to achieve a dense column packing as the only known practical way of packing reproducible, stable, and efficient columns, so this porosity tends to be similar for all columns, with a value close to 0.41.

The internal porosity depends on the nature and structure of the particles used to pack the column. It varies considerably from one stationary phase to another, ranging from zero (nonporous beads) to ca. 60 % for certain porous silicas.

The total porosity is the fraction of the bed accessible to the mobile phase; i.e., it is the sum of the internal and external porosities.

10.5.1.2. Column Permeability

The *permeability* of a column depends on the characteristics of the particular packing material used, and especially on the column external porosity, but it is the same regardless of whether the mobile phase is a gas, a liquid, or a dense gas. This permeability is inversely proportional to the square of the particle diameter:

$$k_0 = \frac{d_p^2}{\Phi} \quad (26)$$

For packed columns the numerical coefficient Φ is of the order of 1000, with reported values ranging between 500 and 1200. It tends to be smaller for spherical than for irregular particles. For empty cylindrical tubes, such as those used for open tubular columns, it is equal to 32.

10.5.2. Viscosity of the Mobile Phase

The viscous properties of gases and liquids differ significantly [36].

10.5.2.1. Viscosity of Gases

The viscosity of a gas increases slowly with increasing temperature, and is nearly independent of the pressure below 2 MPa. It is approximately proportional to $T^{5/6}$. The rare gases (He, Ar) have higher viscosities than diatomic gases (N_2, H_2) and polyatomic molecules. Hydrogen has the lowest viscosity of all possible carrier gases, and is often selected for this reason. The viscosity of the sample vapor is unlikely to exceed that of the carrier gas.

10.5.2.2. Viscosity of Liquids

The viscosity of a liquid decreases with increasing temperature, following an Antoine-type equation:

$$\log \eta_{\mathrm{liq}} = A + \frac{B}{T + C} \quad (27)$$

In most cases, however, C is negligible, and if T is expressed in kelvins B is of the order of 1 to 3×10^{-3}. For example, the viscosity of benzene decreases by one-half with an increase in temperature of 120 °C. The viscosity increases slowly and linearly with increasing pressure below 200 MPa, and may increase by 10 – 25 % when the pressure rises from 0 to 20 MPa. The viscosity of a liquid also increases with the complexity of the molecule and its ability to engage in hydrogen bonding. The viscosity of a sample may be much higher than that of the corresponding mobile phase, which presents a serious potential problem in preparative chromatography [37].

10.6. The Thermodynamics of Phase Equilibria and Retention

Chromatography separates substances on the basis of their equilibrium distributions between the two phases constituting the system. Without retention there can be no separation. Retention data are directly related to the nature of the phase equilibrium and to its isotherm. Although the details depend very much on the nature of the mobile and stationary phases used, a few generalizations are possible.

10.6.1. Retention Data

It is easy to relate retention times to the thermodynamics of phase equilibria. In practice, however, it is difficult to measure many of the relevant thermodynamic parameters with sufficient accuracy, while retention times can be measured with great precision. Furthermore, it should be emphasized at the outset that even though very precise measurements can indeed be made in chromatography [38], the available stationary phases are often so poorly reproducible that such precision is unwarranted. Column-to-column reproducibility for silica-based phases is reasonable only for columns packed with material from a given supplier. For this reason, chromatographers prefer whenever possible to use relative retention data.

10.6.1.1. Absolute Data

The experimental data acquired directly in chromatography are retention time, column temperature, and the mobile-phase flow rate. All other data are derived from these primary parameters. The retention time is inversely proportional to the flow rate, so it is often preferable instead to report the retention volume and the retention factor (both of which are independent of the flow rate).

The *retention factor* k' is the ratio of the fraction of component molecules in the stationary phase n_s to the fraction in the mobile phase n_m at equilibrium. The retention factor is thus

$$k' = \frac{n_s}{n_m} = \frac{1-R}{R} \quad (28)$$

This also corresponds to the ratio of the average times spent by a typical molecule in the two phases:

$$k' = \frac{t_R - t_0}{t_0} \quad (29)$$

Accordingly, k' is proportional to the thermodynamic constant for the phase equilibrium at infinite dilution.

The *retention volume* V_R (see Eq. 4) is the volume of mobile phase required to elute a component peak. For compressible phases it must be referred to a standard pressure:

$$V_R = t_R F_v f(P) \quad (30)$$

where F_v is the mobile-phase flow rate and $f(P)$ is derived from the equation of state for the mobile phase. For the ideal gases used in GC,

$$f(P) = j = \frac{3(P^2-1)}{2(P^3-1)}$$

(Eq. 24); for liquids, $f(P) = 1$; for dense gases it is a complex function [35].

10.6.1.2. Relative Data

Because absolute retention data depend on so many controlled (temperature, pressure) and uncontrollable (activity of the stationary phase) factors, and because their quantification requires independent determinations that are time-consuming (e.g., density of the stationary liquid phase in the case of GC) or not very accurate (flow-rate measurements), and also because chromatographers are more often concerned with practical separation problems than with phase thermodynamics, it is more convenient to work with relative retention data. The parameter of choice is the *relative retention* α, which is also the *separation factor:*

$$\alpha_{2,1} = \frac{t_{R,2} - t_0}{t_{R,1} - t_0} = \frac{k'_2}{k'_1} \quad (31)$$

Relative retention data are more convenient than absolute data, but the choice of an appropriate reference may prove difficult. An ideal reference compound

1) Should not be present in mixtures subject to routine analysis
2) Should be introduced into the sample as a standard (though it might also be used as an external standard in quantitative analysis)
3) Should be nontoxic and inexpensive
4) Should elute near the middle of the chromatogram so the relative retention data will be neither too large nor too small
5) Should have chemical properties similar to those of most of the components in the sample mixture, thereby ensuring that relative retention volumes will be insensitive to changes in the level of activation of the stationary phase

For complex mixtures, especially those analyzed with temperature programming (in GC) or gradient elution (in LC), it is impossible to select a single reference compound. The Kovats retention-index system [39], based on homologous series of reference compounds, provides an elegant solution in GC, one which has been widely accepted. In LC, the absence of suitable homologous series, and the fact that retention depends more than in the case of GC on the polarities of compounds and less on molecular weight makes the use of an index system impractical.

Finally, because of its definition, the separation factor for two compounds is related to the difference between the corresponding Gibbs free energies of transfer from one phase to the other:

$$RT \ln\alpha = \Delta\left(\Delta G^0\right) \quad (32)$$

When α is close to unity [40],

$$\alpha - 1 = \frac{\Delta(\Delta G^0)}{RT}.$$

10.6.2. Partition Coefficients and Isotherms

From Equation 29 it can be shown that for gas–liquid chromatography:

$$k' = \frac{RT\rho}{\gamma^\infty P^0 M_s} \frac{V_L}{V_G} = K \frac{V_L}{V_G} \quad (33)$$

where P^0 is the vapor pressure of the analyte, γ^∞ is the activity coefficient at infinite dilution, ρ is the density, and K is the thermodynamic equilibrium constant, or the initial slope of the isotherm. Alternately, the constant K can be derived from measurement of the specific retention volume. The retention volume, $V_R = F_v t_R$, is the volume of carrier gas required to elute a particular component peak. The *corrected* retention volume V_N is the retention volume corrected for gas compressibility (see Eq. 24) and for the hold-up time in the mobile phase:

$$V_N = j(t_R - t_0)F_v = j k' t_0 F_v \quad (34)$$

The *specific* retention volume is the corrected retention volume under STP conditions, reported for a unit mass of liquid phase:

$$V_g = \frac{V_N}{m_L} \frac{273}{T_c} \frac{p_0}{P_n} = \frac{K}{\varrho} \frac{273}{T_c} \quad (35)$$

where T_c is the column temperature. In practice, derivation of the constant K from k', V_L, and V_G or from V_g requires the same measurements, and the accuracy and precision are the same.

Since V_g is proportional to a thermodynamic constant of equilibrium, it varies with the temperature according to

$$\frac{d(\ln V_g)}{d(1/T)} = -\frac{\Delta H_S}{R} \quad (36)$$

Since the molar enthalpy of vaporization of solute ΔH_S from an infinitely dilute solution is negative, retention volumes decrease with increasing temperature.

In liquid–liquid chromatography the retention volume depends on the ratio of activities of the component in the two phases at equilibrium; i.e., on the ratio of the activity coefficients at infinite dilution. As liquid–liquid chromatography has been little used so far (because of instability of the chromatographic system), and activity coefficients are not easily accessible, this relationship has not been extensively explored.

10.6.3. Gas–Solid Adsorption Equilibria

Gas–solid chromatograhy is used mainly for the separation of gases. In spite of major advantages, such as a high degree of selectivity for geometrical isomers and the high thermal stability of many adsorbents, it is not used for the analysis of organic mixtures.

Two conditions are required for successful analysis by gas–solid chromatography: The isotherm must be close to its initial tangent so that peaks migrate under conditions of linear equilibrium, and the surface of the adsorbent must be highly homogeneous. Gases are analyzed at temperatures much above their boiling points, so their partial pressures relative to their vapor pressures are very low, and equilibrium isotherms deviate little from their initial tangents. This is impractical with heavier organic vapors that decompose thermally at moderate temperatures. Graphitized carbon black, with a highly homogeneous surface, has been used successfully for the separation of many terpenes and sesquiterpenes, especially closely related geometrical isomers [41]. A number of adsorbents developed recently for reversed-phase LC have never been tried in gas chromatography despite their apparent potential advantages.

At low partial pressure (more precisely: low values of the ratio p/P^0 of the partial pressure of the analyte to its vapor pressure) the amount of a substance adsorbed on a surface at equilibrium with the vapor is proportional to the partial pressure:

$$m = K_H p \quad (37)$$

where the constant K_H is the Henry constant [41]. Retention data are sometimes reported in terms of mass of adsorbent in the column, analogous to the specific retention volume, although the adsorbent mass itself is not actually related to the adsorption equilibrium. Retention data should rather be reported as a function of area S_a of adsorbent:

$$K_H = \frac{V_R}{m_a S_a} \quad (38)$$

However, the determination of S_a is complex and relatively inaccurate. The Henry constant of adsorption depends on the geometrical structure of the adsorbent as well as on the interaction energy between the surface and chemical groups in the adsorbate molecule. The Henry constant of adsorption decreases rapidly with increasing temperature. Equation 36 applies in gas–solid chromatography as well.

10.6.4. Liquid–Solid Adsorption Equilibria

Liquid–solid chromatography is the most widely applied method of liquid chromatography. Other modes such as hydrophobic-interaction chromatography are in fact variants of liquid–solid chromatography based on adsorption from complex solutions. The two most important methods in liquid–solid chromatography are *normal phase*, which uses a polar adsorbent (e.g., silica) and a low-polarity mobile phase (e.g., methylene chloride), and *reversed phase*, which uses a nonpolar adsorbent (e.g., chemically bonded C_{18} silica) and a polar eluent (e.g., methanol–water mixtures).

10.6.4.1. Normal-Phase Chromatography

Retention volumes in normal-phase chromatography depend much on the polarity and polarizability of the analyte molecules, and little on their molecular masses [42]. Thus, all alkyl benzenes, or all alkyl esters, will be eluted in a narrow range of retention volumes. On the other hand, structural changes affecting the polarity of a molecule are easily seen. The amount of water adsorbed by the stationary phase controls its activity, hence the resolution between analytes, and this is in turn a sensitive function of water dissolved in the mobile phase. Thus, control of the water content of the eluent is a primary concern. The influence of the composition of the mobile phase on retention has been discussed by SNYDER, who introduced the concept of *eluotropic strength* [42].

Strong solvents (i.e., those that are retained in a pure weak solvent) compete with components of the sample for adsorption. For this reason, a pulse of a weak solvent generates as many peaks as there are components in the mobile phase [43], [44]. These peaks are called *system peaks* [44]. This property has been used for detecting compounds that themselves produce no response from the detector. Thus, if a compound that does produce a response with the detector is introduced at a constant concentration in the mobile phase, injection of a sample is equivalent to injection of a negative amount of the background compound, leading to a negative peak in the chromatogram.

10.6.4.2. Reversed-Phase Chromatography

In reversed-phase chromatography the retention volume depends very much on the molecular mass of the analyte. Methylene selectivity (i.e., the relative retention of two successive homologues) varies over a much wider range in reversed-phase LC than in GC, and can be quite large [45].

Water is the weakest of all solvents in reversed-phase chromatography. The mechanism by which a strong solvent reduces the retention volume of an analyte is quite different in reversed-phase chromatography from that in normal-phase mode. In reversed-phase mode there is no competition. Methanol, acetonitrile, and other organic solvents miscible in water are poorly retained on reversed-phase columns eluted with pure water. Addition of organic solvent instead increases analyte solubility in the mobile phase, thereby decreasing retention. The phenomenon of system peaks is nearly negligible.

10.7. Band Broadening

As discussed previously (Chap. 10.4), the band profile in linear chromatography is Gaussian. The relative standard deviation of this Gaussian profile is used as a way of characterizing column efficiency. By definition

$$N = \frac{t_{\mathrm{R}}}{\sigma_t} = 2k^2 Ln\,2\,\frac{t_{\mathrm{R}}}{w_{1/k}} = 5.64\,\frac{t_{\mathrm{R}}}{w_{1/2}} \qquad (39)$$

where $w_{1/k}$ is the peak width at the fraction $1/k$ of its height, and L is the length of the column. The experimental height equivalent to a theoretical plate is

$$\bar{H} = \frac{L}{N} \qquad (40)$$

where N and $\bar{H}$ depend on the experimental conditions. Experimental conditions must be so established that N is at a maximum and H at a minimum to ensure that the peaks are as narrow as possible, thereby maximizing the resolution between successive bands.

10.7.1. Sources of Band Broadening

Various phenomena contribute to the broadening of peaks during their migration along a column [18]. The most important are:

1) Axial diffusion along the concentration gradient of the band profile
2) Dispersion due to inhomogeneous flow distribution between the packing particles
3) Resistance to mass transfer back and forth from the mobile phase stream passing between the particles to the stagnant mobile phase inside the particles

4) Resistance to mass transfer through pores inside the particles
5) The kinetics of the retention mechanism (e.g., of adsorption – desorption)

Calculations of the contributions from these several phenomena have been accomplished by integration of the mass-balance equation (e.g., Eq. 17), or by a statistical approach.

The simplest approach involves the random-walk model, derived first for the study of Brownian motion. The trajectory of a given molecule through the column can be represented as a succession of steps in which the molecule either moves forward at the mobile-phase velocity (u) or stays in the stationary phase, where it is immobile (i.e., it moves backward with respect to the mass-center of the peak, which moves at the velocity $R\,u$). If one can identify the nature of the steps made by the molecule, and then calculate their average length l and number n, it is possible to derive the contribution to the variance from this particular source of band broadening; the result is in fact $n\,l^2$ [18].

10.7.2. The Plate-Height Equation

Because the plate number is related to particle size, and contributions from axial dispersion and resistance to mass transfer in the mobile phase are related to both the velocity of this phase and the molecular diffusion coefficient, it is convenient to express plate height on the basis of reduced, dimensionless parameters, in the form of a *reduced plate height*, $h = H/d_p$ together with a *reduced velocity*, $v = u\,d_p/D_m$ [46].

Contributions from all the various sources of band broadening listed in the previous section have been calculated [18]. If we assume that these sources are independent, the band variance is the sum of each contribution to the variance, and the column HETP is given by the equation [46]

$$h = \frac{B}{v} + Av^{1/3} + Cv \tag{41}$$

where the coefficients A, B, and C are dimensionless functions of the experimental conditions, but velocity independent. The use of reduced plate height and reduced velocity has become standard practice in LC, but not in GC, probably because the concept arose only in the mid 1970s—although this is not a valid reason for ignoring it.

Depending on the type of chromatography under consideration, the plate-height coefficients are obtained from different relationships. The Golay equation [27] is typical. This describes the plate height for an open tubular column. It is characterized by $A = 0$ and C a simple function of k'. In a packed GC column, anastomosis of the flow channels causes the axial dispersion to be more complex, and results in the $A\,v^{1/3}$ contribution in Equation 41 as well as a more complicated equation for the C coefficient than in the Golay equation. It is striking, however, that GC produces $h = f\,(v)$ (i.e., HETP versus velocity) curves that are quite similar to those from LC, with minimum values of h between 2 and 3, and corresponding values of v between 3 and 5. These values are excellent approximations for order-of-magnitude calculations. The only marked difference between experimental results reflects the distinction between open-tubular and packed columns. For the former, $A = 0$, and the curve is a hyperbola. For the latter, the curve has an inflection point; a tangent drawn from the origin has its contact point at a finite velocity.

Identifying those sources of band broadening that contribute significantly to increased peak width, and then calculating their contributions, makes it possible to derive the value of the local plate height; i.e., the coefficient relating an incremental increase in band variance to an incremental distance of migration [12]:

$$d(\sigma^2) = H\,dz \tag{42}$$

The *apparent plate height*, $\bar{H}$, given by Equation 40, is the column length average of the local plate heights

$$\bar{H} = \frac{\int_0^L H\,dz}{L} \tag{43}$$

This equation can be used to calculate apparent HETPs in both GC and TLC, which are types of chromatography in which the mobile-phase velocity varies quite markedly during the course of peak migration [47].

10.7.3. Resolution Between Peaks

The *resolution* (R) between two successive Gaussian bands is defined as

$$R = 2\frac{t_{R,2} - t_{R,1}}{w_1 + w_2} \qquad (44)$$

Assuming the two bands are close (since otherwise the concept of resolution loses most of its relevance), the relationship $t_{R,2}+t_{R,1} \simeq 2\,t_{R,2}$ represents a good approximation, and

$$R = \frac{\sqrt{N}}{4}\frac{\alpha - 1}{\alpha}\frac{k'}{1+k'} \qquad (45)$$

This equation [48] shows, in the first place, that there is no separation if there is no retention. Separations become very difficult when k' tends toward 0 even if α is large. A case in point is the separation of nitrogen from argon by gas chromatography on molecular sieves (0.5 nm) at room temperature.

Furthermore, solving Equation (45) for N gives the column efficiency required for separating two compounds with a given degree of resolution. This shows that the required column length increases rapidly as α approaches 1. With a 100-plate column it is possible to separate with unit resolution two compounds with $\alpha = 2$ and $k' = 3$; a 1 000 000-plate column is required to separate two compounds with $\alpha = 1.005$.

10.7.4. Optimization of Experimental Conditions

It is quite easy to find the conditions corresponding to maximum column efficiency. The column must be operated at the velocity producing the minimum plate height, and if the efficiency is still insufficient, a longer column must be prepared. In practice, certain constraints are applicable, however. First, there is a maximum pressure at which any column can be operated, a pressure that depends on the equipment available. Furthermore, analysts are often more interested in performing separations rapidly than in achieving the highest possible resolution. It should be emphasized that the less time a band spends in a column, the narrower it will be, hence the higher its maximum concentration and the lower the detection limit. Optimization for minimum analysis time becomes an important consideration. It is easily shown that minimum analysis time is achieved by operating a column at the reduced velocity for which the ratio h/v is a minimum, with a particle size such that this velocity is achieved at the maximum acceptable column inlet pressure [49] – [51].

A considerable amount of work has been published on optimizing the experimental conditions for minimum analysis time under various constraints [52]. One complication arises from the definition of reduced mobile-phase velocity. The actual mobile-phase velocity depends largely on the molecular-diffusion coefficient of the analyte. Thus, very small particles can be used for the analysis of high molecular mass compounds, which have low D_m values. The actual flow rate required will remain compatible with pressure constraints despite the resulting high pneumatic or hydraulic resistance. Detailed results obviously depend greatly on the mode of chromatography used.

10.7.5. Instrumental Requirements

A column alone cannot give the information the analyst requires. A complete instrument is needed, one with a mobile-phase delivery system, an injection device, a detector, and a data station. At all stages the requirements and properties of the modules force the analyst to accept compromises and lose resolution. Contributions of the injector and detector to band broadening should be of special concern. The lucid analysis of these problems with respect to GC by STERNBERG [53] has become a classic. Most of the conclusions are easily extended to LC.

Band broadening in the equipment results from axial dispersion in valves and connecting tubes, as well as from time delays associated with injection of the sample and detection of the separated peaks. The contributions to band broadening from these sources can be expressed as additional terms in the equation for apparent column HETP:

$$H_c = \frac{\sum \tau_j^2 u^2}{(1+k')^2 L} \qquad (46)$$

where τ^2 is the time variance caused by any of the band-broadening sources indicated above.

10.7.5.1. Injection Systems

There is a dearth of data regarding actual injection profiles for every type of column chromatography. Dependence of the instrument variance on the mobile-phase flow velocity has been studied in liquid chromatography [32]. Probably in part because of the onset of eddy turbulence at various flow rates in different parts of the instrument, this variation is not always simple.

If accurate efficiency data are required, the variance of the injection profile should be measured as a function of the flow rate, with a correction applied by subtracting the first moment of the injection profile from the peak retention time and the variance of the injection profile from the band variance. In actual analytical practice it is often sufficient to minimize this contribution by making sure that the volume of the injection device is much smaller than the volume of the column, that the device is properly swept by the mobile phase, and that actual injection is rapid. The use of a bypass at the column inlet permits a higher flow rate through the injection device than through the column, effectively reducing the band-broadening contribution of the injection system.

10.7.5.2. Connecting Tubes

STERNBERG [53] and GOLAY and ATWOOD [32], [54] have studied this particular contribution, showing that for short tubes it is smaller than predicted by application of the Golay equation (Eq. 17) to an empty tube. On the other hand, dead volumes, spaces unswept by the mobile-phase stream but to which the analytes have access by diffusion, act as mixing chambers, and they may have a highly detrimental effect on efficiency. The relative importance of the problem increases with decreasing column diameter.

10.7.5.3. Detectors

The detector is a concentration sensor appropriate to the analysis in question. As in the case of all sensors, detectors do not react immediately, but instead have a time constant. Thus, they contribute to band broadening through both their cell volume and their response time. The cell-volume contribution can be handled in the same way as the volume of the injection device. The response-time contribution is given by Equation 46, and it is proportional to the square of the mobile-phase velocity. The response time should be less than one-tenth of the standard deviation of a Gaussian peak in order to have a negligible effect on the retention time and profile of the peak [31].

10.7.5.4. Instrument Specifications

If a fractional loss of efficiency equal to Φ is acceptable, it follows that:

$$\left(\frac{\sigma_{\mathrm{eq}}}{\sigma_{\mathrm{col}}}\right)^2 \leq \Phi \tag{47}$$

where σ_{eq}^2 and σ_{col}^2 are the variance contributions of the equipment and the column, respectively. Application of this relationship permits easy derivation of equipment specifications as a function of the particular system used and the analysis of interest [52].

10.8. Qualitative Analysis

Although chromatography is a most powerful method of separation, it affords few clues regarding the identity of the separated eluates. Only a retention time, volume, or factor is available for each peak, and the method cannot even establish whether a given peak corresponds to a pure compound or to a mixture. Supplying information that is both meager and not very accurate, chromatography alone is poorly suited to the identification of single components, although previously identified patterns simplify the matter (e.g., alkane series in petroleum derivatives, or phytane–pristane in crude oils). Innumerable attempts at processing retention data for identification purposes have been reported, including the use of the most refined chemometric methods, but success has been limited. In practice, a combination of chromatographic separation and spectrometric identification has become the preferred method of identification for unknown chemicals.

10.8.1. Comparisons of Retention Data

The most efficient identification procedure involves the determination in rapid succession of retention data (retention volumes, or preferably retention factors) for the components of a mixture and a series of authentic compounds. This method permits straightforward identification of components known to be present in the mixture, and it can also rule out the presence of other possible components. However, it cannot positively identify the presence of a given, isolated component. It is recommended that such determinations be carried out using similar amounts of compounds, and that a number of spike analyses be performed, where small amounts of authentic compounds are added to the mixture in order to approximately double the sizes of unknown peaks. Comparison of the efficiencies of an unknown peak in the original sample and in the spiked mixtures increases the probability of recognizing differences between the analyte and an authentic compound.

GIDDINGS and DAVIS [55] have shown that, for complex mixtures, analyte retention data tend to reflect a Poisson distribution. This result permits a simple calculation of the probability of finding a certain number of singlets, doublets, and higher-order multiplets in the course of analyzing a mixture of m components on a column with a peak capacity n. If the ratio m/n is not small, the probability is low. This confirms that there is little chance of separating a complex mixture on the first phase selected [56]. Method development is a long and onerous process because the probability of random success is low.

In practice, qualitative analysis is made easier by the use of retention indices I [39]:

$$I = 100z + \frac{\log\left(t_{R,X} - t_0\right) - \log\left(t_{R,P_z} - t_0\right)}{\log\left(t_{R,P_z+1} - t_0\right)} \tag{48}$$

where $t_{R,X}$ is the retention time of component X, and P_z is the n-alkane containing z carbon atoms. The retention index varies little with temperature, usually following an Antoine-type equation. The retention index increment

$$\Delta I = I^{P} - I^{A} \tag{49}$$

is the difference between the retention indices for a compound on two stationary phases: A polar phase, P, and a nonpolar phase, A. Squalane, Apiezon L (a heavy hydrocarbon grease), and various polymethyl siloxanes have been used as nonpolar phases. KOVATS and WEHRLI [57] have shown that the retention index increment for a compound can be related to its structure.

10.8.2. Precision and Accuracy in the Measurement of Retention Data

The *precision* of retention data can be quite high [38], [58]. This is related to the precision with which the peak maximum can be located; i.e., it depends on column efficiency and on the signal-to-noise ratio. *Reproducibility* depends on the stability of the experimental parameters (i.e., the mobile-phase flow rate and the temperature). Reproducibility also depends on the stability of the stationary phase, which is, unfortunately, much less satisfactory than often assumed. In gas–liquid chromatography, the stationary liquid phase tends to oxidize, decompose, and become more polar unless extreme precautions are taken to avoid oxygen, which diffuses into the carrier gas through septa and flow-rate controller membranes. In liquid–solid chromatography, impurities in the mobile phase (e.g., water in the case of normal phase LC) are adsorbed, and these slowly modify the activity of the stationary phase; alternatively, the aqueous mobile phase may slowly react with a chemically bonded phase and hydrolyze it.

The *accuracy* of retention data is still more questionable than their reproducibility. Poor batch-to-batch reproducibility for silica-based phases has plagued LC for a generation. For this reason, retention data cannot be compiled in the same way as spectra. Such data collections have limited usefulness, providing orders of magnitude for relative retention rather than accurate information on which an analyst can rely for the identification of unknowns. Precision and accuracy with respect to absolute and relative retention data have been discussed in great detail [58].

10.8.3. Retention and Chemical Properties

Considerable effort has been invested in elucidating the relationships between chemical structure and retention data in all forms of chromatography. A wide compendium of results has been published in the literature. These studies are much more detailed for GC than LC, probably because work was begun earlier in the former case, before the sterility of the approach for purposes of qualitative analysis was recognized.

A plot of the logarithm of the retention volumes for homologous compounds versus the number of carbon atoms in their chains produces a nearly straight line. Exceptions are found for the first few members of a series and in other rare cases. Over a wide range of carbon-atom numbers, it also appears that such a plot exhibits definite curvature [59]. This result has been observed in all modes of chromatography capable of separating homologues. It should be noted, however, that the approximate linearity of the plots depends on a proper estimate of the hold-up volume, which is not readily defined in liquid chromatography [60]. Lines observed for different functional groups are parallel, or nearly so.

More generally, the Gibbs free energy of retention can be written as a sum of group contributions [61], and increments can be ascribed to different groups or substituents. This permits the approximate calculation of retention data from tables of increments. Unfortunately, multifunctional compounds are subject to interactions between the respective groups, and interaction coefficients must be introduced, so the system rapidly becomes too complex for practical application.

In GC, the retention volumes of nonpolar or slightly polar compounds obey the following correlation [62]:

$$\log V_g = \log \frac{273R}{M_S} + 0.22k\left(\frac{T_{Eb}}{T} - 1\right) - 2.9 - \log\gamma^{\infty} \quad (50)$$

where M_S is the molecular mass of the stationary phase, k is Trouton's constant (ca. 88 J mol^{-1} K^{-1}), and T_{Eb} is the boiling point of the compound. More detailed relationships linking retention-index increments in gas chromatography with the structures of the corresponding components have been suggested [57]. These correlations apply well within groups of related, monofunctional compounds, but difficulties arise in the attempt to extend the relations to multifunctional compounds. However, the identification of complex compounds is the only real challenge remaining in this area. Applying such relationships to a series of standards has facilitated the classification of gas-chromatographic stationary phases [63].

10.8.4. Selective Detectors

The use of selective detectors is the most promising route to the identification of unknown components in a complex mixture. Chromatography can separate complex mixtures into individual components, but it cannot positively identify them. Most spectrometric methods, on the other hand, supply sufficient information to identify complex compounds, but only if they are so pure that all the spectral clues can be ascribed safely to the same compound. Among the possible combinations, coupling with mass spectrometry, and to a lesser degree IR spectrometry, have been the most fruitful, because the amount of sample easily handled in classical chromatography is well suited to these techniques. Less sophisticated detectors have also been used to advantage.

10.8.4.1. Simple Selective Detectors

Paradoxically, the gas-density detector, which can supply the molecular mass of an unknown, can be considered to be a selective detector in gas chromatography [58]. This comment effectively illustrates the dearth of convenient selective detectors. At best such detectors help to identify compounds that contain halogen atoms (electron-capture or thermoionic detectors), sulfur, nitrogen, or phosphorus (flame-photometric or thermoionic detectors). Physiological detectors have also been used in certain rare cases (insects that react to sexual pheromones, for example, or the chemist's nose, a dangerous and hazardous application).

The choice of selective detectors in liquid chromatography is still more limited. Electrochemical detectors and the UV-spectrophotometric detector can provide clues, but many types of compounds resist oxidation or reduction in an electrochemical cell, and the paucity of clues provided by a UV spectrum has been a source of disappointment to many chromatographers.

Selective detectors are much more useful for the selective quantitation of a small number of chemicals known to be present in a complex matrix than they are for identification purposes. The use of on-line chemical reactions can be very selective for certain chemical families (e.g., amino acids, sugars). This technique has become widespread in column chromatography, both gas and liquid, and is even the principle underlying many detection schemes in thin layer chromatography [64]. The identification of an unknown presupposes methods supplying far more information and consistent with the ready association of elements of this information with structural details of the corresponding compound.

10.8.4.2. Chromatography – Mass Spectrometry (→ Mass Spectrometry)

The coupling of gas chromatography and mass spectrometry (GC/MS) was developed during the 1970s, and it has become the generally accepted method for identifying the components of complex mixtures. Its use has considerably influenced the development of modern environmental, clinical, pharmacological, and toxicological analysis. Preferably implemented with open tubular columns and associated with selective derivatization of the sample components, gas chromatography permits the separation and elution of all compounds that have significant vapor pressures at 400 °C and do not decompose, or have derivatives fulfilling these conditions. Study of the resulting mass spectra in conjunction with a rapid computer search in a large spectrum library permits the identification of separated compounds. Quantitation can be accomplished with the aid of the total ion current, currents for selected masses, or the signal from an associated chromatographic detector.

The coupling of liquid chromatography with mass spectrometry (LC/MS) has also been intensively investigated since the early 1980s, but so far without complete success. The new ionization techniques, especially ion spray, made the coupling of HPLC with MS to a routine technique. The decreasing prices(e.g. of ion trap MS instruments open the use of these instruments in routine HPLC analysis. For more details see→ Mass Spectrometry. In GC/MS it is relatively easy to remove the mobile phase, or to limit and control its effects on the ionization source, but this has proven to be much more difficult in LC/MS. A variety of interfaces has been developed for that purpose;

1) A direct liquid interface, where the mobile phase is nebulized into the source, vaporized, and condensed onto a cold surface
2) Thermo-spray, in which the mobile phase is rapidly heated and vaporized into a spray of fast-moving droplets passing the mass spectrometer inlet, which allows ions to enter the analyzer, but not the unionized solvent
3) Various spraying methods
4) Fast-atom bombardment of liquid flowing out of the column
5) Electro-spray

The variety of these methods, as well as the fact that they produce widely differing spectra and that, for reasons still unclear, one method often appears better suited to a particular group of chemicals, demonstrates that the method is still not a fully developed analytical technique. It nevertheless holds great promise, and has already proven quite useful despite difficulties more serious than those encountered in GC/MS.

10.8.4.3. Chromatography – IR Spectrometry

(→ Infrared and Raman Spectroscopy)

Infrared (IR) spectrometry is less sensitive than MS, and it therefore requires cells with a long optical path length, but IR spectrometry has also proven quite useful as an identification method in chromatography. Fourier-transform IR (FTIR) spectrometry has become the method of choice because of its sensitivity and the possibility of subtracting most of the contributions due to solvent (in liquid chromatography). It is possible to obtain useful spectra with as little as a few nanograms of a compound, provided it has been chromatographically separated from interfering components. The only remaining source of technical difficulty in coupling a chromatographic separation with an FTIR spectrometer is the size of the sample cell, which is generally excessively large relative to the volume of a band. As a consequence, the resolution between separated components must be greater than in the case of coupling to a mass spectrometer. Columns wider than the conventional 4.6 mm i.d., packed columns could be used with advantage.

10.9. Quantitative Analysis

Once a mixture has been separated into its components it becomes possible to determine the amount of each [58]. This requires a determination of the size of the concentration profile. Two possible methods are available: Use of either the *peak height* or the *peak area*. The former is more sensitive in the case of GC to fluctuations in details of the injection procedure and the column temperature, and in the mobile-phase composition and column temperature in LC. Peak areas are more sensitive to flow-rate fluctuations in both GC and LC, at least so long as the detector is concentration sensitive. In practice, however, only the peak-area method has been implemented in data packages currently available for chromatographic computers.

With respect to the most frequent case of a concentration sensitive detector, the mass of a particular compound (i) eluted from a chromatographic column is given by [58]

$$m_i = \int_{\infty} 0 C_i \,\mathrm{d}V \simeq \int_{t_\mathrm{i}}^{t_\mathrm{f}} y_i S_i F_\mathrm{v} \,\mathrm{d}t \simeq F_\mathrm{v} S_i \int_{t_\mathrm{i}}^{t_\mathrm{f}} y_i \,\mathrm{d}t \qquad (51\mathrm{a})$$

where t_i and t_f are the initial and final times between which integration of the detector profile, $y(t)$, is calculated, and S_i is the appropriate response factor. Thus, the peak area $\int_{t_\mathrm{i}}^{t_\mathrm{f}} y_\mathrm{i}\,\mathrm{d}t$ is inversely proportional to the mobile-phase flow rate. For a mass-flow sensitive detector (e.g., mass spectrometer, flame-ionization detector), the result is simpler:

$$m_i = \int_0^\infty \frac{\mathrm{d}m_i}{\mathrm{d}t}\,\mathrm{d}t \simeq \int_{t_i}^{t_f} \mathrm{d}m_i = S_i \int_{t_i}^{t_f} y_i\,\mathrm{d}t \qquad (51b)$$

and the area is independent of the flow rate.

The main problems encountered in quantitative analysis are associated with sampling and sample introduction, and with measurement of the peak area and the calibration factors [58]. Chromatographic measurements are usually carried out with a precision of a few percent, but can be much more precise when the sources of error are properly recognized and controlled [65].

10.9.1. Sampling Problems in Chromatography

As in all analytical techniques, the accuracy of a chromatographic analysis depends on the quality of the sample studied. The techniques involved in ensuring that a representative sample is selected from the system under investigation are discussed elsewhere (→ Sampling). The sample is usually introduced into the mobile-phase stream ahead of the column through a sampling valve, or with a syringe into an injection port. Calibration of the volume of sample injected is difficult, so most analytical work is accomplished by determining only relative concentrations from the actual experimental chromatograms, with internal standards added to permit absolute determinations. A discussion of the repeatability associated with sampling valves can be found in [58]. The approximate volume of an injected sample is most easily measured by injecting a standard acid solution into a flask with the value to be calibrated and titrating.

10.9.2. Measurement of Peak Area

Peak areas are now measured with the aid of automatic integrators or appropriately programmed computers. These devices incorporate A/D converters and noise filters, and make logical decisions based on the value of the observed signal. One of the major problems is the fact that the devices are closed, and often not transparent to the end user. They are typically programmed by electronic engineers or programmers, who are often poorly informed with respect to the analysts' problems. It is necessary that chemists always have access to raw data when needed, and they should demand that. The nature of the algorithms used to correct for baseline drift, to detect the beginning and end of a peak, and to allocate peak areas between incompletely resolved bands should be clearly indicated, and it should be easy to select the particular algorithm required. Various methods used in automatic integration of chromatographic signals and the related problems have been discussed in detail in [58].

Of special concern is the algorithm used for peak-area allocation when two or more bands are not sufficiently well resolved [58]. In most cases, and if the two peaks are comparable in size, a vertical line is dropped from the bottom of the valley between the peaks. This may cause an error ranging from one to a few percent. When the relative concentration is very different from unity, as in the case of trace analysis, and if the minor component is eluted second in the tail of a major component, an artificial baseline is assumed by establishing a tangent to the major-component profile before and after the minor peak, or the profile of the tail may be interpolated and subtracted on the assumption of an exponential decay. Such a procedure can never be as satisfactory as achieving complete resolution of the peaks, and can in fact cause quite significant errors [58].

10.9.3. Calibration Techniques

Quantitative analysis requires the determination of individual response factors for the various components of interest in the analyzed mixture [58]. No detector gives the same response factor for two different compounds unless they are enantiomers. In practice, quantitative analysis requires that the detector be calibrated by injection of mixtures of known compositions containing one or two internal standards. Response factors with respect to the internal standards are then determined in the concentration range expected for the analytes. Response factors should never be extrapolated, even if there are good reasons to believe that the detector provides a linear response. The concentration of the analyte C_{an} is

$$C_{an} = C_{is} S_{an/is} \frac{A_{an}}{A_{is}} \qquad (52)$$

where $S_{an/is}$ is the relative response factor for the analyte and A_{an} and A_{is} are the observed peak areas for analyte and internal standard. The need to calibrate the detector means that no quantitative analysis is possible for compounds that have not been identified. If no source of authentic compound of sufficient purity is available, calibration becomes difficult, but it is not impossible [58].

Use of an external standard, injected separately from the sample after a predetermined delay selected so as to avoid interferences between peaks of the standard and those of the mixture components, represents an extremely useful calibration tool, and it also serves as a control for monitoring proper performance of the analytical instrument, especially in process-control analysis [58], [66]. The peak area of the external, deferred standard should remain constant for as long as the chromatograph is in good working order.

10.9.4. Sources of Error in Chromatography

Any fluctuation of an operating parameter of the chromatograph, including those that control the retention times and band widths of peaks and those that determine response factors, constitutes a source of error in chromatographic analysis [58]. Such parameters include the column temperature and the mobile-phase flow rate, as well as various detector settings [67].

10.10. Theory of Nonlinear Chromatography

Equilibrium isotherms are not linear as they are assumed to be in classical treatments of analytical chromatography. In the low concentration range used for analysis, observed band profiles are in good agreement with predictions made on the basis of replacing an isotherm by its initial tangent. When large samples are used, however, and the concentrations of mixture components are high, the nonlinear character of the thermodynamics of phase equilibria causes band profiles to deviate significantly from the classical Gaussian shape. Furthermore, the equilibrium composition of the stationary phase at high concentrations depends on the concentrations of all the components. In other words, the concentration of one particular compound in the stationary phase at equilibrium is a function of the mobile-phase concentrations of all the components. Thus, the elution profiles interact. In preparative chromatography, unlike analytical chromatography, the chromatogram obtained for a mixture is not simply the sum of those obtained when equivalent amounts of each component are injected separately [23].

The prediction of such band profiles requires the solution of several problems:

1) How does the nonlinear behavior of isotherms influence band profiles?
2) How do bands of incompletely resolved components interact?
3) What is the relative importance of thermodynamic effects compared to the kinetic sources of band broadening already known to exist from studies in linear chromatography?
4) How should one model competitive isotherms?

The answers to all these questions are now well known. The ideal model of chromatography responds to the first and second questions [68]. The equilibrium – dispersive model speaks to the first three questions, at least so long as the column efficiency under linear conditions exceeds 30 – 50 plates [22]. The last question is outside the scope of this review, but has been abundantly discussed in the literature [23].

Band profiles in chromatography are obtained as solutions to the differential mass-balance equations for the mixture components (Eq. 11). The models differ in the way such balance equations are simplified and solved.

10.10.1. The Ideal Model of Chromatography

In this model we assume that the column has infinite efficiency [69]. Thus, this model makes it possible to investigate the influence of the nonlinear thermodynamics of phase equilibria independently of the blurring, dispersive effects of axial dispersion and the finite rate of mass transfer. Neglecting one of the two major phenomena controlling band profiles in chromatography, it is as "theoretical" as linear chromatography. The results must be convoluted with those of linear chromatography to lead to a realistic description of actual chromatograms. Note that in linear, ideal chromatography, band profiles are identical to injection profiles, and the bands move at the same velocity as in linear chromatography.

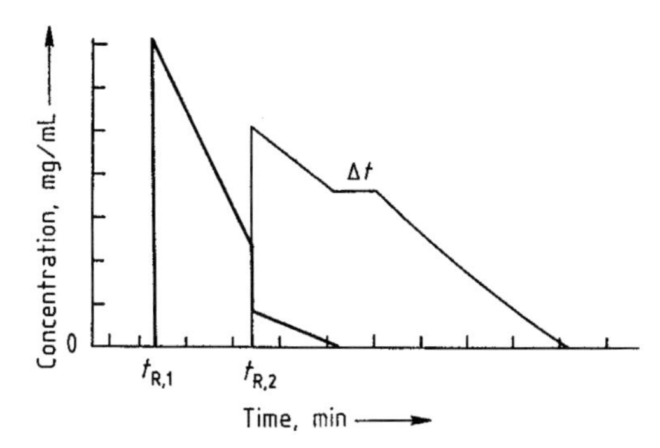

Figure 1. Solution of the ideal model for the elution of a narrow rectangular pulse of a binary mixture
Component 1; — Component 2

For single-component bands it can be shown that, in the ideal model, a particular migration velocity can be associated with each concentration [22], [69]. However, this assigned velocity is a function of the concentration. Therefore, each concentration leads to a well-defined retention time, which is a linear function of the differential of the isotherm q:

$$t_{\mathrm{R,C}} = t_0\left(1 + F\frac{\mathrm{d}q}{\mathrm{d}C}\right) \tag{53}$$

where F is the phase ratio [see Eq. (11)].

If the isotherm is convex upward, as in most cases found in liquid chromatography, the retention time decreases with increasing concentration. High concentrations cannot pass lower ones, however, as there can be only one concentration at a time at any given point. The concentrations thus pile up at the front, creating a concentration discontinuity or shock. This shock migrates at its own velocity, which is a function of its height. It can be shown [22] that the maximum concentration of the shock is the solution of the equation

$$\left|q(C) - C\frac{\partial q}{\partial C}\right| = \frac{N}{Ft_0F_v} \tag{54}$$

The retention time is obtained by introducing the maximum concentration into Equation 53.

In gas chromatography the isotherms, and especially the gas–liquid isotherms, tend to be convex downward. In this case the concentration shock forms at the rear of the band, which appears to be fronting [33]. However, another effect becomes important. The partial molar volumes of the sample components in the two phases, which are nearly equal in LC, are very different in GC. A sorption effect arises from the much higher velocity of the mobile phase in the band, where the component vapor adds to the mobile phase, relative to the region outside the band [33]. The higher the partial pressure of the vapor, the faster the band travels. Thus, bands eluted early display a front shock, later ones a rear shock. An intermediate band has a shock on both sides [70].

In the case of a binary mixture, an analytical solution exists only if the competitive equilibrium behavior is accounted for by the Langmuir isotherm

$$q_i = \frac{a_iC_i}{1 + b_1C_1 + b_2C_2} \tag{55}$$

where a_i and b_i are numerical coefficients characterizing the phase system and the compound.

This solution has been discussed in detail [71]. An example is provided in Figure 1. Characteristics of this solution include:

1) The existence of two concentration shocks at $t_{\mathrm{R,1}}$ and $t_{\mathrm{R,2}}$, one for each of the two components
2) The fact that the positive shock of the second component concentration is accompanied by a partial, negative shock from the first component concentration (at $t_{\mathrm{R,2}}$)
3) A concentration plateau of length Δt for the second component following the end of the elution profile of the first component
4) The fact that the profile for the end of the second component is the same as the profile of a pure band of that compound

The front of the second component displaces the first component (*displacement effect*), while the tail of the first component drags forward the front part of the second-component profile (*tag-along effect*). These effects have important consequences in preparative chromatography [22], [23].

10.10.2. The Equilibrium – Dispersive Model

In this model, constant equilibrium is still assumed between the two phases of the system, but finite column efficiency is accounted for by introducing a dispersion term in the mass-balance equation [22]. This term remains independent of the concentration. Solutions of the model must be obtained by numerical calculation. Several possible algorithms have been discussed. Excellent agreement has been reported between such solutions and experimental band profiles [72].

The solutions of the equilibrium–dispersive model exhibit the same features as those of the ideal model at high concentrations, where thermodynamic effects are dominant and dispersion due to finite column efficiency merely smoothes the edges. When concentrations decrease, the solutions tend toward the Gaussian profiles of linear chromatography.

10.10.3. Moderate Column Overloading

At moderate sample sizes, band profiles depend on the initial curvature of the isotherm, while retention times are functions of both initial slope and curvature [73]. Accordingly, retention varies linearly with increasing sample size [74]. The widespread belief that there is a range of sample sizes over which retention time is independent of sample size is erroneous. What happens is simply that, below a certain size, variation in the retention time is less than the precision of measurement. As the retention time decreases the band front or tail becomes steeper and the band becomes unsymmetrical.

If an isotherm can be reduced to the first two terms of a power expansion (i.e., to a parabolic isotherm), an approximate solution can be obtained by the method of HAARHOFF and VAN DER LINDE [75] in the case of a single-component band.

10.10.4. Preparative Chromatography

Preparative chromatography consists in using chromatography to purify a substance, or to extract one component from a mixture. Unlike analytical chromatography it is not devoted to the direct collection of information. This difference in objective results in marked differences in operation [23]. Production of finite amounts of purified compounds requires the injection of large samples, hence column operation under overloaded conditions. Resolution between individual bands in the chromatogram is no longer necessary provided mixed bands between the components are narrow. Instead, product recovery is the goal, demanding efficient condensers in the case of GC and rotary evaporators for LC. However, equivalent column performance is required in both analytical and preparative applications, and the packing materials must be of comparable quality.

10.11. Reference Material

10.11.1. Journals

The literature of chromatography is enormous. Several journals are dedicated entirely to this field, including the *Journal of Chromatography*, *Journal of Chromatographic Sciences*, and *Chromatographia*, while a larger number specialize in various subfields of chromatography (e.g., *Journal of L iquid Chromatography*, *Journal of Planar Chromatography*, *Journal of Microcolumn Separation*, *High Resolution Chromatography*). Several other journals publish one or several papers on chromatography in every issue. Important papers on chromatography are found in journals important in other scientific fields as well, ranging from *Biochemistry Journal* to *Chemical Engineering Science*.

One journal provides only abstracts of chromatographic papers appearing in other journals: *Chromatography Abstracts*, published by Elsevier for the Chromatographic Society. Chemical Abstracts also issues several *CA Selects* dedicated to specific aspects of chromatography.

10.11.2. Books

Probably the oldest general reference book still available is *Theoretische Grundlagen der Chromatographie* by G. SCHAY [76]. In the theoretical area, *Dynamics of Chromatography* by J. C. GIDDINGS [12] is still a mine of useful discussions, comments, and ideas. A number of excellent books have been written about gas chromatography. *Gas Chromatography* by A. B. LITTLEWOOD [77], though obsolete in many respects, contains excellent fundamental discussions on GC and GC detectors. More practical, *The Practice of Gas Chromatography* by L. S. ETTRE and A. ZLATKIS [78] supplies much valuable information. Similarly, *Modern Practice of Liquid Chromatography* by J. J. KIRKLAND and L. R. SNYDER [79] has long been the primary reference book for LC.

Among recent books, the latest editions of *Chromatography*, edited by E. HEFTMAN [80], and *Chromatography Today* by POOLE and

POOLE [81] contain excellent practical discussions, covering almost all aspects of implementing chromatographic separations. However, two important areas were virtually ignored: quantitative measurement and preparative separations. With respect to the former, *Quantitative Gas Chromatography* [58] is the only book that discusses in detail the sources of errors, their control, and problems of data acquisition and handling, which should not be left exclusively in the hands of computer specialists. For the latter, *Fundamentals of Preparative and Nonlinear Chromatography* [82] presents all aspects of the theory of chromatography, and discusses the optimization of preparative separations.

10.12. References

[1] D. H. Everett, *Trans. Faraday Soc.* **61** (1965) 1637.
[2] T. Tsuda, K. Nomura, G. Nakagawa, *J. Chromatogr.* **248** (1982) 241.
[3] K. Sakodynski, *J. Chromatogr.* **73** (1972) 303.
[4] L. S. Ettre in Cs. Horváth (ed.): *High Performance Liquid Chromatography: Advances and Perspectives,* **vol. 1,** Academic Press, New York 1980, p. 25.
[5] M. S. Tswett, *Ber. Dtsch. Bot. Ges.* **24** (1906) 316.
[6] M. S. Tswett, *Ber. Dtsch. Bot. Ges.* **24** (1906) 384.
[7] M. S. Tswett: *Khromofilly v Rastitel'nom Zhivotnom Mire (Chromophylls in the Plant and Animal World),* Izd. Karbasnikov, Warzaw, Poland, 1910 (reprinted in part in 1946 by the publishing house of the Soviet Academy of Science, A. A. Rikhter, T. A. Krasnosel'skaya, eds.).
[8] V. Berezkin: *M. S. Tswett,* Academic Press, New York 1992.
[9] A. J. P. Martin, R. L. M. Synge, *Biochem. J.* **35** (1941) 1358.
[10] R. Consden, A. M. Gordon, A. J. P. Martin, *Biochem. J.* **38** (1944) 224.
[11] A. J. P. Martin, A. T. James, *Biochem. J.* **50** (1952) 679.
[12] J. C. Giddings: *Dynamics of Chromatography,* Marcel Dekker, New York 1965.
[13] K. K. Unger: *Porous Silica, Its Properties and Use as Support in Column Liquid Chromatography,* Elsevier, Amsterdam 1979.
[14] A. Tiselius, S. Claeson, *Ark. Kemi Mineral. Geol.* **16** (1943) 18.
[15] J. Frenz, Cs. Horváth in Cs. Horváth (ed.): *High-Performance Liquid Chromatography, Advances and Perspectives,* vol. 5, Academic Press, New York 1988.
[16] S. Golshan-Shirazi, G. Guiochon in F. Dondi, G. Guiochon (eds.): "Theoretical Advancement in Chromatography and Related Separation Techniques," *NATO Adv. Study Inst. Ser. Ser. C* **383** (1992) 61.
[17] L. C. Craig, *J. Biol. Chem.* **155** (1944) 519.
[18] J. C. Giddings in E. Heftmann (ed.): *Chromatography,* Van Nostrand, New York 1975, p. 27.
[19] J. C. Giddings, H. Eyring, *J. Phys. Chem.* **59** (1955) 416.
[20] D. A. McQuarrie, *J. Chem. Phys.* **38** (1963) 437.
[21] F. Dondi, M. Remelli, *J. Phys. Chem.* **90** (1986) 1885.
[22] S. Golshan-Shirazi, G. Guiochon: "Theoretical Advancement in Chromatography and Related Separation Techniques," in F. Dondi, G. Guiochon (eds.): *NATO Adv. Study Inst. Ser. Ser. C* **383** (1992) 35.
[23] A. M. Katti, G. Guiochon in J. C. Giddings, E. Grushka, P. R. Brown (eds.), *Adv. Chromatogr. (N.Y.)* **31** (1992) 1.
[24] S. Golshan-Shirazi, G. Guiochon, *J. Chromatogr.* **603** (1992) 1.
[25] L. Lapidus, N. L. Amundson, *J. Phys. Chem.* **56** (1952) 984.
[26] J. J. Van Deemter, F. J. Zuiderweg, A. Klinkenberg, *Chem. Eng. Sci.* **5** (1956) 271.
[27] M. J. E. Golay in D. H. Desty (ed.): *Gas Chromatography 1958,* Butterworths, London 1959, p. 36.
[28] M. Kubin, *Collect. Czech. Chem. Commun.* **30** (1965) 2900.
[29] E. Kucera, *J. Chromatogr.* **19** (1965) 237.
[30] E. Grushka, *J. Phys. Chem.* **76** (1972) 2586.
[31] L. J. Schmauch, *Anal. Chem.* **31** (1959) 225.
[32] J. G. Atwood, M. J. E. Golay, *J. Chromatogr.* **218** (1981) 97.
[33] G. Guiochon, L. Jacob, *Chromatogr. Rev.* **14** (1971) 77.
[34] H. Darcy: *Les Fontaines Publiques de la Ville de Dijon,* Dalmont, Paris 1856.
[35] D. E. Martire, R. E. Boehm, *J. Phys. Chem.* **91** (1987) 2433.
[36] R. C. Reid, J. M. Prausnitz, B. E. Poling: *The Properties of Gases and Liquids,* 4th ed., McGraw-Hill, New York 1987.
[37] A. Felinger, G. Guiochon, *Biotechnol. Progr.* **9** (1993) 421.
[38] C. Vidal-Madjar, M. F. Gonnord, M. Goedert, G. Guiochon, *J. Phys. Chem.* **79** (1975) 732.
[39] E. sz Kovats, *Helv. Chim. Acta* **41** (1958) 1915.
[40] B. L. Karger, *Anal. Chem.* **39** (1967) no. 8, 24 A.
[41] A. V. Kiselev, Ya. I. Yashin: *Gas Adsorption Chromatography,* Plenum Press, New York 1970.
[42] L. R. Snyder in C. Horváth (ed.): *High Performance Liquid Chromatography, Advances and Perspectives,* **vol. 1,** Academic Press, New York 1980, p. 280.
[43] M. Denkert, L. Hackzell, G. Schill, E. Sjögren, *J. Chromatogr.* **218** (1981) 31.
[44] S. Golshan-Shirazi, G. Guiochon, *Anal. Chem.* **62** (1990) 923.
[45] W. Melander, C. Horváth in C. Horváth (ed.): *High Performance Liquid Chromatography, Advances and Perspectives,* **vol. 2,** Wiley, New York 1980, p. 114.
[46] J. H. Knox, *J. Chromatogr. Sci.* **15** (1977) 352.
[47] G. Guiochon, A. Siouffi, *J. Chromatogr. Sci.* **16** (1978) 470.
[48] J. H. Purnell, *J. Chem. Soc.* 1960, 1268.
[49] B. L. Karger, W. D. Cooke, *Anal. Chem.* **36** (1964) 985.
[50] G. Guiochon, *Anal. Chem.* **38** (1966) 1020.
[51] G. Guiochon, E. Grushka, *J. Chromatogr. Sci.* **10** (1972) 649.
[52] G. Guiochon in C. Horváth (ed.): *High Performance Liquid Chromatography, Advances and Perspectives,* **vol. 2,** Wiley, New York 1980, p. 1.
[53] J. C. Sternberg in J. C. Giddings, R. A. Keller (eds.), *Adv. Chromatogr. (N.Y.)* **2** (1966) 205.
[54] M. J. E. Golay, J. G. Atwood, *J. Chromatogr.* **186** (1979) 353.
[55] J. M. Davis, J. C. Giddings, *Anal. Chem.* **57** (1985) 2178.
[56] J. C. Giddings: *Unified Separation Science,* Wiley-Interscience, New York 1991.

[57] A. Wehrli, E. sz Kovats, *Helv. Chim. Acta* **42** (1959) 2709.
[58] G. Guiochon, C. Guillemin: *Quantitative Gas Chromatography,* Elsevier, Amsterdam 1988.
[59] R. V. Golovnya, D. N. Grigoryeva, *Chromatographia* **17** (1983) 613.
[60] H. Colin, E. Grushka, G. Guiochon, *J. Liq. Chromatogr.* **5** (1982) 1391.
[61] A. J. P. Martin, *Discuss. Faraday Soc.* **7** (1949) 332.
[62] D. A. Leathard, B. C. Shurlock: *Identification Techniques in Gas Chromatography,* Wiley-Interscience, New York 1970.
[63] L. Rohrschneider, *J. Gas Chromatogr.* **6** (1968) 5.
[64] R. W. Frei, J. F. Lawrence: *Chemical Derivatization in Analytical Chemistry, I. Chromatography,* Plenum Press, New York 1981.
[65] M. Goedert, G. Guiochon, *J. Chromatogr. Sci.* **7** (1969) 323.
[66] C. L. Guillemin et al., *J. Chromatogr. Sci.* **9** (1971) 155.
[67] G. Guiochon, M. Goedert, L. Jacob in R. Stock (ed.): *Gas Chromatography 1970,* The Institute of Petroleum, London 1971, p. 160.
[68] S. Golshan-Shirazi, G. Guiochon: "Theoretical Advancement in Chromatography and Related Separation Techniques," in F. Dondi, G. Guiochon (eds.): *NATO Adv. Study Inst. Ser. Ser. C* **383** (1992) 1.
[69] D. DeVault, *J. Am. Chem. Soc.* **65** (1943) 532.
[70] P. Valentin, G. Guiochon, *J. Chromatogr. Sci.* **10** (1975) 271.
[71] S. Golshan-Shirazi, G. Guiochon, *J. Phys. Chem.* **93** (1989) 4143.
[72] G. Guiochon et al., *Acc. Chem. Res.* **25** (1992) 366.
[73] S. Golshan-Shirazi, G. Guiochon, *J. Chromatogr.* **506** (1990) 495.
[74] A. Jaulmes, C. Vidal-Madjar, M. Gaspar, G. Guiochon, *J. Phys. Chem.* **88** (1984) 5385.
[75] P. C. Haarhof, H. J. Van der Linde, *Anal. Chem.* **38** (1966) 573.
[76] G. Schay: *Grundlagen der Chromatographie,* Springer Verlag, Berlin.
[77] A. B. Littlewood: *Gas Chromatography,* 2nd ed., Academic Press, New York 1970.
[78] L. S. Ettre, A. Zlatkis: *The Practice of Gas Chromatography,* Interscience, New York 1967.
[79] J. J. Kirkland, L. R. Snyder: *Modern Practice of Liquid Chromatography,* Wiley-Interscience, New York 1971.
[80] E. Heftman (ed.): *Chromatography,* 5th ed., Elsevier, Amsterdam 1993.
[81] C. Poole, S. K. Poole: *Chromatography Today,* Elsevier, Amsterdam 1992.
[82] G. Guiochon, S. Golshan-Shirazi, A. Katti: *Fundamentals of Preparative and Nonlinear Chromatography,* Academic Press, Boston 1994.

11. Gas Chromatography

PAT J. F. SANDRA, University of Stellenbosch, Matieland, South Africa, Universiteit Gent, Vakgroep Organische Chemie, Gent, Belgium

Abbreviations

AED atomic emission detection
CGC capillary gas chromatography
CI chemical ionization
CpSi cyanopropylsilicone
d_f film thickness
D_M diffusion coefficient of the mobile phase
d_p particle diameter
ECD electron capture detector
EI electron ionization
ELCD Hall eletrolytic conductivity detector
EPC electronic pressure control
FFAP free fatty acid phase
FID flame ionization detector
FPD flame photometric detector
FSOT fused silica open tubular
GALP good automated laboratory practice
GLC gas–liquid chromatography
GLP good laboratory practice
GSC gas–solid chromatography
H column efficiency, plate height
HPLC high performance liquid chromatography
HS head space
i.d. internal diameter
K Kelvin temperature
k retention factor
L column length
MAOT maximum allowable operating temperature
MDCGC multidimensional capillary chromatography
MeSi methylsilicone
MiAOT minimum allowable operating temperature
N plate number
NPD nitrogen phosphorus detector
PCGC packed column gas chromatography
PID photoionization detector
PLOT porous layer open tubular
PT purge and trap
PTV programmed-temperature vaporization
r column radius
RCD redox chemiluminescence detector
R_s peak resolution
SFC supercritical fluid chromatography
SFE supercritical fluid extraction
SIM selected ion monitoring
TCD thermal conductivity detector
TD thermal desorption
TEA thermal energy analyzer
TID thermionic detector
t_M retention time of solvent
t_R retention time of analyte
u mobile phase velocity
WCOT wall-coated open tubular
α separation factor

11.1. Introduction

Gas chromatography (GC) is the separation technique that is based on the multiplicative distribution of the compounds to be separated between the two-phase system solid or liquid (*stationary phase*) and gas (*mobile phase*). Gas chromatography by definition thus comprises all separation methods in which the moving phase is gaseous. Contrary to the other chromatographic techniques, i.e., high-performance liquid chromatography (HPLC) and supercritical fluid chromatography (SFC), the role of the gaseous mobile phase—quasi-ideal inert gases such as nitrogen, helium, or hydrogen—is purely mechanical: they just serve for the transport of solutes along the column axis. The residence time (retention) of solutes is affected only by their vapor pressure, which depends on the temperature and on the intermolecular interaction between the solutes and the stationary phase.

Currently, gas chromatography is one of the most important and definitely the most economic of all separation methods. Its applicability ranges from the analysis of permanent gases and natural gas to heavy petroleum products (up to 130 carbon atoms, simulated distillation), oligosaccharides, lipids, etc. Moreover, as far as chromatographic efficiency and GC system selectivity is concerned, no other separation technique can compete with gas chromatography. The dictum "If the separation problem can be solved by gas chromatography, no other technique has to be tried out" is now generally accepted. As illustration, more than 80% of the priority pollutants on the lists of the EC and of the EPA in the United States, are amenable to GC analysis.

Gas chromatography was developed in the early 1950s [8]. Tens of thousands of publications and more than 250 textbooks on theory and practice have been published. The purpose of this article is not to write a new textbook on GC but rather to help newcomers in the field to find their way in the present state of the art of modern GC. For the basic aspects of GC, see → Basic Principles of Chromatography. Fundamental equations are given or repeated only to clarify some important statements. The terms, symbols, and nomenclature used are those that have been advised by IUPAC [9].

At the end of the references, some general and specialized textbooks are listed, together with the most important journals dealing with gas chromatography. The reader is advised to consult these books and journals for a more detailed study.

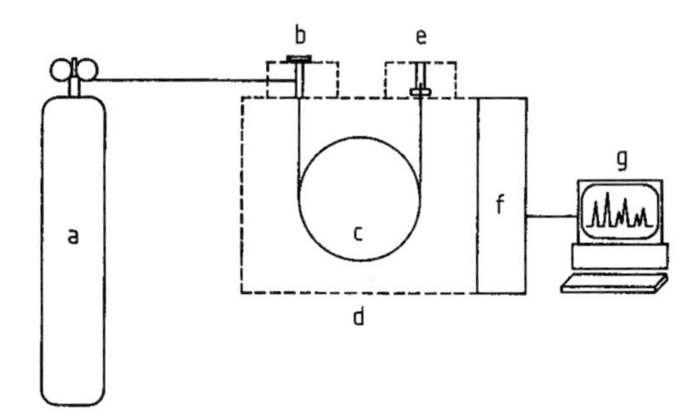

Figure 1. Basic components of a modern GC system
a) Carrier gas supply; b) Injector; c) Column; d) Column oven; e) Detector; f) Controller; g) Recorder

11.2. Instrumental Modules

A schematic drawing of a modern gas chromatographic system is shown in Figure 1. The basic parts are the carrier gas supply (a), the injector (b), the column (c), and the detector (e). The dashed lines indicate thermostatted regions. The carrier gas is usually supplied from a high-pressure cylinder equipped with a two-stage pressure regulator. Via the controller of the GC instrument (f), the gas flow can be fine-tuned, and preheated carrier gas is delivered to the column in the constant-pressure, constant-flow, or pressure-programmed mode. The sample is introduced into the carrier gas stream via the injector, and the vaporized components are transported into the column, the heart of the system, where separation occurs. The column is placed in an oven, the temperature of which can be kept constant (*isothermal operation*) or programmed (*temperature-programmed operation*). Some applications are performed at subambient temperature and instruments can be equipped with cryogenic units. After separation, the component bands leave the column and are recorded (g) as a function of time (chromatogram) by a detection device (e). The chromatogram provides two types of data. The residence time or retention time of the components in the column is characteristic for the solute–stationary phase interaction and can be used for qualitative interpretation or component identification. The detector's response is proportional to the amount of separated sample component and gives quantitative information on the composition of the mixture.

11.3. The Separation System

11.3.1. Modes of Gas Chromatography

Two types of gas chromatography exist: gas–liquid chromatography (GLC) and gas–solid chromatography (GSC). Other classification schemes such as GSC, GLC plus capillary gas chromatography (CGC) are outdated because nowadays GLC and GSC can be performed both in packed columns and in capillary or open tubular columns.

11.3.1.1. Gas–Liquid Chromatography

In GLC the stationary phase is a liquid acting as a solvent for the substances (solutes) to be separated. The liquid can be distributed in the form of a thin film on the surface of a solid support, which is then packed in a tube (packed column GC, PCGC) or on the wall of an open tube or capillary column (capillary GC, CGC). The term WCOT (*wall-coated open tubular*) is used only for capillary columns coated with a thin film of a liquid. In GLC the separation is based on partition of the components between the two phases, the stationary phase and the mobile phase, hence the term partition chromatography. This mode of GC is the most popular and most powerful one.

11.3.1.2. Gas–Solid Chromatography

Gas–solid chromatography comprises the techniques with an active solid as the stationary phase. Separation depends on differences in adsorption of the sample components on inorganic adsorbents (i.e., silica, alumina, carbon black) or on organic adsorbents such as styrene–divinylbenzene copolymers. Separation can also occur by a size exclusion mechanism, such as the separation of gases on synthetic zeolites or molecular sieves. GSC is performed on packed columns or on open tubular columns on the walls of which a thin layer of the porous material is deposited [*porous layer open tubular* (PLOT) columns]. GSC nowadays is used only for special separation problems, and GSC columns are, therefore, referred to as tailor-made columns.

11.3.2. Selection of the Carrier Gas

Common carrier gases are nitrogen, helium, and hydrogen. The choice of carrier gas depends

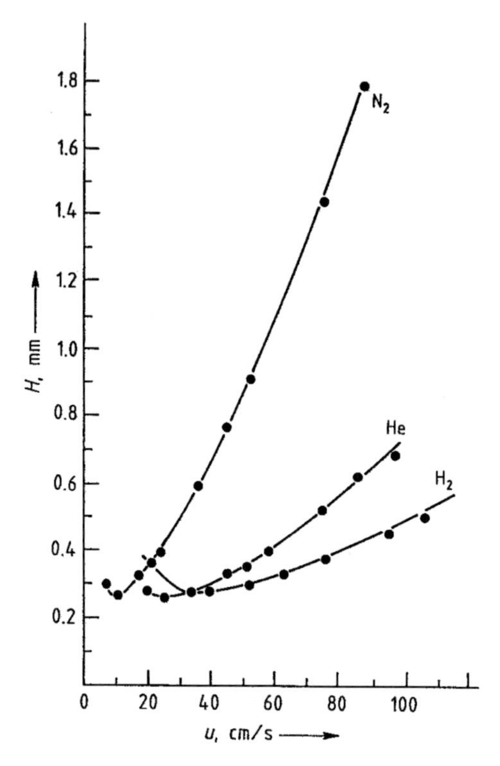

Figure 2. Experimental $H-u$ curves as function of carrier gas

on the column and the detector used but, above all, on the required speed of analysis and detectability. The influence of the mobile-phase velocity u on column efficiency H has been detailed in → Basic Principles of Chromatography. The relation $H-u$ was derived for packed column GC by VAN DEEMTER, ZUIDERWEG, and KLINKENBERG [10] and adapted to capillary columns by GOLAY [11]. Practical consequences of the carrier gas selection in capillary GC have been described in [12], [13]. Due to the relatively high density of nitrogen compared to helium and hydrogen, longitudinal diffusional spreading of the solutes in nitrogen is small, but the resistance to mass transfer in the mobile phase is high. In contrast, for low-density gases such as helium and especially hydrogen, longitudinal diffusional spreading is large but resistance to mass transfer small. For columns with small amounts of stationary phases, the resistance to mass transfer in the stationary phase can be neglected, and the minimum plate height H_{min} is nearly independent of the nature of the carrier gas. The optimum mobile-phase velocity u_{opt}, corresponding to H_{min} is proportional to the diffusion coefficient of the mobile phase D_M; u_{opt} is 10 cm/s for nitrogen, 25 cm/s for helium, and 40 cm/s for hydrogen. In practical GC on thin-film columns, hydrogen is the best choice because the analysis time is reduced by a factor of 4 compared to nitrogen, while the detectability is increased by the same order. Some experimental $H-u$ curves for a *capillary column* 20 m in length and 0.27 mm internal diameter (*i.d.*) coated with a 0.22-μm film of methylsilicone (MeSi) are shown in Figure 2. The experimental data fit the theory perfectly. The fact that the slope (i.e., the increase of plate height with velocity) for nitrogen is much steeper than for the two other gases is noteworthy. Working close to u_{opt} is, therefore, a must with nitrogen, whereas for helium and hydrogen the velocity can be increased to values greater than u_{opt} without losing too much resolution power. For thin-film columns, hydrogen is definitely the best choice. When hydrogen is not allowed for safety reasons—most GC systems nowadays are equipped with hydrogen sensors—helium should be chosen. Because of the important contribution of the resistance to mass transfer in *thick-film columns*, this situation is more complicated. The selection depends on what is considered most important, efficiency or speed. This is dictated by the separation problem at hand. For a given column, the selection of carrier gas and optimum flow rate is best determined experimentally by making $H-u$ plots. This is not difficult and can be carried out in 3–4 h in the following way:

1) A solution of a hydrocarbon (i.e., tridecane) in a highly volatile solvent (i.e., hexane) is injected at an inlet pressure P_x of carrier gas A and a column temperature is selected to give a retention factor k of ca. 5.
2) The plate number N is calculated according to

$$N = 5.54\,(t_R/w_h)^2 \tag{3.1}$$

   where t_R is the retention time of the hydrocarbon and w_h is the peak width at half height.
3) The H value is calculated from

$$H = L/N \tag{3.2}$$

   where L denotes the column length.
4) The corresponding u value is given by

$$u = L/t_M \tag{3.3}$$

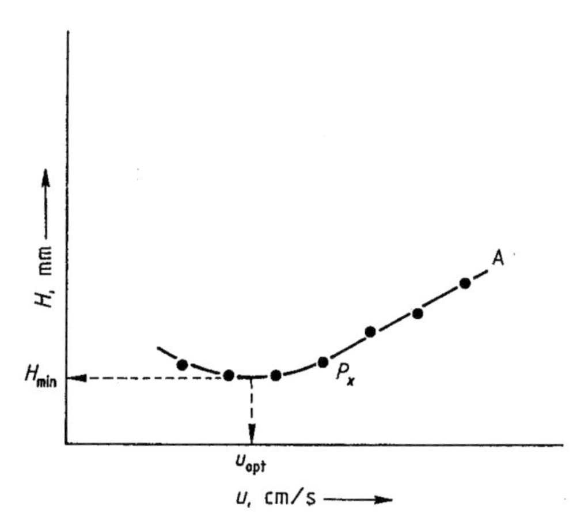

Figure 3. Making a $H-u$ plot

where t_M is the retention time of the solvent; for thick-film columns, methane is injected instead of hexane to give t_M.

5) The H and u values corresponding to P_x of carrier gas A are marked in a $H-u$ plot (Fig. 3).
6) The inlet pressure P is set at values P_x^{-3}, P_x^{-2}, P_x^{-1}, P_x^{+1}, P_x^{+2}, P_x^{+3}, and the corresponding H and u values are drawn in the plot.
7) H_{min} and u_{opt} are deduced from the plot.
8) The same procedure is repeated for carrier gases B and C.

High-purity gases should be used in GC. Most gas suppliers have special GC-grade gases. If lower-quality carrier gases are used, purification through a molecular sieve trap to remove moisture and low molecular mass hydrocarbons and through an oxy-trap to remove oxygen is recommended.

11.3.3. Selection of the Gas Chromatographic Column

Two column types are in use, the packed and the capillary column. The selection is often dictated by the nature and complexity of the sample. The present trend is to replace PCGC by CGC whenever possible, because data obtained with the latter technique are much more reliable. The fundamental difference between the two methods is reflected in the resolution equation, the key equation for separation optimization:

$$R_S = \frac{\sqrt{N}}{4} \frac{\alpha - 1}{\alpha} \frac{k}{k+1} \qquad (3.4)$$

Anything that increases the column efficiency N, the column selectivity α, or the retention factor k will enhance the separation power of the column. Packed columns are characterized by low plate numbers and PCGC is therefore a low-resolution technique. The lower efficiency is compensated by the high selectivity α of the stationary phase, and this is the main reason why so many different stationary phases have been developed for PCGC. Capillary columns on the other hand have very high plate numbers and, therefore, the number of stationary phases can be restricted because the selectivity is less important. In fact, most separation problems can be handled with four basic stationary phases and half a dozen tailor-made stationary phases. Other important features of capillary columns are their inertness and compatibility with spectroscopic detectors. In the framework of this discussion, emphasis is, therefore, on capillary columns.

11.3.3.1. Packed Columns

A packed column is defined by the material of which the tube is made, by its length L and internal diameter *i.d.*, by the nature and diameter d_p of the particles, and of course by the film thickness d_f and the nature of the stationary phase. The latter is most important because the stationary phase controls column selectivity (see Section 11.3.4). All other column parameters represent column efficiency. In packed column GC, H_{min} is controlled mainly by the particle diameter and is always $>2d_p$. A column 1 m long filled with particles of 100–120 mesh (150–125 μm) can never generate more than 3000 plates. Because of the packed bed, the resistance to the mobile flow is high, and columns longer than 3 m exhibit low permeability and are not practical. This means that a total plate number of 10 000 is the maximum that can be reached with packed columns. Most of the columns for routine analysis in fact have only 3000 to 5000 plates. The column efficiency is independent of column diameter. The column tube can be made of copper, stainless steel, nickel, glass, and polytetrafluoroethylene. Glass and polytetrafluoroethylene are the most inert materials and are preferred for the analysis of polar and thermolabile compounds. In *gas–liquid chromatography*, the support material has a silicium oxide network, such as diatomaceous earth, synthetic spherical silica, or glass beads. The support materials are purified by acid washing and deactivated by silanization, which caps the active sites. However, even the most highly deactivated surface will still exhibit adsorption of polar compounds such as organic acids and bases. A variety of support materials are available commercially, the catalogues of the column manufacturers (Agilent, Chrompack, Supelco, Alltech, etc.) should be consulted for detailed information. The stationary-phase film is homogeneously coated on the particles by dissolving the phase in a suitable solvent and adding the support material to the mixture. The solvent is then removed on a rotavapor (rotary evaporator), and after complete drying under a purified nitrogen flow, the column is packed tightly with the coated support. In PCGC, the contents of liquid coatings are expressed as a percentage of the support material: 1, 3, 5, and 10 wt% are commonly used.

The dimensions of columns used in *gas – solid chromatography* are very similar to those of gas–liquid chromatography. Packing materials (silica, alumina, carbon blacks, zeolites, porous polymers) in proper mesh size are available commercially to fill the columns. GSC columns give lower plate numbers than GLC columns but possess very high selectivities for some typical applications. Prepacked GLC and GSC columns can be purchased commercially and are ready to be installed in the GC instrument.

11.3.3.2. Capillary Columns

Major progress in gas chromatography, particularly in the area of applications, has been made by the introduction of capillary columns. Invented by M. Golay in 1957 [11], the real breakthrough came only in 1979 with the development of flexible fused silica as tubing material for capillary columns [14] (*fused silica open tubular columns*, FSOT). The inherent strength and flexibility of fused silica made use of capillary GC easy, compared to the glass capillary columns generally used previously. In addition, no other column material can offer the same inertness. A capillary column is characterized by its dimensions (*L, i.d.*) and by the nature of the stationary phase. For *GLC*, a relatively thin stationary (liquid) phase film, with film thickness d_f, is coated directly on the inner wall (WCOT columns). The film thickness controls the retention factor k and the sample capacity of a capillary column. In the case of GSC, the adsorbent with particle diameter d_p is deposited in a thin layer d_l on the capillary wall (PLOT columns). The main advantage of PLOT columns compared to packed GSC columns is the speed of analysis (see Section 11.3.4.2). Most of the applications (> 90 %) in capillary GC are performed on WCOT columns. These columns can be subdivided in three types according to their internal diameter. Columns with *i.d.* of 0.18 – 0.32 mm are called *conventional capillary columns* because this is the column type most often used for the analysis of complex samples. Columns with an *i.d.* of 0.53 mm are known as *widebore* or *megabore columns*. They have been introduced as an alternative to packed columns for the analysis of less complex mixtures. *Narrowbore columns* with *i.d.*'s ranging from 0.1 to 0.05 mm provide very high efficiencies or short analysis times. This column type is not often used nowadays, but this situation is expected to change in the next few years (see Section 11.10.1).

The selection of a capillary column depends on the complexity of the sample to be analyzed. The column length, the internal diameter, the stationary phase, and its film thickness determine the separation power (resolution), the sample capacity, the speed of analysis, and the detectability or sensitivity. Theoretical considerations [12], [13] indicate that for capillary columns with thin films (< 1 µm), the H_{min} value is roughly equal to the column diameter. This is illustrated in Figure 4, which shows experimental $H - u$ curves for columns varying in internal diameter. H was calculated for dodecane at 100 °C with hydrogen as carrier gas. H_{min} measured experimentally is indeed very close to H_{min} deduced theoretically. By knowing this, the maximum plate number that a capillary column can provide may be calculated without performing any analysis:

$$N = L/H = L/H_{min} = L/d_c \qquad (3.5)$$

The H_{min} values, the maximum number of plates per meter, and the minimum column length needed for 100 000 plates for different *i.d.*'s of commercially available FSOT columns are listed in Table 1. In contrast to packed columns, the length of a capillary column is not restricted because the permeability of an open tube is very high. To a great extent, the length and internal diameter of a capillary column define the efficiency. The selection of the film thickness for a given stationary phase is also of upmost importance. The resolution equation (Eq. 3.4) shows that low k values result in poor resolution. For the same column efficiency N and selectivity α, increasing the k value from 0.25 to 5 corresponds to a resolution gain of 378 % (R_s increases from 0.37 to 1.40).

Table 1. Characteristics of FSOT columns coated with thin films

d_c, mm	H_{min}, mm	N, m^{-1}	L for 100 000 plates, m
0.05	0.05	20 000	5
0.10	0.10	10 000	10
0.18	0.18	5 556	18
0.22	0.22	4 545	22
0.25	0.25	4 000	25
0.32	0.32	3 125	32
0.53	0.53	1 887	53

Increasing the retention factor can be accomplished by increasing the film thickness

$$k = \frac{K\,2\,d_f}{r_c} \tag{3.6}$$

For a fixed column radius (r_c = constant), the retention factor of a solute at a given temperature (K = constant), doubles when the film thickness is increased by a factor of 2. Guidelines for the selection of film thickness for the stationary phase methylsilicone are listed below (values depend on column length and stationary phase):

≈0.1 µm	high molecular mass compounds	500 – 1500
≈0.2 µm	$M_r \approx 300 - 500$	
≈0.3 µm	$M_r \approx 150 - 400$	
≈0.5 µm	$M_r \approx 50 - 200$	
≈1 µm	$M_r \approx 30 - 150$	
2 – 5 µm	widebore – high-capacity narrowbore – high volatiles	

Conventional Capillary Columns. Capillary columns with *i.d.*'s of 0.18 – 0.32 mm and lengths of 10 – 60 m are the "workhorses" for high-resolution separation. Long columns are preferred for the analysis of very complex samples. The less complex the sample, the shorter the column can be. Figure 5 shows the analysis of the oxygen fraction of the essential oil of black pepper [15]. The profile is extremely complex (more than 300 compounds), and the column efficiency must be high to unravel all solutes. Long columns, of course, imply long analysis times. Figure 6 shows the carbon number separation of the neutral lipids in palm oil. The separation is less challenging and the column length can be reduced to 5 m, which results in an analysis time of only 3 min [16]. For samples of unknown complexity, column lengths of 25 – 30 m offer a good compromise. Column dimensions are often adapted to the requirements

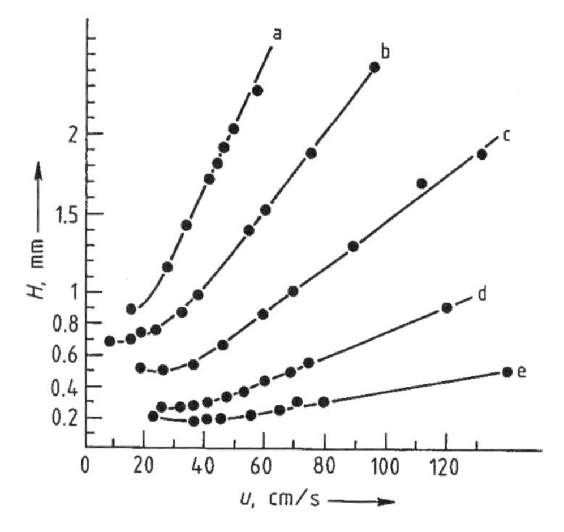

Figure 4. Experimental $H-u$ curves for columns with different *i.d.*'s
a) 0.88 mm; b) 0.70 mm; c) 0.512 mm; d) 0.27 mm; e) 0.18 mm

of pre- or postcolumn devices. As an example, for the combination with mass spectroscopy, an internal diameter of 0.18 – 0.25 mm is preferred because in this case the flow of the mobile phase is of the order of 1 mL/min. This flow rate fits well with the pump capacity of the spectrometer so that special interfacing is not required.

Widebore Capillary Columns. FSOT capillary columns with *i.d.*'s of 0.53 mm were introduced as alternatives to packed columns, with the inertness of fused silica as their main advantage. Widebore or megabore columns have higher sample capacities and are applicable for simple packed column-like separations. A serious limitation of small *i.d.* capillary columns indeed is the low sample capacity (25 m L×0.25 mm *i.d.*; d_f = 0.2 µm results in a capacity of roughly 20 ng per compound). Sample capacity is the ability of a column to tolerate high concentrations of solutes. If the capacity is exceeded, the chromatographic performance is degraded as evidenced by leading (overloaded) peaks. Sample capacity is related to the film thickness and column radius. A column of 10 m×0.53 mm, d_f = 1 µm, has a sample capacity >200 ng per compound. Simple inlet devices can be used with widebore columns in the low-resolution mode. In this mode, the carrier gas velocity is far above optimum values (flows of 8 – 12 mL/min), and a 10-m column generates only 5000 – 8000 plates. For simple mixtures, these efficiencies are sufficient. Megabore columns are

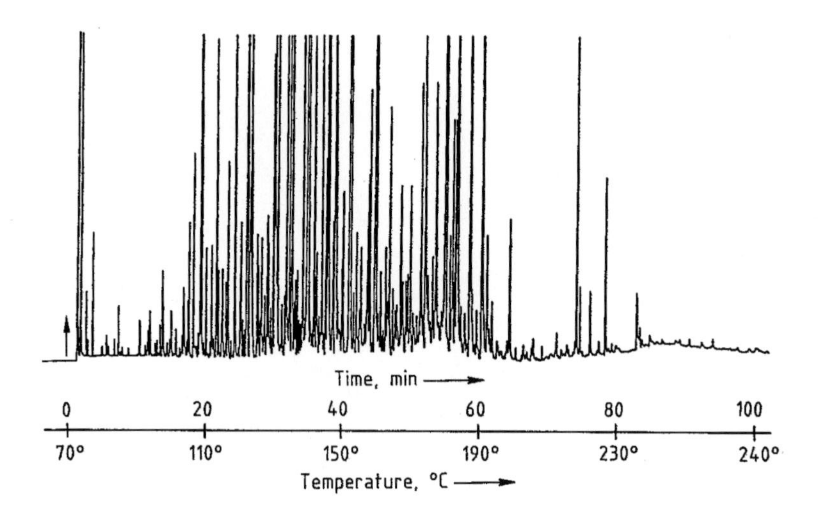

Figure 5. Analysis of oxygen-containing compounds of essential oil of pepper
Column: 40 m L×0.25 mm *i.d.*; 0.25 μm d_f high molecular mass poly(ethylene glycol); temperature as indicated; split injection 1/50; carrier gas hydrogen at 15 psi (≈1 bar); FID

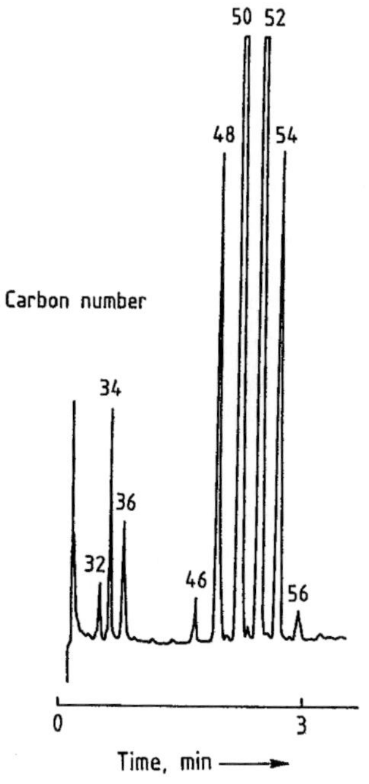

Figure 6. Fast separation of palm oil
Column: 5 m L×0.25 mm *i.d.*; 0.1 μm d_f methylsilicone; temperature: 290–350 °C at 30 °C/min; cool on-column injection; carrier gas hydrogen at 15 psi (≈1 bar); FIDPeaks: 32–36 diglycerides, 46–56 triglycerides

also available in lengths of 30–60 m. At optimum velocities (flows of 1–3 mL/min) they offer 50 000 and 100 000 plates, respectively, which is considered high-resolution GC.

Narrowbore Capillary Columns. Narrowbore columns offer very high efficiencies (100 m L×0.1 mm *i.d.* = 1 000 000 plates) or ultrahigh speed at 100 000 plates but at the expense of sample capacity. This is illustrated in Figure 7 for the analysis of fatty acid methyl esters on a 10 m L×0.1 mm *i.d.*, d_f 0.1 μm, column of cyanopropylsilicone (CpSi). Total analysis time is only 90 s. Sample injection is critical because the sample capacity is <1 ng. This is the reason for the slow acceptance of narrowbore columns. Normal sample size injection is expected to be possible with the development of new sample inlets such as programmed temperature vaporization in the solvent venting mode.

11.3.4. Stationary Phases in Gas Chromatography

11.3.4.1. General Considerations

The importance of the stationary phase in GC can, once again, be deduced from the resolution equation (Eq. 3.4). In Figure 8, the influence of the plate number N, the separation or selectivity factor α, and the retention factor k on peak resolution R_s is represented graphically. The peak resolution of two solutes ($k = 5$) with $\alpha = 1.05$ on a column with a plate number of 20 000 (e.g., 10 m×0.5 mm coated with 0.2 μm methylsilicone) is 1.4. This value, the crossing point of the curves, is taken as reference. The curves are obtained by changing

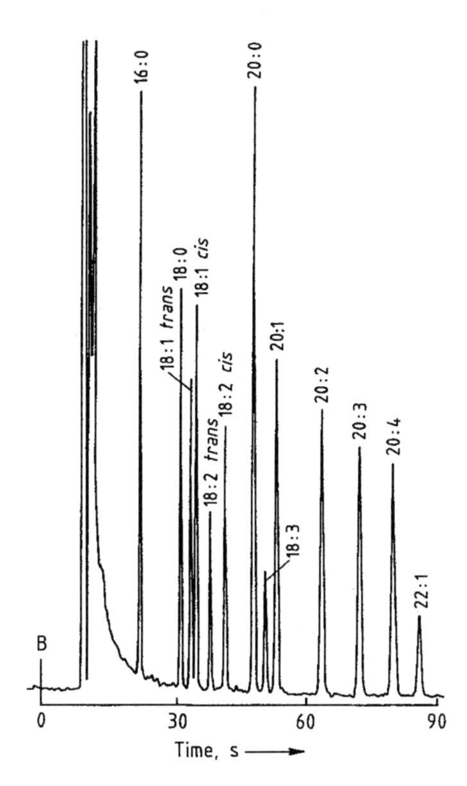

Figure 7. Fast separation of fatty acid methyl esters
Column: 10 m L×0.1 mm *i.d.*; 0.1 µm d_f cyanopropylsilicone; temperature: 180 °C; split injection 1/500; carrier gas hydrogen at 60 psi (≈ 4 bar); FID
Peaks: as indicated (the figure before the colon denotes the number of carbon atoms; that after the colon, the number of double bonds per molecule)

one of the variables while the other two remain constant. From the curves, the following conclusions can be drawn:

Plate Number N. Peak resolution is proportional to the square root of the plate number. Increasing the plate number by a factor of 2 ($N=20\,000$ to $N=40\,000$) improves peak resolution by only 1.4 ($\sqrt{2}$ gives $R_s=1.40 \rightarrow 1.96$).

Retention Factor k. For values >5, the influence of k on R_s is small. Increasing k from 5 to 20 increases resolution by only 14 % ($R_s=1.40 \rightarrow 1.60$).

Separation Factor α. Optimization of the separation factor has the most important impact on peak resolution. Increasing α from 1.05 to 1.10 nearly doubles the peak resolution ($R_s=1.40$

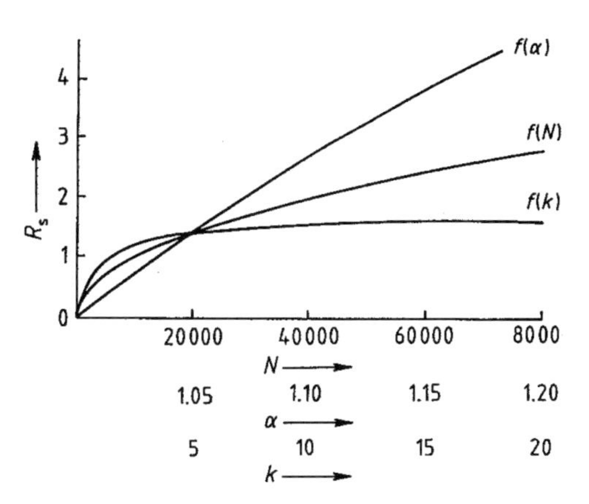

Figure 8. Effect of N, α, and k on R_s

$\rightarrow 2.68$), while a slight decrease ($\alpha=1.03$) causes a drastic loss of resolution ($R_s=1.40 \rightarrow 0.86$). The selection of a "liquid phase" that maximizes the relative retention of solutes is thus of upmost importance. The x-scales in Figure 8 have not been selected arbitrarily. Varying plate number and retention factor, respectively, from $N=20\,000$ to 40 000 and from $k=5$ to 20 is not so difficult to realize. In contrast, increasing the separation factor is not easy to achieve, especially in gas chromatography. First of all, as already mentioned, the mobile phase is an inert gas and its role is purely mechanical. Second, optimization of the selectivity of the stationary phase for a particular sample means that the nature of the solutes must be known. Optimization then requires good chemical insight and intuition. Several separations have been achieved more by luck than by skill. In 1991 a book was published with very detailed discussion of practically every stationary phase used since the inception of gas chromatography [17]. The reader is referred to this book for more information.

The selectivity of a stationary phase is commonly expressed in terms of the relative retention of a compound pair:

$$\alpha = k_2/k_1 \tag{3.7}$$

The thermodynamic description of Equation (3.7) is

$$\alpha = p^{\circ}{}_1/p^{\circ}{}_2 \cdot \gamma^{\circ}{}_1/\gamma^{\circ}{}_2 \tag{3.8}$$

where p° refers to the saturated vapor pressure of the analyte and γ° is the activity of the analyte at

Table 2. Resolution R_s for different α values

α	R_s	
	GC	CGC
1.2	2.9	12.5
1.1	1.6	6.8
1.05	0.8	3.6
1.02	↓	1.5
1.01	other stationary phase	0.8
		↓ other stationary phase or increase of column length

infinite dilution in the stationary phase. Equation (3.8) is split in two contributions: the first term is related to the solutes and depends only on temperature, whereas the second term depends on the selective properties of the stationary phases. Solute – stationary phase interactions are very complex and may include nonspecific interactions (dispersion interactions); specific interactions such as dipole – dipole, dipole – induced dipole, and hydrogen bonds; and chemical interactions such as acid – base, proton donor – proton acceptor, electron donor – electron acceptor, and inclusion. The terms *selectivity* and *polarity* are commonly used to describe a stationary phase. A polarity and selectivity scale of more than 200 stationary phases has been established by McReynolds [18]. The Kovats retention indices (see Chap. 11.7) of benzene, butanol, 2-pentanone, nitropropane, and pyridine on phase X are measured, and the differences in indices in phase X, relative to the stationary phase squalane as typical apolar phase, are listed individually (selectivity) and as a sum (polarity). Although the polarity scale is discussed in all textbooks on GC, it has little practical value. The following are by far the most common methods used for selection of a stationary phase: trying a phase with good chromatographic properties, looking for a similar application that has already been published, or asking a colleague with experience (trial and error) for advice. At this stage the difference between PCGC and CGC should be restressed. Consider a packed column and a capillary column coated with the same stationary phase of the following characteristics: *packed column,* 2 m L×4 mm *i.d.*, packed with 100 – 120 mesh diatomaceous earth particles, 3 % methylsilicone, plate number 5000; *capillary column,* 25 m L×0.25 mm *i.d.*, coated with 0.25 µm methylsilicone, plate number 100 000. By neglecting the k contribution in the resolution equation, the resolution for two solutes with different α values can be calculated (Table 2). For an α value of 1.05, the resolution on the packed column is insufficient and another stationary phase must be selected. On the capillary column for the same solute – stationary phase interaction the resolution is still 3.6 because of the very high efficiency of a capillary column. Even for α values of 1.02, baseline separation is obtained. In the case of α value 1.01, the resolution is only 0.8, but due to the high permeability of a capillary column, its length can be increased to provide the required plate number N_{req} on the same stationary phase. From the resolution equation follows:

$$N_{req} = 16R_s^2\,(\alpha/\alpha - 1)^2 = 160\,000\ \text{plates} \qquad (3.9)$$

and as $N = L/d_c$, the required length for $R_s = 1$ is 40 m. The analysis of a fatty acid methyl ester mixture on a *packed column* with a highly selective phase for these solutes (ethylene glycol succinate) is shown in Figure 9; several peaks are incompletely separated. The same analysis was performed on a capillary column coated with poly(ethylene glycol), a stationary phase with no particular selectivity for the solutes (Fig. 10): all peaks are baseline separated.

11.3.4.2. Solid Phases

For certain applications, GSC enjoys some advantages over GLC. Adsorbents are stable over awide temperature range, and column bleed is virtually nonexistent especially for inorganic adsorbents and molecular sieves. *Silica* and *alumina* adsorbents give excellent separation of saturated and unsaturated hydrocarbons with low molar mass. Medium-polarity and polar solutes interact too strongly with the highly adsorptive surface of these adsorbents. *Graphitized carbon blacks* are well suited for separation of structural and geometric isomers. The disadvantage of carbon blacks is that their retention characteristics differ from batch to batch and they seem difficult to prepare in a reproducible way. Inorganic adsorption columns can be modified by coating the material with a small amount of nonvolatile liquid or inorganic salt. The most adsorptive sites will bind preferentially to these modifiers. As a result, the sites are unavailable to participate in retention, and retention times generally decrease with increased reproducibility. *Molecular sieves* (zeolites) are aluminosilicates of alkali metals. The most common

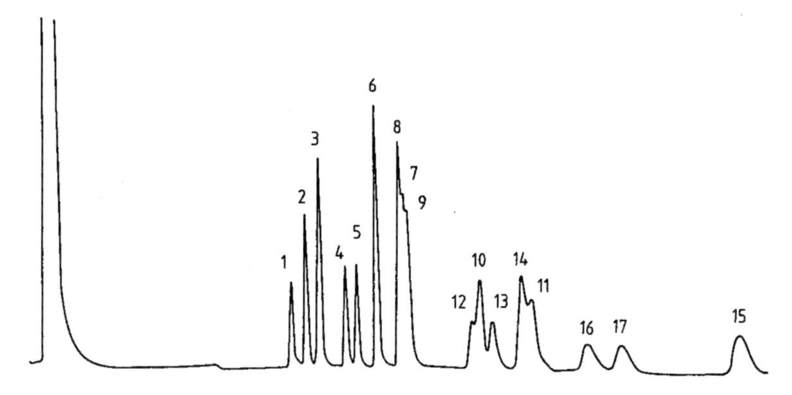

Figure 9. Isothermal analysis (200 °C) of fatty acid methyl esters on a packed column with stationary phase ethylene glycol succinate (EGS-X)
1. $C_{16}:0$; 2. $C_{16}:1$; 3. $C_{17}:0$; 4. $C_{18}:0$; 5. $C_{18}:1$; 6. $C_{18}:2$; 7. $C_{18}:3$; 8. $C_{20}:0$; 9. $C_{20}:1$; 10. $C_{20}:4$; 11. $C_{20}:5$; 12. $C_{22}:0$; 13. $C_{22}:1$; 14. $C_{23}:0$; 15. $C_{22}:6$; 16. $C_{24}:0$; 17. $C_{24}:1$

types for GC are molecular sieves 5A and 13X. Separation is based mainly on the size of the solutes and the porous structure of the molecular sieve, although adsorptive interactions inside and outside the pores may also contribute. Molecular sieves are used primarily for the separation of permanent and inert gases including low molar mass hydrocarbons. *Porous polymer beads* of different properties have found many applications in the analysis of volatile inorganic and organic compounds. The properties of the commercially available materials vary by chemical composition, pore structure, and surface area. Compared to inorganic adsorbents, polymer beads are stable to water injections, and on some of the materials, polar solutes give perfect peak shapes in reasonable analysis times. The best-known adsorbents are the Porapak series, the Chromosorb 101-108 series, and Tenax. The range of applications on those polymers can be found in catalogues of the column and instrument manufacturers. Tenax (a linear polymer of *para*-2,6-diphenylphenylene oxide) is somewhat unique because of its thermal stability up to 380 °C. Relatively high molecular mass polar compounds such as diols, phenols, or ethanolamines can be analyzed successfully. Some doubt exists as to the mechanism of retention on Tenax. At low temperature, adsorption appears to be the principal retention mechanism, but at higher temperature the surface structure becomes liquid-like and partitioning does occur.

Solid Phases for PLOT Columns. At the moment, GSC is enjoying a revival because of the introduction and commercialization of PLOT columns. The advantages of these columns are higher efficiency, speed of analysis, and better stability and reproducibility over time [19]. To date, PLOT columns are available in different lengths and internal diameters, with aluminum oxide – KCl for the separation of saturated and unsaturated hydrocarbons including low-volatile aromatics; with molecular sieve 5A for the analysis of permanent and inert gases; with porous polymers of the Porapak-type (Q,S,U) (PoraPlot columns) for the analysis of polar solutes and water; and with carbon black (CarboPlot columns) for volatile apolar solutes in petroleum products and environmental samples. Figure 11 shows the determination of benzene and toluene in natural gas on an Al_2O_3 – KCl PLOT column. The elution of benzene from the C_6 group and of toluene from the C_7 group illustrates the high selectivity.

11.3.4.3. Liquid Phases

In the past, numerous phases were used in PCGC, leading to what has been called "stationary-phase pollution" [20]. In the beginning, the many phases were a blessing; later, a curse. Today most separations are performed on a dozen preferred phases, which are all specially synthesized for GLC purposes. Some popular PCGC phases, listed according to increasing polarity, are

Methylsilicone (gum)
Methylsilicone (fluid)
Methylphenyl(5 %)silicone
Methylphenyl(50 %)silicone
Methyltrifluoropropyl(50 %)silicone

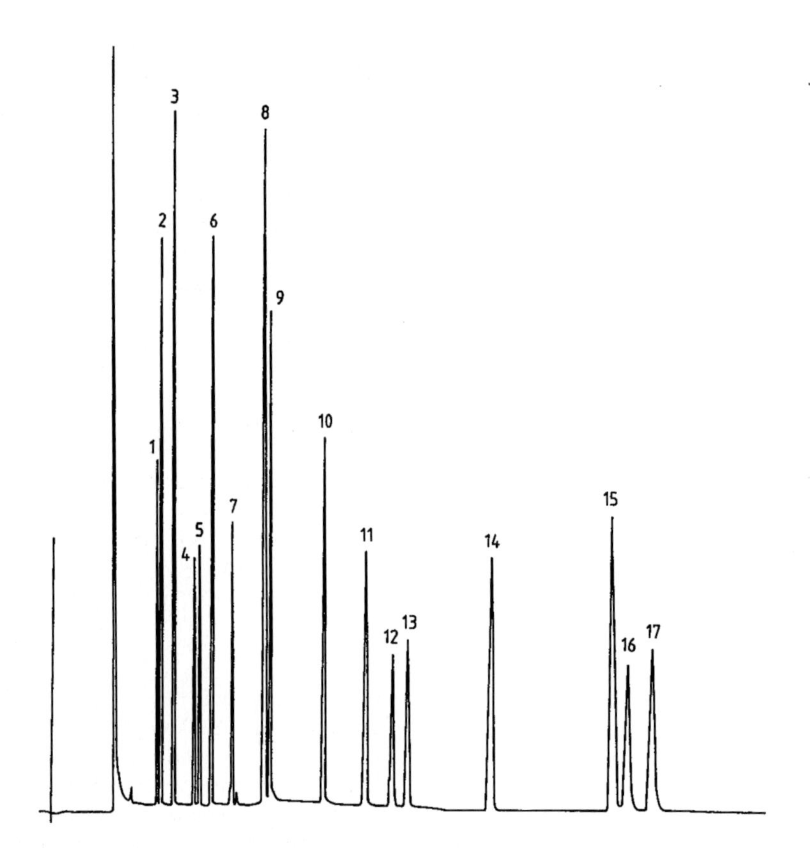

Figure 10. Isothermal analysis (200 °C) of fatty acid methyl esters on a capillary column with high-M_r poly(ethylene glycol) as stationary phase (for designation of peaks, see Fig. 9)

Methylphenyl(25 %)cyanopropyl(25 %)silicone
Poly(ethylene glycol), M_r > 40 000
Cyanopropyl(50 %)phenylsilicone
Poly(ethylene glycol) esterified with 2-nitroterephthalic acid
Diethylene glycol succinate
Cyanopropyl(100 %)silicone
Ethylene glycol succinate
1,2,3-Tris(2-cyanoethoxy)propane

Of this list, by far the most important group are the silicones. The reasons can be summarized as follows: Their diffusion coefficients are high, and thus resistance to mass transfer is low, silicones have a high thermal stability (methylsilicone gum up to 400 °C) and are liquids or gums at low temperature, the structure is well defined, and the polymers are very pure. For practical GC this results in a broad temperature range, which is defined by the minimum allowable operating temperature range (MiAOT) and by the maximum allowable operating temperature range (MAOT), and low column bleeding. Poly(ethylene glycol) with high molecular mass, which has a unique selectivity, is a solid below 60 °C (MiAOT = 60 °C) and bleeds at 240 – 250 °C as a result of the higher vapor pressure compared to silicones.

11.3.4.3.1. Basic Phases for Capillary GC

As already mentioned, the number of phases in CGC can be reduced further due to the high efficiency those columns offer. Stark et al. [21] studied stationary-phase characteristics in depth and concluded that the order of optimized phases for CGC, all higher molecular mass crosslinkable gums, is

Methylsilicone
Methylphenyl(50 – 70 %)silicone
Methylcyanopropylsilicones with medium (25 – 50 %) and high (70 – 90 %) cyanopropyl content

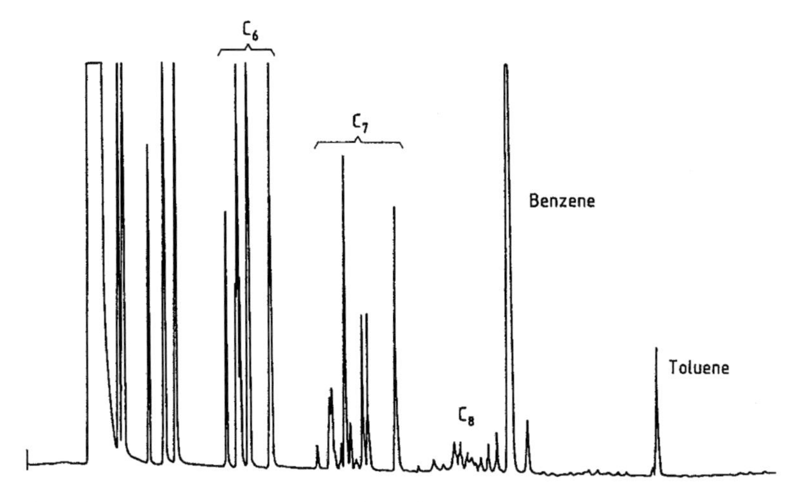

Figure 11. Natural gas analysis on an Al_2O_3 PLOT column
Column: 25 m *L*×0.25 mm *i.d.*; Al_2O_3 – KCl; temperature: 75 – 200 °C at 3 °C/min; split injection; carrier gas nitrogen at 5 psi (≈3.5×10^4 Pa); FID

Methyltrifluoropropylsilicone
High molecular mass poly(ethylene glycol) or a silicone substitute having the same polarity and selectivity

The functionalities in those phases are the methyl, phenyl, cyano, trifluoro, and hydroxy–ether group, respectively. The similarity with HPLC phases is remarkable. Based on the above, SANDRA et al. [22] proposed five basic phases for capillary GC. The formulas are given in Figure 12. For the majority of high-resolution separations, these stationary phases provide more than adequate performance. For some applications, the functionalities are combined on the same siloxane backbone, or columns are coupled (selectivity tuning). Functionalities can also be modified to provide specific interactions. Optical phases and liquid crystals complete the set of preferred CGC phases. These tailor-made phases are discussed in Section 11.3.4.3.3.

The most important phase is *methylsilicone*. Separations occur according to the boiling point differences of the solutes. A typical chromatogram is shown in Figure 13. Fatty acid methyl esters are separated according to their boiling point or molar mass. Hardly any separation occurs among oleic (C_{18}: 1 *cis*), elaidic (C_{18}: 1 *trans*), linoleic (C_{18}: 2) and linolenic (C_{18}: 3) acids. *Methylphenylsilicone* is the second most widely used stationary phase. Selectivity is increased due to the polarizable phenyl group. *Methyltrifluoropropylsilicone* possess rather unique selectivity for electron donor solutes such as ketones and nitro groups. Incorporating the *cyano group* with its large dipole moment into the silicone backbone provides strong interactions with dipolar solutes and unsaturated bonds. Cyano phases are widely used for the separation of unsaturated fatty acid methylesters and for the resolution of dioxins and furans. Figure 14 represents the analysis of a sample whose composition is very similar to that in Figure 13. Separation according to the number of double bonds in the alkyl chains is due to $\pi - \pi$ interactions. Methyl oleate (*cis* C_{18}: 1) and methyl elaidate (*trans* C_{18}: 1) differing only in configuration are baseline separated as well. High *molecular mass poly*(*ethylene glycol*) has a unique selectivity and polarity for the separation of medium-polarity and polar compounds. The most popular of these phases is Carbowax 20M with average M_r of 20 000.

11.3.4.3.2. Selectivity Tuning

A solvent with a selectivity intermediate to the selectivities of the five basic phases can, for some applications, offer a better separation or a shorter analysis time. This is illustrated by the following example. On stationary phase *X*, compounds A and B can be separated from C, but they cannot be separated from each other. On another stationary phase *Y*, A and B are separated, but one of them coelutes with C. By using a chromatographic system with selectivity between *X* and *Y*, an opti-

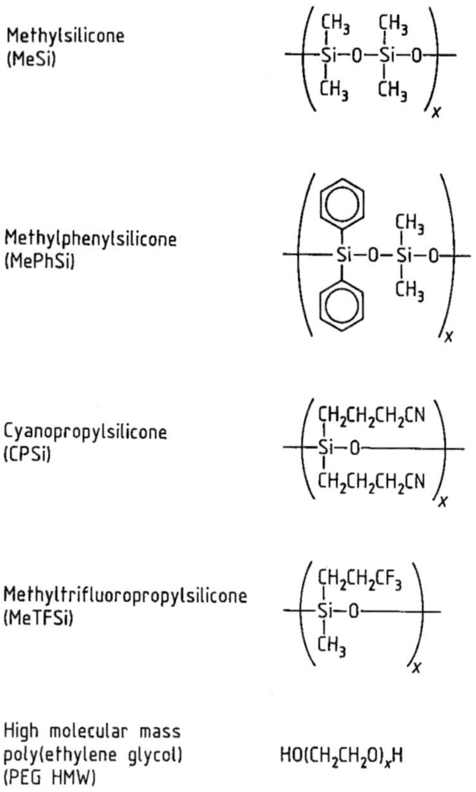

Methylsilicone (MeSi)

Methylphenylsilicone (MePhSi)

Cyanopropylsilicone (CPSi)

Methyltrifluoropropylsilicone (MeTFSi)

High molecular mass poly(ethylene glycol) (PEG HMW) $HO(CH_2CH_2O)_xH$

Figure 12. Structures of the basic phases for capillary GC

mum can be found to separate the three compounds. Intermediate selectivities and polarities can be obtained by selectivity tuning [23]. Selectivity tuning means that the selectivity is adapted to the analytical need by creating a selectivity between two (or more) extreme (basic phase) selectivities. In principle, this can be performed in three different ways: (1) by synthesizing a tuned phase with predetermined amounts of monomers containing the required functional groups; (2) by mixing, in different ratios, two or three basic phases in one column; and (3) by coupling two or more columns of extreme selectivities. Approach (1) and (2) have been commercially used, and columns are available for specific applications. Columns tuned for the analysis of classes of priority pollutants are best known. Method (3) is detailed in Section 11.8.1.

11.3.4.3.3. Tailor-Made Phases

The most widely used tailor-made phases are the free fatty acid phase (FFAP), liquid crystals, and the optical phases Chirasil-Val and derivatized cyclodextrins. *FFAP* is produced by condensing 2-nitroterephthalic acid with Carbowax 20M. The phase is recommended for the analysis of acidic compounds such as organic acids and phenols. The phase is not suitable for the analysis of alkaline compounds and aldehydes with which it reacts. *Liquid crystals* can separate isomers as a function of the solute length-to-breadth ratios. High selectivities have been noted for the separation of polycyclic aromatic compounds and dioxin isomers [24]. The most powerful involves the use of a liquid-crystal column as second column in a multidimensional CGC system. Optical isomers (enantiomers) can be separated principally by two different approaches. Enantiomeric mixtures containing a reactive group can be derivatized with an optically pure reagent to produce a mixture of diastereomers that can be separated on nonoptically active stationary phases such as methylsilicone. The direct separation on optically active stationary phases is, however, the current method of choice because it has certain advantages. Derivatization agents of 100 % enantiomeric purity are not needed, and the method can be applied to enantiomers lacking reactive functional groups. In Chirasil-Val a diamide is incorporated into a silicone backbone, and separation occurs via a triple hydrogen interaction. A number of derivatized α-, β-, and γ-cyclodextrin stationary phases have recently been introduced for the separation of enantiomeric pairs. Separation is based on the formation of inclusion complexes with the chiral cavity of the cyclodextrins. However, none of the phases is universally applicable, and enantiomer separation is still based on trial and error. The separation of γ- and δ-lactones on a 2,3-*O*-acetyl-6-*tert*-butyldimethylsilyl-β-cyclodextrin stationary phase is shown in Figure 15. Full details on the separation of enantiomers by capillary GC can be found in [25], [26].

11.4. Choice of Conditions of Analysis

Under isothermal and constant-flow conditions, for a homologous series of hydrocarbons, an exponential correlation exists between the residence time in the stationary phase (adjusted re-

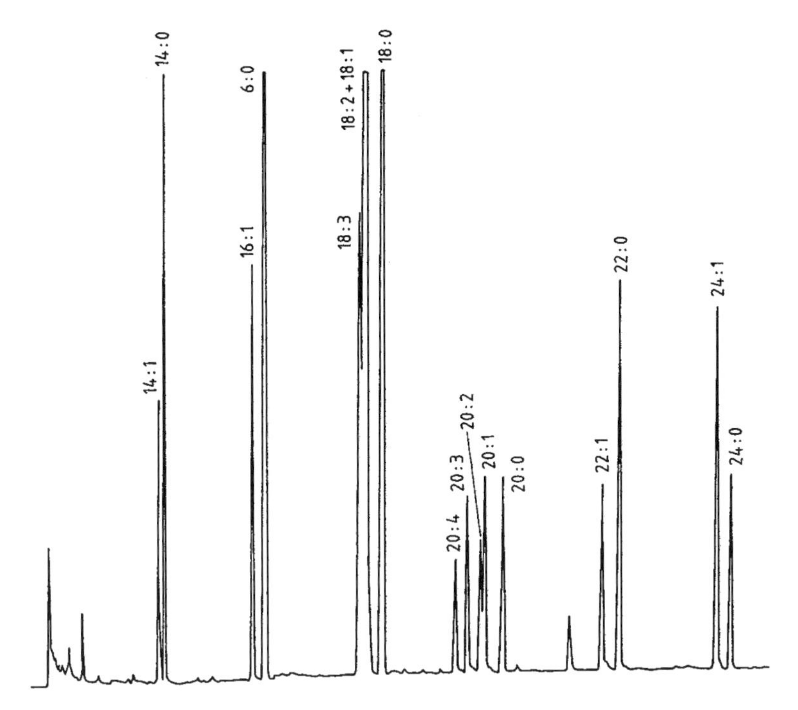

Figure 13. Analysis of fatty acid methyl esters on methylsilicone
Column: 25 m L×0.25 mm *i.d.*; 0.25 μm d_f MeSi; temperature: 100–220 °C at 3 °C/min; split injection 1/100; carrier gas hydrogen at 8 psi (≈5.6×10^4 Pa); FID
(For designation of peaks, see Fig. 7)

tention time) and the solute vapor pressure, boiling point, or number of methylene groups (carbon number).

$$\log t'_R = a\,C_n + b \qquad (4.1)$$

where t'_R is the adjusted retention time; C_n is the number of carbons in the individual homologues; and a, b are constants. For a mixture with boiling point range of the solutes less than 100 °C, an optimum isothermal temperature can be selected. For samples in which the boiling point difference of the solutes exceeds ca. 100 °C, working in the isothermal mode is impractical. Usually, early eluting peaks will show poor resolution (low k values), whereas late eluting peaks will be broad when a compromise temperature has been chosen. The problem can be solved by temperature or flow programming.

11.4.1. Temperature-Programmed Analysis

In temperature-programmed analysis, a controlled change of the column temperature occurs as a function of time. The initial temperature, heating rate, and terminal temperature must be adjusted to the particular separation problem. The initial temperature is chosen so that the low-boiling compounds are optimally separated (k value >3). Selection of the program rate depends on the nature of the solutes and the complexity of the sample. The final temperature is adjusted to give reasonable total analysis times. Temperature programming has been applied since the beginning of gas chromatography (1960), but only since the end of the 1980s have the possibilities been fully exploited. This might be, in part, because of the time-consuming nature of temperature optimization by trial and error, but above all because of ignorance of the importance of the temperature programming rate to selectivity. This is especially so for polar columns where $\Delta t_R/\Delta T$ is much more pronounced than on apolar columns. Today, software is available to optimize a temperature program for a particular separation on the basis of only two experimental programmed runs. The theory of temperature-programmed optimization is treated in detail in [27], [28]. Figure 16 shows the computer predicted (A) and experimental (B)

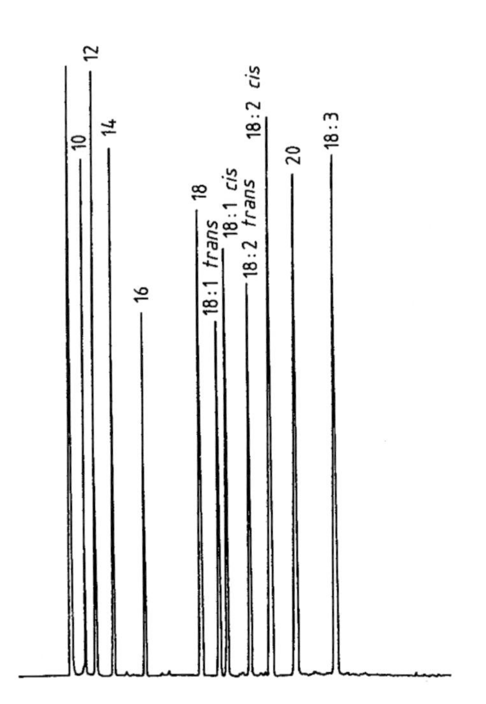

Figure 14. Analysis of fatty acid methyl esters on a cyanopropylsilicone
Column: 25 m L×0.25 mm $i.d.$; 0.25 μm d_f CPSi; temperature: 187 °C; split injection 1/100; carrier gas hydrogen at 8 psi (≈ 5.6×10^4 Pa); FID

chromatograms for a complex fatty acid methyl ester mixture (Table 3) at a programmed temperature increase of 40 °C/min [29]. The column was 55 m in length, and a remarkably good separation was achieved in < 8 min. Without the aid of computer simulation, such fast programming rates would never have been considered for this separation.

11.4.2. Constant Pressure, Constant Flow, and Pressure Programming

Flow programming offers advantages such as reduced separation time, less column bleed, and lower eluting temperature for labile compounds but has not been widely used in gas chromatography because temperature programming is experimentally easier to perform. In 1990, electronic pressure control (EPC) [30] was developed, which allows operation in the constant-pressure, constant-flow, and pressure-programming modes. Whether temperature programming will be replaced by flow programming is questionable, although the features of EPC open new perspectives. Gases play a vital role in GC and are even more important in CGC. The carrier gas is the force moving the solutes through the column, and its velocity or flow controls the chromatographic band broadening and thus the efficiency. Keeping the velocity or flow under control is a prerequisite for good GC practice. Besides this, the carrier gas is often the transport medium of solutes from injection devices such as split, splitless, programmed-temperature vaporizing inlets, purge and trap, thermal desorption units, and pyrolysis units. Often the flows required in the precolumn devices do not correspond to the optimal carrier gas flows, and compromises must be accepted (i.e., split in purge and trap, long residence times of solutes in a splitless liner). On the other hand, optimization of total GC performance also requires precise control of the flows of the detector gases responsible for detector sensitivity and baseline stability. The quality of data generated by the hyphenated spectroscopic methods (see Section 11.6.4) also depends strongly on the stability of the flows of reagent gases (e.g., in chemical ionization mass spectroscopy and atomic emission detection). In the past, very little attention was paid to the mode of operation of the different gases in a GC system, and all gas supplies were more or less operated in the constant-pressure mode. In the constant-flow mode, all parts of a GC system run under optimum conditions. EPC is likely to become a standard mode of operation in the near future. The CGC applications with EPC published until now all show improved chromatographic performance going from reduced discrimination and large volume introduction in splitless injection by a pressure pulse shortening the residence time in the inlet, to increased sensitivity by EPC control of detector gases independently or in concert with the carrier gas flow. These are just a few examples, and more features are detailed in [31]. Yet more is to be expected because the possibilities of EPC have not yet been fully evaluated and exploited. A typical example can be found in multidimensional CGC (see Section 11.8.1). One of the reasons for slow acceptance of this extremely powerful technique is the necessary adjustment of the midpoint pressures, which can require several days of fine tuning with conventional pressure control devices. Flow control at the midpoint should solve this shortcoming. Nevertheless, EPC introduces a new era in good laboratory practice (GLP) and good automated laboratory practice (GALP). An-

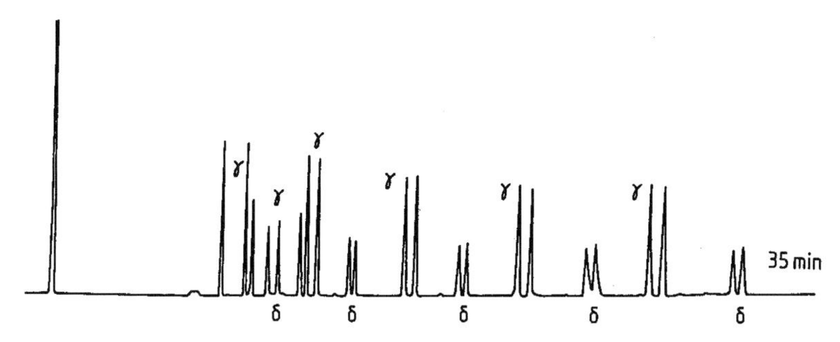

Figure 15. Enantioselective separation of γ- and δ-lactones on 2,3-*O*-acetyl-6-*tert*-butyldimethylsilyl-β-cyclodextrin
Column: 25 m *L*×0.25 mm *i.d.*; 0.25 μm d_f; temperature: 150 – 185 °C at 1 °C/min; split injection 1/100; carrier gas nitrogen at 18 psi (≈1.3 bar); FID
Peaks: γ (6 – 11) and δ (8 – 12) lactones

Table 3. Retention times (n = 5) in initial runs, and predicted and experimental retention times at the optimum programming rate

Compound		Symbol	Initial runs		Optimized program	
No.	Name		1 °C/min, min	12 °C/min, min	Predicted, min	Experimental, min
1	Methyl butanoate	C_4 : 0	2.95	2.49	2.07	2.21
2	Methyl hexanoate	C_6 : 0	6.30	3.74	2.50	2.73
3	Methyl octadecanoate	C_8 : 0	14.65	5.38	3.01	3.32
4	Methyl decanoate	C_{10} : 0	28.64	7.09	3.52	3.88
5	Methyl dodecanoate	C_{12} : 0	45.02	8.69	4.00	4.41
6	Methyl phenylacetate	I.S.*	49.51	9.68	4.37	4.85
7	Methyl tetradecanoate	C_{14} : 0	61.15	10.14	4.44	4.88
8	Methyl hexadecanoate	C_{16} : 0	76.20	11.45	4.83	5.32
9	Methyl *cis*-9-hexadecanoate	C_{16} : 1 *cis* 9	79.38	11.85	4.96	5.47
10	Methyl octadecanoate	C_{18} : 0	90.08	12.65	5.19	5.72
11	Methyl *cis*-9-octadecanoate	C_{18} : 1 *cis* 9	92.43	12.97	5.31	5.85
12	Methyl *cis*-9,12-octadecadienoate	C_{18} : 2	97.23	13.47	5.48	6.04
13	Methyl *cis*-6,9,12-octadecatrienoate	C_{18} : 3 γ	100.56	13.82	5.61	6.19
14	Methyl eicosanoate	C_{20} : 0	102.95	13.75	5.55	6.11
15	Methyl *cis*-9,12,15-octadecatrienoate	C_{18} : 3 α	102.95	14.03	5.68	6.27
16	Methyl docosanoate	C_{22} : 0	114.64	14.77	5.92	6.49
17	Methyl *cis*-5,8,11,14-eicosatetraenoate	C_{20} : 4	114.90	15.10	6.11	6.71
18	Methyl *cis*-4,7,10,13,16-eicosa-pentaenoate	C_{20} : 5	120.20	15.64	6.39	6.89
19	Methyl tetracosanoate	C_{24} : 0	126.08	15.72	6.34	6.97
20	Methyl *cis*-4,7,10,13,16,19-docosa-hexaenoate	C_{22} : 6	133.60	16.83	7.15	7.64

* I.S. = Internal standard.

alytical methods can now be stored completely, providing exactly the same flow characteristics of all gases and tremendously enhancing repeatability and accuracy.

11.5. Sample Inlet Systems

11.5.1. General Considerations

"If the column is described as the heart of chromatography, then sample introduction may,

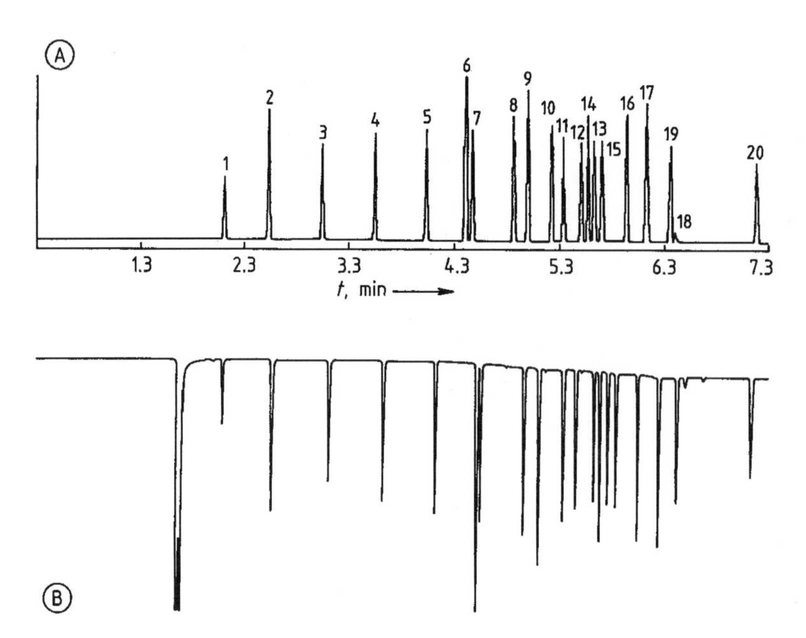

Figure 16. Predicted (A) and experimental (B) chromatograms for fatty acid methyl ester mixture with temperature-programmed analysis at 40 °C/min

with some justification, be referred to as the Achilles heel." This statement [32] clearly illustrates that sample introduction is of primary importance in GC and especially capillary GC. The performance of the sample introduction system is crucial for overall chromatographic performance of the complete system. Much progress has been made in recent years, and the understanding of injection phenomena has increased tremendously. Because of the diversity of samples that can be analyzed with modern GC (wide range of component concentrations; from highly volatile to less volatile components; different thermal stabilities; etc.), several injection methods have been developed. A final universal answer to sample introduction, however, has not been provided yet, and whether such an answer can be provided at all is questionable: "There does not now and probably never will exist injector hardware or methodology that is suitable for all samples, under all conditions" [33]. "There is still no such thing as a universal injection system and there probably never will be" [34]. This does not mean that precise and accurate results cannot be generated with the different injection systems that have been optimized over the years. The possibilities and the limitations of available systems, however, must be known exactly. In combination with a sufficient knowledge of the composition of the sample to be analyzed, proper injector choice will guarantee good results.

A qualitative and quantitative analysis means that the chromatographically determined composition of the sample corresponds to its real composition. Possible difficulties encountered in obtaining such data have to be attributed to the sampling technique, various column effects, or a combination of both. A sample introduction system can be "discriminative," meaning that certain components are not introduced quantitatively into the column. However, the column itself can also act to discriminate (reversible and/or irreversible adsorption), the degree of which can, moreover, be injection dependent. The characteristics of the column in use, therefore, must be well established before any conclusion can be drawn with regard to the sampling system.

The basic prerequisite of a sampling system for gas chromatography is that the sample must be introduced into the column as a narrow band, whose composition is identical to the original composition of the sample. The width of the inlet band must be such that its variance does not contribute significantly, or contributes only slightly, to the chromatographic broadening process. The total measured variance σ_{total} is the sum of all contributions to the peak variance:

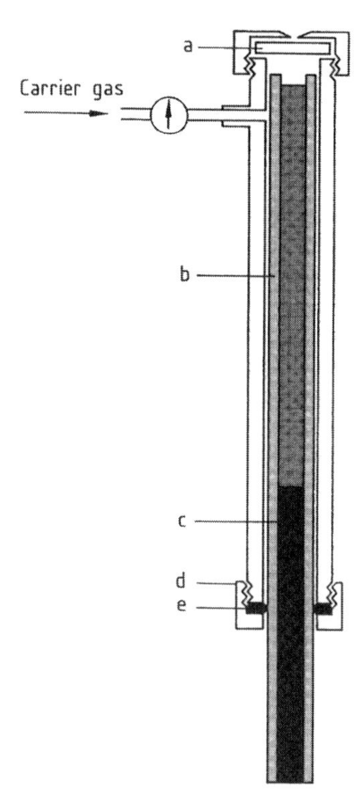

Figure 17. Schematic of a packed column injector
a) Septum; b) Glass wool; c) Packing; d) Nut; e) Ferrule

$$\sigma^2_{\text{total}} = \sigma^2_{\text{column}} + \sigma^2_{\text{injection}} + \sigma^2_{\text{extra}} \quad (5.1)$$

With modern GC instrumentation, the extra peak variance contribution (electrometer, data system, recording) approaches zero. Because of the occurrence of the squared variances in this equation, the contribution of the initial bandwidth to the total peak bandwidth is less dramatic than might be expected, at least for GC on packed columns. For capillary columns, injection bandwidths can be too large compared to the chromatographic band broadening, and narrow initial bands have to be obtained by:

1) Introducing very small sample volumes (1–50 nL) into the column. A sample stream splitting device (splitter) accomplishes this task. The sample vapor, formed on injection at elevated temperature, is split into two sample streams of different flow rates.
2) The entire sample is introduced into the column, but the broad initial band is immediately focused into a narrow initial band by sharpening mechanisms such as solvent, thermal, or stationary-phase focusing. In practice, these effects are used with great advantage to focus the solute bandwidths produced by splitless and on-column injection.

In subsequent sections, the most important sample introduction systems will be discussed in some detail with emphasis on capillary inlets. Because of the much larger σ_{column} in packed columns, injection is less stringent. Inlets are also divided into universal and selective inlets. The aim of universal injection is introduction of the complete sample into the column, whereas selective injection means that only a well-defined fraction enters the column.

11.5.2. Universal Inlets

11.5.2.1. Packed Column Inlets

Packed column inlets are simple to design and use [35]. Injection is performed directly in the column "on-column" (Fig. 17) or in an in-sert placed before the head of the column. Different inserts are used, depending on the type of packed column (1/8" = 0.32 cm or 1/4" = 0.63 cm). Polar and thermolabile compounds are most effectively analyzed by on-column injection on a glass column. The injector zone is heated, and upon manual or automated injection, usually with a microliter syringe, hot mobile-phase gas carries the vaporized sample down to the packing material. Packed column inlets are vaporizing so they can cause sample degradation and needle discrimination. This is discussed in detail for split injection. Sampling valves are often used to sample gases or volatile liquids in constant-flowing streams. The valve is substituted for the syringe and can be used both with packed and with capillary columns. The construction of GC valves is similar to HPLC valves.

11.5.2.2. Capillary Column Inlets

In the context of this contribution, it is impossible to treat every aspect of sample introduction in capillary GC in depth. Full details on the different sampling methods can be found in [36], [37]. A survey of sampling systems is also given in [34], [38].

11.5.2.2.1. Split Injection

Split sampling was the first sample introduction system developed for capillary GC [39]. The conventional split injector is a flash vaporization device. The liquid plug, introduced with a syringe, is volatilized immediately, and a small fraction enters the column, while the major portion is vented to waste. This technique guarantees narrow inlet bands. A schematic diagram of a split injector is shown in Figure 18. Preheated carrier gas enters the injector and the flow is divided in two streams. One stream of carrier gas flows upward and purges the septum (a). The septum purge flow is controlled by a needle valve (b). Septum purge flow rates are usually between 3 and 5 mL/min. A high flow of carrier gas enters the vaporization chamber (c), which is a glass or quartz liner, where the vaporized sample is mixed with carrier gas. The mixed stream is split at the column inlet, and only a small fraction enters the column (d). A needle valve controls the split ratio. Split ratios (i.e., column flow : inlet flow) range typically from 1 : 50 to 1 : 500 for conventional capillary columns (0.22 – 0.32 mm *i.d.*). Lower split ratios can be used in combination with focusing effects. For columns with a high sample capacity such as widebore or thick-film columns, low split ratios (1 : 5 – 1 : 50) are common. In high-speed capillary gas chromatography, applying 0.1 to 0.05 mm *i.d.* columns, split ratios can exceed 1 : 1000. In split injection, initial bandwidths are only milliseconds. Because split injection is a flash vaporization technique, sample discrimination is difficult to avoid. This is especially the case if the sample contains components in different concentrations and with different volatilities and polarities. Sample discrimination in split injection is caused by inlet-related parameters as well as by operational parameters such as syringe handling. *Inlet-related discrimination* is often referred to as nonlinearity of the splitter device. Linearity in this respect means that the split ratio at the point of splitting is equal to the preset split ratio and equal for all the components in the sample. Linear splitting against varying sample components, be it in concentration, volatility, or polarity, is impossible to achieve, even when a sample is introduced in a nondiscriminative manner into the vaporization chamber. Different mechanisms can cause nonlinear splitting: different diffusion speed of the sample components, incomplete evaporation, and fluctuating split ratio. The different mechanisms and their respective contribution are discussed in [36]. Nonlinearity can be minimized by complete vaporization of the sample, followed by homogeneous mixing with the carrier gas before the sample enters the column. As evident as this may seem, the sample introduced by split injection often arrives at the point of splitting as a mixture of vapor and a nonuniform mist of droplets. Two approaches can be used to minimize this phenomenon: (1) increased injection temperature and (2) optimization of the inlet configuration and glass liners. Different glass liners have been proposed (i.e., empty tube, short glass-wool plug in the splitting region, long and tight glass-wool plug, packing with chromatographic support or glass beads, deformation of cross section, Jennings tube, etc.) with the aim of enhancing efficient heat transfer to the injected sample and thoroughly mixing the vaporized sample. Still, whereas such modifications demonstrate an improvement in linearity for some applications, the same setup may give bad results for others. As a general rule, packed liners should not be used unless bad results have been obtained with an open liner. When discussing nonlinearity, it was taken for granted that sample introduction with a syringe into the vaporization chamber occurs without any alteration to the sample — in other words, no discrimination is caused by the syringe introduction. However, most of the discrimination problems encountered with vaporizing injectors are related to *syringe needle effects*. Upon introducing the syringe needle through the septum, volatiles immediately start to evaporate inside the needle itself, which is heated by the injector. Also after pushing down the plunger, solvent and volatile solutes are evaporated more readily than high-boiling solute material, which partly remains on the needle wall. On removing the needle from the injector body, the nonvolatile components are taken out of the vaporization chamber as well, resulting in a mass discrimination according to volatility. Different methods of syringe manipulation have been studied (e.g., filled needle, cold needle, hot needle, solvent flush, air flush, sandwich method) in combination with fast or slow injection. These different procedures are not discussed here, but rather the best method — "manual hot-needle, fast-sample introduction" — is described. This method guarantees minimal syringe discrimination, although complete avoidance is impossible when dealing with solutes having a large volatility difference. In the hot-needle method, the sample is taken into the syringe barrel, for example up to the 5-μL mark when a 10-μL syringe is used, without

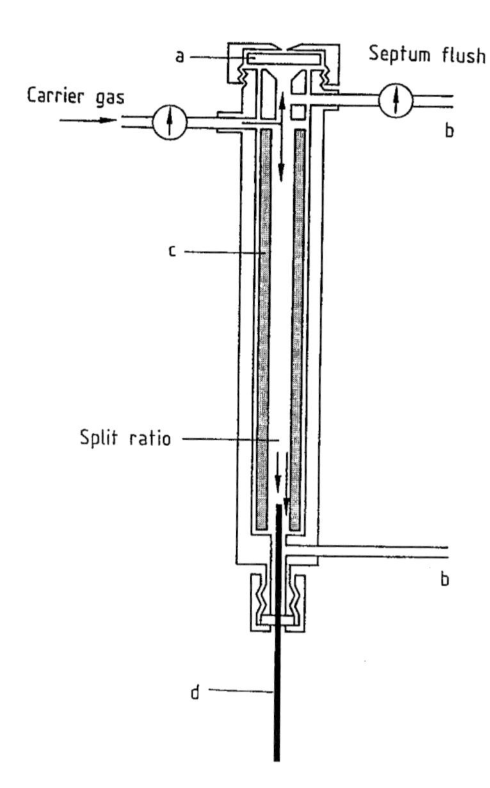

Figure 18. Schematic of a split injector
a) Septum; b) Needle valve; c) Vaporization chamber; d) Capillary column

leaving an air plug between sample and plunger. After insertion into the injection zone, the needle is allowed to heat up for 3 – 5 s. This time is sufficient for the needle to be heated to the injector temperature. The sample is then injected by rapidly pushing the plunger down (fast injection), after which the needle is withdrawn from the injector within 1 s. Syringe discrimination is related to the warming up of the needle in the vaporization chamber. Systems can be worked out in which syringe discrimination is reduced or avoided. The best technique is very fast injection. A prerequisite here is *automated cold-needle* sample introduction since all steps of the injection sequence are identical for each injection. The objective of very fast injection is to get the needle into the injection port, inject the sample, and withdraw the needle so quickly that it has no time to warm up, in this way avoiding selective sample evaporation. Fast injection, moreover, implies that the delivered volume of sample equals the preset volume. The effect of the needle *dwell time* in the injection port on sample discrimination has been

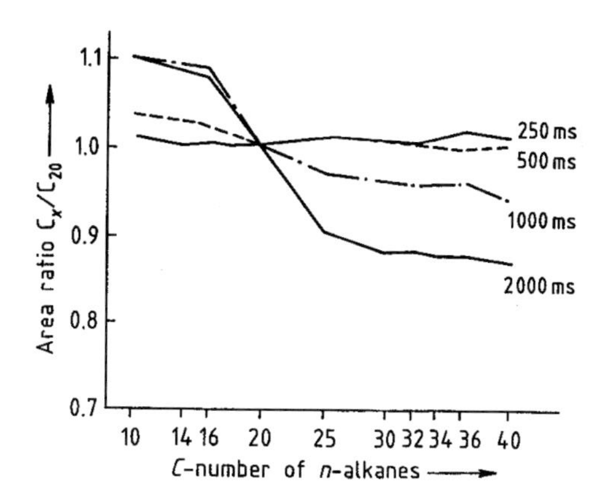

Figure 19. Effect of needle dwell time on sample discrimination

studied [40]. Dwell time is defined as the interval between the needle tip piercing the septum on the way in and reaching the same point on the way out. Figure 19 shows a plot of C_X: C_{20} area ratios (X= 10 – 40) as a function of C-number for different dwell times, with hexane as the solvent. Although these data have been obtained for direct injection, they are also valid for split injection. A dwell time of 500 ms or less shows no noticeable fractionation. Modern autosamplers inject in 100 ms. The previous discussion indicates that obtaining quantitative data by using a split injector can be problematical, but not impossible. For samples in which the solutes do not differ too much in volatility, precision and accuracy can be acceptable if the injection technique has been optimized. Some summarizing guidelines are given. In quantitative analysis, the standard addition or the internal standard method is preferred for manual injections. External standardization in which absolute peak areas are compared can be applied only with fast automated injection. Reproducibility will be enhanced by not varying the injected volume, which typically should be 0.5 – 2 μL. Injector temperature should be adapted to the problem at hand (i.e., close to the boiling point of the last eluting compound). Excessively high injector temperatures should be avoided. The hot-needle, fast-injection method is preferred for manual injection. The use of highly volatile solvents should be avoided whenever possible. If an open liner does not do the job, lose packing with deactivated glass wool or glass beads can provide a solution. However, adsorption and decomposition can occur. One of the main problems associated with split injection is syringe handling. The application of an autosampling system can overcome this difficulty, enhancing both precision and repeatability. For several applications, however, split injection will not give high quantitative accuracy.

11.5.2.2.2. Splitless Injection

In splitless injection a conventional split injector is operated in a nonsplitting mode by closing the split valve during injection. The sample is flash vaporized in the vaporizing chamber from which the sample vapors are carried to the column by the flow of the mobile phase. Since this transfer takes several tens of seconds, broad initial bandwidths would be anticipated. However, through optimal use of focusing effects, such as solvent, thermal, and stationary-phase focusing, initial peak broadening can be suppressed. The main feature of splitless injection lies in the fact that the total injected sample is introduced in the column, thus resulting in much higher sensitivity compared to split injection. For a long time, splitless injection was the only sample introduction technique in capillary GC for trace analysis. The pneumatic configuration is similar to the classical split injector (see Fig. 18). The septum is purged continuously (flow rate 2 mL/min) to maintain the system free of contamination, while a flow of 20 – 50 mL/min passes the split outlet. Prior to injection, a solenoid valve is activated, so that the splitline is closed off while the septum purge is maintained. After waiting a sufficient time for the solvent – solute vapor to be transferred onto the column (i.e., 30 – 80 s), the solenoid valve is deactivated. Residual vapors in the vaporizing chamber are vented to waste via the split line. For this reason, in splitless injection the split line is often referred to as the purge line. The time interval between the point of injection and the activation of the split (purge) line is a function of the characteristics of solvent and solutes, the volume of the vaporizing chamber, the sample size, the injection speed, and the carrier gas velocity. The sample transfer from the vaporizing chamber into the column is a slow process. Solvent vapors especially tend to remain in the inlet for a long time. Purging the insert after injection removes the last traces of vapor from the vaporizing chamber. An important aspect in the design of a splitless injector is the dimension of the vaporizing chamber. Long and narrow inserts are preferred to obtain minimal sample dilution. Internal volumes vary between 0.5 and 1 mL. The column is installed 0.5 cm in the insert, and syringes with long needles are used, creating a distance of 1 – 1.5 cm between the needle tip and the column inlet. Overfilling of the vaporizing chamber is suppressed, and fast sample transfer is achieved. If sample volumes larger than 2 μL are injected, a widebore liner should be used. Because of the relatively long residence time of the solutes in the vaporizing chamber, lower injection temperatures can be used compared to split injection. This can be helpful in minimizing sample degradation. For the same reason, the use of packed inserts should be avoided in splitless injection. An illustration of the effect of glass or fused silica wool in the liner is presented in Figure 20. The organochloropesticide endrin (peak 1) tends to decompose into endrin aldehyde (peak 2) and endrin ketone (peak 3) when active sites are present in the liner. The first chromatogram shows that with an empty liner, 92 % of the endrin is detected, compared to cool on-column injection, which provides 100 % elution. Filling the liner with a 1-cm plug of glass wool or fused silica wool results in the elution of only 83 % and 48 % endrin, respectively. Taking into account the fact that endrin is not the most thermolabile compound, what will happen if polar compounds (i.e., carbamates, barbiturates, etc.) are injected with filled liners can easily be imagined. Recently, electronic pressure control was introduced to reduce the residence time of the solutes in the liner. With EPC, the carrier gas pressure or flow can be programmed. During injection a very high flow rate is applied, which speeds up the transfer liner – column. After injection, the flow is reset to the optimal value for the column in use [41]. Syringe discrimination, which is one of the largest sources of error when applying split injection, also occurs for splitless injection because both are vaporizing inlets. The performance of a splitless injector for a particular application depends on the optimization of experimental variables, the most important of which are sample size, injection speed, purge time, injection temperature, initial column temperature, carrier gas selection, and flow rate. General guidelines cannot be advanced, and some of the variables can be only tuned by trial and error. Whatever the application, however, refocusing of the solutes in the inlet section of the capillary column is necessary. In splitless injection, the initial bandwidths are broadened by two mechanisms: band broadening in time and band broadening in space. *Band broadening in time* is caused by the slow transfer from the vaporizing chamber to the column inlet section, which takes several tens of sec-

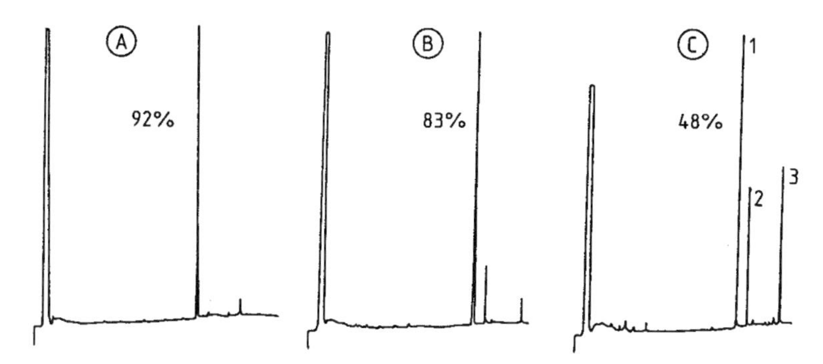

Figure 20. Decomposition of endrin on different splitless liners
A) Empty liner; B) 1-cm plug of glass wool; C) 1-cm plug of fused silica wool

onds. Upon their arrival in the column, the solutes spread over a certain column length, mainly by flooding of the sample liquid, thus causing *band broadening in space*. The fundamental difference between band broadening in time and band broadening in space is that, in the first case, solutes are spread equally with respect to gas chromatographic retention time, whereas in the second case, solutes are spread equally with respect to column length. Both phenomena cause distorted elution profiles if the solutes are not refocused before starting the chromatographic distribution process.

Solvent Focusing, Thermal Focusing, and Stationary-Phase Focusing. Splitless injection most often is combined with column temperature programming, starting with an oven temperature 25 – 30 °C below the boiling point of the solvent. Upon condensation of solvent and solutes, the droplet formed becomes too thick to be stable. The carrier gas pushes the plug further into the column, creating a "flooded zone" (Fig. 21). The solutes now are spread over the full length of the flooded zone, thus creating a solute bandwidth that equals the length of the flooded zone. For 1-μL injections, the length of the flooded zone roughly is 20 cm, provided the stationary phase is perfectly wettable by the solvent [i.e., isooctane solutions on apolar methylsilicone phases, ethyl acetate solutions on poly(ethylene glycol) phases]. On conventional capillary columns, 25 – 30 m in length, *i.d.* 0.32 mm, band broadening will hardly be observed for sample volumes of 1 μL. Only for sample sizes larger than 2 – 3 μL will peak deformation be noted. On the other hand, for polar solvents on apolar stationary phases, the length of the flooded zone can be as large as 1 m, resulting in distorted peaks. Stationary phase focusing then becomes necessary. After creation of the flooded zone, the temperature is increased and the solvent starts to evaporate from the rear of the flooded zone (Fig. 22). Highly volatile solutes evaporate as well and the last traces of solvent act as a barrier since the residual solvent behaves as a thick stationary-phase film for those solutes. High volatiles are thus focused by the *solvent effect*. The low-volatility solutes are cold trapped since the column temperature is still too low to evaporate them. They are *thermally focused* over the length of the flooded zone. When the temperature is high enough to evaporate the low-volatility solutes, they start the chromatographic process at different points. If the flooded zone is homogeneous and short, peak distortion is hardly observed, but when the flooded zone is inhomogeneous and long, which is the case when polar solvents are injected on apolar capillary columns, broad distorted peaks are detected. The only reason for this is that the chromatographic process starts with the solutes distributed over the length of the flooded zone. This can be avoided by removing the stationary phase over the length of the flooded zone or, in other words, by creating a retention gap. Band broadening in space can thus be suppressed by *stationary-phase focusing* via a retention gap (Fig. 23). A retention gap is a definite length of the separation system that is uncoated. All solutes that are spread over the flooded zone are carried with no retention onto the stationary phase where they are retained. In practice, a retention gap is a separate piece of deactivated fused silica, connected to the analytical column through a coupling device such as a press fit connector. The length of the retention gap is a function of the length of the flooded zone, and depends on the sample volume and the nature of the solvent in use. Typical

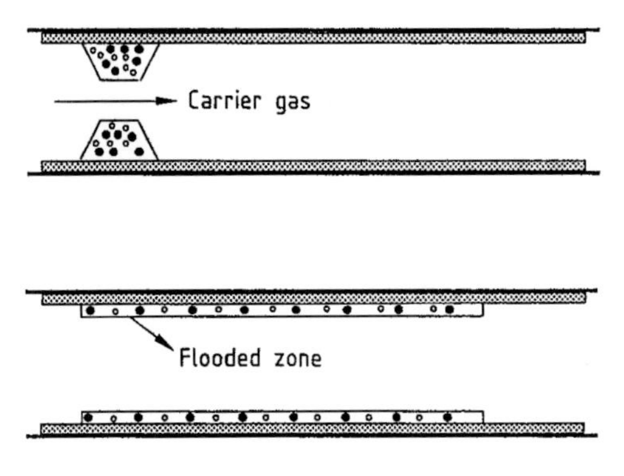

Figure 21. Creation of the flooded zone
o = High-volatile solute; • = Low-volatile solute

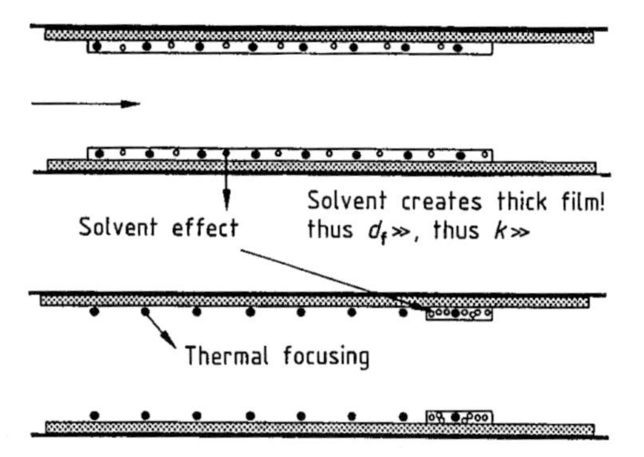

Figure 22. The solvent and thermal focusing effect
o = High-volatile solute; • = Low-volatile solute

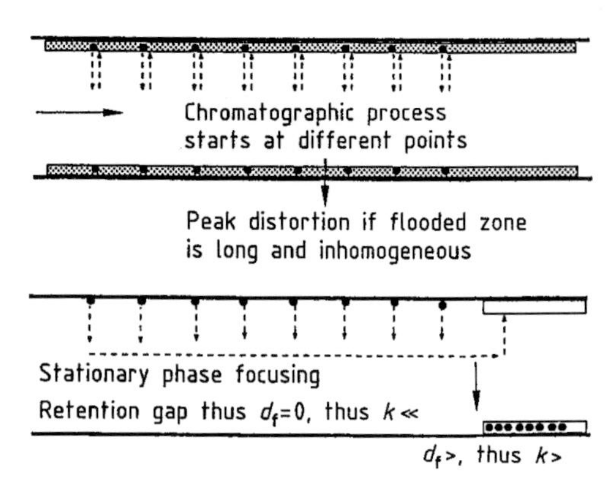

Figure 23. Stationary-phase focusing and the principle of a retention gap
• = Low-volatile solute

lengths are 0.5 – 1 m for injections of 1 to 2 µL. Longer lengths, however, do not harm the separation, and pieces of 3 m are standard and commercially available. Band broadening in space and the use of a retention gap are also required for cool on-column injection (see Section 11.5.2.2.3). In recent years, splitless injection has been overshadowed by cool on-column injection. Without any doubt, the most accurate and precise data are now provided with on-column injection. Notwithstanding this, splitless injection is still used for many routine determinations (i.e., in environmental analysis, pesticide monitoring, drug screening, etc.). In these fields, sample preparation is of primary concern, and cleaning up a sample to such an extent that it is accessible to on-column injection is not always possible or economically justifiable. Traces of nonvolatile or high-boiling components often remain in the sample. Splitless injection is an easy solution to such problems. The "dirty" components remain mainly in the vaporizing inlet, which is easily accessible for cleaning. An illustration of the determination of polychlorinated biphenyls (PCBs) in waste oil via splitless injection is shown in Figure 24. 5 µL of an isooctane solution of the PCBs extracted from waste oil, was injected in the splitless mode on an apolar column equipped with a retention gap of 5 m. A pressure pulse was applied to minimize the residence time of the solutes in the liner [42]. Without a retention gap the flooded zone would have been too large to guarantee good peak shapes. When properly performed, reproducibility in splitless injection can be as high as 1 – 2 % relative standard deviation (RSD). Standard addition or internal standard quantitation is the method of choice for manual injection. The external standard method can be adapted for automated splitless injection. Many capillary GC applications need the best attainable precision and accuracy. Cool on-column injection (see Section 11.5.2.2.4) then is the method of choice.

11.5.2.2.3. Direct Injection

The nomenclature of different injection techiques has been created as new developments occur. This often causes confusion with regard to the interpretation of some terms, which do not always stand for what they really mean. Direct injection is an example of such an expression, often confused or even identified with on-column injection. The understanding of the terminology given in this article, based on the chronological order in which techniques were developed, is as

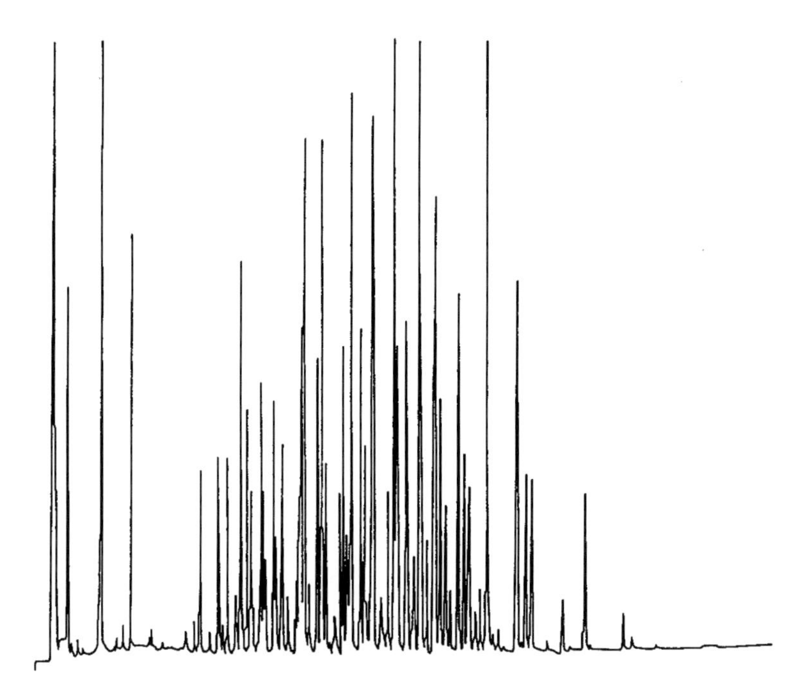

Figure 24. Optimized splitless injection of a PCB sample
Column: 25 m *L*×0.25 mm *i.d.*; 0.25 μm d_f MeSi; retention gap: 5 m *L*×0.25 mm *i.d.*; temperature: 150-280 °C at 5 °C/min; splitless injection with pressure pulse; carrier gas hydrogen; FID

follows: *Direct injection* is a flash vaporizing injection method. The inlet system is heated separately and independently from the column oven, and evaporation occurs in the inlet. This inlet can be a glass liner (out-column evaporation) or a part of the column (in-column evaporation). Different direct injection devices are shown in Figure 25. *On-column injection* is a "cold" injection technique. The sample is injected as a liquid, directly on the column. During injection, the injection zone is cooled to avoid needle discrimination. Injection peak broadening is caused only by band broadening in space. The solutes are focused in the inlet section of the column, from where evaporation gradually starts as the oven temperature is raised. Whereas on-column injection can be applied for all conventional capillary columns (*i.d.* 0.25, 0.32, 0.53 mm), direct injection is restricted to widebore or megabore columns.

Direct injection into widebore columns was described in 1959 [43]. At that time, column efficiency was all important. Through the introduction of widebore fused-silica capillary columns with *i.d.*'s of 0.53 mm, the interest in direct sampling has been renewed. Injection on a 0.53-mm *i.d.* FSOT column, operated in the high-resolution (flow rates 1 – 3 mL/min) or the low-resolution (flow rates 8 – 12 mL/min) mode, as a packed column alternative, is said to be as easy as installing a simple glass liner in the packed column inlet and connecting the column to it. This view is slightly oversimplified since the direct injection technique is based on the evaporation of the sample at elevated injection temperature, all previous statements about syringe discrimination and band broadening effects remain valid. Direct injection must definitely be optimized according to the problem at hand. Some experience-based guidelines on direct injection into widebore columns are advanced. However, when widebore capillary columns are used in the high-resolution mode, they should be connected to conventional capillary injectors (split, splitless, on-column, programmed temperature vaporization). For direct injection on widebore columns operated at high flow rates, the sample size should be kept as small as possible, and the sample should preferably be injected with an autosampler. For manual injections, the fast hot-needle injection technique is preferred to minimize syringe discrimination. To avoid sample

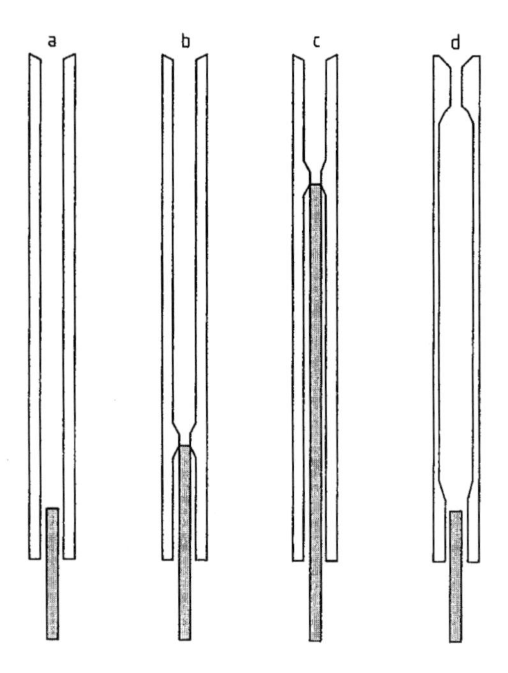

Figure 25. Different injection devices for direct injection
a) Open liner; b) Open liner with conical contraction; c) Direct on-column liner; d) Open liner with expansion volume

losses due to overfilling the insert with sample vapors, the point of injection should be close to the column entrance. Long and narrow inserts are preferred to obtain minimum sample dilution (band broadening in time). Band broadening should be compensated by refocusing the solutes in the inlet section of the column. The principle of the solvent effect and of thermal focusing have been discussed in Section 11.5.2.2.2. Injections of microliter volumes at oven temperatures (far) above the boiling point of the solvent should be avoided, unless the sample is known to contain heavy compounds only, which will be refocused by cold trapping at the chosen column temperature. Figure 26 shows the analysis of the oxygenated fraction of hop oil dissolved in dichloromethane on a narrowbore column with split injection (Fig. 26 A) and on a widebore column with direct injection (Fig. 26 B) [44]. Both columns (25 m×0.25 mm and 50 m× 0.50 mm) offer the same efficiency. Peak broadening did not occur on the widebore column with direct injection, for which only 0.2 μL was injected.

11.5.2.2.4. Cool On-Column Injection

Syringe on-column injection with small-diameter capillary columns was first described in 1978 by the Grobs [45], [46]. Special on-column devices (a micro and a macro version) were introduced in 1977 by Schomburg et al. [47], but the technical requirements were stringent and the devices lacked practical flexibility. Since this pioneering work, all instrument manufacturers have introduced cool on-column injectors. As an example, the cool on-column injector developed by Hewlett–Packard is shown in Figure 27. It can be used for manual and automated injection. The injector has a low thermal mass, which facilitates cooling. A key part of the injector for manual operation is the duck bill valve (b). The duck bill valve, made out of a soft elastomer, is a passive element in that it has no moving parts. It consists simply of two surfaces pressed together by the column inlet pressure. During injection, the syringe needle slips between the two surfaces to maintain the seal. For automated cool on-column injection, the system is modified with a disk septum (h). Today, cool on-column injection systems can be operated in the constant-pressure, constant-flow, or pressure-programmed mode. These features allow reduction of analysis time and elution of high molar mass compounds. The advent of cool on-column injection has extended the range of applicability of capillary GC to many classes of compounds that heretofore were difficult if not impossible to analyze. The technique of introducing the sample directly on the column without prior vaporization offers many advantages, such as elimination of sample discrimination and sample alteration, high analytical precision, and data operator independence. As with all other sampling techniques, the operational parameters strongly affect the chromatographic data. Parameters such as initial column temperature, solvent nature, injection rate, injected volume, and boiling point range of the sample components are interrelated. Once again, all aspects cannot be treated in detail in this article. *On-column Injection in Capillary GC* by K. Grob contains 590 pages [37]! The characteristics of the technique are highlighted here, and some important aspects for daily practical use are discussed. For sample sizes of the order of 0.5–2 μL, injection should be performed as quickly as possible, with the column oven temperature below or equal to the boiling point of the solvent. The oven temperature in on-column injection may be higher than in splitless injection. In

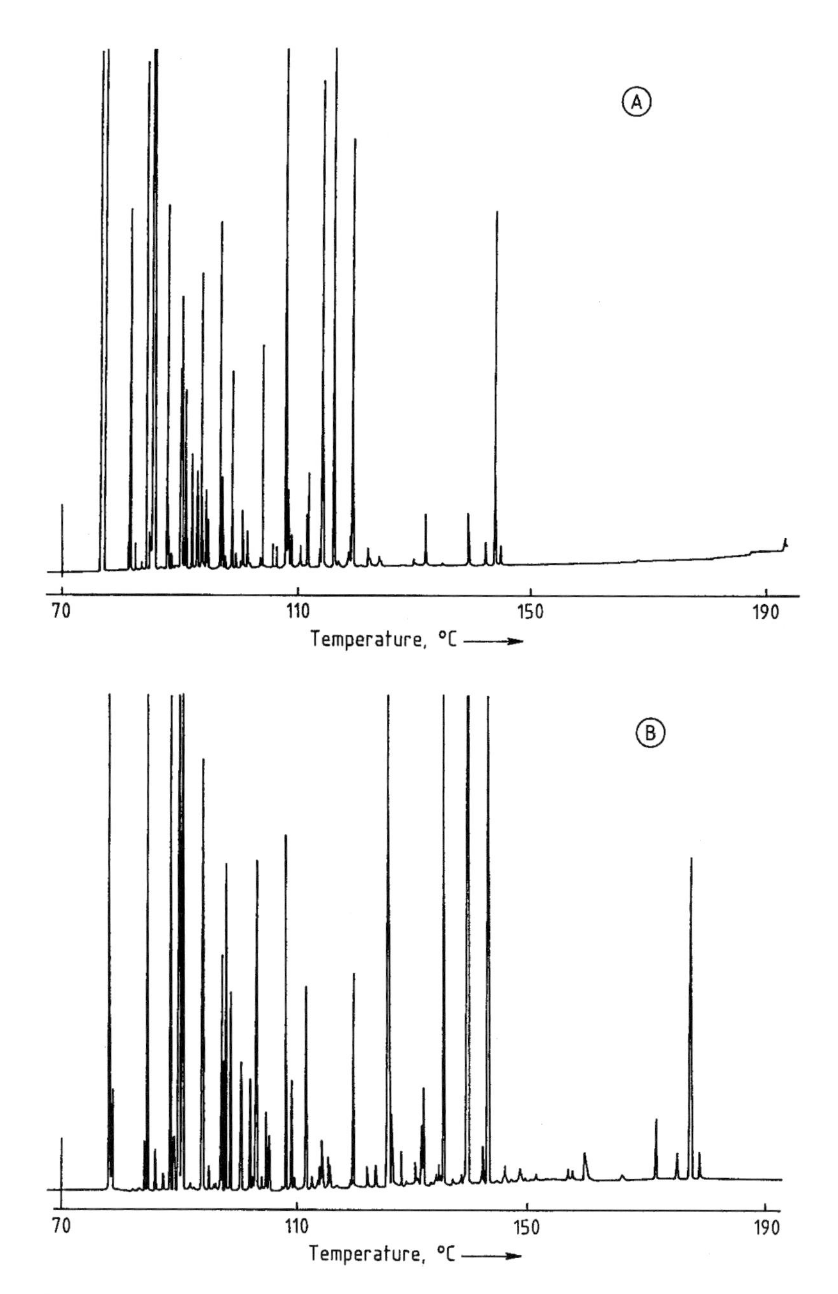

Figure 26. Analysis of the oxygenated fraction of hop essential oil on a conventional and a widebore capillary column
A) Column: 25 m L×0.25 mm *i.d.*; 0.25 μm d_f high-M_r poly(ethylene glycol); temperature: 70–190 °C at 2 °C/min; split injection 1/20; carrier gas hydrogen; FID; B) Column: 50 m L×0.5 mm *i.d.*; 0.25 μm d_f high-M_r poly(ethylene glycol); temperature: 70–190 °C at 2 °C/min; direct injection; carrier gas hydrogen; FID

splitless injection, the solvent has to recondense; whereas in on-column injection, the liquid is introduced directly into the column. As in splitless injection, the length of the flooded zone must be reduced because solutes are distributed over its entire distance. The width of the initial band equals the length of the flooded zone, i.e., band broadening in space (band broadening in time does not occur in on-column injection). The length of the flooded zone is a function of the column di-

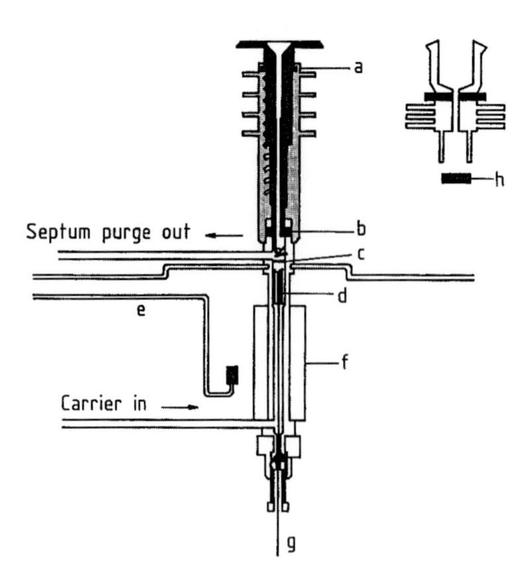

Figure 27. Schematic of a cool on-column injector
a) Cool tower, needle guide; b) Duck bill valve (isolation valve); c) Spring; d) Insert; e) Cryogenic cooling (optional); f) Heater block; g) Column; h) Septum

ameter, the volume injected, the actual inlet temperature, the stationary-phase thickness, and, most important, the affinity of the solvent for the stationary phase. For perfect wettability (i.e., apolar solvents on apolar silicone phases), the length of the flooded zone is about 20 cm per microliter injected. This length easily increases by a factor of five or ten if the wettability is poor, such as for methanol injections on apolar silicone phases. If the solutes are not refocused, band broadening in space makes qualitative and quantitative analysis impossible if large sample volumes are injected or if the solvent does not properly wet the stationary phase. The latter phenomenon is illustrated in Figure 28 with the analysis of a fraction collected from a step-elution HPLC analysis of the essential oil of *Valeriana celtica* L., with hexane and methanol. The fraction of the polar compounds was contained in 1 mL of hexane and 6 mL of methanol in a two-phase system. Both phases were injected on-column (1 μL) onto a 20 m×0.3 mm *i.d.* methylsilicone column with a film thickness of 0.3 μm. Figure 28 A shows the chromatogram of the hexane layer and Figure 28 B that of the methanol layer. Methanol causes peak splitting, whereas hexane does not [48]. The poor wettability (polar–apolar) in the case of methanol results in a long, nonhomogeneous flooded zone. This artifact can be bypassed via a retention gap (see Section 11.5.2.2.2). On-column injection provides the most accurate and precise results because syringe discrimination, which is one of the main sources of error in the quantitative analysis of samples covering a wide range of molar masses, is completely avoided. Moreover, inlet-related discrimination does not occur either, because the liquid is introduced directly onto the column. The automated on-column analysis of free fatty acids from acetic to decanoic acid is shown in Figure 29. Relative standard deviations on absolute peak areas are $< 1\,\%$ and on relative peak areas $< 0.4\,\%$, for 20 injections ($n = 20$). The analysis of polymer additives covering a broad range of functionalities and molecular masses is shown in Figure 30 [49]. To elute Irganox 1010 (M_r 1176) in a reasonable time, cool on-column injection was performed in the constant-flow mode. Both the external standard and the internal standard method can be applied for quantitation. Current literature data contain overwhelming evidence for the superior features of cool on-column injection in quantitative work. A second important feature of cold on-column injection is the elimination of sample alterations. Thermally labile components are not exposed to thermal stress: they begin the chromatographic process at relatively low temperature. Decomposition and rearrangement reactions are nearly completely eliminated. This feature allows analyses that heretofore were impossible with gas chromatography. Figure 31 shows the analysis of some retinoids [50]. Quantitative analysis in plasma samples was possible only with cool on-colum injection. The only disadvantage of cool on-column injection is that since the sample is introduced directly onto the column, relatively "clean" samples have to be prepared. Nonvolatile and less volatile material collects at the head of the column. The presence of this material causes a loss of separation efficiency or introduces adsorptive sites in the column inlet. Proper sample cleanup is thus very important. The use of a precolumn of 3 m, which at the same time acts as a retention gap, is therefore recommended. When the chromatographic performance decreases, 5 to 10 cm of the inlet section of the pre-column is cut off, restoring the original performance of the system.

11.5.2.2.5. Programmed-Temperature Injection

In 1979, Vogt and coworkers [51] described an injection method for the introduction of sample

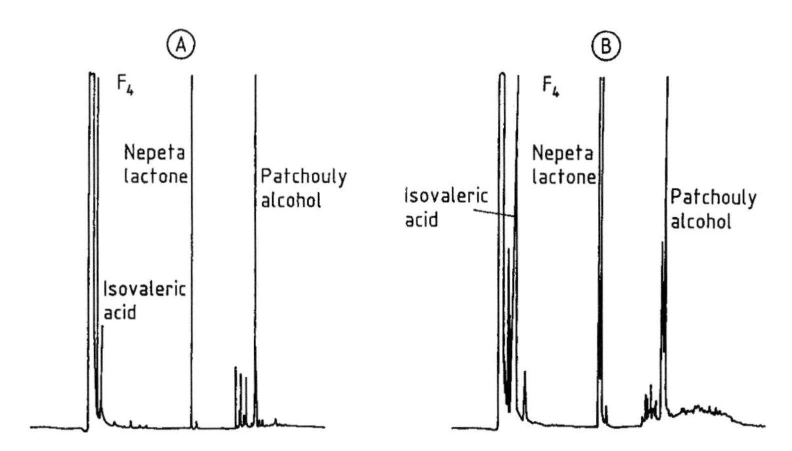

Figure 28. Illustration of peak splitting in cool on-column injection
A) Hexane as solvent; B) Methanol as solvent

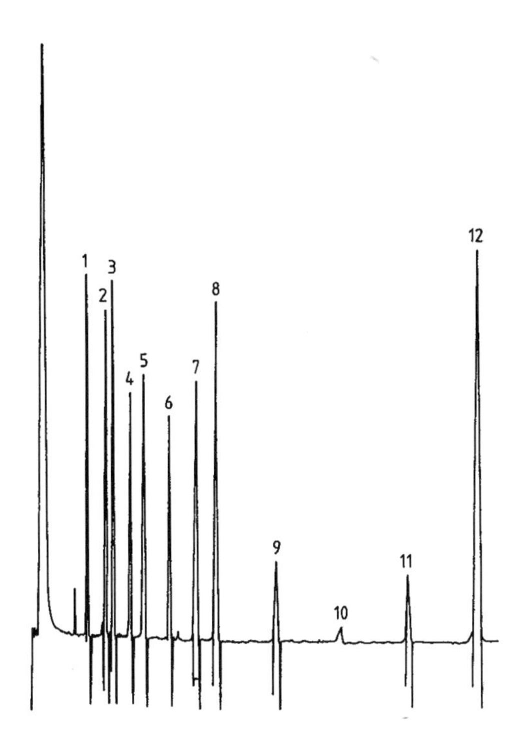

Figure 29. Analysis of free fatty acids on a widebore capillary column
Column: 10 m L×0.53 mm *i.d.*; 0.25 μm d_f FFAP; temperature: 60–100 °C balistically, to 140 °C at 2 °C/min; cool on-column injection; carrier gas hydrogen 4 mL/min; FID-Peaks: (1) acetic acid, (2) propionic acid, (3) isobutyric acid, (4) butyric acid, (5) isovaleric acid, (6) valeric acid, (7) isocaproic acid, (8) caproic acid, (9) heptanoic acid, (10) octanoic acid, (11) nonanoic acid, (12) decanoic acid

volumes up to 250 μL. The sample was introduced slowly into a cold glass insert packed with glass wool. The low-boiling solvent was evaporated continuously and vented through the split exit. The solutes, which remained in the inlet, were transferred to the capillary column by rapidly heating (30 °C/min) the inlet. During this transfer, the split line was closed (splitless injection). Based on this idea, the groups of SCHOMBURG [52] and POY [53] almost simultaneously developed programmed-temperature injection. Their aim, however, was not the injection of large sample volumes but rather the elimination of syringe needle discrimination, which at that time was the subject of interest. Different programmed-temperature vaporization (PTV) injection devices are now commercially available, offering a broad range of possibilities (i.e., hot or cold split injection, hot or cold splitless injection, cool on-column injection, direct injection, etc.). Through these possibilities, together with features such as the injection of large sample volumes, concentration by multiple injection, solvent venting, etc., PTV injection is claimed to be the most universal sample introduction system. It is, however, not the best choice for all applications. The upper part of a PTV inlet resembles a classical split–splitless device, including carrier gas inlet and septum flush. The sample is injected into a glass liner with a low thermal mass while the injector body is cold. After withdrawal of the syringe needle, the vaporizing tube is heated rapidly to volatilize the solvent and the solutes. The heat can be provided electrically or by means of preheated compressed air. Depend-

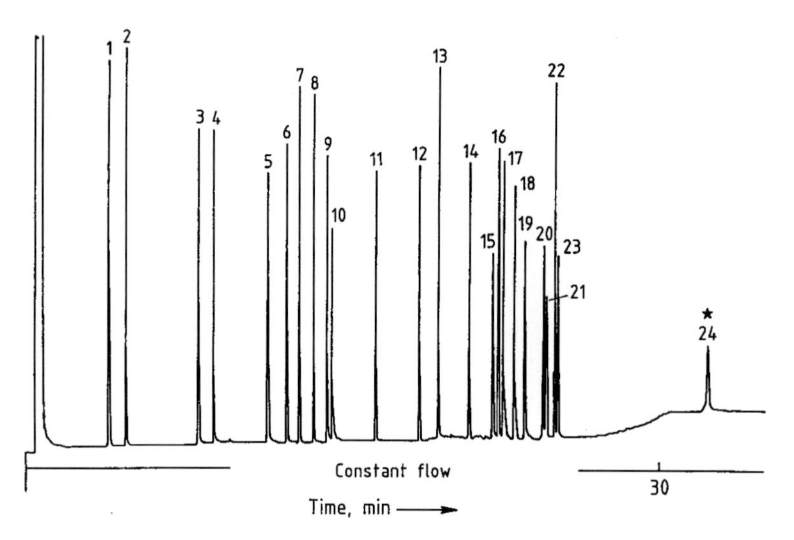

Figure 30. Analysis of polymer additives by high-temperature capillary GC
Column: 25 m L×0.32 mm *i.d.*; 0.17 μm d_f MeSi; temperature: 80–380 °C at 10 °C/min; cool on-column injection; carrier gas hydrogen 50 kPa constant flow; FID
Peaks: (1) butyl-4-methoxyphenol, (2) diethyl phthalate, (3) dibutyl phthalate, (4) Tinuvin P, (5) triphenyl phosphate, (6) dicyclohexyl phthalate, (7) dioctyl phthalate, (8) Tinuvin 327, (9) benzophenone UV, (10) erucamide, (11) Tinuvin 770, (12) Irgaphos 168, (13) Irganox 1076, (14) Tinuvin 144, (15) Irganox 245, (16) Irganox 259, (17) Irganox 1035, (18) Irganox 565, (19) crodamide, (20) Irganox 1098, (21) Irganox 3114, (22) Irganox 1330, (23) Irganox PS 802, (24) Irganox 1010

ing on its construction, heating of the device can be performed ballistically or linearly at selected rates (for example 2–12 °C/s) and even to consecutive temperature levels. These facilities allow optimization of conditions for the analysis of thermally labile compounds, operating in the solvent vent mode when working with specific detection such as electron capture detection (ECD) or mass spectrometry, concentration through multiple injection, etc. Split or splitless injection can be achieved through regulation of the split valve. During or after the chromatographic run, the vaporizing chamber is cooled by air or carbon dioxide, in preparation for the next injection. In an interesting mode of operation (i.e., *solvent elimination injection*), the sample is introduced into the cold injector with the split valve open. Conditions are selected such that only the solvent evaporates while the solutes of interest remain in the liner. The applicability of this technique is, therefore, restricted to the analysis of relatively high-boiling compounds. In this way, large sample volumes can be introduced to enrich traces. The principle has been applied to detect dioxins and furans down to the 100-fg/μL level with capillary GC–low-resolution mass spectroscopy [54].

11.5.3. Selective Inlets

A number of selective inlets are available in GC. Best known are static and dynamic head space (HS) injection (selectivity based on vapor pressure), purge and trap (PT) samplers (selectivity based on solubility of the solutes in water), thermal desorption (TD) units (selectivity based on characteristics of the adsorbent), and pyrolysis units (selectivity based on the pyrolysis or cracking temperature). These inlets are discussed in Sections 11.5.3.1–11.5.3.4. More sophisticated systems, which can also be considered as sample introduction systems, such as solid-phase extraction (SPE; selectivity based on the polarity of both adsorbent and desorbing liquid), supercritical fluid extraction (SFE; selectivity based on the solubility in the supercritical medium, the variables being density, temperature, pressure, and polarity), multidimensional capillary GC (MDCGC), the combinations of capillary GC with liquid chromatography (HPLC–CGC) or with supercritical fluid chromatography (SFC–CGC) allowing preseparation according to molar mass (size-exclusion chromatography, SEC), functionality (normal phase LC), hydrophobicity (reversed phase LC), or solubility (SFC) are treated in Chapter 11.8. In addition, *on-line sample preparation units* became

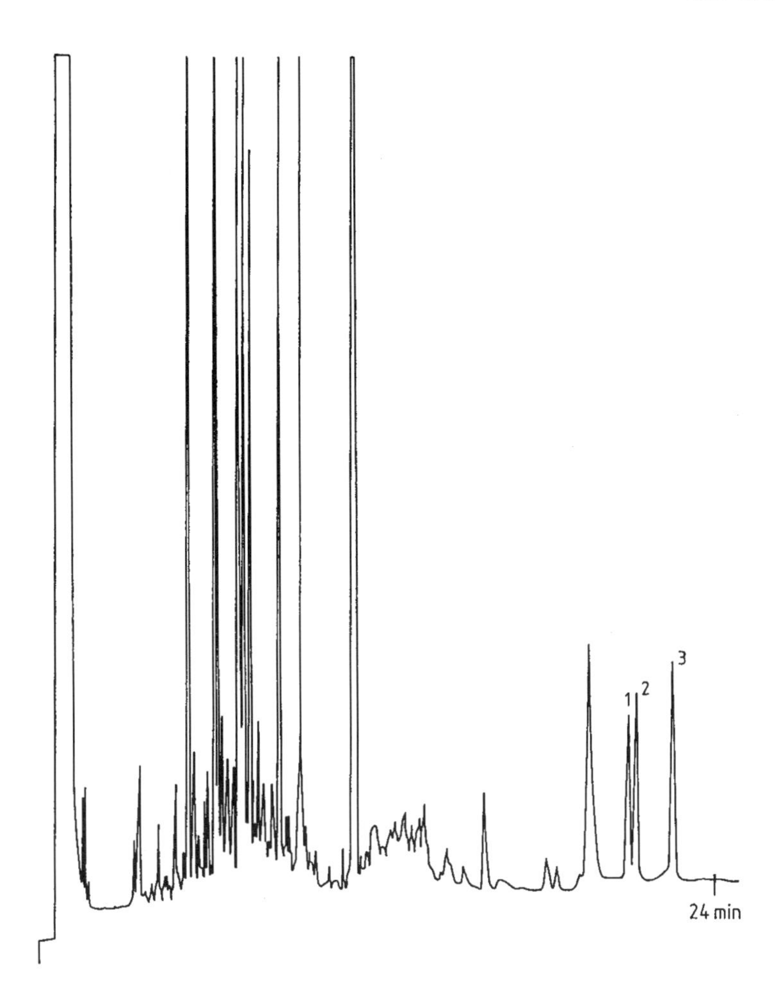

Figure 31. Illustration of the application of cool on-column injection to the analysis of thermolabile compounds
Column: 12.5 m L×0.25 mm *i.d.*; 0.25 μm d_f CPSi; temperature: 60 – 190 °C at 17 °C/min, to 220 °C at 6 °C/min and to 235 °C at 1.5 °C/min; cool on-column injection; carrier gas hydrogen; FIDPeaks: (1) *trans*-acitretin methyl ester, (2) *cis*-acitretin methyl ester, (3) *cis*-acitretin propyl ester

available in 1993, providing functions such as dispensing (addition of reagents or internal standards), diluting (concentration adjustments and multipoint calibration standards), heating (derivatization, solubilization), evaporating (solvent exchange, concentration), solid-phase extraction, and filtering.

11.5.3.1. Static and Dynamic Head Space Analysis

Head space analysis is used to analyze volatiles in samples for which the matrix is of no interest (i.e., water, soil, polymers, etc.). Typical applications include monitoring of volatiles in soil and water (environmental); determination of monomers in polymers, aromas in food and beverages, residual solvents in drugs (pharmaceuticals), etc. Different head space autosamplers are commercially available, and they are based on the principle of static or dynamic head space. The difference between these modes is depicted in Figure 32. In *static head space* (Fig. 32 A), the sample (soil, water, solid) is transferred in a head space vial that is sealed and placed in a thermostat to drive the desirable components into the head space for sampling. Via a gas-tight syringe (a) or a sample loop of a gas sampling valve, an aliquot of the vapor phase is introduced in the GC system, which can be equipped with a packed or a capillary column. Because of the much lower sample loadability of a capillary column, split injection is

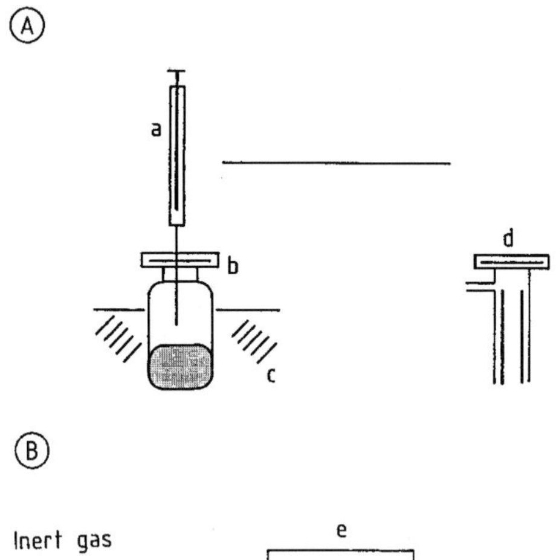

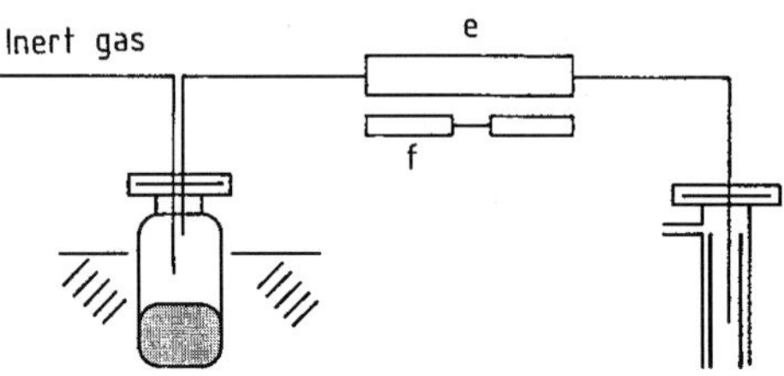

Figure 32. Principle of static (A) and dynamic head space (B)
a) Syringe; b) Head-space bottle; c) Thermostat; d) Injector; e) Cold trap; f) Solid-phase extraction

usually applied in this case. Static head space implies that the sample is taken from one phase equilibrium. To increase detectability, *dynamic head space analysis* has been developed. The phase equilibrium is continuously displaced by driving the head space out of the vial via an inert gas. The solutes are focused by cold trap (e) or an adsorbent such as Tenax and after enrichment the trap is heated and the solutes are introduced into the GC system. Whereas static head space is applied to analyze ppm amounts (1 part in 10^6), dynamic headspace allows determination of ppb amounds (1 part in 10^9) of volatiles. Sample pretreatment can often help to increase sensitivity and enhance reproducibility. The best-known methods are salting out of water samples with sodium sulfate, adding water to solid samples, and adjusting the pH to drive organic acids or bases into the head space.

11.5.3.2. Purge and Trap Systems

Purge and trap samplers have been developed for analysis of apolar and medium-polarity pollutants in water samples. The commercially available systems are all based on the same principle. Helium is purged through the sample placed in a sealed system, and the volatiles are swept continuously through an adsorbent trap where they are concentrated. After a selected time (10 – 20 min), purging is stopped, the carrier gas is directed through the trap via a six-way valve, and the trap is heated rapidly to desorb the solutes. The method has proven to be highly accurate for environmental monitoring of ppb amounts of volatile aromatics, trihalomethanes, etc.

11.5.3.3. Thermal Desorption Units

Thermal desorption is the method of choice to monitor air pollution. A known amount of air is drawn through an adsorbent tube filled with activated carbon, Tenax, silicagel, or mixtures of these. Analytes of interest are trapped and concentrated on the adsorbent. The adsorbent tube is sealed and transported to the laboratory where it is installed in the thermal desorption unit. An example of a thermal desorption unit is shown in Figure 33. After the adsorbent tube is placed in the desorption module, the carrier gas is directed over the adsorbent, which is heated from room temperature to 250 – 300 °C. Since thermal desorption is a slow process, the solutes are focused by cold trapping on a fused silica trap. The cryogenic trap is then heated rapidly to transfer the sample to the column. The method works well for apolar and medium-polarity components in air. Highly polar solutes are very hard to desorb, and other sampling methods must be selected.

11.5.3.4. Pyrolysis Gas Chromatography

Pyrolysis involves the thermal cleavage (cracking) of large molecules, which are not amenable to GC, into small volatile fragments. Pyrolysis GC is an excellent technique for identifying polymers and microorganisms because the pyrolysis profiles are very typical fingerprints. Most important in this respect is the pyrolysis temperature, typically between 400 and 1000 °C. The higher the temperature, the greater is the degree of fragmentation. Several types of analytical pyrolyzers are available. The most common types are the *platinum resistively heated* and the *Curie point pyrolyzers*. Figure 34 shows an example of a resistively heated pyrolyzer. The pyroprobe (c) consists of an injector cap (a), a quartz tube (d), and a platinum wire (f). The sample is placed in the quartz tube, and

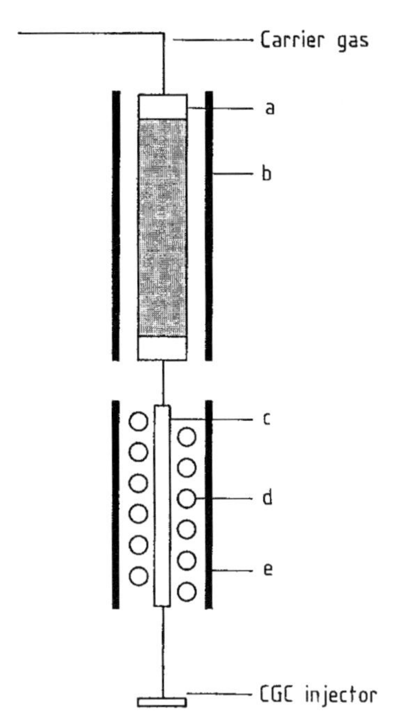

Figure 33. Schematic of a thermal desorption unit
a) Adsorbent tube; b) Heating device; c) FSOT trap; d) Cooling; e) Heating

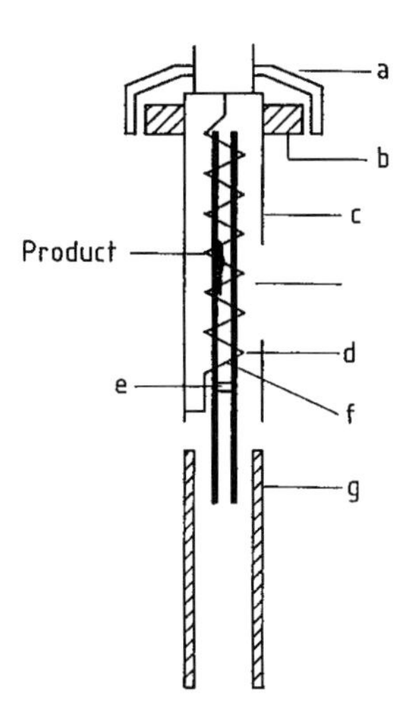

Figure 34. Principle of pyrolysis GC
a) Injector cap; b) Septum; c) Pyroprobe; d) Quartz tube; e) Glass wool; f) Platinum wire; g) Inlet liner

the pyroprobe is introduced in the injection port of the GC. The carrier gas enters the system, and after equilibration, the platinum wire is heated to a preselected temperature. The material decomposes and the volatiles are analyzed.

11.6. Detectors

Detectors in gas chromatography sense the presence of analytes in the carrier gas and convert this information in an electrical signal, which is recorded. Numerous detection devices have been developed in GC [55]; here, only the currently most common detectors are dealt with. An excellent book describing detectors for capillary chromatography has been published [56].

11.6.1. Classification

GC detectors are classified mainly on the basis of response or of detector selectivity. Detectors whose response is proportional to the concentration of a sample component in the mobile phase (g/mL) are called *concentration-sensitive detectors*, whereas detectors whose response is proportional to the amount of sample component reaching the detector in unit time (g/s) are called *mass-flow detectors. Universal detectors* respond to every component in the mobile phase; *selective detectors* respond only to a related group of substances; and *specific detectors* respond to a single sample component or to a limited number of components with similar chemical characteristics. The differentiation of selective versus specific is confusing, and therefore in this article, selective is used for both to distinguish them from universal. Another classification is destructive versus nondestructive detectors. Any detector can be coupled in series with a nondestructive detector. The most important characteristics for practical work are sensitivity, dynamic range, linear range, detector response factors, and selectivity.

Sensitivity is the signal output per unit concentration or unit mass of a substance in the mobile phase and is best expressed by the minimum detectable quantity, which is defined as the concentration or the amount of analyte for which the peak height is four times the intrinsic noise height of the detector (signal-to-noise ratio $S/N=4$). The *dynamic range* of a detector is that range of concentration or quantity over which an incremental change produces an incremental change in detector signal. Figure 35 represents a plot used for the determination of the dynamic range of a detector. That part of the curve for which the response

Table 4. Characteristics of common GC detectors

Name	Type	Selectivity	Minimum detectability	Linear range
Flame ionization detector (FID)	universal [c]	[a]	10 pg C/s	10^7
Thermal conductivity detector (TCD)	universal [c]	[b]	1 ng/mL mobile phase	10^6
Electron capture detector (ECD)	selective	compounds capturing electrons, e.g., halogens	0.2 pg Cl/s	10^4
Nitrogen phosporus detector (NPD)	selective	N and P	1 pg N/s 0.5 pg P/s	10^4
Flame photometric detector (FPD)	selective	P and S	50 pg S/s 2 pg P/s	10^3 10^4
Photoionization detector (PID)	selective	aromatics	5 pg C/s	10^7
Electrolytic conductivity detector (ELCD)	selective	halogens and S	1 pg Cl/s 5 pg S/s	10^6 10^4
Fourier transform infrared spectroscopy (FTIR)	universal [c]	molecular vibrations	1 ng (strong absorber)	10^3
Mass spectroscopy (MS)	universal [c]	characteristic ions	1 ng full-scan mode 1 pg ion-monitoring mode	10^5
Atomic emission detection (AED)	universal [c]	any element	0.2–50 pg/s depending on element	10^4

[a] FID responds to all organic compounds that ionize in a flame.
[b] TCD responds if thermal conductivity is different from carrier gas.
[c] Spectroscopic detectors that can be operated in a universal or a selective mode.

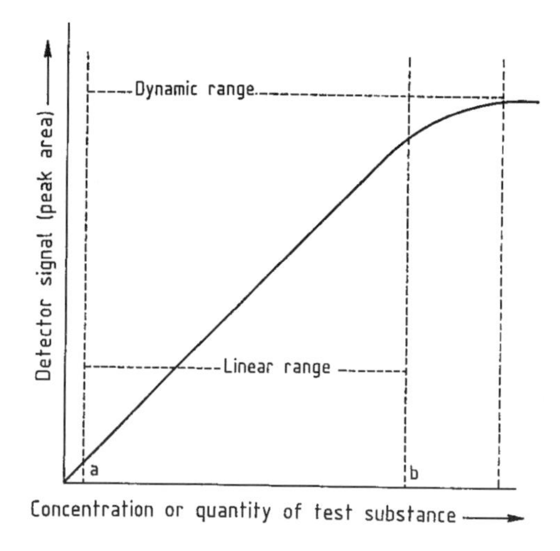

Figure 35. Dynamic range of a GC detector
a) Minimum detectability; b) Upper limit of linearity

increases proportionally with increased concentration or quantity is called the *linear range.* For quantitation, work in the linear range is preferable, it is expressed numerically as the ratio of the upper limit of linearity (b) to the minimum detectability (a), both measured for the same substance. The *detector response factor* f_i is a relative term expressing the sensitivity of a detector to a given compound *i*, relative to its sensitivity to a standard compound:

$$f_i = f_{st}\,(A_i/A_{st}) \qquad (6.1)$$

where A_i and A_{st} are the respective peak areas and f_{st} is the detector response factor for the standard; f_{st} is usually assigned a value of 1.000.

Detector selectivity is the ratio of the signal obtained for a compound to which the detector has special selectivity versus the signal obtained for a compound to which the detector is not particularly sensitive. Both signals must be measured for equal amounts. A compilation of selectivity, minimum detectability, and linear range of the most common GC detectors is given in Table 4.

11.6.2. Universal Detectors

11.6.2.1. Flame Ionization Detector

The flame ionization detector (FID) is the most widely used GC detector. Its operation is simple, sensitivity is of the order of 10 pg carbon per second with a linear range of 10^7, the response is fast, and detector stability is excellent. The carrier gas is mixed with hydrogen, and this mixture is combusted in air at the exit of a flame jet. Ions are formed that are collected at an electrode producing a current that is proportional to the amount of sample compound in the flame. Analytes such as permanent gases, nitrogen oxides, sulfur oxides, carbon oxides, carbon disulfide, water, formic acid, formaldehyde, etc., do not provide a significant FID response. The flows of carrier and combustion gas should be set properly for optimal FID operation. Typical flow rates are 1:1:12

(30 : 30 : 360 mL/min), respectively, for nitrogen as carrier gas, and the combustion gases hydrogen and air. If hydrogen is applied as carrier gas in packed column GC, the hydrogen combustion gas is replaced by nitrogen. For capillary GC where the carrier gas flows are much lower (e.g., 1 – 3 mL/min), hydrogen combustion gas is adjusted to a total hydrogen flow of 30 mL/min. Nitrogen can be added as make-up gas, which increases sensitivity and stability slightly. Detector temperature must always be kept above 150 °C to prevent condensation of water produced in the combustion process. Flame tip and electrodes should be cleaned regularly because certain compounds can cause deposits that must be removed mechanically to prevent loss in sensitivity and detector stability. Bleeding of stationary phases and the use of derivatization agents are important sources of sensitivity loss and spikes. The flame shape for capillary columns should be optimized, and therefore flame tips with different bores are available. Small-bore jets produce the greatest signal for capillary GC.

11.6.2.2. Thermal Conductivity Detector

The thermal conductivity detector (TCD), also called the *hot-wire detector* or *katharometer*, measures the change in the thermal conductivity of the mobile-phase stream. Helium is the recommended carrier gas because of its high thermal conductivity. When an analyte is present in the carrier gas, the thermal conductivity will drop and less heat is lost to the cavity wall. A filament in the detector cell operated under constant voltage will heat up and its resistance will increase. This resistance is measured and recorded.

In the *original TCD* design the resistance is incorporated in a Wheatstone bridge circuit. The difference in thermal conductivity between the column effluent R_1 and a reference flow R_2 is recorded. This original design is still used in packed column GC (e.g., in process control). Nowadays, *single-filament,* flow-modulated thermal conductivity detectors are available for capillary work. The cell size is reduced to 3.5 μL to minimize detector band broadening, and the single filament measures the reduction in voltage required to maintain a constant filament temperature. A diagram of a commercially available TCD cell is presented in Figure 36. In Figure 36 A, the switching flow pushes the column effluent through the filament channel. On changing the switching flow (Fig. 36 B), the column effluent passes through the empty channel while the filament channel fills with the pure reference gas. Switching between column effluent and reference gas occurs every 100 ms. An important parameter in practical work is the detector temperature. The lower the temperature, the higher is the sensitivity and the longer the lifetime of the filament. The temperature must, however, be high enough to avoid condensation of the solutes in the detector cell. Oxygen-contaminated carrier gas, column bleeding, and active components such as strong acids and bases all decrease sensitivity and reduce the lifetime of the filament. Note also that if helium is the carrier gas, and hydrogen together with some other solutes must be measured, the hydrogen peak will be negative because its thermal conductivity is higher than that of helium, while the other solutes give positive peaks.

11.6.3. Selective Detectors

11.6.3.1. Electron Capture Detector

One of the most popular selective GC detectors in use today is the electron capture detector (ECD). The ECD is an extremely powerful tool in environmental and biomedical studies. The principle is based on the fact that electronegative organic species (RX) react with low-energy thermal electrons to form negatively charged ions. The loss of electrons from the electron stream that is measured is related to the quantity of analyte in the carrier gas stream. The capturable electrons originate from ionization of the carrier gas or a make-up gas (usually nitrogen or an argon – methane mixture 95 : 5 or 90 : 10) by β particles from a radioactive source such as ^{63}Ni placed in the cell. The electron flow results in a current that is measured. Electronegative species in the carrier gas capture electrons and decrease the current. The highest sensitivity is noted for organic polyhalides such as trihalomethanes, organochloropesticides, PCBs, dioxins, and furans. Sensitivity increases in the order $F < Cl < Br < I$. The selectivity depends strongly on the compound's affinity for electrons and for polyhalides it is as high as 10^6 compared to hydrocarbons, and 10^4 compared to ketones, anhydrides, amines, etc. By chemical derivatization, the applicability of ECD can be expanded for trace analysis of species that do not or only weakly capture electrons. The best-known reagents are pentafluorobenzoic acid anhydride and pentafluorobenzoyl chloride for the derivati-

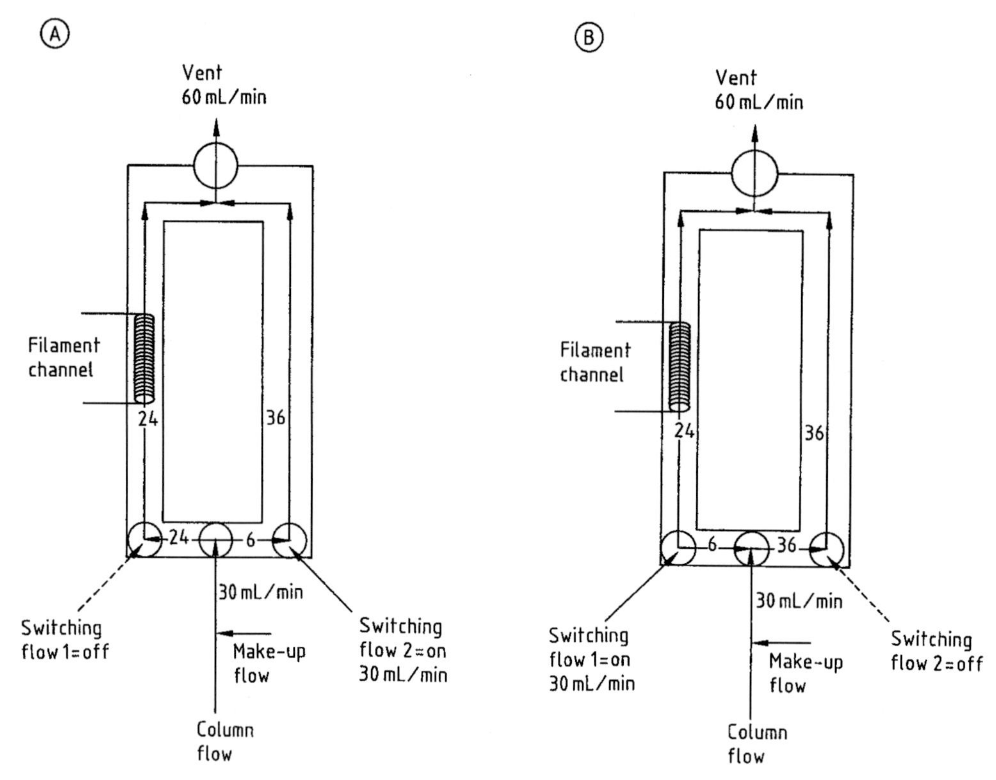

Figure 36. Diagram of a miniaturized TCD cell
A) Measurement of column effluent flow; B) Measurement of reference flow

zation of alcohols, amines, and phenols. (The fluorocompounds are chosen because of their high volatility, even though the detector is less sensitive for fluorine.) The derivatives formed are generally stable and have good chromatographic properties. As an example, the EPA method for the determination of chlorophenols in water samples includes a derivatization step with pentafluorbenzoyl chloride followed by ECD detection, in this way enhancing both sensitivity and selectivity. For proper use of the ECD detector, the carrier gas and the make-up gas must be very clean and dry. In packed column GC, nitrogen or argon – methane is used as carrier gas and no make-up gas is required at the column outlet. In capillary GC, helium or hydrogen is the preferred carrier gas because of the speed of analysis, and 25 – 30 mL/min nitrogen or argon – methane is added as make-up gas to produce the thermal electrons. The column should also be preconditioned before connection to the detector. High stability and sensitivity are guaranteed only with the detector in constant operation. Contamination is normally corrected for by baking out the detector at 350 °C over a 24-h period. Some typical ECD chromatograms are shown in Figure 37 for the direct analysis of trihalomethanes in drinking water [57] and in Figure 38 for the determination of organochloropesticides.

11.6.3.2. Nitrogen Phosphorus Detector

The nitrogen phosphorus detector (NPD), also called the *thermionic detector* (TID) is similar in construction to the FID. The detector is made element specific for N and P because a source of an alkali salt — salts of potassium and rubidium are more commonly used — is positioned above the flame jet. By careful control of the combustion gases hydrogen and air, whose flow rates are less than those for the normal FID, the ionization of compounds containing no nitrogen or phosphorus is minimized, while the ionization of N- and P-containing compounds is increased. The exact mechanism is still unknown. Small changes in detector gas flows have a significant effect on sensitivity and selectivity. Optimal gas flows can

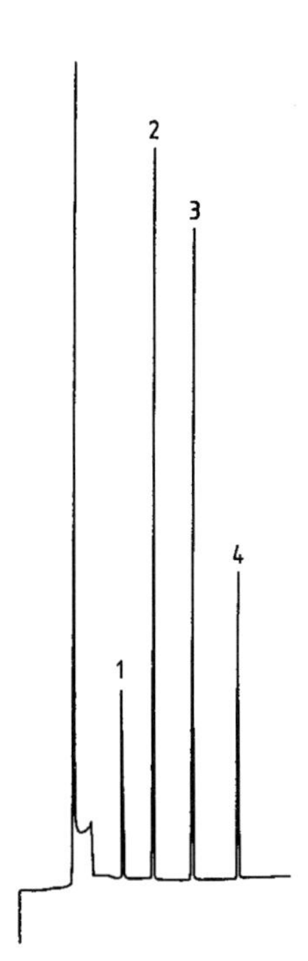

Figure 37. Direct analysis of trihalomethanes in drinking water
Column: 30 m L×0.53 mm *i.d.*; 2.65 μm d_f MeSi; retention gap 1.5 m L×0.53 mm *i.d.*; temperature: 50 – 150 °C at 10 °C/min; cool on-column injection; carrier gas 10 kPa hydrogen; ECD
Peaks: (1) chloroform, (2) dichlorobromoform, (3) dibromochloroform, (4) bromoform

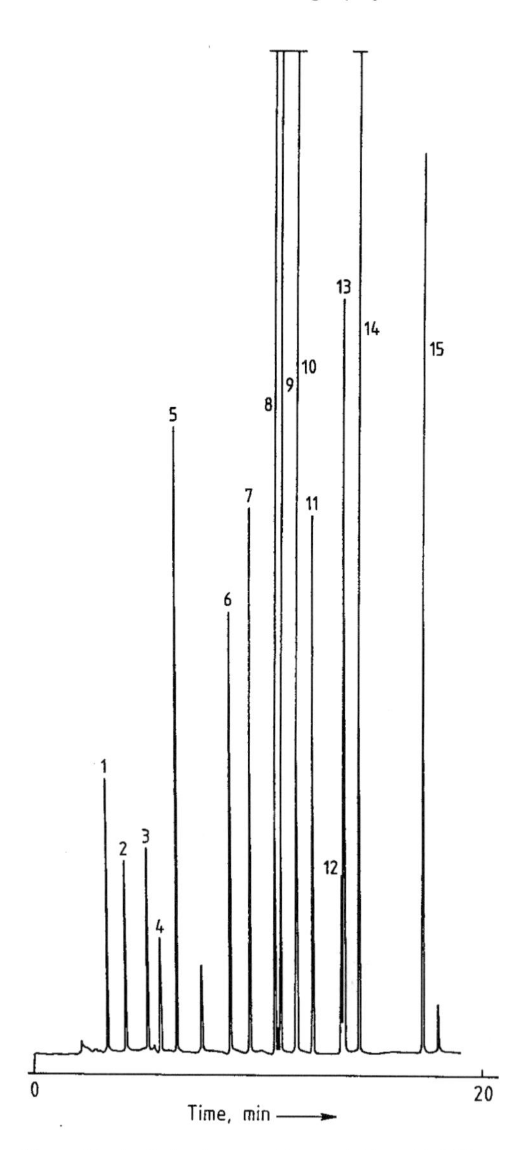

Figure 38. Analysis of organochloropesticides with ECD detection
Column: 20 m L×0.32 mm *i.d.*; 0.25 μm d_f 7 % CPSi, 7 % PhSi, 84 % MeSi (tailor-made called 1701); temperature: 180 – 240 °C at 4 °C/min; carrier gas hydrogen 8 psi (≈ 5.6×10^4 Pa); ECD
Peaks: (1) HCB, (2) hexachlorocyclohexane, (3) lindane, (4) heptachlor, (5) aldrin, (6) heptachlorepoxide, (7) α-endosulfan, (8) 1,1-dichloroethenylidenbis(4-chlorobenzene), (9) dieldrin, (10) endrin, (11) 2,4-DDT, (12) β-endosulfan, (13)1,1′-(2,2-dichloroethylidene)bis(4-chlorobenzene), (14) 4,4′-DDT, (15) methoxychlor

be found by analyzing a mixture containing azobenzene, heptadecane, and malathion in the ratio 1 : 1000 : 2 (100 pg/μL, 100 ng/μL, and 200 pg/μL, respectively). Prominent responses should be obtained for azobenzene and malathion, whereas the hydrocarbon should hardly be detected. Optimal flow rates on the order of 4 – 10 mL/min for hydrogen and 60 –100 mL/min for air are typical. For some versions of the NPD detector the use of a make-up gas such as helium increases the N or P response. The NPD detector is a valuable tool in clinical and pharmaceutical analysis, and especially in environmental monitoring of organophosphorus- and organonitrogen-containing pesticides.

11.6.3.3. Flame Photometric Detector

The flame photometric detector (FPD) uses the principle that sulfur- and phosphorus-containing compounds emit light by chemiluminescence when burned in a hydrogen-rich flame. For sulfur, for example, electronically excited diatomic sulfur (S^*_2) is formed, which upon relaxation emits light over the wavelength region 320 – 450 nm, with an emission maximum at 394 nm. An optical filter at

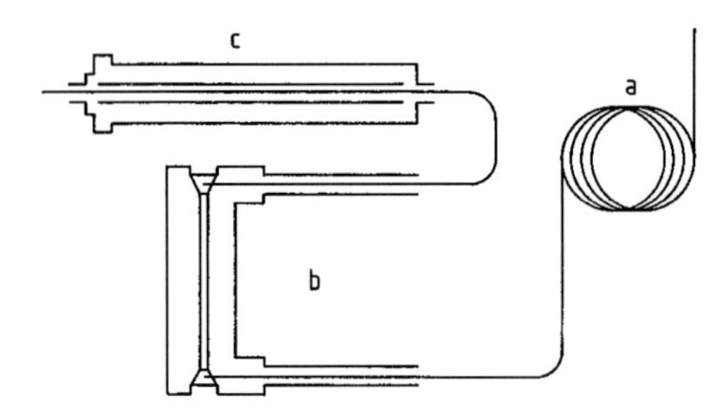

Figure 39. Diagram of a capillary GC – FTIR – MS system
a) Capillary GC; b) FTIR; c) MS

394 nm permits the light to enter a photomultiplier, and a signal is produced. For phosphorus the interference filter is 526 nm. Although widely used mainly as a sulfur-selective detector in the petroleum industry, for the analysis of flavors and fragrances, in environmental analysis, etc., the FPD detector has far from ideal characteristics as a GC detector. It has many shortcomings; the most important one is that the intensity of light emitted is not linear with concentration, but approximately proportional to the square of the sulfur atom concentration. Careful calibration is a must when quantitation is aimed at. Other negative aspects are that the flame extinguishes when high amounts of solvents are introduced (this problem is eliminated when a dual-flame burner is used), hydrocarbon quenching occurs, and self-quenching occurs at high concentrations of the heteroatom species. Moreover, optimization of the different flows is very critical for optimal sensitivity and selectivity. Therefore, for the analysis of sulfur-containing compounds the FPD detector is often replaced by the atomic emission detector, the Hall electrolytic conductivity detector, or the chemiluminescence detector based on ozone reactions, whereas for organophosphorus compounds the NPD is preferred.

11.6.3.4. Overview of Other Selective Detectors

Two of the more recently developed detectors, namely, the Hall electrolytic conductivity detector (ELCD) and the photoionization detector (PID) are recommended by the EPA for the analysis of volatile and semivolatile halogenated organic compounds and low molar mass aromatics. Chemical emission based detectors, such as the thermal energy analyzer (TEA) for its determination of nitrosamines and the redox chemiluminescence detector (RCD) for the quantitation of oxygen- and sulfur-containing compounds, are becoming more and more important for some typical applications.

11.6.4. Detectors Allowing Selective Recognition

The selective detectors discussed in the previous sections often do not provide enough information to elucidate with 100 % probability the nature of the eluting solutes. For this reason, data with selective detectors can be erratic. The future in this respect definitely belongs to the spectroscopic detectors that allow selective recognition of the separated compounds. Today, the hyphenated techniques CGC – mass spectroscopy (CGC – MS), CGC – Fourier transform infrared spectroscopy (CGC – FTIR), and CGC – atomic emission detection (CGC – AED) are the most powerful analytical techniques available. They provide sensitive and selective quantitation of target compounds and structural elucidation or identification of unknowns. The applicability and ease of use of the hyphenated techniques were greatly increased by the introduction of fused silica wall coated open tubular columns. The main reason for this is that because of the low flows of capillary columns, no special interfaces are required and columns are connected directly to the different spectrometers. The introduction of relatively inexpensive benchtop hyphenated systems has enabled many laboratories to acquire such instrumentation, which in turn has expanded their applicability ever further.

The hyphenated techniques CGC – MS, CGC – FTIR, and CGC – AED are generally used as stand-alone units. Due to the nondestructive character of FTIR, CGC – FTIR – MS units (Fig. 39) are possible and have been commercialized. The software then allows simultaneous recording of the infrared and mass spectra of the eluting compounds. In principle, CGC – FTIR – MS – AED is also possible if an open split interface is applied for the CGC – FTIR – MS combination and the split-line is directed into the AED detector. The fundamental aspects of CGC – MS, CGC – FTIR and CGC –AED are discussed in [58], [59], and [60].

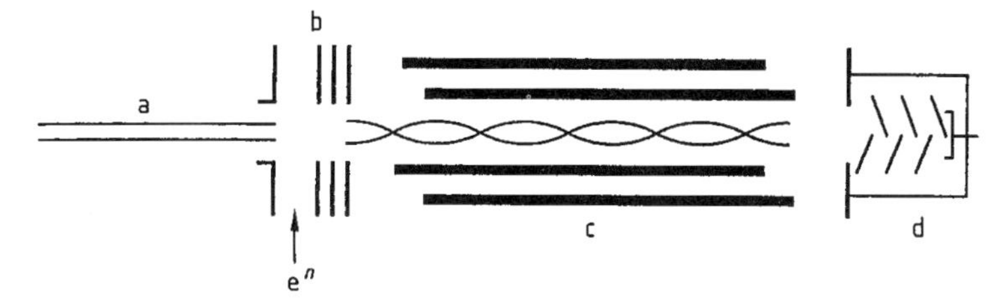

Figure 40. Schematic of a capillary GC – MS system
a) Capillary column; b) Ion source; c) Quadrupole mass filter; d) Multiplier

11.6.4.1. Mass Spectroscopy

The principle of CGC – MS is illustrated with a quadrupole mass analyzer as an example. As sample molecules exit the chromatographic column, they are introduced in the ion source of the mass spectrometer and ionized. Depending on the ionization method, molecular ions or fragment ions are formed, which are accelerated and separated from each other according to their mass-to-charge ratios (m/z) by a mass analyzer such as a quadrupole (c) (Fig. 40). Other mass analyzers are magnetic sector instruments, ion trap detectors, and time-of-flight mass spectrometers. The separated ions are then detected by a electron multiplier (d), and the resulting mass spectrum is a line spectrum of intensity of ions (y-axis) versus their m/z ratio (x-axis).

Two different ionization methods are available in CGC – MS, namely, electron ionization (EI) and chemical ionization (CI). In *electron ionization*, the molecules are bombarded with electrons of 70 eV emitted from a rhenium or tungsten filament. A molecular radical ion is formed ($M^{+}\cdot$) with sufficient amount of energy accumulated in its bonds to dissociate into typical fragment ions, radicals, and neutral species.

$$\begin{array}{l} ABCD \longrightarrow ABCD^{+\bullet}(M^{+\bullet}) + e^- \\ \qquad\qquad \longrightarrow AB^{+\bullet} + CD \\ \qquad\qquad \longrightarrow AB^+ + CD^\bullet \\ \qquad\qquad \longrightarrow ABC^+ + D^\bullet \\ \qquad\qquad\qquad \longrightarrow AB^+ + C \\ \qquad\qquad\qquad \longrightarrow A^+ + BC \end{array}$$

Since low ion source pressures are employed (10^{-5} torr $\approx 1.33\times10^{-2}$ Pa), reactions in the ion source are unimolecular, and association between molecule and fragment ions does not occur. A disadvantage of electron ionization is that energy in the molecule is so large that complete fragmentation can occur and the molecular ion is absent. Less fragmentation is obtained with *chemical ionization*, which is a soft technique of ionizing molecules through gas-phase ion – molecule reactions. In chemical ionization, a reagent gas such as methane is introduced into the ion source and a relatively high source pressure is maintained (1 torr $\approx$ 133 Pa). The reagent gas is ionized by the electrons, and the ions formed interact with the sample molecules. Since these reactions are low in energy, abundant quasi-molecular ions, most often $M^{+}\cdot + 1$ are observed. Figures 41 and 42 show the electron-impact and the chemical-ionization spectra of trimethylsilylated clenbuterol. Chemical ionization is very useful to elucidate the molecular ion of unknowns and is often the ionization method of choice for quantitative mass spectroscopy via ion monitoring. After ionization, the positive ions are accelerated and separated in the mass analyzer by tuning the voltage on the quadrupole rods. An ion of unit mass will pass the rods and reach the detector only if the voltages are properly adjusted. To detect all ions, the voltage is varied in time, and the time to detect all ions is called a scan. This means that if all ions between mass 20 and 400 are to be detected in 1 s, the voltages are on the exact value for each ion to pass during 1/380 of a second. This mode, called the *full-scan mode*, is necessary to identify compounds via their fragmentation pattern. Data are acquired by continuous repetitive scanning of the column eluate. The reconstructed chromatogram of a mass spectrometer is a plot of the total ion current as a function of time or scan number. All recorded spectra or scans are stored in the computer and can be recalled for identification. Nowadays this is performed mostly by comparison with a spectral library. The information in the stored mass scans can provide, in a simple way, valuable information on specific compounds or compound classes. This is illustrated in Figure 43, which shows the recon-

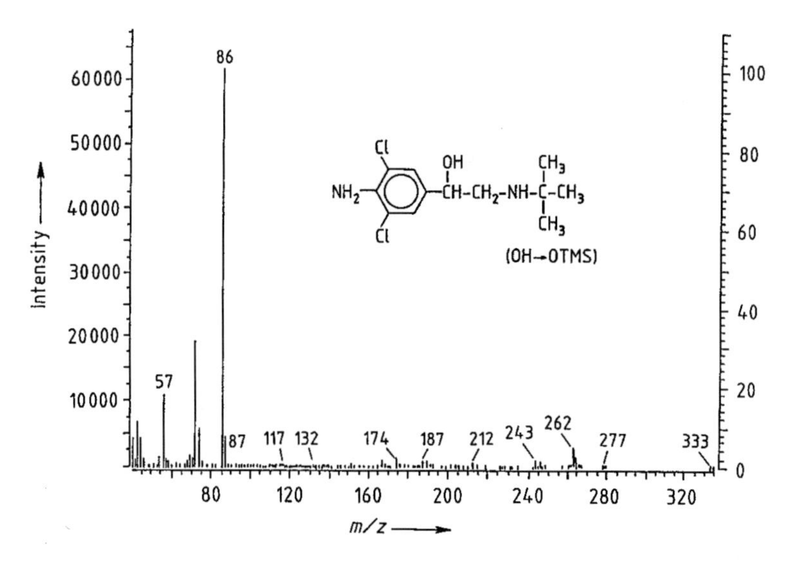

Figure 41. Electron ionization spectrum of TMS – clenbuterol

structed total ion chromatogram of a solvent mixture (Fig. 43 A) and the reconstructed chromatogram of three specific ions typical for C_3 (trimethyl or methylethyl), C_4 (tetramethyl, dimethylethyl, diethyl, etc.), C_5 (pentamethyl, trimethylethyl, dimethylpropyl, etc.) benzenes (Fig. 43 B – D). This is called *mass chromatography*, and only data stored in the computer of the full-scan mode are used to provide selectivity. *Mass fragmentography* or *selected ion monitoring* (SIM), on the other hand, is a method to quantify target compounds with high selectivity and sensitivity. In this mode of operation, the voltages on the quadrupole rods are adjusted stepwise to detect only two, three, or four ions. The time the rods are on a particular voltage to let an ion pass is much longer than in the case of the full-scan mode, which results in enhanced sensitivity. As an illustration, in the previous case for the full-scan mode, each ion was seen for only 1/380 of a second. Suppose clenbuterol has to be quantified. By selecting the ions 259 and 349 only (Fig. 42), each ion is detected in a 1-s scan for 1/2 s. Compared to the full-scan mode, the residence time is 190 times higher; thus the sensitivity increases by a factor 190. With modern CGC – MS systems, picogram amounts can easily be quantified when operated in the ion-monitoring mode. An interesting feature of SIM is that internal standards can be selected that chemically, physically, and chromatographically behave exactly the same as the compound to be measured. In the case of clenbuterol, for example, the best internal standard is D_6-clenbuterol with typical ions at *m/z* 265 and 355. In routine analysis, D_6-clenbuterol is added to the sample (i.e., urine) before any pretreatment. Sample preparation and cleanup are then performed, and the four ions 259 and 349 for clenbuterol and 265 and 355 for the internal standard D_6-clenbuterol are recorded. Quantitation is very easy.

11.6.4.2. Fourier Transform Infrared Spectroscopy

In the combination CGC – FTIR, IR spectra of the eluting peaks are recorded as they emerge sequentially from the column outlet. The eluate is introduced in a light pipe where compounds absorb radiation at well-defined frequencies (i.e., frequencies characteristic of the bonds within a molecule). The absorption is sensed and recorded. The sensitivity of FTIR depends on the functionalities within a molecule, and for strong IR absorbers, 1 ng yields a good spectrum. Current FTIRs include computers, and the recorded spectra can be compared with a library of spectra to aid in compound identification. Monitoring of specific wavelengths allows the elucidation of classes of compounds (e.g., aldehydes, ketones, alcohols, ethers). On-line coupling of CGC–FTIR and CGC – MS can be realized in series since the FTIR is nondestructive. The IR spectra are complementary in nature to MS spectra, especially for the elucidation of isomers for which MS is not very

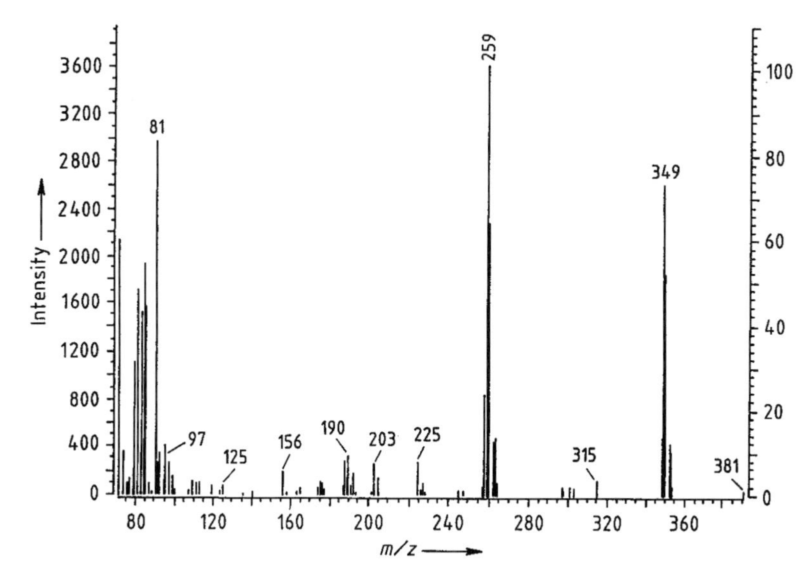

Figure 42. Chemical ionization spectrum (reagent gas isobutane) of TMS – clenbuterol

informative. A typical example is the determination of impurities in α-angelica lactone. Figure 44 shows the total response chromatogram obtained with FTIR and the total ion chromatogram, recorded by the mass spectrometer. Both chromatograms correspond very well, and at any time MS and IR data can be recalled and compared. The mass spectra of peak 1 and 2 are very similar and do not allow the differentiation of structural differences. The infrared spectra of peak 1 and 2 are compared in Figure 45. Some important differences are noted. The shift of the carbonyl band from 1833 cm^{-1} (peak 1) to 1806 cm^{-1} (peak 2) indicates that peak 2 corresponds to β-angelica lactone (α, β-unsaturated instead of β, γ-unsaturated).

11.6.4.3. Atomic Emission Detection

One of the latest developments in hyphenated techniques is the atomic emission detector. The availability of a bench-top model CGC – AED enables the use of this powerful technique in routine analysis. Detectabilities of this element-specific detector are of the order of 0.1 pg/s for organometallic compounds, 0.2 pg/s for carbon (more sensitive than FID), 1 pg/s for sulfur, and 15 pg/s for nitrogen, to mention only a few. The power of the technique lies in its supreme selectivity: all elements can be detected selectively. As opposed to ECD, AED allows differentiation between various halogenated organic compounds (e.g., fluoro, chloro, and bromo components). Multielement analysis can be carried out by simple preselection of the atoms to be monitored. In AED, the solutes eluting from the column are atomized in a high-energy source; the resulting atoms are in the excited state and emit light as they return to the ground state. The various wavelengths of the emitted light are then dispersed in a spectrometer and measured by using a photodiode array. Each element has a typical emission spectrum, with emission lines commonly occurring in clusters. The relative intensities of lines within these clusters are constant. Figure 46 shows the emission spectrum of a sulfur-containing compound, and the 180.7-, 182.0-, and 182.6-nm cluster serves as conclusive proof of the presence of sulfur in a peak.

Multiwavelength detection became possible through the introduction of diode array technology. Quantitative treatment of the data obtained in a multielement analysis moreover allows the calculation of empirical formulas, yielding information that is complementary to and helpful in the interpretation of mass spectral data. Quantitation is simplified considerably, since calibration of the detector no longer depends on the type of components to be quantified. Nontoxic compounds can thus be used to quantify harmful chemicals. Figure 47 illustrates the sensitivity of the detector for organometallic compounds. Organolead halides

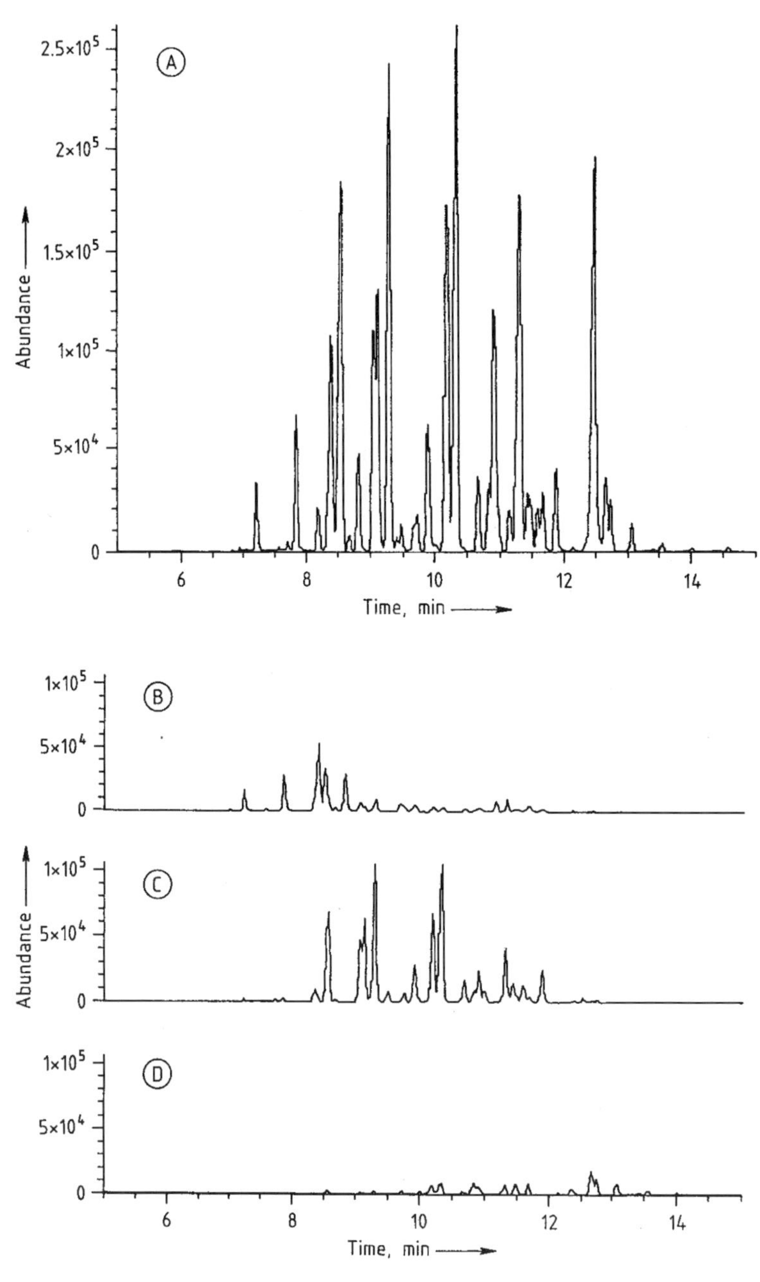

Figure 43. Illustration of mass chromatography
A) Solvent mixture; B) C_3-Benzene; C) C_4-Benzene; D) C_5-Benzene

are derivatized with butylmagnesium bromide into tetraalkylated solutes, which are amenable to CGC analysis. By selecting a wavelength of 406 nm, high selectivity and sensitivity (10 pg per compound) are guaranteed. Another important application concerns direct monitoring of oxygenates in petroleum products. Figure 48 shows the direct analysis of a light naphtha containing low concentrations (ppm) of oxygenated compounds. The concentrations for methanol and ethanol are 100 and 50 ppm, respectively.

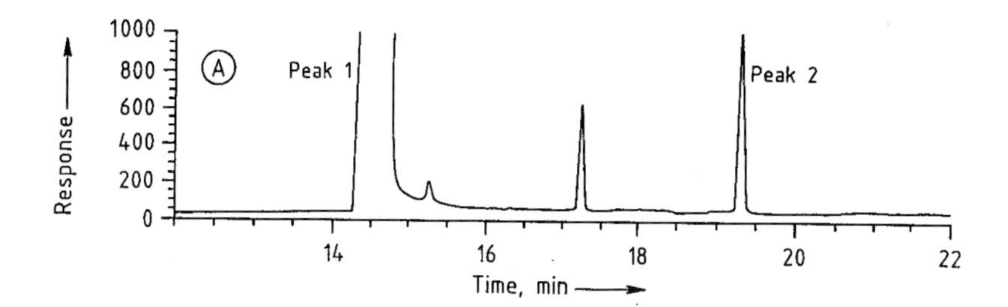

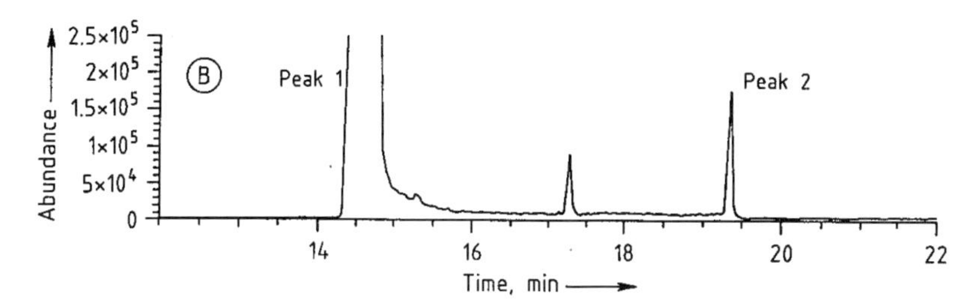

Figure 44. Capillary GC – FTIR – MS chromatograms of α-angelica lactone
A) Total response chromatogram (FTIR); B) Total ion chromatogram (MS)

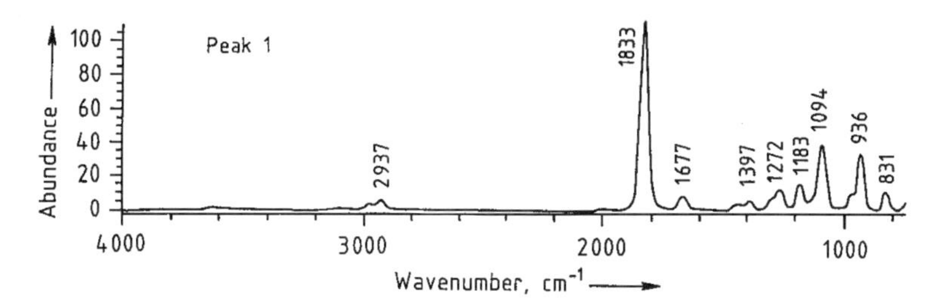

Figure 45. Infrared spectra of peak 1 and 2 of chromatogram in Figure 44

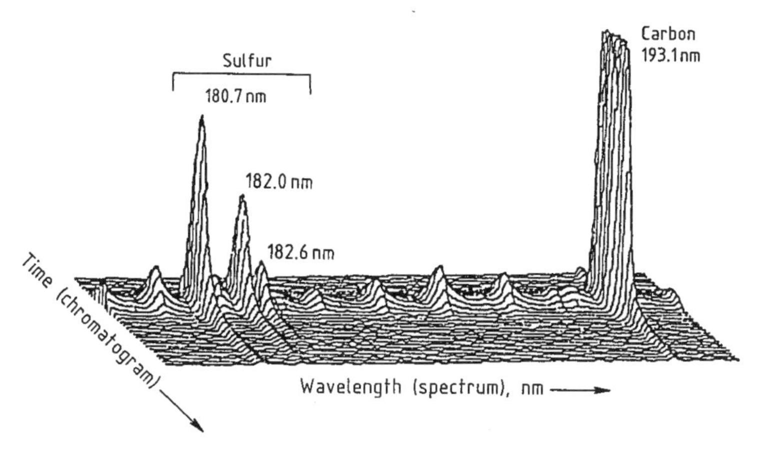

Figure 46. Emission spectrum of a sulfur-containing compound

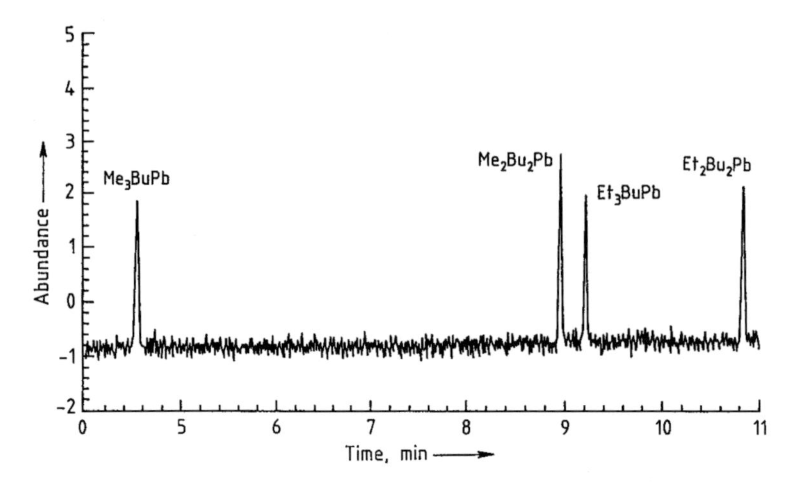

Figure 47. AED analysis of organolead compounds

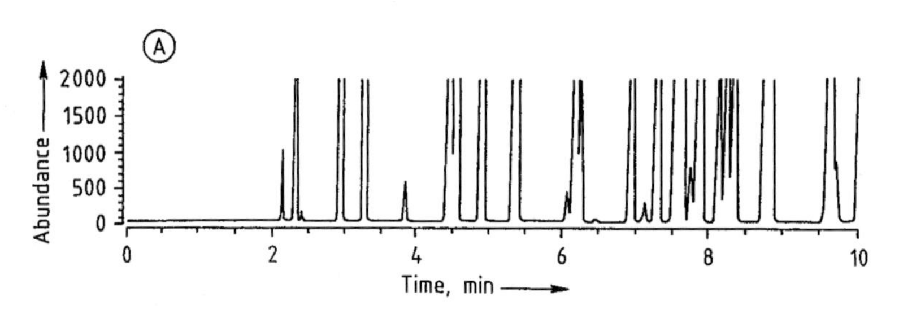

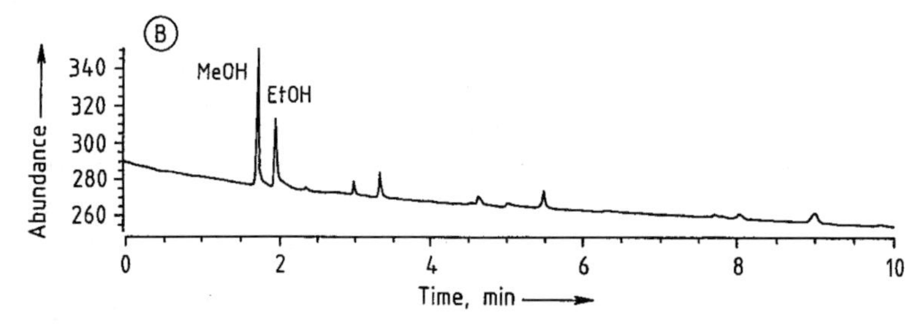

Figure 48. AED analysis of light naphtha in the C (A) and the O (B) emission line

11.7. Practical Considerations in Qualitative and Quantitative Analysis

A chromatogram provides information on the complexity (number of peaks), identity (retention time), and quantity (peak area or height) of the components in a mixture. This information can be considered suspect if the quality of separation is not optimal. Capillary columns have greatly enhanced the use of gas chromatography as a qualitative and quantitative tool.

11.7.1. Qualitative Analysis

Modern capillary GC is characterized by very good precision in retention time, and this allows the use of retention indices for peak identification. In a retention index system, the retention behavior of a particular solute is expressed in a uniform scale determined by a series of closely related standard substances. In the retention index scale developed by Kovats [61] for isothermal separations and by Van den Dool and Kratz [62] for temperature-programmed analyses, the fixed

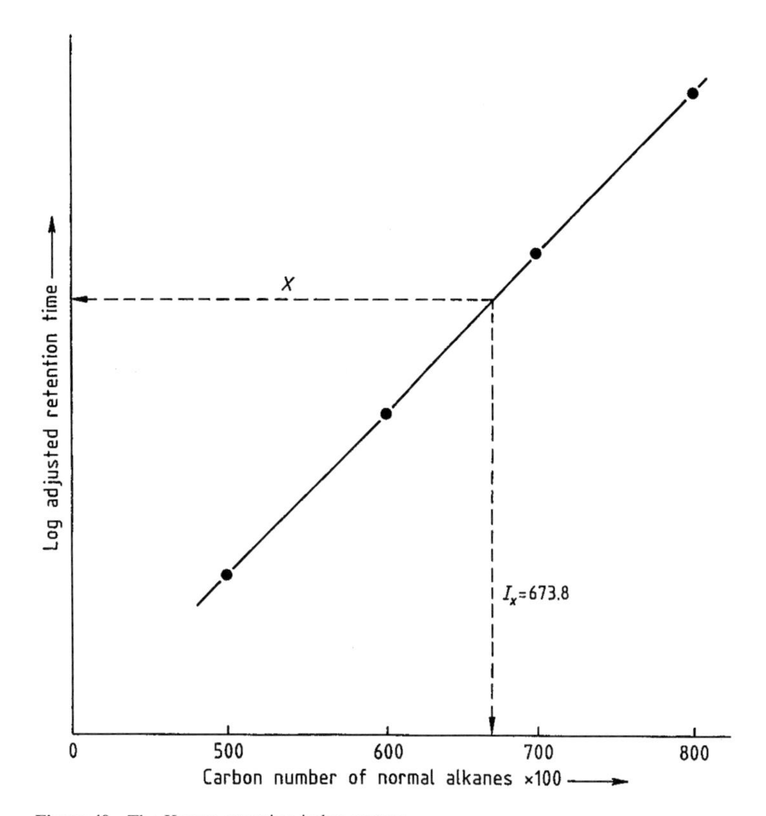

Figure 49. The Kovats retention index system

points of the scale are normal alkanes. The Kovats retention index system is illustrated in Figure 49. Compound X elutes between the hydrocarbons hexane and heptane with carbon numbers 6 and 7, respectively. For convenience, the carbon numbers are multiplied by 100, thus 600 and 700. The retention index I_x of compound X is 673.8. Indices are not graphically measured but calculated by Equation (7.1)

$$I_x = 100z + 100 \frac{\log t'_{Rx} - \log t'_{Rz}}{\log t'_{Rz+1} - \log t'_{Rz}} \qquad (7:1)$$

where z is the carbon number of the n-alkane eluting before X; $z+1$ is the carbon number of the n-alkane eluting after X; and t'_R is the adjusted retention time.

To perform the interpolation, the adjusted retention times of the n-alkanes must be known. This is often done by analyzing a separate mixture containing n-alkanes. This requires, however, automated injection. For manual injection, n-alkanes are added to the sample. In Equation (7.1), logarithmic interpolation is necessary because the adjusted retention times increase exponentially with carbon number (Eq. 4.1). In linear temperature-programmed runs, the retention times of a homologous series (i.e., the n-alkanes) increase linearly with carbon number, and the equation for temperature-programmed analysis is simplified to

$$I_x = 100z + 100 \frac{t_{Rx} - t_{Rz}}{t_{Rz+1} - t_{Rz}} \qquad (:2)$$

Unadjusted retention times may be used. The isothermal indices depend on the stationary phase and film thickness and on the temperature. Specification of those variables is a must. The programmed indices, often called linear retention indices, require specification of the temperature profile as well. Once determined, the indices can be used to identify compounds from libraries with tabulated indices [63], [64]. The general utility of those libraries for identification is questionable. Indices are best used within a laboratory for specific samples (i.e., petroleum products, essential oils, steroids, etc.). Greater confidence in identification is obtained by measuring indices on more than one stationary phase. Unequivocal compound elucidation requires spectroscopic characterization as provided by mass spectroscopy, Fourier transform infrared spectroscopy, and atomic emission detection.

Recently the concept of retention time locking was introduced and this opens new perspectives in capillary GC for solute elucidation (see Section 11.10.2).

11.7.2. Quantitative Analysis

Four techniques are commonly used in quantitative analysis: the normalization method, the external standard method, the internal standard method, and the standard addition method. Whatever method is used, the accuracy often depends on the sample preparation and on the injection technique. Nowadays these are two main sources of error in quantitative analysis. The quantitative results produced by PCGC and CGC are comparable.

11.8. Coupled Systems

Coupled systems include multidimensional and multimodal systems. *Multidimensional chromatography* involves two columns in series preferably two capillary columns, with different selectivity or sample capacity, to optimize the selectivity of some compounds of interest in complex profiles or to provide an enrichment of relevant fractions. In *multimodal systems*, two chromatographic methods or eventually a sample preparation unit and a chromatographic method are coupled in series. Coupled systems that received much interest in recent years are multidimensional CGC (MDCGC), the combination of high-performance liquid chromatography with CGC (HPLC – CGC) and the on- or off-line combination of supercritical fluid extraction with CGC (SFE – CGC). Multidimensional and multimodal techniques in chromatography are described in detail in [65].

11.8.1. Multidimensional Capillary GC

For highly complex samples, the separation power of a single capillary column is insufficient to achieve complete resolution for the compounds of interest. Even after optimization of the selectivity, important compounds will still coelute, since the better separation of one pair of compounds is likely to be counteracted by the overlapping of another pair of compounds present in the sample. The only solution is the combination of more than one column system. In multidimensional CGC, a group of compounds that has not been separated on a first column is transferred (heartcut) to a second column, where complete resolution is achieved. Two operational modes can be distinguished. If the transferred fraction is coldtrapped before analysis on the second column (intermediate trapping), the selectivities of both columns are decoupled. The elution of the transferred solutes will be independent of the selectivity of the first column and will be the same as if the pure sample compounds were analyzed on the second column only. Alternatively, when no intermediate trapping is applied, the elution pattern observed after analysis on the second column will be the same as the elution pattern for analysis on a mixed-phase column or a coupled column system with an intermediate selectivity between the selectivities of both columns. Both principles are illustrated in Figure 50. Peppermint oil was analyzed on a methylsilicone capillary column, and the separation of the menthone – menthol fraction (peak 1 – 7) is incomplete (Fig. 50 A). This fraction was heartcut (Fig. 50 B) and further analyzed on a poly(ethylene glycol) capillary column. With intermediate trapping (Fig. 50 C) the separation is drastically improved, but the highest resolution, in this case, is without intermediate trapping (Fig. 50 D). This is an illustration of the power of multidimensional CGC in combination with selectivity tuning.

Developments in multidimensional CGC diverge in two main directions: the coupling of narrow to narrowbore capillaries to enhance resolution, on the one hand, and the coupling of widebore precolumns to narrowbore capillaries as a selective sample introduction method, on the other. Some inlet devices, such as the programmed temperature vaporization injector operated in the solvent vent mode, can also be classified as multidimensional systems. The solvent is removed through the split vent, while the solutes of interest remain in the liner. The multidimensional approach with selective sampling is of considerable interest in routine analysis. It allows preseparation of relatively large amounts of sample on the precolumn operated under constant-flow conditions. Fine-tuning of the separation can be achieved on the analytical column, operated in the constant-

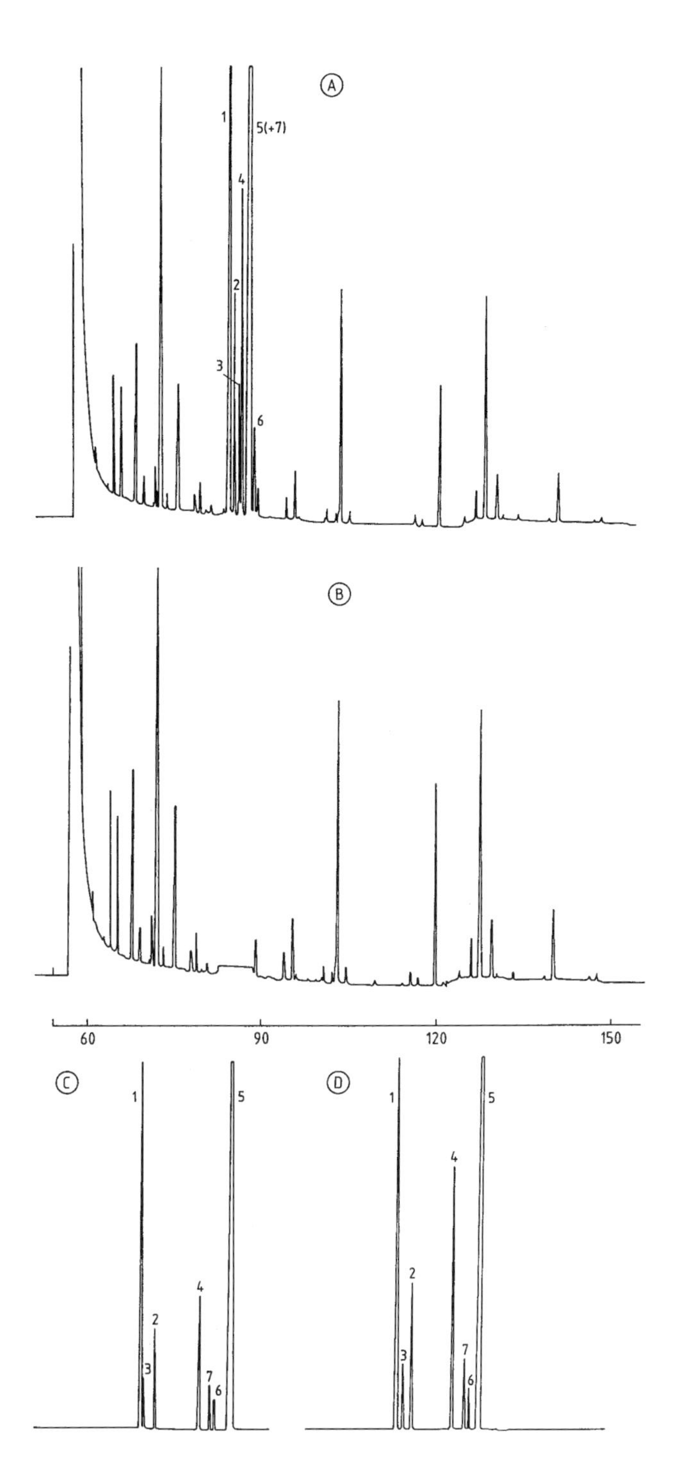

Figure 50. Multidimensional CGC analysis of pepermint oil, two column system constant pressure – constant pressure
A) Separation on MeSi; B) Analysis on MeSi after heartcut ; C) Analysis of heartcut fraction on column B with intermediate trapping; D) Analysis of heartcut fraction on column B without intermediate trapping (selectivity tuning)
Column A: 20 m L×0.32 mm *i.d.*; 0.24 µm d_f MeSi; Column B: 20 m L×0.32 mm *i.d.*; 0.24 µm d_f high-M_r PEG
Temperature: 60 °C (2 min) to 180 °C at 3 °C/min; split injection 1/50; carrier gas hydrogen; FID

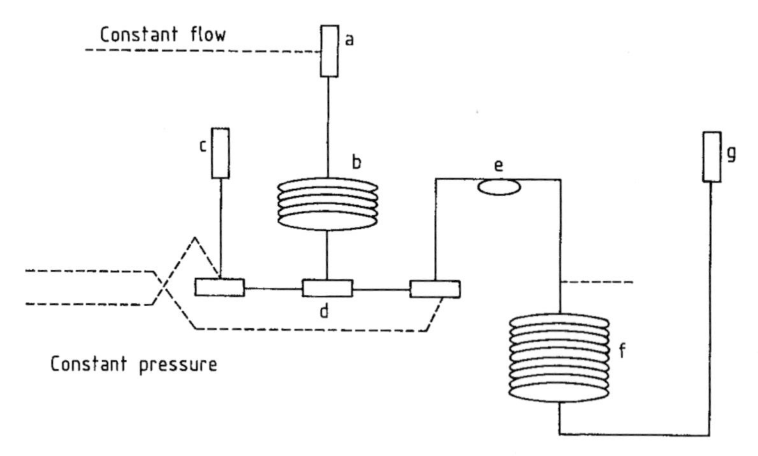

Figure 51. MDCGC system constant flow – constant pressure
a) Injector; b) Precolumn; c) Detector 1; d) Switching device; e) Cold trap; f) Analytical column; g) Detector 2

pressure mode. The system is easy to operate and can be installed without problems in every GC instrument. Systems based on constant pressure – constant pressure and constant flow – constant flow offer more flexibility because column dimensions are less critical than in the constant flow – constant pressure system; however, they are expensive and require time-consuming optimization procedures when new applications are carried out. The possibilities of multidimensional capillary GC as a sampling method are illustrated with the analysis of volatiles emitted by living plants [66]. The atmosphere surrounding a plant (headspace) is sucked in an open tubular trap (3 m in length, 530 µm *i.d.*) coated with a 15-µm film of methylsilicone where the volatiles are retained by the normal chromatographic partitioning process at ambient temperature. The advantages over traps filled with adsorbents are that breakthrough volumes can be calculated, the inertness is high, and thermal desorption of polar compounds is feasible. After sampling, the trap is installed as precolumn in a multidimensional CGC system (Fig. 51). The volatiles are thermally desorbed, reconcentrated in a cold trap (e), and then analyzed on-line on the analytical capillary column (f). Figure 52 shows the headspace CGC – MS analyses of a healthy and a fungus-infected *Mentha arvensis* plant. Menthol (peak 15) is the main compound in the healthy plant (Fig. 52 A), and the highly volatile fraction is relatively poor. In the infected plant (Fig. 52 B), menthyl acetate (peak 21) is the main compound, while the relative percentage of menthone (peak 13) is increased and the highly volatile fraction is more abundant. This hyphenated combination live headspace sampling – multidimensional capillary GC – MS provides interesting results in phytochemical research. In 1992, a more sophisticated system, PTV injection—MDCGC with three columns—MS was described for the analysis of dioxins and furans [54].

11.8.2. Multimodal High-Performance Liquid Chromatography – Capillary GC

The on-line coupling of HPLC to capillary GC has received much attention since the mid-1980s. Pioneering work was performed by K. Grob, Jr. [67]. The coupling of HPLC to capillary GC can be considered as an on-line sample preparation and enrichment step. Preseparation can be carried out according to molecular mass (SEC), functionality (straight-phase LC), or hydrophobicity (reversed-phase LC). Similar results are obtained with off-line techniques (i.e., solid-phase extraction or LC off-line fractionation), but advantages of the on-line approach include easier automation, fewer artifacts, and higher sensitivity because the total elution volume of the fraction of interest is transferred to the capillary GC column.

Recently a fully automated and flexible on-line LC – GC system has been described [68]. The system is based on standard LC and capillary GC instrumentation. The on-line LC – GC system was evaluated for the determination of pesticides in orange oil. The essential oil was fractionated by normal phase LC, resulting in a separation according to polarity. The fraction containing pesticides was transferred and analyzed by capillary GC –

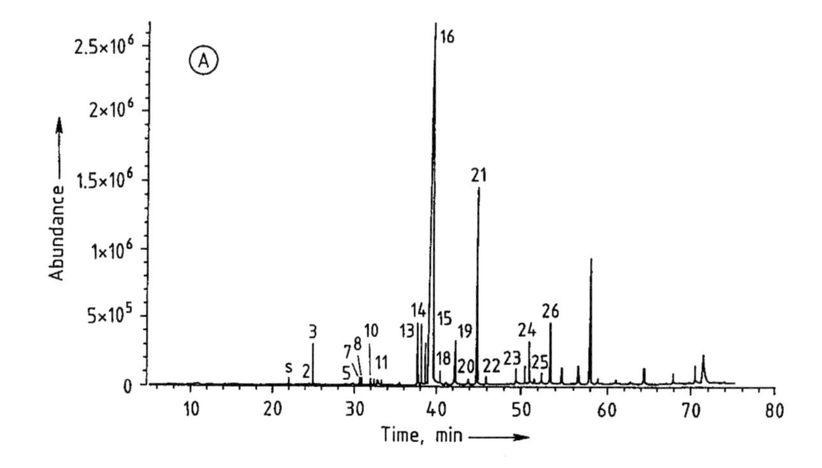

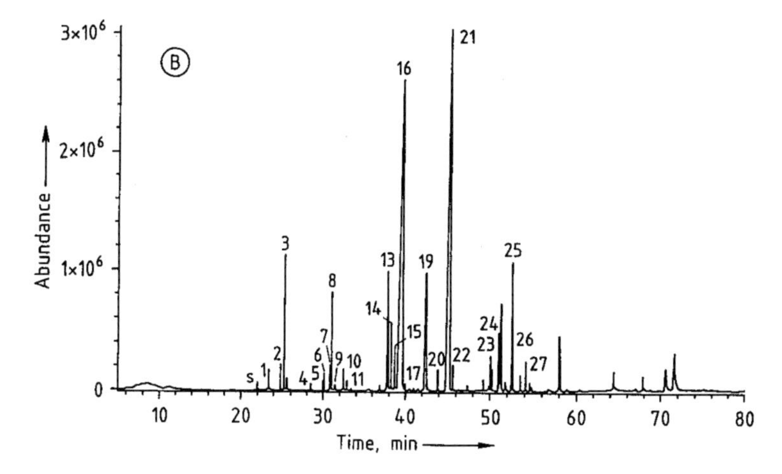

Figure 52. CGC pattern of the head space of a healthy (A) and an infected (B) *Mentha arvensis* plant
Thermal desorption and cryofocusing

NPD. The LC–GC interface is a modified large volume sampler (Gerstel, Mulheim a/d Ruhr, Germany). The heart of the system is a specially designed flow cell that replaces a normal vial. The mobile phase leaving the LC detector is directed via a capillary tube with well-defined dimensions, in a T-shaped flow cell. The cell is equipped with a septumless sampling head through which a syringe needle can be introduced. The sampler is completely computer controlled. To transfer a selected LC fraction, the transfer start and stop times, measured on the LC chromatogram, are introduced into the controller software. The time delay between the LC detector and the flow cell is automatically calculated from the connecting tubing dimensions and the LC flow rate. At the time the LC fraction of interest passes through the flow cell, the syringe needle penetrates the septumless head and samples the LC fraction at a speed equal to the LC flow rate. Volumes of up to 2 mL can be collected in the syringe. After collection, the needle is withdrawn from the flow cell which rotates away from the PTV inlet and a large volume injection is made using the PTV in the solvent vent mode. Depending on the fraction volume and solvent type, the sample introduction parameters (inlet temperature, vent flow, vent time, injection speed,...) are calculated by the PTV calculator program. For fractionation of the orange oil sample, the following HPLC parameters were used: column 250 mm × 4.6 mm i.d. × 10 μm Lichrospher 100 DIOL, injection volume 20 μL, mobile phase in a gradient from 100 % hexane for 10 min, to 40 % isopropanol at 20 min (2 min hold) at a

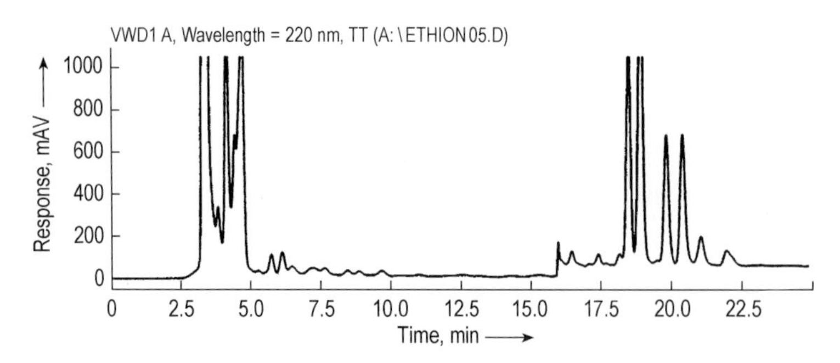

Figure 53. LC fractionation of orange oil on Lichrosorb 100 DIOL

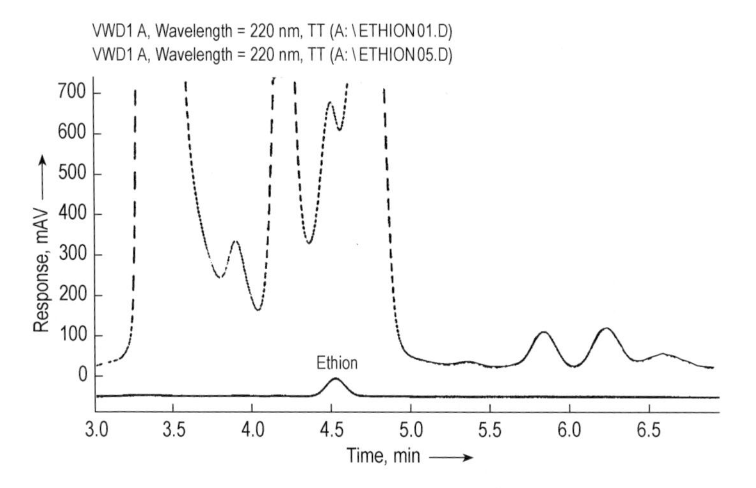

Figure 54. Overlay of relevant parts of LC chromatograms of a 1 ppm ethion standard (bottom trace) and orange oil (top trace)

flow rate of 1 mL/min and UV detection at 220 nm. The fraction eluting from 4.4–4.9 min (volume = 0.5 mL) was automatically transferred to the GC inlet. The LC–GC interface was programmed to take the sample at a 1000 μL/min sampling speed (the same as the LC flow rate). The complete fraction of 500 μL was injected in the PTV inlet at 250 μL/min. This injection speed corresponds to the injection speed calculated by the PTV software program. Capillary GC–NPD analysis was performed on a 30 m × 0.25 mm i.d. × 0.25 μm MeSi column and the oven was programmed from 70 °C (2 min) to 150 °C at 25 °C/min and then to 280 °C at 8 °C/min (10 min). The detector was set at 320 °C with 3 mL/min hydrogen, 80 mL/min air, and 30 mL/min helium make-up flow.

The LC profile for the orange essential is shown in Figure 53. The apolar mono- and sesquiterpenes elute first, followed by the terpenoids and after 16 min also the flavanoids are eluted. These last compounds in particular give most interference in GC analysis because they have similar molecular weights to the pesticides. Using the same conditions, ethion and chlorpyriphos elute at 4.6–4.7 min. Figure 54 shows a comparison of the analysis of a 1 ppm ethion reference sample and the essential oil analysis.

The fraction eluting from 4.4 to 4.9 min was transferred to the CGC-NPD and analyzed. Figure 55 shows the resulting GC–NPD chromatogram and both ethion and chlorpyriphos (t_R 15.79 min) are detected. The chromatograms are very clean and no interferences are present. This demonstrates the excellent selectivity of the LC–GC combination. The concentrations of ethion and chlorpyriphos were calculated by external standard analysis and are 1.9 ppm for chlorpyriphos and 0.8 ppm for ethion.

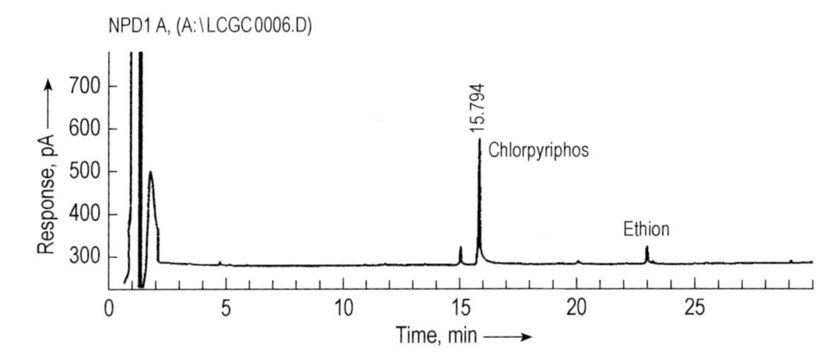

Figure 55. LC – Capillary GC – NPD analysis of orange oil

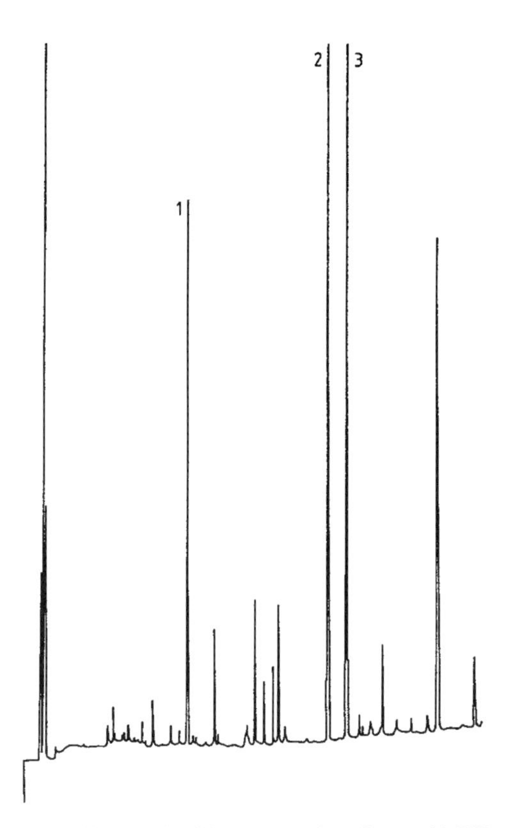

Figure 56. Analysis of SFE extract of a sediment with ECD detection
Peaks: (1) aldrin, (2) dieldrin, (3) endrin

11.8.3. Multimodal Supercritical Fluid Extraction – Capillary GC

Because supercritical fluids have excellent characteristics for the rapid extraction of organic compounds, combining supercritical fluid extraction with CGC is an interesting approach to the enrichment of organic solvents [69], [70]. Since the density of supercritical fluids—usually carbon dioxide—can be varied (pressure, temperature), a high degree of selectivity can be introduced in the sample preparation procedure (i.e., prior to the analytical system). The selectivity can be further enhanced by making use of modifiers that may be polar (e.g., CO_2 – MeOH) or nonpolar (CO_2 – hexane). The main applications of SFE – CGC are related to solid samples, such as soil or sediment samples, and plant materials. For aqueous solutions (e.g., wastewater, urine), samples are enriched by solid-phase extraction. The extraction cartridges are then submitted to SFE – CGC, thus providing all the advantages of SFE (i.e., selective extraction and ease of automation). There is some controversy in the literature whether on-line SFE – CGC is the best approach to the application of analytical supercritical fluid extraction. Indeed off-line SFE has some advantages, and the use of robotics allows automated transfer of the sample to the capillary GC instrument. Selection of on-line or off-line depends mainly on the specific problem. Off-line is preferred for method development and on-line for routine analysis. On-line can, however, be recommended only when the matrix from which the solutes have to be extracted is always the same (e.g., in chemical composition, particle size, pore size). Moreover, the literature on SFE extraction times and recoveries is often incorrect because spiked samples are analyzed. Applied to real contaminated samples, SFE is found to be much more difficult. Figure 56 shows the optimized off-line SFE – CGC analysis of pesticides from sediment: 500 mg sediment was placed in the extraction vessel, and 250 µL of methanol was added on top of the sediment. SF extraction was performed at 60 °C with pure CO_2

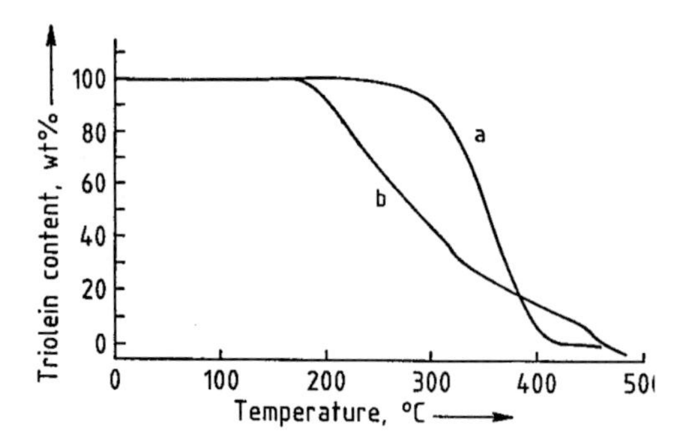

Figure 57. Thermogravimetric analysis of triolein
Ramp from 40 – 500 °C at 5 °C/min
a) Pure nitrogen; b) Air

at 35 MPa for 30 min. The concentration of the main compound dieldrin is 20 ppb. Classical Soxhlet extraction gave the same concentration.

11.9. Applicability

11.9.1. Solute Thermal Stability

Solutes subjected to GC must be thermally stable. The thermostability of organic compounds depends strongly on their nature, on the activity of the environment in which thermal stability is measured, and on the thermal stress given to the solutes. Capillary GC nowadays is performed in a completely inert system, i.e., highly pure carrier gas, purified stationary phases, fused silica with less than 0.1 ppm metal traces and specially deactivated, etc. Moreover, thermal stress can be avoided by applying cool on-column injection. Lipids serve as a good illustration. When oils or fats are used in food preparation, they decompose (formation of aldehydes etc.) or polymerize (formation of dimers, trimers, etc.) as a function of time, which make them no longer useful for cooking. These alterations are caused by the presence of water and oxygen. When heated under inert conditions, fats and oils are stable and evaporate. Figure 57 compares the thermogravimetric profiles recorded for triolein ($M_r = 886$) under a stream of pure nitrogen gas (a) and air (b). In present state-of-the-art GC, the systems are operated under circumstances shown in curve a.

11.9.2. Solute Volatility

Volatility is related to the vapor pressure (boiling point) of the compounds. Polydimethylsiloxanes with molecular masses as high as 5000 are volatile enough to be analyzed by GC, whereas poly(ethylene glycols) with M_r of ca. 1500 are not volatile at all. Volatility is thus related not only to the molecular mass, but to the polarity of the functional groups. Polydimethylsiloxane (M_r 5000) has approximately 68 apolar dimethylsiloxane units, whereas poly(ethylene glycol) (M_r 1500) has 114 polar ethoxy units besides the terminating hydroxyl groups. Some thermolabile compounds, although able to be volatilized, undergo partial decomposition at high temperature. Derivatization can be employed to impart volatility and to yield a more stable product, thereby improving chromatographic performance and peak shape. Three general types of analytical derivatization reactions are used in GC: alkylation, acylation, and silylation [71], [72]. The ideal derivatization procedure must be simple, fast, and complete. Alkylation, acylation, and silylation are used to modify compounds containing functional groups with active hydrogens such as –COOH, –OH, and $–NH_2$. The tendency of these functional groups to form intermolecular hydrogen bonds affects the inherent volatility of compounds containing them, their interaction with the stationary phase, and their stability. Figure 58 shows the analysis of sucrose in molasses with capillary GC. Trehalose is used as internal standard, and the sample is derivatized into the oxime – trimethylsilyl derivatives.

11.9.3. Comparison of Gas Chromatography, Liquid Chromatography, and Supercritical Fluid Chromatography

When a scientist is faced with deciding which separation technique to use for a specific analytical problem, the primary question is, which chromatographic technique best suits this particular separation. The initial answer is dictated by the chemical nature of the analytes themselves. For the analysis of volatile components, gas chromatography is preferred. For analyzing strongly polar or ionic compounds, liquid chromatography may be the best choice. In some cases, supercritical fluid chromatography has distinct advantages over GC and LC. However, for many applications—

Table 5. Calculated percent composition of steroid esters *

Compound	HPLC		CSFC	CGC on-column	Composition in mixture, %
	UV	MD **			
Testosterone propionate	15.13	11.93	10.98	11.98	12.00
Testosterone isocaproate	27.77	23.68	23.57	25.20	24.00
Testosterone decanoate	33.25	40.43	44.33	39.99	40.00
Testosterone phenylpropionate	23.38	23.96	21.11	23.83	24.00
Mean RSD, %	1.43	1.08	2.58	0.58	
Analysis time, min	22.50	25.00	21.50	21.50	

* All values calculated from uncorrected peak areas with response factors = 1.0; mean values of five experiments.
** MD = mass detector.

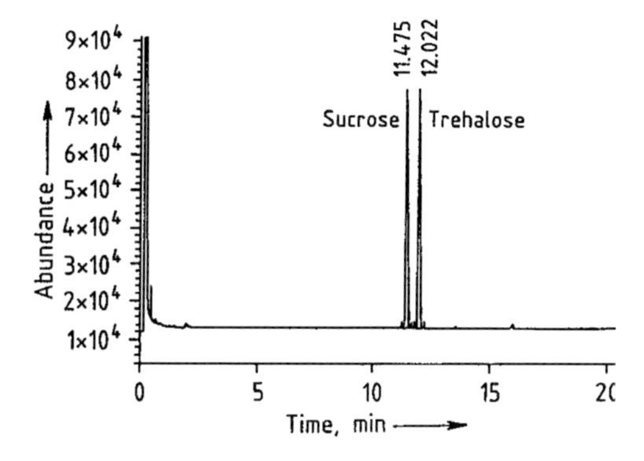

Figure 58. Capillary GC analysis of sugar derivatives (oxime–TMS)
Column: 10 m L×0.53 mm *i.d.*; 0.1 μm d_f MeSi; temperature: 80–300 °C at 10 °C/min; cool on-column injection; carrier gas hydrogen 4 mL/min; FID

such as analyses of polycyclic aromatic hydrocarbons, polychlorinated biphenyls, lipids, fatty acids, steroids, fat-soluble vitamins, mycotoxins, etc.,—several chromatographic techniques can be considered.

In such cases, strange as it may seem, the choice of a chromatographic technique is often biased on secondary considerations such as the analyst's know-how and skill or the availability of a particular instrument. This choice is usually justified with supplementary arguments, especially those concerning the simplicity of a technique. Recent technical developments in nearly all branches of chromatography, however, have largely overcome any "advantage of simplicity" (to some, the "disadvantage of complexity") of various techniques. The extended possibilities of automation further weaken arguments of this kind.

More important, especially when the analytical method must be applied routinely, is its cost. This will be governed mainly by the price of the instrument, columns, and mobile phases, as well as maintenance costs. In this respect, the chromatographic techniques can be ranked in increasing order of the prices of analysis: GC < LC < SFC.

From a practical viewpoint that disregards costs, the choice of method must be based on three fundamental aspects: separation efficiency, accuracy of the analytical method, and total required analysis time. For some applications, the specificity and sensitivity (detectability) of a technique are also important.

A study was conducted in which the performance of three techniques—high-performance LC (HPLC), capillary GC (CGC), and capillary SFC (CSFC)—was compared in the analysis of the percent composition of steroid esters in a pharmaceutical hormone formulation [73]. The study used standard chromatographic conditions, applicable in every laboratory, and the conditions were optimized so as to yield comparable analysis times of <30 min.

High-Performance Liquid Chromatography. In the pharmaceutical industry, analysis of steroid ester formulations is performed routinely by using HPLC. Figure 59 A shows the HPLC chromatogram of the sample mixture applying UV detection at 254 nm. The steroid esters are baseline separated in 24 min. The column plate number, calculated for testosterone decanoate, is 4500, which corresponds to 45 000 plates per meter. With UV detection, allowance must be made for different response factors. Calibration graphs, however, can be recorded only if pure compounds are used, which was not the case in this study. A response factor of 1.0 was, therefore, applied for the various steroid esters, giving the quantitative data listed in Table 5. Very often, in practice, unknown compounds are present in HPLC chromatograms, and quantitation is impossible using UV detection.

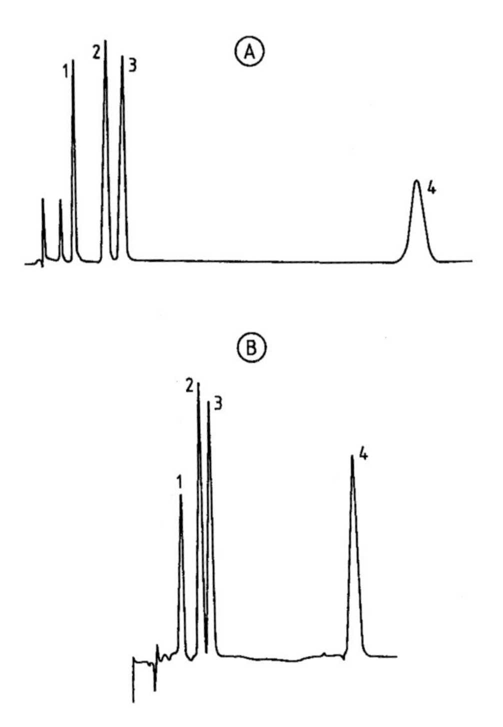

Figure 59. HPLC analysis of steroid esters
A) Column: 10 cm L×0.3 cm *i.d.*, 5 μm ODS, flow 0.5 mL/min methanol–water 85:15, UV at 254 nm; B) see A, light scattering detection
Peaks: (1) testosterone propionate, (2) testosterone isocaproate, (3) testosterone phenylpropionate, (4) testosterone decanoate

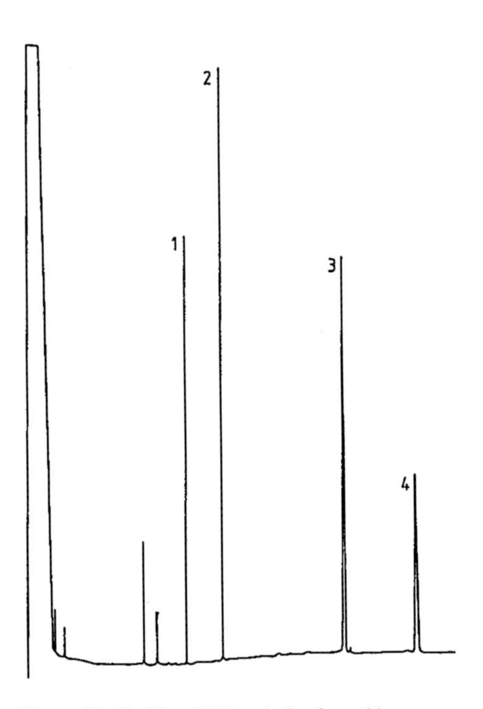

Figure 60. Capillary SFC analysis of steroid esters
Column: 20 m L×0.1 mm *i.d.*; 0.1 μm d_f MeSi; temperature: 130 °C; pressure 200 bar CO_2; FID
Peaks: (1) testosterone propionate, (2) testosterone isocaproate, (3) testosterone decanoate, (4) testosterone phenylpropionate

Calibration can be avoided by using universal HPLC detection systems, such as refractive index detection or light-scattering detection. Compared to UV, however, both of these detectors are less sensitive. Figure 59 B shows the analysis using light-scattering detection; the sample is five times more concentrated than in the UV trace in Figure 59 A. The chromatographic efficiency was reduced by 20 %, to 3600 plates, because of a detector contribution to band broadening. The quantitative data, calculated using response factors of 1, are listed in Table 5 in the column labeled MD (mass detector). The calculated composition corresponds well to the known sample composition.

Capillary Gas Chromatography. Steroid esters are thermally stable. They can be analyzed easily by CGC. Figure 60 shows the analysis of the sample with cold on-column injection. The plate number for this conventional CGC column operated at 130 °C was 85 000 plates; analysis time was 22 min. The quantitative results (response factors = 1.0) are given in Table 5. The data for cold on-column injection correspond very well with the known composition of the sample, and % RSD is < 0.8 % ($n = 5$) for all the steroid esters.

Capillary Supercritical Fluid Chromatography. In the mid-1980s, considerable progress has been made in CSFC, and the technique is now accessible for routine chromatography. CSFC is claimed to have advantages compared to GC. Because of the high solvating power of CO_2, high molar mass compounds elute at much lower temperature, allowing better analysis of thermally labile compounds. Compared with HPLC, for the same column, the speed of analysis for SFC is roughly four times faster. In addition, because of the lower viscosity of the mobile phase, much longer columns can be used, increasing the effective plate number. Universal GC detectors can be applied, and combination with hyphenated techniques such as mass spectrometry or Fourier transform infrared spectrometry is easier. Analysis of the steroid ester sample on a 20 m×0.1 mm

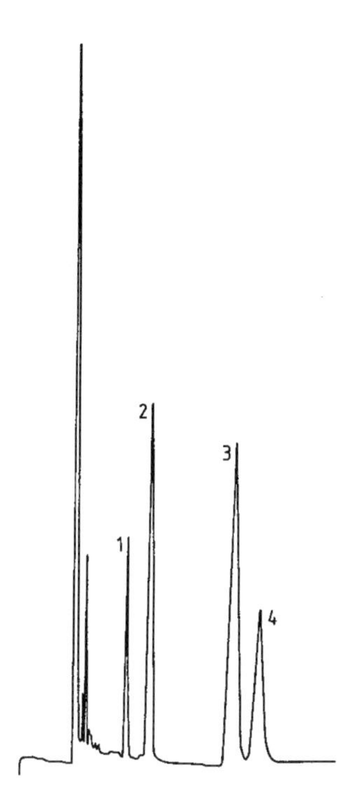

Figure 61. Capillary GC analysis of steroid esters
Column: 25 m L×0.25 mm *i.d.*; 0.15 μm d_f MePhSi; temperature: 60–300 °C at 10 °C/min; cool on-column injection; carrier gas hydrogen 15 psi (≈1 bar); FID
Peaks: see Figure 60

column is shown in Figure 61. In this run at constant density, the plate number is only 5000. This illustrates one of the disadvantages of CSFC. Theoretically the column should offer 150 000 plates, but at a mobile phase velocity of only 0.2–0.3 cm/s. This corresponds to an unretained peak time of 2.5 h, and an analysis time for a compound eluting with a k value of 3.5 (peak 3, testosterone decanoate) of 7.3 h. Such analysis times are, of course, unacceptable. Therefore the column is operated far above the optimal velocity (Figure 61, at 6 cm/s). The slope of the $H-u$ curve is very steep for the 20-m column, and the plate height at 6 cm/s is of the order of 6 mm. Much faster separations can be obtained on short capillary columns with internal diameters of ≈20 μm, but because of technical problems (especially with sample introduction), quantitation is very difficult. Quantitation of the steroid mixture with CSFC (Table 5) yields higher % RSD values and less accurate results than either CGC or HPLC. The reason for this lies mainly in the injection method (valve split discrimination). Compared to conven-

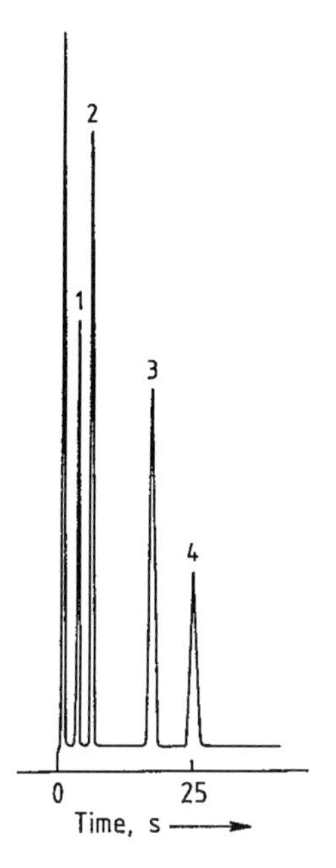

Figure 62. High-speed capillary GC analysis of steroid esters
Column: 2 m L×0.25 mm *i.d.*; 0.1 μm d_f MePhSi; temperature: 260 °C; split injection 1/100; carrier gas hydrogen 0.1 psi (≈700 Pa); FID

tional HPLC, the primary advantage of SFC is the use of the universal FID detector. The results presented in Table 5 show that all three techniques can be applied to analysis of the steroid sample. The difference in chromatographic behavior among HPLC, CGC, and CSFC is determined mainly by the mobile-phase properties such as density, diffusivity, viscosity, and polarity. When using dimensionless parameters, equations can be derived that are valid for the three different types of mobile phases and independent of particle size (packed columns) or column diameter (capillary columns) [74]. The reduced parameters will not be discussed in detail but show clearly, as far as efficiency and speed of analysis are concerned, no other chromatographic methods can compete with GC. Baseline separation of the steroid esters can be obtained on a 2-m capillary column offering 7000 plates in 25 s (Figure 62). This is not a plea for gas chromatography. It is just an example to illustrate the possibilities of different chromatographic techniques for the application selected. Often, however, one might be surprised to see the enormous effort spent in trying a separation with a particular technique, while the use of another technique seems straightforward. Each chromatographic technique has its own possibili-

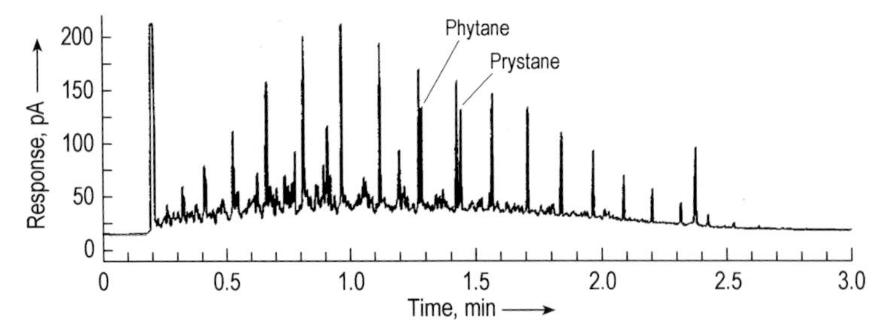

Figure 63. Separation of diesel oil in 2.5 min on standard GC instrumentation

ties, and a good understanding of their performance is a prerequisite before selection for a particular application can be justified.

11.10. Recent and Future Developments

11.10.1. Fast High Resolution Capillary GC

Since the introduction of gas chromatography, there has been an ongoing interest in improving analysis speed. Increasing analysis speed in capillary GC in the first instance is dictated by the problem at hand, the primary objective being the complete separation of the compounds in a mixture. In a large number of cases, the plate number of a capillary column is too high for a given separation problem and the resolution may be impaired. Typical ways to shorten the analysis time in this case are decreasing the column length or increasing the carrier gas flow rate far above the optimum. Recent alternatives include the use of multichannel columns or multicapillaries and flash GC.

Increasing analysis speed for complex profiles without impairing resolution can only be realized by reduction of the internal diameter and the length of the capillary column. A 10 m × 0.1 mm i.d. column offers the same resolution as a 25 m × 0.25 mm i.d. column. Because the column is 2.5 times shorter, the analysis time is reduced drastically. Moreover, since the optimum carrier gas velocity is higher and the H-u curves are flatter for narrow bore columns, higher average carrier gas velocities can be used without loss of resolution. Presently, capillary columns with internal diameters in the order of 100 μm are in the picture for routine operation because state-of-the-art capillary GC instrumentation allows installation of such small diameter columns. This is illustrated with the analysis of a diesel sample with baseline separation of the biomarkers pristane and phytane from the preceeding normal hydrocarbons on a standard GC instrument (Fig. 63). The diesel sample was diluted 10 times in cyclohexane and 1 μL was injected in the split mode (1/1000 split ratio). The column was a 10 m × 0.1 mm i.d. × 0.1 μm MeSi. The oven was programmed from 100 °C to 325 °C at 75 °C/min and held for 1 min (4 min program). FID detection with a signal data acquisition rate of 100 Hz was used.

An important obstacle for the use of narrow bore columns, however, is method development and validation. Using narrow bore columns, different operational conditions (inlet pressure, split ratio, temperature program) have to be used. Since little information is yet available on the use of fast capillary GC, the transfer of standard validated operating procedures developed for conventional capillary columns into operating procedures for narrow bore columns might be difficult and will definitely hamper their use in a routine environment. In this respect, the development of method translation software [75], [76] was very helpful for translating a standard operating procedure for a conventional column (whatever its dimensions and stationary phase film thickness) to an operating procedure for a narrow bore column (coated with the same stationary phase). After performing the analyses on the standard column, the optimized conditions are introduced in the method translation program and all operational conditions for the new column are calculated in order to obtain the same resolution. The gain in analysis time is also predicted. The method translation software is available free of charge from the Agilent website. The principle is illustrated with the analysis of an essential oil. Quality control of essential oil samples is routinely done on a

Table 6. Experimental conditions for essential oil analysis on a conventional and on a narrow bore column. The conditions in italic for the narrow bore column have been calculated by the method translation software. The standard operating conditions (SOP) on the conventional column apply broader conditions than needed for the essential oil analysed here

Column	60 m × 0.25 mm i.d. × 1 μm HP-1	20 m × 0.1 mm i.d. × 0.4 μm HP-1
Injection	split, 1 μL, 1/50 split ratio, 250 °C	split, 1 μL, *1/500* split ratio, 250 °C
Carrier type	helium	hydrogen
Carrier pressure	209 kPa, constant pressure	*411 kPa*, constant pressure
Carrier flow	1.74 mL/min at 50 °C	*0.87 mL/min* at 50 °C
Velocity	29.5 cm/s (average at 50 °C)	*58 cm/s* (average at 50 °C)
Hold-up time	3.39 min	*0.57 min*
Oven program	50 °C – 2 °C/min – 275 °C – 40 min	*50 °C – 11.88 °C/min – 275 °C – 7 min*
Detection	FID	FID
Signal data rate	10 Hz	50 Hz
Analysis time	152.5 min	*25.94 min*

50 – 60 m × 0.25 mm i.d. column using a slow (2 °C/min) temperature program, which results in analysis times in the order of 2 – 3 h. The same peak capacity can be obtained on shorter narrow bore capillary columns, but, although analysis times can be reduced drastically, quality control laboratories hesitate to use this approach because changing the column dimensions implies different operational conditions, which results in different selectivities. Details of well-known fingerprints can be lost. The application of the method translation software allows the translation from conventional to narrow bore columns with hardly any change in resolution, selectivity, and thus fingerprint. This is illustrated with the analysis of nutmeg oil.

The analysis was first performed on a "standard" column used for detailed essential oil profiling (60 m × 0.25 mm i.d. × 1 μm MeSi). The operational conditions optimized for routine QC were applied. Secondly, the analyses were repeated on a 20 m × 0.1 mm i.d. × 0.4 μm MeSi. The operational conditions for the narrow bore column were calculated by using the method translation software. The most important operational conditions are summarized in Table 6. From the method translation software program, a speed gain factor of 5.9 is predicted. Note that the carrier gas in the analyses are helium and hydrogen for the 0.25 mm i.d. and the 0.1 mm i.d. columns, respectively.

The chromatograms obtained on the respective columns are compared in Figure 64 top (Nutmeg oil, standard column) and bottom (Nutmeg oil, narrow bore column). From these chromatograms, it is obvious that the resolution is exactly the same on both columns. The analysis on the narrow bore column is much faster. From the last eluting peak, a speed gain factor of 5.7 is measured, which is close to the predicted speed gain of 5.9. Since for the calculation, the nominal column lengths were used, it might be expected that the correlation could be even better if actual column lengths are measured and used in the method translation software.

11.10.2. The Concept of Retention Time Locking

Recent developments in capillary GC, i.e., electronic pneumatic control (EPC) of the carrier gas, improved oven temperature stability, and excellent reproducibility in column making have led to the concept of retention time locking (RTL) [77], [78]. With retention time locked data bases, absolute retention times instead of retention indices can be used to elucidate the structures of eluting solutes. Moreover, retention time locking can be used in combination with different injectors and detectors. Exact scaling of capillary GC – FID, capillary GC – MS, and capillary GC – AED chromatograms is feasible.

The concept is illustrated with the analysis of some organochlorine pesticides in wine. Figure 65 shows the capillary GC – AED trace in the chloor emission line (837 nm) of a wine extract. RTL conditions were adapted to the analysis and the identities were searched via a pesticide RTL database. A retention time window of 0.5 (± 0.25 min from the measured retention time) and the presence of chloor were introduced in the RTL dialogue box. From the search for the main peak eluting at 21.93 min (Table 7) it is clear that the smallest deviation is obtained for procymidone. The presence of this compound could be confirmed using RTL conditions in capillary GC –

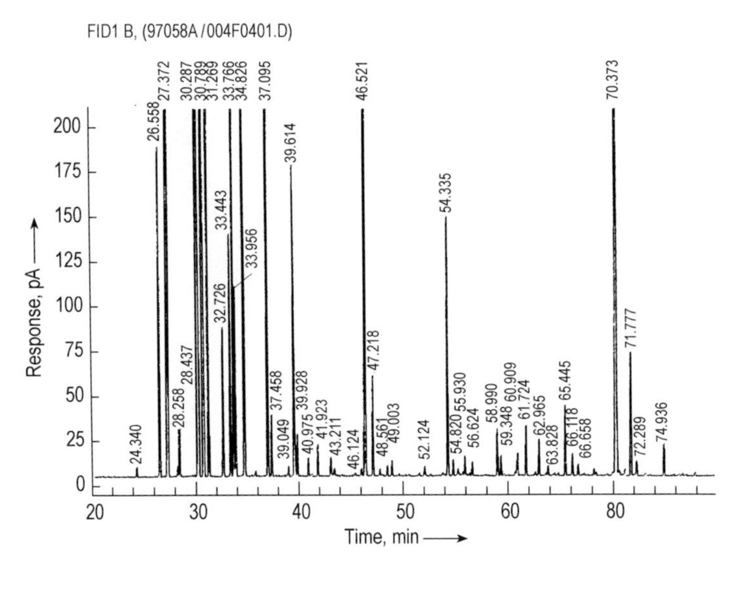

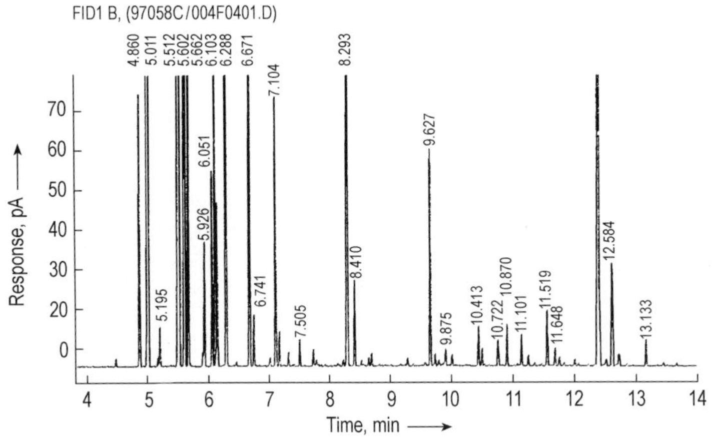

Figure 64. Analyses of Nutmeg oil on a 60 m × 0.25 mm i.d. × 1 μm MeSi column (top trace) and on a 20 m × 0.1 mm × 0.4 μm MeSi column (bottom trace)

MS. The other chlorine compounds, detected by capillary GC–AED analysis, could not be identified directly by capillary GC–MS. Owing to the complexity of the chromatogram no clear spectra were obtained. For these compounds, the result screener was used. Hereby, the TIC is searched for specific ions for all pesticides present in the database (567 pesticides) in their elution window. The peak at 25.64 min in the AED trace was identified as p,p′-DDD. The peak at 28.40 min was identified as iprodione. The peak at 29.76 min could not be identified since no ions corresponding to a pesticide present in the database and eluting in this time window could be detected.

This example clearly demonstrates the complementary nature of capillary GC–AED and capillary GC–MS. The selectivity of the AED results in simple profiles and allows fast screening of pesticides in samples. Capillary GC–MS in combination with the result screener allows confirmation of the detected solutes. Analysis of the samples with capillary GC–MS only results in complex chromatograms and more time consuming data interpretation is needed. Retention time locking also allows the correct matching of the capillary GC–AED chromatograms with the capillary GC–MS chromatograms. The corresponding peaks elute at virtually the same retention times.

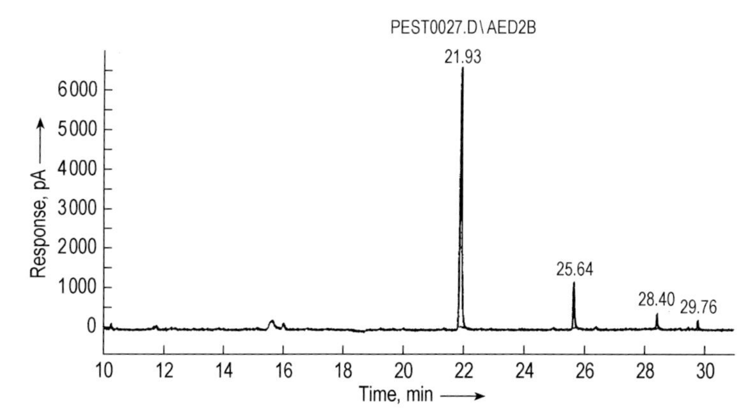

Figure 65. Capillary GC – AED chromatogram at the 837 nm emission line of a white wine extract recorded under RTL conditions

Table 7. Results of retention time locked pesticide data base search

Search results for 21.939 +/– 0.250 min
Contains elements: Cl
Does not contain elements: {no restriction}
RTT file searched: C:\HPCHEM®TL\HPGCPST.RTT

FID_RT	Compound name	MW	Formula	Δ *RT* (min)
21.825	Chlorbenside	269.19	C:13,H:10,Cl:2,S:1,	−0.114
21.971	Procymidone	284.14	C:13,H:11,Cl:2,N:1,O:2,	+0.032
22.037	trans-Chlordane	409.78	C:10,1,Cl:6,H:8,	+0.098
22.95	Chlorflurecol-Me- ester	274.70	C:15,H:11,Cl:1,O:3,	+0.156

Search results for 25.649 +/– 0.250 min
Contains elements: Cl
Does not contain elements: {no restriction}
RTT file searched: C:\HPCHEM®TL\HPGCPST.RTT

FID_RT	Compound name	MW	Formula	Δ *RT* (min)
25.409	Chlorobenzilate	325.19	C:16,H:14,Cl:2,O:3,	−0.240
25.434	Chloropropylate	339.22	C:17,H:16,Cl:2,O:3,	−0.215
25.581	Diniconazole	326.23	C:15,H:17,Cl:2,N:3,O:1,	−0.068
25.654	Cyprofuram	279.72	C:14,H:14,Cl:1,N:1,O:3,	+0.005
25.703	*p,p'*-DDD	320.05	C:14,H:10,Cl:4,	+0.054
25.763	Etaconazole	328.20	C:14,H:15,Cl:2,N:3,O:2,	+0.114
25.786	*o,p'*-DDT	354.49	C:14,H:9,Cl:5,	+0.137
25.823	Flamprop-isopropyl	363.82	C:19,H:19,Cl:1,F:1,N:1,O:3,	+0.174

Search results for 28.404 +/– 0.250 min
Contains elements: Cl
Does not contain elements: {no restriction}
RTT file searched: C:\HPCHEM®TL\HPGCPST.RTT

FID_RT	Compound name	MW	Formula	Δ *RT* (min)
28.239	Endrin ketone	380.91	C:12,H:8,Cl:6,O:1,	−0.165
28.362	Benzoylprop ethyl	366.24	C:18,H:17,Cl:2,N:1,O:3,	−0.042
28.424	Iprodione	330.17	C:13,H:13,Cl:2,N:3,O:3,	+0.020
28.436	Dichlorophen	269.13	C:13,H:10,Cl:2,O:2,	+0.032
28.545	Leptophos oxon	396.00	C:13,H:10,Br:1,Cl:2,O:3,P:1,	+0.141
28.572	Chlorthiophos sulfox-ide	377.24	C:11,H:15,Cl:2,O:4,P:1,S:2,	+0.188

Search results for 29.768 +/– 0.250 min
Contains elements: Cl
Does not contain elements: {no restriction}
RTT file searched: C:\HPCHEM®TL\HPGCPST.RTT

FID_RT	Compound name	MW	Formula	Δ *RT* (min)
29.699	Phosalone	367.80	C:12,H:15,Cl:1,N:1,O:4,P:1,S:2,	−0.069
29.754	Leptophos	412.06	C:13,H:10,Br:1,Cl:2,O:2,P:1,S:1,	−0.014
29.848	Mirex	545.55	C:10,Cl:12,	+0.080

11.10.3. Towards Black Boxes

State-of-the-art GC offers column efficiencies ranging from 5000 to 500 000 plates (packed or short megabore columns to long narrowbore columns), with sample capacities from microgram to subnanogram amounts. Selectivity can be introduced in all parts of the GC system. Column selectivity can be adapted to the specific need by selecting the most suitable *stationary phase*. To enhance resolution of critical pairs, column selectivity can be optimized by *selectivity tuning* between two columns coated with different stationary phases or by *two-dimensional CGC*. *Multidimensional* CGC, HPLC – CGC, SFE – CGC, etc., provide preseparation and enrichment of relevant fractions. In addition to universal inlets a number of *selective inlets* are available in GC. GC enjoys universal detection and sensitive, selective detection. Selective recognition can be performed routinely by CGC – mass spectroscopy, CGC – Fourier transform infrared spectroscopy, and CGC – atomic emission detection, to provide sensitive and selective quantitation of target compounds and structural elucidation or identification of unknowns.

There is much more to come, such as comprehensive capillary GC, hyphenation with time-of-flight MS (capillary GC – TOFMS) and inductively coupled plasma MS (capillary GC – ICP-MS) ... to mention a few recent technological realizations.

Notwithstanding all these capabilities of GC, the data are often disappointing, as evidenced by results from round-robin tests. Deviations can be as high as 100 ppm for BTX measurements in air in the 100 ppm level, as high as 30 ppb for trihalomethanes at the 50 ppb level in drinking water, as high as 60 ppt (ppt = 1 part in 10^{12}) for organochloropesticides in the 70 ppt level in wastewater, etc. How should data such as 3.34 ppt 2,3,7,8 TCDD be interpreted?

What are the reasons for this? Two main aspects can be identified. First of all, there is an educational problem. Technicians in the laboratory often lack experience and expertise, and moreover time and money for training are unavailable. On the other hand, shift of good staff now occurs very quickly without time to transfer the know-how. The second aspect is precolumn errors (e.g., sample introduction and sample preparation), and both parts of the total analytical procedure are interrelated. Future developments in GC are therefore expected in the direction of fully automated sample preparation and sample introduction systems. Some instrumentation nowadays is already tailor-made for a given application. More can be expected in the future and development may result in specific instruments, such as a pesticide analyzer, a PCB analyzer, or a sugar analyzer.

11.11. References

General textbooks on chromatography and gas chromatography

[1] J. C. Giddings: *Unified Separation Science,* J. Wiley & Sons, New York 1991.
[2] C. F. Poole, S. K. Poole: *Chromatography Today,* Elsevier, Amsterdam 1991.
[3] W. Jennings: *Analytical Gas Chromatography,* Academic Press, San Diego 1987.
[4] G. Schomburg: *Gas Chromatography, A Pratical Course,* VCH Verlagsgesellschaft, Weinheim 1990.
[5] D. Rood: *A Practical Guide to the Care, Maintenance and Troubleshooting of Capillary Gas Chromatographic Systems,* Hüthig Verlag, Heidelberg 1991.
[6] M. L. Lee, F. J. Yang, K. D. Bartle: *Open Tubular Column Gas Chromatography, Theory and Practice,* J. Wiley & Sons, New York 1984.
[7] A. Van Es: *High Speed, Narrow-Bore Capillary Gas Chromatography,* Hüthig Verlag, Heidelberg 1992.

Specific References

[8] A. T. James, A. J. P. Martin, *Biochem. J.* **50** (1952) 679.
[9] L. S. Ettre, *Pure Appl. Chem.* **65** (1993) no. 4, 819.
[10] J. J. Van Deemter, F. J. Zuiderberg, A. Klinkenberg, *Chem. Eng. Sci.* **5** (1956) 271.
[11] M. J. E. Golay in D. Detsy (ed.): *Gas Chromatography,* Butterworths, London 1958, p. 36.
[12] P. Sandra, *J. High Resolut. Chromatogr.* **12** (1989) 82.
[13] P. Sandra, *J. High Resolut. Chromatogr.* **12** (1989) 273.
[14] R. D. Dandeneau, E. H. Zerenner, *J. High Resolut. Chromatogr. Comm.* **2** (1979) 351.
[15] M. Verzele, P. Sandra, *J. Chromatogr.* **158** (1978) 111.
[16] P. Sandra, *LC.GC* **5** (1987) 236.
[17] H. Rotzsche: *Stationary Phases in Gas Chromatography,* Elsevier, Amsterdam 1991.
[18] W. O. McReynolds, *J. Chromatogr. Sci.* **8** (1970) 685.
[19] J. de Zeeuw, R. de Nijs, L. Henrich, *J. Chromatogr. Sci.* **25** (1987) 71.
[20] J. R. Mann, S. T. Preston, *J. Chromatogr. Sci.* **11** (1973) 216.
[21] T. J. Stark, P. A. Larson, R. D. Dandaneau, *J. Chromatogr.* **279** (1983) 31.
[22] P. Sandra et al., *J. High Resolut. Chromatogr. Chromatogr. Commun.* **8** (1985) 782.
[23] P. Sandra, F. David in H. J. Cortes (ed.): *Multidimensional Chromatography, Techniques and Applications,* Marcel Dekker, New York 1990, p. 145.
[24] K. E. Markides et al., *Anal. Chem.* **57** (1985) 1296.
[25] W. A. König: *The Practice of Enantiomer Separation by Capillary Gas Chromatography,* Hüthig Verlag, Heidelberg 1987.

[26] W. A. König: *Gas Chromatographic Enantiomer Separation with Modified Cyclodextrins,* Hüthig Verlag, Heidelberg 1992.
[27] D. E. Bautz, J. W. Dolan, W. D. Raddatz, L. R. Snyder, *Anal. Chem.* **62** (1990) 1560.
[28] G. N. Abbay et al., *LC.GC* Int. 4 (1991) 28.
[29] E. Sippola, F. David, P. Sandra, *J. High Resolut. Chromatogr.* **16** (1993) 95.
[30] B. W. Hermann et al., *J. High Resolut. Chromatogr.* **13** (1990) 361.
[31] S. S. Stafford (ed.): *Electronic Pressure Control in Gas Chromatography,* Hewlett-Packard, Little Falls 1993.
[32] V. Pretorius, W. Bertsch, *J. High Resolut. Chromatogr. Chromatogr. Commun.* **6** (1983) 64.
[33] R. Jenkins, W. Jennings, *J. High Resolut. Chromatogr. Chromatogr. Commun.* **6** (1983) 228.
[34] P. Sandra in P. Sandra (ed.): *Sample Introduction in Capillary Gas Chromatography,* Hüthig Verlag, Heidelberg 1985, p. 1.
[35] M. S. Klee: *GC Inlets, An Introduction,* Hewlett-Packard, Little Falls 1990.
[36] K. Grob: *Classical Split and Splitless Injection in Capillary GC,* Hüthig Verlag, Heidelberg 1986.
[37] K. Grob: *On-Column Injection in Capillary GC,* Hüthig Verlag, Heidelberg 1987.
[38] P. Sandra in K. Hyver (ed.): *High Resolution Gas Chromatography,* 3rd ed., Hewlett-Packard, Little Falls 1987.
[39] D. Desty, A. Goldup, H. Whyman, *J. Inst. Pet.* **45** (1959) 287.
[40] W. D. Snyder, Technical Paper no. 108, Hewlett-Packard, Little Falls 1987.
[41] P. Wyllie, K. J. Klein, M. Q. Thompson, B. W. Hermann, *J. High Resolut. Chromatogr.* **14** (1991) 361.
[42] F. David, P. Sandra, S. S. Stafford, B. Slavica, Hewlett-Packard Application Note 228 – 223, Little Falls 1993.
[43] A. Zlatkis, H. R. Kaufman, *Nature (London)* **184** (1959) 4010.
[44] P. Sandra in F. Bruner (ed.): *The Science of Chromatography,* Elsevier, Amsterdam 1985, p. 381.
[45] K. Grob, K. Grob Jr., *J. Chromatogr.* **151** (1978) 311.
[46] K. Grob, *J. High Resolut. Chromatogr. Chromatogr. Commun.* **1** (1978) 263.
[47] G. Schomburg et al., *J. Chromatogr.* **142** (1977) 87.
[48] P. Sandra, M. Van Roelenbosch, M. Verzele, C. Bicchi, *J. Chromatogr.* **279** (1983) 287.
[49] F. David, L. Vanderroost, P. Sandra, S. Stafford, *Int. Labmate* **17** (1992) no. 3, 13.
[50] E. Meyer, A. P. De Leenheer, P. Sandra, *J. High Resolut. Chromatogr.* **15** (1992) 637.
[51] W. Vogt, K. Jacob, H. W. Obwexer, *J. Chromatogr.* **174** (1979) 437.
[52] G. Schomburg in R. E. Kaiser (ed.), *Proc. Int. Symp. Capillary Chromatogr. 4th* 1981, 371, 921 A.
[53] F. Poy, S. Visani, F. Terrosi, *J. Chromatogr.* **217** (1981) 81.
[54] F. David, P. Sandra, A. Hoffmann, J. Gerstel, *Chromatographia* **34** (1992) no. 5 – 8, 259.
[55] R. Buffington, M. K. Wilson: *Detectors for Gas Chromatography, A Practical Primer,* Hewlett-Packard, Little Falls 1987.
[56] H. H. Hill, D. G. McMinn (eds.): *Detectors for Capillary Chromatography,* J. Wiley & Sons, New York 1992.
[57] I. Temmermann, F. David, P. Sandra, R. Soniassy, Hewlett-Packard Co., Application Note 228 – 135, Little Falls 1991.
[58] F. W. Karasek, R. E. Clement: *Basic Gas Chromatography—Mass Spectroscopy,* Elsevier, Amsterdam 1988.
[59] W. Herres: *Capillary Gas Chromatography—Fourier Transform Infrared Spectroscopy,* Hüthig Verlag, Heidelberg 1987.
[60] R. Buffington: *GC-Atomic Emission Spectroscopy Using Microwave Plasmas,* Hewlett-Packard, Little Falls 1988.
[61] E. Kovats, J. C. Giddings, R. A. Keller in: *Advances in Chromatography,* vol. 1, Marcel Dekker, New York 1965, chap. 7.
[62] H. Van den Dool, P. D. Kratz, *J. Chromatogr.* **11** (1963) 463.
[63] W. Jennings, T. Shibamoto: *Qualitative Analysis of Flavor and Fragrance Volatiles by Gas Chromatography,* Academic Press, New York 1980.
[64] J. F. Sprouse, A. Varano, *Am. Lab. (Fairfield, Conn.)* **16** (1984) 54.
[65] H. J. Cortes (ed.): *Multidimensional Chromatography, Techniques and Applications,* Marcel Dekker, New York 1990, p. 145.
[66] C. Biichi, A. D'Amato, F. David, P. Sandra, *J. High Resolut. Chromatogr.* **12** (1989) 316.
[67] K. Grob: *On-Line Coupled LC-GC,* Hüthig Verlag, Heidelberg 1991.
[68] F. David, P. Sandra, D. Bremer, R. Bremer, F. Rogies, A. Hoffmann, *Labor Praxis* **21** (1997) 5.
[69] M. Verschuere, P. Sandra, F. David, *J. Chromatogr. Sci.* **30** (1992) 388.
[70] F. David, M. Verschuere, P. Sandra, *Fresenius J. Anal. Chem.* **344** (1992) 479.
[71] K. Blau, G. S. King: *Handbook of Derivatization for Chromatography,* Heyden (Wiley & Sons), New York 1977.
[72] J. Drozd: *Chemical Derivatization in Gas Chromatography,* Elsevier, Amsterdam 1981.
[73] P. Sandra, F. David, *LC.GC* **7** (1989) 746.
[74] L. S. Ettre, J. V. Hinshaw: *Basic Relationships of Gas Chromatography,* Advanstar, Cleveland 1993.
[75] B. D. Quimby, V. Giarrocco, M. S. Klee, Hewlett-Packard Application Note 228-294, February 1995.
[76] F. David, D. R. Gere, F. Scanlan, P. Sandra, *J. Chromatogr. A* **842** (1999) 309.
[77] V. Giarrocco, B. D. Quimby, M. S. Klee, Hewlett-Packard Application Note 228-392, December 1997.
[78] B. D. Quimby, L. M. Blumberg, M. S. Klee, P. L. Wylie, Hewlett-Packard Application Note 228-401, May 1998.

12. Liquid Chromatography

PETER LEMBKE, Belovo-Spain, S.L., Reus, Spain (Chaps. 12, 12.3 – 12.10, Sections 12.2.1 – 12.2.6.7)

GÜNTER HENZE, Institut für Anorganische und Analytische Chemie der Technischen Universität Clausthal (Section 12.2.6.8)

KARIN CABRERA, Merck KGaA, Darmstadt, Federal Republic of Germany (Chap. 12.11)

WOLFGANG BRÜNNER, formerly Merck KGaA, Darmstadt, Federal Republic of Germany (Chap. 12.11)

EGBERT MÜLLER, Merck KGaA, Darmstadt, Federal Republic of Germany (Chap. 12.11)

Abbreviations

2′,5′-ADP	adenosine 2′,5′-diphosphate
5′-AMP	adenosine 5′-monophosphate
A_{ref}	area or height of the reference peak in the standard solution
A_{si}	area or height of component *i* in the sample
AUFS	bsorption units full scale
C_C	cell capacitance
C_D	double-layer capacitance
CRD	chemical reaction detector
DAD	diode-array detector
D_m	diffusion coefficient
dp	particle size
E	molecular extinction coefficient
ECD	electron capture detector
ELSD	evaporative light-scattering detector
ES	electrospray
F	flow rate
FAB	fast ion bombardment
FD	field desorption
FID	flame ionization detector
FSD	fast-scanning detector
FTIR	Fourier transform IR spectroscopy
GFC	gel filtration chromatography
GPC	gel permeation chromatography
H	plate height, band height
HPIC	high-performance ion chromatography
HPSEC	high-performance size exclusion chromatography
I^*	radiation
I_0/I	transmittance
IEC	ion exclusion chromatography
IEX	ion-exchange chromatography
IMAC	immobilized metal ion affinity chromatography
ISE	ion-selective electrode
K	distribution coefficient
k'	capacity ratio
K_{ij}^{Pot}	selectivity factor
L	length
LC	liquid chromatography
LIF	laser-induced fluorescence
LLC	liquid – liquid chromatography
LSC	liquid – solid chromatography
LSD	light-scattering detector
MSD	mass-selective detector
MWD	multiwavelength detector
N	plate number
NAD	nicotinamide – adenine dinucleotide
NADP	nicotinamide – adenine dinucleotide phosphate
NARP	nonaqueous reversed phase
NP	normal phase
PC	paper chromatography
RI	refractive index
RIG	refractive index gradient
RIU	refractive index units
RP	reversed phase
R_S	electrolyte resistance
Rs	resolution
SAX	strong anion exchanger
SCX	strong cation exchanger
SEC	size exclusion chromatography
SFC	supercritical fluid chromatography
SFE	supercritical fluid extraction
t_0	time analyte spends in the mobile phase
TID	thermoionic detector
TOF	time of flight
t_R	retention time
$t_{R'}$	time analyte spends in the stationary phase
u	velocity
W	peak width
WAX	weak anion exchanger
WCX	weak cation exchanger
WE	working electrode
z_i	atomic number of ion *i*
z_j	atomic number of interfering ion *j*
α	selectivity
Γ	correction factor
λ	correction factor, wavelength
κ	conductivity
ω	angular frequency, correction factor

12.1. General

Liquid chromatography is a separation technique based on a different distribution rate of sample components between a stationary and a liquid mobile phase. Depending on the stationary phase, a distinction is made between thin layer chromatography (TLC), paper chromatography (PC), and liquid column chromatography (LC). Liquid column chromatography can be divided into open and closed systems. Generally an *open system* consists of a glass column packed with large-sized ($\approx$30 μm and larger) particles of the stationary phase onto which the mobile phase is poured. Because of the large particles, the mobile phase can pass the stationary phase only by means of gravity, sometimes assisted by suction. Therefore, no solvent pump is necessary.

However, to increase separation efficiency, high velocities and small-diameter ($\leq$10 μm) particles are essential as the stationary phase. Unfortunately, this leads to an increase in back-pressure. To overcome this back-pressure, a pump is necessary for transport of the mobile phase (high-pressure- or high-performance liquid chromatography, HPLC). Separation is performed in a *closed system.*

HPLC is the LC technique with the highest efficiency (ability to separate different sample compounds in a given time with a given resolution). The selectivity of this technique is based not only on the different types of stationary phases but also on the mobile phase. An advantage of gas chromatography (GC) is that selectivity in this technique is based only on the stationary phase. (In GC, interactions occur only between the solute and the stationary phase. In LC, additional interactions between the solute and the liquid mobile phase are found.) Furthermore GC can be used only for samples that can be volatized without decomposition at temperatures up to ca. 400 °C. This limits the use of GC to about 20 % of the samples of interest [1]. The limit of LC is set by the solubility of the sample in the mobile phase. Since nearly every substance is soluble in some kind of a solvent, the application range of LC is very broad. In contrast, capillary GC (→Gas Chromatography) has a far higher separation efficiency than HPLC.

Neither technique can be said to be better than the other. Both GC and HPLC have their advantages for particular separation problems.

In LC, two sample introduction techniques can be distinguished the continuous and the discontinuous mode. *Continuous sample introduction*, also known as front analysis or adsorptive filtration, is often used for solvent cleanup. In this technique the sample solution is introduced continuously onto the column; thus only the least-retained compound is obtained in the pure state.

Discontinuous Sample Introduction. Discontinuous sample introduction can divided into elution analysis and displacement analysis. In *elution analysis*, the sample is injected into a continuous mobile phase. If the composition of the mobile phase remains unchanged during the analysis this is called the *isocratic mode*. In *gradient elution* the composition of the mobile phase is changed in such a way that the elution strength is increased (or sometimes also decreased) during analysis. In the discontinuous sample introduction mode, an elution diagram (chromatogram) is obtained in which, ideally, each compound is separated by a pure solvent zone from the next eluting compound.

*Displacement analysis*is often used in preparative chromatography. After sample injection, the composition of the mobile phase is changed, and the interaction between the mobile and the stationary phases becomes so dominant that the retained sample compounds are displaced by solvent molecules and move with identical velocity (i.e., that of the eluent).

12.1.1. History

In 1905, TSWETT, a Russian botanist, published the first paper on liquid chromatography [2], in which he described the separation of plant pigments by means of open-column liquid chromatography. The importance of this new technique was recognized only after KUHN (Nobel prize, 1938) and LEDERER used it for the preparative separation of α- and β-carotin from carrots in 1931 [3], [4]. In 1941 MARTIN and SYNGE [5] (Nobel prize, 1951) described the technique of partition chromatography (liquid–liquid chromatography, LLC), and in 1949 SPEDDING [6] and TOMPKINS [7] published the first paper on ion-exchange chromatography (IEX). In 1959, PORATH and FLODIN [8] introduced size exclusion chromatography (SEC). One of the first papers on high-performance liquid chromatography was published by PIEL [9] in 1966. Since then, HPLC has become one of the most frequently applied analytical separation methods. Especially because of the rapid development of instrumentation (pumps, detectors, stationary phase, gradient elution, etc.) and column packings during the 1970s and 1980s, very high selectivity and efficiency can be obtained. Today, HPLC is one of the most versatile techniques in the analytical (e.g., trace analysis) and preparative fields.

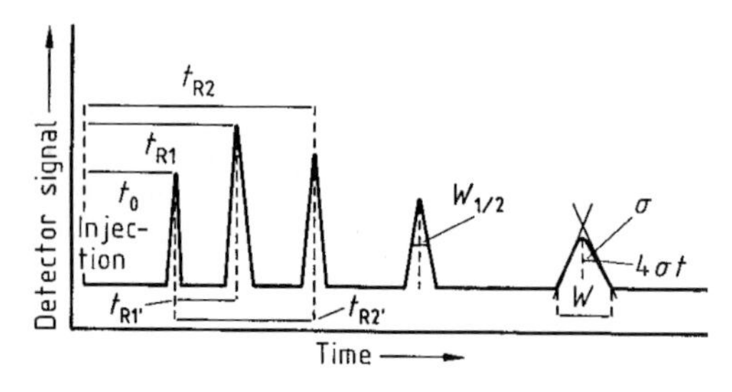

Figure 1. Characteristic information obtainable from a chromatogram

12.1.2. Definition and Theoretical Background (→Basic Principles of Chromatography)

In a chromatographic column, two processes occur simultaneously:

1) The separation process, which is based mainly on the interactions between solute and stationary phase
2) Dilution of the solute in the mobile phase due to diffusion processes; this effect counteracts separation

To understand these effects, several important chromatographic definitions are discussed in this section. Figure 1 shows a typical chromatogram with the most important parameters characterizing a chromatographic separation.

The *retention time* t_R of a particular component is defined by:

$$t_R = t_0 + t_{R'}$$

where t_0 is the time the analyte spends in the mobile phase and $t_{R'}$ is the time the analyte spends in the stationary phase. For given HPLC equipment and a constant flow rate (milliliters per minute), all analytes reside in the mobile phase for the same time t_0. They differ only in the time $t_{R'}$ they spend in the stationary phase. To become independent of column dimension and mobile phase flow the *capacity ratio* k' was introduced:

$$k' = \frac{t_R - t_0}{t_0} = \frac{t_{R'}}{t_0} \frac{\text{time spent in stationary phase}}{\text{time spent in mobile phase}}$$

Another important chromatographic quantity is the relative retention or selectivity of a chromatographic system. The *selectivity* α is defined by equation:

$$\alpha = \frac{t_{R2} - t_0}{t_{R1} - t_0} = \frac{k_{2'}}{k_{1'}} = \frac{t_{R2'}}{t_{R1'}}$$

This represents the ability of a given chromatographic system to separate two components. If $\alpha = 1$, no separation is possible, no matter how high the separation efficiency may be. As in GC, the selectivity can be influenced by the choice of the stationary phase, but in LC the selection of the mobile phase is also a factor. In preparative chromatography the selectivity α is the most important chromatographic parameter to be optimized.

A further quantitative measure for determining the separation of two components in a given sample is the *resolution Rs*.

$$Rs = \frac{2\,(t_{R2} - t_{R1})}{(W_1 + W_2)}$$

where W is the peak width. A resolution of $Rs = 1.5$ (baseline resolution) is ideal for quantification of particular sample peaks. Yet even with a resolution of $Rs = 1.0$, quantification is possible because only ca. 2 % of the peak areas overlap. Chromatograms with resolutions of less than 1.0 should not be used for quantitative analysis. On the other hand, an $Rs > 1.5$ is not necessary, since no advantage is achieved in quantification, but analysis time is prolonged. The above values for Rs are valid only for Gaussian peaks having the same size. Nevertheless, they can assist and guide the chromatographer.

Other important parameters in HPLC are the plate height H and the plate number N. *Plate height* is calculated from a chromatogram by

$$H = \frac{L}{5.54}\left(\frac{W_{0.5}}{t_R}\right)^2$$

where $W_{0.5}$ is the peak width at half maximum peak height; t_R the retention time; and L the length of the column.

Well-packed HPLC columns should have a plate height of 2 – 4 dp (dp = stationary phase particle size).

The *plate number* is an indicator of column quality. The higher it is, the better is the peak capacity (number of peaks separated in a given time) of the column. It is calculated by:

$$N = \frac{L}{H}$$

The plate number can be increased either by reducing the particle size of the stationary phase

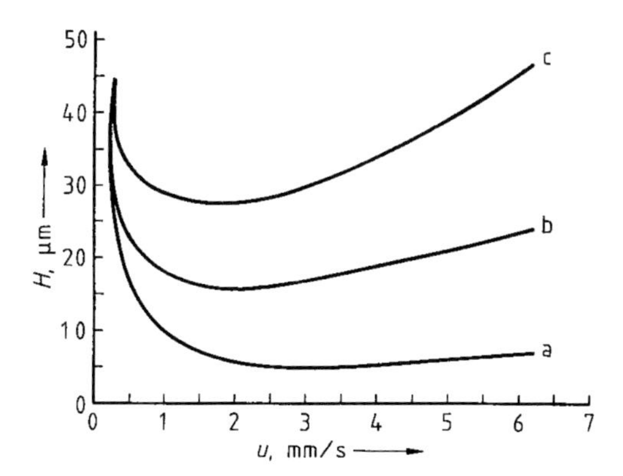

Figure 2. Typical Van Deemter *H/u* curves
a) 3 µm; b) 5 µm; c) 10 µm

dp and thus the achievable plate height or by lengthening the column. Both have the disadvantage of increasing the pressure drop over the column as well. Unfortunately, the equipment in most HPLC pumps, does not work at back-pressures exceeding 40 – 50 MPa ("high-pressure shutdown").

The relationship among most of the aforementioned chromatographic parameters is given by the following equation:

$$Rs = \frac{1}{4}\frac{\alpha - 1}{\alpha}\frac{k_2'}{1 + k_2'}\sqrt{N}$$

This is the most important equation in LC because it combines the chromatographic separation parameters selectivity, capacity ratio, and plate number. Since these three terms are more or less independent, one term can be optimized after the other. The selectivity term can be optimized by changing the stationary or the mobile phase; k' can be influenced by the solvent strength of the mobile phase, and N can be adjusted by changing the column length, particle diameter, or solvent linear velocity u.

$$u = \frac{L}{t_0}$$

The different processes contributing to the overall band broadening are described by the Van Deemter equation [10]

$$H = A + B/u + Cu$$

where A is the eddy diffusion term ($\approx 2\,\lambda dp$), B the longitudinal (axial) diffusion term ($\approx 2\gamma D_m$), C the mass transfer resistance term ($\approx \omega dp^2/D_m$), and u the linear mobile phase velocity.

$$H = 2\lambda dp + 2\gamma D_m/u + \left(dp^2\omega/D_m\right)u$$

where dp is the mean particle diameter; D_m the diffusion coefficient in the mobile phase, and λ, γ, ω are correction factors.

For liquid chromatography the following conclusions can be drawn from the Van Deemter equation: In contrast to gas chromatography, the C term (which increases tremendously with the linear velocity of the mobile phase) is very important in liquid chromatography because the diffusion coefficients in the fluid phase are much lower than in GC. A good HPLC column should have as low a C as possible (i.e., a small slope in the H versus u curve; see Fig. 2). In LC the B term is not as important as in GC because of the low diffusion coefficients. The B term decreases with increasing linear velocity u. The A term is independent of linear velocity; it depends only on particle size dp. Particle size is very important in LC (the A and C terms). The smaller the particle size of the stationary phase, the smaller is the achievable plate height and the higher is the plate number of a particular column. The development of uniform small particles for stationary phases in LC is one reason for the booming development of HPLC since the 1970s. The smallest particle size currently available is 1.5 µm (standard sizes: 5, 7, and 10 µm). Particles < 3 µm are not so popular because the back-pressure of the column increases dramatically with decreasing particle diameter, while packing the column becomes very difficult. As a rule, reducing the particle diameter by half increases the back-pressure of the column to the second power whereas the resolution Rs improves by only a factor of 1.4. Nevertheless, short columns (2 cm) packed with 1.5 µm material do allow extremely fast analysis combined with low consumption of mobile phase and can be very useful for laboratories that have to deal with a large number of samples every day. Because of their nonporous nature, these small particles have a very low capacity, which may occasionally result in column-overload problems (peak tailing, poor resolution, etc.).

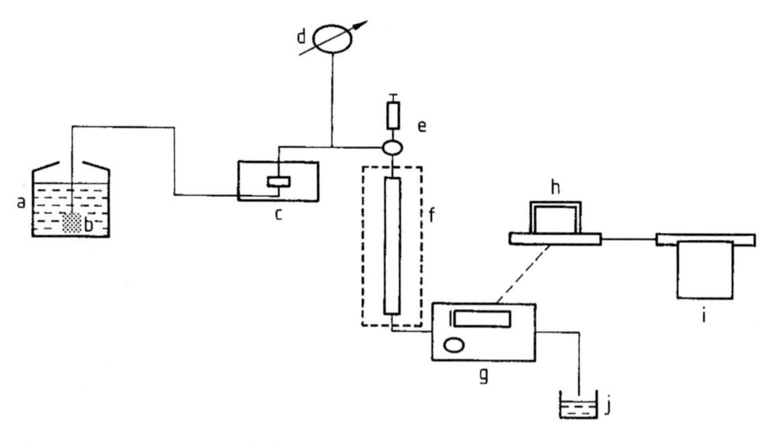

Figure 3. Diagram of a simple HPLC system
a) Solvent reservoir; b) Frit; c) Pump; d) Manometer; e) Injection port; f) Separation column with optional oven; g) Detector; h) Data processing unit; i) Integrator; j) Waste

12.2. Equipment

To obtain separations sufficient for quantitative analysis in liquid chromatography, good equipment is essential. Only the most important parameters are discussed in this chapter. For further information, see [11] – [15] and other monographs on LC. Figure 3 shows a schematic of a simple HPLC system.

12.2.1. Filters and Connecting Tubing

The connecting tubing between solvent reservoir and pump is usually made of polytetrafluoroethylene (PTFE) with an inner diameter i.d. of 2 – 4 mm. The end of each pump – inlet tube should be fitted with an inlet-line filter (frit), which keeps out large, inadvertent contaminants and holds the inlet line at the bottom of the reservoir so that no air (bubbles) is drawn into the pump head. These frits, which are normally made of stainless steel (recently also of glass or PTFE), usually have a pore diameter of 5 – 10 μm and can be cleaned easily with 5 % nitric acid.

The tubing connecting the sample introduction unit, column, and detector consists of small-volume stainless steel capillaries with an inside diameter of 0.25 mm and an outer diameter of 1.25 mm (1/16 inch). For bioseparations that are sensitive to stainless steel, tubing made from polyetheretherketone (PEEK) is recommended. All connecting tubing, especially between sample introduction system and column, should have a small volume (small inside diameter and shortest length possible), to minimize extra column band broadening and loss in separation efficiency.

12.2.2. Containment of Mobile Phase

Depending on the physicochemical properties of the mobile phase (eluent), different containers are used. The most common are brown laboratory bottles for the protection of light-sensitive eluents. If inorganic ions are to be monitored, polyethylene containers should be used because additional inorganic ions may be dessolved from the glass, thus leading to false results. Safety regulations concerning toxic solvents, fire prevention, etc., must also be taken into account [16].

12.2.3. Pumps

The pump is one of the most important parts of an HPLC system. It is responsible for constant flow of the mobile phase, which is necessary for the quantification and reproducibility of an analysis. The retention time t_R can be correlated with the retention volume V_R and the flow rate F:

$$V_R = F\, t_R$$

Therefore a change in the flow rate or strong pulsation changes the retention time of an analyte, making identification by t_R impossible, because the results are not reproducible. Two types of pump pulsation occur that should be kept as low

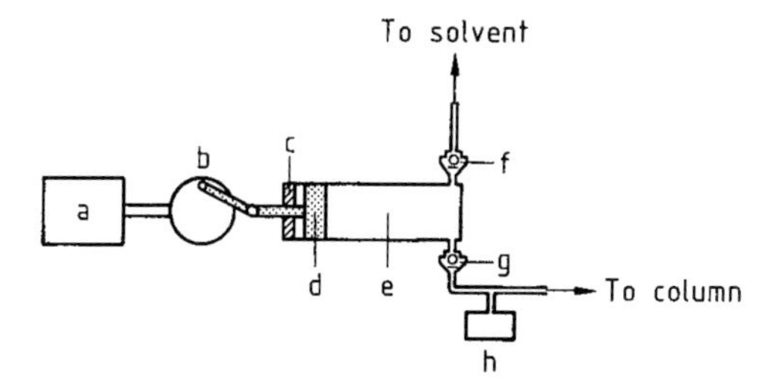

Figure 4. Diagram of a simple reciprocating pump [19]
a) Motor; b) Camshaft; c) Seal; d) Piston; e) Chamber; f) Inlet check valve; g) Outlet check valve; h) Damper

as possible: The *long-time pulsation* (causes irreproducible retention times) and the *short-time pulsation* (causes irreproducible peak areas). The effects of flow instabilities due to pump pulsation and their measurement are discussed in [17].

Good HPLC pumps should be able to deliver a flow rate of 0.5 – 10 mL/min, with a maximum pressure of ca. 40 – 50 MPa, a constant flow rate with variation of $< 0.5\,\%$ (see also [18]), and a rest pulsation of $< 0.1\,\%$. In addition, they should be equipped with adjustable high- and low-pressure limits to avoid damage to the pump or column; should have a solvent compressibility control; must be easy to maintain; and should be chemically inert. The pump heads should not warm up (evaporation of solvent and bubbles), and the pumps should, preferably, not be too noisy.

Basically, two types of pumps exist: Constant-pressure pumps and constant-flow pumps. Pressure-constant pumps are often used to pack HPLC columns.

Modern HPLC systems work with pumps that deliver a constant flow independent of the back-pressure of the column. These *constant-flow pumps* can be divided into continuous and discontinuous systems. In principle, discontinuous pumps work like a huge syringe and have lost their importance in HPLC. Currently, most HPLC pumps work in the continuous mode to deliver a constant flow. In these reciprocating pumps a plunger piston causes suction of the eluent during the back stroke and delivers the solvent under pressure during the forward stroke. To ensure proper flow direction, the pump head is usually equipped with ball check valves (see Fig. 4 f, g). Single-head, one-and-a-half-head, double-head and triple-head reciprocating pumps are available commercially. To reduce the pulsation caused by reciprocating pumps, many different techniques

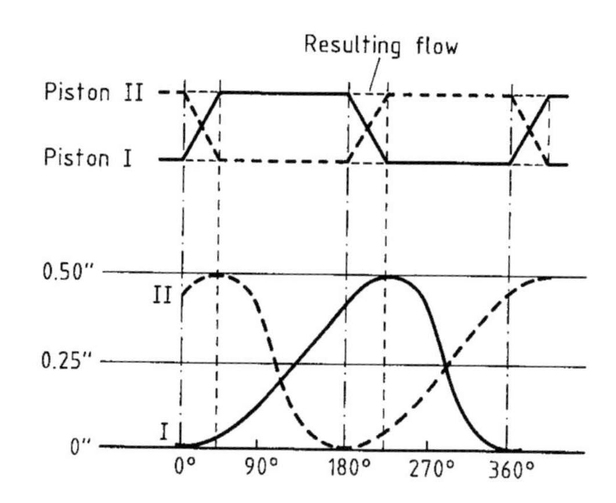

Figure 5. Flow profile of a double-head reciprocating pump (phase shift = 180°)
Courtesy of Laboratory Control/Milton Roy

are applied. For double- and triple-head reciprocating pumps, two or three pistons are used at the same time with a 180° or 120° phase shift between each. The resulting mobile phase flow is more or less periodic (see Fig. 5). The rest pulsation remaining can be decreased further if piston movement is not sinusoidal. Because of an eccentric disk, the second piston starts to deliver before the first has reached its end position. At the end position, the second piston has reached its maximum delivery output, while the first piston is refilled during its back stroke.

12.2.4. Sample Introduction Units

Good column performance is correlated directly with proper sample introduction onto the column in form of a narrow plug. Basically, the sample can be introduced into the high-pressure system in two ways: By syringe injectors or by sample valves.

12.2.4.1. Syringe Injectors

Syringe injection is the simplest way to introduce a sample to the system. The technique is very similar to that applied in GC. The sample is transferred with a syringe, through a self-sealing elastomeric septum, into the pressurized column. A specially designed pressure-resistant microsyringe must be used. One disadvantage of this technique is that column inlet plugging through particles of the septum can occur and the reproducibility of

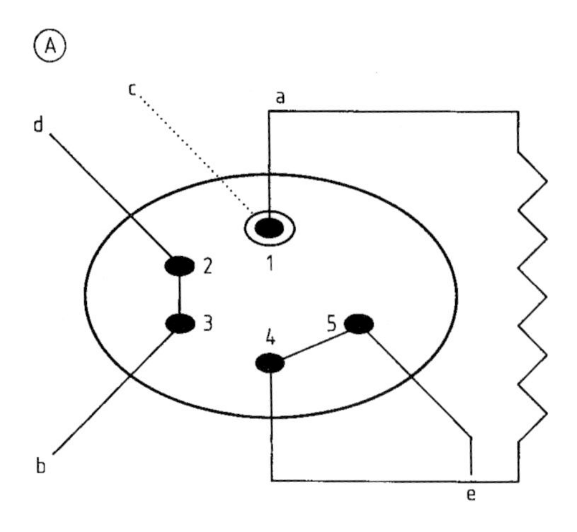

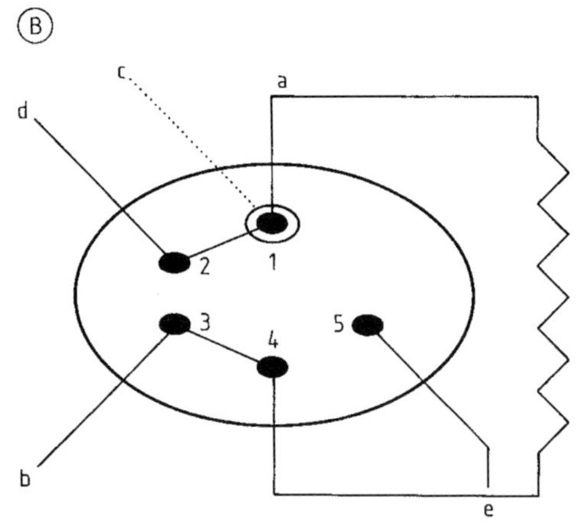

Figure 6. Diagram of a sample loop injection system [20]
A) Load position; B) Inject position
a) Loop; b) Column; c) Injection port; d) Pump; e) Waste

sample injection is often less than 2 % [14]. Therefore, this technique is used rarely, and most LC systems are equipped with sample valves.

12.2.4.2. Sample Valve Injectors

Today, the most common sample valves operate in the loop filling mode and are called *high-pressure sample injectors*. This means that the sample is transferred at atmospheric pressure from a syringe into a sample loop. By means of valving action the sample loop is then connected to the high-pressure mobile phase stream, which carries the sample onto the column (Fig. 6). The valving action can be carried out either manually or automatically by electric or pneumatic actuators.

Commercial sample loops are available from 0.2 μL (for microcolumns) to 5 mL or more. On the other hand, a sample loop can be made easily by cutting an appropriate length of capillary (with known i.d.) and connecting it to the sample injector.

Normally the complete loop filling method is preferred to the partial filling method because the former has excellent volumetric precision. When loop volumes exceed 5 – 10 μL, this can be better than 0.1 % RSD. Nevertheless, the partial filling method still has a volumetric precision (dependent on syringe repeatability) better than 1 % relative standard deviation (RSD) [20].

12.2.4.3. Automatic Introduction Systems

Automatic sample introduction units (auto samplers) are often used for routine analysis (e.g., product quality control). A modern, commercially available HPLC auto sampler can handle ca. 100 samples; has a variable injection volume (e.g., from 1 to 1000 μL); can heat or cool the samples (e.g., 4 – 40 °C); and can be programmed while they are waiting to be injected; and can be programmed individually for each sample. The programmed run of the auto sampler should be able to be interrupted for an urgent analysis and continued later without any problems.

The principle of injection is the same as with sample loop injectors (see Section 12.2.4.2).

With an automatic sample introduction system, very efficient work is possible because sample cleanup, pretreatment (e.g., precolumn derivatization), and sample programming can performed during the day, while the actual analyses are run automatically at night.

12.2.5. Columns

Column Construction. HPLC columns (Fig. 7) are usually manufactured of stainless steel. Current standard HPLC columns have a 4.0- or 4.6-mm inner diameter and a length of 60 – 250 mm. Columns with smaller inner diameter (e.g., 1 or 2 mm — microbore columns) are also available. Although eluent consumption can be reduced with these columns, sensitivity is usually not increased and undesirable "wall effects" and dead volume effects become more noticeable with decreasing column diameter. Microbore columns are preferred only if the sample volume is very limited.

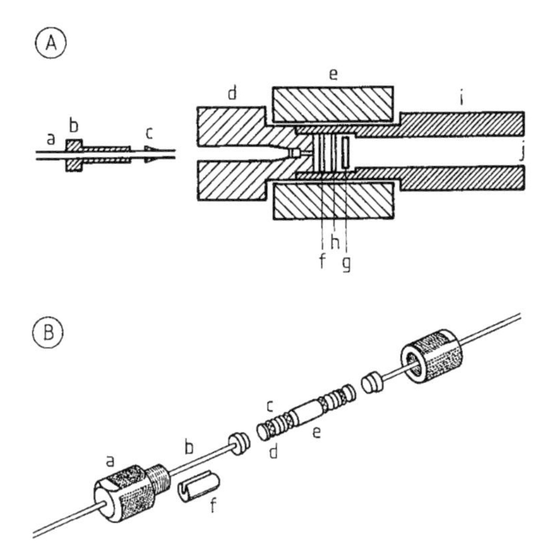

Figure 7. Diagram of the two most common column types for HPLC
A) Conventional HPLC column terminator for 1/16" fittings
a) Capillary; b) Nut; c) Ferrule; d) End fitting with internal thread; e) Coupling body to column; f) Metal frit; g) Filter paper; h) Sealing ring; i) Stainless steel tubing; j) Column packing
B) Cartrige system column
a) Compression nut; b) Column fitting adapter; c) Sieve sandwich; d) PTFE soft seal; e) Column of precolumn; f) Spacer (required when working with precolumns)
Courtesy of Bischoff-Analysentechnik, Leonberg, Germany

For preparative applications, columns with much larger dimensions (i.d. up to 900 mm and more) are available. Generally all parts of the column that do not contain the packing material, should be as small as possible to minimize dead volume. Often, unnecessarily large dead volumes result from false connection of the end fittings.

To ensure good separation, the inner column wall must be well polished and inert to the various mobile phases. Normal glass columns that are inert and have a very smooth inner wall can unfortunately be operated only at pressures less than ca. 5 MPa. To overcome this drawback, glass columns with a surrounding stainless steel jacket were developed that ensure stability at higher operating pressure. Recently, pressure-stable columns made of PEEK which show the same positive properties as glass columns, have come onto the market. These columns are gaining interest for metal-free bioseparations.

12.2.6. Detectors

Another important instrument required in modern LC is a sensitive or selective detector for continuous monitoring of the column effluent. In GC, the differences in physical properties of the mobile phase (carrier gas) and the sample are great enough for universal detectors with good *sensitivity* to be used (e.g., flame ionization detector, thermal conductivity detector, → Gas Chromatography). The problem in LC is that the physical properties of the mobile phase and the sample are often very similar, which makes the use of a universal detector impossible. Nevertheless, presently available LC detectors are very sensitive, are generally selective, and have a relatively wide range of applications (see Table 1).

To compare detectors from different manufacturers, characteristics must be found that determine the performance of the equipment. Besides instrument specifications such as cell geometry, cell volume, dimension of the connecting capillary, or additional special outfitting (e.g., integration marker, auto zero), analytical characteristics such as noise, linearity, sensitivity, and response must be considered. A recently published extensive survey of the different commercially available detectors is given in [21]. A good LC detector should have the following characteristics:

1) High sensitivity and detection limit
2) Good selectivity
3) Fast response
4) Wide range of linearity
5) No contribution to column band broadening (small dwell volume)
6) Reliability and convenience

Detector Sensitivity and Detector Noise. *Detector sensitivity* is the quantitative response of the detector signal to the amount of solute introduced (slope of the plot sample input M versus signal output A: sensitivity $= \mathrm{d}A/\mathrm{d}M$).

Noise originates from the electronics associated with the detector and from fluctuations of the physical parameters in the detector environment (e.g., flow pulsation or temperature change in refraction detectors, or light intensity changes in UV detectors). For further information, see [11].

Detector Linearity. Detectors differ in their linear dynamic range. The response of an ideal detector should be proportional to either the concentration or the mass flow rate of the solute (see

Table 1. Summary of the most important HPLC detectors

Detector	Type	Gradient possible	Maximum linear range	Maximum sensitivity, g/mL	Sensitive to changes in
UV–VIS	selective	yes	10^3-10^5	$10^{-9}-10^{-10}$	pressure (temperature)
Diode-array detector (DAD)	selective	yes	10^3-10^4	$<2\times10^{-5}$	flow, pressure (temperature)
Refractive index (RI) detector	universal	no	10^3-10^4	10^{-7}	temperature (10^{-4} RIU °C), pressure
Fluorescence–phosphorescence detector	selective	yes	10^2-10^4	$<10^{-9}$	flow (temperature)
Fast-scanning detector	selective	yes	10^3	$<10^{-5}$	flow, temperature
Evaporative light scattering detector (ELSD)	universal*	yes	10^2-10^4	10^{-8}	flow
Amperometric detector	selective	yes	10^6	10^{-12}	flow, temperature (1.5 %/°C)
FTIR detector	selective	yes	10^4	10^{-4}	flow
Mass selective detector (MS)	universal	yes	10^2-10^4	$<10^{-9}$ **	flow

* Only universal for nonvolatile solutes.
** Grams per second.

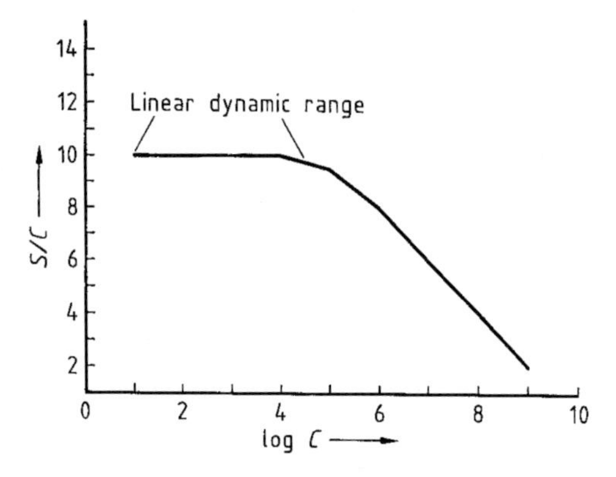

Figure 8. Plot used to determine detector linearity

Section 12.2.6.1). In reality, deviations from linearity usually occur at high concentrations (saturation effect). An easy way to determine the linear dynamic range of a detector is to plot the signal-to-concentration (or sample size) ratio versus the logarithm of concentration (sample size) (see Fig. 8).

12.2.6.1. Concentration-Sensitive and Mass-Flow-Sensitive Detectors

Two kinds of detectors exist, those that respond to solute concentration in the mobile phase and those that respond to changes of the solute mass flow rate into the detector cell. *Concentration-sensitive detectors* are used most frequently in LC. They are independent of the mass flow rate of the solute and are generally nondestructive (e.g., UV and differential detectors). (An easy way to find out whether a detector is concentration sensitive or not is to stop the solvent flow during a run. If the signal remains constant, it is a concentration-sensitive detector.)

The signal of *mass flow-rate-sensitive detectors* is proportional to the product of the solute concentration and the mobile phase flow rate. Consequently, if the flow rate is zero the signal is also zero, whatever the concentration may be. Detectors of this group are usually destructive (e.g., electrochemical detectors and mass spectrometers).

12.2.6.2. Differential Refractive Index (RI) Detector

The RI detector was developed in 1942 by Tiselius [22] and was one of the first commercially available on-line detectors. It is concentration sensitive and nondestructive, which makes it interesting for preparative chromatography. The detection limit of 10^{-6} g/mL is not as good as that of light absorption detectors (e.g., fluorescence or UV detectors), but the RI detector (RID) is a universal detector. To enhance the detection limit, a mobile phase should be chosen that differs as much as possible in its refraction index from the sample. A great disadvantage of the RID is that no eluent gradient can be run and the detector is very sensitive to even slight temperature variations [1 °C causes a change in the refractive index of

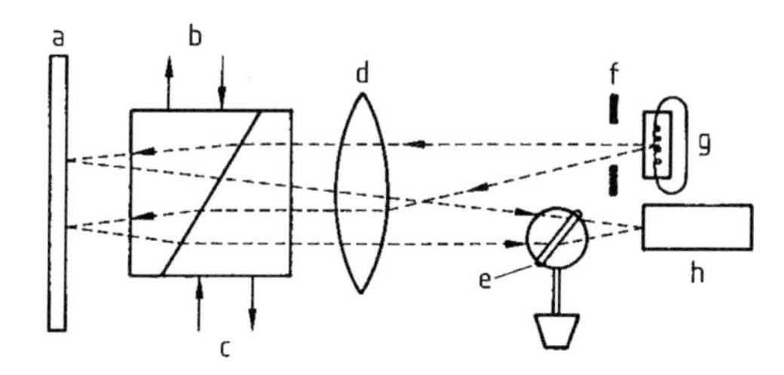

Figure 9. Diagram of a beam deflection (refractive index) detector [24]
a) Mirror; b) Sample cell; c) Reference cell; d) Lens; e) Optical zero-point adjustment; f) Mask; g) Lamp; h) Detector

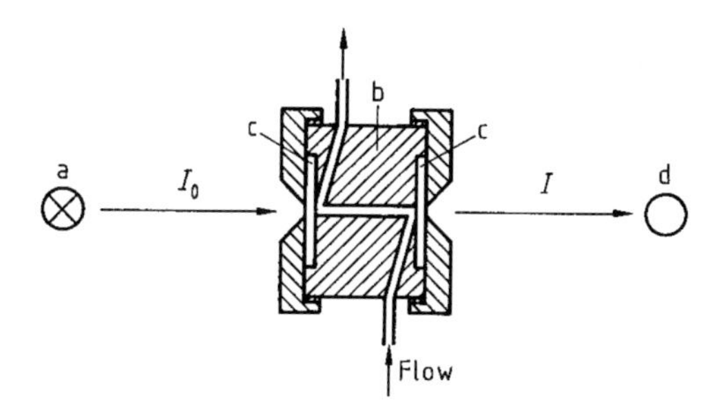

Figure 10. Diagram of a simple UV detector Z-pattern flow cell
a) Lamp; b) Flow cell; c) Quartz window; d) Photo detector

ca. 10^{-4} refractive index units (RIU)]. Therefore, very good thermostating must be provided.

The four general methods of measuring refractive index are the Christiansen, the Frensel, the beam deflection, and the interferometric methods. Today, the most commonly applied technique in HPLC is the beam deflection method.

Beam Deflection Refractive Index Detector. A schematic of the beam deflection refractive index detector is shown in Figure 9. Here, the deflection of a light beam is measured in a single compact cell. A light beam from the incandescent lamp (g) passes through the optical mask (f), which confines the beam within the region of the sample (b) and the reference cell (c). The lens (d) collimates the light beam, and the parallel beam passes through the cells to a mirror (a). The mirror reflects the beam back through the sample and reference cell to the lens, which focuses it onto a photo cell (h). The location of the focused beam, rather than its intensity, is determined by the angle of deflection resulting from the difference in refractive index between the two parts of the cell. As the beam changes location on the photo cell, an output signal is generated that is amplified and sent to the recorder.

12.2.6.3. UV – VIS Absorption Detector

The UV – VIS detector is the most frequently used detector in HPLC (Fig. 10). It is simple to handle, concentration sensitive, selective, and nondestructive. The latter also makes it suitable for preparative chromatography.

When light passes through a liquid (mobile phase), the intensity of absorption is proportional to the concentration of the analyte in the mobile phase C and the optical path length L (Lambert – Beer law):

$$\log I_0/I = E\,L\,C$$

where I_0 is the intensity of the light beam passing through the detector cell without the sample; I the intensity of the light beam passing through the detector cell with the sample; I_0/I the transmittance; E the molar extinction coefficient; L the optical path length; and C the concentration of analyte in the mobile phase.

The Lambert – Beer law is theoretically valid only for measurements at the absorption peak maximum and for monochromatic light; it assumes that only one species present, absorbs light of the wavelength used. Only then does the Lambert – Beer law predict a linear relationship between the concentration and the response of the detector. The chromatographer should be aware of the fact that these optimal conditions are usually not fulfilled with real samples and the linearity of the Lambert – Beer law is not always obeyed. This is often the case when working with highly concentrated sample solutions. Generally, a sample peak should not be quantified if the absorption exceeds 1 AUFS (absorption units full scale). In working with UV – VIS detectors, many solvents also absorb — especially in the UV range (see eluotropic order) — which decrease the sensitivity significantly.

The analytes to be monitored must contain a chromophore; otherwise, UV – VIS detection is impossible. If no chromophore is present, one can often be introduced into the molecule by pre- or postcolumn derivatization (see Section 12.8.2).

Besides fixed-wavelength UV detectors, which usually operate at 254 nm, modern UV – VIS de-

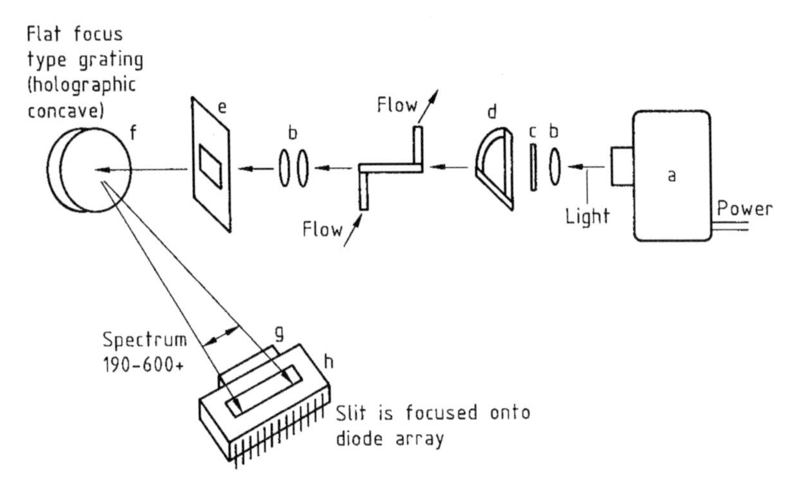

Figure 11. Diagram of a diode-array detector (DAD)
a) D_2 lamp; b) Lens; c) Aperture; d) Shutter; e) Slit; f) Flat focus-type grating (holographic concave); g) Second-order filter; h) Diode array
Courtesy of Waters-Millipore

tectors have an adjustable wavelength from 190 to ca. 600 nm. This enables individual adjustment of the measuring wavelength according to the specific absorption characteristic of the solute. An additional advantage of the adjustable wavelength UV – VIS detector is that a UV – VIS spectrum of the solute can be obtained by simply stopping the mobile phase flow while the solute is inside the detector cell and scanning the sample over a specific range of wavelength. This can be very helpful in determining optimum wavelength at which a particular sample can be detected or identifying an analyte by its characteristic spectrum.

Some UV – VIS detectors, the so-called *multiwavelength detectors* (*MWDs*), are capable of monitoring several different wavelengths simultaneously, thereby enabling the sensitive detection of various analytes that differ in their adsorption maximum.

12.2.6.4. Diode-Array Detector

The diode-array detector (DAD) enables the simultaneous measurement of absorption versus analysis time (chromatogram) and absorption versus wavelength (spectrum), without having to stop the mobile phase flow. Diode-array detectors can record spectra simultaneously every few milliseconds at different points of an elution band (peak) during a chromatographic run. Usually the spectra are taken at the beginning, the maximum, and the end of an eluting peak. Comparison of the spectra obtained supplies important information concerning the identity and purity of the monitored peak.

Like modern UV – VIS detectors, the DAD can also monitor a sample at more than one wavelength at a time, thus enhancing sensitivity dramatically if the absorption maxima of the test compounds differ. Figure 11 shows a schematic diagram of a typical DAD. Here, an achromatic lens system focuses polychromatic light into the detector flow cell. The beam then disperses on the surface of a diffraction grating and falls onto the photo diode array.

At present, good commercially available DADs have 512 diodes to cover a 600-nm wavelength range (ca. 190 – 800 nm). With decreasing number of diodes (for a set wavelength range) the resolution of the obtained spectra decreases as well, which results in a significant loss of information.

The drawback of a DAD is that because the light is split, up in front of the diodes, the resulting sensitivity is not as good as that obtained with classical UV – VIS detectors. Modern DADs are approximately a factor of 2, and older DADs a factor of 10 or more, less sensitive than normal UV – VIS detectors.

Fast-Scanning Detector (FSD). The fast-scanning detector is a relatively inexpensive alternative to the diode-array detector in LC. Compared

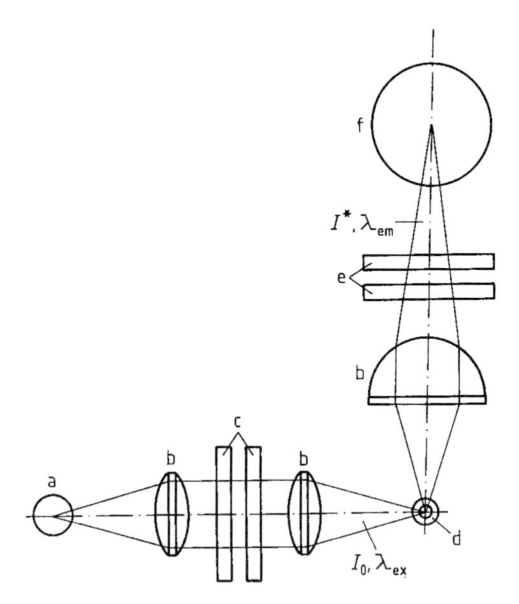

Figure 12. Diagram of a fluorescence detector
a) Lamp; b) Lens; c) Excitation filter; d) Flow cell; e) Emission filter; f) Photomultiplier

to the DAD it has only one diode (the reason why it is cheaper) and functions in principle as follows: Polychromatic light incident on a rotating grated mirror is split up into single wavelengths, which are then registered one after another by the diode. The required scan time is longer (ca. 1 s) than with a DAD (millisecond range). For liquid chromatography, this does not present a major problem because an average peak width is ca. 20 s, so enough time is available to monitor several spectra per peak for purity control, for example. In capillary electrophoresis (CE) this slowscan time can become a problem because peak width can be extreme short.

Indirect UV Detection. At the end of the 1980s, a new variation of UV detection was developed. This so-called indirect UV detection uses the same equipment as the classical DAD or UV–VIS detector. The difference between these methods is that in the classical mode the analyte contains a chromophore. In indirect UV detection a chromophore is added to the mobile phase so that a continuous, positive baseline signal is generated. If an analyte that has no chromophore passes the detector cell, the adsorption of the mobile phase is decreased and a negative peak is recorded. The advantage of this method is that analytes without chromophores (e.g., alkali and alkaline-earth ions) can be detected; the present disadvantage is its relatively poor detection limit (between 0.1 and 1 mg/L).

12.2.6.5. Fluorescence Detector

Fluorescence detection provides high selectivity and sensitivity, which is the reason for its use in trace analysis. The high selectivity results from the fact that only a few analytes are able to fluoresce. Typically, two types of spectra are obtained from fluorescing molecules: *excitation spectra* and *emission spectra*. Both spectra are characteristic for each analyte and therefore a great help in identification.

The high selectivity is due to the different detection technique applied (Fig. 12) compared to absorption spectroscopy (e.g., UV detector). The sample that passes the detector cell is irradiated by a light beam with a certain intensity I_0 and excitation wavelength λ_{ex}. Fluorescing solutes are able to emit a radiation I^* with a characteristically higher wavelength λ_{em} which is recorded and converted into an electric signal.

Since the emited radiation is not monochromatic, selectivity can be enhanced by inserting a monochromator into the beam. In this case the increase in selectivity is paid for by a decrease in sensitivity.

With help of pre- or postcolumn derivatization (Section 12.8.2), nonfluorescing analytes can be transformed into detectable ones, with very high selectivity and detection limits.

Since 1990, *laser-induced fluorescence* (*LIF*) detection has become increasingly important. A good review of the principles and applications of LIF detection is given in [25]. For more detailed information on fluorescence detection, see [26] [27].

12.2.6.6. Evaporative Light-Scattering Detector

The evaporative light-scattering detector (ELSD), sometimes wrongly called a mass detector, is influenced by many variables. Schulz and Engelhardt [28] showed that its specific response in addition to the analyte's molecular size also depends on the linear flow rate of the mobile phase; consequently, this detector is mass sensitive. Furthermore, they found that the ELSD behaves in certain situations more like a selective and not a universal detector. For example, it can detect only those solutes that are nonvolatile under

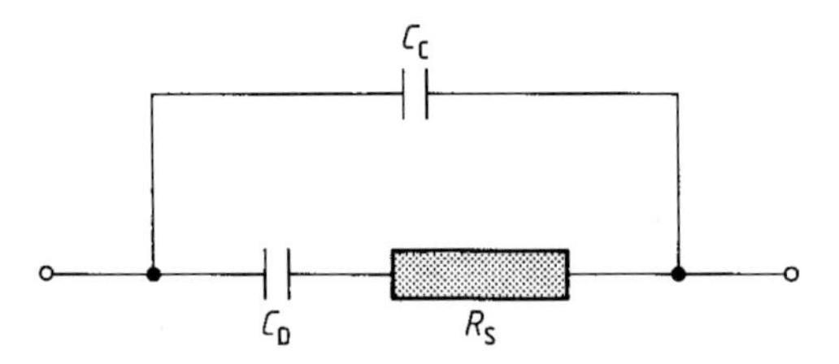

Figure 13. Equivalent circuit of conductivity measuring cell

the applied detection conditions. Nevertheless, ELSD can be a good alternative to the RI detector. Its sensitivity is better by one to two orders of magnitude, and a gradient can be run that would be impossible when measuring the refractive index. Another advantage of the ELSD compared to the RI and UV (< 200 nm) detectors, is the much better signal-to-noise ratio and a reduced baseline drift tendency. Typical applications of the ELSD are in the analysis of sugars and other carbohydrates, polymers, resins, steroids, and bile acids.

12.2.6.7. Infrared Detectors—Fourier Transform Infrared Spectroscopy

Infrared detectors offer the possibility of scanning IR spectra during a chromatographic run. IR detection is limited to a mobile phase that does not absorb in the same range as the analyte.

Modern interfaces joining LC and Fourier transform infrared spectroscopy (FTIR) function in two ways: The first technique is based on a flow cell ("on-line"), in which sensitivity is limited by the dilution of the analyte in the mobile phase. In the second technique the mobile phase is evaporated, leaving the analyte as a small spot that is scanned ("off-line") by FTIR. The advantage of this approach is a significant enhancement in sensitivity. Compared to the refractive index detector the IR detector has a lower sensitivity but theoretically offers the possibility of gradient runs.

12.2.6.8. Electrochemical Detection

In liquid chromatography, electrochemical detection is a superior alternative to optical detection for many analytical problems [29]. Electrochemical detectors monitor currents, electrode potentials, or electrolytic conductivities. In all cases the signal is directly proportional to the concentration or mass flow rate over relatively wide linear ranges. With electronic amplification and simple signal processing, electrical quantities can be measured with high sensitivity.

Conductometric Detectors. Conductometric detectors are sensitive to ionic solutes in a medium with low conductivity, for instance water or another polar solvent [30]. The response is relatively nonspecific since the conductivity of all ions present in the eluate is determined.

The reciprocal of the resistance is the conductance G (Ω^{-1}); it is expressed in the unit siemens (S). The conductivity κ is the product of the conductance G times the quotient of the cell electrode spacing D and area A (the cell constant):

$$\kappa = G(D/A)$$

The conductivity κ has the dimension microsiemens per centimeter.

The cell constant is important only for absolute measurements; it is determined by calibration with solutions of known conductivity. For relative measurements (i.e., to track eluate conductivity changes on a chromatographic column), the cell constant of the detector need not be known.

Conductivity detectors are important mainly for ion-exchange chromatography. In analytical practice, ion-exchange chromatography with the background conductivity suppressed is used chiefly for the highly sensitive determination of inorganic and organic anions and of alkali and alkaline-earth cations (see Section 12.5.3). The calibration curves (response peak height or area versus concentration) are linear over a wide range, up to three orders of magnitude; the limits of detection are in the nanogram-per-gram range [12].

Conductivity is measured in a cell containing two plates as electrodes (usually made of platinum) through which the column effluent flows. The electrodes have a known surface area and are located in the cell at a fixed distance. The cell is connected in an a.c. circuit and is characterized by the double-layer capacitance C_D, the cell capacitance C_C, and the electrolyte resistance R_S of the sample solution (see the equivalent circuit diagram in Fig. 13).

For the electrolyte resistance R_S to be determined, the effect of C_C must be negligible:

$$\frac{1}{2 \cdot \pi \cdot f \cdot C_C} \gg R_S$$

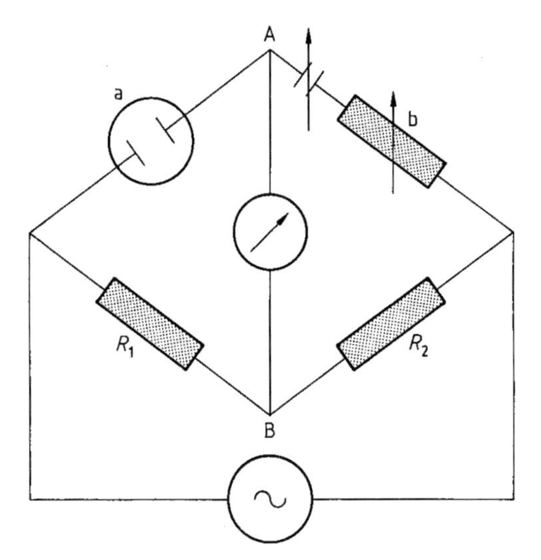

Figure 14. Wheatstone bridge circuit
a) Measuring cell; b) Balancing leg

where f is the frequency in Hz of the alternating current.

In practice, this condition is satisfied whenever very dilute solutions or solutions of weak electrolytes are measured with low-frequency alternating currents ($f = 40 - 80$ Hz) or when solutions having conductivities of about 100 µS/cm are measured with frequencies in the kilohertz range. In this way, R_S can be determined fairly accurately with a Wheatstone bridge circuit, shown in simplified form in Figure 14.

Because the impedance of the measuring cell has a capacitive component, the balancing leg in the bridge circuit must include not only the variable resistance but also a variable capacitance. The bridge is balanced or nulled when the a.c. voltage between points A and B is zero. Direct-reading instruments are used to check the balance; for higher measurement accuracy the signals may be amplified first if necessary.

If the conductivity cell is used in combination with a chromatographic column, the measuring electrodes have the form of capillaries. This makes it possible to keep the cell volume very small, as required; the volume is ca. 1 – 2 µL. Figure 15 illustrates the principle of an HPLC conductivity detector with two measuring electrodes. The cell must be thermally insulated because the conductivity is temperature dependent.

Four- and six-electrode cells have been designed for precision measurements [13], [14].

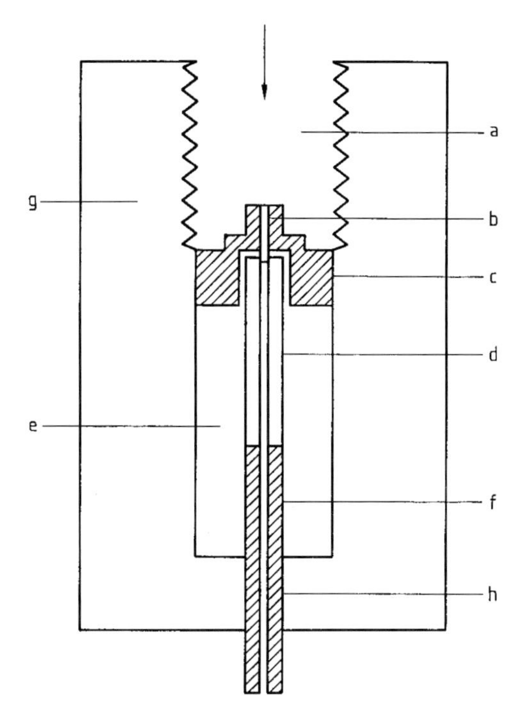

Figure 15. HPLC conductivity detector [31]
a) Connection to column; b) Platinum capillary; c) Electrical connection; d) Polytetrafluoroethylene capillary; e) Polytetrafluoroethylene; f) Electrical connection; g) Plastic detector housing; h) Stainless steel capillary
Measurement is performed between (b) and (h)

Potentiometric Detectors. Essentially, potentiometric detectors consist of ion-selective electrodes (ISEs) to monitor ionic species in chromatographic effluents. Examples of ISEs are glass, solid-state, liquid-membrane, and enzyme electrodes.

The detector response is the potential of the electrode, which depends on the activity of the analyte. This potential is measured with respect to a reference electrode. The response can be described by [32], [33]

$$E = E_0 + \frac{s}{z_i} \log \left(a_i + \sum K_{ij}^{\mathrm{Pot}} a_j^{z_i/z_j} \right)$$

where E is the electromotive force of the cell assembly in volts; z_i and a_i are the charge number of the primary ion i and its activity in the sample solution; z_j and a_j are the charge number of the interfering ion j and its activity in the sample solution; K_{ij}^{Pot} is the selectivity factor, a measure of the sensor's preference for interfering ion j rel-

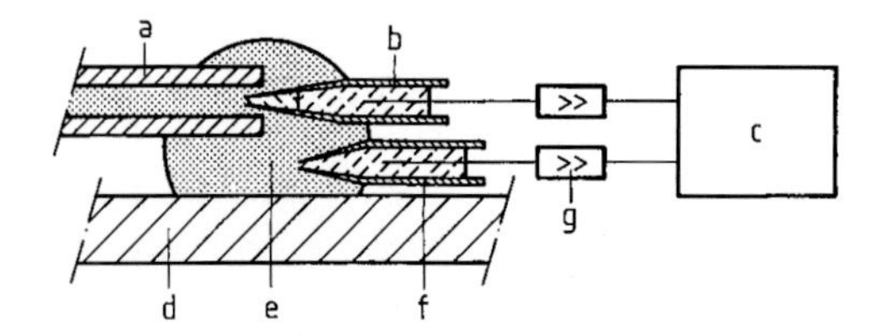

Figure 16. Potentiometric detector with ion-selective microelectrodes [33]
a) Capillary column; b) Ion-selective microelectrode; c) Voltmeter-recorder; d) Glass plate; e) Eluent drop; f) Reference electrode; g) Impedance converter

ative to measured ion i (an ideally selective membrane electrode would have all $K_{ij}^{\mathrm{Pot}} = 0$); E_0 is the potential difference equal to a constant (for a given temperature) plus the liquid junction potential; and $s = 2.303\,RT/z_iF = 59.16\ \mathrm{mV}/z_i$ (at 25 °C for the ideal case).

The activity of an ion is related to its concentration c_i (moles per liter) by $a_\mathrm{i} = c_\mathrm{i} \cdot \gamma_\mathrm{i}$, where γ_i is the activity coefficient of the ion.

Ion-selective electrodes are less important for monitoring chromatographic eluates than for automated flow injection analysis (FIA).

Major characteristics of potentiometric detectors are high specificity, low sensitivity, and slow response.

Only a few examples can be found in the literature in which ISEs have been used as detectors for ion chromatography. They have been employed to detect nitrate in the presence of nitrite [32] and to detect alkali cations [33], [34] and a variety of anions [35], [36] after isolation by ion-exchange liquid chromatography.

Potentiometric detectors vary widely in cell design. One example worth noting is a microtype membrane cell with two compartments separated by an ion-exchange membrane. The column eluate is passed through one compartment of the detector, and the reference solution with the same components of the eluate is passed through the other. This device is useful for monitoring ion-exchange as well as reversed-phase ion-pair chromatography; it offers detection limits in the nanomole range, together with a linear response over about two to three orders of magnitude [34].

The design of a picoliter-volume cell potentiometric detector for open tubular column liquid chromatography is shown in Figure 16. The microelectrode has a tip diameter of 1 μm and is inserted in the open tubular column. A cation-exchange solution can be used as membrane. The reference electrode, located in a micropipette, is a chlorinated silver wire in 3-mol/L KCl. The response time of the microdetector is less than 10 ms [33].

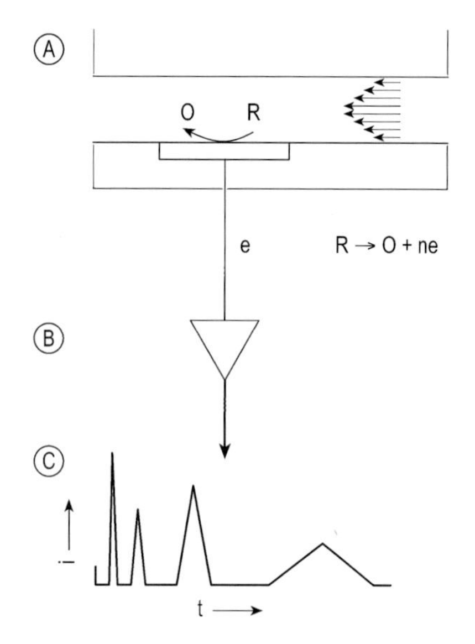

Figure 17. Schematic view of thin-layer amperometric detection (conversion of reactant (R) to product (O) on the surface of the working electrode)
A Detector Cell; B Amplifier; C Chromatogram

Amperometric Detectors. Amperometric detectors are the most commonly used electrochemical detectors for highly sensitive and selective determinations in HPLC [30], [31]. The frequently used synonym "electrochemical detector" is not very precise in relation to the detection mode.

The signal current is the result of an electrochemical conversion of the analyte by oxidation or reduction at the working electrode (Fig. 17). The currents are in the pA – μA range and are measured by the three-electrode technique (→ Analytical Voltammetry and Polarography). As a result extremely low detection limits at the 0.1-pmol level can be achieved.

The simplest and most common detection mode is current measurement at constant potential. The reference and counter electrodes should be located downstream of the working electrode so that reaction products at the counter electrode or leakage from the reference electrode do not interfere with the working electrode. The current is amplified and plotted versus time to yield a chromatogram.

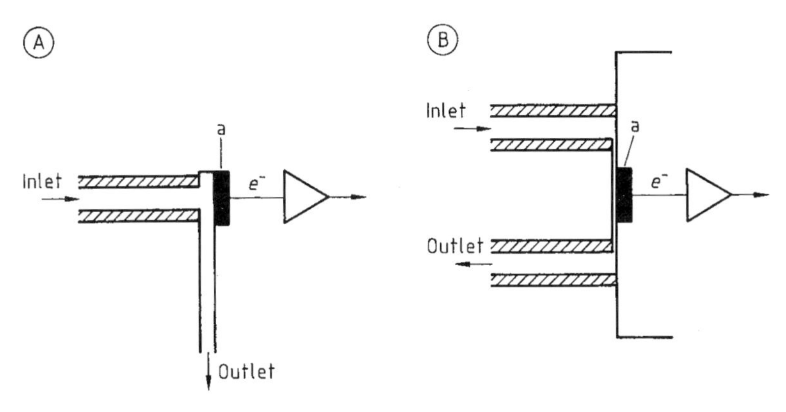

Figure 18. Schematics of the most common amperometric detector cells
A) Wall-jet cell; B) Thin-layer cell
a) Working electrode

Because the electrode process takes place solely at the surface of the working electrode, amperometric flow cells can determine only local (interfacial) concentrations. Conductometric and potentiometric measuring cells, on the other hand, can track the average volumetric concentrations in the eluate. The same holds for UV detection, which is preferred for routine HPLC studies (see Section 12.2.6.3).

In amperometric detector cells, the analyte is transported to the working electrode surface by diffusion as well as convection; migration is suppressed by the supporting electrolyte. Electrolyte concentrations in the eluent of 0.01 – 0.1 mol/L are sufficient. Several cell geometries employing solid electrodes have been designed and tested. The two types shown in Figure 18 are the most commonly used.

In the *thin-layer cell*, the eluate is forced through a thin, rectangular channel, normally 0.05 – 0.5 mm thick. The working electrode (usually a disk electrode, 2.4 mm in diameter) is embedded in the channel wall. In the "wall-jet" design, the solution is introduced into the cell through a small diameter jet or nozzle and impinges perpendicularly onto the planar electrode. In both cases, the volume is ca. 1 – 2 μL. The advantages of these cells in comparison with other detector designs are good hydrodynamic properties and fast response [37], [38].

The thin-layer cell for HPLC was originally described by KISSINGER and coworkers in 1973 [39]. The wall-jet principle has been known since 1967 and was originally introduced for HPLC in 1974 [40]. The development is based on KEMULA'S work with a mercury drop as detector for column chromatography [41].

When mercury is used as the electrode material, detection is possible over a particularly wide cathodic potential range. The anodic range, however, is limited. It is thus commonly used for the detection of reducible species. Detectors which utilize a static mercury drop are particularly attractive as the drop (working electrode) is easily renewed for each chromatogram; this is quite an advantage for the reproducibility of response [42]. The construction and operation of detectors based on mercury drop electrodes have been reviewed [42], [43].

Solid electrodes, which have a larger anodic range than mercury-based electrodes, include different types of carbon and noble-metal electrodes. The preferred electrode material for commercial detectors is glassy carbon. Its electrochemical properties allow measurements in a wide potential range of about – 0.8 to +1.2 V. The effective range (chiefly the cathodic voltage range) depends on the composition, mainly on the pH value, of the eluate. In contrast to carbon-paste electrodes, which have been proposed for amperometric detection — particularly because of the low background response and the good signal-to-noise ratio — glassy carbon electrodes offer better chemical resistance. They are stable in all solvents employed, such as acetonitrile and methanol as mobile phase.

Noble-metal (Au, Pt) electrodes have been employed for special studies, and gold amalgam electrodes as well as mercury-film electrodes have been used for reductive determinations [44] – [46].

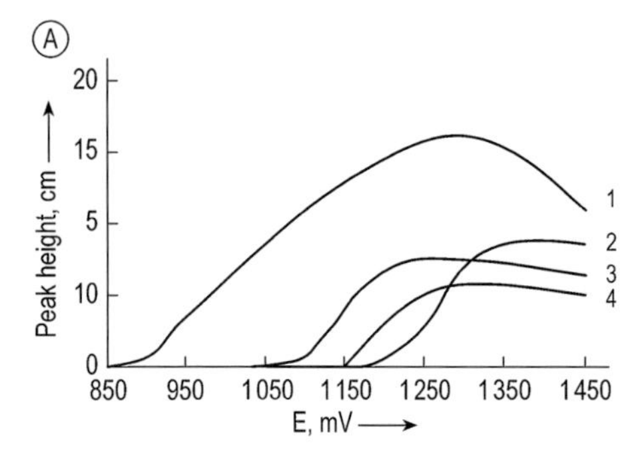

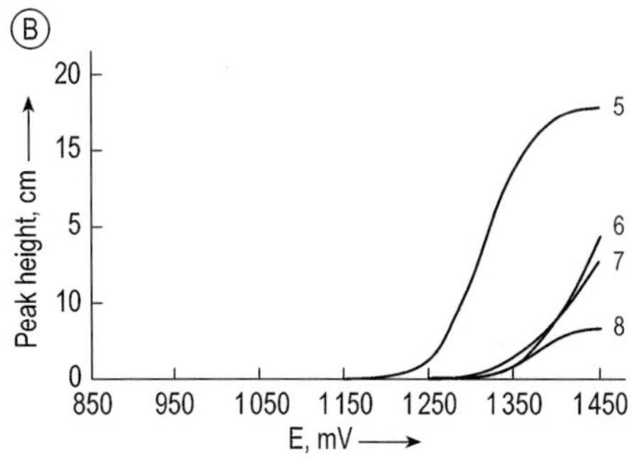

Figure 19. Hydrodynamic voltammograms of polycyclic aromatic hydrocarbons (Eluent: methanol/water 85/15) [47], [67].
1) Benzo[*a*]pyrene; 2) Benzo[*k*]fluoranthene; 3) Benzo[*ghi*]perylene; 4) Indeno[1,2,3-*cd*]pyrene; 5) Acenaphthene; 6) Fluoranthene; 7) Acenaphthalene; 8) Benzo[*b*]fluoranthene.

The basis of the good selectivity of amperometric detection is that only compounds with electroactive functional groups (reducible or oxidizable) can undergo reaction at the electrode. The potential at which these electrode reactions occur depends on the redox behavior of the compounds of interest and the working conditions (working and reference electrodes, composition and pH of the mobile phase). The most suitable detection potential is best determined experimentally from the cyclic voltammogram or the hydrodynamic voltammogram of the analyte.

A *hydrodynamic voltammogram* is a current–potential curve which shows the dependence of the chromatographic peak height on the detection potential. The technique used to obtain the necessary information is voltammetric flow injection analysis, in which an aliquot of the analyte is injected into the flowing eluent prior to the detector and the peak current recorded. This is repeated many times, the detector potential being changed after each injection, until the peak current–potential plot reaches a plateau or a maximum, as shown for some polycyclic aromatic hydrocarbons (PAHs) in Figure 19.

Whereas indeno[1,2,3-*cd*]pyrene, benzo[*ghi*]perylene, benzo[*k*]fluoranthene, and benzo[*a*]pyrene give a maximum detection signal at potentials between + 1.30 and + 1.35 V (Figure 19 A), the current–potential curves of acenaphthalene, acenaphthene, fluoranthene and benzo[*b*]fluoranthene do not reach a maximum even at + 1.45 V (Figure 19 B). A measurable signal is first obtained at + 1.35 V for acenaphthalene benzo[*b*]fluoranthene, and fluoranthene, whereas the other PAHs are oxidized at less positive potentials. Based on these results, it is possible to selectively determine benzo[*a*]pyrene in a mixture of PAHs by using a detection potential of + 1.0 V. At more positive potentials, other peaks appear in the chromatogram until finally, at a potential of + 1.45 V, peaks for all the PAHs are recorded, as shown in the chromatograms in Figure 20.

When using amperometric detectors, the composition of the eluent should remain as constant as possible (isocratic elution). If gradient elution is used to reduce the analysis time, then base line drift is observed, resulting in a loss of sensitivity. This is particularly so with glassy carbon working electrodes and is due to the slow adjustment of the electrochemical equilibrium at the electrode surface to the changing eluent composition.

Working electrodes made of other materials such as the ultratrace graphite electrode (porous graphite structure impregnated with epoxy resin, produced by Metrohm, Switzerland), have much shorter lead-in times. The relatively rapid response of this kind of electrode arises from its semimicro electrode properties [48].

When used as the working electrode in amperometric detection following gradient elution, e.g., in the analysis of beno[*a*]pyrene (BaP) and its metabolites, the hydroxybenzo[*a*]pyrenes, which result from the enzymatic degradation of BaP in animals and humans after uptake in food or from the air, rapid adjustment to steady state conditions is an important requirement and leads to a more stable base line (Figure 21 B). For comparison, the chromatogram using the much slower isocratic elution is shown in Figure 21 A. The detection limits for the analysis of benzo[*a*]pyrene and its metabolites lie between 0.2 ng and 0.4 ng and the determination limits between 0.3 ng and 0.6 ng when either the glassy carbon electrode or the ultratrace electrode is used.

Amperometric detection after HPLC separation has been used to detect easily oxidizable

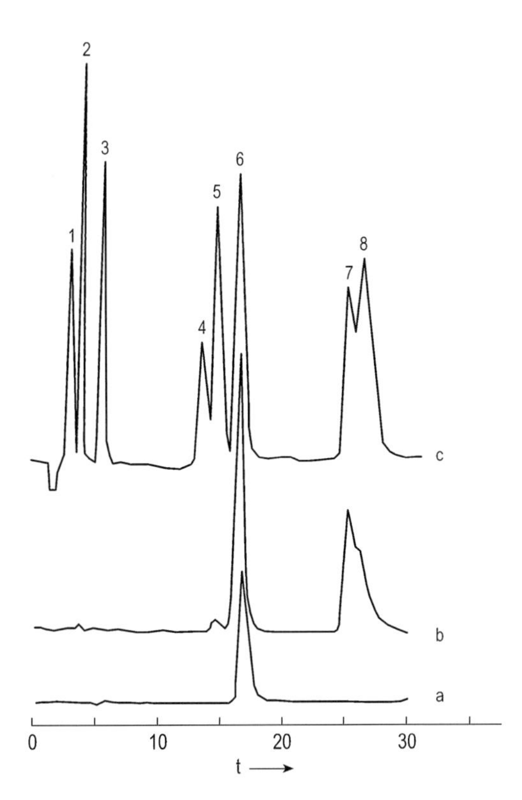

Figure 20. Determination of polycyclic hydrocarbons by HPLC with amperometric detection.
a) Detection potential + 1.0 V : selective determination of 10 ng benzo[*a*]pyrene (6). b) Detemination of 10 ng benzo[*a*]pyrene (6) and 10,5 ng benzo[*ghi*]perylene (7) at + 1.2 V. c) Determination of 8.6 ng acenaphthalene (1), 11.4 ng acenaphthene (2), 51,5 ng fluoranthene (3), 9.5 ng benzo[*a*]fluoranthene (4), 10.5 ng benzo[*k*]fluoranthene (5), 10 ng benzo[*a*]pyrene (6), 10.5 ng benzo[*ghi*]perylene (7) and 10 ng indeno[1,2,3-*cd*]pyrene at a detection potential of + 1.45 V.
Eluent : methanol/water 85/15 with 2 g/L trichloracetic acid as supporting electrolyte. Working electrode: glassy carbon; reference electrode: Ag/AgCl in methanolic KCI. The detection and determination limits were in the range 100 – 300 pg per injection [47], [67].

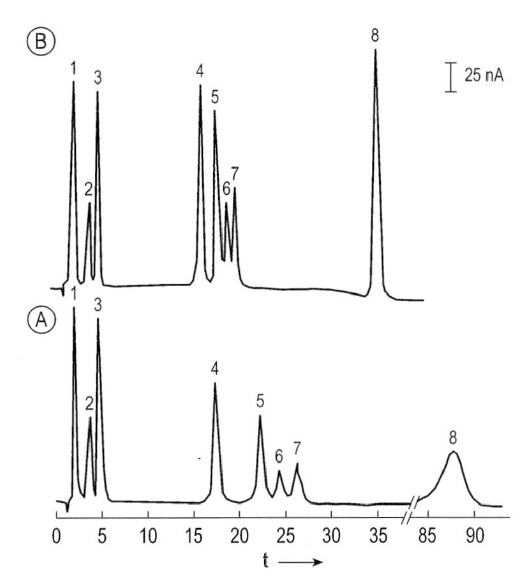

Figure 21. Chromatograms of BaP and metabolites (each 90 ng). Amperometric detection at + 1.35 V (vs Ag/AgCl, 3 M KCl) using an ultratrace electrode with isocratic elution (A) and a glassy carbon electrode with gradient elution (B) [49].
1) Benzo[*a*]pyrene-*trans*-9,10-dihydrodiol; 2) Benzo[*a*]pyrene-*trans*-4,5-dihydrodiol; 3) Benzo[*a*]pyrene-*trans*-7,8-dihydrodiol; 4) 9-Hydroxybenzo[*a*]pyrene; 5) 7-Hydroxybenzo[*a*]pyrene; 6) 1-Hydroxybenzo[*a*]pyrene; 7) 3-Hydroxybenzo[*a*]pyrene; 8) Benzo[*a*]pyrene.

phenols, aromatic amines, indoles, phenothiazines, thiols, and polyaromatic hydrocarbons (Table 2) and easily reducible nitro compounds and quinones (Table 3). These methods have been used in environmental, pharmaceutical, biochemical, and food analysis, as well as for determining hormone residues in soil and water samples and in plant and animal products. The main application is in the analysis of phenols and aromatic amines such as the catecholamines, which can be detected with high sensitivity in the lower picogram region in body fluids and biological samples [51] – [55].

While the current is measured at a constant voltage in amperometric detectors, in voltammetric detectors it is measured while the voltage is changing rapidly. Either linear scan, staircase, square-wave, pulse or a.c. voltammetry can be used for the measurement. When compared to stationary methods, nonstationary methods with superimposed voltage pulses often result in higher selectivity and sensitivity and fewer problems from the adsorption of impurities and electrode reaction products on the electrode surface [53] – [58].

A technique for cleaning the surface of solid working electrodes to improve the reproducibility of the measurements (removal of electrochemical reaction products) is to alternate the potential repeatedly [59].

For the "fast-scan" technique in HPLC, in which the potential is varied rapidly across its full range, the method of square-wave voltammetry (→ Analytical Voltammetry and Polarography) has provided the basis for the development of a "rapid-scan square-wave voltammetric detector"

Table 2. Organic analytes detected amperometrically in the oxidative mode [50]

Analyte with electroactive group	Structural formula	Operating potential, mV	Examples
Aromatic amines			
Anilines		+1000	phenylenediamine, 3-chloroaniline, aniline, methylaniline, naphthylamine, diphenylamine
Benzidines		+ 600	3,3′-dichlorobenzidine
Sulfonamides		1200	
Aromatic hydroxyls			
Phenols, halogenated phenols		+1200	salicylic acid (and derivatives) benzofuranols, chlorophenols (and metabolites), paracetamol, hesperidin, tyrosine
Flavones		+1100	flavonoles, procyanidins
Catecholamines		+ 800	DOPA, dopamine norepinephrine, epinephrine, and their metabolites
Estrogens		+1000	estriol, catecholestrogens
Antioxidants		+1000 (200)	*tert*-butylhydroxyanisole, *tert*-butylhydroxytoluene
Hydroxycumarins		+1000	esculin
Aliphatic amines (*sec*- and *tert*-)		+1200	clonidine, diphenhydramine
Indoles		+1000 + 450	tryptophan, pindolol, bopindolol metabolit
Thiols	R–SH	+ 800	cysteine, glutathione, penicillamine
Phenothiazones		+1000	thioridazine, chlorpromazine
Pyrimidines		+1200	trimethoprim
Purines		+1000	uric acid
Hydrazines	R–NH–NH_2	+ 600	hydralazine
Miscellaneous			
Vitamines		+ 800–1000	vitamins A, E, C
Abusive drugs		+ 700–1200	morphine and derivatives, benzodiazepines, cocaine, hallucinogens, tetrahydrocannabinols, tricyclic antidepressants, neuroleptics
Pesticides, herbicides		+1100–1300	phenylurea derivatives, carbamates

Table 3. Organic analytes detected amperometrically in the reductive mode [50]

Analyte with electroactive group	Structural formula	Operating potential, mV	Examples
Benzodiazepines (and -4-oxides)	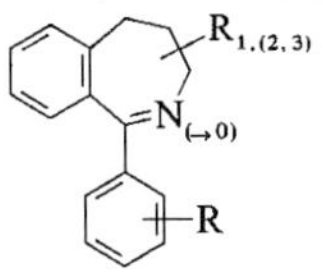 $R_{1,(2,3)}$; $=N_{(\rightarrow O)}$; $-R$	−500 to −1000	chlordiazepoxide, nitrazepam, diazepam
Steroid hormones	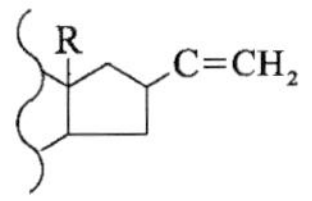 R; $C=CH_2$	−500	ethinylestradiol, levonorgestrel
Aromatic and aliphatic nitro compounds	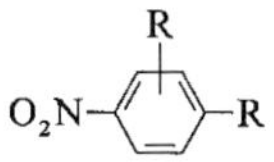 R; O_2N-; $-R$	−400 to −800	explosives (TNT, nitroglycerin), chloramphenicol, nitropolyaromatics, parathion derivatives, dinitrophenol pesticides
Quinones	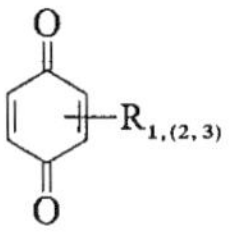 O; $-R_{1,(2,3)}$; O		vitamin K
Miscellaneous Nitrosamines, thioamides, organometallic compounds, peroxides, azo compounds			prothionamide, benzoyl peroxides, triphenyllead, pyridylazo resorcinol

[60]. In addition, a scanning microvoltammetric detector for open tubular liquid chromatography has been described [61]. This detector consists of a 9-μm (outer diameter) carbon fiber as electrode, inserted into the end of a 15-μm (inner diameter) capillary column. The potential can be scanned at rates up to 1 V/s over a range of 0.0 to +1.5 V. A distinct advantage of this detection method is its three-dimensional resolving power.

Figure 22 shows a conventional chromatogram of five compounds (ascorbic acid, epinephrine, tyrosine, dopamine, and hydroquinone) detected at +1.0 V. The third peak in the chromatogram consists of both tyrosine and dopamine, because

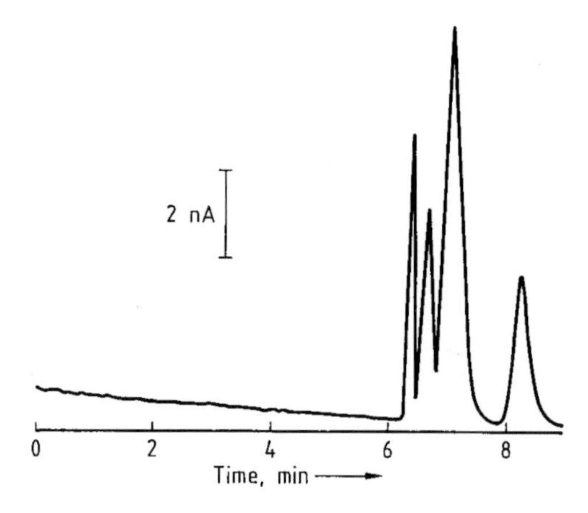

Figure 22. Single-potential chromatogram at +1.0 V versus Ag – AgCl of (in order of elution) 0.1-mmol/L ascorbic acid, 0.1-mmol/L DL-epinephrine, 0.2-mmol/L tyrosine, 0.1-mmol/L dopamine, 0.1-mmol/L hydroquinone
Scan rate was 1 V/s, flow rate 0.59 cm/s; electrode length, 1 mm [61]

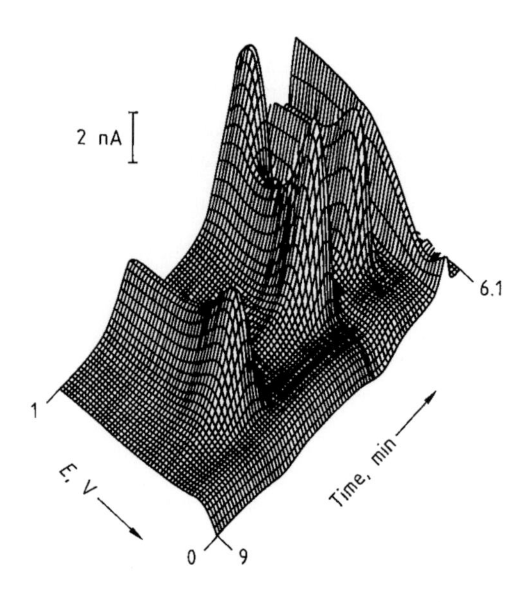

Figure 23. Partial threedimensional chromatovoltammogram of mixture shown in Figure 22, [61]

these are not resolved chromatographically. This is revealed by the three-dimensional chromatogram of Figure 23.

No single potential can be used to detect the presence of the two overlapping peaks. However, a potential scan yields a three-dimensional chromatogram of the mixture, in which the two peaks are quite obvious.

Coulometric Detectors. Coulometric detectors are based on potentiostatic coulometry [30]. The signal of the constant-potential measurements, as in amperometric detection, is the current resulting from an electron-transfer process (reduction or oxidation of the analyte arriving with the eluent) while the working electrode is held at constant potential.

In both flow-through detection techniques, the instantaneous measured current i_t is given by

$$i_t = n \cdot F \frac{dN}{dt}$$

where n is the number of electrons per mol exchanged in the reaction, F is the Faraday constant, and dN/dt is the number of molecules per unit time reacting at the working electrode.

The difference between coulometric and amperometric detection lies in the percentage of analyte converted at the working electrode. Of the molecules located in the measuring cell at a given time, only ca. 1 – 10 % take part in the electrode reaction in amperometric detection, whereas this conversion should be > 95 % in coulometric detection.

To achieve nearly 100 % conversion, the ratio of electrode area to cell volume must be very large in coulometric detectors. The working electrodes employed in coulometric detectors must therefore have a much greater surface area (ca. 5 cm^2) than those used for amperometric detectors. Large electrode areas, however, lower the signal-to-noise ratio and thus increase the detection limits [62]. For this reason, and also because of problems with electrode passivation, coulometric detection has not become an important technique in HPLC, even though the same classes of compounds can be determined as with amperometric detection.

A coulometric detector with two working electrodes and separate reference electrodes, configured one downstream of the other, is called a *dual-electrode cell* [15], [63], [64]. If different potentials are applied to the two working electrodes, two separate chromatograms can be recorded for the same eluate. If the cell is appropriately sized and large-area electrodes (e.g., porous graphite) are employed, current (conversion) efficiencies close to 100 % can be attained. The design of a dual-electrode coulometric detector is illustrated in Figure 24.

Substances with reversible electrode behavior can be determined more sensitively with dual de-

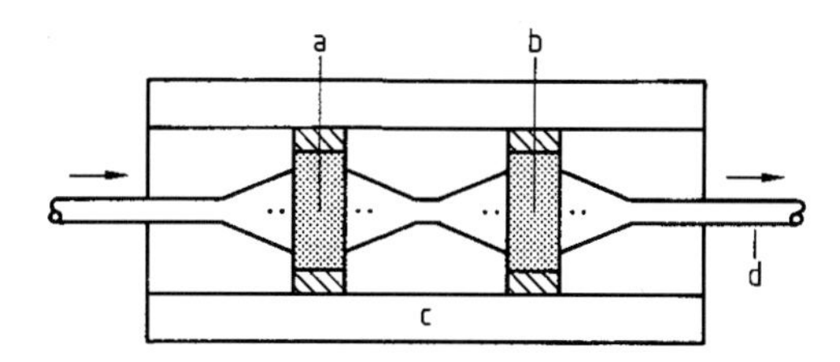

Figure 24. Coulometric detector with dual porous electrodes and total cell volume < 5 μL (Coulochem 5100 A electrochemical detector)
a) Test electrode 1; b) Test electrode 2; c) Stainless steel tubing

tectors than with conventional coulometric detectors. One example is the anodic and cathodic detection (redox mode) of catecholamines [65]. The determination can also be made more sensitive if the two detector signals are fed to a difference circuit so as to eliminate the background that is due chiefly to contaminants in the solvent (differential mode). Dual detectors operated at appropriate electrode potentials can also block out the signals of undesirable compounds (screen-mode detection).

12.3. Solvents (Mobile Phase)

12.3.1. Purification and Pretreatment of LC Solvents

Especially in trace analysis, purification and pretreatment—even of commercially available HPLC-grade solvents—are often necessary (e.g., by filtration or distillation). Occasionally, organic solvents have to be dried as well (normal phase chromatography), for instance, with a molecular sieve. Dissolved atmospheric oxygen in the mobile phase is a major source of trouble in LC. Oxygen increases the UV absorption below 230 nm, thus disturbing sensitive UV detection at lower wavelengths significantly. Dissolved air furthermore, leads to air bubbles inside the chromatographic system (often noticed in "low-pressure" gradient systems), which are responsible for pulsation of the pump or can cause "ghost" peaks when entering the detector cell (e.g., refractive index detector or UV detector). To prevent these problems, the mobile phase has to be degassed. The best method for degassing the eluent is to rinse it with an inert gas (e.g., helium). Often, however, applying vacuum degassing or simply degassing the eluent by ultrasonic treatment for several minutes before use is sufficient.

12.3.2. Elutropic Order

The more strongly a solvent is adsorbed to the stationary phase, the higher is its elution power. Because the solvent molecules outnumber the analyte molecules with respect to the available adsorption sites of the stationary phase, a relative weak solvent (mobile phase) is still strong enough to elute even quite polar sample compounds. In classical normal phase (NP) adsorption chromatography, the different solvents are listed according to increasing elution power (elution strength), resulting in their elutropic order (see Table 4). The elutropic order was developed for normal phase chromatography; consequently, water is the strongest solvent. Today, most separations are carried out in the reversed-phase (RP) chromatography, so the elutropic order is inversed, turning water into the weakest mobile phase (see Section 12.4.3).

With the help of the elutropic order, appropriate solvents can be chosen, or various solvent mixtures created, that have the same elution strength but different selectivity.

12.3.3. Selection of Mobile Phase

The mobile phase in LC consists of one or more solvents, which additionally may contain modifiers. Generally, the composition of the eluent should be kept as simple as possible. When working with gradients, solvated mobile phase modifiers (e.g., salts) might precipitate in the chromatographic system because of the dramatic change in eluent composition. The following factors must be considered in choosing an appropriate mobile phase for a specific LC problem:

1) Solubility of the sample in the eluent
2) Polarity of the eluent
3) Light transmission of the eluent
4) Refractive index of the eluent
5) Viscosity of the eluent
6) Boiling point of the eluent
7) Purity and stability of the eluent
8) Safety regulations
9) pH

Items 1 and 2 must be seen in the context of the separation system applied (see Chap. 12.5).

Table 4. Important characteristics of the most frequently used mobile phases in LC*

Solvent	Polarity	Solvent strength	Dielectric constant	UV cutoff, nm	Refractive index	Viscosity at 25 °C, mPa · s
Heptane	0.2	0.01	1.92	195	1.385	0.40
Hexane	0.1	0.01	1.88	190	1.372	0.30
Pentane	0.0	0.00	1.84	195	1.355	0.22
Cyclohexane	−0.2	0.04	2.02	200	1.423	0.90
Toluene	2.4	0.29	2.40	285	1.494	0.55
Ethyl ether	2.8	0.38	4.30	218	1.350	0.24
Benzene	2.7	0.52	2.30	280	1.498	0.60
Methylene chloride	3.1	0.42	8.90	233	1.421	0.41
1,2-Dichloroethane	3.5	0.44	10.40	228	1.442	0.78
Butanol	3.9	0.70	17.50	210	1.397	2.6
Propanol	4.0	0.82	20.30	240	1.385	1.9
Tetrahydrofuran	4.0	0.57	7.60	212	1.405	0.46
Ethylacetate	4.4	0.58	6.00	256	1.370	0.43
Isopropanol	3.9	0.82	20.30	205	1.384	1.9
Chloroform	4.1	0.40	4.80	245	1.443	0.53
Dioxane	4.8	0.56	2.20	215	1.420	1.20
Acetone	5.1	0.56		330	1.356	0.30
Ethanol	4.3	0.88	24.60	210	1.359	1.08
Acetic acid	6.0		6.20		1.370	1.10
Acetonitrile	5.8	0.65	37.50	190	1.341	0.34
Methanol	5.1	0.95	32.70	205	1.326	0.54
Nitromethane	6.0	0.64		380	1.380	0.61
Water	10.2		80.00		1.333	0.89

* Data taken from elutropic order in [19].

The *light transmission* of the eluent is of importance in using a UV detector. When monitoring a sample at a wavelength of 230 nm, acetone, for instance, (which absorbs up to 330 nm) cannot be used as a mobile phase. As long as the absorption of the eluent does not exceed 0.5 AUFS, it can be tolerated. Table 4 shows some important characteristics and the UV cutoff from often used eluents. Organic or inorganic trace compounds (impurities) in the mobile phase often shift the UV absorption to higher wavelengths.

When working with a refractive index detector, the difference in the *refractive index* of the sample and the mobile phase should be as great as possible to achieve better sensitivity and lower detection limits.

The *viscosity* of the mobile phase is responsible for the back-pressure of a given LC system. In addition, the higher viscosity of the eluent results in a lower diffusion coefficient for the sample, which leads to reduced mass transfer and peak broadening. Especially during gradient elution with two or more relatively polar organic solvents, unexpected changes in viscosity can occur [66].

The possibilities for influencing the viscosity of the mobile phase are limited. Working at higher temperatures can be an advantage in some cases (e.g., high-temperature liquid chromatography).

The *boiling point* of the solvents should also be considered. For preparative chromatography, a high-purity and low-boiling mobile phase is preferred. On the other hand, the boiling point should not be too low because that could cause gas bubbles in the pump or detector.

The pH of the mobile phase often has a dramatic influence on the selectivity and resolution of investigated samples. For many pharmaceuticals, the optimal pH of the mobile phase is between pH 4 and 5. During method development and optimization, it is recommended to run a sequence of samples between pH 2.5 and 7.5 in pH steps of 0.5. Plotting resolution and selectivity against mobile-phase pH then allows the optimum eluent pH to be identified.

Of course, the eluent must also be sufficiently stable toward thermal or chemical influence. Most organic solvents are easily flammable and have a high vapor pressure at room temperature. Therefore, LC instrumentation should always be placed in a well-ventilated area.

The pH of the mobile phase has a strong influence on the solute retention and the peak shape of eluting ionogenic compounds. It is a critical factor for the reproducibility and ruggedness of a HPLC method (see also [67]).

12.4. Column Packing (Stationary Phase)

A very important factor for successful HPLC separation is the correct choice of column packing. In the past, many different packing materials were tested, for instance, powdered sugar, diatomaceous earth, aluminum oxide, calcium carbonate, calcium hydroxide, magnesium oxide, Fuller's earth, and silica. Today, silica is generally used as the basis for adsorbent chromatography; aluminum oxide is also used to a certain extent. Both materials are pressure stable and are unaffected by pH over a wide range.

Particle Size, Particle-Size Distribution. Standard particle size for analytical HPLC columns is 3, 5 or 10 μm. The smallest commercially available particle size is 1.5 μm. The particle-size distribution should be as narrow as possible so that "good" packings and therefore good separations are obtainable. The smaller the particle-size distribution of the stationary phase, the higher is its quality and therefore its price. However, if the size of stationary phase particles is absolutely identical, this can cause problems in packing the column.

Pore Diameter, Specific Surface Area. In adsorption chromatography a packing material with a very high specific surface area is required. The larger the surface area, the more active sites are available for (adsorptive) interaction. In the case of spherical particles (10-μm i.d.) < 1 % of the geometrical surface contributes to the total surface area. The dominant remainder of the surface area originates from the pores of the particles. A normal silica phase has a specific surface area of ca. 300 m^2/g, a mean pore diameter of 10 nm, and a mean pore volume of ca. 0.5 – 1.0 mL/g.

Most analytes have no access to pores with a diameter smaller than 3 – 4 nm. Therefore the packing material should show a narrow pore-size distribution so that the total number of very small pores (to which the analytes have no access) is as low as possible. Generally, the pore diameter should be about 3 – 4 times larger than the diameter of the analyte in solution, to ensure sufficient accessibility.

Nonporous packings with very small particle diameters have developed into useful tools for fast and economic separations. Due to the reduced resistance to mass transfer of the nonporous material, separation efficiency is increased at high flow rates and consequently analysis times are reduced. To overcome the high back-pressure at increased mobile phase velocities, it is recommended to increase the mobile phase operating temperature. This lowers the mobile phase viscosity and helps to reduce the back-pressure generated over the separation column. However, the boiling point of the mobile phase at the corresponding operating pressure should always be taken into consideration to avoid gas bubbles in the pump or in the detector (see also Chapter 12.3.3). More detailed information regarding nonporous column packing material can be found in [68] – [71].

Short analysis times can be achieved by using short conventional columns (4.5 mm i.d. and in 20 to 100 mm in length) combined with higher mobile phase flow rates (e.g., 3 mL/min) and, if possible, by applying sharp gradient profiles, as described in [72].

12.4.1. Silica

Silica is the traditional and most common material used in normal phase chromatography and serves as the basis for many chemically bonded reverse phases. It is a highly porous, amorphous silicic acid in the form of small, rigid particles. A distinction is made between spherical and irregular particles. Traditionally, *irregular silica* was, and to a certain extent still is, prepared by precipitation of water glass with sulfuric acid. Modern manufacturing techniques for *spherical silica* are, for example, sol – gel processing or spray drying of a silica sol; the latter is apparently the most common and economical technique today [73].

Commercially available silicas differ in their physical (specific surface area, mean pore diameter, pore volume, etc.) and chemical (contamination of different metals such as Al, Mg, Fe) characteristics. These metal contaminations can cause secondary retention effects (e.g., strong tailing of basic solutes). Additionally, commercial silicas differ widely in their Na content. The higher the Na content of the silica, the more alkaline its behavior in contact with water. This often acts as a catalyst for the hydrolysis of the Si – C bonds of chemically modified silica phases (see Section 12.4.3). Good silicas have an Na content of < 30 ppm. Excessively high contents of metal contaminants could easily be removed by the manufacturer by means of a thorough acid-washing

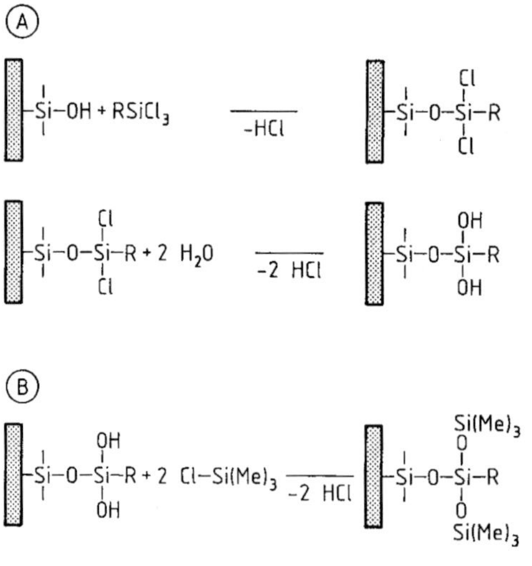

Figure 25. Example (A) derivatization of silica and (B) end capping

step, which would, however, slightly increase the production costs of the material.

The chemical properties of the silica surface are determined by silanol (SiOH) and siloxane (Si–O–Si) groups. There are ca. 7 – 8 μmol SiOH groups per square meter of surface area (i.e., the average distance between silanol groups is 0.4 – 0.5 nm). These silanol groups are responsible for adsorptive interaction with the solute and are the active sites for the derivatization into chemically bonded phases (see Section 12.4.3). Silica is mildly acidic, with a pH of ca. 5. Due to its mildly acidic behavior, problems may occur with very sensitive solutes (acid-catalyzed reaction). At pH > 8, silica goes into solution and the column bed is destroyed.

12.4.2. Alumina

Although alumina is rarely used as the stationary phase in HPLC, in certain applications it could be preferred to silica (e.g., separation of isomeric aromatic hydrocarbons). In contrast to silica, natural alumina is a basic material (basic alumina) with a pH of ca. 12 and a specific surface area of ca. 200 m^2/g (depending on the pore-size distribution). Because of the high pH of the packing, acidic analytes (e.g., carboxylic acids) may show strong peak tailing and are sometimes retained irreversibly. Problems with base-sensitive samples on alumina columns have been reported [74]. Through special treatment, basic alumina can be converted into acidic or neutral alumina; both are commercially available. For adsorption chromatography, neutral alumina is preferred. Acidic alumina can be used as a weak anion exchanger; basic alumina, as a weak cation exchanger.

12.4.3. Chemically Bonded Phases

Chemically bonded phases are usually made by derivatization of the silanol groups of silica (see Fig. 25 A). Remaining unreacted SiOH groups can be removed by end capping with trimethylchlorosilane (TMCS) or hexamethyldisilazane (HMDS) (Fig. 25 B). The positive effect of end capping on the separation of basic solutes on a reversed-phase column is shown in Figure 26. A further advantage of end-capped material is that it is often more stable and therefore has a prolonged service life.

Depending on the size and reactivity of the silica matrix, a maximum of 50 % of all silanol groups present reacts with the silane. This is very important for RP packings because the interactions of basic analytes and the remaining silanol groups can result in severe peak tailing (see Fig. 26 A). Peak tailing can be suppressed by end capping or by the addition of a relatively basic component or a salt to the mobile phase.

A variety of different chemically bonded phases is available (see Table 5). All of them have a stable Si–C bonding. If their organic residue is bonded to a function group (e.g., –OH, $-NH_2$, –CN, or $-NO_2$) the phases are often called selective polar phases or chemically bonded phases, and are used in both the normal phase and the reversed-phase chromatographic modes.

To ensure stable bonding, especially Si–C bonds, the polar functional group, which is responsible for the selectivity of the material, is connected to a spacer [$-(CH_2)_3$ group] that is fixed onto the silica. The result is a nonpolar stationary phase with a relatively polar surface. If the organic residue of a chemically bonded phase consists only of a hydrocarbon chain ($C_1 - C_{18}$), the stationary phase is very nonpolar or hydrophobic. Here, the elution order of hydrocarbon samples is reversed compared to normal phase chromatography (where retention decreases with increasing number of carbon atoms). Therefore, these phases are called reversed phase and the chromatographic mode *reversed-phase chromatography*. Commercially available packing materials for RP columns

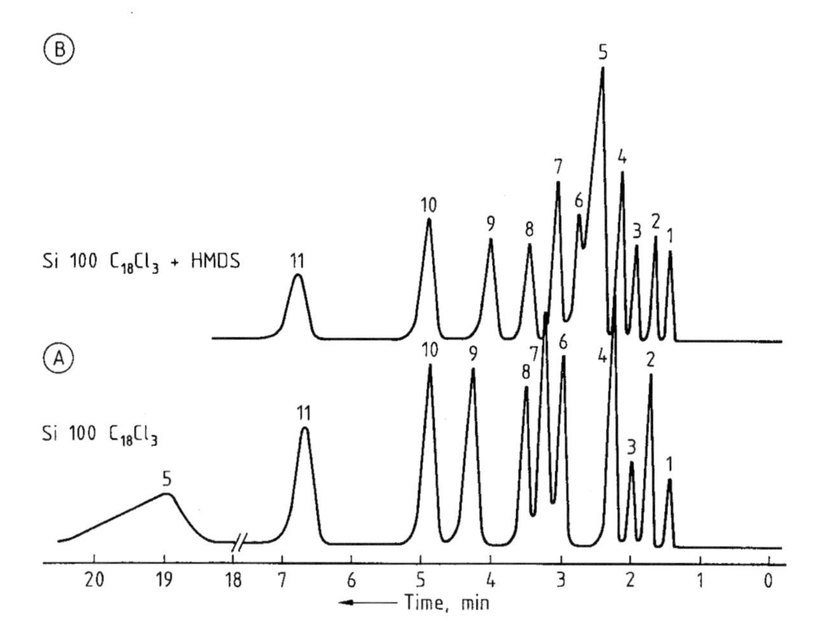

Figure 26. Separation of column performance test on a reversed-phase column
(A) Without end capping; (B) After end capping or basic component or salt to mobile phase
1) H_2O; 2) Benzamide; 3) Phenol; 4) 2-Phenylethanol; 5) Aniline; 6) Nitrobenzene; 7) Methyl benzoate; 8) Benzene; 9) Ethyl benzoate; 10) Toluene; 11) Ethylbenzene
Eluent: H_2O–CH_3OH 34.6 : 65.4 wt %

have a chain length between C_1 and C_{18}. Most common is RP-8 and RP-18 material. RP columns of the same type (e.g., RP-18) from different manufactures, or even from different production batches of the same manufacturer, can differ dramatically in their chromatographic behavior because a different packing material (silica) is often used. For instance, a higher natural alumina contamination of the silica increases the acidity of the Si–OH groups, which again strongly influences the retention of basic analytes (peak tailing).

The development of RP columns by HALASZ and SEBASTIAN [75] was the breakthrough for modern HPLC. Through that development, application of this technique was extended significantly (especially for biological materials), and because of the fast re-equilibration of the column, gradient runs could be performed easily, even in routine analysis.

12.4.4. Column Packing Techniques

Even if excellent equipment is used, but the sample is passed over a "badly packed" column, separation efficiency is poor. Therfore, the goal is to pack the chromatographic bed as densely and uniformly as possible. Care must be taken that no cracks or channels develop and that no fractionation of particles occurs during the packing process. Because experience and know-how are required to pack columns well, especially with small particles (<10 μm), purchase of a prepacked column is recommended.

If columns with relatively coarse particles (>30 μm) have to be packed, the *dry-fill packing process* [19] can be applied. Here vertical tapping and simultaneous rapping of the column during the packing process are necessary for well-packed chromatographic beds.

All analytical HPLC columns with particles < 20 μm (usually 5 or 10 μm) are packed by the *slurry technique* [19], which can be further classified into the high-viscosity technique and the balanced-density technique. In the latter, the particles are suspended in a fluid that has a density similar to theirs, so particle segregation by sedimentation decreases. In both techniques, the suspending fluid must be chosen in such a way that particle dispersion is maintained without aggregation and particle agglomeration is avoided by proper selection of the polarity of the slurry [19], [24].

Table 5. Examples of chemically bonded phases and their application range

Chemical modification	Functional group*	Application range
Butyl	$-(CH_2)_3-CH_3$	RP and ion-pairing chromatography
Octyl	$-(CH_2)_7-CH_3$	RP and ion-pairing chromatography
Octadecyl	$-(CH_2)_{17}-CH_3$	RP and ion-pairing chromatography
Phenyl–diphenyl	$-C_6H_5-(C_6H_5)_2$	RP and ion-pairing chromatography, additional selectivity for $\pi-\pi$ interactions
Nitro	$-NO_2$	analytes containing double bonds
Nitrile (cyano)	$-CN$	NP and RP chromatography
Diol	$-CH_2-CH(OH)-CH_2-OH$	NP and RP chromatography
Amino	$-NH_2$	NP, RP, and weak anion exchanger
Dimethylamino	$-N(CH_3)_2$	NP, RP, and weak anion exchanger
Quaternary ammonium	$-\overset{+}{N}(CH_3)_3$	strong anion exchanger
Sulfonic acid	$-SO_3^{-+}$	strong cation exchanger

* Usually a spacer; for example, a propyl group [$-(CH_2)_3-$] is placed between the silica surface and the functional group.

The suspension is pressed into the empty column, which is fitted with a frit on its far side, by means of a constant-pressure pump (pressure filtration). The usual packing pressure for a standard analytical column (250×4 mm i.d.) is between 40 and 60 MPa. In this case, 1.7 – 2.0 g of stationary phase has to be suspended in ca. 30 – 50 mL of suspension fluid. Detailed discussion of column packing techniques is given in [73], [19], [76].

12.4.5. Column Specifications and Column Diagnosis

The plate number N is often considered the most important column performance specification. A typical plate number for a 250×4 mm i.d. column with 10-μm particles is ca. $N = 10\,000$. The plate number required for a specific separation can be achieved in two ways: By using either a short column with small particles or a long column with larger particle diameter. As an example, the same separation efficiency can be obtained with 10-μm particles in a 250-mm-long column or with 3-μm particles in 50 – 100-mm-long columns.

Even though the plate number is a very useful column specification, it is certainly not sufficient to evaluate the suitability of a tested column for a given separation problem. Especially for RP columns, a variety of column performance tests have been developed [77]. These tests give important information about the quality of the column bed (well packed or not), the kinetic and thermodynamic performance, basic and acidic behavior, and the selectivity and stability (lifetime) of the column (see Table 6)

Especially in routine analysis, an individual test mixture, consisting of characteristic compounds of everyday samples, is often sufficient. This test mixture should be injected onto every new column (to see if selectivity and k' are unchanged) and again from time to time to follow the aging process of the column (peaks get broader, resolution gets worse, k' get lower, peaks become asymmetric, etc.). If one or more of these characteristics are observed, the old column should be replaced by a new one.

12.5. Separation Processes

12.5.1. Adsorption Chromatography

Adsorption chromatography or liquid – solid chromatography (LSC) is the most frequently applied LC mode. Here, specific interactions occur among the solute, the stationary phase, and the mobile phase. Examples are hydrophobic, dipole – dipole, hydrogen bond, charge-transfer, and $\pi-\pi$ interactions.

Every chromatographic separation should give sufficient resolution Rs and good reproducibility, which have great importance for qualitative and quantitative analysis. Good reproducibility in LSC can be achieved only in the range of a linear sorption isotherm. In this case the retention time t_R is independent of sample size and sample vol-

Table 6. Typical reversed-phase column performance tests

Test name	Characterization of	Test compounds	Eluent	pH	Detection
DMD* test	C_8-C_{18} packing hydrophobic interactions	diphenylhydramine, 5-(*p*-methylphenyl)-5-phenylhydantoin, diazepam	acetonitrile – potassium dihydrogenphosphate/phosphoric acid butter, 156 : 340 wt %	2.3	UV (220 nm)
Engelhardt – Jungheim test	C_8-C_{18} packing hydrophobic interaction, behavior of basic and acidic analytes (silanophilic interactions)	aniline, phenol, *o-,m-,p*-toluidine, *N,N*-dimethylaniline, toluene, ethylbenzene, methyl benzoate	methanol – water, 50 : 50 wt %		UV (254 nm)
Aromatics test	C_8-C_{18} packing hydrophobic interaction, selectivity	benzene, naphthalene, fluorene, anthracene, benzanthracene	acetonitrile – water, 75 : 25 wt %		UV (254 nm)

* DMD = Diphenylhydramine,5-(*p*-methylphenyl)-5-phenylhydantoin, diazepam

ume. Generally, a sample size of < 0.1 mg of analyte per gram of stationary phase ensures a linear sorption isotherm and, therefore, reproducible results.

Two groups of stationary phases are used in adsorption chromatography: Polar and nonpolar adsorbents.

12.5.1.1. Polar Adsorbents and Their Properties

The most frequent used polar adsorbents in LSC are silica and alumina (see Sections 12.4.1 and 12.4.2). Polar stationary phases are very selective with respect to the separation of configurational isomers. Figure 27 shows the separation of β-carotin isomers on an alumina column in NP chromatography.

Generally, polar adsorbents show a greater selectivity (higher retention) toward polar solutes, whereas the retention k' of nonpolar analytes decreases with increasing hydrophobicity of the analyte.

Working with polar stationary phases and nonpolar (hydrophobic) eluents is the classical form of LC (which Tswett [2] used in 1903) and is called *normal phase chromatography.* Currently, only ca. 20 % of analytical separations are solved by NP chromatography — for instance, chiral analyses — because re-equilibration of the column can be very time consuming (up to 12 h), and in certain cases, exact water-content control is necessary to obtain reproducible results. In NP chromatography, even traces of water (< 100 mg/L) in the solvent cause a dramatic decrease in k'.

12.5.1.2. Polar Chemically Bonded Phases

Diol phases are often used as "polar" column packings. They belong to the group of polar chemically bonded phases, which also includes nitrile, nitro, amino, and similar polar groups. The diol phase is less polar than silica but is therefore easily wettable with water. Because of their rapid equilibration, these packings are much more suitable for gradient elution than unmodified silica. Diol columns are typically used for the separation of peptides, proteins, and generally components that can build up hydrogen bonds.

12.5.1.3. Nonpolar Adsorbents – Reversed Phase

In contrast to NP, in RP chromatography a nonpolar stationary phase and a polar solvent are used. In the reversed-phase mode the elution-strength order of the solvents is the opposite of that of the elutropic row (see Section 12.4.3 and Table 4).

Some separation problems (e.g., triglyceride separation [76]) require a hydrophobic stationary phase and a nonpolar (nonaqueous) mobile phase. This chromatographic mode is called *nonaqueous reversed-phase (NARP) chromatography.* The most common commercially available reversed-phase materials are C_2, C_4, C_8, C_{18}, and phenyl phases.

As shown in Figure 28 A the retention of a particular hydrophobic analyte increases with growing chain length of the RP material. The selectivity, on the other hand, reaches a plateau near

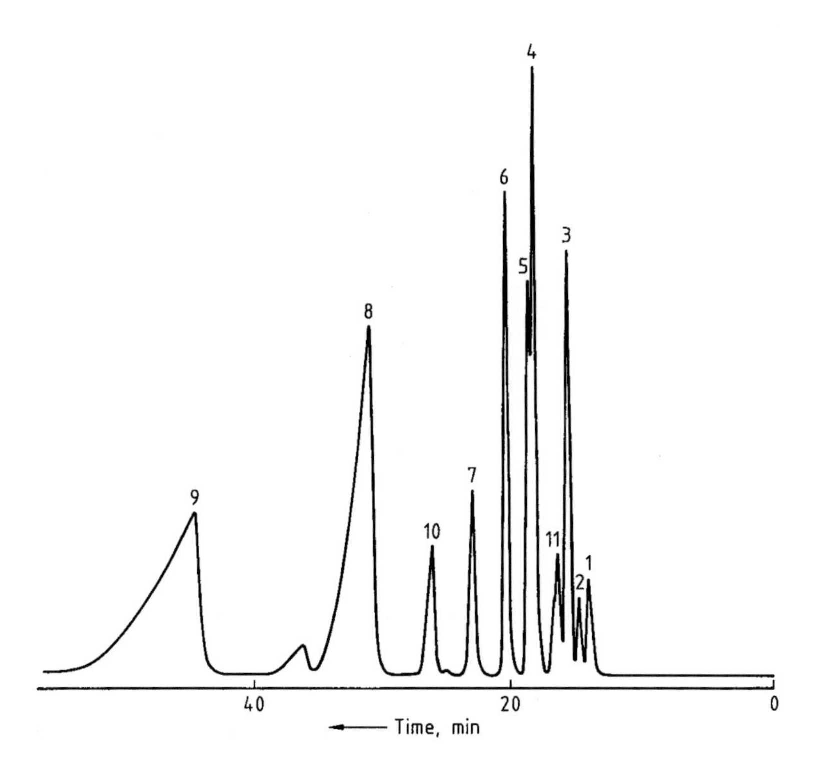

Figure 27. Separation of β-carotin isomers on an alumina column in normal phase chromatography [78]
1) 13,13′-di-*cis*; 2) 9,13,13′-tri-*cis*; 3) 9,13′-di-*cis*; 4) 15-*cis*; 5) 9,13-di-*cis*; 6) 13-*cis*; 7) 9,9′-di-*cis*; 8) all-*trans*; 9) 9-*cis*; 10) α-Carotin; 11) 3 a, b, c

the C_8 column (Fig. 28 B). This means that under identical conditions, the selectivity for the same sample is more or less the same on an RP-8 as on an RP-18 column. The only difference is that the compounds on the RP-18 column are more strongly retained than those on the RP-8 material. Therefore, many separations, which can be performed on an RP-18 column can also be achieved on an RP-8 column by adjusting the elution strength of the mobile phase (in this case by adding more water to the eluent) and vice versa.

Generally, packing materials with shorter alkyl chains (e.g., RP-4 or RP-8) are more suitable for the separation of moderately to highly polar solutes, whereas those with longer alkyl chains (RP-18) are preferred for the separation of moderately polar to nonpolar analytes. Reversed-phase columns are also used for ion-pairing chromatography.

Today, between 50 and 80 % of HPLC applications are run in the RP mode and only 10 to 20 % in the NP mode [79]. This can be explained by the following reasons:

1) The reproducibility of the elution strength of the mobile phase is easier in RP chromatography (more reproducible separations are possible)
2) Re-equilibration between the mobile and stationary phases is much faster in the RP mode (after three to five column volumes) than in NP chromatography, especially if water was added
3) Because of item 2, gradient elution is often not used in NP chromatography
4) RP columns combined with gradient elution are able to separate a wide range of analytes (from hydrophobic to polar components)

Other modes, e.g., IEX (Section 12.5.3) and SEC (Section 12.5.4) are used to a much lesser extent.

The application range of RP chromatography is so large that it is impossible to mention it all. In principle, the technique is applied for nearly all separations in the pharmaceutical, environmental, and biological fields.

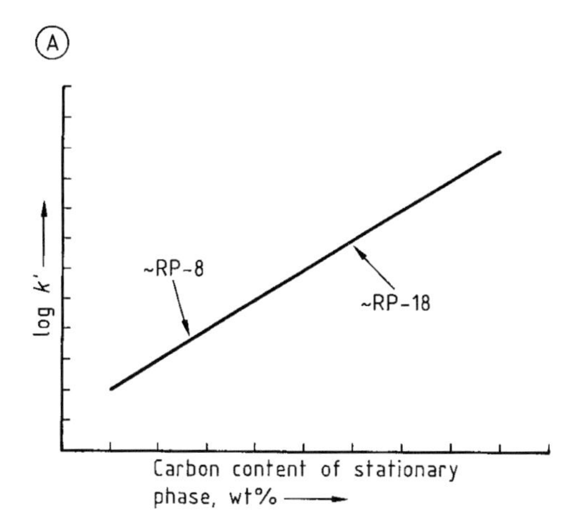

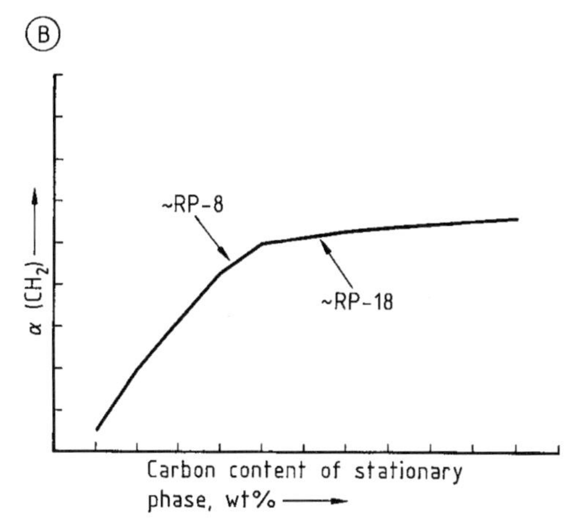

Figure 28. Retention behavior of hydrophobic analytes in RP chromatography
A) Retention; B) Selectivity

12.5.1.4. Other Adsorbents

Polyamide as a stationary phase is used mainly for the separation of compounds that can form hydrogen bonds to the amide group. A typical example is the separation of phenolic compounds on polyamide, due to hydrogen bonds between the phenolic hydroxyl and the amide groups [80]. The drawback of these polyamide packings is that they are not as pressure stable as silica and that undesired swelling of the phase can occur with certain solvents.

Magnesia and magnesia – silica (florisil) are also sometimes used in adsorption chromatography. Magnesia – silica has the disadvantage of retaining many solutes irreversibly (e.g., aromatics, amines, and ester). Possible applications are in the separation of lipids or chlorinated pesticides [80].

Titanium dioxide is also an interesting packing material because of its good mechanical and chemical (pH) stability. Its main drawback is the relatively small surface area (ca. 50 m^2/g). However, this is partially compensated by its high packing density of 1.5 g/mL, which compares to only 0.6 g/mL for silica.

12.5.2. Partition Chromatography (LLC)

Partition chromatography, or liquid – liquid chromatography (LLC), was first mentioned in 1941 in a paper by Martin and Synge [5]. This is a well-known technique in gas chromatography, where the separation is based on distribution of the analyte between a "liquid" stationary phases and a gaseous mobile phase. In liquid chromatography, the distribution takes place between two liquids: The liquid stationary phase, which is dispersed on an inert support (e.g., silica), and the liquid mobile phase. As in the distribution of solutes between two mobile phases in a separatory funnel, the stationary and mobile phases in LLC must be immiscible. However, liquids are always miscible to a certain extent, which results in a washing out of the stationary phase form the column ("column bleeding").

A major difference between adsorption and partition chromatography is that in LLC the separation of a sample is based not on different adsorption energies of the analytes but on different solubilities of the sample components in the liquid mobile and stationary phases. This can be a disadvantage if binary systems (stationary and only one solvent as mobile phase) are used. If the solute is much more soluble in the liquid stationary phase, the retention will be strong, resulting in increased dilution and band broadening; thus, the sensitivity is poor. On the other side, if the solute is more soluble in the mobile phase, the retention will be very weak and the solute peak will elute shortly after the inert peak. To reduce these relatively large differences in polarity (solubility) between the two liquids, a third liquid ("modifier") can be introduced to form a ternary system. In this case the mobile phase is a mixture of two solvents. Now, the more strongly retained solutes elute earlier, with better peak symmetry and smaller elution volumes, thereby increasing the sensitivity significantly. A drawback of this ternary system technique is that the stationary phase also becomes more "soluble" in a solvent containing a modifier, which reduces the column lifetime. Furthermore, even a slight temperature increase can cause total miscibility of all three liquids, which destroys the chromatographic system instantly. Such well-thermostated chromatographic equipment (which also includes the solvent reservoirs) is essential (temperature variations 0.1 – 0.2 °C). If a UV detector is used, neither the mobile nor the liquid stationary phase must absorb at the detector wavelength. A further disadvantage of LLC is that gradient elution is impossible. So if a sample contains components with widely different k' ranges, adsorption chromatography is preferred.

The advantages of partition chromatography compared to adsorption chromatography are as follows:

1) Easy exchange of the stationary phase (in situ) and longer column life
2) High loadability (10 times higher than LSC, ca. $10^{-3} - 10^{-2}$ g of analyte per gram of liquid stationary phase)
3) Reproducible results because the supporting material of the column is not, or at least should not be, involved in the separation process
4) Broad range of selectivity due to the large number of different mobile and stationary phase liquids available (with the condition that the liquids are immiscible)

When selecting the stationary phase in LLC, liquids with lower viscosity should be preferred because the use of high-viscosity liquids leads to poorer separation efficiencies due to imparted diffusion and thus sample transport (mass transfer). In addition, the mobile phase must be saturated with stationary phase to reduce column bleeding. Column life can be further extended by the use of a precolumn for proper equilibration and exclusion of oxygen from both phases.

Column Packings and Preparation. Inert materials with relatively large pores (10 – 50 nm) are used (usually silica) as liquid stationary phase supports. Material with larger pores is not suitable because, unlike in narrow pores, the stationary liquid phase is fixed less strongly to the support by capillary forces. Wide-pore supports start to bleed at low flow rates (< 1 mL/min). Materials with small pores (< 50 nm) are usually stable up to flow rates of 5 mL/min.

Typical stationary phases are, for instance, β,β′-oxydipropionitrile (BOP) or different types of poly(ethylene glycol). Other examples of stationary and mobile phases for LLC are given below:

Stationary phase	Mobile phase
Normal phase LLC	
β,β′-Oxydipropionitrile	pentane, cyclopentane, hexane, heptane, . . .
Carbowax	modified with up to 10 – 20 % chloroform, THF
Triethylene glycol	dichloromethane, acetonitrile, . . .
Ethylene glycol	nitromethane
Ethylenediamine	hexane
Water	butanol
Reversed-phase LLC	
Polydimethylsiloxane	acetonitrile – water
Heptane	methanol – water

The stationary phase in LLC can be prepared in different ways, depending on the support particle size chosen. An often-applied preparation technique is the in situ coating method. In this case the stationary phase liquid is pumped through the dry packed (slurry packing) column. Later, the column is flushed with mobile phase to displace excess stationary phase liquid between the particles.

Separation Modes. Like adsorption chromatography, partition chromatography can also be used in the normal and reversed-phase mode.

In *normal phase LLC* the stationary phase is polar and the mobile phase consists of nonpolar solvents. This mode is suited for more polar, water-soluble analytes. The elution order is similar to that in normal phase adsorption chromatography because nonpolar solutes prefer the moving phase and elute with low k' values. Solutes preferring the polar stationary phase elute later with high k'.

In *reversed-phase LLC* the two phases are interchanged and the elution order reversed. This mode is more suitable for hydrophobic analytes that have poor water solubility. For reversed-phase LLC, the support should be hydrophobic to ensure appropriate fixing of the stationary phase liquid. This can be achieved, for instance, through silanization of the silica.

Applications. LLC can be a good alternative to LSC for analytes that are sensitive to catalytic decomposition on the active surface of the column packings. Typical applications of LLC are in the separation of compounds containing metal ions, radioactive metals, and free and underivatized steroids, as well as in preparative chromatography.

Because of the high loadability of the column ($10^{-3} - 10^{-2}$ g of solute per gram of stationary liquid), LLC is often used for preparative chromatography. Since the mobile phase is saturated with stationary phase, high-boiling stationary phases may be difficult to remove from separated fractions and volatile stationary phases are preferred.

To ensure good chromatographic behavior (as in adsorption chromatography), samples should not be injected in solvents with a higher solution power than the mobile phase. Further, the injected sample volume should not exceed one-third the elution volume of th first peak of interest.

12.5.3. Ion-Exchange Chromatography (IEX)

Ion-exchange chromatography (IEX) can be regarded as a special form of adsorption chromatography resulting from the ionic interactions taking place during the separation process. It was first described by SMALL et al. [81] in 1975. Even in the 1980s, IEX was not very well accepted in HPLC because of unstable column packings and detection problems. These drawbacks have been overcome, and today IEX is a very powerful tool in water analysis, amino acid analysis, and the separation cleanup of proteins, for example. Generally all charged solutes (inorganic ions, organic acids, bases, salts, etc.) can be separated by this chromatographic method.

The following abbreviations are often used in ion chromatography (IC):

SCX	strong cation exchanger
WCX	weak cation exchanger
SAX	strong anion exchanger
WAX	weak anion exchanger
IEX	ion-exchange chromatography
IEC	ion exclusion chromatography
IPC	ion-pair chromatography
HPIC	high-performance ion chromatography

The principle of ion-exchange chromatography is as follows: Ions differing in their charges interact with the functional groups of the exchanger with different intensities, and can be selectively eluted from the matrix by changing the elution strength of the mobile phase (variation of ion concentration, i.e., "salt gradient" or pH gradient).

During the ion-exchange process an equilibrium between, for example, the cation in solution and the anion fixed on the matrix is formed. For a particular ion this equilibrium is described by the equilibrium constant K_D. The equilibrium constant depends on the ion size, its charge, and its polarizability; i.e., the affinity for the stationary phase (ion exchanger) increases with

1) Increasing charge
2) Increasing polarizability
3) Decreasing size of the ion investigated

Usually this equilibrium is shifted toward the side of the fixed ions on the exchanger material. Therefore the effluent in IEX always contains ions (buffers, acids, bases) that compete with the analyte for active sites on the matrix so that the analyte can be desorbed from the matrix. The ionic strength of a mobile phase in IEX is a measure of its elution strength. In addition, the pH and dissociation rate of the analyte also have a great influence on retention. Furthermore, the technique shows a much better selectivity for cations differing in their charges than for anions.

Instrumentation. A schematic of an ion-exchange chromatograph is given in Figure 29. In principle, this is the same setup as for a normal HPLC system consisting of a pump (b), sample introduction system (c), separation column (d), and detector (e). IEX differs from normal HPLC in the "suppressor column" that is installed between the separation column and the detector (e.g., conductivity detector). This suppressor column has two tasks: First, it reduces the high background conductivity of the effluent; second, it converts the analytes into species with higher conductivity. The function of the suppressor column can best be explained by a practical example:

The Cl^- content of a water sample is to be determined. The mobile phase consists of an $Na_2CO_3 - NaHCO_3$ buffer solution. The reactions taking place in the suppressor column — which is a strong cation exchanger (SCX) in the protonated form — are the following:

1) Background conductivity is reduced by exchanging the sodium ions of the buffer with protons. The resulting H_2CO_3 is nonconducting
2) At the same time, NaCl is converted into the strong contucting HCl:

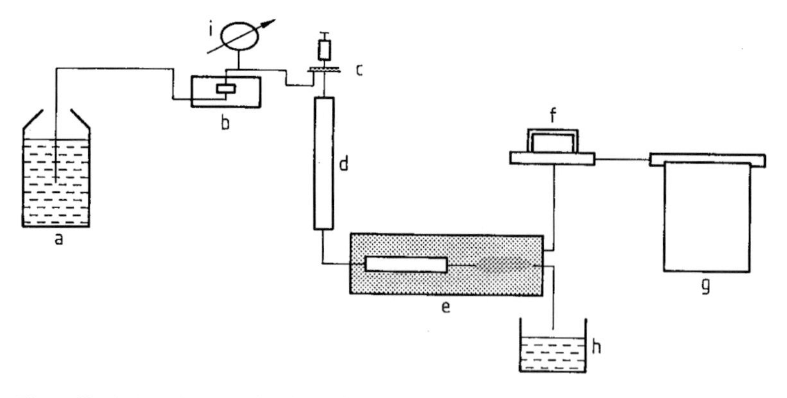

Figure 29. Schematic setup of a simple ion-exchange chromatographic system
a) Solvent; b) Pump; c) Injection port; d) Separation column; e) Detector with suppressor column; f) Control unit; g) Recorder; h) Waste; i) Manometer

$$\text{Resin–SO}_3^-\text{H}^+ + \text{NaHCO}_3 \longrightarrow \text{Resin–SO}_3^-\text{Na}^+ + \text{H}_2\text{CO}_3$$

$$\text{Resin–SO}_3^-\text{H}^+ + \text{NaCl} \longrightarrow \text{Resin–SO}_3^-\text{Na}^+ + \text{HCl}$$

This technique is often referred to as *suppressed conductivity detection.*

Because of the corrosive solvents used in IEX, conventional stainless steel HPLC instrumentation is unsuitable. The pump-head, capillaries, and column should be made of inert material such as PEEK, PTFE, tetrafluoroethylene – ethylene copolymer, polypropylene, ruby and sapphire. Otherwise, contamination problems (especially iron contamination) can occur in the mobile phase, decreasing the separation efficiency and capacity, and increasing the noise of the amperometic detectors.

Detectors. In ion chromatography, suppressed conductivity detection is still the most frequently used technique. Indirect UV detection (Section 12.2.6.3) can also be a powerful tool for the detection of ions, although with respect to sensitivity, conductivity detection (detection power 10 – 100 ng/g) cannot be surpassed. In amino acid analysis, UV and fluorescence detection (after post- or precolumn derivatization) is often applied.

Column Packings. Modern column packings for ion-exchange chromatography are made of a polystyrene matrix with different functional groups, depending on whether it is a cation or an anion-exchange resin.

The following functional groups are used:

SCX: $-SO_3^-$
WCX: $-COO^-$
SAX: $-CH_2NR_3^+$
WCX: $-CH_2–NHR_2$ or $-CH_2–NHR$

Polystyrene column packings are also available with small particle sizes (< 10 μm). Compared to inorganic ion exchangers (silicates or zeolites), they have the advantage of being pH resistant over the entire pH range. Many protein separations must be carried out at a pH of ca. 10. A silica-based exchanger would be destroyed at this basic pH, but polystyrene-based exchangers are not.

A disadvantage of the polystyrene matrix for high-performance ion chromatography is the fact that it tends to swell or shrink if certain solvents are used.

An important characteristic of every ion-exchange material is its "total exchange capacity," expressed in equivalents per unit weight (microequivalents per gram). One equivalent corresponds to 1 mol charge. For weak ion exchangers the capacity is a function of pH, whereas the capacity of strong ion exchangers is independent of pH. Figure 30 shows a simultaneous isocratic separation of alkali and alkaline-earth cations obtained by IEX.

12.5.4. Size Exclusion Chromatography

Although separation due to size exclusion has been known for a long time [82], systematic in-

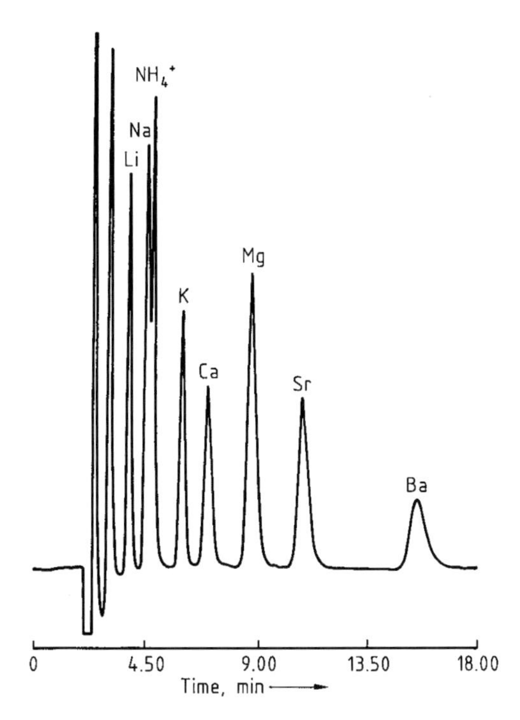

Figure 30. Simultaneous isocratic analysis of alkali and alkaline-earth cations with ion-exchange chromatography
Column: Polyencap WCX; eluent: 4 mmol/L citric acid; 0.7 mmol/L dipicolinic acid in water, 1 mL/min; standard solution: 2.5 – 40 μL/L in water; detection: conductivity

vestigation of this method began only after the development of synthetic matrices [83], [84]. If the mobile phase consists of or contains water, size exclusion chromatography (SEC) is often called *gel filtration chromatography* (GFC). If the mobile phase is an organic solvent, the term *gel permeation chromatography* (GPC) is often used. Another frequently used abbreviation is HPSEC, which stands for high-performance size exclusion chromatography.

In SEC an analyte is, by definition, separated only according to its molecular size, not its molecular mass. This is important to remember since the same solute (with its specific molecular mass) can have a different structure, and therefore different sample size, in different solvents. If all other interactions (ionic interactions, adsorption, ect.) with the stationary phase are suppressed, the selectivity of the system depends entirely on the different accessibility of the pores of the matrix for the solutes. Very small molecules (usually solvent molecules) have access to the entire intra particle volume V_i of the stationary phase. In contrast to the moving interparticle volume V_m (which is the effluent volume outsite the particle pores) the intra particle volume V_i is called the stagnant mobile phase. The total solvent volume V_0 inside the column is therefore

$$V_0 = V_m + V_i$$

Since solvent molecules usually have the smallest size, they elute at the end of an SEC chromatogram with a retention time t_0. In other words, they completely permeate the stationary phase ("total permeation"). In all other chromatographic modes the solvent elutes first, if no exclusion has appeared. In SEC, molecules that are so large that they cannot get into the pores stay only in the faster-moving interparticle volume V_m and elute as the first peaks of the chromatogram. For these large analytes the column behaves as if it were filled with nonporous glass beads. The difference between the retention volumes of the largest (excluded) and smallest (solvent peak) molecules is the pore volume or interparticle volume V_i of the stationary phase. Medium-sized molecules permeate the stationary phase only partially, and consequently elute between the total permeation t_0 and the totally excluded solutes (dead or interparticle volume). Generally, the smaller a molecule, the higher is the accessible pore volume and therefore the longer is the time required to pass through the column.

In SEC the retention volume V_R is used instead of the retention time t_R. V_R is defined as the sum of the interparticle volume V_m and the accessible pore volume V_p for the particular solute:

$$V_R = V_m + V_p$$

For very small molecules (e.g., solvent) V_R is

$$V_R = V_m + V_p = V_0$$

For large and therefore excluded particles, V_R is

$$V_R = V_m$$

To become independent of various packings, column volume, and pore volume, and to enable a comparison of data from different packings, a distribution coefficient K_0 was introduced

$$K_0 = \frac{V_R - V_m}{V_i}$$

The above equation shows that the solvent (last eluting peak) has a K_0 of 1. Consequently, the entire sample must elute with $K_0 < 1$. Because all components of a sample elute so early, band broadening is reduced, which results in high and symmetric peak shapes. Thus, even relatively nonsensitive but therefore universal detectors such as the refractive index detector can be used. Since the solvent peak usually elutes at the end of an chromatogram the analysis time is predictable.

Column Packings in SEC. Two types of packings exist: Rigid packings, which are essential for HPSEC because of their stability, and nonrigid gels. The latter are not resistant to higher pressure and are therefore used in open-column SEC.

In *HPSEC* the most common column packings consist of pure or chemically modified silica or, recently, of highly cross-linked macroporous styrene – divinylbenzene copolymer. Due to the differing production processes for these polymer phases (e.g., amount of inserted cross-links), various pore sizes are available with small (5 and 10 µm), pressure-stable (maximum pressure 10 – 15 MPa), spherical particles [85]. The advantage of these polymer-based columns compared to silica-based columns is that under certain conditions (e.g., pH > 4 – 5), silica can act as a weakly acidic ion exchanger. Interactions between the sample and the active silanol groups (e.g., adsorption) are also possible and must be suppressed by proper choice of effluent. Ion exchange, for instance, is suppressed by increasing the ionic strength of the eluent (e.g., inserting a salt or buffer, 0.1 – 0.5 mol/L). Adsorption effects resulting from silica can be reduced by modifying the silica with chlorotrimethylsilane [86]. An interaction between the analytes and the column packing is indicated either by peak tailing (uncertain) or by the elution of any components after the solvent peak ($t_R > t_0$ or $K_0 > 1$). Nevertheless, silica-based SEC column also have their advantages. Here, pore size and pore volume are independent of the effluent. A silica column can be calibrated and characterized, for instance, with a polystyrene standard in an organic solvent (e.g., CH_2Cl_2); then, the organic solvent is replaced by a water-containing effluent, water-soluble polymers can be investigated without recalibration. In contrast, the pore size and thus the pore volume of polymer-based materials often change with the mobile phase applied. In some cases, a change of solvent can lead to a collapse of the particles, which results in a significant decrease of performance. When polystyrene gels are used the mobile phase must be degassed very well, because air bubbles can damage the packing. In addition, aromatic sample compounds may also form $\pi - \pi$ interactions with the styrene – divinylbenzene copolymer.

To obtain high separation efficiency, similar to other LC techniques, small particles (≤5 µm) and a narrow pore-size distribution are necessary. Common pore sizes in SEC for low molar mass samples are between 6 – 12 nm, whereas high molecular mass samples (e.g., proteins) require a pore diameter from 30 – 100 nm or wider (e.g., 400 nm). Since separation capacity is correlated directly with pore size and therefore pore volume, the latter should be as large as possible. To enlarge the pore volume of an SEC system, either the sample can be reintroduced into the same column after the first separation (recycle SEC) or several columns with the same packing can be connected in series. Doubling the column length doubles the elution volume and the analysis time. If the sample contains molecules with a molecular mass that exceed the separation range of the column, several columns with differing sep-aration ranges can be connected for the "first analysis" of an virtually unknown sample, to obtain maximum information about the molar mass distribution. Two or more column packings differing in pore size should not be inserted into the same column because particle sizing can occur, which decreases the separation efficiency. A problem with columns connected in series is that band broadening may occur. This can be reduced by avoiding any unnecessary dwell volumes, by using small particles and solvents with low viscosities (higher diffusion), and by working at higher temperature (high-temperature gel permeation chromatography).

For choice of the appropriate solvent in SEC the following rules apply:

1) The sample must be soluble in the mobile phase
2) The stationary phase must be resistant to the mobile phase
3) Possible interactions between solute and column packing should be suppressed by appropriate additives to the mobile phase or by control of its pH, ionic strength, etc.
4) The mobile phase should be more strongly adsorbed to the column packing than the solute
5) The maximum permitted solute concentration in SEC should not exceed 0.25 wt %; a concentration of 0.1 wt % is sufficient in most cases

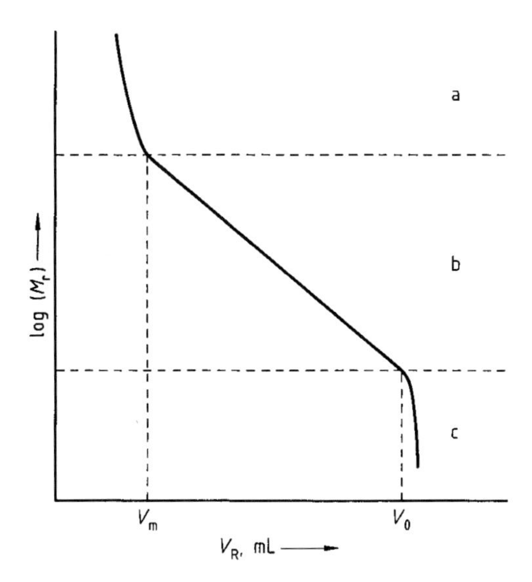

Figure 31. Typical calibration curve in SEC
a) Molecules are too large and elute with interparticle volume; b) Linear range for separation according to the molecular mass of the sample; c) Molecules are too small and penetrate the entire packing

Applications. If all interfering interactions between solute and stationary phase are suppressed, SEC is a very simple, reliable, and fast technique for the separation of many samples, especially oligomers and polymers. The only limitation is that the solutes to be separated must differ by at least 10 % in their molecular masses. Typical applications of SEC:

1) Determination of molar mass distribution of polymers
2) Determination of pore-size distribution of column packings
3) Cleanup technique prior to other high-resolution methods
4) Fingerprint analysis
5) Biopolymer separations (e.g., proteins)

To determine the elution (retention) volume V_R that corresponds to a certain molecular mass, the column must be calibrated with a standard mixture of known molar mass distribution. These standard, high-purity mixtures are available from various column suppliers. Remember that molecule size can change in different solvents and that different solutes (even with comparable molecular masses) can show differing molecular structure and size in the same solvent. In both cases, the retention volume obtained gives an incorrect molar mass. Therefore, the standard must be chosen in such a way that it corresponds as much as possible to the sample structure. Figure 31 shows a typical schematic calibration curve obtained in SEC.

Because retention volume is the most important factor in SEC (it is related to the logarithm of the molecular mass, even small inaccuracies in the determination of V_R are magnified), the pump must deliver a very constant flow which can be controlled by a continuous flow measurement system, for example. Further information on SEC can be found in [19], [24], [87], [88].

12.6. Gradient Elution Technique

In gradient elution the mobile phase is usually composed of two solvents, A and B. The characteristic of gradient elution is that the composition of the mobile phase is changed during the chromatographic run. Usually the concentration of the stronger solvent B (percent) is increased during the run. In the reversed-phase systems that are generally used, solvent A is normally water.

The three main reasons for using gradient elution are:

1) Separation of sample components showing a wide range of k'
2) Fast evaluation of optimum eluent composition for subsequent isocratic analysis
3) Enhancement of detection limit

In addition, the technique can be applied to remove strongly retained solutes or impurities from the column, thus enhancing sensitivity and column life.

Gradient elution is usually applied in reversed-phase chromatography, although for certain separation problems in normal phase chromatography it can also be a very useful tool. Ideally, the solvent strength of the mobile phase should be adjusted in such a way that the k' range of the sample compounds lies between 2 and 10 (although k' values between 1 and 20 are often acceptable). For small molecules in reversed-phase liquid chromatography a 10 % increase in the percentage of B (organic solvent) decreases the k' value of every solute by a factor of ca. 2.

Compared to the isocratic elution mode, the equipment necessary for gradient elution is extended by a second solvent reservoir, an additional pump (in the case of a high-pressure-mixing gradient system), a mixing chamber, and a propor-

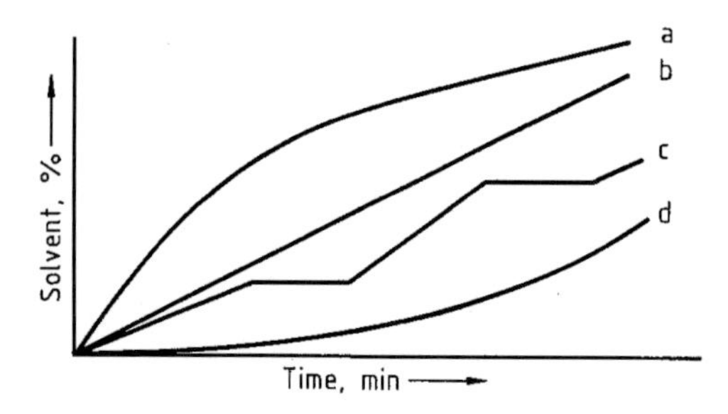

Figure 33. Different possible gradient profiles
a) Convex; b) Linear; c) Step; d) Concave

tioning valve that is regulated by a gradient control system.

Two types of gradient systems are possible (see Fig. 32): The low-pressure-mixing gradient system and the high-pressure-mixing gradient system. In the first case, only one HPLC pump is necessary because the two solvents (A and B) are mixed before the pump (i.e., in the low-pressure area of the system). With the high-pressure gradient system, two pumps are required because solvents A (e.g., water) and B (e.g., organic solvent) are mixed after leaving the pump in the high-pressure area of the system.

Many different gradient profiles are possible (see Fig. 33); the most frequent and useful are the linear and segmented (multilinear) profiles. Besides binary gradients (two solvents), some separations require ternary gradients in which the mobile phase consists of three solvents (A, B, C). With these ternary gradients, special selectivity effects can be achieved [89].

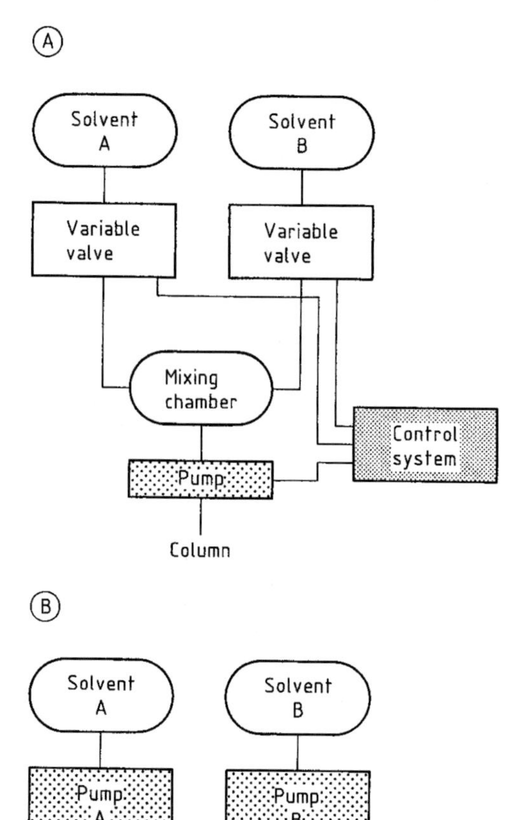

Figure 32. Gradient elution technique
A) Low-pressure system, B) High-pressure system

12.7. Quantitative Analysis

Quantitative analysis should deliver reproducible and reliable information about the composition of the sample investigated. Since end of the 1980s, quantitative analysis has become extremely important for product quality control (e.g., food industry, pharmaceutical industry) and environmental analysis. It is of great economic importance, and an erroneous quantitative analysis can result in fatal consequences. Therefore, this chapter introduces different quantification methods for HPLC and describes some often-occurring sources of error.

Correct sampling and sample storage are essential to obtain exact and reliable analytical results. After a possible cleanup step, the sample is injected into the chromatographic system and a strip chart is obtained from the recorder or electronic integrator, on which the intensity of the detector signal is plotted against analysis time (chromatogram). To be able to make any quantitative statements with the help of this chromatogram, the system must have been previously calibrated. The two most important methods are the external standard method and the internal standard method.

12.7.1. External Standard Method

The external standard method is the technique used most frequently to gather quantitative information from a chromatogram. In this case, a pure reference substance (ideally the same compound as the one to be determined in the sample) is injected in increasing concentrations and the peak areas or peak heights obtained are plotted versus the concentration (calibration curve; →Chemometrics). These calibration curves should show a constant slope (linear curves), and the intercept should be as close to zero as possible. Since the calibration curves usually show nonlinear behavior and flatten off at higher concentrations (see Section 12.2.6), the quantification should be carried out only within the linear part of the curve.

A calibration curve must be recorded for each component of interest. The peak areas/heights of the sample chromatogram are then compared with those in the calibration curve. All chromatographic conditions must remain absolutely constant, for both sample and reference substances. If the volume of injected sample and injected reference solution is constant (which is usually the case if a sample loop is applied), the calculation can be carried out as follows:

$$\frac{A_{si}C_{ref}}{A_{ref}} = C_{si} \quad (1)$$

where A_{si} is the area or height of component i in the sample; A_{ref} the area or height of the reference peak in the standard solution; C_{ref} the concentration of reference substance in the standard solution in milligrams per milliliter; and C_{si} the concentration of component i in the sample solution (milligrams per milliliter).

The amount (percent) of component i in the sample is:

$$\frac{A_{si}C_{ref}V_s}{A_{ref}M_s} \cdot 100 = \%C_{si} \quad (2)$$

where V_s is the volume of the sample solution in milliliter; M_s the mass of sample in volume V_s in milligrams; and $\% C_{si}$ is the percentage component of i in the sample.

12.7.2. Internal Standard Method

The internal standard method is well known from GC (→ Gas Chromatography) and should always be used if exact control of the injection volume is not possible. A defined amount of a chosen compound (internal standard) is added to the sample, so that two peaks (that of the internal standard and that of the compound of interest) must be measured to obtain one result. An internal standard must fulfill several requirements: It must not already be present in the sample; it should elute close to the peak of interest; and the two peak sizes should not differ too much. If possible, peaks should not overlap, and no chemical interaction between the internal standard and other sample components should occur. The internal standard must also have good storage stability and be available in high purity. This method is slightly less accurate than the external standard, but it has several significant advantages:

1) Compensation of varying injection volumes
2) Compensation of recovery of sample during sample cleanup, especially in trace analysis
3) Subjection of the standard to the same chemical reaction (e.g., amino acid analysis) in post-column reactions

Calculations for the internal standard method are similar to those mentioned above for the external standard. More detailed information concerning the external and internal standard methods, as well as other methods of evaluation [e.g., area normalization method (100 % method), addition method] is given in [11].

12.7.3. Error Sources in HPLC (→Chemometrics)

Error sources in HPLC are so manifold that they cannot all be discussed completely. Nevertheless, one can distinguish between random errors and systematic errors. Whereas random errors are relatively easy to recognize and avoid, systematic errors often lead to false quantitative interpretation of chromatograms. Two simple ways of recognizing a systematic error in a newly developed (or newly applied) analytical method are either to test this new method with certified materials or—if an "old" (i.e., validated) method already exists—to compare the result from the old and the new methods. If the new method shows good reproducibility, a correction factor can be calculated easily by

$$C_f = \frac{A_{certified}}{A_{new}} \quad (3)$$

where C_f is the quantitative correction factor; $A_{certified}$ the peak area/height or concentration of component i in the certified reference sample or the "old" method; and A_{new} the peak area/height or concentration of component i obtained with the "new" method. Multiplying the calculated amount of component i (see Eq. 1) in the sample by the correction factor C_f compensates for the difference that can be caused by one or more systematic errors. (If possible, however, the systematic error should be identified and eliminated.)

Sample Injection. For every quantitative HPLC analysis, especially in the external standard mode, a constant volume of injected sample solution is essential. This can be achieved only by using sample loops (5, 10, 20 μL, etc.) or automatic sample introduction system (see Section 12.2.4). The constancy in injection volume of these systems has to be checked from time to time. This is done by injecting the same sample several times and comparing the peak areas and peak heights generated. If a constant injected sample volume cannot be ensured, the internal standard method should be applied to compensate for these variations. A variation of the peak area can also result from short-time pulsation of the pump.

Another error in quantification can be caused by overloading the column (rule of thumb: A maximum of 1 mg of analyte per gram of stationary phase). The overloading of a chromatographic column is shown by the appearance of decreasing peak symmetry (fronting or tailing), peak broadening, and decreasing retention time. Obviously, loss of the sample through leakages or strong adsorption to the stationary phase must be eliminated. The latter is often the reason for a baseline rise or for nonlinear behavior of the calibration curve in the lower concentration range.

Pump. A constant flow rate of the mobile phase is another essential condition for quantitative analysis. Especially for peak area quantification, even small variations in flow can have a disastrous effect because the recorder or integrator plots the concentration (signal intensity) versus analysis time (not versus flow rate). Therefore a variation in flow automatically influences the recorded peak area. Quantification over peak height is not so strongly influenced by fluctuations in the flow of the mobile phase.

As a rule of thumb, peak area calculations are sensitive to baseline variations, and therefore many trace analyses are quantified over the peak height. On the other hand, peak height is influenced strongly by retention time (the longer the retention time, the greater is the dilution of the sample and therefore the weaker is the detector output signal). Generally every variation of flow (even very slow variations—often recognizable by a baseline drift) must be avoided for quantitative analysis.

Detector. If possible, the use of highly selective detectors (e.g., fluorescence detector) should be preferred in quantitative analysis because sensitivity is usually enhanced and, due to the high selectivity, peak overlapping is often avoided. However, most HPLC detectors (even the selective ones) are photometric detectors that convert the light absorption of an analyte in the mobile phase into an electrical signal, which is recorded as a function of time. Therefore, their linearity is limited by the Lambert–Beer law (see Section 12.2.6). The influence of temperature on detection (UV, fluorescence, refractive index, etc.) should not be neglected. The solute concentration in the detector cell is indirectly a function of temperature because the density of the mobile phase is a function of temperature [90]. Additionally, the position of the absorption maxima changes with temperature. Generally, for quantitative analysis, temperature variations must be avoided (in hot countries, daily temperature variations of up to 10 °C are quite common) by thermostating the chromatographic system including the detector cell.

Processing of Data. The chromatogram obtained contains qualitative (t_R and k') and quantitative (peak area or height) information. The quantitative information can be evaluated manually or automatically (by an integrator or computer system).

If the baseline shows any drift, *peak heights* are often more correctly determined manually than by an automatic system (important for trace analysis). The *peak area* is very seldom calculated manually, and if it is, a peak area equivalent is used, which is the product of peak height and peak width at baseline. In older publications, the peak area was often calculated by cutting out the peak and weighing it [88]. This method also gave very good results.

12.8. Sample Preparation and Derivatization

12.8.1. Sample Cleanup and Enrichment

Usually, before a sample is injected into the chromatographic system it has to undergo a cleanup procedure. In the simplest case, cleanup consists of only a filtration step. Each sample solution should be filtered routinely before injection onto the column. This prevents sample loop, capillary, or column plugging by small particles in the sample solution and also prolongs column life significantly. Besides filtration, more complex samples require further cleanup and often an additional enrichment step. No general rule for sample cleanup or sample enrichment can be given; each sample matrix is different, and the particular cleanup must be adjusted to the sample in question. Nevertheless, water-soluble (polar) samples can be cleaned up or enriched simply by liquid – solid extractions (e.g., small silica columns) or liquid – liquid extractions in a separatory funnel. The latter is also very important for the isolation of hydrophobic analytes [e.g., polyaromatic hydrocarbons (PAHs), pesticides] from a polar matrix (i.e., water). Isolation of hydrophobic solutes from a polar matrix is done easily by pumping the solution over a lipophilic stationary phase (e.g., RP-18). Here, the hydrophobic analytes are retained while the more polar components do not interact with the stationary phase. After a certain amount of sample solution (depending on the separation problem and the capacity of the solid phase) has been pumped over the stationary phase (either an open-column "solid-phase extraction" or a small HPLC column), the mobile phase is changed to a hydrophobic (organic) solvent, which elutes the retained analytes. After evaporation of the organic solvent, the analytes can be dissolved in a known amount of solvent for analysis.

Another very interesting and promising cleanup technique is supercritical fluid extraction (SFE). This method, which usually employs supercritical carbon dioxide as an extraction medium, shows a high selectivity toward nonpolar solutes. As a rule of thumb, if a solute is soluble in hydrophobic solvents such as heptane or sometimes even in more polar ones such as dichloromethane, it should also be soluble in supercritical carbon dioxide. Through addition of modifiers (e.g., methanol, water, formic acid), even relatively polar molecules are extractable, although the increase in solvent strength is paid for by a loss in selectivity. The selectivity and thus the solvent strength of the supercritical fluid can also be adjusted by variation of the extraction pressure, density of the extractant, or temperature.

SFE with carbon dioxide offers several advantages compared to classical solvent extraction methods: (1) it is very selective; (2) it is fast, because of lower viscosity and approximately 10 times higher diffusion coefficients; and (3) CO_2 is neither flammable nor explosive and is nontoxic. SFE is suitable for oxidizable and temperature-sensitive samples. In addition, no reactions occur between extraction medium and sample compounds and on-line coupling is possible with both supercritical fluid chromatography (SFC) and gas chromatography.

Nevertheless, this technique also has some disadvantages, especially in collecting the extracted analytes. High molecular mass analytes tend to precipitate inside the restrictor (see Section 12.10.4.5) and cause it to plug, during the expansion of the supercritical fluid (expansion of supercritical fluids results in a decrease of density and therefore solvent power). Low molecular mass samples (volatile samples) can be carried away in form of an "analyte fog" to a certain extent during the expansion procedure, resulting in insufficient recovery. Further information on this promising sample cleanup method is given in [91], [92].

12.8.2. Pre- and Postcolumn Derivatization

Sample derivatization is often applied to enhance detection selectivity or sensitivity. This technique is also applied in GC and TLC. Although derivatization is a very helpful tool in chromatography it has the disadvantage that derivatization reactions make the individual sample compounds more similar in adsorption behavior, which can decrease separation selectivity.

In LC the derivatization can be performed either before the chromatographic separation (precolumn derivatization) or after the analytes have passed the column (postcolumn derivatization).

12.8.2.1. Precolumn Derivatization

Precolumn derivatization is the most frequently applied derivatization technique for many reasons. One very important factor is that the chromatographic equipment need not be

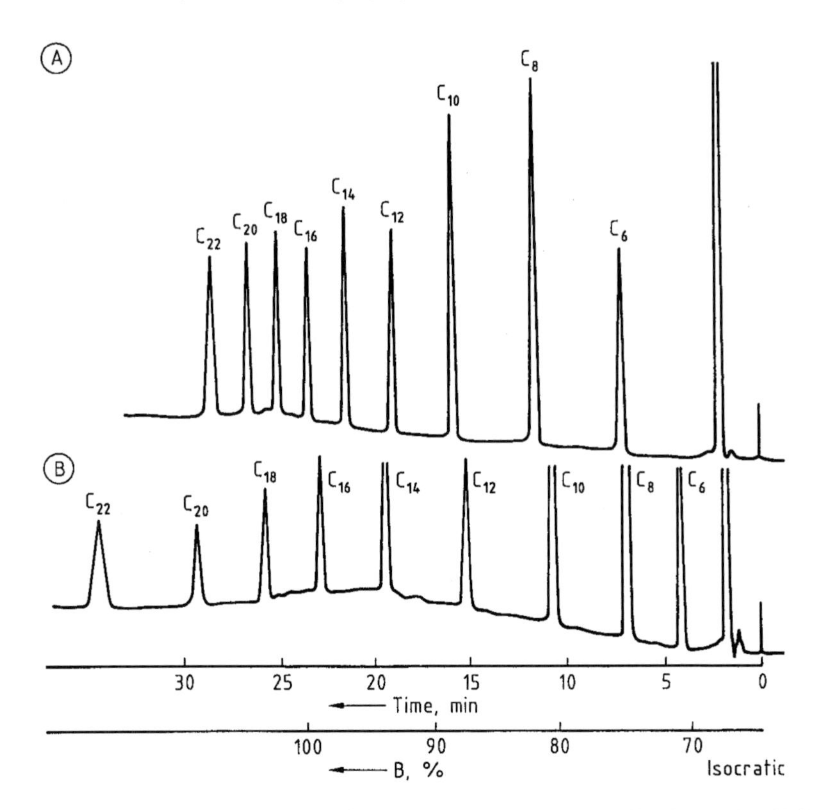

Figure 34. Separation of fatty acid phenacyl esters on an RP-18 column (300×4.2 mm i.d.) with flow = 2 mL/min and t_g= 20 min linear [95]
A) 70 % Methanol in water to 100 % methanol; B) 70 % Acetonitrile in water to 100 % Acetonitrile

changed because precolumn derivatization is performed off-line. Furthermore, the reaction time is of minor importance. Nevertheless, the derivatization should be quantitative, and in certain cases, excess derivatization reagents can lead to poor chromatographic separation.

The most frequent application of precolumn derivatization is in amino acid analysis. Here, the amino acids are derivatized for instance with 9-fluorenyl methoxycarbonyl chloride (FMOC) or *o*-phthalaldehyde (OPA) to form strongly fluorescing compounds, which can be detected easily with a fluorescence detector [93], [94]. Another typical application for precolumn derivatization is in the analysis of fatty acids. Fatty acids have no chromophore, so UV detection is impossible. Even highly unsaturated fatty acids can be detected only in the very far UV range (< 200 nm) with poor sensitivity. Other detectors, such as the refractive index detector, have a disadvantage because they cannot be used in gradient elution, which is essential for fatty acid analysis. Therefore, the fatty acids are reacted with UV-absorbing reagents (e.g., phenacyl bromide) to form fatty acid phenacyl esters, which are easily detectable at 254 nm [95] (see Fig. 34).

12.8.2.2. Postcolumn Derivatization

In postcolumn derivatization, also known as chemical reaction detector (CRD), the sample compounds are derivatized on-line after separation. Derivatization takes place in a reactor situated between the column outlet and the detector (see Fig. 35). The chromatographic equipment is therefore extended by an additional reagent pump, which delivers the reagent into a zero-void-volume mixing chamber (Fig. 36). Here the reagent is mixed thoroughly with the mobile phase containing the solute. (An ordinary three-way unit is not sufficient as a mixing chamber in postcolumn derivatization.) The mixture is then pumped through a reactor where actual derivatization occurs before the sample reaches the detector. The

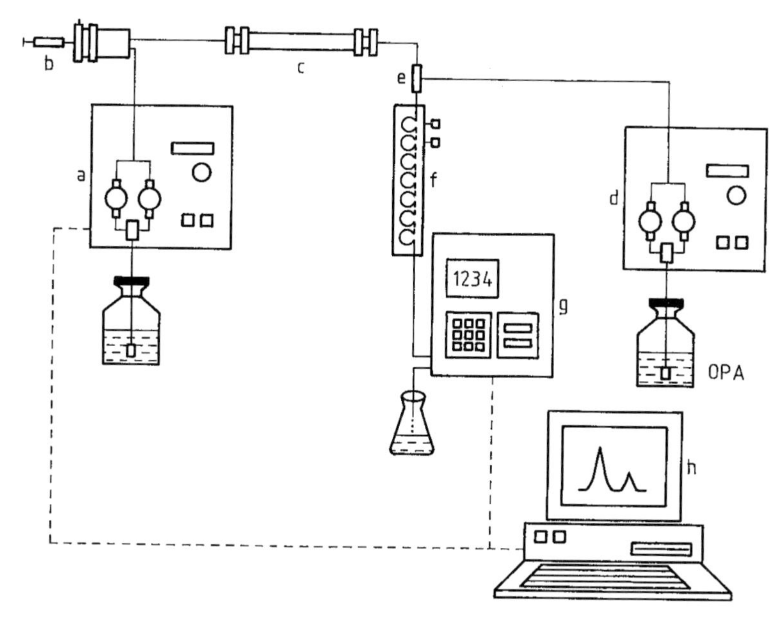

Figure 35. Diagram of postcolumn derivatization system
a) Pump; b) Sample introduction; c) Separation column; d) Reagent pump; e) Mixing chamber; f) Reactor; g) Detector; h) Data control unit

derivatization reagent must be delivered continuously, with as little pulsation as possible.

The reaction time required for a particular derivatization is the limiting factor. It must be short enough so that derivatization can take place during passage of the solutes through the thermostated reactor. Of course, the required reaction time can be prolonged to a certain extent by enlarging the reactor, but this is paid for by increasing backpressure and additional peak dispersion. Even if the required reaction time for a particular derivatization (i.e., for 100% conversion) cannot be achieved with a reactor, this is often no problem in postcolumn derivatization since the derivatization need not to be quantitative, merely reproducible.

The general rule of avoiding any additional void volume between the column outlet and the detector cell cannot be fulfilled in postcolumn derivatization. Therefore, loss of separation efficiency to a certain extent, due to peak dispersion, is unavoidable. To reduce this drawback, several reactor types have been developed. The most important ones are:

1) Packed-bed reactors
2) Open tubes with optimal geometric orientation
3) Open tubes with liquid or gas segmentation
4) Coiled open tubes

Many excellent papers have been published that compare these different types of reactors [94], [96]; therefore, only the most important reactors (open tubes with optimal geographic orientation) are discussed here in more detail.

Coiled Open Tubes. Initially, WALKLING [97] and HALASZ [98] showed that systematic deformation of open tubes can significantly reduce peak dispersion, compared to a simply coiled tube with uniform i.d. A detailed discussion of the different parameters influencing peak dispersion (e.g., relationship of axis, coiling diameter, knot distance, kinematic viscosity of the mobile phase, inner surface of the tube) is given in [99]. The breakthrough in the optimal geometric orientation of open tubes came with the introduction of PTFE capillaries (i.d. = 0.2 – 0.5 mm). With this easily deformable material (compared to glass or stainless steel capillaries), not only could coiled and waved reactors be made [99], or figure eights formed around two rods [100], but the capillary could be knitted [100], [101]. Capillaries are knitted in such a way that each right loop is followed by a left one (with small coiling diameter) so that each loop is bent out of the plane of the preceding one (Fig. 37). Because of the small coiling diameter and continuous change in direction,

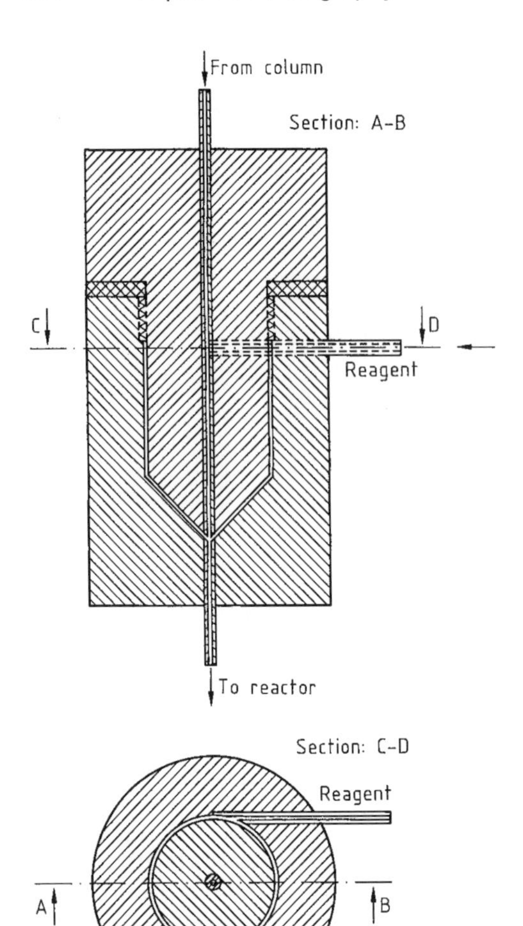

Figure 36. Example of a postcolumn derivatization mixing chamber (cyclon type) [94]

Figure 37. Knitted open tube (KOT) reactor

centrifugal forces are effective even at low flow rates, initiating a secondary flow, that is ultimately responsible for the reduced band broadening. These so-called knitted open tubes (KOTs) have

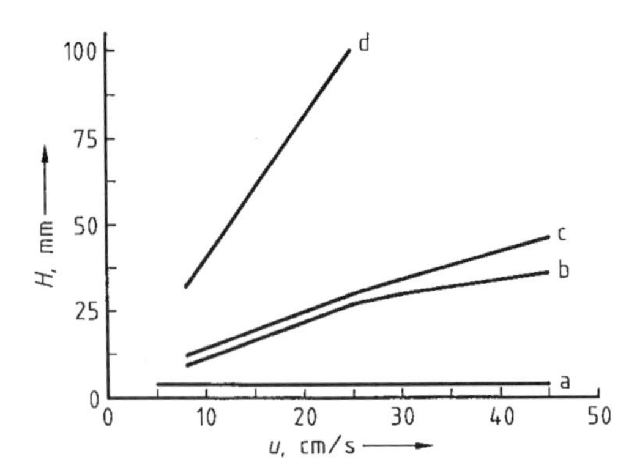

Figure 38. Peak dispersion in open tubes
a) Straight capillary; b) $R = 2.6$ cm; c) $R = 0.35$ cm; d) Knitted open tube (KOT)

the best performance in terms of peak dispersion. With these KOTs, peak dispersion could be made independent of linear velocity over a large velocity range (even at very low flow rates) (Fig. 38).

Packed-Bed Reactors. Packed-bed reactors are packed preferably with small glass beads (20 – 45 μm). The reactor length and therefore the reaction time are limited by the generated back-pressure of the packed-bed reactor. This is the reason it is generally used only for fast derivatization reactions.

Open Tubes with Liquid or Gas Segmentation. Another often-used reactor type (especially for relatively long-lasting derivatization reactions) is the open tube with liquid or gas segmentation [94]. Segmentation prevents the formation of a parabolic flow profile (axial dispersion), and peak dispersion is due mainly to the "phase separator," which is responsible for removal of the segments prior to detection. The principle of segmentation was first used by MOORE and STEIN [102], who developed an amino acid autoanalyzer (ninhydrin reaction) in 1958.

During the development of a new postcolumn derivatization for a particular separation problem, the following points should be considered:

1) An appropriate derivatization reaction should be sought; these are often described in the literature for classical off-line derivatization
2) The reagent concentration should be optimized: Highly concentrated reagents should be delivered with a low flow rate to prevent any unnecessary sample dilution in the mobile phase

Table 7. Examples of postcolumn derivatization techniques in HPLC

Compound	Reagent,* reaction type	Chromatography	Reference
Amino acids	ninhydrin	IEX	[103]
	fluorescamine	IEX	[104]
	OPA	IEX	[105]
Peptides	fluorescamine	RP	[106]
Mono-, oligosaccharides	3,5-dihydroxytoluene	IEX	[107]
	catalytic hydrolysis/ABAH	IEX	[108]
Carbamate pesticides	alkaline hydrolysis/OPA	RP	[109]
Glyphosate (herbicide)	hypochlorite oxidation/OPA	RP	[110]
Vitamin B_1	thiochrome	RP	[111]
Vitamin B_6	semicarbazide	RP	[112]
Vitamin C	oxidation/*o*-phenylenediamine	RP	[113]

* OPA = *o*-phthaladehyde; ABAH = 4-aminobenzoic acid hydrazide.

3) Reaction time should be as short as possible; the longer the required reaction time is, the longer must the reactor be, which results in increased back-pressure and peak dispersion
4) To shorten the required reaction time, the reactor temperature should be as high as possible (an increase of 10 °C doubles the reaction speed)
5) An ca. 80 % conversion of the original solute into the derivatized compound is sufficient to achieve maximum detection sensitivity even for quantitative analysis

Table 7 shows a few derivatization reagents and their typical applications.

12.9. Coupling Techniques

Many different coupling techniques in HPLC were applied since the 1970s but only a few proved to be a significant advantage for the analytical chemist. Even simple HPLC equipment can pose severe handling and reproducibility problems for the beginner. Each additional coupling technique naturally increases these problems, making sophisticated coupling techniques more interesting for academic use than for routine analysis. Nevertheless, coupling techniques such as HPLC – GC and HPLC – MS (mass spectrometry) have become very important tools for analytical chemistry. Another often-applied coupling technique is column switching.

12.9.1. Column Switching

Column switching was described in the late 1960s by Deans [114] for multidimensional GC (→ Gas Chromatography). In principle, it is the coupling of different columns (usually with different selectivities) by means of a manual or automatically driven high-pressure valve. The first application of column switching in HPLC was described by Huber et al. [115] in 1973. They used this technique as a simple and inexpensive alternative to gradient elution. Gradient systems were very expensive in those days. However, in the following years, good and reasonably priced gradient systems came onto the market, making column switching unnecessary as an alternative.

New applications for this technique were found especially in the analysis of very complex samples (e.g., drug analysis in blood serum), where the analyte is selectively "cut out" of the sample matrix. According to the position of the "cut" in a chromatogram, front-cut, heart-cut, and end-cut techniques can be distinguished. Other applications of column switching are on-line sample cleanup, backflush technique, and multidimensional chromatography. Figure 39 shows different examples of switching arrangements.

The high-pressure switching valve is the most important part of the instrumentation in this technique. It must work reliably even after several thousand switches under high pressure, must have a very small dwell volume, and must be chemically inert to both the sample and the mobile phase. Normal switching valves have six ports, but four- and up to ten-port valves are commercially available as well. Their setup is very similar to the usual sample valve injectors, e.g., "rheodyne valves" (Section 12.2.4.2). They can be driven by hand or automatically either by pneumatic force or electronically. Pneumatically or electronically driven valves have the advantage that the entire analysis can be automated.

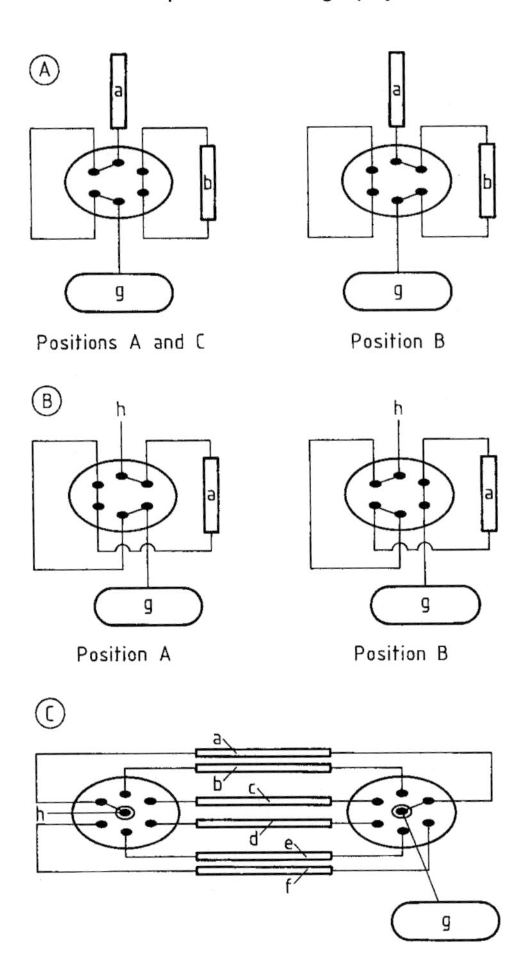

Figure 39. Arrangements for column switching technique
A) Heart-, front-, and end-cut techniques; B) Backflush technique; C) Multidimensional chromatography
a)–f) Column 1–6; g) Detector; h) Injection valve

12.9.1.1. Front-, Heart-, and End-Cut Techniques

A sample is usually preseparated on a short precolumn and then directed onto an analytical column (with the same or different selectivity) where actual separation occurs. The *heart-cut technique* is most often used when the solute co-elutes with another sample or when it is situated on the flank of another peak. In this case the interesting part of the chromatogram is cut out and "parked" on the analytical column, while the rest of the sample is eluted from the precolumn into the waste. Thereafter, the valve is switched back to the analytical column position, and the parked sample can now be baseline separated (Fig. 39 A). The *front- and end-cut techniques* are basically the same, differing only in the position of the cut.

12.9.1.2. On-Line Sample Cleanup and Trace Enrichment

Column switching is also very suitable for sample cleanup and can be combined with trace enrichment. A typical example of this technique is the on-line sample cleanup and enrichment of hydrophobic pesticides from aqueous solutions. Here the sample (water, blood, etc.) is pumped over a small RP column, where only the hydrophobic components are retained. Since the amount of sample can be chosen in any size or volume, hydrophobic trace enrichment takes place. After the entire sample volume has passed the precolumn, the high-pressure valve is switched to the analytical column position and an eluent (having a stronger elution strength for hydrophobic analytes e.g., dichloromethane) is passed through the RP precolumn, eluting the hydrophobic components onto the analytical column where separation occurs.

12.9.1.3. Backflush Technique

The switching arrangement for the backflush technique is shown in Figure 39 B. This technique is used mainly if the components of interest in a particular sample are very strongly retained by the stationary phase. Mobile phase flow is maintained in one direction until all early-eluting components have left the column. At this stage, the more strongly retained components are still near the head of the column. Then, the flow direction is reversed (backflush) to allow faster elution of these analytes. The switching arrangement for the backflush technique can be used in the routine analysis of very complex samples that contain components having a high affinity for the stationary phase. In this case, after a certain number of chromatographic runs the flow direction is reversed to flush the analytical column and elute these strongly retained components [116]. This lengthens column life significantly.

12.9.1.4. Multidimensional HPLC

In multidimensional HPLC, several columns with different selectivities are switched parallel to each other with the help of two high-pressure switching valves (see Fig. 39 C). This technique is used for samples that contain components with a wide range of polarity (from low to high k' values). In this case, one should determine whether gradient elution would be more efficient.

All of the aforementioned column switching techniques can be combined with each other in various ways to provide analytical chemists with important tools for their investigations.

12.9.1.5. Working with Precolumns

In column switching, often at least one column is a precolumn. In the majority of cases the precolumn does not require high separation efficiency. Its purpose is to retain or enrich a certain part of a sample. The actual separation is performed on a second column (the analytical column), which possesses the appropriate separation efficiency. Therefore, the precolumn can be packed with inexpensive, coarse particles. Even a well-packed column bed is unnecessary. If the analyte of interest is retained on the precolumn in a relatively wide zone, this large elution band is reconcentrated as soon as it reaches the head of the analytical column (as long as the analytical column has a better selectivity and higher separation efficiency than the precolumn). Precolumns are usually short stainless steel columns (ca. 20×4 mm i.d.), packed with coarse material (30 – 60 µm), which need not have the small particle-size distribution required for analytical columns. Therefore, these precolumns can easily be homemade, and if purchased, they are inexpensive. If the precolumn is used as a "protection" for the analytical column (its original purpose), it should be well packed and the packing material should be the same as that in the analytical column.

12.9.2. HPLC – Mass Spectrometry

The coupling of high-performance liquid chromatography with mass spectrometry (HPLC – MS) has become a very important tool for the reliable identification and quantification of complex samples. In the pharmaceutical industry, for instance, the metabolic pathways or pharmacological effects of new products are often studied with this system.

The main problem in coupling the LC system with the mass spectrometer is the fact that the mass spectrometer is under a high vacuum and the mobile phase from the HPLC system must be removed before the analyte can enter the MS. This is the reason why HPLC – MS coupling is usually done with microbore or even capillary columns. In using the "normal" 4.0 – 4.6-mm i.d. columns, too much solvent would have to be removed. Initially, solvent removal was performed by the *moving belt* technique, for example. Today, the *direct liquid inlet* technique is often utilized. In this case, solvent eluting from the LC system is sent through an ca. 5-µm hole into the vacuum of the MS. The resulting aerosol comes into a "desolvation chamber," where the solvent is completely removed. With the *thermospray techniqe*, the analyte containing the mobile phase is sent through a heated (60 – 200 °C), narrow pipe (ca. 0.1-mm i.d.) directly into the MS vacuum of a specially designed ion source. This technique has the advantage that no split is necessary after the separation column, so the entire mobile phase can be sent into the MS system, which enhances detection limits significantly. Further, "normal" HPLC columns (4.0 – 4.6-mm i.d.) can be used in this technique, with flow rates between 0.8 and 1.5 mL/min.

For the HPLC – MS systems, many different ionization techniques have been described in the past. Various interface, ionization methods, and operating techniques applicable to LC – MS are discussed in [117] for instance: Thermospray, particle beam, electrospray (ES), field desorption (FD), fast atom bombardment (FAB), time of flight (TOF), etc. The electrospray technique produces a soft ionization for thermally labile compounds, while FAB has the advantage that higher molecular mass samples can be introduced into the mass spectrometer. Table 8 offers a rough guide to the applicability of various LC – MS interfaces. For more detailed information on LC – MS, see [118].

In HPLC – MS coupling, careful sample preparation is recommended. The "dilute-and-shoot" approach is not possible, especially when handling real-world samples such as environmental and biological samples. In these cases sample preparation techniques (see also Chaps. 12.8 and 12.9) such as liquid – liquid extraction, solid-phase extraction [119], or even affinity techniques (see also Chap. 12.11) directly coupled with LC – MS [120] can be useful.

Table 8. Overview of advantages and disadvantages of various LC–MS interfaces * [117]

LC–MS type	LC flow range	Detection limit	Solvent type range	Nonvolatile thermally labile samples	Semivolatile samples	Molecular mass range
Moving belt	+++++	++	+++++	+	+++++	+
Particle beam	+++	++	++++	+	+++++	+
Thermospray	+++++	++	++++	+++	+++++	+++
LC–FAB	+	++	+++	+++++	++++	++++
Electrospray	+	++++	++	+++++	++++	+++++

* Increasing numbers of "+" indicate improved performance.

12.10. Supercritical Fluid Chromatography

12.10.1. History

In 1822, Cagniard de la Tour first described the critical state [121]; In 1869, Andrew [122] published the first paper about a gas–liquid critical point. In 1879, Hannay and Hogarth described the solubility of cobalt and iron chlorides in supercritical ethanol [123]. Lovelock (1958) first suggested the use of supercritical fluids as a mobile phase in chromatography [124], and in 1962, Klesper et al. [125] published the first paper on supercritical fluid chromatography. In this paper they described the separation of nickel porphyrins using supercritical chlorofluoromethanes as the mobile phase. This work was done before HPLC was introduced by Piel in 1966 [9]. In the following years, Sie, Rijnders, and Giddings [126]–[134], through their practical and theoretical studies, made significant contributions to understanding of the separation process with supercritical fluids. Sie and Rijnders at the Shell Laboratories in Amsterdam were the first to use the term supercritical fluid chromatography in 1967 [135].

In the 1970s, SFC declined in popularity, partly because of the very fast development of HPLC and because, in those days, working with supercritical fluids was not so easy due to several technical problems. Another reason SFC declined in popularity was that expectations had been set far too high in the early use of this technique.

Nevertheless, the development of SFC continued. In 1969–1970, Jentoft and Gouw [136], [137] were the first to work with pressure programming and to use modifiers to increase the elution strength of the mobile phase. Schneider and coworkers [138]–[140] investigated the physicochemical aspects of supercritical fluids and thereby contributed much to the understanding of the technique.

The introduction of capillary columns in SFC by Novotny and Lee in 1981 [141] led to a revival of this technique, which was followed by a large number of papers on new applications. At the same time (1981) the first commercial packed-column SFC instrument came on the market. Four years later, in 1985, the first capillary column SFC instrument was available commercially.

Since the late 1980s, capillary and packed-column SFC have become more and more popular, as shown by the growing number of published papers, new books, and international meetings devoted exclusively to SFC and SFE. The reason for this recent gain in popularity is that significant progress has been made in improving the instrumentation, which expanded the applications to many problems such as quantitative environmental analysis (although generally the sensitivity is not as good as in GC, for instance). Another reason is that modern chromatographers have begun to realize that the various analytical separation techniques (HPLC, GC, TLC, SFC, CE, etc.) should not compete against each other. Each technique has its advantages, and which technique is preferred should be decided on a case-to-case basis.

12.10.2. Theoretical Background

If a gas is compressed above its critical pressure p_c at temperatures exceeding the critical temperature t_c, then neither a gas nor a liquid is present but a so-called supercritical fluid, which combines properties of both. Table 9 shows the critical parameters of the most commonly used fluids in SFC.

Table 10 compares some important physicochemical properties of a liquid (HPLC), a gas (GC), and a supercritical fluid. The density—and thus the solvation characteristics of the supercritical fluid—are very similar to those of a liquid.

Table 9. Critical parameters of the most frequently used supercritical fluids [142]

Supercritical fluid	p_c, MPa	t_c, °C	ϱ_c, g/cm^3
CO_2	7.39	31.3	0.47
N_2O	7.34	36.5	0.45
SF_6	3.76	45.5	0.74
Xe	5.92	16.6	1.10
NH_3	11.4	132.5	0.24
CCl_2F_2	4.12	111.8	0.56

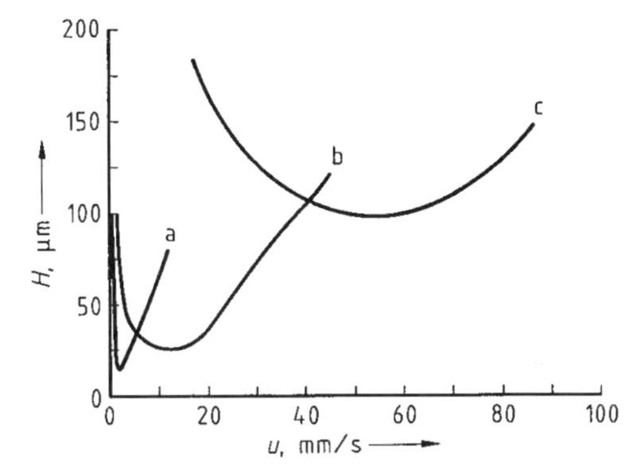

Figure 40. Comparison of VAN DEEMTER curves in GC, HPLC, and SFC
All packed columns (particle size 10 µm); mobile phase in HPLC: methanol, in GC: N_2, in SFC: CO_2; samples in HPLC: benzene, in GC and SFC: methanea) HPLC; b) SFC; c) GC

On the other hand, the viscosity is comparable to a gas, and the diffusion coefficient is about ten times higher than that of a liquid. Therefore in SFC (especially packed-column SFC), analyses can be performed at much higher flow rates than in HPLC without loss of too much efficiency (see Fig. 40), which results in a significantly shorter analysis time. Because of the low viscosity of supercritical fluids, several packed columns can be combined in series, even with small particles (5 µm), without generating too high back-pressure. If several columns packed with the same stationary phase are connected in series, the separation efficiency can be improved dramatically. The combination of different stationary phases (columns) and the variation of their column length result in totally new and interesting selectivity ("selectivity tuning").

In general, a mobile phase for SFC should exhibit the following characteristics:

1) The critical parameters (temperature and pressure) should be achievable with commercial equipment
2) The fluid should be available in high purity and not be too expensive

Table 10. Comparison of physical properties of a liquid, a supercritical fluid, and a gas

	Liquid	Supercritical fluid	Gas
Viscosity η, Pa · s	$<10^{-1}$	$10^{-4}-10^{-3}$	10^{-4}
Density ϱ, g/cm^3	0.8 – 1.0	0.2 – 0.9	10^{-3}
Diffusion coefficient D_m, cm^2/s	$<10^{-5}$	$10^{-3}-10^{-4}$	10^{-1}

3) The fluid should be chemically compatible with the sample and SFC equipment
4) The fluid should be miscible with many organic solvents (modifiers)

Because of the low critical temperature and pressure of carbon dioxide and its compatibility with much equipment, especially with the flame ionization detector (FID), supercritical carbon dioxide is the most frequently used mobile phase in SFC.

The elution strength of a supercritical fluid is often described by its dielectric constant *DC*. Figure 41 shows that the influence of pressure on the *DC* at different temperatures is very similar its influence on the density, which is presented in Figure 42. This diagram is one of the most important for understanding supercritical fluid chromatography and extraction. Generally, the solubility of a solute in a supercritical fluid increases with increasing density of the fluid. Simultaneously, the elution strength of the fluid increases as well, which results in shorter retention times. Figure 42 shows that a small variation of pressure slightly above the critical parameters (p_c and t_c) causes a large change in density and therefore in the elution strength of the fluid. On the other hand, a temperature increase at constant pressure causes a decrease in the density and consequently in the elution strength. In other words, in contrary to gas chromatography, an increase in temperature at constant pressure results in increased retention and therefore analysis time. This explains why negative temperature gradients are often run in SFC.

Further discussions concerning the theory and basics of SFC are given, for example, in [124], [142], [143].

12.10.3. SFC Compared to GC and LC

As mentioned above, SFC should not be regarded as competing with GC or HPLC. SFC is complementary to these two analytical separation

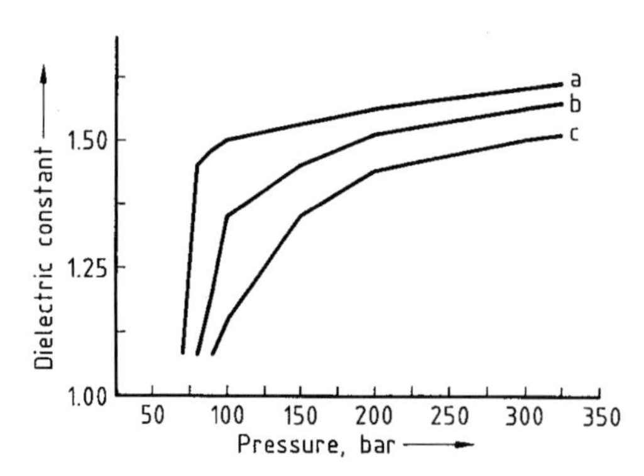

Figure 41. Influence of pressure on the dielectric constant of carbon dioxide at different temperatures
a) $t = 25$ °C; b) $t = 40$ °C; c) $t = 60$ °C

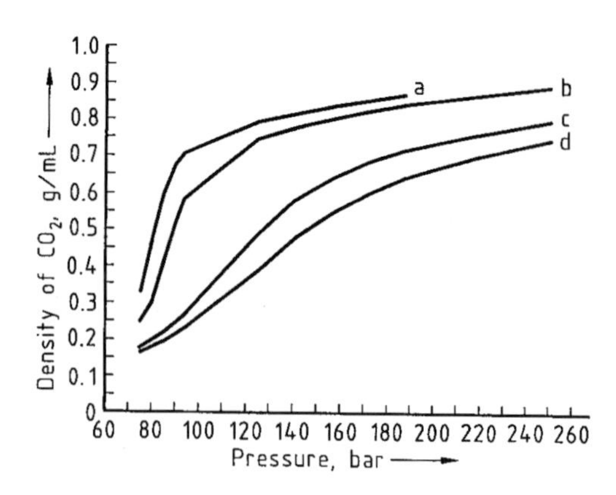

Figure 42. Influence of pressure on the density of carbon dioxide at different temperatures
a) $t = 35$ °C; b) $t = 40$ °C; c) $t = 60$ °C, d) $t = 70$ °C

techniques, and one should decide from case to case which method is preferred.

Because of its intermediate position, SFC combines several advantages of both GC and HPLC. One significant advantage of SFC compared to HPLC is the fact that besides the normal LC detectors (e.g., UV detector), sensitive universal or selective GC detectors (e.g., FID, electron capture detector) are applicable as long as supercritical carbon dioxide is utilized as mobile phase. Furthermore in packed-column SFC, the same variety of stationary phases (selectivities) can be used as in LC (in contrast to GC), with the additional advantage that the analysis time is significantly shorter.

SFC is superior to GC if thermolabile solutes must be separated. However, this is true only as long as the supercritical fluid has a relatively low critical temperature like carbon dioxide (31 °C), for example.

Nevertheless, SFC also has its drawbacks; for example, only a limited number of samples are soluble in supercritical fluids. In supercritical carbon dioxide, for instance, only relatively nonpolar solutes are soluble. As a rule of thumb, solutes that are soluble in organic solvents having a polarity less than or equal to that of *n*-heptane are usually also soluble in supercritical carbon dioxide. Samples that are soluble in water are insoluble in CO_2.

If the elution strength of a supercritical fluid is not sufficient, it can be increased by adding an organic modifier (see Section 12.8.1). In this way, especially in combination with capillary columns, the applicability of SFC is extended to more polar analytes. Methanol is the most commonly used modifier because it shows the widest solubility range for samples that are insoluble in CO_2. If selectivity enhancement is required, a more nonpolar solvent should be chosen. The most common modifiers for SFC are discussed in [144]. Several papers have been published on the mechanism of modifiers in SFC systems [142]. Apparently, the modifier has at least two effects: First, it dramatically alters the solvation properties of the fluid (usually increases the solubility of the sample compound); second, it blocks the most active retention sites of the column packing so that the retention strength of an analyte is decreased as is its analysis time. Even though modifiers have significant advantages, they limit the choice of suitable detectors (e.g., methanol as modifier interferes in FID). Whereas in packed-column SFC, modifier concentrations of 1 – 5 % are usually sufficient, in capillary SFC, modifier concentrations up to 20 % can be necessary. For the separation of acidic or basic samples an additive to the modifier or mobile phase can often be helpful. This improves peak shape and often shortens analysis time considerably. A typical additive for acidic samples is trifluoroacetic acid (TA) whereas for basic samples, primary amines are often used. The function of an additive is not yet clear, although it seems to homogenize the stationary phase (effectively blocking off silanol groups), to improve the solubility of the solute in both phases, and to suppress solute ionization.

A disadvantage of SFC is the fact that even though the same detectors are used as in GC, the detector sensitivity is not as good. In capillary SFC, this results from the very small sample volumes (10 – 250 nL) that must be injected into the column. In the case of packed columns, larger

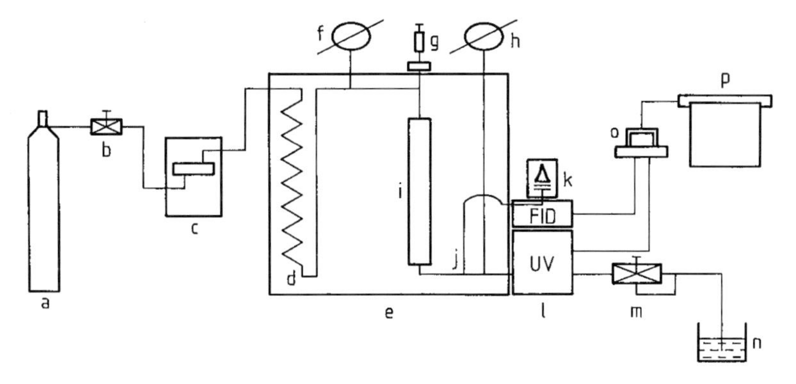

Figure 43. Schematic diagram of an SFC apparatus
a) Gas supply; b) Valve; c) Pump; d) Equilibration coil; e) Oven; f) Manometer; g) Sample introduction unit; h) Manometer; i) Separation column; j) T-unit; k) Flame ionization detector (FID); l) UV detector; m) Back-pressure valve, restrictor, or pressure and flow control unit; n) Fraction collector; o) Data control unit; p) Recorder

sample volumes can be injected but the column outlet going into the combustion detector (e.g., FID, NPD) often has to be split. If not split is applied, the extremely high flow rates, caused by the expanding supercritical fluid, tend to extinguish the flame of the combustion detector.

12.10.4. Equipment

The equipment for packed-column SFC is very similar to that for HPLC, whereas the instrumentation for capillary SFC is more reminiscent of gas chromatography. In Figure 43 a schematic diagram of packed-column SFC instrumentation is shown. It consists of a high-pressure pump, sample introduction system, oven, detector, restrictor, and recorder. In addition, a microcomputer is installed that controls temperature, flow, pressure, and density programming.

12.10.4.1. Pumps

The pump systems used most frequently in open tubular (capillary) and packed-column SFC are the reciprocating piston pumps and the syringe pumps (see Section 12.2.3). The syringe pumps have the advantage of pulseless flow and easy pressure control, while the reciprocating pumps have the advantage of continuous flow. The main difference between the LC pumps described in Section 12.2.3 and SFC pumps is that the latter need a pump head cooling system (temperature of pump head < 5 °C). Reciprocating pumps can deliver only liquids efficiently, not highly compressible supercritical fluids. A further difference is that in SFC a very effective pulse dampener is required to ensure reproducible separations.

The maximum pressure of a modern SFC pump should be as high as possible, at least 40 – 45 MPa, with a flow range from several microliters per minute for open tubular columns (capillary columns) to ca. 5 – 10 mL/min for packed columns.

12.10.4.2. Oven

The oven of an SFC system should meet the same requirements as a normal GC oven. A constant temperature (variation ±0.1 °C) must prevail in the entire oven at any time of a positive or negative temperature gradient. This is very important for reproducible capillary column SFC analysis. These columns are very sensitive to even slight variations in temperature, which can result in peak shape deformation, peak splitting, or irreproducible retention times.

12.10.4.3. Sample Introduction

For packed columns and capillary columns, different sample introduction systems are used.

In the case of capillary columns, finding appropriate solutions for sample introduction is much more challenging than with packed columns. Since packed columns have dimensions similar to normal HPLC, optimal sample introduction can be achieved with ordinary sample loop injectors (see Section 12.2.4.2).

Onto a normal capillary column (10 m × 50 µm i.d.), < 100 mL should be injected to prevent

Table 11. Detection possibilities with SFC *

Detector	Type	Capillary SFC	Packed-column SFC	Modifier possible	Preferred mobile phase
FID	U/D	++	+**	yes	CO_2, SF_6, N_2O
NPD	S/D	++	+**	yes	CO_2, SF_6
UV	S/ND	+(–)	++	yes	CO_2
ECD	S/D	++	+**	yes	CO_2, Xe
LSD	U/(N)	+	++	yes	CO_2
FTIR	U/N	+	++	no	Xe, CO_2, SF_6
MS	U/D	++	+**	yes	CO_2

* U = universal detection; S = selective detection; D = destructive; N = nondestructive; + = detection possible; ++ = preferred column type for the particular detector; – = column type not recommended.
** Split usually necessary.

more than 1 % loss in resolution [142]. To achieve this, different techniques of sample introduction have been used. In 1988, Greibrokk et al. [145] described a technique that involves separation of the solvent from the solutes on a precolumn. The solutes are then transferred to the analytical column where they are focused at the column head. Other solvent elimination techniques based on gas purging and solvent backflush were reported by Lee et al. [146]. The most common reproducible techniques to inject such small volumes of sample onto capillary columns are the dynamic split and the time split technique.

In the *dynamic split mode*, which is the most frequently applied technique in open tubular column SFC, the split ratio is determined by the ratio of the flow out of the column to the flow out of the split restrictor. Thus, if the flow out of the split is 500 times faster than the flow out of the column, the split ratio is 1 : 500. The disadvantage of the dynamic split technique is that obtaining reproducible results can be difficult with certain samples.

The *time split injection* technique is based on early work done by J. C. Moore [147] for LC. In 1986, Richter [148] was the first to report on this injection technique in SFC. A standard injection valve, which is pneumatically driven by a low-viscosity gas such as helium, is connected directly to the analytical column (packed column or capillary column) The valve is electronically driven by high-speed pneumatics to ensure very fast switching times (from load position to inject position and back in the millisecond range). The shorter the switching time is, the smaller is the injected sample volume. Many papers have been published comparing the reproducibility of the time split and the dynamic split techniques, some of them differing significantly in their results. Whereas Richter et al. [149] showed that time split is superior to dynamic split, Schomburg et al. [150] saw no significant difference between the precision of the two techniques. Because of problems in reproducibility with these two injection techniques, the use of an internal standard is strongly recommended for quantitative analysis.

12.10.4.4. Detectors

As mentioned before, a unique advantage of SFC is the fact that a wide variety of detection methods can be applied (see Table 11). Besides the traditional LC detection (e.g., UV) the use of GC detectors, especially, enables relatively sensitive, universal (FID), as well as selective [e.g., electron capture detectors (ECD), thermoionic detectors (TID)] methods of detection.

As in LC, both destructive and nondestructive detectors are used in SFC. The detector most commonly used for this technique is the flame ionization detector. It is a destructive, sensitive, and universal detector, which is well known from gas chromatography. Even though Richter et al. [151] described, in 1989, a new FID detector with a detection limit (5 pg of carbon per second) and a linear dynamic range (10^6) very similar to that achieved in FID – GC systems, the ordinary FID – SFC system is still a factor of 10 – 1000 poorer in sensitivity, which consequently calls for relatively concentrated sample solutions. If very sensitive detection is essential, the FID – SFC detection limit can be increased by a factor of ca. 1000 by replacing carbon dioxide with nitrous oxide, which has similar physicochemical characteristics (see Table 9). However, one should be aware of the fact that N_2O can explode under higher pressure and temperature.

A further drawback of the SFC – FID system is the fact that the amount of organic modifier that can be added to the fluid is limited (methanol ca. 3 – 5 vol %) because the FID responds to organic

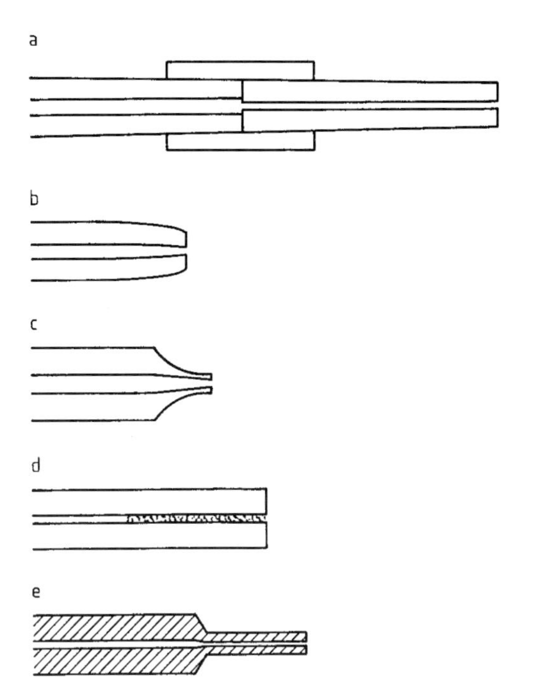

Figure 44. Fixed restrictors
a) Linear capillary restrictor; b) Polished integral restrictor; c) Fast-drawn capillary restrictor; d) Porous frit restrictor; e) Pinched restrictor

molecules. Water can be an alternative, but its solubility in supercritical carbon dioxide is low (< 1 %), and the elution strength does not increase too much.

The UV detector is also very frequently used in SFC, mainly in packed column SFC, because of problems in sensitivity with capillary columns. Since pure carbon dioxide shows little absorbance between 200 and 800 nm, UV detection in packed-column SFC is very sensitive. In general, the characteristics of UV detectors discussed in Section 12.2.6.3 are also valid for UV–SFC detectors, except that the latter must be equipped with a pressure-resistant flow cell. A normal UV detector flow cell in LC can resist a maximum pressure of ca. 7–10 MPa. In SFC, the UV detector cell must be under the same separation conditions (temperature and pressure) as the column (if the pressure in the detector cell were lower, the fluid would loose its solvation power), so it should resist pressure ≤ 40 MPa.

A nondestructive and universal detector in SFC is the FTIR detector (see Section 12.2.6.7). Although its sensitivity is not too high, this technique is highly informative and often helps to identify analytes by their characteristic functional groups or to solve molecular structure problems.

Compared to LC, SFC–FTIR systems have the significant advantage that carbon dioxide is transparent in a large part of the IR region, allowing FTIR flow cell detection [152]. In LC, the mobile phase (especially organic solvents) must usually be removed prior to FTIR detection, because it absorbs in the required IR region. Consequently, organic modifiers used in SFC–FTIR cause the same problem [153], [154], making flow cell detection impossible. The "ideal" fluid for SFC–FTIR flow cell detection is supercritical xenon. It is transparent over the entire IR range and shows sufficient solvation power [155]. Unfortunately, the high price of xenon limits its application in routine analysis. Many other detection techniques are described for SFC. An excellent discussion of the various detection possibilities in SFC is given in [142].

12.10.4.5. Restrictors

The restrictor is an important part of all SFC equipment. It maintains constant pressure in the system and ensures, or at least should ensure, uniform, pulse-free flow of the supercritical fluid; thus, it is also responsible for reproducible analysis. Two types of restrictors exist: Fixed restrictors (see Fig. 44) and variable restrictors. *Fixed restrictors* have the disadvantage that pressure programming automatically also changes the flow. Therefore, independent pressure and flow programming can be achieved only with computer-driven variable restrictors, which often function in principle like a needle valve. Usually the restrictor must be heated since, due to the expansion of the supercritical fluid to atmospheric pressure, strong cooling and formation of solid carbon dioxide result (Joule–Thomson effect), which causes restrictor plugging and therefore irregularities in the flow and reproducibility of the analysis.

12.10.5. Packed-Column SFC (pc-SFC) and Capillary SFC (c-SFC)

The question of whether packed-column or capillary SFC is preferred, is very dificult to answer. Both techniques have their advantages and disadvantages, depending on the particular separation problem. Lately the trend is more toward the packed-column technique because reproducibility and higher sensitivity are ensured. In addi-

Table 12. Column dimensions and characteristics in SFC

Column type	Internal diameter,* mm	Length,* m	Preferred particle size, µm	Supercritical fluid flow (0.4 g/mL), mL/min	Efficiency per meter,* N/m	Efficiency per time,* N/s
Open tubular capillary	0.025 – 0.1	1 – 35	50 **	0.0005 – 0.002	2000 – 22 000	13 – 33
Packed capillary	0.1 – 0.5	0.05 – 0.5	3, 5, 10			
Packed columns	2.0 – 4.6	0.03 – 0.25	3, 5, 10	0.008 – 2.7	33 000 – 51 000	31 – 83
Microbore packed	0.5 – 2.0	0.03 – 0.25	3, 5, 10			

* Data from [142].
** 50-µm-i.d. open tubular (capillary) column.

tion, independent pressure and flow control is possible only in packed-column SFC.

12.10.5.1. Capillary SFC (c-SFC)

Capillary SFC columns are usually manufactured from fused silica, with an internal diameter of 25 – 100 µm and a total column length between 1 and 35 m. The most commonly used are 3 – 10-m, 50-µm-i.d., fused-silica columns. A column like this has a very low supercritical fluid flow rate, from several microliters per minute to ca. 0.01 mL/min which corresponds to a gas flow rate of about 1 mL/min (see Table 12).

c-SFC has a wider application range than the packed-column technique because the phase ratio is much more advantageous. In other words, in c-SFC the absolute amount of stationary phase interacting with the solute (and therefore also the retention strength) is much lower than in pc-SFC. Thus, the separation in c-SFC (as in GC) is dominated by the volatility of the solutes and their solubility in the mobile phase and not so much by the selectivity gained from the interaction between the analyte and the stationary phase. Nevertheless, because of the very low linear velocity (millimeters per second) and therefore low flow rates (millimeter per minute) in c-SFC, the analysis time is usually much longer than in pc-SFC. Even though the separation efficiency is theoretically higher in capillary SFC than in packed-column SFC, in practice, this difference is often not so pronounced. The coupling of c-SFC with mass spectrometry (c-SFC – MS) is much easier than for LC, and many articles on this hyphenated technique have been published [156], [157].

The disadvantages of capillary SFC are that the number of commercially available, stable stationary phases showing different selectivities is limited. Furthermore, problems with reproducibility can occur because of technical difficulties with quantitative sample introduction. Finally, because of the very low flow rates in c-SFC, analysis time is relatively long and independent pressure programming is impossible; the latter can be done only in combination with a flow program.

12.10.5.2. Packed-Column SFC (pc-SFC)

Typical dimensions of packed columns and flow rates in pc-SFC are given in Table 12. This technique is becoming more and more popular compared to capillary SFC. A wide spectrum of stable stationary phases, showing different selectivities, is available commercially. Reproducible, quantitative sample introduction by means of the HPLC sample loop technique is no problem, nor is independent pressure or flow control.

Very high separation efficiencies (more than 100 000 plates) can be achieved by coupling several packed columns in series (e.g., Fig. 45 shows a separation of triazine pesticides with ten 200×4.6 mm i.d. hypersil silica (5 µm) columns in series — note that the pressure drop over all these columns is only 12.4 MPa. This would not be possible in HPLC since, because of the higher viscosity of the solvents, the back-pressure over the column would become too high (high-pressure shutdown).

Several columns with different lengths and selectivities can also be connected in series. This enables individual adjustment of the total selectivity of the system depending on the separation problem (selectivity tuning). Figure 46 shows the separation of fatty acid ethyl esters from human blood serum. The phytanic acid ester (ethyl 3,7,11,15-tetramethylhexadecanate), an indicator for a rarely occurring inherited disease (Refsums syndrome) can be separated and quantified from

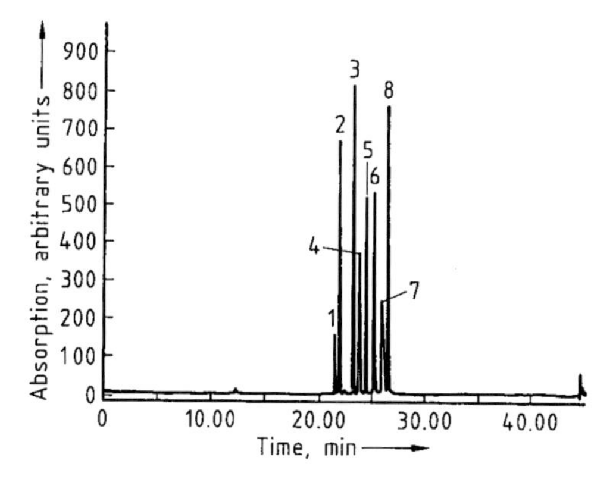

Figure 45. pc-SFC separation of triazine pesticides
Ten 200×4.6 mm i.d. columns hypersil silica columns (particle size 5 µm) in series: program: 2 % methanol, 8 MPa for 5 min, 500 kPa/min + 1 % MeOH/min to 13 MPa and 12 MeOH %; 60 °C, 2.0 mL/min; $\delta p = 12.4$ MPa detector: FID
1) Propazine; 2) Atrazine; 3) Anilazine; 4) Simazine (4-doublet); 5) Prometryne; 6) Ametryne; 7) Bladex; 8) Prometon

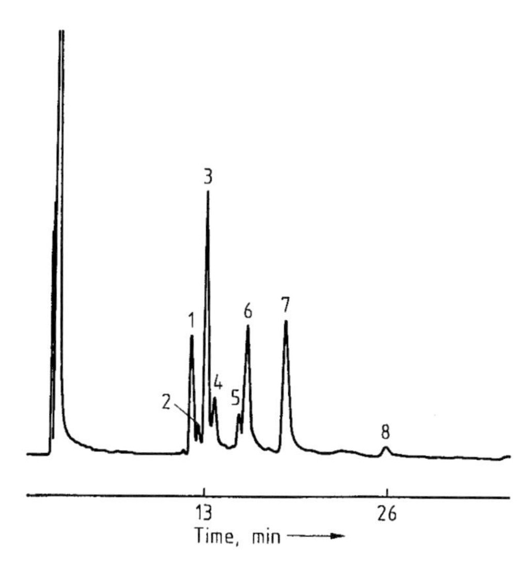

Figure 46. pc-SFC separation of lipids in human blood serum
Columns: 250×4.6 mm ICN silica (3–6 µm); 200× 4.6 mm Nucleosil NH_2 (5 µm); 125×4.6 mm Select B (RP-8) column (5 µm) in series; pure CO_2, 40 °C, 14.5 MPa; detector: FID
1) Phytanic acid; 2) 14:0; 3) 16:3; 4) 18:0; 5) 16:1; 6) 18:1; 7) 18:2; 8) 20:4 (the figure before the colon denotes the number of carbon atoms in the chain, that after the colon, the number of double bonds in the molecule)

other coeluting fatty acid esters only with the column sequence shown.

The pc-SFC separation technique is rapid (ca. five times faster than HPLC with the same column dimensions). Preparative application and coupling with a mass spectrometer (pc-SFC–MS) are possible. However, capillary SFC, or at least narrow-bore columns are preferred for MS coupling because pc-SFC using conventional 100–250 mm × 4–4.6 mm i.d. columns usually has flow rates that are too high.

The retention mechanisms of capillary and packed-column SFC differ from one another. Whereas packed column SFC with supercritical carbon dioxide shows a similar retention mechanism to HPLC (separation due mainly to specific interactions between the analyte and the stationary phase), capillary SFC behaves more like capillary GC (separation in pc-SFC according to "solubility" and "volatility" of solute; separation in c-SFC according to "volatility" of solute). Investigations concerning the retention mechanism in reversed-phase packed-column SFC showed [158] that both a hydrophobic and a polar retention mechanism apply. This can be of significant advantage, as shown in Figure 47. Here, a mixture of salmon oil fatty acid ethyl esters is separated by pc-SFC on an aminopropyl column [159]. A combination of an RP mechanism (retention increases with increasing number of carbon atoms of the esters) and an NP mechanism (retention increases with the growing number of double bonds of the fatty acid esters) can be observed. This separation is an example of the potential of pc-SFC to obtain very fast separation of complex sample mixtures. If plain silica columns are used, the retention mechanism corresponds to the normal phase mechanism in LC, with the advantage that pc-SFC is significantly faster than the NP–HPLC technique.

12.10.6. Applications

Although the application of SFC is limited to those solutes that are soluble in the particular supercritical fluid (generally carbon dioxide), many interesting separations have been reported in various fields. Polymer analysis (M_r up to ca. 5000–8000; even higher with modifier) is certainly one of the most important [160], [161]. Figure 48 shows a typical application in polymer analysis. In the petroleum industry, SFC is becoming more and more important in simulated distillation under mild, nondegradative conditions [142], [162] for analysis of hydrocarbons and of the carcinogenic polycyclic aromatic compounds found in fossil fuels [162] (see Fig. 49). Other

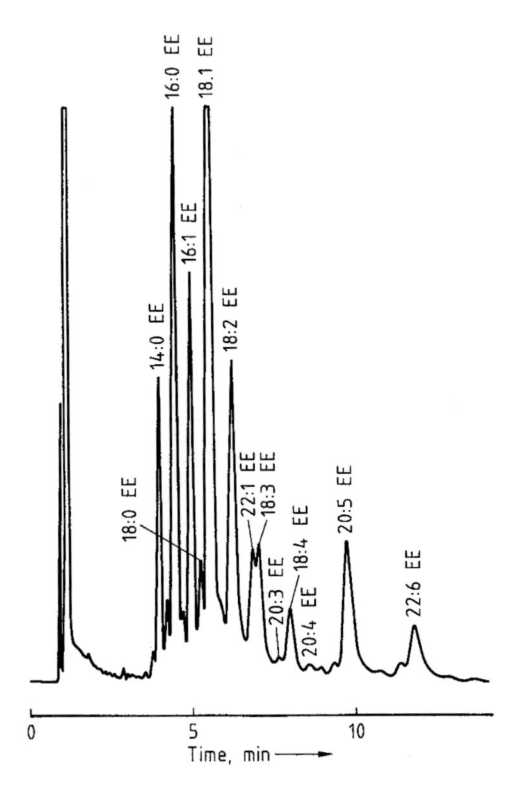

Figure 47. Separation of salmon oil fatty acid ethyl esters with pc-SFC
Column: aminopropyl (particle size 5 µm), 200×4.6 mm i.d.; 34 °C, 1.05 GPa; detector: FID [159]

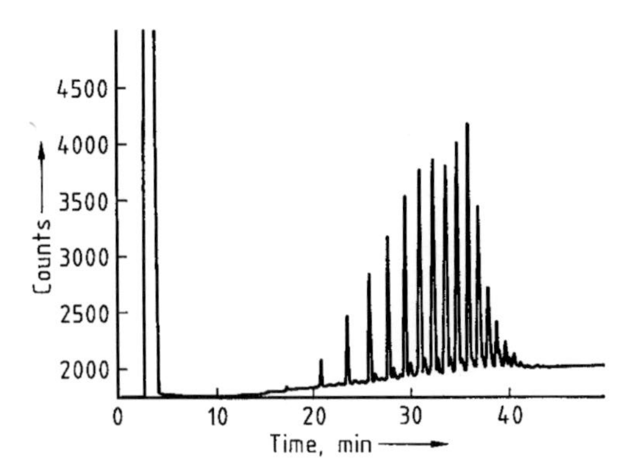

Figure 48. Capillary column SFC separation of Triton X-100 (surfactant)
Column: 50 µm×10 m SB-methyl-1000 (0.35 µm); mobile phase: pure CO_2; temperature: 120 °C; inlet pressure 10 MPa for 10 min, 1 MPa/min to 35 MPa

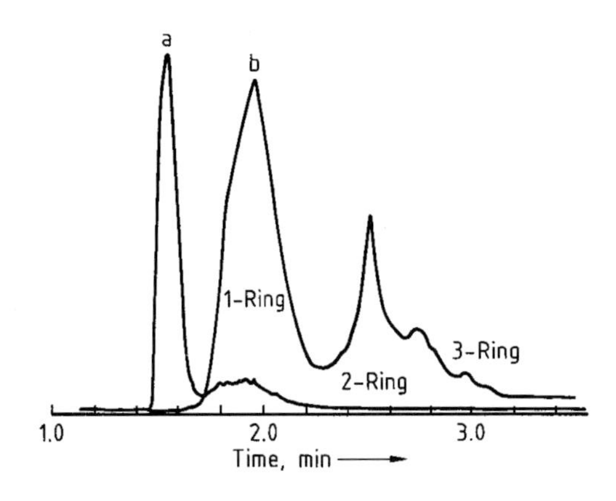

Figure 49. Determination of aromatics in diesel fuel
a) FID detection for saturated compounds and olefins; b) UV detection (at 254 nm) for aromatics
Column: 250×4.6 mm Lichrosorb Si-60; CO_2 isobaric, 15 MPa, 28 °C

typical applications for pc- and c-SFC are the separation of explosives [163], pesticides [164], drugs [165], and pharmaceuticals [166].

Figure 50 shows the potential fo packed-column SFC for an efficient and fast control of the chiral purity of pharmaceuticals. Because of the excellent solubility of lipids in supercritical carbon dioxide, SFC in combination with the universal FID detector is a very helpful tool in lipid analysis.

Although all the applications of pc- and c-SFC would be impossible to mention, their potential in preparative applications should be noted. The latter will certainly become more and more interesting in the future, especially for the food and pharmaceutical industries because very pure extracts are obtainable that are not contaminated with any toxic organic solvents.

In 1997, the worlds first industrial-scale SFC production plant, equipped with separation columns with an internal diameter of over 50 cm and a length of 2 m was designed and started up [167] for the production of high-purity (>95 %) polyunsaturated fatty acids from fish oil [167].

12.11. Affinity Chromatography

Affinity chromatography is a special type of adsorption chromatography for the isolation and purification of biologically active macromolecules. It was used successfully for the first time in 1968 for the purification of enzymes [168]. Since then, innumerable proteins (e.g., enzymes,

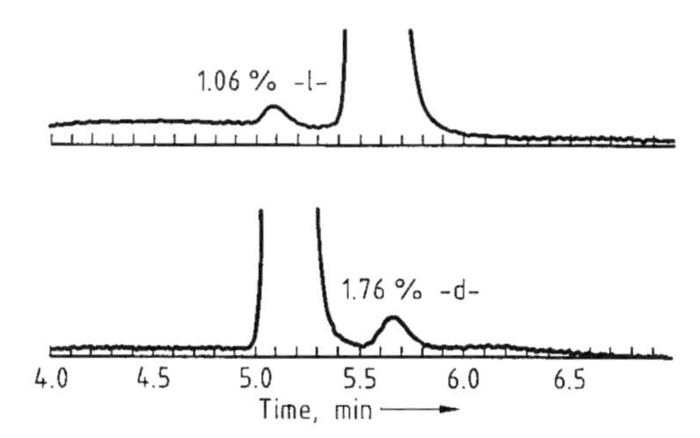

Figure 50. Chiral purity analysis of carbobenzyloxyd,l-alanine at 220 nm with packed-column SFC
Column: 250×4.6 mm Ciralcel-OD; 20 % ethanol (+0.2 % TPA/CO_2) 1 mL/min

receptors, or antibodies) have been isolated with high enrichment (100 – 10 000 fold) from complex raw extracts in a single chromatographic run. The proteins still possess their full biological activity after separation. The routine application of affinity chromatography has been described in many review articles [169] – [176], [177].

The high selectivity of affinity chromatography is based on biospecific interactions, such as those occurring between two molecules in natural biological processes. One interaction partner (ligand) is covalently attached (immobilization) to an insoluble carrier, while the corresponding partner (frequently a protein) is reversibly adsorbed by the ligand because of its complementary biospecific properties.

In laboratory practice, an affinity matrix (stationary phase) is tailor-made for the protein to be purified and filled into a chromatography column. For process applications many different affinity gels are commercially available. The raw extract is then passed through the column by using a physiological buffer as the mobile phase. In this process, the desired product is adsorbed selectively by the ligand and all unwanted components are washed away.

The chemical composition of the mobile phase is then changed to permit desorption and isolation (elution) of the protein (Fig. 51).

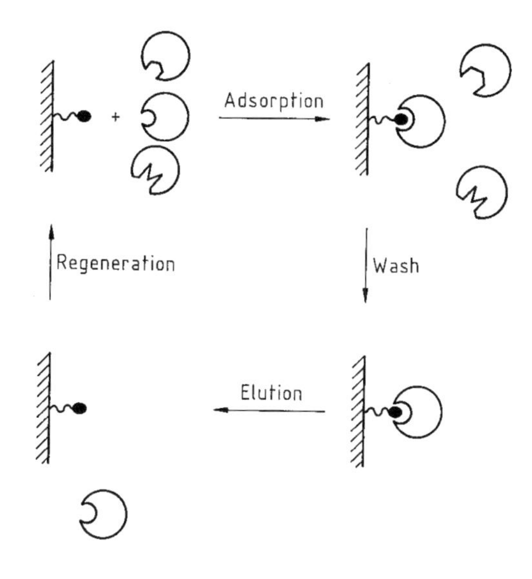

Figure 51. Schematic representation of affinity chromatography

12.11.1. Preparation of an Affinity Matrix

12.11.1.1. Selection of a Carrier [178], [179]

The preparation of an affinity matrix requires a suitable carrier that covalently attaches the selected ligand (immobilization). An ideal carrier should have the following properties:

1) A hydrophilic neutral surface — avoidance of nonspecific adsorption
2) Macroporosity — good permeability
3) Mechanical stability — high flow rates, stability to pressure
4) Chemical and biological stability — insolubility in various solvents, resistance to microbial attack
5) Functional groups on the surface — covalent binding of the ligand

The carriers used can be divided into three groups:

Polysaccharides	agarose dextran cellulose
Synthetic polymers	polyacrylamides polyacrylates poly(vinyl alcohols)
Inorganic materials	SiO_2, ZrO_2, TiO_2 magnetic particles

Agarose, a polysaccharide consisting of galactose units, is so far the carrier of choice because of its favorable chemical and physical properties. However, mechanical instability sometimes results in short column lifetime. In comparison, synthetic polymers and inorganic materials are mechanically more stable.

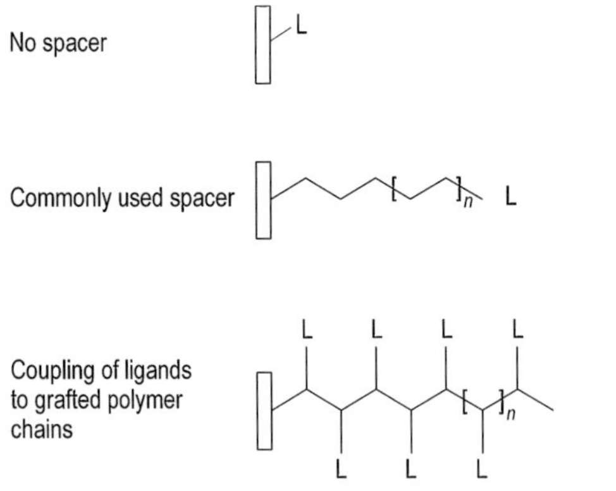

No spacer

Commonly used spacer

Coupling of ligands to grafted polymer chains

Figure 52. Three general possibilities for the covalent fixation of ligands to a matrix surface: direct coupling, coupling to spacers, and coupling to grafted polymer chains. L = ligand.

The spatial availability of immobilized ligands is often substantially improved by coupling ligands to spacers that keep them at a distinct distance from the matrix [180]. Basically, there are three general possibilities for the covalent fixation of ligands to the surface of appropriate support materials (Fig. 52):

1) Direct coupling to the surface (i.e., with no spacers)
2) Commonly used spacer arms : linear aliphatic hydrocarbons, 6-aminohexanoic acid, hexamethylenediamine, poly(ethylene glycol) [181]. One of the groups (often a primary amine, $-NH_2$) is attached to the matrix, while the group at the other end is selected on the basis of the ligand to be bound. The latter group which also is called the terminal group is usually a carboxyl (–COOH) or amino group.
3) Coupling of ligands to grafted polymer chains which freely rotate (i.e., in a "tentacle" arrangement) [182].

12.11.1.2. Immobilization of the Ligand

Covalent binding of a ligand to a carrier requires the presence of functional groups on both binding partners.

Generally, the following groups are available for immobilization of the ligand: $-NH_2$ (amino), –COOH (carboxyl), –CHO (aldehyde), –SH (thiol), or –OH (hydroxyl). High-molecular ligands (e.g., proteins) are usually bound through ε-NH_2 groups of lysine residues or occasionally through carboxyl groups. The functional group of the ligand must be mainly located in the nonspecific region of the molecule that is of no significance for reversible complex formation with the complementary molecule to be purified.

The functional groups of the carrier are either already present in the natural structure (e.g., OH groups of polysaccharides) or introduced later by the chemical functionalization of the surface:

$-CH-CH_2$ (with O bridge)	epoxide (a)
$-CHO$	aldehyde (b)
$-NCS$	isothiocyanate (c)
$-SO_2-CH=CH_2$	vinyl sulfone (d)
$-OH$	hydroxyl (e)
$-COOH$	carboxyl (f)
$-NH_2$	amine (g)

A carrier with reactive functional groups (a) – (d) can be used directly for the immobilization of ligands (e.g., via their NH_2 groups). In the case of epoxides (a), however, the coupling yields obtained are low. In the case of aldehydes (b), chemically instable ligand bonds (Schiff base) are formed that require further reduction (e.g., with boron hydride). Isothiocyanates (c) and vinyl sulfones (d) have the disadvantage that after immobilization, chargeable ligand bonds (thiourea and amine derivatives) are formed. These groups can act as weak ion exchangers and cause additional nonspecific adsorptions.

Functional groups (e) – (g) must be converted to a chemically reactive form (activation) before coupling to the ligand can occur. After immobilization has been carried out, any unreacted reactive groups on the carrier must be deactivated to prevent chemical reactions with the raw extract. This is frequently effected by additional coupling to a low-molecular ligand (e.g., ethanolamine) or by hydrolysis of the remaining activated groups.

Methods of Activation. Functional groups such as hydroxyl, carboxyl, or amino groups (e) – (g) must be converted to an activated form before they can be attached covalently to the ligand.

The classical method for activating *hydroxyl groups* involves reaction with cyanogen bromide [171], [183] to give a reactive cyanate ester (Fig. 53 A). However, the coupling of ligands (e.g., via ε-NH_2 groups of lysine residues) results in chemically unstable isourea linkages that can cause bleeding of the ligand during affinity chro-

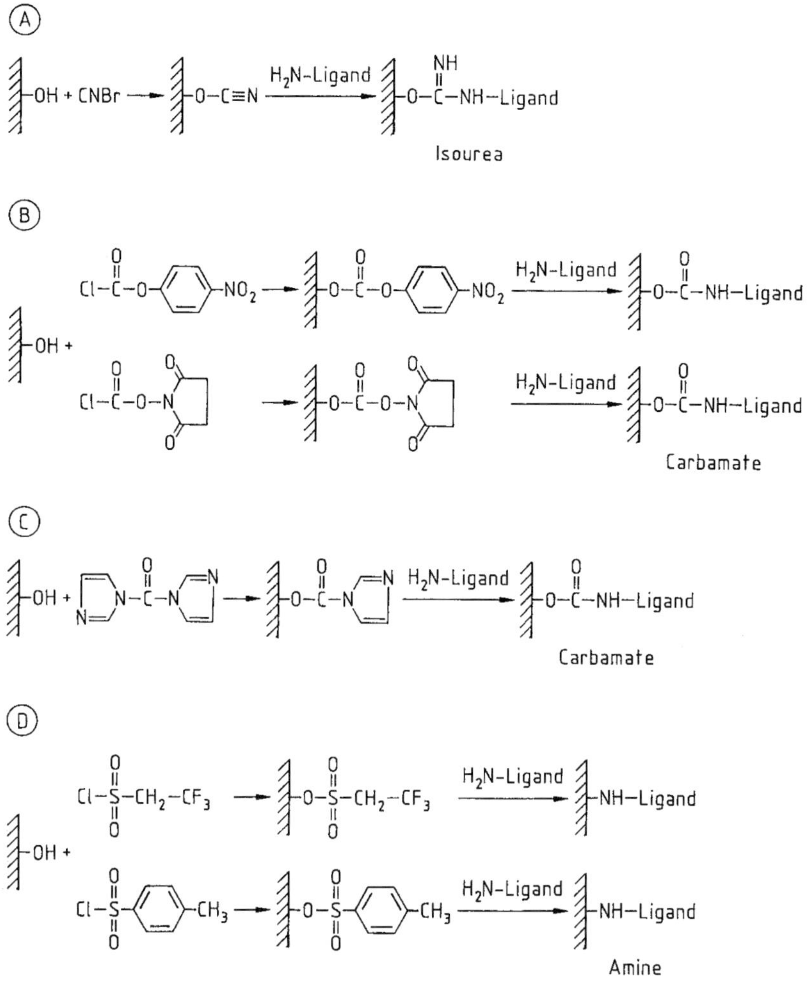

Figure 53. Activation and coupling of ligands to base supports with hydroxyl functions
A) Cyanogen bromide; B) Chloroformates; C) Carbonyldiimidazole; D) Sulfonyl chlorides

matography. This disadvantage can be avoided by activation with other reagents such as 1,1′-carbonyldiimidazole (Fig. 53 C) [184], *p*-nitrophenyl and *N*-hydroxysuccinimide chloroformates (Fig. 53 B) [185], [186], or sulfonyl chlorides [187], such as 4-toluenesulfonyl chloride and 2,2,2-trifluoroethanesulfonyl chloride (Fig. 53 D). The resulting imidazoyl, carbonate, and sulfonyl derivatives react, for example, with the amino groups of the ligand, with high coupling yields, and the ligand is attached to the carrier through chemically stable carbamate or amine bonds. The carbamate bond also has the advantage of being neutral and, consequently, causing no undesirable, nonspecific interactions. Unreacted activated groups can be hydrolyzed easily to reform the original neutral hydroxyl groups.

Carboxyl groups can be converted to active esters (e.g., with *p*-nitrophenol or *N*-hydroxysuccinimide), which react with the ligand with high coupling yields. Furthermore, carboxyl groups can be activated with water-soluble carbodiimide.

However, unreacted carboxyl groups act as weak ion exchangers, leading to unwanted nonspecific adsorptions. This disadvantage is also encountered with base matrices that are functionalized with amino groups.

12.11.2. Ligands

A ligand suited to the production of an affinity matrix should have the following properties:

1) Ability to form reversible complexes with the protein to be isolated and purified
2) Suitable specificity for the planned application
3) Large enough complexation constant to obtain stable complexes or sufficient delays during the chromatographic process
4) Complex that can be dissociated easily by simple changes in the mobile phase without irreversible loss of the biological activity of the protein

A distinction is made between low-molecular and high-molecular ligands as well as between group-specific and monospecific ligands. Group-specific ligands exhibit binding affinity for a range of substances that have similar properties with regard to function or structure. Typical examples are the triazine dyes Cibacron blue F 3 G-A and Procion red HE-3 B, which are used as ligands for many dehydrogenases, kinases, phosphatases, and other proteins. Monospecific ligands (e.g., antibodies) are "tailor-made" and can be used for only one application. In general, the binding of monospecific ligands to their complementary substance is substantially stronger (i.e., unfavorable elution conditions are occasionally required).

12.11.2.1. Group-Specific Ligands

12.11.2.1.1. Low-Molecular Ligands

The low-molecular, group-specific ligands belong to the largest class of ligands used in affinity chromatography:

Ligand	*Target molecule*
Cibacron blue F 3 G-A	kinases, phosphatases, dehydrogenases albumin, clotting factors II and IX, interferon, lipoproteins
Procion red HE-3 B	dehydrogenases, carboxypeptidase G, interferon, plasminogen, dopamine β-monooxygenase, inhibin
5′-AMP	NAD^+-dependent dehydrogenases, ATP (adenosine 5′-triphosphate)-dependent kinases
2′,5′-ADP	$NADP^+$-dependent dehydrogenases
NAD^+ and $NADP^+$	dehydrogenases
Phenylboric acid	nucleosides, nucleotides, carbohydrates, glycoproteins
Benzamidine	trypsin, urokinase, pronase E, glu- and lys-plasminogen

The coenzymes nicotinamide – adenine dinucleotide (NAD) and nicotinamide – adenine dinucleotide phosphate (NADP), their reduced forms, and adenosine 5′-monophosphate (5′-AMP) and adenosine 2′,5′-diphosphate (2′,5′-ADP) represent well-known examples of this class. They can be considered for the purification of ca. 25 % of the known enzymes (especially dehydrogenases).

Moreover, the triazine dyes (e.g., Cibacron blue F 3 G-A and Procion red HE-3 B) [188], [189] mentioned above are also used as ligands for the purification of these proteins. Triazine dyes can replace ligands of the NAD group because they have similar spatial structures. In addition, they bind a series of blood proteins (e.g., albumins, lipoproteins, clotting factors, and interferon).

Thiophilic ligands [190], [191] form a separate class of group-specific ligands that possess high selectivity with respect to immunoglobulins. They are low-molecular compounds, and contain sulfone and thioether groupings (see below), which show a strong affinity for mono- and polyclonal antibodies in the presence of ammonium sulfate (ca. 1 mol/L).

$$--CH_2-S-CH_2-CH_2-\overset{\overset{O}{\uparrow}}{\underset{\underset{O}{\downarrow}}{S}}-CH_2-CH_2-S-CH_2-CH_2-OH$$

Immobilized metal ion affinity chromatography (IMAC) uses gels with group-specific ligands which display selectivity for certain amino residues (histidine, cysteine, tryptophan) exposed at the outer surface of proteins [192]. The ligands are immobilized chelating agents [e.g., iminodiacetic acid (IDA) or tris (carboxymethyl)ethylenediamine (TED)] with bound transition metal ions. The metal ions are mostly by Cu^{2+}, Ni^{2+}, Zn^{2+}, and Co^{2+}, with decreasing complex stability and affinity [193]. The amino acids at the protein surface are bound to a free coordination site of the immobilized complex (Fig. 54) [194].

Adsorbed proteins can be recovered by lowering the pH to 3 or 4 or by competitive elution with an increasing concentration of glycine, imidazole, or histidine [195]. Through advances in molecular biology, histidine residues can be inserted into the

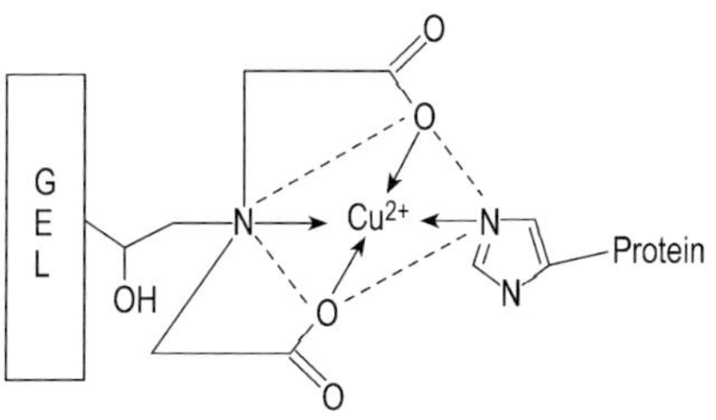

Figure 54. Schematic of binding of a histidine protein residue to an immobilized IDA copper(II) complex

protein sequence (fusion protein), resluting in improved separation in IMAC. The modified protein has a high affinity to the immobilized metal ions due to the addition of the histidine residues in the primary sequence. The fusion part can cleaved chemically or with a specific proteinase after purification.

12.11.2.1.2. Macromolecular Ligands

Macromolecular group-specific ligands include proteins that are used widely in the purification of biologically relevant macromolecules. Important macromolecular ligands are lectins (for the isolation of glycoproteins), protein A and protein G (for the purification of immunoglobulins), calmodulin (for Ca-dependent enzymes), and the polysaccharide heparin (for the purification of clotting factors and other plasma proteins).

Lectins [196]. Lectins exhibit a marked affinity for sugar molecules. They are perfectly suited to the isolation and purification of soluble and membrane-bound glycoproteins and polysaccharides (e.g., enzymes, hormones, plasma proteins, antigens, antibodies, and blood group substances). Among the available lectins, Con A (from *Canavalia ensiformis*) is the most widely used because of its specificity with respect to the frequently occurring α-D-glucose and α-D-mannose structural units. The desorption of the component to be purified from the ligand elution is often achieved by the addition of these low-molecular sugars to the mobile phase (see Section 12.11.3).

Lentil lectin (from *Lens culinaris*) has the same specificity as Con A, but a lower binding constant. For this reason, lentil lectin is suited to the purification of membrane-bound glycoproteins, which often have an exceedingly strong binding affinity for Con A.

Proteins A and G [197], [198]. Apart from the thiophilic ligands mentioned above, proteins A and G from bacterial cell walls are also suited to the purification of immunoglobulins. Protein A binds different subclasses of immunoglobulins to different extents. Accordingly, elution is often achieved with a decreasing stepwise, pH gradient. Mouse immunoglobulins (Igs) are eluted, for example, with 0.1 mol/L citrate buffer at the following pH values:

	IgG 1	IgG 2a	IgG 2b	IgG 3
pH	6.0	5.0	4.0	4.5

Protein G in its natural form can bind not only immunoglobulins, but also albumin. For this reason, it is applied in a genetically engineered form in which the albumin binding domain is eliminated (Fig. 55). Immunoglobulins generally display a higher binding affinity for protein G than for protein A.

12.11.2.2. Monospecific Ligands

Highly specific protein–protein interactions play a large role in many natural biological processes. For instance, the transport and transmission of information proceed via highly specific receptors that can frequently interact with only one molecular species (e.g., hormone or transmitter). In spite of this, the affinity chromatographic isolation of receptors and other highly specific proteins is rather an exception. However, antibodies occupy a special position as monospecific macromolecular ligands.

Antibodies and Antigens (Immunoaffinity Chromatography) [199], [200]. Immunoaffinity chromatography refers to the immobilization of antibodies (monoclonal or polyclonal) or antigens on an insoluble carrier (immunosorbent) for isolation and purification of the corresponding partner.

Antibodies are ligands of special interest because they show high biospecificity and strong binding to the molecules that elicited their synthesis, the antigens. The antibody–antigen interaction takes place between a particular part of the antibody (the antigen-binding site) and its complementary part on the antigen (the antigenic determinant or epitope).

Polyclonal antibodies represent a wide spectrum of antibody molecules that all have specificities with respect to different epitopes and, therefore, different avidities. The association constants

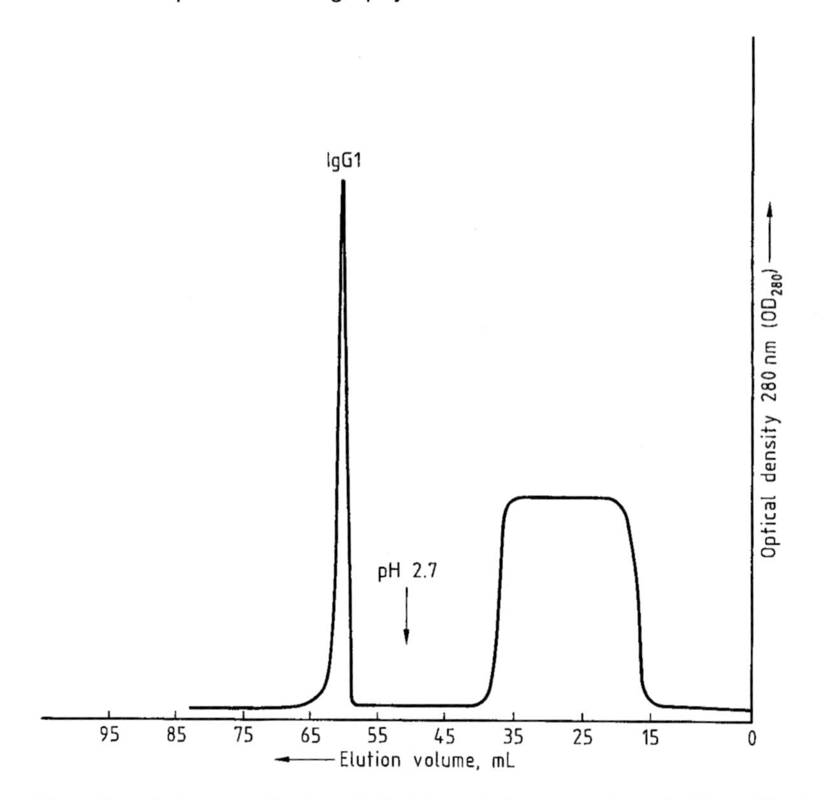

Figure 55. Isolation and purification of IgGl with genetically engineered protein G immobilized onto Sepharose

of an antigen and its corresponding antibody may range from 10^4 to 10^{11} L/mol. As a result, multipoint binding of the antigen to two or more different antibodies can occur on an immunosorbent prepared from polyclonal antibodies, resulting in such tight binding that dissociation and recovery of the antigen may be difficult.

In contrast to polyclonal antibodies, *monoclonal antibodies* show absolute specificity for a single epitope, the smallest immunologically recognized submolecular group of an antigen. Therefore, from a group of monoclonal antibodies the one most suited to antigen purification can be selected. Consequently, immunoaffinity chromatography has gained importance because monoclonal antibodies can be produced in large quantities by a hybridoma technology developed by KÖHLER and MILSTEIN [201]. This allows the use of immunoaffinity chromatography even for industrial-scale applications. Several proteins of potential therapeutic value are now produced by bioengineering techniques and purified with immunoaffinity chromatography (Table 13).

12.11.3. Elution [176]

Above all, successful affinity chromatography requires suitable desorption conditions (i.e., conditions that permit elution of the adsorbed substance with full biological activity). In principle, desorption is achieved by cleaving the binding sites responsible for ligand–protein interaction. These are mainly electrostatic, hydrophobic, and hydrogen bonds that can be specifically or nonspecifically disrupted. A *nonspecific method* of eluting an adsorbed substance is by altering the pH or ionic strength of the mobile phase. A pH decrease to pH 2–4 or an increase in ionic strength through a NaCl gradient, for instance, typically causes the dissociation of mainly electrostatic complex binding site. Ligand–protein complexes with high binding constants, such as those encountered in immunoaffinity chromatography, can often be eluted only under drastic conditions that can lead to denaturation. For this purpose, chaotropic salts (e.g., KSCN, KI) are used in concentrations of 1–3 mol/L. Urea or guanidin-

Table 13. Proteins of potential therapeutic value purified by immunoaffinity chromatography

Protein	Potential use
Interferons	
α-Interferon	antiviral, antitumor drug
β-Interferon	
γ-Interferon	
Interleukin 2	antitumor drug; also used against immuno-deficiency syndrome
Anticlotting factor	
Urokinase	dissolving blood clots
Urokinase-type plasminogen activator	
Zymogen of plasminogen activator	
Plasminogen activator	
Tissue plasminogen activator	
Human protein C	
Clotting factor	
Factor VIII	blood coagulant
Factor IX	
α- and β-Kallikrein	
Insulin-like growth hormone	growth hormone
Vaccines	
Human hepatitis B surface antigen	vaccine
Rubella virus antigen	
Tetanus toxin	
Polio virus	

ium chloride (4 – 6 mol/L) is also employed for the elution of strongly adsorbed substances.

In the case of *specific elution*, low-molecular substances with a certain affinity for the ligand or the adsorbed protein are added to the mobile phase. They displace the adsorbed protein from the ligand by competitive binding. This method of elution is frequently used with group-specific ligands (e.g., lectins). Glycoproteins can be eluted gently from immobilized lectins by the addition of low concentrations (10 – 100 mmol/L) of a monosaccharide to the mobile phase. Another example is the elution of dehydrogenases from coenzyme affinity supports by using NAD and its derivatives. In general, specific elution is a gentle method that usually ensures retention of biological activity.

12.12. References

[1] J. J. Kirkland, *Anal. Chem.* **43** (1971) no. 12, 36 A.

[2] M. Tswett, *Tr. Protok. Varshav. Obshch. Estestvoistpyt. Otd. Biol.* **14** (1903, Publ. 1905) 20 (Reprinted and translated in G. Hesse, H. Weil: *Michael Tswett's erste chromatographische Schrift,* Woelm, Eschwege 1954, pp. 37.

[3] R. Kuhn, E. Lederer, *Naturwissenschaften* **19** (1931) 306.

[4] R. Kuhn, E. Lederer, *Ber. Dtsch. Chem. Ges.* **64** (1931) 1349 – 1357.

[5] A. Martin, R. Synge, *Biochem. J.* **35** (1941) 1358 – 1368.

[6] F. Spedding, *Discuss. Faraday Soc.* **7** (1949) 214 – 231.

[7] E. Tompkins, *Discuss. Faraday Soc.* **7** (1949) 232 – 237.

[8] J. Porath, P. Flodin, *Nature (London)* **183** (1959) 1657 – 1659.

[9] E. Piel, *Anal. Chem.* **38** (1966) 370 – 372.

[10] J. J. Van Deemter, F. J. Zuiderweg, A. Klinkenberg, *Chem. Eng. Sci.* **5** (1956) 271.

[11] H. Engelhardt (ed.): *Practice of High Performance Analysis,* Springer Verlag, Berlin 1986.

[12] F. C. Smith, R. C. Chang, *CRC Crit. Rev. Anal. Chem.* **9** (1980) 197.

[13] P. L. Bailey, *Anal. Chem.* **50** (1978) 698 A.

[14] K. Stulik, V. Pacakova, *CRC Crit. Rev. Anal. Chem.* **14** (1982) 297.

[15] C. J. Blank, *J. Chromatogr.* **117** (1976) 35.

[16] L. Roth: *Sicherheitsdaten/MAK-Werte,* Verlag moderne Industrie, W. Dummer & Co., München 1978.

[17] H. Schrenker, *Int. Lab.* (1978) 67.

[18] I. Halasz, P. Vogtel, *J. Chromatogr.* **142** (1977) 241.

[19] L. R. Snyder, J. K. Kirkland: *Introduction to Modern Liquid Chromatography,* 2nd ed., John Wiley and Sons, New York 1979.

[20] *Rheodyne, Products for Liquid Chromatography,* Catalog no. 3, Cotati, Calif., USA, 1986.

[21] J. Meister, H. Engelhardt, *Nachr. Chem. Tech. Lab.* **40** (1992) no. 10.

[22] A. Tiselius, D. Claesson, *Ark. Kemi Mineral. Geol.* **15** (1942) no. 22.

[23] Refractive gradient detector

[24] H. Engelhardt: *Hochdruck-Flüssigkeits-Chromatographie,* Springer Verlag, Berlin 1975.

[25] U. A. T. Brinkmann, G. J. De Jong, C. Gooijer, *Methodol. Surv. Biochem. Anal.* **18** (1988) 321 – 338.

[26] T. M. Vickrey (ed.): *Liquid Chromatography Detectors, Chromatographic Science,* vol. 23, Marcel Dekker, New York 1983.

[27] R. P. W. Scott, *J. Chromatogr. Libr. Liquid Chromatography Detectors* **11** (1977).

[28] R. Schulz, H. Engelhardt, *Chromatographia* **29** (1990) nos. 11/12, 517 – 522.

[29] M. Varadi, J. Balla, E. Pungor, *Pure Appl. Chem.* **51** (1979) 1177.

[30] G. Henze, R. Neeb: *Elektrochemische Analytik,* Springer Verlag, Berlin 1986.

[31] G. Henze: *Polarographie und Voltammetrie; Grundlagen and analytische Praxis,* Springer Verlag, Berlin 2000.

[32] F. A. Schultz, D. E. Mathis, *Anal. Chem.* **46** (1974) 2253.

[33] A. Manz, W. Simon, *J. Chromatogr. Sci.* **21** (1983) 326.

[34] R. S. Deelder, H. A. J. Linssen, J. G. Koen, A. J. B. Beeren, *J. Chromatogr.* **203** (1981) 153.

[35] H. Akaiwa, H. Kawamoto, M. Osumi, *Talanta* **29** (1982) 689.

[36] T. Deguchi, T. Kuma, H. Nagai, *J. Chromatogr.* **152** (1978) 379.

[37] T. H. Ryan: *Electrochemical Detectors,* Plenum Press, New York 1984.

[38] K. Brunt: *Electrochemical Detectors in Trace Analysis,* vol. 1, Academic Press, New York 1981.

[39] P. T. Kissinger, C. Refshauge, R. Dreiling, R. N. Adams, *Anal. Lett.* **6** (1973) 465.

[40] B. Fleet, C. J. Little, *J. Chromatogr. Sci.* **12** (1974) 747.

[41] W. Kemula, *Rocz. Chem.* **26** (1952) 281.

[42] P. Just, M. Karakaplan, G. Henze, F. Scholz, *Fresenius J. Anal. Chem.* **32** (1993) 345.

[43] W. W. Kubiak, *Electroanalysis NBS* (1989) 379.

[44] J. Frank, *Chimia* **35** (1981) 24.

[45] P. T. Kissinger, C. S. Brunett, K. Bratin, J. R. Rice, *Spec. Publ. (U.S.)* **519** (1979) 705.

[46] S. Yao, A. Meyer, G. Henze, *Fresenius J. Anal. Chem.* **339** (1991) 207.

[47] H.-P. Nirmaier, E. Fischer, G. Henze, *GIT Labor-Fachzeitschrift* **42** (1998) 12.

[48] J. Schiewe, A. M. Bond, G. Henze, in G. Henze, M. Köhler, J. P. Lay (eds.): *Voltammetrische Spurenanalyse mit Mikroelektroden in Umweltdiagnostik mit Mikrosystemen,* Wiley-VCH Verlag, Weinheim 1999, p. 121.

[49] H.-P. Nirmaier, E. Fischer, G. Henze, *Electroanalysis,* **10** (1998) 187.

[50] R. Wintersteiger, G. Berlitz, *GIT Suppl. 3* (1989) 19.

[51] J. P. Hart, *Electroanalysis of Biologically Important Compounds,* Ellis Horwood, Chichester 1990.

[52] D. M. Radzik, S. M. Lunte: "Application of Liquid Chromatography/Electrochemistry in Pharmaceutical and Biochemical Analysis," *CRC Critical Reviews in Analytical Chemistry* **20** (1989) no. 5, 317.

[53] K. Štulik, V. Pacáková: "Electrochemical Detection in High-Performance Liquid Chromatography," *CRC Critical Reviews in Analytical Chemistry* **14** (1982) no. 4, 297.

[54] K. Štulik, V. Pacáková: "Electrochemical Detection Techniques in High-Performance Liquid Chromatography," *J. Electroanal. Chem.* **129** (1981) 1.

[55] G. Patonay, HPLC Detection, Newer Methods, VCH **1992.**

[56] K. Štulik, V. Pacáková: "Electrochemical Detection Techniques in High-Performance Liquid Chromatography," *J. Electroanal. Chem.* **129** (1991) 1.

[57] R. J. Rucki: "Electrochemical Detectors for flowing Liquid Systems," *Talanta* **27** (1980) 147.

[58] D. C. Johnson, S. G. Weber, A. M. Bond, R. M. Wightman, R. E. Shoup, J. S. Krull: "Electroanalytical Voltammetry in Flowing Solutions," *Anal. Chim. Acta* **180** (1986) 187.

[59] D. S. Austin-Harrison, D. C. Johnson: "Pulsed Amperometric Detection Based on Direct and Indirect Anodic Reactions," *Electroanalysis* **1** (1989) 189.

[60] R. Samuelsson, J. O'Dea, J. Osteryoung, *Anal. Chem.* **52** (1980) 2215.

[61] J. G. White, R. L. St. Claire, J. W. Jorgenson, *Anal. Chem.* **58** (1986) 293.

[62] P. T. Kissinger, *Anal. Chem.* **49** (1977) 447 A.

[63] M. Goto et al., *J. Chromatogr.* **226** (1981) 33.

[64] E. L. Craig, P. T. Kissinger, *Anal. Chem.* **55** (1983) 1458.

[65] J. Dutrieu, Y. A. Delmotte, *Fresenius J. Anal. Chem.* **314** (1983) 416.

[66] C. L. Putzig et al., *Anal. Chem.* **64** (1992) 270 R – 302 R.

[67] H.-P. Nirmaier, E. Fischer, A. Meyer, G. Henze, *J. Chromatography A,* 1996, 730, 169.

[68] K. K. Unger, G. Gilge, J. N. Kinkel, M. T. W. Hearn, J. Chromatogr. 359, 61 (1986).

[69] H. Chen, C. Horvath, J. Chromatogr. 705, 3 (1995).

[70] K. Stulik, V. Pacáková, J. Suchánková, H. Claessens, Anal. Chim. Acta 352, 1 (1997).

[71] R. Ohnmacht, B. Boros, I. Kiss, L. Jelinek, *Chromatographia* **50** (1999) no. 1/2.

[72] I. M. Mutton, *Chromatographia* **47** (1998) no. 5/6, 291 – 298.

[73] K. K. Unger (ed.): "Packings and Stationary Phases in Chromatographic Techniques," *Chromatographic Science Series,* vol. 47, Marcel Dekker, New York 1990.

[74] J. Persek, H. Lin, *Chromatographia* **28** (1989) 565.

[75] I. Halasz, I. Sebastian, *Angew. Chem. Int. Ed. Engl.* **8** (1969) 453.

[76] B. G. Herslöf, *HRC & CC, J. High Resolut. Chromatogr. Chromatogr. Commun.* **4** (1981) 471 – 473.

[77] A. Sándy et al., *Chromatographia* **45** (1997) 206 – 214.

[78] M. Vecchi, G. Englert, R. Maurer, V. Meduna, *Helv. Chim. Acta* **64** (1981) 2746.

[79] E. R. Majors, *LC-GC Int.,* **8** (1995) no. 7, 368 – 374.

[80] K. Aitzetmüller, *J. Chromatogr.* **113** (1975) 231 – 266.

[81] H. Small, T. S. Stevens, W. C. Baumann, *Anal. Chem.* **47** (1975) 1801.

[82] J. W. BcBain, *Kolloid Z.* **40** (1926) 1.

[83] I. Porath, P. Flodin, *Nature (London)* **183** (1959) 1657.

[84] J. C. Moore, *J. Polymer Sci. Part A* **2** (1964) 835.

[85] Macherey & Nagel & Co.: "HPLC", Düren 1991.

[86] K. K. Unger: "Adsorbents in Column Liquid Chromatography," in [24]

[87] Z. Deyl, K. Macek, J. Janak (eds.): "Liquid Column Chromatography," *J. Chromatogr. Libr.* **3** (1975).

[88] V. Meyer: *Praxis der Hochdruck-Flüssigkeits-Chromatographie,* Diesterweg-Salle-Sauerländer, Frankfurt 1979.

[89] K. Aitzetmüller, J. Koch, *J. Chromatogr.* **145** (1987) 195.

[90] J. E. Campbell, M. Hewins, R. J. Lych, D. D. Shrewsbury, *Chromatographia* **16** (1982) 162 – 165.

[91] S. A. Westwood (ed.): *Supercritical Fluid Extraction and its Use in Sample Preparation,* Blackie Academic & Professional, London 1993.
[92] K. Jinno (ed.): "Hyphenated Techniques in Supercritical Fluid Chromatography and Extraction," *J. Chromatogr. Libr.* **53** (1992).
[93] H. Godel, Th. Graser, P. Földi, *J. Chromatogr.* **297** (1984) 49.
[94] I. S. Krull (ed.): "Reaction Detection in Liquid Chromatography," *Chromatographic Science Series,* **vol. 34,** Marcel Dekker, New York 1986.
[95] H. Engelhardt, H. Elgass, *Chromatographia* **22** (1986) no. 1 – 6, June.
[96] A. H. M. T. Scholten, U. A. T. Brinkman, R. W. Frei, *J. Chromatogr.* **205** (1981) 229.
[97] P. Walkling, Ph. D. Thesis, Frankfurt 1968.
[98] I. Halasz, P. Walkling, *Ber. Bunsenges. Phys. Chem.* **74** (1970) 66.
[99] K. Hofmann, I. Halasz, *J. Chromatogr.* **199** (1980) 3.
[100] H. Engelhardt, U. D. Neue, *Chromatographia* **15** (1982) 403.
[101] U. D. Neue, Ph. D. Thesis, Saarbrücken 1976.
[102] D. Moore, W. H. Stein, D. H. Speckman, *Anal. Chem.* **30** (1958) 1190.
[103] D. H. Speckman, W. H. Stein, S. Moore, *Anal. Chem.* **30** (1958) 1190.
[104] M. Weigele, S. L. De Bernardo, J. P. Tengi, W. Leimgruber, *J. Am. Chem. Soc.* **94** (1972) 5927.
[105] M. Roth, *Anal. Chem.* **43** (1971) 880.
[106] S. Stein et al., *Arch. Biochem. Biophys.* **155** (1973) 202.
[107] D. Gottschalk, H. Körner, J. Puls in H. Engelhardt, K.-P. Hupe (eds.): *Kopplungsverfahren in der HPLC,* GIT Verlag, Darmstadt 1985.
[108] P. Vratny, U. A. T. Brinkman, R. W. Frei, *Anal. Chem.* **57** (1985) 224.
[109] L. Nondek, R. W. Frei, U. A. T. Brinkman, *J. Chromatogr.* **282** (1983) 141.
[110] H. A. Moye, C. J. Miles, S. J. Scherer, *J. Agric. Food Chem.* **31** (1983) 69.
[111] J. Schrijver et al., *Ann. Clin. Biochem.* **19** (1982) 52.
[112] J. Schrijver, A. J. Speek, W. H. P. Schreurs, *Int. J. Vitam. Nutr. Res.* **51** (1981) 216.
[113] J. T. Vanderslice, D. J. Higgs, *J. Chromatogr. Sci.* **22** (1984) 485.
[114] Deans-säulenschalten GC.
[115] J. F. K. Huber et al., *J. Chromatogr.* **83** (1973) 267.
[116] S. Wielinski, A. Olszanowski, *Chromatographia* **50** (1999) no. 1/2, 109 – 112.
[117] A. P. Bruins, *Adv. Mass Spectrom.* **11 A** (1989) 23 – 31.
[118] A. L. Burlingame, D. S. Millington, D. L. Norwood, D. H. Russel, *Anal. Chem.* **62** (1990) 268 R – 303 R.
[119] H. Kataoka, J. Pawliszyn, *Chromatographia* **50** (1999) no. 9/10, 532 – 538.
[120] M L. Nedved, S. Habbi-Gondarzi, B. Ganem, J.-D. Henion, *Anal. Chem.* **68** (1996) 4228 – 4236.
[121] Cagnaird de la Tour, *Ann. Chim. Phys.* **21** (1822) 127.
[122] T. Anrews, *Philos. Trans. R. Soc. London* **159** (1869) 575.
[123] J. B. Hannay, J. Hogarth, *Proc. R. Soc. London* **30** (1880) 178.
[124] C. M. White: *Modern Supercritical Fluid Chromatography,* Hüthig, Heidelberg 1988.
[125] E. Klesper, A. H. Corwin, D. A. Turner, *J. Org. Chem.* **27** (1962) 700.
[126] S. T. Sie, W. van Beersum, G. W. A. Rijnders, *Sep. Sci.* **1** (1966) 459.
[127] S. T. Sie, G. W. A. Rijnders, *Sep. Sci.* **2** (1967) 699.
[128] S. T. Sie, G. W. A. Rijnders, *Sep. Sci.* **2** (1967) 729.
[129] S. T. Sie, G. W. A. Rijnders, *Sep. Sci.* **2** (1967) 755.
[130] S. T. Sie, J. P. A. Bleumer, G. W. A. Rijnders in C. L. A. Harbourn (ed.): *Gas Chromatography 1968,* The Institute of Petroleum, London 1969, p. 235.
[131] J. C. Giddings, *Sep. Sci.* **1** (1966) 73.
[132] J. C. Giddings, M. N. Myers, L. McLaren, R. A. Keller; *Science (Washington, D.C. 1883)* **162** (1968) 67.
[133] J. C. Giddings, M. N. Myers, J. W. King, *J. Chromatogr. Sci.* **7** (1969) 276.
[134] J. C. Giddings, L. M. Bowman, N. M. Myers, *Anal. Chem.* **49** (1977) 243.
[135] S. T. Sie, G. W. A. Rijnders, *Anal. Chim. Acta* **38** (1967) 31.
[136] R. E. Jentoft, T. H. Gouw, *J. Chromatogr. Sci.* **8** (1970) 138.
[137] R. E. Jentoft, T. H. Gouw, *J. Polym. Sci. Polym. Lett. Ed.* **7** (1969) 811.
[138] D. Bartmann, G. M. Schneider, *J. Chromatogr.* **83** (1973) 135.
[139] U. van Wasen, G. M. Schneider, *Chromatographia* **8** (1975) 274.
[140] U. van Wasen, G. M. Schneider, *J. Phys. Chem.* **84** (1980) 229.
[141] M. Novotny et al., *Anal. Chem.* **53** (1981) 407 A.
[142] M. L. Lee, K. E. Markides (eds.): *Analytical Supercritical Fluid Chromatography and Extraction,* Chromatography Conferences, Inc., Provo, Utah, 1990.
[143] K. P. Johnston, J. M. L. Penninger (eds.): "Supercritical Fluid Science and Technology," *ACS Symp. Ser.* **406** (1989).
[144] M. L. Lee, *J. Microcol. Sep.* **4** (1992) 91 – 122.
[145] A. F. Buskhe, B. E. Berg, O. Gyllenhaal, T. Greibrokk, *HRC & CC, J. High Resolut. Chromatogr.* **11** (1988) 16.
[146] M. L. Lee et al., *J. Microcol. Sep.* **1** (1989) 7.
[147] J. C. Moore, *J. Polym. Sci. Part* **A 2** (1964) 835.
[148] B. E. Richter: *1986 Pittsburgh Conference and Exposition on Analytical Chemistry and Applied Spectroscopy,* Atlantic City, NJ, March 10 – 14, 1986, paper 514.
[149] B. E. Richter et al., *HRC & CC, J. High Resolut. Chromatogr.* **11** (1988) 29.
[150] G. Schomburg et al., *HRC & CC, J. High Resolut. Chromatogr.* **12** (1989) 142.
[151] B. E. Richter et al., *J. Chromatogr. Sci.* **27** (1989) 303.
[152] P. Morin, M. Caude, H. Richard, R. Rosset, *Chromatographia* **21** (1986) 523.
[153] J. W. Jordan, L. T. Taylor, *J. Chromatogr. Sci.* **24** (1980) 82.
[154] R. C. Wieboldt, G. E. Adams, D. W. Later, *Anal. Chem.* **60** (1988) 2422.
[155] S. B. French, M. Novotny, *Anal. Chem.* **58** (1986) 164.
[156] J. D. Pinkston, D. J. Bowling in [43]
[157] M. Smith (ed.): "Supercritical Fluid Chromatography," *RSC Chromatography Monographs,* 1989.
[158] J. Zapp, Ph. D. Thesis, Saarbrücken 1990.
[159] K. D. Pharma GmbH: DE P 42 065 39.9, 1992 (P. Lembke, H. Engelhardt, R. Krumbholz).
[160] Y. Hirata, F. Nakata, *J. Chromatogr.* **295** (1984) 315.
[161] H. H. Hill, M. A. Morrissey in G. M. White (ed.): *Modern Supercritical Fluid Chromatography,* Huethig, Heidelberg 1988.

[162] P. Morin, M. Caude, H. Richard, R. Rossel, *Analysis* **15** (1987) 117.
[163] J. M. F. Douse, *J. Chromatogr.* **445** (1988) 244.
[164] B. W. Wright, R. D. Smith, *J. High Resolut. Chromatogr.* **9** (1986) 73.
[165] J. B. Crowther, J. D. Henion, *Anal. Chem.* **57** (1985) 2711.
[166] R. M. Smith, *J. Chromatogr.* **505** (1990) 147.
[167] P. Lembke, in K. Anton, C. Berger (eds.): *Supercritical Fluid Chromatography with Packed Columns,* Marcel Dekker, New York 1998, pp. 429–443.
[168] P. Cuatrecasas, M. Wilchek, C. Anfinsen, *Proc. Natl. Acad. Sci. U.S.A.* **61** (1968) 636.
[169] W. H. Scouten: *Affinity Chromatography,* John Wiley & Sons, New York 1981.
[170] W. Brümmer in R. Bock, W. Fresenius, H. Günzler, W. Huber, G. Tölg (eds.): *Analytiker-Taschenbuch,* **vol. 2,** Springer Verlag, Berlin 1981, pp. 63–96.
[171] M. Wilchek, T. Miron, J. Kohn in W. B. Jacoby (ed.): *Methods of Enzymology,* **vol. 104,** Academic Press, New York 1984, pp. 3–55.
[172] I. R. Birch, C. R. Hill, A. C. Kenney in P. N. Cheremisinoss, R. P. Ovellette (eds.): *Biotechnology Applications and Research,* Techcromic Publishing Company Inc., Lancaster 1985, p. 594.
[173] P. Mohr, K. Pommering: "Äffinity Chromatography, Practical and Theoretical Aspects," in J. Cazes (ed.): *Chromatography Science,* vol. 33: Marcel Dekker, New York 1986.
[174] K. Ernst-Cabrera, M. Wilchek in R. Burgess (ed.): *Protein Purification: Micro to Macro, UCLA Symposia on Molecular and Cellular Biology, new series,* **vol. 68,** A.R. Liss Inc., New York 1987, pp. 163–175.
[175] St. Ostrove in J. N. Abelson, M. I. Simon (eds.): *Methods of Enzymology,* **vol. 182,** Academic Press, New York 1990, pp. 357–371.
[176] J. Carlsson, J. C. Janson, M. Sparrmann in J. C. Janson, L. Ryden (eds.): *Protein Purification,* VCH Publishers, New York 1989, pp. 275–329.
[177] G. T. Hermanson, A. K. Mallia, P. K. Smith (eds.): *Immobilized Affinity Ligand Techniques,* Academic Press, New York 1992.
[178] R. F. Taylor, *Anal. Chim. Acta* **172** (1985) 241.
[179] K. Ernst-Cabrera, M. Wilchek, *Makromol. Chem., Makromol. Symp.* **19** (1988) 145–154.
[180] C. R. Lowe, M. J. Harvey, D. B. Craven, P. D. G. Dean, *Biochem. J.* **133** (1973) 499–506.
[181] P. Cuatrecasas, *Nature* **228** (1970) 1327.
[182] E. Müller, "Coupling Reactions" in M. Kastner (eds.): *Protein Liquid Chromatography,* Elsevier, Amsterdam 2000.
[183] J. Kohn, M. Wilchek, *Enzyme Microb. Technol.* **4** (1982) 161–163.
[184] G. S. Bethell, J. S. Ayers, M. T. W. Hearn, W. S. Hancock, *J. Chromatogr.* **219** (1981) 361.
[185] T. Miron, M. Wilchek, *Appl. Biochem. Biotechnol.* **11** (1985) 445–456.
[186] K. Ernst-Cabrera, M. Wilchek, *J. Chromatogr.* **397** (1987) 187.
[187] K. Nilsson, K. Mosbach in W. B. Jacoby (ed.): *Methods of Enzymology,* **vol. 104,** Academic Press, New York 1984, p. 56.
[188] E. Stellwagen in J. N. Abelson, M. I. Simon (eds.): *Methods of Enzymology,* **vol. 182,** Academic Press, New York 1990, pp. 343–357.
[189] St. J. Burton in A. Kenney, S. Fowell (eds.): *Methods in Molecular Biology,* **vol. 11,** The Humana Press Inc., Totowa, N.Y., 1992, pp. 91–103.
[190] J. Porath, F. Maisano, M. Belew, *FEBS Lett.* **185** (1985) 306–310.
[191] B. Nopper, F. Kohen, M. Wilchek, *Anal. Biochem.* **180** (1989) 66–71.
[192] J. Porath, J. Carlson, I. Olsson et al., *Nature* **258** (1975) 598–599.
[193] Z. Horvath, G. Nagydiosi, *J. Inorg. Nucl. Chem.* **37** (1975) 767–769.
[194] V. A. Davankov, A. V. Semechkin, *J. Chrom.* **141** (1977) 313–353..
[195] E. Sulkowski, *Trends Biotechn.* **3** (1985) 1–7.
[196] P. J. Hogg, D. J. Winzor, *Anal. Biochem.* **163** (1987) 331–338.
[197] J. J. Langone, *Adv. Immunol.* **32** (1982) 157–252.
[198] A. Jungbauer et al., *J. Chromatogr.* **476** (1989) 257–268.
[199] K. Ernst-Cabrera, M. Wilchek, *Med. Sci. Res.* **16** (1988) 305–310.
[200] M. L. Yarmush et al., *Biotechnol. Adv.* **10** (1992) 413–446.
[201] G. Köhler, C. Milstein, *Nature (London)* **256** (1975) 495–497.

13. Thin Layer Chromatography

JOSEPH C. TOUCHSTONE, School of Medicine, University of Pennsylvania, Philadelphia, PA 19104-6080, United States

13.1. Introduction

The following constitutes a general guide to the use of thin layer chromatography (TLC). Pertinent reference sources on the subject include [1]–[3]. The discussion here is focused on precoated plates, since these have largely supplanted homemade layers. For further information on the relationship between TLC and other types of chromatography → Basic Principles of Chromatography.

13.2. Choice of the Sorbent Layer

Precoated plates for TLC have been commercially available since 1961. The sorbents may be coated on glass, plastic, or aluminum supports. Sorbents with and without binder, and with and without UV indicators, are available in a variety of layer thicknesses, ranging from 100 μm in the case of plastic plates and high-performance layers to 2311 μm for preparative layers. The largest selection is presented by glass plates, coated to a thickness of 250 μm in the case of "analytical layers."

Important considerations with respect to all TLC plates are that the layers be uniform in thickness and surface, and that the plates yield reproducible results from package to package. These are difficult characteristics to duplicate with homemade plates, and are of prime concern in analytical work. The availability of high-quality commercial precoated plates is in fact one of the reasons why quantitative thin layer chromatography has been so successful in many applications.

Plates with a plastic support have an additional distinguishing feature relative to glass-backed plates: They can be cut with scissors. This is advantageous, for example, when small plates are desired but only 20×20 cm plates are on hand. With plastic-backed plates it is also easy to elute separated substances from the developed layer by cutting out an appropriate region and immersing the cut piece in a suitable extraction solvent. This is an especially important consideration when working with radioactive substances (→ Radionuclides in Analytical Chemistry). Measuring the radioactivity of such a substance is facilitated by simply immersing the portion of plate containing the radioactive zone in a scintillation vial containing counting fluid. Because the layer need not be scraped, potential contamination and losses caused by flying dust are minimized.

Considering only the cost of the sorbent and plates, precoated plates are obviously more expensive than homemade plates. However, the total expense of homemade plates also includes the cost of the coating apparatus and drying – storage racks, as well as the time involved. After these factors are taken into account it will be seen that precoated plates are highly desirable.

13.2.1. Properties of Presently Available Sorbents

Silica Gel. Silica gel is prepared by the precipitation of a silicate solution with acid, or by hydrolysis of silicon derivatives. The surface area and diameter of the silica gel particles depend on the method of precipitation. Variations in pH during precipitation can produce silica gels with surface areas ranging from 200 – 800 m^2/g. It has been shown [4] that silica gel provides three types of surface hydroxyl groups: Bound, reactive, and free. Relative reactivity and adsorption follow the order bound > free > reactive. Thus, control of the distribution of surface functions can have a significant effect on the chromatographic properties of a silica.

Special methods have been developed for preparing spherical particles with specific pore characteristics. Thermal treatment can also affect the pore size. Many of the silicas currently in use have a pore size of 6 nm.

Alumina. Alumina has seen limited use as a sorbent in TLC. Three types of alumina are available: Acidic, basic, and neutral. The amount of water present greatly affects the chromatographic behavior of alumina. For control purposes the plates may be heated to a specific temperature prior to use.

Kieselguhr. Kieselguhr is a purified diatomaceous earth usually characterized by wide pores. It is frequently used as a *preadsorbent layer* on other types of TLC plates. Commercial kieselguhr preparations include Celite, Filter Cel, and Hyflo Super Cel.

Cellulose. Cellulose has been used as a sorbent since the earliest days of chromatography, originally in the form of paper chromatography. Most methods of preparing celluloses for TLC are based on procedures for preparation of this medium for other uses. The types used for TLC are powdered (fibrous) or microcrystalline celluloses. The fiber length is shorter with TLC cellulose than in paper chromatography, which leads to less diffusion of the spots relative to paper chromatography.

Polyamides. Polyamides, previously used almost exclusively in column chromatography, have seen increasing application as sorbents for TLC. Certain separations are accomplished much more readily on polyamide than on conventional layers, since this medium lends itself to the analysis of water-soluble samples.

Ion Exchange. Ion exchange provides greater selectivity than most other forms of TLC, and recent developments have broadened its applicability. Ion exchange can be utilized with either organic or inorganic substances. Within these classifications, a further division can be made into anionic or cationic processes according to the nature of the ions subject to exchange. Several sorbents are now available for ion-exchange TLC, and the number will grow as development continues.

Reversed-Phase Sorbents. Reversed-phase (RP) sorbents can be classified into (1) those in which a liquid is impregnated in a solid support, and (2) chemically bonded sorbents, where the liquid phase is bonded chemically instead of physically. Chemical bonding provides layers that are more stable, preventing the stationary phase from being removed by solvent. Three types of reactions are used for bonding: Esterification, reaction of a chlorosilane with a selected organic substituent, and chlorination of the support followed by reaction with an organolithium compound to produce Si–C bonds. These bonding procedures can lead to various types of structures depending on the mode of attachment to the silica and the nature of substituents present at the other end of the alkyl side chain. The major bonding types have the structures Si–O–C, Si–O–Si–C, and Si–C–O–Si–N. The length of commercially available chains ranges from 2 to 18 carbon atoms.

Gels. Gels provide an entirely different type of medium for separation on thin layers. Sephadex was one of the first gels used as a stationary phase in the resulting *steric exclusion* TLC.

Cyclodextrin. More recently, cyclodextrin-modified layers have been introduced for use in the separation of enantiomers. The increasing importance of isomer separation in pharmaceutical applications has led to a demand for these specially modified layers. Chiral layers can take several different forms. Most are C_{18} layers impregnated with a chiral-selecting protein derivative together with copper ions. Others consist of layers in which a sorbent silica has been chemically bonded to a cyclodextrin derivative.

13.2.2. Binders

Sorbents are prepared both with and without binding substances. A binder changes the properties of a sorbent such that the latter will adhere to its support and not readily flake off during handling, chromatography, visualization, or documentation. Nevertheless, binding agents are not always necessary, nor are they always desirable. Cellulose, for example, does not usually require a binder, since it has good inherent properties of adhesion and strength. Binders are obviously undesirable if they interfere with development or visualization. Calcium sulfate, a commonly used binder, is slightly soluble in water, and it is therefore not recommended in conjunction with aqueous solvent systems. Layers employing starch or polymer binders (so-called organic binders) should not be used when very high-temperature (>150 °C) visualization techniques are contemplated. They are also not suitable with universal visualization reagents such as sulfuric acid spray. Starch is appropriate only when it does not interfere, and when tough, strong layers are essential.

Calcium sulfate binder concentrations range from ca. 5 wt % in some silica gel preparations to 9 wt % in aluminum oxide G sorbent, 13 wt % in the widely used silica gel G, and up to 30 wt % in certain silica gel preparations for preparative (high capacity) TLC, where a "G" in a sorbent name denotes the presence of calcium sulfate (gypsum).

Commercial precoated plates are sometimes made with a small amount of organic polymeric binder, which makes the layer very hard and resistant to abrasion. This binder does not dissolve in relatively strong, polar solvents or in organic solvents, which increases the range of possible applications for separation but precludes certain detection procedures. Such binders, which are of a proprietary nature, absorb less moisture from the atmosphere than calcium sulfate, and for this reason the plates usually do not require activation prior to use.

13.2.3. Suppliers

A number of distributers and manufacturers supply precoated TLC plates, but not all carry a complete selection of sorbent/backing-plate combinations in the four major sizes (5×20, 10×20, 20×20, and 20×40 cm):

Analtech Inc., Newark, Del, United States	full line
J. T. Baker Chemical Co., Phillipsburg, N.J., United States	full line
Brinkmann Instrument Inc., Westbury, N.Y., United States	full line
Desaga GmbH, Heidelberg, Germany	relatively broad line
Eastman Kodak, Rochester, N.Y., United States	mylar backed layers
EM Science, Darmstadt, Germany	full line
Ingenur F. Heidenreith, Oslo, Norway	relatively broad line
Macherey-Nagel GmbH, Düren, Germany	sorbents and layers
Schleicher & Schuell GmbH, Kassel, Germany	specialty layers
Whatman Chemical Separations, Clifton, N.J., United States	full line

Some commercial 20×20 cm plates have prescored lines on the back at 5, 10, and 15 cm so they can easily be broken into smaller plates.

There is little literature available comparing plate-layer characteristics. To the benefit of the method, only two or three major manufacturers produce each sorbent type, so diversity with respect to a given sorbent is kept to a minimum. However, different manufacturers of the plates themselves employ different coating techniques, which in turn has an effect on the characteristics of sorbent that has been coated on the plates. Thus, even though two precoated-plate companies buy their sorbent from the same manufacturer, the resulting plates may not display identical separation characteristics.

The recently introduced *hydrophilic modified* precoated plates have broadened the spectrum of selectivity available for TLC. These plates are referred to as "HPTLC-NH_2 precoated plates," in which γ-aminopropyl groups are bonded to the surface of a silica-gel skeleton, and "HPTLC-CN precoated plates," with γ-cyanopropyl moieties as the functional groups. In terms of polarity the two hydrophilic modifications can be classified as follows: Normal $SiO_2 > NH_2 > CN > RP$. Since the surface NH_2 and CN groups are of medium polarity, no wetting problems are encountered with the corresponding plates even at the maximum level of modification. Any mobile phase can be used, without restriction.

The presence of the functional group means that an NH_2-modified support can act as a weakly basic ion exchanger, useful, for example, in separating charged substances. Thus, various nucleotide building-blocks can be distinguished on the basis of differing charge numbers, with substances retained more strongly as the charge numbers increase. Other retention mechanisms can also be involved in sample retention on an NH_2 plate. In the case of oligonucleotides, the observed selectivity arises through differences in hydrophobicity among the various nucleobases.

A CN-modified layer can be used for either normal-phase or reversed-phase chromatography depending on the polarity of the solvent system.

The *hardness* of a layer varies as a function of any binder present, and this in turn affects the speed of development and the resulting R_f for a given substance. Factors such as activation and developing-chamber saturation are also important influences in this regard.

Generalizations regarding cellulose layers are easier to formulate. As noted previously, the two most popular forms of cellulose are fibrous and microcrystalline. The major supplier of fibrous cellulose is Macherey, Nagel and Co., whose main product is MN 300 cellulose. The major supplier of microcrystalline cellulose is FMC Corporation, under the trade name Avicel. Microcrystalline cellulose layers usually develop more rapidly and produce zones with lower R_f values that are more compact in size relative to fibrous cellulose layers.

13.3. Sample Cleanup

Proper preparation of the sample is the most important preliminary step in conducting any analysis (→ Sample Preparation for Trace Analysis), and this is especially true in chromatography. Of particular concern is the amount of extraneous material present in the sample solution. The greater the amount of "junk," the more sample preparation that will be required.

The Sample. Before undertaking the preparation of any sample, an aliquot of material representative of that sample must be obtained. Otherwise the analysis itself may be meaningless. The best way to assay a solid sample is to collect random aliquots from the whole and then mix them together before removing the extraction aliquot. If a package of tablets or capsules is to be analyzed, random portions from the whole should similarly be mixed together prior to selecting an aliquot for analysis.

The Solvent for the Sample. An appropriate solvent must of course be selected for dissolving the sample. When a sample must be derivatized, solubility characteristics may change drastically, or there may not be enough sample remaining for performing general solubility tests. The solubility of a crystalline material can be readily determined with a microscope. A few crystals of the sample are placed on a slide in the field of the microscope and carefully observed as a drop of solvent is added. If the crystals do not dissolve, the solvent can be evaporated and a new one selected for further testing. Testing with a number of solvents greatly expedites selection of the one that is optimal for experimental analysis.

Extraction of the compounds of interest from a complex sample and subsequent dissolution in the proper solvents requires different strategies depending on the compound class under consideration. For example, polar compounds do not dissolve well in nonpolar solvents, and nonpolar compounds will not dissolve in polar solvents. Ionic compounds dissolve best in a polar solvent to which the counter ion has been added. Satisfactory sample application on a TLC layer requires that the solvent be highly volatile and as nonpolar as possible. Nonvolatile solvents spread through the plate-layer matrix during the period of contact, taking sample with them. Polar solvents likewise spread solute through the matrix as a consequence of both high solvent power and low volatility.

Sample Extraction. Several alternatives are available for extracting the sample from its solvent. These include:

1) *Direct application* of the solution, with extraction taking place on the TLC layer
2) Application of a solution of the sample on a *preadsorbent layer,* followed by direct development

   With both of these techniques proper selection of the mobile phase can ensure that extraneous materials are left at the chromatographic origin. In some cases the layer can be predeveloped in such a way that analyte is deposited on a line which can be regarded as an origin for sub-

Table 1. Sorbents for use in microcolumns

Nonpolar	Polar	Ion exchange
C_{18}	cyanopropyl	carboxylic acid
C_8	diol	propylsulfonic acid
C_2	silica gel	benzenesulfonic acid
Phenyl	aminopropyl	diethylaminopropyl
Cyclohexyl	propyl	quaternary or secondary amine

sequent development with the correct mobile phase for separation.

3) *Liquid Extraction of the Sample.* Some samples may be too impure for simple extraction by one of the first two methods, but control of the pH often improves the situation. For example, organic acids are converted into water-soluble salts at high pH. Lowering the pH restores the unionized acids, which are then extractable into organic solvents. At high pH, basic compounds are extracted into organic solvents, returning to the water phase as their salts at low pH.
4) *Solid Phase Extraction.* Columns and microcolumns (also referred to as cartridges) prepacked with sorbents can be used for *solid phase extraction* (*SPE*) of samples. Such columns are available commercially, but one can easily pack a simple tube with the required sorbent, such as those in Table 1. Some extraction columns are made to fit onto a syringe, whereby the column becomes an extension of the syringe barrel, and the plunger is used to force the sample and eluting solvents through the packing. Table 2 lists a number of United States suppliers of packed SPE columns.
SPE columns require minimal amounts of solvent. Eluants of increasing strength can be used to elute analytes with different polarities in separate fractions. Thus, silica-gel columns release low-polarity substances upon elution with a nonpolar solvent, whereas the analyte is retained as a consequence of its polarity. The analyte is then eluted with a more polar solvent. The least polar solvent capable of eluting the analyte should always be used to ensure that very polar impurities remain behind. The eluant is subsequently evaporated; alternatively, if an analyte concentration is sufficiently high, an aliquot can be taken for application to the TLC plate. This methodology presents unlimited possibilities for analyte isolation.
5) *Liquid–Liquid Extraction Using Microcolumns.* Special microcolumns buffered at pH 4–9 and packed with diatomaceous earth have been developed for liquid–liquid partitioning. These can be utilized for sample volumes ranging from one to several hundred milliliters. The principle of liquid–liquid partitioning is also the basis of a device known as an Extube. This is a disposable plastic syringe packed with an inert fibrous matrix presenting a large surface area. A sample and an extraction solvent are poured into the tube, where liquid–liquid partitioning occurs. Analyte is extracted into the organic phase, which passes through the matrix, while impurities and other polar components, including water, are retained. The eluate is then concentrated by evaporation prior to TLC.

General Guidelines for the Use of SPE Columns [5]. The SPE cartridge must always be *conditioned* with the particular solvent that will be used. This is necessary both to activate the sorbent and to clean the column. Conditioning procedures vary depending on whether normal or reversed-phase sorbents are employed. Ion-exchange conditioning is a function of the polarity of the sample solvent.

For normal-phase packings the first conditioning wash is often conducted in the solvent in which the analyte is dissolved. With most SPEs 2 mL of solvent is sufficient. An SPE can also be conditioned with 3–5 mL of a nonpolar solvent such as hexane. This can be adjusted by adding methylene chloride, but the final mix should always be miscible with the sample solvent.

With reversed-phase sorbents, methanol is used for the first wash. A typical cartridge requires 3–5 mL of methanol, which is followed by the solvent used to prepare the sample. A film of water-miscible solvent remains on the sorbent after the aqueous solvent has been added, improving the interaction between aqueous sample and hydrophobic sorbent.

For ion-exchange sorbents, deionized water is usually used first, but if the sample is in a nonpolar solvent, the SPE should be conditioned with 2–5 mL of this solvent instead. Aqueous samples require use of 2 mL of methanol followed by 2 mL of water. The sorbent is then adjusted to the organic solvent content and salt concentration of the sample solution. Sorbent must not be allowed to

Table 2. Suppliers of SPE columns in the United States

Company	Column designation	Accessories available
Alltech Inc., Deerfield, Ill.	"Extract-Clean"	manifold
Analtech Inc., Newark, Del.	"Spice" cartridges	manifold
J. T. Baker Chem. Co., Phillipsburg, N.J.	"Baker-Bond"	manifold
Applied Separations, Bethlehem, Pa.	"Spe-ed" cartridges	various accessories
Bodman Chemicals, Aston, Pa.	"Pure-Sep" cartridge	various accessories
Isolab Inc., Akron, Ohio	"Quick-Sep"	manifold
J & W Scientific, Folsom, Ca.	"Accuford" tubes	manifold
Versa-Prep Lab Systems Inc., Nanuet, N.Y.	"Tef-Elutor"	various accessories
Varian Inc., Sunnyvale, Ca.	"Bond-Elute"	various accessories
Waters Associates, Milford, Mass.	"Sep-Pak"	various accessories
Worldwide Monitoring Corp., Hatfield, Pa.	"Cleanscreen" columns	full line of sorbents

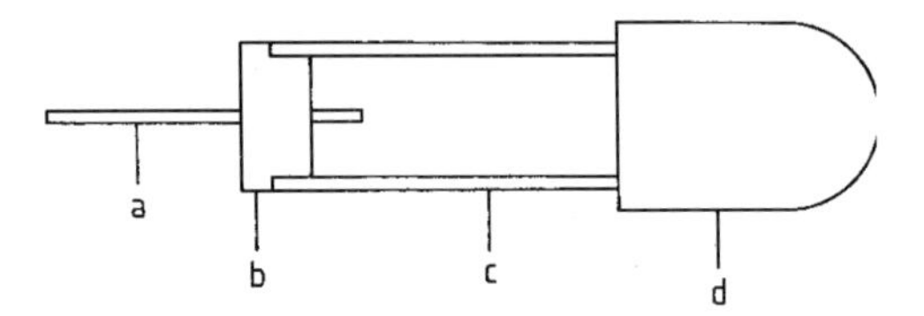

Figure 1. Microcapillary spotting device for TLC
a) Capillary tube; b) Septum; c) Support tube; d) Rubber bulb

dry between conditioning and addition of the sample.

The sample solution is subsequently added to the top of the cartridge and drawn through the sorbent bed by application of a vacuum. Special vacuum manifolds are available for this purpose, some of which handle 12 or more tubes, thus increasing sample throughput.

The next step is *washing* of the column (and sample). For reversed-phase work, washing with a polar solvent generally ensures that the analyte will be retained. Normal-phase applications require a wash with some nonpolar solvent that leaves the analyte on the sorbent. An ion-exchange column is washed with water, again so that the analyte will be retained.

Washing is followed by *elution* of the analyte. In reversed-phase mode a nonpolar solvent is used to remove the analyte; in normal phase, it is a polar solvent that removes the analyte. Adjustment of the ionic strength or modification of the eluting solvent is used to free the solute from an ion-exchange column. In each case, two volumes of 100 – 500 μL should suffice for complete elution.

The eluant is then *evaporated,* and the residue (extract) is *reconstituted* in a solvent suitable for introduction into the required chromatographic system. Typically, 5 – 200 μL of solvent is used for this purpose, 2 – 10 μL of which is taken for analysis. If the concentration is suitable, an aliquot of the eluant can be analyzed directly without the need for evaporation and reconstitution.

13.4. Sample Application

Numerous devices are available for applying samples to TLC plates. *Microliter syringes*, such as those manufactured by Hamilton, have frequently been used for applying 5 to 10 μL amounts of samples. Application of small volumes of sample with a syringe results in relatively narrow or small application areas, since little solvent is present to cause diffusion. Sucessful use of these devices takes practice, however, and it necessitates washing of the delivery device between sample applications. *Microcapillaries* for this purpose are available from many suppliers as an alternative, and these offer the best volume-delivery precision. Capillaries are also disposable, ruling out contamination between samples. These capillaries with holders, as illustrated in Figure 1, are available in a wide range of fixed volumes, and they have proven to be ideal for TLC applications.

Automated or *semiautomated sample applicators* are available as well. These devices apply consistent and reproducible sample spots, but it is a misconception that they are necessary for quantitative work. With proper technique, manual methods of sample application can provide results entirely comparable to those from automatic devices. Commercial automated units employ syringes or rows of microcapillaries to apply a spot or band of sample, and some actually "spray" the sample onto the layer. Many are designed for preparative separations, applying large amounts of sample as streaks across the sorbent layer.

The Sample. In the case of silica-gel layers the sample should be dissolved in the least polar solvent in which it is soluble. With a solution of the proper concentration, 1 – 5 µL should suffice to permit detection of the analyte of interest after chromatography. Concentrating may prove to be the best procedure with a dilute solution. However, dilute solutions can also be applied in the form of several successive aliquots. If the sample is too dilute, spreading of the application area can be prevented by accelerating evaporation of solvent with a stream of air blown over the area in question. A hair dryer is often used for this purpose. Correct application ensures that the separated zones will be both symmetrical and compact.

Sample Amounts. The amounts of the sample and the analyte are critical factors in achieving good results, since applied load affects the shapes and positions of developed zones. Application of a sample that is too concentrated can result in overloading. Overloaded zones produce comet-like vertical streaks after development, which can cause overlap of previously separated components.

TLC is a sensitive technique, and it is important to "think small" with respect to sample size. The highest resolution will always be obtained with the smallest sample commensurate with the ability to visualize the analyte. This is also one of the prerequisites for reproducibility, especially if densitometry is to be used for quantitation (see Chap. 13.8). Zones that are too concentrated result in integrated curve areas that behave in a nonlinear way on regression plots. Depending on the nature of the analyte, sample sizes as small as picogram amounts are not uncommon, especially in high-performance thin layer chromatography (HPTLC). This is especially true in the analysis of fluorescent substances. Thus, for ultimate sensitivity it is sometimes desirable that a fluorescent derivative of the analyte be prepared.

With regard to preadsorbent plates, the applied sample volume has an effect on preadsorbent efficiency. Up to certain levels, increasing the sample volume appears to increase plate efficiency, since a larger volume prevents compacted zones, and uniformity of the starting line after predevelopment is maximized.

Using Micropipettes for Manual Application of Analytical Samples. Care should be taken to ensure that separate spots are placed on a straight line horizontally across the plate. With hard-layer plates a guideline can be drawn on the plate surface itself with a soft pencil to act as a spotting guide. A white sheet of paper with a dark line drawn on it can also serve as a guide, since such a line will be visible through a layer positioned over it. Use of a guideline makes it easier to keep the sample origins evenly spaced across the plate and parallel to the edge. In the case of plates with a preadsorbent zone no guide or guideline is required, because the sample can be applied anywhere within the preadsorbent zone. This zone is inert, so it will not retain the analyte, permitting samples to move with the solvent front until they reach the adsorbent zone. A preadsorbent provides a convenient means of applying a dilute sample without the need for evaporation. No matter how the sample is applied, the result after predevelopment will be a fine line at the juncture between the adsorbent and the preadsorbent material.

Samples are applied to a 20×20 cm TLC plate 2 cm from the bottom. The two outermost samples should be 1.5 to 2 cm from the edges. Normally, individual "lanes" are kept about 1 cm apart. The origin should be located 2 cm from the bottom to ensure that sample will not be eluted inadvertently into the mobile-phase reservoir. Standards are interspersed at every third or fourth position. A single plate can provide space for 20 or more samples.

Usually 2 – 10 µL of sample solution is applied to the TLC plate. A fixed-volume micropipette is completely filled with the solution, and the end of the capillary is then touched to the surface of the TLC adsorbent, causing solution to be "wicked" into the adsorbent, and producing a round "spot." The capillary is most conveniently held by thrusting it through a hole in a septum, which is in turn inserted into a larger piece of glass tubing. Some manufacturers provide a rubber bulb (similar to the top of a medicine dropper) that fits over this larger tube. The bulb has a hole in it to equilibrate pressure while the capillary is draining (see Fig. 1).

Since it is known that small samples lead to improved separation, a method known as *overspotting* is sometimes used for sample application. Thus, a 10 µL sample might be applied to the plate as five spots of 2 µL each, all delivered at the same place. Successive spots must not be applied until solvent from the previous application has completely evaporated. Failure to observe this precaution can result in distorted zones. Sample application in this way can be used to form a single spot, or a series of spots can be joined to form a band. Application of a sample as a band gives greater resolution [6]. Treatment between applications with a hair dryer or a stream of filtered air facilitates solvent evaporation. The zone of application should be kept as narrow as possible compatible with the degree of concentration required. Spots or zones that are too concentrated result in poor separation, since the mobile phase tends to travel around the concentrated center, which leads to distorted zones or tailing.

Applying Preparative Samples. The preferred method for introducing a sample onto a preparative layer is to apply it as a narrow streak across the entire width of the plate. Both manual and automatic means are available for applying such a streak. The streak should be made as straight (horizontally) and as narrow as possible. Most commercial streakers are capable of producing a sample zone only 1 – 3 mm wide. A layer 1000 μm (1 mm) thick can readily handle 100 mg of sample.

Some successful preparative work has been reported based on the same manual techniques developed for applying analytical spots. In these cases a small (i.e., 2 – 4 μL) pipette is used to apply successive round spots in direct contact with each other across a plate, but the method is time-consuming and tedious, and it requires great care. Any spotting irregularities are greatly magnified in the separation process, and this often makes separation of closely spaced components very difficult. However, if the separated components are widely spaced, then satisfactory results are achievable in this way.

The use of preparative plates with a preadsorbent zone reduces the amount of care necessary to produce a narrow band. Sample in this case is applied to the relatively inert preadsorbent layer as a crude "streak" across the plate. Even so, it is best to exercise some care and apply the sample evenly across the plate, because a high local sample concentration can cause an overloading effect in the corresponding region of the plate.

The Importance of Sample Drying. Thorough drying of applied samples is a necessary step that is frequently ignored. Errors can result if a sample carrier solvent is not allowed to evaporate completely from the TLC plate. Many chromatographers erroneously assume that the sample spot is sufficiently dry when solvent is no longer visible on the plate. The preferred practice is to allow a drying time of five to ten minutes, perhaps assisted by the air stream from a hair dryer.

13.5. The Mobile Phase

Proper selection of the mobile phase is essential to achieving optimal TLC results. Indeed, the great strength of TLC lies in the wide range of applicable mobile-phase solvents, extending from the extremely nonpolar to the extremely polar. A highly polar mobile phase is generally used when the separation involves very polar substances, whereas a less polar mobile phase is best for less polar components. A very polar substance is strongly attracted to silica gel, so a polar solvent is required to displace it. Chromatographic conditions cited in references may be useful as starting points, but some knowledge of solvent – solute characteristics is required for optimal separation.

Selecting a Mobile Phase. It is quite rare for nothing whatsoever to be known about the chemical nature and solubility characteristics of a sample. If one knows something about what constitutes a "good" solvent for the sample, this may imply what will be a "poor" solvent. When a "good" solvent is used as the carrier, adsorbed materials will move in the solvent front, while a "poor" solvent will probably not move anything. What is required for effective TLC is a "mediocre" solvent, one that will cause partial movement and some measure of selective discrimination. Usually the first step is to find a solvent or mobile phase that will move the analyte with an R_f of 0.5. In difficult separations, a mobile phase with a changing composition may be necessary.

Interaction forces involved in TLC include: Intramolecular forces, inductive forces, hydrogen bonding, charge transfer, and covalent bonding [7]. Understanding selectivity thus requires that something be known about the nature of these various types of interaction.

Selecting a Mobile Phase Based on Solvent – Solute Characteristics. The mobile phase in TLC is the source of mobility, but more importantly it produces selectivity. Selectivity — the degree of interaction between components of the mobile phase and analytes in the solute — determines the extent of any resolution that will be achieved. Solvent strength is the factor that determines mobility. Knowing the characteristics of a solute makes it easier to decide which solvents to try during the search for a mobile phase that will produce the desired separation.

Important general characteristics of mobile-phase solvents important in TLC include:

1) *Water content;* this can vary widely, and it is a common source of reproducibility problems, since traces of water may lead to large changes in solvent strength and selectivity
2) *Autoxidation* sometimes occurs during solvent storage, especially over prolonged periods
3) Solvents kept in metal containers may be subject to *decomposition* and unknown chemical reactions, which may change the solvent characteristics
4) Solvent *impurities* may differ from batch to batch, and such impurities may interact with the solute, the solvent, or the sorbent
5) *Chloroform* often contains ethanol as a preservative, and since ethanol has high solvent strength, small amounts may have a dramatic effect

The optimal composition for a mobile phase depends on the nature and number of functional groups in the material to be analyzed. A molecule containing many oxygen functions will display high interactivity, whereas a hydrocarbon with no functional groups will show little interaction. On the other hand, a hydrocarbon solvent will dissolve other hydrocarbons, but not molecules characterized by numerous polar functional groups. Functional groups affect the solvent–solute interaction in the order $RH < ROCH_3 < RN(CH_3) < ROR < RNH_2 < ROH < RCONH_2 < RCO_2H$. This sequence also applies to components of the mobile phase [8].

The level of interaction between the solute and the mobile phase as opposed to the sorbent determines the distribution of the solute between these two phases. The mobile phase is generally less reactive than the sorbent. If the analyte interacts more strongly with the mobile phase it will prefer to remain in the mobile phase, and the separation will be poor. If the solute has a more powerful interaction with the sorbent the separation potential will be greater. Chromatography thus consists of an interplay between solute–sorbent and solute–mobile-phase interactions (→ Basic Principles of Chromatography).

The *solvent strength parameter* (see below) provides a quantitative measure of the relationship between solvent strength and the type of sorbent used, although solvent behavior differs greatly in normal- vs. reversed-phase chromatography. The interaction of a solvent with the sorbent is thus a function of the type of sorbent used. Methanol is more polar than chloroform. As polarity increases, so does elution strength with respect to a normal-phase sorbent such as silica gel. Depending on the analyte, methanol will lead to more rapid analyte migration up the layer relative to a less polar solvent. The opposite is the case with a reversed-phase system. Here, the more polar the mobile phase the greater the tendency toward retention by the sorbent. Thus, water will not readily move analytes in a reversed-phase system, but will do so easily under normal-phase conditions. Even so, if the analyte does not dissolve, nothing will happen. Hexane, a nonpolar solvent, is at the opposite end of the scale from methanol.

Perhaps the most widely recognized characteristic solvent property is the *solvent strength parameter* $E°$ [7]. This was defined by Snyder as the adsorption energy for a solvent per unit of standard sorbent. The higher the solvent strength, the greater will be the R_f value of an analyte in normal-phase TLC. The opposite is true in the case of reversed-phase sorbents. Solvent strength parameters are listed below for a variety of common solvents [7]. The values cited are based on heats of adsorption with alumina, but a similar sequence and similar values apply to silica gel. The same principles also hold true for reversed phases, but the corresponding solvent-strength values are reversed.

Solvent	$E°$, Al_2O_3
Fluoroalkanes	−0.25
n-Pentane	0.00
Hexane	0.00
Isooctane	0.01
Petroleum ether, Skellysolve B, etc.	0.01
n-Decane	0.04
Cyclohexane	0.04
Cyclopentane	0.05
Diisobutene	0.06
1-Pentene	0.08
Carbon disulfide	0.15
Carbon tetrachloride	0.18
Amyl chloride	0.26
Butyl chloride	0.26
Xylene	0.26
Diisopropyl ether	0.28
2-Chloropropane	0.29
Toluene	0.29
1-Chloropropane	0.30
Chlorobenzene	0.30
Benzene	0.32
Ethyl bromide	0.37
Ethyl ether	0.38
Ethyl sulfide	0.38
Chloroform	0.40
Methylene chloride	0.42

Solvent	$E°$, Al_2O_3
Methyl isobutyl ketone	0.43
Tetrahydrofuran	0.45
Ethylene dichloride	0.49
Methyl ethyl ketone	0.51
1-Nitropropane	0.53
Acetone	0.56
Dioxane	0.56
Ethyl acetate	0.58
Methyl acetate	0.60
Amyl alcohol	0.61
Dimethyl sulfoxide	0.62
Aniline	0.62
Diethyl amine	0.63
Nitromethane	0.64
Acetonitrile	0.65
Pyridine	0.71
Butyl cellosolve	0.74
Isopropyl alcohol, *n*-propyl alcohol	0.82
Ethanol	0.88
Methanol	0.95
Ethylene glycol	1.11
Acetic acid	large
Water	larger
Salts and buffers	very large

Solvent *selectivity* refers to the ability of a solvent to establish interaction differences for two different analytes. Resolution is dependent on three factors:

1) The nature of the sorbent
2) The migration rate of the two solutes, which in turn depends on the solvent strength
3) Separation factors, which are determined by interactions between the solvent and the mobile phase

If two solutes cannot be separated by a particular mobile phase of predetermined strength, it may still be possible to achieve separation with a different mobile phase of the same strength. The above list of solvent strengths can be used to illustrate application of the solvent selectivity concept. Selectivity arises from the fact that a given solvent will interact differently with two solutes of different functional makeup. The solvent-strength list includes several examples of solvents with identical strength values. Ethyl acetate and diethylamine for example, have approximately the same solvent strength, but entirely different chemical structures, and thus different reactivities with respect to other molecules. In particular, two solutes that cannot be separated with a given mobile-phase system containing ethyl acetate may separate when diethylamine is substituted for the ethyl acetate despite a minimal change in overall mobility.

The mechanisms of retention on chemically bonded reversed phases are not clearly understood, but observed R_f values for a series of solutes usually turn out to be reversed compared to the corresponding sequence on silica gel provided water constitutes a large fraction of the mobile phase. It is also possible to achieve separations on RP plates using an entirely organic mixture as the mobile phase.

Most RP chromatography is conducted with two-solvent mixtures composed of water together with an alcohol, acetonitrile, acetone, or dioxane. Methanol – water (8 : 2) is a convenient starting point for work with RP layers. The solvent strength is then varied by changing the water–organic modifier ratio, and selectivity is improved by using a different modifier. More polar samples require relatively more water. Small changes in mobile-phase strength have less effect on R_f values in RPTLC than in normal-phase TLC, making mobile-phase selection easier in the former case.

Mobile phases for gel TLC must be capable of dissolving the sample and at the same time causing the gel network to swell. Organic solvents such as tetrahydrofuran (THF) are common for gel permeation TLC on organic polymer layers (e.g., styrene – divinylbenzene), while aqueous buffers are employed for gel filtration chromatography on layers such as dextran.

Useful general rules related to selectivity and solvent strength include the following:

1) Increasing the solvent strength will decrease separation and increase R_f.
2) Decreasing the solvent strength will increase separation, but decrease R_f.
3) Selectivity is greatest when the concentration of the stronger component of the mobile phase is greater than 5 vol % and less than 50 vol %. In other words, more favorable solute – solvent interactions result when the solvent of greater strength is not dominant.
4) Introduction of a solvent with greater hydrogen-bonding potential may increase resolution.
5) With respect to (3), it is generally best to examine the effect of small changes in the amount of the stronger modifying solvent.
6) If the observed R_f value is too high, a primary solvent of lower solvent strength should be tried rather than drastically changing the proportions of a mixture, and vice versa.

13.6. Development

13.6.1. Apparatus

Most TLC development is accomplished by allowing the mobile phase to ascend a plate standing in a suitable-sized chamber or tank containing the appropriate mobile phase. Glass chambers are by far the most common. Simple chambers can be made from wide-mouthed screw-cap, battery, or museum jars. Chambers specifically made for TLC are available in a variety of sizes, both rectangular and circular, accommodating all sizes and formats of TLC plates. Aluminum racks are available for supporting and developing several conventional plates at one time in a single tank. One company has introduced a special tank with a raised glass hump running lengthwise on the inside bottom to act as a tank divider. The latter requires only 20 mL of solution per side, and makes it possible for two large plates to be developed simultaneously. The divided bottom also allows the tank to be saturated with two mobile-phase systems simultaneously, placed on opposite sides of the divider, so that a plate can be introduced on either side for development. Small chambers have been designed specifically for use with 10×10 cm or HPTLC plates. These small chambers occupy much less bench space, and require considerably less solvent.

A *sandwich chamber* for use with 10×10 cm plates offers several advantages, including small solvent volume (6 mL), rapid chamber saturation, and rapid development. A glass-backed TLC plate forms one wall of the chamber, with the adsorbent facing inward toward a second glass plate. The sides are clamped together with metal clips and the open end of the chamber is inserted in a solvent trough containing the developing solvent. Other commercially available sandwich chambers accommodate larger plates (20×20 and 10×20 cm). Such a chamber can be purchased commercially from a number of suppliers for about the same price as a normal glass TLC tank.

Ascending or Linear Development. The separation process in TLC occurs in a three-phase system consisting of stationary, mobile, and gas phases, all of which interact with each other until equilibrium is reached. If a plate is developed in a chamber whose atmosphere is not saturated with the mobile phase, flow up the plate may occur in an unsymmetrical way because some mobile phase evaporates from the plate into the chamber atmosphere in an attempt to establish equilibrium. The entire chromatographic process then depends on the volatility of the mobile phase, the temperature, and the rate of development. It is for this reason that the development chamber should be made as small as conveniently possible based on the plate size and the number of plates to be developed at one time.

Reproducible results are achieved only with a constant environment. Laboratory benches exposed to the sun during part of the day should be avoided because of temperature variation. Most development chambers are of such a size that they may conveniently be placed in an environmentally controlled space, room, or bath as a way of maintaining consistency. It is advisable to weigh down the lid of the chamber so that it cannot be dislodged by vapor pressure or heat produced by the mobile phase, which would disturb the equilibrium within the system. Some newer chambers include latches that actually "lock down" the lids.

Chamber Saturation [9]. Saturation of the developing chamber atmosphere is facilitated by lining the inner walls with blotting paper or thick chromatography paper (e.g., Whatman 3 MM), or by using saturation pads specially designed for TLC, although some separations can be accomplished in unlined tanks. The individual components of the mobile phase should be mixed thoroughly before they are introduced into the developing chamber. Reported solvent proportions are assumed to be by volume unless stated otherwise. It is advantageous to pour the mixed mobile phase over the lining paper as it is added to the developing chamber, since this hastens the process of saturating the paper and the chamber. Enough time should be allowed for complete equilibration of the chamber.

Plate development may be described in terms of either time or distance. With timed development a plate is allowed to develop for a specific period of time, after which it is carefully lifted out of the chamber. The location of the solvent front is then marked with a pencil or a sharp object such as a syringe needle. If mobile phase is instead expected to travel a fixed distance (e.g., 10 cm), this distance should be marked on the plate at the time of sample application. It then becomes easy to observe when the mobile phase has reached the desired point so the plate can be removed promptly from the chamber. A fixed development distance is used to standardize R_f values; because

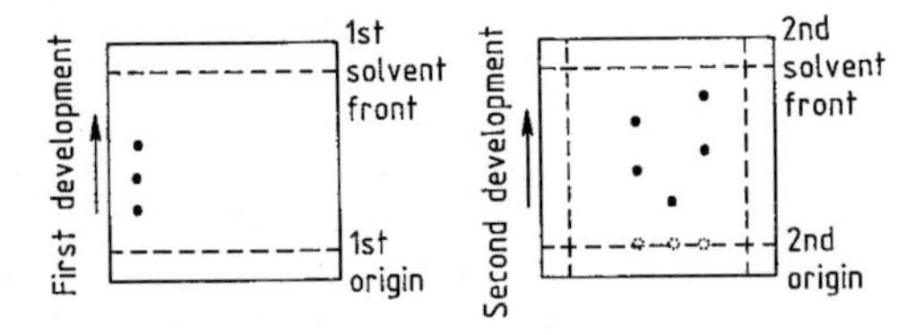

Figure 2. An example of two-dimensional development

R_f values by definition fall between 0.0 and 1.0, 10 cm is a convenient standard distance.

13.6.2. Development Techniques

Standard Development. The standard method of development has already been described in the preceding paragraphs. If the TLC characteristics of the analyte are unknown, a simple single-component mobile phase is preferable for preliminary experiments aimed at establishing separation behavior. If no single-component mobile phase provides the separation desired, however, multicomponent systems should be examined.

Multiple Development. If a single development of a plate suggests that a partial separation has been achieved, the plate should be allowed to dry, after which it can be reinserted into the developing chamber for redevelopment. This procedure may be repeated as many times as necessary to achieve a satisfactory separation. Repeated development increases the effective length of the layer, thereby increasing the likelihood of separation, because the various molecules will have an opportunity to interact over a greater distance, perhaps culminating in complete resolution.

With a mixture containing compounds that differ considerably in their polarities, development with a single mobile phase may not provide the desired separation. However, separation may result if such a plate is developed successively with various mobile phases that have different selectivities or strengths. Multiple mobile phases can be used in a number of ways. For example the least polar phase might be introduced first, followed by a more polar phase, or vice versa.More than two developments can also be tried, and the development distance on the plate might be changed with each development. For example, a very polar mobile phase, such as methanol, could be used first with a given sample mixture in order first to separate the polar substances. After the methanol had developed part way up the layer the plate would be removed from the chromatography tank and dried. Then a second mobile phase, a nonpolar solvent such as cyclohexane, would be allowed to develop the entire length of the plate in an attempt to resolve in the upper portion those nonpolar substances that were not resolved in the previous (polar-phase) development. Depending on the circumstances, the same procedure might also be carried out with the reverse sequence.

Continuous Development. Continuous development involves a continuous flow of mobile phase along the length of the plate, with the mobile phase being permitted to evaporate as it reaches the end. The chromatography tank in this case is not as tall as a normal tank, which means that a normal-sized TLC plate extends beyond the top. Mobile phase then evaporates from the exposed end of the plate, causing a continuous flow to be maintained.

Continuous development offers the advantage of greater separation without the need for changing the mobile phase. Solvent selection is simplified because one can exploit the high separation power of a low-polarity solvent without suffering the usual consequence of reduced R_f. Sensitivity may be increased as well, because zone diffusion is reduced as a result of the shorter distance traveled by the analyte. The technique is essentially equivalent to development on a very long TLC plate without the requirement for special equipment.

Two-Dimensional Development. This developing technique involves the application of normal, multiple, or stepwise development in two successive dimensions. A very versatile method, it has been widely used in paper chromatography as well as in certain TLC separations. A single sample is applied at one corner of a plate and developed in the usual way. The plate is then removed from the developing chamber and dried, after which it is rotated by 90° and placed in a chamber containing a different mobile phase so that it will develop again, this time in a direction perpendicular to that of the first development. The line of partially resolved sample components from the first development becomes the origin for the second development (see Fig. 2). It is important that initial sample application not be too close to the edge of the plate, since otherwise the resulting line of developed spots might fall below the level of solvent in the second developing chamber. A

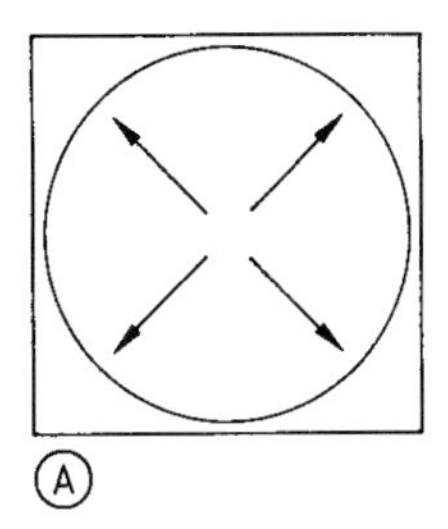

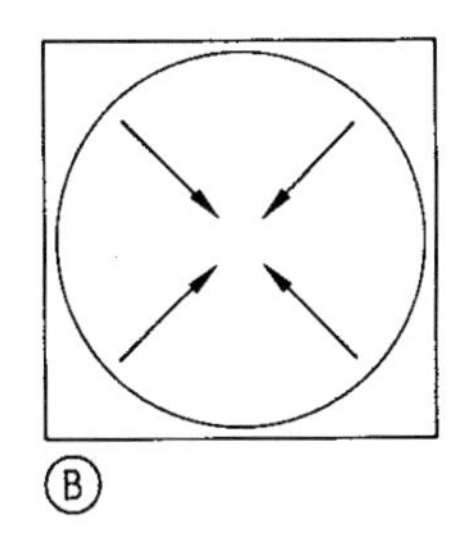

Figure 3. Circular and anticircular TLC development
A) Circular development; a wick is used to apply mobile phase to the center, causing radial flow outward; B) Anticircular development, in which solvent is applied with a wick to the outer ring, causing solvent migration inward toward the center

distance of 2.5 – 3 cm from the edge is usually sufficient. The versatility of the technique arises not only from the second development with a different mobile phase, but also from the opportunity provided for treating or modifying the layer and the sample prior to the second development. Standards can be applied to the second origin before the second development to permit direct comparisons and calculation of R_f values after development and visualization. The technique is also applicable to plates containing two different sorbents, such as silica gel on one half and RP_{18} on the other.

Circular and Anticircular Development. Circular and anticircular development have been accomplished in a number of ways. Because of the need for solvent-delivery control in such systems, chromatography is best carried out in equipment specially designed for the purpose. It is also important that the chamber be kept level. Camag, Analtech, and Anspec all offer chambers for circular development. Samples are applied at the center, and mobile phase is wicked to the surface. Development causes the analyte mixture to separate into a series of concentric rings. Figure 3 illustrates these devices.

Reversed-Phase Partition Development. Reversed-phase development employs a relatively nonpolar stationary phase, which means that the layer is hydrophobic rather than hydrophilic as in normal partition TLC. The corresponding mobile phase is relatively polar, and may be water or an alcohol. Most of the common TLC sorbents have been used in conjunction with partition development, including cellulose, silica gel, and kieselguhr. This technique is generally applicable to the separation of polar compounds, which often remain at the origin during conventional development. With reversed-phase development, nonpolar compounds are the ones that usually remain near the origin. The need for polar eluents results in development times longer than those normally encountered with ordinary silica gel and less polar eluents.

Horizontal Development. Linear development can also be conducted from two directions simultaneously, with mobile phase entering from two opposing sides of a single plate. Such development requires the use of a horizontal developing chamber. The main advantage of the procedure, which was originally designed for use with 10×10 HPTLC layers, is that the number of samples per plate can be doubled. In closed systems and with small samples it is possible to separate in this way as many as 72 samples at the same time.

13.7. Visualization

Once a sample has been transferred to a TLC plate, all the components of that sample will be located somewhere on the layer. If the analyte is colored, localization is simple. Most organic compounds are not colored, however, so special techniques are required to change the materials in such a way that some optical contrast is established in the visible or the ultraviolet region. Reagents useful in color formation have been listed in [1].

A satisfactory visualization technique must take account of the following considerations:

1) Significant contrast is essential between the separated zone and the background.
2) A high degree of sensitivity is important to ensure detection of minute amounts of analyte.
3) Many compounds are not stable once separated on a sorbent; fluorescent compounds in particular are often unstable. This may not be a serious concern however, if only qualitative information is required. Nonetheless, the stability characteristics of the solutes should be ascertained.
4) A universal reagent capable of detecting all the solutes present is most desirable.
5) Satisfactory detection methods always result in well-defined zones of separation. Spraying or

dipping procedures can be problematic as a result of dispersion of the separated solutes.

6) An ideal detection method would permit quantitative analysis based on proportional absorbance intensities (see Chap. 13.8).

Methods for inducing detectability may be either *nondestructive* or *destructive*, where nondestructive methods are ones that do not destroy the analyte.

13.7.1. Nondestructive Visualization Methods

Ultraviolet absorbance can be utilized for detection purposes in several ways. For example, ultraviolet light often makes it possible to locate analytes separated on a fluorescent TLC plate. Thus, a UV-absorbing component may quench the background fluorescence, appearing as a dark spot against a luminescent field. Zinc silicate is a common phosphor in plates marked "F" for fluorescent, which fluoresce under the influence of light with a wavelength of 254 nm. Plastic fluorescent plates sometimes contain the fluorescent dye Rhodamine 6 G.

However, this fluorescence quenching is not a very sensitive detection method. As a rule it is recommended that visualization of the thin layer chromatogram take place in a dark room under the influence of lamps with emission maxima at 254 and 366 nm.

Inherent fluorescence can also be used for detecting separated zones, although the commonly available 254- and 366-nm lamps are suitable only for detecting compounds that fluoresce within the output range of the lamps themselves. Inherent fluorescence is the most sensitive aid to detection, occasionally revealing as little as 1 pg of an analyte.

Nondestructive detection is sometimes possible simply by spraying a plate with water. Free polar adsorbents (silica gel and alumina) are able to adsorb large amounts of water, whereas regions on the plate occupied by spots of organic material adsorb much less moisture, and appear as fatty zones against an almost translucent white background.

Detection with iodine vapor is one of the most widely applicable visualization techniques. If iodine crystals are placed in the bottom of a covered glass development chamber, within a few hours the atmosphere in the chamber will be saturated with the vapors. Analyte components on a thin layer chromatogram placed in such an iodine chamber tend to adsorb iodine, subsequently appearing as dark brown spots against a lighter tan background [10]. This method is characterized by either destructive or nondestructive detection depending on the analyte. After the developed layer in an iodine tank has been examined for evidence of reactive zones, the plate can be removed and areas of interest can be marked. Once the iodine has evaporated, analyte can sometimes be recovered for use in another procedure—provided it has not reacted with the iodine. Some analytes do undergo reaction with iodine, however, in which case the detection process is nonreversible. For example, phenolic compounds are known to react irreversibly under these conditions.

The iodine reaction can be surprisingly sensitive in some cases. Thus, morphine and codeine have been detected in this way at the 0.1-μg level. Iodine reagent can also be applied by spraying with a 1 % solution of iodine in an inert, volatile solvent such as hexane.

13.7.2. Destructive Visualization Methods

Charring is a destructive method that can be regarded as essentially universal. When an organic material on a TLC plate is heated in the presence of acid, an easily distinguishable black or brownish-black spot is formed. In many cases, quantitation of such a spot is possible by densitometry (see Chap. 13.8).

Acid solutions frequently used for TLC charring include:

sulfuric acid : acetic anhydride (1 : 3)
sulfuric acid : sodium dichromate (the well known "chromic acid cleaning agent")
sulfuric acid : nitric acid (1 : 1)
sulfuric acid : copper acetate or copper sulfate
phosphoric acid : copper sulfate or copper acetate

The plate under investigation is first treated with the acid solution and then heated. Charring is accomplished at 120 – 280 °C over a period of 1 – 40 min. Initial experiments should be conducted at 120 °C for 1 min, since this may produce dehydration products that fluoresce with different colors. For example, steroids generally produce fluorescent products when sprayed with 1 % sulfuric acid in ethanol followed by heating to as little as 100 °C. As noted previously, this type of reagent has the inherent advantage of offering universal detection; most analytes produce some type of reaction, and if no reaction is initially apparent the sample on the layer can be heated to increasingly higher temperatures until charring occurs. The obvious disadvantage is destruction of the analyte (to say nothing of surrounding structures and clothing if care is not exercised!).

13.7.3. Reactive Visualization Methods

This category includes the preparation of derivatives, either before or after development of the chromatogram. It is sometimes feasible to improve the detectability of an analyte by subjecting it to a chemical reaction prior to chromatography. The separation that follows such a reaction is obviously not of the original materials, but rather of the reaction products. Confirmatory identification should be based on R_f values for both the unchanged and the reacted product. Reactive derivatization often has the additional advantage of improving the resolution of compounds that are otherwise difficult to separate.

13.7.4. Application of Visualization Reagents

The most common approach to application of a visualization reagent is probably spraying, using a spray gun powered either by compressed air from a laboratory line or compressed gas in a canister. The usual device incorporates a propellant container mounted on a screw cap that fits the top of a glass reagent jar. Reagent is drawn out of the jar by the venturi effect, and is dispersed as fine droplets in the emerging gas "cone." Spraying should be conducted in a fume hood to eliminate or minimize the exposure of personnel to reactive and toxic materials. Many distributers offer a wide variety of specially designed plate supports, spraying racks, and disposable or washable spray booths.

A TLC plate should be sprayed as evenly as possible, combining side-to-side and up-and-down motions of the spray equipment. Careful work is required to ensure uniform distribution of the sprayed reactant. Quantitation becomes particularly uncertain if spot intensity is dependent upon both the concentration of the spot component and the amount of reagent locally applied.

Application of reactive solutions is also possible by immersion of the TLC plate in some type of "dipping tank." The chief disadvantage of dipping techniques, especially with highly reactive agents, is that the back of the plate also becomes wet, and must be wiped in order to remove excess reagent. Care must also be taken when attempting to dip "soft" plates, since a soft coating is easily damaged by the forces associated with moving the plate in a surrounding liquid. One alternative is allowing the reactive reagent to ascend the plate by capillary action, just as in development.

If the need arises to spray a large number of plates it may be advantageous to consider setting up a *vapor charring* system, in which a plate is heated in a chamber through which SO_3 gas is passed, which in turn induces charring. The gaseous approach has the advantage that it distributes charring reagent evenly over the entire plate. The procedure utilizes a square Pyroceram dish approximately 10×10×25 cm in size and equipped with a curved glass cover. These are readily available as accessories for microwave cooking. The TLC plate is placed in the bottom of the vessel, adsorbent side up. A small amount of fuming sulfuric acid is then swabbed around the glass cover about 2.5 – 5 cm from the edge. The prepared cover is placed over the base, and the entire assembly is transferred to a hot plate or an oven. Appropriate conditions of time and temperature must be precisely established for each required analysis. It must also be emphasized that extreme precautions are necessary withrespect to the fuming acid, particularly when opening the vessel after charring. The advantages, of course, are very even charring and highly reproducible results.

Both Camag and Analtech also distribute sealed chambers commercially, utilizing ammonium hydrogencarbonate for visualization purposes. Vapors formed as the salt is heated tend to char or otherwise cause visual changes in a plate which is heated simultaneously. Many compounds form fluorescent derivatives under these conditions.

13.8. Quantitation

Modern approaches to *quantitative* TLC require the use of densitometry. Alternative visual quantitation techniques will not be discussed here, nor will elution techniques and the associated spectrophotometric methodology. For a discussion in this area one should consult the literature, for instance the "Practice of Thin Layer Chromatography" by TOUCHSTONE [1] now in its third edition. The focus here will be densitometric methodology, which is clearly becoming the method of choice. For a detailed discussion of theoretical aspects of densitometry see POLLAK [11].

In situ methods based on densitometers and spectrodensitometers are now available as well, and these can be very precise, with relative standard deviations of less than 5 %. Instruments currently on the market (1994) for in situ scanning of TLC layers include the following:

Ambias Inc., San Diego, Ca., USA	scanner for combined densitometry and isotope detection
Analtech Inc., Newark, Del., USA	one of the only video scanners designed for TLC
Camag, Muttenz, Switzerland	linear scanner used in many laboratories; equipped with a monochromator as a standard feature
Desaga GmbH, Heidelberg, FRG	linear scanner with monochromator
Helena Laboratories, Beaumont, Tex., USA	densitometer originally designed for electrophoresis scanning
Shimadzu Scientific Instruments, Columbia, Md. USA	linear scanner with monochromator

Video Scanners. Video scanners (such as the Uniscan device from Analtech) permit very rapid data acquisition. An entire TLC plate can beanalyzed at once with this device. Three detector modes are available: transmittance, reflectance, and fluorescence. The scanner incorporates its own computer with a video digitizer, light sources (visible and UV), and an imaging detector. Band-pass filters in front of the lens are used to select specific emission wavelengths, which can extend the usefulness of the instrument. In the case of fluorescent compounds separated by HPTLC, quantities in the low picogram range are readily assayed. For rapid quantitation in many regions, multichannel detector instruments offer a great advantage over conventional detection systems. Inexpensive instruments now permit digital storage of large amounts of data, opening the way to mathematical methods of data analysis. The most satisfactory instruments take advantage of a charge-coupled discharge video or standard camera, which can capture a complete image of a TLC layer in fractions of a second. The Analtech Uniscan comes with an enclosed view box featuring visible and short- or long-wave UV illumination. The image is projected on a monitor, and it is processed with the aid of a cursor for setting scan limits. Signals are processed in such a way as to provide both a chromatogram and tabulated quantitative data in the form of a printout. The quantitative information is calibrated on the basis of data from serial applications of standards that have been separated on the same chromatogram.

Linear Scanners. Linear scanners usually rely on a fixed light source under which the TLC layer must be moved. Changes in density are recorded by a computer programmed to convert the signal into a chromatogram, which is than evaluated on the basis of parameters so adjusted as to provide optimal scanning. Both Shimadzu and Camag market linear scanners of this type, which are also capable of determining absorption spectra of separated materials suitably positioned unter the light beam from a monochromator. The same instruments can be used to scan fluorescent compounds as well. Judicious use of filters can enhance the specificity of an analysis. For example, a 400-nm cut-off filter provides useful screening in the case of fluorescence analyses.

In *double-beam* operation the primary signals are amplified and fed directly into logarithmic converters. At the output, the reference signal is substracted from the sample signal, and the result is fed into an integrator. From there the signal passes to a conventional chart recorder that records the ratio of the two beam signals.

Important properties to consider in evaluating an instrument include the nature of the light source, the lines of highest emission energy, and the response curve of the detecting element, which may be either a photomultiplier or a cadmium sulfide diode. For example, the following spectral characteristics were observed with one instrument based on a xenon – mercury lamp (Hanovia 901.B-11) and photomultiplier detectors: Mercury bands from the xenon – mercury lamp were apparent at 312 – 313 nm and 365 –366 nm, with the band at 365 – 366 nm by far the most intense. The lamp had an intensity curve rising gradually from 200 nm to a hump at ca. 500 nm, after which it leveled off. The photomultipliers showed a maximum response to light at 380 nm. This wavelength was preceded by a shoulder at the end of a steady rise. Above 380 nm the response declined gradually, becoming almost zero at 570 nm. The peaks of highest emission of the lamp coincided with the bands of greatest activation for the fluorescence of quinine sulfate separated on a silica-gel layer. This is significant, because it is usually possible to obtain greater sensitivity by using light with a wavelength close to the lines of maximum emission of the lamp.

It is important to scan at the wavelength of maximum absorption of the solute in the sorbent, where the absorption maximum for a compound adsorbed on silica gel particles is typically not the same as that for the same compound in an alcohol solution. Precise and sensitive determination requires that the absorption maximum be determined in situ on the plate. This is possible with an instrument equipped with sources of variable wavelength or monochromators. Narrow-bandpass filters are available for instruments based instead on filters.

Single-beam recordings show the expected background variations, together with an increased noise level in both transmission and reflectance measurements. (The background is much smoother with a double-beam system.) Over short distances near the center of the plate the background variations and noise experienced in the course of reflectance measurements in single-beam operation may appear acceptable, but in reflectance mode the peak response for a given spot is substantially lower than in transmission mode, so the S/N value (i.e., precision) should be expected to decrease. Peak signals obtained in transmission are about five times as large as the corresponding signals in reflectance, whereas the noise level is only twice as great.

Errors in sample application are among the most significant sources of inaccuracy. The technique of spotting can have a marked effect on delivery volumes between 1 and 5 μL. A constant *starting zone size* should be maintained so that a given amount of substance applied in the form of different volumes will yield spots with identical areas, whereas different amounts spotted at constant volume will exhibit a linear relationship in a plot of peak area (from recordings of densitometer scans) versus mass.

Errors introduced during movement of the solute through the sorbent layer include lateral diffusion during development, and problems can also arise from structural variations for different plates or variations between tanks. Lateral diffusion is a function of the relative rates of diffusion of solute and mobile phase (→ Basic Principles of Chromatography). Errors due to lateral diffusion when substantially different amounts of solute are applied next to each other can be prevented by scoring the plate into strips ca. 1 cm wide prior to spotting. In this way two dissimilar samples can safely be developed side by side. Variation in structure from one layer to another leads to incorrect extrapolation of data obtained when a known quantity of standard is compared to the quantity of an unknown. The ratio between the quantity in a final spot and the quantity spotted varies for different layers, so it is recommended that a standard be applied to each test plate. Factors such as the spot size before and after development and the R_f value also become important in this context.

There are several possible sources of error in quantifying the results obtained from a densitometer or radiochromatogram scan. Important considerations include sampling methods, separation variables, variations in detector response, and calibration and measurement techniques.

Many of the errors in quantitation originate in the way a sample is applied to the layer, since this affects the shape of the spot after chromatogram development. The chromatographic process itself determines whether there will be tailing or overlap of the spots. High R_f values result in broad peaks that are not compatible with reproducible results from scanning in situ.

Another source of error is sample decomposition during chromatography. Evidence of decomposition can be obtained by developing chromatograms with standards of known serial dilutions. If the peak area does not increase linearly as a function of concentration, sample decomposition should be suspected. Errors due to tailing and overlap are quite common. These can be eliminated by changing the mobile phase to one that produces better separation.

The peak-area calibration procedure is usually less subject to influence from changes in instrumental parameters than calibration based on peak height. In addition, it offers improved results with peaks whose shapes are not Gaussian. However, results obtained with the peak-area quantitative method are adversely affected by neighboring peaks. Peak areas should always be used when precision is important.

The electronic integrators used in GC analysis are generally satisfactory for measuring peak areas in TLC scans as well. An electronic digital integrator measures peak areas automatically and converts them into numerical data, which are then printed. Sophisticated versions of such devices also correct for baseline drift.

Thin layer chromatography peak areas derived from scans can also be measured conveniently with computer systems developed for GC applications, often without any change. Most devices of this type are programmed to print out a complete report, including compound names, retention times, peak areas, area correction factors, and weight-percent composition with respect to the various sample components.

13.9. References

[1] J. C. Touchstone: *Practice of Thin Layer Chromatography,* 3rd ed., Wiley, New York 1992.
[2] J. Sherma, B. Fried: *Handbook of Thin-Layer Chromatography,* Marcel Dekker, New York 1991.
[3] F. Geiss: *Fundamentals of Thin Layer Chromatography,* Hüthig-Verlag, New York 1987.
[4] T. E. Beesley in J. C. Touchstone, D. Rogers (eds.): *Thin Layer Chromatography, Quantitative Environmental and Clinical Applications,* J. Wiley, New York 1980, p. 10.
[5] M. Zief, R. Riser, *Am. Lab. (Fairfield, Conn.)* **6** (1990) 70.
[6] D. E. Jaenchen, H. Issaq, *J. Liq. Chromatogr.* **11** (1988) 1941.
[7] L. R. Snyder: *Principles of Adsorption Chromatography,* Marcel Dekker, New York 1968.
[8] L. R. Snyder, J. J. Kirkland: *Introduction to Modern Liquid Chromatography,* Wiley, New York 1979.
[9] R. A. de Zeeuw, *J. Chromatogr.* **33** (1968) 222.
[10] G. C. Barrett in J. C. Giddings, R. A. Keller (eds): *Advances in Chromatography,* **vol. 11,** Marcel Dekker, New York 1974, p. 151.
[11] V. Pollak in J. C. Giddings et al. (eds.): *Advances in Chromatography,* Marcel Dekker, New York 1979, pp. 1 – 50.

14. Electrophoresis

Pier Giorgio Righetti, Department of Agricultural and Industrial Biotechnologies, University of Verona, I-37134 Verona, Italy

Cecilia Gelfi, ITBA, CNR, L.I.T.A., Segrate 20090 (Milano), Italy

14.1. Introduction

The theory of electrophoresis has been adequately covered in the excellent textbooks of Giddings [1] and Andrews [2] as well as in specific manuals [3], [4]. For discussion on electrophoresis in free liquid media, e.g., curtain, free-flow, endless belt, field-flow-fractionation, particle, and cell electrophoresis the reader is referred to a comprehensive review by Van Oss [5] and to a book largely devoted to continuous-flow electrophoresis [6]. Here the focus is mostly on electrophoresis in a capillary support, i.e., in gel-stabilized media, and discussion is limited to protein applications.

Electrophoresis is based on the differential migration of electrically charged particles in an electric field. As such, the method is applicable only to ionic or ionogenic materials, i.e., substances convertible to ionic species (a classical example: neutral sugars, which form negatively charged complexes with borate). In fact, with the advent of capillary zone electrophoresis (CZE) it has been found that a host of neutral substances can be induced to migrate in an electric field by inclusion in charged micelles, e.g., of anionic (sodium dodecyl sulfate, SDS) or cationic (cetyltrimethylammonium bromide, CTAB) surfactants [7]. Even compounds that are not ionic, ionogenic, or complexable can often be analyzed by CZE as they are transported past the detector by the strong electrosmotic flow on the capillary walls [8].

Electrophoretic techniques can be divided into four main types: zone electrophoresis (ZE), moving-boundary electrophoresis (MBE), isotachophoresis (ITP), and isoelectric focusing (IEF). This classification is based on two criteria: (1) initial component distribution (uniform vs. sharply discontinuous); and (2) boundary permeability (to all ions vs. only to H^+ and OH^-). Alternatively, electrophoretic techniques may be enumerated in chronological order, as follows: moving-boundary electrophoresis, zone electrophoresis, disc electrophoresis, isoelectric focusing, sodium dodecyl sulfate – polyacrylamide gel electrophoresis (SDS – PAGE), two-dimensional (2-D) maps, isotachophoresis (ITP), staining techniques, immobilized pH gradients (IPG), and capillary zone electrophoresis.

Zone electrophoresis became practicable when hydrophilic gels (acting as an anticonvective support) were discovered. GRABAR and WILLIAMS in 1953 first proposed the use of an agar matrix (later abandoned in favor of a highly purified agar fraction, agarose) [9]. They also combined, for the first time, electrophoresis on a hydrophilic support with biospecific detection (immunoelectrophoresis). Barely two years after that, SMITHIES [10] applied another gel, potato starch. The starch blocks were highly concentrated matrices (12 – 14 % solids) and introduced a new parameter in electrophoretic separation: molecular sieving. Human sera, which in cellulose acetate or paper electrophoresis were resolved into barely five bands, now produced a spectrum of 15 zones.

14.2. Basic Principles

In a free buffer solution, in the absence of molecular sieving, the velocity $\boldsymbol{v}$ of a particle is proportional to the field strength $\boldsymbol{E}$, multiplied by its electrophoretic mobility μ:

$$\boldsymbol{v} = \mu \boldsymbol{E}$$

The velocity and $\boldsymbol{E}$ are both vectors, while μ is a scalar, being positive for cations and negative for anions. A distinction is made between absolute mobility (the μ value of species at infinite dilution and complete ionization) and effective mobility (the μ value of species under the actual experimental conditions, e.g., as a result of sieving in gel matrices, or in the presence of a strong electrosmotic flow). In deriving an expression for μ which takes into account parameters such as charge, size, and solvent effects, one can consider the different forces acting on a species. At infinite dilution, the electric field is exerting a force on the charge Q of the particle and this force is opposed by the frictional force, given by the friction factor f_c multiplied by the velocity v. At steady state:

$$Q\boldsymbol{E} = f_c \boldsymbol{v}$$

For a spherical particle obeying Stokes' law, f_c is given by:

$$f_c = 6\pi\eta\, r$$

where η is the solvent viscosity and r the particle radius. By substituting and rearranging one can derive the expression for the absolute mobility:

$$\mu = \boldsymbol{v}/\boldsymbol{E} = Q/(6\pi\eta\, r)$$

According to the Debye – Hückel theory, the total charge Q of a particle is correlated to the dielectric constant ε and to the zeta potential by the expression:

$$Q = \varepsilon\, r \zeta$$

whereby the expression for μ follows:

$$\mu = \varepsilon\zeta/(6\pi\eta)$$

The mobility is directly proportional to the dielectric constant and zeta potential and inversely related to the solvent viscosity. Thus, it follows that μ is strongly reduced in organic solvents, is zero at the isoelectric point of an amphoteric species (where the net charge is zero), and is greatly reduced in any form of gel media, which can be viewed as highly viscous solutions.

14.3. Electrophoretic Matrices

Only three matrices are in use today: cellulose acetate, agarose, and polyacrylamide.

14.3.1. Cellulose Acetate

Cellulose acetate is a cellulose derivative in which each hexose ring of the polysaccharide chain has two hydroxyl groups esterified to acetate (in general, in the C-3 and C-6 positions). This medium was developed for electrophoresis largely through the work of KOHN [11], who also designed the original tank, produced by Shandon Southern

Instruments as Kohn's Electrophoresis Apparatus Model U77 and imitated throughout the world.

Cellulose acetate is still very popular in clinical chemistry, as it offers a ready-made support that can be equilibrated with buffer in a few seconds and produces good separations of proteins from biological fluids in ca. 20 min. Completely automatic systems (for sample loading, staining, destaining, and densitometry) have been built around it, so that this system has become the first example of combined electrophoresis and robotics, an integration which had its testing ground and underwent development in clinical chemistry and is now spreading to basic research laboratories.

Cellulose acetate allows migration of even large serum proteins, such as pre-β-lipoprotein ($M_r > 5 \times 10^6$) and is amenable to immunofixation in situ [12], [13] without any need for prior transfer by blotting (as required, e.g., after SDS-PAGE). In addition, the immunoprecipitate, obtained directly on the cellulose acetate, can be enhanced by use of a second antibody, conjugated with, e.g., alkaline phosphatase [14]. Moreover, owing to their large porosity and their availability as thin films, cellulose acetate matrices can be stained with gold micelles, increasing the detection sensitivity to the subnanogram range [15], whereas agarose and polyacrylamide are incompatible with this gilding process [16]. However, the use of cellulose acetate is limited to clinical electrophoresis and is not much in vogue in basic research for high-resolution runs, even though focusing and two-dimensional techniques on cellulose acetate membranes have been described [17].

14.3.2. Agarose Gels

Agarose is a purified linear galactan hydrocolloid, isolated from agar or recovered directly from agar-bearing marine algae, such as the Rhodophyta. Genera from which agarose is extracted include *Gelidium, Gracilaria, Acanthopeltis, Ceramium, Pterocladia*, and *Campylaephora*. Agar is extracted from red algae by boiling in water and is then isolated by filtering off the particulates, gelling and freeze-thawing the colloid to remove water-soluble impurities, and precipitation with ethanol [18]. The resultant product is a mixture of polysaccharides, which are alternating copolymers of 1,4-linked 3,6-anhydro-α-L-galactose and 1,3-linked β-D-galactose. The repeating disaccharide is referred to as agarobiose [19]. Of course, this is an idealized structure, since even purified agarose contains appreciable amounts of the following substituents: sulfate, pyruvate, and methoxyl groups. Sulfate is generally esterified to the hydroxyl at C-4 of β-D-galactopyranose, and C-2 and C-6 of 3,6-anhydro-α-L-galactose. Pyruvate is linked to both C-4 and C-6 of some residues of β-D-galactopyranose. The resulting compound is referred to as 4,6-*O*-carboxyethylidene [20]. In addition, the hydroxyl group on C-6 of some residues of β-D-galactopyranose can also be methylated.

Agarose was first used to form gels for electrophoresis by HJERTÉN [21] and also for immunodiffusion [22] and chromatography [23]. Agarose can be separated from agaropectin by several procedures, including preferential precipitation of agaropectin by a quaternary ammonium salt, followed by fractionation of agarose with polyethylene glycol. By using a combination of polyethylene glycol precipitation and anion-exchange chromatography, an agarose with no detectable pyruvate and 0.02 % sulfate has been prepared [24]. A further treatment consists of alkaline desulfation in the presence of $NaBH_4$ followed by reduction with $LiAlH_4$ in dioxane [25]. When an agarose solution is cooled, during gelation, single strands of agarose (average molecular mass 120 kDa or ca. 392 agarobiose units) [26] dimerize to form a double helix. The double helix has a pitch of 1.9 nm, and each strand has threefold helical symmetry [27]. However, formation of the double helix is not sufficient to produce a gel; it is assumed that these helices aggregate laterally during gelation to form suprafibers (Fig. 1). In fact, the width of an agarose double helix is no more than 1.4 nm, whereas in gelled agarose, fibers one order of magnitude wider than this have been observed by electron microscopy and light scattering. The greater the number of aggregated double helices (i.e., the thicker the diameter of the "pillar"), the larger is the "pore" defined by the supporting beam [28]. Thus, it is reasonable to assume that it is the presence of these suprafibers that causes agarose gels to have strengths and pore sizes greater than those of gels that do not form suprafibers, such as starch and 2.5 – 5 % cross-linked polyacrylamide gels. In electrophoresis, agarose gels are used for fractionation of several types of particles, including serum lipoproteins [29], nucleic acids [30], virus and related particles [31], and subcellular organelles, such as microsomes. However, a suspension of 0.14 % agarose particles is used for the latter, rather than a coherent gel mass [32]. More recently, by ex-

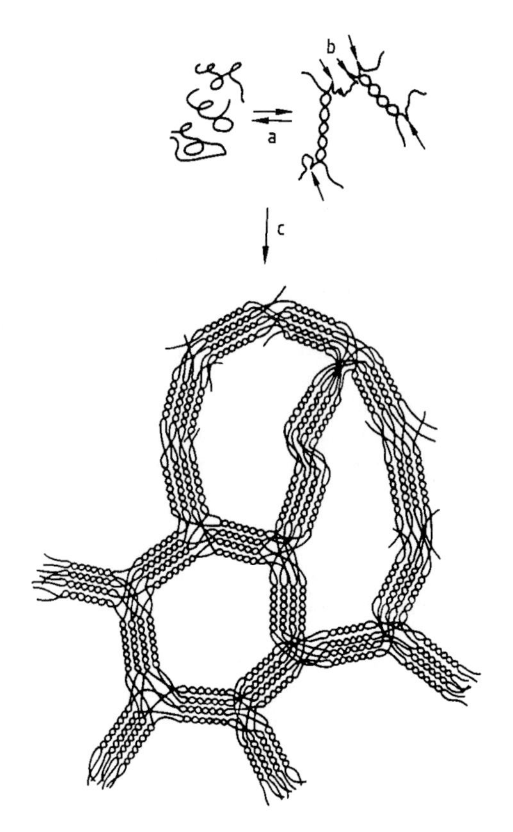

Figure 1. Formation of an agarose gel
Strands of agarose join to form double helices (a). At kinks (b) the strands exchange partners. Higher-order aggregation results in suprafibers (c). In hydroxyethylated agarose, it appears that the pillar looses its structure and more of the double helices are dispersed into the surrounding space: this results in a decrease in average pore size, i.e., higher sieving for the same % matrix as compared with underivatized agarose [31].

ploiting ultradilute gels (barely 0.03 % agarose, a special high-strength gel, Seakem LE, from Marine Colloids, Rockwille, ME, United States) SERWER et al. [33] have even been able to fractionate *E. coli* cells. Previously, this had been thought to be impossible, and in fact RIGHETTI et al. [34] reported a maximum pore diameter, for even the most dilute agaroses (0.16 %) of 500 – 600 nm. According to SERWER et al. [33] the effective pore radius P_E of a 0.05 % agarose gel is 1.2 μm; hence, the average pore of a 0.03 % gel should be several micrometers wide.

FMC BioProducts (formerly Marine Colloids, a division of FMC Corporation) of Rockland, ME, United States, has produced a number of highly sophisticated agaroses to meet the diverse requirements of electrophoresis users; the reader is referred to an extensive review on this topic [35]. Mixed-bed gels, containing 0.8 % agarose and variable amounts (2.5 – 9 %T) of acrylamide (%T indicates the total amount of monomers) have also been produced [36], [37]. Such gels were quite popular in the 1970s for nucleic acid fractionation, as they allowed fine resolution of lower M_r DNAs and RNAs, but they are seldom used in modern electrokinetic methods. However, these gels (under the trade name Ultrogels) in a spherical form are still popular for chromatographic purposes. Five types of bead exist, with a fractionation range of 1 – 1200 kDa [38].

14.3.3. Polyacrylamide Gels

The most versatile of all matrices are the polyacrylamide gels. Their popularity stems from several fundamental properties: (1) optical clarity, including ultraviolet (280 nm) transparency; (2) electrical neutrality, due to the absence of charged groups; and (3) availability in a wide range of pore sizes. Their chemical formula, as commonly polymerized from acrylamide and *N,N′*-methylenebisacrylamide (Bis), is shown in Figure 2, together with that of the two most widely employed catalysts, persulfate (ammonium or potassium) and *N,N,N′N′*-tetramethylethylenediamine (TEMED) [39]. Four groups simultaneously and independently developed polyacrylamide gel electrophoresis: RAYMOND and WEINTRAUB [40], DAVIS [41], ORNSTEIN [42], and HJERTÉN [43]. In addition, HJERTÉN [44] introduced polyacrylamide beads as a matrix for chromatography. Normal polyacrylamide gels (i.e., cross-linked with standard amounts of 3 – 5 % Bis) have porosities that decrease linearly with the monomer content (%T). Their porosities are substantially smaller than those of corresponding agarose gels. However, highly cross-linked gels (> 30 %C) can have porosities comparable to or even greater than those of agarose matrices (%C is the mass in grams of cross-linker per 100 g total monomers) [45], [46]. However, high-C polyacrylamide matrices are brittle and opaque, and tend to collapse and exude water above 30 %C [34]. Thus, at present their use is limited.

Hydrophilic gels are considered to be a network of flexible polymer chains with interstices into which macromolecules are forced to migrate by an applied potential difference, according to a partition governed by steric factors. Large mole-

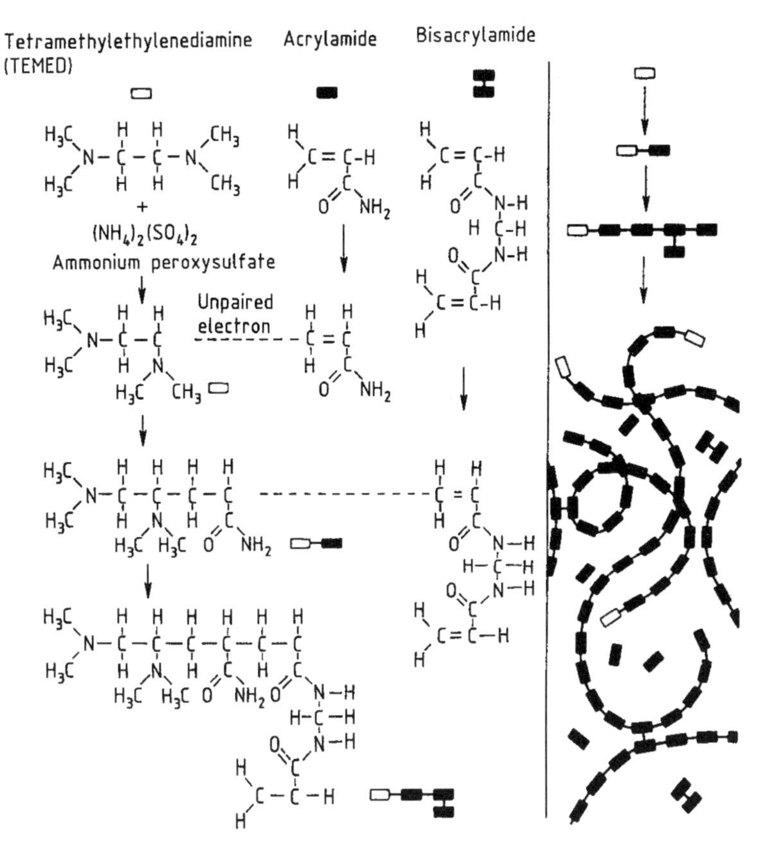

Figure 2. The polymerization reaction of acrylamide
The structure of acrylamide, *N,N'*-methylenebisacrylamide (Bis), and of a representative segment of cross-linked polyacrylamide are shown. Initiators shown are peroxysulfate and *N,N,N', N'*-tetramethylethylendiamine (TEMED).

cules can only penetrate into regions where the meshes in the net are large, while small molecules find their way into tightly knit regions of the network. Perhaps the best way to envision a hydrophilic matrix is to consider it as being composed of two interlaced fluid compartments, with different frictional coefficients [47]. In general, agaroses are more porous than polyacrylamides and these two matrices are used to complement each other.

The versatility of polyacrylamide gels is also shown by the large number of cross-linkers, besides Bis, that can be used to cast gels with particular properties for different fractionation purposes. *N,N'*-(1,2-dihydroxyethylene)bisacrylamide (DHEBA) can be used for casting reversible gels (i.e., gels that can be liquefied), since the 1,2-diol structure of DHEBA renders them susceptible to cleavage by oxidation with periodic acid [48]. The same principle should also apply to *N,N'*-diallyltartardiamide (DATD) gels [49]. Alternatively, ethylenediacrylate (EDA) gels [50] may be used, since this cross-linker contains ester bonds that undergo base-catalyzed hydrolytic cleavage. The poly(ethylene glycol) diacrylate cross-link belongs to the same series of ester derivatives [34]. As a further addition to the series, *N,N'*-bisacrylylcystamine (BAC) gels, which contain a disulfide bridge cleavable by thiols, have been described [51]. Practically any cross-linker can be used, but DATD and *N,N',N''*-triallylcitrictriamide (TACT) should be avoided, since, being allyl derivatives, they are inhibitors of gel polymerization when mixed with compounds containing acrylic double bonds. Their use at high %C is simply disastrous [52], [53].

Other monomers, in addition to acrylamide, have been described. One of them is Trisacryl, *N*-acryloyl-2-amino-2-hydroxymethyl-1,3-propanediol. The Trisacryl monomer creates a microenvironment that favors the approach of hydrophilic solutes (proteins) to the gel polymer surface, since the polyethylene backbone is buried underneath a layer of hydroxymethyl groups. In addition, owing to the much larger M_r of the monomer, as compared with acrylamide, gels made with the same nominal matrix content (%T) should be more porous, since they would have on average thicker fibers than the corresponding polyacrylamide matrix. This type of gel is extensively used by IBF Reactifs (VilleneuveLa-Garenne, France) for preparing gel filtration media or ion exchangers and also for electrophoresis [54], [55]. However, this monomer degrades with zero-order kinetics and hence seems to be of limited use in electrophoresis [56].

In 1984, ARTONI et al. [57] described two novel monomers that impart peculiar properties to polyacrylamide-type gels: acryloylmorpholine (ACM) and bisacrylylpiperazine (BAP). These gels exhibited some interesting features: owing to the presence of the morpholino ring on the nitrogen atom involved in the amido bond, the latter was rendered stable to alkaline hydrolysis, which bedevils conventional polyacrylamide matrices. In addition, such matrices are fully compatible with a host of hydro-organic solvents, thus allowing electrophoresis in mixed phases. The use of the BAP cross-linker has been reported to have beneficial effects when silver staining polyacrylamide matrices.

More recently, novel monomers combining high hydrophilicity with extreme hydrolytic stability have been described. One of them is *N*-acryloylaminoethoxyethanol (AAEE). As a free monomer, it has ten times higher resistance to hydrolysis than plain acrylamide, but as a polymer its stability (in 0.1 N NaOH, 70 °C) is 500 times higher [58]. The strategy: the ω-OH group in the *N*-substituent was sufficiently remote from the amido bond so as to impede formation of hydrogen bonds with the amido carbonyl group; if at all, the oxygen in the ethoxy moiety of the *N*-substituent would act as a preferential partner for hydrogen-bond formation with the ω-OH group. The second is acryloylaminopropanol (AAP), in which the ether group on the *N*-substituent was removed and the chain shortened [59]. As a result, this monomer was even more hydrophilic than AAEE and highly resistant to hydrolysis as compared to free acrylamide (hydrolysis constant of AAP $0.008\ L\,mol^{-1}\,min^{-1}$ versus $0.05\ L\,mol^{-1}\,min^{-1}$ for acrylamide, in an alkaline milieu). The AAP monomer was found to have a unique performance in DNA separations when used as a sieving liquid polymer in capillary zone electrophoresis (CZE) [60]. In addition, poly(AAP) has excellent performance in CZE at high temperatures (60 °C and higher), which are necessary for screening for DNA point mutations in temperature-programmed CZE [61]. It is known that acrylamide is a neurotoxin and possibly even (at high doses) a carcinogen. Now the situation could have dramatically changed: if we assess toxicity in terms of the ability of these monomers to alkylate free SH groups in proteins (thus forming a cysteinyl-*S*-β-propionamide adduct) and evaluate this reaction by high-resolution, delayed extraction MALDI-TOF MS (matrix-assisted laser desorption ionization time of flight mass spectrometry) both in the linear and reflectron mode, this picture will change substantially: at one end of the scale the major offenders remain indeed acrylamide and DMA (strongly alkylating, hence highly toxic agents), whereas at the opposed end (minimally alkylating) one can locate only AAP [62]. Tables 1 and 2 list a number of monomers and cross-linkers, respectively, in use today.

14.4. Discontinuous Electrophoresis

In 1959 RAYMOND and WEINTRAUB [40] described the use of polyacrylamide gels (PAG) in ZE, offering UV and visible transparency (starch gels are opalescent) and the ability to sieve macromolecules over a wide range of sizes. In 1964, ORNSTEIN [42] and DAVIS [41] created discontinuous (disc) electrophoresis by applying a series of discontinuities to PAG (leading and terminating ions, pH, conductivity, and porosity), thus further increasing the resolving power of the technique. In disc electrophoresis (Fig. 3) the proteins are separated on the basis of two parameters: surface charge and molecular mass. FERGUSON [63] showed that one can distinguish between the two by plotting the results of a series of experiments with polyacrylamide gels of varying porosity. For each protein under analysis, the slope of the curve log m_T (electrophoretic mobility) versus gel density (%T) is proportional to molecular mass, while the *y*-intercept is a measure of surface charge [64]. Nonlinear Ferguson plots have been reported [65], related to the reptation mode of DNA in sieving media.

Table 1. Monomers utilized for preparing polyacrylamide gels

Name	Chemical formula	M_r
N,*N*-Dimethylacrylamide (DMA)		99
N-Acryloyl-2-amino-2-hydroxymethyl-1,3-propane (Trisacryl)		175
N-Acryloylmorpholine		141
N-Acryloyl-1-amino-1-deoxy-D-glucitol		235
N-Acryloylaminoethoxyethanol		159
N-Acryloylaminopropanol		129
N-Acryloylaminobutanol		143

14.5. Isoelectric Focusing

New routes were explored toward a simpler assessment of charge and mass of proteins. As early as 1961 SVENSSON [66] experimented with isoelectric focusing (IEF), an electrophoretic technique by which amphoteric compounds are fractionated according to their isoelectric point (pI) along a continuous pH gradient. In contrast to ZE, where the constant pH of the separation medium establishes a constant charge density at the surface of the molecule and causes it to migrate with constant mobility (in the absence of sieving), the surface charge of an amphoteric compound in IEF keeps changing and decreasing according to its titration curve, as it moves along the pH gradient toward its steady-state position, i.e., the region where the pH matches its pI. There its mobility is zero and the molecule comes to a stop.

The gradient is created and maintained by the passage of current through a solution of amphoteric compounds with closely spaced pI values, encompassing a given pH range. The electrophoretic transport causes these buffers (carrier ampho-

Table 2. Cross-linkers described for gelling polyacrylamide matrices

Name	Chemical formula	M_r	n*
N,N'-Methylene bisacrylamide (Bis)	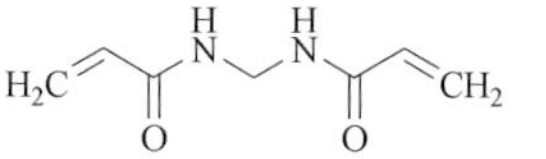	154	9
Ethylene diacrylate (EDA)	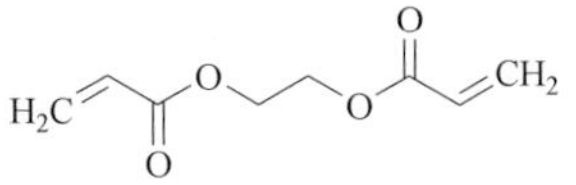	170	10
N,N'-(1,2-Dihydroxyethylene) bisacrylamide (DHEBA)	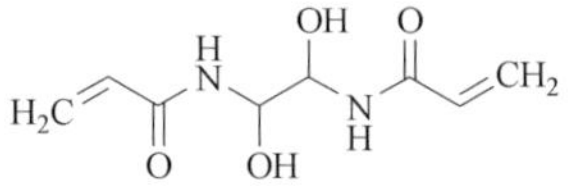	200	10
N,N'-Diallyltartardiamide (DATD)	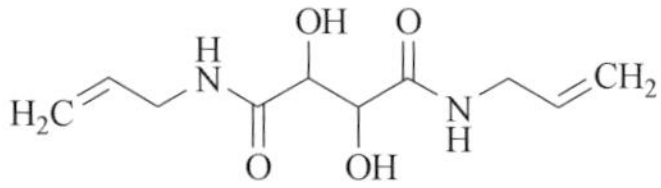	228	12
N,N',N''-Triallyl citric triamide (TACT)	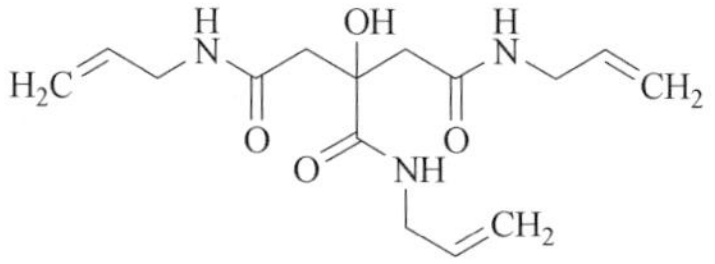	309	12–13
Polyethylene glycol diacrylate 200 ($PEGDA_{200}$)	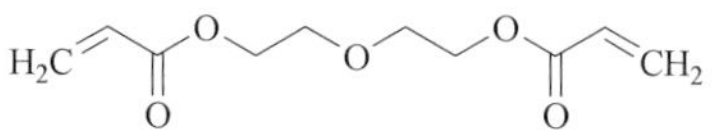	214	13
N,N'-Bisacrylylcystamine (BAC)	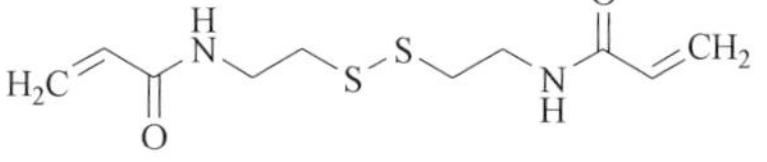	260	14
Polyethylene glycol diacrylate 400 ($PEGDA_{400}$)	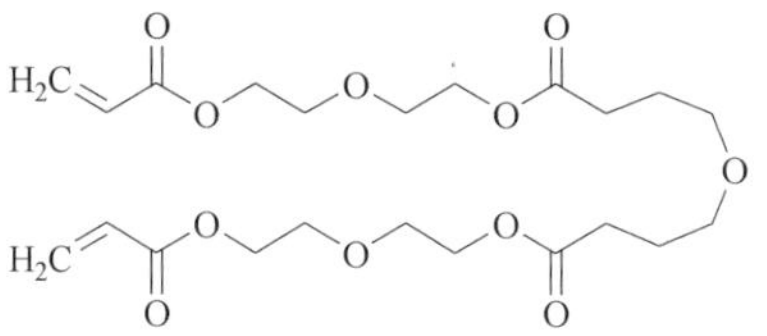	400	25

Table 2. (continued)

Name	Chemical formula	M_r	n*
N,N'-Bis-acrylylpiperazine (BAP)	$H_2C{=}CH{-}CO{-}N(CH_2CH_2)_2N{-}CO{-}CH{=}CH_2$	194	10

* Chain length.

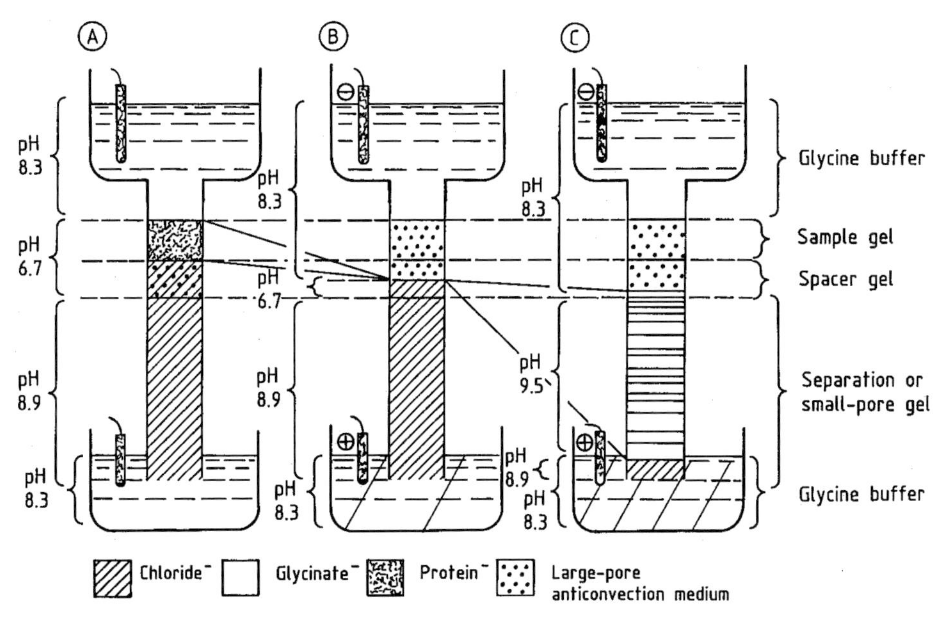

Figure 3. Principle of discontinuous electrophoresis
A) Sample in sample gel; B) Sample concentration in stacking gel; C) Sample separation in small-pore, separation gel [42]

lytes, CA) to stack according to their pI values, and a pH gradient, increasing from anode to cathode, is established. This process is shown in Figure 4, which gives the formulae of the CA buffers and portions of their titration curves in the proximity of the pI value. When the system has achieved a steady state, maintained by the electric field, no further mass transport is expected except from symmetric to-and-fro micromovements of each species about its pI, generated by the action of two opposite forces, diffusion and voltage gradient, acting on each focused component. Thus, this diffusion–electrophoresis process is the primary cause of the residual current under isoelectric steady-state conditions.

The technique applies only to amphoteric compounds and, more precisely, to good ampholytes with small $|pI\text{-}pK_1|$ values, i.e., with a steep titration curve around their pI (Fig. 4); the essential condition for any compound to focus in a narrow band. In practice, notwithstanding the availability of CAs covering the pH range 2.5–11, the limit of CA-IEF is in the pH 3.5–10 range, due to solvent conductivity outside this interval. When a restrictive support, such as polyacrylamide, is used, a limit also exists for the size of the largest molecules exhibiting an acceptable mobility through the gel. A conservative evaluation sets an upper limit around 750 kDa for standard techniques. The molecular form in which the proteins are separated

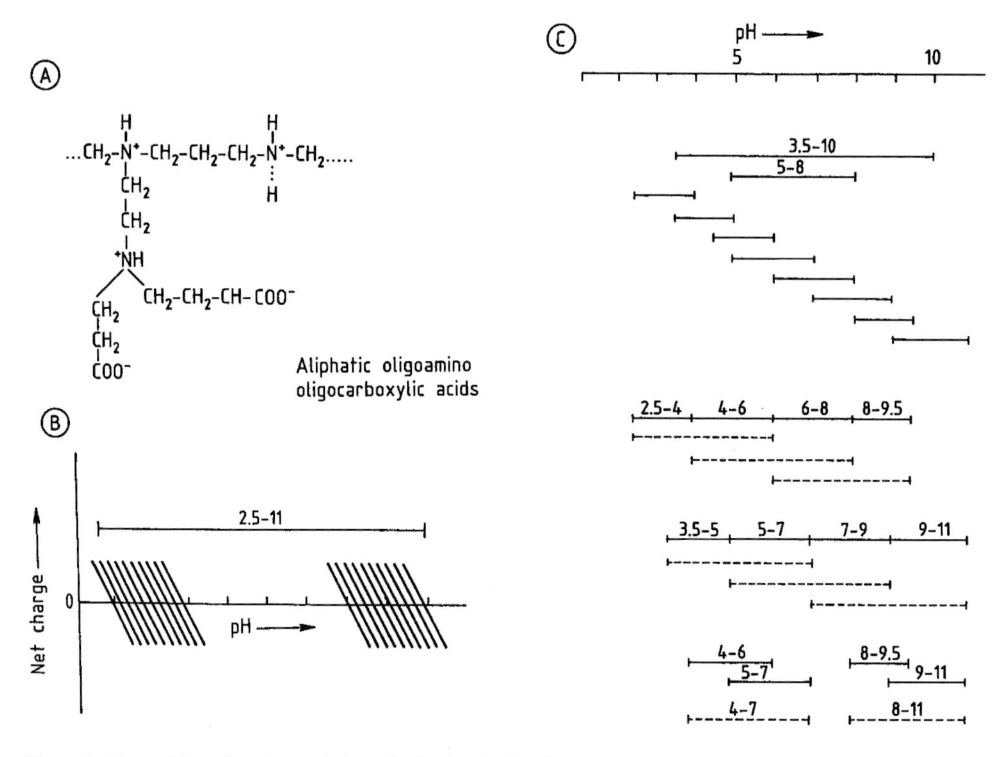

Figure 4. Composition of carrier ampholytes for isoelectric focusing
A) Representative chemical formula (oligoamino oligocarboxylic acids); B) Portions of titration curves of carrier ampholytes near pI; C) Narrow pH intervals, obtained by subfractionation of the wide pH 3.5 – 10 gradient components

depends strongly on the presence of additives, such as urea and/or detergents. Moreover, supramolecular aggregates or complexes with charged ligands can be focused only if their K_d is lower than 1 μmol/L and if the complex is stable at pH = pI. Aggregates with higher K_d are easily split by the pulling force of current, whereas most chromatographic procedures are unable to modify the native molecular form.

The general properties of the carrier ampholytes, i.e., of the amphoteric buffers used to generate and stabilize the pH gradient in IEF, can be summarized as follows:

1) Fundamental “classical” properties: (a) zero buffering ion mobility at pI; (b) good conductance; (c) good buffering capacity
2) Performance properties: (a) good solubility; (b) no influence on detection system; (c) no influence on sample; (d) separable from sample
3) “Phenomena” properties: (a) plateau effect, drift of the pH gradient; (b) chemical change in the sample; (c) complex formation

In chemical terms, CAs are oligoamino, oligocarboxylic acids, available from different suppliers under different trade names (Ampholine from LKB Produkter, Pharmalyte from Pharmacia Fine Chemicals, Biolyte from Bio-Rad, Servalyte from Serva, Resolyte from BDH). There are two basic synthetic approaches: (1) the Vesterberg approach [67], allowing different oligoamines (tetra-, penta-, and hexa-amines) to react with acrylic acid; and (2) the Pharmacia synthetic process [68], which involves the copolymerization of amines, amino acids, and dipeptides with epichlorohydrin. The wide-range synthetic mixture (pH 3 – 10) seems to contain hundreds, possibly thousands of different amphoteric chemicals, having pI values evenly distributed along the pH scale. CAs, from any source, should have an average molecular mass around 750 Da (range 600 – 900 Da, the higher values referring to the more acidic CA species). Thus, they should be easily separable from macromolecules by gel filtration, unless they are hydrophobically complexed to proteins. A further complication arises from the chelating effect of acidic CAs, especially toward Cu^{2+} ions, which could inactivate metalloenzymes. In addition, focused CAs represent a medium of very low ionic strength (less than 1 mequiv./L at steady state) [3], [69]. Since the isoelectric state involves a minimum of solvation and thus of solubility for the protein macroion, there could be a tendency for some proteins (e.g., globulins) to precipitate during IEF close to the pI. This is a severe problem in preparative IEF, but in analytical procedures it can be lessened by reducing the total amount of sample applied.

14.6. Sodium Dodecyl Sulfate Electrophoresis

Sodium dodecyl sulfate (SDS) electrophoresis fractionates polypeptide chains essentially on the basis of their size. It is therefore a simple, yet powerful and reliable method for molecular mass determination. In 1967 SHAPIRO et al. [70] first reported that electrophoretic migration in SDS is proportional to the effective molecular radius and, thus, to the M_r of the polypeptide chain. This means that SDS must bind to proteins and cancel out differences in molecular charge, so that all components migrate solely according to size. Surprisingly large amounts of SDS appear to be bound (an average of 1.4 g/g protein). This means that the number of SDS molecules bound is approximately half the number of amino acid residues in a polypeptide chain. This amount of highly charged surfactant molecules is sufficient to overwhelm effectively the intrinsic charges of the polymer coil, so that their net charge per unit mass becomes approximately constant. If migration in SDS (and disulfide reducing agents, such as 2-mercaptoethanol, in the denaturing step, for proper unfolding of the proteins) is proportional only to M_r, then, in addition to canceling out charge differences, SDS also equalizes molecular shape differences (e.g., globular vs. rod-shaped molecules). This seems to be the case for protein – SDS mixed micelles. These complexes can be assumed to behave as ellipsoids of constant minor axis (ca. 1.8 nm) and a major axis proportional to the length of the amino acid chain (i.e., molecular mass) of the protein. The rod length for the 1.4 g/g protein complex is of the order of 0.074 nm per amino acid residue. For further information on detergent properties, see HELENIUS and SIMONS [71].

In SDS electrophoresis, the proteins can be prelabeled with dyes that covalently bind to their NH_2 residues. The dyes can be conventional (e.g., the blue dye Remazol) or fluorescent, such as dansyl chloride, fluorescamine, *O*-phthaldialdehyde, and MDPF (2-methoxy-2,4-diphenyl-3[2*H*]-furanone). Prelabeling is compatible with SDS electrophoresis, as the size increase is minimal, but would be disastrous in disc electrophoresis or IEF, as it would generate a series of bands of slightly altered mobility or pI from an otherwise homogeneous protein.

For data treatment the sample and M_r standards are electrophoresed side-by-side in a gel slab. After detection of the polypeptide zones, the migration distance (or R_F) is plotted against log M_r to produce a calibration curve [72] from which the M_r of the sample can be calculated. Note that, in a gel of constant %T, linearity is obtained only in a certain range of molecular size. Outside this range a new gel matrix of appropriate porosity should be used. Two classes of protein show anomalous behavior in SDS electrophoresis: glycoproteins (because their hydrophilic oligosaccharide units prevent hydrophobic binding of SDS micelles) and strongly basic proteins, e.g., histones (because of electrostatic binding of SDS micelles through their sulfate groups). The first can be partially alleviated by using alkaline Tris/borate buffers [73], which will increase the net negative charge on the glycoprotein and thus produce migration rates well correlated with molecular size. The migration of histones can be improved by using pore gradient gels and allowing the polypeptide chains to approach the pore limit [74].

14.7. Porosity Gradient Gels

When macromolecules are electrophoresed in a continuously varying matrix concentration (which results in a porosity gradient), rather than in a gel of uniform concentration, the protein zones are compacted along their track, because the band front is, at any given time, at a gel concentration somewhat higher than at the rear of the band, so that the former is decelerated continuously. A progressive band sharpening thus results. There are other reasons for resorting to gels of graded porosity. Disc electrophoresis separates macromolecules on the basis of both size and charge differences. If the influence of molecular charge could be eliminated, then clearly the method could be used, with a suitable calibration, for measuring molecular size. This has been accomplished by overcoming charge effects in two main ways. In one, a relatively large amount of charged ligand, such as SDS, is bound to the protein, effectively swamping the initial charges present on the protein molecules and giving a quasi-constant charge/mass ratio. However, in SDS electrophoresis proteins are generally dissociated into their constituent polypeptide subunits, and the concomitant loss of functional integrity and antigenic properties cannot be prevented. Therefore, the size of the original, native molecule must be evaluated in the absence of denaturing substances.

The second method for M_r measurements relies on a mathematical canceling of charge effects, following measurements of the mobility of native proteins in gels of different concentrations. This is the so-called Ferguson plot [63], discussed in Chapter 14.4. A third method for molecular size measurements uses gels of graded porosity. This method is characterized by high resolving power and relative insensitivity to variations in experimental conditions. Under appropriate conditions (at least 10 kV · h), the mobility of most proteins becomes constant and eventually ceases as each constituent reaches a gel density region in which the average pore size approaches the diameter of the protein (pore limit) [75]. Thus, the ratio of the migration distance of a given protein to that of any other becomes constant after the proteins have all entered a gel region in which they are subjected to drastic sieving conditions. This causes the electrophoretic pattern to become constant after prolonged migration in a gel gradient. The gel concentration at which the migration rate for a given protein becomes constant is called the "pore limit": If this porosity is properly mapped with the aid of a suitable set of marker proteins, it is possible to correlate the migration distance to the molecular mass of any constituent in the mixture.

After electrophoresis is over and the proper experimental data are gathered, they can be handled by two-step or one-step methods. Among the former, the most promising approach appears to be that of LAMBIN and FINE [76], who observed that there is a linear relationship between the migration distance of proteins and the square root of electrophoresis time, provided that time is kept between 1 and 8 h. The slopes of the regression lines of each protein are an indication of molecular size. When the slopes of the various regression lines thus obtained are plotted against the respective molecular masses, a good linear fit is obtained, which allows molecular mass measurements of proteins between 2×10^5 and 10^6 Da. The shape of the proteins (globular or fibrillar), their carbohydrate content (up to 45 %), and their free electrophoretic mobilities (between 2.1 and 5.9×10^{-5} cm^2 V^{-1} s^{-1}) do not seem to be critical for proper M_r measurements by this procedure. One-step methods have been described by ROTHE and PURKHANBABA [77] who found that, when $\log M_r$ is plotted against either D (distance migrated) or %T (acrylamide + Bis), a nonlinear correlation is always obtained. However, when $\log M_r$ is plotted against $\sqrt{\%T}$ or $\sqrt{D}$, a linear regression line is obtained, which allows the accurate determination of M_r values of proteins (standard deviation ± 3.7 %; Fig. 5). The correlations $\log M_r - \sqrt{\%T}$ or $\log M_r - \sqrt{D}$ are not significantly altered by the duration of electrophoresis. Therefore, a constant M_r value should be obtained for a stable protein, no matter how long electrophoresis has been going on. ROTHE and MAURER [78] have demonstrated that the relationship $\log M_r - \sqrt{D}$ is also applicable to SDS electrophoresis in linear polyacrylamide gel gradients.

14.8. Two-Dimensional Maps (Proteome Analysis)

By sequentially coupling pure charge (IEF) to pure size fractionation (SDS-PAGE; the latter orthogonal to the first), one can distribute the polypeptide chains two-dimensionally, with charge and mass as coordinates (IEF-SDS or ISO-DALT, according to ANDERSON's nomenclature: ISOelectric for charge and DALT for mass separation) [79]. When the first dimension is performed in immobilized pH gradients, the technique is called IPG-DALT [4]. The technique was first reported by BARRETT and GOULD [80] and described in more detail by O'FARRELL [81], KLOSE [82], and SCHEELE [83]. Large gels (e.g., 30 × 40 cm) [84] and prolonged exposure to radiolabeled material (up to two months) have allowed resolution of as many as 12 000 labeled peptides in a total mammalian cell lysate. Thus, there is a good chance that, in a properly prepared 2-D map, a spot will represent an individual polypeptide chain, uncontaminated by other material. On this assumption, and provided that enough material is present in an individual spot (about 1 μg), it is possible to blot it

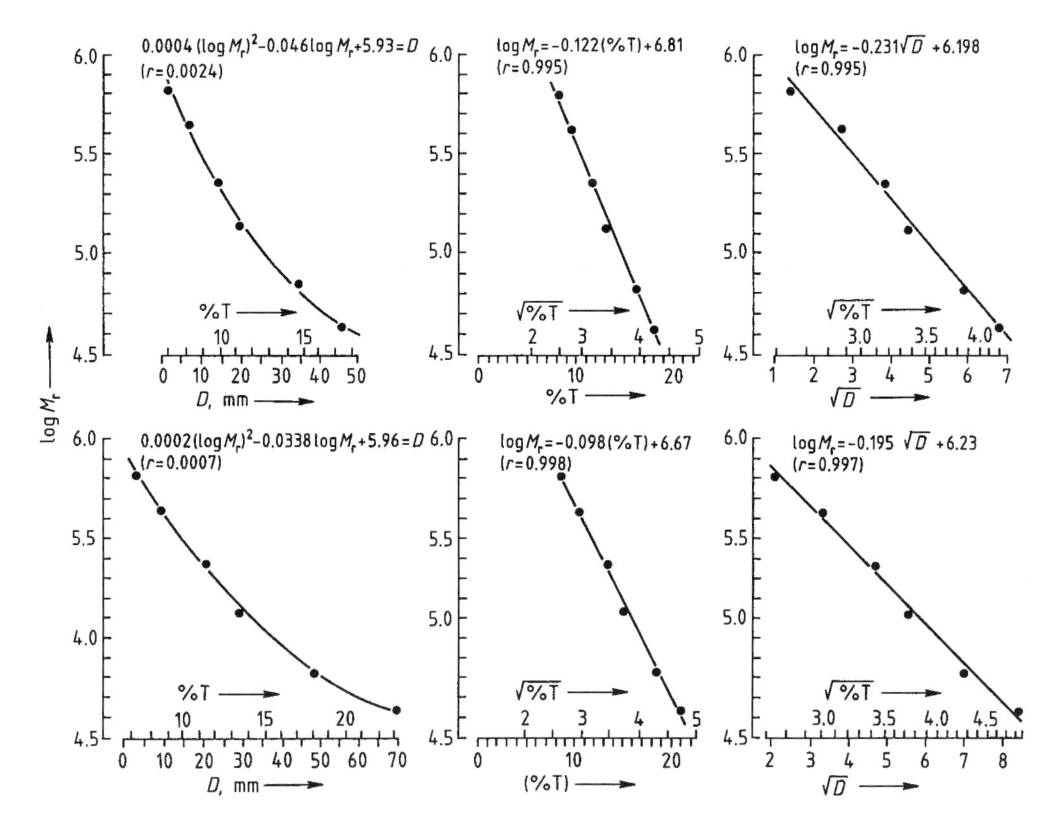

Figure 5. Typical plots of log M_r against migration distance D or gel composition %T after pore gradient electrophoresis Note that these plots are nonlinear, whereas when log M_r is plotted against $\sqrt{D}$ or $\sqrt{\%T}$ a linear relationship is obtained [77].

onto a glass fiber filter and perform sequencing on it [85].

Two-dimensional electrophoresis (2-DE) excited grandiose projects, like the Human Protein Index System of Anderson & Anderson [79], who initiated the far-reaching goal of mapping all possible phenotypes expressed by all the different cells of the human organism. An Herculean task, if one considers that there may be close to 75 000 such phenotypes (assuming a total of ca. 250 differentiated cells, each expressing > 3000 polypeptides, of which 300 are specific to a given cell). This started a series of meetings of the 2-DE group, particularly strong in the field of clinical chemistry [86], [87]. A number of books were devoted to this 2-DE issue [88], [89], and a series of meetings was started [90]. Today, 2-DE has found a proper forum in the journal *Electrophoresis*, which began hosting individual papers dealing with variagate topics in 2-D maps. Starting in 1988, *Electrophoresis* launched special issues devoted to 2-D maps, not only in clinical chemistry and human molecular anatomy, but in fact in every possible living organism and tissue, the first being dedicated to plant proteins [91]. Soon, a host of such "Paper Symposia" appeared, collecting databases on any new spots of which a sequence and a function could be elucidated, under the following editorships: Celis [92]–[101]; Dunn [102]–[105]; Lottspeich [106]; Tümmler [107]; Williams [108]; Humphery-Smith [109]; Appel et al. [110], [111] and Cash [112]. This collection of "Paper Symposia" is a gold mine of new information and novel evolutionary steps on the IEF technique (such as new solubilizers, new staining procedures, sample pretreatment before IEF and the like). The world "proteome" is a recent neologism that refers to PROTEins expressed by the genOME or tissue. Just as genome has become a generic term for "big-science" molecular biology, so proteome is becoming a synonym of bioinformatics, since this project requires the building and continuous updating of a vast body of information, such as the amino acid sequence and biological function of the newly discovered proteins. An idea of the complexity of this field is given in [113], [114]. A partial list of the many sites available on the WWW for consulting databases and for visualizing codified 2-D maps is given in the following:

1) A 2D PAGE protein database of *Drosophila melanogaster*
   http://tyr.cmb.ki.se/
2) ExPASy Server (2D liver, plasma, etc. SWISS-(PROT, 2DPAGE, 3DIMAGE), BIOSCI, Melanie software)
   http://expasy.ch/
3) CSH QUEST Protein Database Center (2D REF52 rat, mouse embryo, yeast, Quest software)
   http://siva.cshl.org/
4) NCI/FCRDC LMMB Image Processing Section (GELLAB software)
   http://www-lmmb.ncifcrf.gov/lemkin/gellab.html
5) 2-D Images Meta-Database
   http://www-lmmb.ncifcrf.gov/2dwgDB/
6) *E. coli* Gene-Protein Database Project-ECO2DBASE (in NCBI repository)
   ftp://ncbi.nlm.nih.gov/repository/ECO2DBASE/
7) Argonne Protein Mapping Group Server (mouse liver, human breast cell, etc.)
   http://www.anl.gov/CMB/PMG/
8) Cambridge 2D PAGE (a rat neuronal database)
   http://sunspot.bioc.cam.uk/Neuron.html
9) Heart Science Centre, Harefield Hospital (Human Heart 2D Gel Protein DB)
   http://www.harefield.mthames.nhs.uk/nhli/protein/
10) Berlin Human Myocardial 2D Electrophoresis Protein Database
   http://www.chemie.fu-berlin.de/user/pleiss/
11) Max Delbruck Ctr. for Molecular Medicine-Myocardial 2D Gel DB
   http://www.mdc-berlin.de/~emu/heart/
12) The World of Electrophoresis (EP literature, ElphoFit)
   http://www.uni-giessen.de/~gh43/electrophoresis.html
13) Human Colon Carcinoma Protein Database (Joint Protein Structure Lab)
   http://www.ludwig.edu.au/www/jpsl/jpslhome.html
14) Large Scale Biology Corp (2D maps: rat, mouse, and human liver)
   http://www.lsbc.com/
15) Yeast 2D PAGE
   http://yeast-2dpage.gmm.gu.se/
16) Proteome Inc. YPD Yeast Protein Database
   http://www.proteome.com/YPDhome.html
17) Keratinocyte, cDNA Database (Danish Centre for Human Genome Research)
   http://biobase.dk/cgi-bin/celis/
18) Institut de Biochemie et Genetique-Yeast 2D gel DB
   http://www.ibgc.u-bordeaux2.fr/YPM/
19) Swiss-Flash newsletter
   http://www.expasy.ch/swiss-flash/
20) Phosphoprotein Database
   http://www-lmmb.ncifcrf.gov/phosphoDB/

Figure 6 shows the complexity of one such 2-D map, downloaded from the ExPASy site, representing a partial panorama (limited to the pH gradient 5.5 – 8.5) of the thousands of polypeptide spots present in the human liver.

14.9. Isotachophoresis

The theoretical foundations of isotachophoresis (ITP) were laid in 1897 by KOHLRAUSCH [115], who showed that, at a migrating boundary between two salt solutions, the concentrations of ions were related to their effective mobilities (Kohlrausch autoregulating function) [116]. The term isotachophoresis underlines the most important aspect of this technique, namely the identical velocities of the sample zones at equilibrium [117]. ITP will take place when an electric field is applied to a system of electrolytes, consisting of:

1) A leading electrolyte, which must contain only one ionic species, the leading ion L^-, having the same sign as the sample ions to be separated, and an effective mobility higher than that of any sample ions
2) A second, terminating electrolyte, which contains one ion species, the terminating ion T^-, having the same sign as the sample ions to be separated, and an effective mobility lower than that of any sample ions
3) An intermediate zone of sample ions, having the same sign as the leading and terminating ions and intermediate mobilities

The three zones are juxtaposed, with the proviso that sharp boundaries must be created at the start of the experiment. The polarity of the electric field must be such that the leading ion migrates in front

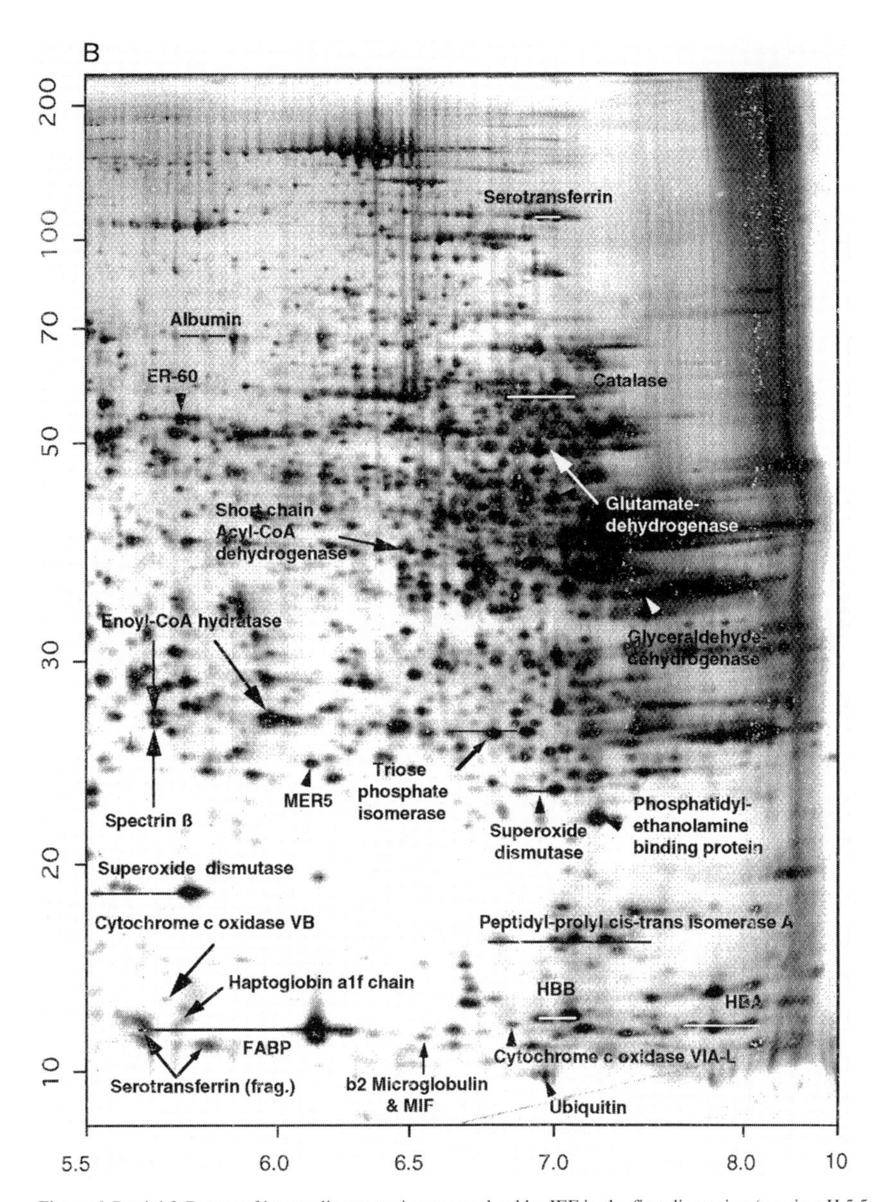

Figure 6. Partial 2-D map of human liver proteins, as resolved by IEF in the first dimension (*x*-axis, pH 5.5 – 8.5) and SDS-PAGE in the second dimension (*y*-axis, molecular mass).

of the ITP train at all times. When the system has reached the steady state, all ions move with the same speed, individually separated into a number of consecutive zones, in immediate contact with each other, and arranged in order of effective mobilities. Once the ITP train is formed, the ionic concentration in a separated sample zone adapts itself to the concentration of the preceding zone. The Kohlrausch function, which is given at the leading/terminating ion boundary, in fact gives the conditions at any boundary between two adjacent ions A^-, B^-, with one common counterion R^+ when the boundary migrates in the electric field.

There are two fundamental properties of ITP built into the autoregulating function: the concentrating effect and the zone-sharpening effect

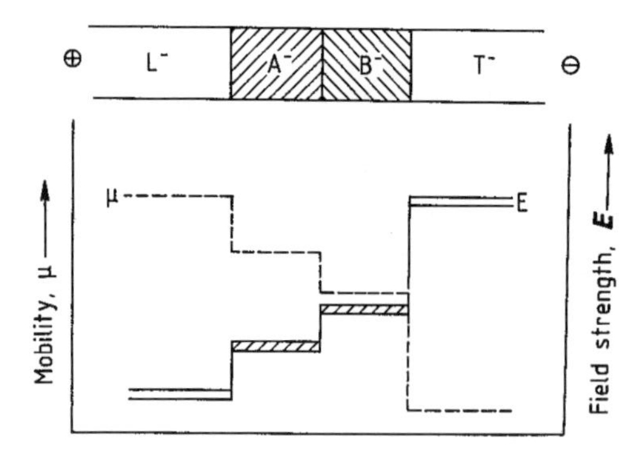

Figure 7. Scheme of an isotachophoretic process, with adjacent zones of leading ion L^-, samples A^- and B^- and terminating ion T^-
The respective mobilities (dotted line) and voltage gradients (solid line) in each zone are represented.

(Fig. 7). Suppose component A^- is introduced at very low concentration, even lower than that of the terminating ion T^-. Since the mobility of A^- is intermediate between that of L^- and T^-, its concentration must also be intermediate between those of L^- and T^-. This will result in A^- being concentrated (decrease in zone length) until it reaches the theoretically defined concentration. Conversely, if the concentration of A^- is too high (even higher than that of L^-), the A^- zone will be diluted (increase in zone length) until the correct equilibrium concentration is reached. This is an example of the zone-sharpening effect: if A^- diffuses into the B^- zone, it is accelerated, since it enters a zone of higher voltage gradient, therefore, it automatically migrates back into its zone. Conversely, if it enters the L^- zone, it finds a region of lower voltage, is decelerated, and thus falls back into its zone (Fig. 7). This applies to all ions in the system. Note that the first few minutes of disc electrophoresis, the period of migration in the sample and stacking gels, represent in fact an ITP migration. This is what produces the spectacular effects of disc electrophoresis: stacking into extremely thin starting zones (barely a few micrometers thick, from a zone up to 1 cm in thickness, a concentration factor of 1000 – 10 000) and sharpening of the zone boundaries. A characteristic of ITP is that the peaks, unlike those in other separation techniques (except for displacement chromatography), are square-shaped as a result of the Kohlrausch autoregulating function, i.e., the concentration of the substance within a homogeneous zone is constant from front to rear boundary [118]. In fact, under ideal conditions, the diffusive forces, which cause an eluted peak to spread into a Gaussian shape, are effectively counteracted. Although ITP is not much in vogue today, it is now used for stacking purposes as a transient step in CZE, for concentrating and sharpening dilute sample zones [119].

14.10. Immunoelectrophoresis

The basic technique of immunoelectrophoresis was first described by GRABAR and WILLIAMS in 1953 [9]. The antigen sample is applied to a hole in the middle of a glass plate, coated with a 1-mm layer of agarose gel. Electrophoresis separates the antigen mixture into various zones. A longitudinal trough, parallel to the long edge of the plate, is then made in the gel and filled with the antiserum to the antigen mixture. A passive, double-diffusion takes place; the antiserum diffuses into the gel (advancing with a linear front) while the antigens diffuse radially in all directions from the electrophoresis zones. The antigen – antibody complexes are formed at equivalence points, the number of precipitates formed corresponding to the number of independent antigens present. The precipitates form a system of arcs, resulting from the combination of linear and circular fronts. A number of variants have been described over the years, but in this chapter only some of the most popular immunoelectrophoretic techniques are reviewed. For a more extensive treatise, see [126] – [128].

14.10.1. Rocket Immunoelectrophoresis

In rocket immunoelectrophoresis, the antigen – antibody reaction occurs during the electrophoresis of an antigen in an antibody-containing medium. Both antigen and antibody move according to their electrophoretic mobilities, and they also react with each other, resulting in flame-shaped precipitation zones of antigen – antibody complexes (LAURELL's rocket technique) [120]. Under the influence of the electric field, unbound antigen within the peak of the flame-shaped precipitate migrates into the precipitate, which redissolves in the excess antigen. Thus, the leading edge of the flame is gradually displaced in the direction in which the antigen is migrating. The amount of antigen within the leading boundary edge is successively diminished because of the formation of soluble antigen – antibody complexes. When the antigen is diminished to equivalence with the antibody, the complexes can no longer be dissolved and a stable precipitate forms at the leading edge, which is thereafter stationary. The distance finally traveled by the peak depends on the relative excess of antigen over antibody and can be used as a measure of the amount of antigen present. Every precipitation band featured as a flame represents an individual antigen [120].

14.10.2. Crossed Immunoelectrophoresis

In the case of polydisperse samples, crossed immunoelectrophoresis (CIE) as described by CLARKE and FREEMAN [121] is employed. First, a mixture of protein antigens is subjected to conventional electrophoresis in agarose. Then the agarose gel is cut into strips, which are transferred to a second glass plate. Melted agarose (1 % w/v) containing the antiserum is then cast to form a gel of the same thickness as that of the first-dimension gel, and a secure junction between the two layers is achieved by melting to fuse their edges. During the second-dimension electrophoresis, performed perpendicular to the first, each antigen migrates independently, and precipitation zones are formed which resemble the Laurell rockets but have a wider base. Remarkably, two antigens present in a single zone, i.e., possessing equal electrophoretic mobilities, can be distinguished by the second-dimension electroimmunoassay as the shape and height of the respective rockets generally do not coincide exactly. In a typical product of the Clarke and Freeman method, ca. 50 rockets can be counted in human serum, and their relative abundance can be assessed by measuring the areas of the respective peaks [121]. Peak assessment and evaluation have been greatly facilitated by the development of a computer system for the specific analysis of CIE patterns [122].

14.10.3. Tandem Crossed Immunoelectrophoresis

It would be quite difficult to identify a single antigen in the complex patterns of CIE. One of the proposed methods for complex mixtures is the tandem CIE technique. Before the electrophoresis in the first dimension, two wells rather than one are cut into the gel strip, positioned one after the other in the direction of electrophoretic migration. One well is loaded with a mixture of the antigens to be analyzed, while the pure antigen whose peak is to be identified in the mixture is introduced into the other. The remaining manipulations are performed according to classical CIE. In the final pattern there will be two peaks which fuse smoothly and are separated by exactly the distance between the two wells. These two peaks may be of different heights (owing to difference in antigen concentration in the sample and reference wells), whereas fusion between them is indicative of the fact that they are caused by the same protein antigen. It is this canceling of the inner flanks of the two rockets and the fusion process that allow the unknown antigen to be located. By repeating this process with different purified antigens, it is possible to map, one by one, the components of a heterogeneous mixture [123].

14.10.4. Intermediate Gel Crossed Immunoelectrophoresis

Tandem CIE aimed at the identification of a single antigen in a mixture requires the availability of a pure marker antigen (see Section 14.10.3). In an alternative approach, identification can also be made when monospecific antiserum to the antigen is available. This approach is called intermediate gel CIE. After the first-dimension electrophoresis, the slab is prepared as for conventional CIE. Then a 1- to 2-cm strip of polyspecific antiserum-containing agarose gel nearest to the first-dimension gel is excised and a monospecific antiserum-containing agarose is cast instead. In the course of the second-dimension electrophoresis, the antigen to be detected will precipitate with the specific antiserum just as it enters the intermediate gel, while other antigens will pass through this gel without being retarded. The bases of their precipitation peaks, formed in the polyspecific antiserum-containing agarose, will be positioned on the borderline between the two latter gels. The antigen under study is thus distinguishable from the others [124].

14.10.5. Fused-Rocket Crossed Immunoelectrophoresis

In contrast to the Laurell rocket technique, this method is not used for quantitative determinations. Rather, it is applied to detect heterogeneity in seemingly homogeneous protein fractions, obtained by gel chromatography or ion exchange chromatography. A set of wells is punched in a checkerboard pattern in a strip of antibody-free agarose gel. Aliquots of each fraction eluted from the column are placed in the wells and the gel is left in a humid chamber for 30 – 45 min to allow the proteins in the wells to migrate and fuse, thus reproducing the continuous elution profile on an extended scale (hence the term "fused rockets"). On the remainder of the plate a thin layer of agarose is cast containing a polyvalent antiserum, and electrophoresis is performed as usual. If the collected chromatographic peak is indeed homogeneous, a single fused rocket will appear, while inhomogeneity will be revealed by an envelope of subpeaks within the main eluate fraction [125].

14.11. Staining Techniques and Blotting

MERRIL et al. [129] described a silver-staining procedure in which the sensitivity — which in Coomassie Blue is merely on the order of a few micrograms per zone — is increased to a few nanograms of protein per zone, thus approaching the sensitivity of radioisotope labeling. In the gilding technique of MOEREMANS et al. [130] polypeptide chains are coated with 20-nm particles of colloidal gold and detected with a sensitivity of < 1 ng/mm^2. Proteins can also be stained with micelles of Fe^{3+}, although with a sensitivity about one order of magnitude lower than with gold micelles [131]. These last two staining techniques became a reality only after the discovery of yet another electrophoretic method, the so-called Southern [132] and Western [133] blots, in which nucleic acids or proteins are transferred from hydrophilic gels to nitrocellulose or any of a number of other membranes, where they are immobilized by hydrophobic adsorption or covalent bonding. The very large porosity of these membranes makes them accessible to colloidal dyes. In addition, transfer of proteins to thin membranes greatly facilitates detection by immunological methods. This has resulted in new, high-sensitivity methods called "immunoblotting." After saturation of potential binding sites, the antigens transferred to the membrane are first made to react with a primary antibody, and then the precipitate is detected with a secondary antibody, tagged with horseradish peroxidase, alkaline phosphatase, gold particles, or biotin, which is then allowed to react with enzyme-linked avidin [134]. In all cases the sensitivity is greatly augmented. It appears that immunoblotting will pose a serious threat to quite a few of the standard immunoelectrophoretic techniques [135]. In terms of colloidal staining, a direct staining method for polyacrylamide gels with colloidal particles of Coomassie Blue G-250 is said to be as sensitive as silver staining [136]. For an extensive review covering all aspects of staining techniques, together with a vast bibliography on polypeptide detection methods, see [37].

14.12. Immobilized pH Gradients

In 1982, immobilized pH gradients (IPG) were introduced, resulting in an increase in resolution by one order of magnitude compared with conventional IEF [138]. Conventional IEF was besieged by several problems: (1) very low and unknown ionic strength; (2) uneven buffering capacity; (3) uneven conductivity; (4) unknown chemical environment; (5) nonamenable to pH gradient engineering; (6) cathodic drift (pH gradient instability); and (7) low sample loadability. In particular, a most vexing phenomenon was the near-isoelectric precipitation of samples of low solubility at the isoelectric point or of components present in large amounts in heterogeneous samples. The inability to reach stable steady-state conditions (resulting in a slow pH gradient loss at the cathodic gel end) and to obtain narrow and ultranarrow pH gradients, aggravated matters. Perhaps, most annoying was the lack of reproducibility and linearity of pH gradients produced by the so-called carrier ampholyte buffers [3]. IPGs proved able to solve all these problems.

IPGs are based on the principle that the pH gradient, which exists prior to IEF itself, is copolymerized, and thus rendered insoluble, within the fibers of a polyacrylamide matrix (Fig. 8). This is achieved by using, as buffers, a set of eight commercial chemicals (known as Immobilines, by analogy with Ampholine, produced by Pharmacia-LKB Biotechnologies, Uppsala, Sweden) having pK values distributed over the pH range 3.1 – 10.3. They are acrylamido derivatives with the general formula $CH_2=CH–CO–NH–R$, where R denotes one of three weak carboxyl groups, with pK values of 3.1, 3.6, and 4.6, for the acidic compounds, or one of five tertiary amino groups, with pK values of 6.2, 7.0, 8.5, 9.3, and 10.3, for the basic buffers [139]. Two additional compounds are needed: a strongly acidic (pK 1.0) and a strongly basic (p$K > 12$) titrant for producing linear pH gradients

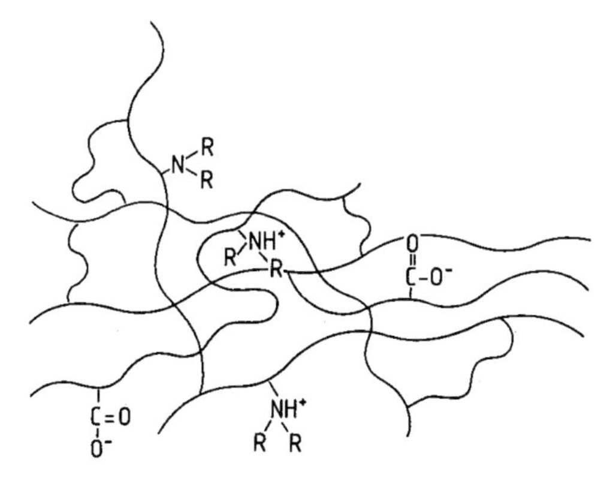

Figure 8. Scheme of a polyacrylamide matrix with a grafted pH gradient, depicted as bound negative and positive charges The ratio of such charges in each gel layer between anode and cathode defines a unique pH value.

covering the entire pH range 3–10 [140]. Computer simulations had shown that, in the absence of these two titrants, extended pH intervals would exhibit strong deviations from linearity at the two extremes, as the most acidic and most basic of the commercial Immobilines would act simultaneously as buffers and titrants [141]. The 2-acrylamidoglycolic acid (p*K* 3.1) species [142] is useful for separating strongly acidic proteins, since it extends the pH gradient to as low as pH 2.5. *N,N'*-Diethylaminopropylacrylamide (p*K* 10.3) has been used for analysis of strongly alkaline proteins [143]. Given the fairly evenly spacing of the p*K* values along the pH scale, it is clear that the set of ten chemicals proposed here (eight buffers and two titrants) is quite adequate to ensure linear pH gradients in the range pH 2.5–11 (the ideal ΔpK for linearity would be 1 pH unit between two adjacent buffers). The rule $\Delta pK = 1$ is fairly well obeyed, except for two "holes" between p*K* 4.6 and 6.2 and between p*K* 7.0 and 8.5. For a more detailed treatise on how to use an IPG gel, and IPG recipes, the reader is referred to an extensive manual [4] and to reviews [144]–[148]. Owing to the much increased resolution of IPG, quite a number of so-called electrophoretically silent mutations (bearing amino acid replacements with no ionizable groups in the side chains) have now been fully resolved.

14.13. Capillary Zone Electrophoresis

Capillary zone electrophoresis (CZE) appears to be a most powerful technique, perhaps equaling the resolving power of IPG. If one assumes that longitudinal diffusion is the only significant source of zone broadening, then the number of theoretical plates *N* in CZE is given by [149]:

$$N = \mu V / 2D$$

where μ and D are the electrophoretic mobility and diffusion coefficient, respectively, of the analyte, and V is the applied voltage. This equation shows that high voltage gradients are the most direct way to high separation efficiencies. For nucleic acids, it has been calculated that N could be as high as 10^6 theoretical plates. Figure 9 is a schematic drawing of a CZE system, illustrating one of the possible injection systems, namely gravity feed [150]. The fused-silica capillary has a diameter of 50–100 μm and a length up to 1 m. It is suspended between two reservoirs, connected to a power supply that is able to deliver up to 30 kV (typical operating currents 10–100 μA). One of the simplest ways to introduce the sample into the capillary is by electromigration, i.e., by dipping the capillary extremity into the sample reservoir, under voltage, for a few seconds. Detection is usually accomplished by on-column fluorescence and/or UV absorption. Conductivity and thermal detectors, as usually employed in ITP, exhibit too low a sensitivity in CZE. The reason for this stems from the fact that the flow cell where sample monitoring occurs has a volume of barely 0.5 nL, allowing sensitivities down to the femtomole level. In fact, with laser-induced fluorescence detection, a sensitivity on the order of 10^{-21} mol is claimed [151]. By forming a chiral complex with a component of the background electrolyte (copper aspartame) it is possible to resolve racemates of amino acids [8]. Even neutral organic molecules can be made to migrate in CZE by complexing them with charged ligands, such as SDS. This introduces a new parameter, a hydrophobicity scale, in electrokinetic migrations. For more on CZE, readers are referred to the Proceedings of the International Symposia on High-Performance Capillary Electrophoresis [152]–[162], which give an account of the evolution of and new developments in the CZE technique, spanning a ten-year period. Other special issues dedicated to CZE can be found in *J. Chromatogr. B* [163] and *J. Biochem. Biophys. Methods* [164]. In addition, the

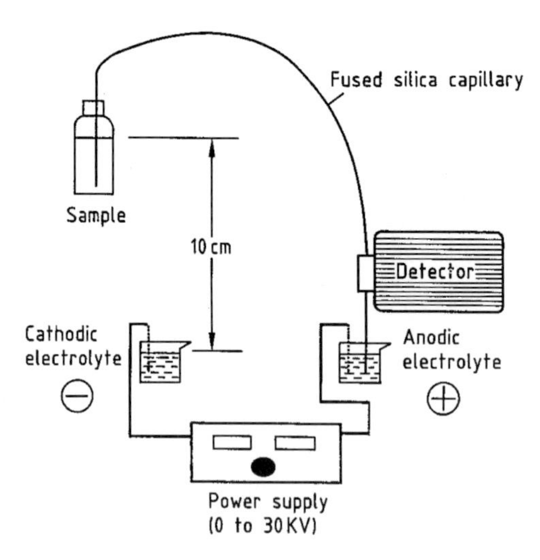

Figure 9. Scheme of a capillary zone electrophoresis apparatus
The detector includes a high-pressure mercury – xenon arc lamp, oriented perpendicular to the migration path, at the end of the capillary. The sample signal is measured with a photomultiplier and a photometer connected to the analog/digital converter of a multifunction interface board, connected to a microcomputer.

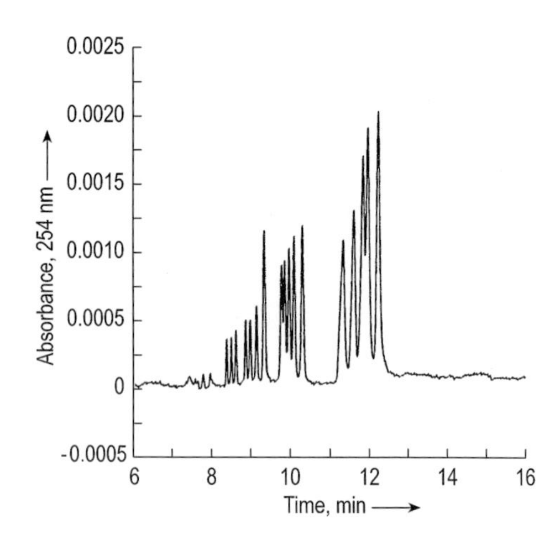

Figure 10. CZE of DNA marker V in isoelectric buffers. Conditions: 150 mM His buffer, pH=pI=7.6 in presence of 1.5 % poly(DMA); poly(AAP)-coated capillary, i.d. 100 μm, total length 30 cm (25.4 cm to the detector). UV absorbance detection at 254 nm. Run at 25 °C and 100 V/cm [174].

journal Electrophoresis has devoted a number of Paper Symposia to various aspects of CZE [165] – [171]. Two books which offer a broad coverage of the variegated aspects of CZE are [172] and [173].

It is of interest to note here some aspects of the CZE technique, which offers a unique performance, unrivalled by other electrokinetic methods, except for 2-D maps. These aspects regard the use of sieving liquid polymers instead of true gels for separation of macromolecules, notably proteins and nucleic acids. Any polymer above the entanglement threshold can exhibit sieving and separate macro-analytes almost as efficiently as true gels. Such polymers embrace the classical polyacrylamides, as well as a number of others, such as celluloses, poly(vinyl alcohol), and poly(vinyl pyrrolidone). The advantages: after each run, the polymers can be expelled from the lumen of the capillary and thus each new run can be performed under highly controlled and reproducible starting conditions, in the absence of carryover from previous runs. In addition, all celluloses are transparent to low wavelengths, down to 200 nm, thus permitting on-line detection of proteins and peptides via the adsorption of the peptide bond, with much increased sensitivity. Other interesting, recent developments regard the use of isoelectric buffers, i.e., amphoteric species as the sole buffering ion at pH = pI. Such buffers, due to their very low conductivity, allow delivery of high voltage gradients, with much increased resolution and very short analysis times. Examples on separations of nucleic acids in isoelectric His [174] and of peptides in isoelectric Asp [175] are given in Figures 10 and 11, respectively.

14.14. Preparative Electrophoresis

Several variants of electrophoresis have been described for preparative runs. They utilize either completely fluid beds or gel phases. This chapter restricts itself to those techniques exploiting isoelectric focusing principles, both in carrier ampholytes or in immobilized, nonamphoteric buffers, since they couple, in general, high loading capacities to a high resolving power. A classical example involves vertical columns filled with a sucrose density gradient, which is then emptied and collected in fractions at the end of the IEF run. This was Rilbe's original idea for performing IEF [176]; however, even though LKB has sold more than 3000 columns over the years, they are seldom used today. The following sections describe a few methods which appear to be more accessible; a complete survey of all other variants is available [3].

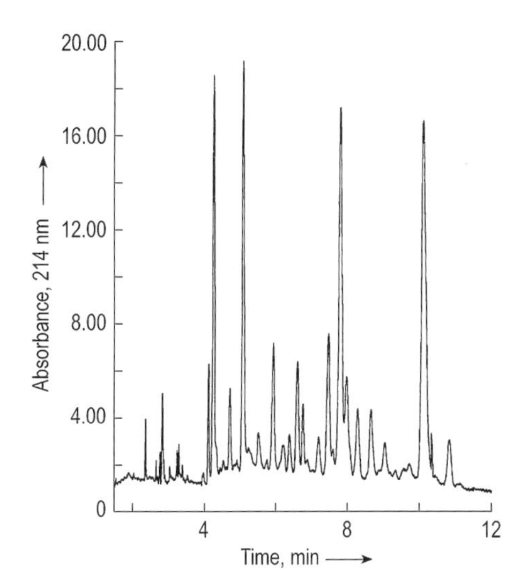

Figure 11. CZE of a tryptic digest of β-casein in a 37 × 100 μm i.d. capillary, bathed in 50 mM isoelectric aspartic acid (pH approximating the pI value of 2.77) added with 0.5 % hydroxyethyl cellulose. Run: 600 V (current 58 μA). The three major peptides (pI 6.1, 6.93, and 3.95) were eluted and analyzed by mass spectrometry [175].

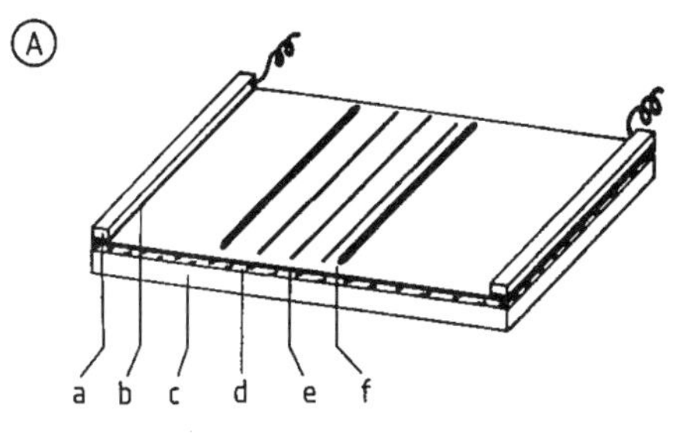

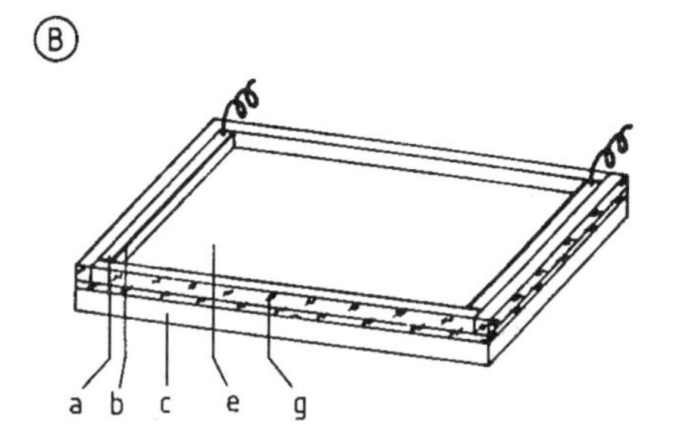

Figure 12. Preparative IEF in granulated Sephadex layers
A) Small-scale separation; B) Large-scale separation
a) Electrode; b) Filter paper pad soaked in electrode solution; c) Cooling block; d) Glass plate; e) Gel layer; f) Focused proteins; g) Trough [178].

14.14.1. Preparative Isoelectric Focusing in Granulated Gel Layers

RADOLA [177], [178] described a method for large-scale preparative IEF in troughs coated with a suspension of granular gels, such as Sephadex G-25 superfine (7.5 g/100 mL), Sephadex G-200 superfine (4 g/100 mL), or Bio Gel P-60 (4 g/100 mL). The trough consists of a glass or quartz plate at the bottom of a Lucite frame (Fig. 12). Various sizes of trough may be used; typical dimensions are 40 × 20 cm or 20 × 20 cm, with a gel layer thickness of up to 10 mm. The total gel volume in the trough varies from 300 to 800 mL. A slurry of Sephadex (containing 1 % Ampholine) is poured into the trough and allowed to dry in air to the correct consistency. Achieving the correct consistency of the slurry before focusing seems to be the key to successful results. Best results seem to be obtained when 25 % of the water in the gel has evaporated and the gel does not move when the plate is inclined at 45°. The plate is run on a cooling block maintained at 2 – 10°C. The electric field is applied via flat electrodes or platinum bands which make contact with the gel through absorbent paper pads soaked in 1 mol/L sulfuric acid at the anode, and 2 mol/L ethylenediamine at the cathode. In most preparative experiments, initial voltages of 10 – 15 V/cm and terminal voltages of 20 – 40 V/cm are used. Power gradients as high as 0.05 W/cm are well tolerated in 1-cm thick layers. Samples may be mixed with the gel suspension or added to the surface of preformed gels, either as a streak or from the edge of a glass slide. Larger sample volumes may be mixed with the dry gel (ca. 60 mg Sephadex per mL) and poured into a slot in the gel slab. Samples may be applied at any position between the electrodes. After focusing, proteins are located by the paper print technique. Focused proteins in the surface gel layer are absorbed onto a strip of filter paper which is then dried at 110°C. The proteins may then be stained directly with dyes of low sensitivity, such as Light Green SF or Coomassie Violet R-150, after removing ampholytes by washing in acid. The pH gradient in the gel can be measured in situ with a combination microelectrode sliding on a calibrated ruler. Radola's technique offers the advantage of combining high resolution, high sample load, and easy recovery of focused components. As much as 5 – 10 mg/mL protein per milliliter gel suspension can be fractionated over wide pH ranges. Purifi-

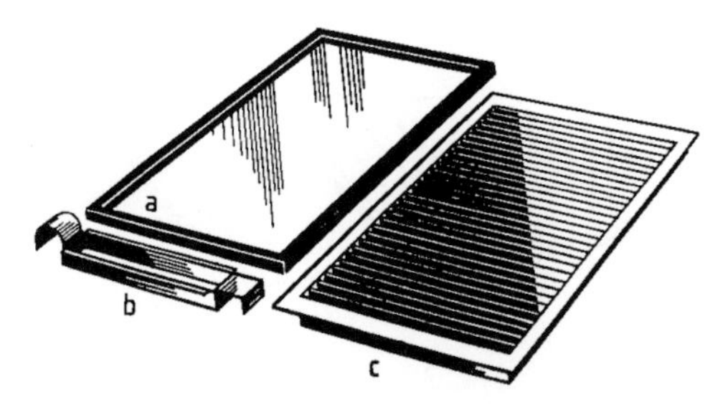

Figure 13. Trough (a), sample applicator (b), and fractionation grid (c) for preparative IEF in granulated gel layers (LKB Multiphor 2117 apparatus)

cation of 10 g pronase E in 800 mL gel suspension has been reported. At these high protein loads, even colorless samples can be easily detected, since they appear in the gel as translucent zones. As there is no sieving effect for macromolecules above the exclusion limits of the Sephadex, high molecular mass substances, such as virus particles, can be focused without steric hindrance. The system has a high flexibility, since it allows analytical and small- and large-scale preparative runs in the same trough, merely by varying the gel thickness.

For recovery of protein bands at the end of the IEF run, a fractionation grid pressed onto the gel layer allows recovery of 30 fractions, i.e., one fraction each 8 mm along a separation axis of 25 cm (Fig. 13). After scraping off each fraction with a spatula, the protein is eluted by placing the gel into a syringe equipped with glass wool as bottom filter, adding a buffer volume, and ejecting the eluate with the syringe piston.

14.14.2. Continuous-Flow, Recycling Isoelectric Focusing

Another interesting approach, called recycling IEF, has been described by BIER's group [179], [180]. It is well known that continuous-flow techniques, which appear essential for large-scale preparative work, are disturbed by parabolic and electrosmotic flows, as well as by convective flows due to thermal gradients. BIER et al. have improved this system by separating the actual flow-through focusing cell, which is miniaturized, from the sample and heat-exchange reservoir, which can be built up to any size. Minimization of parabolic flow, electrosmosis, and convective liquid flow is achieved by flowing the sample to be separated through a thin focusing cell (the actual distance from anode to cathode is only 3 cm) built of an array of closely spaced filter elements oriented parallel to the electrodes and parallel to the direction of flow. Increased sample load is achieved by recirculating the process fluid through external heat-exchange reservoirs, where the Joule heat is dissipated. During each passage through the focusing cell, only small sample migrations toward their eventual pl are obtained, but through recycling a final steady state is achieved. The IEF cell has ten input and output ports for sample flow-through, monitored by an array of ten miniaturized pH electrodes and ten UV sensors. The entire system is controlled and operated by a computer. A scheme of the entire apparatus can be seen in Figure 14. By activating pumps at two extreme channels, the computer can alter the slope of the pH gradient, counteracting any effect of cathodic drift, which results in a net migration of the sample zones towards the cathode.

14.14.3. The Rotofor

The Rotofor cell is a preparative-scale, free-solution, isoelectric focusing apparatus that provides a useful alternative to the recycling system described above. This cell [181] consists of a cylindrical focusing chamber that rotates around its axis, achieving liquid stabilization, not by recycling, but by gravity stabilization, as reported in 1967 by HJERTÉN [182]. The inside of the cylindrical focusing apparatus is divided by 19 parallel, monofilament, polyester membrane screens into 20 discrete compartments. Figure 15 gives a schematic view of the focusing chamber with the assembled Nylon screens supported by a cooling tube. There are several advantages of such a horizontal, rotating system. First of all, in the classical vertical density gradient IEF column of VESTERBERG et al. [176] isoelectric protein precipitates (not at all uncommon at the high protein loads of preparative runs) used to sediment along the supporting density gradient, thus contaminating other focused zones along the separation path. The Rotofor cell overcomes this problem, since, by focusing in a horizontal axis, potential protein precipitates stay focused in their compartment and do not disturb the remainder of the gradient. Secondly, the vertical density gradient column for IEF had a weak point in the elution funnel: all the liquid in the column was emptied via a bottom outlet, and this resulted in severe remixing of the separated protein zones and in contamination

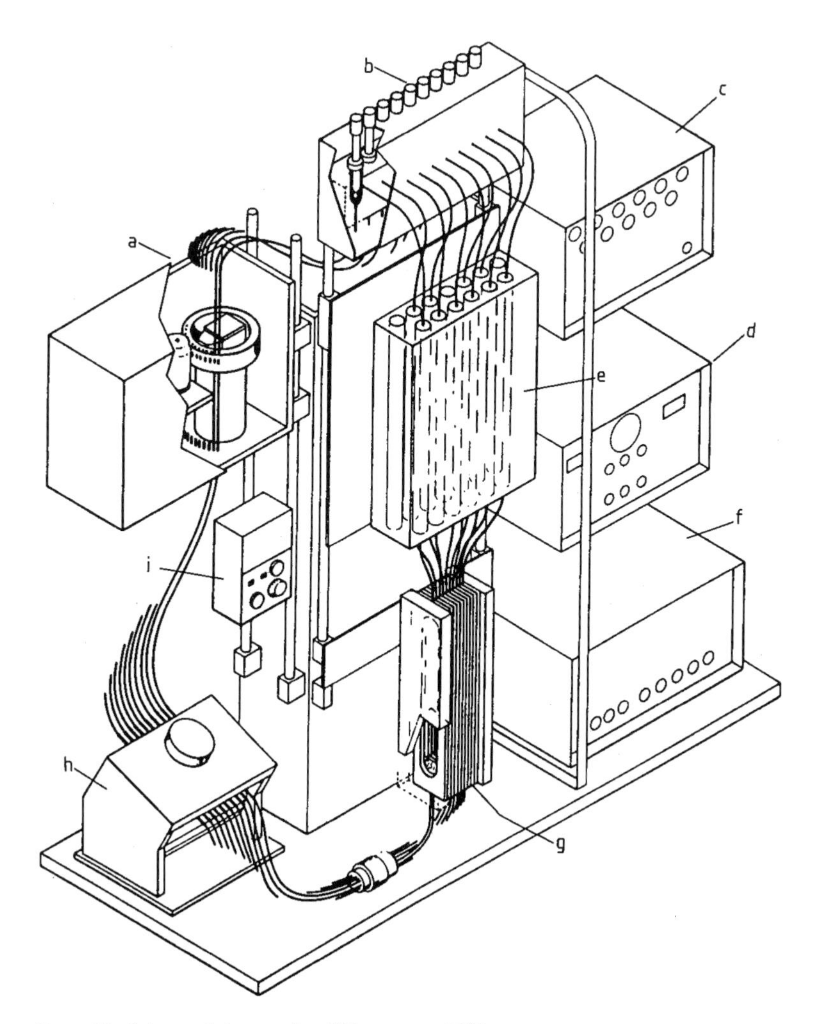

Figure 14. Scheme of the recycling IEF apparatus [180]
a) UV monitor; b) pH meter; c) Data interface; d) Power supply; e) Heat-exchange reservoir; f) UV control; g) Focusing cell; h) Pump; i) Pump control

by diffusion, owing to the long times required for harvesting the column content (usually 30 – 60 min, depending on column size). These problems are eliminated in the Rotofor apparatus, since the 20 chambers inside the column are emptied simultaneously and instantaneously by an array of 20 collection needles. The cylindrical focusing cell holds a total of 55 ml of solution for preparative protein purification; during operation, the chamber is rotated at 10 rpm around its axis to stabilize against convective disturbances. The run is usually performed at 12 W constant power (up to 2000 V maximum voltage); focusing is in general accomplished in as little as 4 h. Once the system is at steady state, collection is quickly performed by using an array of 20 collection needles that lead to 20 test tubes nested in a specially designed vacuum collection box. The needles simultaneously pierce a tape sealing 20 small holes, each at the bottom of an individual chamber in the Rotofor. By applying vacuum to the collection box, the 20 compartments are simultaneously emptied. In common with other preparative IEF techniques, the Rotofor can be used in a cascade set-up, that is, the contents of a single cell, containing the protein of interest, can be spread over the entire 20-cell assembly, and focused a second time over a shallower pH gradient, upon dilution with an appropriate range of carrier ampholytes.

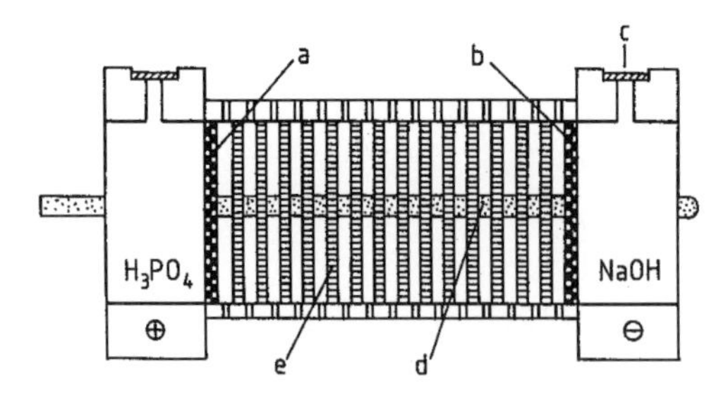

Figure 15. Scheme of the Rotofor cell
a) Cation-exchange membrane; b) Anion-exchange membrane; c) Vent caps; d) Ceramic cold finger; e) Screening material (6-μm polyester)

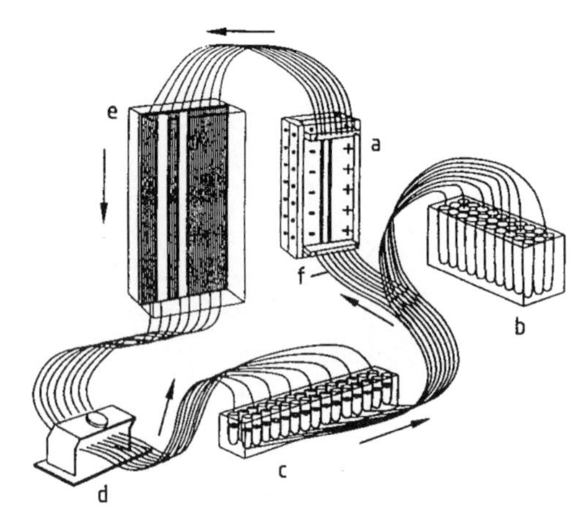

Figure 16. Scheme of the recycling free-flow focusing (RF3) apparatus
The arrows indicate the fluid flow pattern. At the end of the process, the flow pattern is switched to the harvesting position toward the fraction collector [183]
a) Focusing cell; b) Fraction collector; c) Bubble trap; d) Peristaltic pump; e) Heat exchanger; f) 30-Channel manifold

14.14.4. Recycling Free-Flow Focusing (RF3)

In the recycling IEF apparatus, stabilization of fluid flow is performed by screen elements; in the Rotofor, stabilization is achieved by screen elements and by rotation. In the RF3 apparatus shown schematically in Figure 16, fluid stabilization is achieved by means of rapid flow through narrow gaps, with continuous recycling. The key to stability of laminar flow through the cavity is its shallow depth (0.75 mm), combined with the rapid flow of process fluid (the latter necessary for avoiding electrohydrodynamic distortion). The lateral electrodes are separated from the focusing cavity by ion-permselective membranes. This instrument can also be used in the electrodialysis mode, by having the cation-selective membrane facing the cathodic compartment and the anion-selective membrane in front of the anodic chamber [183].

14.14.5. Multicompartment Electrolyzers with Isoelectric Membranes

Perhaps the final evolution of preparative IEF is the concept of a membrane apparatus, in which the entire IPG gel is reduced to isoelectric membranes delimiting a series of flow chambers. Figure 17 gives an exploded view of this novel apparatus [184], [185]: it consists of a stack of chambers sandwiched between an anodic and a cathodic reservoir. The apparatus is modular and in the present version can accommodate up to eight flow chambers. Figure 17 shows a stack of three chambers already assembled to the left, a central compartment, and a thinner chamber to the right for connection to the other electrode. All flow chambers are provided with inlet and outlet for sample or electrolyte recycling, an O-ring for ensuring flow-tight connections, and four holes for threading four long metal rods which can be tightened by hand-driven butterfly nuts for assembling the apparatus. Several versions of these cells have been built, capable of housing Immobiline membranes from 4.7 cm (the present apparatus) up to 9 cm diameter. The pH-controlling membranes are housed in the central depression between two 1-cm-wide caoutchouc rings. After assembling and tightening the apparatus, each compartment is flow-tight, so that no net liquid bulk flow ensues (except, when applicable, that generated by electrosmosis). The Pt electrodes are housed in two rectangular Perspex mountings, which also act as legs on which the electrolyzer stands. The distance between adjacent cells is only 10 mm, so that each chamber holds ca. 5 ml of liquid. The reason why this system works is shown in Figure 18: two isoelectric membranes facing each flow chamber act by continuously titrating the protein of interest to its isoelectric point. They can be envisaged as highly selective membranes, which retain any protein having pIs in between their limiting values, and which allow transmigration of any nonamphoteric, nonisoelectric species. The only condition required is that $pI_{cm} > pI_p > pI_{am}$, where the subscripts cm and am denote cathodic and anodic membranes, and p is the protein

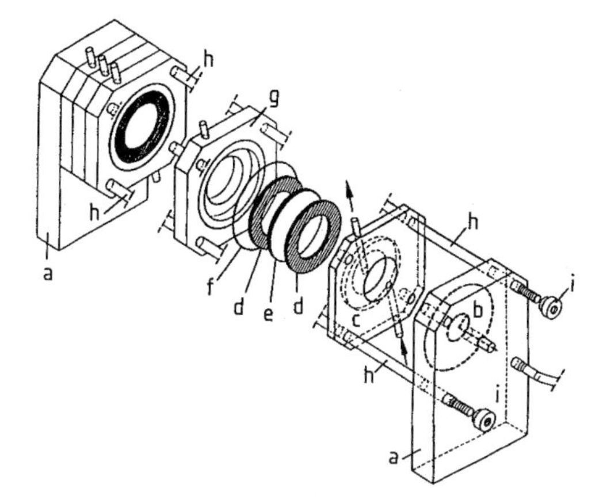

Figure 17. Exploded view of the multicompartment electrolyzer
a) Rectangular supporting legs; b) Pt electrode; c) Thin terminal flow chamber; d) Rubber rings for supporting the membrane; e) Isoelectric Immobiline membrane cast onto the glass-fiber filter; f) O-ring; g) One of the sample flow chambers; h) Threaded metal rods for assembling the apparatus; i) Nuts for fastening the metal bolts [128]

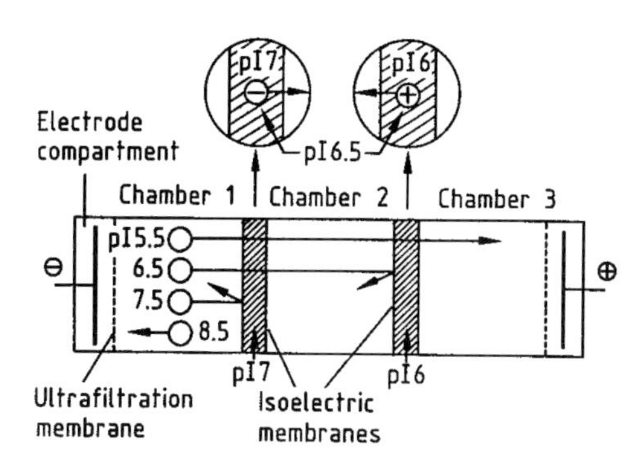

Figure 18. Mechanism of the purification process in the multicompartment electrolyzer
In the case shown, a protein with pI 6.5 is trapped between the pI 6 anodic and pI 7 cathodic membranes, whereas proteins with higher pI move toward the cathode and a species with lower pI (5.5) crosses the two membranes and collects at the anode [128].

having a given isoelectric point between the two membranes [184]. For this mechanism to be operative, the two isoelectric membranes must possess good conductivity and good buffering capacity, so as to effectively titrate the protein present in the flow chamber to its pI, while ensuring good current flow through the system. WENGER et al. [186] had in fact synthesized amphoteric, isoelectric Immobiline membranes and demonstrated that they are good conductors and good buffers at their pI.

14.15. References

[1] J. C. Giddings: *Unified Separation Science,* Wiley-Interscience, New York 1991, chap. 8.
[2] A. T. Andrews: *Electrophoresis: Theory, Techniques and Biochemical and Clinical Applications,* Clarendon Press, Oxford 1986.
[3] P. G. Righetti: *Isoelectric Focusing: Theory, Methodology and Applications,* Elsevier, Amsterdam 1983.
[4] P. G. Righetti: *Immobilized pH Gradients: Theory and Methodology,* Elsevier, Amsterdam 1990.
[5] C. J. Van Oss, *Sep. Purif. Methods* **8** (1979) 119–198.
[6] P. G. Righetti, C. J. Van Oss, J. W. Vanderhoff (eds.): *Electrokinetic Separation Methods,* Elsevier, Amsterdam 1979.
[7] B. L. Karger, *Nature (London)* **339** (1989) 641–642.
[8] P. Gozel, E. Gassmann, H. Michelsen, R. N. Zare, *Anal. Chem.* **59** (1987) 44–49.
[9] P. Grabar, C. A. Williams, *Biochim. Biophys. Acta* **10** (1953) 193–201.
[10] O. Smithies, *Biochem. J.* **61** (1955) 629–636.
[11] J. Kohn, *Nature (London)* **180** (1957) 986–987.
[12] L. P. Cawley, B J. Minard, W. W. Tourtellotte, P. I. Ma, C. Clelle, *Clin. Chem. (Winston Salem N.C.)* **22** (1976) 1262–1268.
[13] A. A. Keshgegian, P. Peiffer, *Clin. Chim. Acta* **108** (1981) 337–340.
[14] J. Kohn, P. Priches, J. C. Raymond, *J. Immunol. Methods* **76** (1985) 11–16.
[15] P. G. Righetti, P. Casero, G. B. Del Campo, *Clin. Chim. Acta* **157** (1986) 167–174.
[16] P. Casero, G. B. Del Campo, P. G. Righetti, *Electrophoresis (Weinheim Fed. Repub. Ger.)* **6** (1985) 373–376.
[17] T. Toda, T. Fujita, M. Ohashi in R. C. Allen, P. Arnaud (eds.): *Electrophoresis '81,* De Gruyter, Berlin 1981, pp. 271–280.
[18] P. Grabar, *Methods Biochem. Anal.* **7** (1957) 1–38.
[19] C. Araki, *Proceedings 5th International Seaweed Symposium,* Halifax, 1965, pp. 3–17.
[20] M. Duckworth, W. Yaphe, *Carbohydr. Res.* **16** (1971) 189–197.
[21] S. Hjertén, *Biochim. Biophys. Acta* **53** (1961) 514–517.
[22] S. Brishammar, S. Hjertén, B. Van Hofsten, *Biochim. Biophys. Acta* **53** (1961) 518–521.
[23] S. Hjertén, *Arch. Biochem. Biophys.* **99** (1962) 466–475.
[24] M. Duckworth, W. Yaphe, *Anal. Biochem.* **44** (1971) 636–641.
[25] T. J. Låås, *J. Chromatogr.* **66** (1972) 347–355.
[26] T. G. L. Hickson, A. Polson, *Biochim. Biophys. Acta* **168** (1965) 43–58.
[27] S. Arnott, A. Fulmer, W. E. Scott, I. C. M. Dea, R. Moorhourse, D. A. Rees, *J. Mol. Biol.* **90** (1974) 269–284.
[28] P. G. Righetti in R. C. Allen, R. Arnaud (eds.): *Electrophoresis '81,* De Gruyter, Berlin 1981, pp. 3–16.

[29] J. J. Opplt in L. A. Lewis, J. J. Opplt (eds.): *Handbook of Electrophoresis,* CRC Press, Boca Raton 1980, pp. 151 – 180.
[30] N. C. Stellwagen, *Adv. Electrophor.* **1** (1987) 177 – 228.
[31] P. Serwer, *Electrophoresis (Weinheim Fed. Repub. Ger.)* **4** (1983) 375 – 382.
[32] S. Hjertén, *J. Chromatogr.* **12** (1963) 510 – 526.
[33] P. Serwer, E. T. Moreno, G. A. Griess in C. Schafer-Nielsen (ed.): *Electrophoresis '88,* VCH, Weinheim 1988, pp. 216 – 222.
[34] P. G. Righetti, B. C. W. Brost, R. S. Snyder, *J. Biochem. Biophys. Methods* **4** (1981) 347 – 363.
[35] P. G. Righetti, *J. Biochem. Biophys. Methods* **19** (1989) 1 – 20.
[36] J. Uriel, J. Berges, *C. R. Hebd. Seances Acad. Sci. Ser. 3* **262** (1966) 164 – 170.
[37] A. C. Peacock, C. W. Dingman, *Biochemistry* **7** (1968) 668 – 673.
[38] E. Boschetti in P. G. D. Dean, W. S. Johnson, F. A. Middle (eds): *Affinity Chromatography,* IRL Press, Oxford 1985, pp. 11 – 15.
[39] A. Chrambach, D. Rodbard, *Science (Washington D.C.)* **172** (1971) 440 – 451.
[40] S. Raymond, L. Weintraub, *Science (Washington D.C.)* **130** (1959) 711 – 712.
[41] B. J. Davis, *Ann. N.Y. Acad. Sci.* **121** (1964) 404 – 427.
[42] L. Ornstein, *Ann. N.Y. Acad. Sci.* **121** (1964) 321 – 349.
[43] S. Hjertén, *J. Chromatogr.* **11** (1963) 66 – 70.
[44] S. Hjertén, *Anal. Biochem.* **3** (1962) 109 – 118.
[45] S. Hjertén, *Arch. Biochem. Biophys. Suppl. 1* (1962) 276 – 282.
[46] S. Hjertén, S. Jersted, A. Tiselius, *Anal. Biochem.* **27** (1969) 108 – 129.
[47] J. H. Bode in B. J. Radola (ed.): *Electrophoresis '79,* De Gruyter, Berlin 1980, pp. 39 – 52.
[48] P. B. H. O'Connell, C. J. Brady, *Anal. Biochem.* **76** (1976) 63 – 76.
[49] H. S. Anker, *FEBS Lett.* **7** (1970) 293 – 296.
[50] P. N. Paus, *Anal. Biochem.* **42** (1971) 327 – 376.
[51] J. N. Hansen, *Anal. Biochem.* **76** (1976) 37 – 44.
[52] C. Gelfi, P. G. Righetti, *Electrophoresis (Weinheim Fed. Repub. Ger.)* **2** (1981) 213 – 219.
[53] A. Bianchi-Bosisio, C. Loherlein, R. S. Snyder, P. G. Righetti, *J. Chromatogr.* **189** (1980) 317 – 330.
[54] P. G. Righetti, C. Gelfi, M. L. Bossi, E. Boschetti, *Electrophoresis (Weinheim Fed. Repub. Ger.)* **8** (1987) 62 – 70.
[55] B. Kozulic, K. Mosbach, M. Pietrzak, *Anal. Biochem.* **170** (1988) 478 – 484.
[56] C. Gelfi, P. De Besi, A. Alloni, P. G. Righetti, *J. Chromatogr.* **608** (1992) 333 – 341.
[57] G. Artoni, E. Gianazza, M. Zanoni, C. Gelfi, M. C. Tanzi, C. Barozzi, P. Ferruti, P. G. Righetti, *Anal. Biochem.* **137** (1984) 420 – 428.
[58] M. Chiari, C. Micheletti, M. Nesi, M. Fazio, P. G. Righetti, *Electrophoresis* **15** (1994) 177 – 186.
[59] E. Simò-Alfonso, C. Gelfi, R. Sebastiano, A. Cittero, P. G. Righetti, *Electrophoresis* **17** (1996) 723 – 731; **17** (1996) 732 – 737; **17** (1996) 738 – 743.
[60] C. Gelfi, M. Perego, F. Libbra, P. G. Righetti, *Electrophoresis* **17** (1996) 1342 – 1347.
[61] E. Simò-Alfonso, C. Gelfi, M. Lucisano, P. G. Righetti, *J. Chromatogr. A* **756** (1996) 255 – 262.
[62] E. Bordini, M. Hamdan, P. G. Righetti, *Rapid Commun. Mass Spectrom.* **13** (1999) 2209 – 2215.
[63] K. A. Ferguson, *Metab. Clin. Exp.* **13** (1964) 985 –995.
[64] H. R. Maurer: *Disc Electrophoresis,* De Gruyter, Berlin 1972.
[65] A. Chrambach in C. Schafer-Nielsen (ed.): *Electrophoresis '88,* VCH, Weinheim 1988, pp. 28 – 40.
[66] H. Svensson, *Acta Chem. Scand.* **15** (1961) 325 – 341;**16** (1962) 456 – 466.
[67] O. Vesterberg, *Acta Chem. Scand.* **23** (1969) 2653 –2666.
[68] L. Söderberg, D. Buckley, G. Hagström, *Protides Biol. Fluids* **27** (1980) 687 – 691.
[69] C. Schafer-Nielsen in C. Schafer-Nielsen (ed.): *Electrophoresis '88,* VCH, Weinheim 1988, pp. 41 – 48.
[70] A. L. Shapiro, E. Vinuela, J. V. Maizel, Jr., *Biochem. Biophys. Res. Commun.* **28** (1967) 815 – 822.
[71] A. Helenius, K. Simons, *Biochim. Biophys. Acta* **415** (1975) 29 – 79.
[72] J. L. Neff, N. Munez, J. L. Colburn, A. F. de Castro in R. C. Allen, P. Arnauds (eds.): *Electrophoresis '81,* De Gruyter, Berlin 1981, pp. 49 – 63.
[73] J. F. Poduslo, *Anal. Biochem.* **114** (1981) 131 – 140.
[74] P. Lambin, *Anal. Biochem.* **85** (1978) 114 – 124.
[75] J. Margolis, K. G. Kenrick, *Anal. Biochem.* **25** (1968) 347 – 358.
[76] P. Lambin, J. M. Fine, *Anal. Biochem.* **98** (1979) 160 – 168.
[77] G. M. Rothe, M. Purkhanbaba, *Electrophoresis (Weinheim Fed. Repub. Ger.)* **3** (1982) 33 – 42.
[78] G. M. Rothe, W. D. Maurer in M. J. Dunn (ed.): *Gel Electrophoresis of Proteins,* Wright, Bristol 1986, pp. 37 – 140.
[79] N. G. Anderson, N. L. Anderson, *Clin. Chem. (Winston Salem N.C.)* **28** (1982) 739 – 748.
[80] T. Barrett, H. J. Gould, *Biochim. Biophys. Acta* **294** (1973) 165 – 170.
[81] P. O'Farrell, *J. Biol. Chem.* **250** (1975) 4007 – 4021.
[82] J. Klose, *Humangenetik* **26** (1975) 231 – 243.
[83] G. A. Scheele, *J. Biol. Chem.* **250** (1975) 5375 – 5385.
[84] R. A. Colbert, J. M. Amatruda, D. S. Young, *Clin. Chem. (Winston Salem N.C.)* **30** (1984) 2053 – 2058.
[85] R. H. Aebersold, J. Leavitt, L. E. Hood, S. B. H. Kent in K. Walsh (ed.): *Methods in Protein Sequence Analysis,* Humana Press, Clifton 1987, pp. 277 – 294.
[86] D. S. Young, N. G. Anderson (eds.): Special Issue on Two Dimensional Electrophoresis, *Clin. Chem.* **28** (1982) 737 – 1092.
[87] J. S. King (ed.): Special Issue on Two Dimensional Electrophoresis, *Clin. Chem.* **30** (1984) 1897 – 2108.
[88] J. E. Celis, R. Bravo (eds.): *Two Dimensional Gel Electrophoresis of Proteins,* Academic Press, Orlando 1984, pp. 1 – 487.
[89] B. S. Dunbar: *Two-Dimensional Electrophoresis and Immunological Techniques,* Plenum Press, New York 1987, pp. 1 – 372.
[90] M. J. Dunn (ed.): *2-D PAGE '91,* Zebra Printing, Perivale 1991, pp. 1 – 325.
[91] C. Damerval, D. de Vienne (eds.): Paper Symposium: Two Dimensional Electrophoresis of Plant Proteins, *Electrophoresis* **9** (1988) 679 – 796.
[92] J. E. Celis (ed.): Paper Symposium: Protein Databases in Two Dimensional Electrophoresis, *Electrophoresis* **10** (1989) 71 – 164.
[93] J. E. Celis (ed.): Paper Symposium: Cell Biology, *Electrophoresis* **11** (1990) 189 – 280.
[94] J. E. Celis (ed.): Paper Symposium: Two Dimensional Gel Protein Databases, *Electrophoresis* **11** (1990) 987 – 1168.

[95] J. E. Celis (ed.): Paper Symposium: Two Dimensional Gel Protein Databases, *Electrophoresis* **12** (1991) 763 – 996.
[96] J. E. Celis (ed.): Paper Symposium: Two Dimensional Gel Protein Databases, *Electrophoresis* **13** (1992) 891 – 1062.
[97] J. E. Celis (ed.): Paper Symposium: Electrophoresis in Cancer Research, *Electrophoresis* **15** (1994) 307 – 556.
[98] J. E. Celis (ed.): Paper Symposium: Two Dimensional Gel Protein Databases, *Electrophoresis* **15** (1994) 1347 – 1492.
[99] J. E. Celis (ed.): Paper Symposium: Two Dimensional Gel Protein Databases, *Electrophoresis* **16** (1995) 2175 – 2264.
[100] J. E. Celis (ed.): Paper Symposium: Two Dimensional Gel Protein Databases, *Electrophoresis* **17** (1996) 1653 – 1798.
[101] J. E. Celis (ed.): Genomics and Proteomics of Cancer, *Electrophoresis* **20** (1999) 223 – 243.
[102] M. J. Dunn (ed.): Paper Symposium: Biomedical Applications of Two-Dimensional Gel Electrophoresis, *Electrophoresis* **12** (1991) 459 – 606.
[103] M. J. Dunn (ed.): 2D Electrophoresis: from Protein Maps to Genomes, *Electrophoresis* **16** (1995) 1077 – 1326.
[104] M. J. Dunn (ed.): From Protein Maps to Genomes, Proceedings of the Second Siena Two-Dimensional Electrophoresis Meeting, *Electrophoresis* **18** (1997) 305 – 662.
[105] M. J. Dunn (ed.): From Genome to Proteome: Proceedings of the Third Siena Two-Dimensional Electrophoresis Meeting, *Electrophoresis* **20** (1999) 643 – 845.
[106] F. Lottspeich (ed.): Paper Symposium: Electrophoresis and Amino Acid Sequencing, *Electrophoresis* **17** (1996) 811 – 966.
[107] B. Tümmler (ed.): Microbial Genomes: Biology and Technology, *Electrophoresis* **19** (1998) 467 – 624.
[108] K. L. Williams (ed.): Strategies in Proteome Research, *Electrophoresis* **19** (1998) 1853 – 2050.
[109] I. Humphrey-Smith (ed.): Paper Symposium: Microbial Proteomes, *Electrophoresis* **18** (1997) 1207 – 1497.
[110] R. D. Appel, M. J. Dunn, D. F. Hochstrasser (eds): Paper Symposium: Biomedicin and Bioinformatics, *Electrophoresis* **18** (1997) 2703 – 2842.
[111] R. D. Appel, M. J. Dunn, D. F. Hochstrasser (eds): Paper Symposium: Biomedicin and Bioinformatics, *Electrophoresis* **20** (1999) 3481 – 3686.
[112] P. Cash (ed.): Paper Symposium: Microbial Proteomes, *Electrophoresis* **20** (1999) 2149 – 2285.
[113] M. R. Wilkins, K. L. Williams, R. D. Appel, D. F. Hochstrasser (eds.): *Proteome Research: New Frontiers in Functional Genomics,* Springer, Berlin 1997.
[114] S. M. Hanash, in B. D. Hames (ed.): *Gel Electrophoresis of Proteins: a Practical Approach,* Oxford University Press, Oxford 1998, pp. 189 – 211.
[115] F. Kohlrausch, *Ann. Phys. Chem.* **62** (1897) 209 – 234.
[116] F. M. Everaerts, J. L. Becker, T. P. E. M. Verheggen: *Isotachophoresis: Theory, Instrumentation and Applications,* Elsevier, Amsterdam 1976.
[117] P. Gebauer, V. Dolnik, M. Deml, P. Bocek, *Adv. Electrophor.* **1** (1987) 281 – 359.
[118] S. G. Hjalmarsson, A. Baldesten, *CRC Crit. Rev. Anal. Chem.* **18** (1981) 261 – 352.
[119] L. Krivankovà, P. Bocek, in M. G. Khaledi (ed.): *High Performance Capillary Electrophoresis,* John Wiley & Sons, New York 1998, pp. 251 – 275.
[120] C. B. Laurell, *Anal. Biochem.* **15** (1966) 45 – 52.
[121] H. G. M. Clarke, T. Freeman, *Protides Biol. Fluids* **14** (1967) 503 – 509.
[122] I. Sondergaard, L. K. Poulsen, M. Hagerup, K. Conradsen, *Anal. Biochem.* **165** (1987) 384 – 391.
[123] J. Kroll in N. H. Axelsen, J. Kroll, B. Weeke (eds.): *A Manual of Quantitative Immunoelectrophoresis,* Universitetsforlaget, Oslo 1973, pp. 61 – 67.
[124] P. J. Svendsen, N. H. Axelsen, *J. Immunol. Methods* **2** (1972) 169 – 176.
[125] P. J. Svendsen, C. Rose, *Sci. Tools* **17** (1970) 13 – 17.
[126] N. H. Axelsen, J. Kroll, B. Weeke (eds.): "A Manual of Quantitative Immunoelectrophoresis," *Scand. J. Immunol. Suppl.* **1** (1973) no. 2, pp. 1 – 169.
[127] N. H. Axelsen (ed.): "Quantitative Immunoelectrophoresis," *Scand. J. Immunol. Suppl.* **2** (1975) pp. 1 – 230.
[128] N. H. Axelsen (ed.), *Scand. J. Immunol. Suppl.* **10** (1983) no. 17, pp. 1 – 280.
[129] C. R. Merril, R. C. Switzer, M. L. Van Keuren, *Proc. Natl. Acad. Sci. U.S.A.* **76** (1979) 4335 – 4339.
[130] M. Moeremans, G. Daneels, J. De Mey, *Anal. Biochem.* **145** (1985) 315 – 321.
[131] M. Moeremans, D. De Raeymaeker, G. Daneels, J. De Mey, *Anal. Biochem.* **153** (1986) 18 – 22.
[132] E. M. Southern, *J. Mol. Biol.* **98** (1975) 503 – 510.
[133] H. Towbin, T. Staehelin, J. Gordon. *Proc. Natl. Acad. Sci. U.S.A.* **76** (1979) 4350 – 4355.
[134] H. Towbin, J. Gordon, *J. Immunol. Methods* **72** (1984) 313 – 340.
[135] O. J. Bjerrum, N. H. H. Heegaard, *J. Chromatogr.* **470** (1989) 351 – 367.
[136] V. Neuhoff, R. Stamm, H. Eibl, *Electrophoresis* **6** (1985) 427 – 448.
[137] C. R. Merril, K. M. Washart, in: *Gel Electrophoresis of Proteins, a Practical Approach,* Oxford University Press, Oxford 1998, pp. 53 – 92, 319 – 343.
[138] B. Bjellqvist, K. Ek, P. G. Righetti, E. Gianazza, A. Görg, W. Postel, R. Westermeier, *J. Biochem. Biophys. Methods* **6** (1982) 317 – 339.
[139] M. Chiari, E. Casale, E. Santaniello, P. G. Righetti, *Appl. Theor. Electrophor.* **1** (1989) 99 – 102;**ibid. 1** (1989) 103 – 107.
[140] E. Gianazza, F. Celentano, G. Dossi, B. Bjellqvist, P. G. Righetti, *Electrophoresis* **5** (1984) 88 – 97.
[141] G. Dossi, F. Celentano, E. Gianazza, P. G. Righetti, *J. Biochem. Biophys. Methods* **7** (1983) 123 – 142.
[142] P. G. Righetti, M. Chiari, P. K. Sinha, E. Santaniello, *J. Biochem. Biophys. Methods* **16** (1988) 185 –192.
[143] C. Gelfi, M. L. Bossi, B. Bjellqvist, P. G. Righetti, *J. Biochem. Biophys. Methods* **15** (1987) 41 – 48;P. K. Sinha, P. G. Righetti, **ibid. 15** (1987) 199 – 206.
[144] P. G. Righetti, E. Gianazza, C. Gelfi, M. Chiari, P. K. Sinha, *Anal. Chem.* **61** (1989) 1602 – 1612.
[145] P. G. Righetti, C. Gelfi, M. Chiari, in B. L. Karger, W. S. Hancock (eds.): *Methods in Enzymology: High Resolution Separation and Analysis of Biological Macromolecules, Part A: Fundamentals,* **vol. 270,** Academic Press, San Diego 1996, pp. 235 – 255.
[146] P. G. Righetti, A. Bossi, *Anal. Biochem.* **247** (1997) 1 – 10.
[147] P. G. Righetti, A. Bossi, *J. Chromatogr. B* **699** (1997) 77 – 89.

[148] P. G. Righetti, A. Bossi, C. Gelfi, in B. D. Hames (ed.): *Gel Electrophoresis of Proteins,* 3rd ed., Oxford University Press, Oxford 1998, pp. 127 – 187.

[149] J. W. Jorgenson in J. W. Jorgenson, M. Phillips (eds.): "New Directions in Electrophoretic Methods," *ACS Symp. Ser. 335,* Am. Chem. Soc., Washington, 1987, pp. 70 – 93.

[150] D. J. Rose, Jr., J. W. Jorgenson, *J. Chromatogr.* **447** (1988) 117 – 131.

[151] F. Foret, P. Bocek, *Advances Electrophor.* **3** (1989) 271 – 347.

[152] B. L. Karger (ed.): "Proceedings of the 1st International Symposium on High-Performance Capillary Electrophoresis," *J. Chromatogr.* **480** (1989) 1 – 435.

[153] B. L. Karger (ed.): "Proceedings of the 2nd International Symposium on High-Performance Capillary Electrophoresis," *J. Chromatogr.* **516** (1990) 1 –298.

[154] J. W. Jorgenson (ed.): "Proceedings of the 3rd International Symposium on High-Performance Capillary Electrophoresis," *J. Chromatogr.* **559** (1991) 1 –561.

[155] F. M. Everaerts, T. P. E. M. Verheggen (eds.): "Proceedings of the 4th International Symposium on High-Performance Capillary Electrophoresis," *J. Chromatogr.* **608** (1992) 1 – 429.

[156] B. L. Karger (ed.): "Proceedings of the 5th International Symposium on High-Performance Capillary Electrophoresis," *J. Chromatogr. A* **652** (1993) 1 – 574.

[157] B. L. Karger, S. Terabe (eds.): "Proceedings of the 6th International Symposium on High-Performance Capillary Electrophoresis," *J. Chromatogr. A* **680** (1994) 1 – 689.

[158] H. Engelhardt (ed.): "Proceedings of the 7th International Symposium on High-Performance Capillary Electrophoresis," *J. Chromatogr. A* **716** (1995) 1 – 412; **717** (1995) 1 – 431.

[159] B. L. Karger (ed.): "Proceedings of the 8th International Symposium on High-Performance Capillary Electrophoresis," *J. Chromatogr. A* **744** (1996) 1 – 354; **745** (1996) 1 – 303.

[160] W. S. Hancock (ed.): "Proceedings of the 9th International Symposium on High-Performance Capillary Electrophoresis," *J. Chromatogr. A* **781** (1997) 1 – 568.

[161] S. Fanali, B. L. Karger (eds.): "Proceedings of the 10th International Symposium on High-Performance Capillary Electrophoresis," *J. Chromatogr. A* **817** (1998) 1 – 382.

[162] E. S. Yeung (ed.): "Proceedings of the 11th International Symposium on High-Performance Capillary Electrophoresis," *J. Chromatogr. A* **853** (1999) 1 – 576.

[163] A. M. Krstulovic (ed.): "Capillary Electrophoresis in the Life Sciences," *J. Chromatogr. B* **697** (1997) 1 – 290.

[164] Y. Baba (ed.): "Analysis of DNA by Capillary Electrophoresis," *J. Biochem. Biophys. Methods* **41** (1999) 75 – 165.

[165] B. L. Karger (ed.): Paper Symposium: Capillary Electrophoresis, *Electrophoresis* **14** (1993) 373 – 558.

[166] N. J. Dovichi (ed.): Paper Symposium: Nucleic Acid Electrophoresis, *Electrophoresis* **17** (1996) 1407 – 1517.

[167] F. Foret, P. Bocek (eds.): Paper Symposium: Capillary Electrophoresis: Instrumentation and Methodology, *Electrophoresis* **17** (1996) 1801 – 1963.

[168] J. P. Landers (ed.): Paper Symposium: Capillary Electrophoresis in the Clinical Sciences, *Electrophoresis* **18** (1997) 1709 – 1905.

[169] Z. El Rassi (ed.): "Capillary Electrophoresis and Electrochromatography," *Electrophoresis* **18** (1997) 2123 – 2501.

[170] B. R. McCord (ed.): Paper Symposium: Capillary Electrophoresis in Forensic Science, *Electrophoresis* **19** (1998) 3 – 124.

[171] Z. El Rassi (ed.): Paper Symposium: Capillary Electrophoresis and Electrochromatography Reviews, *Electrophoresis* **20** (1999) 2989 – 3328.

[172] P. G. Righetti (ed.): *Capillary Electrophoresis: an Analytical Tool in Biotechnology,* CRC Press, Boca Raton, FL 1996.

[173] M. G. Khaledi (ed.): *High Performance Capillary Electrophoresis: Theory, Techniques and Applications,* John Wiley & Sons, New York 1998.

[174] P. G. Righetti, C. Gelfi, *J. Biochem. Biophys. Methods* **41** (1999) 75 – 90.

[175] P. G. Righetti, A. Bossi, E. Olivieri, C. Gelfi, *J. Biochem. Biophys. Methods* **40** (1999) 1 – 16.

[176] O. Vesterberg, T. Waldström, K. Vesterberg, H. Svensson, B. Malmgren, *Biochim. Biophys. Acta* **133** (1967) 435 – 445.

[177] B. J. Radola, *Ann. N.Y. Acad. Sci.* **209** (1973) 127 –143.

[178] B. J. Radola, *Biochim. Biophys. Acta* **295** (1973) 412–428.

[179] M. Bier, N. Egen in H. Haglund, J. C. Westerfeld, J. T. Ball Jr. (eds.): *Electrofocus '78,* Elsevier, Amsterdam 1979, pp. 35 – 48.

[180] M. Bier, N. Egen, T. T. Allgyer, G. E. Twitty, R. A. Mosher in E. Gross, J. Meienhofer (eds.): *Peptides: Structure and Biological Function,* Pierce Chem. Co., Rockford 1979, pp. 79 – 89.

[181] N. Egen. W. Thormann, G. E. Twitty, M. Bier in H. Hirai (ed.): *Electrophoresis '83,* De Gruyter, Berlin 1984, pp. 547 – 550.

[182] M. Bier, T. Long, *J. Chromatogr.* **604** (1992) 73 –83.

[183] P. G. Righetti, E. Wenisch, M. Faupel, *J. Chromatogr.* **475** (1989) 293 – 309.

[184] P. G. Righetti, E. Wenisch, A. Jungbauer, H. Katinger, M. Faupel, *J. Chromatogr.* **500** (1990) 681 –696.

[185] P. G. Righetti, E. Wenisch, M. Faupel, *Advances Electrophor.* **5** (1992) 159 – 200.

[186] P. Wenger, M. de Zuanni, P. Javet, P. G. Righetti, *J. Biochem. Biophys. Methods* **14** (1987) 29 – 43.

15. Structure Analysis by Diffraction

Erich F. Paulus, Institute of Mineralogy and Crystallography of the University of Frankfurt, Frankfurt/Main, Federal Republic of Germany

Alfred Gieren, formerly Institute for Analytical Chemistry and Radiochemistry, University of Innsbruck, Innsbruck, Austria

15.1. General Principles

When electromagnetic or corpuscular radiation is diffracted by matter, i.e., when it is involved in an elastic interaction with matter, a scattering diagram or diffraction pattern is produced; this pattern is the Fourier transform of the diffracting object. If the inverse transform is applied, going from "reciprocal" to "direct" space, an image of the object is obtained. This inverse transformation is called Fourier synthesis; the transformation from direct to reciprocal space is called Fourier analysis. The principle of Fourier analysis (diffraction) followed by Fourier synthesis (imaging) controls the classical process of magnification in an optical microscope. Here, an object is irradiated, and elastic scattering takes place. The objective lens causes the scattered rays to interfere, generating a magnified image. To apply this principle at atomic resolution (ca. 10^{-10} m), the first thing needed is radiation with wavelength of the same order of magnitude (maximum resolution is $\lambda/2$)

[55]. A further condition is that the destruction caused by the interaction of the radiation with the object should be as slight as possible, so that the image is not substantially degraded by radiation damage.

The chief kinds of radiation used are X rays, neutrons, and electrons. X rays and neutrons cause relatively little damage, but electrons have a scattering power several orders of magnitude greater, because of their charge, which leads to more radiation damage. Electrons have the advantage that they can be imaged with lenses in a microscope configuration. The use of a lens for imaging (Fourier synthesis) is not, however, essential. Rapid advances in electronic computers, make it possible to replace the lens by a computer. However, this is possible only if all the relevant parameters of the scattering diagram, including the phases of the scattered waves, are known. As a rule, the phases of X ray and neutron radiation are not accessible to direct measurement, so that computer-intensive techniques are required to obtain them.

X-ray and neutron structure analysis generally deal with crystalline matter, either single crystals or crystalline powders. Nonperiodic objects, such as glasses, play only a secondary role.

15.2. Structure Analysis of Solids

15.2.1. Diffraction by Crystal Lattices

With a few exceptions, this article deals with the crystalline solid state; other aggregations such as liquid crystals are not covered.

Naturally occurring crystals differ from ideal crystals in two main aspects:

1) They are finite with surfaces
2) They contain structural defects and vacancies, e.g., the mosaic structure of real crystals

The crystals available in practice thus lie somewhere between amorphous and ideally crystalline, but nearer the ideal end of the scale ("ideal imperfect crystals"). A crystalline specimen may be a single crystal (smallest dimension ca. 0.05 – 0.5 mm) or a polycrystalline material with crystallite size ca. 0.04 mm.

High-polymer plastics occupy an intermediate position. According to the two-phase model [56] the amorphous phase is present along with the crystalline phase. The unit cells in the crystalline portion may be oriented in some preferred direction (as in fibers), or there may be no preferred direction.

15.2.1.1. Materials and Methods

"Single crystals" are the classical objects investigated by crystallography. The diffraction of X rays by crystals was discovered in 1912 by von Laue, Friedrich, and Knipping in a study of copper sulfate pentahydrate crystals. It was later found that diffraction also takes place from crystalline powders. The methodological principles are explained here for single crystals and to a certain extent also for crystalline powder.

Single-crystal structure analysis, the determination of the structure of the asymmetric unit of a crystal, is so far the only analytical method that can yield direct and unambiguous information about the three-dimensional atomic structure of a substance. Crystalline powder structure analysis needs in the average also a model.

Simple Inorganic Compounds. The class of simple inorganic compounds includes substances such as metals and salts, that contain very few atoms in the asymmetric unit. In such cases, the complete crystal structure can often be derived as soon as the crystal symmetry (space group) has been determined. If several possibilities exist, the correct structure can be arrived at by trial and error. When this approach also fails, a Patterson synthesis (see Eq. 27) can be used to solve the structure.

Complex Inorganic and Organic Compounds. Two methods can be used for the solution of a more complex molecular and crystal structure problem.

1) Patterson synthesis followed by successive Fourier syntheses of the electron density
2) "Direct" determination of the phases of the structure factors, possibly followed by successive Fourier syntheses of the electron density (direct methods)

These methods complement each other in many cases. When there are several possible solutions for the Patterson function, some may be eliminated if the direct methods yield a possible solution by which the Patterson synthesis can be interpreted. The contrary can also occur, when a Patterson synthesis is used to decide which phase set is correct.

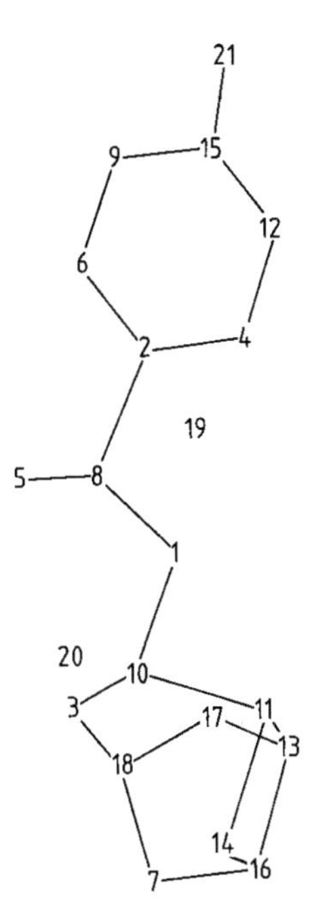

Figure 1. Computer-generated molecular structure diagram after solution of a crystal structure by direct methods

A prerequisite for the use of Patterson synthesis is that concrete assumptions about the complete molecular structure exist, or at least that parts of the structure are known. The position and orientation of the known molecular part can then be determined, for example with the aid of the convolution molecule method [57] or minimum search programs. In the case of heavy-atom derivatives, the known molecular part consists of one or several heavy atoms [58].

Direct methods are most commonly used to solve crystal structures. A number of advanced program systems exist and, in many cases, make solution of the phase problem a routine matter for the expert. The proposed structure, which often proves to be correct and complete, can be output directly in graphical form (Fig. 1). The numbers represent atomic positions; a small numerical value indicates a larger electron density maximum and a large value indicates a smaller one. The program determines which electron density maxima should be connected to each other to obtain the structure; tables are generated that show the positions of the atoms in the cluster (see also Table 3). However, if this method does not succeed immediately, generally much experience, including an understanding of the principles of direct methods, is required to solve the phase problem. Given sufficient time, virtually all structure problems (ca. 95 %) satisfying certain minima criteria are solvable; these include structures with up to 100 and many more nonhydrogen atoms in the asymmetric unit (the atomic cluster from which the entire unit cell can be generated by symmetry operations), and correspond to those structures of greatest interest to the synthetic chemist. If the data set is improved by cooling the crystal (below−100 °C) to reduce the thermal vibrations of the atoms, this limit can be extended in special cases to more than 300.

Proteins. In these much larger molecules, it is often possible to determine the phases by the method of isomorphous replacement. Most commonly, this means allowing ions of heavy atoms to diffuse into the solvent-filled vacancies in protein crystals, where they adopt well-defined positions. Thus, this represents the incorporation of atoms in completely new positions, rather than an atom-replacement process. One exception is insulin, in which zinc ions are replaced by lead or mercury ions. Newer methods allow the structure to be solved by direct methods when excellent protein crystals are available.

A number of techniques have been devised for determining the positions of heavy atoms in the unit cells of proteins [6, pp. 337−371], but these are not discussed here.

In the noncentrosymmetric case, which always holds for proteins, phase determination of the structure factors requires at least two heavy-atom derivatives to estimate the phases for the native protein from the heavy atom positions [6, pp. 337−371]. As a rule, three or four derivatives are investigated in order to reduce phase errors.

All reflections whose phases are to be determined have to be measured on the native protein and the isomorphous crystals. A computer-controlled diffractometer is used. Alternatively, photographic data and automatic photometers can be employed; this has the advantage that the crystals do not have to be exposed to X rays for such a long time, reducing radiation damage. Recently, films have been replaced by area detectors which produce the intensities in digital form. If a synchrotron is used as radiation source, and the pattern is recorded by the Laue method (see Section 15.2.1.5), data acquisition takes only a few minutes.

The extent to which the Fourier synthesis of the electron density can be interpreted depends on the resolution with which the data could be measured. Resolution is said to be low when the smallest lattice plane spacings to which measured reflections can be assigned are $>5\times10^{-10}$ m; aside from the dimensions and overall shape of the molecule, no detail can be extracted from the Fourier synthesis. At a resolution of 3×10^{-10} m, the shape of polypeptide chains can be identified; at 1.5×10^{-10} m, single atoms can be distinguished and side chains can be identified without knowledge of the amino acid sequence. The atomic positions cannot be refined by conventional least-squares methods, because the computational problem becomes immense and the number of observed data seldom exceeds the number of parameters. For special methods, see [6, pp. 420–442].

Supplementary Structure Investigations by Neutron Diffraction. Single-crystal X-ray structure analysis cannot determine the positions of hydrogen atoms in many cases. Even with error-free measurements, there are essentially two reasons:

1) Because hydrogen atoms generally occupy peripheral locations in the molecule, they are smeared out by static and/or dynamic disorder (small conformational differences), and their scattering contribution at higher scattering angles is not significantly higher than the background.
2) The compound under study may contain not only light atoms, but also heavy atoms, such as bromine and iodine. The intensity measurements are not accurate enough to detect the small relative intensity differences due to hydrogen atoms.

In most such cases, well-known stereochemical rules make it possible to calculate the positions of most hydrogen atoms from those of the other atoms. But if one wishes to investigate, say, hydrogen bonds more closely, the use of neutron diffraction is indicated, because H or, better, D atoms (because of the smaller incoherent scattering) are strong neutron scatterers.

Neutron diffraction analysis may also be necessary when elements near one another in the periodic table need to be distinguished. A well-known example is differentiation between the isoelectronic ions Si^{4+} and Al^{3+}.

By virtue of their magnetic moment, neutrons can also be used to investigate magnetic structures [5, pp. 395–511], [59].

Chemical bonds cause a displacement of the bonding electrons. Because X rays are also scattered by these electrons, a slight distortion of the apparent atomic positions is unavoidable. For example, X-ray diffraction always gives carbon–hydrogen bond lengths that are too short, because the hydrogen electron is pulled closer to its bonding partner. Neutron diffraction analysis, in contrast, determines the position of the atomic nucleus, not the maximum of the electron density distribution. The deformation density resulting from chemical bonding may be directly revealed by difference Fourier syntheses of electron densities from X-ray diffraction data, and the structure factors calculated from the atomic positions obtained by neutron diffraction analysis (X–N method).

The measurements must be performed with the utmost care, because the effects are comparable to the standard deviations. X-ray and neutron diffraction patterns should be recorded at the lowest possible temperatures, to minimize thermal motion.

For reasons of cost, neutron diffraction is virtually always preceded by single-crystal X-ray structure analysis. In many cases, the deformation densities can be determined by X-ray measurements alone, in view of the fact that the scattering contribution of the bonding electrons vanishes at large scattering angles (X–X method).

15.2.1.2. Introduction to X-Ray Crystallography; History

Democritos of Abdera (born ca. 465 B.C.) wrote, "It is customary to say that there is color, sweetness, bitterness; in actuality, however, there are atoms and void" [60]. In this sentence, the atom, as the fundamental building block of matter, was introduced into the discussion of the structure of matter. In 1807, Dalton had only to invoke this hypothesis in order to impose order on the experimental facts known at the time. For example, a crystalline structure consisting of identical particles was also in agreement with these principles.

Nikolaus Steno, as early as 1669, formulated the law of constant angles, which states that two like edges of crystals of the same substance always make equal angles with each other. In 1823, Franz Ernst Neumann discovered the law of simple rational indices. Planes drawn through faces of a crystal intersect three coordinate axes, constructed parallel to crystal edges, at minima lengths a, b, c that are in a constant ratio to one another. The other faces of the crystal can be

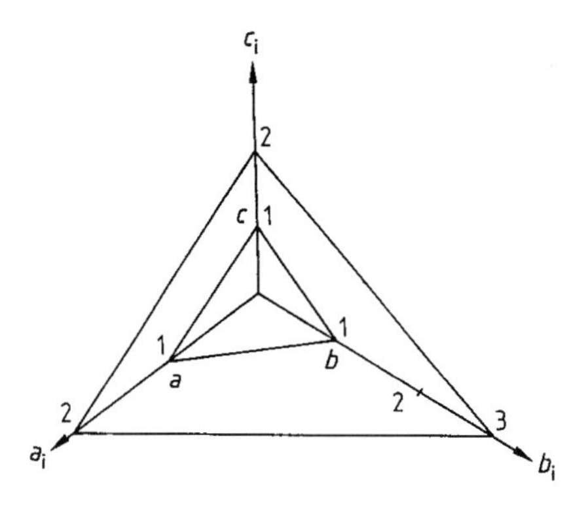

Figure 2. The law of simple rational indices
Planes shown are: $u=v=w=1$; and $u=2$, $v=3$, $w=2$

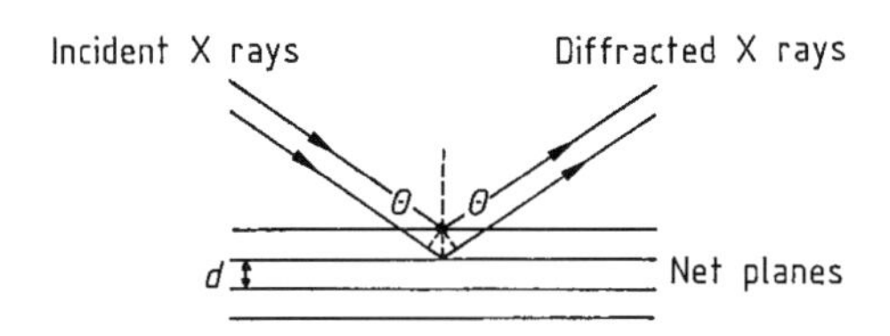

Figure 3. Derivation of the Bragg equation

characterized by planes drawn through them that cut the axes at lengths a_i, b_i, c_i which are in whole-number (possibly negative) ratios to the intercepts a, b, c $(a_i{:}\,b_j{:}\,c_i = u \cdot a : v \cdot b : w \cdot c)$ (Fig. 2).

About 1890, FEDOROV, SCHOENFLIES, and BARLOW more or less simultaneously showed that for crystals built up from discrete particles in a three-dimensionally ordered manner, there can be no more than 230 different combinations of elements, the 230 space groups. In 1912, on the basis of their key diffraction experiment, which yielded discrete X-ray reflections from a crystal, LAUE, FRIEDRICH, and KNIPPING demonstrated both the periodic construction of crystals from atoms and the wave nature of X rays.

W. H. and W. L. BRAGG were the first to show that the models proposed by BARLOW for simple compounds, such as NaCl, CsCl, and ZnS are in agreement with the scattered X-ray intensities. PAUL NIGGLI found that space groups could be determined by X-ray methods via the systematic extinction laws.

The Bragg equation:

$$\sin\theta = n\frac{\lambda}{2d} \tag{1}$$

$$n\lambda = 2d\sin\theta \tag{2}$$

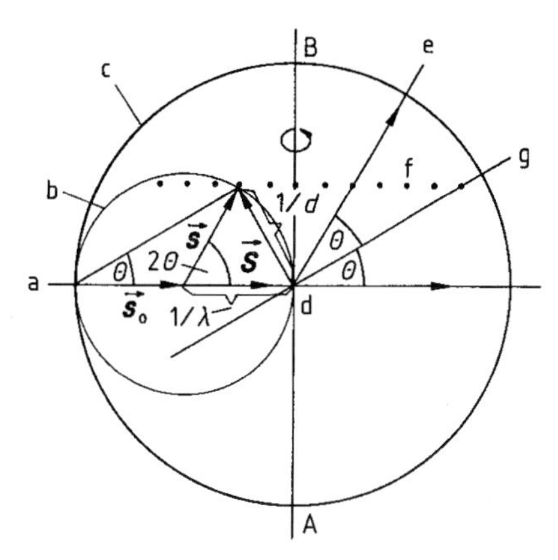

Figure 4. Ewald construction
a) Incident beam; b) Reflection sphere; c) Limiting sphere; d) Crystal; e) Diffracted beam; f) Lattice points on a plane of the reciprocal lattice perpendicular to the plane of the diagram; g) Reflecting lattice plane

describes the diffraction of X rays by a crystal as reflection from the net planes. Constructive interference occurs only if the difference in path length between two rays is an integer multiple of the wavelength λ (Fig. 3). Here θ is the glancing angle ($2\,\theta$ is the diffraction angle), and d is the net plane spacing.

A construction due to EWALD illustrates the importance of the reciprocal lattice in X-ray crystallography. As Figure 4 shows, the Bragg equation is satisfied where the reflection sphere is cut by a lattice point of the reciprocal lattice constructed around the center of the crystal. Rotating the crystal together with the reciprocal lattice around a few different directions in the crystal fulfills the reflection condition for all points of the reciprocal space within the limiting sphere. The reciprocal lattice vector $\boldsymbol{S}$ is perpendicular to the set of net planes, and has the absolute magnitude $1/d$. In vector notation:

$$\boldsymbol{S} = \boldsymbol{s} - \boldsymbol{s}_0 \tag{3}$$

Vectors $\boldsymbol{s}$ and $\boldsymbol{s}_0$ each have modulus $1/\lambda$.

The reciprocal lattice provides a simple way to illustrate the positions of diffraction points in space. Figure 4 also shows that the Ewald limiting sphere determines how many reflections can be measured. All reflections inside the limiting sphere with radius $2/\lambda$ are obtained only if the crystal axis AB can take up any orientation. This is not so in many data acquisition methods, e.g., the rotating crystal method. If the crystal is rotated

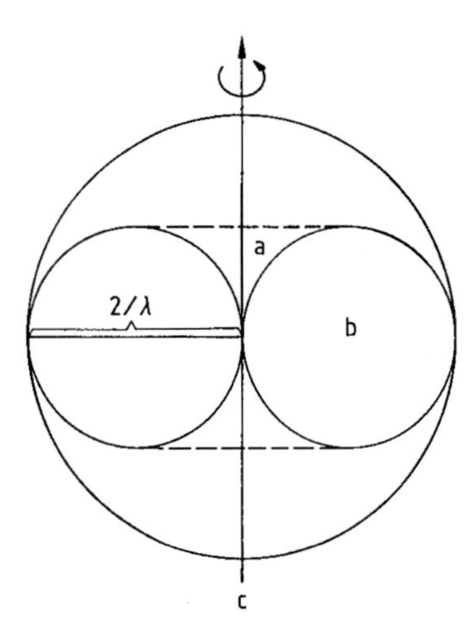

Figure 5. The region of reciprocal space that can be measured for a single setting of a rotating crystal
a) Blind spot; b) Toroidal region of reciprocal space covered for a single setting; c) Crystal axis AB

about AB, reciprocal lattice points lying on reciprocal lattice planes perpendicular to AB follow circular paths, cutting the reflection sphere. Every such cut corresponds to a reflection.

The portion of reciprocal space inside the Ewald limiting sphere that can be measured with a single crystal setting is shown in Figure 5. Almost all lattice points of the limiting sphere are registered when measurements are done with two crystal settings about axes forming an angle ca. 90°.

It is generally true that a vector of the reciprocal lattice is perpendicular to a plane in the direct lattice and, conversely, a vector of the direct lattice is perpendicular to a plane of the reciprocal lattice.

The deeper connection between the real and reciprocal lattices becomes clear when it is shown (see Eqs. 13, 18) that the absolute magnitudes of the functional values of the Fourier transform of the entire crystal are equal to the square roots of the measured intensities. The position coordinates of these functional values are coordinates in reciprocal space (i.e., the indices of the reflections) if nonzero functional values are allowed only at reciprocal lattice points. A reflection can then be defined completely generally as a reciprocal lattice point. Further details are presented in the textbooks cited at the end of this article.

15.2.1.3. Experimental Principles, Applications

If suitable single crystals of a compound are available, it is a routine matter to determine the cell constants. The crystalline specimen should not be much smaller than 0.3 – 0.5 mm. A crystal volume of 0.005 mm^3 is adequate for measurements on organic compounds; much smaller crystals can be used in the case of inorganic compounds with strongly scattering atoms. Air-sensitive substances can be sealed in borosilicate capillaries.

The following method has rapidly gained acceptance where it is possible or desirable to measure the crystals at low temperature (ca. – 100 °C). The crystalline specimen is embedded in an oil that is liquid at the embedding temperature, but solidifies at low temperature [61]. Measurements can be performed on extremely air-sensitive substances, and on those that are liquid at room temperature. The method has the drawback that the crystal employed for data collection is normally not recoverable at the end of the measurement.

Before the introduction of laboratory computers, virtually the only way of determining cell constants was the photographic method. With the Weissenberg and precession cameras, as well as the Space Explorer, a "mixture" of cell constants in both reciprocal and direct space is obtained. The relationships between these cell constants are expressed by the following equations:

$$a = \frac{1}{V^*} b^* c^* \sin\alpha^* \tag{4}$$

$$b = \frac{1}{V^*} c^* a^* \sin\beta^* \tag{5}$$

$$c = \frac{1}{V^*} a^* b^* \sin\gamma^* \tag{6}$$

$$\cos\alpha = (\cos\beta^* \cos\gamma^* - \cos\alpha^*)/(\sin\beta^* \sin\gamma^*) \tag{7}$$

$$\cos\beta = (\cos\alpha^* \cos\gamma^* - \cos\beta^*)/(\sin\alpha^* \sin\gamma^*) \tag{8}$$

$$\cos\gamma = (\cos\alpha^* \cos\beta^* - \cos\gamma^*)/(\sin\alpha^* \sin\beta^*) \tag{9}$$

$$V^* = a^* b^* c^* \sqrt{1 + 2\cos\alpha^* \cos\beta^* \cos\gamma^* - \cos^2\alpha^* - \cos^2\beta^* - \cos^2\gamma^*} \tag{10}$$

Symbols with asterisks relate to the reciprocal lattice; a, b, c are lengths, α, β, γ are angles, and V is the cell volume. If the left-hand sides of these equations are replaced by the cell constants of the reciprocal lattice, the right-hand sides contain the corresponding lattice values in direct space.

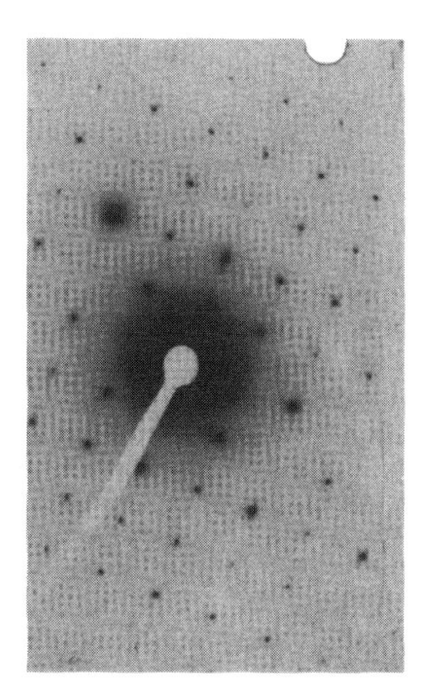

Figure 6. Part of the electron diffraction pattern of a single crystal (MoO_3)

The determination of cell constants is particularly convenient with the precession camera, in which the crystal has to be set on a goniometer head; one reciprocal lattice axis must lie parallel to the goniometer head axis. Two settings of the goniometer spindle yield two principal planes in the reciprocal lattice. The third angle of the reciprocal lattice, obtained in terms of the corresponding angle in direct space, is the difference between the two spindle settings. A photographic cone record (see also Fig. 25) represents a complete projection of the reciprocal lattice; this avoids mistakes in the choice of the cell dimensions due to missing planes of the reciprocal lattice. The first two layer lines supply the remainder of the data needed for space group determination.

Equations 4–10 make it possible to calculate the cell constants in direct and reciprocal space. From the symmetries present and the systematic extinction of reflections, the crystal system, the lattice type (Bravais lattice), and the space group can be derived [4, pp. 124–159]. Because the diffraction pattern always has a center of symmetry (Friedel's law, which holds when anomalous scattering is neglected), ambiguities in the space group determination are still possible (11 Laue groups instead of 32 crystal classes). In principle, the space group is not definite until the structure has been completely solved, although in most cases there is a high prior probability for a particular space group.

If the slightest doubt exists whether the crystal system is right, no time should be lost in carrying out a cell reduction by the Azaroff–Buerger method [4]; many computer programs are on the market for this operation, but it is also reasonably easy to perform "manually".

If a computer-controlled single-crystal diffractometer is available, the cell constants, and space group can be determined by the computer. Three methods exist, differing only in the means of determining the reflection coordinates in reciprocal space. In "peak-hunting" programs, the region of reciprocal space within specified boundaries is systematically searched for reflections [62]; the Acedeva program makes a "systematically unsystematic" search for reflections [63], with a random number generator determining the directions in which to drive the diffractometer. Normally, either of these two methods yields a number of usable reflections in reciprocal space, from which the three shortest noncoplanar vectors can be extracted and used to index all the reflections. Finally, the reduced cell and the Niggli matrix are calculated, and thus the crystal system and Bravais lattice can be derived [4]. The third method employs an oscillation photograph, made on the diffractometer by using a polaroid film, from which two-dimensional coordinates of selected reflections are obtained; the search for about ten reflections is based on these coordinates [64].

A diffractometer which performs only point measurements, does not give the cell constants unambiguously, since it deals only with discrete points in the reciprocal space. "Continuous" measurement in the reciprocal space is necessary to make certain that no essential points for unit cell determination have been missed. This fact, together with pedagogical considerations, is the reason that film techniques are still used in the X-ray diffraction laboratory. A modern four-circle diffractometer is equipped to perform the necessary photographic measurements which a three-circle diffractometer is unable to do.

Electron diffraction studies on tiny crystals ($6 \leq 10^{-8}$ m) are possible in the electron microscope. The resulting single-crystal diffraction patterns can be indexed similarly to precession photographs (Fig. 6), provided multiple orientations of the same specimen relative to the electron beam can be set up. The results should always be combined with those of X-ray powder patterns. This technique has the advantage that it does not require the laborious growth of single crystals, but its greatest disadvantage is that the intensities are

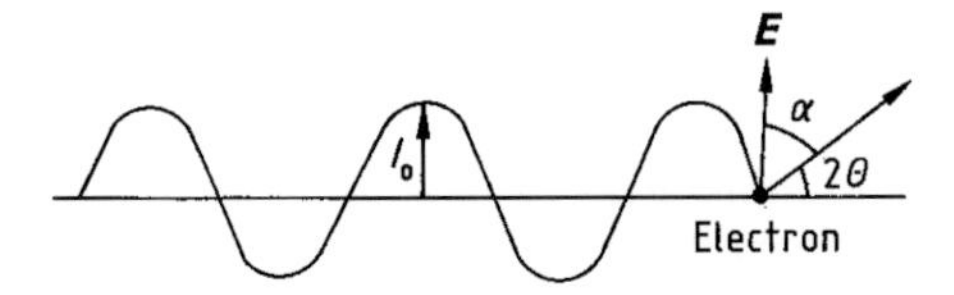

Figure 7. Scattering of X rays by an electron

often totally falsified (this is, however, of no importance for unit cell determination).

Up to this point, only the geometry of diffraction has been considered. However, the intensity ratios of the reflections are the basis for determining the structure of the unit cell contents.

For the intensity ratios to be quantified, scattering must be discussed in more detail. Consider the scattering of X rays by electrons in the atomic electron shells. The scattering power I_e of an electron at point A is given by Maxwell's theory as:

$$I_e = \frac{I_0 e^4}{r^2 m^2 c^4} \sin^2\alpha \tag{11}$$

where I_0 is the incident X-ray intensity, e is the electron charge, m is the electron mass, r is the distance from the scattering electron to the observer, c is the speed of light, and α is the angle between the electric field vector $\boldsymbol{E}$ and the direction to the observer (Fig. 7).

If the incident X rays are unpolarized, $\sin^2\alpha$ must be replaced by the factor $(\frac{1}{2}+\frac{1}{2}\cos^2 2\theta)$, the polarization factor. Here, 2θ is the deflection angle.

X rays employed in structure analysis normally have wavelengths $0.5-2.3\times10^{-10}$ m, the same order of magnitude as the size of atoms and their electron shells. Only at a scattering angle of 0° can e and m in Equation (11) be replaced by Ze and Zm, for an atom with atomic number Z. At other angles, the X rays scattered by the electrons interfere, and the total intensity is reduced as a function of scattering angle. The loss of intensity is greater, the higher the glancing angle θ and the larger the electron shell are.

The best-known example demonstrating the effect of atomic radius is the case of O^{2-} and Si^{4+} (important in the analysis of silicates). As θ increases, the scattered intensity falls off much faster for O^{2-} than it does for the smaller Si^{4+}. In this way, it is possible to distinguish between these two atomic species (Fig. 8). Atomic form factors f for atoms and ions [29, vol. III, pp. 201 – 246] are a function of the scattering power of the corresponding atom, and are expressed in units of a free electron.

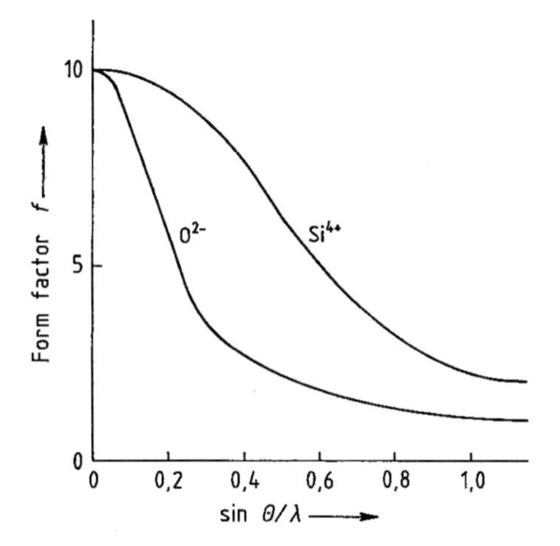

Figure 8. Atomic form factors f for O^{2-} and Si^{4+}

If the frequency of the incident X rays is close to the frequency of the K absorption edge of the irradiated atom, scattering can no longer be regarded as elastic; some of the incident energy is used to excite the electrons in the atom. The atomic form factors have to be corrected for anomalous dispersion; the tables in [29] include this correction. The significance of these corrections for the determination of the absolute configuration is discussed in Section 15.2.4.

Among the key properties of electron beams relating to their scattering by atoms, the electron charge is particularly important. Electrons are scattered by inhomogeneities in the electrostatic potential distribution, chiefly at atomic nuclei. The process is described by the equation:

$$f^e = (Z - f)\frac{e^2 m}{2h^2} \cdot \frac{\lambda^2}{\sin^2\theta} \tag{12}$$

where Z is the atomic number, f is the atomic form factor for X rays, f^e is the atomic form factor for electron beams, and e and m are the electron charge and mass, respectively.

Neutrons are scattered only by atomic nuclei (provided there is no magnetic scattering due to unpaired $3d$ or $4f$ electrons). The size of the atomic nucleus is orders of magnitude smaller than the wavelength of thermal neutrons, and so no dependence of scattered amplitude on glancing angle can be observed.

The X-ray intensities I of reflections produced by lattice planes having Miller indices (hkl) are measured. These are related in the following way to the absolute magnitudes of the structure factors $|F|$, i.e., the structure amplitudes:

$$\|F(hkl)\| = k\sqrt{I(hkl)} \tag{13}$$

Similarly to Equation (11), the following can be derived: Given a crystal of volume V, z unit cells per unit volume, X rays with wavelength λ, and rotation of the crystal through the reflection position at rotational frequency ω; the intensity I of a reflection having structure factor F is:

$$I = \frac{e^4}{m^2c^4} I_0 \|F\|^2 \frac{\lambda^3 z^2 V}{\omega} \frac{1 + \cos^2 2\theta}{2} \frac{1}{\sin 2\theta} \tag{14}$$

where c is the speed of light. The last two factors depend on the recording technique, and are commonly combined in the Lorentz-polarization (LP) factor. Strictly speaking, Equation (14) holds only if the coherent interaction between incident and diffracted X rays is neglected (primary extinction), and if one ignores the fact that the incident X-ray beam is attenuated by diffraction as it propagates through the crystal (secondary extinction), so that not all portions of the crystal contribute equally to diffraction.

All these falsifications become negligibly small if the crystal has an "ideal mosaic" structure, and the "mosaic particles" are smaller than 1 μm. Generally:

$$I \sim \|F\|^n \tag{15}$$

where n is a number between 1 and 2. If $n = 1$, Equation (14) gives:

$$I = \frac{8}{3\pi} \frac{e^2}{mc^2} I_0 \|F\| \frac{\lambda^2 z}{\omega} \frac{1 + \|\cos^2 2\theta\|}{2 \sin 2\theta} \tag{16}$$

Diffraction is controlled by Equation (14) if the kinematic theory is viewed as completely valid ($n = 2$), and by Equation (16) if the dynamic theory ($n = 1$) is applied. Normally, neither of these extremes is exactly fulfilled, and there is a combination of both. The kinematic theory is usually adopted, with "dynamic" contributions being handled by an extinction correction, although this (like the absorption correction) often proves unnecessary for measurements of standard level.

The structure factor F is needed to determine the electron density ϱ at point x, y, z (in fractional units of the cell constants) and thus to determine the crystal structure:

$$\varrho(xyz) = \frac{1}{V} \sum_{h,k,l=-\infty}^{+\infty} F(hkl) \exp\left[-2\pi i(hx + ky + lz)\right] \tag{17}$$

$$F(hkl) = \sum_{n=1}^{N} f_n \exp\left[2\pi i(hx_n + ky_n + lz_n)\right] \tag{18}$$

The summation in Equation (17) is over all measurable reflections (hkl), and that in Equation (18) is over all atoms N in the unit cell; V is the cell volume; x_n, y_n, z_n are the coordinates of atom n; and f_n is its atomic form factor.

The significance of systematic extinctions for space-group determination can be seen from Equation (18). In space group $P2_1$, for example, for an atom at x, y, z there is another at $-x, y+½, -z$. The effect on the ($0k0$) reflections is that the contributions of these two atoms to the structure factor differ in phase by 180°. If k is odd they therefore cancel each other, but are additive if k is even. The following equation represents the situation for space group $P2_1$:

$$F(0k0) = \sum_{n=1}^{N/2} f_n \left[\exp(2\pi i k y_n) + \exp\left(2\pi i k\left(y_n + \frac{1}{2}\right)\right)\right] \tag{19}$$

Corresponding relations have been tabulated for all space groups [29], [30].

Equations (17) and (18) further show a relationship between real and reciprocal space. The function $F(hkl)$ is the Fourier transform of the unit cell contents, expressed in the reciprocal space coordinates h, k, and l. Because the symmetry operation of translation holds for all three spatial directions in crystals, the Fourier transform of the entire crystal is zero, except at reciprocal lattice points.

15.2.1.4. Crystal Growth for X-Ray Structure Analysis

While the growing of single crystals from the melt or the gas phase may sometimes be important for industrial purposes (as in the crucible pulling method used for semiconductors such as Si and Ge), in X-ray structure analysis such methods are secondary to crystal growth from solution, and they are employed only when other techniques are not possible.

In principle, suitable crystals can be grown in either of two ways:

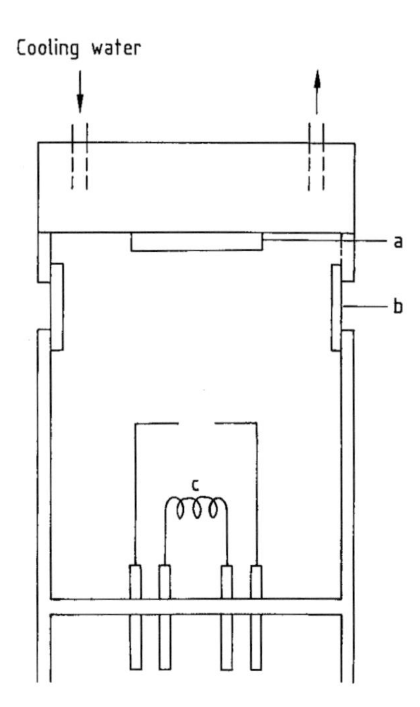

Figure 9. X-ray tube
a) Anode; b) Be window; c) Hot cathode

1) The first approach uses the temperature dependence of solubility. In accordance with the maxim "like dissolves like", the solvent employed is a compound with the same or similar parent structure, and/or having the same or similar functional groups. Normally, compounds are more soluble at higher temperature. Slow cooling, which can be effected under programmed control, yields crystals suitable for structure analysis in most instances (maximum dimension 0.3 – 0.8 mm). A simple way to bring about slow cooling is to place the solution in a Dewar flask filled with a liquid that can be heated to the boiling point of the solvent without itself vaporizing to any great extent.
2) The second approach depends on raising the concentration. The simplest method is to evaporate the solvent; generally, the more slowly this is done, the better are the crystals. A straightforward way of doing this is to pierce one or more small holes in the stopper of the crystallization flask.

Evaporation is not the only way of raising the concentration. Another possibility is to use the diffusion of two liquids into each other; the liquids should be miscible, and the substance to be crystallized should have greatly different solubilities in them. The diffusion can take place in the liquid or the gas phase. In the first case, a layer of the "weaker" solvent (lower solubility) is formed on top of the solution, and the first crystal nuclei will form at the interface between the two solvents. In the simplest example of diffusion via the gas phase, an open vessel containing the solution with the "stronger" solvent (higher solubility) is placed inside a vessel containing the "weaker", and the whole assembly is hermetically sealed.

In principle, the handling and growing of crystals at low temperature can be done by the same methods, with appropriate modifications [61]. It is even possible to grow a crystal in the measuring instrument itself [65].

15.2.1.5. X Rays, Neutrons, and Electrons

X rays of sufficient intensity for diffraction studies are obtained from an X-ray generator, which essentially consists of two portions, the high-voltage generator and the tube housing. The generator transforms 220 V alternating current to a d.c. voltage of ≤ 60 kV. The total power is generally in the range 3 – 6 kW, but the output of the sealed X-ray tubes (Fig. 9) used for diffraction purposes is preferably 2.5 kW. In general, there is no improvement in luminance with a 3 kW tube; the only difference is that the focus is longer.

In "normal-focus" tubes, the focal spot measures 1×10 mm. If this spot is viewed at a 6° angle in the direction of its longest dimension, the effective spot size is 1×1 mm (point focus); in the perpendicular direction, the effective size is 0.1×10 mm (line focus). A typical X-ray tube has four windows, two with point focus and two with line focus. Specimens extended in two dimensions are irradiated with the line focus; point, sphere, or needle-shaped specimens, with the point focus. Fine-focus tubes (0.4×8 mm) are commonly used for single-crystal studies; these offer markedly higher luminance at the same energy output.

The residual gas pressure in the envelope of an X-ray tube (where the hot cathode and the anode are located) is $< 10^{-1}$ Pa. The appropriate voltage for the anode material is applied between the cathode and the anode (Table 1). Electrons emitted by the hot cathode are accelerated by the applied voltage, then braked when they strike the anode. If the electrons have sufficient kinetic energy to cause internal ionization of the atoms in the anode, a small part of the energy released (ca. 1 %) is consumed in ejecting electrons from the K shell

Table 1. Data on X-ray generation

Anode material	Excitation voltage, kV	Operating voltage, kV	$K\alpha_1$, nm	$K\alpha_2$, nm	$K\alpha$, nm	$K\beta_1$, nm	$K\beta$ filter
Cr	6.0	20–25	0.22896	0.22935	0.22909	0.20848	V
Fe	7.5	25–30	0.19360	0.19399	0.19373	0.17565	Mn
Co	7.8	30–35	0.17889	0.17928	0.17902	0.16207	Fe
Ni	8.5	30–35	0.16578	0.16617	0.16591	0.15001	Co
Cu	9	35–40	0.15405	0.15443	0.15418	0.13922	Ni
Mo	20	50–60	0.07093	0.07135	0.07107	0.06323	Zr
Ag	25.5	60	0.05594	0.05638	0.05609	0.04970	Pd

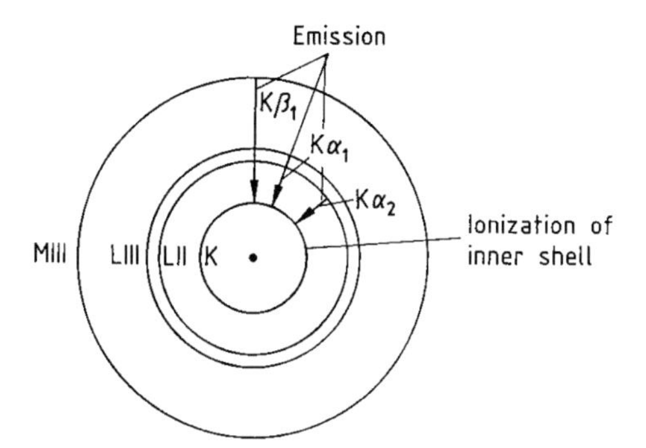

Figure 10. Generation of X rays

of the atoms. The vacant energy levels are filled, chiefly by electrons from the L shell (L III and L II levels) and also from the M shell (M III level). The energy thus gained is emitted as X rays (Fig. 10). The quantum transitions differ in probability, which can be varied to some extent through the operating voltage (see Table 1).

Given the proper tube voltage, the intensities of the principal components are in the ratio:

$$K\alpha_1 : K\alpha_2 : K\beta_1 = 10 : 5 : 2 \qquad (20)$$

A mixture of $K\alpha_1$ and $K\alpha_2$ radiation can be used for most diffraction studies; the $K\beta_1$ is removed by a filter or a crystal monochromator. The main constituent of the filter material is an element with an atomic number one less than the anode material; the X-ray spectrum of the filter material then has an absorption edge at the wavelength of the $K\beta$ radiation to be removed. When a monochromator is used, the various wavelengths are separated by diffraction through a crystal, usually a graphite or α-quartz single crystal. The Bragg equation then gives different glancing angles θ for the several wavelengths, so that the unwanted radiation components can be eliminated.

The most common anode materials for X-ray tubes are copper and molybdenum; silver, iron, and chromium are also used. Compounds containing only light atoms (C, N, and O) show only small absorption for molybdenum radiation, while heavier elements have correspondingly higher absorption, but a much more reduced one compared to Cu K_α radiation. Because the scattered intensity is directly proportional to λ^3 (Eq. 14), a copper anode has to be used for extremely small and weakly scattering specimens. The opposite is the case for larger specimens: The low absorption of Mo radiation shifts the intensity ratios in favor of molybdenum. X-ray tubes with chromium anodes generally have low output, air scattering is very high, and absorption problems are difficult or impossible to cope with, at least in single-crystal studies.

Contrasted with these sealed X-ray tubes are rotating anode tubes. The tube is not sealed and the vacuum must be maintained by a pump. The rotation of the anode and the more intensive cooling of the anode material mean that the specific energy loading of the anode can be made far higher; e.g., a generator of power 1.2 kW can yield a specific output of 12 kW/mm^2 with focal spot 0.1×1 mm. For comparison, a fine-focus tube at 90 % of maximum loading has a specific output of only 0.56 kW/mm^2. Rotating anode tubes have now become so reliable that sealed X-ray tubes may disappear from practical use in the near future.

Another kind of X-ray generators are the microfocus X-ray sources. The focus has an area of only 30 μm^2, and with an operating power of 24 W it attains the same brightness as a 5 kW rotating anode generator. The future will show whether its promise holds for very small crystals. A two orders of magnitude higher X-ray intensity is obtained by using synchrotron radiation. This highly monochromatic, parallel, and brilliant X-ray radiation should only be used when the experiments demand this very expensive X-ray source (small crystals, protein crystals, high scattering angle resolution, very special experiments).

Neutron Beams. The neutron was discovered by CHADWICK in 1932. The equation:

$$m v = \frac{h}{\lambda} \tag{21}$$

where h is Planck's constant, m is the mass of the particle, λ is the wavelength, and v is the particle velocity, makes it possible to regard any moving particle as a wave. The neutron has not only mass but also magnetic moment; accordingly, it can interact with atomic nuclei and with unpaired electrons. Practical neutron sources are high-flux nuclear reactors such as those at Grenoble, Oak Ridge, and Brookhaven, which offer neutron fluxes of roughly $1.5\times10^{15}\ cm^{-2}\, s^{-1}$. Reactors available in Germany have fluxes about an order of magnitude lower (BERII in HMI Berlin: $1.2\times10^{14}\ cm^{-2}\, s^{-1}$).

The wavelength of neutron radiation is controlled by the absolute ambient temperature T:

$$\frac{1}{2} m v^2 = \frac{2}{3} k T \tag{22}$$

where k is Boltzmann's constant, and Equation (21) gives the value of λ. Neutrons in equilibrium at room temperature (thermal neutrons) have a wavelength of ca. 10^{-10} m. Monochromators for neutrons are generally diffracting crystals. Time-of-flight spectrometers make use of the fact that neutrons with a certain velocity correspond to neutron radiation with a certain wavelength.

Electron Beams. Electrons, like neutrons, are a form of particle radiation. If matter absorbs neutrons far less than X rays, exactly the opposite holds for electrons. They are so strongly absorbed by air that electron diffraction by the solid phase must generally be carried out in the high vacuum of the electron microscope. The wavelength, and thus the energy of electron radiation are controlled by the voltage drop V through which the electrons pass:

$$\frac{1}{2} m v^2 = e V \tag{23}$$

where e is the elementary charge, m is the electron mass, and v is the electron velocity. This, in combination with Equation (21), gives the wavelength:

$$\lambda = 12.236 V^{-1/2}\left(10^{-10}\,\text{m}\right) \tag{24}$$

For example, if the voltage drop is 100 kV, the electron beam has a wavelength of 0.039×10^{-10} m. Because electrons do not penetrate to a great depth, they are particularly suited to the study of thin films and surface structures. The scattering length is four orders of magnitude larger than for X rays and neutrons. This leads not only to a correspondingly higher absorption, but also to an extremely high scattering power of the atoms.

15.2.1.6. Cameras and Diffractometers

Three types of detector are available for measuring the X rays diffracted by a crystal: Point detectors, linear and area detectors. The film method involves a two-dimensional detector, which can provide simultaneous information about every point in the region of reciprocal space investigated. The measurements are not immediately available in digital form, but powerful computer-controlled photometers can digitize them.

This technology is sometimes still used in protein crystallography, but the electronically controlled area counter is being introduced in this field. This type of position-sensitive counter directly supplies digital values as a function of position in reciprocal space; it also has a dynamic range several orders of magnitude wider than the film. In conjunction with a rotating anode X-ray source, it forms a data acquisition system with which not just proteins, but also other large molecules with poorly scattering atoms or very small crystals can be measured in an optimal way [66] [67]. One-dimensional position-sensitive detectors are available for powder diffractometry [68].

Point measurements with scintillation or proportional counters yield equally precise values. However, important parts of the information can be lost if it is not borne in mind that only those points can be measured that are moved into reflection position with complete prior knowledge of that position.

Recording methods also differ with which clompleteness a region of reciprocal space can be measured. In the analysis of crystalline powders, where the orientations of the lattice planes are expected to show a statistical distribution, the largest measurable glancing angle controls the ex-

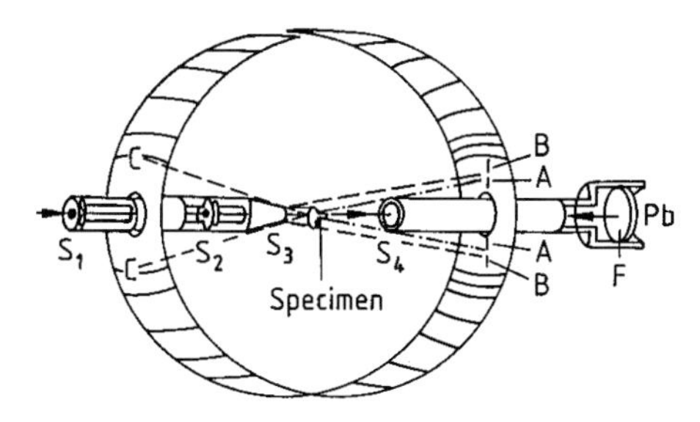

Figure 11. Debye – Scherrer camera

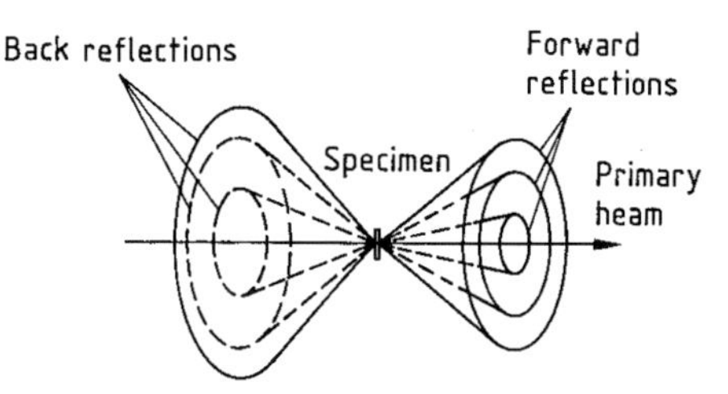

Figure 12. Diffraction cones in the crystalline powder method

Figure 13. Debye – Scherrer diffraction pattern of LiF

tent of reciprocal space that can be covered. The situation is similar with the four-circle single-crystal diffractometer; every net plane can be brought to reflection position, provided a maximum glancing angle (characteristic of the instrument) is not exceeded. In all other single-crystal goniometers, there are "blind regions" in the Ewald sphere that cannot be measured.

Film Methods. The prerequisite for using the film method is a suitable film material. The Committee for Crystallographic Apparatus of the International Union of Crystallography has tested a variety of X-ray films [69]. The properties of a usable X-ray film can be characterized as: Fine grain and homogeneous emulsion; low sensitivity to temperature and moisture; low susceptibility to fogging; wide range of linearity between film blackening and logarithm of exposure time; and high sensitivity to the wavelengths produced by X-ray tubes.

Nowadays it can be difficult to obtain film material of such quality because the medical purposes for which it was primarily intended have mostly vanished. However, the conditions can be relaxed because quantitative intensity measurements are no longer made by using films.

Debye – Scherrer Camera. Figure 11 illustrates the principle of the Debye – Scherrer camera, while Figure 12 shows schematically how the diffraction pattern is generated. A specimen of crystalline powder (ca. 0.1 mg) is placed in a borate glass capillary of diameter 0.3 – 1 mm and length ca. 1 cm, and the capillary is rotated about a specified axis. The film fits snugly inside the camera. In a camera with internal diameter 114.6 mm, 1 mm on the film corresponds exactly to a diffraction angle of 1°. The radiation passes through the collimator (with three diaphragms S_1, S_2, S_3) to reach the specimen; S_1 and S_2 have the same diameter, while S_3 is somewhat larger. The primary beam is absorbed in the primary beam trap, which carries diaphragm S_4, and can be looked at through the lead-glass window, Pb, on the fluorescent screen F for setting purposes. The position and size of screen S_4 limit the smallest measurable diffraction angle 2θ, and thus the resolution power of the camera (A – A) in the small angle region. The position of S_3 relative to S_2 determines the range of angles within which scattered radiation from S_2 is observed (B – B). S_3 also masks backscattering, so that no diffracted X rays can reach the region C – C. Figure 13 shows a Debye – Scherrer pattern. The intersections of the diffraction cones in Figure 12 with the film give the Debye – Scherrer rings.

The camera should be evacuated while the pattern is being recorded, to minimize the background due to air scattering.

This method has the advantages that a small specimen can be used, operation is simple, and there is only slight falsification of the intensities due to nonstatistical distribution of crystallite orientations in the specimen. Disadvantages of the method are that exposure times are quite long (up to and beyond 12 h) and interpretation of measurements is inconvenient, especially when there are many specimens, because the results are not in digital form.

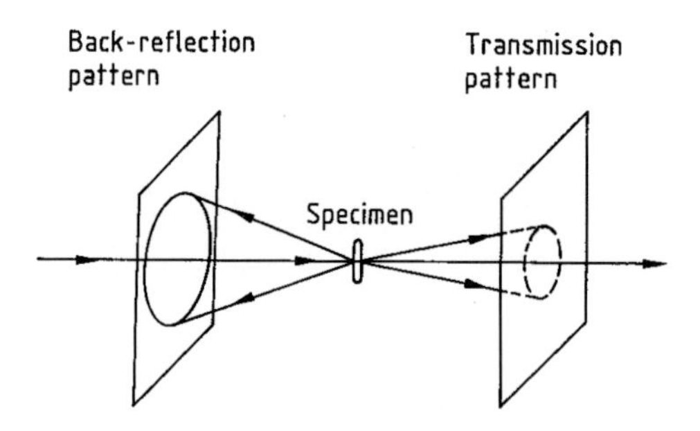

Figure 14. Laue camera

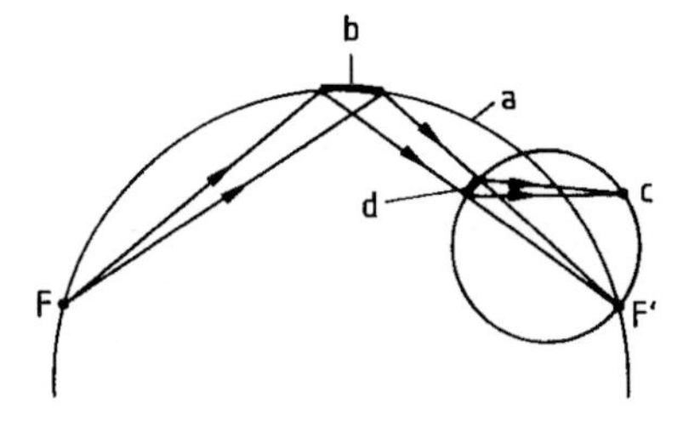

Figure 16. Principle of the Guinier method
a) Seemann – Bohlin circle; b) Monochromator; c) Reflection; d) Specimen

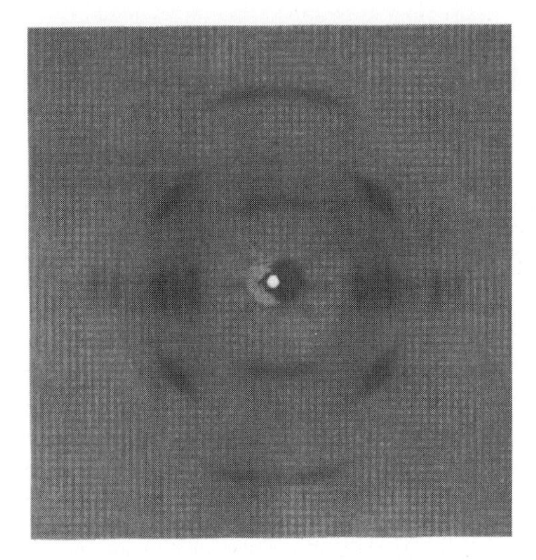

Figure 15. Flat-film diffraction pattern of polypropylene fibers

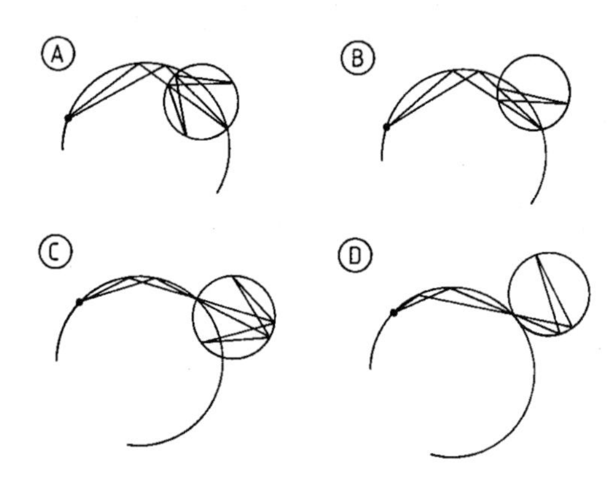

Figure 17. Guinier method, showing the four possible relative positions of specimen and monochromator
A) Symmetrical transmission; B) Asymmetrical transmission; C) Symmetrical reflection; D) Asymmetrical reflection

Laue Camera. This flat-film camera can record both transmission and back-reflection patterns. The principle is illustrated in Figure 14; Figure 15 shows a flat-film pattern. Transmission patterns with polychromatic radiation yield information about the symmetry of a crystal (Laue symmetry) or with monochromatic radiation the orientation of the crystallites (preferred orientation, texture); such patterns are recorded, for example, from fibers. Back-reflection patterns permit the nondestructive examination of workpieces.

In the Laue method, single crystals are irradiated with polychromatic X rays, so that the crystal need not be rotated. All net planes have wavelengths for which the reflection conditions are satisfied. The patterns are employed for the alignment of single crystals and the identification of their symmetries.

Guinier Camera. The Guinier camera is a focusing goniometer; for a certain geometric configuration of specimen and X-ray tube focus F, the X rays are focused, so that higher angular resolution is achieved (Fig. 16) . The radiation is reflected from a bent quartz monochromator (Johansson principle). The focussing circle (Seemann – Bohlin circle) is coupled with it in such a way that both circles have the same focal axis [70]. In addition to the symmetrical transmission configuration shown, three other relative positions of the monochromator and Seemann – Bohlin circles are possible, so that the entire X-ray diffraction spectrum can be recorded; these are the asymmetric transmission position, the symmetric reflection position, and the asymmetric reflection position (Fig. 17).

The specimen, in the form of a crystalline powder, is applied to the substrate in the thinnest possible layer. With strongly scattering inorganic

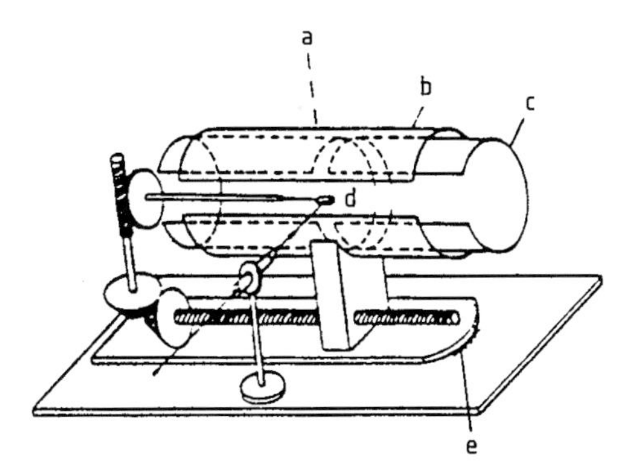

Figure 18. Principle of the Weissenberg camera
a) Equi-inclination axis; b) Film; c) Layer-line screen; d) Crystal; e) Equi-inclination angle setting

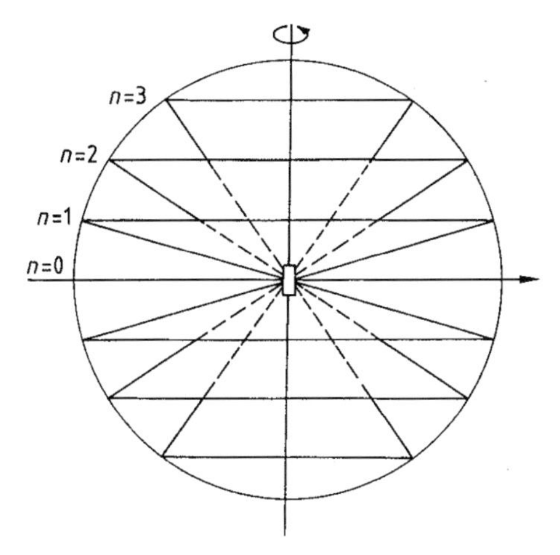

Figure 19. Diffraction cones in the rotating-crystal method (n = number of layer line)

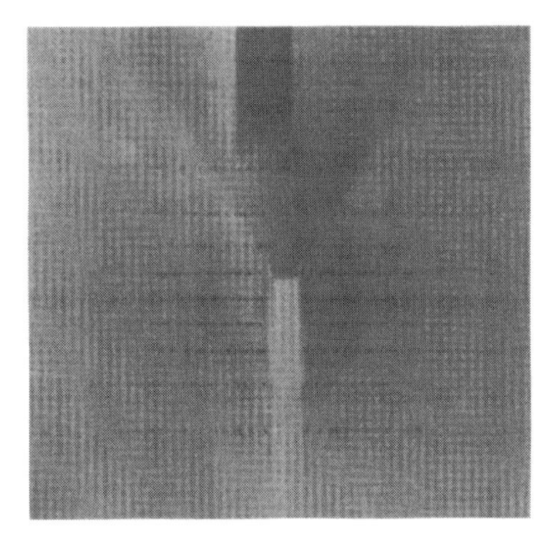

Figure 20. Rotating crystal pattern

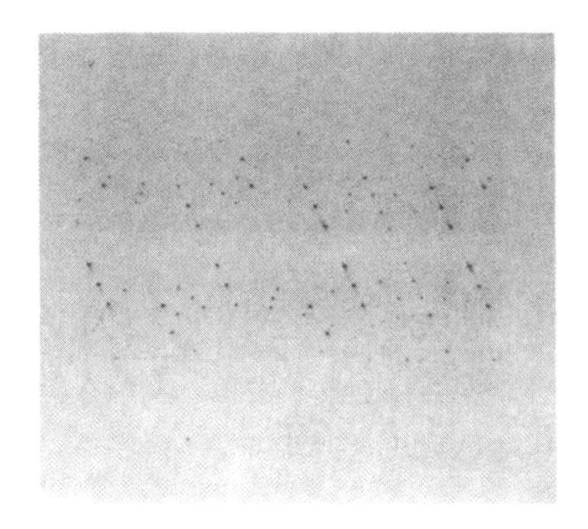

Figure 21. Weissenberg pattern

substances, the specimen can be dusted onto a film of grease. With organic substances, it is necessary to work with a layer at least 0.1 – 0.3 mm thick, which may be sandwiched between foils, with the result, that the resolution of the system is somewhat less.

Weissenberg Camera. Figure 18 shows the camera schematically, while Figure 19 illustrates the registration of a diffraction pattern. This camera is used in recording diffraction data from single crystals [10, pp. 221 – 225]. If the layer-line screens are removed, an overall record of the reciprocal lattice is obtained (rotating crystal pattern; Fig. 20). With these screens in place, the individual diffraction cones of Figure 19 (the layer lines) can be isolated. Coupling the crystal rotation with a translation of the film holder in the direction of the rotation axis spreads the one-dimensional layer line of the rotating crystal pattern into a layer plane of the reciprocal space. If the crystal is rotated about one of the crystallographic axes a, b, c, the number of the layer line gives the index of the corresponding reciprocal axis. The diffraction points can be uniquely indexed, even though the reciprocal lattice is recorded in distorted form (Fig. 21). The method is superbly suited to intensity measurements, since an entire film pack can be placed in the film cassette and exposed to the X-ray beam; the films are then exposed simultaneously, but exposure varies with the penetration depth of the X rays.

The normal technique is the equi-inclination method, in which the direct and diffracted X-ray

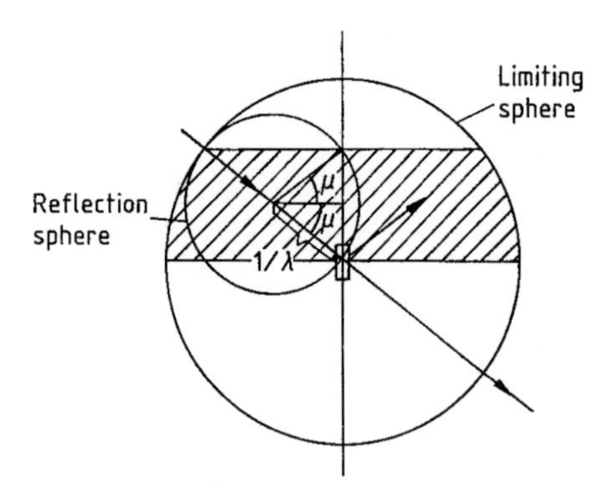

Figure 22. The portion of the limiting sphere that is accessible with the equi-inclination method

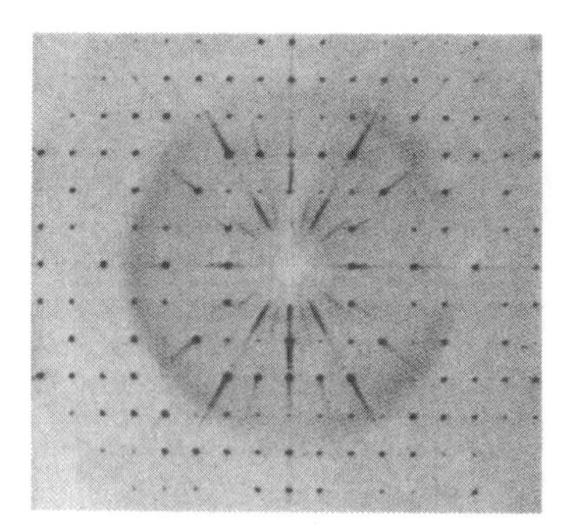

Figure 24. Precession pattern

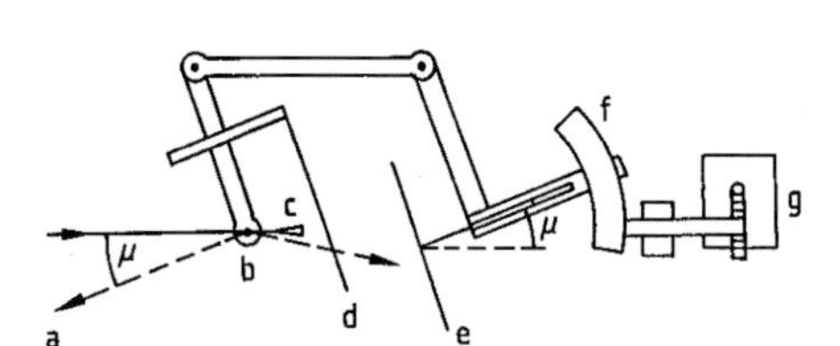

Figure 23. Precession camera
a) Vector in real space; b) Crystal; c) Primary beam stop; d) Layer-line screen; e) Film plane; f) Precession angle setting; g) Motor

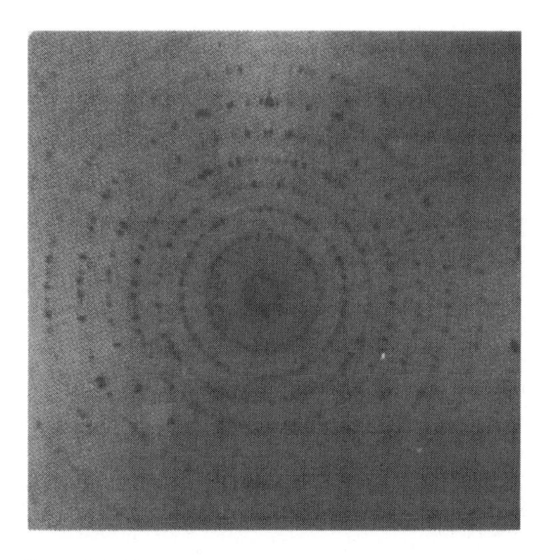

Figure 25. Cone pattern

beams make equal angles with the layer line being measured. In addition to increasing the number of measurable reciprocal lattice points, this also simplifies the necessary intensity corrections. Figure 22 shows that the maximum length of the reciprocal lattice vector parallel to the rotation axis is:

$$\zeta_{max} = \frac{2}{\lambda} \sin\mu_{max} \tag{25}$$

The maximum equi-inclination angle μ_{max} is normally 30° ($\zeta_{max} = 1/\lambda$), so that ζ_{max} is the same as in the normal rotating crystal pattern, except that there is no blind region near the rotation axis (Figs. 5 and 22).

Precession (Buerger) Camera. In contrast to the Weissenberg camera, the crystal carrier is now coupled to the film cassette so that both move in exactly the same way while the pattern is being recorded (Fig. 23). A direction in the crystal precesses (precession angle μ) about the incident X-ray beam, so that reciprocal lattice points lying in a plane perpendicular to this crystal direction cut the reflection sphere. A precession pattern represents the reciprocal lattice in undistorted form (Fig. 24); the reflections are easily indexed and the pattern is easily interpreted [9, pp. 15–68]. The cone pattern of Figure 25, which shows the layer lines measurable with a given crystal setting, corresponds to the rotating crystal pattern.

Figure 26 illustrates the region of reciprocal space whose lattice points can be measured with a single spindle setting (shaded area). The maximum ζ is $2/\lambda \cos\mu$. One further spindle setting, which should be as close as possible to 90° away from the first, will eliminate most of the blind spot about the ζ-axis.

Space Explorer. The Weissenberg camera records only reciprocal lattice planes perpendicular to the crystal axis that is parallel to the goniometer

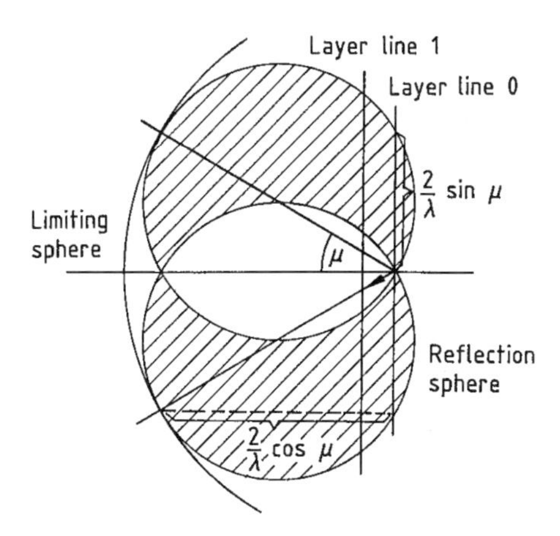

Figure 26. Region of the limiting sphere covered by the precession method

head axis, and the precession camera can measure only lattice planes that have a common reciprocal lattice axis parallel to the goniometer head axis and lattice planes which are parallel to these [24, pp. 98–167]. With the Space Explorer both types of series can be recorded. This facilitates the determination of cell constants and space groups in many cases.

These film methods may appear old fashioned, but this is wrong. Any serious X-ray laboratory will have a few of these machines, which are always necessary to give a first impression of the diffraction problem on a relatively cheap apparatus, especially when the expensive computer-controlled diffractometers are not immediately available or when there are problems and a three-dimensional inspection of the reciprocal space is necessary to gain an impression of the difficulties.

Counter Tube Method. The term "counter tube" means a detector that moves on a circle, the equator of the diffractometer. Only point measurements along this circle are possible. If one wishes to measure reflections that do not lie on this plane, the specimen must be oriented to position the reflections to be measured on the equator.

The use of this type of counter tube demands complicated electronics that make it possible to count the absorbed X-ray quanta. In the proportional counter, every X-ray quantum initiates a gas discharge; the voltage surge is electronically amplified and recorded as a pulse. In the scintillation counter, the X-ray quanta absorbed by the scintillator crystal are transformed into light (fluorescence), and amplified by a photomultiplier. The two types of counter tube have very similar properties, and the pulses, once amplified, can be processed by the same electronics. A device called an electronic discriminator can block a given energy range ("window") and thus render the beam more monochromatic, greatly increasing the ratio of peak to background. Unfortunately, the use of a filter or a monochromator crystal to separate the $K\alpha$ and $K\beta$ radiation cannot be dispensed with, because the resolution of an ordinary discriminator is not sufficient to do so.

The situation is different if a semiconductor detector (lithium-doped silicon crystal) and the proper electronics are available. This setup can separate $K\alpha$ and $K\beta$ radiation, so that the beam does not have to be passed through a monochromator. The counter tube itself has nowadays become easier to operate, for the semiconductor crystal no longer has to be maintained at liquid nitrogen temperature—the problem has been solved electronically. Efforts continue to find broader applications for this type of detector, since replacement of the monochromator by electronics means that there is no longer a 50% loss of primary intensity through diffraction by the monochromator crystal or absorption by the β filter.

The diffractometer must have a precision mechanical system for the positioning of specimen and detector, with a reproducibility of at least 0.01°. Digital evaluation makes it relatively easy to perform a profile analysis of the reflections.

In comparison with film cameras, diffractometers are roughly a factor of 10 more costly. They are still more expensive if the point counter is replaced by a one- or two-dimensional position-sensitive detector. The data acquisition time is, however, shortened by a factor of 50–100, or even more, depending on the geometry of the crystal lattice measured.

There are four types of position-sensitive detectors: Wire detectors (one- and two-dimensional), video detectors, CCD (charge coupled device chip), and image-plate detectors. The wire detector is based on the proportional counter principle; voltage pulses produced by quanta incident at different points along the wire take different times to reach its two ends. The travel times are derived from the profile of the sum of the two pulses. In the video counter, a phosphor layer generates light flashes where X-ray quanta are incident, and the flashes are picked up by a television camera. The CCD chip is covered by a scintillation foil, which

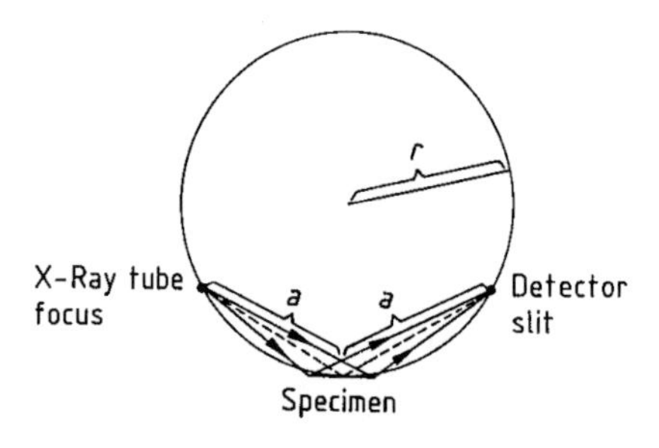

Figure 27. Beam path of the Bragg – Brentano two-circle powder diffractometer

is specific for a certain X-ray wavelength. The generated light flashes are counted by the CCD. The image-plate has a coating of rare-earth oxides, in which metastable states are produced by X-ray quanta. Transitions back to the ground state, induced by laser light, can be observed and quantified.

These position-sensitive detectors permit the use of simpler mechanical systems in diffractometers. A one-dimensional counter makes it possible to leave out one circle; a two-dimensional counter, two circles. In principle, then, a single-crystal diffractometer equipped with a two-dimensional counter needs only one circle, while a powder diffractometer equipped with a one-dimensional counter does not require a circle at all. Experience has shown that the greatest flexibility is achieved by incorporating these counters without reducing the number of circles.

A direct oscillation photograph, for example, can only be made by a four-circle diffractometer equipped with a two-dimensional detector.

Two-circle powder diffractometer fitted with a point counter. There are three powder diffractometer geometries with focusing optics, whereby the monochromator, the specimen, or both, can have a focusing characteristics. Usually, they are two-circle diffractometers.

The Bragg – Brentano two-circle diffractometer has a Seemann – Bohlin focusing circle, on which the X-ray tube anode, the counter tube window, and the specimen are located (Fig. 27). Only the specimen is focusing.

The planar specimen is placed tangentially to the focusing circle. Because theoretically this can be the case only along a line parallel to the line focus, there are residual geometric beam-spreading effects. The radius r of the focusing circle decreases with increasing θ, from infinity at 0° to $a/2$ at 90°. In order to maintain the focusing

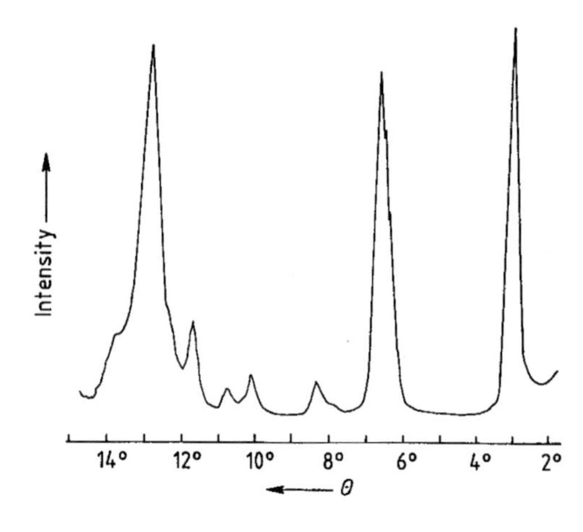

Figure 28. X-ray diffraction pattern of the γ-phase of quinacridone, recorded in a Bragg – Brentano diffractometer with Cu K α radiation

conditions while a diffraction pattern is being recorded, the detector is moved at twice the angular velocity of the specimen carrier. Because the specimen extends in two dimensions, it is possible, and necessary, to work with the line focus of the X-ray tube (perpendicular to the plane of Fig. 27); this makes very effective use of the X-ray tube anode surface.

As an alternative to a β filter, beam monochromatization can be achieved by a graphite monochromator between specimen and detector; this arrangement removes not only K β radiation, but also fluorescence emitted by the specimen. The effect is particularly beneficial in the study of iron-containing specimens with Cu K α radiation.

An $x-y$ plotter connected to the instrument (if it is not inherently computer-controlled) makes it possible to output the scattered X-ray intensity in analog form as a function of diffraction angle (Fig. 28). The intensity of the diffracted beam can also be output in digital form. An especially economical approach is to fit the instrument with an automated specimen changer, record the measured X-ray intensities on a mass storage medium (magnetic tape, diskette, or preferably yet a hard disk), and control the operation of the diffractometer by computer. Data measured, say, overnight or over the weekend can then be interpreted during regular working hours with the same computer.

In the *Debye – Scherrer diffractometer* with monochromator (Fig. 29), the line focus of the X-ray tube, the monochromator crystal, and the detector window all lie on the focusing circle of

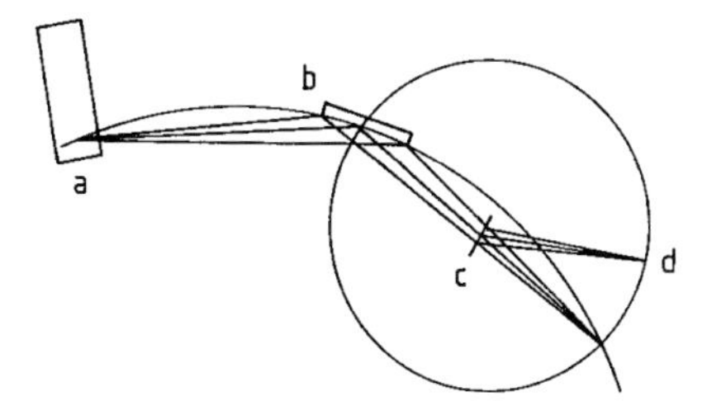

Figure 29. Beam path of the Debye–Scherrer diffractometer
a) X-ray tube; b) Monochromator; c) Specimen; d) Detector and detector slit

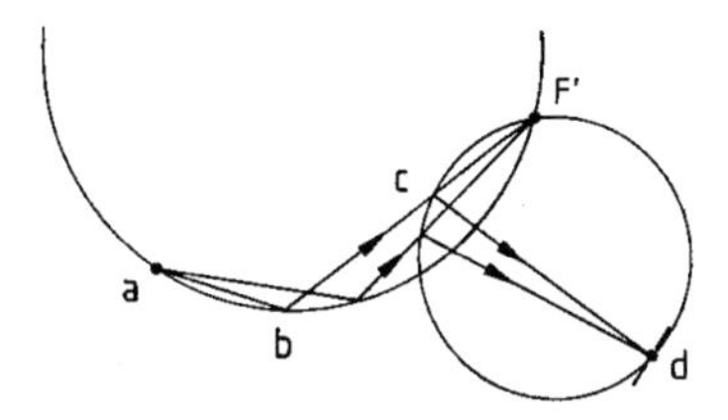

Figure 30. Beam path of the Guinier diffractometer
a) Line focus of X-ray tube; b) Monochromator; c) Specimen; d) Detector slit

the curved monochromator crystal (zero setting). The specimen is measured in transmission. For inorganic substances with relatively high scattering power, a specimen dusted onto a foil with the thinnest possible coating of grease is sufficient. This technique is generally not adequate for organic specimens; instead, the specimen is contained between two thin foils, separated by a spacer. It is useful to do a transmittance measurement in the primary beam before performing the diffraction measurement itself. A transmittance of about 70 % should be obtained. For substances that do not absorb too strongly, the specimen can be placed in a capillary; among the advantages of this technique is that the measurement can be carried out in a controlled atmosphere.

Figure 29 further shows that the specimen position does not contribute to focusing. In a fixed specimen position, only absorption varies somewhat with diffraction angle. This setup is thus suitable when a curved one-dimensional position-sensitive detector is employed. There are detectors on the market that can measure up to $2\theta = 120°$ at the same time. When such a detector is used, no diffractometer is needed—only a base plate support. This technique has proven itself useful, particularly with programmed heating of specimens, and with in situ recording of diffraction diagrams.

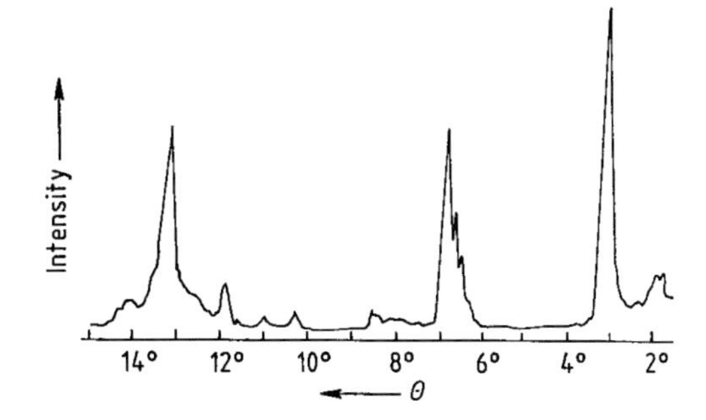

Figure 31. X-ray diffraction pattern of the γ phase of quinacridone, recorded by the Guinier method with Cu Kα_1 radiation

The principle of the Guinier camera, described in section p. 385, has been developed by replacing the film by a counter tube to give the *Guinier diffractometer* (Fig. 30). The reflection intensities can be digitally evaluated; this is a major advantage for quantitative phase and profile analyses. The resolution obtained is much better than with the Bragg–Brentano method (compare Figs. 28 and 31), but specimen preparation is more complicated, and no automatic sample changer is available.

Single-Crystal Diffractometer. Most intensity measurements are now carried out with a counter tube; measurements are more accurate, and the results are obtained directly in digital form.

A relatively simple and inexpensive diffractometer is the single-crystal instrument with Weissenberg geometry [24, pp. 98–167] (see Fig. 18). Measurements can be automatically controlled by computer, but the counter tube must be realigned manually for each layer line. The instrument has the advantage of low cost, but frequent realignment is needed (semiautomatic instrument), and the individual reflections are not all measured in the same geometrical arrangements (Weissenberg geometry).

Four-circle diffractometers (Figs. 32 and 33) are free of these disadvantages. The ϕ and χ circles set the crystal in such a way that the reciprocal lattice vector of the reflection being measured lies in the equatorial plane of the instrument, while the counter tube moves along this circle (2θ circle). The θ and 2θ circles make it possible to set up the Bragg relation. The χ circle, called the Eulerian cradle, can be either closed or open (as in Fig. 32). A closed Eulerian cradle has the advantage that any reciprocal lattice vector can, in principle, be

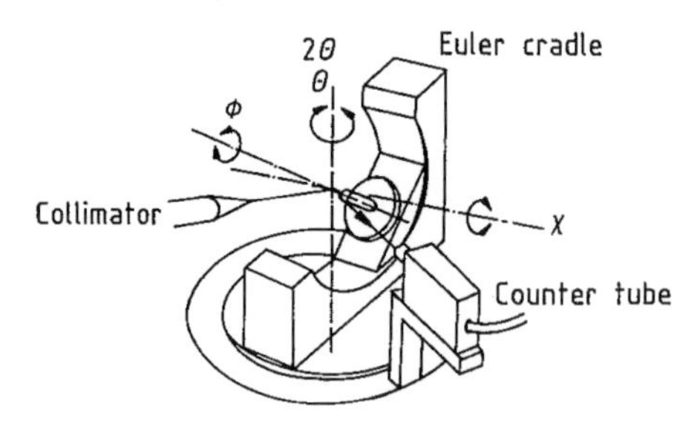

Figure 32. Four-circle diffractometer with open Eulerian cradle (χ circle)

brought into the equatorial plane; with an open 90° Eulerian cradle, this is possible only for half the Ewald sphere, though normally this is sufficient. The open Eulerian cradle permits a much sturdier construction, and "shading" by the closed Eulerian cradle ceases to be a problem. Complete control and monitoring by computer is possible with either form.

An important advantage of the closed Eulerian cradle is that in low-temperature measurements the cold nitrogen stream can keep its lamellar characteristics when it is passing the crystal on the rod, provided the nitrogen stream and the axis of the goniometer head remain always coaxial, what can be easily arranged with a closed Eulerian cradle.

An elegant version of the four-circle diffractometer is the CAD 4 diffractometer made by Nonius. Here the Eulerian cradle or χ circle is replaced by the κ circle, whose axis intersects that of the θ circle at an angle of 50° (Fig. 33). The axis of the ϕ circle likewise makes an angle of 50° with that of the κ circle, so that the ϕ axis can lie anywhere on a 100° cone.

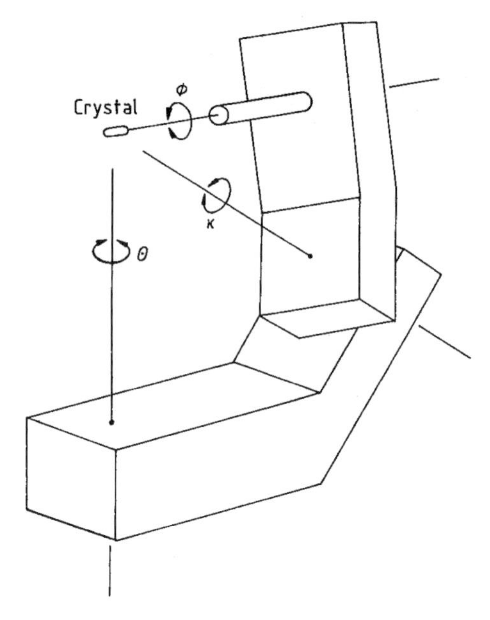

Figure 33. CAD 4 diffractometer

Data Processing by Computer. Because minicomputers have become so inexpensive, it is an obvious step to use the computer not only to control the diffractometer, but also to solve the structure problem. In fact, it is common to use one computer for control, and a second for data processing. While the diffractometer, for example, is at work measuring the diffracted intensities of the computer-adjusted crystal of a compound, another structure can be solved, by either the same minicomputer or a second one. Nowadays, a personal computer is used for the least-squares refinement of the full matrix of the atomic parameters even for large structures. The same holds for the solution of the phase problem.

The combination of the rotating anode X-ray generators which are nowadays very stable, the robust one-, two-, three-, or four-circle single-crystal diffractometers, the area detectors (image plate, wire counters, or CCD), and powerful PCs make solving the crystal structure of a medium-sized molecule a job of a few hours, of a few days, or exceptionally of a week for crystals that are relatively small ($< 50\ \mu m$ in one or all dimensions) and/or show greater disorder.

15.2.1.7. Safety

For electron radiation, the very high absorption in air means that no special safety rules have to be followed; but the X rays produced by electrons do call for precautions. The manufacturers of most electron microscopes have solved this problem. Neutron diffraction experiments are not performed in the chemist's own lab, and the host institute therefore has the responsibility of complying with safety regulations. In this section, only practices relating to X rays are discussed.

The regulations, for example, currently in force in Germany are set forth in the "X-Ray Regulation" of 8, January 1987 [71], which supersedes the similarly titled regulation of 1, March 1973. The new regulation is based on recommendations issued in 1958 by the International Committee for

Radiation Protection [72], [73]. The code prescribes, in part, as follows:

1) A radiation protection supervisor must be designated for the operation of X-ray instruments other than fully shielded instruments. If necessary, the radiation protection supervisor designates radiation protection officers, who must demonstrate familiarity with the requisite specialist knowledge.
2) X-ray equipment other than a fully shielded instrument may be operated only in a room enclosed on all sides and designated as an X-ray room. The controlled access area is defined as an area where persons can receive more than 15 mSv (1.5 rem) in a calendar year as whole-body exposure; the corresponding figure for the monitored area is 5 mSv.
3) Workers occupationally exposed to radiation may not exceed a whole-body dose of 50 or 15 mSv in a calendar year. In the first case, much more stringent provisions apply to the medical supervision of the worker. Special provisions apply for pregnant women and persons under 18 years of age.
4) The individual dose is measured by two independent methods. The first method is performed with dosimeters, which are to be submitted, at intervals not to exceed 1 month, to the government agency having jurisdiction. The second measurement must permit the determination of dose at any time, and may be put into use at the direction of the radiation protection supervisor or of the radiation protection officer or at the request of the worker.

Exposure to the primary beam from an X-ray tube (ca. 10^5 R/min) must be avoided under all circumstances.

15.2.1.8. Instrument Manufacturers

What follows is a list of companies trading in Germany, together with the instruments they market. No claim for completeness is made.

Nonius B. V., Röntgenweg 1, P. O. Box 811, 2600 AV Delft, The Netherlands. X-ray sources: Sealed tubes or rotating anode generators, microsource generators; goniometer heads; high- and low-temperature devices; four-circle K-axis goniometers; one- and two-dimensional detector systems, CCD and imaging plate detectors.

Freiberger Präzisionsmechanik GmbH, Hainichener Strasse 2 a, 09599 Freiberg, Germany. X-ray sources; X-ray tubes; Debye–Scherrer cameras; two-circle powder diffractometers; precession cameras; Weissenberg cameras.

Huber Diffraktionstechnik GmbH, Sommerstraße 4, 83253 Rimsting/Chiemsee, Germany. X-ray diffraction equipment, goniometer heads: Laue, Debye-Scherrer, Buerger-precession, and Weissenberg cameras; imaging-plate Guinier cameras; X-ray diffractometers; high- and low-temperature attachments; monochromators; Eulerian cradles, rotation stages, linear stages, multiaxis goniometers for synchrotron and neutron facilities.

Philips GmbH, Miramstrasse 87, 34123 Kassel, Germany. X-ray sources; X-ray tubes; goniometer heads; high- and low-temperature stages; position-sensitive detectors; texture goniometers; two-circle powder diffractometers; thin-film high-resolution diffractometers.

Molecular Structure Corporation Europe, Unit D2, Chaucer Business Park, Watery Lane, Kemsing, Sevenoaks, Kent TN15 6YU, United Kingdom. Imaging plate detectors; CCD detectors; partial-Chi circle; low-temperature devices; rotating anode generators; peripheral devices.

Bruker AXS GmbH, Östliche Rheinbrückenstrasse 50, 76187 Karlsruhe, Germany. Rotating-anode and sealed-tube sources; ceramic X-ray tubes; multilayer optics; mono/poly capillaries; two-circle powder diffractometer; reflectometer; texture/stress diffractometer; attachments for high/low temperature, high pressure, humidity control, reactive gas; micro diffractometer; 2D SAXS camera; process automation diffractometer; chemical and biological 2D-CCD single-crystal systems.

Rich. Seifert & Co GmbH Röntgenwerk, Freiberger Präzisionsmechanik GmbH, Bogenstraße 41, 22926 Ahrensburg, Germany. Four-circle diffractometer, Theta/Theta powder diffractometer, low- and high-temperature sample stages, phase-texture-stress-diffractometer for laboratory and automotive industry, thin-film texture/stress Eta-diffractometer, reflectometer, high-resolution diffractometer, wafer ingot orientation systems for industrial applications, Rayflex Windows Software, X-ray generators, X-ray tubes, X-ray optics, parabolic multilayers, monochromators, film cameras, position-sensitive detectors PSD, energy dispersive detectors, electroluminescence detectors LUX.

Stoe & Cie GmbH, P.O. Box 101302, Hilpertstrasse 10, 64295 Darmstadt, Germany. Goniometer heads; Debye–Scherrer cameras; Guinier cameras; Weissenberg cameras; precession cameras; Reciprocal Lattice Explorers; powder diffrac-

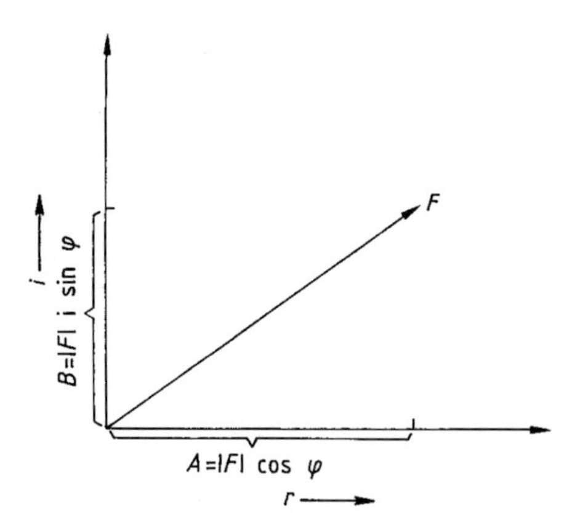

Figure 34. Gaussian plane
The real part of the complex number F is plotted along the r-axis; the imaginary part along the i-axis

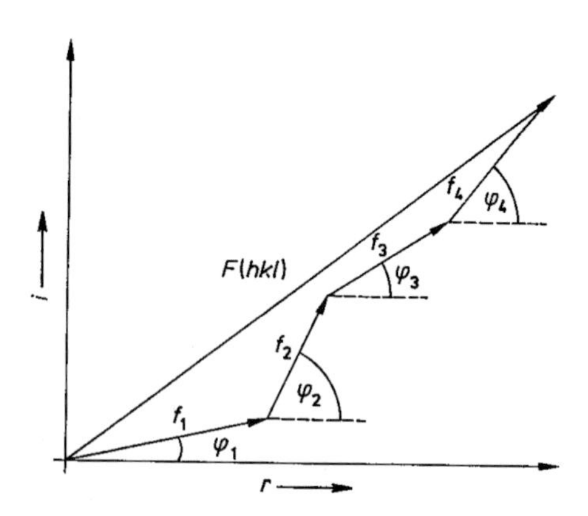

Figure 35. Addition of structure factor contributions in the complex plane

tometers with scintillation counter, position-sensitive detectors or imaging plates; special sample holders; high-temperature and texture sample stages; four-circle single-crystal diffractometers also with CCD detector; one- and two-circle single-crystal diffractometers with imaging plates; crystallographic software.

15.2.2. The Phase Problem

In general, $F(hkl)$ is a complex number; when plotted in the Gaussian plane (Fig. 34), it has not only a modulus $|F|$ but also a phase angle φ.

By the Euler–Moivre theorem

$$\|F\| \exp(i\varphi) = \|F\| \cos\varphi + i\,\|F\| \sin\varphi = A + iB \tag{26}$$

A structure factor is obtained not by simple scalar addition of the scattering contributions atom-by-atom, but by vector addition, as the plot in the Gaussian plane makes clear (Fig. 35). The angle between a scattering contribution and the real axis is given by $2\pi(hx_n + ky_n + lz_n)$ (Eq. 18). The phase angles of the structure factors give the angles by which the sine waves of the individual structure factors in the Fourier synthesis must be shifted with respect to the origin in order to yield the electron density.

An important special case is where B in Equation (26) vanishes, and the only two phase angles possible for $F(h,k,l)$ are 0° and 180°. The phase problem then becomes a matter of adding signed quantities. According to Equations (18) and (26), this occurs whenever an atom with coordinates x, y, z always has an atom of the same species at coordinates $-x$, $-y$, $-z$ corresponding to it. In other words, the phase problem becomes a signed addition problem in all centrosymmetric space groups.

In the methods now commonly used for measuring diffracted X-ray intensities, it is generally not possible to measure the phase angle of a structure factor. The phase is said to be "lost" in the intensity measurement. In the early days of X-ray structure analysis, it was thought that phases could not be deduced from measured intensities.

15.2.2.1. Patterson (Vector) Methods

Not until A. L. Patterson's work [74] did it gradually become clear that the measured intensities also contain phase information.The function:

$$P(uvw) = \frac{1}{V} \sum_{hkl=-\infty}^{+\infty} \|F(hkl)\|^2 \exp[2\pi i(hu + kv + lw)] \tag{27}$$

does not give the electron density at point x, y, z but does yield maxima in a vector space. Each such maximum has a position (relative to the origin) corresponding to the vector between two atoms; its weight is proportional to the product of the electron numbers of the two atoms. The vector space equation requires knowledge only of the structure amplitudes $|F|$, not of the phases. It is possible to solve a structure that is not too complicated, if all the maxima of a Patterson synthesis are known.

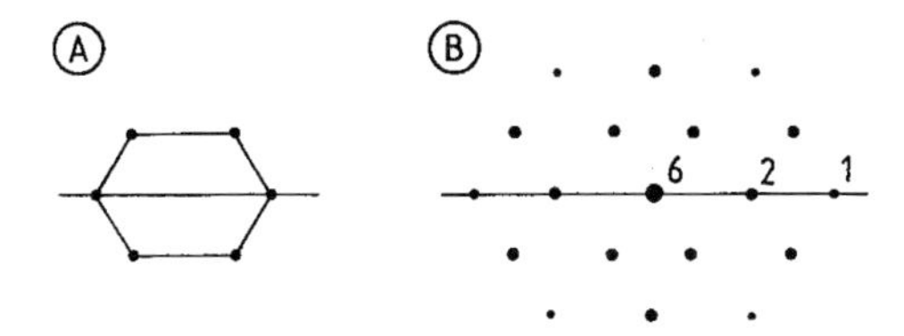

Figure 36. A) Crystal space; B) Vector space

Multiple solutions may exist for one and the same set of vectors in extremely rare cases (homometric structures). No case is known in which two homometric structures are both chemically meaningful.

The complete interpretation of a Patterson synthesis is generally made much more difficult because the maxima are not resolved, and cannot be explicitly identified. Figure 36 illustrates this situation for the two-dimensional case. Corresponding to the 6 maxima in the crystal space, there are 19 maxima in the vector space; these must be placed in the same unit cell volume that accommodates the 6 atoms. What is more, the maxima in the vector space require more volume than atoms in the crystal space, because the indeterminacies of each pair of atoms add at the Patterson maximum. The half-width values of the maxima in the vector space are thus roughly twice those in the crystal space. As a consequence, the peaks overlap strongly. The relative weights of the maxima are indicated in Figure 36.

Patterson synthesis by itself cannot solve even moderately large molecular structures (five or six equally heavy atoms) unless concrete structure information already exists so that the vector space can be systematically searched for a presumed structure or substructure. A search of this kind might be done with a computer, perhaps by the method of pattern-seeking functions, or the convolution molecule method [57], [75]. These techniques, however, are not so often used today, having been supplanted by direct phase-determination methods (Section 15.2.2.2), which make it possible to reach the goal quickly on the basis of relatively little prior knowledge [76]. Patterson synthesis may return to prominence when used to identify sets of starting phases (Section 15.2.2.2).

An exception of that is PATSEE [77], which uses one or two known fragments of the structure, whose orientation is found by a real-space Patterson rotation search and its translation in the cell by direct methods (Section 15.2.2.2), starting from random positions. Figures of merit related to those of the direct methods are used to refine only promising phase sets.

Heavy Atom Method. The principal use of Patterson synthesis at present is in the heavy atom method. If a few heavy atoms, such as bromine or iodine, are present in a molecule along with light atoms, such as carbon, nitrogen, or oxygen, the Patterson technique makes it possible to analyze the X-ray diffraction pattern into heavy- and light-atom components. For a structure to be solved reliably, however, the light and heavy atoms must have sufficiently different scattering powers, but the sum of the heavy-atom scattering powers (proportional to the sum of the squares of the atomic numbers Z) must not be much different from that of the light atoms:

$$\sum Z_{\mathrm{H}}^2 \geq \sum Z_{\mathrm{L}}^2 \tag{28}$$

Given these conditions, the heavy-atom vectors in the Patterson synthesis have markedly higher maxima associated with them than the heavy/light- or light/light-atom vectors. In centrosymmetric structures, this holds even when the ratio of heavy to light atoms is less favorable, since in this case the knowledge of a small structural component is enough to establish with fair confidence the sign trend of a large structure factor.

If, say, a bromine atom has been incorporated into an organic molecule to facilitate X-ray structure analysis, the weights of bromine–bromine, bromine–carbon, and carbon–carbon vectors are in the ratio:

$$Z_{\mathrm{Br}}^2 : Z_{\mathrm{Br}} Z_{\mathrm{C}} : Z_{\mathrm{C}}^2 = 1225 : 210 : 36 \tag{29}$$

Thus, it should not be hard to identify the bromine–bromine vectors, given the sizes of the maxima. Further, heavy-atom maxima are not expected to overlap, because they have virtually the entire cell available, while carbon–carbon maxima and, to a great extent, bromine–carbon maxima will recede into the background noise.

Example. Vectors between heavy atoms are especially easy to identify if just one heavy atom is located in the asymmetric unit. This is shown here for the centrosymmetric space group $P\,2_1/c$. From the coordinates in the crystal space:

$$x, y, z; x, \frac{1}{2} - y, \frac{1}{2} + z; \; -x, \; -y, \; -z; \; -x, \frac{1}{2} + y, \; \frac{1}{2} - z$$

Table 2. Patterson maxima* for the space group $P\,2_1/c$ with one heavy atom in the asymmetric unit

Atomic coordinates	x	x	$-x$	$-x$
	y	$\frac{1}{2}-x$	$-y$	$\frac{1}{2}+y$
	z	$\frac{1}{2}+z$	$-z$	$\frac{1}{2}-z$
x	0	0	$2x$	$2x$
y	0	$\frac{1}{2}+2y$	$2y$	$\frac{1}{2}$
z	0	$\frac{1}{2}$	$2z$	$\frac{1}{2}+2z$
x	0	0	$2x$	$2x$
$\frac{1}{2}-y$	$\frac{1}{2}-2y$	0	$\frac{1}{2}$	$-2y$
$\frac{1}{2}+z$	$\frac{1}{2}$	0	$\frac{1}{2}+2z$	$2z$
$-x$	$-2x$	$-2x$	0	0
$-y$	$-2y$	$-\frac{1}{2}$	0	$-\frac{1}{2}-2y$
$-z$	$-2z$	$-\frac{1}{2}-2z$	0	$-\frac{1}{2}$
$-x$	$-2x$	$-2x$	0	0
$\frac{1}{2}+y$	$\frac{1}{2}$	$2y$	$\frac{1}{2}+2y$	0
$\frac{1}{2}-z$	$\frac{1}{2}-2z$	$-2z$	$\frac{1}{2}$	0

* Vector space coordinates are differences between pairs of atomic coordinates.

the coordinates in the vector space can be found by forming the 16 possible vectors, as shown in Table 2. Such a table should always be prepared for difficult cases.

The space group in the vector space is $P\,2/m$, which is the Laue group of $P\,2_1/c$. Only 21 such "Patterson groups" are possible with conservation of the Bravais lattice. The space group of the vector space (Patterson group) is derived from that of the crystal by adding a center of symmetry (if none is present) and removing translation from the group of symmetry elements.

Four maxima are located at the origin; four at $0, \frac{1}{2}\pm 2y, \frac{1}{2}$; four at $\pm 2x, \frac{1}{2}, \frac{1}{2}\pm 2z$; two at $\pm 2x, \pm 2y, \pm 2z$; and two at $2x, \pm 2y, 2z$. For the $0, \frac{1}{2}\pm 2y, \frac{1}{2}$ maximum, only y is variable; the maximum lies on a "Harker line". At $\pm 2x, \frac{1}{2}, \frac{1}{2}\pm 2z$, the section cuts through the cell at $y=\frac{1}{2}$ (Harker section). The first condition yields the y coordinate; the second yields the x and z coordinates of the heavy atom. The maxima at $2x$, $2y$, $2z$ (and the symmetry-related positions) simply provide confirmation of the values found.

Maxima on Harker lines and sections are always doubly occupied, so that they are still easier to find.

If there is more than one heavy atom in the asymmetric unit, there are also vectors between the heavy atoms in the asymmetric unit. A table should be prepared in order to make clear the weighting of individual maxima. The convolution technique can lead to a solution in a very direct way [58].

All of this work is nowadays carried out by computer programs (e.g., SHELX-97 [49]). It is advisable, however, to do at least part of it by hand, so that one is familiar with the theory when there are difficulties.

From the heavy-atom positions obtained by this procedure, the structure factors are calculated from Equation (18), and the signs or phases are assigned to the measured structure amplitudes. The result is a Fourier synthesis of electron density. Because the phases depend only on the positions of the heavy atoms, this first Fourier synthesis often fails to reveal the whole structure of the molecule; instead, it yields only a partial structure, but this is generally enough to lead to better phases for the next Fourier synthesis. The structure should be solved after two or three (at most six or seven) of these successive Fourier syntheses of electron density. This step can be automated. The maxima of a Fourier synthesis can be analyzed by computer; the only maxima accepted as atoms are those that form a cluster with chemically meaningful geometries [73]. The combination of this technique with direct methods is discussed further on.

Special difficulties arise when the heavy-atom structure, taken separately, contains one more symmetry element (usually a center of symmetry) than the overall structure. The first Fourier synthesis of the electron density then shows two light-atom structures along with the heavy-atom structure. In order to calculate the second Fourier synthesis, the heavy-atom structure is then used together with only that part of the light-atom structure that can be assigned with high confidence to just one of the two light-atom structures indicated in the Fourier synthesis. The additional symmetry element vanishes in the next Fourier synthesis, and the structure analysis proceeds in the usual way.

15.2.2.2. Direct Methods

As Equation (17) implies, the distribution of the electron density in a crystal structure is unambiguously determined once the structure factors have been obtained. A structure factor comprises the structure amplitude and the phase angle; as has already been shown, all that can be determined by experiment is generally the structure amplitude, not the phase angle φ. The preceding section discussed a fairly straightforward way of using the heavy-atom method to gain knowledge of a substructure, and using this to determine the φ values indirectly from the Patterson synthesis.

As the theory of Patterson synthesis shows, it must be generally possible to infer the phases, given the associated structure amplitudes.

The best-known equation for direct phase determination is that of David Sayre [78]:

$$F(h) = \frac{1}{V}\frac{f}{g} \sum_{h'} F(h')F(h-h') \tag{30}$$

where $(1/V)\,(f/g)$ can be regarded as merely a kind of scaling factor and h, h' and $h-h'$ are vectors in reciprocal space, i.e., each is a triplet of indices (h, k, l). Equation (30) is not a probability relation, but a true equation; all atoms are assumed identical, and all combinations of structure factors are required. Sayre used this formula to elucidate a centrosymmetric projection of the crystal structure of hydroxyproline [78].

Multiplying both sides of Equation (30) by $F(-h)$ yields:

$$\|F(h)\|^2 = \frac{1}{V}\frac{f}{g}\sum_{h'} F(-h)F(h')F(h-h') \tag{31}$$

For large $F(h)$, the right-hand side is also large, real, and positive. If $F(h)$, $F(h')$, and $F(h-h')$ are large, the following probability statements then hold:

$$s(h)s(h')s(h-h') \approx +1 \tag{32}$$

in the centrosymmetric case, and:

$$\varphi(-h) + \varphi(h') + \varphi(h-h') \approx 0 \quad (\text{mod } 2\pi) \tag{33}$$

in the noncentrosymmetric case. Here, $s(h)$ denotes the sign and $\varphi(h)$ the phase of $F(h)$, and similarly for the other terms. Given the signs or phases of two structure factors, it is thus possible to infer the sign or phase of a third.

The form in which the Sayre equation is commonly used today is due to HUGHES [79]:

$$E(h) = N^{1/2}\left(E(h')E(h-h')\right)_{h'} \tag{34}$$

where N is the number of atoms in the unit cell and $\langle\,\rangle_{h'}$ denotes the average over all h'; $E(h)$ is the normalized structure factor defined as:

$$\|E(h)\|^2 = \frac{\|F(h)\|^2}{\sum_{j=1}^{N} f_j^2} = \frac{I(h)}{\bar{I}} \tag{35}$$

where the sum of the squares of the atomic form factors f is taken over all N atoms in the unit cell. This is the expected value of the intensity of a reflection; $|F(h)|$ is the structure amplitude after absolute scaling is applied (see Eq. 58); $I(h)$ is the intensity of the reflection (h) and I the average intensity at the glancing angle of (h).

Equation (32) is a probability relation. The probability that the positive sign is correct is:

$$P_+(h) = \frac{1}{2} + \frac{1}{2}\tanh\left(N^{-1/2}\,\|E(h)E(h')E(h-h')\|\right) \tag{36}$$

It should be noted that the probability of the calculated phases depends not on the absolute intensities, but on the relative intensities (Eq. 35) that describe the scattering power of a structure with point atoms. Basically, the highest E values are the reflections containing the largest amount of information. The more atoms in the unit cell, the lower the probability that relation (32) holds. For structures where the unit cell contains atoms differing in atomic number:

$$P_+(h) = \frac{1}{2} + \frac{1}{2}\tanh\left(\sigma_3\sigma_2^{-3/2}\,\|E(h)\|\sum_{h'} E(h')E(h-h')\right) \tag{37}$$

for all products $E(h')E(h-h')$ available for $E(h)$, and:

$$\sigma_n = \sum_{j=1}^{N} Z_j^n \tag{38}$$

Using Equation (37) for the probability of relation (32) takes care of the fact that not all products required for the Sayre equation are available. A strict equation has become a probability relation. One of the corresponding equations for the probability that relation (33) is correct in the noncentrosymmetric case reads as follows:

$$P(\varphi(-h) + \varphi(h-h') + \varphi(h')) = \frac{\exp\left[2\sigma_3\sigma_2^{-3/2}\,\|E(-h)E(h')E(h-h')\|\cos(\varphi(-h)+\varphi(h-h')+\varphi(h'))\right]}{2\pi I_0 2\sigma_3\sigma_2^{-3/2}\,\|E(-h)E(h')E(h-h')\|} \tag{39}$$

where I_0 is a Bessel function of the second kind [80].

The Hughes form of the Sayre equation (34) is for the centrosymmetric case; in the noncentrosymmetric case, at least as far as physical information is concerned, it corresponds to the tangent formula. In contrast to Equation (33), here the sum is weighted with the normalized structure factors, and all pairs of structure factors possible in a special case are taken care of:

$$\tan\varphi(h) = \frac{\sum_{h'} K(hh') \sin\left[\varphi(h') + \varphi(h - h')\right]}{\sum_{h'} K(hh') \cos\left[\varphi(h') + \varphi(h - h')\right]} \quad (40)$$

where:

$$K(hh') = 2\sigma_3\sigma_2^{-3/2} \left\| E(-h)E(h')E(h - h') \right\| \quad (41)$$

A measure of the reliability of a phase obtained in this way is $\alpha(h)$, where:

$$[\alpha(h)]^2 = \left[\sum_{h'} K(hh') \cos\left(\varphi(h') + \varphi(h - h')\right)\right]^2 + \left[\sum_{h'} K(hh') \sin\left(\varphi(h') + \varphi(h - h')\right)\right]^2 \quad (42)$$

These equations are also used in the centrosymmetric case for determining the signs of structure factors (only the cosine terms are required).

In summary, it is probable that the sum of the phases for two reflections equals the phase of a third reflection, if the sums of the indices of the first two reflections are the indices of the third, and if all three reflections are relatively strong. The probability that this holds can be calculated. It is also possible to determine the probability for one phase when many phase combinations exist.

As Equation (34) shows, phases must be known or assumed at the start of a phase determination, before the equations can be used at all. Depending on the space group or Bravais lattice type, the phases of up to three reflections can be freely chosen; this fixes the origin of the coordinate system [81]–[84]. In the case of noncentrosymmetric space groups or those without mirror symmetry, the phase of one further reflection must be restricted to a 180° angular range in order to fix one of the two enantiomers. Other phases can be represented by letter symbols (symbolic addition) [85], [86] or simply by angle values (multisolution) [87]–[89]. In the case of a center of symmetry, for every additional reflection with freely selected phase, only two possibilities (0° and 180°) must be taken into account; this is done by letting each resulting phase combination serve as the starting point for determination of a new phase set. In noncentrosymmetric space groups, on the other hand, an infinite number of possibilities between 0° and 360° would have to be examined, but experience has shown that four values per structure factor (45°, 135°, 225°, 315°) are enough to place an adequate restriction on one phase for the determination of other phases. If there are n additional reflections with freely selected phases, the number of phase sets will be 2^n in the centrosymmetric case and 4^n in the noncentrosymmetric case. If the phases have been calculated in symbolic form, the stated possibilities must be permuted over all the symbols to yield the same number of phase sets as in the latter cases (unless relations can be established between the symbols, as is very often the case).

In another direct phasing method, a random number generator is used to supply phases for a relatively large number of reflections. "Figures of merit" are calculated in order to decide whether the calculation of a phase set should be completed. With present-day fast computers, 20 000 or more such phase sets can be calculated. Because one starting phase set contains a great number of reflections, there is no longer any need to fix the origin, or select one enantiomer through the phases of some reflections [90].

Ultimately, the only way to decide whether a phase set is the right one is to look for a chemically meaningful interpretation of the Fourier synthesis of the electron density. But it would be rather tedious to check, say, 64 Fourier syntheses in this way (256 or more in troublesome cases). An attempt is therefore made to find figures of merit [91] that will make one phase set more probable than the others, even before the Fourier synthesis is calculated. Such a figure of merit might refer to the internal consistency of a phase set. The determination of phases or signs by direct methods then involves the following steps:

1) Calculate the magnitudes $|E|$ of the normalized structure factors
2) Rank the E values (decreasing order)
3) List pairs of reflections that can figure in a Sayre relation with a given reflection

4) For the starting phase set, select the reflections defining the origin and a few additional reflections
5) Determine the phases of each phase set
6) Select the phase sets that appear best under given criteria (figures of merit)
7) Calculate the Fourier synthesis of the electron density (E map) and interpret it

Different systems of weighting for phase determination and an almost countless number of figures of merit have been developed [92]–[94].

H is now possible to solve rather large crystal structures, including small proteins with more than 1000 non-hydrogen atoms, by real/reciprocal space Fourier recycling methods [95]–[99]. These are very lenghtly procedures, and a powerful PC may require four weeks or more without breaks. These major problems usually involve measuring synchrotron data.

15.2.2.3. Trial-and-Error Method, R Test

The trial-and-error method used to be employed for simple inorganic structures, but it is no longer important.

The R value:

$$R_1 = \frac{\sum \|F(hkl)\|_{obs} - \|F(hkl)\|_{calc}}{\sum \|F(hkl)\|_{obs}} 100\,[\%] \tag{43}$$

indicates how close the solution approaches the correct structure. The sums are taken over all measured structure factor magnitudes. The $|F|_{obs}$ are observed structure amplitudes, while the $|F|_{calc}$ are the amplitudes calculated from Equation (18). If a least-squares fit is also done, a second R test becomes important; this includes the weights w, a function based on the counting statistics of the measurements, assigned to the squares of the differences between observed and calculated structure amplitudes:

$$R_2 = \left[\frac{\sum w\left(\|F(hkl)\|_{obs} - \|F(hkl)\|_{calc}\right)^2}{\sum w\,\|F(hkl)\|^2_{obs}}\right]^{1/2} 100\,[\%] \tag{44}$$

A value of $R < 50\,\%$ generally indicates that one is on the right path toward a solution of the structure. In particularly simple cases, space-filling considerations and the requirements of a highly symmetric space group may limit the number of possible solutions.

15.2.2.4. Experimental Phase Determination

High-quality crystals, together with intense X-ray sources, such as synchrotron radiation (and, in especially simple cases, rotating anode generators), make it possible to find the phase angles of structure invariants (triple products or reflections with even indices) by experimental means. This supplemental information can be derived from intensity fluctuations that occur when a crystal is rotated about the reciprocal lattice vector of a net plane in reflection position, and another reciprocal lattice point additionally passes through the Ewald sphere. The results can no longer be interpreted by kinematic lattice theory; the dynamic theory must be employed [100], [101].

15.2.3. Least-Squares Refinement

The least-squares method is used to achieve the best possible fit of the structure model, found after phase determination, to the measured data. For every atom, in general, the least-squares calculation refines three positional parameters as well as one (isotropic) or six (anisotropic) temperature coefficients. There are also "overall" parameters such as the absolute scaling factor. Special cases may require further parameters, for example a factor for enantiomorphism.

Formerly the general practice was to minimize the quantity:

$$\sum \left(\|F_{obs}\| - \|F_{calc}\|\right)^2 \tag{45}$$

This approach has the disadvantage that no significance is accorded to "negative" intensities [103], which means that the zero point was too high.

Modern least-squares computer programs minimize:

$$\sum (F_{\text{obs}}^2 - F_{\text{calc}}^2)^2 \quad (46)$$

Among other advantages of this procedure is that even low-intensity data sets can result in useful parameter refinements [50].

Least-squares refinement yields not only the structure parameters, but also their standard deviations. The R values defined above can serve as figures of merit if w represents the weighting function employed in the refinement. A more important point for assessing the quality of a structure analysis, however, is that the results fit with the wealth of experience of crystallographers and chemists, and deviate from prior knowledge only in justifiable cases.

Spectroscopic data must also be taken into account.

Constraints (fixed parameters) and restraints (partly fixed parameters with a specified allowance of deviation) have become important in the last few years [51]. In this way it is possible to extract as much "truth" as possible out of relatively bad data (originating from the "best" crystal which was available).

Another important problem is twinning, which usually gets apparent during parameter refinement. This "model" is very often more "true" than any model of disorder [102].

15.2.4. Determination of Absolute Configuration

Friedel's law states that:

$$\|F(hkl)\| = \|F(-h-k-l)\| \quad (47)$$

even for noncentrosymmetric space groups. When there is anomalous dispersion, however, Friedel's law no longer holds rigorously. The atomic form factors adopt complex values:

$$f = f_0 + f' + if'' \quad (48)$$

where f_0 is the atomic form factor with no allowance for anomalous scattering, f' is the real part of the anomalous scattering, and f'' is its imaginary part.

For the noncentrosymmetric case, the calculated magnitudes of $F(h,k,l)$ and $F(-h,-k,-l)$ are different, so that the absolute configuration of a molecule or atomic group can be determined. The effect is the greater the more strongly an atom scatters and the closer (on the long-wavelength side) the absorption edge of an atom is to the X-ray wavelength. Normally, when extremely accurate intensity measurements are not possible, an atom heavier than oxygen must be present in the molecule (or must be incorporated synthetically) for the configuration of the molecule to be unambiguously determinable. Naturally, the synthesis must exclude racemization.

There are several methods for determining the absolute configuration. After the structure has been solved, one can calculate which pairs of reflections $(h,k,l/-h,-k,-l)$ differ most in intensity, and then measure these reflections more accurately. This approach makes it relatively easy to decide between enantiomers. D. W. ENGEL has used the technique to determine the absolute configuration for compounds that do not contain atoms heavier than oxygen [104].

A more convenient method that does not call for further measurements is to obtain the best possible least-squares fits of both enantiomers to the measurements. The R_2 values (Eq. 44) are subjected to a χ^2 test, which yields the statistical confidence level for one of the two configurations. The confidences are normally well over 99 %. This method was proposed and developed by W. C. HAMILTON [105].

Another method is to attach a factor to the imaginary part in Equation (48) and to introduce this factor as a parameter in the least-squares fit. If it refines to + 1, the configuration is correct; to – 1, the other configuration is [106]. A new formalism has been introduced [107].

15.2.5. Example of a Single-Crystal X-Ray Structure Analysis Solved by Direct Methods

To illustrate the technique, consider a compound for which two structures have been proposed (Fig. 37). Spectroscopic measurements (NMR, mass spectroscopy) have not yielded decisive results. The coupling constants and chemical shifts are too unusual, especially for proposed structure **1**, which further involves bond angles so far from the norm that no Dreiding model could be built. At this point in the structure analysis, if not earlier, X-ray examination becomes essential.

A glass-clear crystal measuring 0.45×0.45×0.2 mm was mounted on the inside wall of an capillary with a tiny amount of grease, and the capillary was sealed. The capillary was glued to the glass fiber of a goniometer head in such a way that the primary beam avoided the fiber. Alignment patterns of a precession camera showed a very high probability that the specimen was a single crystal, because it yielded sharp reflections.

It is important to consider what type of X rays to use. Photographic recording of patterns was done with Cu Kα radiation; diffractometer measurements were performed with Mo Kα radiation. The absorption coefficient μ was calculated as:

$$\mu = \left[0.68\,(\mu/\varrho)_C + 0.06\,(\mu/\varrho)_H + 0.26\,(\mu/\varrho)_O\right]\varrho = 0.788\varrho \tag{49}$$

where the weighting corresponds to the mass fractions of the atoms in the compound.

Values of μ/ϱ by element and wavelength are given in the International Tables [29, vol. III, pp. 157–200]. The density ϱ is calculated from the molecular mass M, the number of molecules Z in the cell, Avogadro's number (6.022×10^{23}), and the cell volume V (in 10^{-30} m^3):

$$\varrho = \frac{10 M Z}{6.022 V}\,(\mathrm{g\,cm^{-3}}) \tag{50}$$

With a computer-controlled single-crystal diffractometer, the cell constants and the setting matrix for the diffractometer were automatically determined on 98 reflections [63]; an appropriate program for controlling the diffractometer was employed, and the measurements were subjected to least-squares refinement. The result was the triclinic space group P$\bar{1}$ with cell constants:

$a = 10.561\,(5)$; $b = 6.134\,(3)$; $c = 10.418\,(6)$ (10^{-10} m)
$\alpha = 89.37°\,(4)$; $\beta = 64.08°\,(4)$; $\gamma = 89.46°$

The numbers in parentheses are the standard deviations derived from the least-squares calculations. The density, determined by suspension in aqueous K_2HgI_4, was 1.35 g/cm^3; this implies that the unit cell contains two molecules, i.e., the X-ray measurements give a density of 1.347 g/cm^3. Accordingly, the absorption coefficient is 1.06 cm^{-1}. The fraction of the radiation not absorbed can be calculated:

$$I/I_0 = \exp(-\mu d) \tag{51}$$

It is 93 % for the maximum path length d (0.67 mm) in the crystal, and 98 % for the minimal path length (0.2 mm). Thus, no absorption correction is required with Mo Kα radiation; the lower intensity of the reflections compared with Cu Kα radiation must, however, be taken into consideration (Eq. 14).

Diffracted X-ray intensities were measured on a computer-controlled diffractometer with a $\theta/2\theta$ scan mode [108]. The reference reflection was remeasured at the beginning and end of each cycle of 20 reflections. This reflection had to be carefully selected because it formed the basis for the relative scaling. At a maximum glancing angle of 28°, the resolution was 0.76×10^{-10} m [$=\lambda/(2\sin\theta)$].

Of the 2930 reflections to be measured, 2391 had X-ray intensities significantly above the background.

The five-point measurement used here (Fig. 38) is not so often employed today. Large mass storage devices make it possible to store all the measured points for a reflection profile, so that laborious corrections do not have to be applied during the measurements.

The standard deviation σ of the intensity of a reflection is derived from the counting statistics:

$$\sigma(hkl) = \sqrt{\mathrm{meas}_1 + \mathrm{meas}_3 + \mathrm{meas}_5 + 4\,\mathrm{meas}_2 + 4\,\mathrm{meas}_4} \tag{52}$$

The relative error $r(hkl)$ is given by

$$r(hkl) = \frac{\sigma(hkl)}{I(hkl)_{\mathrm{obs}}} \tag{53}$$

In the least-squares calculations, the reciprocal of $r(hkl)$ is needed for the weight function $w(hkl)$ in order to fit the structure parameters to the measured data.

The Lorentz–polarization factor correction had to be applied to the diffracted intensities:

$$I(hkl) = (\mathrm{LP})^{-1} I(hkl)_{\mathrm{obs}} \tag{54}$$

For the case under consideration:

$$\mathrm{LP} = \frac{1 + \cos^2 2\theta}{2\sin 2\theta} \tag{55}$$

The absolute structure amplitude is then calculated as:

$$\|F(hkl)\| = c\sqrt{I(hkl)} \tag{56}$$

The use of direct methods for phase determination requires not just the relative structure amplitudes, but also the absolute structure amplitudes (Eq. 56). These can be calculated with the aid of Wilson statistics [109]:

$$\bar{I} = \exp\left(-2B\sin^2\theta/\lambda^2\right)k\sum_{n=1}^{N} f_n^2 \tag{57}$$

$$\ln\left(\bar{I}\Big/\sum_{n=1}^{N} f_n^2\right) = -2B\sin^2\theta/\lambda^2 + \ln k \tag{58}$$

The theory says that the average $\bar{I}$ of the diffracted X-ray intensities for a given glancing angle equals the sum of the squares of the atomic form factors of the N atoms in the unit cell. If the left-hand side of Equation (58) is plotted against $\sin^2\theta/\lambda^2$, the slope of the regression line is twice the overall temperature coefficient B, and the vertical-axis intercept is the natural logarithm of the scaling factor k (Fig. 39). The resulting values were $B = 4.3\times10^{-20}$ m^2 and $k = 0.32$. After parameter refinement, a scaling factor of 0.28 and an overall temperature coefficient of 4.8×10^{-20} m^2 were obtained. Factor B can be described by:

$$B = 8\pi^2\bar{u}^2 \tag{59}$$

where $\bar{u}^2$ is the mean-square displacement of atoms from their average positions perpendicular to the diffracting net planes.

From Equation (35), the normalized structure amplitudes $|E|$ were obtained. These provided the basis for direct phase determination.

Table 3. Maxima in the E map plotted by computer

Peak	Peak height	Initial coordinates			Interpretation	Coordinates after least-squares refinement		
		x	y	z		x	y	z
1	999	0.0914	0.2955	0.1963	O(3)	0.0881	0.3031	0.2044
2	873	0.3008	0.4824	0.1738	C(9)	0.2904	0.4925	0.1766
3	871	−0.1507	0.4079	0.2851	O(2)	−0.1478	0.4049	0.2865
4	844	0.3264	0.3231	0.2409	C(14)	0.3287	0.3208	0.2401
5	842	0.1360	0.6191	0.0820	O(4)	0.1401	0.6100	0.0727
6	790	0.3636	0.6667	0.1563	C(10)	0.3715	0.6809	0.1391
7	771	−0.3413	0.1716	0.4126	O(1)	−0.3412	0.1781	0.4128
8	747	0.1515	0.5000	0.1563	C(8)	0.1671	0.4802	0.1443
9	743	0.4954	0.6820	0.1616	C(11)	0.4886	0.6936	0.1655
10	728	−0.0462	0.2833	0.1889	C(7)	−0.0403	0.2748	0.1869
11	703	−0.0724	0.0370	0.2020	C(1)	−0.0691	0.0334	0.2044
12	672	0.4485	0.3440	0.2635	C(13)	0.4467	0.3348	0.2647
13	561	−0.1418	−0.0400	0.3657	C(2)	−0.1393	−0.0384	0.3634
14	554	−0.2104	−0.0157	0.1946	C(4)	0.2051	−0.0301	0.1941
15	550	0.5288	0.5165	0.2304	C(12)	0.5265	0.5216	0.2272
16	536	−0.2746	−0.0438	0.3560	C(3)	−0.2813	−0.0325	0.3561
17	453	−0.1515	0.1667	0.4688	C(5)	−0.1499	0.1433	0.4641
18	426	−0.2326	0.3038	0.4336	C(6)	−0.2334	0.3031	0.4214
19	322	0.2147	0.3104	0.2263				
20	309	−0.0538	0.4083	0.0974				
21	225	0.6667	0.5556	0.2188				

The subsequent calculations were done with the MULTAN 77 program system [110], which comprises six programs linked through data files. From the corrected intensities $I(hkl)$, NORMAL generates normalized structure amplitudes $|E(hkl)|$. Wilson statistics (or Debye statistics [111] if parts of the molecule are already known) are used to determine the scaling factor and the overall temperature factor, and to work out the statistics of the scattered X-ray intensities by reflection groups. The actual phase determination takes place in the MULTAN program. The 393 largest E values were used for phase determination, the maximum E being 3.5 and the minimum 1.0. First, all combinations of two reciprocal lattice vectors for each reflection were found: $H = H' + (H - H')$, where H stands for an index triplet (h, k, l). Processing continued with the 2000 largest triple products of $|E|$ values.

The critical question in phase determination is which reflections are assigned phases in the starting set. The case under study was centrosymmetric, so that only signs had to be allocated. The origin of the coordinate system was fixed by the reflections (7, 3, 3), (1, 1, 4), and (0, 1, 2). The four reflections (1, 1, 2), (8, 2, 2), (5, 0, 2), and (0, 5, 2) gave 16 sets of signs. The reflections were suggested by the program. The basic criteria for including a reflection in the starting set were the number, size, and type of triple-product relations. In more difficult cases this selection must be performed manually.

The program used the tangent formula (Eq. 40) to generate from the initial sets, sign sets with the fewest contradictions. Statistical criteria (figures of merit) were then applied to choose the most probable set of signs. The most probable set was that in which only (1, 1, 2) among all the starting reflections had negative sign. From these results, and the normalized structure amplitudes $|E|$, the programs EXPAND, FFT, and SEARCH were used to calculate an E map (a Fourier synthesis of the electron density of point scatterers) and to interpret it. Figure 1 and Table 3 present the result. Maximum 19 in Table 3 is markedly smaller than those which precede it. Comparison with Figure 37 reveals that strucural formula **1** is in agreement with the computer suggestion. The 18 nonhydrogen atoms of proposed structure **1** were assigned to the 18 strongest maxima of the Fourier synthesis. A structure factor calculation (Eq. 18) gave $R_1 = 28.6\,\%$ (Eq. 43).

The structure model obtained is chemically meaningful. The chemist's experience thus comes into play only when a possible solution to the phase problem has already been found. In contrast to indirect structure analysis methods (such as spectroscopy), the analysis does not start with structure model.

In the least-squares refinement the structure parameters (one isotropic temperature coefficient and three position parameters per atom, one scaling factor) had to be fitted to the 2391 observed reflections; unit weights were assigned to all reflections. When R_1 reached 14.3 %, every atom was assigned six temperature coefficients. The expression for the anisotropic temperature factor is:

$$\exp - \left(b_{11}h^2 + b_{22}k^2 + b_{33}l^2 + 2b_{12}hk + 2b_{13}hl + 2b_{23}kl\right) \quad (60)$$

After further refinement cycles, hydrogen atoms in chemically meaningful positions were assigned to the largest maxima in the difference Fourier synthesis of electron density. This is a Fourier synthesis in which the Fourier coefficients are the differences between the observed and calculated structure factors. When the hydrogen coordinates had been fitted to the measured data, the isotropic temperature coefficients of the hydrogen atoms were also refined. The R_1 was now 5.3 %. Finally, weighting the squared deviations in

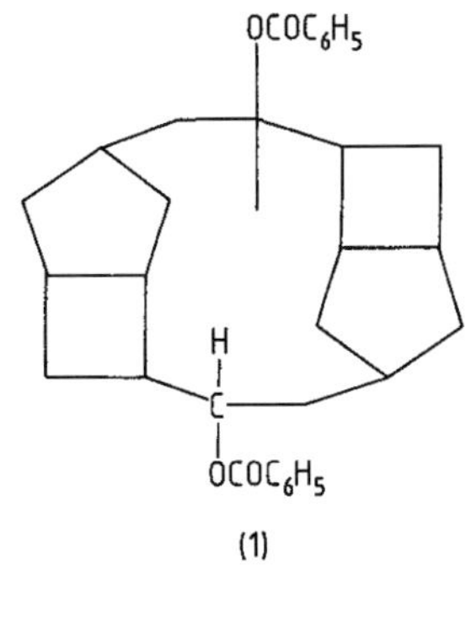

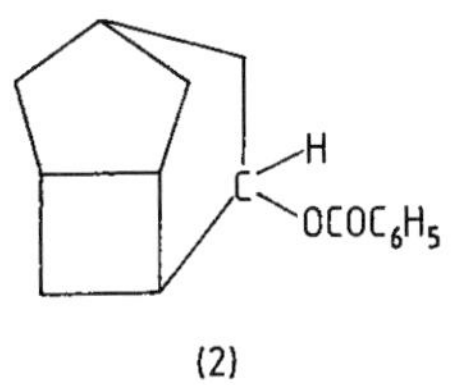

Figure 37. Two suggested structures for the compound under study

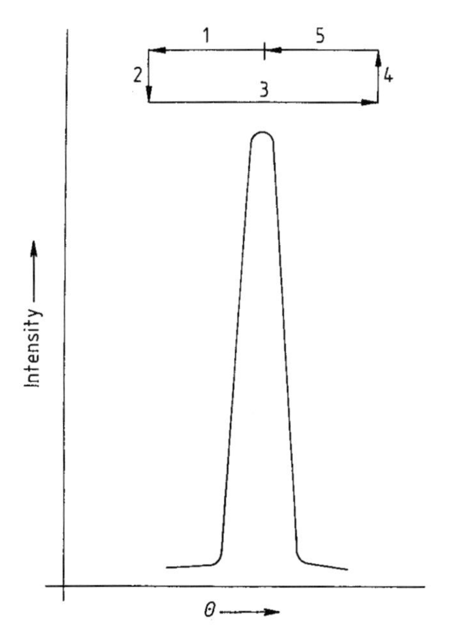

Figure 38. Steps in the five-point measurement

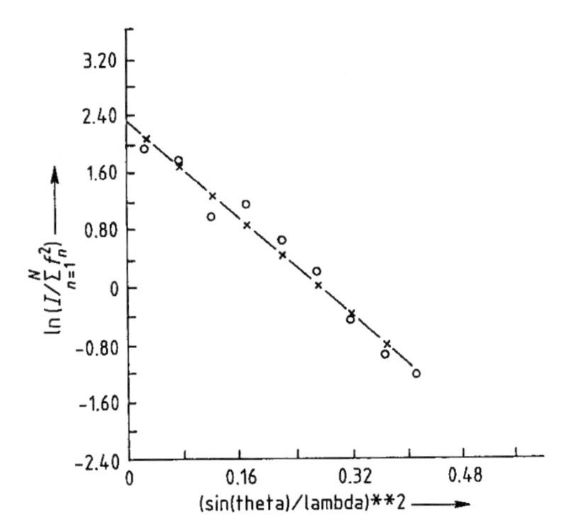

Figure 39. Computer plot of results from Wilson statistics

accordance with the counting statistics gave $R_1 = 4.7\,\%$. The weighted R_2 from Equation (44) was 3.6 %.

At this point, the least-squares refinement could be regarded as complete because the largest change in a parameter was smaller than 10 % of the corresponding standard deviation. In a difference Fourier synthesis of the electron density, the values for the 10 largest maxima were 0.12–0.16 electrons/10^{-30} m^3.

The software used was MULTAN 77 [110]. There are a number of more up-to-date program systems:

SIR 92 [42]
DIRDIF [43]
MULTAN 88 [44]
NRCVAX [46]
XTAL [47]
ORTEP II [48]
SHELXS 86 [49]
SHELXL 92 [50]
SHELX 97 [51]
CRYSTALS [52]

Representations of the molecule are shown in Figures 40 (configuration and bond lengths), 41 (stereoscopic view), and 42 (projection of unit cell content along the *a*-axis). If the bond and torsion angles (not shown here) are considered, along with the bond lengths, it becomes clear that major difficulties arise in determining the molecular structure by spectroscopic means alone. The results of the structure analysis have been published elsewhere [112].

15.2.6. Diffraction by Polycrystalline Specimens

A substance is characterized as polycrystalline if its crystalline particles cannot be distinguished with the naked eye (<0.01 mm). Larger particles, if any, must be ground before the measurement, provided this does not adversely affect the crystal structure of the substance.

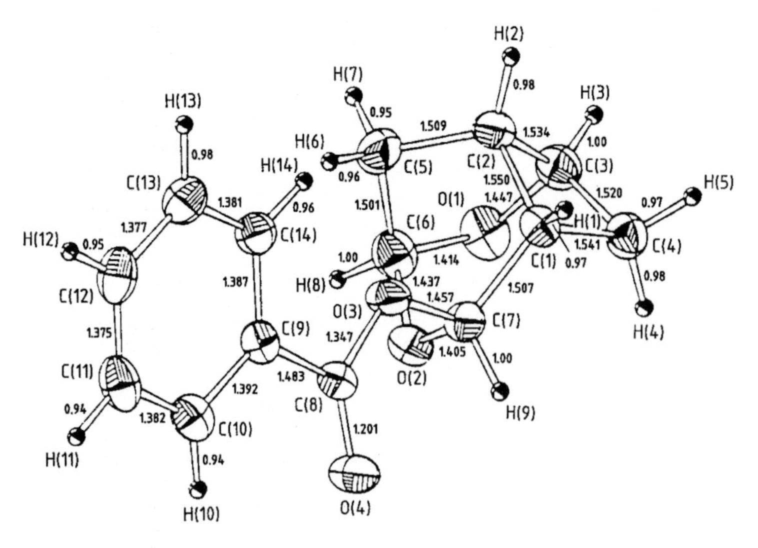

Figure 40. Molecular structure of (1 RS, 3 SR, 4 RS, 6 SR, 8 SR)-5,9-dioxatricyclo[4.2.1.$0^{3,8}$]-non-4-yl benzoate. Lengths in 10^{-10} m. Standard deviations $\bar{\sigma}$: 0.002×10^{-10} m for O – C, 0.002×10^{-10} m for C – C, 0.01×10^{-10} m for C – H

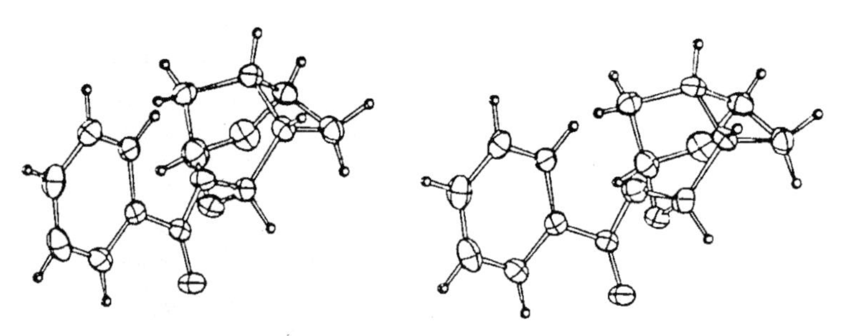

Figure 41. Stereoscopic view

15.2.6.1. General Considerations

If the apparatus ordinarily used to record diffraction patterns of polycrystalline specimens is compared with that used for single-crystal diffractometry, the following essential differences are seen:

1) Instruments for powder diffractometry are much simpler. If an "ideally polycrystalline" specimen is assumed, and if a one- or two-dimensional detector with appropriately large angular aperture is employed, there is no need to rotate either the specimen or the detector. Given an ideal statistical distribution of crystallites in the specimen, together with not too small a specimen size, it can be assumed that every orientation of a crystallite relative to the incident X-ray beam occurs in the specimen (as a statistical average). A rotation about one axis is enough to get a quite substantial increase in the scattered X-ray intensities, because then a large number of crystallites contribute to the intensity for a selected net plane. Single-axis goniometers used for this purpose are the Debye – Scherrer and the transmission diffractometers.
2) The specimen sizes employed are much larger than those in single-crystal studies. Because ideally there must be a statistical distribution of crystallites, only a small fraction of the specimen can contribute to a given reflection.

One of the problems in X-ray diffraction by a polycrystalline powder is that the specimen is not "uniformly polycrystalline". The crystallites vary widely in size and their relative orientations are not random. The predominance of one habitus or

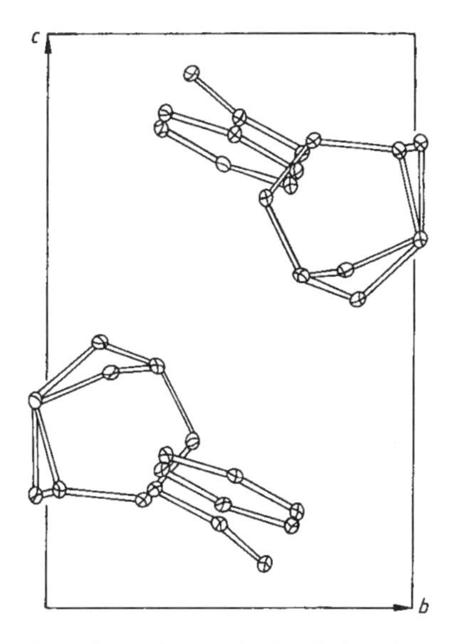

Figure 42. Projection of unit cell along the a-axis

another is crucial in the preparation of the specimen; for example, needles line up parallel to one another, while plates align themselves parallel to the specimen surface. In principle, this effect can be diminished by grinding the specimen and screening out the coarse particles, but there are two disadvantages to this:

1) Organic substances, in particular, are so sensitive that the crystal structure is disrupted by grinding, the specimen undergoes a change in degree of crystallinity, or there may even be a phase transformation
2) The specimen thus treated is no longer in the as-delivered condition that was to be characterized

If one has a phase mixture, demixing is possible, because the components often differ in their grinding characteristics.

Far better than excessive size reduction is to rotate the specimen appropriately and integrate over all directions in space. These ideal conditions can be approximated if an area detector is employed.

15.2.6.2. Qualitative and Quantitative Analyses of Powder Specimens

The X-ray diffraction pattern of a polycrystalline specimen is a plot of diffracted X-ray intensity against diffraction angle. Because the diffracted intensity is a single-valued function of the unit cell contents (Eqs. 15, 18), and the diffraction angle is a single-valued function of the unit-cell dimensions (Eq. 2), an identical unit cell with identical contents must also yield an identical diffraction pattern. The diffraction pattern is thus a genuine fingerprint.

But the fingerprint also contains information on the macroscopic properties of a specimen. The half-value widths of the reflections are inversely proportional to the crystallite sizes, and directly proportional to the concentration and magnitude of structural defects. The coherent scattered component of the background results from amorphous constituents in the specimen, while the incoherent scattering is chiefly due to inelastic atomic scattering.

Determination of Cell Constants, Crystal System, and Space Group. The requirements on specimen preparation are a bit less demanding when a powder pattern is used to determine the cell constants only as long as lines are not fully missing, because preferred orientation is so heavy. Unfortunately, an unknown unit cell cannot always be determined in this simple way. The method is most suitable for the cubic case. Squaring the Bragg equation gives:

$$\sin^2\theta = \frac{\lambda^2}{4d^2} \tag{61}$$

For the cubic crystal system:

$$1/d^2 = \frac{h^2 + k^2 + l^2}{a^2} \tag{62}$$

where a is the cell constant. It follows that:

$$\sin^2\theta = \frac{\lambda^2}{4a^2}\left(h^2 + k^2 + l^2\right) \tag{63}$$

For a cubic substance, the constant $\lambda^2/4\,a^2$ can normally be identified directly if the $\sin^2\theta$ values of the reflections are ranked in ascending order of magnitude. The value of $\lambda^2/4\,a^2$ can be recognized and all $\sin^2\theta$ values can be obtained by multiplying by a sum of at most three squares.

Unambiguous results are, however, obtained only if the errors of measurement are small. It can never be excluded with utter confidence that slight deviations from, say, the cubic system occur. Absolutely certain results come only from single-crystal patterns, which additionally reveal the symmetry.

The powder patterns of substances belonging to other crystal systems can be indexed on the basis of analogous regularities, and thus assigned to a unit cell. The method of T. ITO even makes it possible to index the diffraction spectra of triclinic compounds [113]. The precondition is that many discrete reflections must be present, and their glancing angles must be determinable with great accuracy. Organic compounds commonly do not fulfill these conditions. Large cell constants lead to overlapping of reflections at relatively small glancing angles; many crystal structure defects and large temperature factors have the same effect because they increase the half-value widths of the reflections.

Special difficulties arise in determining the cell constants of high polymers. Using single crystals instead of crystalline powders is not ordinarily an option because in most cases there are principally no single crystals. In most cases, however, filaments can be drawn from the substance; after appropriate heat treatment, such filaments tend to take on a single-crystal character. The reflections are then located on layer lines in flat-film patterns. With some luck, trial and error will lead to not only the translational period in the fiber direction, but also the other cell constants [1, pp. 405–406].

If provisional cell constants have been determined for a specimen, all reflections can be indexed. There must be no common divisor for all the indices h, k, and l (primitive cell). Now the reduced cell (whose edges are the three shortest noncoplanar translational periods of the lattice) can be calculated. From the Niggli matrix:

$$\begin{pmatrix} \boldsymbol{a}\cdot\boldsymbol{a} & \boldsymbol{b}\cdot\boldsymbol{b} & \boldsymbol{c}\cdot\boldsymbol{c} \\ \boldsymbol{b}\cdot\boldsymbol{c} & \boldsymbol{c}\cdot\boldsymbol{a} & \boldsymbol{a}\cdot\boldsymbol{b} \end{pmatrix} \tag{64}$$

which is formed from the scalar products of the cell constants, it is possible to deduce the crystal system and Bravais lattice [4, pp. 124–159].

With the final cell constants, the crystal system derived from them, and the Bravais lattice, the reflections can be indexed. The space group can be identified from systematic absences [19, pp. 21–53].

As the preceding discussion implies, the determination of cell constants from powder patterns involves large uncertainties, even more so for the determination of the crystal system and space group, because the diffraction pattern does not directly lead to the symmetries, which are instead deduced from measured angles which include experimental errors, e.g., 90° for one of the cell constants. The case is similar with systematic extinctions: Lack of resolution can result in uncertainty or a wrong conclusion.

A better procedure is the use of powder patterns for crystalline phases whose cell constants are known. Changes in the cell under given conditions can be tracked in a convenient and reliable way. The cell data can be fitted to the measured data by least-squares methods.

Phase Analysis. If a specimen is crystalline, an obvious way to study and identify it is by using X-ray diffraction patterns. The X-ray diffraction pattern of a crystalline compound functions as a fingerprint that should permit unambiguous identification. Isomorphous crystalline compounds often create major problems; elemental analysis can be of aid here.

The case is even more difficult when the specimen for analysis is a mixture of crystalline phases differing markedly in the content of the individual components. This problem may be so aggravated that one of the crystalline phases is represented solely by its strongest reflection. Other methods of investigation are then needed in order to verify this crystal phase. All that X-ray diffraction can do in such instances is to make the presence of a crystalline phase probable.

If two isomorphous crystalline compounds cannot be distinguished from the X-ray diffraction pattern, neutron diffraction is an option. The atomic form factors for neutron diffraction and X-ray diffraction are completely different, so that the reflections can be expected to differ in intensity.

The successful use of X-ray diffraction to analyze unknown substances and mixtures of substances requires a knowledge of many X-ray diffraction patterns. The International Center for Diffraction Data (ICDD, 1601 Park Lane, Swarthmore, PA 19081, United States) has taken on the job of compiling X-ray diffraction patterns. The tabulated data can be obtained in the form of data cards, microfiches, compact disc (CD) or magnetic tape, together with a search program. In the first two cases, the X-ray diffraction data can be searched by means of tables ordered via lattice plane distances and relative intensities. The continually updated compilation concentrates on inorganic substances. The collection of data for organic compounds is far less complete. Furthermore, analysis is made more difficult by the poorer crystal quality and larger cell constants of organic compounds. Both these factors tend to push the principal reflections to relatively small glancing angles, so that the net plane distances must be assigned relatively large errors.

Quantitative Analysis of Crystalline Phases. The intensity of a reflection is directly proportional to the quantity of the crystalline phase in a mixture, but it is also a function of the crystal structure. Comparing the intensities of two reflections is therefore not enough to determine the mass ratio of two phases in a mixture; the constant of proportionality must also be found.

The following technique can be used in the case of a two-phase mixture in which the percentage content of one phase is to be determined. In a number of specimens of known composition, the intensity ratio for the two crystalline phases is plotted against the mass ratio. The dependence is linear only if both crystalline phases have similar absorption coefficients, which is roughly so for, say, crystal modifications of the same compound.

A method for the absolute quantitative determination of crystalline phases in a multiphase mixture is to mix a standard substance into the specimen. Calibration curves are necessary for mixtures of the standard with each of the phases to be determined. The components can then be determined independently of each other. The critical point is to determine accurately the integral intensity of a sufficiently strong reflection that does not have any other reflection superposed on it. With a diffractometer linked to a computer, it is possible to separate overlapping reflections and determine their intensities, although some idealizations are required.

When a phase mixture contains just a few percent ($< 8\,\%$) or even fractions of a percent of the phase being analyzed a more accurate result can be obtained by recording X-ray diffraction patterns for specimens that lie in the composition range of interest. These allow the content of the strongly deficient phase to be estimated. The results are often distorted because it proves quite difficult to find a suitable standard for the specimen being measured. The standard should have the same crystal quality and crystallite size as the substance to be determined; otherwise, errors in the results are inevitable. These idealized conditions are scarcely ever fully achievable.

15.2.6.3. Solid Solutions

Compounds are normally miscible in all proportions in the gas phase, but in the liquid phase two or more substances must have certain chemical properties in common if they are to be miscible. The situation is much more complicated in the solid phase. Here, miscibility demands not just chemical affinity, but also a similar manner of space filling. For the formation of a continuous series of solid solutions, a molecule or atom of one phase must be arbitrarily replaceable by a molecule or atom of the other one. The formation of solid solutions has a relatively great effect on the cell volume, and this effect can be used for the very exact determination of a compound in trace element analysis (e.g., quantitative determination of trace impurities in semiconductor crystals).

Very often, solid solutions can be formed only over a limited range. The host lattice tries to adapt, more or less, to the guest until a point is reached where the stability limit of the host lattice is passed. Variations of the lattice parameters in the stability range are often controlled by the Vegard rule, which says that variations of parameters are proportional to the content of foreign components. The validity of this rule can, conversely, be regarded as evidence of solid-solution formation. It is generally sufficient if proportionality can be demonstrated for a few net plane distances.

The use of computers to control the diffractometer and interpret the measurements has proven itself here in particular. Not only can glancing angles be determined accurately, but any new reflections that appear can be identified with sufficient statistical reliability.

15.2.6.4. Rietveld Method

Measurements on crystalline powders lead to a solution of only relatively small and simple structures: Most of the reflections must be resolved, so that intensities can be associated with them unequivocally. Even theoretically, this is often impossible, e.g., when several reflections occur at precisely the same glancing angle. In the cubic crystal system this is so when the sum of the squares of the indices does not permit a single indexing; e.g., (221) and (300), where the sum of the squares of the indices is 9 for each case. Measurements on crystalline powders that give unambiguous values for all reflections should be inherently preferable to single-crystal measurements, because many experimental difficulties (e.g., extinction and absorption) either do not occur at all, or are easier to handle.

This is theoretically correct, but this ideal case never occurs. The atomic resolution is always low-

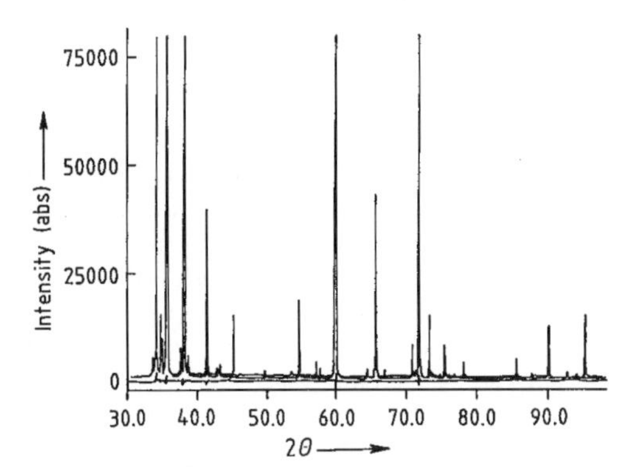

Figure 43. Diffraction pattern of a polytypic mixture of SiC

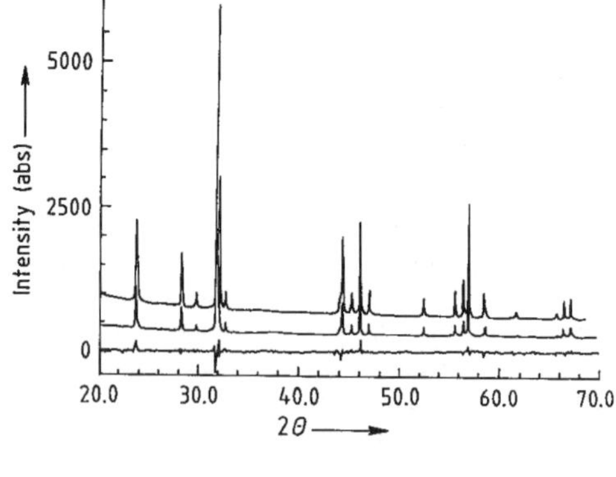

Figure 44. Diffraction pattern of a solid solution and a mixed phase: 12.9 % $Nd_{0.4}Ce_{0.6}O_{1.8}$, 87.1 % $Nd_{1.85}Ce_{0.15}CuO_4$

er. In general a single-crystal structure analysis is always better and gives more significant data, nevertheless the single crystal might be very poor. The powder method is used when it is not possible to obtain sufficiently good single crystals.

The strength of the Rietveld method lies in the refinement of the structure parameters of the whole powder sample not just of one single crystal. RIETVELD [114] recognized that it is by no means necessary to allocate the measured intensities to individual reflections in a one-dimensional powder pattern. Instead, it is better to measure a diffraction pattern as exactly as possible and use it in its entirety as the basis for refinement of the structure parameters. Parameters such as the atom coordinates, temperature coefficients, and scaling factors are fitted to the diffraction data by a least-squares procedure.

This method is particularly important for neutron diffraction analysis, where in general a previous single-crystal X-ray structure analysis has led to a structure model that is satisfactory within certain limits. The powder method can then be used, for example, to examine the magnetic properties of a substance in more detail, without having to grow large single crystals.

Rietveld analysis has recently proved especially valuable for the quantitative determination of components in polytypic mixtures. If the structures of the polytypes are known, the only parameters to be determined are the temperature coefficients, the scaling factors, and the cell constants. Figure 43 shows an analysis of SiC. The specimen contains 11.2 % of polytype 4 H, 85.2 % of 6 H, and 3.6 % of 15 R. The profile R value is ca. 6.7 %. Another very important application is the study of solid-solution systems. As Figure 44 shows, the content of an additional phase can be estimated when there are gaps in the series of solid solutions [115]. The profile R value in this case was 2.67 %.

In the past this method could only be used when the overall crystal structure was known and only special characteristics had to be explored. Now many methods have been developed to perform quasi-a-priori crystal structure analyses of crystal powder samples. Computer programs for single-crystal work often have an option for powder structure analysis (e.g., [42], [50]). They need the successful extraction of intensities of as many as possible single reflections out of the powder diagram, what is often performed by the method of Le Bail [117]. This method is also used by POSIT, in which only low-order reflections are necessary to determine a model from known fragments of the unknown structure for the Rietveld refinement [118]. The computer program Powder Cell is a complete program system for solving crystal structures from powder data [53]. Fully another approach is Monte Carlo/simulated annealing [119]. Only an approximate starting model with a few internal degrees of freedom is necessary. Thousands of diffraction diagrams are calculated and compared with the experimental one. The software package Powder Solve is fully integrated within the Cerius^2 molecular modeling environment [54]. A further approach, quite different from the above, is the ab initio prediction of possible molecular crystal structures by energy minimization. No cell determinations, no indexing, and no extraction of intensities of single reflections are necessary [53], [120] – [122], [124].

15.2.6.5. Texture

Up to this point, two distinct types of specimen, single crystals and polycrystalline powders, have been discussed. In the analysis of a single crystal, it is essential that the specimen contains no polycrystalline material; in polycrystalline powders, one condition for X-ray examination is that the directions of the net planes have a uniform statistical distribution in space, otherwise more or less complicated corrections must be applied, perhaps by preparation or by computer programs. An intelligent use of preferred orientation for structure analysis is recording the diffraction diagram by an area detector. The reflections are partly spread out into two dimensions, many more reflections are resolved, more reflections can be used for determining the cell constants, and overlap of intensities of reflections is diminished [116].

The properties of many substances, however, exhibit a direction dependence, which is similar to a single crystal, but at the same time indicates that the substance has only partial single-crystal character. Such a solid is said to have texture. The individual crystallites are fully or partly aligned; they are said to have preferred orientations.

Texture in a polycrystalline structure can arise in quite varied ways, e.g., deformation texture, casting texture, recrystallization texture, plate texture, or fiber texture.

Texture studies on wires and plates are very important. Initial information can be acquired from transmission and reflection patterns recorded with a flat-film camera (see Fig. 15). The type and degree of texture can be determined more or less exactly, depending on the object of investigation and the type of instrument used. One of the best-developed instruments is the Lücke texture goniometer [125]. With this apparatus, the distribution of given net planes can be directly imaged in a pole figure.

The fully automated, computer-controlled X-ray texture goniometer sold by Seifert allows reflection and transmission measurements; the background is measured and the appropriate correction applied automatically. Other corrections, e.g., absorption, are also performed automatically. The pole figure is output to a terminal. Several reflections can be measured in a fully automated programmed sequence.

15.2.7. Crystallographic Databanks

X-ray studies typically generate large amounts of data. Specimen synthesis and preparation, data acquisition, and interpretation of results are generally so expensive that repeating a structure determination is a waste of resources, unless there are compelling reasons to do so. For this reason, databanks have been set up and made available to scientists worldwide. Unfortunately, using these databanks generally incurs high costs; the importance of structure studies to the advance of the natural sciences suggests that the scientific community ought to be trying to make such data freely available to all. The following databanks deserve mention:

Powder Patterns. PDF (Powder Diffraction File) Databases [38]

Crystal Structures of Organic and Organometallic Compounds. CSDS (Cambridge Structural Database System) [37]

Crystal Structures of Inorganic Compounds. Inorganic Crystal Structure Database [39]; MDAT (minerals) [40]

Crystal Structures of Proteins. Protein Data Bank [41]

15.2.8. Noncrystallinity

The diffraction of X rays by solid matter is only possible when a certain regularity of structure exists. Scattering does not become diffraction unless the material contains equal and repeated distances. These conditions are fulfilled by a crystal lattice. If the crystallites are too small, or the crystals contain structural defects, the coherent scattering domains become smaller and the reflections become broader; this broadening may be so great that it leaves only an amorphous halo, often with a half-value width of several degrees, as is the case with quartz glass, for example. In addition, the halo is often centered near the highest intensity reflection of the corresponding crystalline phase. A similar effect occurs in solids that do not have this continuous gradation from crystalline to amorphous, but in which similar atomic distances recur. Some amorphous substances give no halo at all. If the particles in a uniform host matrix are large ($20-1000\times10^{-10}$ m and above) scattering takes place at small angles ($< 0.5°$), where profile forms can yield information about the particle shape. The precondition for such studies is that the particles and the enclosing matrix differ sufficiently in electron density. Several problem areas are dealt with in somewhat more detail.

15.2.8.1. Crystal Quality Analysis

The relations discussed above hold for an ideal crystal; in practice, crystals are ideal only to a limited extent. In the following section, an attempt is made to indicate where a particular specimen should be placed in the range from ideal crystal to amorphous matter. Along with structural defects in the interior of the crystallites, or at their surfaces, the crystallite size also has to be estimated. A crystal may also contain stresses, or it may be that the single-crystal ordering is only partial, or applies in just one or two dimensions. Each of these factors affects the X-ray diffraction pattern.

15.2.8.2. Crystallite Size

A real crystal is not infinitely large. In the case of X rays and neutrons, size does not begin to affect the diffraction pattern until the crystallites (or more precisely the coherent scattering domains) become smaller than ca. 0.5 μm. When electrons are used, this limit is nearly two orders of magnitude smaller (0.01 μm).

SCHERRER, in 1918, derived the following equation between the "physical half-value width" β (in degrees 2θ), i.e., that part of the broadening of a reflection due solely to inadequate crystallite size, and the crystallite size L:

$$\beta = \frac{K\lambda}{L\cos\theta} \tag{65}$$

Theoretically, L is the dimension of the crystallite perpendicular to the diffracting net planes; K is a constant, close to unity (often taken as 0.9); and λ is the wavelength of the radiation employed. The observed half-value width B is commonly assumed to be the sum of the instrumental broadening b and the physical broadening β. Some authors also give the formula:

$$B = \sqrt{\beta^2 + b^2} \tag{66}$$

It should be kept in mind that the error of determination is 20–30 %.

The instrumental broadening b can be obtained by calibrating with a substance with a large crystallite size (>0.5 μm) [126].

15.2.8.3. Defects

Like excessively small crystallite size, crystal structure defects also increase the half-value width of a reflection. Such defects make the coherent scattering domains smaller. The amount of broadening depends on the type of defect, because in general newly generated scattering domains remain coherent with one another across the new disruptions.

It is very difficult to separate the broadening of a reflection into components due to small crystallite size and to structural defects. Equation (65) implies that $\beta\cos\theta$ should be constant for all glancing angles, if the broadening results entirely from small crystallite size. Because L is the dimension perpendicular to the diffracting net planes, this statement cannot hold rigorously, except for the diffraction orders of one and the same net plane. If line broadening is the consequence of crystal structure defects alone, the width increases with θ.

15.2.8.4. X-Ray Crystallinity

The two-phase model generally accepted for high polymers has amorphous and crystalline domains occurring next to each other. If the crystallinity obtained by X-ray examination isstated as a percentage, this figure should not be regarded as a serious value; it serves only for comparison within a series of experiments. In no case should comparison be made between the X-ray crystallinities of high polymers derived from different monomers.

It has become customary to extend the term X-ray crystallinity to compounds other than high polymers, in which case numerical figures should be treated with even more caution. For example, the properties of organic pigments depend very strongly on the crystal quality of the primary particles, i.e., the coherent scattering domains. Figure 45 presents X-ray diffraction patterns of organic pigments consisting of the same compound. Between A and C, a continuous transition (such as case B) is possible. It therefore appears meaningful to speak of a crystalline component, but numerical values can never be assigned in such a case, as they can for high polymers, where por-

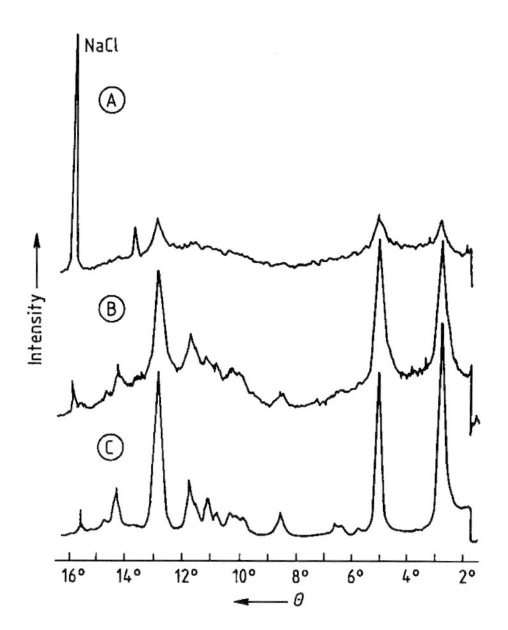

Figure 45. Organic pigments consisting of the same compound, but differing in crystal quality (patterns recorded with Cu K α radiation)
A) "Over-ground" pigment; B) Moderately well-finished pigment; C) Well-finished pigment

tions of the X-ray diffraction pattern can be attributed to amorphous and crystalline parts (Fig. 46).

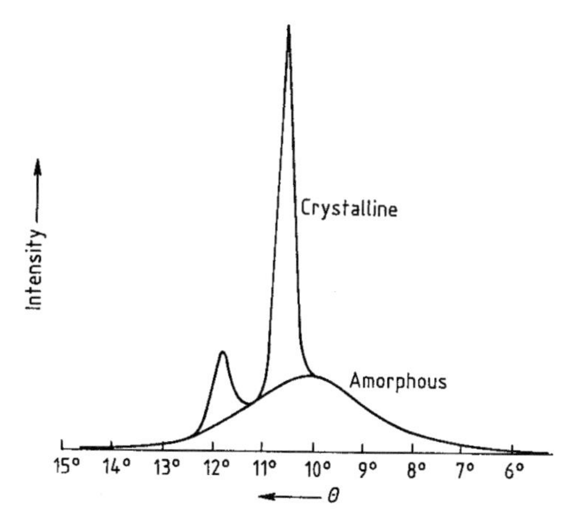

Figure 46. Amorphous and crystalline components in the X-ray diffraction pattern of a high polymer, recorded with Cu K α radiation

15.2.8.5. Elastic Stress

The measurement of elastic stress is especially important in the examination of steels. Elastic stress (residual macroscopic stress) expands the lattice in the direction of the tensile stress and contracts it in the perpendicular direction. The corresponding shifts toward lower and higher glancing angles, respectively, can be measured by X-ray methods. These lattice changes can be correlated with the elastic moduli E:

$$\sigma_{\varphi} = \frac{d_{\psi} - d_0}{d} \frac{E}{1 + \nu} \frac{1}{\sin^2 \psi} \qquad (67)$$

where d is the net-plane distance in the stress-free material; d_{ψ} is the distance in the stressed material, where the lattice planes make an angle of ψ with the surface, and similarly d_0 for 0°; ν is the Poisson ratio (ratio of transverse contraction to longitudinal extension); and σ_{φ} is the stress in direction φ in the specimen surface (the X-ray beam direction projected onto the specimen).

The macrostress should be distinguished from the residual microstress, which is due to internal stresses caused by factors such as the hardening of steels.

In contrast to other methods of measuring elastic stress, the X-ray method is not subject to distortion of the results by plastic deformation. Furthermore, highly inhomogeneous stress fields can be measured, because the irradiated surface can be as small as $1-10\ mm^2$. The technique has the disadvantage that X rays penetrate only some 10^{-2} cm into steels, so that only stresses at the surface can be determined.

The elastic stress in a given direction can be obtained from Equation (67), by making measurements at a range of incident angles. If a photographic method is used, objects such as bridge piers can be tested nondestructively in situ.

15.2.8.6. Radial Electron Density Distribution

The systems discussed up to this point have been crystalline in the common meaning of that term. As a rule of thumb, there must be more than 10 translational periods in every direction in space before a material can be deemed crystalline (a one- or two-dimensional state of order does not qualify). The expression "noncrystalline" here means the same as "X-ray amorphous".

The more noncrystalline a sample becomes, the less possible it is to give precise statements about individual atomic distances. In these cases, only a radial distribution function (RDF) of density $\varrho(r)$ of electrons or atoms can be given, according to the equation [17, p. 795]

$$4\pi r^2 \varrho(r) = 4\pi r^2 \varrho_0 + \frac{2r}{\pi} \int_0^\infty S\, i(S) \sin r S \,\mathrm{d}S \qquad (68)$$

where ϱ_0 is a function of the density of the solid, S equals $4\pi \sin \theta/\lambda$, and $i(S)$ is a function of the diffraction pattern. This equation can only be applied to samples with only one kind of atom.

15.3. Synchrotron Radiation

Synchrotron radiation is generated when electrons travel along curved trajectories at speeds close to that of light. The curvature is produced by magnetic fields. The nature and intensity of these fields determine the spectral distribution of the electromagnetic radiation produced. In general, a spectrum ranging from short-wavelength X rays ($< 0.5\times10^{-10}$ m) up to the ultraviolet (200 nm) is available. In contrast to X rays from rotating anode tubes, the synchrotron source offers high intensity, but also a broad spectrum, only slight divergence, and well-defined state of polarization. The radiated power (whole spectrum) obtained from an electron current of 100 mA is around 50 kW, several orders of magnitude greater than the output of conventional X-ray sources [127].

For elastic interactions at large angles (>5°), scattering by an amorphous material (short-range density fluctuations) is distinguished from diffraction, where interference effects play a key role. Studies in this field employ the powder, rotating crystal, and diffractometer methods, but the principal method and the most developed is also the oldest — the Laue method. Protein crystal structures can be examined in the time-resolved mode (submillisecond range) with this technique. Because of the high degree of polarization of the radiation, magnetic structures can also be investigated. At small angles, the Bragg reflections of structures with very large cells (including long periods) are accompanied by scattering, which yields information about domain sizes, defects, microcracks, and clusters and segregations of all kinds — provided there are regions differing in electron density. Time-resolved studies are possible in all these cases because of the high intensity of the X-ray beam used.

The inelastic interaction between synchrotron radiation and solid matter is also utilized by crystallographers, although spectroscopic transitions rather than diffraction phenomena are involved (EXAFS, XANES, UPS, XPS). These techniques measure the fine structure of absorption edges, to yield information about the bonding of the atom under study.

15.4. Neutron Diffraction

Many experiments can be performed with either X rays or neutrons. Neutron radiation is some orders of magnitude more expensive and so is not a candidate for studies that can be done equally well or better (because of the higher intensity) with X rays [127], [128].

An important application of neutron diffraction is based on the fact that the dependence of scattering power on atomic number is different for neutrons and X rays. For X rays, the atomic form factor is directly proportional to the atomic number, but the relationship is more complicated for neutron radiation, and there is a correlation with the energy states of isotopes. Between hydrogen and deuterium, for example, there is an appreciable difference in scattering power. The fact that neutrons, unlike X-ray quanta, are scattered by atomic nuclei means that the scattering power is virtually independent of the diffraction angle. A further advantage of the use of neutrons is the lower absorption.

By virtue of these factors, diffraction of neutrons by polycrystalline matter yields far more information than that of X rays. Thicker layers of material can be penetrated, and preferred orientations can be eliminated. Useful intensities can be measured out to large diffraction angles.

For most structure problems investigated with neutrons, an X-ray structure analysis already exists. The task is to improve or refine certain parameters. There are four essential problems:

1) To determine the coordinates of hydrogen atoms
2) To distinguish between atoms of similar atomic number
3) To perform Rietveld profile analyses on powder samples to determine special structure parameters or achieve structure refinement
4) To analyze magnetic structures

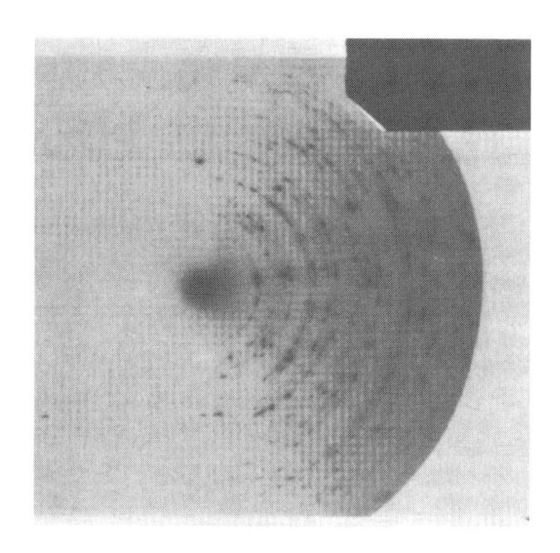

Figure 47. Electron diffraction pattern of titanium sheet, measured in reflection immediately after annealing

15.5. Electron Diffraction

15.5.1. Diffraction in the Electron Microscope

Because of their strong interaction with matter, electrons have little penetrating power, and are thus particularly suited to surface studies. No line broadening is expected up to a crystallite size of 100×10^{-10} m for electrons with a wavelength of 0.05×10^{-10} m (accelerating voltage 60 kV). If the crystallite orientation has a statistical distribution, a diffraction spectrum comparable to the Debye–Scherrer diagram is obtained. This can be utilized for phase analysis (like an X-ray powder pattern); in the case of reflection measurements, the analysis is valid for layers near the surface (Fig. 47).

Surface regions just a few atomic layers thick can be examined by diffraction of electrons that have been accelerated by a low voltage (150 V). Epitaxy can be dealt with in this way. Because of absorption, deeper layers contribute nothing to the diffraction.

Often, the distribution of the vectors of the net planes in a powder specimen is not statistical. Because the curvature of the Ewald sphere is small relative to the dimensions of the reciprocal lattice, many net planes are unequally well in their reflection position, resulting in a correspondingly unequal degree of error in the diffracted electron intensities. Electron diffraction results should therefore be verified by comparative X-ray measurements whenever possible.

If the X-ray diffraction pattern offers a fingerprint with two parameters, the glancing angle and the reflection intensity, the electron diffraction pattern gives only the glancing angle, and one cannot speak of a fingerprint, especially as even strong reflections may be absent altogether. Transmission electron diffraction data, however, can be very similar to the corresponding X-ray patterns, and may contain a similar level of information (see Figs. 48 and 13). This is only the case when the distribution of net plane vectors is statistical.

15.5.2. Electron Diffraction in the Gas Phase

The discussion of this technique has been kept short because the diffraction spectra are very similar to Debye–Scherrer patterns; the method is becoming very important for molecular structure analysis. An electron beam in a high vacuum (0.1 – 10 Pa) collides with a molecular beam, and the electrons are diffracted by the molecules. The film reveals washed-out Debye–Scherrer rings, from which a radial electron density distribution function for the molecule can be derived. Together with spectroscopic data, the distribution makes it possible to infer the molecular structure [129].

15.6. Future Developments

Many of the techniques discussed here have become routine. X-ray diffraction data of organic pigments, for example, are employed for product surveillance in industry. Not just the fingerprint as such, but also the half-value widths of the reflections and the level of the background, are utilized. The situation is similar in the fiber industry, as well as many other areas of chemical production. (Fig. 48)

The most spectacular successes of X-ray methods, however, are in molecular and crystal structure analysis. Examples are the structures of insulin [130], hemoglobin [131], and vitamin B_{12} [132]. Today, single-crystal X-ray structure analyses of relatively complex compounds (up to ca. 40 – 200 nonhydrogen atoms) are performed with computer-controlled diffractometers whose computers can be used simultaneously to solve the phase and structure problem within a few days or just a few hours.. The method has gained parity in investigation time with spectroscopic methods

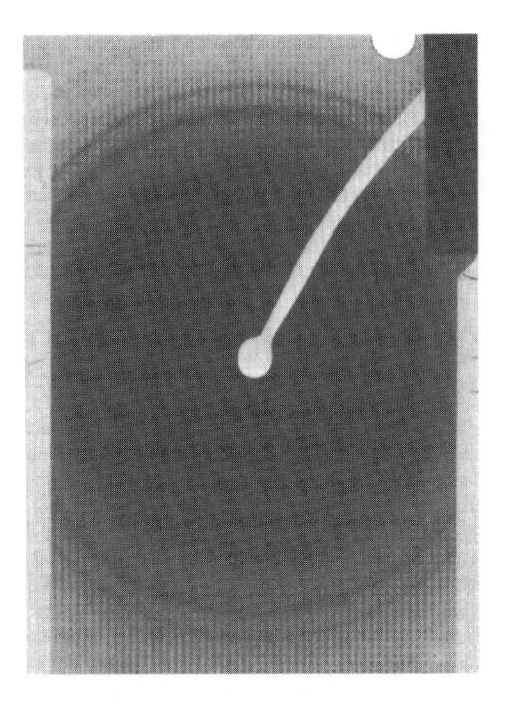

Figure 48. Transmission electron diffraction pattern of LiF

for the solution of relatively complex structure problems as far as the compounds are crystallizable. What is more, X-ray structure analysis is far superior to all other methods with respect to the reliability of the results. A single-crystal X-ray structure analysis of a compound yields everything that one could "see" if it were possible to examine a single molecule under the microscope. But in the end it is very important to remove all the discrepancies between diffraction and spectroscopic methods.

However, it can be quite laborious to solve the phase problem, even for relatively simple crystal structures. There is not yet any general method; a special solution of the phase problem must be found for every structure. The methods have been developed to the point where the solution can often be achieved by a computer alone; if this fails, however, it is usually not clear why, considerable ingenuity may be needed to reach the solution.

The fundamental reason for this situation is that the relationships in direct phase-determination methods are nothing more than probability relations. If these can be turned into true equations, single-crystal X-ray structure analyses would become a routine analytical procedure. H. HAUPTMANN, co-creator of the direct methods of phase determination, expressed the dilemma as "One good idea and we are out of a job".

One attempt to replace the probability relations with true equations for phase determination was that of R. ROTHBAUER [133], based on earlier work by SAYRE [78], and WOOLFSON [134]. The future must show whether the equations derived can be used for practical structure determination [135]. Their use for the determination of the absolute scaling factor and the overall temperature factor has already been tested in practice [136]. A publication by R. ROTHBAUER shows, that solving the phase problem generally is still up to date [0]

15.7. References

General References

Text Books

[1] L. E. Alexander: *X-Ray Diffraction Methods in Polymer Science,* Wiley, New York 1969.

[2] K. W. Andrews, D. J. Dyson, S. R. Keown: *Interpretation of Electron Diffraction Patterns,* Hilger & Watts, London 1967.

[3] U. W. Arndt, B. T. M. Willis: *Single Crystal Diffractometry,* University Press, Cambridge 1966.

[4] L. V. Azaroff, M. J. Buerger: *The Powder Method in X-Ray Crystallography,* McGraw-Hill, New York 1958.

[5] G. E. Bacon: *Neutron Diffraction,* 3rd ed., Clarendon Press, Oxford 1975.

[6] T. L. Blundell, L. N. Johnson: *Protein Crystallography,* Academic Press, New York 1976.

[7] M. J. Buerger: *Vector Space and Its Application in Crystal-Structure Investigation,* Wiley, New York 1959.

[8] M. J. Buerger: *Crystal-Structure Analysis,* Wiley, New York 1960.

[9] M. J. Buerger: *The Precession Method in X-Ray Crystallography,* Wiley, New York 1964.

[10] M. J. Buerger: *X-Ray Crystallography,* 7th ed., Wiley, New York 1966.

[11] M. J. Buerger: *Kristallographie, eine Einführung in die geometrische und röntgenographische Kristallkunde,* De Gruyter, Berlin 1977.

[12] J. D. Dunitz: *X-Ray Analysis and the Structure of Organic Molecules,* Cornell University Press, Ithaca 1979.

[13] C. Giacovazzo et al.: "Fundamentals of Crystallography," in C. Giacovazzo (ed.): *International Union of Crystallography,* Oxford University Press, Oxford 1992.

[14] J. P. Glusker, K. N. Trueblood: *Crystal Structure Analysis, a Primer,* Oxford University Press, New York 1985.

[15] E. F. Kaelble (ed.): *Handbook of X-Rays,* McGraw-Hill, New York 1967.

[16] A. J. Kitaigorodsky: *Molecular Crystals and Molecules,* Academic Press, New York 1973.

[17] H. P. Klug, L. E. Alexander: *X-Ray Diffraction Procedures for Polycrystalline and Amorphous Materials,* J. Wiley & Sons, New York 1974.
[18] H. Krebs: *Grundzüge der anorganischen Kristallchemie,* Enke-Verlag, Stuttgart 1968.
[19] H. Lipson, W. Cochran: *The Determination of Crystal Structures,* 3rd ed., G. Bell & Sons, London 1966.
[20] P. Luger: *Modern X-Ray Analysis on Single Crystals,* De Gruyter, Berlin 1980.
[21] D. McKie, C. McKie: *Essentials of Crystallography,* Blackwell Scientific Publ., Oxford 1986.
[22] H. Neff: *Grundlagen und Anwendungen der Röntgenfeinstrukturanalyse,* 2nd ed., R.-OldenbourgVerlag, München 1962.
[23] G. H. Stout, L. H. Jensen: *X-Ray Structure Determination,* Macmillan, New York 1968.
[24] E. R. Wölfel: *Theorie und Praxis der Röntgenstrukturanalyse,* Vieweg, Braunschweig 1975.
[25] M. M. Woolfson: *An Introduction to X-Ray Crystallography,* University Press, Cambridge 1970.
[26] W. Borchardt-Ott: Kristallographie, eine Einführung für Naturwissenschaftler; 5. Auflage; Springer-Verlag, Berlin 1997.
[27] J. P. Glusker, M. Lewis, M. Rossi: Crystal Structure Analysis for Chemists and Biologists; New York, Weinheim, Cambridge (UK) 1994
[28] J. S. Rollet (ed.): *Computing Methods in Crystallography.* Pergamon Press, Oxford 1965.

General References

Reference Works

[29] *International Tables for X-Ray Crystallography,* Kynoch Press, Birmingham, N. F. M. Henry, K. Lonsdale (eds.): "Symmetry Groups," vol. I, 1952;
J. S. Kasper, K. Lonsdale (eds.): *"Mathematical Tables,"* **vol. II,** 1959;
C. H. MacGillavry, G. D. Rieck (eds.): *"Physical and Chemical Tables,"* **vol. III,** 1962;
J. A. Ibers, W. C. Hamilton (eds.): *"Revised and Supplementary Tables,"* **vol. IV,** 1974.
[30] International Tables for Crystallography, Kluwer Academic Publishers, Dordrecht: Th. Hahn (ed.), Vol. A: Space-Group Symmetry, fifth revised edition, 2000; U. Shmueli (ed.), Vol. B: Reciprocal Space, second edition, 2000; A. J. C. Wilson, E. Prince (eds.), Vol. C: Mathematical, Physical & Chemical Tables, second edition, 1999.

General References

Databanks

[31] J. D. H. Donnay, H. M. Ondik (eds.): *Crystal Data: Determinative Tables,* 3rd. ed., **vol. 1, "Organic Compounds"; vol. 2, "Inorganic Compounds,"** U.S. Dep. of Commerce – Nat. Bur. of Standards – Joint Committee on Powder Diffraction Standards, 1972.
[32] *Landolt-Börnstein,* 7th ed., **III,** 5; III 6; III 7; III 8.
[33] A. R. Hölzel: *Systematics of Minerals, Systematik in der Mineralogie,* Ober-Olm, Germany 1990.
[34] O. Kennard, F. H. Allen, D. G. Watson (eds.): *Molecular Structures and Dimensions, Guide to the Literature 1935–1976, Organic and Organometallic Crystal Structures,* Crystallographic Data Centre, Cambridge 1977.
[35] O. Kennard et al.: *Interatomic Distances, 1960–1965, Organic and Organometallic Crystal Structures,* Crystallographic Data Centre, Cambridge 1972.
[36] R. W. G. Wyckoff: *Crystal Structures,* 2nd ed., **vols. 1–6,** Wiley, New York 1963–1971.
[37] F. H. Allen, et al., *J. Chem. Inf. Comput. Sci.* **31** (1991) 187–204.
[38] L. G. Berry (ed.): *Powder Diffraction File (Organic and Inorganic),* International Centre for Diffraction Data (ICDD), Swarthmore, USA.
[39] G. Bergerhoff et al., ICSD, Inorganic Crystal Structure Database, University Bonn.
[40] A. R. Hölzel: MDAT, Datenbank der Mineralien, Systematik in der Mineralogie, Ober-Olm, Germany 1992.
[41] Protein Data Bank, Brookhaven National Laboratory, USA.

General References

Computer Programs

[42] A. Altomare et al., SIR 92, User's Manual, University of Bari and Perugia, Italy 1992 [138]
[43] P. T. Beurskens et al., DIRDIF, a Computer Program System for Crystal Structure Determination by Patterson Methods and Direct Methods applied to Difference Structure Factors, 1992.
[44] T. Debaerdemaeker et al., MULTAIN 88, a System of Computer Programs for the Automatic Solution of Crystal Structures from X-Ray Diffraction Data, Universities of Ulm (Germany), Louvain (Belgium), and York (England) 1988.
[45] A. L. Spek, PLATON, a Multi-Purpose Crystallographic Tool, University of Utrecht, the Netherlands (1998)
[46] E. J. Gabe, A. C. Larson, F. L. Lee, Y. Lepage: NRCVAX, Crystal Structure System, National Research Council, Ottawa, Canada.
[47] S. R. Hall, J. M. Stewart (eds.): XTAL 3.0, Reference Manual, University of Western Australia, Australia, and Maryland, USA.
[48] C. K. Johnson: ORTEP II, Report ORNL 5138, Oak Ridge National Laboratory, Tennessee, USA.
[49] G. M. Sheldrick: SHELXS 86, Program for the Solution of Crystal Structures, University of Göttingen, Germany 1986.
[50] G. M. Sheldrick: SHELXL 93, Program for the Refinement of Crystal Structures, University of Göttingen, Germany 1993.
[51] G. M. Sheldrick: SHELX-97, Computer program for the solution and refinement of crystal strucutres. University of Göttingen, Germany 1997.
[52] D. J. Watkin, J. R. Carruthers, P. W. Betteridge: Crystals User Guide, Chemical Crystallography Laboratory, University of Oxford, England 1985.

Specific References

[53] R.-G. Kretschmer, G. Reck, POSIT in W. Kraus, G. Noke, Computer program PowderCell, BAM Berlin (1996).
[54] G. E. Engel, S. Wilke, O. König, K. D. M. Harris, F. J. J. Leusen, PowderSolve (1999) in Cerius^2 (Molecular Simulations Inc., 9685 Scranton Road, San Diego, CA 92121-3752, USA).
[55] E. Keller, *Chem. unserer Zeit* **16** (1982) 71–88, 116–123.
[56] R. Bonart, R. Hosemann, *Makromol. Chem.* **39** (1960) 105–118.

[57] W. Hoppe, *Z. Elektrochem.* **61** (1957) 1076–1083.
[58] W. Hoppe, E. F. Paulus, *Acta Crystallogr.* **23** (1967) 339–342.
[59] H. Fuess, *Chem. Br.* **14** (1978) 37–43.
[60] H. Diehls: *Die Fragmente der Vorsokratiker,* 8th ed., **vol. 2,** Weidmannsche Verlagsbuchhandlung, Berlin 1952, p. 97.
[61] T. Kottke, D. Stalke, *J. Appl. Crystallogr.* **26** (1993) 615–619.
[62] *Sci. Ind. (Eindhoven)* **18** (1972) 22–28.
[63] D. Kobelt, E. F. Paulus, *Siemens Power Eng.* **1** (1979) 269–271.
[64] R. A. Sparks, *Abstracts of Internat. Summer School on Crystallography Computing,* Prague, July, 28–Aug. 5, 1975, pp. 452–467.
[65] R. Boese, M. Nussbaumer: "In situ Crystallization Techniques," in D. W. Jones (ed.). *Organic Crystal Chemistry,* IUCR, Cryst. Symposia, vol. VI, Oxford University Press, Oxford 1993.
[66] E. Pohl et al., *Acta Crystallogr.* **D 49** (1993) in press.
[67] G. M. Sheldrick, E. F. Paulus, L. Vértesy, F. Hahn, Acta Crystallogr. B51, 89–98.
[68] B. Baumgartner et al.: *Modern Fast Automatic X-Ray Powder Diffractometry (Stoe & Cie GmbH, company brochure),* Darmstadt 1988.
[69] Commission of Crystallographic Apparatus of the Internat. Union of Crystallography, *Acta Crystallogr.* **9** (1956) 520–525.
[70] Robert Huber Diffraktionstechnik: *Manual Guinier-System 600 (company brochure),* Rimsting, Germany.
[71] Verordnung über den Schutz vor Schäden durch Röntgenstrahlen (Röntgenverordnung) January 8, 1987 (BGBl. I S. 114). Textausgabe mit Anm. von E. Witt. C. Heymanns-Verlag, Köln 1987.
[72] Permissible Dose from External Sources of Ionising Radiations, NBS Handbook 59, pp. 2, 3, addendum, April 15, 1958, Government Printing Office, Washington.
[73] Recommendations of the International Commission on Radiological Protection, Pergamon Press, New York 1959.
[74] A. L. Patterson, *Phys. Rev.* **46** (1934) 372–376.
[75] R. Huber, *Acta Crystallogr.* **19** (1965) 353–356.
[76] J. Karle, *Adv. Struct. Res. Diffr. Methods* **1** (1964) 55–88.
[77] E. Egert, G. M. Sheldrick, *Acta Crystallogr.* **A43** (1985) 262–268.
[78] D. M. Sayre, *Acta Crystallogr.* **5** (1952) 60–65.
[79] E. W. Hughes, *Acta Crystallogr.* **6** (1953) 871.
[80] G. N. Watson: *The Theory of Bessel Functions,* 2nd ed., University Press, Cambridge 1958, p. 77.
[81] H. Hauptmann, J. Karle: "Solution of the Phase Problem. I, The Centrosymmetric Crystal," *Monogr.* no. 3, Polycrystal Book Service, Pittsburgh 1953.
[82] H. Hauptmann, J. Karle, *Acta Crystallogr.* **12** (1959) 93–97.
[83] H. Hauptmann, J. Karle, *Acta Crystallogr.* **9** (1956) 45–55.
[84] J. Karle, H. Hauptmann, *Acta Crystallogr.* **14** (1961) 217–223.
[85] I. L. Karle, J. Karle, *Acta Crystallogr.* **16** (1963) 969–975.
[86] I. L. Karle, J. Karle, *Acta Crystallogr.* **17** (1964) 835–841.
[87] G. Germain, M. M. Woolfson, *Acta Crystallogr.* **B 24** (1968) 91–96.
[88] M. M. Woolfson, *Acta Crystallogr.* **A 33** (1977) 219–225.
[89] P. Main, *Acta Crystallogr.* **A 34** (1978) 31–38.
[90] G. M. Sheldrick, *Acta Crystallogr.* **A 46** (1990) 467–473.
[91] G. Germain, P. Main, M. M. Woolfson, *Acta Crystallogr.* **A 27** (1971) 368–376.
[92] T. Debaerdemaeker, C. Tate, M. M. Woolfson, *Acta Crystallogr.* **A41** (1985) 268–290.
[93] M. M. Woolfson, *Acta Crystallogr.* **A43** (1987) 593–612.
[94] H. Hauptman, *Science* **233** (1986) 178–183.
[95] R. Miller, G. T. DeTitta, R. Jones, D. A. Langs, C. M. Weeks, H. A. Hauptman, *Science* **259** (1993) 1430–1433.
[96] R. Miller, S. M. Gallo, H. G. Khalak, C. M. Weeks, *J. Appl. Cryst.* **27** (1994) 613–621.
[97] G. M. Sheldrick in S. Fortier (ed.): *Direct Methods for Solving Macromolecular Structurs,* Kluwer Academic Publishers, Dordrecht 1998, pp. 401–411.
[98] I. Usón, G. M. Sheldrick, *Curr. Op. Struct. Biol.* **9** (1999) 643–648.
[99] M. Schäfer, Th. R. Schneider, G. M. Sheldrick, *Structure* **4** (1996) 1509–1515.
[100] B. Post, *Acta Crystallogr.* **A 39** (1983).
[101] E. Weckert, W. Schwegle, K. Hümmer, *Proc. R. Soc. London A* **424** (1993) 33–46.
[102] R. Herbst-Irmer, G. M. Sheldrick, *Acta Crystallogr.* **B54** (1998) 443–449.
[103] F. L. Hirshfeld, D. Rabinovich, *Acta Crystallogr.* **A 29** (1973) 510–513.
[104] D. W. Engel, *Acta Crystallogr.* 28 (1972) 1496–1509.
[105] W. C. Hamilton, *Acta Crystallogr.* **18** (1965) 502–510.
[106] D. Rogers, *Acta Crystallogr.* **A 37** (1981) 734–741.
[107] H. D. Flack, *Acta Crystallogr. Sect. A: Found Crystallogr.* **A 39** (1983) 876–881.
[108] W. Hoppe, *Angew. Chem.* **77** (1965) 484–492.
[109] A. J. C. Wilson: *Nature (London)* **150** (1942) 151–152.
[110] P. Main et al.: *MULTAN 77, a System of Computer Programmes for the Automatic Solution of Crystal Structures from X-Ray Diffraction Data,* University of York, England 1977.
[111] P. Debye, *Ann. Phys. (Leipzig)* **46** (1915) 809–823.
[112] D. Reuschling, E. F. Paulus, H. Rehling, *Tetrahedron Lett.* 1979, 517–520.
[113] T. Ito: *X-Ray Studies on Polymorphism,* Maruzen Co., Tokyo 1950, pp. 187–228.
[114] H. M. Rietveld, *Acta Crystallogr.* **22** (1967) 151–152.
[115] E. F. Paulus et al., *Solid State Commun.* **73** (1990) 791–795.
[116] R. Hedel, H. J. Bunge, G. Reck, *Material Science Forum* **157–162** (1994) 2067–2074.
[117] A. Le Bail, Structure Determination from Powder Diffraction-Database (1993–1999). http://www.cristal.Org/iniref.html.
[118] G. Reck, R.-G. Kretschmer, L. Kutschabsky, W. Pritzkow, *Acta crystallogr.* **A44** (1988) 417–421.
[119] G. E. Engel, S. Wilke, O. König, K. D. M. Harris, F. J. J. Leusen, *J. Appl. Cryst.* **32** (1999) 1169–1179.
[120] R. J. Gdanitz in A. Gavezzotti (ed.): *Theoretical Aspects and Computer Modeling,* John Wiley & Sons 1997, pp. 185–201.
[121] M. U. Schmidt, R. E. Dinnebier, *J. Appl. Cryst.* **32** (1999) 178–186.

[122] M. U. Schmidt in D. Braga, G. Orpen (eds.): *Crystal Engineering: From Molecules and Crystals to Materials,* Kluwer Academic Publishers, Dordrecht 1999, pp. 331 – 348.
[123] K. D. M. Harris, M. Tremayne, *Chem. Mater.* **8** (1996) 2554 – 2570.
[124] P. Erk in D. Braga, G. Orpen (eds.): *Crystal Engineering: From Molecules and Crystals to Materials,* Kluwer Academic Publishers, Dordrecht 1999, pp. 143 – 161.
[125] G. Ibe, K. Lücke, *Texture (London)* **1** (1973) 87 – 98.
[126] R. C. Rau, *Adv. X-Ray Anal.* **5** (1962) 104 – 115.
[127] A. K. Cheetham, A. P. Wilkinson, *Angew. Chem.* **104** (1992) 1594 – 1608.
[128] H. Fuess: "Application of Neutron Scattering in Chemistry. Pulsed and Continuous Sources in Comparison," in M. A. Carrondo, G. A. Jeffrey (eds.): *Chemical Crystallography with Pulsed Neutrons and Synchrotron X-Rays,* D. Reidel Publ., Dordrecht 1988, pp. 77 – 115.
[129] L. Schäfer, *Appl. Spectrosc.* **30** (1976) 123 – 149.
[130] T. Blundell, G. Dodson, D. Hodgkin, D. Mercola, *Adv. Protein Chem.* **26** (1972) 279 – 402.
[131] M. F. Perutz, H. Muirhead, J. M. Cox, L. G. C. Goaman, *Nature (London)* **219** (1968) 131 – 139.
[132] C. Brink-Shoemaker et al., *Proc. R. Soc. London A* **278** (1964) 1 – 26.
[133] R. Rothbauer, *Acta Crystallogr.* **A 33** (1977) 365 – 367.
[134] M. M. Woolfson, *Acta Crystallogr.* **11** (1958) 277 –283.
[135] R. Rothbauer, *Z. Kristallogr.* **209** (1994) 578 – 582.
[136] R. Rothbauer, *Acta Crystallogr.* **A 34** (1978) 528 – 533.
[137] R. Rothbauer, *Z. Kristallogr.* **215** (2000) 158 – 168.
[138] A. Altomare, G. Cascarano, C. Giacovazzo, A. Guagliardi, M. C. Burla, G. Polidori, M. Camalli, *J. Appl. Cryst.* **27** (1994) 435 – 436.

16. Ultraviolet and Visible Spectroscopy

GÜNTER GAUGLITZ, Universität Tübingen, Institut für Physikalische und Theoretische Chemie, Tübingen, Federal Republic of Germany

Abbreviations

a	linear (decadic) absorption coefficient, concentration
A	absorption
$\mathbf{b}$	matrix of signals
$\mathbf{B}$	magnetic field vector
c	velocity of light, concentration
d	thickness, pathlength
E	energy, absorbance, extinction
$\mathbf{E}$	electric field vector
f	spectral distribution
F	spontaneous emission
g	grating constant
h	Planck's constant
H	magnetic field
I	fluorescence intensity
k	Boltzmann constant, absorption coefficient
l	path length
n	refractive index
N	number of molecules
p	polarization
P	photochemistry
$\mathbf{Q}$	inverse of the matrix of the absorption coefficient
R	reflectance of the sample
s	scattering coefficient
S	singlet state
T	triplet state
v	velocity
$[\alpha]$	specific rotation
α	angle, absorptance
β	angle
Γ	geometric factor
δ	phase shift
ε	extinction coefficient, molar (decadic) absorption coefficient
η	fluorescence quantum yield
ϑ	angle
Θ	ellipticity
λ	wavelength
μ	dipole moment
ν	frequency
$\tilde{\nu}$	wavenumber
τ	turbidity coefficient
φ	phase shift
Φ	angle
$[\Phi]$	molecular rotation
ω	Verdet constant

16.1. Introduction

16.1.1. Comparison with Other Spectroscopic Methods

Spectroscopy in the ultraviolet and visible regions of the spectrum (UV–VIS spectroscopy) involves observation of the excitation of electrons and is, therefore, often referred to as "electron spectroscopy." The term electron spectroscopy does not mean spectroscopy in which excitation is performed by electrons, but spectroscopy of the electronic states (excitation of the electrons).

An electron is excited if the frequency of the incident electromagnetic radiation matches the difference in energy between two electronic states. This energy difference depends on the electronic structure of the molecule being investigated and its environment. In addition, vibrational and rotational transitions can be excited. Therefore, bands in UV–VIS spectroscopy are generally broad. These bands provide little information about the molecular structure and functional groups present, especially for spectroscopy of liquids and solids [1]–[4].

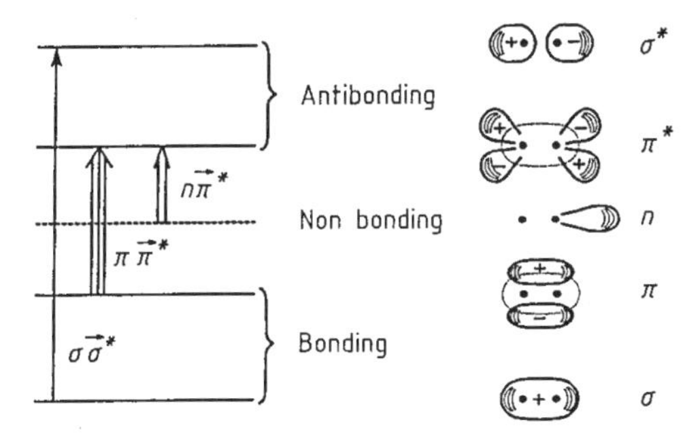

Figure 1. Bonding and antibonding orbitals, showing symmetries (+/−: sign of the wave function) and positions of atomic nuclei (•) for a diatomic molecule (HOMO = π or n; LUMO = π^*)
Thickness of vertical arrows indicates intensity of the transition

16.1.2. Development and Uses

The present state of UV–VIS measurement technology has developed from its starting point *colorimetry*. In colorimetry, the intensity of a color was first taken as a measure of concentration, the sample being compared visually with standards, indicator paper being a typical example. This mainly qualitative procedure was then improved by the use of filter-photometers. These measure the attenuation of light by the sample in a narrow spectral region (photometry) and compare it with solutions containing only pure solvent. Replacement of the filter by a monochromator with a wavelength scanning mechanism then resulted in *spectrometry*.

Whereas IR, NMR, and mass spectroscopy are used mainly for the elucidation of structure and the identification of substances, UV–VIS spectroscopy enables quantitative determinations to be carried out much more precisely and reproducibly. Therefore, its primary areas of application are in quantitative analysis and clinical medicine, in the determination of drug concentrations, in the quantification of pharmaceuticals and as detectors in chromatographic processes (HPLC, TLC) (→ Liquid Chromatography) [4], [5]. Furthermore, mixtures as well as pure substances can be studied and the components determined quantitatively by methods of multicomponent analysis [6]. Since modern spectrometers operate very rapidly and can be constructed in the form of photodiode arrays, they have the advantage over other analytical methods of being usable not only for observing stationary systems, but also for carrying out repeated determinations very rapidly, sometimes within milliseconds [3], [4].

16.2. Theoretical Principles

16.2.1. Electronic States and Orbitals

The energy of a molecule is given by its movement through space (translational energy), its rotary motion (rotational energy), the oscillatory movement of the atoms with respect to each other (vibrational energy), and the distribution of its electron density (electronic energy) [7]–[9]. The corresponding energy states can be calculated as eigenvalues by the Schrödinger equation through use of the electronic, vibrational, and rotational eigenfunctions, and the boundary conditions given by the molecular structure [10].

The energy levels or states in the UV–VIS spectral region correspond to electron density distributions (orbitals). Some typical orbitals for an organic molecule are shown schematically in Figure 1 [3]. Also shown are possible transitions of varying intensity corresponding to the symmetry of the electron density distributions of the initial and final states.

16.2.2. Interaction Between Radiation and Matter

In electromagnetic radiation, either the particle (photons) or the wavelike character is more noticeable. In the ultraviolet and visible wavelength region, the latter is the case. The electric field vector of the radiation induces charge separation (polarization) within the molecules. In the UV–VIS region, the sign of the electric field vector changes at ca. 10^{15} Hz, causing the electron density to be polarized at this frequency. The extent of this polarization depends on the dielectric properties of the molecules, their environment, and the wavelength of the radiation [11]–[13]. The Bohr–Einstein relationship

$$\Delta E = E_2 - E_1 = h\nu$$
$$h\nu = hc\tilde{\nu} = hc/\lambda \qquad (1)$$

connects the discrete atomic or molecular electron energy states with the frequency ν of the exciting electromagnetic radiation, where $\tilde{\nu} = 1/\lambda$ is the

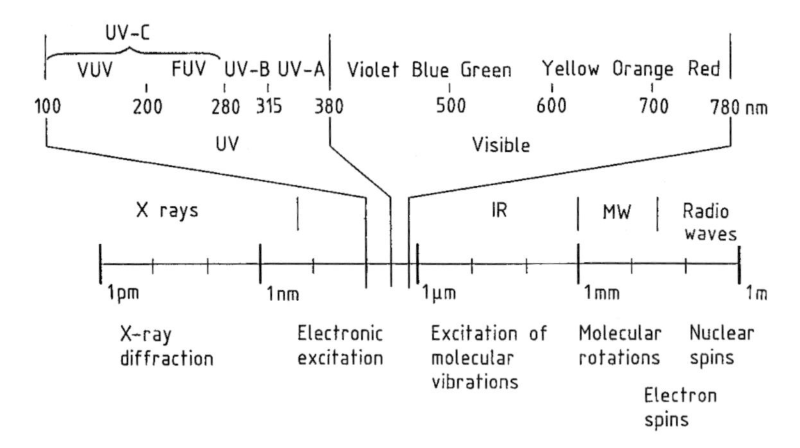

Figure 2. Location of UV and visible regions within the complete electromagnetic radiation spectrum, showing visible colors and types of excitation
VUV = Vacuum UV; FUV = Far UV; MW = Microwaves

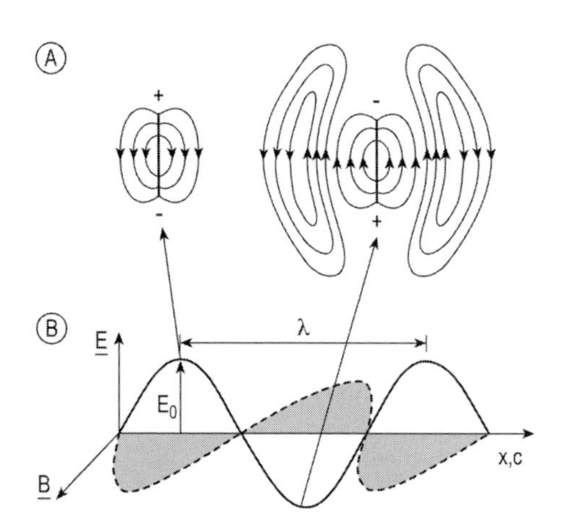

Figure 3. A) Electrical fields for a Hertz dipole (**E** || *dipole moment* μ); B) Electrical and magnetic field vectors in the propagation of electromagnetic radiation

wavenumber, c is the velocity of light in a vacuum ($c = 2.99\times10^8$ m/s), and h is Planck's constant (6.62×10^{-34} J · s). The location of this spectral region within the complete electromagnetic radiation spectrum is shown in Figure 2, which also indicates the colors of the visible range.

16.2.2.1. Dispersion

The simple model of a harmonic oscillator can account for the processes of polarization, dispersion, absorption, and scattering. In *polarization*, interaction of the radiation with matter (dielectric) induces a reversing dipole in each molecule. The effect is proportional to the number of particles, so that polarization is a volume-based quantity. The induced dipole moment represents a Hertz dipole reversing with the excitation frequency, thus behaving as a radiation source. Because of the fluctuation of the charge distribution that fluctuates with the dipole's eigenfrequency. As shown in Figure 3, this radiation is emitted perpendicularly to the axis of polarization.

When radiation propagates through a medium, its frequency remains the same due to the conservation of energy, but its velocity decreases [11], [13]. The ratio of propagation rates is the refractive index n ($n \geq 1$). The variation of refractive index with frequency or wavelength is the dispersion [11], [13]. Two cases must be distinguished – absorption and scattering.

16.2.2.2. Absorption

If the frequency of the radiation corresponds closely to the energy difference of the transition between two energy states, as given by Equation (1), this leads to a resonance excitation, a change in the electron density distribution, and an electronic transition (see Fig. 1) from the highest occupied molecular orbital (HOMO) to the lowest unoccupied molecular orbital (LUMO). In solids and metals, HOMO corresponds to the valence, and LUMO to the conduction band [12], [14], [15]. The energy required for this transition is provided by the radiation. This process is known

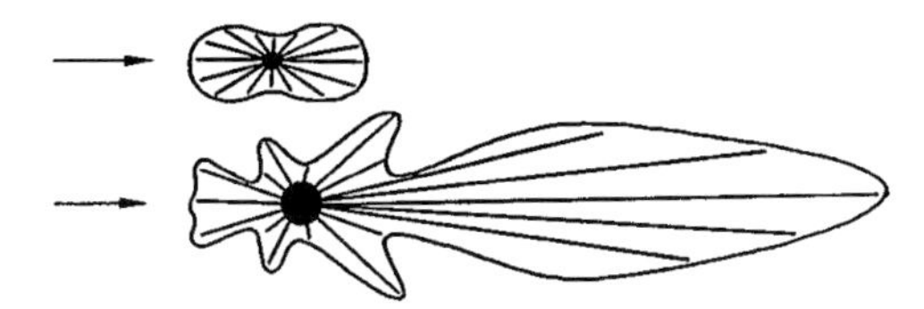

Figure 4. Intensity distribution of Rayleigh (above) and Mie (below) scattered radiation projected onto a plane. Excitation radiation is in the direction of the arrow. The black spot represents the superposition of many molecular dipoles in all spatial directions.

as (induced) absorption. In the range of an absorption band, anomalous dispersion is observed (refractive index increases with wavelength). Depending on the molecular environment and possible pathways of deactivation, the new excited state can exist for 10^{-13} to 10^{-3} s.

Only gaseous atoms give line spectra. Molecules in the gaseous state and in solution produce bands (sometimes with vibrational structure), because the energy for the electronic transition is very high, and this transition is usually accompanied by vibrational and rotational excitation. Line broadening in solution is caused by interaction between the solvent and the molecules, leading to a reduction in the lifetime of the excited state. According to Heisenberg's uncertainty principle

$$\Delta E \Delta \tau \geq h/2\pi \tag{2}$$

the energy uncertainty of the excited state increases at the expense of its lifetime. The relationship between refractive index and absorption coefficient can be expressed by the Kramers–Kronig equations [16].

16.2.2.3. Scattering

If the frequency of incident radiation does not correspond to the energy difference between two electronic levels, the polarization does not lead to a new stable (stationary) electron density distribution. After a very short delay, the molecule emits the energy that it absorbed previously. In this frequency range normal dispersion is observed (i.e., refractive index decreases with wavelength). When refractive index is plotted as a function of wavelength a dispersion curve is obtained. It characterizes the molecule together with the polarizability. With the aid of the dispersion curve, relationships between the molecular structure and the spectrum can be established. This gives good results in the infrared, but only poor results in the UV–VIS region [17].

Since the energy causing the polarization is reemitted immediately in the nonresonant case (see Fig. 3), no energy is absorbed. Even the smallest molecules show scattering effects. However, since the intensity of scattered radiation increases proportionally to the sixth power of the length of the induced dipole moment, the radiation becomes observable only for molecular diameters of $\geq \lambda/10$ (Rayleigh scattering) and/or extremely high radiation intensities (lasers) [18], [19].

In the case of Mie scattering (larger molecules with a diameter of $\lambda/4 - \lambda/2$), a molecule contains several scattering centers. Thus, the different scattering centers emit waves that interfere to varying extents in various spatial directions, depending on the location of the scattering centers. Hence, the interference pattern contains information about the size and shape of the molecule. The intensity distribution of Rayleigh and Mie scattering assuming many molecules with statistical orientations overlap is shown in Figure 4 [18].

16.2.2.4. Reflection

Whereas in gases or liquids the particles are disordered and cause random scattering, atoms or molecules in the surface layer of solids (especially crystals and metals) are very highly ordered. Especially metals exhibit extremely polarizable electron density distributions. The many Hertz dipoles show constructive interference in certain directions. At solid surfaces this leads to directional reflection, which encloses the same angle with the optical normal to the surface as the incident radiation [11], [20] (Fig. 5). Reflectance R is given by the Fresnel equation, which, for the simplest case of normal incidence and a transparent medium, is [11], [13]

$$R = \left[\frac{(n_1 - n_2)}{(n_1 + n_2)}\right]^2 \tag{3}$$

When radiation enters a material with a higher refractive index n, it is refracted toward the optical axis (Snell's law):

$$n_1 \sin\alpha = n_2 \sin\beta \tag{4}$$

For a material with a lower refractive index, the angle between reflected radiation and the optical axis becomes larger. Under these conditions, total internal reflection occurs at the critical angle β_{cr}

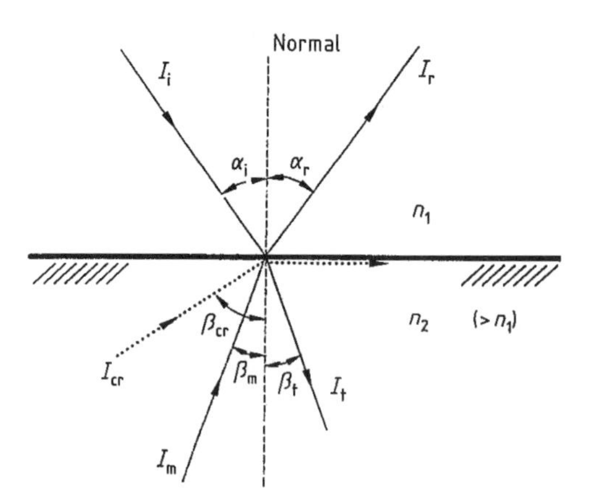

Figure 5. Ray paths for internal, external, and total internal reflection
Indices i, r, t = Incident, reflected, transmitted; I = Intensity; cr = Critical angle; m = Ray out of medium 2 corresponding to incident ray I_i, forms I_r

(see Fig. 5) [20]. In recent years, techniques based on reflection have become very important in analysis.

16.2.2.5. Band Intensity

Regarding transitions from ground to excited state and vice versa, the interaction of the radiation with the molecules are in principle the same. For this reason, the Einstein coefficients, which are a measure of probability of the two transitions, are the same for both (induced) absorption and induced emission [11], [12], [14], [15]. Which of the two transitions is more effective for the interaction with radiation depends only on the relative distribution of the molecules between the two states.

In the normal case, according to Boltzmann's law,

$$N_2 = N_1 e^{-\Delta E/kT} \tag{5}$$

where N_1 is the number of molecules in the ground state and N_2 that in the excited state; the ground state is occupied to a considerably greater extent than the excited state (not strictly true for rotational levels) so absorption mainly occurs (ΔE = energy difference, k = Boltzmann's constant, T = temperature in K). However, if intermediate energy levels are produced by high-intensity radiation (optical pumping), leading to an inversion in the populations of the two energy levels, induced emission predominates. This principle is utilized in lasers [21].

16.2.3. The Lambert – Beer Law

Many years ago, BOUGUER, LAMBERT, and BEER discovered a relationship between the number of particles in a sample, their properties, the path length of the sample, and the observed attenuation of light [22], [23].

16.2.3.1. Definitions

When radiation penetrates a medium, an exponential relationship exists among the "extinction" (attenuation of transmitted radiation), the concentration of absorbing particles in the medium, and path length. The extinction is the sum of the effects of absorption and scattering. The proportionality constant is the *molar* (decadic) *absorption coefficient* ε_λ in the case of absorption and the *turbidity coefficient* in the case of scattering [24], [25]. By taking the logarithm of the ratio of the radiant powers, i.e., of the nonabsorbing reference [$\Phi(\lambda,0)$] to that observed passing a sample [$\Phi(\lambda,d)$], a linear relationship is obtained:

$$\begin{aligned} E(\lambda) &= \log \frac{\Phi(\lambda,0)}{\Phi(\lambda,d)} \\ &= \varepsilon_\lambda c\, d \\ &= a_\lambda d \end{aligned} \tag{6}$$

In the definition of extinction (E, decadic internal absorbance), the sample thickness d is expressed in centimeters, and the concentration c of the solutions in moles per liter. IUPAC definitions of these quantities are given in Table 1, with their meanings and formal relationships [23], [26], [27].

For investigation of solid bodies or thin films, molar concentrations such as those used for solutions cannot be quoted. In these cases, absorption coefficients a, with dimensions of cm^{-1}, are used instead, which correspond to the product of the molar absorption coefficient and the concentration of the particles.

16.2.3.2. Deviations from the Lambert – Beer Law

The Lambert – Beer law assumes that monochromatic radiation falls perpendicularly onto the sample. Since the absorption coefficient is a

Table 1. Physical quantities relating to radiation, with symbols and units in accordance with the IUPAC standard

Quantity	Symbol	Unit
Radiant power (radiant energy Q)	$\Phi_i = dQ/dt$	W
Radiant power transmitted	Φ_t	W
Radiant power reflected	Φ_r	W
Radiant power absorbed	Φ_a	W
Radiant intensity	I	W/sr, W
Reflectance or reflection factor	$\varrho = \Phi_r / \Phi_0$	1
Absorptance or absorbance factor	$\alpha = 1 - \tau$	1
Internal transmittance	τ_i	1
Internal absorptance	α_i	1
(Decadic) internal absorbance	$A = -\log_{10} \tau_i(\lambda)$	1
Linear (decadic) absorption coefficient	$a(\lambda) = A/d$	cm^{-1}
Molar (decadic) absorption coefficient	$\varepsilon = A/(c\,d)$	$L\ mol^{-1}\ cm^{-1}$
Molar concentration of absorber	c	mol/L

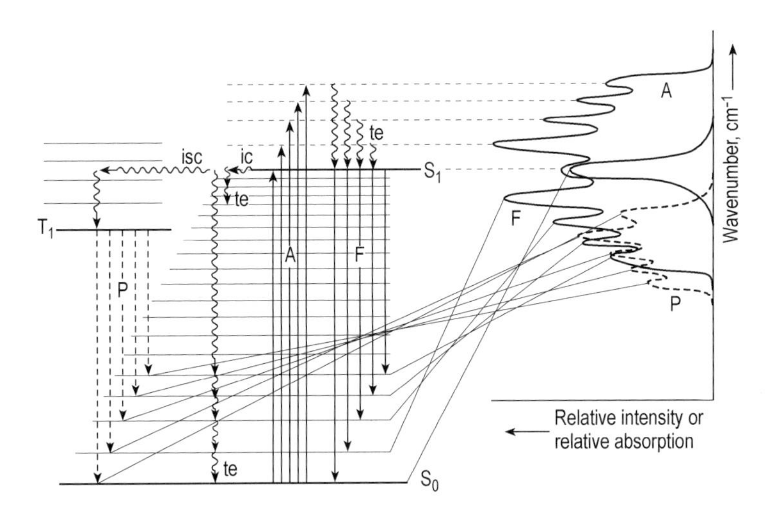

Figure 6. Schematic diagram of energy levels of an organic molecule with singlet ground state (S_0), singlet excited state (S_1), and triplet excited state (T_1) with possible photophysical transitions
The spectra resulting from these transitions are shown on the right. A = Absorption; F = Fluorescence; P = Phosphorescence

molecule-specific quantity, any chemical change of the molecule or interaction with the solvent, with other components or dissociation/association leads to false results when the Lambert – Beer law is applied (chemical deviations). On the other hand physical deviations are observed if the natural bandwidth of the absorption spectrum of the components in the sample is narrower than the spectral bandwidth (see Section 16.3.3) of the "monochromatic" radiation produced by the equipment [3], [22]. If more than one component in the sample absorbs radiation, the total effect is the sum of the molar absorption coefficients of the different components times their concentration.

16.2.4. Photophysics

16.2.4.1. Energy Level Diagram

The energy level diagram for a typical organic molecule is shown in Figure 6, which illustrates the energy levels and transitions described above. In the absorption (A) of electromagnetic radiation at the resonance wavelengths, various vibrational states are reached in the excited electronic state (S_1), and are coupled with the electronic transition (within 10^{-15} s). Many vibrational and rotational levels exist in each electronic state (S_n), but these are not shown for reasons of clarity. Normally, only the first excited electronic state is important in spectroscopy, because all higher states have very short lifetimes (relaxation times).

Table 2. Possible deactivation processes from the vibrational ground state of the first excited singlet state (see also Fig. 6)

Process type	Abbreviation	Name of process	Process description
Radiationless (rd)	te	thermal equilibration	relaxation from high vibrational level in the same electronic state lifetime of excited vibrational level determined by: – external effects (collisions, medium) – inner deactivation (transfer of energy in torsional vibrations to heavy substituents)
	ic	internal conversion	isoelectronic transition within the same energy level system from the vibrational ground state of a higher electronic state into the very high energy vibrational state of a lower electronic state
	isc	intersystem crossing, intercombination	isoelectronic transition into another energy level system ($S \leftrightarrow T$), usually from the vibrational ground state such a radiative transition is forbidden because of the spin inversion prohibition (except for heavy nuclei)
Radiation transfers (spontaneous emission)	F	fluorescence	without spin inversion from S_n (provided that lifetime is 10^{-8} s) within singlet system
	P	phosphorescence	out of triplet into singlet system (provided that lifetime is 10^{-3} s) very low probability, only possible at low temperatures, or in a matrix
Photochemical reactions			photoinduced reaction leading to ionization, cleavage, or a bimolecular step, provided that lifetime in excited state is relatively long

In organic molecules in the ground state, electron spins are usually paired (singlet states). However, in inorganic transition-metal complexes in the ground state, triplets, and other states with a larger number of unpaired electrons occur, in accordance with Hund's rule. Therefore, several energy level systems usually exist side by side as solutions of the Schrödinger equation. A direct excitation of the singlet electronic ground state to the excited triplet state (T_1 parallel unpaired spins) is forbidden in principle because of spin inversion, so that the process of induced absorption from singlet states always leads to excited singlet states (S_n) [3], [5], [7], [28].

16.2.4.2. Deactivation Processes

Figure 6 shows deactivation as well as excitation processes. After excitation to a higher level, the molecule usually relaxes to the vibrational ground state of the S_1 level by thermal equilibration (te). This S_1 level is the starting point for several competitive processes, the preferred one depending on the type of molecule, temperature, and its environment [29]. These include (1) radiationless deactivation (rd), which consists of internal conversion (ic) followed by thermal equilibration (te); (2) intercombination or intersystem crossing (isc); (3) spontaneous emission (F, P); and (4) photochemical processes (see survey in Table 2) [3], [28]–[31]. The vibrational ground states in electronically excited states have a relatively long lifetime (lifetime $S_1 = 10^{-8}$ s, lifetime $T_1 = 10^{-3}$ s).

In competition with radiationless deactivation, energy can also be lost in the form of radiation (F, P) [3], [28]. Fluorescence occurs with molecules that either (1) have extended π-systems (such as polycondensed aromatics); (2) do not permit deactivation by torsional or rotational motion of parts of the molecule; or (3) have no heavy atoms as substituents [32], [33]. In addition to these molecular properties, the environment also plays a part. Thus, the fluorescence intensity increases at low temperature and in solid matrices. This is even more important in phosphorescence, where the $T_1 \rightarrow S_0$ transition is in fact spin-forbidden. If a higher vibrational level ($v' > 0$) is occupied in T_1, in accordance with the Boltzmann equation (Eq. 5), so-called delayed fluorescence [28] can occur by backward intersystem crossing $[T_1 \rightarrow S_1 (v = 0) \rightarrow S_0]$.

Alternatively, during this relatively long lifetime, the molecule can undergo an internal structural change or form new reaction products by an activated collision (photochemistry). This deactivation route is promoted by high light intensities and is used in photochemical synthesis, but it can lead to undesired side effects (e.g., in fluorescence spectroscopy) because photochemical processes

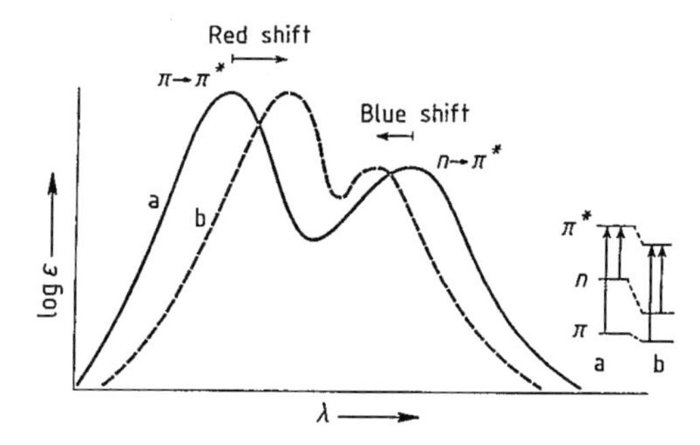

Figure 7. Schematic diagram showing effect of the solvent polarity on positions of the maxima of the $\pi \rightarrow \pi^*$ and $n \rightarrow \pi^*$ transitions
a) Methanol; b) Cyclohexane

reduce the number of molecules that can fluoresce (photobleaching) [5].

16.2.4.3. Transition Probability and Fine Structure of the Bands

A wave function can be assigned to each energy state. In the induced transition between two energy levels, the overlapping of these wave functions is important and strongly influences the intensity. For this reason, all transitions do not have the same intensity; rather, they have varying transition moments, so that absorption spectrum A on the right of Figure 6 shows a vibrational fine structure with individual peaks of varying intensities [14], [31].

Depending on the positions of the nuclei at their equilibrium distances in the molecule in the ground and electronically excited states, these wave functions overlap to varying extents and lead to differing intensity distributions for the vibrational transitions within the entire band. Since atomic nuclei hardly move during the electronic transition (Born – Oppenheimer approximation), the shape of the entire absorption band, according to the Franck – Condon principle, is given mainly by the relative positions of the equilibrium distances of the atoms in the electronic ground and excited states. The lower the degree of interaction between the molecules and their environment, the larger the vibrational splitting becomes. Absorption, fluorescence, and phosphorescence spectra are shown schematically in the right-hand side of Figure 6.

16.2.5. Chromophores

Unlike IR spectra (normal vibrations), electron density distributions in UV – VIS spectra do not depend so much on molecular structure or on interaction with solvent molecules. However, changes in the electronic ground or excited state lead to a shift in the relative position of the energy levels and to a change in polarizability. Therefore, even in the UV – VIS region, molecules may be identifiable by the position and height of the absorption bands. Both are influenced, e.g., by extension of conjugated π-systems by substitution or, in the case of aromatics, by electron-attracting or electron-repelling substituents in the *ortho*, *para*, or *meta* position and by the effects of solvents and of pH. This is especially important in pharmaceutical chemistry [5].

As shown in Figure 7, $\pi \rightarrow \pi^*$- and $n \rightarrow \pi^*$-transitions often behave differently. If cyclohexane (b) is replaced by methanol (a) as the solvent, the polarity increases considerably, so that the $\pi \rightarrow \pi^*$ band is shifted by 10 – 20 nm towards the red (bathochromic) and the $n \rightarrow \pi^*$-band by 10 nm towards the blue (hypsochromic). This shift gives an indication of the effect not only of the solvent polarity on the behavior of the bands, but also of the structure of the system. A change in intensity of the absorption band is described as hyperchromic (intensity increase) or hypochromic (intensity decrease) [5], [31].

16.2.6. Optical Rotatory Dispersion and Circular Dichroism

16.2.6.1. Generation of Polarized Radiation

For substances that are optically active or for molecules that are arranged in thin films or with a given orientation, additional information can be obtained with polarized electromagnetic radiation [11], [34]. Light can be plane, circular, or elliptic polarized. Plane-polarized light can be obtained with either polarizing films or special anisotropic crystals. Other possibilities are shown in Figures 8 and 9 [11], [13], [35]. Plane-polarized light is obtained from two circular-polarized beams that rotate in opposite directions and have the same intensity and phase (see Fig. 8).

As shown in Figure 9 A plane-polarized light is formed from two plane-polarized beams whose phase difference $\Delta \varphi$ equals zero. If a phase shift of 90° ($\Delta \varphi = \pi/2$) is chosen, as shown in Fig-

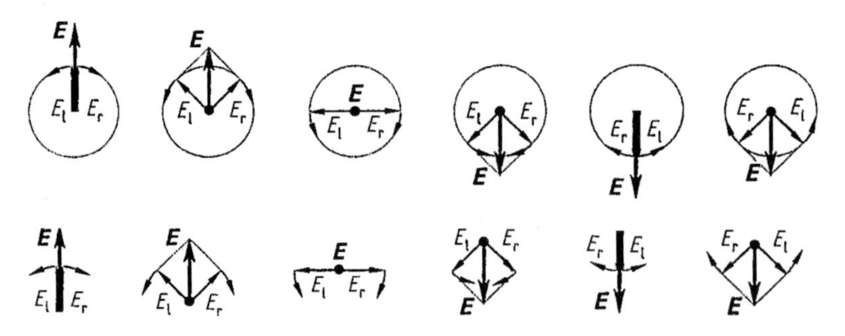

Figure 8. Production of plane-polarized light from two superposed beams of left and right circular-polarized light

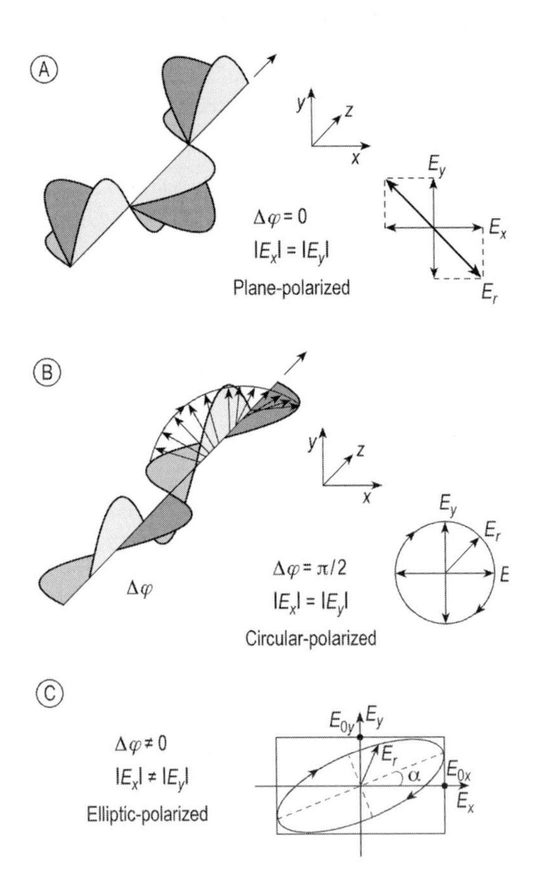

Figure 9. Production of plane- (A), circular- (B), and elliptic- (C) polarized light

ure 9 B, spiral propagation of the resulting field vector is produced, giving circular-polarized light. If the phase difference between two plane-polarized light beams is chosen arbitrarily, or if the amplitudes of the two field vectors are different, elliptic-polarized light (Fig. 9 C) is obtained, which represents the most general form of polarization.

16.2.6.2. Interaction with Polarized Radiation

The plane of polarization of plane-polarized light can be rotated to the left or the right by interaction with optically active substances. By convention, an observed rotation in a counter clockwise direction when viewed against the direction of propagation of radiation is called levorotatory and is given a negative sign (the opposite holds true for clockwise – dextrorotatory – rotation). Optical activity reveals structural properties usually referred to as chirality. These structural properties remain unchanged in molten materials, in solution, and also in complexes. For chiral substances, *quantitative polarimetry* is a very sensitive technique; it is used mainly in pharmacy and medicine to determine concentration.

The two circular-polarized beams of the incident radiation (l, levorotatory; r, dextrorotatory) not only can be influenced with respect to their direction of rotation, but also, in the region of an absorption band, can be absorbed to a different extent by the sample, so that in addition to so-called optical rotatory dispersion (ORD), circular dichroism (CD) or the Cotton effect is observed [35] – [38]. Along with these classical methods of analysis, modern methods for the analytical investigation of surfaces and boundary layers have become very important, including the use of polarized light in ellipsometry (→ Surface Analysis) and surface plasmon resonance. Under the influence of external forces (e.g., a magnetic field), even optically inactive substances can be caused to produce magnetooptic rotation.

16.2.6.3. Optical Rotatory Dispersion

The rotation α caused by optically active bodies is proportional to the thickness d (in millimeters) of the substance through which light

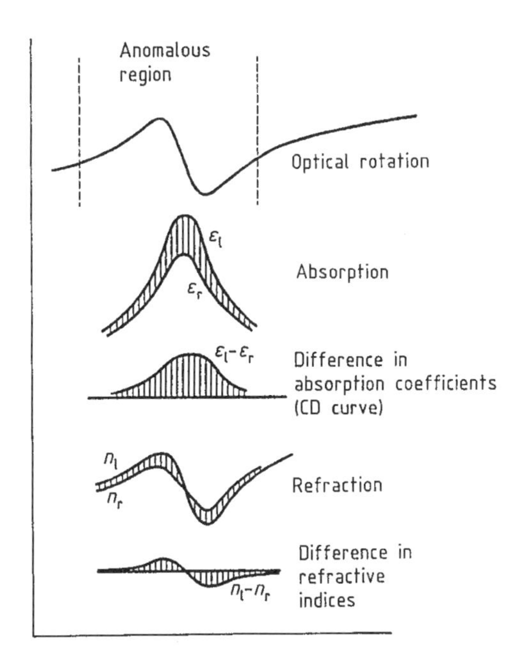

Figure 10. Rotation and circular dichroism optical rotatory dispersion in normal and anomalous spectral regions [38]

passes. The specific rotation $[\alpha]$ for wavelength λ is defined by [36], [38]:

$$[\alpha]^{\lambda} = \frac{\alpha^{\lambda}}{d} \tag{7}$$

If an optically active substance is dissolved in an inactive solvent, α is given by Biot's law:

$$\alpha^{\lambda} = [\alpha]^{\lambda} \frac{c\,d}{100} \tag{8}$$

Here, because of the relatively small rotation of solutions, sample thickness is given in decimeters, and concentration c in grams per 100 mL of solvent. The sodium D line ($\lambda = 589.3$ nm) is often used as incident radiation, and measurement is performed at 25 °C. These parameters are indicated by quoting indices in the form $[\alpha]^{\mathrm{D}}_{25}$. The *molar* or *molecular rotation* $[\Phi]^{\lambda}$ (sometimes $[\alpha]^{\lambda}{}_{\mathrm{M}}$) is given by

$$[\Phi]^{\lambda} = [\alpha]^{\lambda} \frac{M}{100} = \frac{100\alpha^{\lambda}}{d\,c} \tag{9}$$

where M is the molar mass, c concentration in moles per liter, and d path length in centimeters. Specific rotation is a characteristic quantity for a substance, but it also depends on experimental conditions (temperature, concentration, solvent, wavelength of light) and perhaps time. Along with concentration and temperature, intermolecular forces also play a part. A change of solvent can even lead to a reversal of the direction of rotation.

If optically active substances undergo chemical conversion (e.g., sucrose in acid solution), the specific rotation becomes time dependent (*mutarotation*). For example, when dextrorotatory sucrose reacts until chemical equilibrium is reached, a levorotatory mixture of glucose and fructose is formed (inversion of sucrose).

The specific rotation as a function of wavelength is known as *optical rotatory dispersion* [39]. The angle of rotation usually increases as wavelength decreases. This *normal rotatory dispersion* is found in broad spectral regions that are sufficiently remote from absorption bands of the optically active substance. Rotation can be represented by Drude's equation

$$\alpha^{\lambda} = \sum_i \frac{A_i}{\lambda^2 - \lambda_i^2} \tag{10}$$

where A_i is a constant and λ_i is the wavelength of the nearest absorption maximum. This region corresponds to the normal dispersion curve.

16.2.6.4. Circular Dichroism and the Cotton Effect

The different refractive indices n_l and n_r for left- and right-polarized light lead to a specific rotation. In the region of an absorption band the left- and right-hand polarized beam are absorbed to different extents, and the originally plane-polarized light beam becomes ellipticpolarized ($\varepsilon_l \neq \varepsilon_r$). This phenomenon is called circular dichroism. Circular dichroism is coupled with peaks and troughs in the ORD curve (anomalous dispersion) [35], [38]. Both anomalous ORD and CD are grouped under the term *Cotton effect* [34].

Figure 10 shows the relationships between the spectral behavior of the two polarized beams with respect to rotation, absorption, and refraction. Only optically active substances interact differently with two polarized beams, so that this difference leads to observable effects. Whereas in the region of anomalous rotatory dispersion the ORD curve shows a point of inflection, the CD curve has a maximum or minimum [38]. The extreme of the CD curve coincides approximately with the point of inflection of the ORD curve. In analogy to molar rotation, the molar ellipticity of the CD curve is given by

$$[\Theta]^{\lambda} = 100\,\Theta^{\lambda}\,\frac{d}{c} \tag{11}$$

Where Θ is the ellipticity of the transmitted beam in angular degrees, d is expressed in centimeters, and c in moles per liter. In the SI system, molar ellipticity should be expressed in deg m^2 mol^{-1}.

The relationship between rotation and refractive index or between ellipticity and absorption coefficient can be represented, in analogy to dispersion theory, by the model of coupled linear oscillators or by quantum mechanical methods. ORD and CD are related to one another by equations analogous to the Kramers – Kronig equations [34].

16.2.6.5. Magnetooptical Effects

If plane-polarized radiation enters a transparent isotropic body parallel to a magnetic field, the plane of polarization is rotated (magnetooptical rotation, MOR, or Faraday effect) [40]. The angle of rotation α is given by

$$\alpha = \omega l H \tag{12}$$

where H is the magnetic field strength, l the path length, and ω the Verdet constant [38]. The degree of rotation depends on the substance, the temperature, and the wavelength of the radiation. The wavelength dependence is known as *magnetooptical rotatory dispersion* (MORD). Since the Faraday effect is practically instantaneous, it is used in the construction of rapid optical shutters and for the modulation of optical radiation paths. If Zeeman splitting of atomic and molecular terms takes place in the magnetic field, this leads to *magnetic circular dichroism* (MCD) [38], [40].

All of these effects are very important in laser technology for modulation and, as an extension of absorption spectroscopy, for the assignment of electronic transitions.

16.3. Optical Components and Spectrometers

16.3.1. Principles of Spectrometer Construction

Photometers or spectrometers in the ultraviolet and visible spectral region either are built only for the measurement of light absorption, fluorescence, reflection, or scattering, or have a modular construction so that they can be used for several measuring operations. Spectrometers always include a polychromatic light source, a monochromator for the spectral resolution of the beam, a sample holder, and a detector for measuring the radiation that has been modified by its passage through the sample. Depending on the measuring problem, these components can be modified, and the light path changed [3], [4].

16.3.1.1. Sequential Measurement of Absorption

In the measurement of absorption, the radiation is measured against a reference for each wavelength (monochromatic) after attenuation by a sample. By this reference measurement, such effects as reflections at the cell windows or other optical surfaces can be corrected for, if sample and reference differ only with respect to the substance being examined and are otherwise optically identical. The equipment can be of the single- or double-beam type. In *single-beam equipment*, radiation of the selected wavelength is passed alternately through the reference cell and the sample-containing cell before it strikes the detector. In older, more simple equipment, the position of the sample holder in the cell compartment is adjusted manually. Usually, only a small number of wavelengths are measured because a spectrum cannot be recorded automatically.

Photometers can be of the *broadband type* (broad spectral band light source + filter) or the *narrow-band type* (line spectrum light source + filter) [41], [42]. In more expensive equipment, continuous radiation sources (broadband) are used, and the filter is replaced by a monochromator, so that a complete spectrum can be observed point by point. Only a very good monochromator can give resolution (monochromaticity) as good as that from a combination of a line source and filter since in the latter system, the very small width of the

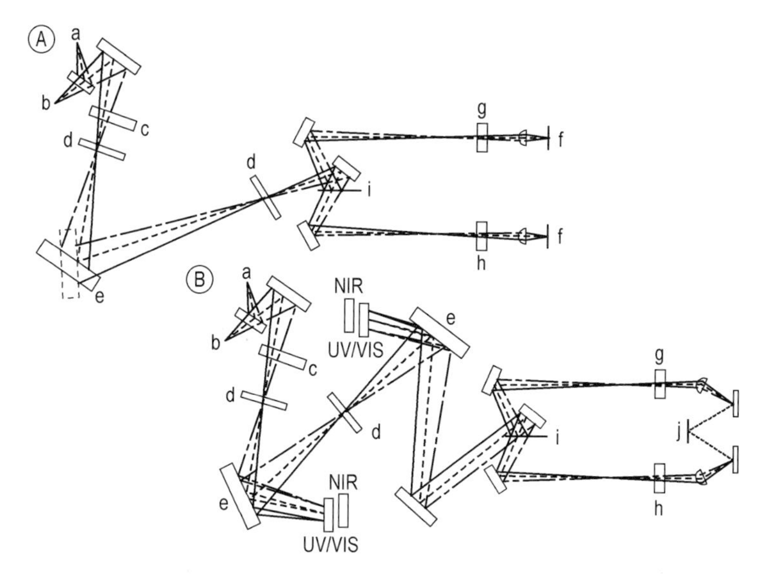

Figure 11. Optical light paths in double beam spectrometers [43]
A) Apparatus with single monochromator and two detectors; B) Research apparatus with double monochromator and beam splitter with one detector
a) Halogen lamp; b) Deuterium lamp; c) Filter wheel; d) Slit; e) Monochromator; f) Photodiode; g) Reference; h) Sample; i) Beam splitter; j) Detector

natural line determines the bandwidth of the radiation used [41], [42].

Single-Beam Equipment. Modern single-beam equipment first records a spectrum of the standard in the reference optical path and then the spectrum of the sample. The internal microprocessor serves not only for control but also for calculating the intensity ratio according to the Lambert – Beer law (see Eq. 6). The optical light path of such equipment is shown in Figure 11 A [43]. This equipment has fewer optical components and is less mechanically complex (no beam splitter) than the double-beam spectrometer, but the time required for each sample measurement is longer (successive measurements of dark, reference, and test spectra) [44]. Good stability of the light source and of the electronics is necessary to avoid any drift between measurements.

Double-Beam Equipment. In double-beam equipment, the two light paths (through the sample and the reference) are automatically continuously interchanged during the wavelength scan. A rotating mirror, oscillating mirror, or half-transparent mirror splits the beam after it emerges from the monochromator. However, for accurate measurement, at each wavelength, the "dark" signal must be observed as well as that for the reference cell and the sample. The optical paths in equipment with single and double monochromators are shown in Figure 11 A and 11 B.

High recording rates lead to errors in the spectral resolving power (which is fixed by the slit width of the monochromator) because sample and reference values are no longer determined at the same wavelength [3]. In many types of equipment, a rotating mirror, which is divided into four segments, is used to measure two dark periods, one reference, and the sample. Usually the mirror wheel rotates at 50 Hz, so that a maximum of 50 points per second is observed. Wavelength scan rates today can be as high as >1200 nm/min or >20 nm/s. Thus, at 50 Hz and this extreme scan rate, the reading is taken only every 0.4 nm (with a very short signal integration time); in addition, the wavelength differs from the reference wavelength by 0.2 nm (restricting spectral resolution to ca. 1 nm). Therefore, a considerably slower recording rate is used for high-resolution spectra. Alternatively, it ispossible to stop at each measured wavelength ("stop and go") [45], [46].

In double-beam spectrometers, fewer problems of stability of the light source and the electronics

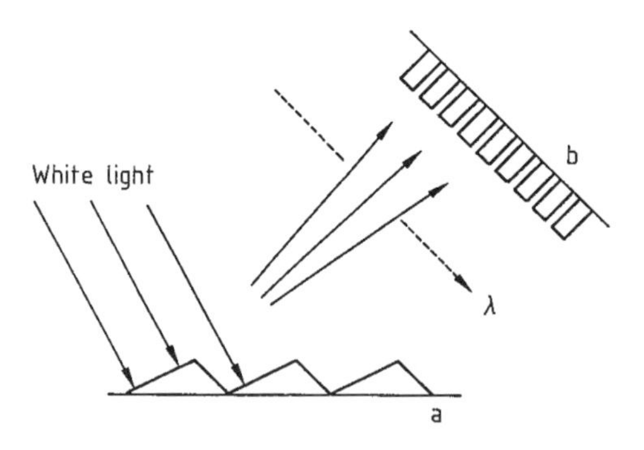

Figure 12. Diagram of multiplex detector (diode array) combined with a grating for spectral dispersion of white light
a) Grating; b) Diode array

arise. Also, a spectrum can be recorded considerably more rapidly than with single-beam equipment. Disadvantages include the greater number of optical components (each optical component – i.e., mirrors and lenses – reduces the available light intensity due to reflection); the polarization of the light beams because of movement of the optical components; and the cost and complexity of construction [3].

Radiation in the UV – VIS region is of high energy. Therefore, in sequential spectrometers the monochromator is positioned before the sample. In IR equipment, in which the sample itself emits radiation, the reverse arrangement is used [3].

16.3.1.2. Multiplex Methods in Absorption Spectroscopy

With multiplex detectors, the radiation of the various wavelengths no longer need to be measured sequentially (one after the other). Instead, by parallel imaging of many wavelengths on a spatially resolving detector they can be measured simultaneously so that a complete spectrum is recorded. This can be done by using linear photodiode arrays or, more recently, so-called charge-coupled devices (CCDs) [2], [3], [47], [48].

For this equipment, the classical single-beam arrangement is used. However, the light striking the sample is not monochromatic, but white. After passing through the sample, light is spectrally dispersed by a monochromator without an exit slit (polychromator) and then strikes an array of diodes. A possible arrangement of such a polychromator with a multiplex detector is shown in Figure 12. The resolution is determined by the specification of the grating and the number of diode elements in the array (256 – 1024).

By this method, complete spectra can be recorded in fractions of a second (1 – 100 ms, depending on the equipment). Within these times, a reference measurement is impossible so, as with modern microprocessor-controlled single-beam equipment, the reference and sample can only be observed consecutively (except when using a double diode array configuration). The sample is exposed to white light, which can result in photoinduced processes. However, this type of equipment has the advantage of recording hundreds of spectra within a few seconds, so many dynamic processes can be observed in real time. This requires that neither the diode array nor the light source drifts within the time necessary for measurement, which is true to only a limited extent. In fact, the problems are the same as those with microprocessor-controlled single-beam equipment. The application of this type of equipment for routine measurements, in kinetics, and as a detector in chromatography is continually increasing [49].

Hadamard Spectroscopy. If, instead of the exit slit, an irregular arrangement of slits of various widths is used, and an arrangement in the form of a comb-like second slit is moved in front of this, for each relative position a different summation of intensities at various wavelengths is obtained on a nonmultiplex detector. By using convolution functions, spectra can be computed rapidly for a cycle of different slit positions. Since for mathematical deconvolution so-called Hadamard matrices are used, the principle is known as Hadamard spectroscopy. It has gained some importance in waste gas analysis [50].

Fourier Transform Spectroscopy. After the development of diode arrays, the use of Fourier transform spectroscopy, which is very successful in the IR, NIR, and FIR regions, has not been very widespread for routine analyis in the UV –VIS region. In Fourier transform UV spectroscopy [11], the polychromatic light beam is split in a Michelson interferometer, 50 % falling on a fixed mirror and 50 % on a second moving mirror, and then recombined by the beam splitter. Each position of the moving mirror gives a phase relationship between the two partial beams. After recombination of the component beams the modulation of intensity caused by the position change depending on time can be transformed by Fourier analysis to spectral information (wavenumber) [51] – [54].

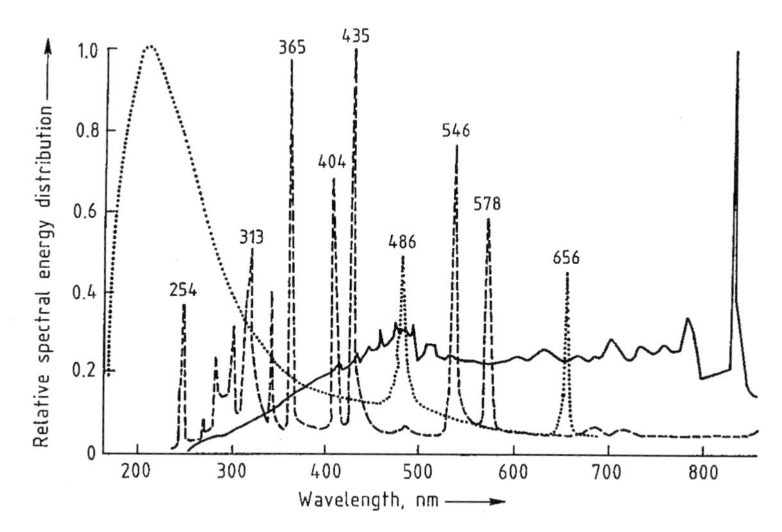

Figure 13. Line and continuous spectra from various light sources
—— = Xenon lamp; – – – = Mercury vapor arcs; · · · · · = Hydrogen lamp

Although modern computers calculate such spectra in fractions of a second, they can give high resolution only if the moving mirror is positioned exactly by using laser interference ($\Delta x \approx 1/\Delta\lambda$).

16.3.2. Light Sources

16.3.2.1. Line Sources

In filter photometers, the spectral bandwidth can be improved considerably if line sources are combined with filters. Mercury line sources [56] emit various wavelengths, as shown in Figure 13. Such equipment is used mainly in clinical analysis, because in general, single wavelengths are only used for photometry or in simple kinetic experiments. Mercury vapor arcs are also employed in polarimeters. By using these, rotations at ca. 10 wavelengths can be measured. In polarimetry, the use of sodium vapor arcs is also common. These emit the doublet 589.0 and 589.6 nm (sodium D line). In the past, only these wavelengths were used to determine optical rotation. The sodium lamp is rarely used today for routine measurements; however, the sodium D line is still indispensable for calibration and reference measurements.

16.3.2.2. Sources of Continuous Radiation

In spectrometers, a light source supplying continuous radiation must be available. Normally, a high-radiant power source that emits in the visible range (tungsten – halogen lamp) is combined with an ultraviolet emitter (hydrogen or deuterium lamp) to provide a wider spectral range.

Tungsten – Halogen Lamps. By addition of iodine to incandescent lamps a higher filament temperature can be reached, and consequently a higher radiation density at the shortwave end of the spectrum. The emission maximum of a tungsten – halogen lamp is in the green part of the spectrum. At wavelengths < 300 nm, the intensity decreases rapidly unless the lamps are operated at high current (high filament temperature), because the intensity maximum is shifted toward shorter wavelengths by an amount proportional to T^4 (Wien's law). However, the lifetime of these light sources is thereby decreased drastically [57], [58].

Hydrogen and Deuterium Lamps. Deuterium lamps, whose emission maximum is at ca. 200 nm, are used mainly in sequential spectrometers. The emission maximum of hydrogen lamps is at a longer wavelength. At >400 nm, both types have rather low intensities, and show the familiar Fraunhofer lines at 486 and 656 nm whereby the lines of the two isotopes differ by fractions of a nanometer [57], [59]. These lines can be used for wavelength calibration. The user must realize that equipment that is fitted only with deuterium lamps can be used only to a limited extent for measurement in the visible range.

This is a very important consideration, particularly for diode array spectrometers, since these, like sequential routine equipment, must often cover the range between 200 and 800 nm. Despite the higher costs and greater complexity of the high-voltage supply, *xenon lamps* are preferable because they cover the entire 200 – 800-nm range with sufficient intensity, as shown in Figure 13. However, in xenon lamps, a marked structure at 490 nm is present in the continuum so that balancing the two optical paths is difficult in this range. These light sources are also used in ORD and CD equipment.

16.3.2.3. Lasers

Lasers represent a special type of light source [16], [21], [60], [61]. They are used in trace analysis by fluorescence measurement or laser-induced fluorescence (LIF) (→ Laser Analytical Spectroscopy) [62] – [64], in high-resolution spectroscopy, and in polarimetry for the detection of very small amounts of materials. Lasers can be of the gas, solid, or dye type [21]. In dye lasers, solutions of dyes are pumped optially by another laser or a flash lamp and then show induced emission in some regions of their fluorescence bands. By tuning the resonator the decoupled dye laser line can be varied to a limited extent, so that what may be termed sequential laser spectrometers can be constructed [65]. In modern semiconductor lasers, pressure and temperature can also be used to “detune” the emission wavelength by 20 – 30 nm [66], [67].

16.3.3. Selection of Wavelengths

In simple equipment, only relatively wide transmission bands are selected by means of colored glass or interference filters. A working beam of narrow bandwidth is obtained by combining this with a mercury source. Interference filters give spectral bandwidths in the 3 – 10-nm range [42], [68]. These operate on the principle of multiple reflections between layers of dielectric material that have been produced in a definite sequence by vacuum deposition. The wave character of light leads to interference of the reflected beams at many interfaces. A suitable choice of thicknesses of the evaporated films enables a narrow band containing a small number of wavelengths to be selected.

16.3.3.1. Prism Monochromators

Prisms are relatively expensive and very difficult to manufacture in the highest optical quality. Furthermore, dispersion by diffraction is not linear, so that so-called cam disks must be used to give a wavelength scaling that is as linear as possible over the entire range. Prism monochromators are mechanically very complex and therefore very expensive. They were at one time used widely in the UV region. Today they are often used as premonochromators (predispersion devices) since the decomposition of white light into rainbow colors by a prism gives only one spectrum order [42].

16.3.3.2. Grating Monochromators

In the dispersion spectrum produced by gratings, several wavelengths of different orders appear at the same place. As shown in Figure 14, incident white light is reflected at a number of grooves that are either ruled mechanically on a plane surface or produced holographically by superposition of two laser rays. The reflected beams interfere in different ways, depending on their wavelength, the distance between lines of the grating, and the angle of reflection, so that constructive or destructive interference can occur. Thus, reinforcement at a given wavelength takes place only in a certain direction. However, several orders are produced (see Fig. 14) so that, e.g., the first-order reflection of 400 nm and the second-order reflection of 200 nm form an image at the same place. The greater the number of grooves, the more do the partial beams interfere, and the more significant is the destructive interference at certain wavelengths [11], [42].

The dispersion of the grating is almost linear. However, as already mentioned, individual orders overlap so that either an order filter must be fitted or a predispersion device (prism) must be used. If the grooves are blazed, reflected intensities of mainly one order can be collected. This is at a maximum for a particular angle (blaze angle) and a particular wavelength (blaze wavelength).

An inlet slit is usually located in front of the prism or grating symmetrical to the exit slit. If the inlet and exit slits are coupled and have the same width, the slit function (see below) forms a trian-

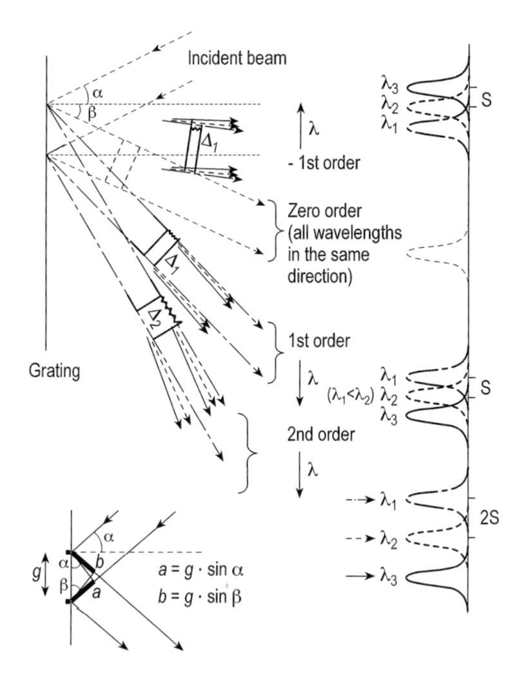

Figure 14. Principle of the grating
Beams reflected at the grating grooves overlap, giving constructive interference at an angle β for each wavelength except the zero-order case. The separation (resolution) of wavelength maxima S on a projection surface for the second order is twice that for the first order.

gular intensity distribution. In high-quality equipment, the first monochromator is followed by a second. These double monochromators have considerably better resolving power, although less light intensity is available since part of the light is cut off by the nonparallel beam through each slit. However, in a qualitatively high-grade double monochromator, a central slit must also be present, adjusted synchronously with the inlet and outlet slits; otherwise slit function is impaired [42], [69].

The slit function can be observed well in the imaging of line sources when the inlet and outlet slits (and the middle slit if present) are moved synchronously. If the slit is too wide the line of a mercury lamp does not appear as the expected Gaussian-shaped curve intensity distribution, but as a triangle. If the inlet and exit slits are different, a trapezoid is obtained. Triangular shapes in the spectrum indicate defective adjustment of slit widths in the equipment. These effects are observed mainly in spectrometers in which only a small number of preset slit widths can be selected. Distortion of a spectral band is negligible only if the spectral bandwidth of the monochromator (slit width) is not more than one-tenth the natural bandwith of the sample [3], [42], [70].

16.3.4. Polarizers and Analyzers

The polarizer and analyzer separate electromagnetic radiation into an ordinary and extraordinary beam. The so-called Glan–Thompson prism is suitable for this. The path of the ordinary beam, whose mode is parallel to the plane of incidence (incident beam normal to prism edge), is refracted at the edge of the prism, while the extraordinary beam continues along the axis [11], [34]. Simple polarization filters (polarizing films) are also often used, in which two mutually perpendicular, plane-polarized beams are absorbed to different extents. Such polarizing films can be obtained by dyeing highly stretched (oriented) plastic film with dichroic dyes. Additional information (e.g., phase dependence) can be obtained by modulation. This technique is very important, especially in equipment for spectral measurements, particularly ellipsometry, because automatic balancing is possible. Modulation can be achieved by either rotational or oscillatory movement of the analyzer or polarizer, or by superposition with the inclusion of a Faraday modulator, which utilizes the Faraday effect described in Section 16.2.6.5 [11], [38]. Since spectral polarimeters (ORD equipment) are also used in the shortwave UV region, prism assemblies that include Canada balsam adhesive (which absorbs in this spectral range) must be excluded. Calcite can also cause problems because it absorbs at wavelengths below ca. 220–250 nm [38]. For these reasons, quartz Rochon prisms are used mainly today [34], [71].

16.3.5. Sample Compartments and Cells

16.3.5.1. Closed Compartments

Light from the monochromator is usually convergent and is focused on a minimum beam diameter at the position of the cell. The less convergent beam, the easier can measurements be performed in cells having long optical paths (10 cm) without the light grazing the sidewall of the cell. To prevent this grazing, the position of the cell in the path of the beam should be checked. In the visible region of the spectrum, a simple sheet of paper is suitable for this (monochromator set at ca. 550 nm and with a very wide slit) [3]. It should be placed

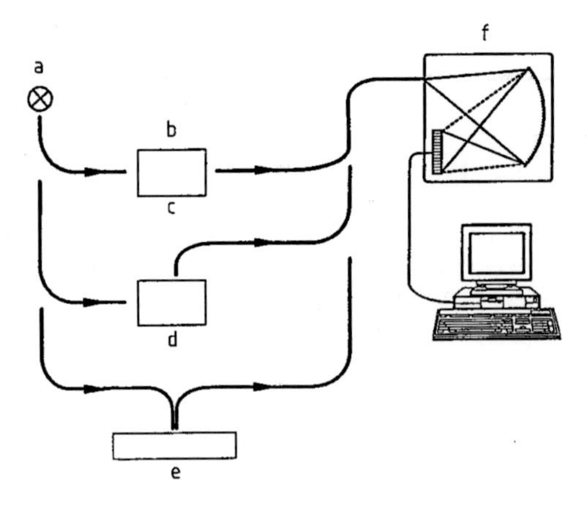

Figure 15. Combination of diode array and glass fibers in various arrangements for measuring absorption, fluorescence, or reflection

perpendicularly to the beam to monitor the green spot with respect to the cell position. Furthermore, the edges of the cell and the regions of adhesive at the edges should be inspected for any radiation. In the UV region, paper containing optical brighteners that emit blue light can be used.

Depending on spectral region, the correct material must be selected from appropriate catalogues, for example, because cells of glass or polymer are not transparent to UV radiation. The plastic cells often used in routine analysis differ in their optical properties. This holds true for cells from the same batch and from different batches [72]. Because of the manufacturing process used, their interior surfaces are not parallel, so the reproducibility and potential applications of these cells are limited.

In the sample compartment, as much space as possible should be available to allow for thermostatically controlled sample holders or other equipment (Ulbricht globes, praying mantis devices, multicell holders, sippers). A secondary cooling circuit is useful to protect the optics and electronics from high temperature. Large sample compartments require well-thought-out, carefully calculated optical paths. Photochemical reactions can be followed directly in a spectrometer by special accessories [45].

16.3.5.2. Modular Arrangements

In modular construction, monochromators, light sources, and detectors can be arranged in various ways adjacent to the sample compartment, so that absorption, fluorescence, and reflection measurements are possible by using the same set of equipment components.

16.3.5.3. Open Compartments

Today, diode arrays combined with optical fiber arrangements are very well suited to this modular concept. The possible variety of such arrangements is illustrated by the combination of a diode array with optical fibers in Figure 15 [55]. The optical path between the light source (a) and the sample (b), and between the sample and the detection system (f), is realized by means of glass or quartz fibers, so that with one experimental arrangement, absorption (c), fluorescence (d), and reflection (e) measurements are possible by a variable arrangement of optical components. This type of equipment is very important for color measurements and becomes increasingly applied in optical sensor technology [55], [73]. In principle, the use of sequential recording equipment with designs based on optical fibers is also possible [74], [75].

16.3.6. Detectors

Detectors are of mainly two types, based on either the external or the internal photoelectric effect [59], [76]–[78].

Photomultipliers. In photomultipliers, the external photoelectric effect is utilized. Photons incident on a photocathode, which have an energy greater than the work function of cathode material expel electrons, which are captured by an anode that is positively charged relative to the cathode. The number of electrons produced is proportional to light intensity. They can be measured by the current flowing between electrodes.

Dynodes are often placed between the photocathode and the anode as intermediate electrodes. High-quality photomultipliers have 10–14 dynodes (see Fig. 16). This internal amplification is critical because in addition to electrons from the light-induced process, other thermally activated electrons give rise to a dark current. Photomultipliers are therefore often cooled for the measurement of low light intensities.

The head-on arrangement (Fig. 16 A) has a larger inlet angle and is extended lengthwise, with a correspondingly greater distance separating the

A

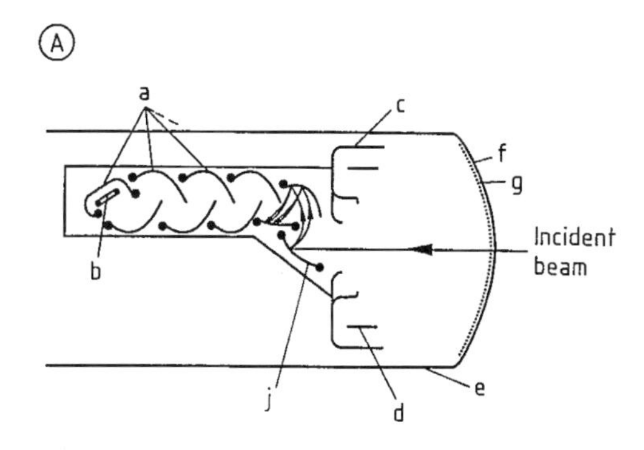

B

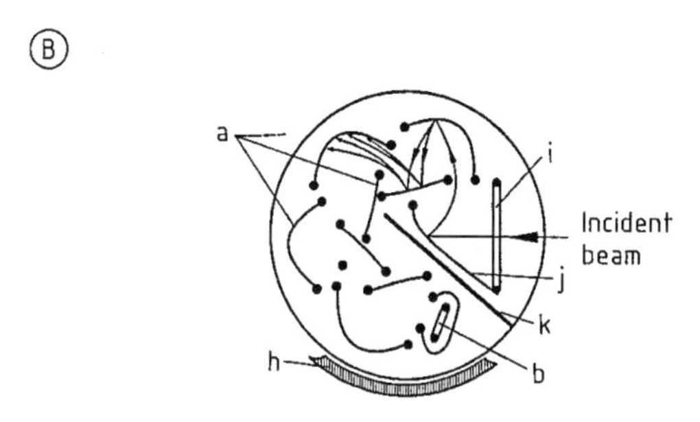

Figure 16. Types of photomultiplier
A) Head-on arrangement; B) Side-on arrangement (on a larger scale)
a) Dynodes; b) Anode; c) Focusing electrode; d) Focusing ring; e) Conductive internal coating; f) Front plate; g) Semitransparent photocathode; h) Envelope; i) Grill; j) Opaque photocathode; k) Shield

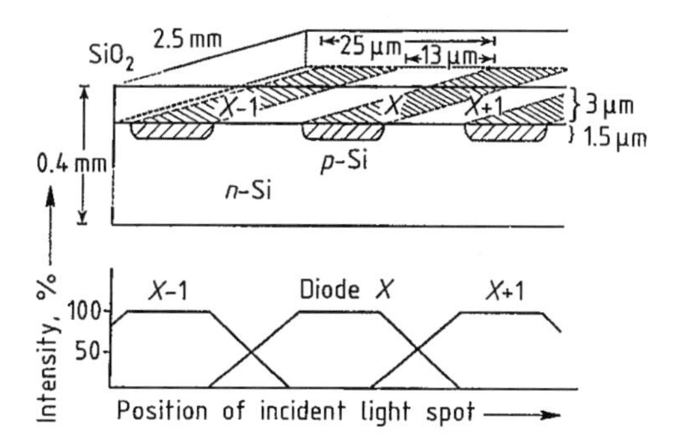

Figure 17. Linear photodiode array [3], [81]
The p-doped islands can be found in the n-silicon under a layer of SiO_2. In the region of these p-islands, photons can produce charges that discharge the capacitors arranged in parallel.

A

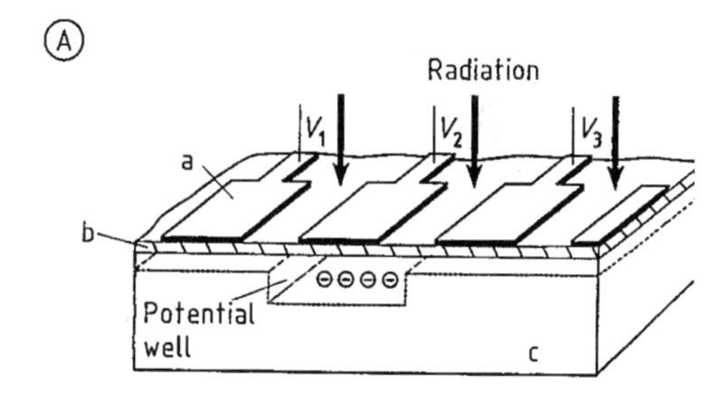

B

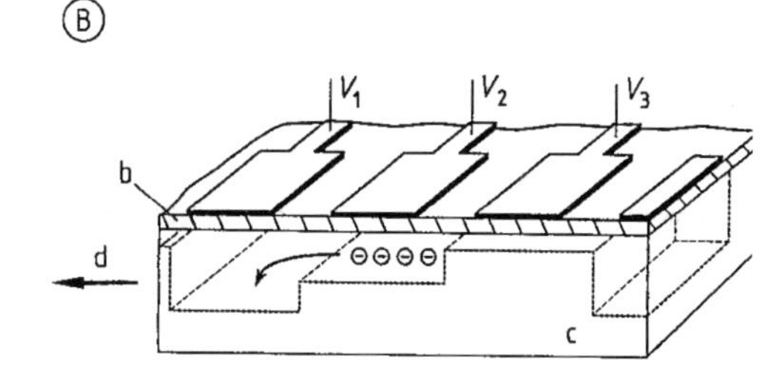

Figure 18. Charge-coupled device (CCD) [55]
A) Free charges are produced by incident light and localized in potential wells (countervoltage); B) Periodic application of different voltages causes charges to be displaced from the potential well like a shift register
a) Metallic electrode (gate); b) SiO_2 (insulator); c) p-Type silicon

dynodes (lower field strengths). It can accommodate more dynodes than the side-on arrangement, and therefore has a greater range of amplification. However, response times are longer. The side-on arrangement (Fig. 16 B) is considerably more compact, the distances between dynodes are shorter and the response times are relatively short. However, the "capture angle" is smaller. For detecting only a small number of photons, the more slowly reacting head-on photomultiplier is preferable [79].

Photodiodes, on the other hand, use the internal photoelectric effect [77]. When these diodes are arranged in arrays, they are known as diode arrays or diode matrices. Multiplex detectors are used in simultaneous spectrometers [80].

An example is the Reticon array shown in Figure 17. Also, CCD arrays (charge coupled devices), which were developed for television cameras and fax machines, have also become important in spectroscopy (Fig. 18) [4], [49], [81] – [83]. Apart from the advantage of being multiplex devices, these charge-transfer components have characteristics more similar to those of photomultipliers than normal photodiodes.

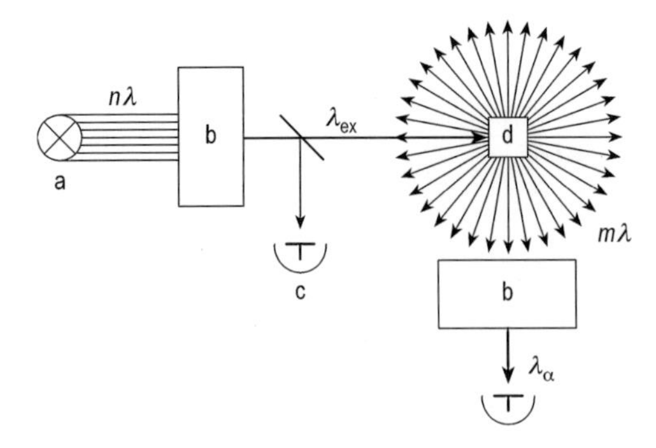

Figure 19. Diagram of optical light paths in a fluorescence experiment
a) Light source; b) Monochromator; c) Photomultiplier; d) Sample

16.3.7. Optical Paths for Special Measuring Requirements

16.3.7.1. Fluorescence Measurement

Unlike transmission spectroscopy, in fluorescence measurement the sample is observed perpendicularly to the incident excitation beam. In modern equipment, monochromators are placed both in the path of the excitation beam and the observation beam (see Fig. 19). Since no direct comparative measurement is possible in fluorescence, the intensity of the excitation light beam is usually controlled at least by a semitransparent mirror and a photodiode. To measure a *fluorescence spectrum*, the wavelength of the excitation beam is kept constant and the wavelength of the observation monochromator is varied, so that the excitation monochromator can be replaced by a filter in simple equipment.

For observation of *fluorescence excitation spectra*, the monochromator at the observation end is adjusted to a wavelength of highest possible fluorescence intensity, and the wavelength of the monochromator on the excitation side is varied. Hence, for these spectra, a shape analogous to an absorption spectrum is obtained [84]. If the absolute fluorescence spectra must be measured, the equipment must be calibrated for all wavelengths by using fluorescence standards [85]. This operation is highly complex and can be performed only with special equipment and appropriate software. Commercial equipment of this type is illustrated in Figure 20 [43].

16.3.7.2. Measuring Equipment for Polarimetry, ORD, and CD

Simple *polarimeters* consist of a polarizer, several slits, a light source (usually the sodium D line), and a detector, which in the simplest case is the eye of the observer. To adjust an optimal minimum of intensity passing the analyzer half-shaded apparatus are used. The measuring field is separated into two adjacent areas which are adjusted to equal brightness for an adequate analyzer position. Equipment with self-equalizing polarimeters is more complex. Self-equalizing polarimeters contain an additional Wollaston prism and a Faraday modulator, as shown in Figure 21. The polarizer (a) and analyzer (e) are crossed with respect to each other. Radiation from the polarizer is split in the Wollaston prism into two beams of plane-polarized light, whose planes of polarization are perpendicular to each other. The prism (c) is positioned such that the plane of polarization of the incident light forms an angle of 45° to the two beams passing the prism (without sample). The two split beams are of equal intensity under such conditions. Passage through Faraday modulator causes the direction of rotation of the two partial beams to oscillate, and they combine to give a constant radiation flux. This situation is disturbed when an optically active substance is placed in front of the Wollaston prism. An additional oscillating radiation flux falls on the photomultiplier. By use of a feedback system, the polarizer is rotated until this signal disappears.

In *spectral polarimeters*, another monochromator is introduced in front of the polarizer. In this type of ORD equipment, the Wollaston prism and polarizer are often combined by pivoting the polarizer to and from in a defined angle. CD equipment is often provided as an additional device for classical absorption spectrometry. With these devices, absorption of left and right circular-polarized light can be measured independently. For this, combinations of Rochon prisms (as polarizers) and Fresnel quartz rhombic prisms (for production of circular-polarized light) are used [34]. Equipment of this type is shown in Figure 22.

16.3.7.3. Reflection Measurement

Reflection can be either regular (at smooth surfaces) or diffuse (by material whose small particle size causes scattering). *Diffuse reflection* is used mainly in thin-film chromatography

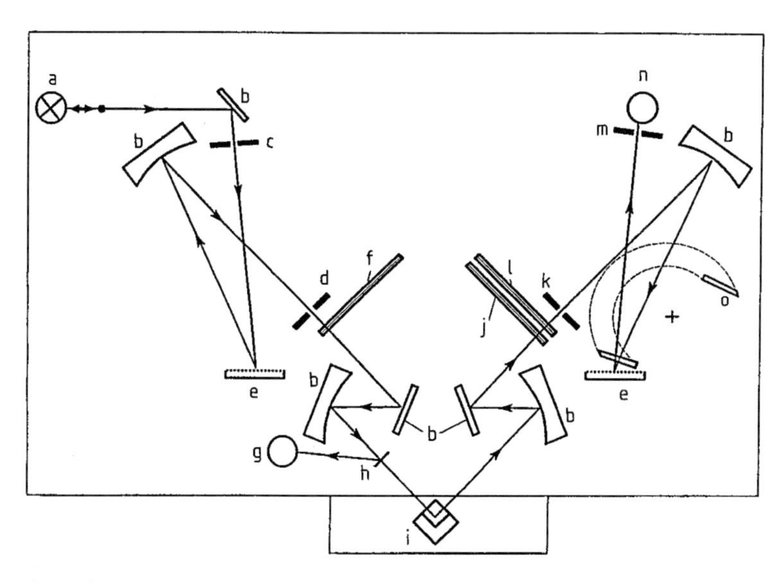

Figure 20. Light paths in commercial fluorimeter from Perkin–Elmer [43]
a) Light source; b) Mirror; c) Slit excitation entry; d) Slit excitation exit; e) Grating; f) Excitation filter wheel with excitation polarizer accessory; g) Reference photomultiplier; h) Beam splitter; i) Sample; j) Emission polarizer accessory; k) Slit emission entry; l) Emission filter wheel; m) Slit emission exit; n) Sample photomultiplier; o) Zero-order mirror accessory

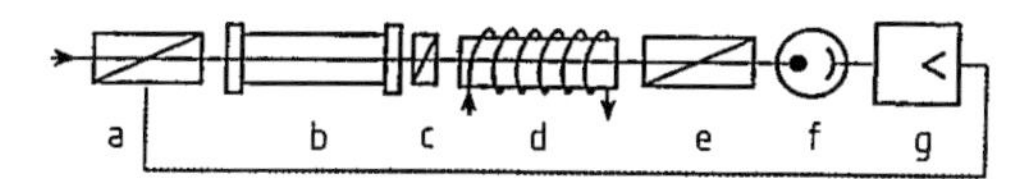

Figure 21. Schematic arrangement of self-balancing polarimeter [38]
a) Polarizer (rotatable); b) Sample cell; c) Wollaston prism; d) Faraday modulator; e) Analyzer; f) Photomultiplier; g) Amplifier
— = Feedback between detector and polarizer

[87]–[89], to evaluate the Kubelka–Munk function at an angle of $< 60°$ from the Lambert cosine squared law [4], [90]. Diode arrays can also be used as detector [91].

To observe *regular (specular) reflection*, goniometer equipment is used in combination with optical fibers and spectrometers. A modular system of this type is shown in Figure 15.

16.3.7.4. Ellipsometry

If measuring equipment for ORD and CD (Section 16.3.7.2) is combined with a reflection system, either ellipsometry at one wavelength or spectral ellipsometry can be carried out at defined angles (Fig. 23). In this apparatus, a rotating polarizer is used to produce rotating plane-polarized light. After reflection, an analyzer is used to measure the phase and attenuated relative intensity of the two beams (i.e., ellipticity). From the modulated signal, Hadamard matrices can be used to calculate the refractive index and absorption coefficient of the sample as a function of wavelength [93], [94], [95].

16.3.8. Effect of Equipment Parameters

In many spectrometers, some parameters (e.g., slit width, time constant, and amplification) cannot be chosen freely. Hence, their effects on measurement must be known. These settings depend on the type of spectrum being observed (e.g., a spectrum for routine analysis; a rapid scan spectrum, or high-resolution spectrum).

The following properties of the sample or the equipment can affect measurement [3], [5], [70], [96], [97]:

1) After being switched on, all equipment requires time to warm up until the electronics, light sources, and optical components are in thermal equilibrium.
2) The number of digits that can be read correctly from the display is affected by noise. The number of correct digits does not necessarily correspond with those displayed.

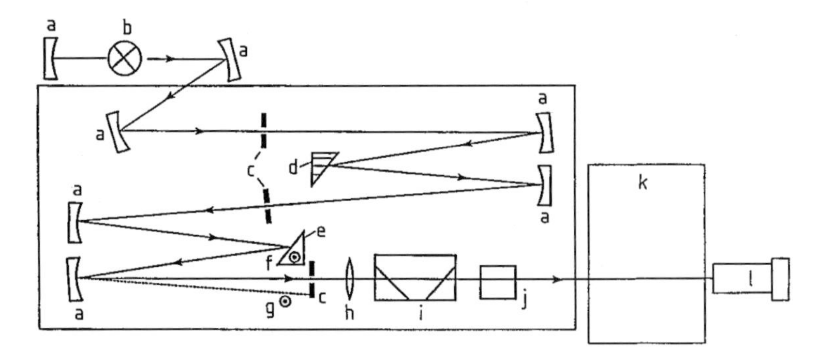

Figure 22. Light path of CD apparatus from Jasco company [86]
a) Mirror; b) Light source; c) Slit; d) Prism (vertical optical axis); e) Prism (horizontal optical axis); f) Ordinary ray; g) Extraordinary ray; h) Lens; i) Filter; j) CD modulator; k) Sample compartment; l) Photomultiplier

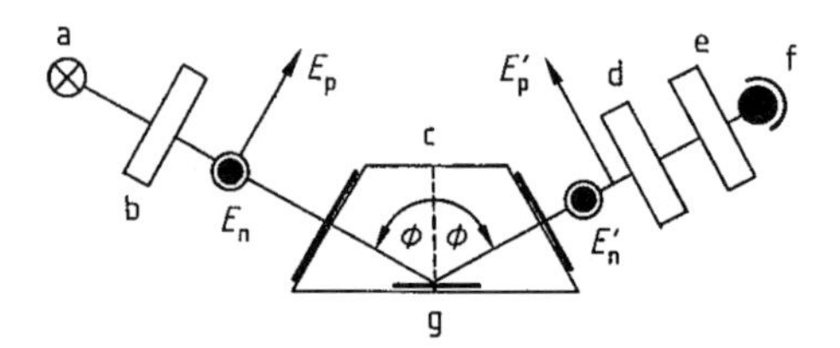

Figure 23. Design of MOSS ES4G ellipsometer from SOPRA [92]
a) Light source; b) Rotating polarizer; c) Measuring cell; d) Analzyer; e) Monochromator; f) Detector; g) Sample

3) The cell compartment as well as the cell holder should be temperature controllable to protect the electronics and optics from high measuring temperature.
4) To prevent any effects of fluctuation in voltage supply on the "bright value" (100 % signal), such fluctuations should be controllable even under load.
5) Measurements in the spectral range < 190 nm can be performed accurately, only if the entire optical light path and sample compartment can be spilled by nitrogen.
6) Blaze wavelengths of ca. 320 nm ensure good measurements down to 200 nm even if the transmission of the optical components and sensitivity of the detector decrease in this range. However, this short blaze wavelength cannot usually be compensated for by higher detector sensitivity if wavelengths of 800 – 1000 nm are measured.
7) If cells with a long path length are used, they must be filled to the minimum level, and the geometry of the measuring beam must be correct.
8) If the intensity of the measuring beam in the UV is too high, it can lead to photochemical reactions in the sample.
9) In principle, the wavelength scan should be stopped for each measurement to prevent integration of the changing signal with the wavelength. Some equipment includes this facility, but the overall scan rate is reduced [45], [46].

Damping (time constant), amplification, spectral bandwidth, and scan rate have a considerable effect on the signal obtained and are interdependent. For these reasons, depending on the spectral characteristics of the sample and the purpose of the experiment (exploratory, routine, or high-resolution spectrum), parameters must be optimized relative to each other. Some criteria are given below. Other information on standards and calibration methods can be found in the literature [98], [99].

Reproducibility and Accuracy of Wavelengths. In both grating and prism equipment, mechanical wandering of the wavelength scale can occur over time. Therefore, the accuracy of the wavelength must be monitored, and adjusted if necessary, using holmium perchlorate solution, holmium chloride solution, or filters, or by using the lines of a mercury high pressure arc. The simplest method is to use the lines of the hydrogen or deuterium lamp that is part of the equipment.

Wavelength reproducibility can be simply checked by an overlay of repetitively recorded line emitter or holmium perchlorate solution spectra. Since wavelength reproducibility is especially important in kinetic investigations, a chemical reaction can also be used as standard, for which the

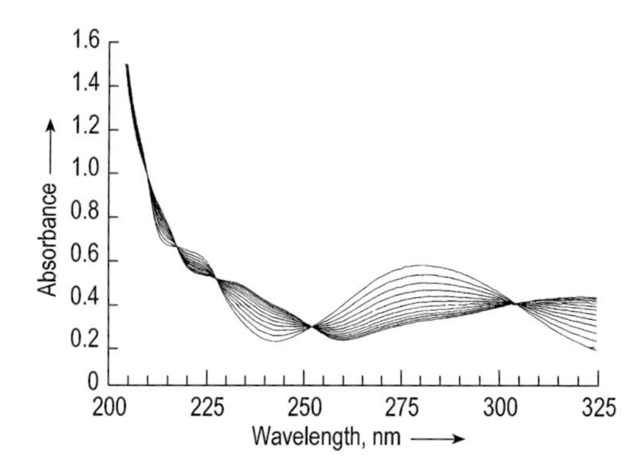

Figure 24. Sulfone reaction spectrum for characterization of wavelength reproducibility of a spectrometer

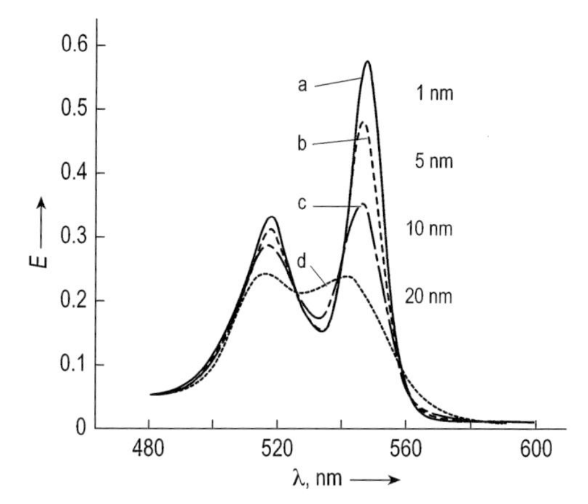

Figure 25. Effect of slit width on spectrum for various ratios of spectral slit width to natural bandwidth
a) 1 : 20 (1 nm); b) 1 : 4 (5 nm); c) 1 : 2 (10 nm); d) 1 : 1 (20 nm)

absorbance at certain wavelengths does not change during reaction (isosbestic points). An example is shown in Figure 24, in which repeated measurements were carried out on a thermal reaction of 2-hydroxy-5-nitrobenzylsulfonic acid [3].

Photometric Reproducibility and Accuracy. Standards for absolute determination of photometric accuracy are available from, for example, national standards offices (Physikalisch-Technische Bundesanstalt, PTB; NBS). However, the accuracy of these standards (both filters and standard solutions) is not greater than ca. 0.3 %, and even that can be achieved only by expert operators.

A simpler determination of photometric reproducibility is often carried out with the aid of a

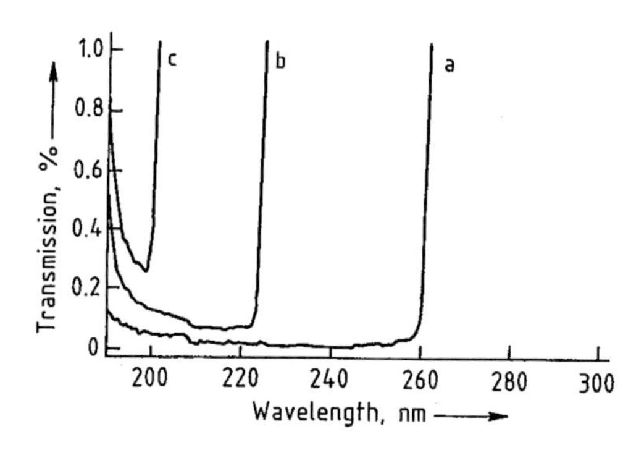

Figure 26. Cutoff filter properties for various totally absorbing salt solutions
a) NaI (cutoff at 260 nm); b) NaBr (cutoff at 220 nm); c) KCl (cutoff at 200 nm)

solution of potassium dichromate in 0.005 mol/L sulfuric acid [100], [101]. The Lambert – Beer law requires a linear relationship between the signal and concentration. Deviations can be due to several reasons provided chemical effects can be excluded: (1) the recording rate or time constant may be too high; (2) the spectral bandwidth chosen may be too large compared to the natural bandwidth of the sample; (3) too much stray light may be present. As shown in Figure 25, if the spectral bandwidth used is too large, this leads not only to a reduction in the peak height, but also to a displacement of the maximum. A good rule of thumb is that the spectral bandwidth of the equipment should be only one-tenth the natural bandwidth of the substance [3]; otherwise, peak height will be > 0.5 % less than the true value. Therefore, equipment with fixed slit widths is of limited use.

Stray Light. In addition to radiation of the desired wavelength, every monochromator passes a proportion of so-called stray light (light of different wavelength). If radiation at the extremes of the wavelength range (< 250 nm or > 500 nm) is strongly attenuated owing to high absorption properties of the sample, even stray light of low intensity can lead to severe signal distortion because it usually has a wide bandwidth, is scarcely attenuated by the sample, and strikes the detector, which is not wavelength selective. This effect is most noticeable at absorbances > 2. The transmission of the sample appears higher, a false value of absorbance is recorded, and the absorption maxima are reduced. As shown in Figure 26, totally absorbing salt solutions with cutoff filter properties can be used to reveal stray light [3], [102].

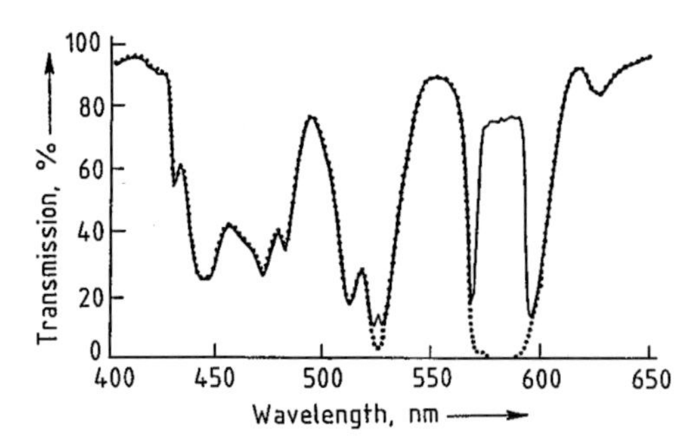

Figure 27. Determination of amount of stray light by a difference measurement of colored glass of various thicknesses
····· = Colored glass 1 mm thick with respect to silica glass 1 mm thick (without chromophore); —— = Colored glass 4 mm thick measured against colored glass 3 mm thick

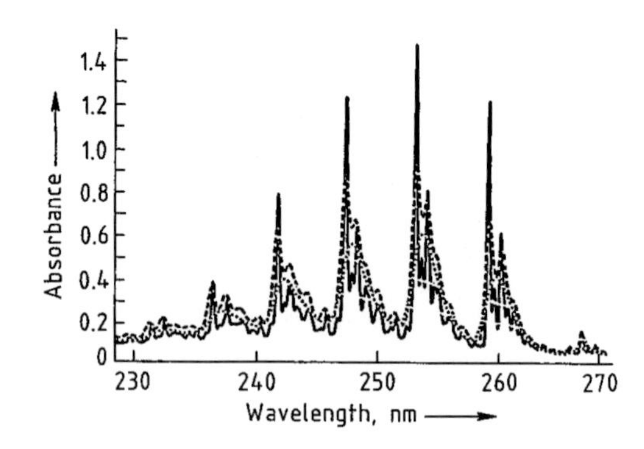

Figure 28. Benzene vapor spectrum recorded with the double monochromator equipment Specord M-500 (C. Zeiss) at various settings of spectral slit widths
··· = > 2.0 mm; --- = 0.5 mm; —— = 0.2 mm

Another possibility, which is successful mainly in the visible region, is provided by comparative measurements of colored glasses of different thicknesses. As shown in Figure 27, the bands have a tendency to even out in the high-absorption region. In an extreme case, this can be manifest by an intermediate minimum in the resulting spectrum which can lead to misleading interpretations [3], [103].

In principle, the following rules of thumb can be given for recording various types of spectra:

1) For exploratory spectra (recorded at high speed), high amplification and small time constants are used, with a large spectral slit width.
2) For routine spectra, a medium-sized spectral slit width is used, depending on the natural bandwidth, and the scan rate selected should not be too high. A medium time constant with a medium degree of amplification avoids noisy spectra.
3) For high-resolution spectra, a very low recording rate with large time constants, high amplification, and an extremely small slit width is used.

The difference between these three types of spectra is illustrated in Figure 28, which shows results obtained on benzene vapor under various conditions by using a spectrometer capable of high resolution.

Cleaning the Optical Cell. The type of material used for the cell and the cleaning method are very important for measurement quality. Standard cleaning procedures recommended by cell manufacturer should be used [104]. Under no circumstances should finger marks be left on the cell because this seriously impairs the refractive properties of the cell walls. Alteration of the optical path and the refractive properties of a cell with scratched walls are revealed by a change in the signal of an empty or solution-filled cell when the cell is rotated through 180°.

Since plastic cells generally do not have two plane-parallel walls, fluctuations can occur even in different cells from the same batch [72]. Cells with polytetrafluoroethane stoppers should be used for quantitative measurements because, with other types of cells, the organic solvent can creep over the edges or evaporate.

16.3.9. Connection to Electronic Systems and Computers

Modern spectrometers have digital readouts and integrated microprocessors for purposes of calibration, control of the wavelength drive, slit adjustment, and sometimes for the interchange of the light sources. They also include interfaces to personal computers sometimes called data stations. The trend is now toward spectrometers with a limited amount of built-in intelligence, which are controlled by personal computers that can also readout data. For this purpose, software for control, data collection, and evaluation is available, operating under commonly used operating systems. In this way, available additional development costs are reduced. Built-in analog-to-digital converters should be provided with preamplifiers to ensure that a wide dynamic range is available.

In the UV – VIS range, personal computers are used in data evaluation, especially in multicomponent analysis. When connected to detection systems in chromatography, they are also used for peak detection, peak integration, and eluent optimization [105]. Although in UV – VIS spectrometry, libraries of spectra do not play an important role, standard spectra are used in chromatography for identification [106].

In larger laboratories, personal computers are linked via a local area network (LAN) to a laboratory information and management system (LIMS) [107].

16.4. Uses of UV – VIS Spectroscopy in Absorption, Fluorescence, and Reflection

16.4.1. Identification of Substances and Determination of Structures

Although the main strength of UV – VIS spectroscopy does not lie in the identification of substances, it can often be used as a quick, inexpensive method for identifying certain classes of materials [108] – [111]. This is especially true in the area of pharmaceutical chemistry, where many otherwise difficult-to-identify substances can be identified with certainty, especially at different pH values and via color reactions. Other examples are aromatic carboxylic acids [111], where the position of the absorption maxima is characteristic and depends on the substituents, for example, with benzadiazepines [112]; and also pregnenones and androstenones [108], [109].

Since steroids contain α, β-unsaturated keto groups absorption maxima can even be predicted by increment rules [108]. In this case, identification is easier by UV – VIS spectroscopy than by the more sophisticated NMR method. The position of methyl or hydrogen substituents relative to the keto groups in the ring enables the wavelength of absorption peaks to be calculated. The spectra of antidiabetics such as carbutamide, tolbutamide, etc. [111], can be calculated and classified readily, although they sometimes differ only in the *para*-substituents on the aromatic sulfonamide ring. Since carbutamide contains an amino group, a significant pH effect occurs that is not observed with the other compounds.

16.4.2. Quantitative Analysis

16.4.2.1. Determination of Concentration by Calibration Curves (→ Chemometrics)

Molar decadic absorption coefficients can be determined from the slope of the absorbance-versus-concentration curve. This slope is known as the sensitivity. Any intersections with axes or nonlinearities can be caused either by impurities inside or outside the cell, by limited photometric linearity of the instrument, or by deviations from the Lambert – Beer law. Influences on the spectrum of a pure substance are shown in Figure 29 (A – C). These changes are caused by the presence of an impurity, fluorescence of the absorbing substance, or scattering. Three measuring wavelengths (λ_1, λ_2, λ_3) are also indicated in the figure. *Scattering* is apparent at all three measuring wavelengths, although to different extents. The *impurity* affects only the spectrum of the pure substance at wavelengths λ_2 and λ_3, and fluorescence appears only at wavelength λ_3. By drawing extinction – concentration diagrams for the three cases (see Fig. 30), instead of the "true" absorbance marked by the symbol dot in circle, a false signal is used. Calibration at one wavelength can lead to incorrect determination. The measured absorbance of a sample can lead to various apparent concentrations, depending on what is causing the false reading.

To ensure correctness of calibration, this operation must always be carried out at several wavelengths that are carefully selected from the spectrum. A simple graphical method of detecting errors in calibration and measurement, and also of recognizing their causes, is given in Figure 31 (A – C). Absorbances are compared for various wavelength combinations by using the same sample. These show a pattern that depends on the type of influence, thereby enabling various types of disturbance to be distinguished. The diagram obtained (see Fig. 31) is known as an extinction diagram [3], [30]. If the influence is due not to the presence of an additional substance in solution, but to an impurity on the cell wall (e.g., finger marks or cells that are not equally clean), this appears in the calibration diagram as an intersection with an axis (i.e., a parallel shift of the cal-

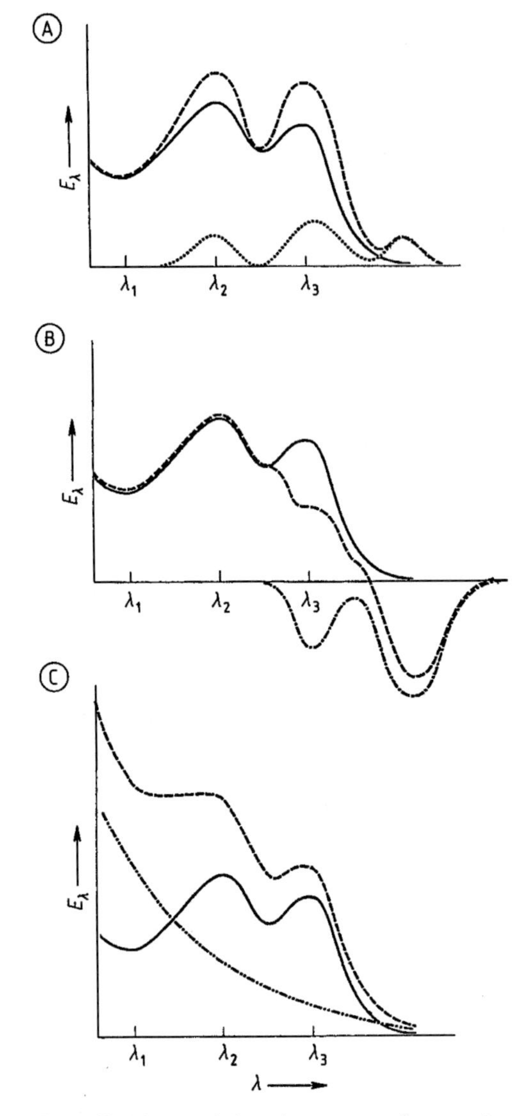

Figure 29. Diagram of absorption spectrum of a pure substance (—)
A) Superimposed on spectrum of an impurity (· · ·); B) With a fluorescent component (–·–·–·–); C) With a scattering substance (- - - - - - - - - - - -)
– – – = Measured spectrum

ibration line with offset), as shown by curve e in Figure 30 [3].

16.4.2.2. Classical Multicomponent Analysis

Calibration at several wavelengths also enables multicomponent analysis to be carried out in mixtures of substances by using the spectra of pure

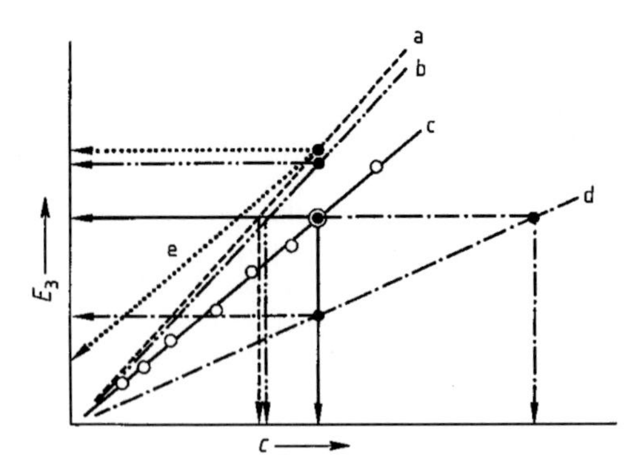

Figure 30. Absorbance plotted as a function of concentration to give a calibration line
Various signals are observed for the same concentration. Same absorbance value corresponds to different concentrations.
a) Substance + impurity; b) Substance + scattering; c) Pure substance; d) Substance + fluorescence; e) Calibration curve if, e.g., finger marks occur on outside faces of cell
○○○ = Measured points

substances or defined mixtures for calibration (see Fig. 32) [113]. In this case, the Lambert – Beer law for a pure substance is modified, giving the relationship [3]

$$E_\lambda = d \sum_i \varepsilon_{\lambda,i}\, c_i \tag{13}$$

Thus, for three components A, B, C, a linear equation system is obtained, which could take the form

$$E_{500} = d(\varepsilon_{500,\mathrm{A}}\, a + \varepsilon_{500,\mathrm{B}}\, b + \varepsilon_{500,\mathrm{C}}\, c)$$
$$E_{400} = d(\varepsilon_{400,\mathrm{A}}\, a + \varepsilon_{400,\mathrm{B}}\, b + \varepsilon_{400,\mathrm{C}}\, c)$$
$$E_{300} = d(\varepsilon_{300,\mathrm{A}}\, a + \varepsilon_{300,\mathrm{B}}\, b + \varepsilon_{300,\mathrm{C}}\, c)$$

$$\boldsymbol{E} = \varepsilon\, \boldsymbol{c}\, d \tag{14}$$

if three wavelengths are being measured. In classical multicomponent analysis, this linear equation system is solved by matrix operations and normal equations. However, in this so-called K-matrix method [6], [114], errors that arise in calibration and measurement must be taken into account. These are represented by the error vector $\boldsymbol{e}$, so that the following matrix equation is obtained

$$\boldsymbol{E} = \varepsilon\, \boldsymbol{c}\, d + \boldsymbol{e} \tag{15}$$

The number of components that can be determined independently in such multicomponent analysis is normally limited to three or four if individual

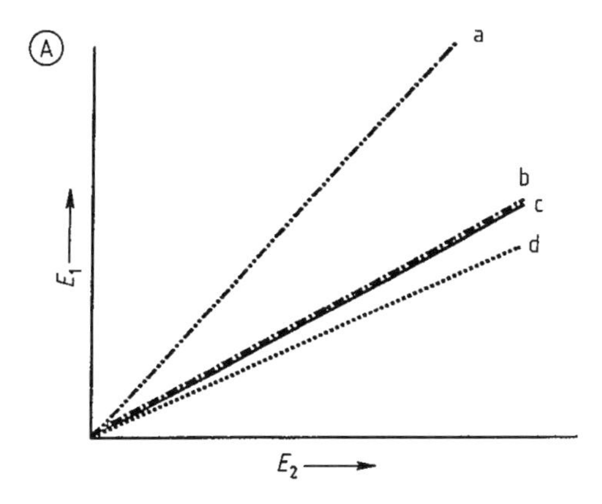

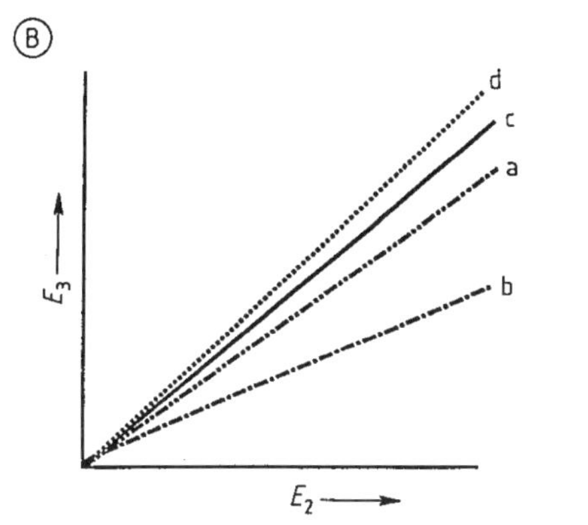

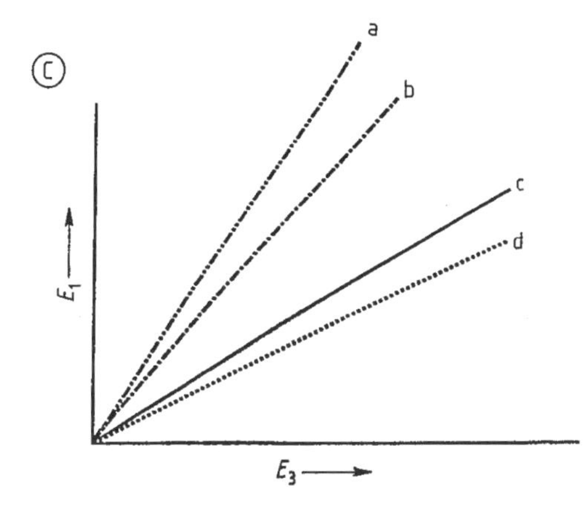

Figure 31. Extinction diagram in various combinations for detection of errors in calibration measurement
a) Substance + scattering; b) Substance + fluorescence; c) Pure substance; d) Substance + impurity

components are not situated favorably in the spectrum with respect to their wavelengths and bandwidths. The ε-matrix is obtained from the calibration measurements. Inversion of this matrix often leads to numerical problems [3], [6], [115].

16.4.2.3. Multivariate Data Analysis

To take account of interactions between individual components (association, nonlinearities), calibration using multivariate data analysis is often also carried out with mixtures rather than pure substances. Despite this fact, limitations to this method of assessment are encountered quickly. Therefore, the so-called inverse method using the $\boldsymbol{Q}$-matrix is employed, and either principal component regression (PCR) or the partial least squares (PLS) method is used [6], [114], [116]. In both methods, calibration is carried out not with pure substances, but with various mixtures, which must cover the expected concentration range of all components. Within limits, this can allow for nonlinearities:

$$c = \boldsymbol{Q}\boldsymbol{b} \tag{16}$$

Principal component analysis first gives the number of significant components, which avoids a typical error of classical multicomponent analysis, which distributes an unconsidered component among the others. However, principal component analysis reveals it as an additional "impurity." Then a large number of regression steps is carried out with respect to each of the mixtures. In PCR, all possible combinations of factors must be considered since the algorithm does not produce these factors in systematic order whichis the intended aim (quantitative multiple component determination with minimal error). By using the Lanczos bidiagonalization method (PLS), this validation is simplified considerably, since only combinations of the first sequence of factors need be considered [113], [117].

Another method is the so-called Kalman filter [118], [119]. Iterative calculations enable the spectrum of an unknown component to be determined using a sufficiently large number of calibrating wavelengths. Methods of multicomponent analysis by fuzzy logic have not proved very successful [106]. In contrast, a method using Fourier transforms followed by multicomponent analysis is available commercially [120]. The most recent methods, using neuronal networks [6], [121], are used much less in classical multicomponent analysis [122] than in sensor arrays with relatively low selectivity, where they are employed to determine relative concentrations in gas mixtures or liquids [123], [124]. Another wide field of application, in the infrared, is in foodstuffs analysis and pharmaceutical chemistry [125].

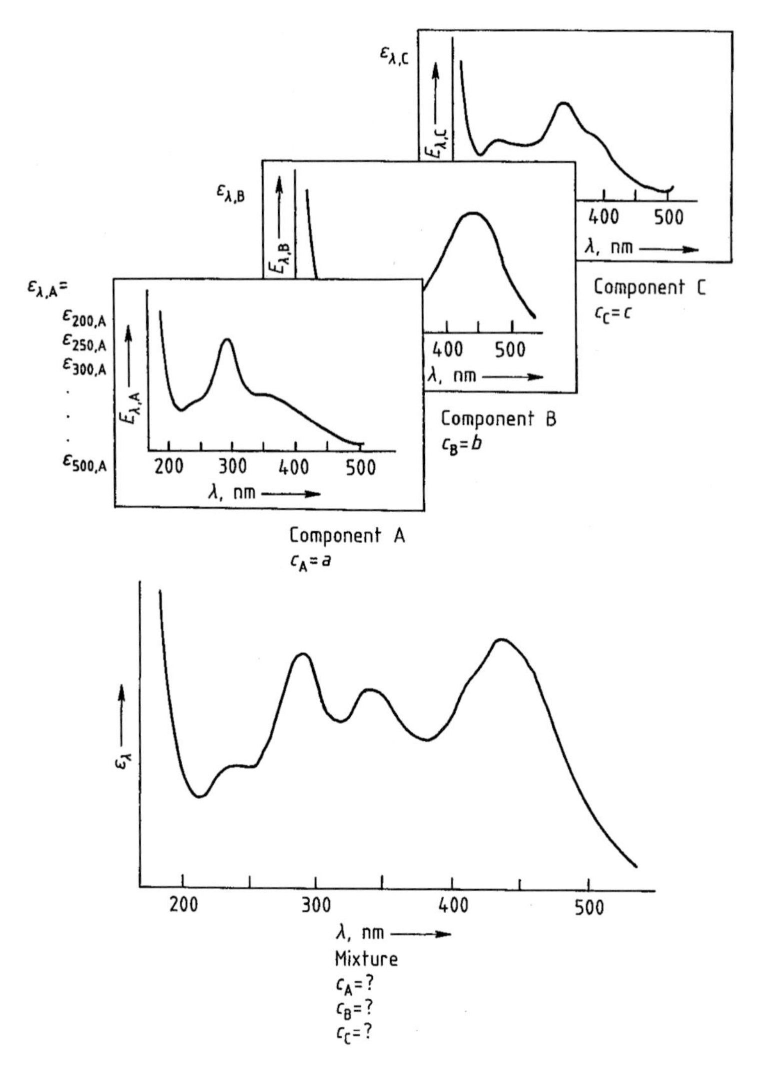

Figure 32. Diagram of pure spectra of three components A, B, and C for calibration, and spectrum of an unknown mixture

16.4.2.4. Use in Chromatography

In principle, multicomponent analysis can also be used in chromatography if peaks cannot be separated completely. By using a diode array while recording a chromatogram, components hidden within a peak can be detected by factor analysis if they are spectrally distinguishable [126]. If their calibration spectra are known, their relative concentrations can be determined.

Every multicomponent analysis should be assessed critically by the analyst, and different algorithms should be used on the same data to improve the level of significance of the evaluation.

16.4.3. Fluorimetry

As explained in Section 16.2.4, fluorescence takes place in competition with radiationless deactivation. Fluorescent emission is normally spherically isotropic and of longer wavelength than the exciting radiation. It is extinguished by heavy atoms in the molecule, by the excitation of free rotation about chemical bonds and torsional vibrations, by high concentrations because of increased collision probabilities, or by the presence of paramagnetic oxygen in the sample solution. This can affect reproducibility in quantitative measurements [28], [84], [127].

The observed fluorescence intensity I_α^F depends not only on the absorbing power of the molecule at the excitation wavelength $\varepsilon(\lambda)$, but also on the observed wavelength α within the fluorescence spectrum, depending on its spectral distribution f_α. Measuring fluorescence does not involve any reference measurement, so any fluctuations in the lamps or the electronics, especially long-term drifts, distort the measured signal. An advantage of fluorescence measurement is that by increasing the intensity of the excitation light source, trace analysis can be carried out even with samples that fluoresce only slightly [28], [84], [127]. However, high intensities of the excitation light source can also cause photochemical decomposition reactions in the sample. Even laser dyes have such poor photo stability that two fluorescence spectra of these dyes observed consecutively can give different results [128].

16.4.3.1. Inner Filter Effects

At high analyte concentrations, inner filter effects can occur. These lower fluorescence intensity by self-absorption if the observed wavelength is incorrectly chosen (within the absorption band). To prevent this quenching due to high concentrations, measurements are taken at the lowest possible concentrations (usually $< 10^{-7}$ mol/L). A rule of thumb is that absorbance at the excitation wavelength should be < 0.02 to prevent errors $> 2.5\,\%$. Under these conditions, fluorescence intensity I_α^F is proportional to the concentration c of the fluorophor [3]:

$$I_\alpha^F = \Gamma I_0(\lambda)\eta f_\alpha c\varepsilon(\lambda)\frac{1-10^{-E(\lambda)}}{E(\lambda)} \tag{17}$$

Here, η is the fluorescence quantum yield, $E(\lambda)$ the absorbance, and Γ a geometric factor. In practice (cuboid-shaped cell), this factor corrects for non-isotropic intensity distribution. The factor $(1-10^{-E(\lambda)})/E(\lambda)$ describes the finite absorption and becomes E^{-1}, if $E \ll 1$ [3], [28]. A linear relationship exists between observed fluorescence intensity and concentration [3], [28].

16.4.3.2. Fluorescene and Scattering

If fluorescence is measured in the direction of excitation radiation and analyte concentration is low, much unattenuated excitation radiation falls on the detector. Therefore, measurements are usually made at 90° to the path of the excitation beam. Despite this, the fluorescence signal is usually mixed with scattered light, especially if the solution is cloudy and the substances show only weak fluorescence.

Particularly in trace analysis, artifacts are recorded that occur because of the so-called Raman bands of the solvent. These result in incoherent scattering by the solvent at high excitation intensities (→ Infrared and Raman Spectroscopy). To avoid this, in good-quality equipment, excitation radiation is produced with the aid of a monochromator rather than a filter, so that these additional Raman lines can be cut off by the second (observation) monochromator. In fluorescence measurements, the spectrum of pure solvent should always be recorded as a "blank" before the sample is investigated.

16.4.3.3. Excitation Spectra

Whereas absorption spectra describe the relative position of the vibrational level of the first excited state, fluorescence spectra give the position of the vibrational level of the ground state.

If the observation monochromator is kept at the wavelength of the fluorescence maximum and the wavelength of the excitation monochromator is varied, a so-called excitation spectrum is obtained that resembles the absorption spectrum in its spectral distribution. If a mixture of substances contains only one component capable of fluorescence, a fluorescence excitation spectrum can be used to obtain spectral information about this specific component [28], [33].

16.4.3.4. Applications

Fluorometry is used to determine elements such as boron, silicon, aluminum, beryllium, and zirconium [7], [129], [130], as well as organic compounds [131] (e.g., vitamins [132]). An important application in the field of trace analysis is the determination of aromatic hydrocarbons in wastewater [133]. Here, the limits of detection can be improved considerably [64] if fluorescence is excited by a laser (LIF). This is very important for the measurement of polycyclic aromatic hydrocarbons [134].

Other applications, especially those involving the use of sensors, have been reviewed recently [135], [136].

16.4.4. Reflectometry

16.4.4.1. Diffuse Reflection

Depending on the surface of the substrate, incident electromagnetic radiation can be reflected either diffusely or regularly (see Fig. 5). If the surface is rough or consists of many small crystallites, diffuse reflection is observed. For a pulverized sample with a very large layer thickness (so that no appreciable reflection of substrate material on which the powder is deposited is observable), the reflectance R_∞ can be related to both the scattering coefficient s and the absorption coefficient k. The reflection behavior of a powdered sample can be described by the so-called Kubelka–Munk function [90], [137]:

$$F(R_\infty) = \frac{k}{s} = \frac{(1 - R_\infty)^2}{2R_\infty} \tag{18}$$

Diffuse reflection has a wide range of uses, e.g., in research, where it is used in the observation of molecules adsorbed on a surface [138] (especially in catalysis), and also in analysis (e.g., thin-film chromatography [89], [139]). To "capture" as much of the diffusely reflected radiation as possible, an Ulbricht sphere is often used. This type of arrangement, in conjunction with optical fibers, is currently used in production lines in the automobile industry for quality control of bodywork paint [140].

16.4.4.2. Color Measurement

An important use of reflection spectroscopy is in color measurement [141]. Visual comparisons are replaced by methods based on spectral quantities and the C.I.E. standard colorimetric system (DIN 5033). These techniques are very important in the paint industry [142]. Subtractive and additive mixing of several components can be represented in the color space by means of vectors. Proportions of standard color values based on standard spectral curves, together with the physiological color stimulus specification, give a color stimulus corresponding to the color equation [143]. This method of measuring paint colors is very important, especially in the automobile manufacturing industry.

16.4.4.3. Regular Reflection

Depending on the change in refractive index at an interface, internal and external reflection can occur. In *external reflection* (transition to a medium of higher refractive index, i.e., into an optically denser medium), reflection occurs at all angles of incidence. The reflection can be calculated from Fresnel's laws (see Eq. 3).

In *internal reflection*, at angles of incidence larger than the critical angle, electromagnetic radiation is totally reflected (attenuated total reflectance, ATR, see Section 16.2.2.4 and Fig. 5). This special case is very important in analysis for two approaches. First, simple transportation of radiation within the fiber (or a waveguide). Second, in total reflection, an evanescent field appears in which the electrical field vector decays exponentially in the optically less dense medium. Every change within the medium with lower refractive index influences the field vector coupled to the field in the optically denser medium. Therefore, the totally reflected radiation contains information about effects on the other side of the phase boundary (the medium with lower refractive index) [20], [144]. Various principles to interogate this effect are known and used in evanescent field sensors.

If ATR experiments are combined with dichroitic measurements, additional information on orientation and lateral spacing can be obtained. This is especially important if molecules are oriented perpendicularly or parallel at interfaces owing to self-organizing effects [145].

Both approaches of internal reflection (transportation of radiation, application of evanescent fields) are discussed in the following.

16.4.4.4. Determination of Film Thickness

If reflection occurs at not just one interface but at least two adjacent boundaries, the regularly reflected beams can superimpose and interfere. Interference occurs in certain preferred directions, at certain wavelengths, and with certain substrate and superstrate materials (see Fig. 33 A). If the partial beams extinguish each other, it is called destructive interference; if they reinforce each other, constructive interference [11]. If white light is used, well-resolved interference spectra are obtained with film thicknesses of 1–5 µm [146]. Bifurcated fibers are used for incident radiation and its collection after reflection at the phase boundaries.

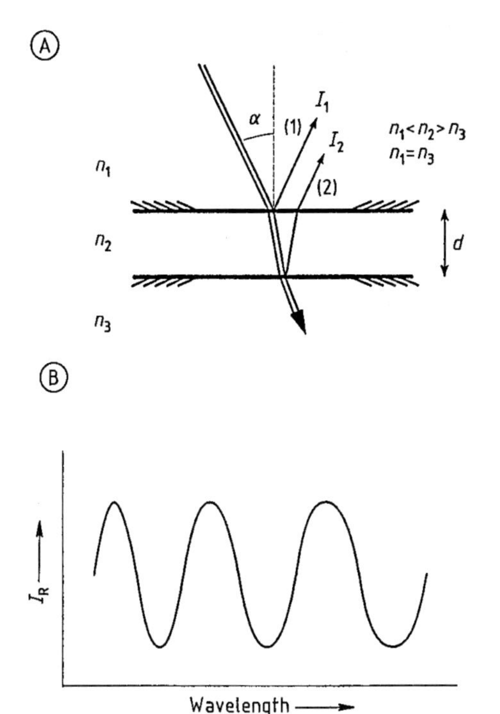

Figure 33. A) Reflected beams at a multilayer system (upper diagram) in which two beams (1) and (2) interfere constructively or destructively; B) Resulting interference spectrum (lower diagram)

An interference pattern of this type is shown in Figure 33 B; the modulation is given by a $\cos(1/\lambda)$ function. Reflection at the phase boundaries leads to reflected intensities I_1 and I_2 which are of the order of a few percent of the total intensity. According to wave theory, these intensities superimpose as follows [11]:

$$I_R = I_1 + I_2 + 2\sqrt{I_1 I_2}\cos\left(2\pi\frac{2n_2 d}{\lambda} + \delta\right) \tag{19}$$

where δ corrects for a phase shift on reflection at the optically denser medium; n_2 is the refractive index of the film; and d is the thickness of the film at whose phase boundaries the reflections take place. According to Equation (19), maxima in the interference spectrum occur where $m = 0, 1, 2, \ldots$

$$2\pi\frac{2n_2 d}{\lambda} = 2\pi m \tag{20}$$

Whereas absolute film thicknesses can be determined only by complete curve fitting of the many distances between maxima of the calculated dispersion curve [124], relative film thickness changes can be determined simply and accurately. According to [11] and [147], from the equation ($\Delta m = 1$)

$$2n_2 d = \frac{\lambda_1 \lambda_2}{\lambda_1 - \lambda_2} \tag{21}$$

the optical film thicknesses can be determined from the spacing of the extrema at the two wavelengths. Minimal changes in refractive index lead to a shift of extrema in the interference pattern.

Improvement of the signal-to-noise ratio and refinement of the evaluation algorithm [148] enable film thickness changes of ca. 1 pm to be observed. The method is dynamic and has response times of < 1 s if diode arrays are used to obtain interference spectra [149]. The method is called reflectometric interference spectroscopy (RIfS) [147]. Similar procedures are also followed in the IR region [116].

16.4.4.5. Ellipsometry

The method described in Section 16.3.7.4 for performing ellipsometric measurements using polarized light (see Fig. 23) [150] is very important in microchip production [151]. This method is also used in the characterization of alloys and the investigation of biological films or membranes [152].

The disadvantage of ellipsometry is undoubtedly that it does not represent a direct analytical method, but attempts to match functional relationships as well as possible to the experimental data by the development of models [93]. On the other hand, ellipsometry is one of the few nondestructive analytical methods that allows measurement of the monolayer or submonolayer. However, even when a "microspot" of measuring light is used, it represents several square millimeters in area. In this surface area, the product nd (optical film thickness) is averaged laterally.

Compared to analyses carried out with monochromatic light at a fixed angle, the result is improved by measurement either as a function of angle or as a function of wavelength at a fixed angle [94]. Samples with weak absorbing properties cause problems [153]. Complex refractive indices must then be used for calculations.

However, in contrast to simple reflection measurements or by surface plasmon spectroscopy at one angle or wavelength [153], spectral meas-

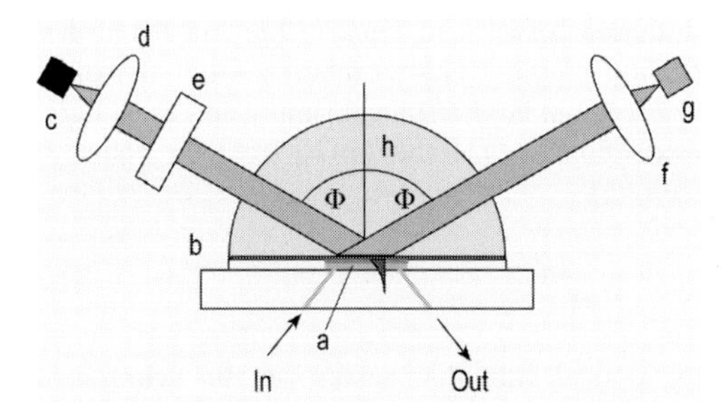

Figure 34. Diagram of system for measuring spectral surface plasmon resonance with a flow cell and resonance point
a) Flow cell; b) Silver layer; c) Light source; d) Collimator lens; e) Polarizer; f) Lens; g) Optical fiber (to the diode array); h) Glass cylinder

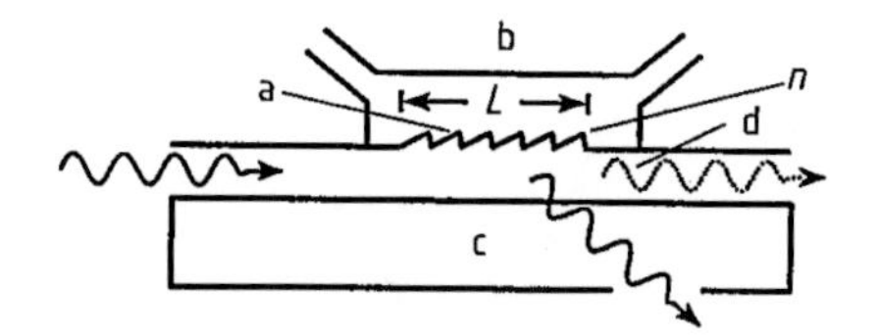

Figure 35. Principle of grating coupler [160]
a) Grating; b) Flow cell; c) Carrier; d) Waveguide

urements with ellipsometry enable n and d to be determined independently.

16.4.5. Resonance Methods

16.4.5.1. Surface Plasmon Resonance

Surface plasmon resonance (SPR) is a typical method to interrogate evanescent field effects. The measuring system consists of a thin metal film produced by vacuum deposition onto a glass prism (or half cylinder). Depending on wavelength, angle of incidence, and refractive index at the phase boundary, resonance conditions are produced if electromagnetic radiation leaving the prism excites plasmons in the metal to coupled vibrations [153] – [156].

Generally, monochromatic radiation is used and the angle of incidence is varied. Resonance leads to attenuation of the reflected beam because some of its energy is used for plasmon excitation. The second possibility, spectral excitation by white light at a preset angle, is also of interest. Here, the resonance wavelength is detected by a diode array [94]. An apparatus of this type, with a flow cell in the path of the signal, is shown in Figure 34. Gold, silver, and occasionally copper are used to form the films.

SPR is used to characterize biological membranes and Langmuir – Blodget films, as well as a sensor in bioanalysis [157], [158]. SPR application to sensors is reviewed in [159].

16.4.5.2. Grating Couplers

Another resonance technique involves the use of grating couplers, introduced by LUKOSZ [160], [161]. These can be obtained from the ASI company in Zürich. Grating patterns are imprinted on thin films of highly refractive material that is supported by a carrier material. Depending on the refractive index ratios at the grating, a resonance condition is produced for the film waveguide under the grating, leading to coupling or decoupling of the interfering beams at the grooves of the grating. The principle is shown in Figure 35. Both modes of electromagnetic radiation (transversal electric, TE, and transversal magnetic, TM) can be used, so that a difference is produced [162]. A large number of uses have been published in the field of gas sensors and for rapid analysis with biosensors [163].

16.4.6. On-Line Process Control

16.4.6.1. Process Analysis

By use of flow cells, normal spectrometers can be employed for routine measurement. Since recording a spectrum by sequential spectrometers required a certain amount of time, the absorption was usually observed continuously at one wavelength. The development of diode arrays has enabled this system of measurement using photometers to be superseded. Also, these new types of measuring equipment use optical fibers, enabling measuring cells to be installed at a considerable distance from the spectrometer, so that continuous measurements are possible in dangerous or explosion-protected spaces. Equipment designs such as those shown in Figure 15 are very suitable for such tasks, enabling absorption, reflection, and fluorescence measurements to be carried out with the aid of glass fibers.

By using optical fibers, biotechnological processes in fermentors can be monitored continuously with spectral measurements. In simple cases, transmission measurements are performed [164]; or for monitoring reactions, scattering phenomena are employed. These techniques are also used in wastewater analysis by continuous measurement of turbidity. If multicomponent systems are to be observed (e.g., in the analysis of foods), methods using the near IR are often more suitable [125].

Many analytical problems can be solved by the use of flow cells. If these are combined with flow injection analysis equipment [165], the limits of detection in water analysis can be improved [166] or the scope of biochemical analysis increased [147].

16.4.6.2. Measurement of Film Thicknesses

When beams of radiation reflected from the two surfaces of a film superimpose, constructive or destructive interference occurs when visible light is used if the film thickness is in the region of a few micrometers, since one partial beam has a longer path length to travel than the other. The resulting phase difference depends on the physical distance between film interfaces and the refractive index of the film (see also Fig. 33). From the interference pattern, the so-called optical film thickness $n\,d$ can be calculated. Use of the diode arrays previously described, in combination with optical fibers, enables rapid measurements of optical film thicknesses to be performed even on films produced at high speed in coating machines. The possibility of monitoring coating thickness at various parts of the strip by means of glass fibers, and of reacting to production problems within a few seconds, makes this technique extremely useful in process control.

16.4.6.3. Optical Sensors

For a general definition of sensors, see → Chemical and Biochemical Sensors. The term sensor denotes a small specialized device that operates selectively on several analytes [167]. Optical sensors are certainly competitive with ion-selective electrodes because they are now less complex than in the past and miniaturization promises further success in the future. Furthermore, optical sensors have advantages [168] in remote sensing and by utilization of the spectral information, especially in the case of sensor arrays.

Along with simple optical sensors that measure transmission, so-called optodes [169] play an important role. In these, either the color change of an indicator dye in a membrane is monitored by using diffuse reflection [170], or the effect of an analyte on the fluorescence spectrum is observed. In addition, many other effects are used, such as surface plasmon resonance (see Section 16.4.5.1), and grating couplers (see Section 16.4.5.2). In most cases, interaction of the analyte with the evanescence field of a glass fiber, a slab or strip waveguide plays the important part, influencing the wave propagation of the coupled electromagnetic field by changing the refractive index of the superstrate [171]. Furthermore, determination of the thickness of thin films by interferometric methods (see Section 16.4.4.4) can be used for analysis of both gases [146], [172], [173] and liquids [147], [174]. Here, the amount of swelling of polymer film due to the permeation by the analyte is determined. Accumulations of material caused by antigen – antibody interactions [147] can also be observed by methods sensitive to changes in layer thickness. Label-free detection of biomolecular interaction by optical sensors has been reviewed recently [175].

Direct optical detection of DNA hybridisation by SPR [176] or RIfS [177] has been achieved. In routine sensing, assays with labeled compounds are preferred using different fluorescent techniques [159].

Optical sensors have become very important, are used more and more widely, and have been reviewed comprehensively in several monographs [178].

16.4.7. Measuring Methods Based on Deviations from the Lambert – Beer Law

Effects that lead to deviations from the Lambert – Beer law are given below. On the left-hand side, the factors that influence the spectrum are listed, on the right the chemical and physical aspects are given:

Changes to the spectrum due to	Caused by
Concentration	self-association, dissociation
pH change	acids/bases proteolytic reaction, equilibrium
Solvent	interaction between solvent and solute
Presence of additional component B	complex formation
Time	thermal chemical reaction (including solvolysis)
Incident light	photochemical reaction, photochromism
Temperature	thermochromism

The most important of these are the temperature dependence of the molar absorption coefficient (due to density change); the effect of association

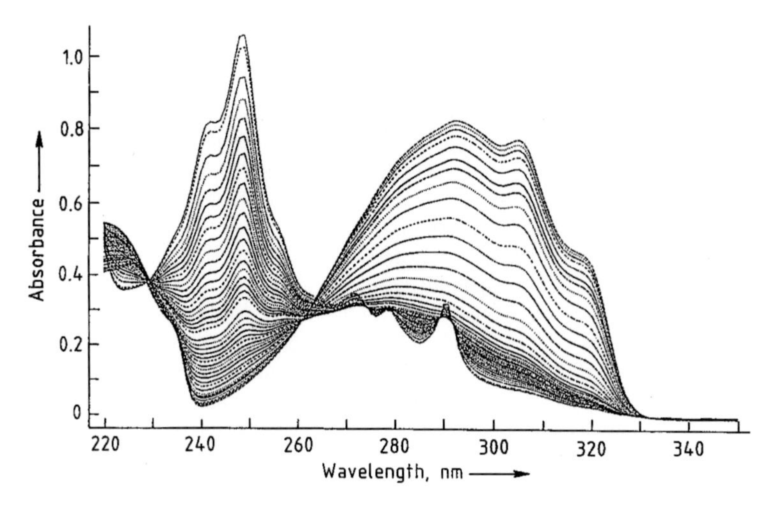

Figure 36. Reaction spectrum of the photoreaction of *trans*-stilbene [183]

and dissociation equilibria, which change with temperature, concentration, and pH; and, not least, concentration changes caused by chemical and photochemical reactions.

Like typical titrations in which electrodes are used, photometric titrations can be carried out by using indicator dyes. Here, the pH is plotted not against the concentration of anions or dissociated components, but against the absorbance at one or more wavelengths [179]. From these diagrams, p*K* values of even polyvalent acids and bases or of complex equilibria can be determined. Furthermore, it is possible to linearize the diagram, which allows better control of the model of the number of steps of titration [180], [181].

UV–VIS spectroscopy has also found wide use in monitoring thermal and photochemical reactions. Absorbance is measured at several wavelengths during the reaction, and dynamic multicomponent analysis is carried out by using the Lambert–Beer law (see Section 16.4.2). In the simplest case, wavelengths can be found at which just one component absorbs. The absorbance signal is then directly proportional to the concentration of one reactant and can be evaluated by the usual kinetic equations [182], [183]. An example is given in Figure 24, in which case the thermal reaction has also been used to check the reproducibility of the wavelength of a spectrometer.

The interpretation of photochemical reactions is more complicated if they have to be followed spectroscopically. An example is given in Figure 36, which illustrates the photochemical isomerization of *trans*-stilbene to *cis*-stilbene followed by ring closure to form the end product phenanthrene via dihydrophenanthrene, an intermediate that is not detectable by conventional methods. The diagram shows a photochemical reaction spectrum. The measured data can be evaluated by using various methods [184]. When investigating such reactions, a process-controlled system is usually employed in which the irradiation equipment and the spectrometer are combined [185].

16.5. Special Methods

16.5.1. Derivative Spectroscopy

If absorption bands of two compounds overlap (one broad and the other narrow), derivative spectroscopy can be used for resolution. Herewith even trace amounts can be analyzed quantitatively [186]. Derivative spectroscopy is also used to investigate turbid solutions such as industrial wastewaters or biological samples. This method is also suitable for the analysis of shoulder-shaped spectra and overlapping absorption bands. The derivative of transmission with respect to wavelength is not proportional to concentration, but the derivative of absorbance with respect to wavelength is. Generally, the first and second derivatives are used [115]. Higher derivatives present noise problems, as the amount of noise increases with each derivative. Higher derivatives have been claimed to be useful for special applications [187].

Table 3. Various methods of obtaining derivatives, with their advantages and disadvantages

Method	Details of method	Advantages	Disadvantages
Optical	frequency modulation of the light beam by oscillating gratings, slits, or mirrors determination of Fourier coefficients equipment with constant wavelength distance	no effect on the spectrum of variations in the intensity of the light source very good signal to noise ratio	higher derivatives not obtainable special optical construction necessary
Electronic	derivative obtained from the analog signal by differentiation and attenuation circuits	direct on-line spectra higher derivatives possible	derivative as a function of time and not of wavelength
Computed	the digitalized data are stored and the derivative is calculated	direct derivative as a function of wavelength original spectrum is not lost; use of various algorithms optimization of the selected parameters	necessity for an integrated computer or data station

Table 4. Various mathematical algorithms for smoothing curves and producing their derivatives, with their advantages and disadvantages

Numerical algorithm	Advantages	Disadvantages
Differential coefficient $\Delta E/\Delta\lambda$	simple, quick	signal-to-noise ratio is made worse by subtraction of two measured values that are subject to error
Polynomial smoothing	derivative obtained directly from polynomial coefficients number of restart points determines the degree of smoothing	loss of measured points at the beginning and end of the spectrum problems with sharp peaks
Cubic spline	treatment of the entire spectrum in one step good linear dependence on concentration	large influence of the parameters

Optical, electronic, and mathematical methods are available [3], [188], whose advantages and disadvantages are listed in Table 3. The algorithms generally used for derivation are listed in Table 4, along with their advantages and disadvantages. For three overlapping Gaussian peaks, the normal spectra and the first, second, third, and seventh derivatives are shown schematically in Figure 37. For both the first and the second derivative, by forming the difference fora neighboring minimum and maximum, a linear relationship can be found between these values and concentration. On differentiation, the peaks become narrower and the slopes steeper. Thus, in the second derivative, a steep slope occurs in the calibration curve, and the extreme values are further apart. The derivative signals are less marked if the distance between supporting points is com-parable to, or wider than, the width at half peak value.

However, Figure 37 shows not only that the number of peaks increases with higher derivatives, but that when several bands overlap, the number of minima and maxima increases so that the shape of the curve becomes unclear. If real, noisy, raw data are used instead of simulated data, interpretation with the aid of derivatives becomes much more critical (see Fig. 37 B). If too much smoothing is used, information is lost. If too little smoothing is employed, significantly higher derivatives are impossible.

Each differentiation step leads in principle to an increase in noise, so that no additional information can be obtained by this mathematical procedure, although in some cases this form of "spectral manipulation" offers a possible way of revealing information hidden in raw data. However, as implied by the word "manipulation," the final result is affected by the procedure used.

The potential and limitations of derivative spectroscopy are described in many literature references, which also include several examples of its use [7], [186], [189].

Derivative spectroscopy can be applied to fluorescence and fluorescence excitation spectra [188]. It is also used in chromatography in the search for peak maxima, and in interferometric measurements for determining minima and maxima in interference spectra [124]. A typical application is the determination of phenol in water (Fig. 38) [190].

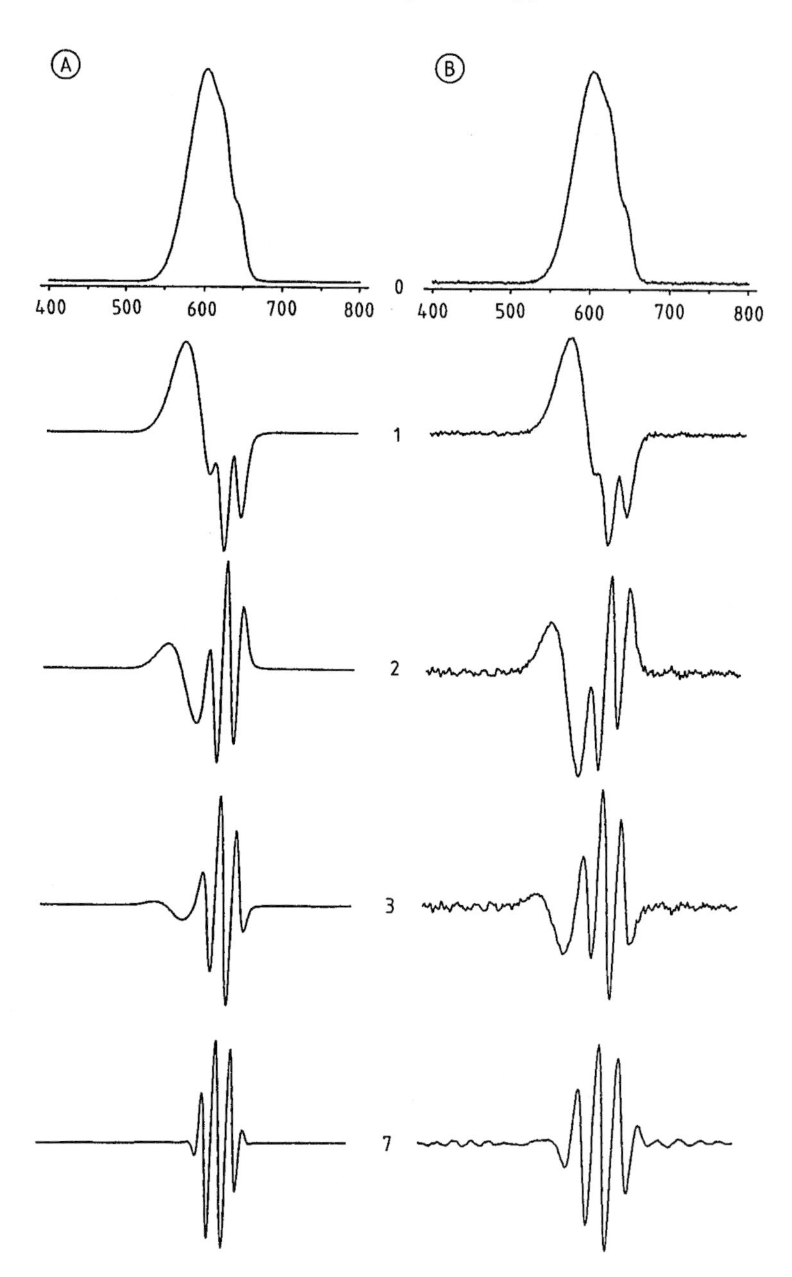

Figure 37. Schematic diagram of various derivatives of three overlapping Gaussian peaks
A) Simulated spectra; B) Spectra with 0.3 % errors superimposed

16.5.2. Dual-Wavelength Spectroscopy

Another possible method of investigating biological samples that cause light scattering is by dual-wavelength spectroscopy [191], [192], although this method is limited to very special applications. The number of types of equipment used is therefore also limited. First, normal spectrometers can be used, the signals for two wavelengths being taken from the observed spectra. Second, so-called bifrequency equipment can be used, which operates with two beams at different wavelength at the same time. This is possible either with two monochromators, which cause light of

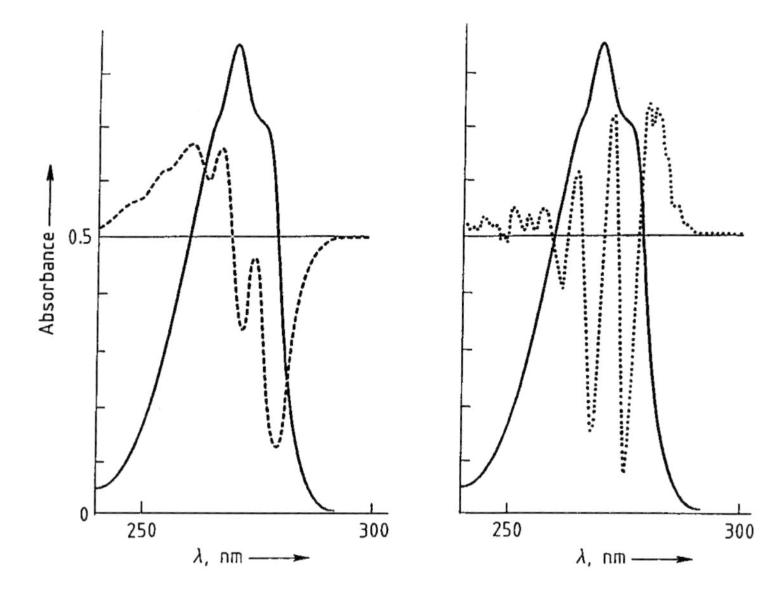

Figure 38. Absorption spectrum of phenol in water (—), showing first (– – –) and second (· · · ·) derivatives of the spectrum [190]

two wavelengths to fall on the sample simultaneously, or with one monochromator and a modulated slit, which provide light of two wavelengths that alternate continuously.

This method has the advantage of avoiding problems from differences in cell positions, differences in the cells themselves due to spurious absorptions or impurities inside or outside the cell, scattering effects, or turbidity. Two techniques are used, first the so-called *equiabsorption method* in which two wavelengths are chosen for which one of the components has the same absorption. The second component must absorb at only one of the two wavelengths. This enables sodium nitrite to be determined, for example, in the presence of sodium nitrate (see also Fig. 39).

The *signal amplification method* is of general application. It is used either if no region occurs in which only one component absorbs or if the interfering component has no absorption maximum. With this type of equipment, the absorption can be corrected by a factor at one wavelength to give the same value as that obtained at the other wavelength (the uncharacteristic spectrum is "evened out"). This enables the concentration of the desired component to be determined from the absorbance difference at the two wavelengths if its absorbance coefficients at the two measuring wavelengths are known. This method has been used in medicine and for biological samples [192]. However, it has recently decreased in importance.

16.5.3. Scattering

As described in Section 16.2.2.3, scattering of aqueous solutions increases considerably when larger suspended particles such as soot, mist droplets, fat droplets, or bacteria are present. In analogy to the Lambert–Beer law, attenuation of the incident light by a scattering medium is described in the ideal case by

$$I = I_0 e^{-\tau l} \quad (22)$$

where I is the intensity after passage through the medium, I_0 is the incident intensity, l is the optical path length, and τ is the so-called turbidity coefficient. This equation is valid only for scattering particles whose diameter is small compared to the wavelength of the light. The intensity of scattered light depends on the number of scattering particles per unit of volume, the angle at which the scattered light is observed, the size and shape of the scattering particles, and the wavelength of the light [18], [193]. These various relationships can be utilized in analysis [194].

16.5.3.1. Turbidimetry

Measurement of the attenuation of an incident light beam on passing through a scattering medium is known as turbidimetry and can be carried out with simple colorimeters or photometers.

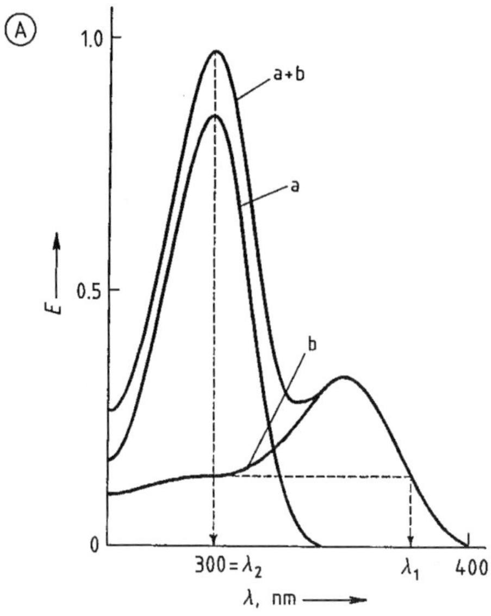

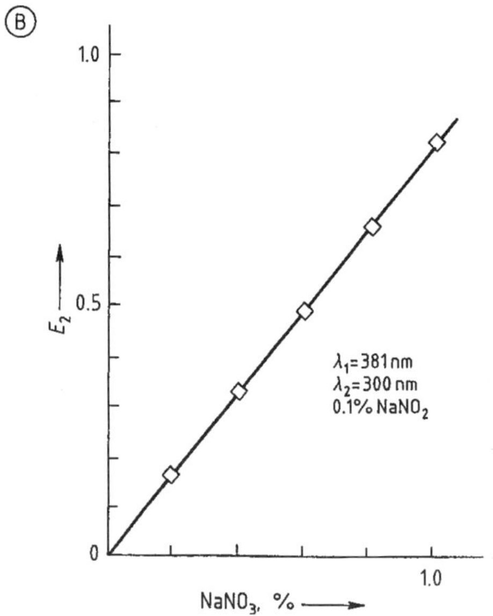

Figure 39. Determination of sodium nitrate and sodium nitrite in the presence of each other [192]
A) Spectra of 1 % $NaNO_3$ (a) and 0.1 % $NaNO_2$ (b);
B) Calibration curve

However, spectrometers in which the wavelengths of the incident radiation can be chosen are used for more exact determinations. These are preferable for measuring either relatively low or very high turbidity. Measuring errors can become very large in both cases.

16.5.3.2. Nephelometry

Nephelometry is used if turbidity is slight. Here, the intensity of scattered radiation is measured, not the difference between incident and attenuated radiation. Under these conditions, errors due to absorption are less likely to occur in nephelometry. The intensity of light deflected to the side is determined; this can be carried out by using fluorometers. Since measurement at an angle of 90° to the incident light direction does not always give the best results (even with particles having a diameter of 1 μm, ca. 90 % of the scattered light is in a cone at an angle of only 10° to the incident light beam), the equipment must include a goniometer so that measurements can be carried out at various angles [195], [196].

Synthetic turbidity standards are generally used in the quantitative measurement of turbidity. Alternatively, calibration curves can be obtained by precipitating known amounts of the substance to be measured. However, grain size depends not only on precipitation conditions, but also on the time delay between precipitation and measurement, so conditions must be standardized carefully to ensure reproducible results. Nephelometric and turbidimetric methods have also become important in recent years in continuous automatic on-line process control (e.g., of wastewater).

16.5.3.3. Photon Correlation Spectroscopy

Turbidimetry and nephelometry give no information on particle shape. However, if the principle of nephelometry is combined with time-dependent measurement, photon correlation spectroscopy can provide information on the shape of the scattering particle. This dynamic light scattering has become very important in colloid chemistry [197], [198].

16.5.4. Luminescence, Excitation, and Depolarization Spectroscopy, and Measurement of Lifetimes

If the molecules and their direction of polarization are not distributed statistically in the sample (anisotropy), a preferred spatial direction for the fluorescence exists. If polarized light is used for excitation [199], only those molecules whose axis of polarization is parallel to the axis of polarization of incident light are excited substantially. This results in a preferred direction of polarization for the fluorescence radiation. However, if molecules

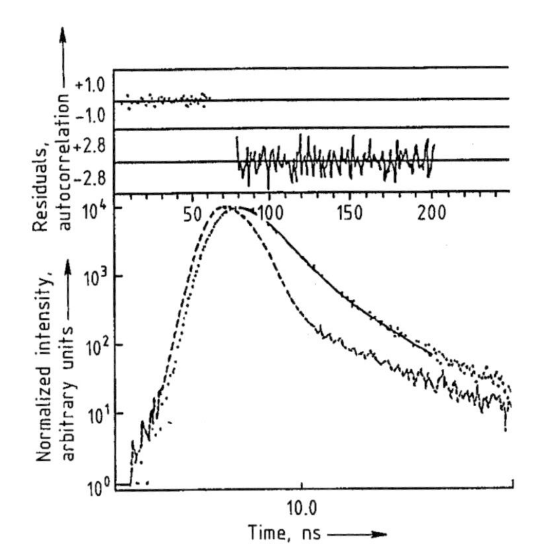

Figure 40. Logarithmic plot of measured intensity of fluorescence against time in a single photon counting experiment
Top diagram shows autocorrelation and residuals between fluorescence intensity decay curve (· · ·) and excitation light pulse (– – –) [138]

can move or rotate freely within the lifetime of the excited state, this preferred direction of fluorescence polarization is disturbed (depolarization effects) [200]. These effects enable statements to be made about the environment of the molecules. The degree of polarization is given by

$$p = \frac{I_{\parallel} - I_{\perp}}{I_{\parallel} + I_{\perp}} \quad p = \frac{3\cos^2\alpha - 1}{\cos^2\alpha + 3} \tag{23}$$

where α is the angle between the transition moments during absorption and emission processes.

Since the lifetime of the excited state is also characteristic of individual substances, and particularly of the environment of the molecule, time-dependent fluorescence measurements with polarized light, with variation of the excitation and polarization wavelength, can be performed. They give important information about the orientation and dynamics of the sample. Many of these methods are successfully applied in biosensing and bioanalytics [159].

The method of *single photon counting* [201], [202] is generally used for such lifetime determinations. A weak pulsed light source causes excitation of a small number of molecules. Spread over a period of time (spontaneous fluorescence is a statistical process), single fluorescence photons are emitted and are "counted" only in a small time window which proceeds between pulses by a ramp technique. By following thousands of pulses, the statistically emitted photons are counted in the channels (time windows), giving the decay curve shown in Figure 40. The time constants are determined by curve fitting.

If several deactivation processes occur at the same time, an incorrectly chosen number of overlapping exponential functions for fitting can lead to physical misinterpretation. Figure 40 shows the intensity of the excitation light source, the decay of fluorescence intensity, and curve fitting. If application of the residuals leads to systematic deviation from the mean zero value, this can indicate a poor curve fit or an incorrectly chosen number of exponential curves.

16.5.5. Polarimetry

16.5.5.1. Sugar Analysis

Saccharimetry is very important in the sugar industry. An aqueous solution of sugar is analyzed by determining the optical rotation α, using the green mercury line (546 nm) or yellow sodium line (589 nm). However, the effects of evaporation, the unstable nature of sugar solutions, and temperature effects can easily lead to errors in the determination of optical rotation.

In sugar manufacture, automatic quality control is important, and in yeast production the molasses stream must be determined to have optimum sugar content. Here, too, the optical rotation of molasses is measured automatically in a continuous flow cell [38].

16.5.5.2. Cellulose Determination

Whereas conventional gravimetric cellulose determination is complicated and time-consuming, cellulose concentrations can be determined rapidly in ammoniacal copper solution by measuring the optical rotation of the levorotatory solution. However, the solutions must be completely free of turbidity [138].

16.5.5.3. Stereochemical Structural Analysis

Optical rotatory dispersion can be used as a routine procedure, especially for steroids, since the asymmetry of spatial orientation of a molecule is also affected by its conformation. Cholestan-

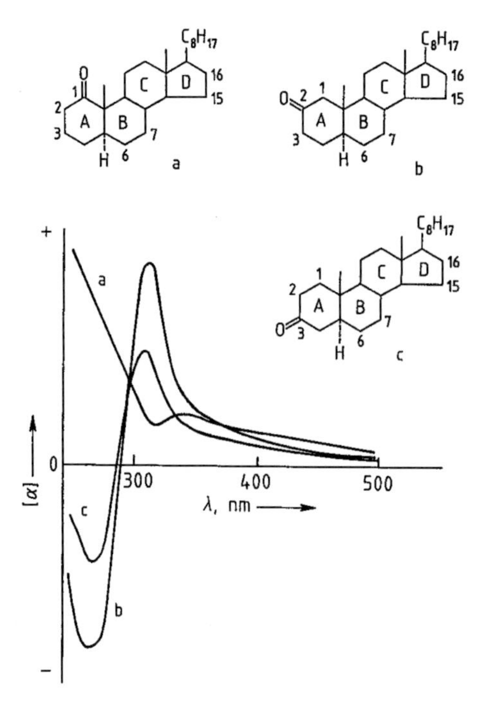

Figure 41. ORD curves of cholestan-1-one (a), cholestan-2-one (b), and cholestan-3-one (c), all dissolved in methanol [38]

1-one, cholestan-2-one, and cholestan-3-one, for example, can be distinguished because of their different ORD curves (see Fig. 41). Molecular asymmetry (dissymmetry) is a prerequisite of optical activity. A dissymmetric disturbance of the electron system of the keto group caused by a change in the stereochemical arrangement is the case of other ketosteroids (such as androsterone). Here, ORD and CD curves often give information only on relative configuration.

Such ORD and CD curves are used for purity control of optically active substances in pharmaceutical chemistry.

16.5.5.4. Use of Optical Activity Induced by a Magnetic Field

In solid-state inorganic chemistry, MCD is an important technique for the determination of color centers in alkali-metal halide crystals, rare earths, and highly symmetric compounds such as hexahalides of platinum or ruthenium. In organic chemistry, MCD is observed with, e.g., metal porphyrins, pyrimidine bases, and nucleoside derivatives.

16.5.6. Photoacoustic Spectroscopy (PAS)

Photophysical processes can take place in combination with the excitation of molecules to more highly excited electronic states (see Section 16.2.4.2). In the subsequent radiationless transitions, the energy of excited molecules is taken up by the environment (neighboring molecules). Thermal equilibrium is thus reestablished and the temperature of the environment is increased. This is associated with a volume change, which is appreciable for a gas but less pronounced for liquids and solids.

If periodically modulated visible radiation is used, the periodic volume changes cause pressure variations and sound waves that can be detected (e.g., by a microphone). The photoacoustic or optoacoustic effect is based on this principle [203], [204]. Although this has been known for a long time, it has become more important since the late 1970s as an adjunct to reflection spectroscopy if samples are either very strongly or very weakly absorbing. The "heating" of the sample or its environment depends on the extent to which the modulated electromagnetic radiation can be absorbed. This corresponds to the spectral distribution of the molar absorption coefficients. If the wavelength of excitation radiation is varied, the amplitudes of pressure fluctuations are proportional to dynamic changes in the temperature of the system, and thus to the absorption spectrum [205].

Photoacoustic spectra can be observed with a simple microphone and optical apparatus. A diagram of a modern piece of equipment is given in Figure 42, in which there is a blackbody sample at the second input of a look-in amplifier. Corrected photoacoustic spectra can be recorded with this type of monochromator equipment. Frequency-modulated xenon lamps are used as white light sources, so that a chopper is not needed for modulation.

Photoacoustic spectroscopy is used with gases [206] – [208], liquids [209], and solid samples [210] – [212]. Simple solutions of the thermal diffusion equation can be used only under certain conditions. Areas of use [213] include investigation of waste gases (see Fig. 43), absolute measurements of fluorescence quantum yields [214], determination of dye concentrations in films, investigation of thin films in catalysis [210], [215], and biological and physiological problems such as the investigation of leaves and samples of skin or

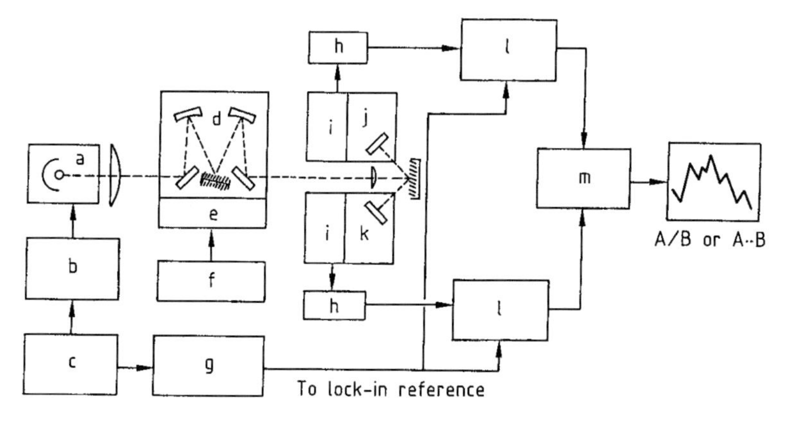

Figure 42. Block diagram of commercial photoacoustic equipment
a) 300-W Xenon lamp; b) Lamp power supply; c) Sine-wave modulator; d) Monochromator (two gratings); e) Wavelength drive stepper motor; f) Monochromator controls; g) Modulation frequency display; h) Preamplifier; i) Microphone; j) Cell A; k) Cell B; l) Lock-in amplifier; m) Ratiometer/difference amplifier

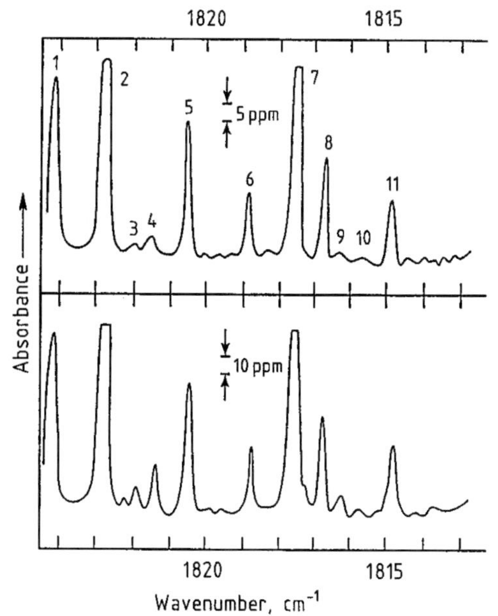

Figure 43. Absorption spectra of gas samples
Upper diagram shows calibration spectra for 20 ppm NO in N_2. Lines 1, 5, 6, 8, and 11 are due to NO, and other lines to water vapor; lower diagram shows spectrum of automobile exhaust gas

blood. Despite the wide area of application, use of this equipment is very limited so few types of equipment are available on the market. However, individual instances of the application of PAS in sensor technology have been described [216].

16.6. References

[1] J. D. Ingle, Jr., S. R. Crouch: *Analytical Spectroscopy*, Prentice-Hall, Englewood Cliffs, NJ, 1988.

[2] G. Svehla (ed.): "Analytical Visible and Ultraviolet Spectrometry," in: *Comprehensive Analytical Chemistry*, vol. XIX, Elsevier, Amsterdam 1986.

[3] G. Gauglitz: *Praktische Spektroskopie*, Attempto Verlag, Tübingen 1983.

[4] D. A. Skoog, J. J. Leary: *Principles of Instrumental Analysis*, Saunders College Publishing, Fort Worth 1992.

[5] G. Gauglitz, K.-A. Kovar in E. Nürnberg, P. Surmann (eds.): *Hagers Handbuch der pharmazeutischen Praxis*, 5th ed., **vol. 2**, Methoden, Springer Verlag, Berlin 1991, p. 157.

[6] D. L. Massart, B. G. M. Vandeginste, S. N. Demmig, Y. Michotte, L. Kaufman: Data Handling in Science and Technology 2: *Chemometrics: a Textbook*, Elsevier, Amsterdam 1988.

[7] H.-H. Perkampus: *UV-VIS Spectroscopy and its Applications*, Springer Verlag, Berlin 1992.

[8] H. H. Jaffé, M. Orchin: *Theory and Applications of Ultraviolet Spectroscopy*, Wiley, New York 1962.

[9] M. Klesinger, J. Michl: *Lichtabsorption und Photochemie organischer Moleküle*, VCH Verlagsgesellschaft, Weinheim 1989.

[10] P. W. Atkins: *Molecular Quantum Mechanics*, 2nd ed., Oxford University Press, Oxford 1983.

[11] L. Bergmann, C. Schaefer: *Lehrbuch der Experimentalphysik*, vol. 6: Optik, De Gruyter, Berlin 1987.

[12] M. Born, E. Wolf: *Principles of Optics*, Pergamon, New York 1980.

[13] E. Hecht, A. Zajac: *Optics*, Addison-Wesley, Reading 1974.

[14] P. W. Atkins: *Physikalische Chemie*, VCH Verlagsgesellschaft, Weinheim 1990.

[15] R. P. Feynman, R. B. Leighton, M. Sands: *The Feynman Lectures on Physics*, Bilingna Edition, R. Oldenbourg, München, Addison-Wesley, Reading 1977.

[16] W. Demtröder, "Laser Spectroscopy," Springer Series in Chemical Physics 5, Springer Verlag, Berlin 1981.
[17] H. A. Stuart: *Molekülstruktur,* Springer Verlag, Berlin 1967.
[18] H. C. van der Hulst: *Light Scattering by Small Particles,* Wiley, New York 1957.
[19] K. A. Stacey: *Light-Scattering in Physical Chemistry,* Butterworths, London 1956.
[20] N. J. Harrick: *Internal Reflection Spectroscopy,* Harrick Scientific Corporation, New York 1979.
[21] H. Weber, G. Herziger: *Laser,* Physik-Verlag, Weinheim 1972.
[22] G. Kortüm: *Kolorimetrie, Photometrie und Spektrometrie,* Springer Verlag, Berlin 1962.
[23] M. Krystek in W. Erb (ed.): *Leitfaden der Spektroradiometrie,* **INSTAND vol. 6,** Springer Verlag 1989, p. 321.
[24] H.-H. Perkampus: *Lexikon der Spektroskopie,* VCH Verlagsgesellschaft, Weinheim 1993.
[25] K. Laqua, W. H. Melhuish, M. Zander: "Molecular Absorption Spectroscopy, Ultraviolet and Visible (UV/VIS)," *Pure Appl. Chem.* **60** (1988) 1449–1460.
[26] V. A. Fassel et al., *Pure Appl. Chem.* **45** (1976) 105.
[27] J. Inczedy, T. Lengyel, N. Ure: *Compendium of Analytical Nomenclature,* Blackwell Science, Oxford 1998.
[28] C. A. Parker: *Photoluminescence of Solutions,* Elsevier, Amsterdam 1969.
[29] W. H. Melhuish, *Pure Appl. Chem.* **56** (1984) 231.
[30] H. Mauer, G. Gauglitz: "Photokinetics; Theoretical Fundamentals and Applications," in R. G. Compton, G. Hancock (eds.): *Comprehensive Chemical Kinetics,* **vol. 36,** Elsevier, Amsterdam 1998.
[31] H. A. Staab: *Einführung in die theoretische organische Chemie,* Verlag Chemie, Weinheim 1966.
[32] G. G. Guilbault: *Practical Fluorescence; Theory, Methods, and Techniques,* Marcel Dekker, New York 1973.
[33] Th. Förster: *Floureszenz organischer Verbindungen,* Vandenhoek, Göttingen 1951.
[34] J. Michl, E. W. Thulstrup: *Spectroscopy with Polarized Light,* VCH Verlagsgesellschaft, Weinheim 1986.
[35] H. Rau in H. Naumer, W. Heller (eds.): *Untersuchungsmethoden in der Chemie; Optische Aktivität und Polarimetrie,* Thieme-Verlag, Stuttgart 1990.
[36] C. Djerassi: *Optical Rotatory Dispersion Applications to Organic Chemistry,* McGraw Hill, New York 1960.
[37] G. Snatzke: *Optical Rotatory Dispersion and Circular Dichroism in Organic Chemistry,* Heyden & Son, London 1967.
[38] *Ullmann,* 4th ed., **5,** 628.
[39] F. Ciardelli, P. Salvadori (eds.): *Fundamental Aspects and Recent Developments in Optical Rotatory Dispersion and Circular Dichroism,* Heyden & Son, London 1973.
[40] J. Michl: "Magnetic Circular Dichroism of Cyclic π-Electron Systems," *J. Am. Chem. Soc.* **100** (1978) 6801, 6812, 6819.
[41] DIN Norm 5030 Part 3 (1983), DIN Norm 32 635, Beuth-Verlag, Berlin 1982.
[42] A. Reule in W. Erb (ed.): *Leitfaden der Spektroradiometrie,* INSTAND vol. 6, Springer Verlag, Berlin 1989, p. 179.
[43] Perkin-Elmer, company information.
[44] W. Kaye, D. Barber, R. Marasco: "Design of a Microcomputer-Controlled UV–VIS Spectrophotometer," *Anal. Chem.* **52** (1980) 437A.
[45] G. Gauglitz, T. Klink, A. Lorch, *Fresenius Z. Anal. Chem.* **319** (1984) 364.
[46] Application information by Varian, Darmstadt, Germany.
[47] P. M. Epperson, J. V. Sweedler, R. B. Bilhorn, G. R. Sims, M. B. Denton, *Anal. Chem.* **60** (1988) 327A.
[48] J. V. Sweedler, R. B. Bilhorn, P. M. Epperson, M. B. Denton, *Anal. Chem.* **60** (1988) 282A.
[49] D. G. Jones, *Anal. Chem.* **57** (1985) 1057A.
[50] J. A. Decker: "Hadamard Transform Spectrometry: a New Analytical Technique," *Anal. Chem.* **46** (1974) 1803.
[51] R. Geick, *Chem. Labor Betr.* **23** (1972) 193, 250, 300.
[52] L. Genzel, *Fresenius Z. Anal. Chem.* **273** (1975) 391.
[53] A. G. Marshall, F. R. Verdon: *Fourier Transform in NMR, Optical, and Mass Spectrometry,* Elsevier, New York 1990.
[54] P. R. Griffiths: *Chemical Fourier Transform Spectroscopy,* Wiley, New York 1975.
[55] O. S. Wolfbeis, G. Boisdé, G. Gauglitz in W. Göpel (ed.): *Sensors and Biosensors,* VCH Verlagsgesellschaft, Weinheim 1992, p. 575.
[56] J. Kiefer (ed.): *Ultraviolette Strahlen,* De Gruyter, Berlin 1977.
[57] L. Endres, H. Frietz in W. Erb (ed.): *Leitfaden der Spektroradiometrie;* Strahler, INSTAND vol. 6, Springer Verlag, Berlin 1989, p. 23.
[58] H. A. Strobel, W. R. Heinemann: *Chemical Instrumentation: a Systematic Approach,* Wiley & Sons, New York 1989.
[59] J. F. Rabeck: *Experimental Methods in Photochemistry and Photophysics,* Wiley, Chichester 1982.
[60] J. Wilson, J. F. B. Hawkes: *Lasers, Principles and Applications,* Prentice-Hall, Englewood Cliffs, NJ, 1987.
[61] D. Engelage: *Lichtwellenleiter in Energie und Automatisierungsanlagen,* Hüthig-Verlag, Heidelberg 1986.
[62] J. Pfab, *Anal. Proc. (London)* **28** (1991) 415.
[63] K. Niemax, *Fresenius J. Anal. Chem.* **337** (1990) 551.
[64] U. Panne, F. Lewitzka, R. Nießner, *Analusis* **20** (1992) 533.
[65] F. P. Schäfer in F. P. Schäfer (ed.): *Dye Lasers, Topics in Applied Physics,* **vol. 1,** Springer Verlag, Berlin 1973, pp. 1–90.
[66] F. K. Kneubühl, M. W. Sigrist: *Laser,* Teubner Studienbücher Physik, Stuttgart 1991, p. 331.
[67] J. Eichler, H.-J. Eichler: *Laser,* Springer Verlag, Berlin 19990, p. 173.
[68] H. A. Macleod: *Thin-film Optical Filters,* Adam Hilger Ltd., Bristol 1986.
[69] A. G. Reule: "Errors in Spectrometry and Calibration, Procedures to Avoid Them," *J. Res. Nat. Bur. Stand. Sect. A* **80** (1976) 609.
[70] Optimum Parameters for Spectrophotometry, Varian Instruments, OPT-720A.
[71] L. Velluz, M. Legrand, M. Grosjean: *Optical Circular Dichroism,* Verlag Chemie, Weinheim 1965.
[72] U. B. Seiffert, D. Janson: "Zur optischen Qualität von Photometerküvetten aus Plastik," *J. Clin. Chem. Clin. Biochem.* **19** (1981) 41.
[73] G. Gauglitz: "Opto-Chemical and Opto-Immuno Sensors," in H. Baltes, W. Göpel, J. Hesse (eds.): *Sensors Update,* **vol. 1,** VCH Verlagsgesellschaft, Weinheim 1996.
[74] W. Böhme, H. W. Müller, W. Liekmeier, *Int. Lab.* (1991) Jan/Feb.
[75] W. Böhme, K. Horn, D. Meissner, *LaborPraxis* **11** (1987) 628.

[76] E. L. Dereniak, D. G. Crowe: *Optical Radiation Detectors,* Wiley, New York 1984.
[77] K. Möstl in W. Erb (ed.): *Leitfaden der Spektroradiometrie,* INSTAND vol. 6, Springer Verlag, Heidelberg 1989, p. 101.
[78] F. Grum, C. J. Bartleson: *Optical Radiation Measurements,* vol. 4: Physical Detectors of Optical Radiation, Academic Press, New York 1983.
[79] Photomultiplier Handbook, Burle Ind., Lancaster 1989.
[80] C. Zeiss: Produktbeschreibung Simultanspektrometer, Oberkochen.
[81] Firma Reticon, München, data sheets.
[82] Y. Talmi: *Multichannel Image Detectors,* **vols. 1, 2,** Am. Chem. Soc., Washington, DC, 1979, 1982.
[83] J. V. Sweedler, K. L. Ratzlaff, M. B. Denton (eds.): *Charge-Transfer Devices in Spectroscopy,* VCH Verlagsgesellschaft, Weinheim 1994.
[84] E. L. Wehry (ed.): *Modern Fluorescence Spectroscopy,* Plenum, New York 1981.
[85] E. Lippert, W. Nägele, I. Seibold-Blankenstein, U. Steigert, W. Voss, *Fresenius Z. Anal. Chem.* **170** (1959) 1.
[86] Operation manual CD-spectrometer Jasco.
[87] R. Hamilton, S. Hamilton: *Thin Layer Chromatography,* Wiley, New York 1987.
[88] D. Oelkrug, A. Erbse, M. Plauschinat, *Z. Phys Chem.* **96** (1975) 283.
[89] U. Hezel: "Direkte quantitative Photometrie an Dünnschichtchromatogrammen," *Angew. Chem.* **85** (1973) 334;*GIT Fachz. Lab.* **21** (1977) 694.
[90] P. Kubelka, *J. Opt. Soc. Am.* **38** (1948) part I, 448.
[91] S. Bayerbach, G. Gauglitz, *Fresenius J. Anal. Chem.* **335** (1989) 370.
[92] Operation manual of MOSS ES4G ellipsometer, SOPRA Paris, France.
[93] R. M. A. Azzam, N. M. Bashera: *Ellipsometry and Polarized Light,* North Holland, New Amsterdam 1977.
[94] Ch. Striebel, A. Brecht, G. Gauglitz: "Characterization of Biomembranes by Spectral Ellipsometry, Surface Plasmon Resonance and Interferometry with regard to Biosensor Application," in *Biosen. Bioelectr.,* **9** (1994) 139.
[95] H. Arwin: "Ellipsometry," in A. Baszkin, W. Norde (eds.): *Physical Chemistry of Biological Interfaces,* Marcel Dekker, New York 2000.
[96] "Maintaining Optimum Spectrophotometer Performance," Perkin-Elmer, Application Data Bulletin; *Am. Soc. Clin. Pathol. Techn. Improv. Ser.* **27.**
[97] Recommended Practices for General Techniques of Ultraviolet Quantitative Analysis, E169-63, ASTM in Manual of Recommended Practices in Spectroscopy, Philadelphia: American Society of Testing Materials Committee E13, 1969, 10 – 34.
[98] C. Burgess, K. D. Mielenz (eds.): *Advances in Standards and Methodology in Spectrophotometry,* Elsevier, Amsterdam 1987.
[99] A. Knowless, C. Burgess: *Practical Absorption Spectrometry,* Chapman and Hall, London 1984.
[100] C. Burgess, A. Knowless: *Standards in Absorption Spectrometry,* Ultraviolet Spectrometry Group, vol. 1, Chapman and Hall, London 1981.
[101] R. W. Burke, E. R. Deardorff, O. Menis: "Liquid Absorbance Standards," *J. Res. Nat. Bur. Stand. Sect. A* **76** (1972) 469.
[102] R. E. Poulson: "Test Methods in Spectrophotometry: Stray Light Determination," *Appl. Opt.* **3** (1964) 99.
[103] W. Luck, *Z. Elektrochem.* **64** (1960) 676.
[104] Instruction manual, Hellma, Mühlheim/Baden.
[105] S. Tesch, K. H. Rentrop, M. Otto, *Fresenius J. Anal. Chem.* **344** (1992) 206.
[106] H. Bandemer, M. Otto, *Mikrochim. Acta* 1986, no. 2, 93.
[107] K. Schuchardt, *Nachr. Chem. Tech. Lab.* **40** (1992) M1 – M20;H. J. Majer, *GIT Fachz. Lab.* **37** (1993) 881.
[108] D. H. Williams, I. Flemming: *Strukturaufklärung in der organischen Chemie,* Thieme-Verlag, Stuttgart 1985.
[109] G. Rücker, M. Neugebauer, G. G. Williams: *Instrumentelle pharmazeutische Analytik,* Wissenschaftliche Verlagsgesellschaft, Stuttgart 1988.
[110] M. Hesse, H. Meier, B. Zeeh: *Spektroskopische Methoden in der organischen Chemie,* Thieme Verlag, Stuttgart 1985.
[111] G. Gauglitz, K.-A. Kovar in E. Nürnberg, P. Surmann (eds.): *Hagers Handbuch der pharmazeutischen Praxis,* 5th ed., **vol. 2,** Methoden, Springer Verlag, Berlin 1991, p. 471.
[112] I. Barret, W. F. Smyth, I. E. Davidson, *J. Pharm. Pharmacol.* **25** (1973) 387.
[113] H. Martens, T. Naes: *Multivariate Calibration,* J. Wiley & Sons, Chichester 1993.
[114] E. R. Malinowski, D. G. Howery: *Factor Analysis in Chemistry,* Wiley, New York 1980.
[115] S. Ebel, E. Glaser, S. Abdulla, U. Steffens, V. Walter, *Fresenius Z. Anal. Chem.* **313** (1982) 24.
[116] H. M. Heise, *Fresenius J. Anal. Chem.* **345** (1993) 604.
[117] R. Marbach, H. M. Heise, *Chemometr. Intel. Lab. Syst.* **9** (1990) 45.
[118] S. D. Brown, *Chemometr. Intel. Lab. Syst.* **10** (1991) 87.
[119] M. Mettler, G. Gauglitz, *TrAc Trends Anal. Chem.* **11** (1992) 203.
[120] J. A. Weismüller, A. Chanady, *TrAc Trends Anal. Chem. (Pers. Ed.)* **11** (1992) 86.
[121] T. Kohonen: *Self-Organization and Associative Memory,* Springer Verlag, Berlin 1989.
[122] C. Schierle, M. Otto, *Fresenius J. Anal. Chem.* **344** (1992) 190.
[123] J. Göppert et al., Proc. Neuro Nimes 1992.
[124] G. Kraus, G. Gauglitz, *Fresenius J. Anal. Chem.* **346** (1993) 572.
[125] K. Molt, M. Egelkraut, *Fresenius J. Anal. Chem.* **327** (1987) 77.
[126] W. Windig, *Chemometr. Intel. Lab. Syst.* **16** (1992) 1.
[127] J. Eisenbrand: *Fluorimetrie,* Wissenschaftliche Verlagsgesellschaft, Stuttgart 1966.
[128] G. Gauglitz, R. Goes, W. Stooss, R. Raue, *Z. Naturforsch. A Phys. Phys. Chem. Kosmophys.* **40 A** (1985) 317.
[129] G. Elliott, A. Radley, *Analyst (London)* **86** (1961) 62.
[130] S. Udenfried: *Fluorescence Assay in Biology and Medicine,* Academic Press, New York, vol. I, 1962, vol. II, 1969.
[131] I. B. Berlman: *Handbook of Fluorescence Spectra of Aromatic Compounds,* Academic Press, New York 1965.
[132] R. Strohecker, H. M. Hemming: *Vitaminbestimmungen,* Verlag Chemie, Weinheim 1963.

[133] F. P. Schwarz, St. P. Wasik, *Anal. Chem.* **48** (1976) 524.
[134] U. Panne, R. Nießner, *Vom Wasser* **79** (1992) 89.
[135] O. S. Wolfbeis (ed.): *Fluorescence Spectroscopy, New Methods and Applications,* Springer Verlag, Berlin 1993.
[136] U. Schobel, C. Barzen, G. Gauglitz: "Immunoanalytical techniques for pesticide monitoring based on fluorescence detection," *Fresenius J. Anal. Chem.* **366** (2000) 646 – 658.
[137] G. Kortüm: *Reflexionsspektroskopie,* Springer Verlag, Berlin 1969.
[138] H. Wilsing, *Leitz-Mitt. Wiss. Tech.* **1** (1960) 133.
[139] A. Zlatkis, R. E. Kaiser, *J. Chromatogr. Lib.* **9** (1977).
[140] *Accessory Lambda-Series* Perkin-Elmer, Überlingen.
[141] H. Arens: *Farbmetrik,* Akademie Verlag, Berlin 1957.
[142] M. Richter: *Einführung in die Farbmetrik,* De Gruyter, Berlin 1981.
[143] H. Naumer, W. Heller (eds.): *Untersuchungsmethoden in der Chemie,* Thieme-Verlag, Stuttgart 1990, p. 348.
[144] R. Th. Kersten: *Einführung in die optische Nachrichtentechnik,* Springer Verlag, Berlin 1983.
[145] V. Hoffmann et al., *J. Mol. Struct.* **293** (1993) 253.
[146] G. Gauglitz, A. Brecht, G. Kraus, W. Nahm, *Sens. Actuators B* **11** (1993) 21.
[147] A. Brecht, G. Gauglitz, W. Nahm, *Analusis* **20** (1992) 135.
[148] G. Kraus, G. Gauglitz, *Fresenius J. Anal. Chem.* **344** (1992) 153.
[149] G. Gauglitz, J. Krause-Bonte, H. Schlemmer, A. Matthes, *Anal. Chem.* **60** (1988) 2609.
[150] D. E. Aspnes: "The Accurate Determination of Optical Properties by Ellipsometry," in: E. D. Palik (ed.): *Handbook of Optical Constants of Solids,* Academic Press, London 1986, pp. 89 ff.
[151] D. E. Aspnes, A. A. Studna, *Phys. Rev. B Condens. Matter* **27** (1983) 985.
[152] H. Arwin, D. E. Aspnes, *Thin Solid Films* **138** (1986) 195.
[153] G. Kraus, A. Brecht, G. Gauglitz, Ch. Striebel, Lecture held at the Bunsentagung, Leipzig 1993.
[154] H. Raether: Surface Plasmons on Smooth and Rough Surfaces and on Gratings, Springer Tracts in Modern Physics, vol. 111, Springer Verlag, Berlin 1988.
[155] E. Kretchmann, H. Raether, *Z. Naturforsch. A Astrophys. Phys. Phys. Chem.* **23A** (1968) 2135.
[156] A. Otto, *Z. Phys.* **216** (1968) 398.
[157] B. Liedberg, C. Nylander, I. Lundström: "Surface Plasmon Resonance for Gas Detection and Biosensing," *Sens. Actuators* **4** (1983) 299.
[158] P. B. Daniels, J. K. Deacon, M. J. Edowes, D. G. Pedley: "Surface Plasmon Resonance Applied to Immunosensing," *Sens. Actuators* **15** (1988) 11.
[159] J. Homola, S. S. Yee, G. Gauglitz: "Surface plasmon resonance sensors: review," *Sensors & Actuators* **B 54** (1999) 3 – 15.
[160] W. Lukosz, K. Tiefenthaler, *Opt. Lett.* **8** (1983) 537.
[161] W. Lukosz, Th. Brenner, V. Briguet, Ph. Nellen, P. Zeller, *Proc. SPIE Int. Soc. Opt. Eng.* **1141** (1989) 192.
[162] Ch. Stamm, W. Lukosz, *Sens. Actuators* **B11** (1993) 177.
[163] D. Clerc, W. Lukosz, *Sens. Actuators* **B11** (1993) 177.
[164] W. Böhme, P. Pospisil, K. Horn, H. Morr, *Spectrosc. Int.* **3** (1991) 36.
[165] U. Ruzicka, E. H. Hansen: *Flow Injection Analysis,* John Wiley, New York 1981.
[166] H. Müller, B. Frey, W. Böhme, *Fresenius J. Anal. Chem.* **341** (1991) 647.
[167] W. Göpel et al. (eds.): *Sensors,* VCH Verlagsgesellschaft, Weinheim 1992.
[168] O. S. Wolfbeis, *Fresenius J. Anal. Chem.* **325** (1986) 387.
[169] D. W. Lübbers, N. Opitz, *Z. Naturforsch. C. Biosci.* **C 30** (1975) 532.
[170] G. Gauglitz, M. Reichert, *Sens. Actuators* **B6** (1992) 83.
[171] G. Gauglitz, J. Ingenhoff, *Ber. Bunsen-Ges. Phys. Chem.* **95** (1991) 1558.
[172] M. A. Butler, R. Buss, A. Galuska, *J. Appl. Phys.* **70** (1991) 2326.
[173] W. Nahm, G. Gauglitz, *GIT Fachz. Lab.* **34** (1990) 889.
[174] G. Kraus, A. Brecht, V. Vasic, G. Gauglitz, *Fresenius J. Anal. Chem.* **348** (1994) 598.
[175] H.-M. Haake, A. Schütz, G. Gauglitz, "Label-free detection of biomolecular interaction by optical sensors," *Fresenius J. Anal. Chem.* **366** (2000) 576 – 585.
[176] C. Mischiati et al.: "Interaction of the human NF-kappaB p52 transcription factor with DNA-PNA hybrids mimicking the NF-kappaB binding sites of the human immunodeficiency virus type 1 promoter," *J. Biol. Chem.* **274** (1999) 33114 – 33122.
[177] M. Sauer et al.: "Interaction of Chemically Modified Antisense Oligonucleotides with Sense DNA: A Label-Free Interaction Study with Reflectometric Interference Spectroscopy," *Analytical Chemistry* **71** (1999) 2850 – 2857.
[178] W. Göpel, J. Hesse, J. N. Zemel (eds.): "Optical Sensors," in *Chemical and Biochemical Sensors,* VCH Verlagsgesellschaft, Weinheim 1992.
[179] J. Polster, H. Lachmann: *Spectrometric Titrations,* VCH Verlagsgesellschaft, Weinheim 1989.
[180] R. Blume, H. Lachmann, H. Mauser, F. Schneider, *Z. Naturforsch. B Anorg. Chem. Org. Chem.* **B 29** (1974) 500.
[181] H. Lachmann, *Z. Anal. Chem.* **290** (1978) 117.
[182] G. Gauglitz: "Physical Properties of Photochromic Substances," in H. Dürr, H. Bois-Laurant (eds.): *Photochromic Systems,* Elsevier, Amsterdam 1990.
[183] G. Gauglitz in H. G. Seiler, A. Sigel, H. Sigel (eds.): *Handbook on Metals in Clinical Chemistry,* Marcel Dekker, New York, in press.
[184] H. Mauser, G. Gauglitz: *Principles and Applications of Photokinetics,* Elsevier, Amsterdam, in preparation.
[185] G. Gauglitz, E. Lüddecke, *Z. Anal. Chem.* **280** (1976) 105.
[186] T. C. O'Haver, G. L. Green, *Anal. Chem.* **48** (1976) 312, *Am. Lab. (Fairfield, Conn.)* March (1975) 15.
[187] G. Talsky, L. Mayring, H. Kreuzer, *Angew. Chem.* **90** (1978) 840.
[188] G. L. Green, T. C. O'Haver, *Anal. Chem.* **46** (1974) 2191.
[189] S. Shibata, M. Furukawa, T. Honkawa, *Anal. Chim. Acta* **81** (1976) 206.
[190] *Ullmann,* 4th ed., **5,** 269.
[191] S. Shibata, *Angew. Chem.* **88** (1976) 750.
[192] A. Schmitt: *Labor-Praxis,* vol. 9, Vogel-Verlag, Würzburg 1979.
[193] G. Stella, *Chem. Phys. Lett.* **39** (1976) 176.
[194] H. P. Engelbert, R. Lawaczek, *Chem. Phys. Lipids* **38** (1985) 365.

[195] J. C. Sternberg, *Clin. Chem.* **23** (1977) 1456.
[196] G. Träxler, *Medizintechnik (Stuttgart)* **99** (1979) 79.
[197] K. Oka et al., *Int. Lab.* March (1986) 24.
[198] E. Lüddecke, D. Horn, *Chem. Ing. Tech.* **54** (1982) 266.
[199] F. Dörr: "Polarized Light in Spectroscopy and Photochemistry," in A. A. Lamola (ed.): *Creation and Detection of the Excited State,* vol. 1, Part A, Dekker Inc., New York 1971, pp. 53 ff.
[200] Th. Förster: *Fluoreszenz organischer Verbindungen,* Vandenhoeck G. Ruprecht, Göttingen 1982.
[201] D. V. O'Connor, D. Phillips: *Time-Correlated Single Photon Counting,* Academic Press, New York 1984.
[202] A. E. W. Knight, B. K. Selinger, *Aust. J. Chem.* **26** (1973) 1.
[203] R. B. Somoano, *Angew. Chem.* **90** (1978) 250.
[204] Yok-Han Pao (ed.): *Optoacoustic Spectroscopy and Detection,* Academic Press, New York 1977.
[205] L. B. Kreuzer, *J. Appl. Phys.* **42** (1971) 2934.
[206] L. G. Rosengreen, *Appl. Opt.* **14** (1975) 1960.
[207] C. F. Dewey, R. D. Kamm, C. E. Hackett, *Appl. Phys. Lett.* **23** (1973) 633.
[208] K. Kaya et al., *J. Am. Chem. Soc.* **97** (1975) 2153.
[209] J. F. McClelland, R. N. Knisely, *Appl. Phys. Lett.* **28** (1976) 467;*Appl. Opt.* **15** (1976) 2658.
[210] A. Rosencwaig: *Optoacoustic Spectroscopy and Detection,* Academic Press, New York 1977, p. 193.
[211] J. C. Murphy, L. C. Aamodt, *Appl. Phys. Lett.* **31** (1977) 728.
[212] A. Rosencwaig, A. Gersho, *J. Appl. Phys.* **47** (1976) 64.
[213] W. R. Harshberger, M. B. Robin, *Acc. Chem. Res.* **6** (1973) 329.
[214] M. G. Rockley, K. M. Wangle, *Chem. Phys. Lett.* **54** (1978) 597.
[215] A. Rosencwaig, *Science (Washington, D.C.)* **181** (1973) 657.
[216] A. Petzold, R. Niessner, SPIE 1716 (1992) 510.

17. Infrared and Raman Spectroscopy

HANS-ULRICH GREMLICH, Novartis AG, Basel, Switzerland

17.1. Introduction

The roots of infrared spectroscopy go back to the year 1800, when WILLIAM HERSCHEL discovered the infrared region of the electromagnetic spectrum. Since 1905, when WILLIAM W. COBLENTZ ran the first infrared spectrum [1], vibrational spectroscopy has become an important analytical tool in research and in technical fields. In the late 1960s, infrared spectrometry was generally believed to be a dying instrumental technique that was being superseded by nuclear magnetic resonance and mass spectrometry for structural determination, and by gas and liquid chromatography for quantitative analysis.

However, the appearance of the first research-grade Fourier transform infrared (FT-IR) spectrometers in the early 1970s initiated a renaissance of infrared spectrometry. After analytical instruments (since the late 1970s) and routine instruments (since the mid 1980s), dedicated instruments are now available at reasonable prices. With its fundamental multiplex or Fellgett's advantage and throughput or Jacquinot's advantage, FT-IR offers a versatility of approach to measurement problems often superior to other techniques. Furthermore, FT-IR is capable of extracting from samples information that is difficult to obtain or even inaccessible for nuclear magnetic resonance and mass spectrometry. Applications of modern FT-IR spectrometry include simple, routine identity and purity examinations (quality control) as well as quantitative analysis, process measurements, the identification of unknown compounds, and the investigation of biological materials.

Raman and infrared spectra give images of molecular vibrations which complement each other; i.e., the combined evaluation of both spectra yields more information about molecular structure than when they are evaluated separately. The 1990s witnessed a tremendous growth in the capabilities and utilization of Raman scattering measurements. The parallel growth of Fourier transform Raman (FT-Raman) with charge-coupled device (CCD) based dispersive instrumentation has provided the spectroscopist with a wide range of choice in instrumentation. There are clear strengths and weaknesses that accompany both types of instrumentation. While FT-Raman systems virtually eliminate fluorescence by using near-IR laser sources, this advantage is paid for with limited sensitivity. On the other hand, fluorescence interferences always play a role in dispersive measurements with higher sensitivity.

As an alternative to infrared spectroscopy, Raman spectroscopy can be easier to use in some cases; for example, whereas water and glass are strong infrared absorbers they are weak Raman scatterers, so that it is easy to produce a good-quality Raman spectrum of an aqueous sample in a glass container.

General References. Details of vibrational theory are given in the books by HERZBERG [2], [3]; WILSON, DECIUS, and CROSS [4]; STEELE [5]; KONINGSTEIN [6]; and LONG [7]. The technique and applications of FT-IR spectroscopy are discussed in detail by GRIFFITHS and DE HASETH [8], and by MACKENZIE [9]. An introduction to infrared and Raman spectroscopy, dealing with theoretical as well as experimental and interpretational aspects, is provided by COLTHUP, DALY, and WIBERLEY [10] and by SCHRADER [11]. Books about Raman spectroscopy in particular are those of KOHLRAUSCH [12]; BARANSKA, LABUDZINSKA, and TERPINSKI [13]; and HENDRA, JONES, and WARNES [14]. Practical FT-IR and Raman spectroscopy is covered by FERRARO and KRISHNAN [15]; PELLETIER [16]; and GRASSELLI and BULKIN [17]. For the interpretation of IR and Raman spectra of organic compounds, see SOCRATES [18], BELLAMY [19], [20], DOLLISH, FATELEY, and BENTLEY [21], and LIN-VIEN, COLTHUP, FATELEY, and GRASSELLI [22]. Vibrational spectroscopy of inorganic and coordination compounds is described by NAKAMOTO [23]. Biological applications of IR and Raman spectroscopy are covered by GREMLICH and YAN [24]. Various collections of vibrational spectra are available. Both Raman and IR spectra of about 1000 organic compounds are offered as an atlas by SCHRADER [25]. The most comprehensive reference-spectra libraries, digital and hardcopy, can be obtained from SADTLER [26]. A book on modern techniques in applied optical spectroscopy was edited by MIRABELLA [27].

17.2. Techniques

17.2.1. Fourier Transform Infrared Technique

The basis of Fourier transform infrared (FT-IR) spectroscopy is the two-beam interferometer, designed by MICHELSON in 1891 [28], and shown

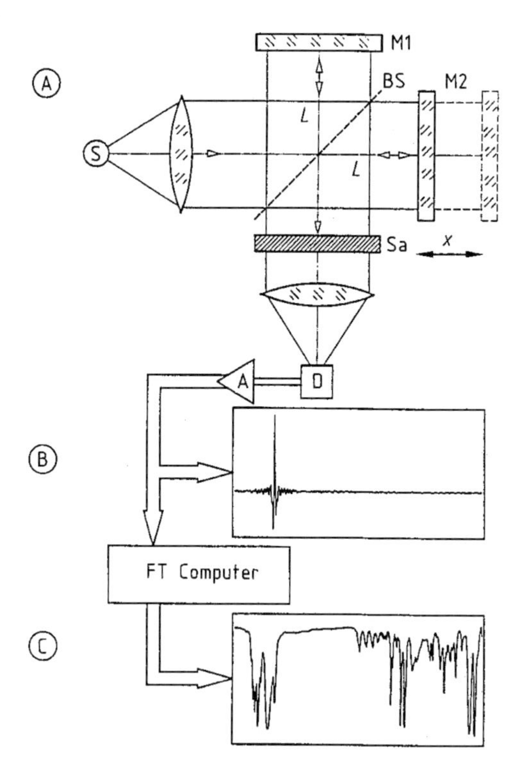

Figure 1. A) Schematic diagram of a Michelson interferometer; B) Signal registered by the detector D, the interferogram; C) Spectrum obtained by Fourier transform (FT) of the interferogram
S = Radiation source; Sa = Sample cell; D = Detector; A = Amplifier; M1 = Fixed mirror; M2 = Movable mirror; BS = Beam splitter; x = Mirror displacement

schematically in Figure 1 A. Broadband infrared radiation is emitted by a thermal source (globar, metal strips, Nernst glower) and falls onto a beam splitter which, in the ideal case, transmits half the radiation and reflects the other half. The reflected half, after traversing a distance L, falls onto a fixed mirror M1. The radiation is reflected by M1 and, after traversing back along distance L, falls onto the beam splitter again. The transmitted radiation follows a similar path and also traverses distance L; however, the mirror M2 of the interferometer can be moved very precisely along the optical axis by an additional distance x. Hence, the total path length of the transmitted radiation is $2(L+x)$. On recombination at the beam splitter the two beams possess an optical path difference of $\Delta = 2x$. Since they are spatially coherent, the two beams interfere on recombination. The beam, modulated by movement of the mirror, leaves the interferometer, passes through the sample cell and is finally focussed on the detector. The signal registered by the detector, the interferogram, is thus the radiation intensity $I(x)$, measured as a function of the displacement x of the moving mirror M2 from the distance L (Fig. 1 B). The mathematical transformation, a Fourier transform [8], of the interferogram, performed by computer, initially provides a so-called single-beam spectrum. This is compared with a reference spectrum measured without the sample to obtain a spectrum analogous to that measured by conventional dispersive methods. This spectrum is stored digitally in the computer, from which it can be retrieved for further use (Fig. 1 C). In addition to Michelson interferometers, which often have air bearings for the moving mirror and thus require a source of dry compressed air, mechanical interferometer concepts such as frictionless electromagnetic drive mechanisms or refractively scanned interferometers have been developed by instrument manufacturers.

Two major kinds of detector are used in mid-infrared FT-IR spectrometers, the deuterated triglycine sulfate (DTGS) pyroelectric detector and the mercury cadmium telluride (MCT) photodetector. The DTGS type, which operates at room temperature and has a wide frequency range, is the most popular detector for FT-IR instruments. The MCT detector responds more quickly and is more sensitive than the DTGS detector, but it operates at liquid-nitrogen temperature and has both a limited frequency range and a limited dynamic range; therefore, it is used only for special applications [8].

The resolution of an interferometer depends mainly on two factors: firstly, the distance x the moving mirror travels or the maximum optical path difference $\Delta = 2x$ (Fig. 1), and secondly, the apodization function used in computing the spectrum.

The first point is derived from the Rayleigh criterion of resolution, which states that to resolve two spectral lines separated by a distance d, the interferogram must be measured to an optical path length of at least $1/d$ [29].

The second factor is related to the fact that in real life the interferogram is truncated at finite optical path difference. In addition, in the fast Fourier transform (FFT) algorithm, according to Cooley and Tukey [30], which is used to perform the Fourier transform faster than the classical method, certain assumptions and simplifications are made. The result is that the FFT of a monochromatic source is not an infinitely narrow line,

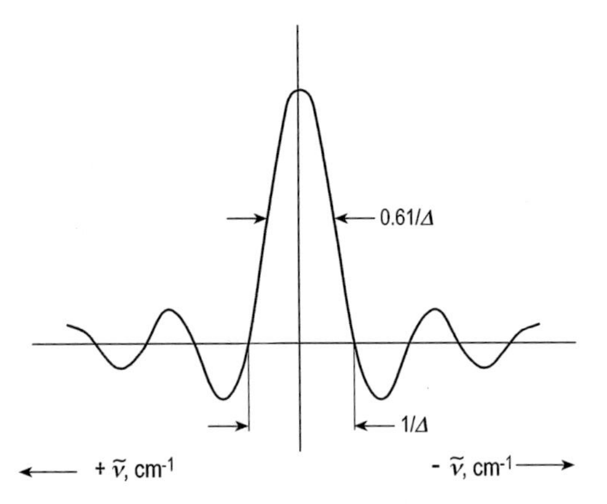

Figure 2. The sinc x function: instrumental line shape function of a perfectly aligned Michelson interferometer with no apodization
$\tilde{\nu}$ = wavenumber, cm^{-1}; Δ = maximum retardation, cm

corresponding to infinite optical path difference and hence to infinite resolution (see above), but a definable shape known as instrumental line shape (ILS) function [8]. The ILS of a perfectly aligned Michelson interferometer has the shape of a $(\sin x)/x$ or sinc x function (Fig. 2), exhibiting strong side lobes or feet, which may even have negative intensities. To attenuate these spurious feet, so-called apodization functions are used as weighting functions for the interferogram. Numerous such functions exist (e.g., triangular, Happ–Genzel, or Blackman–Harris) which produce instrumental line shapes with lower intensity side lobes, but also with broader main lobes than the sinc function. Hence, side-lobe suppression is only possible at the cost of resolution, because the full width at half-height of the ILS defines the best resolution achievable with a given apodization function [31]. In contrast to dispersive spectroscopy, where the triangular apodization function is determined by the slit of the grating spectrometer, in FT-IR spectroscopy, a choice between highest resolution (no apodization) and best side lobe suppression (Blackman–Harris apodization function) is possible.

In comparison with conventional spectroscopy, the FT-IR method possesses significant advantages. In grating spectrometers, the spectrum is measured directly by registering the radiation intensity as a function of the continuously changing wavelength given by a monochromator [10]. Depending on the spectral resolution chosen, only a very small part (in practice less than 0.1 %) of the radiation present in the monochromator reaches the IR detector. In the FT-IR spectrometer, all frequencies emitted by the IR source reach the detector simultaneously, which results in considerable time saving and a large signal-to-noise ratio advantage over dispersive instruments; this is the most important advantage, known as the multiplex or Fellgett's advantage [8]. A further advantage stems from the fact that the circular apertures employed in FT-IR spectrometers have larger surface areas than the linear slits of grating spectrometers and hence permit radiation throughputs at least six times greater; this is known as the throughput or Jacquinot's advantage [8]. In FT-IR spectroscopy, the measuring time is the time required to move the mirror M2 the distance necessary to give the desired resolution; since the mirror can be moved very rapidly, complete spectra can be obtained within fractions of a second. By contrast, the measuring time for a conventional spectrum is on the order of minutes. In an FT spectrum, the accuracy of each wavenumber is coupled to the accuracy with which the position of the moving mirror is determined; by using an auxiliary HeNe laser interferometer the position of the mirror can be determined to an accuracy better than 0.005 μm. This means that the wavenumbers of an FT-IR spectrum can be determined with high accuracy ($< 0.01\ cm^{-1}$). In other words, FT-IR spectrometers have a built-in wavenumber calibration of high precision; this advantage is known as accuracy or Connes' advantage [29]. As a result, it is possible to carry out highly precise spectral subtractions and thus very reliably detect slight spectral differences between samples.

17.2.2. Dispersive and Fourier Transform Raman Techniques

In modern dispersive Raman spectrometers, arrays of detector elements such as photodiode arrays or charge-coupled devices (CCDs) are arranged in a polychromator. Here, each element records a different spectral band at the same time, thus making use of the multichannel advantage. In dispersive instruments, the very strong radiation at and near the exciting line, the Rayleigh radiation, may produce stray radiation in the entire spectrum with an intensity much higher than that of the Raman lines. In order to remove this unwanted radiation, line filters in the form of so-called notch filters or holographic filters are used which specifically reflect the Rayleigh line. In conventional, dispersive Raman spectroscopy visible lasers are

Ⓐ

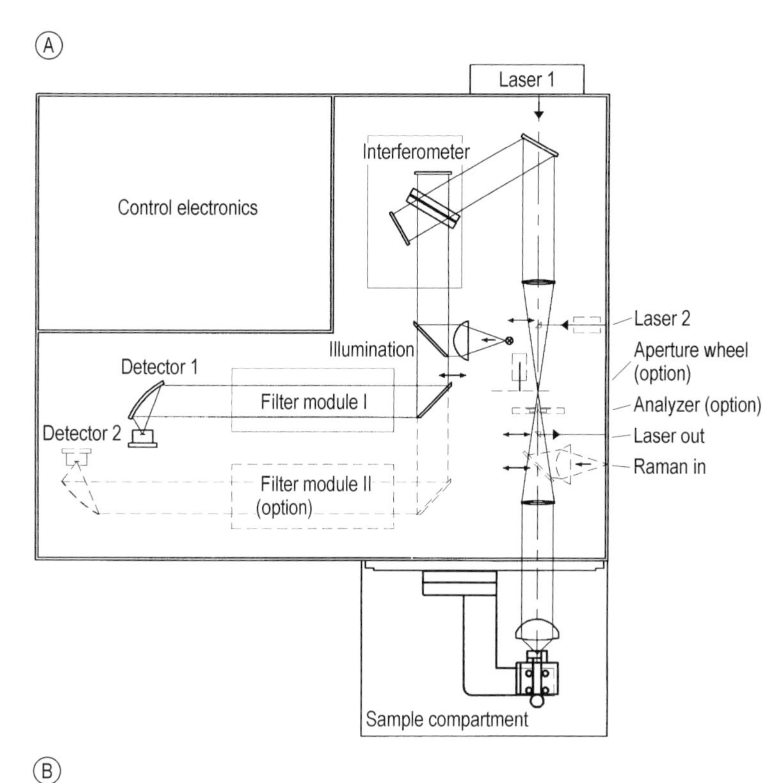

Ⓑ

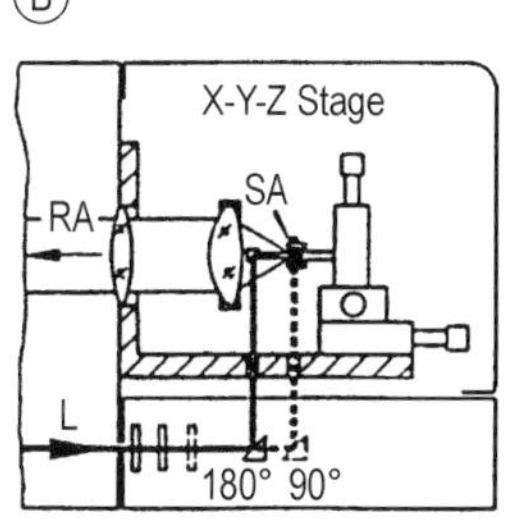

Figure 3. A) Schematic diagram of an FT-Raman spectrometer (Bruker RFS 100); B) Schematic diagram of the sample compartment
L = laser; RA = Raman; SA = Sample
(Reproduced by permission of Bruker Optik GmbH, D-76275 Ettlingen, Germany)

usually used to excite the Raman effect (see Section 17.3.1), while in FT-Raman spectroscopy near-IR lasers are employed. Excitation with visible lasers often produces strongly interfering fluorescence originating from the sample or impurities; however, this is eliminated by using a neodymium yttrium aluminum garnet (Nd : YAG) laser providing a discrete line excitation at 1.064 μm, as at this wavelength in the near-IR region no electronic transitions are induced.

An FT-Raman spectrometer is shown in schematic form in Figure 3 A. The beam of the laser source passes to the sample compartment where it is focused, typically to a spot of 100 μm diameter, onto the sample, either in 90° or 180° geometry (Fig. 3 B). The scattered Raman light is collected by an aspherical lens and passed through a filter module to remove the Rayleigh line and Rayleigh wings [32]. It is then directed through the infrared input port into the interferometer which is optimized for the near-IR region as is the liquid-nitrogen-cooled germanium or room-temperature indium gallium arsenide (InGaAs) detector [33]. The procedure to obtain the FT-Raman spectrum from the interferogram, the signal registered by the detector, is essentially the same as that described above for an FT-IR spectrum.

Before FT-Raman spectrometers could be successfully used several difficulties had to be overcome. As the intensity of Raman scattering is proportional to the fourth power of the frequency of excitation a loss in sensitivity by a factor between 20 and 90 must be compensated for when the exciting radiation is shifted from the visible to the near-IR region [32]. Moreover, the noise equivalent power (NEP) of near-IR detectors is greater by about one to two orders of magnitude than that of a photomultiplier tube, the usual detector for conventional Raman spectroscopy. However, both disadvantages are compensated for by the throughput (Jacquinot) and multiplex (Fellgett) advantages of Fourier transform spectroscopy (see Section 17.2.1). Another major problem is the fact that Raman scattering is always accompanied by very strong radiation, the Rayleigh line and Rayleigh wings, occurring at and near the frequency of excitation with six to ten orders of magnitude higher intensity than the Raman signal. The shot noise of the Rayleigh line and Rayleigh wings produces a background noise which would completely mask all Raman lines. Therefore, an effective filter that removes the Rayleigh line and Rayleigh wings in the range 750 – 100 cm^{-1} is one of the most important components of a powerful FT-Raman instrument [32]. In addition, optimization of the sampling technique is indispensable in order to convert the flux of exciting radiation into a maximum flux of Raman radiation to be received by the optical system of the spectrometer [34]. The 180° backscattering arrangement (Fig. 3 B), rather than the 90° arrangement used in visible, dispersive Raman technique, is undoubtedly the most efficient way of collecting the Raman signal and provides the best quality data in the near-IR FT-Raman experiment.

17.3. Basic Principles of Vibrational Spectroscopy

17.3.1. The Vibrational Spectrum

To explain the origins of a vibrational spectrum the vibration of a diatomic molecule is considered first. This can be illustrated by a molecular model in which the atomic nuclei are represented by two point masses m_1 and m_2, and the interatomic bond by a massless spring. According to Hooke's law, the vibrational frequency ν (in s^{-1}) determined by classical methods is given by

$$\nu = \frac{1}{2\pi}\sqrt{\frac{f}{\mu}} \tag{1}$$

where f is the force constant of the spring in N/m, and μ is the reduced mass in kg:

$$\mu = \frac{m_1 m_2}{m_1 + m_2} \tag{2}$$

Thus, the vibrational frequency is higher when the force constant is higher, i.e., when the bonding of the two atoms is stronger. Conversely, the heavier the vibrating masses are, the smaller is the frequency ν. The frequency of fundamental vibrations is on the order of magnitude of 10^{13} s^{-1}. Quantum-mechanical methods give the following expression for the vibrational energy of a Hooke's oscillator in a state characterized by the vibrational quantum number v:

$$E_{\text{vib}} = \frac{h}{2\pi}\sqrt{\frac{f}{\mu}}\left(v + \frac{1}{2}\right) = h\nu_0\left(v + \frac{1}{2}\right) \tag{3}$$

where f and μ have the same definition as in Equation (1) and ν_0 is the vibrational frequency of the ground state. This relation is valid with good accuracy for vibrational transitions from the ground state ($v = 0$) to the first excited state ($v = 1$). In free molecules (gas state), vibrational transitions are always accompanied by rotational transitions. The corresponding energies are given by:

$$E_{\text{vib,rot}} = h\nu_0\left(v + \frac{1}{2}\right) + BhJ(J+1) \tag{4}$$

where J is the rotational quantum number and B is the rotational constant:

$$B = \frac{h}{8\pi^2 I} \tag{5}$$

where I is the moment of inertia. In the real world, molecular oscillations are inharmonic. Therefore, the transition energy decreases with increasing vibrational quantum number until the molecule finally dissociates. The quantum mechanical energy of a diatomic inharmonic oscillator is given in good approximation by:

Table 1. Infrared and microwave regions of the electromagnetic spectrum

Region	Wavelength λ, μm	Wavenumber $\tilde{\nu}$, cm^{-1}	Frequency ν, s^{-1}
Infrared			
near	7.8×10^{-1} to 2.5	12 800 to 4000	3.8×10^{14} to 1.2×10^{14}
mid	2.5 to 5.0×10^{1}	4000 to 200	1.2×10^{14} to 6.0×10^{12}
far	5.0×10^{1} to 1.0×10^{3}	200 to 10	6.0×10^{12} to 3.0×10^{11}
Microwave	1.0×10^{3} to 1.0×10^{6}	10 to 0.01	1.0×10^{12} to 3.0×10^{8}

$$E_{\mathrm{vib}} = h\nu_0\left[\left(v+\frac{1}{2}\right) - x_0\left(v+\frac{1}{2}\right)^2\right] \tag{6}$$

where x_0 is the inharmonicity constant.

A complex molecule is considered as a system of coupled inharmonic oscillators. If there are N atomic nuclei in the molecule, there will be a total of $3N$ degrees of freedom of motion for all the nuclear masses in the molecule. Subtracting the pure translations and rotations of the entire molecule leaves $3N-6$ internal degrees of freedom for a nonlinear molecule and $3N-5$ internal degrees of freedom for a linear molecule. The internal degrees of freedom correspond to the number of independent normal modes of vibration; their forms and frequencies must be calculated mathematically [10]. In each normal mode of vibration all the atoms of the molecule vibrate with the same frequency and pass through their equilibrium positions simultaneously. The true internal motions of the molecule, which constitute its vibrational spectrum, are composed of the normal vibrations as a coupled system of these independent inharmonic oscillators. Thus, a molecule is unambiguously characterized by its vibrational spectrum. Real-world molecules are best described by the model of inharmonic oscillators because here transitions are allowed which are forbidden to the harmonic oscillator: transitions from the ground state ($v=0$) to states with $v=2$, 3,... (overtones), transitions from an excited level of a vibration (difference bands), and transitions between states belonging to different types of normal mode (combination bands).

In vibrational spectroscopy, instead of the frequency ν (in s^{-1}) the so-called wavenumber $\tilde{\nu}$ expressed in cm^{-1} (reciprocal centimeters) is mostly used; this signifies the number of waves in a 1 cm wavetrain:

$$\tilde{\nu} = \frac{\nu}{(c/n)} = \frac{1}{\lambda} \tag{7}$$

where c is the velocity of light in a vacuum (2.997925×10^{10} cm/s), c/n is the velocity of light in a medium with refractive index n in which the wavenumber is measured (the refractive index of air is 1.0003), and λ is the wavelength in centimeters. In the infrared region of the electromagnetic spectrum the practical unit of wavelength is 10^{-4} cm or 10^{-6} m (i.e., μm), and wavenumber and wavelength are related as follows:

$$\tilde{\nu}/\mathrm{cm}^{-1} = \frac{10^4}{\lambda/\mu\mathrm{m}} \tag{8}$$

In infrared spectroscopy, both wavenumber and wavelength are used, whereas in Raman spectroscopy only wavenumber is employed.

The infrared and microwave regions of the electromagnetic spectrum are shown in Table 1. In the near-infrared region, overtones and combination bands are observed. Since these transitions are only possible because of inharmonicity, their probability is diminished. A good rule of thumb is that the intensity of the first overtone is about one-tenth that of the corresponding fundamental vibration. The fundamentals occur in the mid-infrared range, which is the most important infrared region. In the far-infrared region, the fundamentals of heavy, single-bonded atoms and the absorptions of inorganic coordination compounds [23] are found.

Infrared spectroscopy is based on the interaction of an oscillating electromagnetic field with a molecule, and it is only possible if the dipole moment of the molecule changes as a result of a molecular vibration. While the absorption frequency depends on the molecular vibrational frequency, the absorption intensity depends on how effectively the infrared photon energy is transferred to the molecule. It can be shown [3] that the intensity of infrared absorption is proportional to the square of the change in the dipole moment μ, with respect to the change in the normal coordinate q describing the corresponding molecular vibration:

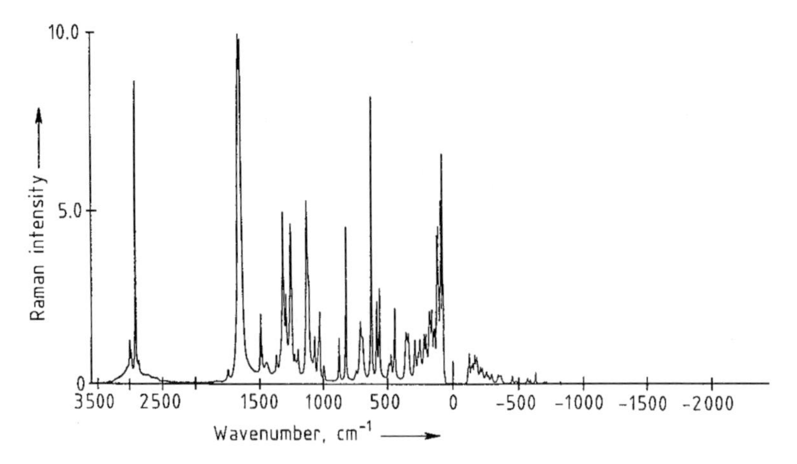

Figure 4. FT-Raman spectrum of ascorbic acid recorded on a Bruker FT-Raman accessory FRA 106 directly interfaced to a Bruker FT-IR spectrometer IFS 66
The accessory was equipped with a germanium diode cooled by liquid nitrogen. Resolution was 2 cm^{-1} and laser output power was 300 mW. Stokes shift: 3600 to 50 cm^{-1}, anti-Stokes shift: – 100 to – 2000 cm^{-1}.

$$I_{IR} \sim \left(\frac{\partial \mu}{\partial q}\right)^2 \quad (9)$$

A second way to induce molecular vibrations is to irradiate a sample with an intense source of monochromatic radiation, usually in the visible or near-infrared region, this leads to the Raman effect, which can be regarded as an inelastic collision between the incident photon and the molecule. As a result of the collision, the vibrational or rotational energy of the molecule is changed by an amount ΔE_m. For energy to be conserved, the energy of the scattered photon $h\nu_s$ must differ from that of the incident photon $h\nu_i$ by an amount equal to ΔE_m:

$$h\nu_i - h\nu_s = \Delta E_m = h\nu_m \quad (10)$$

If the molecule gains energy, then ΔE_m is positive and ν_s is smaller than ν_i, giving rise to so-called Stokes lines in the Raman spectrum. If the molecule loses energy, ΔE_m is negative and ν_s is larger than ν_i, giving rise to so-called anti-Stokes lines in the Raman spectrum. The dipole moment μ induced by the Raman effect can be related to the electric field E of the incident electromagnetic radiation as follows:

$$\mu = \alpha E \quad (11)$$

where α is the polarizability of the molecule. It can be derived from classical Raman theory [3] that for a molecular vibration to be Raman active it must be accompanied by a change in the polarizability of the molecule. The intensity of Raman absorption is proportional to the square of the change in polarizability α, with respect to the corresponding normal coordinate q:

$$I_{RA} \sim \left(\frac{\partial \alpha}{\partial q}\right)^2 \quad (12)$$

Although it follows from classical theory that Stokes and anti-Stokes lines should have the same intensity, according to quantum theory and in agreement with experiment (Fig. 4) anti-Stokes lines are a much less intense since the number of molecules in the initial state $v = 1$ of anti-Stokes lines is only $e^{-(h\nu_m/kT)}$ times the number of molecules in the initial state $v = 0$ of the Stokes lines [3]. The Raman shift in cm^{-1}, i.e., the difference $\Delta\tilde{\nu}$ between the wavenumber $\tilde{\nu}_e$ of the exciting laser and the wavenumber $\tilde{\nu}_s$ of the scattered Raman light, is indicated on the abscissa of each Raman spectrum.

17.3.2. Quantitative Analysis

Quantitative analysis is well established not only in ultraviolet/visible and in near-infrared spectroscopy, but it is also very important in mid-infrared measurements. The general prerequisite for spectrometric quantitative analysis is defined as follows [35]: information derived from the spectrum of a sample is related in mathematical terms to changes in the level(s) of an individual component, or several components within the sample or series of samples, i.e., the spectral response of an analyte can be related by a mathematical function to changes in concentration of the analyte. In the ideal situation, the measured spectroscopic feature varies linearly with concentration. In reality, however, true linearity is not always obtained, but this is not important provided the measured function is reproducible. Most practical analyses are not absolute measurements, and normally measurements are made on a given instrument within a fixed working environment; under such set circumstances, reproducibility and consistency of the measurement are the most important factors.

If I_0 is the intensity of monochromatic radiation entering a sample and I is the intensity transmitted by the sample, then the ratio I/I_0 is the transmittance T of the sample. The percent transmittance ($\%T$) is $100T$. If the sample cell has thickness b and the absorbing component has concentration c, then the fundamental relation governing the absorption of radiation as a function of transmittance is:

$$T = \frac{I}{I_0} = 10^{-abc} \tag{13}$$

The constant a is called the absorptivity and is characteristic of a specific sample at a specific wavelength. As the transmittance T does not vary linearly with the concentration of an absorbing species the equation above is usually transformed by taking the logarithms of both sides of the equation and replacing I/I_0 with I_0/I to eliminate the negative sign:

$$\log \frac{I_0}{I} = abc \tag{14}$$

The term $\log(I_0/I)$ is the absorbance A, which changes linearly with changes in concentration of an absorbing species:

$$A = abc \tag{15}$$

This fundamental equation for spectrometric quantitative analysis is known as Beer–Lambert–Bouguer law, sometimes shortened to Beer's law or Beer–Lambert law.

In practice, several steps are involved in the development of a quantitative method [35]. The first and probably most crucial for the ultimate success of the analytical method is to understand the system, i.e., to obtain reference spectra of the analyte and all other components. In the second step, the best method of sampling should be determined. Then, with standards having been prepared, the system must be calibrated. The last step before analyzing samples is to prepare validation samples and to evaluate the method.

With modern sampling techniques, good quantitative infrared analysis with virtually every type of sample is practicable; however, liquids are ideal for this purpose, being measured in a liquid cell of fixed thickness, either as 100 % sample or diluted with solvent. In this connection it should be taken into account that there are no ideal solvents for infrared spectroscopy [35]. In addition, because absorption bands and path length can be influenced by the temperature of the transmission cell, it is advisable to control the temperature.

Polymers are usually analyzed as pressed films, and solids are often difficult to examine quantitatively by infrared methods; in these cases, absorption band ratios [10] give the best results. In fact, the main application of multicomponent quantitative infrared analysis is gas analysis. If the difficulty in handling the gases is overcome, multicomponent analysis can be readily performed.

A simple chemical system can consist of a pure single component, or of a single component or more than one component in a mixture with no spectral interference; it is assumed that the radiation absorption by one component is not affected by the presence of other components. In simple systems, absorbance peak height measurements, directly or by using a selected baseline [35], are often employed for calibration and analysis. Because of intrinsic instrumental errors the practical limit for usable absorbance values is about three. Peak height measurements are also sensitive to changes in instrumental resolution and can vary considerably from instrument to instrument. To circumvent these problems, an alternative method is the use of integrated absorbance or peak area [10].

Complex chemical systems are composed of one or more components in a mixture with a significant degree of spectral interference, or of several components with a large amount of mutual physical and/or chemical interaction. In these cases, quantitative analysis is best performed by statistical methods such as principal component regression (PCR) or partial least squares (PLS) [36]; these are offered in the software packages of instrument manufacturers and software suppliers. Artificial neural networks (ANNs) should be primarily used when a data set is nonlinear [37].

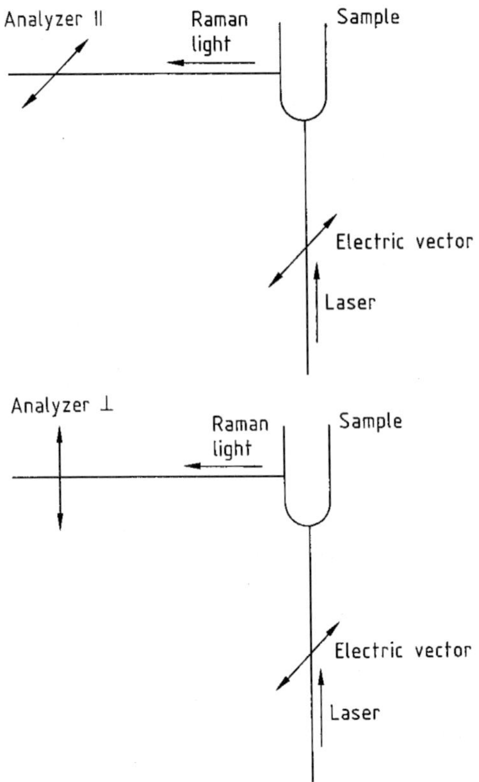

Figure 5. Orientation of the analyzer in relation to the direction of the electrical field of the exciting laser

17.3.3. The Symmetry of Molecules and Molecular Vibrations; Selection Rules

All molecules show symmetry properties and they all possess at least one (trivial) symmetry element, the identity. The symmetry of a molecule is important in spectroscopy because changes in symmetry during molecular vibration determine whether a vibrational dipole moment μ occurs or not. As mentioned, a vibration is infrared active if μ changes, if not, it is infrared inactive; analogously, this is true of the polarizability α and Raman activity.

In infrared and Raman spectroscopy, the symmetry of molecules is usually discussed in terms of five symmetry elements and their corresponding five symmetry operations [38]. If a wide variety of molecules is investigated, it will be found, as can be proved by mathematical group theory [39], that only certain combinations of symmetry elements are possible. Such restricted combinations of symmetry elements, known as point groups, are used for the classification of molecules. Each point group is associated with a set of normal vibrations. These in turn are classified according to the symmetry of vibration. From this it is possible to predict whether a normal vibration is infrared or Raman active or neither. These are the so-called selection rules. Many vibrations of molecules with low symmetry are both infrared and Raman active, and it is chiefly in the band intensities that the two types of spectra differ, sometimes markedly so. On the other hand, when a center of symmetry is present, i.e., for molecules with a high degree of symmetry, bands that are infrared active are Raman inactive and vice versa. By using laser excitation (i.e., linearly polarized light) (see Section 17.2.2), Raman spectroscopy also provides a means of recognizing totally symmetrical vibrations [3]. The intensity of a Raman line depends on the direction of the exciting electric field in relation to the orientation of the analyzer (Fig. 5). The latter can be either perpendicular or parallel to the direction of the electric field. The depolarization ratio ϱ is the ratio between the intensities of scattered light measured at each of these two positions:

$$\varrho = \frac{I_{\perp}}{I_{\parallel}} \qquad (16)$$

The depolarization ratio of a Raman line depends on the symmetry of the molecular vibration involved (i.e., the change in molecular symmetry induced by the corresponding vibration); the maximum value of depolarization observed with linearly polarized light is $\varrho_l^{max} = 3/4$. If a Raman line shows this extent of depolarization, it is said to be depolarized, whereas, if the degree of depolarization is smaller, the line is polarized. It can be shown [3] that only Raman lines corresponding to totally symmetric vibrations can have a degree of depolarization smaller than the maximum value, that is, can be polarized.

17.4. Interpretation of Infrared and Raman Spectra of Organic Compounds

17.4.1. The Concept of Group Frequencies

A complex molecule can be considered as a system of coupled inharmonic oscillators. Empir-

Table 2. Commonly used symbols and descriptions or different vibrational forms

Symbol	Designation	Example	Illustration
s	symmetric	ν_s	
as	antisymmetric (asymmetric)	ν_{as}	
ip	in-plane (ip)	δ	
oop	out-of-plane (oop)	γ	
ν	stretching		
δ	ip deformation		
γ	oop deformation		
ω	wagging		
τ	twisting		
ϱ	rocking		

ically it is found that vibrational coupling is restricted to certain submolecular groups of atoms. This coupling is relatively constant from molecule to molecule, so that the submolecular groups produce bands in a characteristic frequency region of the vibrational spectrum. These bands — the characteristic group frequencies — are readily predictable and so form the empirical basis for the interpretation of vibrational spectra. Their position, intensity, and width are decisive for the correlation of an absorption band with a certain submolecular group. For example, the vibrational spectra of *n*-heptane, *n*-octane, and *n*-nonane have a number of bands in common; these are the group frequencies for normal alkanes. This concept of group frequencies corresponds to the concept of chemical shift in nuclear magnetic resonance spectroscopy. These spectra also have a number of bands which are not in common; these so-called fingerprint bands are characteristic of the individual chemical compound and are used to distinguish one compound from another. In Table 2, commonly used symbols and descriptions of different vibrational forms are given, and Table 3 lists usual abbreviations for the classification of vibrational absorption bands. For practical applications, spectral interpretation is mainly based on personal experience.

Table 3. Usual abbreviations for the classification of vibrational absorption bands

Abbreviation	Signification
vs	very strong
s	strong
m	medium
w	weak
vw	very weak
sh	shoulder
b	broad
sr	sharp
v	variable

In the following sections, dealing with spectrum – structure correlations, characteristic group frequencies having diagnostic value are given in the form of charts. These correlation charts show the positions of the characteristic group frequencies which are the same in infrared and Raman spectra. The two types of spectra mainly differ in the band intensities, which, however, are not indicated in the correlation charts. To provide this information, examples of infrared and Raman spectra of most of the functional groups discussed are shown. Fourier transform Raman spectra (Fig. 4) are usually recorded between 3600 and 50 cm^{-1} (Stokes shift) and −100 to −2000 cm^{-1} (anti-Stokes shift). For comparison with mid-infrared spectra, in the following sections the Fourier transform Raman spectra are presented in the range between 4000 and 400 cm^{-1}. The relative intensities of vibrational absorption bands, which are evident in the model spectra presented, are important for appropriate spectral interpretation. Specific differences between infrared and Raman spectra are given special mention.

In the figures in this chapter (Figs. 8 – 44), the upper curve is the infrared spectrum, with intensity increasing from the bottom to the top of the diagram. The lower curve is the Raman spectrum with an ordinate linear in relative intensity units increasing from the bottom to the top of the diagram. The Raman spectra have not been corrected for fluctuations in the sensitivity of the spectrometer. The infrared spectra were recorded with a Bruker IFS 66 FT-IR spectrometer; for recording of the Raman spectra a Bruker FRA 106 FT-Raman accessory was used.

17.4.2. Methyl and Methylene Groups

Figure 6 shows characteristic group frequencies, and Figure 7 illustrates typical vibrations of

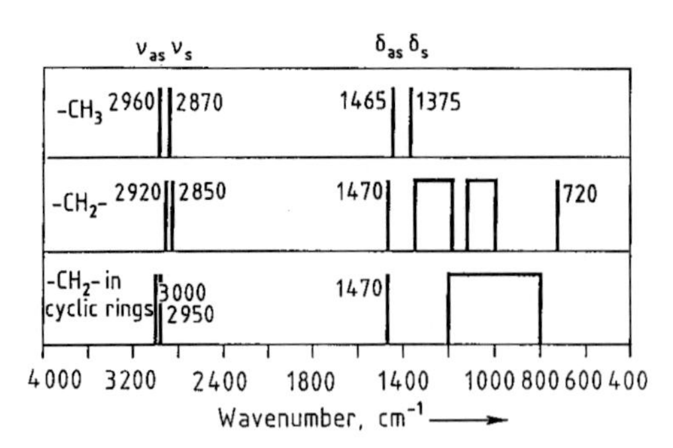

Figure 6. Characteristic group frequencies of alkanes

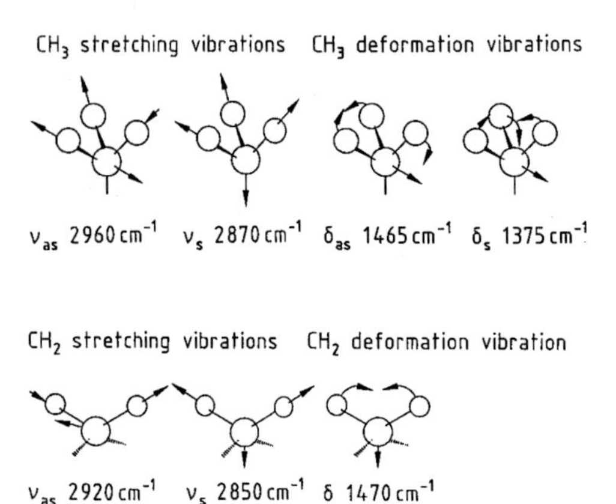

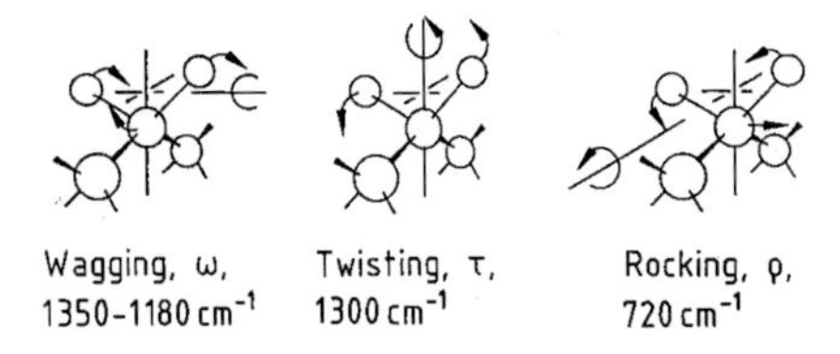

Figure 7. Typical vibrations of alkanes

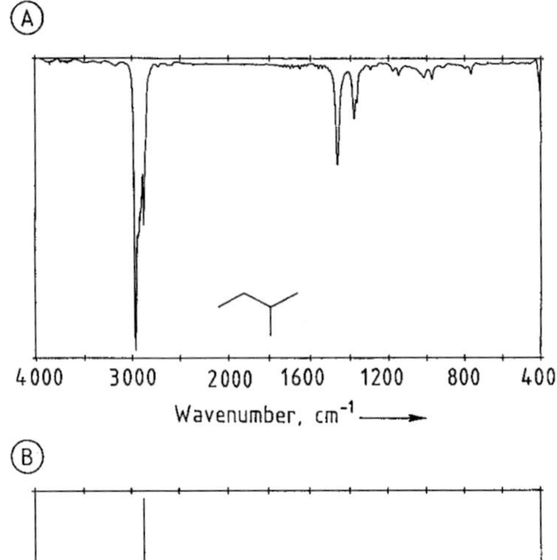

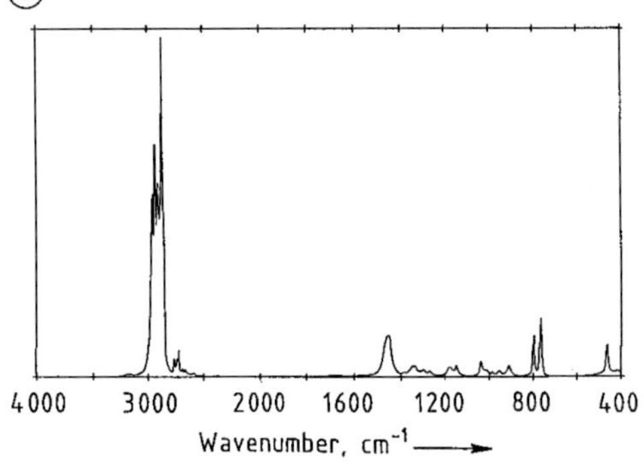

Figure 8. FT-infrared (A) and FT-Raman (B) spectra of 2-methylbutane

alkanes. In Figure 8, infrared and Raman spectra of 2-methylbutane (isopentane) are given as an example of alkane spectra. The most intense bands are those of the antisymmetric and symmetric CH_3 or CH_2 stretching vibrations between 2960 and 2850 cm^{-1}. Typical of isopropyl and tertiary butyl groups is that the symmetric CH_3 deformation band, usually observed at 1375 cm^{-1}, is split into two bands near 1390 and 1370 cm^{-1}. In the Raman spectra, the symmetric CH_3 deformation band near 1375 cm^{-1} is generally relatively weak in alkanes, but when the CH_3 group is next to a double or triple bond or an aromatic ring the Raman intensity is noticeably enhanced [10]. This is demonstrated in Figure 9, which shows the spectra of 2-methyl-2-butene.

The CH_2 wagging bands are spread over a region between 1350 and 1180 cm^{-1}, occurring as a characteristic progression of weak bands. They are best seen in the solid-phase spectra of long straight-chain compounds such as fatty acids [10].

The CH_2 twisting vibrations in CH_2 chains have frequencies dispersed over the same region as the wagging bands, as can be seen in the spectra of polyethylene [25], for example. The infrared intensity of these bands is weak, whereas in the Raman spectrum at about 1300 cm^{-1} the in-phase CH_2 twist vibration is a useful group frequency.

The C–C stretching vibration of CH_2 chains is seen only in the Raman spectrum, at 1131 and 1061 cm^{-1} [40], while the CH_2 rocking vibration at 720 cm^{-1}, which is characteristic for longer $(CH_2)_n$ chains with $n \geq 4$, is observed only in the infrared spectrum (e.g., polyethylene) [25].

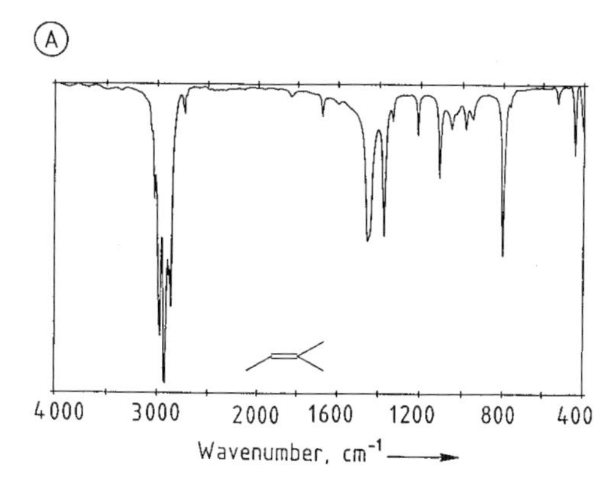

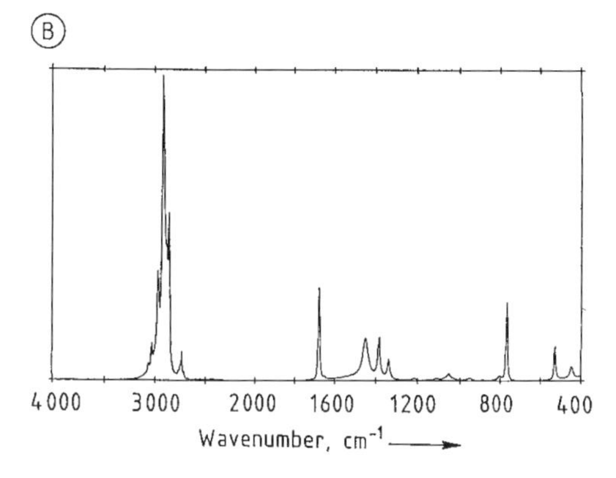

Figure 9. FT-infrared (A) and FT-Raman (B) spectra of 2-methyl-2-butene

Aliphatic ring compounds are best characterized by their Raman spectra, since in the infrared only very weak characteristic ring frequencies due to C–C ring stretching vibrations are observed near 1000 cm^{-1} (Fig. 10). Raman spectra, in contrast, show prominent lines due to ring stretching vibrations: cyclopropane 1188 cm^{-1} [25], cyclobutane 1001 cm^{-1} [40], cyclopentane 886 cm^{-1} [25], and cyclohexane 802 cm^{-1} (Fig. 10).

17.4.3. Alkene Groups

Characteristic group frequencies of alkenes are shown in Figure 11. In alkenes, the =CH stretching vibration generally occurs above 3000 cm^{-1}. In symmetrical *trans* or symmetrical tetrasubstituted double-bond compounds, the C=C stretching frequency near 1640 cm^{-1}, usually a medium intensity band, is infrared inactive because, in this case, no change of dipole moment occurs as a result of the vibration. However, the C=C stretching vibration gives rise to a strong Raman signal in the region between 1680 and 1630 cm^{-1} in all types of alkenes (Fig. 12).

In vinyl and vinylidine groups, the $=CH_2$ in-plane deformation gives rise to a medium intensity band in both the infrared and Raman spectrum near 1415 cm^{-1}. In the case of the vinyl group, an additional infrared and Raman deformation band is observed near 1300 cm^{-1} [10]. In *cis*-1,2-dialkyl ethylenes, the in-plane deformation band appears near 1405 cm^{-1} in the infrared and at ca. 1265 cm^{-1} in the Raman spectrum. In the corresponding *trans* compounds, the in-plane bending vibrations occur at 1305 cm^{-1} in the Raman and at 1295 cm^{-1} in the infrared spectrum. In these *trans* compounds, the =CH out-of-plane bending vibration gives rise to a strong infrared band, of high diagnostic value, occurring between 980 and 965 cm^{-1}.

17.4.4. Aromatic Rings

Figure 13 shows characteristic group frequencies of aromatic rings. The bands between 1600 and 1460 cm^{-1} are ring modes involving C–C partial double bonds of the aromatic ring. The band at 1500 cm^{-1} is usually the strongest of these bands, which generally show medium to strong infrared but weak Raman intensities. The use of both infrared and Raman spectra gives reliable information about the type of substitution of an aromatic ring: in infrared spectra, intense bands in the 850–675 cm^{-1} region, due to out-of-plane CH wagging and out-of-plane sextant ring bending vibrations, are indicative of the type of substitution [10]. In the Raman spectra, there are very strong lines at ca. 1000 cm^{-1} involving ring stretching and ring bending vibrations. As shown in Figures 14–17, a very strong signal at 1000 cm^{-1} occurs for mono and *meta* substitution, a strong line is observed between 1055 and 1015 cm^{-1} for *ortho* substitution, while no signal is observed around 1000 cm^{-1} for *para* substitution. In infrared spectra, the out-of-plane CH wagging vibrations give rise to relatively prominent summation bands in the 2000–1650 cm^{-1} region. As the summation band patterns are approximately constant for a particular ring substitution, they can also be used to determine the type of substitution [10].

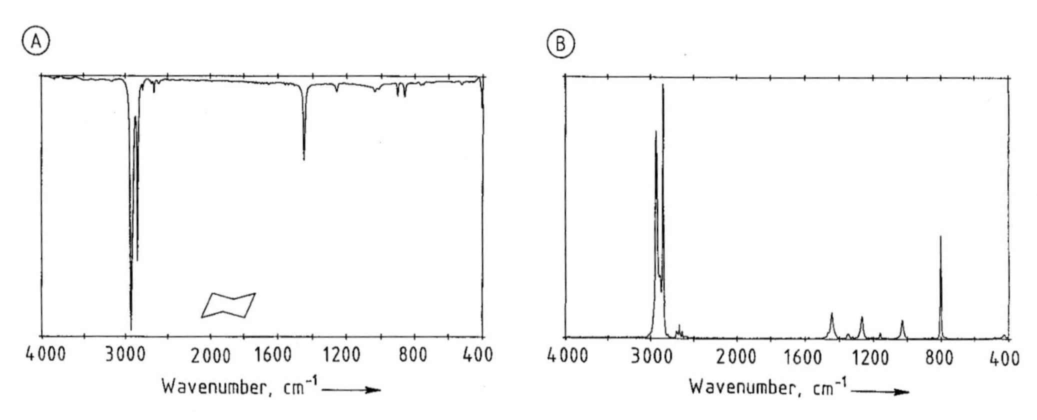

Figure 10. FT-infrared (A) and FT-Raman (B) spectra of cyclohexane

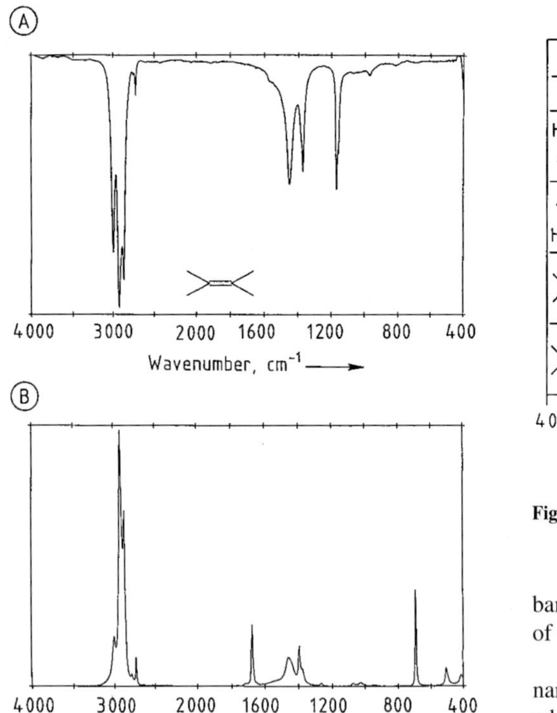

Figure 12. FT-infrared (A) and FT-Raman (B) spectra of 2,3-dimethyl-2-butene

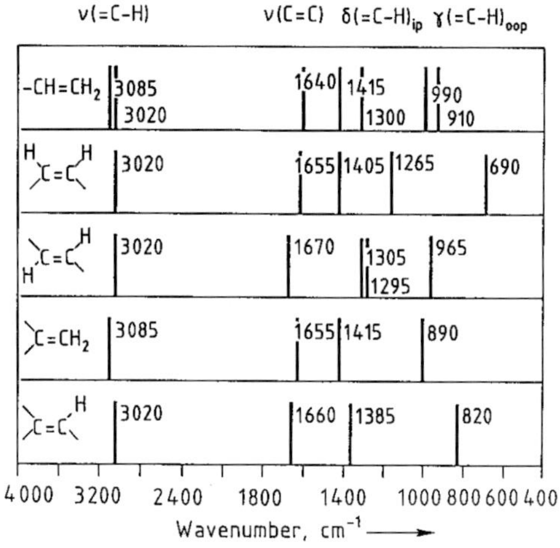

Figure 11. Characteristic group frequencies of alkenes

17.4.5. Triple Bonds and Cumulated Double Bonds

Triple bonds (X≡Y) and cumulated double bonds (X=Y=Z) absorb roughly in the same region, namely, 2300–2000 cm^{-1} (Table 4). The bands are very diagnostic because in this region of the spectrum no other strong bands occur.

Monosubstituted acetylenes have a relatively narrow ≡CH stretching band near 3300 cm^{-1} which is strong in the infrared and weaker in the Raman. In monoalkyl acetylenes, the C≡C stretching vibration shows a weak infrared absorption near 2120 cm^{-1}, while no infrared C≡C absorption is observed in symmetrically substituted acetylenes because of the center of symmetry. In Raman spectra, this vibration always occurs as a strong line. Nitriles are characterized by a strong C≡N stretching frequency at 2260–2200 cm^{-1} in the infrared and in the Raman (Fig. 18). Upon electronegative substitution at the α-carbon atom, the infrared intensity of the C≡N stretching vi-

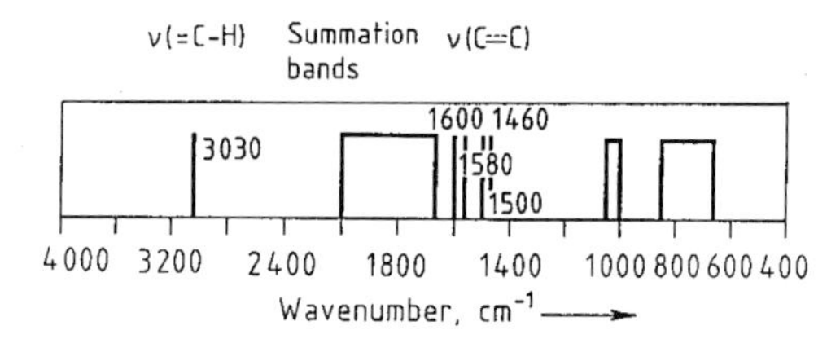

Figure 13. Characteristic group frequencies of aromatic rings

bration is considerably weakened so that only the Raman spectrum, where a strong signal is always observed, can be used for identification (Fig. 19).

17.4.6. Ethers, Alcohols, and Phenols

17.4.6.1. Ethers

Characteristic group frequencies of ethers are shown in Figure 20. The key band for noncyclic ethers is the strong antisymmetric C–O–C stretching vibration between 1140 and 1085 cm^{-1} for aliphatic, between 1275 and 1200 cm^{-1} for aromatic, and between 1225 and 1200 cm^{-1} for vinyl ethers. With the exception of aliphatic – aromatic ethers, the symmetric C–O–C stretching band at ca. 1050 cm^{-1} is too weak to have diagnostic value. In cyclic ethers, both the antisymmetric (1180 – 800 cm^{-1}) and the symmetric (1270 – 800 cm^{-1}) C–O–C stretching bands are characteristic. The antisymmetric C–O–C stretch is usually a strong infrared band and a weak Raman signal, the symmetric C–O–C stretching vibration generally being a strong Raman line and a weaker infrared band (Fig. 21).

17.4.6.2. Alcohols and Phenols

Characteristic group frequencies of alcohols and phenols are shown in Figure 22. Because of the strongly polarized OH group, alcohols generally show a weak Raman effect (and therefore are excellent solvents for Raman spectroscopy). In the liquid or solid state, alcohols and phenols exhibit strong infrared bands due to O–H stretching vibrations. Due to hydrogen bonding of the OH groups, these bands are very broad in the pure liquid or solid, as well as in concentrated solutions or mixtures. The position of the similarly intense C–O stretching band permits primary, secondary, and tertiary alcohols, as well as phenols to be distinguished between.

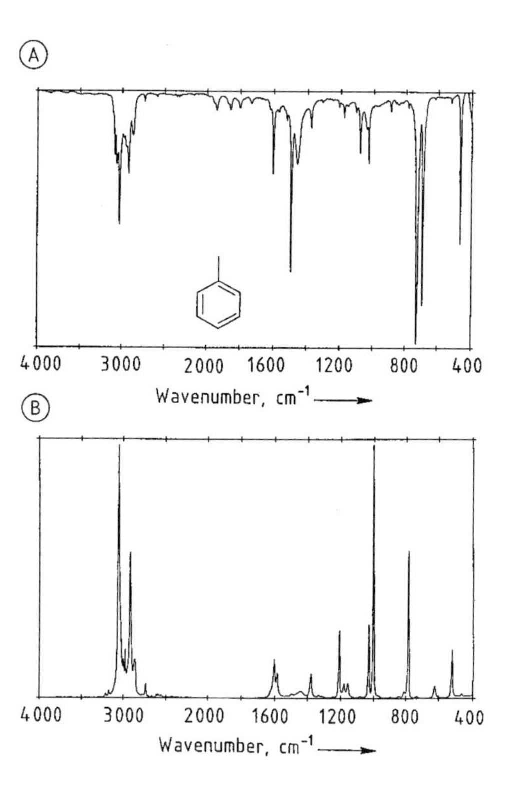

Figure 14. FT-infrared (A) and FT-Raman (B) spectra of toluene

17.4.7. Amines, Azo, and Nitro Compounds

17.4.7.1. Amines

Characteristic group frequencies of amines are shown in Figure 23. In the infrared and Raman, key bands are the NH stretching bands, although they are often not very intense. Primary amines exhibit two bands due to antisymmetric (3550 – 3330 cm^{-1}) and symmetric (3450 – 3250 cm^{-1}) stretching, while for secondary amines only one NH stretching band is observed. The intensities of NH and NH_2 stretching bands are generally weaker in aliphatic than in aromatic amines. The C–N stretching vibration is not necessarily diagnostic, because in aliphatic amines it gives rise to absorption bands of only weak to medium intensity between 1240 and 1000 cm^{-1}. In aromatic amines, strong C–N stretching bands are observed in the range 1380 – 1250 cm^{-1}.

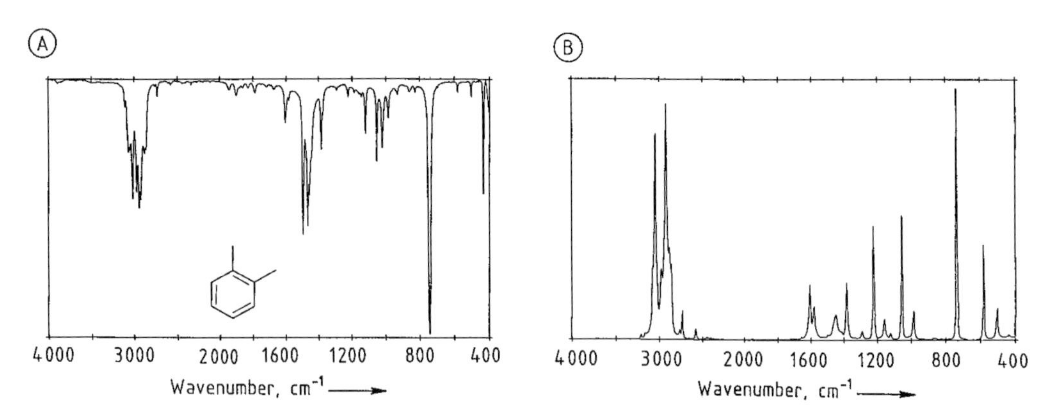

Figure 15. FT-infrared (A) and FT-Raman (B) spectra of *o*-xylene

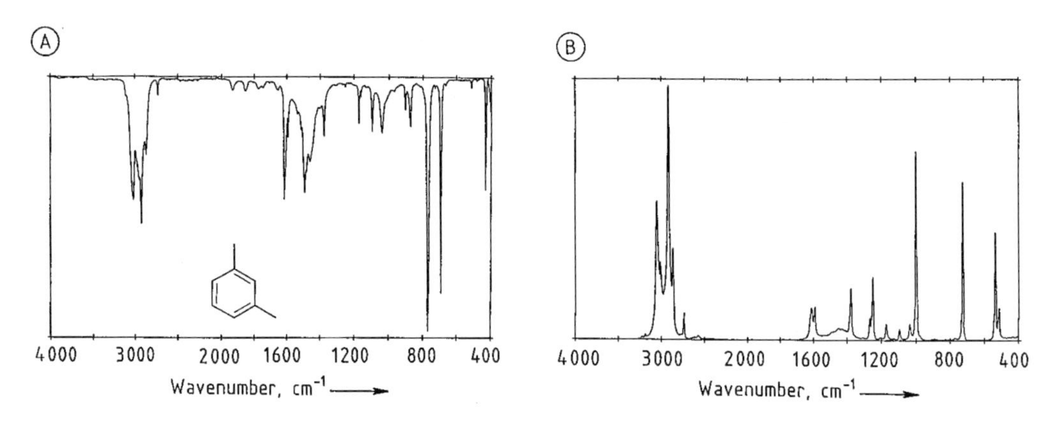

Figure 16. FT-infrared (A) and FT-Raman (B) spectra of *m*-xylene

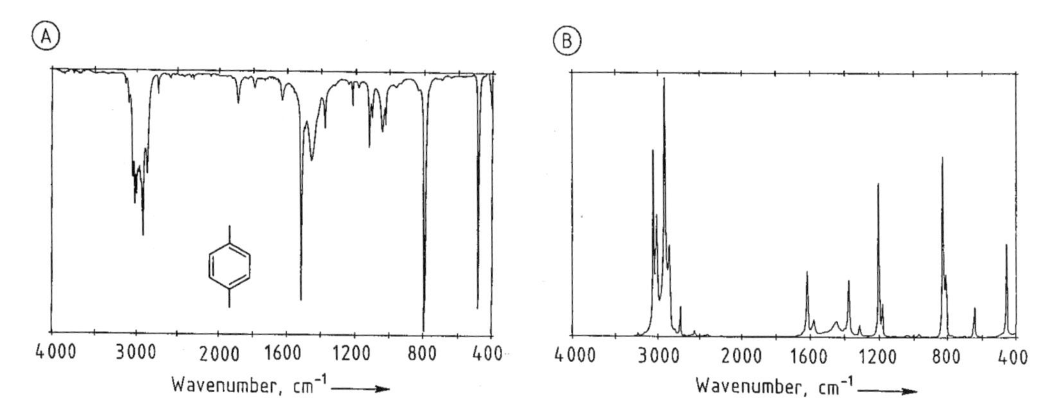

Figure 17. FT-infrared (A) and FT-Raman (B) spectra of *p*-xylene

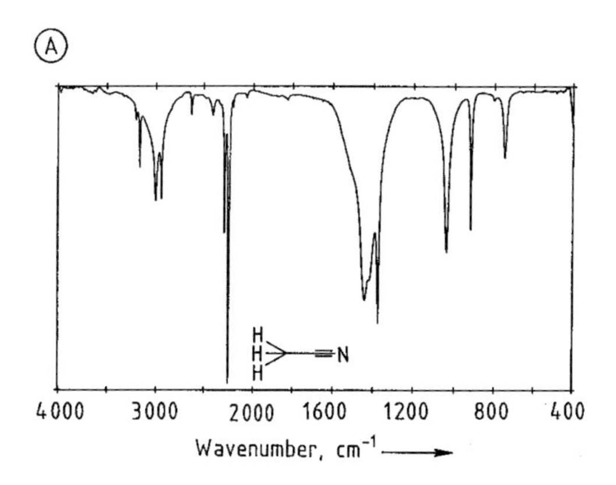

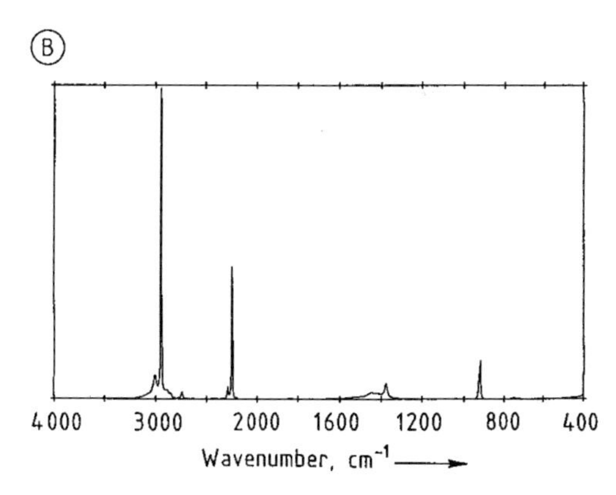

Figure 18. FT-infrared (A) and FT-Raman (B) spectra of acetonitrile

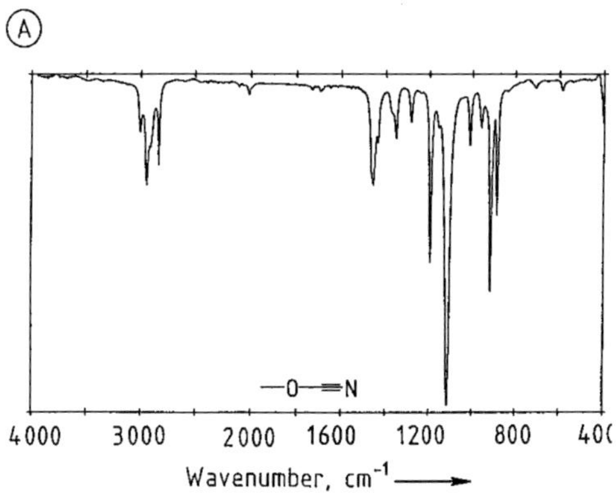

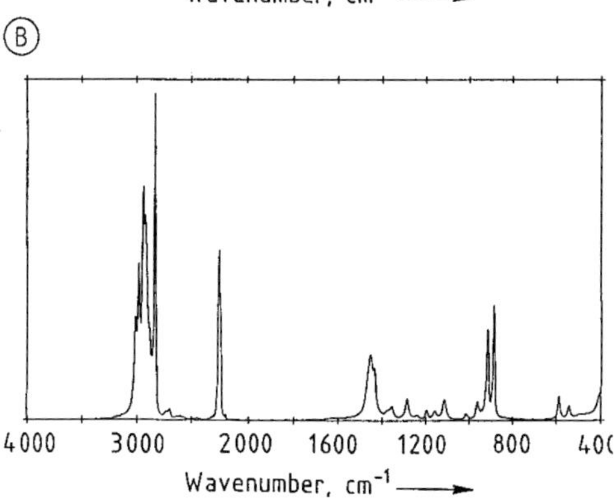

Figure 19. FT-infrared (A) and FT-Raman (B) spectra of methoxy-acetonitrile

Table 4. Characteristic group frequencies of triple bonds and cumulated double bonds

Bond	$\nu(\equiv C{-}H)$, cm^{-1}	$\nu(X\equiv Y)$ or $\nu_{as}(X{=}Y{=}Z)$, cm^{-1}
$-C\equiv C{-}H$	3300	2260 – 2100
$-C\equiv N$		2260 – 2200
$-N\equiv C$		2165 – 2110
$-C\equiv N\rightarrow O$		2300 – 2290
$-N^+\equiv N$		2300 – 2140
$-S{-}C\equiv N$		2180 – 2140
$-N{=}C{=}S$		2150 – 2000
$-N{=}C{=}O$		2275 – 2230
$-N{=}N^+{=}N^-$		2250 – 2100
$-N{=}C{=}N-$		2150 – 2100
$-C{=}C{=}O$		2155 – 2130
$-C{=}N^+{=}N^-$		2100 – 2010
$-CO{-}C{=}N^+{=}N^-$		2100 – 2050

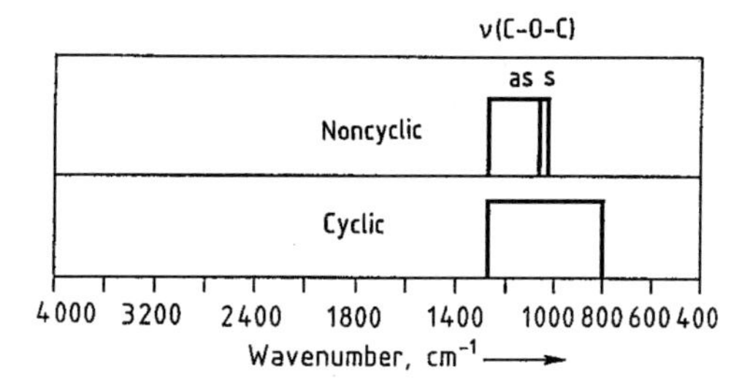

Figure 20. Characteristic group frequencies of ethers

17.4.7.2. Azo Compounds

Characteristic group frequencies of azo compounds are shown in Figure 24. In the infrared spectrum, similarly to symmetrical *trans*-substituted ethylenes, the N=N stretching vibration of the *trans* symmetrically substituted azo group does not occur, whereas all azo compounds show a medium (aliphatic compounds) or strong (aromatic compounds) N=N stretching Raman line. This band occurs between 1580 and 1520 cm^{-1} for purely aliphatic or aliphatic – aromatic, and

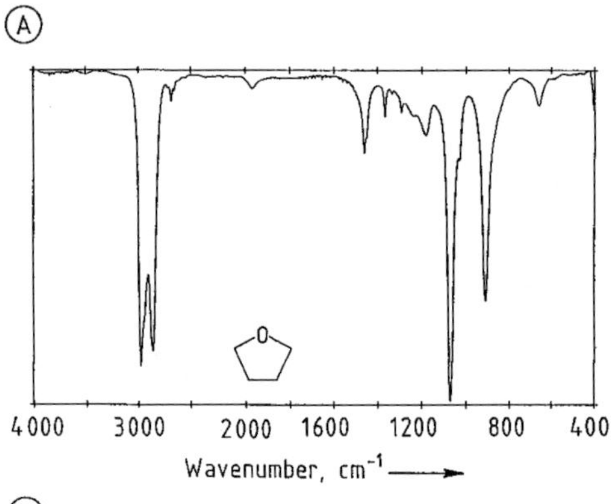

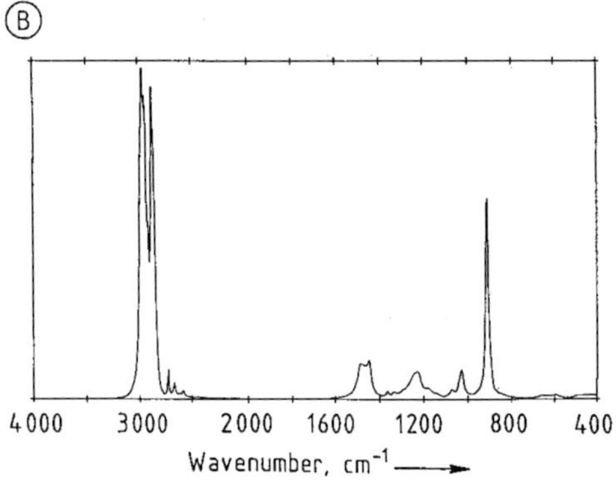

Figure 21. FT-infrared (A) and FT-Raman (B) spectra of tetrahydrofuran

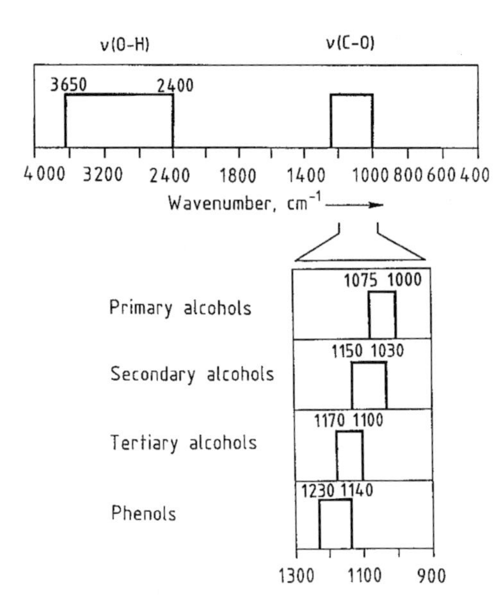

Figure 22. Characteristic group frequencies of alcohols and phenols

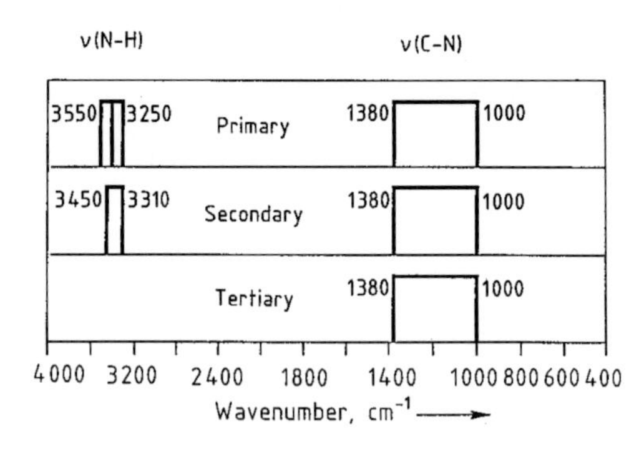

Figure 23. Characteristic group frequencies of amines

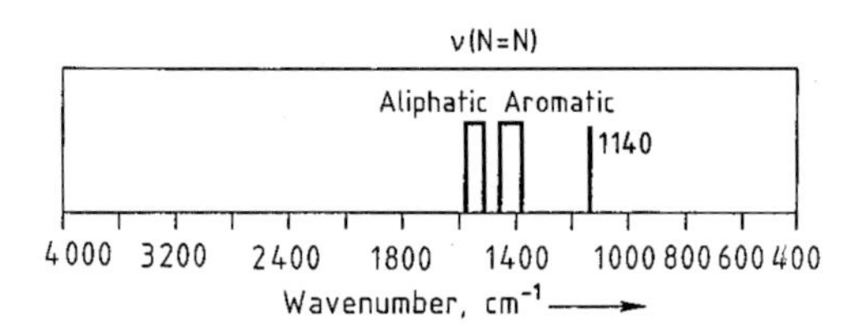

Figure 24. Characteristic group frequencies of azo compounds

between 1460 and 1380 cm^{-1} for purely aromatic azo compounds. In azoaryls, an additional strong Raman line involving aryl – N stretch is observed near 1140 cm^{-1}. As an example, the spectra of azobenzene are shown in Figure 25.

17.4.7.3. Nitro Compounds

Characteristic group frequencies of nitro compounds are shown in Figure 26. The nitro group, with its two identical N–O partial double bonds, gives rise to two bands due to antisymmetric and symmetric stretching vibrations. In the infrared, the antisymmetric vibration causes a strong absorption at 1590 – 1545 cm^{-1} in aliphatic, and at 1545 – 1500 cm^{-1} in aromatic nitro compounds. In comparison, the infrared intensities of the symmetric stretching vibration between 1390 and 1355 cm^{-1} for aliphatic, and 1370 and 1330 cm^{-1} for aromatic nitro compounds are somewhat weaker. However, a very strong Raman line due to symmetric vibration is observed in this region (see Fig. 27).

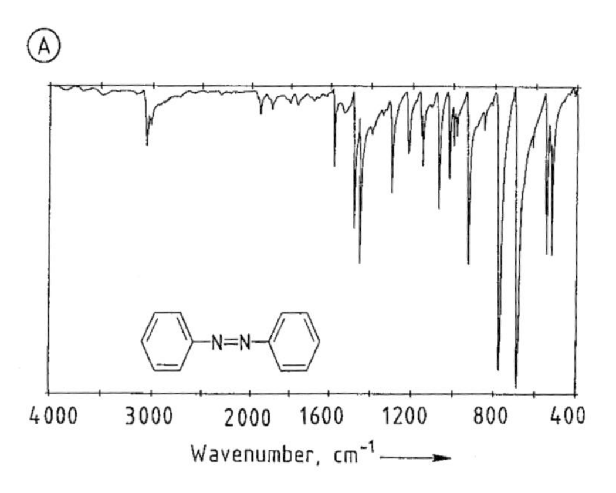

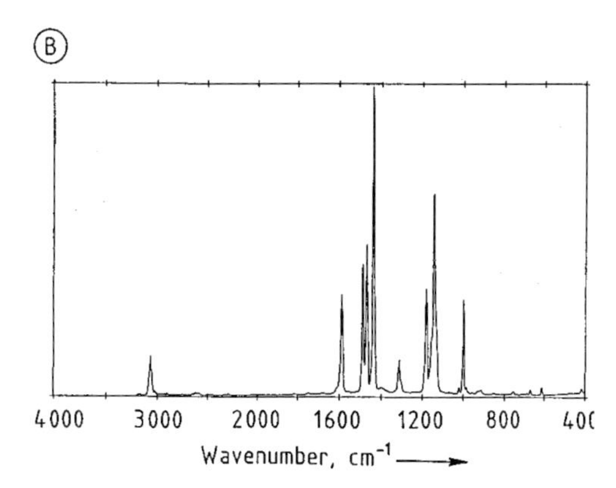

Figure 25. FT-infrared (A) and FT-Raman (B) spectra of azobenzene

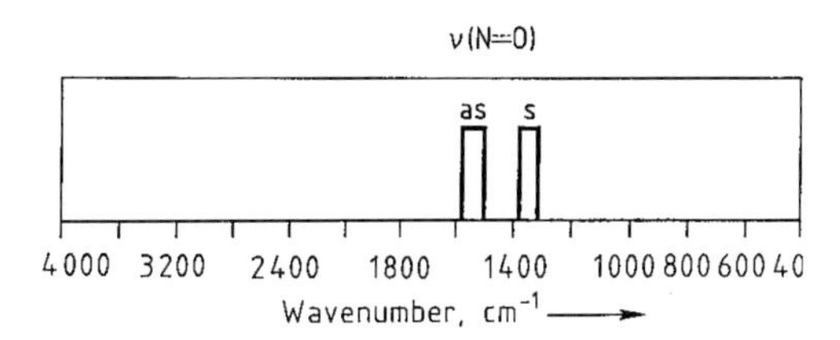

Figure 26. Characteristic group frequencies of nitro compounds

17.4.8. Carbonyl Compounds

Because of the stretching of the C=O bond, carbonyl compounds generally give rise to a very strong infrared and a medium to strong Raman line between 1900 and 1550 cm^{-1}. The position of the carbonyl band is influenced by inductive, mesomeric, mass, and bond-angle effects [10].

17.4.8.1. Ketones

Characteristic group frequencies of ketones are shown in Figure 28. Ketones are characterized by a strong C=O band near 1715 cm^{-1}, often accompanied by an overtone at ca. 3450 cm^{-1}. Ketones in strained rings show considerably higher C=O frequencies, whereas in di-*tert*-butyl ketone, for example, the carbonyl frequency is observed at 1687 cm^{-1}; a steric increase in the C–C–C angle causes this lower value.

17.4.8.2. Aldehydes

Characteristic group frequencies of aldehydes are shown in Figure 29. Together with the carbonyl band, the most intense band in the infrared spectrum, aldehydes are characterized by two CH bands at 2900 – 2800 cm^{-1} and 2775 –2680 cm^{-1}. These bands are due to Fermi resonance [3], i.e., interaction of the CH stretch fundamental with the overtone of the O=C–H bending vibration near 1390 cm^{-1}.

17.4.8.3. Carboxylic Acids

Characteristic group frequencies of carboxylic acids are shown in Figure 30. Because carboxylic acids have a pronounced tendency to form hydrogen-bonded dimers, in the condensed state they are characterized by a very broad infrared OH stretching band centered near 3000 cm^{-1}. As the dimer has a center of symmetry, the antisymmetric C=O stretch at 1720 – 1680 cm^{-1} is infrared active only, whereas the symmetric C=O stretch at 1680 – 1640 cm^{-1} is Raman active only; this is shown in Figure 31 using acetic acid as an example. In the infrared, a broad band at 960 – 880 cm^{-1} due to out-of-plane OH···O hydrogen bending is diagnostic of carboxyl dimers, as is the medium intense C–O stretching band at 1315 – 1280 cm^{-1}. On the other hand, monomeric acids have a weak, sharp OH absorption band at 3580 – 3500 cm^{-1} in the infrared, and the C=O stretching band appears at 1800 – 1740 cm^{-1} in both infrared and Raman.

17.4.8.4. Carboxylate Salts

Characteristic group frequencies of carboxylate salts are shown in Figure 32. In carboxylate salts, the two identical C–O partial double bonds give rise to two bands: the antisymmetric CO_2

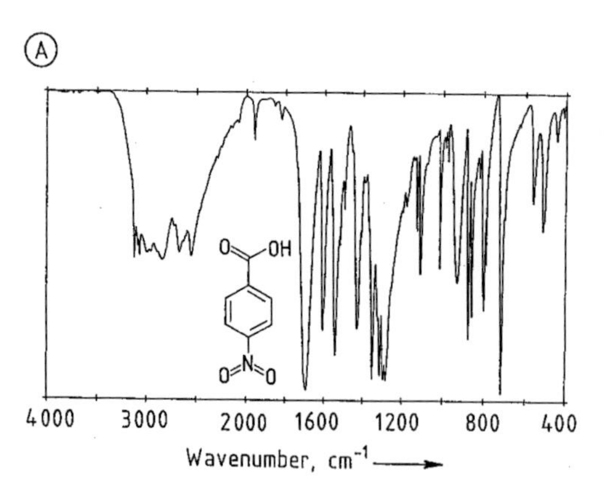

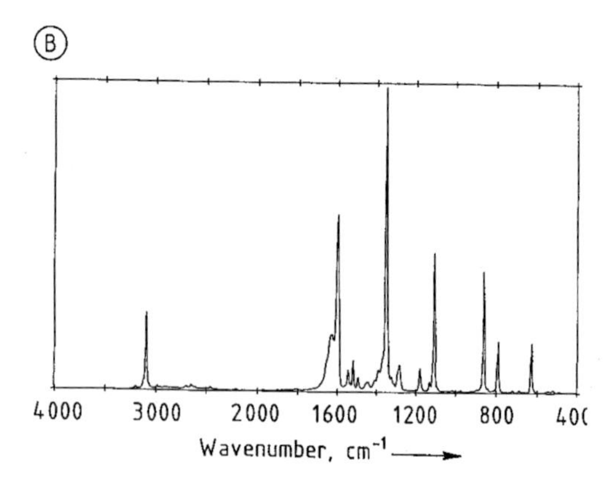

Figure 27. FT-infrared (A) and FT-Raman (B) spectra of 4-nitrobenzoic acid

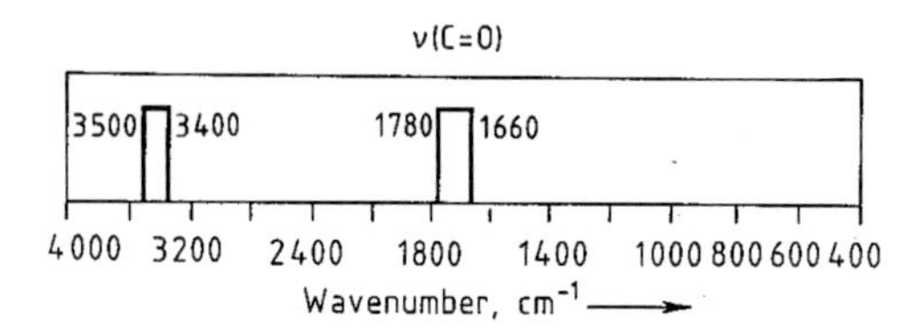

Figure 28. Characteristic group frequencies of ketones

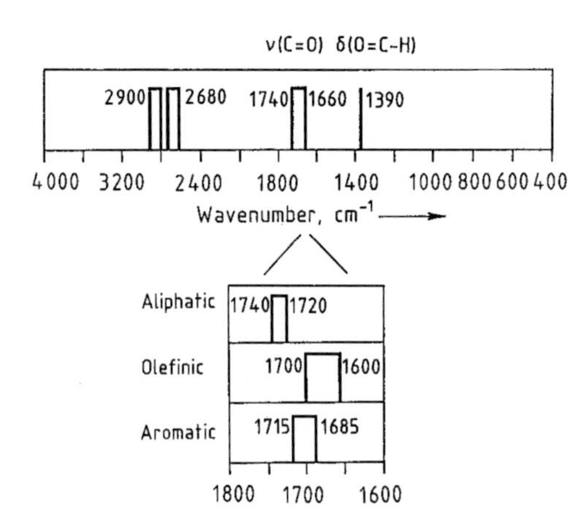

Figure 29. Characteristic group frequencies of aldehydes

stretching vibration at 1650 – 1550 cm^{-1} is very strong in the infrared and weak in the Raman, while the corresponding symmetric vibration at 1450 – 1350 cm^{-1} is somewhat weaker in the infrared but strong in the Raman.

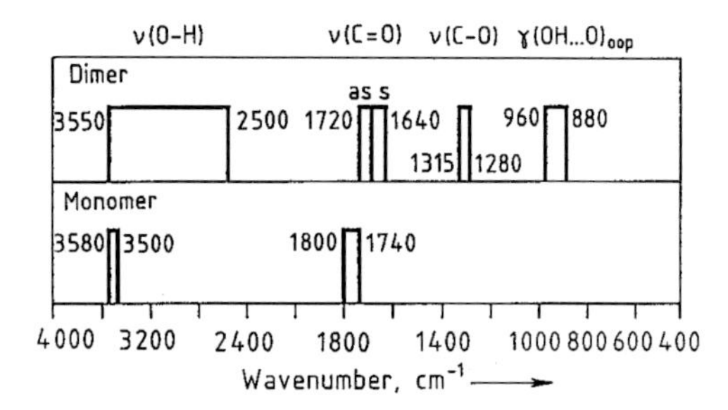

Figure 30. Characteristic group frequencies of carboxylic acids

17.4.8.5. Anhydrides

Characteristic group frequencies of anhydrides are shown in Figure 33. Anhydrides are characterized by two C=O stretching vibration bands, one of which occurs above 1800 cm^{-1}. These bands are strong in the infrared but weak in the Raman. In noncyclic anhydrides, the infrared C=O band at higher wavenumber is usually more intense than the C=O band at lower wavenumber, whereas in cyclic anhydrides the opposite is observed. In the Raman, the C=O line at higher wavenumber is generally stronger. Unconjugated straight-chain anhydrides have a strong infrared absorption, due to the C–O–C stretching vibration, at 1050 – 1040 cm^{-1} (except for acetic anhydride: 1125 cm^{-1}), while cyclic anhydrides exhibit this infrared band at 955 –895 cm^{-1}. In Figure 34, the spectra of acetic anhydride are shown as an example.

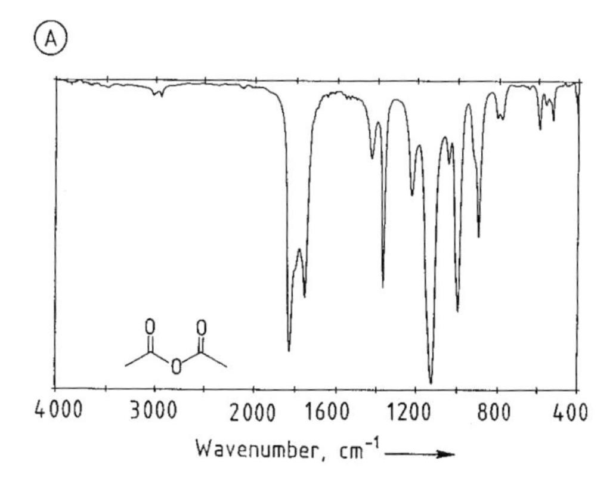

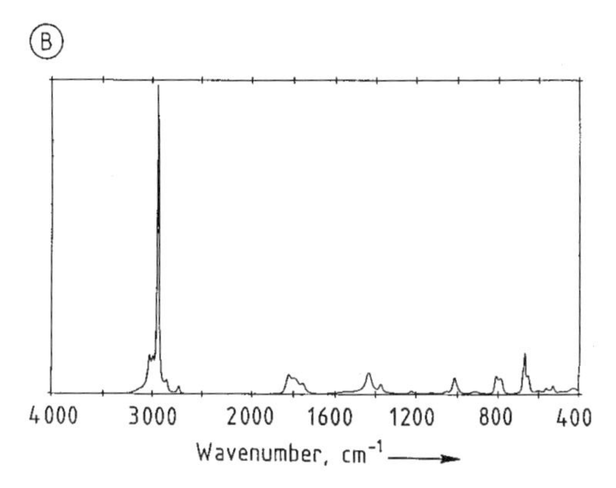

Figure 34. FT-infrared (A) and FT-Raman (B) spectra of acetic anhydride

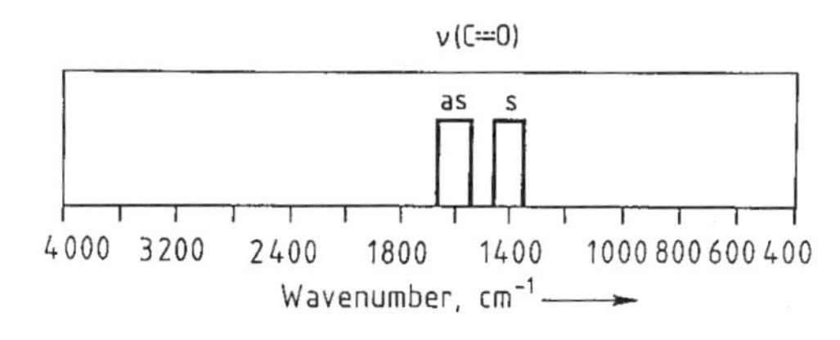

Figure 32. Characteristic group frequencies of carboxylate salts

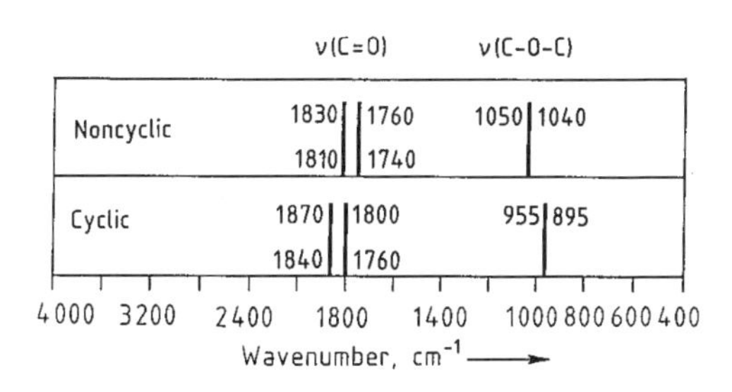

Figure 33. Characteristic group frequencies of anhydrides

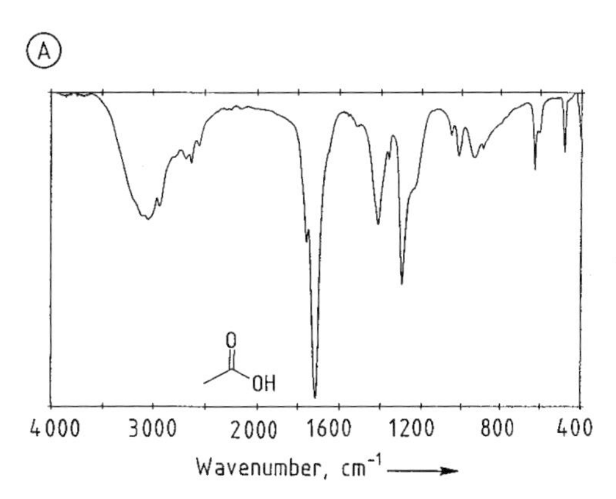

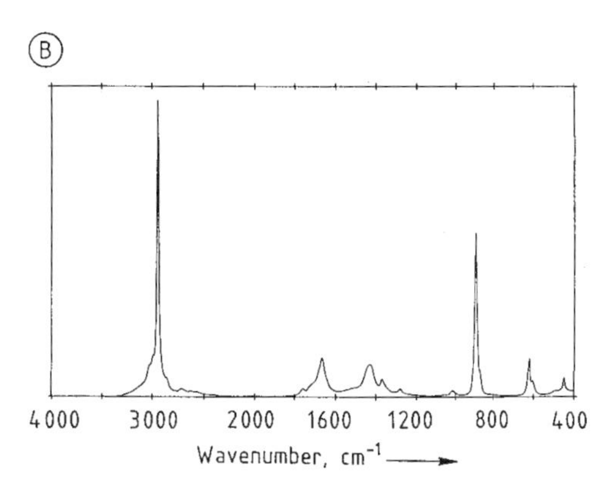

Figure 31. FT-infrared (A) and FT-Raman (B) spectra of acetic acid

17.4.8.6. Esters

Figure 35 shows the characteristic group frequencies of esters. In esters, the C=O stretching vibration gives rise to an absorption band at 1740 cm^{-1} which is very strong in the infrared and medium in the Raman. Its position is strongly influenced by the groups adjacent to the ester group. The C–O stretching band at 1300 – 1000 cm^{-1} often shows an infrared intensity similar to the C=O band.

17.4.8.7. Lactones

In lactones, the position of the C=O band is strongly dependent on the size of the ring (Table 5). In lactones with unstrained, six-membered and larger rings the position is similar to that of noncyclic esters.

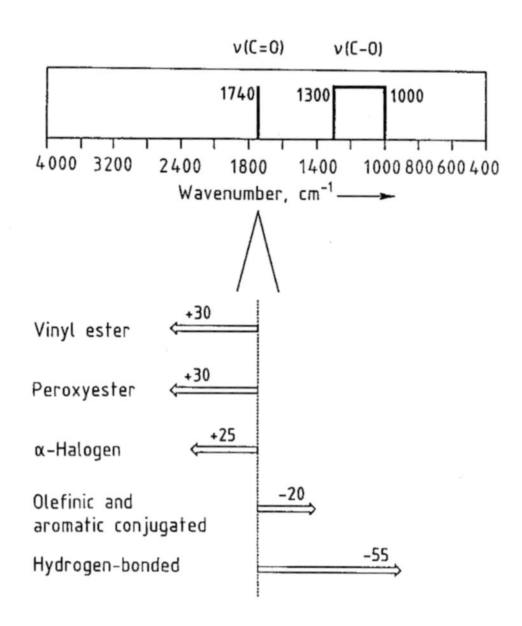

Figure 35. Characteristic group frequencies of esters

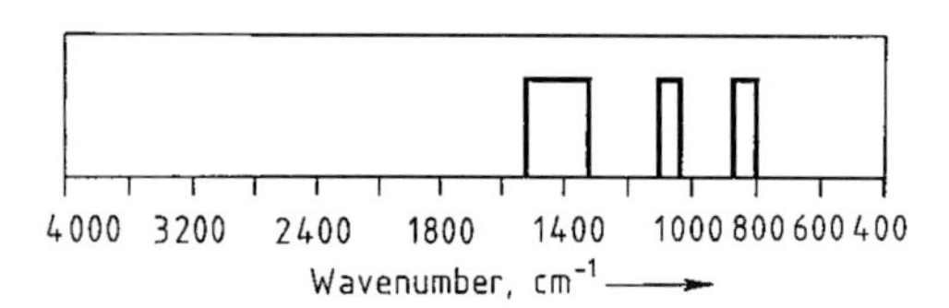

Figure 36. Characteristic group frequencies of carbonate salts

17.4.8.8. Carbonate Salts

Characteristic group frequencies of carbonate salts are shown in Figure 36. In inorganic carbonate salts, the antisymmetric CO_3 stretching vibration gives rise to a very broad absorption band at 1520 – 1320 cm^{-1}, which is very strong in the infrared but only weak in the Raman. The symmetric CO_3 stretching vibration, on the other hand, gives rise to a very strong Raman line at 1100 – 1030 cm^{-1}, which is normally inactive in the infrared. Another characteristic medium sharp infrared band at 880 – 800 cm^{-1} is caused by out-of-plane deformation of the CO_3^{2-} ion (Fig. 37).

17.4.8.9. Amides

Figure 38 shows characteristic group frequencies of amides. In amides, the C=O stretching vibration gives rise to an absorption at

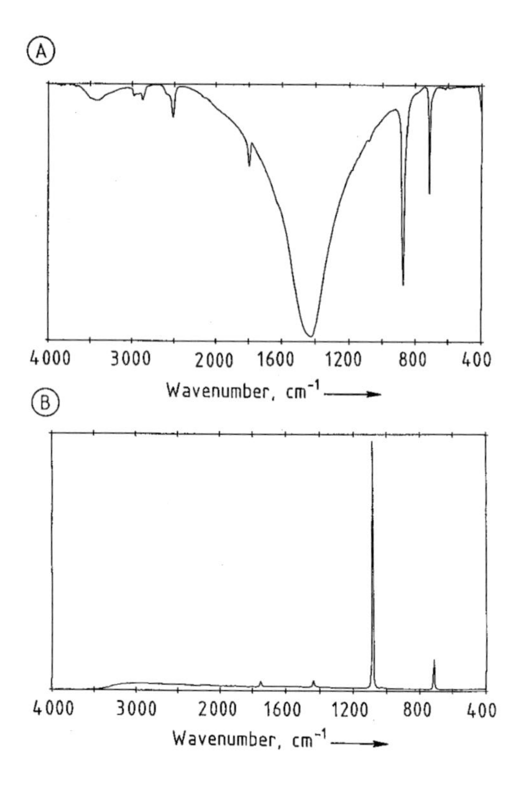

Figure 37. FT-infrared (A) and FT-Raman (B) spectra of calcium carbonate

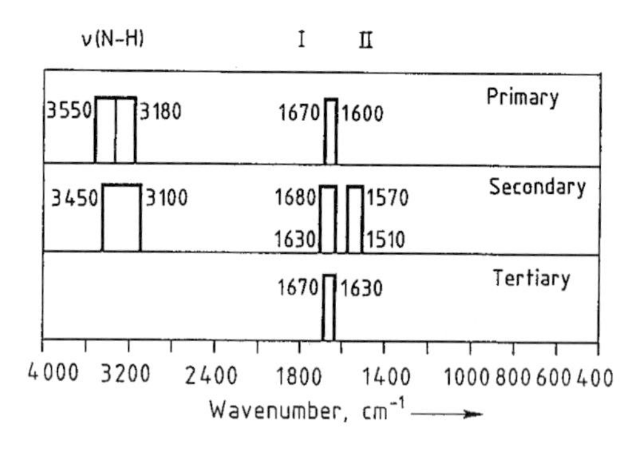

Figure 38. Characteristic group frequencies of amides

1680 – 1600 cm^{-1} known as the amide I band; it is strong in the infrared and medium to strong in the Raman. In primary amides, a doublet is usually observed in the amide I region, the second band being caused by NH_2 deformation. A characteristic band for secondary amides, which mainly exist with the NH and C=O in the *trans* configu-

Table 5. Characteristic group frequencies of lactones

Lactone		Structure	ν(C=O), cm^{-1}
β-Lactones		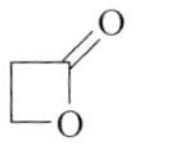	1840
γ-Lactones:	saturated	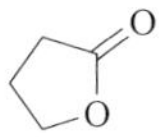	1795 – 1775
	α, β-unsaturated	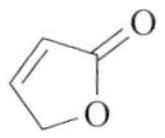	1785 – 1775 and 1765 – 1740
	β, γ-unsaturated	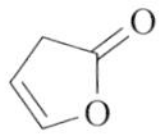	1820 – 1795
δ-Lactones:	saturated	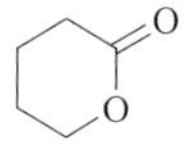	1750 – 1735
	α, β-unsaturated	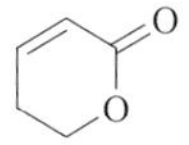	
	β, γ-unsaturated	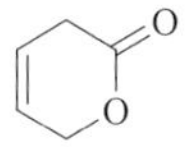	1760 – 1750 and 1740 – 1715

ration, is the amide II band at 1570 – 1510 cm^{-1}, involving in-plane NH bending and C – N stretching. This band is less intense than the amide I band.

As in amines, the NH stretching vibration gives rise to two bands at 3550 – 3180 cm^{-1} in primary amides and one near 3300 cm^{-1} in secondary amides; these are strong both in infrared and Raman. In secondary amides a weaker band appears near 3100 cm^{-1} due to an overtone of the amide II band.

17.4.8.10. Lactams

Characteristic group frequencies of lactams are shown in Figure 39. As in lactones, the size of the ring influences the C=O frequency in lactams. In six- or seven-membered rings the carbonyl vibration absorbs at 1680 – 1630 cm^{-1}, the same as the noncyclic *trans* case. Lactams in five-membered rings absorb at 1750 – 1700 cm^{-1}, while *β*-lactams absorb at 1780 – 1730 cm^{-1}. These bands are strong in the infrared and medium in the Raman.

Because of the cyclic structure the NH and C=O groups are forced into the *cis* configuration so that no band comparable to the amide II band in *trans* secondary amides occurs in lactams. A characteristically strong NH stretching absorption near 3200 cm^{-1} occurs in the infrared, which is only weak to medium in the Raman. A weaker infrared band near 3100 cm^{-1} is due to a combination band of the C=O stretching and NH bending vibrations.

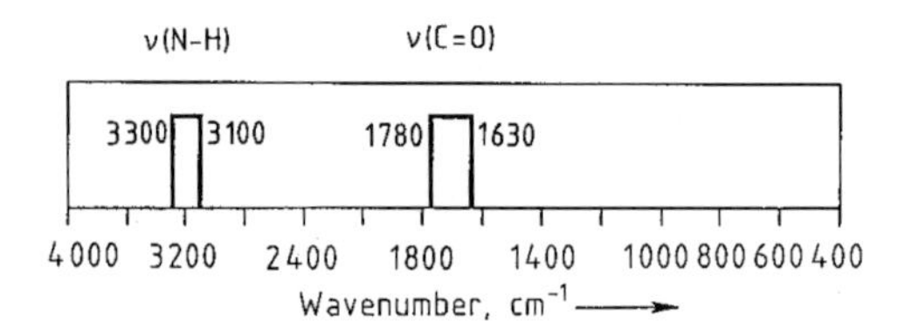

Figure 39. Characteristic group frequencies of lactams

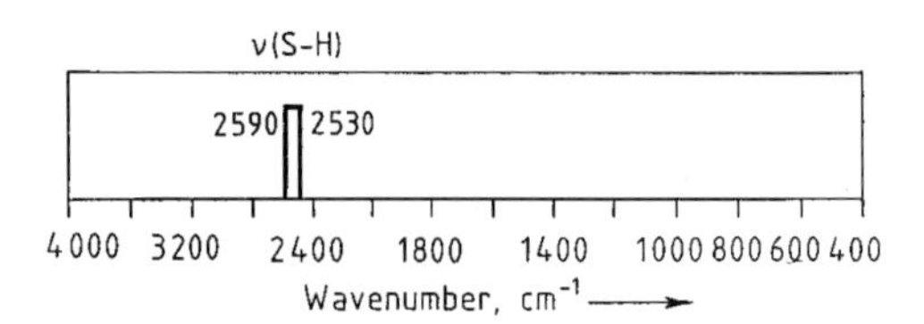

Figure 40. Characteristic group frequencies of thiols

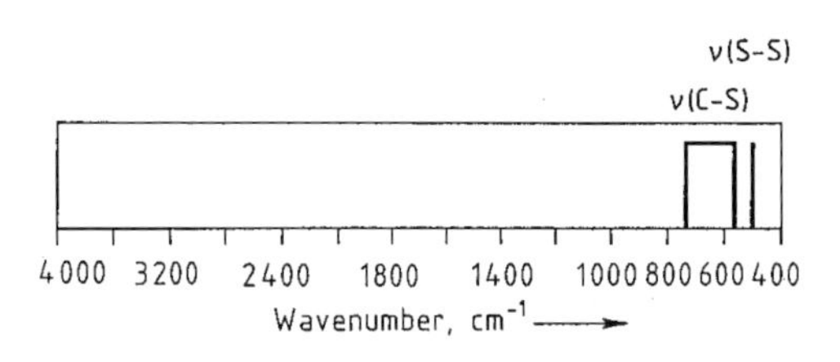

Figure 41. Characteristic group frequencies of sulfides and disulfides

17.4.9. Sulfur-Containing Compounds

Sulfur-containing functional groups generally show a strong Raman effect, and groups such as C–S and S–S, for example, are extremely weak infrared absorbers. Hence, Raman spectra of sulfur-containing compounds have a much greater diagnostic value than their infrared spectra.

17.4.9.1. Thiols

Characteristic group frequencies of thiols are shown in Figure 40. The S–H stretching vibration of aliphatic and aromatic thiols gives rise to a medium to strong Raman line at 2590 – 2530 cm^{-1}; this band is weak or very weak in the infrared.

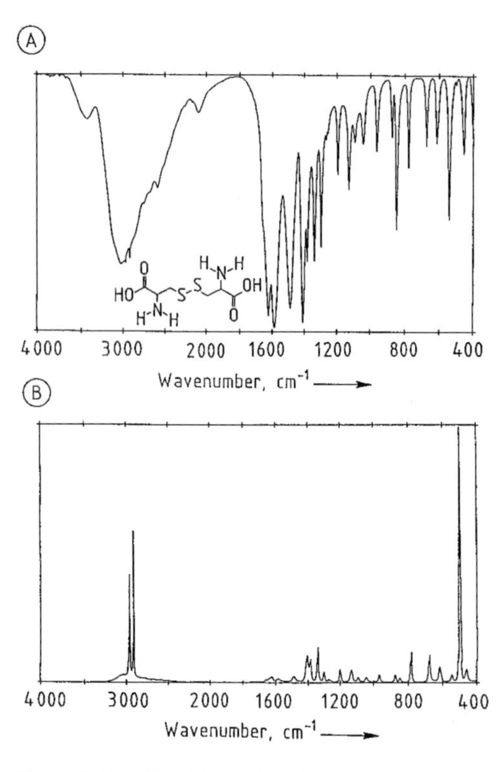

Figure 42. FT-infrared (A) and FT-Raman (B) spectra of L-cystine

17.4.9.2. Sulfides and Disulfides

Figure 41 shows characteristic group frequencies of sulfides and disulfides. The stretching vibration of C–S bonds gives rise to a weak infrared but a strong Raman signal at 730 – 570 cm^{-1}. Similarly, the S–S stretching vibration at 500 cm^{-1} is a very strong Raman line but very weak infrared band (Fig. 42). Both Raman signals are very diagnostic in the conformational analysis of disulfide bridges in proteins.

17.4.9.3. Sulfones

Similar to nitro groups and carboxyl salts, sulfones are characterized by two bands due to antisymmetric and symmetric stretching of the SO_2 group, the former at 1350 – 1290 cm^{-1} and the latter at 1160 – 1120 cm^{-1} (Fig. 43). Both bands are strong in the infrared, while only the symmetric vibration results in a strong Raman line (Fig. 44).

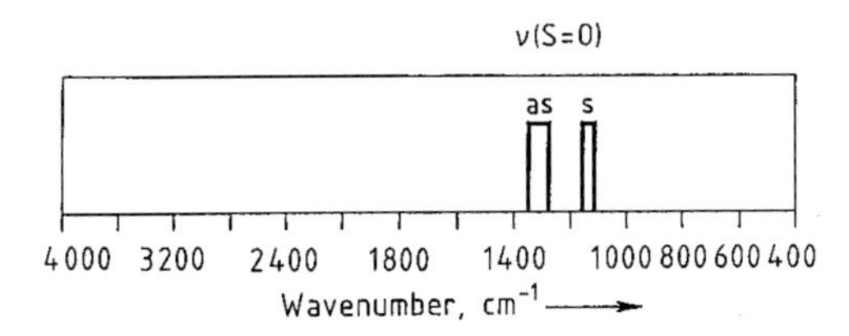

Figure 43. Characteristic group frequencies of sulfones

17.4.10. Computer-Aided Spectral Interpretation

After his first experiences an interpreting vibrational spectra the spectroscopist soon realizes the limited capacity of the human brain to store and selectively retrieve all required spectral data. Powerful micro- and personal computers are now available together with fast and efficient software systems to help the spectroscopist to identify unknown compounds. IR Mentor [41] is a program that resembles an interactive book or chart of functional group frequencies. Although the final interpretation remains in human hands, this program saves the spectroscopist time by making tabular correlation information available in computer form, and, moreover, it is also an ideal teaching tool. To facilitate the automated identification of unknown compounds by spectral comparison numerous systems based on library search, e.g., SPECTACLE [42], GRAMS/32 [43], or those offered by spectrometer manufacturers, are used. These systems employ various algorithms for spectral search [44], and the one that uses full spectra according to the criteria of LOWRY and HUPPLER [45] is the most popular. The central hypothesis behind library search is that if spectra are similar then chemical structures are similar [46]. In principle, library search is separated into identity and similarity search systems. The identity search is expected to identify the sample with only one of the reference compounds in the library. If the sample is not identical to one of the reference compounds the similarity search presents a set of model compounds similar to the unknown one and an estimate of structural similarity. The size and contents of the library are crucial for a successful library search system, especially a similarity search system. A smaller library with carefully chosen spectra of high quality is more useful than a comprehensive library containing spectra of all known chemical compounds or as many as are available because, in similarity search, the retrieval of an excessive

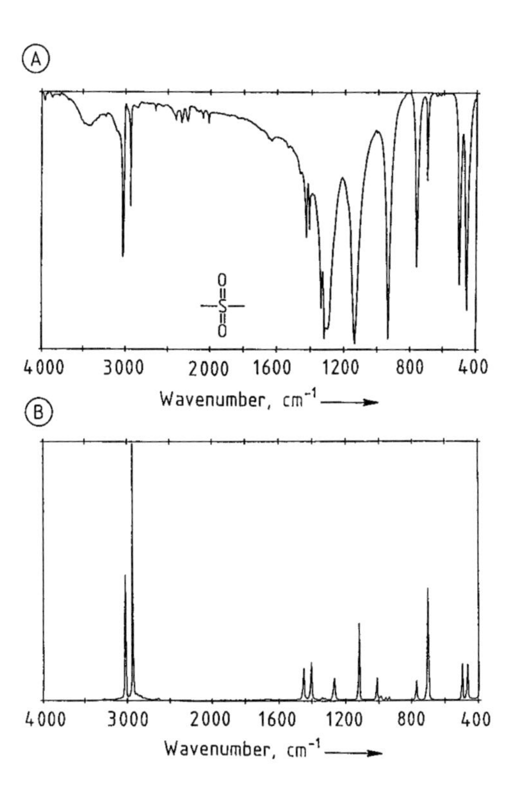

Figure 44. FT-infrared (A) and FT-Raman (B) spectra of dimethyl sulfone

number of closely similar references for a particular sample only increases output volume without providing additional information [46]. A critical discussion of the performance of library search systems is presented by CLERC [46]. While library search systems are well-established, invaluable tools in daily analytical work, expert systems (i.e., computer programs that can interpret spectral data) based on artificial neural networks (ANNs) are more and more emerging [47], [48].

17.5. Applications of Vibrational Spectroscopy

17.5.1. Infrared Spectroscopy

17.5.1.1. Transmission Spectroscopy

Transmission spectroscopy is the simplest sampling technique in infrared spectroscopy, and is generally used for routine spectral measure-

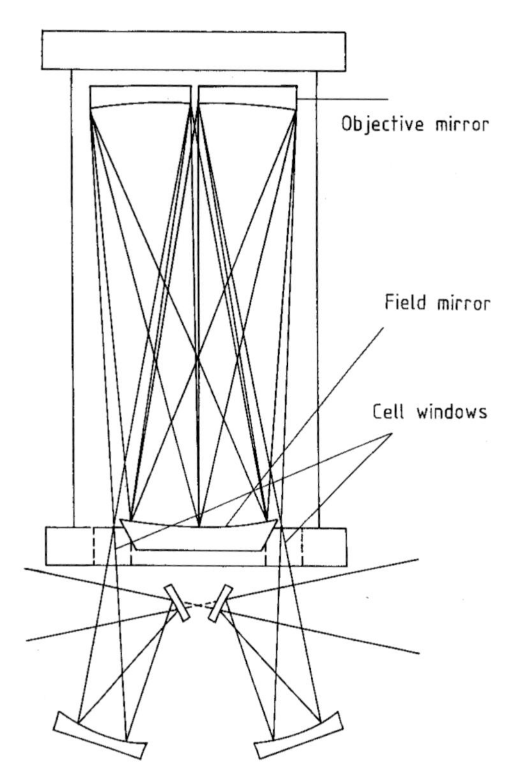

Figure 45. Optical diagram of a White gas cell (A mode) showing eight passes, to change the path length only one objective mirror is rotated, the rotation is nonlinear and is very small at long path lengths. For B mode operation, the light beam enters and exits on one and the same side of the field mirror, the objective mirrors are coupled and are rotated together linearly over the entire operational range (Reproduced by permission of Harrick Scientific Corporation, Ossining, NY 10562)

ments on all kinds of samples. It requires only simple accessories, such as infrared-transparent windows [e.g., potassium bromide, sodium chloride, or thallium bromoiodide (KRS-5)] for gases and liquids, or a sample holder for solids. The sample is placed in the light beam of an infrared spectrometer, and the intensity of the incident beam is compared with that transmitted by the sample. According to the fundamental relation governing the absorption of radiation as a function of transmittance (see Section 17.3.2), parameters that can be determined in transmission spectroscopy are the thickness or the concentration of the sample. The thickness can range from micrometers for solids and liquids to even kilometers for gas samples.

Apart from general laboratory gas analysis, gas-phase transmission spectroscopy is used for the characterization of air pollutants as well as the monitoring of stack gas and of air quality in work places. As air quality has become a major concern for environmentalists, industrial manufacturers and many others, spectroscopic measurement in the gas phase is receiving increasing attention. However, two major problems must be overcome: absorptions in the gas phase are many times weaker than those in the liquid or solid phase, and the species to be measured usually makes up only a small fraction of the total gas volume. The most popular way of surmounting these problems is to increase the path length by reflecting the beam several times from precisely controlled mirrors. Such multiple-pass gas cells are all based on the White design [49] or variations thereof. Figure 45 shows an optical diagram of a White cell aligned for eight passes. The length of the gas cell and the number of passes employed determines the optical path length, which may vary from centimeters to kilometers. The number of passes is adjusted in four pass increments by rotating one of the objective mirrors. In the A mode of operation, only one objective mirror is rotated, while in the B mode the objective mirrors are coupled and rotated together linearly over the entire operational range to change the path length. To prepare a sample, the gas cell is evacuated, the sample is bled into the cell, and the cell is sealed.

Liquid samples are easily studied with the aid of a wide variety of liquid cells, including heatable, flow through, and variable path length cells. These cells are constructed of two infrared transparent windows with a spacer between them, thereby forming a cavity for the sample. As the samples are often strong infrared absorbers, for this spectral range the liquid transmission cell must be constructed with short optical path lengths (0.025 – 1 mm).

Polycrystalline or powdered samples can be prepared as a suspension in mineral oil (Nujol mull), as a potassium bromide disk (pellet), or as thin films deposited on infrared-transparent substrates. The potassium bromide pellet is the most common way of preparing powder samples; in this method a small amount, usually 1 mg, of finely-ground solid sample is mixed with powdered potassium bromide, usually 300 mg, and then pressed in an evacuated die under high pressure. The resulting disks are transparent and yield excellent spectra. The only infrared absorption in the potassium bromide matrix is due to small amounts of adsorbed water, which can, however, be confused with OH-containing impurities in the sam-

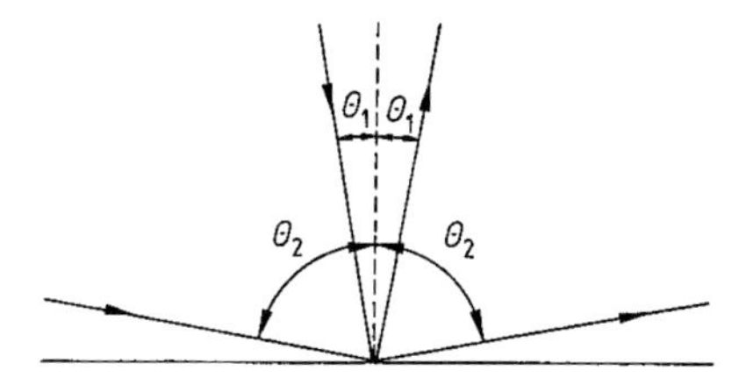

Figure 46. Near-normal (θ_1) and grazing angle (θ_2) incidence

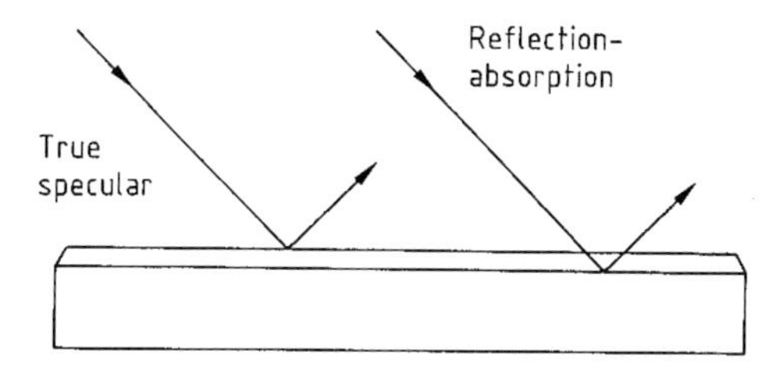

Figure 47. Types of specular reflection

ple. To minimize adsorbed water the potassium bromide is ground as little as possible. The sample is ground separately and then mixed with potassium bromide, after which the mixture is ground only so far as is necessary to achieve good mixing.

17.5.1.2. External Reflection Spectroscopy

In external reflection spectroscopy, electromagnetic radiation is reflected at the interface of two materials differing in refractive index or from a smooth surface, a conductor such as metal, a semiconductor, a dielectric, or a liquid. The intensity of light reflected is measured to obtain information about the surface chemistry and structure. The angle of collection is the same as the angle of incidence and is measured from the normal to the surface (Fig. 46). Small-angle incidence (θ_1, usually 10°) is called near-normal incidence while that at large angles (θ_2, usually 80°) is known as grazing angle incidence. A nondestructive method requiring no sample preparation, external reflection spectroscopy comprises two different types of measurements, both being termed specular reflectance. In the first type, the reflection from a smooth mirrorlike surface, e.g., free-standing film, single crystal face, or surface of an organic material, is referred to as true specular reflection or simply specular reflection (Fig. 47). In the second, a film of sample is deposited on a highly reflecting substrate, usually a metal, from which the infrared beam is reflected, the radiation is transmitted through the sample layer, reflected from the substrate surface, and finally transmitted back through the sample. This technique is called reflection – absorption (RA), infrared reflection-absorption spectroscopy (IRRAS), or transflectance.

A material's object properties are characterized by its complex refractive index, $n' = n + ik$, where n is the refractive index and k is the absorption index, both of which are dependent on the frequency of incident light. Unlike normal transmission spectra, true specular reflection spectra often have distorted, derivative bands because specular reflectance from a surface frequently has a large refractive index component. However, by using a mathematical operation known as the Kramers – Kronig transformation [50] a transmission-like spectrum showing the expected absorptivity information can be calculated from a measured, distorted spectrum. As true specular reflectance is often measured at or near normal incidence, the reflected energy is small, usually between 0 and 12 % for most organic materials. However, when this low signal-to-noise ratio is overcome by using more sensitive cooled MCT detectors the true specular reflection technique has the important advantage of permitting the noncontact analysis of bulk solids and liquids. Depending on the thickness of the film of sample, two types of reflection – absorption measurement are used: near-normal and grazing angle. In near-normal reflection – absorption, the thickness of the sample usually ranges between 0.5 and 20 µm and the angle of incidence varies between 10 and 30°. Hence, the amount of energy reflected from the substrate is much greater than the amount reflected from the front surface of the sample film, and a completely transmission-like reflection – absorption spectrum is obtained. Grazing angle reflection – absorption measurements are more sensitive than near-normal reflectance and are used to analyze sample films less than 1 µm thick, and even monolayers, on highly reflective materials. The improved sensitivity is due to the enhanced electromagnetic field that can be produced at the reflection surface by using polarized light at large angles of incidence [51], [52]. As the plane of incidence contains the incident and reflected rays, as well as the normal to the reflection surface, the electrical field of p-polarized light is parallel to it whereas that of s-polarized light is perpendicular to it. For s-polarized light, the phase of the reflected beam is shifted 180° from the incident

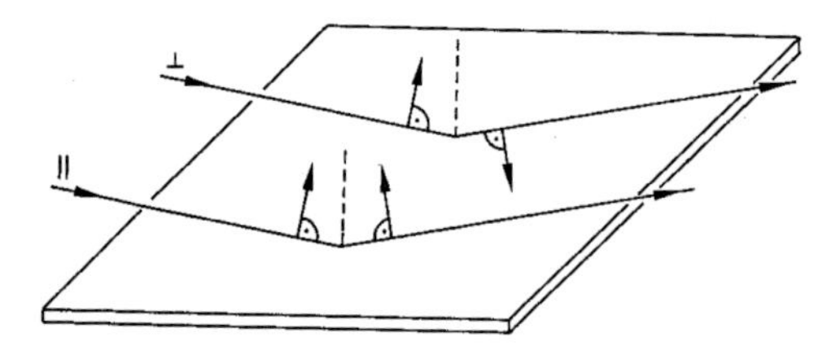

Figure 48. Phase shift of the reflected beam occurring at grazing angle incidence with perpendicular (180° shift) and parallel (90° shift) polarized light

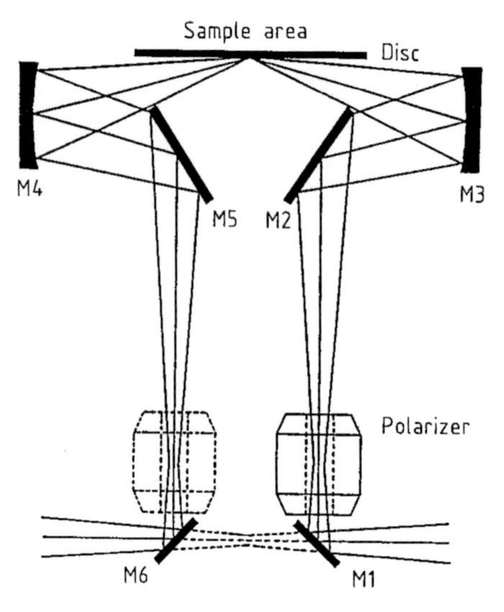

Figure 49. Ray diagram of the wafer/disk checker attachment for recording spectra of thin films on large samples, e.g., lubricants on hard magnetic disks;
$M_1 - M_6$ = mirrors
(Reproduced by permission of Harrick Scientific Corporation, Ossining, NY 10562)

beam at all angles of incidence (Fig. 48), and the two beams interfere at the surface to produce a node. Hence little interaction occurs between the infrared radiation and films thinner than one-quarter of the wavelength. However, for p-polarized light, the phase shift is approximately 90° at large angles of incidence, and the combination of incident and reflected beams at the surface produces a standing wave, giving an intense electrical field oriented normal to the surface. For metal substrates the greatest enhancement of the electrical field is produced by using p-polarization at large angles of incidence (e.g., 80°).

Reflection – absorption measurements are used in the analysis of a wide range of industrial products, including polymer coatings on metals, lubricant films on hard magnetic disks (Fig. 49), semiconductor surfaces, or the inner coating of food and beverage containers. In addition, grazing angle reflectance is especially suited to the study of molecular orientations at surfaces [53]. External reflectance at different angles of incidence can be employed to determine both the thickness and refractive index of a film of sample by measuring the interference fringe separation [54]; Figure 50 shows one of the various external reflection attachments which are available.

17.5.1.3. Internal Reflection Spectroscopy

Internal reflection spectroscopy, also known as attenuated total reflectance (ATR) or multiple internal reflectance (MIR), is a versatile, nondestructive technique for obtaining the infrared spectrum of the surface of a material or the spectrum of materials either too thick or too strongly absorbing to be analyzed by standard transmission spectroscopy.

The technique goes back to Newton [55] who, in studies of the total reflection of light at the interface between two media of different refractive indices, discovered that an evanescent wave in the less dense medium extends beyond the reflecting interface. Internal reflection spectroscopy has been developed since 1959, when it was reported that optical absorption spectra could conveniently be obtained by measuring the interaction of the evanescent wave with the external less dense medium [56], [57]. In this technique, the sample is placed in contact with the internal reflection element (IRE), the light is totally reflected, generally several times, and the sample interacts with the evanescent wave (Fig. 51) resulting in the absorption of radiation by the sample at each point of reflection. The internal reflection element is made from a material with a high refractive index; zinc selenide (ZnSe), thallium iodide – thallium bromide (KRS-5), and germanium (Ge) are the most commonly used. To obtain total internal reflection the angle of the incident radiation θ must exceed the critical angle θ_c [58]. The critical angle is defined as:

$$\theta_c = \sin^{-1} \frac{n_2}{n_1} \qquad (17)$$

where n_1 is the refractive index of the internal reflection element and n_2 is the refractive index of the sample. What makes ATR a powerful technique is the fact that the intensity of the evanescent

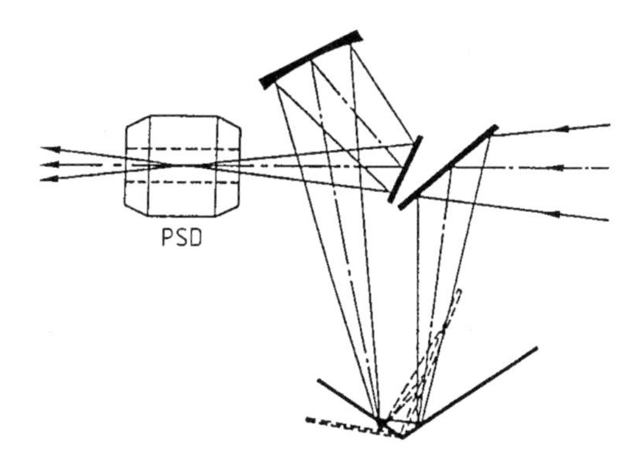

Figure 50. Ray diagram of the versatile reflection attachment (VRA) with single diamond polarizer (PSD)
(Reproduced by permission of Harrick Scientific Corporation, Ossining, NY 10562)

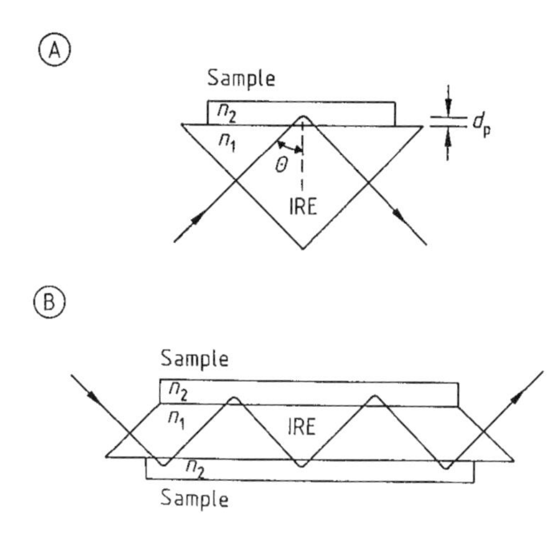

Figure 51. Schematic representation of total internal reflection with: A) Single reflection; B) Multiple reflection
IRE (internal reflection element)
n_1 = Refractive index of the internal reflection element; n_2 = Refractive index of the sample with $n_2 < n_1$; θ = Angle of incidence; d_p = Depth of penetration

wave decays exponentially with the distance from the surface of the internal reflection element. As the effective penetration depth is usually a fraction of a wavelength, total internal reflectance is generally insensitive to sample thickness and so permits thick or strongly absorbing samples to be analyzed. The depth of penetration d_p, defined as the distance required for the electrical field amplitude to fall to e^{-1} of its value at the interface, is given by:

$$d_p = \frac{\lambda_1}{2\pi\left(\sin^2\theta - n_{21}^2\right)^{1/2}} \tag{18}$$

where $\lambda_1 = \lambda / n_1$ is the wavelength in the denser medium, and $n_{21} = n_2 / n_1$ is the ratio of the refractive index of the less dense medium divided by that of the denser [56].

Although ATR and transmission spectra of the same sample closely resemble each other, differences are observed because of the dependency of the penetration depth on wavelength: longer wavelength radiation penetrates further into the sample, so that in an ATR spectrum bands at longer wavelengths are more intense than those at shorter ones.

The depth of penetration also depends on the angle of incidence; hence, an angle of 45°, which allows a large penetration depth, is generally used to analyze organic substances, rather than an angle of 60°, which results in a substantially weaker spectrum due to the decreased depth of penetration.

The degree of physical contact between sample and internal reflection element determines the sensitivity of an ATR spectrum. To achieve this, a horizontal ATR accessory such as FastIR [58], in which the top plate is the sampling surface, is used [59]; reproducible contact is ensured by a special sample clamp or powder press. Good quality spectra are thus obtained for many materials that present problems of analysis with routine transmission methods, e.g., powders, pastes, adhesives, coatings, rubbers, fibers, thick films, textiles, papers, greases, foams, and viscous liquids. Possible methods of obtaining a spectrum from a variety of samples are discussed in [60]; in situ ATR spectroscopy of membranes is described in [61], [62]. Liquid samples are also well suited to ATR analysis. Most liquids require a very short path length; aqueous samples, for instance, are measured at path lengths of no more than ca. 15 μm, which makes the design of transmission cells difficult because flow of liquids is hindered; they also exhibit interference fringes because of the small spacing between the high refractive index infrared windows. These problems are eliminated by using liquid ATR cells, a variant of solid ATR, in which the internal reflection element is surrounded by a vessel into which the liquid is poured. Various such liquid cells are available, e.g., the Circle [63], the Prism Liquid Cell (PLC) [58], the Squarecol [64] and the Tunnel Cell [65]. With optimized optical design and fixed internal reflection elements these cells provide a highly reproducible path length which permits the quantitative analysis of liquids and aqueous solutions [66]. Liquid ATR

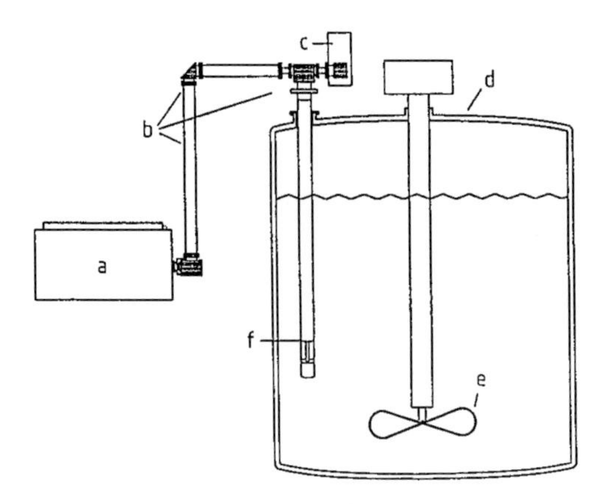

Figure 52. Schematic drawing of an FT-IR measurement system utilizing the Deep Immersion Probe Model DPR-124 mounted in a batch reaction vessel [68]
a) FT-IR spectrometer; b) Optical transfer elements; c) Detector assembly; d) Reaction vessel; e) Mixing blade; f) ATR sensing head
(Reproduced by permission of Axiom Analytical, Inc., Irvine, CA 92614)

cells are uniquely suited to fully automatable, on-line process-monitoring of liquids and viscous fluids. In addition, with suitable optical transfer modules such as the Axiot system [67], it is possible to operate liquid cells outside the sample compartment of the FT-IR spectrometer. This is important for on-line applications in which it is not practicable to pipe the sample to the spectrometer because the analysis can be carried out at the most desirable location, even in a fume hood or on a process line. With an ATR sensing head, immersion probes [68] offer the possibility of in situ FT-IR spectroscopy, for monitoring batch process reactions (Fig. 52), laboratory process development, the identification of hazardous waste, the verification of the contents of storage drums, and the inspection of incoming raw materials.

17.5.1.4. Diffuse Reflection Spectroscopy

Diffuse reflection spectroscopy, also known as diffuse reflectance infrared Fourier transform (DRIFT) spectroscopy, enables the analysis of many samples for which traditional techniques fail, to be made with little or no sample preparation. Many substances in their natural state, especially powders but also any solid with a rough surface, such as dyed textiles and printed papers, exhibit diffuse reflection, i.e., incident light is scattered in all directions, unlike specular reflection (see Section 17.5.1.2) where the angle of incidence equals the angle of reflection. Diffuse reflectance spectra are in practice complex, because of spectral distortions caused by the mixing of both absorbance and reflectance features owing to contributions from transmission, internal, and specular components in the measured radiation. Hence, DRIFT spectra are affected by particle size, packing, and dilution of the sample.

In diffuse reflectance spectra large glossy particle faces or crystal surfaces cause specular reflection which produces inverted bands, known as reststrahlen bands; this effect is reduced by fine grinding of the sample. Reststrahlen bands may also occur in neat samples with a rough surface, such as dyed textiles, printed papers, and agricultural and botanical specimens [69], measured by direct nondestructive, DRIFT spectroscopy. In this case, roughening the surface with emery paper reduces the disturbing bands because with smaller, thinner particles on the roughened surface the transmission component of the collected light increases, thus decreasing the reststrahlen effect. In strongly absorbing samples, the slight penetration of the incident beam produces mainly specular reflection at the surface, which also results in strong reststrahlen bands. By diluting the sample with a nonabsorbing powder such as KBr the effect is minimized or even eliminated, as this ensures deeper penetration of the incident beam and, therefore, an increased contribution of transmission and internal reflection components to the DRIFT spectrum. The diffuse reflectance spectrum is usually calculated as the ratio of sample and reference reflectance, with the pure diluent being taken as reference. Typically, a mixture of 90–95 % diluent and 10–5 % sample is used. Although the diluted spectra and the bulk reflectance spectra of weakly absorbing compounds have a similar appearance, they show enhanced intensity at lower wavenumbers compared to transmission spectra.

The interpretation of diffuse reflectance is based on a theory developed by Kubelka and Munk [70] and extended by Kortüm, Braun, and Herzog [71] for diluted samples in nonabsorbing matrices. The diffuse reflectance R_∞ of a diluted sample of infinite thickness (i.e., a sample for which an increase in thickness does not appreciably change the spectrum) is linearly related to the concentration c of the sample, as given by the Kubelka–Munk (K–M) equation:

$$f(R_\infty) = \frac{(1 - R_\infty)^2}{2R_\infty} = \frac{2.303ac}{s} \quad (19)$$

where a is the absorptivity and s is the scattering coefficient. Because the scattering coefficient depends on both particle size and degree of sample packing, the linearity of the K–M function can only be assured if particle size and packing method are strictly controlled. If these conditions are fulfilled, accurate quantitative diffuse reflectance measurements can be obtained [72], [73].

Initiated by the work of GRIFFITHS et al. [74], diffuse reflectance accessories have been developed that pay special attention to the optimal reduction of the disturbing specular component. For example, the Praying Mantis diffuse reflectance attachment (Fig. 53) uses an off-line collection angle incorporating two 6 : 1 90° off-axis ellipsoidal mirrors. One ellipsoidal mirror focuses the incident beam on the sample, while the radiation diffusely reflected by it is collected by the other. Both ellipsoidal mirrors are tilted forward; therefore, the specular component is deflected behind the collecting ellipsoid, permitting the diffusely reflected component to be collected. As diffuse reflectance is a very sensitive technique for the analysis of trace quantities of materials, all available attachments allow for microsampling [75]. The analysis of thin-layer chromatograms by DRIFT spectroscopy, in situ or after transfer, is described in [76], [77]. Catalytic reactions can be monitored by using a controlled reaction chamber attached to a DRIFT accessory [78], [79]. Intractable samples, such as paint on automobile panels, can be analyzed by using silicon carbide abrasives [80]: a silicon carbide disk is rubbed against the sample, a small amount of which adheres to the disk. The DRIFT spectrum of the sample is then quickly obtained by placing the disk in the diffuse reflectance attachment.

17.5.1.5. Emission Spectroscopy

Infrared emission spectroscopy is a useful and effective technique for studying the surface of liquid and solid organic, inorganic, and polymeric materials, and for the in situ process-monitoring of high-temperature reactions and thermal transformations [81], [82]. In contrast to the conventional transmission technique, in emission spectroscopy the sample itself is the infrared source. The temperature of the sample is raised, which increases the Boltzmann population of the vibrational energy states. Infrared radiation is emitted on return to the ground state. In theory, emission spectra can always be measured provided the sample is at a different temperature to the detector. Like transmission spectra, emission spectra are obtained by dividing the emission of the sample by that of the reference, which is often a blackbody source at the same temperature as the sample. The main reason for the lack of any extensive work in this technique before 1970 was the low intensity of infrared emission, which results in spectra with poor signal-to-noise ratio. However, the availability of FT-IR spectrometers with the potential for low-energy sampling and for the development of highly efficient FT-IR emission spectroscopy accessories [83], [84] has led to a resurgence of interest in infrared emission spectroscopy.

Because of the phenomenon of self-absorption the ideal sample for conventional emission studies is a thin layer (e.g., a polymer film), on both metal and semiconductor surfaces [81]. A sample is usually heated from below the emitting surface, the lower surface thus having a higher temperature than the upper one. Therefore, radiation emitted from below the upper surface is absorbed before it reaches the surface, and this self-absorption of previously emitted light severely truncates and alters features in the emission spectra of optically thick samples. This problem is overcome by using a laser for controlled heat generation within a thin surface layer of the sample, self-absorption of radiation thus being minimized. These methods, known as laser-induced thermal emission (LITE) spectroscopy [85], [86] and transient infrared emission spectroscopy (TIRES) [87], [88] can produce analytically useful emission spectra from optically thick samples. Quantitative applications of infrared emission spectroscopy are described in [89]–[91].

17.5.1.6. Photoacoustic Spectroscopy

Photoacoustic spectroscopy (PAS) [92]–[95] is a fast, nondestructive method for analyzing various materials in the gas, liquid, or solid state with virtually no sample preparation. In cases where samples are insoluble, difficult to grind into a powder, or of irregular shape—such as coal, carbon-filled and conducting polymers, pharmaceutical preparations, spin coatings on fibers, and coatings on irregular surfaces—photoacoustic spectroscopy complements other established techniques such as attenuated total reflectance (ATR) and diffuse re-

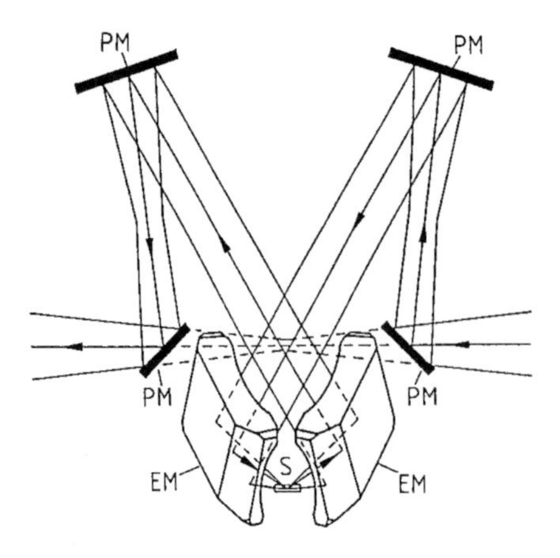

Figure 53. Ray diagram of the Praying Mantis diffuse reflectance attachment
EM = Ellipsoidal mirror; PM = Planar mirror; S = Sample
(Reproduced by permission of Harrick Scientific Corporation, Ossining, NY 10562)

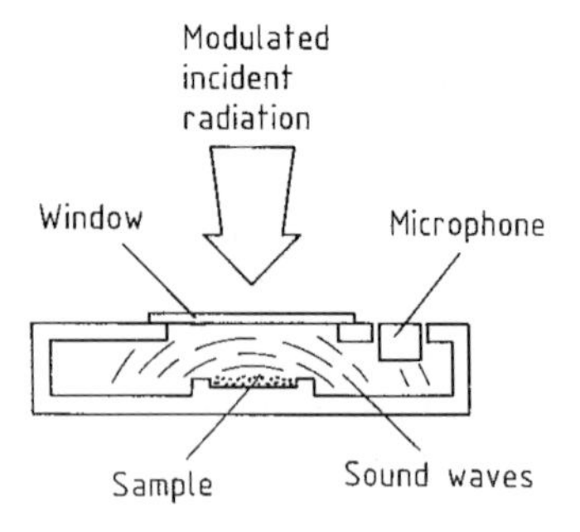

Figure 54. Schematic of a photoacoustic cell

flectance (DRIFT). Moreover, measurements on samples that show the Christiansen effect [96] when prepared as KBr pellets for standard transmission spectroscopy can be better performed by the photoacoustic method. As most photoacoustic measurements are made with solid or semisolid samples, the sample is sealed in a small-volume cell containing a nonabsorbing gas such as helium and having windows for optical transmission (Fig. 54). The sample is illuminated with modulated radiation from the monochromator of a dispersive instrument or the interferometer of a Fourier transform infrared spectrometer. At wavelengths where the sample absorbs a fraction of the incident radiation, a modulated temperature fluctuation at the same frequency, but not necessarily with the same phase, as that of the incident radiation is generated in the sample, and the surrounding inert gas produces periodic pressure waves in the sealed cell. These fluctuations in pressure can be detected by a microphone because the modulation frequency of the incident beam usually lies in the acoustic range, e.g., between 100 and 1600 Hz. The low power available from dispersive instruments results in spectra with a rather poor signal-to-noise ratio; hence, photoacoustic mid-infrared spectra of condensed substances are preferably obtained by using an FT-IR spectrometer with a photoacoustic cell accessory. In a rapid-scanning Michelson interferometer with a mirror velocity of v (in cm/s), radiation of wavenumber $\tilde{\nu}$ (in cm^{-1}) is modulated with a frequency of $2 v \tilde{\nu}$ Hz [8]. Therefore, FT-IR spectrometers are ideal for photoacoustic measurements in the mid-infrared, where mirror velocities on the order of $0.05 - 0.2$ cm s^{-1} provide modulation frequencies in the acoustic range. As carbon black absorbs all incident radiation, the sample single-beam spectrum is usually ratioed against a carbon black single-beam spectrum to obtain an absorbance-like spectrum.

Before the first Fourier transform infrared photoacoustic spectrum was reported in 1978 [97], most infrared photoacoustic experiments were performed by using high-intensity tunable laser sources. However, as the source is monochromatic, this type of experiment is only useful for monitoring a single line in the spectrum. Highly-developed photoacoustic accessories are now available from FT-IR manufacturers or other companies [98], which have a high sensitivity also for microsamples. As the signal is inversely proportional to the cell volume, cells optimized for the measurement of condensed samples have a small volume, typically well under 1 cm^3, to enhance the signal amplitude [99]. Water vapor or carbon dioxide generate good photoacoustic signals, so even traces of these impurities cause serious interference in the photoacoustic spectroscopy signal; therefore, the purity of the gas in the cell is essential for good-quality photoacoustic spectra.

Important applications of photoacoustic spectroscopy are near-surface analysis [100] and depth profiling of polymers and polymer layers [101] – [103]. The theory of depth profiling is described in detail in [92]. It is based on the fact that thermal diffusion length (the distance a thermal wave travels in the sample before its intensity is reduced by 1/e in magnitude) and hence intensity of the photoacoustic signal are increased by lowering the modulation frequency. In a rapid-scanning interferometer this is achieved by decreasing the mirror velocity. To date, rapid-scanning interferometers that provide modulation frequencies which continuously change with wavenumber have generally been used. However, step-scan interferometers accommodating any desired modulation frequency, constant over the entire wavelength range, produce improved photoacoustic spectra and are thus more effective in depth profiling [104] – [107].

Saturation effects, which can occur for stronger absorption bands, are the main limitation of photoacoustic spectroscopy. These distortions can be overcome by using thin samples [93] or by analyzing the phase of the photoacoustic signal [108]. The latter method is particularly useful for quantitative analysis.

17.5.1.7. Chromatography/Fourier Transform Infrared Spectroscopy

The combination of chromatographic separation with Fourier transform infrared spectroscopy has significantly improved the analysis of complex mixtures [109]. In chromatography/FT-IR systems, an infrared detector providing information on the structure of separated species is used instead of standard bulk chromatography detectors such as thermal conductivity, flame ionization, ultraviolet, or fluorescence; these detectors can be used for quantitative analysis when the identities of the mixture components are known.

In these systems the following factors must be taken into consideration: the physical state of the matter to be analyzed, the optimization of the interface between chromatograph and FT-IR, and the specific requirements and constraints of all three parts. In gas chromatography (GC)/FT-IR spectroscopy [110], one type of interface is composed of temperature-controlled, long, thin flow-through gas cells or "light pipes" coated with gold for high reflectivity. As the light-pipe techniques have a detection limit in the nanogram range, subnanogram limits can be routinely attained with the matrix-isolation technique (GC/MI/FT-IR) [109], and even better detection limits are achieved with the subambient trapping GC/FT-IR technique [109], [111]. Rapid-scanning interferometers able to record the entire spectrum in less than one second at low resolution (usually 4 or 8 cm^{-1}), together with highly sensitive detectors (e.g., cooled MCT detectors), are essential in all column chromatography/FT-IR spectroscopic techniques, with capillary GC/FT-IR requiring the most rapid data acquisition. In comparison to vapor phase species, rotations of matrix-isolated species are hindered, which results in sharper and more intense bands. Hence, for accurate spectra representation, GC/MI/FT-IR spectra are usually measured at 1 cm^{-1} resolution.

The Gram – Schmidt vector orthogonalization method [112] is the most sensitive and computationally most efficient and, hence, the most popular procedure for the reconstruction of the total chromatographic trace from interferometric infrared data. Selective detection is obtained by functional-group-specific chromatograms (chemigrams [113]), which are produced by computing absorbance spectra during chromatographic separation and plotting integrated absorbance for prespecified wavelength windows as a function of separation time, representing the quantity of a specific functional group that elutes as a function time. In chromatography/FT-IR spectral evaluation, in order to classify unknowns as to compound type, library searching (see Section 17.4.10) has proved to be very useful.

The addition of a complementary detector significantly increases the power of GC/FT-IR analysis. Mass spectrometry (MS) is an ideal choice [114] because the disadvantages of each method — infrared spectroscopy often cannot distinguish between long-chain homologues, and mass spectrometry frequently fails to distinguish isomeric species — are offset by the other.

The crucial point in liquid chromatography (LC)/FT-IR methods, however, is the interferences caused by the liquid phase. Actually, two viable LC-IR systems are commercially available: the LC-Transform [115] and the InfraRed Chromatograph (IRC) [116], [117]. While the LC-Transform is a solvent-elimination interface, the IRC provides full FT-IR spectra of the eluent stream from an LC column. The LC-Transform system utilizes a high-performance ultrasonic nebulizer or a pneumatic linear capillary nozzle to remove the mobile phase from samples as they elute from the chromatograph. The capillary is surrounded by hot, flowing sheath gas which provides sufficient thermal energy to evaporate the mobile phase and to contain the nebulized spray in a tight, focused cone as the spray emerges from the nozzle. The sample is deposited on a germanium sample-collection disk as spots with diameters between 0.5 and 1 mm. In a consecutive, uncoupled step, good-quality infrared spectra can be obtained from these spots, which contain microgram to sub-microgram amounts, by the use of a specially designed optics module, comprising a beam condenser. While this optics module can be used in any FT-IR spectrometer uncoupled from the LC-Transform interface, the IRC includes an FT-IR spectrometer, a data station for driving the spectrometer and for data-processing, and a vacuum interface for depositing and scanning the sample. Being a real-time direct deposition interface, the IRC uses an ultrasonic nebulizer, a desolvation tube and a vacuum chamber to evaporate the solvent from the LC stream. The sample residue is then deposited onto a moving zinc selenide window. The data station collects a continuous array of spectra as the dry sample track on the window moves through the focused beam of the spectrometer.

LC-IR provides powerful identification possibilities, based on molecular structure. A high degree of confidence can be placed on substance identification that simultaneously matches infrared spectral characteristics and chromatographic elution time. This is especially so when molecules that have various conformational isomers are to be distinguished.

Supercritical fluid chromatography (SFC)/FT-IR spectroscopy, generally with carbon dioxide as the mobile phase, bridges the gap between GC/FT-IR and LC/FT-IR, and is particularly useful for separating nonvolatile or thermally labile materials not amenable to gas chromatographic separation [109]. Flow cells, mobile phase elimination and matrix-isolation techniques are used as SFC/FT-IR interfaces.

Applications of chromatography/FT-IR spectroscopy involve toxins and carcinogens, wastewater constituents, sediment extracts, airborne species, pesticides and their degradation products, fuels and fuel feedstocks, flavors and fragrances, natural products, pharmaceuticals, biomedicine, and polymers.

17.5.1.8. Thermogravimetry/Fourier Transform Infrared Spectroscopy

Thermogravimetry (TG), also called thermogravimetric analysis (TGA), measures changes in sample mass as a function of increasing temperature and has become an important tool for the analysis and characterization of materials [118], [119]. To correlate weight changes with chemical changes, it is necessary to identify gases evolved from the thermogravimetric analyzer furnace. This is made possible by coupling thermogravimetry with FT-IR spectroscopy, which provides the necessary speed and sensitivity.

A system has been described in which the infrared beam of the FT-IR spectrometer is led directly into the TG equipment [120], but in commercially available systems [121], [122] the evolved gases are conducted from the TG system to the spectrometer by carrier gas flow via a transfer line. The interface is heated to prevent recondensation and incorporates the transfer line and a flow-through gas cell with nitrogen or helium as carrier gas.

After infrared spectra at low resolution (usually 4 cm^{-1}) have been recorded as a function of time and temperature, the further procedure is similar to GC/FT-IR (see Section 17.5.1.7) and, in fact, GC/FT-IR software, such as Gram – Schmidt thermogram reconstruction and library search with a vapor-phase spectra database is commonly used. Helpful for the identification in the vapor phase is the virtual absence of interaction between molecules, and therefore the nondestructive IR detection enables in many cases an easy and fast interpretation of single and overlapping absorption bands.

Providing structural information for sample components down to nanogram levels, TG/FT-IR applications include decomposition studies of polymers and laminates [123], the analysis of coal, oil shales, and petroleum source rocks [124], [125], and the determination of activation energies [126] and thermal decomposition of energetic materials [127].

The IR detection of evolved gases opens new possibilities in the quantification of thermogravimetric analysis. Besides measuring the mass loss of a TG step, the total amount can also be calculated after calibration of a single component and integration over time [128]. A further improvement in quantitative TG/FT-IR is achieved by PulseTA [129]. Here, a certain amount of gas or liquid is injected into the system and used as internal calibration standard for the quantification of the unknown amount of evolved gas.

17.5.1.9. Vibrational Circular Dichroism

Vibrational circular dichroism (VCD) extends the functionality of electronic circular dichroism (CD) into the infrared spectral region, with VCD bands of enantiomeric pairs of chiral molecules having opposite signs. Early VCD spectrometers were dispersive instruments equipped with a CaF_2 photoelastic modulator for creating left and right circularly polarized radiation and a liquid-nitrogen-cooled InSb detector. FT-IR-VCD measurements are carried out by using the so-called double modulation technique [130] in which the VCD interferogram is carried by the polarization modulation frequency in the region of 50 kHz, and the IR transmission interferogram occurs in the usual range of Fourier frequencies, typically below 5 kHz.

Combining the structural specificity of FT-IR spectroscopy with the stereosensitivity of circular dichroism, VCD allows the determination of optical purity without enantiomeric separation or derivatization, as well as of absolute configuration without crystallization. Simultaneous monitoring of the optical purity of multiple chiral species, such as reactant and product molecules in a reaction process is also possible, as is the determination of conformations of biological molecules such as proteins, peptides, and DNA [131], [132].

17.5.2. Raman Spectroscopy

In mid-infrared spectroscopy, Fourier transform instruments are used almost exclusively. However, in Raman spectroscopy both conventional dispersive and Fourier transform techniques have their applications, the choice being governed by several factors [133], [134]. Consequently, a modern Raman laboratory is equipped with both Fourier transform and CCD-based dispersive instruments. For a routine fingerprint analysis, the FT system is generally used, because it requires less operator skill and is quicker to set up; the FT system is also be tried first if samples are highly fluorescent or light sensitive. However, if the utmost sensitivity is required, or if Raman lines with a shift smaller than 100 cm^{-1} are to be recorded, conventional spectrometers are usually preferred.

Raman spectroscopy offers a high degree of versatility, and sampling is easy because no special preparation is required. As it is a scattering technique, samples are simply placed in the laser beam and the backscattered radiation is analyzed.

Glass gives only a weak Raman signal, so that FT-Raman spectra from liquids are recorded by using a capillary cell [34], a sapphire sphere [34], or glass cuvettes with the back surface silvered for the favored 180° backscattering arrangement (see Section 17.2.2). Moreover, the study of aqueous samples in glass or quartz vessels, which is difficult in the mid-infrared region owing to strong infrared absorption bands of water and quartz, is possible in Raman spectroscopy because there are no significant bands causing interference in the solution spectrum. Powders are either measured by using microspherical cells [33] or are filled into cells consisting of a brass cylinder with a 3 mm hole drilled through, into which a reflective metal rod is placed to compress the powder [14]. The Raman intensity is related to the sample thickness, but once the thickness exceeds 2 mm the intensity does not increase. Optical fibers are used for remote sampling, and for FT-Raman spectroscopy this is especially advantageous as the transmission by the fibers, excited by a near-infrared Nd : YAG laser [33], [135], can be at its highest just in the range of the Raman spectrum. Guidelines for optimizing FT-Raman recording conditions are given in [34]. Conventional dispersive Raman spectroscopy has not been widely used for quantitative measurements, mainly because of difficulties with band reproducibility. However, by using an FT-Raman spectrometer, spectra with excellent reproducibility, both in band position and intensity, are obtained, so that this technique has great potential for quantitative analysis [136] – [138].

Along with general reviews [139] – [142] and surveys of major areas of Raman spectroscopy [16], [17], various reviews covering special applications of Raman spectroscopy, such as polarization measurements [143], polymers [144], proteins and nucleic acids [145], pharmaceuticals and biomaterials [146] – [150], paint systems [151], [152], inorganic and organometallic materials [153], and thin-layer chromatography [154], have been published. For further applications and special Raman techniques, such as resonance Raman (RR), surface-enhanced Raman spectroscopy (SERS), coherent anti-Stokes Raman scattering (CARS), and stimulated Raman scattering (SRS), see [14], [155].

17.5.3. Fourier Transform Infrared and Raman Microspectroscopy

In conventional infrared microsampling, beam size is reduced four to six times by using beam condensers. These accessories, which permit submicrogram samples of between 2 and 0.5 mm in diameter to be measured, have, however, drawbacks such as heating effects, since the entire power of the infrared beam is condensed on the sample, and the fact that the sample cannot be visibly located, so that a skilled operator is needed for this time-consuming work. These disadvantages are largely overcome by coupling an optical microscope to an FT-IR spectrometer. As infrared lenses are poor in performance, infrared microscopes use reflecting optics such as Cassegrain objectives instead of the glass or quartz lenses

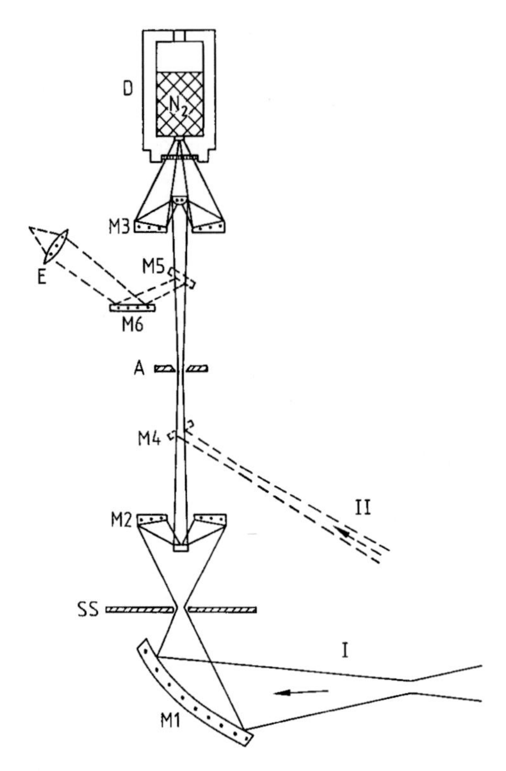

Figure 55. Beam path of a typical infrared microscope
I = Infrared transmittance beam; II = Infrared reflectance beam; M1 = Condensing mirror; M2 and M3 = Cassegrain objectives; M4 = Movable, semi-transparent mirror; M5 = Movable mirror; M6 = Mirror; SS = sample stage; A = Aperture; E = Eye-piece; D = MCT-Detector, liquid N_2 cooled
(Reproduced by permission of Bruker Optik GmbH, D-76275 Ettlingen, Germany)

employed in conventional optical microscopes. Figure 55 shows the beam path of a typical infrared microscope with both transmission and reflection modes. The infrared beam from the FT-IR instrument is focussed onto the sample placed in a standard microscope, and the beam that passes through the sample is collected by a Cassegrain objective with 15- or 36-fold magnification that produces an image of the sample within the barrel of the microscope. The radiation passes through the variable aperture placed in the image plane and is focussed by another Cassegrain objective on a small area (250×250 μm or less) of the specially matched MCT detector. As the visible optical train is collinear with the infrared light path, it is possible to position the sample and to isolate and aperture the area for analysis visually. This is the main advantage of an FT-IR microscope over a beam condenser, and it enables the measurement of, for example, microcontaminants that cannot be removed from the sample, or the individual layers of a polymer laminate. FT-IR microscopy requires virtually no sample preparation. A further advantage of the technique is its unsurpassed sensitivity, which permits samples in the picogram range to be analyzed, with the diffraction limit of infrared radiation (10 – 20 μm) as the limiting condition for the size of measurement spot. To minimize stray light due to diffraction being collected by the microscope, particularly when the size of the area investigated is approximately equal to the wavelength employed, a second aperture is introduced into the optics that focus the radiation onto the sample [156].

Uses of FT-IR microspectroscopy include general characterization of particulate matter, dichroic measurements with polarized light, polymer characterization, semiconductor measurements, the identification of contaminants, as well as forensic, biological, and pharmaceutical applications [157] – [159].

By using diamond cells with the beam condenser or microscope, the thickness of a sample (e.g., paint chips) can be adjusted by squeezing. An alternative to this is an ultrasmall sample analyzer [160], which allows nondestructive internal reflectance studies of microgram and nanogram samples to be performed.

In principle, Raman spectroscopy is a microtechnique [161] since, for a given light flux of a laser source, the flux of Raman radiation is inversely proportional to the diameter of the laser-beam focus at the sample, i.e., an optimized Raman sample is a microsample. However, Raman microspectroscopy able to obtain spatially resolved vibrational spectra to ca. 1 μm spatial resolution and using a conventional optical microscope system has only recently been more widely appreciated. For Raman microspectroscopy both conventional [162] and FT-Raman spectrometers [163], [164] are employed, the latter being coupled by near-infrared fiber optics to the microscope.

Applications of Raman microspectroscopy include the analysis of a wide variety of organic and inorganic materials, e.g., semiconductors, polymers, single fibers, molecular crystals, and minerals [165] – [168].

17.5.4. Time-Resolved FT-IR and FT-Raman Spectroscopy

Time-resolved FT-IR and FT-Raman techniques allow studies of molecular reaction mechanisms with time resolutions up to nanoseconds [169], which is a prerequisite for the understanding of biological processes at the molecular level. The major problem in measuring biological reactions consists in selecting small absorption bands of the molecular groups undergoing reactions from the large background absorption of water and the entire biological molecule. This difficulty is met by obtaining difference spectra of two reaction states of the molecules under investigation. The intrinsic advantages of FT spectroscopy, i.e., the multiplex, the throughput, and the accuracy (see Section 17.2.1) enable such small absorption changes to be reliably detected. With operation of the FT spectrometer in the step-scan mode [169], a time resolution of a few nanoseconds is achieved. Biological applications of time-resolved FT-IR and FT-Raman spectroscopy are described in [169].

17.5.5. Vibrational Spectroscopic Imaging

Although vibrational spectroscopic imaging methods are relative newcomers to the panoply of vibrational techniques, they inherit the versatility of the traditional single-point infrared and Raman approaches and, in combination with digital imaging technology, allow for new perspectives in the interpretation of the spectra. Imaging advantages are derived, in part, from the ability of an investigator to process and discern effectively a complex two-dimensional spatial representation of a sample in terms of a variety of spectrally related molecular parameters. In particular, a single spectroscopic image, or chemical map, is capable of summarizing and conveying a wealth of spatial, chemical, and molecular data of benefit to both specialists and nonspecialists. Regardless of the specific method used to collect data during a vibrational spectroscopic imaging experiment, the final representation is typically expressed as a three-dimensional image cube, or hypercube, consisting of two spatial dimensions and one spectral dimension. The image cube concept, shown in Figure 56, is interpreted in one of two ways, either (1) a series of images collected as a function of wavenumber (Figure 56 A), or (2) a vibrational spectrum corresponding to each spatial location, or pixel, within the image plane (Figure 56 B). The ability to merge and recall, within a single analytical technique, information corresponding to the spatial and spectral axes of the image cube provides extraordinary flexibility in probing and extracting structural and compositional information from samples. Thus, a data set may be either summarized as a single image or expressed as a series of images derived from one or more spectroscopic parameters, such as spectral peak intensity, band frequency, or linewidth. The chemical and morphological interpretations of the resulting images are then based on the spatial variations of the intrinsic molecular property being highlighted and their relevance to the biological questions being pursued. In contrast to images generated by conventional point mapping, which generally suffer from low spatial resolution due to relatively few sampling points, direct imaging approaches using the advantages inherent in two-dimensional array detectors derive an enormous benefit from the highly parallel nature of the data collection. That is, a single data set which typically contains many tens of thousands of independent spectral channels may be analyzed and interpreted by using a variety of statistical algorithms, allowing subtle spatial and spectral variations that are often overlooked or misinterpreted in traditional spectroscopic studies to be tested and revealed.

Vibrational spectroscopic imaging has been widely applied to diverse materials, including polymers, semiconductors, pharmaceuticals, cosmetics, consumer products, and biological materials. The strengths and adaptability of vibrational spectroscopic imaging rest not only on the ability to determine chemical compositions and component distribution within a sample, but also on being able to extract localized molecular information relevant to the sample architecture. For example, vibrational spectra are sensitive to alterations in protein secondary and tertiary structures. Spectral shifts could therefore be used to image specifically the distributions of specific structural moieties within a biological sample as opposed to measuring simply an overall distribution of a general class of molecules. Multivariate approaches may also be used to generate a composite metric indicative of either disease progression or biological function. Although these metrics can not be readily interpreted in terms of biochemical changes, they may often be statistically correlated with disease. In addition, by simultaneously recording data on a two-dimensional focal-plane array detector from a myriad of spatial locations

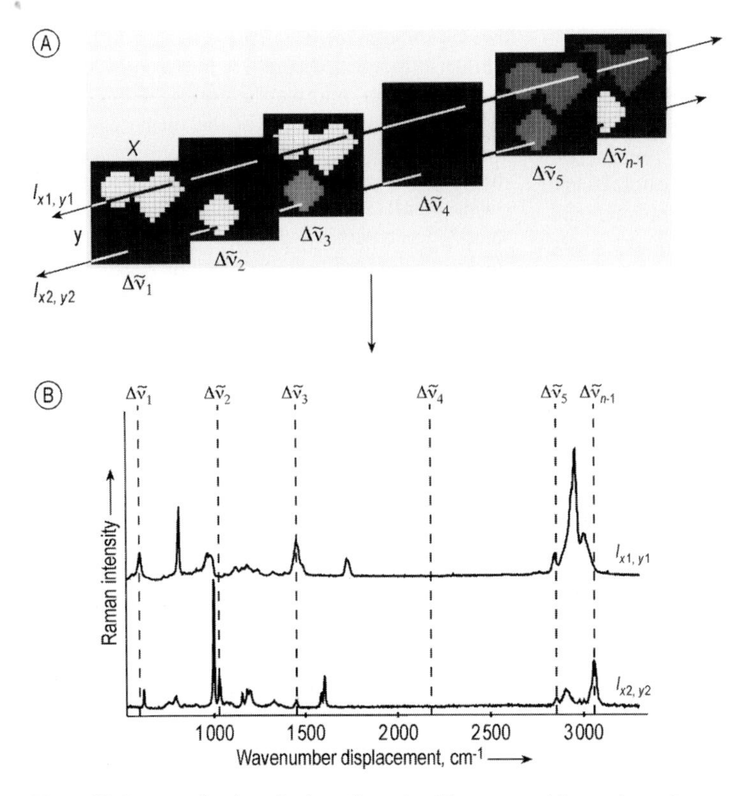

Figure 56. Conceptualization of a three-dimensional hyperspectral image data cube
(A) Images as a function of wavenumber; B) Selected spectra associated with pixels (*i, j*) in the image
(Reproduced by permission of Marcel Dekker, Inc., 270 Madison Avenue, New York, NY 10016-0602)

within a biological matrix, pooled spectra may be treated statistically. In this manner, analysis of variance can be employed to more robustly test the significance of observed spectral changes appearing, for example, because of a perturbed or diseased state.

Raman and infrared imaging techniques as well as their biological applications are described in detail in [170].

17.5.6. Infrared and Raman Spectroscopy of Polymers

As polymer analysis is a discipline of its own beyond the scope of this article, the reader is referred to the comprehensive works by HUMMEL [171], KOENIG [172], and ZERBI et al. [173].

17.6. Near-Infrared Spectroscopy

17.6.1. Comparison of Mid-Infrared and Near-Infrared Spectroscopy

Near-infrared spectroscopy as a valuable tool for qualitative and quantitative analysis [174], [175] has experienced a revival through the development of chemometric methods and the introduction of fiber optics during the 1980s, and high-quality near-infrared spectra are provided by improved conventional and by Fourier transform instruments.

Various aspects differentiate a mid-infrared spectrum from the corresponding near-infrared spectrum (Table 6). In the mid-infrared region fundamental molecular vibrations always occur between 4000 and 200 cm^{-1}, while in the near-infrared region overtones and combination bands of these fundamentals are observed between 12 800 and 4000 cm^{-1}. Near-infrared bands are due primarily to hydrogenic stretches of C–H, N–H, and

Table 6. Comparison of mid-infrared and near-infrared spectroscopy

Mid-Infrared	Near-Infrared
4000 – 200 cm^{-1}	12 800 – 4000 cm^{-1}
Fundamentals	overtones and combinations
High intensity	low intensity
High sensitivity (trace analysis possible)	low sensitivity (not suited for trace analysis)
Sharp, separated bands	strongly overlapping bands
Easy quantitation with isolated bands	quantitation complex, chemometrics necessary
FT-instruments	grating, filter, FT and acousto-optical tunable scanning instruments

O–H bonds whose fundamental vibrations give rise to strong bands between 4000 and 2000 cm^{-1}, so that their overtones and combinations occur above 4000 cm^{-1} as the most intense absorption bands in the near-infrared spectrum. Since the near-infrared absorptions of polyatomic molecules thus mainly reflect vibrational contributions from very few functional groups, near-infrared spectroscopy is less suitable for detailed qualitative analysis than mid-infrared spectroscopy, which shows all (active) fundamentals and the overtones and combinations of low-energy vibrations. The near-infrared overtone and combination bands depend more on their environment than does the fundamental of the same vibration; slight perturbation in the bonding scheme causes only small changes in the fundamental but significant frequency shifts and amplitude changes in the near-infrared. In going from the fundamental to the first overtone the intensity of an absorption band decreases by a factor of about 10 – 100, so that the sensitivity of near-infrared spectroscopy is reduced in comparison with the mid-infrared. This is a disadvantage when gases are to be measured, but for liquids, it is a considerable advantage as regards sample handling because cells with convenient path lengths between 1 mm and 10 cm can be used. Glass or quartz is used as window material. Near-infrared bands of liquids and solids have relatively large bandwidths between 30 and 60 cm^{-1}, so that they strongly overlap and direct assignment of bands is generally not possible for larger, complex molecules. This is why, in the near-infrared, easy quantitative analysis using isolated bands, as in the mid-infrared, is not possible. Chemometrics [37], [176] – [178], however, by the application of mathematical and statistical procedures, generates correlations between experimental data and the chemical composition or physical properties of the sample, and can be used in a general manner to solve quantitative and qualitative analytical problems.

17.6.2. Applications of Near-Infrared Spectroscopy

Applications of near-infrared spectroscopy include chemistry [179], [180], the oil industry [181], clinical analysis [182], biological and medical analysis [183], and the pharmaceutical industry [184], [185].

Since the intensities of characteristic near-infrared bands are independent of or only slightly dependent on the state of the system, near-infrared spectroscopy is widely applicable for the quantitative study of liquid and compressed gaseous systems, including fluid systems, up to high pressures and temperatures.

The application of near-infrared data to routine quantitative analysis was initiated by NORRIS [186], who used diffuse reflectance measurements to quantitatively determine major components, such as moisture, protein and fat, in agricultural commodities. In comparison with mid-infrared, near-infrared diffuse reflectance [187] is able to measure powdered samples with minimal sample preparation, and lends itself extremely well to quantitative analysis because the smaller absorptivities and larger scattering coefficients at shorter wavelengths result in a larger scattering/absorption ratio. Although near-infrared reflectance analysis was initially developed for agricultural and food applications [188], in combination with chemometrics it is now applied to many other areas, e.g., polymers [189], pharmaceuticals [190], organic substances [191], geological samples [192], and thin-layer chromatography [193], [194].

The excellent transmittance of quartz in the near-infrared range has led to a further enhancement of the potential of near-infrared spectroscopy by the introduction of fiber optics. Fiber-optic waveguides are used to transfer light from the spectrometer to the sample and, after transmission or reflection, back to the spectrometer. Most fiber optic cables consist of three concentric components: the inner part, through which the light propagates, is called the core, the middle layer is the cladding, and the outer protective layer is the jacket. Crucial to the light throughput of the fiber optic cable are the optical properties of the core and cladding. The difference in refractive index of the two materials defines the highest angle at which the core/cladding interface exhibits total internal reflection, and hence allows the light to propagate through the cable. This angle is related to the numerical aperture (N.A.) of the fiber by the following equation:

$$\sin\alpha = \text{N.A.} = \left(n_{\text{core}}^2 - n_{\text{cladding}}^2\right)^{1/2} \quad (20)$$

where α is the half-angle of acceptance and n is the refractive index of the material. The greater the difference in refractive index the greater the half-angle of acceptance. Any light entering at an angle greater than the half-angle of acceptance is lost to the cladding through absorption. The core and cladding are usually of quartz (of different refractive index), while the outer protective layer is a polymer; other fiber optic materials are zirconium fluoride (ZrF) and chalcogenide (AsGeSe) [195]. The obvious advantage of the optical fiber technique is that the location of measurement can be separated from the spectrometer by between two and several hundred meters, depending on the length of the waveguide. Since almost all practically relevant substances show characteristic near-infrared absorption bands, quantitative analysis by near-infrared spectroscopy is generally applicable to on-line concentration measurements in connection with chemical reactions, chemical equilibria, and phase equilibria. Various probes can be integrated into different reactors or bypasses and offer numerous remote-sensing and on-line process control applications [196]. In the analysis of hazardous or toxic materials this has proved of outstanding importance. In the refining, petrochemical, polymer, plastics, and chemicals industries, typical applications of near-infrared remote spectroscopy are the determination of the octane number and oxygenates in gasoline, aromatics in fuels, the composition of solvent mixtures [197], [198], the hydroxyl number of polyols, low contents of water in solvents, and for polymer analysis.

17.7. References

[1] W. W. Coblentz: *Investigations of Infrared Spectra,* Carnegie Institution, Washington 1905, re-published, The Coblentz Society, Norwalk 1962.

[2] G. Herzberg: *Molecular Spectra and Molecular Structure I,* Van Nostrand, Princeton 1953.

[3] G. Herzberg: *Molecular Spectra and Molecular Structure II,* Van Nostrand, Princeton 1960.

[4] E. B. Wilson, Jr., J. C. Decius, P. C. Cross: *Molecular Vibrations,* McGraw-Hill, New York 1955, re-published Dover, Mineola 1980.

[5] D. Steele: *Theory of Vibrational Spectroscopy,* Saunders, Philadelphia 1971.

[6] J. A. Koningstein: *Introduction to the Theory of the Raman Effect,* Reidel, Dordrecht 1972.

[7] D. A. Long: *Raman Spectroscopy,* McGraw-Hill, New York 1977.

[8] P. R. Griffiths, J. A. de Haseth: *Fourier Transform Infrared Spectroscopy,* J. Wiley & Sons, New York 1986.

[9] M. W. Mackenzie: *Advances in Applied Fourier Transform Infrared Spectroscopy,* J. Wiley & Sons, New York 1988.

[10] N. B. Colthup, L. H. Daly, S. E. Wiberley: *Introduction to Infrared and Raman Spectroscopy,* 3rd ed., Academic Press, San Diego 1990.

[11] B. Schrader (ed.): *Infrared and Raman Spectroscopy,* VCH Verlagsgesellschaft, Weinheim 1995.

[12] K. W. F. Kohlrausch: *Ramanspektren,* Akademische Verlagsgesellschaft, Leipzig 1943, re-published Heyden & Son, London 1973.

[13] H. Baranska, A. Labudzinska, T. Terpinski: *Laser Raman Spectroscopy,* J. Wiley & Sons, New York 1983.

[14] P. Hendra, C. Jones, G. Warnes: *Fourier Transform Raman Spectroscopy,* Ellis Horwood, Chichester 1991.

[15] J. R. Ferraro, K. Krishnan (eds.): *Practical Fourier Transform Infrared Spectroscopy,* Academic Press, San Diego 1990.

[16] M. J. Pelletier (ed.): *Analytical Applications of Raman Spectroscopy,* Blackwell Science, Oxford 1999.

[17] J. G. Grasselli, B. J. Bulkin (eds.): *Analytical Raman Spectroscopy,* J. Wiley & Sons, New York 1991.

[18] G. Socrates: *Infrared Characteristic Group Frequencies,* J. Wiley & Sons, New York 1980.

[19] L. J. Bellamy: *The Infrared Spectra of Complex Molecules,* 3rd ed., vol. 1, Chapman & Hall, London 1975.

[20] L. J. Bellamy: *The Infrared Spectra of Complex Molecules,* 2nd ed., vol. 2, Chapman & Hall, London 1980.

[21] F. R. Dollish, W. G. Fateley, F. F. Bentley: *Characteristic Raman Frequencies of Organic Compounds,* Wiley Interscience, New York 1974.

[22] D. Lin-Vien, N. B. Colthup, W. G. Fateley, J. G. Grasselli: *The Handbook of Infrared and Raman Characteristic Frequencies of Organic Molecules,* Academic Press, Boston 1991.

[23] K. Nakamoto: *Infrared and Raman Spectroscopy of Inorganic and Coordination Compounds,* 5th ed., J. Wiley & Sons, Chichester 1997.

[24] H.-U. Gremlich, B. Yan (eds.): *Infrared and Raman Spectroscopy of Biological Materials,* Marcel Dekker, New York 2000.

[25] B. Schrader: *Raman/Infrared Atlas of Organic Compounds,* 2nd ed., VCH Verlagsgesellschaft, Weinheim 1989.

[26] Sadtler IR Digital Spectra Libraries, Sadtler Division, Bio-Rad Laboratories, Inc., 3316 Spring Garden Street, Philadelphia, PA 19104.
[27] F. M. Mirabella (ed.): *Modern Techniques in Applied Molecular Spectroscopy,* J. Wiley & Sons, New York 1998.
[28] A. A. Michelson, *Philos. Mag.* **31** (1891) 256 – 259.
[29] W. Herres, J. Gronholz, *Comput. Appl. Lab.* **4** (1984) 216 – 220.
[30] J. W. Cooley, J. W. Tukey, *Math. Comput.* **19** (1965) 297 – 301.
[31] J. Gronholz, W. Herres, *Instrum. Computers* **3** (1985) 10 – 19.
[32] B. Schrader, A. Hoffmann, A. Simon, J. Sawatzki, *Vib. Spectrosc.* **1** (1991) 239 – 250.
[33] B. Schrader in J. R. Ferraro, K. Krishnan (eds.): *Practical Fourier Transform Infrared Spectroscopy,* Academic Press, San Diego 1990, p. 167.
[34] B. Schrader, A. Hoffmann, S. Keller, *Spectrochim. Acta* **47 A** (1991) 1135 – 1148.
[35] J. Coates in W. O. George, H. A. Willis (eds.): *Computer Methods in UV, Visible and IR Spectroscopy,* The Royal Society of Chemistry, Cambridge 1990, p. 95.
[36] P. S. Wilson in W. O. George, H. A. Willis (eds.): *Computer Methods in UV, Visible and IR Spectroscopy,* The Royal Society of Chemistry, Cambridge 1990, p. 143.
[37] F. Despagne, D. L. Massart, *Analyst* **123** (1998) 157R – 178R.
[38] G. M. Barrow: *Introduction to Molecular Spectroscopy,* McGraw-Hill, New York 1962.
[39] S. R. La Paglia: *Introductory Quantum Chemistry,* Harper & Row, New York 1971.
[40] *Ullmann,* 4th ed., **5,** 303.
[41] Available from Sadtler Division, Bio-Rad Laboratories, Inc., 3316 Spring Garden Street, Philadelphia, PA 19104.
[42] Available from LabControl GmbH, D-50858 Köln 40, Max-Planck-Str. 17 a, Germany.
[43] Available from Galactic Industries Corporation, 395 Main Street, Salem NH 03079.
[44] J. P. Coates, R. W. Hannah in T. Theophanides (ed.): *Fourier Transform Infrared Spectroscopy,* D. Reidel Publ., Dordrecht 1984, p. 167.
[45] S. R. Lowry, D. A. Hupplar, *Anal. Chem.* **53** (1981) 889 – 893.
[46] J. T. Clerc in W. O. George, H. A. Willis (eds.): *Computer Methods in UV, Visible and IR Spectroscopy,* The Royal Society of Chemistry, Cambridge 1990, p. 13.
[47] D. A. Cirovic, *Trends Anal. Chem.* **16** (1997) 148 – 155.
[48] H. Yang, P. R. Griffiths, *Anal. Chem.* **71** (1999) 751 – 761.
[49] J. U. White, *J. Opt. Soc. Am.* **32** (1942) 285 – 288.
[50] R. T. Graf, J. L. Koenig, H. Ishida, *Appl. Spectrosc.* **39** (1985) 405 – 412.
[51] R. G. Greenler, *J. Chem. Phys.* **44** (1966) 310 – 315.
[52] R. G. Greenler, *J. Chem. Phys.* **50** (1969) 1963 – 1968.
[53] D. L. Allara, J. D. Swalen, *J. Phys. Chem.* **86** (1982) 2700 – 2704.
[54] N. J. Harrick, *Appl. Opt.* **10** (1971) 2344 – 2349.
[55] I. Newton: *Opticks,* Dover, New York 1952.
[56] N. J. Harrick: *Internal Reflection Spectroscopy,* Harrick Scientific Corporation, Ossining, New York 1979.
[57] J. Fahrenfort in A. Mangini (ed.): *Adv. Mol. Spectrosc. Proc. Int. Meet. 4th 1959,* **vol. 2,** 1962, p. 701.
[58] Available from Harrick Scientific Corporation, 88 Broadway, Ossining, NY 10562.
[59] N. J. Harrick, M. Milosevic, S. L. Berets, *Appl. Spectrosc.* **45** (1991) 944 – 948.
[60] S. Compton, P. Stout, FTS 7/IR Notes no. 1, Bio-Rad, Digilab Division, Cambridge, MA 02139, 1991.
[61] U. P. Fringeli in F. M. Mirabella (ed.): *Internal Reflection Spectroscopy: Theory and Application,* Marcel Dekker, New York 1992, p. 255.
[62] U. P. Fringeli, *Chimia* **46** (1992) 200 – 214.
[63] Available from Spectra-Tech Inc., Stamford, CT 06906.
[64] Available from GRASEBY SPECAC Ltd., Lagoon Road, St. Mary Cray, Orpington, Kent BR5 3QX, U.K.
[65] W. M. Doyle, *Appl. Spectrosc.* **44** (1990) 50 – 69.
[66] B. E. Miller, N. D. Danielson, J. E. Katon, *Appl. Spectrosc.* **42** (1988) 401 – 405.
[67] Available from Axiom Analytical, Inc., 17751-C Sky Park Circle, Irvine, CA 92614.
[68] W. M. Doyle, *Process Control Quality* **2** (1992) 11 –41.
[69] N. L. Owen, D. W. Thomas, *Appl. Spectrosc.* **43** (1989) 451 – 455.
[70] P. Kubelka, F. Munk, *Z. Tech. Phys.* **12** (1931) 593–601.
[71] P. Kortüm, W. Braun, G. Herzog, *Angew. Chem., Int. Ed. Engl.* **2** (1963) 333 – 341.
[72] P. H. Turner, W. Herres, BRUKER FT-IR Application Note 23, Bruker Analytische Meßtechnik GmbH, D-76153 Karlsruhe, Germany.
[73] D. Reinecke, A. Jansen, F. Fister, U. Schernau, *Anal. Chem.* **60** (1988) 1221 – 1224.
[74] M. P. Fuller, P. R. Griffiths, *Anal. Chem.* **50** (1978) 1906 – 1910.
[75] M. Fuller in *Nicolet FT-IR Spectral Lines,* Nicolet Analytical Instruments, Madison, WI 53711, Spring/ Summer 1990, p. 15.
[76] G. Glauninger, K.-A. Kovar, V. Hoffmann, *Fresenius J. Anal. Chem.* **338** (1990) 710 – 716.
[77] H. Yamamoto, K. Wada, T. Tajima, K. Ichimura, *Appl. Spectrosc.* **45** (1991) 253 – 259.
[78] S. A. Johnson, R. M. Rinkus, T. C. Diebold, V. A. Maroni, *Appl. Spectrosc.* **42** (1988) 1369 – 1375.
[79] K. A. Martin, R. F. Zabransky, *Appl. Spectrosc.* **45** (1991) 68 – 72.
[80] Si-Carb Sampler, available from Spectra-Tech Inc., Stamford, CT 06906.
[81] F. J. DeBlase, S. Compton, *Appl. Spectrosc.* **45** (1991) 611 – 618.
[82] A. M. Vassallo, P. A. Cole-Clarke, L. S. K. Pang, A. J. Palmisano, *Appl. Spectrosc.* **46** (1992) 73 – 78.
[83] M. Handke, N. J. Harrick, *Appl. Spectrosc.* **40** (1986) 401 – 405.
[84] R. T. Rewick, R. G. Messerschmidt, *Appl. Spectrosc.* **45** (1991) 297 – 301.
[85] L. T. Lin, D. D. Archibald, D. E. Honigs, *Appl. Spectrosc.* **42** (1988) 477 – 483.
[86] A. Tsuge, Y. Uwamino, T. Ishizuka, *Appl. Spectrosc.* **43** (1989) 1145 – 1149.
[87] R. W. Jones, J. F. McClelland, *Anal. Chem.* **61** (1989) 650 – 656.
[88] R. W. Jones, J. F. McClelland, *Anal. Chem.* **61** (1989) 1810 – 1815.
[89] R. J. Pell, B. C. Erickson, R. W. Hannah, J. B. Callis, B. R. Kowalski, *Anal. Chem.* **60** (1988) 2824 –2827.
[90] J. A. McGuire, B. Wangmaneerat, T. M. Niemczyk, D. M. Haaland, *Appl. Spectrosc.* **46** (1992) 178 – 180.
[91] B. Wangmaneerat et al., *Appl. Spectrosc.* **46** (1992) 340 – 344.

[92] A. Rosencwaig: *Photoacoustics and Photoacoustic Spectroscopy,* J. Wiley & Sons, New York 1980.
[93] J. F. McClelland, *Anal. Chem.* **55** (1983) 89 A – 105 A.
[94] J. A. Graham, W. M. Grim III, W. G. Fateley in J. R. Ferraro, L. J. Basile (eds.): *Fourier Transform Infrared Spectroscopy,* **vol. 4,** Academic Press, Orlando 1985, p. 345.
[95] H. Coufal, J. F. McClelland, *J. Mol. Struct.* **173** (1988) 129 – 140.
[96] G. Laufer, J. T. Huneke, B. S. H. Royce, Y. C. Teng, *Appl. Phys. Lett.* **37** (1980) 517 – 519.
[97] G. Busse, B. Bullemer, *Infrared Phys.* **18** (1978) 631 –634.
[98] MTEC Photoacoustics, Inc., Ames, IA 50010.
[99] R. O. Carter III, S. L. Wright, *Appl. Spectrosc.* **45** (1991) 1101 – 1103.
[100] C. Q. Yang, *Appl. Spectrosc.* **45** (1991) 102 – 108.
[101] J. C. Donini, K. H. Michaelian, *Infrared Phys.* **24** (1984) 157 – 163.
[102] M. W. Urban, J. L. Koenig, *Appl. Spectrosc.* **40** (1986) 994 – 998.
[103] C. Q. Yang, R. R. Bresee, W. G. Fateley, *Appl. Spectrosc.* **41** (1987) 889 – 896.
[104] R. Rubinovitz, J. Seebode, A. Simon, *Bruker Report* **1** (1988) 35 – 38.
[105] M. J. Smith, C. J. Manning, R. A. Palmer, J. L. Chao, *Appl. Spectrosc.* **42** (1988) 546 – 555.
[106] R. A. Crocombe, S. V. Compton, FTS/IR Notes no. 82, Bio-Rad, Digilab Division, Cambridge, MA 02139, 1991.
[107] K. H. Michaelian, *Appl. Spectrosc.* **45** (1991) 302 –304.
[108] L. Bertrand, *Appl. Spectrosc.* **42** (1988) 134 – 138.
[109] R. White: "Chromatography/Fourier Transform Infrared Spectroscopy and Its Applications," *Practical Spectroscopy Series,* vol. 10, Marcel Dekker, New York 1990.
[110] W. Herres: *HRGC-FTIR: Capillary Gas Chromatography-Fourier Transform Infrared Spectroscopy,* Hüthig Verlag, Heidelberg 1987.
[111] S. Bourne et al., FTS/IR Notes no. 75, Bio-Rad, Digilab Division, Cambridge, MA 02139, 1990.
[112] R. L. White, G. N. Giss, G. M. Brissey, C. L. Wilkins, *Anal. Chem.* **53** (1981) 1778 – 1782.
[113] P. Coffey, D. R. Mattson, J. C. Wright, *Am. Lab.* **10** (1978) 126 – 132.
[114] K. S. Chiu, K. Biemann, K. Krishnan, S. L. Hill, *Anal. Chem.* **56** (1984) 1610 – 1615.
[115] Available from Lab Connections, Inc., 201 Forest Street, Marlborough, MA 01752.
[116] Available from Bourne Scientific, Inc., 65 Bridge Street, Lexington, MA 02421-7927.
[117] S. Bourne, *Am. Lab.* **16** (1998) 17F – 17J.
[118] C. M. Earnest, *Anal. Chem.* **56** (1984) 1471 A –1486 A.
[119] D. Dollimore, *Anal. Chem.* **64** (1992) 147 R – 153 R.
[120] J. A. J. Jansen, J. H. van der Maas, A. Posthuma de Boer, *Appl. Spectrosc.* **46** (1992) 88 – 92.
[121] K. Ichimura, H. Ohta, T. Tajima, T. Okino, *Mikrochim. Acta* **1** (1988) 157 – 161.
[122] R. C. Wieboldt, S. R. Lowry, R. J. Rosenthal, *Mikrochim. Acta* **1** (1988) 179 – 182.
[123] J. Mullens et al., *Bull. Soc. Chim. Belg.* **101** (1992) 267 – 277.
[124] R. M. Carangelo, P. R. Solomon, D. J. Gerson, *Fuel* **66** (1987) 960 – 967.
[125] J. K. Whelan, *Energy Fuels* **2** (1988) 65 – 73.
[126] L. A. Sanchez, M. Y. Keating, *Appl. Spectrosc.* **42** (1988) 1253 – 1258.
[127] M. D. Timken, J. K. Chen, T. B. Brill, *Appl. Spectrosc.* **44** (1990) 701 – 706.
[128] E. Post, S. Rahner, H. Möhler, A. Rager, *Thermochim. Acta* **263** (1995) 1 – 6.
[129] M. Maciejewski, F. Eigenmann, A. Baiker, E. Dreser, A. Rager, *Thermochim. Acta* (2000) in press.
[130] L. A. Nafie in M. W. Mackenzie (ed.): *Advances in Applied FTIR Spectroscopy,* John Wiley and Sons, New York 1988, p. 67.
[131] L. A. Nafie, T. B. Freedman in H.-U. Gremlich, B. Yang (eds.): *Infrared and Raman Spectroscopy of Biological Materials,* Marcel Dekker, New York 2000, p. 15.
[132] T. A. Keiderling in H.-U. Gremlich, B. Yang (eds.): *Infrared and Raman Spectroscopy of Biological Materials,* Marcel Dekker, New York 2000, p. 55.
[133] B. Schrader, S. Keller, *Proc. SPIE Int. Soc. Opt. Eng. 1575 (Int. Conf. Fourier Transform Spectrosc., 8th, 1991)* (1992) 30 – 39.
[134] N. J. Everall, J. Lumsdon, *Spectrosc. Eur.* **4** (1992) 10 – 21.
[135] C. G. Zimba, J. F. Rabolt, *Appl. Spectrosc.* **45** (1991) 162 – 165.
[136] T. Jawhari, P. J. Hedra, H. A. Willis, M. Judkins, *Spectrochim. Acta* **46 A** (1990) 161 – 170.
[137] F. T. Walder, M. J. Smith, *Spectrochim. Acta* **47 A** (1991) 1201 – 1216.
[138] H. Sadeghi-Jorabchi, R. H. Wilson, P. S. Belton, J. D. Edwards-Webb, D. T. Coxon, *Spectrochim. Acta* **47 A** (1991) 1449 – 1458.
[139] L. A Lyon et al., *Anal. Chem.* **70** (1998) 341R – 361R.
[140] B. Chase in A. M. C. Davies, C. S. Creaser (eds.): *Anal. Appl. Spectrosc. 2,* The Royal Society of Chemistry, Cambridge 1991, p. 13.
[141] K. P. J. Williams, S. M. Mason, *TRAC* **9** (1990) 119 –127.
[142] P. J. Hendra, *J. Mol. Struct.* **266** (1992) 97 –114.
[143] A. Hoffmann et al., *J. Raman Spectrosc.* **22** (1991) 497 – 503.
[144] P. J. Hendra, C. H. Jones, *Makromol. Chem. Macromol. Symp.* **52** (1991) 41 – 56.
[145] G. J. Thomas, *Ann. Rev. Biophys. Biomol. Struct.* **28** (1999) 1 – 27.
[146] A. M. Tudor et al., *J. Pharm. Biomed. Anal.* **8** (1990) 717 – 720.
[147] P. R. Carey, *J. Biol. Chem.* **274** (1999) 26 625 – 26 628.
[148] J. Góral, V. Zichy, *Spectrochim. Acta* **46 A** (1990) 253 – 275.
[149] I. W. Levin, E. N. Lewis, *Anal. Chem.* **62** (1990) 1101 A – 1111 A.
[150] S. Nie et al., *Spectroscopy* **5** (1990) 24 – 32.
[151] G. Ellis, M. Claybourn, S. E. Richards, *Spectrochim. Acta* **46 A** (1990) 227 – 241.
[152] D. Bourgeois, S. P. Church, *Spectrochim. Acta* **46 A** (1990) 295 – 301.
[153] T. N. Day, P. J. Hendra, A. J. Rest, A. J. Rowlands, *Spectrochim. Acta* **47 A** (1991) 1251 – 1262.
[154] N. J. Everall, J. M. Chalmers, I. D. Newton, *Appl. Spectrosc.* **46** (1992) 597 – 601.
[155] D. L. Gerrard, J. Birnie, *Anal. Chem.* **64** (1992) 502 R – 513 R.
[156] B. Yan, *Acc. Chem. Res.* **31** (1998) 621 – 630.

[157] R. G. Messerschmidt, M. A. Harthcock (eds.): *Infrared Microspectroscopy: Theory and Applications,* Marcel Dekker, New York 1988.
[158] K. Krishnan, S. L. Hill in J. R. Ferraro, K. Krishnan (eds.): *Practical Fourier Transform Spectroscopy,* Academic Press, San Diego 1990, p. 103.
[159] S. Compton, J. Powell, FTS 7/IR Notes no. 2, Bio-Rad, Digilab Division, Cambridge, MA 02139, 1991.
[160] N. J. Harrick, M. Milosevic, S. L. Berets, *Appl. Spectrosc.* **45** (1991) 944 – 948.
[161] B. Schrader, *Fresenius J. Anal. Chem.* **337** (1990) 824 – 829.
[162] J. D. Louden in D. J. Gardiner, P. R. Graves (eds.): *Practical Raman Spectroscopy,* Springer Verlag, Berlin 1989, p. 119.
[163] J. Sawatzki, *Fresenius J. Anal. Chem.* **339** (1991) 267 –270.
[164] A. Hoffmann, B. Schrader, R. Podschadlowski, A. Simon, *Proc. SPIE-Int. Soc. Opt. Eng. 1145 (Int. Conf. Fourier Transform Spectrosc., 7th, 1989)* (1989) 372 – 373.
[165] F. Bergin, *Proc. Annu. Conf. Microbeam Anal. Soc.* **25** (1990) 235 – 239.
[166] B. W. Cook in A. M. C. Davies, C. S. Creaser (eds.): *Anal. Appl. Spectrosc.* **2,** The Royal Society of Chemistry, Cambridge 1991, p. 61.
[167] A. J. Sommer, J. E. Katon, *Appl. Spectrosc.* **45** (1991) 527 – 534.
[168] D. Gernet, W. Kiefer, H. Schmidt, *Appl. Spectrosc.* **46** (1992) 571 – 576.
[169] K. Gerwert in H.-U. Gremlich, B. Yang (eds.): *Infrared and Raman Spectroscopy of Biological Materials,* Marcel Dekker, New York 2000, p. 193.
[170] M. D. Schaeberle, I. W. Levin, E. N. Lewis in H.-U. Gremlich, B. Yang (eds.): *Infrared and Raman Spectroscopy of Biological Materials,* Marcel Dekker, New York 2000, p. 231.
[171] D. O. Hummel: *Atlas of Polymer and Plastics Analysis,* 3rd ed., VCH Verlagsgesellschaft, Weinheim 1991.
[172] J. L. Koenig: *Spectroscopy of Polmers,* American Chemical Society, Washington, D.C. 1992.
[173] G. Zerbi et al.: *Modern Polymer Spectroscopy,* Wiley-VCH, Weinheim 1998.
[174] J. Workman, Jr., *J. Near Infrared Spectrosc.* **1** (1993) 221.
[175] I. Murray, I. A. Cowe (eds.): *Making Light Work: Advances in Near Infrared Spectroscopy,* VCH Verlagsgesellschaft, Weinheim 1992.
[176] P. Geladi, E. Dabbak, *J. Near Infrared Spectrosc.* **3** (1995) 119.
[177] D. L. Massart, B. G. M. Vandeginste, L. M. C. Buydens, S. de Jong, P. J. Lewi, J. Smeyers-Verbecke: *Handbook of Chemometrics and Qualimetrics, Part A,* Elsevier, Amsterdam 1997.
[178] D. L. Massart, B. G. M. Vandeginste, L. M. C. Buydens, S. de Jong, P. J. Lewi, J. Smeyers-Verbecke: *Handbook of Chemometrics and Qualimetrics, Part B,* Elsevier, Amsterdam 1998.
[179] S. M. Donahue, C. W. Brown, B. Caputo, M. D. Modell, *Anal. Chem.* **60** (1988) 1873 – 1878.
[180] J. Lin, C. W. Brown, *Vibr. Spectrosc.* **7** (1994) 117 – 123.
[181] S. R. Westbrook, *SAE Tech. Paper Ser.* **930734** (1993) 1 – 10.
[182] H. M. Heise in H.-U. Gremlich, B. Yan (eds.): *Infrared and Raman Spectroscopy of Biological Materials,* Marcel Dekker, New York 2000, p. 259.
[183] R. J. Dempsey, D. G. Davis, R. G. Buice, Jr., R. A. Lodder, *Appl. Spectrosc.* **50** (1996) 18A – 34A.
[184] J. D. Kirsch, J. K. Drennen, *Appl. Spectrosc. Rev.* **30** (1995) 139 – 174.
[185] W. Plugge, C. van der Vlies, *J. Pharm. Biomed. Anal.* **14** (1996) 891 – 898.
[186] I. Ben-Gera, K. H. Norris, *Israel J. Agric. Res.* **18** (1968) 125 – 132.
[187] D. L. Wetzel, *Anal. Chem.* **55** (1983) 1165 A – 1176 A.
[188] P. Williams, K. Norris (eds.): *Near-Infrared Technology in the Agricultural and Food Industries,* AACC, St. Paul 1987.
[189] C. E. Miller, *Appl. Spectrosc. Rev.* **26** (1991) 277 –339.
[190] B. F. MacDonald, K. A. Prebble, *J. Pharm. Biomed. Anal.* **11** (1993) 1077 – 1085.
[191] L. G. Weyer, *Appl. Spectrosc. Rev.* **21** (1985) 1 – 43.
[192] D. E. Honigs, T. B. Hirschfeld, G. M. Hieftje, *Appl. Spectrosc.* **39** (1985) 1062 – 1065.
[193] E. W. Ciurczak, L. J. Cline-Love, D. M. Mustillo, *Spectrosc. Int.* **3** (1991) no. 5, 39 – 42.
[194] E. W. Ciurczak, W. R. Murphy, D. M. Mustillo, *Spectrosc. Int.* **3** (1991) no. 7, 39 – 44.
[195] M. J. Smith, T. E. May, *Int. Lab.* **22** (1992) 18 – 24.
[196] L. P. McDermott, *Ad. Instrum. Control.* **45** (1990) 669 – 677.
[197] Guided Wave, Inc., Application Note no. A3-987, El Dorado Hills, CA 95630.
[198] Guided Wave, Inc., Application Note no. A4-188, El Dorado Hills, CA 95630.

18. Nuclear Magnetic Resonance and Electron Spin Resonance Spectroscopy

REINHARD MEUSINGER, Institute of Organic Chemsitry, University of Mainz, Mainz, Federal Republic of Germany

A. MARGARET CHIPPENDALE, ZENECA Specialties, Manchester, United Kingdom

SHIRLEY A. FAIRHURST, AFRC Institute for Plant Science Research, Brighton, United Kingdom

Symbols

a ESR isotropic hyperfine coupling constant; ESR total hyperfine interaction
AT acquisition time
B magnetic field
DW dwell time
g electronic or nuclear g factor
h Planck constant
I nuclear spin quantum number
J spin – spin coupling constant
k Boltzmann constant
K intensity of ESR response
M magnetization vector
m_I, m_S magnetic quantum numbers
N Avogadro number
s shielding constant
S electronic spin quantum number
SW spectrum width
T ESR anisotropic hyperfine coupling
T_1 spin – lattice relaxation time
T_2 spin – spin relaxation time
T_d relaxation delay
β electronic Bohr or nuclear magneton
γ magnetogyric ratio
δ chemical shift
μ magnetic dipole moment
τ rotational correlation time, pulse width
ν resonance frequency
ω_0 Larmor frequency

Units

ppm parts per million (unit of the dimensionless δ scale of the chemical shift)
T tesla (unit of magnetic field strength, $1\ T = 10^4$ gauss)

18.1. Introduction

It is now about 50 years since the phenomena of electron spin resonance (ESR) and nuclear magnetic resonance (NMR) were first discovered. During the last thirty years NMR spectroscopy has developed into perhaps the most important instrumental measuring technique within chemistry. This is due to a dramatic increase in both the sensitivity and the resolution of NMR and ESR instruments. NMR and ESR spectroscopy are used today within practically all branches of chemistry and in related sciences, such as physics, biology, and medicine, both at universities and in industrial laboratories. NMR has its most important applications in the determination of molecular structures. It can today be applied to a wide variety of chemical systems, from small molecules to proteins and nucleic acids. ESR has played a key role in the study of free radicals, many transition metal compounds, and other species containing unpaired electrons.

The continuing development of each technique has resulted from a series of step changes. Both methods were first developed using continuous wave methodology. By the late 1960s the pulsed Fourier transform techniques were developed and additionally the NMR spectroscopists began to use magnets based on superconducting material. On grounds of these developments the quality of spectra, expressed both in terms of sensitivity and resolution, improved quickly during the 1970s. The increase in sensitivity has made it possible to study small amounts of material as well as chemically interesting isotopes of low natural occurrence, e.g., carbon-13 (^{13}C NMR), routinely. Pulsed ESR techniques have since followed.

Since the 1970s two-dimensional NMR methods and more recently three- and multi-dimensional varieties have allowed ever more complex molecules to be studied. In the last two decades an additional increase of sensitivity was obtained by inverse detection techniques simultaneously combined with a drastic shortening of measuring time by gradient enhanced spectroscopy. In the area of solid states NMR techniques have been developed that obtain solution-like spectra from solids. Superconducting magnet systems for NMR frequencies up to 900 MHz (21.13 T) allow now the elucidation of the structure of large proteins and make NMR spectroscopy invaluable for the development of new drugs.

Since the mid-1970s NMR and ESR imaging has also been developed. The construction of whole-body magnets with magnetic field strength of up to 8.0 T and with magnet bores of up to 90 cm has paved the way for investigations on humans and NMR imaging today is an indispensable aid for medical diagnosis.

Table 1. Nuclear properties

Mass number	Atomic number	Nuclear spin I
Odd	even or odd	1/2, 3/2, 5/2...
Even	even	0
Even	odd	1, 2, 3,...

18.2. Principles of Magnetic Resonance [1], [92]–[96]

Many of the principles of magnetic resonance are common to both ESR and NMR spectroscopy. To exhibit magnetic resonance the system concerned must have a magnetic moment μ. For NMR this is provided by a nucleus with nonzero nuclear spin, and for ESR by an unpaired electron in an atom, radical, or compound. The two phenomena differ in the signs and magnitudes of the magnetic interactions involved. When matter is placed in a strong magnetic field, these systems (nuclei or electrons) behave like microscopic compass needles. According to the laws of quantum mechanics these tiny compass needles can orient themselves with respect to the magnetic field only in a few ways. These orientations are characterized by different energy levels. If the sample is exposed to electromagnetic waves of an exactly specified frequency the nuclear spins or the unpaired electrons can be forced to jump between different energy levels. In magnetic fields with a field strength of a few teslas, the resonance frequencies for the NMR and ESR are in the radio and microwave frequency regions, respectively. During the experiment the frequency is varied. When it exactly matches the characteristic frequency of the nuclei or the electrons an electric signal is induced in the detector. The strength of the signal is plotted as a function of this resonance frequency in a diagram called the NMR or ESR spectrum, respectively.

18.2.1. Nuclear and Electronic Properties

The nuclear spin I of an atomic nucleus has values of 0, 1/2, 1, 3/2, etc. in units of $h/2\pi$, where h is the Planck constant. The value for a given nucleus is dependent on the mass and atomic numbers of the nucleus (Table 1). Thus some important, common nuclei such as ^{12}C and ^{16}O which have both even mass and atomic numbers have zero spin. An electron has a spin S of 1/2. Nuclear (μ_N) and electronic (μ_e) magnetic moments are proportional to the magnitude of the spin:

$$\mu_N = \gamma_N \frac{hI}{2\pi}$$

where γ_N is the magnetogyric ratio of the nucleus. In SI units μ_N is in amperes meter squared and γ_N in radian per tesla per second. Alternatively the magnetic moment can be expressed as:

$$\mu_N = g_N \beta_N I$$

where g_N is a dimensionless constant (nuclear g factor) and β_N the nuclear magneton:

$$\beta_N = \frac{eh}{4\pi Mc}$$

where e and M are the charge and mass of the proton, and c the velocity of light. The quantities g_N and I distinguish nuclei from one another.

The magnetic moment of an electron is given by an analogous expression:

$$\mu_e = -g\beta S$$

where g is the dimensionless electron g factor, and β the electronic Bohr magneton:

$$\beta = -\frac{eh}{4\pi mc}$$

where $-e$ and m are the charge and mass of the electron. Because the electron is negatively charged, its moment is negative.

18.2.2. Nuclei and Electrons in a Stationary Magnetic Field

When a nucleus with nonzero spin or an electron is placed in a strong magnetic field B the orientation of the spin becomes quantized, with each possible orientation having a different energy. A nucleus of spin I has $2I+1$ possible orientations, with associated magnetic quantum numbers m_I of $-I$, $-I+1, \ldots$ $I-1$, I. Hydrogen nuclei (protons) have a spin quantum number $I = 1/2$. Consequently the proton can exist in only two spin states, which are characterized by their magnetic quantum numbers $m_I = 1/2$ and $m_I = -1/2$. The proton can therefore be pictured as a magnetic dipole, the z component μ_z of which can have a parallel or antiparallel orientation with respect to the z direction of the coordinate system

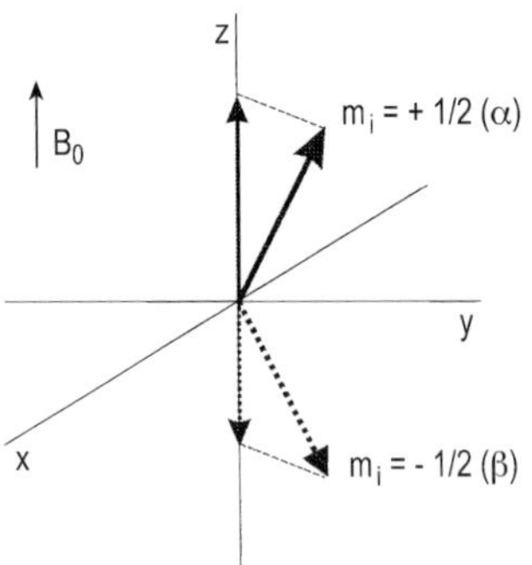

Figure 1. Orientation of the nuclear magnetic moments in the magnetic field B_0.

(Fig. 1). In the following discussion the denotations α and β are used for these two orientation states, where α denotes the lower, more stable energy level and β the higher one. For a single electron with $S=1/2$, allowed values for the magnetic quantum number m_S are also + 1/2 and – 1/2.

In the absence of a magnetic field the α and β states have the same energy. Only in a static magnetic field B_0 is this degeneracy destroyed as a result of the interaction of the nuclear or electronic moment μ with B_0 and an energy difference for the two orientation states results. The energy of interaction is proportional to the nuclear or electronic moment and the applied field. For a nucleus and electron, respectively, this can be written as:

$$E_I = -g_N \beta_N m_I B_0 = \frac{-\gamma_N h m_I B_0}{2\pi}$$

$$E_S = g \beta m_S B_0$$

Energy level diagrams for nuclei with spins of 1/2 and 3/2 are given in Figure 2.

The energy levels diagram for an electron looks exactly like that of the nucleus in Figure 2 A), but with exchanged signs of the magnetic quantum numbers m_S. The selection rule for NMR and ESR allows transitions for which m_I and m_S change by one unit. Thus:

$$\Delta E_I = g_N \beta_N B_0 = \frac{\gamma_N h B_0}{2\pi}$$

$$\Delta E_S = g \beta B_0$$

18.2.3. Basic Principles of the NMR and ESR Experiments

To observe transitions, radiation given by $\Delta E = h\nu$ must be applied. The fundamental resonance conditions for NMR and ESR experiments respectively are:

$$h\nu = g_N \beta_N B_0 = \frac{\gamma_N h B_0}{2\pi} \tag{1}$$

and

$$h\nu = g \beta B_0$$

where ν is the resonance frequency. For a proton in a magnetic field of 2.349 T, ν is 100 MHz. The resonance frequency of a free electron in the same field is 65.827 MHz.

Resonance absorption can only be detected if the spin levels involved differ in population. For a very large number N of protons in thermal equilibrium, the distribution of the spin 1/2 nuclei between the ground (α) state and the excited (β) state is given by the Boltzmann relation:

$$\frac{N_\beta}{N_\alpha} = \exp(-h\nu/kT)$$

The excess of spins in the lower state is only ca. 1 in 10^5 for NMR at a magnetic field strength of 9 T and at room temperature. This is the fundamental reason for the low sensitivity of NMR compared with IR and UV spectroscopy and even ESR.

The magnetic resonance experiment itself is best explained and visualized in classical as opposed to quantum mechanical terms. How is the energy transferred to nuclei or electrons aligned in a magnetic field, and how is the energy absorbed measured? The following description is for nuclei of spin 1/2 but is equally applicable to electrons. In a magnetic field B_0, applied in the z-direction, the nuclei have an orientation which is either aligned or opposed to the direction of B_0. Additionally, the nuclei precess randomly around the z-axis with a frequency ω_0, known as the Larmor frequency, where $\omega_0 = 2\pi\nu$. A macroscopic magnetization vector M along the z-axis but not in the x-y plane results from the individual nuclear magnetic moments μ_z defined in Figure 1. The slight excess of nuclei aligned with the magnetic field as shown above, gives rise to the alignment of this magnetization vector M_z in direction of B_0 (Fig. 3).

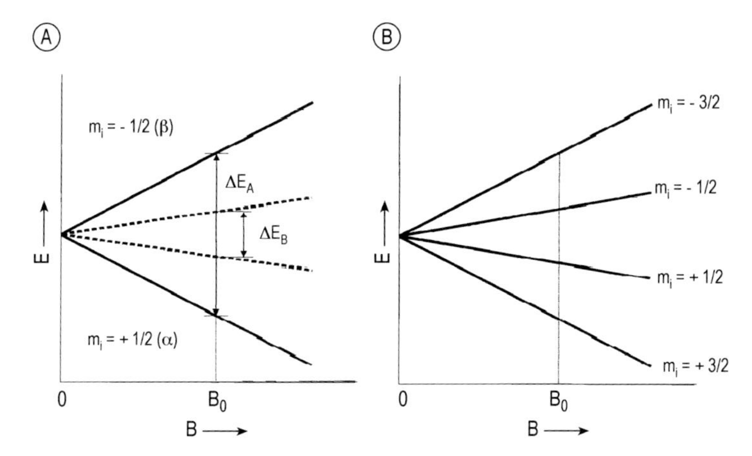

Figure 2. Energy levels for a nucleus with a spin of A) $I = 1/2$, and B) $I = 3/2$. On the left-hand side the levels for two different isotopes A and B with different magnetogyric ratios are represented ($\gamma_A > \gamma_B$). For the nuclei of the isotope B an obviously smaller energy difference ΔE results at the same magnetic field strength B_0. Only the value and not the direction of the magnetic field B_0 is plotted on the abscissas.

The quantum description of the lower and higher energy levels can be equated with the classical description of precessing nuclei aligned with and against the applied field. Up to here the description of the nuclear motions was referred to in the stationary or laboratory frame of reference. If the laboratory could be rotated at the Larmor frequency ω_0, the nuclei would no longer appear to precess but would become stationary and coincident with the magnetic field axis B_0 as shown in Figure 4; x' and y' are the rotating x- and y-axes. The magnetic behavior is now completely described by a magnetization vector M_z acting along B_0. Use of the rotating frame system greatly simplifies the description of pulsed NMR experiments (see Section 18.3.1.2).

In the magnetic resonance experiment the longitudinal net magnetization M_z is tipped toward the $x-y$ plane. Radio-frequency electromagnetic energy is applied such that its magnetic component B_1 is at right angles to B_0 and is rotating with the precessing nuclei. The RF irradiation is applied along the x-axis and, at the resonance frequency, tilts the net magnetization towards the $x-y$ plane. This creates an x-y component, the so-called transverse magnetization, which induces a signal in the detector coil. Conventionally the receiver is located on the y-axis of the Cartesian coordinates (Fig. 4).

To obtain a spectrum, either the RF frequency or the magnetic field B_0 can be scanned over a narrow range. Historically, this mode, known as continuous wave (CW), was employed in both NMR and ESR instruments. For NMR it has been almost completely superceded by the more sensitive and sophisticated pulsed Fourier transform (FT) method. Pulsed methods are now becoming more widely used in ESR.

18.2.4. Relaxation

In the absence of any mechanism enabling the nuclear or electronic spins to return to the ground state, all the excess population in the lower energy state will eventually be raised to the higher energy state, and no more energy will be absorbed. However, such mechanisms do exist, the so-called relaxation phenomena. That is, the magnitudes of both longitudinal and transversal magnetization are time dependent . One is the spin – lattice or longitudinal relaxation process, which occurs by loss of energy from excited nuclear spins to the surrounding molecules. By this process the component of magnetization along the z-axis relaxes back to its original value M_z, with an exponential decay characterized by a time constant T_1, the spin – latice relaxation time.

After application of the RF energy the magnetization has components in the $x-y$ plane, and these also decay at least as fast as the spin – lattice relaxation returns the magnetization to the z-axis. However, $x-y$ magnetization can also be lost by other processes, which cause the $x-y$ magnetism to dephase to give a net x – y magnetization of zero. The rate constant in the $x-y$ plane for this process is T_2, the spin – spin or transverse relaxation time. Such relaxation involves transfer of

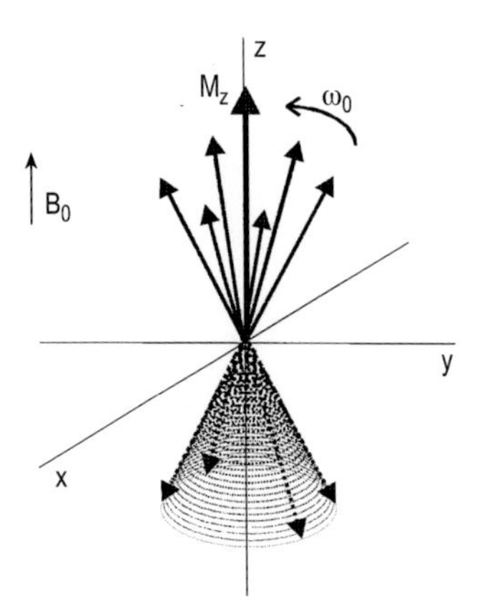

Figure 3. Motion of spin 1/2 nuclei in a magnetic field B_0. Note that more nuclei are arranged parallel than antiparallel to the direction of the static magnetic field B_0.

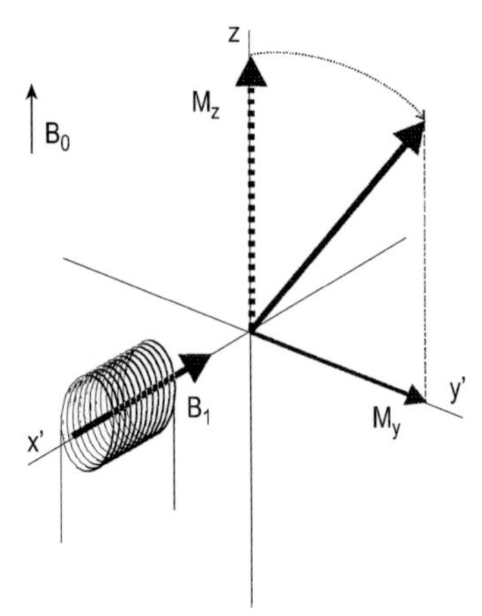

Figure 4. Motion of spin 1/2 nuclei in the rotating frame and generation of transverse magnetization M_y by rotation of the vector of the macroscopic magnetization M_z. The B_1 field caused by the transmitter coil along the x-axis deflects M_z into the $x'-y'$-plane.

energy from one nucleus to another. An important feature of spin – spin relaxation is that it determines the natural linewidths of the signals in a spectrum, which is defined in terms of the half-height linewidth Δ. In hertz one has:

$$\Delta = \frac{1}{\pi T_2}$$

if T_2 is measured in seconds. However, the observed linewidth is always larger because of a contribution from the magnetic field inhomogeneity.

18.3. High-Resolution Solution NMR Spectroscopy [2], [3], [4] – [6], [97] – [109].

Knowing the gyromagnetic ratio γ of a nucleus enables the resonance frequency from Equation (1) to be calculated for any magnetic field strength. Table 2 lists useful, commonly studied nuclei and their magnetic properties. The frequencies are for a 2.35 T applied magnetic field, i.e., relative to ^1H at 100 MHz. Table 2 gives the relative sensitivity for equal numbers of each nucleus and their relative receptivities, i.e., the product of the relative sensitivity and natural abundance.

The abundant nuclei, ^1H, ^{19}F, and ^{31}P, with $I = 1/2$ were the first to be studied by CW NMR. Owing to sensitivity problems, less abundant isotopes of important spin 1/2 nuclei such as ^{13}C and ^{15}N require FT techniques for observation in natural abundance. Nuclei with spin $I \geq 1$ have a quadrupole moment in addition to their magnetic moment. The interaction of the nuclear quadrupole moment with the electric field gradient at the nucleus provides a very efficient process for nuclear relaxation. Often, the quadrupolar mechanism is dominant, the resulting relaxation times are very short, and the signals broad. In extreme cases (e.g., covalently bonded ^{35}Cl or ^{37}Cl), the signals are too broad to detect. However, nuclei with very small quadrupole moments such as ^2H and ^{11}B can be observed in the usual manner as the broadening is much less severe.

18.3.1. The NMR Experiment

The first successful NMR experiments were reported in 1945 by two independent groups in the USA [7], [8]. The discoveries by Bloch at Stanford and by Purcell at Harvard were awarded a Nobel Prize in Physics in 1952. Already in September 1952 the first commercial NMR spectrometer was installed. Early instruments used permanent magnets or electromagnets with fields from 0.94 up to 2.35 T, corresponding to 40 up to 100 MHz for proton resonance. Permanent magnets give very stable magnetic fields and are often used without further stabilization. Electromagnets need additional control mechanisms to provide the necessary stability. The commonest

Table 2. Magnetic properties of some nuclei

Isotope	Spin	Natural abundance, %	NMR frequency,* MHz	Relative sensitivity	Relative receptivity
^{1}H	1/2	99.985	100.000	1.00	1.00
^{2}H	1	0.015	15.351	9.65×10^{-3}	1.45×10^{-6}
^{11}B	3/2	80.42	32.084	0.17	0.13
^{13}C	1/2	1.108	25.144	1.59×10^{-2}	1.76×10^{-4}
^{14}N	1	99.63	7.224	1.01×10^{-3}	1.01×10^{-3}
^{15}N	1/2	0.37	10.133	1.04×10^{-3}	3.85×10^{-6}
^{17}O	5/2	0.037	13.557	2.91×10^{-2}	1.08×10^{-5}
^{19}F	1/2	100	94.077	0.83	0.83
^{23}Na	3/2	100	26.451	9.25×10^{-2}	9.25×10^{-2}
^{29}Si	1/2	4.70	19.865	7.84×10^{-3}	3.69×10^{-4}
^{31}P	1/2	100	40.481	6.63×10^{-2}	6.63×10^{-2}

* At $B = 2.35$ T.

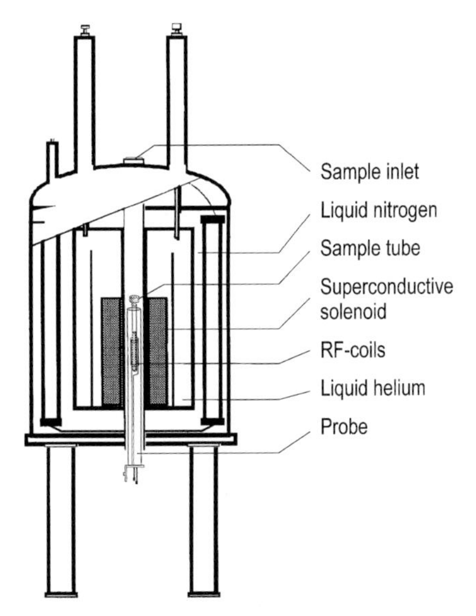

Figure 5. Schematic diagram of a cryomagnet. The superconducting coil is immersed in liquid helium. The sample tube is placed from above into the middle of the solenoid, where the field is strongest and has the best homogeneity. The lines of the static magnetic field are parallel to the long axis of the sample tube.

way to do this is to use the NMR signal of another nucleus in the sample to provide a field-frequency locking mechanism. The other nucleus may be the reference nucleus (homo-lock) or a different nucleus (hetero-lock), e.g., the deuterium resonance of a deuterated solvent. The need for instruments with higher resolution and sensitivity has led to the development of commercially available systems operating at up to 900 MHz for protons. All instruments operating above 100 MHz use helium-cooled superconducting solenoids to provide the magnetic field. High-resolution NMR requires both the magnetic field and the RF source to be homogeneous and stable to better than 1 part in 10^8. To achieve this performance, which allows to obtain the maximum information possible from an NMR experiment, the magnetic field profile has to be made extremely uniform in the region of the sample. The two parameters which quantify these performance characteristics are magnetic field "homogeneity" and field "drift". The magnets are designed to minimize any magnetic field variations in time. Inevitable imperfections are introduced during manufacture of the wire of the solenoid and also caused by different samples, temperature variations, etc. They are eliminated using correction coils and so-called shim coils to remove residual field gradients. Also the magnetic field stability is of major importance. In conventional magnets the superconducting wire is bathed in liquid helium under atmospheric pressure where it has a boiling point of 4.2 K. By reducing the pressure the boiling point of helium can be lowered to 2.3 K. By doing this the performance of the superconducting wire is enhanced and higher magnetic fields can be achieved (cryostabilization). In Figure 5 a schematic diagram of a superconducting magnet is shown.

The sample, normally a solution in a deuterated solvent in a 5 or 10 mm glass tube (smaller tubes have been employed for specific applications for some years as well), is placed in the instrument probe between the poles of an electromagnet/permanent magnet or inside the solenoid of a superconducting magnet. The probe contains the RF transmitter and receiver coils and a spinner to spin the tube about its vertical axis. Sample spinning is used to average out magnetic field inhomogeneity across the sample. Multidimensional NMR spectra are measured in modern instruments without rotating.

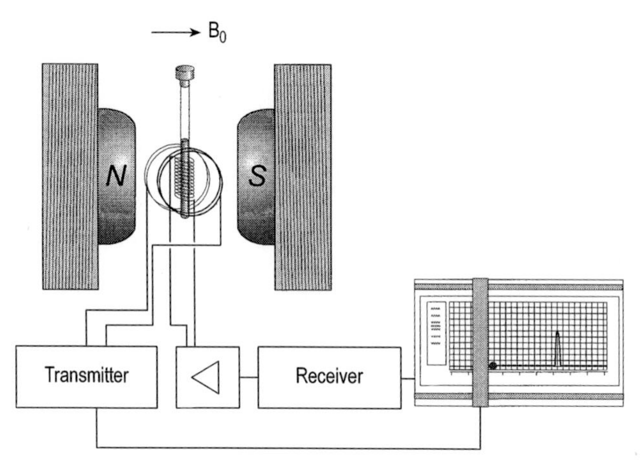

Figure 6. Block diagram of a CW NMR spectrometer.

The ideal NMR solvent should dissolve the sample, be inert, nonpolar, liquid over the temperature range of interest, inexpensive, and not give rise to large interfering peaks in the spectrum being recorded. Deuterated solvents such as deuterochloroform and deuterodimethyl sulfoxide are most commonly used. They offer the advantage of being "transparent" in proton spectra and provide a deuterium signal which can be used to stabilize and lock the spectrometer system. A clear, mobile solution is required for a high-quality spectrum. Solid particles or viscous solutions degrade the resolution. Traces of paramagnetic impurities in a solution can dramatically reduce relaxation times and cause peak broadening. With increasing field strength the demands both on the sample and on the used NMR tubes are usually growing.

Spectra are recorded either by the CW scan or pulsed FT method. The amount of sample required depends on the method of detection (CW or FT) and the receptivity of the nucleus being studied. For most routine CW proton spectrometers 10–20 mg of compound of M_r 200–300, dissolved in 0.4 mL of solvent should give a reasonable spectrum. On a 500 MHz FT instrument it should be possible to obtain a proton spectrum from less than one milligram of such a compound. Even smaller amounts can be measured with specific techniques, e.g., special NMR tube inserts. For specially designed probes amounts of few nanograms are sufficient [110], [144].

To measure nuclei with low receptivity, such as ^{13}C, a more concentrated solution must be used, in order to obtain a spectrum in as short a time as possible. Larger diameter sample tubes (10 or 15 mm) and, consequently, larger sample volumes can be used to increase the amount of sample. Recently the so-called inverse detection techniques were used for observation of nuclei with small sensitivities (Section 18.3.6.4).

18.3.1.1. Continuous Wave Methodology

Until the late 1960s all NMR instruments were of the CW type. Figure 6 shows a block diagram of such a spectrometer. To obtain a spectrum from a sample, either the magnetic field or the RF frequency is slowly varied. When the resonance condition (Eq. 1) is met for the nuclei being observed, the sample absorbs RF energy and the resulting signal is detected by the receiver coils, amplified, and recorded. Spectra are recorded on precalibrated charts relative to a reference compound.

The CW method is suitable for recording spectra of abundant nuclei from relatively large amounts of sample. As a result, early NMR work was essentially limited to the ^{1}H, ^{19}F, and ^{31}P nuclei. Nowadays the use of CW instruments, which are relatively inexpensive, is mainly limited to ^{1}H observation. Such instrumentation is usually to be found in university teaching or synthetic organic chemical laboratories rather than specialist NMR facilities.

One of the main limitations of NMR spectroscopy is its inherent lack of sensitivity. As a result, the CW method is unsuitable for recording spectra from nuclei of low natural abundance or even from abundant nuclei in solutions of very low concentration. One way of improving the signal-to-noise (S/N) ratio is to record several spectra from a sample and add them together. The noise averages

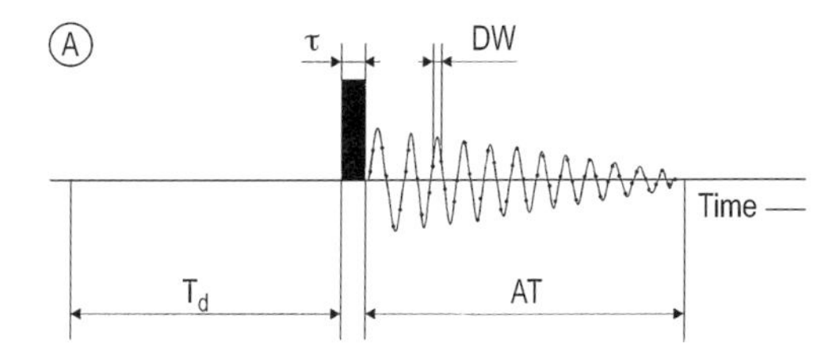

Figure 7. Schematic presentation of a pulse or FT NMR experiment (A); the free induction decay (FID) (B); and the appendant Fourier-transformed signal (NMR spectrum) (C). In A the FID is shown in a simplified manner also in the digital form, which will be generated by an analog-to-digital converter (ADC).

out, and the S/N ratio increases by a factor equal to the square root of the number of spectra accumulated. Using a computer to store and add the spectra allows many thousands of spectra to be accumulated in this way. The main drawback to this method on a CW instrument is the time to acquire each spectrum, typically ca. 4 min, resulting in excessive and often prohibitive total experiment times. A technique was required which could simultaneously excite the whole spectral region of interest. The pulsed FT technique, which first became available on commercial instruments in the late 1960s, provided the answer.

18.3.1.2. Fourier Transform Methodology

In 1966 ERNST and ANDERSON discovered that the sensitivity of NMR spectra could be increased dramatically if the sample was not exposed to the slow radio frequency sweep but to short and intense radio frequency pulses. The intensity of the signal was measured as a function of time after this pulse. The next pulse and signal acquisition were started after a few seconds, and the signals after each pulse were summed in a computer. If the RF frequency ν is turned on and off very rapidly to obtain a pulse τ seconds long, this is equivalent to irradiating the sample with a range of frequencies centered about ν over a frequency range $\nu \pm 1/\tau$. Thus if $\tau = 50$ µs a range of $\nu \pm$ 20 000 Hz is covered.

If an RF pulse, at the resonant frequency ν_0, is applied along the x-axis in the laboratory frame, this is equivalent to applying a static field $\boldsymbol{B}_1$ along the x'-axis of the rotating frame. This drives the sample magnetization about the x'-axis by an angle θ, as shown in Figure 4. Spectrometers are normally designed to detect signals along the y'-axis. The component of magnetization along this direction is given by $\boldsymbol{M}_0 \sin\theta$, and the maximum signal is obtained when $\theta = 90°$. A pulse producing this effect is known as a 90° pulse, and the time for which the pulse must be applied to achieve this is known as the 90° pulse time. On modern FT spectrometers 90° pulse times are usually significantly less than 50 µs. Since the pulse applied is very short it must be powerful enough to excite nuclei in the time available across the whole spectral region of interest. To do this it must satisfy the condition:

$$\gamma \boldsymbol{B}_1 \geq 2\pi(\mathrm{SW})$$

where SW is the spectral width required.

At the end of the RF pulse, the spin system begins to relax back towards its equilibrium condition by means of the two relaxation mechanisms described in Section 18.2.4. After a time $5\,T_1$ the magnetization along y'—the signal detection axis—has decayed essentially to zero. For the remainder of this article the rotating frame is assumed and the axes are referred to simply as x, y, and z.

In the CW experiment, descibed in Section 18.3.1.1, the intensity of the NMR signal is recorded as function of frequency. However, in the pulsed experiment, the intensity of the decaying signal, following the RF pulse, is detected and recorded as a function of time. The signal is known as the free induction decay (FID) and it is not amenable to a simple interpretation. In Figure 7 a pulse experiment is represented as diagram in the usual notation. After the delay period T_d the excitation occurs through a pulse with the pulse width (length) τ and the subsequent acquisition. The resultant FID is also illustrated in Figure 7 together with the conventional NMR spectrum, for a sample which gives a spectrum with a single peak. For a spectrum containing more than one peak, the FID is an interference pattern resulting from the frequencies corresponding to all the individual peaks in the spectrum. The FID is often referred to as the time domain signal, whereas the corresponding NMR spectrum provides a frequency domain signal.

Time and frequency domain data can be interconverted by the mathematical process of Fourier transformation (FT), which can be performed rapidly by the computer. The intensity of the FID is measured at n intervals equally spaced by the so-called dwell time DW and stored in digital form. In other words, the dwell time is the time used to produce a particular data point. The total time to acquire the data is referred to as the acquisition time AT. In order to obtain a spectral width SW the following equation must be satisfied:

$$AT = \frac{n}{2(SW)} = n \cdot DW$$

Thus, typical values for obtaining a 100 MHz ^{13}C NMR spectrum would be:

SW = 20 000 Hz (equivalent to 200 ppm)
N = 16 384 (stored in 16 K computer words)
DW = 25 µs
AT = 0.41 s

Fourier transformation of the 16 K data points gives a spectrum with 8 K points since the transform contains both real and imaginary components, the real component being identical to the conventional CW spectrum.

The principal advantage of the FT method is that full spectral data can be obtained in a few seconds acquisition time as compared with minutes for the CW method. Signal averaging, by adding together the FIDs obtained after each of a series of RF pulses is therefore a much more efficient process than its CW analogue. The introduction of FT instruments made the routine study of less abundant isotopes, and in particular ^{13}C, possible.

FT instruments are available today with magnetic fields from 1.9 to 21.1 T, i.e., 80 to 900 MHz for ^{1}H observation. Most such instruments are available with a range of probeheads for studying different nuclei. Some probeheads are selectively tuned for observation of a single nucleus at maximum sensitivity. Dual- and trial-tuned probeheads for the convenient observation of ^{1}H and other fixed nuclei (e.g., ^{13}C, ^{15}N, ^{31}P) on a single probehead are now widely used, as are tunable probeheads which enable a wide range of nuclei to be studied. However, the most extensive development occurred in the case of the pulse sequences. Several hundred pulse sequences were described in the last twenty years and it is impossible to know all. A very good overview of the 150 most important ones is given in [109]. Furthermore, numerous information about basics and new developments are to be found in the Internet. Some of the most impressive web pages are listed in [111], [112].

The relatively high sensitivity and computing facilities associated with modern FT instruments have led to the development of fully automated systems with robotic sample changers. This enables unattended, round-the-clock use of expensive high-field instrumentation. A typical sample changer can accommodate from 6 up to 100 samples, and data is automatically acquired, processed, and plotted.

18.3.2. Spectral Parameters

Around 1950, it was discovered that nuclear resonance frequencies depend not only on the nature of the atomic nuclei, but also on their chemical environment. The possibility of using NMR as a tool for chemical analysis soon became apparent. All applications of solution NMR spectroscopy make use of one or more of the spectral parameters chemical shift, spin–spin coupling, signal intensity, and relaxation time. In the following sections examples from ^{1}H and ^{13}C NMR are used but the concepts described apply to magnetic nuclei in general.

18.3.2.1. Chemical Shift

Equation 1 might suggest that only a single peak would be obtainable from each magnetic isotope. However, when the molecule under observation is placed in the magnetic field, the electrons in the molecule shield the nuclei from the applied magnetic field to a small extent. For a particular nucleus within a molecule the degree of shielding depends on the density of the electrons circulating about that nucleus. Protons in different chemical environments therefore have different shieldings. Figure 8 shows diagramatically what happens to spherically symmetrical s electrons around a nucleus under the influence of an external magnetic field B_0. They circulate around the nucleus, producing their own magnetic field which opposes the applied field. The effective field B at the nucleus is then given by

$$B = B_0(1 - \sigma)$$

Thus, to obtain the resonant condition (Eq. 1) it is necessary to increase the applied field above that for an isolated nucleus. The shielding constant σ is a small fraction, usually of the order of parts per million. For electrons in p-orbitals where there is no spherical symmetry the situation is more complex since shielding and deshielding effects depend on the orientation of the molecule with respect to the applied field. Such effects are considered in more detail in Section 18.3.3.1.

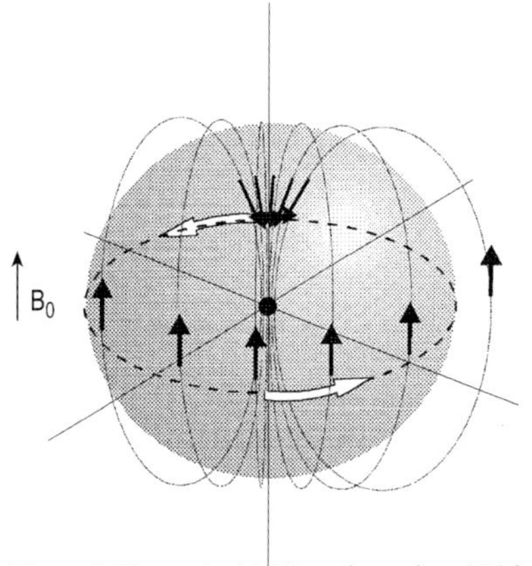

Figure 8. Electronic shielding of a nucleus (●) by circulating *s*-electrons (.......) in a magnetic field B_0

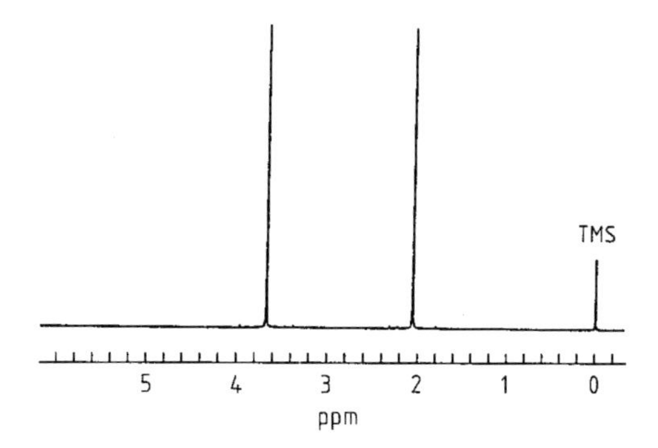

Figure 9. 250 MHz ^{1}H NMR spectrum of methyl acetate

The magnitude of the nuclear shielding is proportional to the applied magnetic field. The difference in the peak position of a particular nucleus from the peak position of a reference nucleus is called the chemical shift δ. In ^{1}H NMR if two peaks are separated by 80 Hz in a magnetic field corresponding to 100 MHz they will be separated by 320 Hz in a field corresponding to 400 MHz. Chemical shifts δ can be expressed in dimensionless units as parts per million (ppm) of the applied frequency in MHz as follows:

$$\delta = \frac{(\nu_{\text{sample}} - \nu_{\text{reference}})}{\text{Applied frequency}}$$

Thus for the example given above

$$\text{At 100 MHz} \quad \delta = \frac{80\,\text{Hz}}{100 \times 10^6\,\text{Hz}} = 0.8\,\text{ppm}$$

$$\text{At 400 MHz} \quad \delta = \frac{320\,\text{Hz}}{400 \times 10^6\,\text{Hz}} = 0.8\,\text{ppm}$$

The chemical shift δ is therefore a molecular parameter that does not depend on the spectrometer frequency. ^{1}H and ^{13}C nuclei in organic molecules give peaks which cover a range of approximately 20 ppm and 220 ppm, respectively.

Tetramethylsilane (TMS) is the most commonly used chemical shift reference compound for ^{1}H and ^{13}C NMR. In both types of spectra TMS gives a single sharp resonance at lower frequency than most other proton or carbon resonances. The TMS is usually added to the sample solution as an internal reference. The accepted sign convention is for chemical shifts to high and low frequency of the reference peak to be positive and negative, respectively. Figure 9 shows the 250 MHz ^{1}H spectrum of methyl acetate referenced to TMS; the peaks at 2.0 δ and 3.6 δ arise from the acetyl and methyl protons, respectively.

18.3.2.2. Spin – Spin Coupling

Chemical shift is one source of information on fine structure in NMR spectra. The second is spin – spin coupling. Figure 10 shows the aromatic region of the proton spectrum of 3-methoxy-6-nitro-*o*-xylene. One might expect two peaks from the two different aromatic protons in the molecule. In fact four peaks are observed. This results from the indirect coupling of proton spins through the intervening bonding electrons. In Figure 11 this is shown schematically for the two diastereotopic protons in a methylene group. The magnetic B_0 field experienced by nucleus A has a small component arising from nucleus B. Nucleus B has two equally probable orientations in the magnetic field, $m_B = +1/2$ or α and $m_B = -1/2$ or β. As a result nucleus A gives two equally intense peaks corresponding to the two orientations of B. By the same reasoning B itself also gives two equally intense peaks with the same separation as those for A. This splitting is called the spin – spin coupling J, where J_{AB} is the coupling between nuclei A and B. The magnitude of the coupling, which is measured in hertz, is independent of the applied magnetic field. Couplings between protons may be observed from nuclei separated by up to five bonds. Their magnitude rarely exceeds 20 Hz.

If more than two nuclei are interacting with each other all possible orientations of the coupling nuclei must be considered. When three nuclei have different chemical shifts, as for the olefinic protons of vinyl acetate, then each nuclear signal is split by coupling to both of the other nuclei. Thus

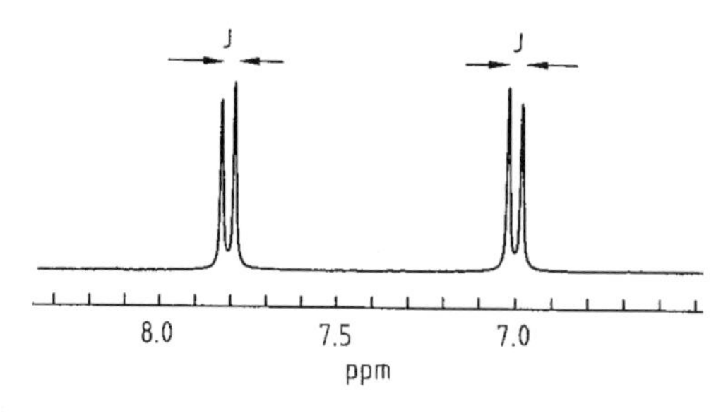

Figure 10. Aromatic region of the 250 MHz ^{1}H NMR spectrum of 3-methoxy-6-nitro-*o*-xylene

each resonance can be envisaged as being split into a doublet by the first coupling, each line of which is further split into a doublet to give a quartet pattern for each nucleus. When two of the three nuclei are equivalent, the total number of lines in the spectrum is reduced because some of the couplings are identical. Chemically equivalent protons (e.g., in a methyl group) do not exhibit splitting from coupling among each other.

Simple splitting patterns, for nuclei of spin 1/2, obey the following general rules:

1) If a nucleus interacts with n other nuclei with a different coupling to each then the signal will consist of 2^n lines of equal intensity
2) If a nucleus A couples to n equivalent nuclei B then the pattern for A comprises $(n+1)$ lines with relative intensities given by the coefficients of the binomial expansion of $(1+x)^n$

These simple rules are only strictly applicable to so-called first-order spectra. For spectra to be first order a single criterion must be obeyed. The chemical shift difference in hertz Δv for the coupled nuclei must be much larger than the coupling between them ($\Delta v/J \geq 10$). If this condition is not met the spectra may show more lines than the rules predict, and their intensities are distorted, with a build up of intensity towards the middle of the coupling pattern. It is beyond the scope of this article to descibe these second-order spectra.

Because the magnitude of the spin–spin coupling interaction is independent of the applied magnetic field, whereas the chemical shift separation is proportional to it, a complex NMR spectrum can be simplified by using higher applied fields.

All the above examples of spin–spin coupling involve homonuclear coupling in proton spectra. In natural abundance ^{13}C spectra homonuclear coupling is not normally observed since the probability of two adjacent ^{13}C atoms is extremely low

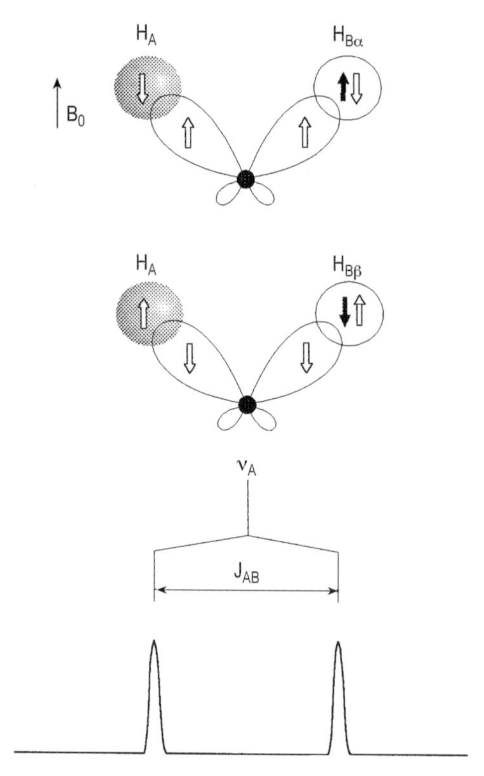

Figure 11. Nuclear spin-spin-interaction through the bonding electrons in a CH_2-group. The low-energy state corresponds to the antiparallel orientation of the magnetic moments of the nuclei (black arrows) and electrons (white arrows). The two possible states of the magnetic moments of hydrogen nucleus B are shown. As a result the resonance signal of nucleus A shows a splitting into a doublet. The effect is only observable in a methylene group in the case of diastereotopic, i.e., chemically inequivalent protons.

(10^{-4}). For organic molecules the most commonly encountered couplings are those to protons. One bond couplings range from about 110 to 250 Hz. Most longer range couplings are < 10 Hz. Proton-coupled ^{13}C spectra often show complex overlapping multiplets. To simplify the spectra and obtain them in a shorter time, routine ^{13}C measurements are usually proton noise decoupled (see Section 18.3.4.2). Thus, in the absence of other coupling nuclei such as ^{31}P or ^{19}F, a single sharp peak is observed for each chemically inequivalent ^{13}C atom. Figure 12 shows the proton-coupled and -decoupled ^{13}C NMR spectra of ethylbenzene.

18.3.2.3. Signal Intensity [113]

NMR is an inherently quantitative technique. For any nucleus the NMR signal obtained, at least

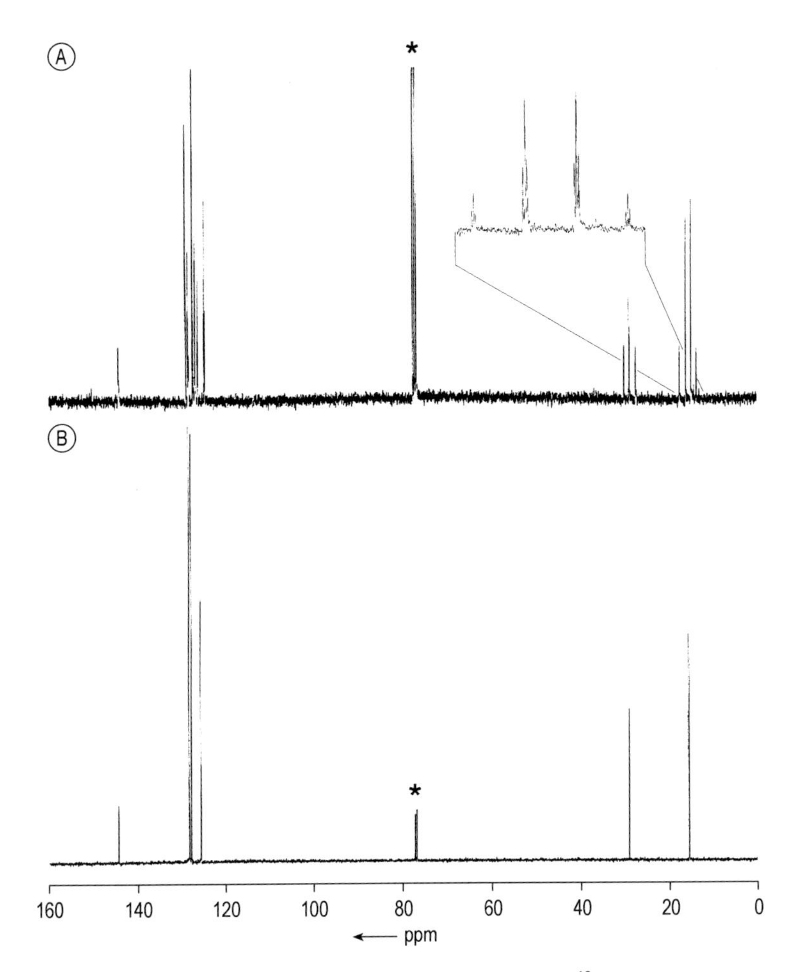

Figure 12. Proton-coupled (A) and proton broadband decoupled (B) ^{13}C NMR spectra of 10 % ethylbenzene in $CDCl_3$ (*). Applied frequency 100.6 MHz; puls width 13.5 μs; repetition time 5 s; number of accumulations 1024 and 128; measuring time 85 min and 11 min, respectively.

for a single-scan CW experiment or a single-pulse FT experiment, is directly proportional to the number of nuclei producing it. For FT experiments in general to be quantitative the spin system must be given sufficient time to relax completely between successive pulses. Most proton spectra are recorded under essentially quantitative conditions and are integrated to give the relative number of protons contributing to each resonance. Figure 13 shows an integrated proton spectrum of camphor with relative peak areas of approximately 1 : 1 : 1 : 1 : 1 : 2 : 3 : 3 : 3 for the CH-protons, the diastereotopic methylene, and the methyl protons, respectively.

^{13}C spectra and those of other low natural abundance nuclei are not normally recorded under quantitative conditions. For structural identification work with such nuclei, conditions are usually selected for optimum sensitivity, not quantification. There is an additional problem in spectra such as ^{13}C where broadband proton decoupling is usually used. Apart from collapsing the multiplet structure, the decoupling irradiation causes an increase in signal intensity due to nuclear Overhauser enhancement (NOE, see Section 18.3.4.2). The enhancement factor is often different for the various ^{13}C resonances in the spectrum. Quantitative ^{13}C work is possible but the experimental conditions must be carefully selected [113].

18.3.2.4. Relaxation Times

Spin – lattice and spin – spin relaxation, the two processes by which a perturbed spin system re-

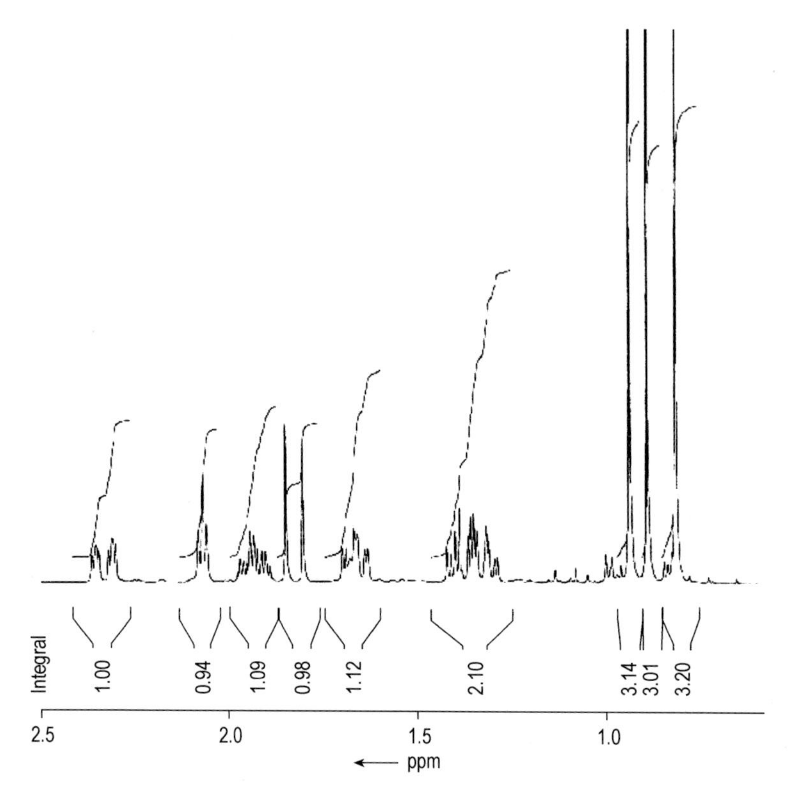

Figure 13. Integrated 400 MHz ^{1}H NMR spectrum of camphor. The reason for the differences between the experimental and the theoretical integral values are impurities and spectrum processing effects.

turns to equilibrium, are outlined in Section 18.2.4. The various mechanisms that contribute to these processes are discussed in [3]. This section is concerned with the effect variations of T_1 and T_2 have on the appearance of a spectrum. Proton spectra are usually little affected by differences in relaxation times for different protons in a molecule because most protons have short relaxation times (a few seconds). For ^{13}C and many other nuclei this may not be the case. ^{13}C relaxation times in organic materials range from a few milliseconds to > 100 s, depending largely on the molecular mass of the molecule and the environment of a particular ^{13}C atom in the molecule.

For spin 1/2 nuclei the principle T_1 mechanism is often dipole – dipole relaxation, which results from interactions between two nuclei with magnetic spins. Its efficiency is inversely proportional to the sixth power of the internuclear distance. In ^{13}C spectra of organic molecules, protons provide the main source of such relaxation. Where a ^{13}C nucleus is directly bonded to a proton, T_1 is usually quite short, whereas a quaternary carbon atom in the same molecule will probably have a significantly longer value. In a FT ^{13}C spectrum this will result in a relatively low intensity for a quaternary resonance unless the interpulse delay is sufficiently long to allow complete relaxation of all nuclei. Such effects can be used as an aid in spectral assignment.

For polymers and other high molecular mass molecules, T_1 can be very short, in the range 10^{-3} to 1 s. For organic molecules with molecular masses < 1000, values for proton-bonded carbons lie typically in the range 0.1 – 10 s, and for non-proton-bonded carbons, 10 – 300 s.

The most important effect of the spin – spin relaxation time T_2 is that it determines the natural width of the lines in the spectrum. However, for most ^{1}H and ^{13}C spectra the observed linewidth is determined by the homogeneity of the applied magnetic field.

Table 3. ^1H chemical shifts of CH_3X compounds

X	δH, ppm
$SiMe_3$	0.0
H	0.13
Me	0.88
CN	1.97
$COCH_3$	2.08
NH_2	2.36
I	2.16
Br	2.68
Cl	3.05
OH	3.38
F	4.26

18.3.3. NMR and Structure

In this section the relationship between chemical structure and the NMR parameters of chemical shift and spin–spin coupling are discussed for some of the more commonly studied nuclei.

18.3.3.1. Hydrogen (^1H and ^2H)

The two naturally occurring isotopes of hydrogen are the proton and the deuteron with natural abundances of 99.985 % and 0.015 %, respectively. Whilst the former with $I = 1/2$ is the most commonly studied of all magnetic nuclei, the latter with $I = 1$ is rarely studied in natural abundance and then only for a few specialist applications. One such application, the determination of deuterium levels in samples, is discussed in Section 18.3.8.2. If small isotope effects are disregarded, corresponding ^1H and ^2H chemical shifts are the same. ^2H spectra are less well dispersed than the corresponding ^1H spectra by a factor of 6.51, the ratio of the magnetogyric ratios. The smaller magnetogyric ratio also results in correspondingly smaller coupling constants. Whilst the following discussion is exemplified by reference to the proton, most of the content is equally applicable to the deuteron. However, since deuterium homonuclear coupling constants are very small, ^2H spectra with broadband ^1H decoupling give a single peak from each chemically inequivalent deuteron even for deuterium enriched samples.

Chemical Shifts. The shielding effect of spherically symmetrical *s* electrons is discussed in Section 18.3.2.1. This diamagnetic upfield shift affects all nuclei since every molecule has *s* electrons. For electrons in *p*-orbitals there is no spherical symmetry and the phenomenon of diamagnetic anisotropy is used to explain some otherwise anomalous chemical shifts. Diamagnetic anisotropy means that shielding and deshielding depend on the orientation of the molecule with respect to the applied magnetic field.

The hydrogen atom is a special case since it has no *p* electrons. A consequence of this is that in proton NMR the direct influence of the diamagnetic term can be seen. For example in substituted methanes CH_3X, as X becomes more electronegative, the electron density round the protons decreases and they resonate at lower fields (Table 3). For the same reason acidic protons in carboxylic acid groups resonate at very low fields. Figure 14 shows the chemical shift ranges for a selection of organic groups.

Whilst the electronegativity concept can be used to explain chemical shifts for saturated aliphatic compounds, the effects of diamagnetic anisotropy must also be considered for some other classes of compounds. For example, aromatic compounds such as benzene are strongly deshielded, the protons of acetylene are much more strongly shielded than those of ethylene (acetylene 1.48 ppm compared with ethylene 5.31 ppm), and aldehydic protons resonate at very low fields. In the case of benzene the deshielding results from the so-called ring current effect. When a benzene molecule is orientated perpendicular to the applied magnetic field B_0, there is a molecular circulation of π electrons about the field, as shown in Figure 15. The resulting ring current produces an additional magnetic field which opposes the applied magnetic field along the sixfold axis of the benzene ring but adds to it at the protons on the benzene ring, giving a low-field shift. The ring current is only induced when the applied magnetic field is perpendicular to the benzene ring. In practice only a proportion of the rapidly tumbling molecules are perpendicular to the magnetic field, but the overall averaged shift is affected by them.

Similar arguments can be invoked to explain the apparently anomalous shifts of acetylenic and aldehydic protons. The free circulation of electrons which gives rise to the diamagnetic effects in spherically symmetric atoms and the benzene ring can also occur around the axis of any linear molecule when the axis is parallel to the applied field. In the case of acetylene the π electrons of the bond can circulate at right angles to the applied field, inducing a magnetic field that opposes the applied field. Since the protons lie along the magnetic axis they are shielded and an upfield shift results. In the case of the aldehydic proton, the effect of the applied magnetic field is greatest

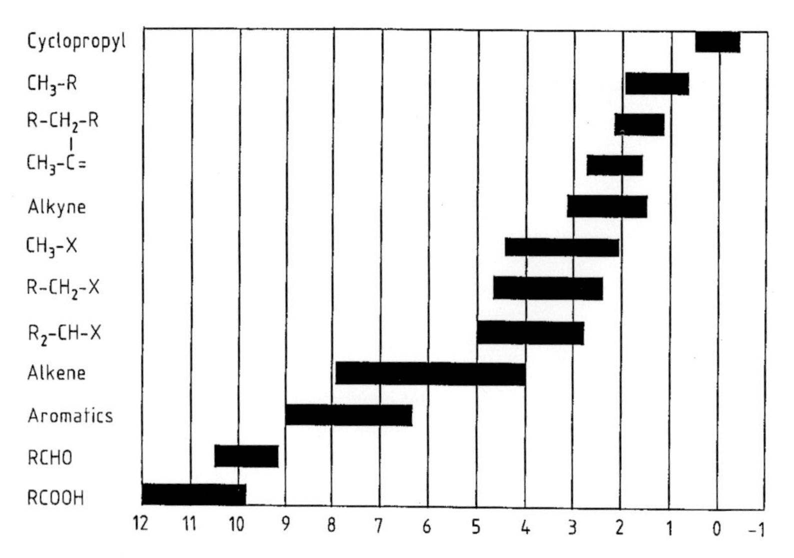

Figure 14. ^{1}H NMR chemical shift ranges
X = Halogen, –N⟨, –S–, –O–; R=Alkyl

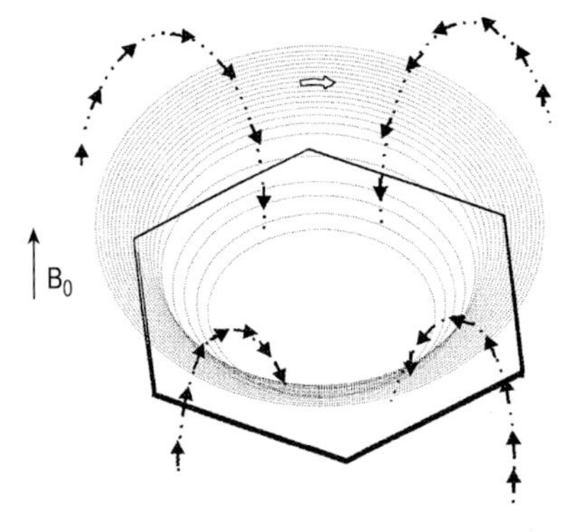

Figure 15. Ring current effects in benzene. The overlapping of the $2p_z$ orbitals leads to two current loops, above and below the *s*-bond plane of the benzene molecule. Only one loop is represented here. A magnetic field (black arrows) is induced by the ring current (white arrows), changing the shielding constant.

along the transverse axis of the C=O bond. This anisotropy causes deshielding of protons lying in a cone whose axis is along the C=O bond, and shielding outside this cone. The aldehydic proton, which lies within the cone, thus experiences a low-field shift and consequently resonates at low fields (9.5 – 10.0 ppm).

The effect of substituents on the proton chemical shifts of hydrocarbons has been extensively investigated. Schoolery's rules [2], [137] enable the prediction of the chemical shift of any CH_2XY and CHXYZ proton. The chemical shift is the shift for methane plus the sum of substituent contributions, which are largely determined by the electronegativity of the substituent. In aromatic compounds strongly electron-withdrawing groups such as nitro and carboxyl deshield all the protons, but the effect is largest at the *ortho* and *para* positions. The opposite is true for strongly electron-donating groups such as NH_2 and OH, while the halogens show smaller effects. For the various types of compounds the substituent increments can be used in an additive manner for compounds with more than one substituent. However, the accuracy of prediction decreases as the number of substituents increases.

Spin – Spin Coupling. The origin of spin – spin coupling and the resulting multiplet structure in proton NMR spectra is introduced in Section 18.3.2.2. The magnitude of couplings plays an important role in chemical structure determination, particularly in conformational analysis, stereochemistry, and in differentiating between isomeric structures. In proton NMR couplings are observed between protons and to other magnetic nuclei in the molecule such as ^{19}F and ^{31}P. Couplings to rare nuclei are only observed in accordance with their small abundance as so-called "satellites" in ^{1}H NMR spectra. A vicinal proton – proton coupling over three bonds is denoted by $^{3}J_{HH}$, whereas $^{1}J_{PH}$ represents a one bond coupling between a proton and a phosphorus atom. Depending

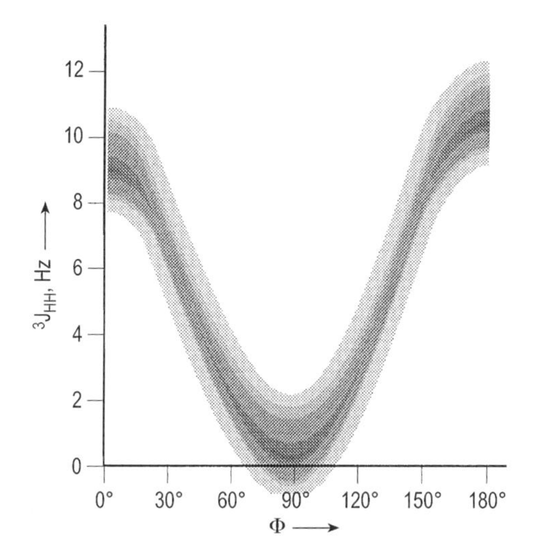

Figure 16. The Karplus curve shows the dependence of vicinal proton-proton coupling $^3J_{HH}$ on the dihedral angle Φ. The shaded area symbolizes the range of the empirical results.

on whether the coupling occurs over two, three, ore more bonds it is designated geminal, vicinal, or long-range coupling.

The magnitude of geminal couplings (^2J) depends to a first approximation on the H–C–H angle between the coupled protons. For tetrahedrally substituted carbon atoms a coupling in the range 12 – 15 Hz is typical. In aliphatic systems where there is free rotation about bonds geminal couplings are only observed where methylene protons are diastereotopic, that is, when the molecule contains a chiral center (C*):

```
   Ha X
   |  |
R—C—C*—Y
   |  |
   Hb Z
```

In such compounds H_a and H_b cannot be interchanged by symmetry operations or by rapid rotation and therefore may have different chemical shifts and exhibit coupling to each other. The most commonly encountered coupling in NMR is the vicinal proton-proton coupling. For protons on vicinal carbon atoms in saturated aliphatic groups the coupling constant depends primarily on the dihedral angle Φ between the CH bonds:

$$^3J_{(\text{H-C-C-H})} = \text{A} + \text{B}\cos\Phi + \text{C}\cos^2\Phi$$

where A, B and C are constants with the values 4.22, – 0.5, and 4.5, respectively. The graph of this function-the so-called Karplus curve-is shown in Figure 16. A series of important regularities is explained by the Karplus curve, e.g., those for couplings in six-membered rings of fixed geometry. An axial-axial coupling J_{aa} in a cyclohexane ring where $\Phi = 180°$ would be ca. 10 Hz according to the Karplus equation. Experimentally, J_{aa} varies from 8 to 13 Hz, J_{ae} from 2 to 4 Hz, and J_{ee} from 1 to 3 Hz, depending on the nature of the substituents present. In acyclic systems the same angular dependence exists, with *gauche* and *trans* couplings typically in the ranges of 2 – 4 and 8 – 12 Hz, respectively. However, in systems with free rotation around the intervening C-C bond, an average coupling of ca. 7 Hz is observed.

In olefinic compounds the $^3J_{HH}$ coupling constant is used to differentiate *cis* and *trans* isomers since $J_{trans} > J_{cis}$. Ranges for *trans* and *cis* couplings are 12 – 18 and 6 – 12 Hz, respectively. Geminal couplings in olefins lie in the range 0 – 3 Hz. In aromatic compounds the observed couplings are also characteristic of the disposition of the hydrogen atoms. For benzene derivatives the following ranges are typical:

$^3J_{ortho} = 6 - 10$ Hz
$^4J_{meta} = 1 - 3$ Hz
$^5J_{para} = 0 - 1$ Hz

Coupling of protons to other nuclei are discussed in the sections concerned with those nuclei.

18.3.3.2. Carbon (^{13}C) [9] – [11], [114] – [116]

Although present in natural abundance at only 1.1 %, the ^{13}C nucleus has, since the advent of FT methodology, been extensively studied. Only the approximately 5700 times more sensitive ^{1}H NMR is more widely used. In organic chemistry the information obtained from the two nuclei is often complementary for solving structural problems. The large ^{13}C chemical shift range and the small line width of the ^{13}C signals increases spectral resolution effectively. So, ^{13}C NMR is the method of choice for structural investigations of complex organic molecules, of complex mixtures of compounds, as well as biological oligomers and macromolecules.

Chemical Shifts. For organic chemicals most shifts are spread over a range of 0 – 200 ppm — about twenty times that for the proton. Figure 17 shows chemical shift ranges for some common organic groups. By comparison with Figure 14 it can be seen that there is an overall similarity in

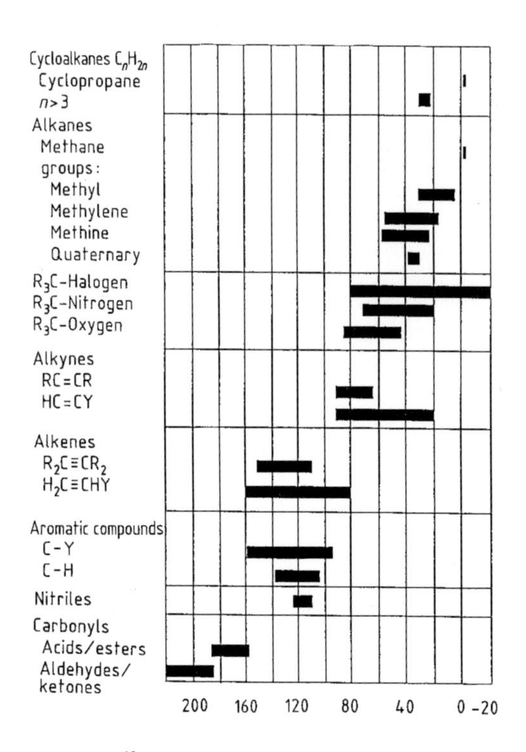

Figure 17. ^{13}C NMR chemical shift ranges
R = H, alkyl; Y = All substituents

ordering on going downfield from TMS to that observed for proton chemical shifts. However, in ^{13}C NMR the effects of substitution are often significant several bonds away from the point of substitution. In simple alkanes α- and β-carbons result in deshielding effects of approximately 9 ppm each whilst a γ-carbon shields by about 2.5 ppm. As a result the C-1 shifts for methane, ethane, and propane are −2.3, 5.7, and 15.4 ppm, respectively. For alkanes and most alkenes the effects are additive and a set of rules have been derived [12]. Similarly, substitution leads to typical stereochemical effects in cycloalkanes as also in open-chain compounds. This is of importance for conformational analysis, where the stereochemistry of different conformers can be assessed through their ^{13}C data.

Polar substituents have large effects as can be seen by comparing the shifts for *n*-pentane and 1-pentanol.

13.7 22-.6 34.-5 22.-6 13.7
$CH_3–CH_2–CH_2–CH_2–CH_3$
13.9 22-.9 28.-5 32.-7 62.0
$CH_3–CH_2–CH_2–CH_2–CH_2–OH$

Substituent chemical shifts in benzene derivatives have been extensively studied [13]. Such shifts can be used with caution in an additive manner to predict shifts and hence probable assignments for multi-substituted aromatics.

Spin – Spin Coupling. Carbon – proton couplings are usually sacrificed by broadband proton decoupling in the interests of spectral simplification and obtaining an acceptable spectrum more quickly. However, such couplings can provide valuable structural information.

One-bond carbon – proton couplings J_{CH} are of particular interest. At the simplest level the resulting multiplet provides information about the type of carbon atom; for example, a methyl group gives a quartet resulting from coupling to three equivalent hydrogen atoms. However, such information can be more conveniently obtained from editing experiments (see Section 18.3.5.3). $^1J_{CH}$ values vary from ca. 110 to 320 Hz, increasing with increasing *s* character of the C–H bond. Thus the couplings for methane (sp^3), ethylene (sp^2), and acetylene (sp) are 125, 156, and 248 Hz, respectively. Polar substituents also have significant effects on the magnitudes of couplings.

$^2J_{CH}$ values in aliphatic hydrocarbons range from 5 to 60 Hz. Couplings are often characteristic of substitution and/or bond hybridization. Thus whilst $^2J_{CH}$ for ethane is 4.5 Hz, the coupling to the methyl carbon of acetaldehyde is 26.7 Hz. Acetylene has a very large $^2J_{CH}$ of 49.3 Hz.

$^3J_{CH}$ couplings show orientational effects analogous to those observed in the corresponding H–H couplings. Thus $^3J_{cis}$ is always less than $^3J_{trans}$ in olefins, and the same Karplus type of dihedral angle dependence exists for saturated aliphatic groups. In aromatic rings $^3J_{CH}$ values are characteristically larger than $^2J_{CH}$ values. For benzene itself, $^3J_{CH} = 7.4$ Hz and $^2J_{CH} = 1.0$ Hz.

Table 4. Representative $^{13}C-X$ coupling constants (Hz)

Compound	X	$^1J_{CX}$	$^2J_{CX}$	$^3J_{CX}$	$^4J_{CX}$
C_6H_5F	F	245.3	21.0	7.7	3.3
n-$C_6H_{13}F$	F	166.6	19.9	5.3	
$(C_6H_5)_3P$	P	12.4	20.0	6.7	
$(n$-$C_4H_9)_3P$	P	10.9	11.7	12.5	
C_6H_6	H	157.5	1.0	7.4	
C_6D_6	D	25.5			
$CHCl_3$	H	209.0			
$CDCl_3$	D	31.5			

Table 5. Representative ^{19}F chemical shifts*

Compound	δ, ppm	Compound	δ, ppm
CH_3F	–271.9	CF_2Cl_2	–6.9
C_2H_5F	–211.5	CF_3CO_2H	–78.5
CF_3CF_3	–88.2	$C_6H_5CF_3$	–63.9
$FCH{=}CH_2$	–114.0	F_2	422.9
$F_2C{=}CH_2$	–81.3	AsF_3	–43.5
$F_2C{=}CF_2$	–135.2	AsF_5	–67.2
C_6F_6	–162.9	SF_6	48.5
C_6H_5F	–113.2	PF_3	–36.2
$C_6H_5SO_2F$	65.5	PF_5	–77.8

* Referenced to $CFCl_3$.

Note that heteronuclear couplings to magnetic nuclei other than ^1H (e.g., ^{19}F, ^{31}P, and ^2H) are observed in proton decoupled ^{13}C NMR spectra. Thus coupling to deuterium is seen in signals from deuterated solvents. Some representative C–X couplings constants are given in Table 4.

18.3.3.3. Fluorine (^{19}F) [14], [16], [17], [106], [127]

With 100 % natural abundance and $I = 1/2$, ^{19}F is ideally suited to NMR investigations. Since the earliest days of NMR spectroscopy ^{19}F applications in organic, inorganic, and organometallic chemistry have been widespread. As in the case of the ^{13}C nucleus broadband proton decoupling is often used to simplify spectra.

Chemical Shifts. ^{19}F NMR spectroscopy has the advantage of a very large chemical shift range. Organofluorine compounds alone cover over 400 ppm, whilst the range is extended to some 700 ppm if inorganic compounds are included. $CFCl_3$ is now the generally adopted internal reference. Most organofluorine resonances are high field shifted to the $CFCl_3$ reference peak. In NMR spectroscopy the accepted convention is for shifts to high field of the reference to be assigned negative values. Table 5 gives some representative ^{19}F chemical shifts. The shifts are sensitive not only to directly attached substituents but also to long range effects and thus are very good probes of changes in molecular structure. The chemical shifts given for the following compound provide an idea of the wide chemical shift range observed for some fluorocarbons.

$$F_3C-CF_2-CF_2-CF_2-CF_2-CF(CF_3)-C(=O)-F$$

Shifts: F_3C –81.6; CF_2 (successive) –126.8, –123.0, –120.4, –117.1; CF –180.4; CF_3 (branch) –72.5; C(O)F +32.3

Spin–Spin Coupling. Large compilations of data are available for both fluorine–fluorine and fluorine–hydrogen coupling constants. The presence of several relatively large couplings can lead to complicated spectra even from relatively simple molecules if they are highly fluorinated. However, the spectra are often amenable to first-order analysis, and in general the coupling constants are very sensitive to structure and substituents.

Three-bond fluorine–hydrogen couplings show orientational effects similar to those observed in the corresponding H–H couplings, i.e., the dihedral angle dependence in saturated aliphat-

Table 6. Selected F–F and F–H coupling constant ranges

Group	Coupling	J_{FX} coupling constant range, Hz	
		X=F	X=H
>C(F)X	^2J	40–370	40–80
–C(F)–C(X)–	^3J	40–80	0–40
>C=C(F)X	^2J	0–110	70–90
F(>C=C<)X (trans)	^3J	100–150	10–50
F(>C=C<)X (cis)	^3J	0–60	0–20
Fluorobenzenes:			
ortho-X	^3J	18–35	7–12
meta-X	^4J	0–15	4–8
para-X	^5J	4–16	0–3

ics and that $^3J_{cis}$ is less than $^3J_{trans}$ in olefins. Typical coupling constant ranges are given in Table 6.

18.3.3.4. Phosphorus (^{31}P) [18], [19], [117]

The ^{31}P nucleus has 100 % natural abundance and $I = 1/2$. However, a relatively small magnetogyric ratio results in the sensitivity of the ^{31}P nucleus being only 0.066 that of the proton. Nevertheless, with Fourier transform methodology, the nucleus is easy to detect even at low levels. There is much data in the literature covering applications involving inorganic compounds and complexes, organophosphorus chemistry, and biological and medical studies. In addition to these, even the correlation between ^{31}P and ^{13}C is observable with modern spectrometers in 2D correlation experiments [118].

Chemical Shifts. The accepted standard for ^{31}P NMR work is 85 % H_3PO_4, which is normally used as an external reference. Phosphorus compounds are often reactive and no suitable internal reference standard has been found for general use. Relative to H_3PO_4, ^{31}P shifts cover the range from ca. 250 ppm to −470 ppm. Phosphorus(III) compounds cover the complete range, showing that the chemical shifts are very substituent dependent, whilst phosphorus(V) compounds cover a much smaller range of −50 ppm to 100 ppm. To a first approximation the substituents directly attached to the phosphorus atom determine the chemical shift. Three variables are important: the bond angles, the electronegativity of the substituent, and the π-bonding character of the substituent. Some representative ^{31}P shifts are given in Table 7.

The phosphorus nuclei are well suited for biological and medical investigations. Relatively simple high resolution ^{31}P NMR spectra are obtained for the chemically different phosphorus atoms of cell metabolism products. The quality of the spectra which can be obtained from living systems (in vivo ^{31}P NMR spectroscopy) allows to detect metabolic changes. Furthermore, the position of the ^{31}P resonance of inorganic phosphate in the cell is a sensitive indicator for the intracellular pH value [119].

Spin–Spin Coupling. Coupling constants have been studied between ^{31}P and a wide variety of other nuclei. Couplings to protons, fluorine, and other phosphorus nuclei have been extensively used for structure elucidation. One-bond phosphorus–hydrogen couplings depend on the oxidation state of the phosphorus atom. Thus for P^{III} compounds $^1J(^{31}P, ^1H)$ values are on the order of 200 Hz, whilst for P^V compounds they are usually in the range 400–1100 Hz. By contrast, one-bond phosphorus–fluorine coupling constants for both P^{III} and P^V compounds have a common range of ca. 800 to 1500 Hz. Phosphines are widely used as ligands in inorganic complexes and therefore phosphorus couplings to some transition metals are important aids to structure determination.

18.3.3.5. Nitrogen (^{14}N and ^{15}N) [20], [21], [106], [127]

There are two magnetic isotopes of nitrogen, ^{14}N and ^{15}N, with natural abundances of 99.63 % and 0.36 %, respectively. In a magnetic field of 9.4 T the ^{14}N resonance frequency is 28.9 MHz, while that for ^{15}N is 40.5 MHz. There are problems associated with observing both nuclei. The ^{14}N nucleus possesses a quadrupole moment since $I = 1$. As a result, although ^{14}N signals can often be detected easily, the lines can be very broad with widths often between 100 and 1000 Hz. With a 9.4 T applied magnetic field this is equivalent to 3.5–34.6 ppm. This can cause resolution problems when a molecule contains more than one type of nitrogen atom.

Table 7. Representative ^{31}P shifts*

P^{III} compounds		P^{V} compounds	
Compound	δ, ppm	Compound	δ, ppm
PF_3	97	PF_5	–35
PCl_3	219	PCl_5	–80
PBr_3	227	PBr_5	–101
$P(OMe)_3$	141	$P(OEt)_5$	–71
$P(OPh)_3$	127	$P(OPh)_5$	–86
PH_3	–240	$(MeO)_3P{=}O$	2
$MePH_2$	–164	$(PhO)_3P{=}O$	–18
Me_2PH	–99	$Me_3P{=}S$	59
Me_3P	–62	$Me_3P{=}O$	36
Ph_3P	–6	$Ph_3P{=}O$	25

* Referenced to 85 % H_3PO_4.

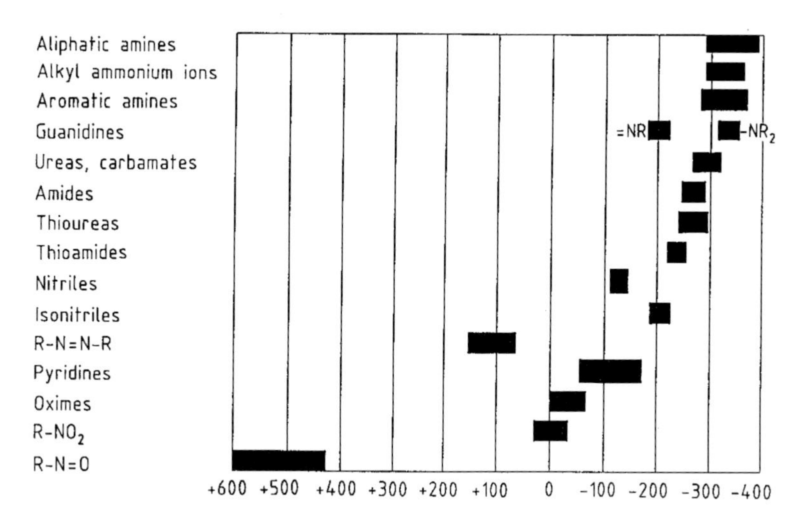

Figure 18. $^{14}N/^{15}N$ NMR chemical shift ranges
R = alkyl, aryl

Although $I = 1/2$ for the ^{15}N nucleus its very low natural abundance results in low receptivity. It also has a negative magnetogyric ratio which leads to a negative NOE when broadband proton decoupling is used. This reduces the intensity of peaks and can result in negative and even null resonances. It has therefore been necessary to use experimental procedures which minimize or eliminate the NOE. The two principal ones are inverse-gated proton decoupling (see Section 18.3.4.2) and the use of a paramagnetic relaxation reagent such as $[Cr(acac)_3]$, which reduces T_1 and consequently the NOE. With modern FT equipment it is now possible to obtain ^{15}N spectra from most small and medium-sized nitrogen-containing molecules at natural abundance level. However, for work with large biological macromolecules ^{15}N enrichment is commonly used.

Chemical Shifts. In principle the use of two isotopes raises the possibility of a primary isotope shift, but in practice ^{14}N and ^{15}N shieldings differ by no more than a few tenths of a ppm and are interchangeable for practical purposes. Nitromethane is used as reference standard for calibration purposes and the convention of assigning negative values to signals occurring high field of the reference is used here also. However, the chemical shift of nitromethane is somewhat solvent and concentration dependent. The total nitrogen chemical shift range covers about 1000 ppm, and ranges for a selection of chemical classes are shown in Figure 18.

Spin–Spin Coupling. It is unusual to observe coupling in ^{14}N spectra where resonances are very broad as a result of quadrupolar relaxation. Cou-

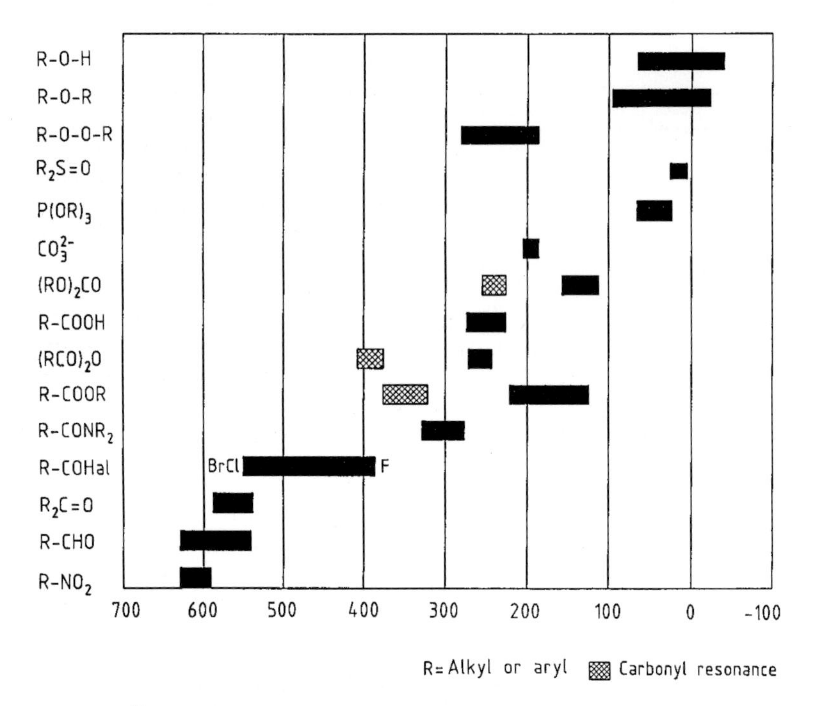

Figure 19. ^{17}O NMR chemical shift ranges
R = Alkyl, aryl

pling is, however, often resolved for ammonium salts. The relative magnitudes of the corresponding ^{14}N and ^{15}N coupling constants are given by the ratio of their magnetogyric ratios. The discussion here is restricted to ^{15}N coupling. The observation of coupling to low-abundance nuclei such as ^{13}C or another ^{15}N nucleus normally requires isotopic enrichment. Couplings to abundant nuclei such as ^{1}H, ^{19}F, and ^{31}P are more readily observed, as in the case of ^{13}C spectroscopy.

One-bond couplings to hydrogen depend on hybridization and electronegativity. Thus the $^{1}J(^{15}N, ^{1}H)$ values for NH_4^+, $O{=}C(NH_2)_2$, and $HC{\equiv}NH^+$ are 73.3, 89.0, and 134.0 Hz, respectively. One-bond couplings to ^{13}C also show a reasonable correlation with hybridization. The magnitudes of one- and two-bond couplings to ^{13}C are often comparable.

18.3.3.6. Oxygen (^{17}O) [120], [121]

The only magnetic isotope of oxygen is ^{17}O. With a natural abundance of only 0.037 %, a receptivity that is 0.61 that of ^{13}C, and a quadrupole moment ($I = 5/2$), ^{17}O is difficult to study. Nevertheless, because of the importance of oxygen in both inorganic and organic chemistry as well as biology, the ^{17}O nucleus has been quite widely studied both in natural abundance and in isotopically enriched samples. For small molecules linewidths are typically in the range of several tens to several hundreds of hertz. With large molecules or viscous solutions much larger linewidths may be encountered.

Chemical Shifts. Water is the usual chemical shift reference, being readily available in enriched form. A chemical shift range of over 1000 ppm affords wide spectral disperison. Figure 19 shows representative chemical shift ranges for some organic groups. There is a clear distinction between single and double carbon – oxygen bonds. However, a single resonance is observed from a carboxylic group at the average position expected for the carbonyl and OH groups as a result of rapid proton exchange facilitated by dimeric species. Primary alcohols and ethers resonate close to the water position, whilst lower field shifts are observed for secondary and tertiary species.

Spin – Spin Coupling. Since ^{17}O resonances are usually broad, the majority of reported values are

Table 8. ^{29}Si chemical shifts for SiX_nMe_{4-n} (ppm)

X	$n = 1$	$n = 2$	$n = 3$	$n = 4$
H	−15.5	−37.3	−65.2	−93.1
Et	1.6	4.6	6.5	8.4
Ph	− 5.1	− 9.4	−11.9	−14.0
OMe	17.2	− 2.5	−41.4	−79.2
NMe_2	5.9	− 1.7	−17.5	−28.1
Cl	30.2	32.3	12.5	−18.5

for couplings between directly bonded nuclei. $^1J(^{17}O, H)$ in water and alcohols are ca. 80 Hz. The $^1J(P, ^{17}O)$ couplings involving P=O groups cover the range 145 – 210 Hz. In contrast, couplings involving P–O groups are usually smaller (90 – 100 Hz). Some two-bond couplings have been reported. Examples of two-bond hydrogen – oxygen couplings are 38, 7.5, and 10.5 Hz for the O–CH, O–CH_3, and HC=O interactions in methyl formate.

18.3.3.7. Silicon (^{29}Si) [22], [122]

^{29}Si, with an abundance of 4.7 %, is the only naturally occurring isotope of silicon with nonzero spin ($I = 1/2$). Whilst its receptivity is 2.1 times that of the ^{13}C nucleus there are problems associated with its detection. In common with the ^{15}N nucleus a negative magnetogyric ratio results in negative NOEs when broadband proton decoupling is employed, as is usually the case. Inverse-gated decoupling and relaxation reagents are used to circumvent the problem. Broad signals from the glass in the NMR probehead or tube can also cause problems. The glass can be replaced by alternative materials such as teflon or the spectrum obtained from a blank run (i.e., without sample) can be subtracted from the spectrum of the sample. ^{29}Si is very important for solid state NMR, e.g., for analysis of glasses and zeolites [128].

Chemical Shifts. The ^{29}Si resonance of TMS is the usual chemical shift reference. ^{29}Si chemical shifts cover a range of ca. 400 ppm. The large dispersion arising from structural effects makes ^{29}Si NMR a valuable tool for molecular structure determination. The nature of the atoms directly bonded to silicon is of particular importance. Table 8 lists chemical shifts for a series of compounds SiX_nMe_{4-n}. ^{29}Si NMR has proved to be an ideal tool for structure determination of polysiloxane macromolecules.

Spin – Spin Coupling. Couplings to abundant nuclei such as the proton and fluorine can be measured either in a coupled ^{29}Si spectrum or by observing the ^{29}Si satellites in the spectrum of the abundant nucleus. The latter are relatively weak compared with peaks from the main ^{28}Si isotopomer. Nevertheless many determinations were made in this way during the 1960s by using CW spectrometers. In a well-resolved ^{1}H NMR spectrum the satellites due to a $^2J_{SiH}$ coupling of 6.8 Hz are readily observed at the base of the TMS reference peak. One-bond proton – silicon coupling constants in silanes are usually in the range 150 – 380 Hz, the magnitude of J increasing with substituent electronegativity. The literature contains data about coupling to a range of other nuclei, including ^{31}P, ^{13}C, ^{15}N, and ^{29}Si itself.

18.3.4. Double Resonance Techniques

In Section 18.3.2.2 reference is made to the use of proton noise (or broadband) decoupling to simplify ^{13}C spectra and improve the signal-to-noise ratio. This is just one example of a large number of double resonance techniques which are available to either simplify spectra or as spectral interpretation tools to aid structure elucidation.

18.3.4.1. Homonuclear Spin Decoupling

The principles of homonuclear double resonance can be illustrated by considering two coupled protons, A and B, as shown in Figure 11. In the double resonance experiment nucleus A is observed in the usual way in a CW or FT experiment whilst simultaneously selectively irradiating nucleus B with a second much stronger RF field at the "decoupling" frequency. This irradiation induces transitions between the two spin states of nucleus B. If sufficient irradiating power is applied, B flips between the α and β spin states so rapidly that nucleus A can no longer distinguish

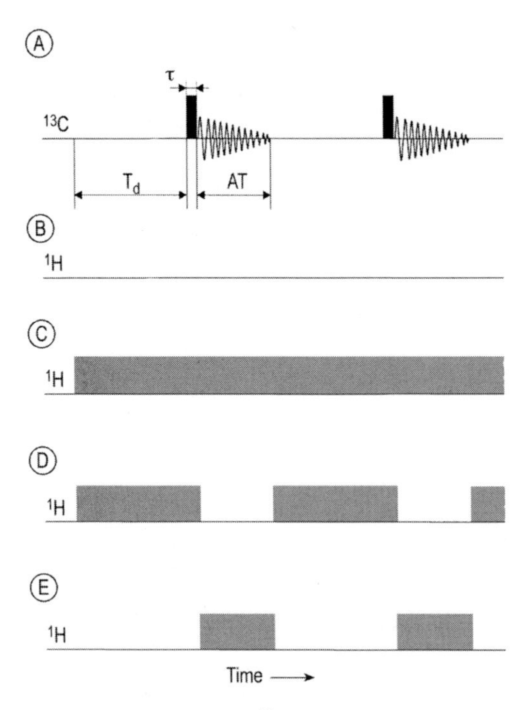

Figure 20. Schematics of an ^{13}C NMR experiment (A) without (B) and with different ^{1}H decoupling techniques: C) broadband decoupling; D) gated decoupling and E) inverse gated (igated) decoupling

between the two orientations of B and perceives an average orientation. The coupling J_{AB} disappears and the A signal doublet collapses to give a single line. In the same way irradiation of an arbitrary signal in a complex NMR spectrum results in the obvious collapsing of all multiplets which are coupled with the irradiated proton. At time before 2D NMR was feasible, this method was often used to prove the connectivities between different protons in a molecule.

18.3.4.2. Heteronuclear Spin Decoupling

In the case of heteronuclear spin decoupling the decoupled and observed nuclei are of different chemical types. The experiments are illustrated with examples in which protons are decoupled and the effect observed on the ^{13}C spectrum. However the methodologies are equally applicable to other heteronuclear spin systems.

Selective spin decoupling is analogous to the homonuclear decoupling experiment described above. In general higher RF power is required than for homonuclear decoupling but lower than is generally used for broadband heteronuclear decoupling. An individual proton signal is selected and irradiated to remove all couplings to that specific proton in the ^{13}C spectrum.

Broadband Spin Decoupling. If the proton decoupler frequency is set to the center of the proton spectrum and modulated by using a noise generator with a bandwidth wide enough to cover the complete proton region, then this is equivalent to simultaneously irradiating every proton frequency. This effectively decouples all the protons in the molecule producing the effect shown in Figure 12 B. Two problems are, however, associated with this type of broadband decoupling. The high RF power required heats the sample and, secondly, decoupling is ineffective if the bandwidth to be irradiated is large. The latter is more of a problem on higher field spectrometers and where broadband decoupling of other nuclei such as ^{19}F or ^{31}P, with their large spectral widths, is required. In recent years alternative methods such as WALTZ-16 [23] based on composite pulse decoupling have largely overcome these problems.

Broadband decoupling also produces a change in intensity, known as nuclear Overhauser enhancement, of the observed signals. For ^{13}C the increase in intensity can be as much as 200 %.

Gated Decoupling. When decoupling is turned on or off, coupling effects appear and disappear instantaneously. The NOE effects however, grow and decay at rates related to some relaxation process. As a result it is possible by computer control of the decoupler to completely separate the two effects.

Sometimes it is necessary to record broadband proton decoupled spectra without any NOE. The NOE may be undesirable either because quantitative information is needed about the relative number of atoms contributing to each peak in, for example, a ^{13}C spectrum, or because the nucleus being studied has a negative NOE (e.g., ^{15}N or ^{29}Si). Such spectra are also used as controls when measuring NOE values. In other instances coupled spectra with NOE are used as a means of improving sensitivity.

Figure 20 shows a FT ^{13}C experiment with three possible broadband decoupling schemes. The repetitive pulse experiment consists of three parts: a relaxation delay T_d, a ^{13}C RF pulse, and the acquisition time AT during which time the FID is sampled. AT and T_d are on the order of seconds, whilst the pulse duration is a few microseconds.

For conventional broadband decoupling (Fig. 20 C) the decoupler is on throughout the experiment. If the decoupler is gated off just before the RF pulse as in Figure 20 D, NOE effects can be observed without any decoupling. Conversely, if the decoupler is switched on when the pulse is applied and off as soon as the acquisition is complete, as shown in (E), the resultant decoupled spectrum does not exhibit an NOE. The latter experiment is referred to as inverse gated decoupling.

Off-Resonance Decoupling. Whilst the broadband proton decoupling described above produces considerable simplification of ^{13}C and other spectra, it removes all the coupling information. As a result the spectra may be difficult to assign. In the early days of ^{13}C NMR a technique known as off-resonance decoupling was used to circumvent this problem. Off-resonance decoupling is achieved by setting the central frequency of the broadband proton decoupler 1000 – 2000 Hz outside the proton spectrum. This results in residual couplings from protons directly bonded to ^{13}C atoms, whereas longer range couplings are lost. The magnitude of the residual coupling, which is smaller than the true coupling J, is determined by the amount of the offset and the power of the decoupler. In such spectra it is usually possible to determine the multiplicities of all the individual carbon atoms, i.e., methyl carbons appear as quartets, methylenes as triplets, etc. Nowadays the use of gated decoupling for multiplicity determination has largely been replaced by spectral editing experiments (Section 18.3.5.3).

18.3.4.3. NOE Difference Spectroscopy [24], [25]

The phenomenon of nuclear Overhauser enhancement was introduced in Section 18.3.2.3 to explain the increase in intensity of ^{13}C peaks when broadband proton decoupling is used. In general terms the NOE is defined as the change in signal intensity of a nucleus when a second nucleus is irradiated. It is beyond the scope of this article to give a theoretical treatment of the origin of the effect, which is caused by through space nuclear relaxation by dipolar interaction. Among other factors the magnitude of the enhancement has a $1/r^6$ dependence on the distance between the two nuclei. Therefore, homonuclear NOEs between protons have been used extensively to obtain information about internuclear distances and thereby to distinguish between possible structures.

The most convenient way of observing such enhancements, which rarely exceed 20 % in the most favorable cases and are sometimes < 1 %, is by a technique known as NOE difference spectroscopy. In such an experiment two spectra are recorded. The first is a conventional ^{1}H spec-trum, while in the second a chosen resonance is selectively irradiated during the interpulse delay of an FT experiment. The second spectrum is then subtracted from the first to give a "difference" spectrum.

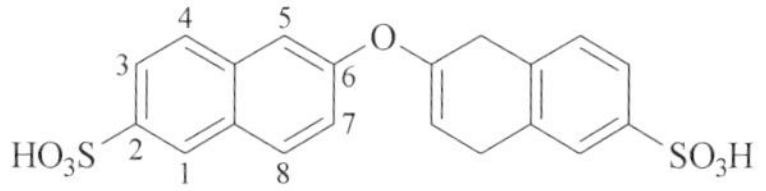

HO_3S 2 1 3 4 5 6 O 7 8 SO_3H

1

Figure 21 A shows the ^{1}H NMR spectrum for **1**. From this spectrum alone it is not obvious whether the ether linkage is at C-6 or C-7. The peak at 8.2 ppm is clearly due to H-1 whilst that at ca. 8.0 ppm is from H-5 or H-8, depending on which isomer is present. Figure 21 B shows the difference spectrum obtained when the resonance at 8.0 ppm is selectively irradiated. The irradiated resonance, which is partially or fully saturated, gives a negative signal in the difference spectrum. Positive signals are obtained from those resonances which exhibit an enhancement. Since a positive enhancement is observed for H-1 this is a good indication that H-1 and H-8 are close together. Measurable effects can be observed between atoms up to about 0.4 nm apart. Since the above description is rather simplistic it should be stressed that because dipolar relaxation of a particular nucleus may involve interaction with more than one other nucleus a basic understanding of the processes involved is a prerequisite to applying the technique successfully. Also spin-spin coupling may complicate NOE measurements. If irradiation is applied to spin multiplets, NOE measurements may fail if selective population transfer is present and special precautions have to be taken in such cases. Furthermore, the NOE effect depends on the molecular correlation time τ. During slow molecular motions frequencies dominate the relaxation process. For macromolecules or small molecules in viscous solvents the NOE may completely vanish or become negative. A positive NOE effect is observable, therefore, only under condition $\omega\tau \ll 1$, which is always true for small

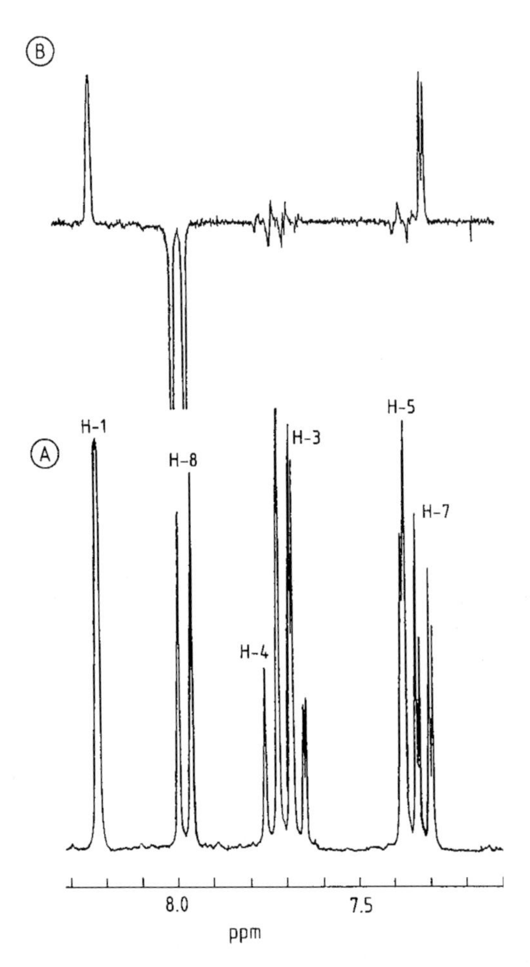

Figure 21. (A) 250 MHz ^{1}H NMR spectrum of **1**; (B) NOE difference spectrum on irradiation of H-8

and isotropic molecules of low molecular weight compounds ($M_r < 500$) in solutions of low viscosity. In the last years the two-dimensional NMR technique has been used more frequently for measurement of the NOE effect (NOESY spectroscopy, see Section 18.3.6.5).

18.3.5. One-Dimensional Multi-Pulse FT Experiments

The FT experiment described in Section 18.3.1.2. and depicted in Figure 7 is a single-pulse experiment. This means that only one RF pulse is applied prior to recording the FID response. A large number of experiments have been developed, for a variety of purposes, in which more than one pulse is applied prior to recording the FID. One of these, the inversion recovery method of determining T_1, is described below in some detail. Others will be mentioned only in so far as their application is concerned.

18.3.5.1. T_1 Measurement

Spin–lattice relaxation is described briefly in Section 18.3.2.4. T_1 values can provide valuable information about a molecule. Each of the magnetic nuclei in a different environment in a molecule has a different T_1. Thus each type of hydrogen or carbon atom will have a different value. A knowledge of these values and their relative sizes can give an insight into such things as the type of atom (e.g., whether it is a quaternary or proton-bonded carbon atom), the mechanism responsible for the relaxation, and the molecular motion of that part of the molecule. A knowledge of T_1 values is also required for using NMR as an accurate quantitative tool (see Section 18.3.8.2.).

The most common method of measuring T_1 is the inversion recovery experiment, which uses the pulse sequence: $[T_d - 180° - \tau - 90°\,(\mathrm{FID})]_n$. The principle is pictured in Figure 22. Whereas a 90° pulse rotates the magnetization vector, which at equilibrium is M_{z0} and lies along the z-axis (Fig. 22 B), to the y-axis of the rotating frame, a 180° pulse which is twice as long inverts the magnetization to the $-z$-axis. Immediately after the pulse the magnetization vector $\boldsymbol{M}_z$ begins to relax back to its equilibrium value M_{z0} according to the following equation:

$$\boldsymbol{M}_z = M_{z0}[1 - 2\exp(-\tau/T_1)] \tag{2}$$

With time, $\boldsymbol{M}_z$ becomes less negative, passes through zero, and eventually relaxes back to M_{z0}. This process proceeds at a different rate for each peak in a spectrum. At a time τ after the 180° pulse, a 90° pulse is applied which tips the magnetization vector onto the $-y$- or $+y$-axis, depending on whether $\boldsymbol{M}_z$ was negative or positive. The delay T_d required before the sequence can be repeated should be at least five times the longest T_1 being measured to enable the system to relax completely. As in the case of the single pulse experiment the sequence is repeated n times and the FID added together. The resulting FID can then be transformed in the usual way. The sign and the amplitude of the resulting peaks depends on the length of the interpulse delay τ and T_1 for each nucleus (Fig. 22 C).

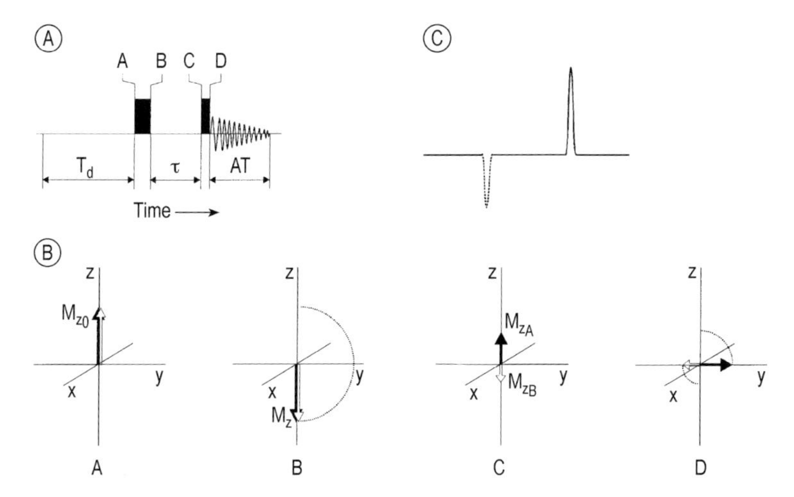

Figure 22. The principle of the inversion-recovery experiment for T_1 measurement: Puls sequence (A) and position of the macroscopic magnetization M_z (B) during the experiment at four selected points (A-D), shown for two nuclei A and B with different T_1 values, resulting in the spectrum (C)

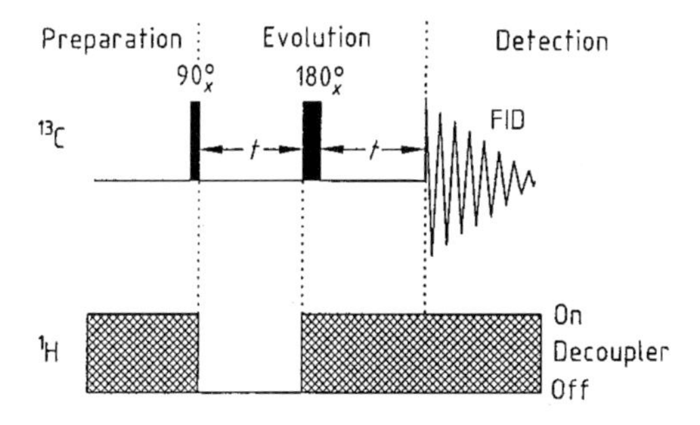

Figure 23. J-modulated spin-echo pulse sequence

Rearranging Equation (2) and taking logarithms gives:

$$\ln(M_z - M_{z0}) = -\ln(2M_{z0}) - \tau/T_1$$

Thus a plot of $\ln(M_z - M_{z0})$ against τ for each peak gives a straight line with a gradient of $-1/T_1$. In practice a series of experiments is carried out with increasing values of τ. The value of M_{z0} is determined by performing an experiment in which τ is very long.

18.3.5.2. T_2 Measurement

It is much more difficult to measure T_2 than T_1. One of the main reasons is that imhomogeneities in the magnetic field can make a significant contribution to the apparent spin–spin relaxation time. Complex, so-called spin-echo pulse sequences [27], [28] have been developed for such measurements.

18.3.5.3. Spectral Editing Experiments [0], [26]

The most frequently used multi-pulse experiments are probably those for multiplicity determination in ^{13}C NMR. Both the *J*-modulated spin-echo [29] and the distortionless enhancement by polarization transfer (DEPT) [30] experiments are extensively used for this purpose.

Figure 23 shows the pulse sequence used for the *J-modulated spin-echo* experiment which includes the three periods characteristic of all modern pulse experiments, namely preparation, evolution, and detection. The preparation period is a relaxation delay ending with a $90°_x$ pulse along the *x*-axis of the rotating frame. The decoupler is switched off, and during the first half of the evolution period the magnetization in the $x-y$ plane evolves under the influence of the proton–carbon couplings. At the beginning of the second period the decoupler is switched on again and a $180°_x$ pulse refocusses all effects other than *J*-modulation. In the detection period a proton-decoupled FID signal with NOE is observed. If $t = 1/J$, signals from quaternary and CH_2 carbons are in anti-phase to those from CH and CH_3 carbons. Figure 24

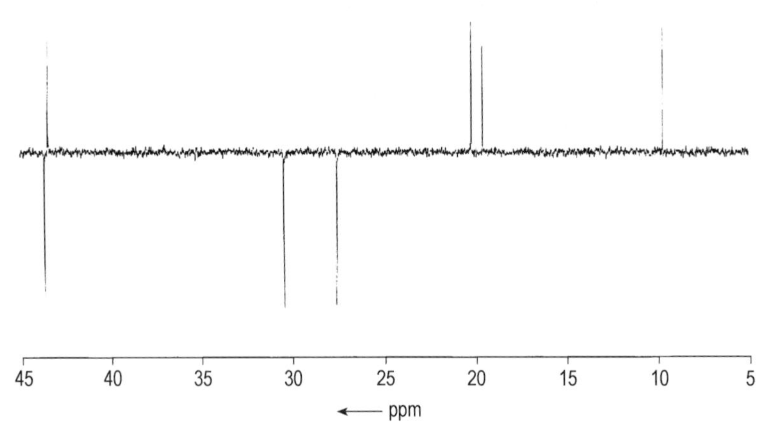

Figure 24. Part of the *J*-modulated ^{13}C NMR spin-echo spectra of camphor. Note the distinction between the C-3 methylene and the C-4 methine carbons with a shift difference of only 0.3 ppm.

shows such a *J*-modulated spin-echo ^{13}C spectrum for camphor (**2**).

2

When $t = 1/2J$ only signals from quaternary carbons are observed. Whilst widely used, this experiment has two shortcomings for multiplicity assignment. Firstly, signals from CH and CH_3 carbons cannot be distinguished, and secondly it is sensitive to variations in J values. One-bond carbon–proton couplings are typically ca. 125 and ca. 160 Hz for saturated aliphatic groups and aromatic/olefinic carbon atoms, respectively. For such compounds use of a compromise value for t works well. However, for carbon atoms with much larger couplings such as alkynyl carbons and some unsaturated heterocyclic carbons it is less successful.

In such instances new methods for ^{13}C assignment are more suitable. Compared with the *J*-modulated spin-echo experiment, the *DEPT* (Distortionless Enhancement by Polarization Transfer) pulse sequence is insensitive to variations in $^{1}J_{CH}$. The method is based on the use of polarization transfer, whereby magnetization is transferred from protons to ^{13}C nuclei, and as a result quaternary carbon atoms are not detected in such experiments. This experiment involves pulsing both the ^{1}H and ^{13}C channels. The final proton pulse is along the y-axis and by performing two experiments with angles for this pulse of 90° and 135°, full multiplicity data can be obtained. The 135° experiment gives negative CH_2, but positive CH and CH_3 carbon signals whilst in the 90° experiment only CH signals are observed.

18.3.6. Multi-Dimensional NMR [123]

A wide range of two-dimensional (2 D) experiments are now available as routine tools, and more recently three- and four-dimensional methods have been developed. In general the experiments are used to extract information from complex spectra (e.g., about which nuclei are J coupled to each other) and to measure J couplings. In this article it is only possible to give a very brief introduction to two-dimensional methodology together with a few examples of its application in ^{1}H and ^{13}C NMR. Further information about this important field is available in [31], [32], [99], [103], [108], [109].

18.3.6.1. Basic Principles

The common feature of the 1 D multipulse experiments described above was the time sequence preparation – evolution – detection, whereby the detected signal is only a function of the detection time t_2. The important difference in 2 D NMR is that the evolution time t_1 is now a variable. In a 2 D experiment n separate experiments are performed with incremented values of t_1. For each experiment a free induction decay $S(t_2)$ is measured. In this way a matrix $S(t_1, t_2)$ is built up.

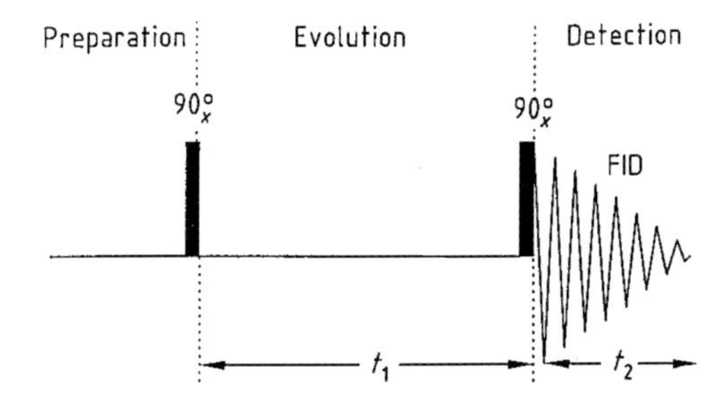

Figure 25. Basic COSY pulse sequence

Fourier transformation of this matrix with respect to t_2 gives a series of spectra $S(t_1, f_2)$. Further transformation with respect to t_1 gives a matrix $S(f_1, f_2)$ which is a spectrum in two independent frequency dimensions. The dispersion of the signals as a function of f_1 depends on the effect of the pulse sequence used on the spin system as t_1 is varied. In general there are two distinct types of 2 D experiments, referred to as *J*-resolved and correlated spectroscopy. In a *J-resolved spectrum* one frequency axis (f_1) contains spin – spin coupling information and the other (f_2) chemical shift information. In *correlated spectroscopy* both axes contain chemical shift information, the connection between the two being via spin – spin coupling, NOE, or exchange effects.

18.3.6.2. *J*-Resolved Spectra

J-Resolved spectra provide the possibility to observe the resonance frequencies δ and the coupling constants J on two distinct frequency axes separately. This is achieved by a pulse sequence $[T_d - 90° - t_1/2 - 180° - t_1/2 - \mathrm{FID}(t_2)]_n$. After excitation by a 90° pulse the evolution time t_1 is divided by a 180° pulse. The signal (FID) is then detected in t_2. This experiment, known as 2 D *J*-resolved or *J*,δ-spectroscopy is mainly used in analysis of crowded spectra. The overlapping absorptions in a 1D proton spectrum can sometimes be resolved by this experiment, in which chemical shifts are presented on one axis and coupling constants on the other. The projection of the 2 D spectrum onto the δ-axis is effectively a "^{1}H broad band decoupled" proton spectrum. However, the relatively long measuring time, which are usually larger than that of other 2 D methods and a number of artifacts, shown in strongly coupled spin systems, caused a decrease of the application frequency.

In heteronuclear *J*-resolved spectra chemical shifts of an arbitrary nucleus X which couples with protons (this is mostly ^{13}C) are presented on one axis and proton-X J couplings on the other. The information content is equivalent to that in a proton-coupled ^{13}C spectrum (Fig. 12) but without the severe overlap of multiplets which is usually encountered in the latter. In common with off-resonance proton decoupling, *J*-modulated spin echo, and DEPT experiments, it facilitates multiplicity determination. In addition, it enables proton-X coupling constants to be measured.

18.3.6.3. Homonuclear Chemical Shift Correlation (COSY)

The most frequently applied 2 D technique is proton homonuclear correlation using one of the many variants of the COSY (COrrelated SpectroscopY) experiment. The basic COSY experiment consists of two $90°_x$ pulses separated by a time t_1, as shown in Figure 25. Figure 26 shows the aliphatic part of the COSY spectrum obtained from camphor **2**.

The spectrum is presented as a contour plot in which intensities above a chosen threshold level are plotted like height contours on a map. Since both frequency domains are proton chemical shifts the matrix is square. Contours along the diagonal of the square correspond to the peaks in a one-dimensional spectrum. Homonuclear couplings give rise to off-diagonal contours or cross peaks. The one-dimensional spectrum is also plotted as a projection along both axes. The cross-peaks provide the same information about proton – proton connectivities as can be obtained from a series of homonuclear decoupling experiments, but all in a single spectrum.

The connectivities for a particular proton can be extracted by drawing a horizontal line from the relevant diagonal peak that intercepts cross peaks corresponding to correlations. In Figure 26 the multiplet at 2.35 ppm is due to H-3_{exo}, and from the cross-peaks it shows correlations to three other multiplets. By drawing a vertical line from each cross-peak back to the diagonal it can be determined with which contour the first diagonal contour is correlated (i.e., coupled). In this case correlations are observed at 2.1, and 1.84 ppm from H-4 and the attached H-3_{endo}, respectively, and a weak one at 1.9 ppm from the H-5_{exo}, four bonds away from the H-3_{exo}. Note that the cross-peaks are found symmetrically on both sides of the diagonal.

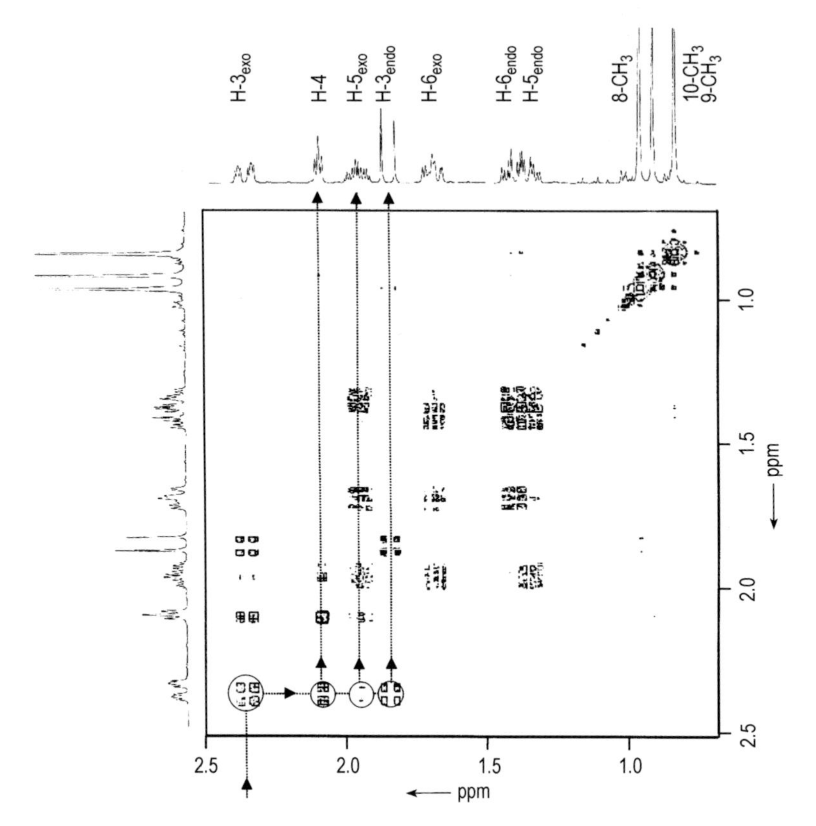

Figure 26. ^{1}H COSY spectrum of camphor

2 D methodology is inherently much less sensitive than 1 D, principally because of the requirement to perform a large number of individual experiments. However, because of the relatively high sensitivity of the proton, a few milligrams of sample are sufficient to yield a good COSY spectrum on a modern high-field instrument. Even shorter measuring time have become possible by the application of gradient accelerated or enhanced COSY spectroscopy, a new technique where the traditional phase cycling techniques will be replaced by the use of pulsed field gradients. Specially designed probe-heads and electronics are necessary to employ the gradients. Depending on the manufacture of the spectrometer, either one or three additional coils are required to generate the *z*- or the *x*-, *y*-, *z*-field gradients within the sample. There exists, however, no real enhancement by field gradients, and the abbreviation *gs-COSY* (gradient selected) become increasingly common. Indeed, gradients select desired and undesired coherences. Simplified, a coherence describes all possible mechanism for the exchange of spin population between different energy levels in a NMR diagram. This task was previously performed by time-consuming phase cycling. The selection of the desired coherence already happens in the probe, and only one single transient is sufficient, provided that enough substance is available. Thus, a typical gradient enhanced COSY experiment with 256 time increments can be recorded in a few minutes.

In addition, several other modifications of the COSY pulse sequence exist: COSY-45 for the reduction of the diagonal signals, long-range COSY to emphasize small couplings, relayed-COSY for the observation of remote protons, and the double quantum filter DQF-COSY for elimination of singlet signals. The last one is very important for suppression of water signals in biological samples [124].

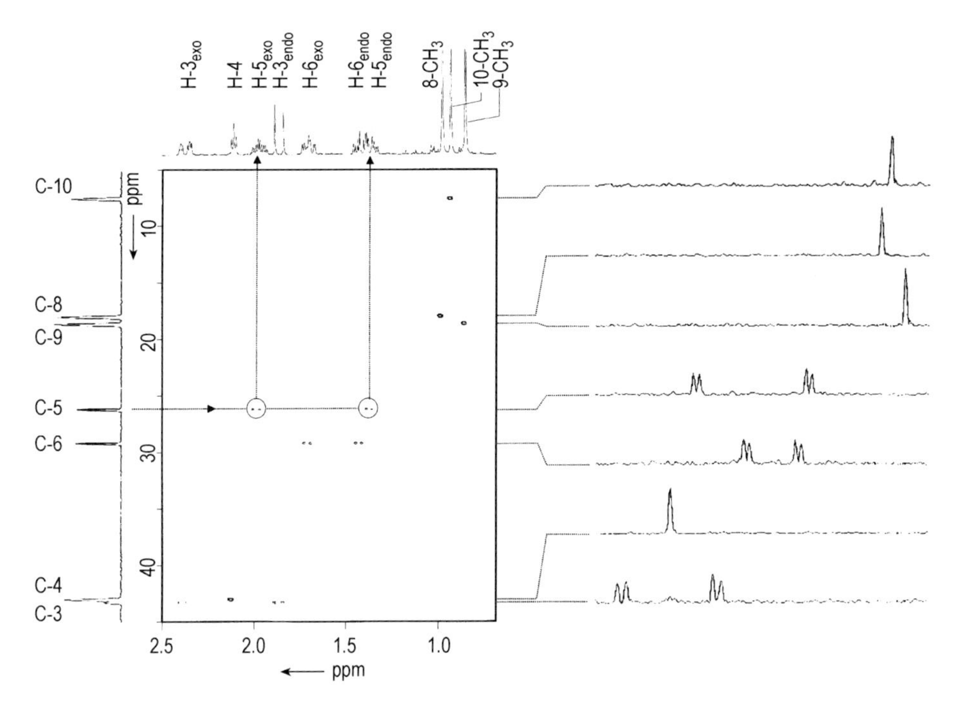

Figure 27. Aliphatic part of the heteronuclear chemical shift correlated (HETCOR) spectrum of camphor. Contour diagram (left) and the traces in the ^1H dimension (right) of all C-H-systems.

18.3.6.4. Heteronuclear Chemical Shift Correlation (HETCOR, HMQC)

In the *HETCOR* experiment the peaks of an insensitive nucleus (^{13}C, ^{15}N) are correlated with those of a sensitive nucleus (^1H, ^{19}F, ^{31}P). In Figure 27 the aliphatic part of the HETCOR spectrum of camphor (**2**) shows the specific resonances of the protons which are attached to each ^{13}C nucleus. The relevant parts of the corresponding 1 D spectra are plotted along the axes. A correlation is observed as a cross-peak at the intersection of two lines drawn from a proton resonance and from a carbon peak, respectively. The three pairs of diastereotopic methylene protons H-$3_{endo/exo}$, H-$5_{endo/exo}$ and H-$6_{endo/exo}$ give individual cross peaks at the same carbon resonance, respectively. Correlations are not observed for quaternary carbon atoms. The technique is an important tool for chemical shift assignment and thus structure elucidation.

The cross peaks in the basic HETCOR experiment are generated by a magnetization transfer from the sensitive (^1H) to the adjacent insensitive nucleus (^{13}C). A number of variants have been developed to increase the sensitivity and to enable observation of long range proton-carbon correlations.

A dramatic increase in sensitivity has been obtained by introduction of the reverse or inverse shift correlation, where the sensitive nucleus (^1H) is used for signal detection [125]. This method is known by the acronym *HMQC* (heteronuclear multiple quantum coherence). In theory, a ^1H-excited and ^1H-detected H,C-correlation experiment is 32 times more sensitive (S/N) compared to a normal ^{13}C detection. In practice, an average S/N-enhancement of 2 – 3 for a HMQC compared to a proton decoupled 1 D ^{13}C spectrum will be found. For detection of long-range connectivities between carbons and protons the *COLOC sequence* (correlation via long -range couplings) and an inverse technique with the highest possible intensity, the *HMBC* (heteronuclear shift correlations via multiple bond connectivities) experiment, were developed [126]. In Figure 28 the gradient selected HMBC of camphor are shown. The corresponding 1D ^1H and ^{13}C spectra are plotted on the axis analogous to the HMQC spectrum, but in contrast cross peaks are shown here only for H,C-couplings over two (^1H-C-^{13}C) or three bonds (^1H-C-C-^{13}C). That is illustrated

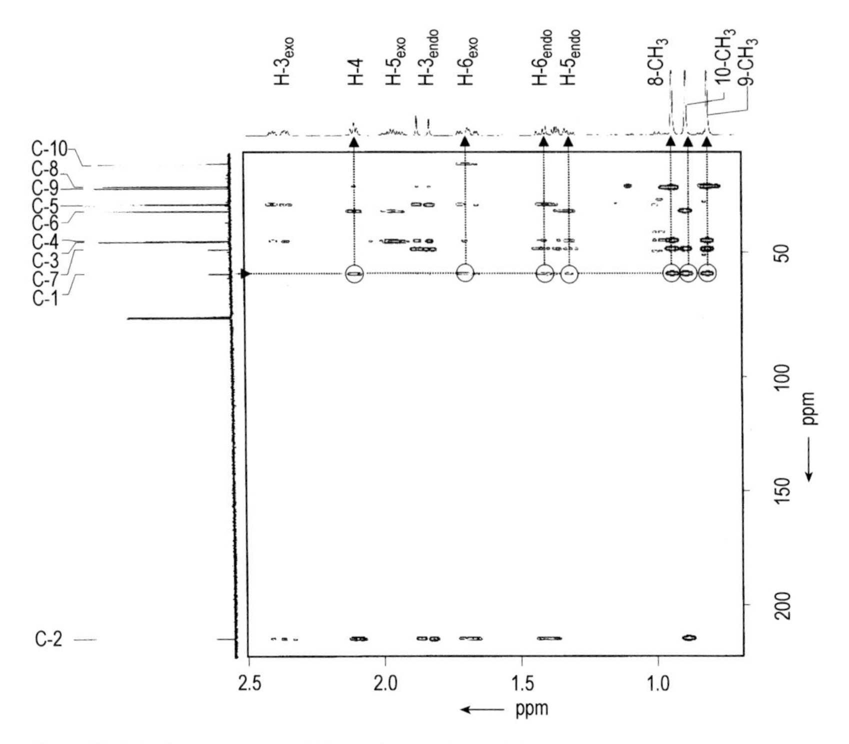

Figure 28. HMBC (*h*eteronuclear shift correlations via *m*ultiple *b*ond *c*onnectivities) spectrum of camphor

clearly for the carbons C-1 and C-7 in Figure 28. Both quaternary carbons show cross peaks to all three methyl groups whereas the couplings over one bond are suppressed, as shown for the C-10 methyl carbon. Therefore, the interpretation of a HMBC spectrum requires the previous analysis of the HETCOR or HMQC spectrum.

However, there are some problems connected with inverse correlation experiments that should be mentioned. In contrast to the HETCOR, in the inverse experiment the ^{13}C resolution does not depend on the acquisition time *AT*, but on the number of time increments, because of the interchange of the f_1 and f_2 frequency axes (see Section 18.3.6.1). Another large problem is the suppression of the undesired signals of protons bound to ^{12}C. In theory, phase cycling should remove all non-^{13}C bonded protons. But a perfect cancellation can never be achieved. Thus, a severe so-called t_1 noise ridge is usually shown at all ^{1}H chemical shift frequencies. In this article it is only possible to name the two methods that help to reduce these problems dramatically: The BIRD (bilinear rotational decoupling) sequence and the pulsed field gradients (PFG, see Section 18.3.6.3). With PFG the homogeneity of the magnetic field can be destroyed in a controlled way. This yields artifact-free correlation spectra in a fraction of the time need previously (e.g., gs-HMQC). Today with modern spectrometers, the 2 D gs-HMQC is a good alternative to the normal 1 D ^{13}C NMR spectrum and there is no reason to run conventional HETCOR experiments anymore.

18.3.6.5. Homonuclear NOE Correlation

The 2 D NOESY (nuclear Overhauser and exchange spectroscopy) technique in principal enables all proton – proton NOE effects to be assembled in a single spectrum. Such spectra are comparable to the respective COSY spectrum. The cross peaks arise from through-space proton-proton interactions between nonbonded protons that are nearby in space. The technique is extensively used in the determination of three-dimensional structures and conformations of molecules. This is shown in Figure 29 with a part of the NOESY spectrum of camphor (**2**). The ^{1}H NMR signals at 0.83 ppm (^{13}C 19.8 ppm) and 0.97 ppm (^{13}C 19.2 ppm) can be unambiguously assigned to the methyl groups 8 and 9 by use of the NOE effects. The high-field proton methyl signal shows an ob-

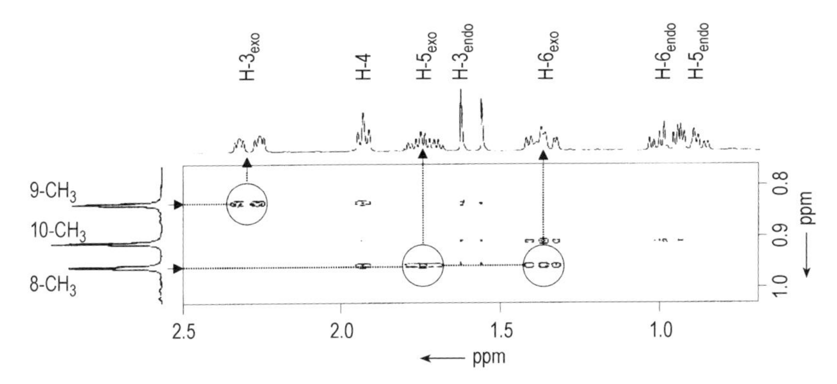

Figure 29. Part of the NOESY spectrum of camphor. Only the NOE effects of the three methylgroups (left) to all other protons (top) are shown here.

vious NOE to the H-3_{exo} proton what is only possible from the 9-position whereas the low-field signal shows effects to the H-5_{exo} and H-6_{exo} protons.

18.3.7. NMR Spectral Collections, Databases, and Expert Systems

NMR is a powerful tool for the determination of structures from first principles and the chemical shift is the most important NMR parameter in structural analysis. For estimating the relationship between chemical structures and chemical shifts three possibilities exist: the calculation of the chemical shift values by empirical methods [137], the computation by quantum chemical procedures, e.g., with the IGLO-method (Individual Gauge for Localized Orbitals [129]), or the use of large compilations of NMR spectra and the associated chemical structures. The access to relevant reference data for identical or similar compounds can facilitate the assignment process enormously. Reference data may assist by reducing the amount of experimental and/or interpretive effort required or increase confidence in the suggested structure.

Many data compilations exist, containing significant amounts of reference data, especially those dedicated to a single nucleus. In addition there are many spectral libraries, some available in hard-copy format only. One of the first ^{1}H NMR libraries was published by Varian Associates always in the early 1960s [33]. Such older hard-copy libraries contain only low-field data and are normally searchable by parameters such as name, molecular formula, chemical shift value, or compound type [34]–[38], [137]. However, state of the art are search softwares in computer-readable form. Computer-searchable libraries, especially those which allow substructure searching of compound data in conjunction with spectral data searching, are much more powerful than hard-copy libraries. In 1973 BREMSER published a procedure for the description of the chemical environment of nuclei by their hierarchically ordered spheres of environment (HOSE code) [130]. Today several databases exist containing hundreds of thousands of chemical shift values in particular for ^{13}C nuclei and the corresponding information about the chemical environment of the individual carbons. Only some examples of electronically stored databases should be mentioned here: SpecInfo [39], CSEARCH [132], and CNMR [133]. Their current contents are 149 000 (500), 265 000 (0) and >67 000 (>82 000) spectra for ^{13}C NMR (^{1}H NMR) spectroscopy, respectively. All three systems listed above contain assigned NMR data which can be searched by molecular formula, name, peak positions/-spectrum similarity, and structure/substructure. Users are also able to add their own spectral data to the libraries. All databases can be used to identify known substances contained within the database, and they are an efficient tool for elucidating the structure of unknown substances.

The SpecInfo system which was originally developed in-house by BREMSER et al. at BASF and was also marketed in a searchable microfiche version [40], is the premier spectroscopic archive and interpretation system for ^{13}C, ^{1}H, hetero-NMR, IR, UV/VIS, Raman, and mass spectra. In a more automated way the structure assignment can be achieved by using a structure generator, which assembles all possible isomers for a given

Table 9. ^{13}C chemical shift estimation of the camphor carbons with different databases

C	CH_n	exp	SpecTools [a] [153]	C_shift [a] [134]	SBSD [135]	ACD [133]
1	C	57.7	64.7	60.3	57.6	57.2
2	C	219.7	213.9	216.5	219.3	217.3
3	CH_2	43.3	33.1	42.9	43.3	43.7
4	CH	43.1	36.3	46.6	43.1	43.3
5	CH_2	27.1	22.1	26.5	27.1	27.3
6	CH_2	30.0	26.3	29.0	29.95	30.1
7	C	46.8	31.8	47.6	46.8	46.6
8	CH_3	19.2	21.4 [b]	20.2 [b]	19.15	18.7
9	CH_3	19.8	21.4 [b]	20.2 [b]	19.8	20.7
10	CH_3	9.3	12.7	11.8	9.25	9.5

[a] In contrast to the databases which fall back on the experimental spectra, the shift values were calculated with this PC programs by use of additional increments (SpecTools) or neural networks (C_shift).
[b] Exchangeable since stereochemistry is not considered here.

molecular formula and performs an automatic ranking of the candidate structures (SPECTACLE [131]). In 2000 a PC program was developed on basis of the SpecInfo database for the calculation of ^{13}C NMR spectra of any proposed organic molecular structure [134]. The spherically encoded chemical environments of more than 500 000 carbons were used here to train artificial neural networks which allow the fast determination of ^{13}C chemical shift values.

CSEARCH spectrum estimation is based on scripts written by R. Bobrovsky, G. Löffler ,and W. Robien and can be accessed via an e-mail based server system at the University of Vienna [132]. Four different versions are available. Optionally the stereochemistry of the molecules can be considered and/or neural networks can be used for the fast estimation of chemical shifts.

The Advanced Chemistry Development (ACD) presents a toolset of PC and web-based software for NMR prediction, processing, and database management also for ^{19}F (> 11 500 spectra), ^{31}P NMR (> 18 500 spectra) and for MS, IR, UV-Vis and chromatographic databases. The operation of an actual NMR spectrometer can be simulated here, allowing to choose among different spectra modes (off-resonance, *J*-modulation, DEPT) just as the operating frequency, the solvent and the concentration of the solute.

Another web-based integrated spectral data base system for organic compounds (SDBS) was developed by the National Institute of Materials and Chemical Research of Tsukuba (Japan) [135]. This system includes six different types of spectra: ^{13}C NMR (ca. 11 000 spectra), ^{1}H NMR (ca. 13 500 spectra), MS, IR, laser-Raman, and ESR (ca. 2000 spectra). The studies on the SDBS started already in early 1970. This system only allows the search for available structures and not the determination of spectra from unknown compounds.

However, spectra databases are statistical tools to establish the relationships between NMR spectral parameters and chemical environment of individual atoms. So, the results can only be offered with a statistical probability, depending on the quantity and quality of the available database entries. In other words, the accuracy of the predicted data cannot be more precise than the stored data. Usually, ^{13}C NMR spectra can be calculated for almost any drawn organic structure to an accuracy of ± 3 ppm or better, apart from stereochemical problems which can not be considered by some databases. Table 9 summarizes the results of the determination of the ^{13}C NMR chemical shifts for the ten carbon atoms in camphor (**2**) obtained with different methods.

Many publications on a variety of expert systems and computer-based structure determinations will be found in the recent literature [136]. In general they are not associated with the large commercially available databases but they are useful for the elucidation of a completely unknown structure. However, in most practical cases the more common type of structure determination is structure verification. Here, the structure information achieved via the chemical shift is usually sufficient.

18.3.8. Applications

Solution NMR spectra usually contain a wealth of information. Previous sections have provided an overview of experimental techniques, the parameters which can be determined from NMR spectra and techniques which can be used for the assignment of signals in spectra. Here some of the applications of solution NMR spectroscopy are briefly summarized.

18.3.8.1. Chemical Structure Determination

NMR spectroscopy is the most powerful tool available for the unambiguous determination of molecular structures of compounds in solution. In a modern chemical research laboratory the principal tools for structure determination are usually mass spectrometry, to obtain molecular mass data, and NMR spectroscopy which provides detailed information about the groups present and how they are assembled. The abundant nuclei ^{1}H, ^{19}F, and ^{31}P have been used for over 40 years to study the structures of organic and inorganic chemicals. In the early 1970s the advent of FT methodology enabled a much wider range of nuclei to be studied and in particular resulted in an explosive growth in the use of ^{13}C NMR as a complementary technique to ^{1}H NMR in the identification of organic compounds. The most useful NMR parameters are, in general, chemical shifts, coupling constants, and integrals from which structural information can be deduced.

In addition to the study of monomeric species, NMR is a long established tool for studying the structures of synthetic polymers [41], [138]. In the case of copolymers the various groups present can be identified, their relative concentrations determined, and often information about sequence distributions obtained. The latter provides a measure of the random/block character of the polymer. Whilst ^{1}H and ^{13}C are the most commonly studied nuclei for such work several others such as ^{19}F and ^{29}Si have been used where appropriate. For vinyl polymers ^{13}C and ^{1}H NMR spectra provide information on tacticity. Chain branching and end groups can also be identified from NMR spectra of polymers.

During the last decade NMR has developed as a powerful tool for biochemical structure determination. The most impressive applications are in the determination of protein structures with molecular masses up to the 25 000 – 30 000 range [42], [139]. Such work has been made possible by the development of multidimensional NMR techniques. 2 D and 3 D techniques are widely used for such work and 4 D experiments are being developed. 3 D and 4 D methodologies are used in conjunction with ^{15}N and/or ^{13}C labelling of the amino acid residues to reduce resonance overlap. To date NMR protein structures have usually been calculated on the basis of interproton distances and dihedral angles derived from NOE and *J* coupling measurements, respectively. Each structure determination necessitates a large number of sophisticated experiments, so that data acquisition and subsequent analysis may take several months. The value of the solution structure lies in the fact that it provides a starting point for the study of protein/substrate interactions, a knowledge of which may then aid drug design.

18.3.8.2. Quantiative Chemical Analysis by NMR [43]

NMR spectroscopy offers several important advantages for quantitative analysis over other techniques, including LC and GC chromatography. Firstly a single technique can be used to unambiguously confirm the identity of the components and quantify them. For chromatographic analysis spectroscopic techniques such as MS and NMR are generally required initially to identify components. Secondly and most importantly pure samples of the compounds of interest are not required to calibrate the response of the instrument in an NMR experiment. This is a result of the fact that, given due attention to the experimental NMR conditions, the integrated resonance intensity is directly proportional to the concentration of nuclei giving rise to the resonance. Relative concentrations can be obtained directly from relative resonance intensities and absolute concentrations by adding a known amount of another compound as an internal intensity standard. The nondestructive nature of the NMR experiment offers an additional advantage. On the negative side, the NMR experiment is inherently much less sensitive than other spectroscopic and chromatographic methods. Therefore the technique is rarely suitable for quantifying components present at very low concentrations even when the more sensitive nuclei such as ^{1}H, ^{19}F, or ^{31}P are used.

In a modern NMR laboratory, pulsed FT techniques are normally used for quantitative analysis. For such work careful selection of the experimental parameters is required to obtain accurate intensity relationships between the resonances in a spectrum. Whilst ^{1}H NMR spectra are normally integrated for structural identification work, the accuracy required to identify the relative numbers of different types of protons in a molecule is much less than for quantitative analysis.

For proton NMR the main consideration is ensuring complete relaxation between successive pulses for all the different types of hydrogen atoms present. This requires the interpulse delay to be at least 5 times the longest T_1. In addition, for nuclei such as ^{13}C, where broadband decoupling is usually required, the inverse gated technique (Section 18.3.4.2) should be used to prevent the occurrence of NOE effects. A further consideration in the case of spectra from nuclei such as ^{13}C and ^{19}F, which may have very wide spectral widths, is whether the RF pulse has sufficient power to irradiate all the nuclei equally effectively. The digital resolution and data processing requirements for a particular application also require careful selection.

The wide range of chemical, biochemical, and clinical applications include strength determinations, mixture analyses, polymer analyses, and reaction monitoring. Choice of nucleus obviously depends on the individual application. For organic chemical applications, the proton would normally be the preferred nucleus because of its relatively high sensitivity. However, proton spectra of mixtures are often highly congested and the greater dispersion afforded by ^{13}C NMR may make it the nucleus of choice. Two specialist applications are described briefly below.

Isotope Content Determination. A highly developed example of isotope content determination is the site-specific natural isotope fractionation (SNIF) NMR of deuterium [44]. The natural abundance of deuterium at different sites in a natural product can vary significantly depending on its biochemical origin. This pattern of natural abundances of deuterium is often quite characteristic of the source of the material, as is the case for alcohol in wine. By using a standard with a known deuterium content, ^{2}H NMR can be used to determine the deuteration levels in the alcohol. This data provides information about the origin of the wine and is a means of detecting watering down or artificial enrichment.

Enantiomeric Purity Determination. NMR has become an increasingly important tool for the determination of enantiomeric purity [45]. Many pharmaceuticals and agrochemicals are chiral, with only one enantiomer having the required effect. At best the other is inactive, at worst it may have undesirable properties. This situation has led to a surge in enantioselective synthesis, the products of which require analysis. Chromatographic methods which use chiral columns are usually used for quality control purposes with an established process, but at the research/development stage NMR is often the method of choice. In an achiral medium, enantiomers cannot be distinguished by NMR because their spectra are identical. Diastereoisomers, however, which have different physical properties, can be distinguished by NMR. Therefore the determination of enantiomeric purity by NMR requires the use of a chiral auxiliary that converts a mixture enantiomers into a mixture of diastereoisomers. Chiral lanthanide shift reagents [46] and chiral solvating agents form diastereomeric complexes in situ and can be used directly for NMR analysis. A third method involves reaction with chiral derivatising agents prior to NMR examination.

Lanthanide shift reagents (LSR) are *tris* (β-diketonate) complexes. Further complexation with organic compounds may result in large shifts in the NMR resonances of the latter. This is caused by the magnetic properties of the lanthanide ions. Compounds containing a wide range of functional groups, including alcohols, amines, and ketones, form complexes with LSRs. Prior to the widespread introduction of high-field instruments such materials were widely used to increase chemical shift dispersion in spectra. If an LSR chelate containing an optically pure ligand interacts with a pair of optical isomers, two diastereoisomers result which can be distinguished by NMR. Many of the ligands used in such reagents are derivatives of camphor.

Chiral solvating agents (CSA) form diastereoisomeric solvation complexes with solute enantiomers via rapidly reversible equilibria in competition with the bulk solvent.

Where quantitative chiral purity determinations are required, the minimum level of detection for the minor component and accuracy achieved depend on factors which include the resolution between the resonances from the diastereoisomers formed and the signal-to-noise ratio in the spectrum.

18.3.8.3. Rate Processes and NMR Spectra [47]

NMR is an important tool for the study of certain types of rate processes. In deciding whether a process is amenable to study by NMR the rate must be compared with the "NMR timescale," which refers to lifetimes of the order of 1 s to 10^{-6} s. An example of a process which can be studied in this way is rotation about the C–N bond in an amide such as *N,N*-dimethylacetamide (3).

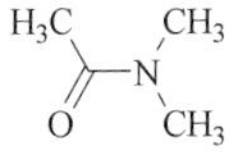

H_3C CH_3 N O CH_3

3

Rotation causes the two *N*-methyl groups to exchange positions. In the room-temperature ^1H NMR spectrum separate peaks are observed for the two *N*-methyl groups. If the temperature is raised, the rate of rotation increases and the *N*-methyl resonances broaden until a temperature is reached when the two peaks coalesce into a single broad resonance. This peak sharpens as the temperature is raised further until finally the fast exchange limit is reached, above which no further change is detected. The temperatures at which slow exchange, coalescence, and fast exchange are observed for a particular process depend on the rate of the process and the frequency difference between the exchanging lines. Thus at higher field coalescence will take place at a higher temperature. Analysis of such spectra enables exchange rates to be determined over a range of temperatures. Examples of other processes which have been studied in this way include keto–enol tautomerism, ring inversion, and proton-exchange equilibria.

18.3.8.4. NMR Methods Utilized in Combinatorial Chemistry and Biochemistry [110], [140]

Combinatorial chemistry provides a powerful means of rapidly generating the large numbers of structurally diverse compounds necessary for the biological screening required by drug discovery. Thus, combinatorial chemistry has become an indispensable tool in pharmaceutical research. The diversity of techniques being employed has caused the adaptation of a number of NMR methods for combinatorial chemistry purposes. These methods include the NMR analysis of reaction intermediates and products in solution or in the gel state and the analysis of ligands interacting with their receptors. High-throughput screening strategies were developed which involve the acquisition of 2D spectra of small organic molecules in a few minutes. Libraries of more than 200 000 compound can be tested in less than one month. There are many advantages of high-throughput NMR-based screening compared to conventional assays, such as the ability to identify high-affinity ligands for protein targets. This suggests that the method will be extremely useful for screening the large number of targets derived from genomics research [145].

Shuker et al. described a NMR method in which small organic molecules that bind to proximal subsites of a protein and produce high-affinity ligands are identified. This approach was called "SAR by NMR" because structure-activity relationships (SAR) are obtained here from NMR. This technique involves a series of 2D spectra of a labeled receptor protein in the presence and absence of potential ligands. The method reduces the amount of chemical syntheses and time required for the discovery of high-affinity ligands and appears particularly useful in target-directed drug research [141].

NMR has become established also as a valuable technique in clinical biochemistry where ^1H NMR in particular is used to study complex biochemical fluids such as plasma, urine, and bile. The major problem associated with such work is the removal or reduction of the very large water signal. One method involves freeze-drying the sample and redissolving it in deuterated water. NMR techniques are also available for selective suppression of the water signal. These methods, such as presaturation of the water, are required to reduce the water signal such that acceptable suppression can be attained from phase cycling. The same effect can be accomplished with magnetic field gradients alone because the removal of the undesirable resonances is accomplished in a single scan.

Under the conditions used, only low molecular mass constituents are observable, the signals from proteins and other large molecules are not resolved. Applications include screening and monitoring for metabolic diseases, the study of biochemical mechanisms associated with disease processes, and the identification of drug metabolites.

Translational diffusion measurements are a useful tool for studying supramolecular complexes and for characterizing the association state of molecules that aggregate at NMR concentrations. NMR spectra quality can provide a direct measure of the mobility of organic polymeric compounds and pulsed field gradients can be used to measure diffusion coefficients. Therefore, diffusion measurements can play an important role in the study of the properties and dynamics of the resins used in combinatorial chemistry. *Diffusion-ordered NMR* (DOSY) is another powerful tool for the analysis of complex mixtures. It allows to resolve the NMR signals of discrete compounds in a mixture on basis of variance of their molecular diffusion coefficients [142].

One of the most powerful methods for performing analysis of complex mixtures represents the redevelopment of HPLC-NMR coupling techniques [143]. HPLC-NMR has been widely applied in drug metabolism and also used for combinatorial chemistry. Data collection usually occurs with the system operating in either on-flow or stopped-flow mode. In the latter, up to three dozen fractions may be collected within storage loops. If the differences in the retention time of individual compounds are smaller than the required NMR measuring time, the storage loop acts like a sample changer.

18.4. NMR of Solids and Heterogeneous Systems [48]

Solids give very broad NMR signals devoid of the fine structure which provides detailed information about chemical structures. The restriction of NMR spectroscopy to the liquid, solution, or gas phase, however, imposes considerable limitations on its applicability since some compounds are insoluble or have very low solubility. Moreover, a compound may experience a change in molecular structure on dissolution (e.g., by tautomerism). Studying samples as solids should provide some insight about how molecules pack as well as about molecular structure. Since the 1970s developments in experimental design and spectrometer hardware have made it possible in some situations to obtain spectra from solids in which the resolution approaches that obtainable in a solution experiment.

This section provides a brief explanation of the origins of the broadening effects, followed by an overview of the techniques employed to obtain high-resolution NMR spectra from solids and examples of their applications. Only spin 1/2 nuclei are considered. There are two principal factors which cause broadening in the NMR spectra of spin 1/2 nuclei obtained from solid samples. The first is chemical shift anisotropy: the chemical shifts are orientation dependent and the different crystallites in a sample have a range of orientations to the applied magnetic field. Broadening resulting from dipole – dipole interactions is the second factor. Both of these interactions are present in solution but are usually averaged out by molecular tumbling. A further problem results from the fact that nuclei in solids generally have very long relaxation times. This means that they take a long time to relax back to the equilibrium magnetization, which in turn affects the amount of time required to obtain a spectrum with a reasonable signal-to-noise ratio.

18.4.1. High-Resolution NMR of Solids

The study of "dilute" nuclear spin systems such as ^{13}C has been more amenable to obtaining high-resolution solid-state spectra than that of abundant nuclei such as the proton. The reason for this is that the problems associated with homonuclear dipolar interactions are essentially eliminated in a dilute spin system. Dilution may be the result of low natural abundance, as in the case of ^{13}C, or of abundant isotopes present in low concentration (e.g., ^{31}P in organophosphates). In either case homonuclear dipolar interactions are negligible, since dipolar interaction has a $1/r^3$ dependence. Heteronuclear dipolar interactions, usually to protons, remain but these can be removed effectively by the use of high-power proton decoupling. A technique known as magic angle spinning (MAS) is used to remove the effects of chemical shift anisotropy. This involves spinning the sample at an angle of 54°44′ to the applied magnetic field. To be effective the rate of rotation must be comparable to the frequency range being observed (i.e., several kilohertz). To overcome the problems associated with low sensitivity a technique known as cross-polarization (CP) is used. This involves the transfer of magnetization from highly polarizable abundant spins such as protons to dilute spins such as ^{13}C.

The combined application of CP, MAS, and high-power proton decoupling can give high-resolution spectra from powders and amorphous solids, comparable in quality to those obtained in solution. The ^{13}C, ^{29}Si, ^{15}N, and ^{31}P nuclei have been widely studied in this way in the last 20 years [128], [146], [147] . There are two principal reasons for obtaining a high-resolution NMR spectrum from a solid. Either it is insoluble or information is being sought about the solid-state structure. In the latter case crystallographic information is sought either because a single-crystal X-ray structure is not available or to enable a comparison of solid and solution state structures/conformations to be made when spectra are available in both phases. Crystallographic effects frequently give rise to splitting of lines in spectra of solids

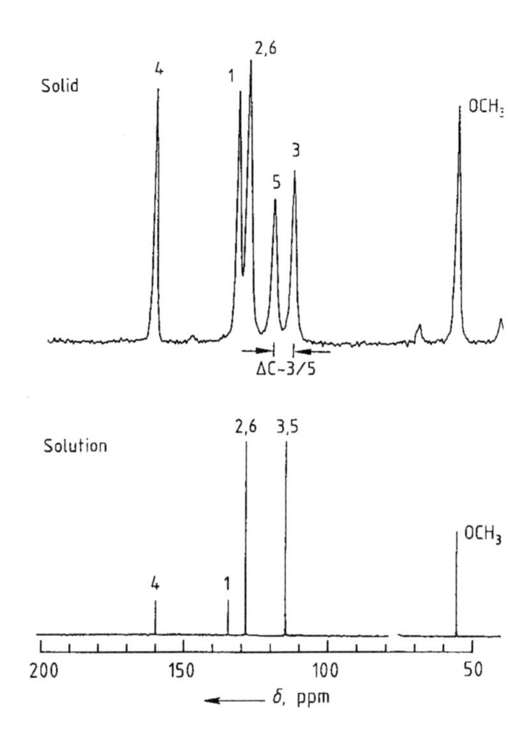

Figure 30. ^{13}C NMR spectra of 4,4′-dimethoxybiphenyl (**4**)

when compared with those obtained in solution. For a particular type of atom to yield a single line in a spectrum obtained from a solid, all the carbons of that type must be related by symmetry in the crystal. Nonequivalence is not an uncommon occurrence and may be intramolecular or intermolecular. Figure 30 shows the solid-state and solution spectra of 4,4′-dimethoxybiphenyl (**4**). In solution a single peak is observed for C-3 and C-5, whereas for the solid two peaks separated by 6.4 ppm are present. This is attributed [49] to the fact that in the solid the compound exists in the locked conformation (**4**) whereas in solution rotation about the C-4–O bond causes an average shift to be observed.

4

Solid-state NMR spectroscopy, particularly ^{13}C, has been widely used to study polymorphism since polymorphs invariably give spectra which are discernably different.

18.4.2. Low Resolution ^1H NMR of Heterogeneous Systems

Thus far the discussion has assumed a requirement for the highest resolution NMR spectra. However, there are important analytical applications where it is not necessary to resolve the signals from all the different hydrogen atoms in a sample. As in the case of high-resolution NMR, early instruments used CW methodology. Pulsed NMR, which is much better suited to and more versatile for this type of measurement, is now used. After excitation the bulk magnetization of a sample is studied. The initial amplitude of the signal is proportional to the total number of hydrogen nuclei in the sample. Each component of the signal decays at a different rate, so, for example, signals from hydrogen atoms in solid phases decay far more rapidly than those from liquids. There are three types of measurements on such instruments. For many years the methodology has been used to measure solid/liquid ratios (e.g., to determine the solid fat content in margarine). An absolute measurement method is also available where the absolute quantity of protons in a sample can be measured and expressed as a percentage by weight. The third more sophisticated method involves the measurement of bulk relaxation times, which can be correlated with a property of the sample. For instance the decrease in the spin – spin relaxation time T_2 has been used to follow the styrene polymerization reaction [50].

18.5. NMR Imaging [148] – [151]

Since 1970 a technique known as magnetic resonance imaging (MRI) has been developed which has revolutionized diagnostic medicine. The technique produces an NMR picture or image. In addition to the NMR methodology described above, MRI also uses magnetic field gradients in the x, y, and z directions to make the resonance frequency a function of the spatial origin of the signal. First, a slice at height z and thickness dz of the object is selected. This is done by means of a selective RF pulse, combined with a field gradient in z direction. Next, the spatial coding for the x direction is obtained by means of an x-gradient. Finally, the spatial coding for the second direction, i.e., the y direction, is obtained by means of a y-gradient. In medical imaging the protons in water are usually detected. While the water content in different types of tissue may show little variation, the T_1 and T_2 relaxation times are different. Pulse sequences are therefore used which produce contrast that reflects the different relaxation times. The technique has become particularly important in the diagnosis of cancer, since cancerous tissue has a longer relaxation time than healthy tissue. In recent years this method acquired special importance in neurochemistry [152].

Equipment is available for a range of applications from whole human body imagers operating at field strengths of up to ca. 8 T to equipment for imaging small samples up to 11.7 T. The achievable resolution depends on the magnitude and duration of the field gradients applied. For small objects the best resolution which has been achieved is of the order of 5 – 10 μm. For whole body in-vivo applications it is much lower.

The techniques developed for biomedical applications to study the liquid state have also been used in materials science. Such applications include the study of solvents ingressed into polymers, liquids absorbed onto porous media, and polymerization reactions [51], all of which can be studied as a function of space and time.

18.6. ESR Spectroscopy

Electron spin resonance (ESR) spectroscopy is also known as electron paramagnetic resonance (EPR) spectroscopy or electron magnetic resonance (EMR) spectroscopy. The main requirement for observation of an ESR response is the presence of unpaired electrons. Organic and inorganic free radicals and many transition metal compounds fulfil this condition, as do electronic triplet state molecules and biradicals, semicon-ductor impurities, electrons in unfilled conduction bands, and electrons trapped in radiation-damaged sites and crystal defect sites.

The principles of ESR and general applications are covered in various monographs [52] – [57]. Reviews of technique developments and applications have been published [58], [59]; industrial applications have been reviewed [60]; and dosimetry [61] and biological applications described [62] – [66].

18.6.1. The ESR Experiment

The resonance condition for ESR, for a system having a spin value of 1/2, is:

$$\Delta E = h\nu = g\beta B \tag{3}$$

where ΔE is the separation of energy levels produced by the application of an external magnetic field B, and β is the Bohr magneton (9.274×10^{24} J/K). Most ESR spectrometers operate at a fixed frequency ν and record an ESR spectrum by sweeping the external field B. The most convenient frequency is ca. 9 GHz, which is in the microwave region. It corresponds to a wavelength of ca. 3 cm and is known as X-band. Spectrometers have been built which operate at other microwave frequencies: L-band (1.0 – 1.2 GHz), S-band (3.8 – 4.2 GHz), K-band (24 GHz), and Q-band (34 GHz); and some even operate at a few megahertz in the earth's field. The lower frequencies are used when “lossy” samples, that is those with high dielectric constants such as aqueous or biological samples, are examined. The higher frequencies are used when greater dispersion is required, e.g., when improved separation of anisotropic g components in solid samples is required to reduce overlap of lines. Unfortunately, at the higher frequencies the sample size is restricted and noisier spectra often result.

18.6.1.1. Continuous Wave ESR

The basic components of a continuous wave (CW) ESR spectrometer are a frequency source (klystron or Gunn diode), an electromagnet (having a field of ca. 350 mT for X-band) and a resonant cavity (or loop gap resonator) in which the sample is placed (Fig. 31). To improve the signal-to-noise ratio, it is customary to modulate the microwave frequency at 100 kHz (other frequencies are also used) and to detect at this frequency by using a phase sensitive detector. This results in the usual recorder tracing from the spectrometer being the first-derivative instead of absorption as found in NMR spectrometers. A simple ESR spectrum is shown in Figure 32.

18.6.1.2. ENDOR and Triple Resonance

Electron – nuclear double resonance (ENDOR) offers enhanced resolution compared with conventional ESR. This is achieved mainly because the technique is ESR-detected NMR spectroscopy.

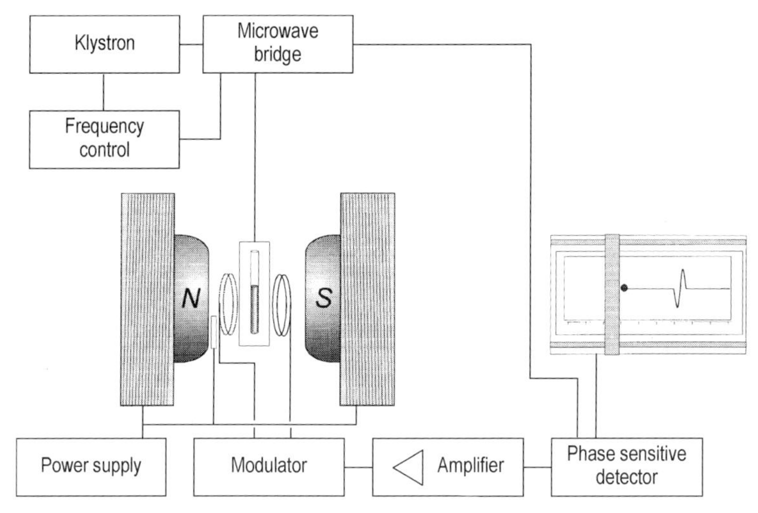

Figure 31. Block diagram of a CW ESR spectrometer

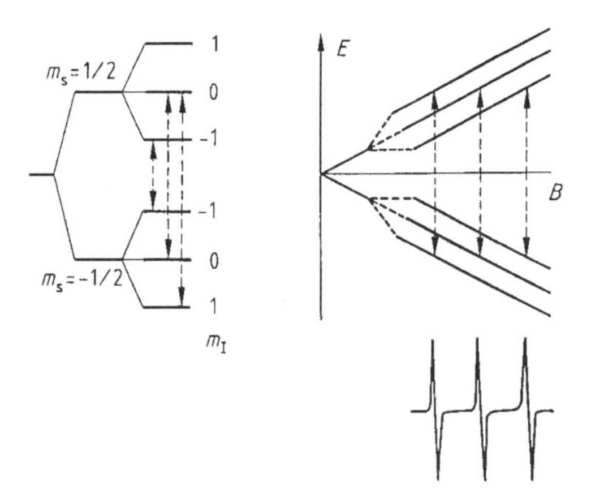

Figure 32. Simple ESR spectrum
P. L. Nordio in [64] (with permission)

Another advantage is that only two ENDOR lines arise from each group of equivalent nuclei, resulting in a simplification in the spectra when compared with ESR. ENDOR has been applied to the study of organic [67] and inorganic radicals and transition metal complexes [62], [68] in both the solid (powdered and crystalline) and liquid states.

The ENDOR experiment involves setting the spectrometer on a chosen line in the ESR spectrum and then increasing the microwave power to saturate the transition while sweeping the selected NMR frequency.

Two NMR transitions are observed at the frequencies:

$$\nu = \|\nu_n \pm a/2\| \qquad (4)$$

where ν_n is the NMR frequency for the applied magnetic field of the ESR spectrometer and a is the ESR hyperfine coupling constant. Equation (4) applies when $\nu_n > |a/2|$; however, when $\nu_n < |a/2|$, the two lines are centred at $|a/2|$ and are separated by $2\,\nu_n$.

Other experiments performed with an ENDOR spectrometer include special and general electron-nuclear-electron triple resonance, where three frequencies are employed. In the special triple experiment, two NMR transitions belonging to one set of equivalent nuclei are irradiated. This results in improved spectral resolution compared with ordinary ENDOR. In the general triple experiment, NMR transitions belonging to different sets of nuclei are irradiated. This allows the relative signs of the hyperfine coupling constants to be determined. As an example the ENDOR spectrum, and the corresponding ESR, general triple and special triple spectra, of bacteriochlorophyll a (**5**) are shown in Figure 33.

Apart from the improvement in resolution already described, ENDOR spectroscopy has several additional advantages over CW ESR, namely:

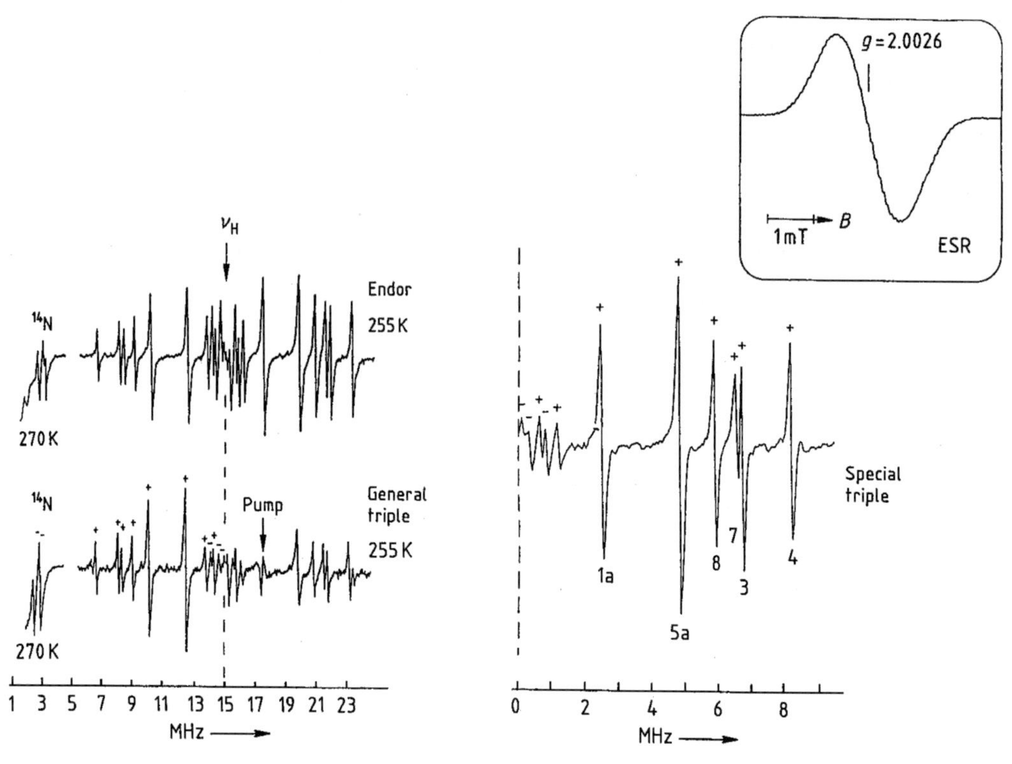

Figure 33. Typical ESR, ENDOR, general triple and special triple spectra of bacteriochlorophyll **a** (**5**) cation radical (iodine oxidation in 6 : 1 CH_2Cl_2 : CH_3OH)
K. Mobius, W. Lubitz and M. Plato in [66] (with permission)

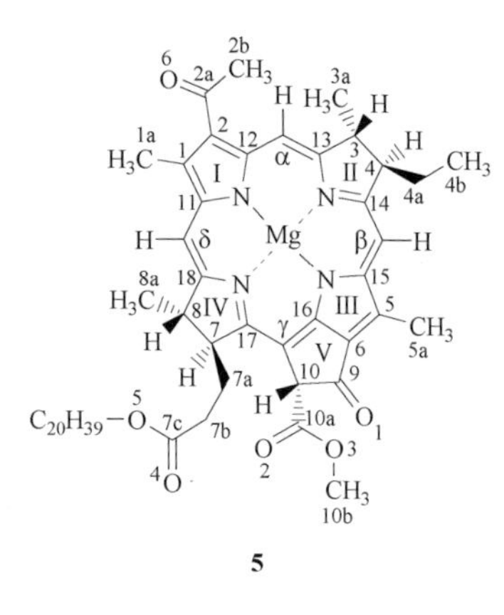

1) Simplification of the spectrum when several equivalent nuclei are present
2) Direct measurement of hyperfine splittings
3) Unambiguous assignment of the hyperfine splittings of magnetic nuclei since the spectrometer is set at the appropriate NMR frequency for that particular nucleus
4) Direct determination of quadrupolar splittings
5) Determination of relative signs of hyperfine and quadrupolar interactions
6) Spectra of individual species in a mixture can be determined
7) Very small hyperfine coupling constants can be measured

18.6.1.3. Pulse ESR [69] – [73]

Although the basic principles of ESR and NMR are similar, practical difficulties mean that pulse methods are less useful in ESR. This is because the pulse power required to produce the frequency span of a typical ESR spectrum would be several kilowatts and the pulse very short (nanoseconds). The pulse or FT (Fourier transform) ESR spectrometer is usually based on a standard CW instrument because it is often useful to record a standard ESR spectrum before carrying out pulse experiments. The pulse microwave source is usually a travelling wave tube; other essentials include a travelling wave tube amplifier, a programmable pulse generator, and fast digitizer.

SCHWEIGER [72] has reviewed pulse ESR and pulse ENDOR methods and HOFF [66] includes many recent applications. Echo methods are similar to those used in NMR. The electron spin echo (ESE) method involves the application of two or three microwave pulses to a spin system in a constant magnetic field. The ESE signal decay due to relaxation is accompanied by periodic rises and falls in the echo amplitude. These modulation effects depend on electron–nuclear and electron–electron couplings in the spin system, and are termed electron spin echo envelope modulation (ESEEM) [70], [71]. ESEEM is used to investigate hyperfine interactions in magnetically diluted solids, and most experiments are performed at temperatures between 4 and 77 K. ESEEM has been applied to the study of surface complexes, coordination of water in metmyoglobin, structures of ligands in copper proteins, and magnetic properties of electronic triplets and solitons in polyacetylene. Pulses and detection need to be on a very short time scale since a typical spin–lattice relaxation time is about 1 μs.

Spin dynamic studies, including saturation and inversion recovery, Hahn echo and stimulated echo decay, as well as Carr–Purcell–Meiboom–Gill sequences, can be performed, yielding relaxation times. When rapid reaction kinetics are being measured, information is also gained on transient phenomena such as chemically induced electron polarization (CIDEP).

Pulsed ENDOR has several advantages over conventional ENDOR for solids [72], [73]. This means that pulsed ENDOR and ESE–ENDOR can be used at any temperature, provided that a spin echo can be detected.

18.6.1.4. ESR Imaging

ESR imaging [74] is used for both biological and materials research and, as with the more commonly used NMR imaging, it is based on the application of field gradients in the *x, y,* and *z* directions, to allow volume elements to be selectively studied. Large field gradients are required because of the wide field range of the ESR spectrum, typically ca. 3 mT.

Imaging has been performed at various irradiation frequencies. X-band (9 GHz) imaging is limited to samples with a maximum diameter of 10 mm and is unsuitable for most biological samples because of their high water content. Larger or biological samples are usually examined at lower frequency. Small-animal studies have been performed at L-band (1.0–1.2 GHz), and the use of even lower frequencies permits samples as large as 100 mm diameter to be examined.

ESR microscopy (at X-band) has been applied to a variety of materials including in situ studies of the oxidation of coals; defects in diamonds; radiometric dosimeters; polymer swelling and to the diffusion of oxygen in model biological systems. The resolution of ESR imaging depends on the field gradient and line width, and the best resolution to date from an X-band imager is 1 μm.

18.6.2. Spectral Parameters

18.6.2.1. g-Factor

The ESR g-factor is also known as the Landé g-factor or spectroscopic splitting factor and depends on the resonance condition for ESR (Eq. 3) and is independent of both applied field and frequency. The g-factor of a free electron g_e is 2.002322, while the g-factors of organic free radicals, defect centers, transition metals, etc. depend on their electronic structure. The g-factors for free radicals are close to the free electron value but may vary from 0 to 9 for transition metal compounds. The most comprehensive compilations of g-factors are those published in [75], [76]. The magnetic moments and hence g-factors of nuclei in crystalline and molecular environments are anisotropic, that is the g-factor (and hyperfine interactions) depend on the orientation of the sample. In general, three principal g-factors are encountered whose orientation dependence is given by:

$$g^2 = (g_{xx})^2(l_{xx})^2 + (g_{yy})^2(l_{yy})^2 + (g_{zz})^2(l_{zz})^2 \tag{5}$$

where l_{xx}, l_{yy}, and l_{zz} are the direction cosines between the direction of B and the principal g-factors, g_{xx}, g_{yy}, and g_{zz}. When the site shows threefold or fourfold symmetry only two g-factors, $g_\parallel$ (g-parallel) and $g_\perp$ (g-perpendicular) are observed, where $g_\perp = (g_{xx} + g_{yy})/2$. For solutions, free molecular tumbling leads to averaging and the isotropic g-factor g_{iso} is observed:

$$g_{iso} = 1/3\,(g_{xx} + g_{yy} + g_{zz}) \tag{6}$$

or

$$g_{iso} = 1/3\left(g_\parallel + 2g_\perp\right) \tag{7}$$

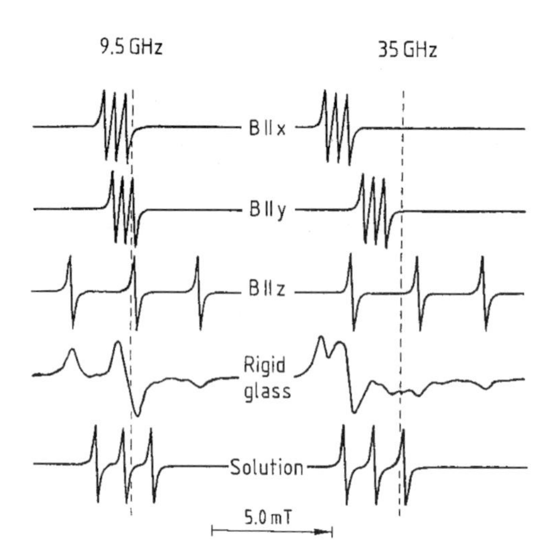

Figure 34. Simple crystal, powder and solution spectrum of a typical aminoxyl radical such as TEMPO (2,2,6,6-tetramethylpiperidine-*N*-oxyl), (**6**).
O. Hayes Griffith, P. Jost in [64] (with permission)

Figure 34 shows the single crystal, powder and solution spectra expected for an aminoxyl free radical (**6**): this sample has anisotropic g and hyperfine interactions (see Section 18.6.2.2).

6

18.6.2.2. Nuclear Hyperfine Interaction

If the interaction of an electron with a magnetic field were the only effect operative, then all ESR spectra of free radicals would consist of one line. When the nuclear spin quantum number I is nonzero a nuclear hyperfine interaction $\boldsymbol{A}$ is observed. When several equivalent nuclei are present (e.g., $\cdot CH_3$, $\cdot C_2H_5$), the number of lines in the spectrum is given by $\Pi_i (2 n_i I_i + 1)$, where n is the number of magnetically equivalent nuclei i.

The isotropic, or Fermi, hyperfine interaction a arises from the presence of nonzero electron density at the nucleus. Interaction between electrons and nuclear dipoles gives rise to the anisotropic, or dipolar, hyperfine coupling $\boldsymbol{T}$. This interaction is orientation dependent, but averages to zero in freely tumbling solution. The total hyperfine interaction $\boldsymbol{A}$ is the sum of a and $\boldsymbol{T}$, where $\boldsymbol{A}$ and $\boldsymbol{T}$ are tensors.

The orientation dependence is given by:

$$A^2 g^2 = (A_{xx})^2 (g_{xx})^2 (l_{xx})^2 + (A_{yy})^2 (g_{yy})^2 (l_{yy})^2 + (A_{zz})^2 (g_{zz})^2 (l_{zz})^2 \quad (8)$$

where A_{xx}, A_{yy}, and A_{zz} are the principal hyperfine constants. As with the g-factor, an axially symmetric spectrum with only $A_{\parallel}$ and $A_{\perp}$ is often observed.

18.6.2.3. Quantitative Measurements

g-Factor and Hyperfine Splitting. The determination of g-factor and hyperfine splitting depends on accurate determination of the external magnetic field at the sample. This can be done directly by using a NMR magnetometer. The microwave frequency is measured directly with a frequency counter, and the g-factor can then be found from Equation (3). Alternatively, the g-factor and magnetic field sweep of the spectrometer can be determined by comparison with secondary standards. Several authors list suitable materials [52], [77].

Spin Concentration. To determine the electron spin concentration it is essential to obtain, as accurately as possible, the intensity K of the ESR response. Preferably, this is done by double integration of the first derivative spectrum using ESR data acquired in a computer; however, if this is not available or if there is considerable overlap with lines from a second species then the relation:

$$K = hw^2 \quad (9)$$

is used, where h is the peak-to-peak amplitude and w is the peak-to-peak line width. The latter relationship works satisfactorily if relative intensities are required for signals with similar line shapes. Generally, concentration determinations are made relative to a concentration standard. Some of the points which must be considered in concentration determinations including the choice of concentration standard are noted in [52], [77], [78]. Examples of ESR concentration standards include solutions of α,α'-diphenylpicrylhydrazyl (DPPH), potassium peroxylamine disulfonate ($K_2NO(SO_3)_2$), nitroxide radicals, copper sulfate pentahydrate ($CuSO_4 \cdot 5\,H_2O$), and manganese sulfate monohydrate ($MnSO_4 \cdot H_2O$). Solid standards include single crystals of ruby and $CuSO_4 \cdot 5\,H_2O$, or powders of weak and strong pitch in potassium

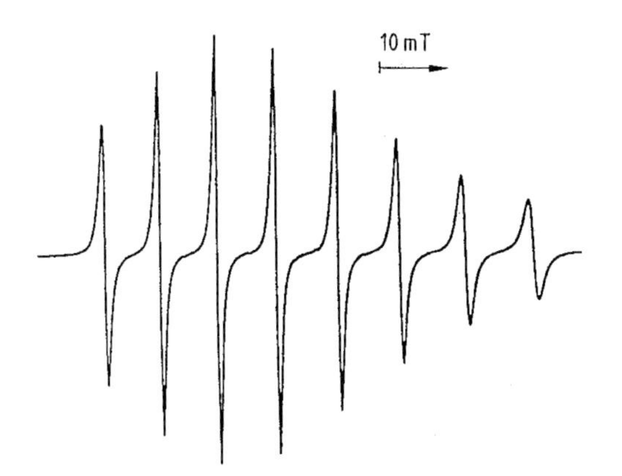

Figure 35. ESR spectrum of vanadyl acetylacetonate (**7**) in 1 : 1 toluene : chloroform, showing asymmetric line broadening

chloride; DPPH; $MnSO_4 \cdot H_2O$; F-centers in alkyl halides and defects in magnesium oxide. Some of the latter are suitable as g-factor standards also.

For magnetically dilute samples the concentration of the unknown C_X is given by:

$$C_x = C_{std} K_x (g_{std})^2 [S(S+1)]_{std} / K_{std} (g_x)^2 [S(S+1)]_x \quad (10)$$

where S is the total spin (e.g., $S = 1/2$ for organic free radicals, 5/2 for manganese(II) which has five d electrons), and the unknown (X) and standard samples (std) have been examined under identical conditions (temperature, spectrometer gain, modulation and scan range), without microwave power saturation. For neat materials the following applies:

$$M_p = g^2 \beta^2 N_0 S(S+1) B_r / 3k(T-\theta) \quad (11)$$

where M_p is the magnetization of a nondilute paramagnetic material, β is the Bohr magneton, N_0 is the Avogadro number, B_r is the applied field at resonance, k is the Boltzmann constant, T is the absolute temperature, and θ is the Curie temperature.

18.6.3. ESR in the Liquid State

In solution, molecular motion leads to a simplification of the spectra due to the averaging of anisotropic interactions to zero, resulting in the observation of isotropic spectra.

There have been many studies on organic and inorganic free radicals in solution, including their role as reaction intermediates, mechanistic studies, and studies of molecular dynamics.

18.6.3.1. Slow Molecular Tumbling

When molecular tumbling leads to incomplete averaging, perhaps because of high solution viscosity, variations are observed in the line width of the hyperfine lines. This variation is shown in Figure 35 for vanadyl acetylacetonate (**7**) dissolved in 1 : 1 toluene : chloroform at room temperature.

7

Asymmetric line broadening is commonly observed for aminoxyl radicals, which are therefore used in a wide variety of spin label and spin probe studies (see Section 18.6.5.2).

The variation of line width $1/T_2$ with m_I has been studied extensively [79], [80]. For isotopic tumbling with correlation times of 10^{-11} to 10^{-9} s the following expression is generally used:

$$[T_2(m_I)]^{-1} = A + B m_I + C m_I^2 + D m_I^3 \quad (12)$$

The parameters A, B, C, and D depend on the sign and magnitudes of the g and hyperfine anisotropies and upon the rate of molecular reorientations in the liquid. D is usually small compared with B and C and only terms up to m_I^2 need normally be retained.

In Equation (12) the A term includes broadening contributions arising from instrumental factors and mechanisms which are independent of the magnetic anisotropies. The term in m_I progressively broadens the line across the spectrum, transitions with the largest negative m_I value being broadened the least.

The term in m_I^2 broadens the lines at the extremities of the spectrum more than those in the middle, but its effect is symmetrical. The B and C parameters are given by:

$$B = 4/15 (b \Delta\gamma B_0 \tau_c) \quad (13)$$

$$C = b^2 \tau_c / 8 \quad (14)$$

where B_0 is the operating field of the spectrometer,

$$b = 4\pi\left[A_{zz} - \left(A_{xx} + A_{yy}\right)/2\right]/3 \qquad (15)$$

$$\Delta\gamma = -\beta_e h^{-1}\left[g_{zz} - \left(g_{xx} + g_{yy}\right)/2\right] \qquad (16)$$

and τ_c is the rotational correlation time.

According to the Debye diffusion model, a measure of the length of time a molecule remains in a given orientation τ_c can be calculated from the expression:

$$\tau_c = 4\pi\eta r^3/3kT \qquad (17)$$

where r is the spherical radius of the molecule and η is the viscosity of the solvent. This method has been used to estimate radical radii and the effects of hindered environments [81].

Correlation times with time scales of 10^{-11} to 10^{-9} s are determined by the above method and are said to be in the fast tumbling region. In the slow tumbling region (10^{-9} to 10^{-7} s) more sophisticated theoretical approaches [79], [80] are used. The very slow region (10^{-6} to 10^{-3} s) cannot be studied by conventional ESR and requires special techniques such as saturation transfer spectroscopy (see Section 18.6.5.4).

18.6.3.2. Exchange Processes

Generally, it is assumed that the electrons in free radicals and paramagnetic materials are completely independent and noninteracting. Several effects become important when electron spins interact magnetically and chemically with each other and with their environment.

Electron spin exchange is a bimolecular reaction in which the unpaired electrons of two free radicals exchange their spin states. At low concentration (ca. 10^{-3} mol/L) the usual isotropic spectrum is observed. As the radical concentration is increased the lines gradually broaden until at high concentration ($> 10^{-1}$ mol/L) the lines coalesce to a single line. This single line sharpens further as the concentration is increased.

The spectrum is said to be exchange narrowed since the electron spins are exchanging so fast that the time average of the hyperfine field is close to zero. Electron spin exchange must be avoided if narrow lines are desired in solution and this is readily achieved if the concentration is maintained below ca. 10^{-3} mol/L.

Exchange narrowed spectra (where hyperfine splittings are coalesced) are observed for many pure solid free radicals, such as DPPH and $CuSO_4 \cdot 5\,H_2O$. For this reason, single crystal spectra are obtained by incorporating about 1 % of the paramagnetic compound into an isomorphous diamagnetic host crystal.

18.6.4. Computer Simulation of Spectra

Isotropic Spectra. Numerous computer programs have been written for the simulation of solution state (isotropic) ESR spectra. Often these include the facility to simulate spectra from mixtures of radicals and to allow for contributions from satellite lines arising from the presence of low-abundance magnetic nuclei such as ^{13}C, ^{15}N, and ^{33}S.

Time-dependent phenomena, including the alternating line width effect [52], [82] and asymmetric line broadening [79], [80], require simulated spectra to obtain the ESR parameters and rates.

Anisotropic Spectra. Programs are available for the simulation of single crystal, partially oriented, and powder simulation (sometimes in the same program) and in several cases allowance for noncoincidence of the axes of the $\boldsymbol{g}$ and $\boldsymbol{A}$ tensors can be included. Various programs are described in detail in [53], [56], [57].

18.6.5. Specialist Techniques

18.6.5.1. Spin Trapping

A limitation to the application of ESR spectroscopy to the study of short-lived transient radicals is the difficulty of producing sufficient free radicals for direct observation. Spin trapping provides a simple and efficient means of overcoming this problem [83], [84].

A diamagnetic compound (spin trap) is introduced into the radical producing system to give a relatively stable ESR-observable free radical (spin adduct). Typical spin traps are nitrone compounds such as phenyl-*N*-*tert*-butylnitrone (PBN) (**8**) and 5,5-dimethyl-1-pyrroline-*N*-oxide (DMPO) (**9**) and nitroso compounds such as 2-methyl-2-nitrosopropane (MNP) (**10**) and 2,4,6-tri-*tert*-butylnitrosobenzene (TBN) (**11**)

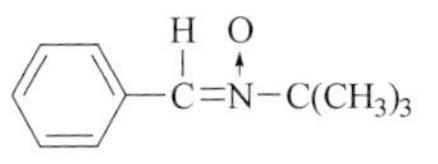

$C_6H_5{-}CH{=}N(\rightarrow O){-}C(CH_3)_3$

PBN (**8**)

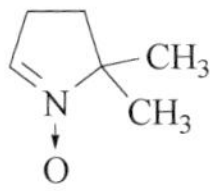

DMPO (**9**)

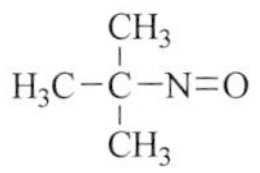

$(CH_3)_3C{-}N{=}O$

MNP (**10**)

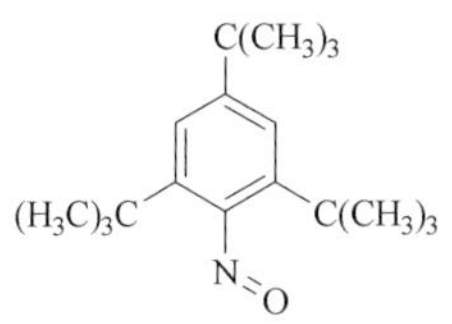

TBN (**11**)

Spin trapping of a radical X with nitrone and nitroso compounds occurs as follows:

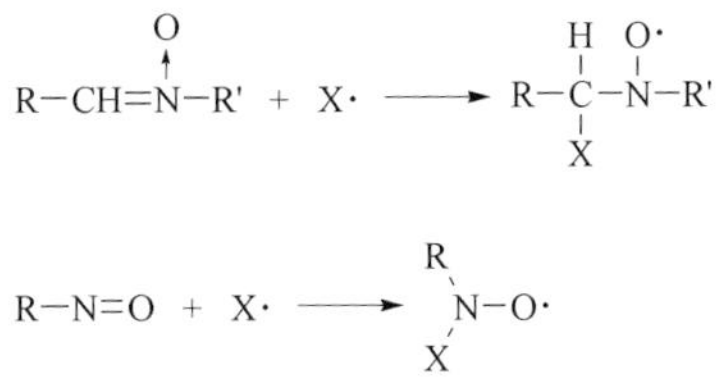

$$R{-}CH{=}\overset{O}{\overset{\uparrow}{N}}{-}R' + X\cdot \longrightarrow R{-}\underset{X}{\overset{H}{C}}{-}\overset{O\cdot}{N}{-}R'$$

$$R{-}N{=}O + X\cdot \longrightarrow \begin{matrix} R \\ \;\; N{-}O\cdot \\ X \end{matrix}$$

In favorable cases, the original free radical can be identified from the g-factor and hyperfine coupling constants of the spin adduct.

When a methyl radical is trapped by MNP (**10**) the resulting ESR spectrum is a 1 : 1 : 1 triplet of 1 : 3 : 3 : 1 quartets. However, if PBN (**8**) is used, the hyperfine structure is not indicative of the attached radical and its identity must be deduced from hyperfine splittings originating from the nitrogen and lone proton already present in PBN.

The main use of spin trapping is in identifying radical intermediates in organic and inorganic reactions. The reactions can be thermally, electrochemically, or radiation induced. Probably the most important application of spin trapping is the study of radicals in biological systems.

18.6.5.2. Spin Labeling [64], [65], [85]

Spin labeling is a powerful means of studying biochemical and biophysical characteristics of lipids, proteins, and amino acids. Many investigations have been devoted to the synthesis of radicals having a suitable reactive functional group for coupling selectively to the biological system whilst causing minimum perturbation of biological activity.

In spin labeling, a stable free radical is chemically attached to a macromolecule, and the increase in molecular mass of the radical reduces its rate of tumbling, which can be easily monitored by ESR spectroscopy, as described in Section 18.6.3.1. The most suitable class of free radicals for the purpose are aminoxyls (also referred to as nitroxides and nitroxyls) because of their stability and because their anisotropic $\boldsymbol{g}$ and $\boldsymbol{A}$ tensors are sufficiently different in the x, y, and z directions to provide high sensitivity to a change in the rate of molecular tumbling. Rates of tumbling can be measured by the change from a completely isotropic spectrum (rates of about $10^{10}\ s^{-1}$) to a near-powder spectrum (rates of about $10^{6}\ s^{-1}$).

Thousands of aminoxyl spin labels have been synthesized and some examples are given below (compounds **12**–**15**) [85].

A technique related to spin labeling is that of using stable free radicals as spin probes to monitor molecular motions. In this technique a radical, usually an aminoxyl, is synthesized that mimics as closely as possible the structure of the molecule under investigation. The major difference compared with spin labeling is that the radical is not chemically attached to the molecule under investigation.

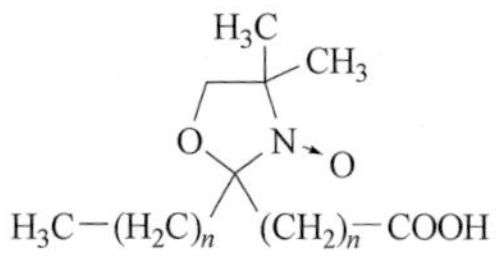

H_3C CH_3 O N O $H_3C-(H_2C)_n$ $(CH_2)_n-COOH$

12

H_3C CH_3 O N O O O R' $H_3C-(H_2C)_n$ $(CH_2)_n$ O O O O–P–O–X O⁻

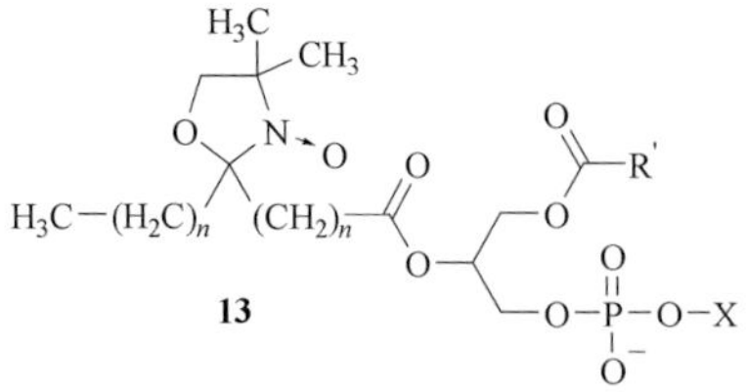

13

X = $(CH_2)_2\overset{+}{N}(CH_3)_3$ (phosphatidylcholine)
X = $(CH_2)_2\overset{+}{N}H_3$ (phosphatidylethanolamine)
X = $CH_2CH(OH)CH_2OH$ (phosphatidylglycerol)
X = $CH_2CH(\overset{+}{N}H_3)COO^-$ (phosphatidylserine)
X = H (phosphatidic acid)

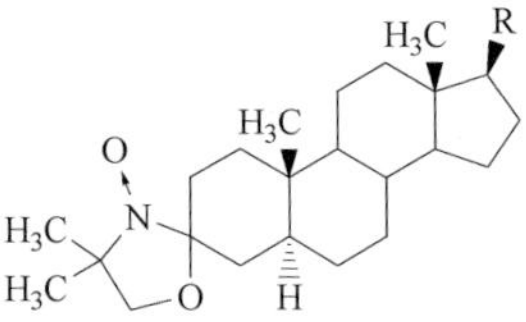

H_3C R H_3C O N H_3C H_3C O H

14

R = CH_3 CH_3 CH_3 (cholestane derivative)
R = —OH (androstanol derivative)

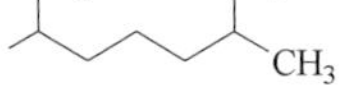

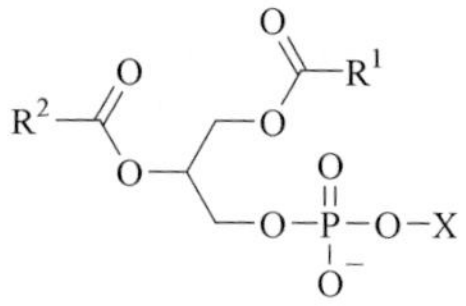

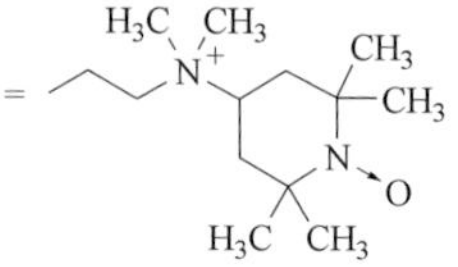

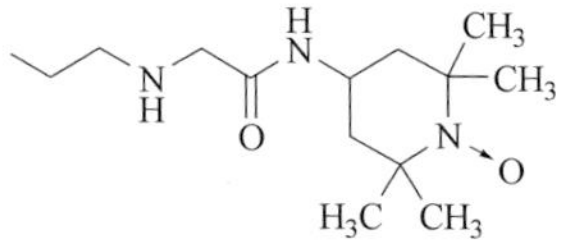

O R^1 R^2 O O O O O O–P–O–X O^-

15

X = H_3C CH_3 N^+ CH_3 CH_3 N O H_3C CH_3

X = N H H N O CH_3 CH_3 N O H_3C CH_3

X = COO^- O N H CH_3 CH_3 H_3C N O H_3C

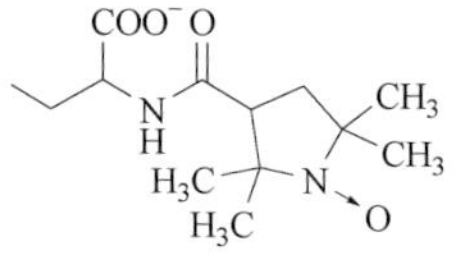

18.6.5.3. Oximetry

Molecular oxygen is paramagnetic and this property is used to provide an indirect method for the detection and measurement of oxygen by oximetry [86]. Bimolecular collisions of oxygen with free radicals results in broadening of the ESR spectrum. By using radicals with sharp lines and measuring the extent of broadening produced by oxygen it is possible to determine the oxygen concentration. In principle, dissolved oxygen concentrations as low as 10^{-7} mol/L can be rapidly measured.

Probably the most important application of this technique is the measurement of oxygen concentrations in cells and animals, both in vivo and in vitro [87]. Aminoxyl radicals can be used, the most commonly used radical being 3-carboxamido-2,2,5,5-tetramethylpyrrolidine-*N*-oxyl, CTPO (**16**) and its deuterated analogue 3-carbamido-2,2,5,5-tetra-[$^2H_3$]methylpyrrolidine-*N*-oxyl:

O NH_2 H_3C CH_3 H_3C N CH_3 O

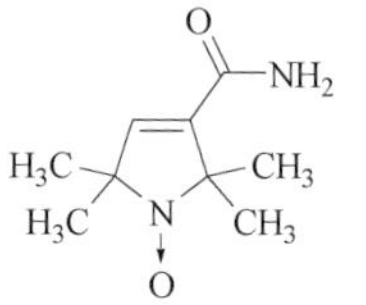

16

Another approach is to use a solid which is pervious to oxygen and whose ESR spectrum has a single very sharp line. This can be achieved with a small crystal of lithium phthalocyanine, which, in the absence of oxygen has a peak-to-peak line width of less than 0.002 mT [88]. A less good but cheaper material is a type of coal called fusinite.

18.6.5.4. Saturation Transfer

Saturation transfer is a technique that is sensitive to very slow motions with correlation times from 10^{-6} to 10^{-3} s — a region where conventional ESR is not sensitive [89]. It has been used extensively in spin label studies of biological systems [90], [91]. Spectra are recorded under saturation conditions and high modulation amplitudes, as second harmonic in-phase and quadrature spectra. The method measures the transfer of saturation throughout a spectrum by reorientational molecular motions. The line shape is sensitive to molecular motions and the saturation properties of the spin label.

18.7. References

[1] A. Abragam: *The Principles of Nuclear Magnetism,* Clarendon Press, Oxford 1961.
[2] R. J. Abraham, J. Fisher, P. Loftus: *Introduction to NMR Spectroscopy,* J. Wiley & Sons, Chichester 1988.
[3] R. K. Harris: *Nuclear Magnetic Resonance Spectroscopy,* Longman Scientific & Technical, Harlow 1986.
[4] J. Mason: *Multinuclear NMR,* Plenum Publishing, New York 1987.
[5] D. Shaw: *Fourier Transform NMR Spectroscopy,* 2nd ed., Elsevier, Amsterdam 1984.
[6] T. C. Farrar: *An Introduction to Pulse NMR Spectroscopy,* Farragut Press, Chicago 1987.
[7] F. Bloch, W. W. Hansen, M. E. Packard, *Phys. Rev.* **69** (1946) 127.
[8] E. M. Purcell, H. C. Torrey, R. V. Pound, *Phys. Rev.* **69** (1946) 37.
[9] G. C. Levy, R. L. Lichter, G. L. Nelson: *Carbon-13 Nuclear Magnetic Resonance for Organic Chemists,* 2nd ed., Wiley, New York 1980.
[10] F. W. Wehrli, A. P. Marchand, S. Wehrli: *Interpretation of Carbon-13 NMR Spectra,* 2nd ed., Wiley, New York 1988.
[11] E. Breitmaier, W. Voelter: *Carbon-13 NMR Spectroscopy,* 3rd ed., VCH Publishers, New York 1987.
[12] L. P. Lindeman, J. Q. Adams, *Anal. Chem.* **43** (1971) 1245.
[13] D. E. Ewing, *Org. Magn. Reson.* **12** (1979) 499.
[14] E. F. Mooney: *An Introduction to ^{19}F NMR Spectroscopy,* Heyden-Sadtler, London 1970.
[15] J. W. Emsley, L. Phillips, *Prog. Nucl. Magn. Reson. Spectrosc.* **7** (1971) 1–526.
[16] V. Wray, *Annu. Rep. NMR Spectrosc.* **10 B** (1980) 1.
[17] V. Wray, *Annu. Rep. NMR Spectrosc.* **14** (1983) 1.
[18] D. G. Gorenstein: *Phosphorus-31 NMR: Principles and Applications,* Academic Press, New York 1984.
[19] J. Emsley, D. Hall: *The Chemistry of Phosphorus,* Harper & Row, New York 1976.
[20] G. C. Levy, R. L. Lichter: *Nitrogen-15 NMR Spectroscopy,* J. Wiley & Sons, New York 1979.
[21] M. Witanowski, L. Stefaniak, G. A. Webb, *Annu. Rep. NMR Spectrosc.* **18** (1986) 1.
[22] E. A. Williams, J. D. Cargioli, *Annu. Rep. NMR Spectrosc.* **9** (1979) 221.
[23] A. J. Shaka, J. Keeler, R. Freeman, *J. Magn. Reson.* **53** (1983) 313.
[24] J. H. Noggle, R. E. Schirmer: *The Nuclear Overhauser Effect – Chemical Applications,* Academic Press, New York 1971.
[25] D. Neuhaus, M. P. Williamson: *The Nuclear Overhauser Effect in Structural and Conformation Analysis,* VCH Publishers, New York 1989.
[26] J. K. M. Sanders, B. K. Hunter: *Modern NMR Spectroscopy,* Oxford University Press, Oxford 1987.
[27] H. Y. Carr, E. M. Purcell, *Phys. Rev.* **94** (1954) 630.
[28] S. Meiboom, D. Gill, *Rev. Sci. Instrum.* **29** (1958) 688.
[29] C. LeCocq, J. Y. Lallemand, *J. Chem. Soc. Chem. Commun.* 1981, 150.
[30] D. M. Doddrell, D. T. Pegg, M. R. Bendall, *J. Magn. Reson.* **48** (1982) 323.
[31] G. A. Morris, *Magn. Reson. Chem.* **24** (1986) 371.
[32] R. Benn, H. Günther, *Angew. Chem. Int. Ed. Engl.* **22** (1983) 350.
[33] Varian Associates, Palo Alto: *High Resolution NMR Spectra Catalogue,* **vol. 1, 1962; vol. 2, 1963.**
[34] The Sadtler Collection of High Resolution Spectra, Sadtler Research Laboratories, Philadelphia.
[35] C. J. Pouchert: *Aldrich Library of NMR Spectra,* vols. 1 and 2, Aldrich, Milwaukee, WI, 1983.
[36] W. Brugel: *Handbook of NMR Spectral Parameters,* vols. 1–3, Heyden, Philadelphia 1979.
[37] L. F. Johnson, W. C. Jankowski: *Carbon-13 NMR Spectra, a Collection of Assigned, Coded, and Indexed Spectra,* Wiley, New York 1972.
[38] E. Breitmaier, G. Haas, W. Voelter: *Atlas of C-13 NMR Data,* vols. 1–3, Heyden, Philadelphia 1979.
[39] SPECINFO, Chemical Concepts, Weinheim.
[40] W. Bremser et al.: *Carbon-13 NMR Spectral Data (microfiche),* 4th ed., VCH Publishers, New York 1987.
[41] J. L. Koenig: *Spectroscopy of Polymers,* American Chemical Society, Washington 1992.
[42] A. Bax, S. Grzesiek, *Acc. Chem. Res.* **26** (1993) 131.
[43] D. L. Rabenstein, D. A. Keire, *Pract. Spectrosc.* **11** (1991) 323.
[44] G. J. Martin, X. Y. Sun, C. Guillou, M. L. Martin, *Tetrahedron* **41** (1985) 3285.
[45] D. Parker, *Chem. Rev.* **91** (1991) 1441.
[46] T. C. Morrill (ed.): *Lanthanide Shift Reagents in Stereochemical Analysis,* VCH Publishers, New York 1986.
[47] J. I. Kaplan, G. Fraenkel: *NMR of Chemically Exchanging Systems,* Academic Press, New York 1980.
[48] C. A. Fyfe: *Solid State NMR for Chemists,* CFC Press, Guelph 1983.
[49] A. M. Chippendale et al., *Magn. Reson. Chem.* **24** (1986) 81.
[50] Bruker Minispec Application Note 30, Bruker Analytische Messtechnik GmbH, Karlsruhe.
[51] P. Jackson et al., *Poly. Int.* **24** (1991) 139.

[52] J. E. Wertz, J. R. Bolton: *Electron Spin Resonance; Elementary Theory and Practical Applications,* Chapman & Hall, New York 1986.
[53] J. A. Weil, J. R. Bolton: *Electron Paramagnetic Resonance Spectroscopy, Theory and Examples,* Wiley, New York 1993.
[54] N. M. Atherton: *Principles of Electron Spin Resonance,* Ellis Horwood, Chichester 1993.
[55] A. Abragam, B. Bleaney: *Electron Paramagnetic Resonance of Transition Ions,* Clarendon Press, Oxford 1970.
[56] J. R. Pilbrow: *Transition Ion Electron Paramagnetic Resonance,* Clarendon Press, Oxford 1990.
[57] F. E. Mabbs, D. Collison: *Electron Paramagnetic Resonance of d Transition Metal Compounds,* Elsevier, Amsterdam 1992.
[58] M. C. R. Symons (ed.): "Specialist Periodical Reports," *Electron Spin Resonance,* **vols. 1 – 13,** Royal Society of Chemistry, London 1965 – 1993.
[59] J. A. Weil (ed.): *Electron Magnetic Resonance of the Solid State,* Canadian Society for Chemistry, Ottawa 1987.
[60] S. A. Fairhurst: "Specialist Periodical Reports," in M. C. R. Symons (ed.): *Electron Spin Resonance,* vol. 13 A, Royal Society of Chemistry, London 1992.
[61] M. I. Ikeya: *New Applications of Electron Spin Resonance Dating, Dosimetry and Microscopy,* World Scientific Publ., New York 1993.
[62] H. Sigel (ed.): "ENDOR, EPR and Electron Spin Echo for Probing Co-ordination Spheres," *Metal Ions in Biological Systems,* vol. 22, Marcel Dekker, New York 1987.
[63] L. J. Berliner, J. Reuben (eds.): "Biological Magnetic Resonance," *EMR of Paramagnetic Molecules,* vol. 13, Plenum Press, New York 1993.
[64] L. J. Berliner (ed.): *Spin Labeling, Theory and Applications,* Academic Press, New York 1976.
[65] L. J. Berliner, J. Reuben (eds.): "Biological Magnetic Resonance," *Spin Labeling Theory and Applications,* **vol. 8,** Plenum Press, New York 1989.
[66] A. J. Hoff (ed.): *Advanced EPR: Applications in Biology and Biochemistry,* Elsevier, Amsterdam 1989.
[67] H. Kurreck, B. Kirste, W. Lubitz: *Electron Nuclear Double Resonance Spectroscopy of Radicals in Solution; Applications to Organic and Biological Chemistry,* VCH Publishers, New York 1988.
[68] L. Kevan, L. D. Kispert: *Electron Spin Double Resonance Spectroscopy,* Wiley, New York 1976.
[69] C. P. Keijzers, E. J. Reijerse, J. Schmidt (eds.): *Pulsed EPR: A New Field of Applications,* North Holland, Amsterdam 1989.
[70] L. Kevan, M. K. Bowman (eds.): *Modern Pulsed and Continuous-Wave Electron Spin Resonance,* Wiley, New York 1990.
[71] S. A. Dikanov, Y. D. Tsvetkov: *Electron Spin Echo Envelope Modulation (ESEEM)* Spectroscopy, CRC Press, Boca Raton 1992.
[72] A. Schweiger, *Angew. Chem. Int. Ed. Engl.* **30** (1991) 265.
[73] A. Schweiger, *Pure Appl. Chem.* **64** (1992) 809.
[74] G. R. Eaton, S. S. Eaton, K. Ohno: *EPR Imaging and In vivo EPR,* CRC Press, Boca Raton 1991.
[75] *Landolt-Börnstein,* **New Series Group II, vols. 1, 9 a, b, c1, c2, d1, d2.**
[76] *Landolt-Börnstein,* **New Series Group II, vols. 2, 8, 10, 11.**
[77] I. B. Goldberg, A. J. Bard: *Treatise on Analytical Chemistry,* part 1, vol. 10, Wiley Interscience, New York 1983.
[78] R. G. Kooser, E. Kirchmann, T. Matkov, *Concepts Magn. Reson.* **4** (1992) 145.
[79] J. H. Freed in L. J. Berliner (ed.): *Spin Labeling Theory and Applications,* Academic Press, New York 1976.
[80] J. H. Freed in L. J. Berliner, J. Reuben (eds.): *Spin Labeling Theory and Applications,* vol. 8: "Biological Magnetic Resonance," Plenum Press, New York 1989.
[81] R. F. Boyer, S. E. Keinath: *Molecular Motions in Polymers by ESR,* Harwood Academic Press, Chur 1980.
[82] P. D. Sullivan, J. R. Bolton in J. S. Waugh (ed.): *Advances in Magnetic Resonance,* vol. 4, Academic Press, New York 1970.
[83] M. J. Perkins in V. Gold, D. Bethell (eds.): *Advances in Physical Organic Chemistry,* vol. 17, Academic Press, London 1980.
[84] E. G. Janzen, D. L. Haire in D. D. Tanner (ed.): *Advances in Free Radical Chemistry,* vol. 1, JAI Press, London 1990.
[85] D. Marsh, *Tech. Life Sci. B4/II* **B 426** (1982) 1 – 44.
[86] J. S. Hyde, W. K. Subczynski in L. J. Berliner, J. Reuben (eds.): "Biological Magnetic Resonance," *Spin Labeling Theory and Applications,* **vol. 8,** Plenum Press, New York 1989.
[87] R. K. Woods et al., *J. Magn. Reson.* **85** (1989) 50.
[88] X.-S. Tang, M. Mousavi, G. C. Dismukes, *J. Am. Chem. Soc.* **113** (1991) 5914.
[89] J. S. Hyde, L. R. Dalton in L. J. Berliner (ed.): *Spin Labeling Theory and Applications,* Academic Press, New York 1979.
[90] M. A. Hemminga, P. A. Jager: "Biological Magnetic Resonance," in L. J. Berliner, J. Reuben (eds.): *Spin Labeling Theory and Applications,* vol. 8, Plenum Press, New York 1989.
[91] P. F. Knowles, D. Marsh, H. W. E. Rattle: *Magnetic Resonance of Biomolecules; An Introduction to the Theory and Practice of NMR and ESR in Biological Systems,* Wiley-Interscience, New York 1976.
[92] D. M. Grant, R. K. Harris (eds.): *Encyclopedia of Nuclear Magnetic Resonance,* J. Wiley, Chichester 1996.
[93] C. P. Slichter: *Principles of Magnetic Resonance,* 3rd ed., Springer Verlag, Berlin 1996.
[94] R. S. Macomber: *A Complete Introduction to Modern NMR Spectroscopy,* Wiley, New York 1998.
[95] R. Freeman: *A Handbook of Nuclear Magnetic Resonance,* 2nd ed., Addison Wesley Longman, Essex 1997.
[96] S. W. Homans: *Dictionary of Concepts in NMR,* Oxford University Press, Oxford 1992.
[97] A. E. Derome: *Modern NMR Techniques for Chemistry Research,* Pergamon Press, Oxford 1987.
[98] J. K. M. Sanders, B. K. Hunter: *Modern NMR Spectroscopy,* Oxford University Press, Oxford 1987.
[99] R. R. Ernst, G. Bodenhausen, A. Wokaun: *Principles of Nuclear Magnetic Resonance in One and Two Dimensions,* Clarendon Press, Oxford 1990.
[100] D. M. S. Bagguley: *Pulsed Magnetic Resonance,* Oxford Univ. Press, Oxford 1992.
[101] J. W. Akitt: *NMR and Chemistry: An Introduction to the Multinuclear Fourier Transform Era,* 3rd ed., Chapman & Hall, London 1992.
[102] P. J. Hore: *Nuclear Magnetic Resonance,* Oxford University Press, Oxford 1995.

[103] H. Günther: *NMR Spectroscopy: Basic Principles,Concepts and Applications in Chemistry,* J. Wiley & Sons, New York 1995.
[104] R. Freeman: *Spin Choreography, Basic Steps in High Resolution NMR,* Freeman, Basingstoke 1996.
[105] D. Canet: *Nuclear Magnetic Resonance: Concepts and Methods,* Wiley, Chichester 1996.
[106] S. Berger, S. Braun, H. O. Kalinowski: *NMR-Spectroscopy of the Non-metallic Elements,* John Wiley, Chichester 1996.
[107] G. Batta, K. Kovir, L. Jossuth, C. Szankey: *Methods of Structure Elucidation by High-Resolution NMR,* Elsevier, London 1997.
[108] H. Friebolin: *Basic One and Two-Dimensional NMR Spectroscopy,* Wiley-VCH, Weinheim 1998.
[109] S. Braun, H.-O. Kalinowski, S. Berger: *150 and More Basic NMR Experiments,* Wiley-VCH, Weinheim 1998.
[110] P. A. Keifer, *Drugs of the Future* **23** (1998) 301.
[111] J. P. Hornak, http://www.cis.rit.edu/htbooks/nmr/inside.htm, Rochester Institute of Technology, Rochester, NY.
[112] P. A. Petillo, http://www-chem430.scs.uiuc.edu/lessons/lessons.htm, University of Illinois, Urbana, IL.
[113] D. J. Cookson, B. E. Smith, *J. Magn. Reson.* **57** (1984) 355.
[114] H.-O. Kalinowski, S. Berger, S. Braun: *Carbon-13 NMR Spectroscopy,* Wiley, Chichester 1988.
[115] K. Pihlaja, E. Kleinpeter: *Carbon-13 NMR Chemical Shifts in Structural and Stereochemical Analysis,* VCH, Weinheim 1994.
[116] Attar-ur-Rahman, V. U. Ahmad: *13C NMR of Natural Products,* **Vol. 1 and 2,** Plenum Press, New York 1992.
[117] L. D. Quin, J. G. Verkade: *Phosphorus-31 NMR Spectral Properties in Compound Characterization and Structural Analysis,* VCH Publishers, New York 1994.
[118] S. Berger, T. Faecke, R. Wagner, *Magn. Reson. Chem.* **34** (1996) 4.
[119] D. G. Gadian: *Nuclear Magnetic Resonance and its Application to Living Systems,* Clarendon Press, Oxford 1982.
[120] D. W. Boykin: *17O NMR Spectroscopy,* CRC Press, Boca Raton 1990.
[121] S. Kuroki: *O-17 NMR in I. Ando, T. Asakura (ed.): Solid State NMR of Polymers,* Elsevier Science Publ, Amsterdam 1998, p. 236.
[122] Y. Takeuchi, T. Takayama: "Si-29 NMR spectroscopy of organosilicon compounds" in Z. Rappoport, Y. Apeloig (eds.): *Chemistry of Organic Silicon Compounds,* **vol. 2,** J. Wiley & Sons, Chichester 1998, p. 267.
[123] F. J. M. van de Veen: *Multidimensional NMR in Liquids: Basic Principles and Experimental Methods,* VCH, Weinheim 1995.
[124] A. E. Derome, M. P. Williamson, *J. Magn. Reson.* **88** (1990) 177.
[125] A. A. Maudsely, L. Müller, R. R. Ernst, *J. Magn. Reson.* **28** (1977) 463.
[126] W. Bermel, C. Griesinger, H. Kessler, K. Wagner, *J. Magn. Reson. Chem.* **25** (1987) 325.
[127] Landolt Bernstein: *III/35B Nuclear Magnetic Resonance (NMR)Data: Chemical Shifts and Coupling Constants for 19F and 15N,* Springer Verlag, Berlin 1998.
[128] G. Engelhardt, D. Michel: *High Resolution Solid State NMR of Silicates and Zeolites,* Wiley, Chichester 1987.
[129] M. Schindler, W. Kuttzelnigg, *J. Chem. Phys.* **76** (1982) 1919.
[130] W. Bremser, *Anal. Chim. Acta* **103** (1973) 355.
[131] LabControl Scientific Consulting and Software Development GmbH, *Spectroscopic Knowledge Management System — SPECTACLE,* Köln, Germany.
[132] V. Purtuc, V. Schütz, S. Felsinger, W. Robien, Institute for Organic Chemistry, University of Vienna, Vienna, Austria, http://felix.orc.univie.ac.at/
[133] Advanced Chemistry Development Inc., *CNMR data base,* Toronto, Canada.
[134] J. Meiler, R. Meusinger, M. Will, *J. Chem. Inf. Comput. Sci.* **40** (2000) 1169.
[135] K. Hayamizu, M. Yanagisawa, O. Yamamoto, *Integrated Spectral Data Base System for Organic Compounds (SDBS),* National Institute of Materials and Chemical Research, Tsukuba, Japan, http://www.aist.go.jp/RIODB/SDBS/menu-e.html
[136] E. M. Munk, *J. Chem. Inf. Comput. Sci.* **38** (1998) 997.
[137] E. Pretsch, J. T. Clerc: *Spectra Interpretation of Organic Compounds,* VCH, Weinheim 1997.
[138] F. A. Bovey, P. A. Mireau: *NMR of Polymers,* Academic Press, San Diego 1996.
[139] D. G. Reid: *Protein NMR Techniques,* Humana Press, Totowa, N.J. 1997.
[140] M. J. Shapiro, J. S. Gounarides, *Prog. NMR Spectr.* **35** (1999) 153.
[141] S. B. Shuker, P. J. Hajduk, R. P. Meadows, S. W. Fesik, *Science* **274** (1996) 1531.
[142] H. Barjat, G. A. Morris, S. Smart, A. G. Swanson, S. C. R. Williams, *J. Magn. Reson. Ser. B.* **108** (1995) 170.
[143] M. Spraul, M. Hoffman, P. Dvortsak, J. K. Nicholson, I. D. Wilson, *Anal. Chem.* **65** (1993) 327.
[144] D. L. Olson, T. L. Peck, A. G. Webb, R. L. Magin, J. V. Sweedler, *Science* **270** (1995) 1967.
[145] P. J. Hajduk, T. Gerfin, J. M. Boehlen, M. Häberli, D. MArek, S. W. Fesik, *J. Med. Chem.* **42** (1999) 2315.
[146] I. Ando, J. Austin: *Solid State NMR Spectroscopy,* Elsevier, Amsterdam 1995.
[147] E. O. Stejskal, J. D. Memory: *High Resolution NMR in the Solid State,* Oxford University Press, New York 1994.
[148] B. Hills: *Magnetic Resonance Imaging in Food Science,* Wiley-VCH, Weinheim 1998.
[149] R. Kimmich (ed.): *NMR Tomography, Diffusometry, Relaxometry,* Springer Verlag, Berlin 1997.
[150] R. R. Edelmann, J. R. Hesselink, M. B. Zlatkin (eds.): *Clinical Magnetic Resonance Imaging,* W. B. Saunders Comp. 1996.
[151] M. T. Vlaardingerbroek, J. A. den Boer: *Magnetic Resonance Imaging,* Springer Verlag, Berlin 1996.
[152] H. Bachelard (ed.): *Magnetic Resonance Spectroscopy and Imaging in Neurochemistry,* Plenum Press, New York 1997.
[153] A. Gloor, M. Cadisch, R. Bürgin-Schaller, M. Farkas, T. Kocsis, J. T. Clerc, E. Pretsch, R. Aeschimann, M. Badertscher, T. Brodmeier, A. Fürst, H.-J. Hediger, M. Junghans, H. Kubinyi, M. E. Munk, H. Schriber, D. Wegmann: *SpecTool 1.1, A Hypermedia Book for Structure Elucidation of Organic Compoundswith Spectroscopic Methods,* Chemical Concepts GmbH, Weinheim 1995.